SOLUTIONS MANUAL

Joseph Topich
Virginia Commonwealth University

CHEMISTRY

Seventh Edition

John E. McMurry
Robert C. Fay
Jill K. Robinson

PEARSON

Editor-in-Chief: Jeanne Zalesky
Acquisitions Editor: Chris Hess
Project Manager: Crissy Dudonis
Program Manager: Lisa Pierce
Program Management Team Lead: Kristen Flatham
Project Management Team Lead: David Zielonka
Compositor: Lumina Datamatics Ltd.
Cover Designer: Elise Lansdon
Executive Marketing Manager: Will Moore
Cover Photo Credit: Dr. Keith Wheeler/ Science Source

PEARSON

www.pearsonhighered.com

ISBN 10: 0-133-89229-8
ISBN 13: 978-0-13-389229-1

Contents

Preface

Chapter 1:	Chemical Tools: Experimentation and Measurement	1
Chapter 2:	Atoms, Molecules, and Ions	21
Chapter 3:	Mass Relationships in Chemical Reactions	41
Chapter 4:	Reactions in Aqueous Solution	73
Chapter 5:	Periodicity and the Electronic Structure of Atoms	109
Chapter 6:	Ionic Compounds: Periodic Trends and Bonding Theory	131
Chapter 7:	Covalent Bonding and Electron-Dot Structures	145
Chapter 8:	Covalent Compounds: Bonding Theories and Molecular Structure	183
Chapter 9:	Thermochemistry: Chemical Energy	205
Chapter 10:	Gases: Their Properties and Behavior	231
Chapter 11:	Liquids, Solids, and Phase Changes	273
Chapter 12:	Solutions and Their Properties	301
Chapter 13:	Chemical Kinetics	345
Chapter 14:	Chemical Equilibrium	393
Chapter 15:	Aqueous Equilibria: Acids and Bases	437
Chapter 16:	Applications of Aqueous Equilibria	489
Chapter 17:	Thermodynamics: Entropy, Free Energy, and Equilibrium	551
Chapter 18:	Electrochemistry	589
Chapter 19:	Nuclear Chemistry	647
Chapter 20:	Transition Elements and Coordination Chemistry	679
Chapter 21:	Metals and Solid-State Materials	721
Chapter 22:	The Main-Group Elements	751
Chapter 23:	Organic and Biological Chemistry	777

Preface

Problem solving is your key to success in chemistry! CHEMISTRY, 7/e by McMurry, Fay, and Robinson contains thousands of questions and problems for you to answer. Develop a problem-solving strategy. Read each problem carefully. List the information contained in the problem. Understand what the problem is asking. Use your knowledge of chemistry principles to identify connections between the information in the problem and the solution that you are seeking. Set up and attempt to solve the problem. Look at your answer. Is your answer reasonable? Are the units correct? Then, and only then, check your answer with the *Solutions Manual*.

The *Solutions Manual* to accompany CHEMISTRY, 7/e by McMurry, Fay, and Robinson contains the solutions to all in-chapter, conceptual, and end-of-chapter questions and problems.

I have worked to ensure that the solutions in this manual are as error free as possible. Solutions have been double-checked and in many cases triple-checked. Small differences in numerical answers between student results and those in the *Solutions Manual* may result because of rounding and significant figure differences. It also should be noted that there is, in many instances, more than one acceptable setup for a problem.

I would like to thank John McMurry, Robert Fay, and Jill Robinson for the opportunity to contribute to their CHEMISTRY package. I also want to thank them for their helpful comments as I worked on this solutions manual. I also want to acknowledge and thank Alton Hassell (accuracy checker) and the entire Pearson staff. Finally, I want to thank in a very special way my wife, Ruth, and our daughter, Judy, for their constant encouragement and support as I worked on this project.

Joseph Topich
Department of Chemistry
Virginia Commonwealth University

1 Chemical Tools: Experimentation and Measurement

1.1 (a) 5×10^{-9} m; 5 nm (b) 4.0075017×10^{7} m; 40.075017 Mm

1.2 (a) 7×10^{-5} m (b) 2×10^{13} kg

1.3 $$^{\circ}C = \frac{5}{9} \times (^{\circ}F - 32) = \frac{5}{9} \times (1474 - 32) = 801\ ^{\circ}C$$

$$K = ^{\circ}C + 273.15 = 801 + 273.15 = 1074.15\ K \text{ or } 1074\ K$$

1.4 The melting point of gallium is converted from 302.91 K to °F for comparison.

$$^{\circ}C = K - 273.15 = 302.91 - 273.15 = 29.76\ ^{\circ}C$$

$$^{\circ}F = (\frac{9}{5} \times ^{\circ}C) + 32 = (\frac{9}{5} \times 29.76) + 32 = 85.57\ ^{\circ}F$$

The temperature in the compartment (88 °F) is above the melting point, so the liquid state exists.

1.5 $$\text{Volume} = 9.37\ g \times \frac{1\ mL}{1.483\ g} = 6.32\ mL$$

1.6 Bracelet mass = 80.0 g

Bracelet volume = 17.61 mL – 10.0 mL = 7.61 mL

$$\text{Bracelet density} = \frac{80.0\ g}{7.61\ mL} = 10.5\ g/mL$$

The density of the bracelet matches the density of silver. Since density is one way to identify an unknown substance, it is likely that the bracelet is made of pure silver.

1.7 (a) $$E_K = \frac{1}{2}mv^2 = \frac{1}{2}(6.6 \times 10^{-27}\ kg)\left(\frac{1.5 \times 10^{7}\ m}{s}\right)^2 = 7.4 \times 10^{-13}\ \frac{kg{\cdot}m^2}{s^2} = 7.4 \times 10^{-13}\ J$$

(b) 0.74 pJ

1.8 $$450\ g = 0.450\ kg;\quad E_K = 406\ J = 406\ \frac{kg \cdot m^2}{s^2}$$

$$E_K = \frac{1}{2}mv^2$$

$$v = \sqrt{\frac{2 \times E_K}{m}} = \sqrt{\frac{2 \times 406\ kg{\cdot}m^2/s^2}{0.450\ kg}} = 42.5\ m/s$$

1.9 (a) 76.600 kJ has 5 significant figures because zeros at the end of a number and after the decimal point are always significant.
(b) 4.502 00 x 10^3 g has 6 significant figures because zeros in the middle of a number are significant and zeros at the end of a number and after the decimal point are always significant.
(c) 3000 nm has 1, 2, 3, or 4 significant figures because zeros at the end of a number and before the decimal point may or may not be significant.
(d) 0.003 00 mL has 3 significant figures because zeros at the beginning of a number are not significant and zeros at the end of a number and after the decimal point are always significant.
(e) 18 students has an infinite number of significant figures because this is an exact number.
(f) 3 x 10^{-5} g has 1 significant figure.
(g) 47.60 mL has 4 significant figures because a zero at the end of a number and after the decimal point is always significant.
(h) 2070 mi has 3 or 4 significant figures because a zero in the middle of a number is significant and a zero at the end of a number and before the decimal point may or may not be significant.

1.10 To indicate the uncertainty in a measurement, the value you record should use all the digits you are sure of plus one additional digit that you estimate. The volume can be read to the tenths place and therefore the hundredths place should be estimated. The volume reported to the correct number of significant figures is 4.55 mL.

1.11 (a) In figure (c) darts are scattered (low precision) and are away from the bull's-eye (low accuracy).
(b) In figure (b) darts are clustered together (high precision) and hit the bull's-eye (high accuracy).

1.12 The three measurements are 0.7783 g, 0.7780 g, and 0.7786 g. There is little variation between the three measurements so they have fairly high precision. However, the measurements are all lower than the true value and therefore, the accuracy is low.

1.13 (a)

24.567 g	
+ 0.044 78 g	
24.611 78 g	

This result should be expressed with 3 decimal places. Because the digit to be dropped (7) is greater than 5, round up. The result is 24.612 g (5 significant figures).

(b) 4.6742 g / 0.003 71 L = 1259.89 g/L
0.003 71 has only 3 significant figures so the result of the division should have only 3 significant figures. Because the digit to be dropped (first 9) is greater than 5, round up. The result is 1260 g/L (3 significant figures), or 1.26 x 10^3 g/L.

(c)

0.378 mL	
+ 42.3 mL	
– 1.5833 mL	
41.0947 mL	

This result should be expressed with 1 decimal place. Because the digit to be dropped (9) is greater than 5, round up. The result is 41.1 mL (3 significant figures).

1.14 NaCl mass = 36.2365 g – 35.6783 g = 0.5582 g
NaCl concentration = 0.5582 g/25.0 mL = 0.0223 g/mL = 2.23×10^{-2} g/mL

1.15 1 carat = 200 mg = 200×10^{-3} g = 0.200 g

Mass of Hope Diamond in grams = $44.4 \text{ carats} \times \dfrac{0.200 \text{ g}}{1 \text{ carat}} = 8.88 \text{ g}$

1 ounce = 28.35 g

Mass of Hope Diamond in ounces = $8.88 \text{ g} \times \dfrac{1 \text{ ounce}}{28.35 \text{ g}} = 0.313 \text{ ounces}$

1.16 Volume of Hope Diamond = $8.88 \text{ g} \times \dfrac{1 \text{ cm}^3}{3.52 \text{ g}} = 2.52 \text{ cm}^3$

1.17 (a) area = $113.112 \text{ in}^2 \times \left(\dfrac{2.54 \text{ cm}}{1 \text{ in}}\right)^2 = 729.753 \text{ cm}^2$

(b) volume = $355 \text{ mL} \times \dfrac{1 \text{ cm}^3}{1 \text{ mL}} \times \left(\dfrac{1 \times 10^{-2} \text{ m}}{1 \text{ cm}}\right)^3 = 3.55 \times 10^{-4} \text{ m}^3$

1.18 Volume of a cylinder = $\pi r^2 h$
Cell radius = 6×10^{-6} m/2 = 3×10^{-6} m
Cell volume = $\pi(3 \times 10^{-6} \text{ m})^2(2 \times 10^{-6} \text{ m}) = 6 \times 10^{-17} \text{ m}^3$

Cell volume = $6 \times 10^{-17} \text{ m}^3 \times \left(\dfrac{1 \text{ cm}}{1 \times 10^{-2} \text{ m}}\right)^3 = 6 \times 10^{-11} \text{ cm}^3 = 6 \times 10^{-11} \text{ mL}$

Cell volume = $6 \times 10^{-11} \text{ mL} \times \dfrac{1 \times 10^{-3} \text{ L}}{1 \text{ mL}} = 6 \times 10^{-14} \text{ L} = 0.06 \times 10^{-12} \text{ L} = 0.06 \text{ pL}$

1.19 The diameter of a human hair (~1×10^{-5} m) is approximately 1,000 times larger than the diameter of a 10 nm nanoparticle. (b) A red blood cell (~1×10^{-6} m) is approximately 10,000 times larger than a glucose molecule (1×10^{-10} m).

1.20 Assume that individual atoms pack as cubes in the nanoparticle.
5.0 nm = 5.0×10^{-9} m; 10.0 nm = 10.0×10^{-9} m; 250 pm = 250×10^{-12} m
Atom volume = $(250 \times 10^{-12} \text{ m})^3 = 1.6 \times 10^{-29} \text{ m}^3$
(a) Particle volume = $(5.0 \times 10^{-9} \text{ m})^3 = 1.3 \times 10^{-25} \text{ m}^3$

Atoms/particle = $\dfrac{\text{particle volume}}{\text{atom volume}} = \dfrac{1.3 \times 10^{-25} \text{ m}^3/\text{particle}}{1.6 \times 10^{-29} \text{ m}^3/\text{atom}} = 8125 \text{ atoms/particle}$

Atom face area = $(250 \times 10^{-12} \text{ m})^2 = 6.25 \times 10^{-20} \text{ m}^2$
Particle face area = $(5.0 \times 10^{-9} \text{ m})^2 = 2.5 \times 10^{-17} \text{ m}^2$

Atoms/particle face = $\dfrac{\text{particle face area}}{\text{atom face area}} = \dfrac{2.5 \times 10^{-17} \text{ m}^2/\text{particle}}{6.25 \times 10^{-20} \text{ m}^2/\text{atom}} = 400 \text{ atoms/particle face}$

% atoms on surface = $\dfrac{(6 \text{ faces})(400 \text{ atoms/face})}{8125 \text{ atoms}} \times 100 = 30\%$

(b) Particle volume = $(10.0 \times 10^{-9}\ m)^3 = 1.0 \times 10^{-24}\ m^3$

$$\text{Atoms/particle} = \frac{\text{particle volume}}{\text{atom volume}} = \frac{1.0 \times 10^{-24}\ m^3/\text{particle}}{1.6 \times 10^{-29}\ m^3/\text{atom}} = 62{,}500 \text{ atoms/particle}$$

Atom face area = $(250 \times 10^{-12}\ m)^2 = 6.25 \times 10^{-20}\ m^2$

Particle face area = $(10.0 \times 10^{-9}\ m)^2 = 1.0 \times 10^{-16}\ m^2$

$$\text{Atoms/particle face} = \frac{\text{particle face area}}{\text{atom face area}} = \frac{1.0 \times 10^{-16}\ m^2/\text{particle}}{6.25 \times 10^{-20}\ m^2/\text{atom}} = 1{,}600 \text{ atoms/particle face}$$

$$\%\text{ atoms on surface} = \frac{(6\text{ faces})(1{,}600\text{ atoms/face})}{62{,}500\text{ atoms}} \times 100\% = 15\%$$

1.21 (a) Diameter = 5.0 nm = 5.0×10^{-9} m; radius = 2.5 nm = 2.5×10^{-9} m

$SA = 4\pi r^2 = 4\pi(2.5 \times 10^{-9}\ m)^2 = 7.9 \times 10^{-17}\ m^2$

$$SA = (7.9 \times 10^{-17}\ m^2)\left(\frac{1\ \mu m}{1 \times 10^{-6}\ m}\right)^2 = 7.9 \times 10^{-5}\ \mu m^2$$

Diameter = 5.0 μm = 5.0×10^{-6} m; radius = 2.5 μm = 2.5×10^{-6} m

$SA = 4\pi r^2 = 4\pi(2.5 \times 10^{-6}\ m)^2 = 7.9 \times 10^{-11}\ m^2$

$$SA = (7.9 \times 10^{-11}\ m^2)\left(\frac{1\ \mu m}{1 \times 10^{-6}\ m}\right)^2 = 79\ \mu m^2$$

(b) Diameter = 5.0 nm = 5.0×10^{-9} m; radius = 2.5 nm = 2.5×10^{-9} m

$$\text{Volume} = \frac{4}{3}\pi r^3 = \frac{4}{3}\pi(2.5 \times 10^{-9}\ m)^3 = 6.5 \times 10^{-26}\ m^3$$

$$\text{Volume} = (6.5 \times 10^{-26}\ m^3)\left(\frac{1\ \mu m}{1 \times 10^{-6}\ m}\right)^3 = 6.5 \times 10^{-8}\ \mu m^3$$

Diameter = 5.0 μm = 5.0×10^{-6} m; radius = 2.5 μm = 2.5×10^{-6} m

$$\text{Volume} = \frac{4}{3}\pi r^3 = \frac{4}{3}\pi(2.5 \times 10^{-6}\ m)^3 = 6.5 \times 10^{-17}\ m^3$$

$$\text{Volume} = (6.5 \times 10^{-17}\ m^3)\left(\frac{1\ \mu m}{1 \times 10^{-6}\ m}\right)^3 = 65\ \mu m^3$$

(c) 5.0 nm particle $\quad \frac{SA}{\text{Volume}} = \frac{7.9 \times 10^{-5}\ \mu m^2}{6.5 \times 10^{-8}\ \mu m^3} = 1{,}200\ \mu m^{-1}$

5.0 μm particle $\quad \frac{SA}{\text{Volume}} = \frac{79\ \mu m^2}{65\ \mu m^3} = 1.2\ \mu m^{-1}$

(d) $1{,}200\ \mu m^{-1}/1.2\ \mu m^{-1} = 1{,}000$ times

1.22 (a) As particle size decreases there is a larger fraction of atoms on the surface and surface atoms are more reactive.

(b) Smaller particles maximize the number of reactive atoms while minimizing the total amount of the expensive metal.

(c) Color, electrical conductivity, or melting point.

Conceptual Problems

1.23 For balance (a), the mass of the red block is greater than the mass of the green block. The volume of the red block is less than the volume of the green block.

Because density $= \frac{\text{mass}}{\text{volume}}$, the red block is more dense.

For balance (b), the mass of the green block is greater than the mass of the red block. The volume of both blocks is the same. Because density $= \frac{\text{mass}}{\text{volume}}$, the green block is more dense.

1.24 The level of the liquid in the thermometer is just past the 32 °C mark on the thermometer. The temperature is 32.2°C (3 significant figures).

1.25 (a) 32.0 mL (3 significant figures) (b) 2.72 cm (3 significant figures)

1.26

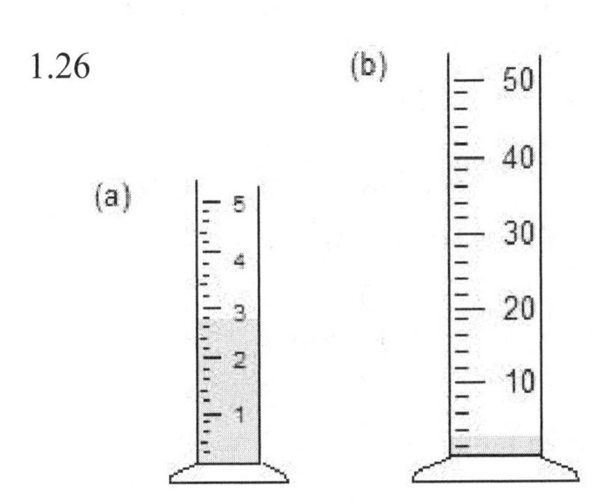

The 5 mL graduated cylinder is marked every 0.2 mL and can be read to ± 0.02 mL. The 50 mL graduated cylinder is marked every 2 mL and can only be read to ± 0.2 mL. The 5 mL graduated cylinder will give more accurate measurements.

1.27 A liquid that is less dense than another will float on top of it. The most dense liquid is mercury, and it is at the bottom of the cylinder. Because water is less dense than mercury but more dense than vegetable oil, it is the middle liquid in the cylinder. Vegetable oil is the least dense of the three liquids and is the top liquid in the cylinder.

Section Problems
Scientific Method (Section 1.1)

1.28 (a) experiment (b) hypothesis (c) observation

1.29 (a) hypothesis (b) observation (c) experiment

1.30 (c) is the correct statement.

1.31 hypothesis

1.32 Molecular models are simplified representations of more complex systems. These models can be used to visualize structure-function relationships that help make theories more concrete.

1.33 (a) qualitative (b) quantitative

1.34 (a), (b) and (d) are quantitative. (c) and (e) are qualitative.

1.35 (b) and (d) are quantitative. (a) and (c) are qualitative.

Units and Significant Figures (Sections 1.2–1.10)

1.36 Mass measures the amount of matter in an object, whereas weight measures the pull of gravity on an object by the earth or other celestial body.

1.37 There are only seven fundamental (base) SI units for scientific measurement. A derived SI unit is some combination of two or more base SI units.
Base SI unit: Mass, kg; Derived SI unit: Density, kg/m^3

1.38 (a) kilogram, kg (b) meter, m (c) kelvin, K
(d) cubic meter, m^3 (e) joule, $(kg \cdot m^2)/s^2$ (f) kg/m^3 or g/cm^3

1.39 (a) kilo, k (b) micro, μ (c) giga, G (d) pico, p (e) centi, c

1.40 A Celsius degree is larger than a Fahrenheit degree by a factor of $\frac{9}{5}$.

1.41 A kelvin and Celsius degree are the same size.

1.42 The volume of a cubic decimeter (dm^3) and a liter (L) are the same.

1.43 The volume of a cubic centimeter (cm^3) and a milliliter (mL) are the same.

1.44 (a) and (b) are exact numbers because they are both definitions.
(c) and (d) are not exact numbers because they result from measurements.

1.45
$$\begin{array}{r} 4.8673\ \text{g} \\ -\ 4.8\ \text{g} \\ \hline 0.0673\ \text{g} \end{array}$$
The result should contain only 1 decimal place. Because the digit to be dropped (6) is greater than 5, round up. The result is 0.1 g.

1.46 cL is centiliter (10^{-2} L)

1.47 (a) dL is deciliter (10^{-1} L) (b) dm is decimeter (10^{-1} m)
(c) μm is micrometer (10^{-6} m) (d) nL is nanoliter (10^{-9} L)
(e) MJ is megajoule (10^{6} J)

1.48 (a) Convert cm to km and compare the two quantities.

$$5.63 \times 10^6 \text{ cm} \times \frac{1 \times 10^{-2} \text{ m}}{1 \text{ cm}} \times \frac{1 \text{ km}}{1{,}000 \text{ m}} = 5.63 \times 10^1 \text{ km}$$

6.02×10^1 km is larger.

(b) Convert μs to ms and compare the two quantities.

$$46\ \mu\text{s} \times \frac{1 \times 10^{-6} \text{ s}}{1\ \mu\text{s}} \times \frac{1 \text{ ms}}{1 \times 10^{-3} \text{ s}} = 4.6 \times 10^{-2} \text{ ms}$$

46 μs is larger.

(c) Convert g to kg and compare the two quantities.

$$200{,}098 \text{ g} \times \frac{1 \text{ kg}}{1000 \text{ g}} = 20.0098 \times 10^1 \text{ kg}$$

200,098 g is larger.

1.49 (a) Convert pm to cm and compare the two quantities.

$$154 \text{ pm} \times \frac{1 \times 10^{-12} \text{ m}}{1 \text{ pm}} \times \frac{1 \text{ cm}}{1 \times 10^{-2} \text{ m}} = 15.4 \times 10^{-9} \text{ cm}$$

7.7×10^{-9} cm is smaller.

(b) Convert μm to km and compare the two quantities.

$$1.86 \times 10^{11}\ \mu\text{m} \times \frac{1 \times 10^{-6} \text{ m}}{1\ \mu\text{m}} \times \frac{1 \text{ km}}{1000 \text{ m}} = 1.86 \times 10^2 \text{ km}$$

1.86×10^{11} μm is smaller.

(c) Convert GA to μA and compare the two quantities.

$$2.9 \text{ GA} \times \frac{1 \times 10^9 \text{ A}}{1 \text{ GA}} \times \frac{1\ \mu\text{A}}{1 \times 10^{-6} \text{ A}} = 2.9 \times 10^{15}\ \mu\text{A}$$

2.9 GA is smaller.

1.50 $1 \text{ mg} = 1 \times 10^{-3} \text{ g}$ and $1 \text{ pg} = 1 \times 10^{-12} \text{ g}$

$$\frac{1 \times 10^{-3} \text{ g}}{1 \text{ mg}} \times \frac{1 \text{ pg}}{1 \times 10^{-12} \text{ g}} = 1 \times 10^9 \text{ pg/mg}$$

$35 \text{ ng} = 35 \times 10^{-9} \text{ g}$

$$\frac{35 \times 10^{-9} \text{ g}}{35 \text{ ng}} \times \frac{1 \text{ pg}}{1 \times 10^{-12} \text{ g}} = 3.5 \times 10^4 \text{ pg/35 ng}$$

1.51 $1\ \mu\text{L} = 10^{-6} \text{ L}$

$$\frac{1\ \mu\text{L}}{10^{-6} \text{ L}} = 10^6\ \mu\text{L/L}$$

$20 \text{ mL} = 20 \times 10^{-3} \text{ L}$

$$\frac{20 \times 10^{-3} \text{ L}}{20 \text{ mL}} \times \frac{1\ \mu\text{L}}{10^{-6} \text{ L}} = 2 \times 10^4\ \mu\text{L/20 mL}$$

1.52 (a) $5 \text{ pm} = 5 \times 10^{-12} \text{ m}$

$$5 \times 10^{-12} \text{ m} \times \frac{1 \text{ cm}}{1 \times 10^{-2} \text{ m}} = 5 \times 10^{-10} \text{ cm}$$

$$5 \times 10^{-12} \text{ m} \times \frac{1 \text{ nm}}{1 \times 10^{-9} \text{ m}} = 5 \times 10^{-3} \text{ nm}$$

(b) $8.5\ \text{cm}^3 \times \left(\frac{1 \times 10^{-2}\ \text{m}}{1\ \text{cm}}\right)^3 = 8.5 \times 10^{-6}\ \text{m}^3$

$8.5\ \text{cm}^3 \times \left(\frac{10\ \text{mm}}{1\ \text{cm}}\right)^3 = 8.5 \times 10^3\ \text{mm}^3$

(c) $65.2\ \text{mg} \times \frac{1 \times 10^{-3}\ \text{g}}{1\ \text{mg}} = 0.0652\ \text{g}$

$65.2\ \text{mg} \times \frac{1 \times 10^{-3}\ \text{g}}{1\ \text{mg}} \times \frac{1\ \text{pg}}{1 \times 10^{-12}\ \text{g}} = 6.52 \times 10^{10}\ \text{pg}$

1.53 (a) A liter is just slightly larger than a quart.
(b) A mile is about twice as long as a kilometer.
(c) An ounce is about 30 times larger than a gram.
(d) An inch is about 2.5 times larger than a centimeter.

1.54 (a) 35.0445 g has 6 significant figures because zeros in the middle of a number are significant.
(b) 59.0001 cm has 6 significant figures because zeros in the middle of a number are significant.
(c) 0.030 03 kg has 4 significant figures because zeros at the beginning of a number are not significant and zeros in the middle of a number are significant.
(d) 0.004 50 m has 3 significant figures because zeros at the beginning of a number are not significant and zeros at the end of a number and after the decimal point are always significant.
(e) 67,000 m^2 has 2, 3, 4, or 5 significant figures because zeros at the end of a number and before the decimal point may or may not be significant.
(f) 3.8200×10^3 L has 5 significant figures because zeros at the end of a number and after the decimal point are always significant.

1.55 (a) $130.95 is an exact number and has an infinite number of significant figures.
(b) 2000.003 has 7 significant figures because zeros in the middle of a number are significant.
(c) The measured quantity, 5 ft 3 in., has 2 significant figures. The 5 ft is certain and the 3 in. is an estimate.
(d) 510 J has 2 or 3 significant figures because zeros at the end of a number and before the decimal point may or may not be significant.
(e) 5.10×10^2 J has 3 significant figures because zeros at the end of a number and after the decimal point are always significant.
(f) 10 students is a count, therefore 10 is an exact number with an infinite number of significant figures.

1.56 To convert 3,666,500 m^3 to scientific notation, move the decimal point 6 places to the left and include an exponent of 10^6. The result is $3.6665 \times 10^6\ \text{m}^3$. Because the digit to be dropped is 5 with nothing following, round down. The result is $3.666 \times 10^6\ \text{m}^3$ (4 significant figures). Because the digit to be dropped (the second 6) is greater than 5, round up. The result is $3.7 \times 10^6\ \text{m}^3$ (2 significant figures).

1.57 Because the digit to be dropped (3) is less than 5, round down. The result to 4 significant figures is 7926 mi or 7.926×10^3 mi.
Because the digit to be dropped (2) is less than 5, round down. The result to 2 significant figures is 7900 mi or 7.9×10^3 mi.

1.58 (a) To convert 453.32 mg to scientific notation, move the decimal point 2 places to the left and include an exponent of 10^2. The result is 4.5332×10^2 mg.
(b) To convert 0.000 042 1 mL to scientific notation, move the decimal point 5 places to the right and include an exponent of 10^{-5}. The result is 4.21×10^{-5} mL.
(c) To convert 667,000 g to scientific notation, move the decimal point 5 places to the left and include an exponent of 10^5. The result is 6.67×10^5 g.

1.59 (a) Because the exponent is a negative 3, move the decimal point 3 places to the left to get 0.003 221 mm.
(b) Because the exponent is a positive 5, move the decimal point 5 places to the right to get 894,000 m.
(c) Because the exponent is a negative 12, move the decimal point 12 places to the left to get 0.000 000 000 001 350 82 m^3.
(d) Because the exponent is a positive 2, move the decimal point 2 places to the right to get 641.00 km.

1.60 (a) Because the digit to be dropped (0) is less than 5, round down. The result is 3.567×10^4 or 35,670 m (4 significant figures).
Because the digit to be dropped (the second 6) is greater than 5, round up. The result is 35,670.1 m (6 significant figures).
(b) Because the digit to be dropped is 5 with nonzero digits following, round up. The result is 69 g (2 significant figures).
Because the digit to be dropped (0) is less than 5, round down. The result is 68.5 g (3 significant figures).
(c) Because the digit to be dropped is 5 with nothing following, round down. The result is 4.99×10^3 cm (3 significant figures).
(d) Because the digit to be dropped is 5 with nothing following, round down. The result is 2.3098×10^{-4} kg (5 significant figures).

1.61 (a) Because the digit to be dropped (1) is less than 5, round down. The result is 7.000 kg.
(b) Because the digit to be dropped is 5 with nothing following, round down. The result is 1.60 km.
(c) Because the digit to be dropped (1) is less than 5, round down. The result is 13.2 g/cm^3.
(d) Because the digit to be dropped (1) is less than 5, round down. The result is 2,300,000. or $2.300\,000 \times 10^6$.

1.62 (a) $4.884 \times 2.05 = 10.012$
The result should contain only 3 significant figures because 2.05 contains 3 significant figures (the smaller number of significant figures of the two). Because the digit to be dropped (1) is less than 5, round down. The result is 10.0.

(b) 94.61 / 3.7 = 25.57
The result should contain only 2 significant figures because 3.7 contains 2 significant figures (the smaller number of significant figures of the two). Because the digit to be dropped (second 5) is 5 with nonzero digits following, round up. The result is 26.
(c) 3.7 / 94.61 = 0.0391
The result should contain only 2 significant figures because 3.7 contains 2 significant figures (the smaller number of significant figures of the two). Because the digit to be dropped (1) is less than 5, round down. The result is 0.039.

(d)
$$\begin{array}{r} 5502.3 \\ 24 \\ +\quad 0.01 \\ \hline 5526.31 \end{array}$$
This result should be expressed with no decimal places. Because the digit to be dropped (3) is less than 5, round down. The result is 5526.

(e)
$$\begin{array}{r} 86.3 \\ +\ 1.42 \\ -\ 0.09 \\ \hline 87.63 \end{array}$$
This result should be expressed with only 1 decimal place. Because the digit to be dropped (3) is less than 5, round down. The result is 87.6.

(f) 5.7 x 2.31 = 13.167
The result should contain only 2 significant figures because 5.7 contains 2 significant figures (the smaller number of significant figures of the two). Because the digit to be dropped (second 1) is less than 5, round down. The result is 13.

1.63 (a) $\dfrac{3.41 - 0.23}{5.233} \times 0.205 = \dfrac{3.18}{5.233} \times 0.205 = 0.12457 = 0.125$

Complete the subtraction first. The result has 2 decimal places and 3 significant figures. The result of the multiplication and division must have 3 significant figures. Because the digit to be dropped is 5 with nonzero digits following, round up.

(b) $\dfrac{5.556 \times 2.3}{4.223 - 0.08} = \dfrac{5.556 \times 2.3}{4.143} = 3.08 = 3.1$

Complete the subtraction first. The result of the subtraction should have 2 decimal places and 3 significant figures (an extra digit is being carried until the calculation is completed). The result of the multiplication and division must have 2 significant figures. Because the digit to be dropped (8) is greater than 5, round up.

1.64 1 mile = 1.6093 km; The time is 1 h, 5 min, and 26.6 s.
Convert the time to seconds and then hours.

$$\text{time} = \left(1 \text{ hr} \times \frac{60 \text{ min}}{1 \text{ h}} \times \frac{60 \text{ s}}{1 \text{ min}}\right) + \left(5 \text{ min} \times \frac{60 \text{ s}}{1 \text{ min}}\right) + 26.6 \text{ s} = 3926.6 \text{ s}$$

$$\text{time} = 3926.6 \text{ s} \times \frac{1 \text{ min}}{60 \text{ s}} \times \frac{1 \text{ h}}{60 \text{ min}} = 1.0907 \text{ h}$$

Convert meters to miles.

$$20{,}000 \text{ m} \times \frac{1 \text{ km}}{1000 \text{ m}} \times \frac{1 \text{ mi}}{1.6093 \text{ km}} = 12.4278 \text{ mi}$$

average speed = $\frac{12.4278 \text{ mi}}{1.0907 \text{ h}}$ = 11.394 mi/h

1.65 1 mile = 1.6093 km

Convert g to mg.

$12.0 \text{ g} \times \frac{1000 \text{ mg}}{1 \text{ g}} = 12{,}000 \text{ mg}$

Convert km to miles.

$1 \text{ km} \times \frac{1 \text{ mi}}{1.6093 \text{ km}} = 0.6214 \text{ mi}$

CO limit = $\frac{1.20 \times 10^4 \text{ mg}}{0.6214 \text{ mi}} = 1.93 \times 10^4 \text{ mg/mi}$

Unit Conversions (Section 1.11)

1.66 (a) $0.25 \text{ lb} \times \frac{453.59 \text{ g}}{1 \text{ lb}} = 113.4 \text{ g} = 110 \text{ g}$

(b) $1454 \text{ ft} \times \frac{12 \text{ in.}}{1 \text{ ft}} \times \frac{2.54 \text{ cm}}{1 \text{ in.}} \times \frac{1 \times 10^{-2} \text{ m}}{1 \text{ cm}} = 443.2 \text{ m}$

(c) $2{,}941{,}526 \text{ mi}^2 \times \left(\frac{1.6093 \text{ km}}{1 \text{ mi}}\right)^2 \times \left(\frac{1000 \text{ m}}{1 \text{ km}}\right)^2 = 7.6181 \times 10^{12} \text{ m}^2$

1.67 (a) $5.4 \text{ in.} \times \frac{2.54 \text{ cm}}{1 \text{ in.}} \times \frac{1 \times 10^{-2} \text{ m}}{1 \text{ cm}} = 0.14 \text{ m}$

(b) $66.31 \text{ lb} \times \frac{1 \text{ kg}}{2.2046 \text{ lb}} = 30.08 \text{ kg}$

(c) $0.5521 \text{ gal} \times \frac{3.7854 \text{ L}}{1 \text{ gal}} \times \frac{1 \times 10^{-3} \text{ m}^3}{1 \text{ L}} = 2.090 \times 10^{-3} \text{ m}^3$

(d) $65 \frac{\text{mi}}{\text{h}} \times \frac{1.6093 \text{ km}}{1 \text{ mi}} \times \frac{1000 \text{ m}}{1 \text{ km}} \times \frac{1 \text{ h}}{60 \text{ min}} \times \frac{1 \text{ min}}{60 \text{ s}} = 29 \frac{\text{m}}{\text{s}}$

(e) $978.3 \text{ yd}^3 \times \left(\frac{1 \text{ m}}{1.0936 \text{ yd}}\right)^3 = 748.0 \text{ m}^3$

(f) $2.380 \text{ mi}^2 \times \left(\frac{1.6093 \text{ km}}{1 \text{ mi}}\right)^2 \times \left(\frac{1000 \text{ m}}{1 \text{ km}}\right)^2 = 6.164 \times 10^6 \text{ m}^2$

1.68 (a) $1 \text{ acre-ft} \times \frac{1 \text{ mi}^2}{640 \text{ acres}} \times \left(\frac{5280 \text{ ft}}{1 \text{ mi}}\right)^2 = 43{,}560 \text{ ft}^3$

(b) $116 \text{ mi}^3 \times \left(\frac{5280 \text{ ft}}{1 \text{ mi}}\right)^3 \times \frac{1 \text{ acre-ft}}{43{,}560 \text{ ft}^3} = 3.92 \times 10^8 \text{ acre-ft}$

1.69 (a) $18.6 \text{ hands} \times \frac{1/3 \text{ ft}}{1 \text{ hand}} \times \frac{12 \text{ in.}}{1 \text{ ft}} \times \frac{2.54 \text{ cm}}{1 \text{ in.}} = 189 \text{ cm}$

(b) $(6 \times 2.5 \times 15) \text{ hands}^3 \times \left(\frac{1/3 \text{ ft}}{1 \text{ hand}}\right)^3 \times \left(\frac{12 \text{ in.}}{1 \text{ ft}}\right)^3 \times \left(\frac{2.54 \text{ cm}}{1 \text{ in.}}\right)^3 \times \left(\frac{1 \times 10^{-2} \text{ m}}{1 \text{ cm}}\right)^3 = 0.2 \text{ m}^3$

1.70 $8.65 \text{ stones} \times \frac{14 \text{ lb}}{1 \text{ stone}} = 121 \text{ lb}$

1.71 (a) $\frac{200 \text{ mg}}{100 \text{ mL}} \times \frac{1 \text{ mL}}{1 \times 10^{-3} \text{ L}} = 2000 \text{ mg/L}$

(b) $\frac{200 \text{ mg}}{100 \text{ mL}} \times \frac{1 \times 10^{-3} \text{ g}}{1 \text{ mg}} \times \frac{1 \, \mu\text{g}}{1 \times 10^{-6} \text{ g}} = 2000 \, \mu\text{g/mL}$

(c) $\frac{200 \text{ mg}}{100 \text{ mL}} \times \frac{1 \times 10^{-3} \text{ g}}{1 \text{ mg}} \times \frac{1 \text{ mL}}{1 \times 10^{-3} \text{ L}} = 2 \text{ g/L}$

(d) $\frac{200 \text{ mg}}{100 \text{ mL}} \times \frac{1 \times 10^{-3} \text{ g}}{1 \text{ mg}} \times \frac{1 \text{ mL}}{1 \times 10^{-3} \text{ L}} \times \frac{1 \text{ ng}}{1 \times 10^{-9} \text{ g}} \times \frac{1 \times 10^{-6} \text{ L}}{1 \, \mu\text{L}} = 2000 \text{ ng}/\mu\text{L}$

(e) $2 \text{ g/L} \times 5 \text{ L} = 10 \text{ g}$

1.72 $160 \text{ lb} \times \frac{1 \text{ kg}}{2.2046 \text{ lb}} = 72.6 \text{ kg}$

$72.6 \text{ kg} \times \frac{20 \, \mu\text{g}}{1 \text{ kg}} \times \frac{1 \text{ mg}}{1 \times 10^3 \, \mu\text{g}} = 1.452 \text{ mg} = 1.5 \text{ mg}$

1.73 $55 \frac{\text{mi}}{\text{h}} \times \frac{5280 \text{ ft}}{1 \text{ mi}} \times \frac{12 \text{ in.}}{1 \text{ ft}} \times \frac{2.54 \text{ cm}}{1 \text{ in.}} \times \frac{1 \text{ h}}{3600 \text{ s}} \times \frac{2.5 \times 10^{-4} \text{ s}}{1 \text{ shake}} = 0.61 \frac{\text{cm}}{\text{shake}}$

Temperature (Section 1.5)

1.74 $^\circ\text{F} = (\frac{9}{5} \times {}^\circ\text{C}) + 32$

$^\circ\text{F} = (\frac{9}{5} \times 39.9^\circ\text{C}) + 32 = 103.8 \, ^\circ\text{F}$ (goat)

$^\circ\text{F} = (\frac{9}{5} \times 22.2^\circ\text{C}) + 32 = 72.0 \, ^\circ\text{F}$ (Australian spiny anteater)

1.75 For Hg: mp is $\left[\frac{9}{5} \times (-38.87)\right] + 32 = -37.97 \, ^\circ\text{F}$

For Br_2: mp is $\left[\frac{9}{5} \times (-7.2)\right] + 32 = 19.0 \, ^\circ\text{F}$

For Cs: mp is $\left[\frac{9}{5} \times (28.40)\right] + 32 = 83.12 \, ^\circ\text{F}$

For Ga: mp is $\left[\frac{9}{5} \text{ x } (29.78)\right] + 32 = 85.60\ ^\circ\text{F}$

1.76 $^\circ\text{F} = (\frac{9}{5} \text{ x } ^\circ\text{C}) + 32 = (\frac{9}{5} \text{ x } 175) + 32 = 347\ ^\circ\text{F}$

1.77 $^\circ\text{C} = \frac{5}{9} \text{ x } (^\circ\text{F} - 32) = \frac{5}{9} \text{ x } (6192 - 32) = 3422\ ^\circ\text{C}$

$\text{K} = ^\circ\text{C} + 273.15 = 3422 + 273.15 = 3695.15 \text{ K or } 3695 \text{ K}$

1.78

Ethanol boiling point	78.5 °C	173.3 °F	200 °E
Ethanol melting point	–117.3 °C	–179.1 °F	0 °E

(a) $\frac{200\ ^\circ\text{E}}{[78.5\ ^\circ\text{C} - (-117.3\ ^\circ\text{C})]} = \frac{200\ ^\circ\text{E}}{195.8\ ^\circ\text{C}} = 1.021\ ^\circ\text{E}/^\circ\text{C}$

(b) $\frac{200\ ^\circ\text{E}}{[173.3\ ^\circ\text{F} - (-179.1\ ^\circ\text{F})]} = \frac{200\ ^\circ\text{E}}{352.4\ ^\circ\text{F}} = 0.5675\ ^\circ\text{E}/^\circ\text{F}$

(c) $^\circ\text{E} = \frac{200}{195.8} \text{ x } (^\circ\text{C} + 117.3)$

H_2O melting point = 0°C; $^\circ\text{E} = \frac{200}{195.8} \text{ x } (0 + 117.3) = 119.8\ ^\circ\text{E}$

H_2O boiling point = 100°C; $^\circ\text{E} = \frac{200}{195.8} \text{ x } (100 + 117.3) = 222.0\ ^\circ\text{E}$

(d) $^\circ\text{E} = \frac{200}{352.4} \text{ x } (^\circ\text{F} + 179.1) = \frac{200}{352.4} \text{ x } (98.6 + 179.1) = 157.6\ ^\circ\text{E}$

(e) $^\circ\text{F} = \left(^\circ\text{E} \text{ x } \frac{352.4}{200}\right) - 179.1 = \left(130 \text{ x } \frac{352.4}{200}\right) - 179.1 = 50.0\ ^\circ\text{F}$

Because the outside temperature is 50.0°F, I would wear a sweater or light jacket.

1.79

NH_3 boiling point	–33.4 °C	–28.1 °F	100 °A
NH_3 melting point	–77.7 °C	–107.9 °F	0° A

(a) $\frac{100\ ^\circ\text{A}}{[-33.4 - (-77.7\ ^\circ\text{C})]} = \frac{100\ ^\circ\text{A}}{44.3\ ^\circ\text{C}} = 2.26\ ^\circ\text{A}/\ ^\circ\text{C}$

(b) $\frac{100\ ^\circ\text{A}}{[-28.1 - (-107.9\ ^\circ\text{F})]} = \frac{100\ ^\circ\text{A}}{79.8\ ^\circ\text{F}} = 1.25\ ^\circ\text{A}/\ ^\circ\text{F}$

(c) $^\circ\text{A} = \frac{100}{44.3} \text{ x } (^\circ\text{C} + 77.7)$

H_2O melting point = 0°C; $^\circ\text{A} = \frac{100}{44.3} \text{ x } (0 + 77.7) = 175\ ^\circ\text{A}$

H_2O boiling point = 100°C; $^\circ\text{A} = \frac{100}{44.3} \text{ x } (100 + 77.7) = 401\ ^\circ\text{A}$

(d) $^{o}A = \frac{100}{79.8} \times (^{o}F + 107.9) = \frac{100}{79.8} \times (98.6 + 107.9) = 259\ ^{o}A$

Density (Section 1.7)

1.80 $d = \frac{m}{V} = \frac{27.43\ g}{12.40\ cm^3} = 2.212\ g/cm^3$

1.81 $d = \frac{m}{V} = \frac{206.77\ g}{15.50\ cm^3} = 13.34\ \frac{g}{cm^3}$

1.82 $3.10\ g/cm^3 = 3.10\ g/mL$

$mass = 3.10\ g/mL \times \frac{1\ kg}{1000\ g} \times \frac{1\ mL}{1 \times 10^{-3}\ L} \times 4.67\ L = 14.5\ kg$

1.83 $250\ mg \times \frac{1 \times 10^{-3}\ g}{1\ mg} = 0.25\ g;\quad V = 0.25\ g \times \frac{1\ cm^3}{1.40\ g} = 0.18\ cm^3$

$500\ lb \times \frac{453.59\ g}{1\ lb} = 226{,}795\ g;\quad V = 226{,}795\ g \times \frac{1\ cm^3}{1.40\ g} = 161{,}996\ cm^3 = 162{,}000\ cm^3$

1.84 For H_2: $V = 1.0078\ g \times \frac{1\ L}{0.0899\ g} = 11.2\ L$

For Cl_2: $V = 35.45\ g \times \frac{1\ L}{3.214\ g} = 11.03\ L$

1.85 $mass = 10.5\ g/cm^3 \times \frac{1\ kg}{1000\ g} \times \left(\frac{1\ cm}{1 \times 10^{-2}\ m}\right)^3 \times (0.62\ m)^3 = 2500\ kg$

1.86 $d = \frac{m}{V} = \frac{220.9\ g}{(0.50 \times 1.55 \times 25.00)\ cm^3} = 11.4\ \frac{g}{cm^3} = 11\ \frac{g}{cm^3}$

1.87 diameter = 2.40 mm = 0.240 cm, r = diameter/2 = 0.120 cm, and $V = \pi r^2 h$

$d = \frac{m}{V} = \frac{0.3624\ g}{(3.1416)(0.120\ cm)^2(15.0\ cm)} = 0.534\ g/cm^3$

1.88 Silverware mass = 80.56 g

Silverware volume = 15.90 mL – 10.00 mL = 5.90 mL

Silverware density = $\frac{80.56\ g}{5.90\ mL} = 13.7\ g/mL$

The density of the silverware and pure silver are different. The silverware is not pure silver.

1.89 Mass of 10 pennies = 24.656 g

Volume of 10 pennies = 12.90 mL – 10.0 mL = 2.90 mL

$$\text{Pennies density} = \frac{24.656\ \text{g}}{2.90\ \text{mL}} = 8.50\ \text{g/mL}$$

The density of the pennies and pure copper are different. The pennies are not pure copper.

Energy (Section 1.8)

1.90 $$\text{Car: } E_K = ½(1400\ \text{kg})\left(\frac{115 \times 10^3\ \text{m}}{3600\ \text{s}}\right)^2 = 7.1 \times 10^5\ \text{J}$$

$$\text{Truck: } E_K = ½(12{,}000\ \text{kg})\left(\frac{38 \times 10^3\ \text{m}}{3600\ \text{s}}\right)^2 = 6.7 \times 10^5\ \text{J}$$

The car has more kinetic energy.

1.91 Heat = q = 7.1×10^5 J (from Problem 1.90)

q = (specific heat) x m x ΔT

$$m = \frac{q}{(\text{specific heat}) \times \Delta T} = \frac{7.1 \times 10^5\ \text{J}}{\left(4.18\ \frac{\text{J}}{\text{g}\cdot{}^\circ\text{C}}\right)(50\ {}^\circ\text{C} - 20\ {}^\circ\text{C})} = 5.7 \times 10^3\ \text{g of water}$$

1.92 1 oz = 28.35 g

$$\text{energy} = 0.450\ \text{oz} \times \frac{28.35\ \text{g}}{1\ \text{oz}} \times \frac{2498\ \text{kJ}}{45.0\ \text{g}} \times \frac{1\ \text{kcal}}{4.184\ \text{kJ}} = 169\ \text{kcal}$$

1.93 $$\text{g Na} = \frac{1.00\ \text{g Na}}{17.9\ \text{kJ}} \times \frac{4.184\ \text{kJ}}{1\ \text{kcal}} \times 171\ \text{kcal} = 40.0\ \text{g Na}$$

$$\text{g Cl} = \frac{1.54\ \text{g Cl}}{1.00\ \text{g Na}} \times 40.0\ \text{g Na} = 61.6\ \text{g Cl}$$

1.94 (a) $$540\ \text{Cal} \times \frac{1000\ \text{cal}}{1\ \text{Cal}} \times \frac{4.184\ \text{J}}{1\ \text{cal}} \times \frac{1\ \text{kJ}}{1000\ \text{J}} = 2259\ \text{kJ} = 2300\ \text{kJ}$$

(b) 100 watts = 100 J/s

$$\text{time} = 2259\ \text{kJ} \times \frac{1000\ \text{J}}{1\ \text{kJ}} \times \frac{1\ \text{s}}{100\ \text{J}} \times \frac{1\ \text{min}}{60\ \text{s}} \times \frac{1\ \text{h}}{60\ \text{min}} = 6.275\ \text{h} = 6.3\ \text{h}$$

1.95 (a) $$238\ \text{Cal} \times \frac{1000\ \text{cal}}{1\ \text{Cal}} \times \frac{4.184\ \text{J}}{1\ \text{cal}} \times \frac{1\ \text{kJ}}{1000\ \text{J}} = 996\ \text{kJ}$$

(b) 75 watts = 75 J/s

$$\text{time} = 996\ \text{kJ} \times \frac{1000\ \text{J}}{1\ \text{kJ}} \times \frac{1\ \text{s}}{75\ \text{J}} \times \frac{1\ \text{min}}{60\ \text{s}} \times \frac{1\ \text{h}}{60\ \text{min}} = 3.69\ \text{h}$$

Chapter Problems

1.96 $d = \frac{m}{V} = \frac{8.763\ g}{(28.76 - 25.00)\ mL} = \frac{8.763\ g}{3.76\ mL} = 2.331\ \frac{g}{cm^3} = 2.33\ \frac{g}{cm^3}$

1.97 volume of sphere = $\frac{4}{3}\pi r^3$; sphere radius = 7.60 cm/2 = 3.80 cm

(a) $d = \frac{m}{V} = \frac{313\ g}{\frac{4}{3}\pi(3.80\ cm)^3} = 1.36\ g/cm^3 = 1.36\ g/mL$

(b) Because the density is greater than 1.0 g/mL, the sphere will sink in water.
(c) Because the density is less than 1.48 g/mL, the sphere will float in chloroform.

1.98 NaCl melting point = 1074 K
°C = K – 273.15 = 1074 – 273.15 = 800.85 °C = 801 °C

$°F = (\frac{9}{5} \times °C) + 32 = (\frac{9}{5} \times 800.85) + 32 = 1473.53\ °F = 1474\ °F$

NaCl boiling point = 1686 K
°C = K – 273.15 = 1686 – 273.15 = 1412.85 °C = 1413 °C

$°F = (\frac{9}{5} \times °C) + 32 = (\frac{9}{5} \times 1412.85) + 32 = 2575.13\ °F = 2575\ °F$

1.99 1 gal = 3.7854 L

(a) volume = $3.4 \times 10^4\ L \times \frac{1\ gal}{3.7854\ L} = 9.0 \times 10^3\ gal$

(b) value = $9.0 \times 10^3\ gal \times \frac{\$3.00}{1\ gal} = \$27{,}000$

1.100 $V = 112.5\ g \times \frac{1\ mL}{1.4832\ g} = 75.85\ mL$

1.101 1 lb = 453.59 g

volume = $3.6 \times 10^{11}\ lb \times \frac{453.59\ g}{1\ lb} \times \frac{1\ cm^3}{1.8302\ g} \times \frac{1\ L}{1000\ cm^3} = 8.9 \times 10^{10}\ L$

1.102 (a) density = $\frac{1\ lb}{1\ pint} \times \frac{8\ pints}{1\ gal} \times \frac{1\ gal}{3.7854\ L} \times \frac{453.59\ g}{1\ lb} \times \frac{1\ L}{1000\ mL} = 0.958\ 61\ g/mL$

(b) area in m^2 =

$$1\ acre \times \frac{1\ mi^2}{640\ acres} \times \left(\frac{5280\ ft}{1\ mi}\right)^2 \times \left(\frac{12\ in.}{1\ ft}\right)^2 \times \left(\frac{2.54\ cm}{1\ in.}\right)^2 \times \left(\frac{1 \times 10^{-2}\ m}{1\ cm}\right)^2 = 4047\ m^2$$

(c) mass of wood =

$$1 \text{ cord} \times \frac{128 \text{ ft}^3}{1 \text{ cord}} \times \left(\frac{12 \text{ in.}}{1 \text{ ft}}\right)^3 \times \left(\frac{2.54 \text{ cm}}{1 \text{ in.}}\right)^3 \times \frac{0.40 \text{ g}}{1 \text{ cm}^3} \times \frac{1 \text{ kg}}{1000 \text{ g}} = 1450 \text{ kg} = 1400 \text{ kg}$$

(d) mass of oil =

$$1 \text{ barrel} \times \frac{42 \text{ gal}}{1 \text{ barrel}} \times \frac{3.7854 \text{ L}}{1 \text{ gal}} \times \frac{1 \text{ mL}}{1 \times 10^{-3} \text{ L}} \times \frac{0.85 \text{ g}}{1 \text{ mL}} \times \frac{1 \text{ kg}}{1000 \text{ g}} = 135.1 \text{ kg} = 140 \text{ kg}$$

(e) fat Calories =

$$0.5 \text{ gal} \times \frac{32 \text{ servings}}{1 \text{ gal}} \times \frac{165 \text{ Calories}}{1 \text{ serving}} \times \frac{30.0 \text{ Cal from fat}}{100 \text{ Cal total}} = 792 \text{ Cal from fat}$$

1.103 amount of chocolate =

$$2.0 \text{ cups coffee} \times \frac{105 \text{ mg caffeine}}{1 \text{ cup coffee}} \times \frac{1.0 \text{ ounce chocolate}}{15 \text{ mg caffeine}} = 14 \text{ ounces of chocolate}$$

14 ounces of chocolate is just under 1 pound.

1.104 (a) number of Hershey's Kisses =

$$2.0 \text{ lb} \times \frac{453.59 \text{ g}}{1 \text{ lb}} \times \frac{1 \text{ serving}}{41 \text{ g}} \times \frac{9 \text{ kisses}}{1 \text{ serving}} = 199 \text{ kisses} = 200 \text{ kisses}$$

(b) Hershey's Kiss volume = $\frac{41 \text{ g}}{1 \text{ serving}} \times \frac{1 \text{ serving}}{9 \text{ kisses}} \times \frac{1 \text{ mL}}{1.4 \text{ g}} = 3.254 \text{ mL} = 3.3 \text{ mL}$

(c) Calories/Hershey's Kiss = $\frac{230 \text{ Cal}}{1 \text{ serving}} \times \frac{1 \text{ serving}}{9 \text{ kisses}} = 25.55 \text{ Cal/kiss} = 26 \text{ Cal/kiss}$

(d) % fat Calories =

$$\frac{13 \text{ g fat}}{1 \text{ serving}} \times \frac{9 \text{ Cal from fat}}{1 \text{ g fat}} \times \frac{1 \text{ serving}}{230 \text{ Cal total}} \times 100\% = 51\% \text{ Calories from fat}$$

1.105 Let Y equal volume of vinegar and ($422.8 \text{ cm}^3 - \text{Y}$) equal the volume of oil.

Mass = volume x density

$397.8 \text{ g} = (\text{Y} \times 1.006 \text{ g/cm}^3) + [(422.8 \text{ cm}^3 - \text{Y}) \times 0.918 \text{ g/cm}^3]$

$397.8 \text{ g} = (1.006 \text{ g/cm}^3)\text{Y} + 388.1 \text{ g} - (0.918 \text{ g/cm}^3)\text{Y}$

$397.8 \text{ g} - 388.1 \text{ g} = (1.006 \text{ g/cm}^3)\text{Y} - (0.918 \text{ g/cm}^3)\text{Y}$

$9.7 \text{ g} = (0.088 \text{ g/cm}^3)\text{Y}$

$\text{Y} = \text{vinegar volume} = \frac{9.7 \text{ g}}{0.088 \text{ g/cm}^3} = 110 \text{ cm}^3$

oil volume $= (422.8 \text{ cm}^3 - \text{Y}) = (422.8 \text{ cm}^3 - 110 \text{ cm}^3) = 313 \text{ cm}^3$

1.106 $^{\circ}\text{C} = \frac{5}{9} \times (^{\circ}\text{F} - 32)$

Set $^{\circ}\text{C} = {}^{\circ}\text{F}$: $^{\circ}\text{C} = \frac{5}{9} \times (^{\circ}\text{C} - 32)$

Solve for °C: $°C \times \frac{9}{5} = °C - 32$

$(°C \times \frac{9}{5}) - °C = -32$

$°C \times \frac{4}{5} = -32$

$°C = \frac{5}{4}(-32) = -40\ °C$

The Celsius and Fahrenheit scales "cross" at −40 °C (−40 °F).

1.107 Cork: volume = 1.30 cm x 5.50 cm x 3.00 cm = 21.45 cm^3

$$\text{mass} = 21.45\ cm^3 \times \frac{0.235\ g}{1\ cm^3} = 5.041\ g$$

Lead: volume = $(1.15\ cm)^3 = 1.521\ cm^3$

$$\text{mass} = 1.521\ cm^3 \times \frac{11.35\ g}{1\ cm^3} = 17.26\ g$$

total mass = 5.041 g + 17.26 g = 22.30 g
total volume = 21.45 cm^3 + 1.521 cm^3 = 22.97 cm^3

$$\text{average density} = \frac{22.30\ g}{22.97\ cm^3} = 0.971\ g/cm^3$$ so the cork and lead will float.

1.108 $d = 0.037\ \frac{lbs}{in^3} \times \frac{453.59\ g}{1\ lb} \times \left(\frac{1\ in}{2.54\ cm}\right)^3 = 1.0\ g/cm^3$

1.109 $d = 0.55\ \frac{oz}{in^3} \times \frac{1\ lb}{16\ oz} \times \frac{453.59\ g}{1\ lb} \times \left(\frac{1\ in}{2.54\ cm}\right)^3 = 0.95\ g/cm^3$

1.110 Convert 8 min, 25 s to s. $8\ min \times \frac{60\ s}{1\ min} + 25\ s = 505\ s$

Convert 293.2 K to °F:

$293.2 - 273.15 = 20.05\ °C$ and $°F = (\frac{9}{5} \times 20.05) + 32 = 68.09\ °F$

$$\text{Final temperature} = 68.09\ °F + 505\ s \times \frac{3.0\ °F}{60\ s} = 93.34\ °F$$

$$°C = \frac{5}{9} \times (93.34 - 32) = 34.1\ °C$$

1.111 Ethyl alcohol density $= \frac{19.7325\ g}{25.00\ mL} = 0.7893\ g/mL$

total mass = metal mass + ethyl alcohol mass = 38.4704 g
ethyl alcohol mass = total mass – metal mass = 38.4704 g – 25.0920 g = 13.3784 g

$$\text{ethyl alcohol volume} = 13.3784 \text{ g} \times \frac{1 \text{ mL}}{0.7893 \text{ g}} = 16.95 \text{ mL}$$

metal volume = total volume – ethyl alcohol volume = 25.00 mL – 16.95 mL = 8.05 mL

$$\text{metal density} = \frac{25.0920 \text{ g}}{8.05 \text{ mL}} = 3.12 \text{ g/mL}$$

1.112 Average brass density = $(0.670)(8.92 \text{ g/cm}^3) + (0.330)(7.14 \text{ g/cm}^3) = 8.333 \text{ g/cm}^3$

$$\text{length} = 1.62 \text{ in.} \times \frac{2.54 \text{ cm}}{1 \text{ in.}} = 4.115 \text{ cm}$$

$$\text{diameter} = 0.514 \text{ in.} \times \frac{2.54 \text{ cm}}{1 \text{ in.}} = 1.306 \text{ cm}$$

$$\text{volume} = \pi r^2 h = (3.1416)[(1.306 \text{ cm})/2]^2(4.115 \text{ cm}) = 5.512 \text{ cm}^3$$

$$\text{mass} = 5.512 \text{ cm}^3 \times \frac{8.333 \text{ g}}{1 \text{ cm}^3} = 45.9 \text{ g}$$

1.113 $35 \text{ sv} = 35 \times 10^9 \ \frac{\text{m}^3}{\text{s}}$

(a) $$\text{gulf stream flow} = \left(35 \times 10^9 \ \frac{\text{m}^3}{\text{s}}\right)\left(\frac{1 \text{ cm}}{1 \times 10^{-2} \text{ m}}\right)^3\left(\frac{1 \text{ mL}}{1 \text{ cm}^3}\right)\left(\frac{60 \text{ s}}{1 \text{ min}}\right) = 2.1 \times 10^{18} \text{ mL/min}$$

(b) $$\text{mass of } H_2O = \left(2.1 \times 10^{18} \ \frac{\text{mL}}{\text{min}}\right)\left(\frac{60 \text{ min}}{1 \text{ h}}\right)(24 \text{ h})\left(\frac{1.025 \text{ g}}{1 \text{ mL}}\right) = 3.1 \times 10^{21} \text{ g} = 3.1 \times 10^{18} \text{ kg}$$

(c) $$\text{time} = \left(1.0 \times 10^{15} \text{ L}\right)\left(\frac{1 \text{ mL}}{1 \times 10^{-3} \text{ L}}\right)\left(\frac{1 \text{ min}}{2.1 \times 10^{18} \text{ mL}}\right) = 0.48 \text{ min}$$

1.114 (a) $$\text{Ga density} = \frac{0.2133 \text{ lb}}{1 \text{ in.}^3} \times \frac{453.59 \text{ g}}{1 \text{ lb}} \times \frac{1 \text{ in.}^3}{(2.54 \text{ cm})^3} = 5.904 \text{ g/cm}^3$$

(b)

Ga boiling point	2204 °C	1000 °G
Ga melting point	29.78 °C	0 °G

$$\frac{1000 \text{ °G} - 0 \text{ °G}}{2204 \text{ °C} - 29.78 \text{ °C}} = \frac{1000 \text{ °G}}{2174.22 \text{ °C}} = 0.4599 \text{ °G/°C}$$

°G = 0.4599 x (°C – 29.78)
°G = 0.4599 x (801 – 29.78) = 355 °G

The melting point of sodium chloride (NaCl) on the gallium scale is 355 °G.

1.115 1 knot = 1 nautical mile per hour
Historically, 1 knot = 47 ft 3 in. per 28 s

(a) $$1 \text{ knot} = \frac{\left(47 \text{ ft x } \frac{12 \text{ in.}}{1 \text{ ft}} + 3 \text{ in.}\right)}{28 \text{ s}} = 20.25 \text{ in/s}$$

Convert knots in in./s to knots in ft/hr.

$$1 \text{ knot} = 20.25 \text{ in./s x } \frac{1 \text{ ft}}{12 \text{ in.}} \text{ x } \frac{60 \text{ s}}{1 \text{ min}} \text{ x } \frac{60 \text{ min}}{1 \text{ hr}} = 6075 \text{ ft/hr}$$

Therefore, 1 nautical mile = 6075 ft

$$1 \text{ nautical mile in meters} = 6075 \text{ ft x } \frac{12 \text{ in.}}{1 \text{ ft}} \text{ x } \frac{2.54 \text{ cm}}{1 \text{ in.}} \text{ x } \frac{1 \text{ x } 10^{-2} \text{ m}}{1 \text{ cm}} = 1851.66 \text{ m} = 1852 \text{ m}$$

(b) $$\text{speed in knots} = 48 \text{ mi/hr x } \frac{5280 \text{ ft}}{1 \text{ mi}} \text{ x } \frac{1 \text{ nautical mi}}{6075 \text{ ft}} = 41.7 \text{ knots} = 42 \text{ knots}$$

(c) $$\text{depth in leagues} = 35{,}798 \text{ ft x } \frac{1 \text{ nautical mi}}{6075 \text{ ft}} \text{ x } \frac{1 \text{ league}}{3 \text{ nautical mi}} = 1.964 \text{ leagues}$$

(d) $$\left|\frac{1851.66 \text{ m} - 1852 \text{ m}}{1851.66 \text{ m}}\right| \text{ x } 100 = 0.0184\%$$

The current definition of the nautical mile is 0.0184% larger than the original definition.

Convert the current definition of the nautical mile in meters to feet.

$$1 \text{ nautical mile in feet} = 1852 \text{ m x } \frac{1 \text{ cm}}{1 \text{ x } 10^{-2} \text{ m}} \text{ x } \frac{1 \text{ in.}}{2.54 \text{ cm}} \text{ x } \frac{1 \text{ ft}}{12 \text{ in.}} = 6076.115 \text{ ft}$$

$$\left|\frac{5280 \text{ ft} - 6076.115 \text{ ft}}{5280 \text{ ft}}\right| \text{ x } 100 = 15.1\%$$

The current definition of the nautical mile is 15.1% larger than the statute mile.

2 Atoms, Molecules, and Ions

2.1 It is a metal, and most likely near the end of the transition metals because it can be found in nature in its pure form. A likely candidate is silver.

2.2 (a) K, potassium, metal (b) shiny, metallic solid
(c) It would deform and not crack since metals are malleable.
(d) It would conduct electricity, but it is not a good choice for wiring because it is reacts when exposed to oxygen and humidity (water) in the atmosphere. Potassium is also a soft metal which, makes it unsuitable for use as a wire.

2.3 First, find the S:O ratio in each compound.
Substance A: S:O mass ratio = (6.00 g S) / (5.99 g O) = 1.00
Substance B: S:O mass ratio = (8.60 g S) / (12.88 g O) = 0.668

$$\frac{\text{S:O mass ratio in substance A}}{\text{S:O mass ratio in substance B}} = \frac{1.00}{0.668} = 1.50 = \frac{3}{2}$$

2.4 In compound A the O/S mass ratio is 1. In compound B, the O/S mass ratio is 3/2. This means there is 3/2 times more O in compound B. To find the formula of Compound B multiply the subscript on O in Compound A by 3/2. Compound B is SO_3.

2.5 $$0.005 \text{ mm} \times \frac{1 \times 10^{-3} \text{ m}}{1 \text{ mm}} \times \frac{1 \text{ Au atom}}{2.9 \times 10^{-10} \text{ m}} = 2 \times 10^{4} \text{ Au atoms}$$

2.6 $$1 \times 10^{19} \text{ C atoms} \times \frac{1.5 \times 10^{-10} \text{ m}}{\text{C atom}} \times \frac{1 \text{ km}}{1000 \text{ m}} \times \frac{1 \text{ time}}{40{,}075 \text{ km}} = 37.4 \text{ times} \sim 40 \text{ times}$$

2.7 $^{75}_{34}Se$ has 34 protons, 34 electrons, and (75 – 34) = 41 neutrons.

2.8 The element with 24 protons is Cr. The mass number is the sum of the protons and the neutrons, 24 + 28 = 52. The isotope symbol is $^{52}_{24}Cr$.

2.9 atomic mass = (0.6915 x 62.93) + (0.3085 x 64.93) = 63.55
(63.546 from periodic table)

2.10 (a) The atomic weight of Ga is 69.7231. This mass is closer to the mass of gallium-69 than to gallium-71; therefore gallium-69 must be more abundant.
(b) The total abundance of both isotopes must be 100.00%. Let Y be the natural abundance of ^{69}Ga and [1 – Y] the natural abundance of ^{71}Ga.
(Y x 68.9256) + ([1 – Y] x 70.9247) = 69.7231

Solve for Y. $Y = \frac{-1.2016}{-1.9991} = 0.6011$

^{69}Ga natural abundance = 60.11% and ^{71}Ga natural abundance = 100.00 – 60.11 = 39.89%

2.11 Pt, 195.078

$$\text{mol Pt} = 9.50 \text{ g Pt} \times \frac{1 \text{ mol Pt}}{195.078 \text{ g Pt}} = 0.0487 \text{ mol Pt}$$

$$\text{atoms Pt} = 0.0487 \text{ mol Pt} \times \frac{6.022 \times 10^{23} \text{ atoms Pt}}{1 \text{ mol Pt}} = 2.93 \times 10^{22} \text{ atoms Pt}$$

2.12 $$\text{atomic mass in g} = \frac{1.50 \text{ g}}{2.26 \times 10^{22} \text{ atoms}} \times 6.022 \times 10^{23} \text{ atoms} = 40.0 \text{ g; Y = Ca}$$

2.13 Figure (b) represents a collection of hydrogen peroxide (H_2O_2) molecules.

2.14 (a) Figures (b) and (d) illustrate pure substances.
(b) Figures (a) and (c) illustrate mixtures.
(c) Figures (b) and (d) illustrate the law of multiple proportions.

2.15

```
    H   H
    |   |
H—C—N—H
    |
    H
```

2.16 adrenaline, $C_9H_{13}NO_3$

2.17 (a) LiBr is composed of a metal (Li) and nonmetal (Br) and is ionic.
(b) $SiCl_4$ is composed of only nonmetals and is molecular.
(c) BF_3 is composed of only nonmetals and is molecular.
(d) CaO is composed of a metal (Ca) and nonmetal (O) and is ionic.

2.18 Figure (a) most likely represents an ionic compound because there are no discrete molecules, only a regular array of two different chemical species (ions). Figure (b) most likely represents a molecular compound because discrete molecules are present.

2.19 (a) magnesium fluoride, MgF_2
(b) tin(IV) oxide, SnO_2
(c) iron(III) sulfide, Fe_2S_3

2.20 red – potassium sulfide, K_2S; green – strontium iodide, SrI_2; blue – gallium oxide, Ga_2O_3

2.21 (a) potassium hypochlorite, KClO
(b) silver(I) chromate, Ag_2CrO_4
(c) iron(III) carbonate, $Fe_2(CO_3)_3$

2.22 Drawing 1 represents ionic compounds with one cation and two anions. Only (c) $CaCl_2$ is consistent with drawing 1.
Drawing 2 represents ionic compounds with one cation and one anion. Both (a) LiBr and (b) $NaNO_2$ are consistent with drawing 2.

2.23 (a) disulfur dichloride, S_2Cl_2
(b) iodine monochloride, ICl
(c) nitrogen triiodide, NI_3

2.24 (a) PCl_5, phosphorus pentachloride (b) N_2O, dinitrogen monoxide

2.25 The green chemistry principle of atom economy accounts for the mass of every atom in a chemical reaction.

2.26 $C_3H_8O \rightarrow C_3H_6 + H_2O$

Reactants:
3 C = (3)(12.0) = 36.0
8 H = (8)(1.0) = 8.0
1 O = (1)(16.0) = 16.0
Σ = 60.0

Desired Product:
3 C = (3)(12.0) = 36.0
6 H = (6)(1.0) = 6.0
Σ = 42.0

$$\% \text{ Atom Economy} = \frac{\Sigma \text{ Atomic Weight}_{(\text{atoms in desired product})}}{\Sigma \text{ Atomic Weight}_{(\text{atoms in all reactants})}} = \frac{42.0}{60.0} \times 100 = 70\%$$

2.27 (a) Reaction 1 has the higher % atom economy because all reactant atoms are used.
(b) rxn 1: $C_2H_4 + Cl_2 \rightarrow C_2H_4Cl_2$
There are no undesired products, therefore the % Atom Economy = 100%.

rxn 2: $CH_3Cl + Br^- \rightarrow CH_3Br + Cl^-$

Reactants:
1 C = (1)(12.0) = 12.0
3 H = (3)(1.0) = 3.0
1 Cl = (1)(35.5) = 35.5
1 Br = (1)(80.0) = 80.0
Σ = 130.5

Desired Product:
1 C = (1)(12.0) = 12.0
3 H = (3)(1.0) = 3.0
1 Br = (1)(80.0) = 80.0
Σ = 93.0

$$\% \text{ Atom Economy} = \frac{\Sigma \text{ Atomic Weight}_{(\text{atoms in desired product})}}{\Sigma \text{ Atomic Weight}_{(\text{atoms in all reactants})}} = \frac{93.0}{130.5} \times 100 = 71\%$$

2.28 Ibuprofen, $C_{13}H_{18}O_2$

2.29 (a) $1 \text{ mol Na} \times \dfrac{23.0 \text{ g Na}}{1 \text{ mol Na}} = 23.0 \text{ g Na}$

$23 \text{ mol H x } \frac{1.0 \text{ g H}}{1 \text{ mol H}} = 23.0 \text{ g H}$

$1 \text{ mol N x } \frac{14.0 \text{ g N}}{1 \text{ mol N}} = 14.0 \text{ g N}$

$7 \text{ mol C x } \frac{12.0 \text{ g C}}{1 \text{ mol C}} = 84.0 \text{ g C}$

$8 \text{ mol O x } \frac{16.0 \text{ g O}}{1 \text{ mol O}} = 128.0 \text{ g O}$

$1 \text{ mol Cl x } \frac{35.5 \text{ g Cl}}{1 \text{ mol Cl}} = 35.5 \text{ g Cl}$

(b) 23.0 g + 23.0 g +14.0 g + 84.0 g + 128.0 g + 35.5 g = 307.5 g

(c) $6.6 \times 10^7 \text{ mol Ibuprofen x } \frac{307.5 \text{ g waste}}{1 \text{ mol Ibuprofen}} \text{ x } \frac{1 \text{ kg}}{1000 \text{ g}} = 2.0 \times 10^7 \text{ kg waste}$

2.30 (a) $4 \text{ mol H x } \frac{1.0 \text{ g H}}{1 \text{ mol H}} = 4.0 \text{ g H}$

$2 \text{ mol C x } \frac{12.0 \text{ g C}}{1 \text{ mol C}} = 24.0 \text{ g C}$

$2 \text{ mol O x } \frac{16.0 \text{ g O}}{1 \text{ mol O}} = 32.0 \text{ g O}$

(b) 4.0 g + 24.0 g + 32.0 g = 60.0 g

(c) $6.6 \times 10^7 \text{ mol Ibuprofen x } \frac{60.0 \text{ g waste}}{1 \text{ mol Ibuprofen}} \text{ x } \frac{1 \text{ kg}}{1000 \text{ g}} = 4.0 \times 10^6 \text{ kg waste}$

(d) savings = 2.0×10^7 kg – 4.0×10^6 kg = 1.6×10^7 kg

Conceptual Problems

2.31

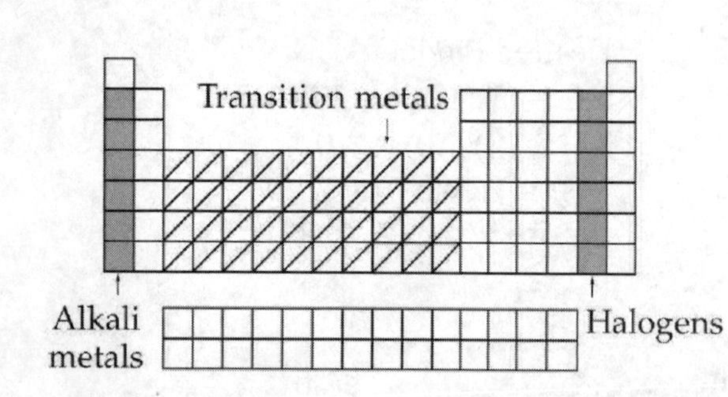

Alkaline-earth metals

H

He

2.32

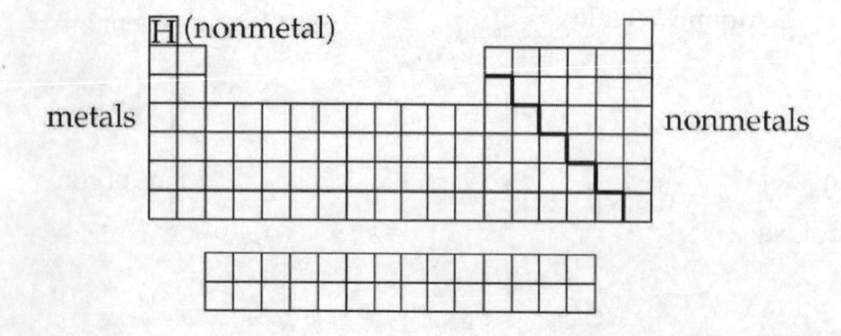

2.33 red – gas; blue – 42;
green – lithium, sodium, potassium or rubidium are possible answers.

2.34 The element is americium (Am) with atomic number = 95. It is in the actinide series.

2.35 The three "coinage metals" are copper (Cu), silver (Ag), and gold (Au).

2.36 Drawing (a) represents a collection of SO_2 units. Drawing (d) represents a mixture of S atoms and O_2 units.

2.37 To obey the law of mass conservation, the correct drawing must have the same number of red and yellow spheres as in drawing (a). The correct drawing is (d).

2.38 Figures (b) and (c) both contain two protons but different numbers of neutrons. They are isotopes of the same element. Figure (a) contains only one proton. It is a different element than (b) and (c).

2.39 methionine, $C_5H_{11}NO_2S$

2.40 thymine, $C_5H_6N_2O_2$

2.41 (a) alanine, $C_3H_7NO_2$ (b) ethylene glycol, $C_2H_6O_2$ (c) acetic acid, $C_2H_4O_2$

2.42 A Na atom has 11 protons and 11 electrons [drawing (b)].
A Ca^{2+} ion has 20 protons and 18 electrons [drawing (c)].
A F^- ion has 9 protons and 10 electrons [drawing (a)].

2.43 (a) $MgSO_4$ (b) Li_2CO_3 (c) $FeCl_2$ (d) $Ca_3(PO_4)_2$

Section Problems
Elements and the Periodic Table (Sections 2.1–2.3)

2.44 118 elements are presently known. About 90 elements occur naturally.

2.45 C is the second most abundant element in the human body. Si is the second most abundant element in the Earth's crust. C and Si have low reactivity but both do form compounds with other elements.

2.46 (a) gadolinium, Gd (b) germanium, Ge (c) technetium, Tc (d) arsenic, As

2.47 (a) cadmium, Cd (b) iridium, Ir (c) beryllium, Be (d) tungsten, W

2.48 (a) Te, tellurium (b) Re, rhenium (c) Be, beryllium (d) Ar, argon
(e) Pu, plutonium

2.49 (a) B, boron (b) Rh, rhodium (c) Cf, californium (d) Os, osmium
(e) Ga, gallium

2.50 (a) Tin is Sn. Ti is titanium. (b) Manganese is Mn. Mg is magnesium.
(c) Potassium is K. Po is polonium.
(d) The symbol for helium is He. The second letter is lowercase.

2.51 (a) The symbol for carbon is C. (b) The symbol for sodium is Na.
(c) The symbol for nitrogen is N. (d) The symbol for chlorine is Cl.

2.52 The rows are called periods, and the columns are called groups.

2.53 There are 18 groups in the periodic table. They are labeled as follows:
1A, 2A, 3B, 4B, 5B, 6B, 7B, 8B (3 groups), 1B, 2B, 3A, 4A, 5A, 6A, 7A, 8A

2.54 Elements within a group have similar chemical properties.

2.55

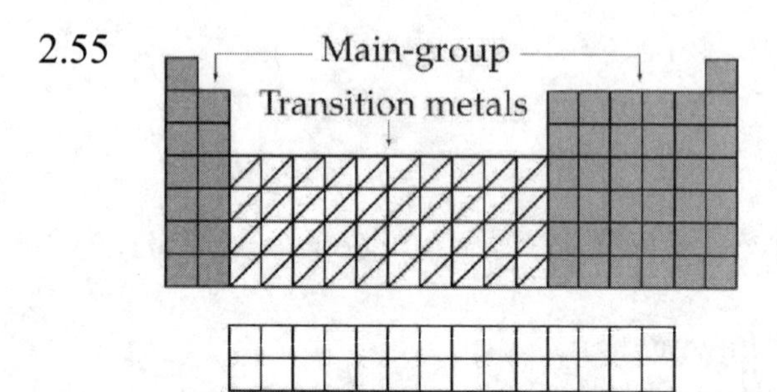

2.56

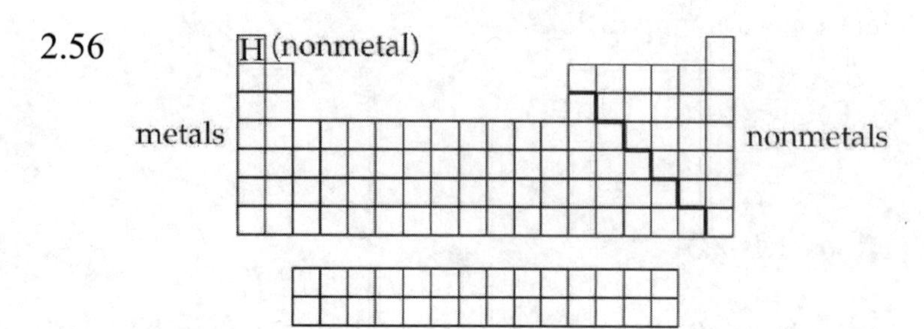

2.57

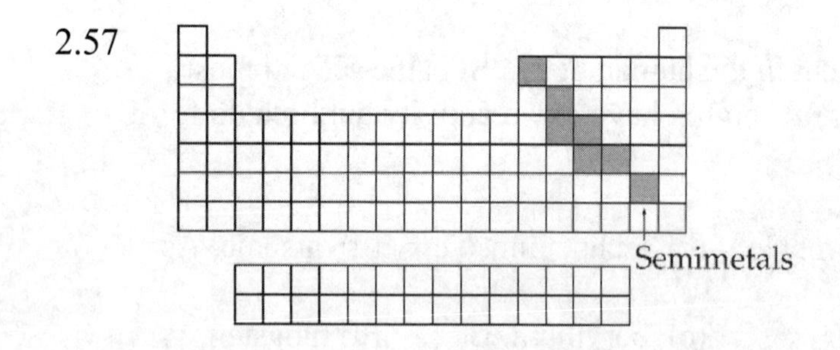

A semimetal is an element with properties that fall between those of metals and nonmetals.

2.58 (a) Ti, metal (b) Te, semimetal (c) Se, nonmetal
(d) Sc, metal (e) Si, semimetal

2.59 (a) Ar, nonmetal (b) Sb, semimetal (c) Mo, metal
(d) Cl, nonmetal (e) N, nonmetal (e) Mg, metal

2.60 (a) The alkali metals are shiny, soft, low-melting metals that react rapidly with water to form products that are alkaline.
(b) The noble gases are gases of very low reactivity.
(c) The halogens are nonmetallic and corrosive. They are found in nature only in combination with other elements.

2.61 (a) Li, Na, K, Rb, and Cs (b) Be, Mg, Ca, Sr, and Ba

2.62 F, Cl, Br, and I

2.63 He, Ne, Ar, Kr, Xe, and Rn

2.64 An element that is a soft, silver-colored solid that reacts violently with water and is a good conductor of electricity has the characteristics of a metal.

2.65 An element that is a brittle, shiny, silver-colored solid that is a poor conductor of electricity has the characteristics of a semimetal.

2.66 An element that is a yellow crystalline solid, does not conduct electricity, and when hit with a hammer it shatters has the characteristics of a nonmetal.

2.67 An element that is a colorless, unreactive gas has the characteristics of a nonmetal.

2.68 All match in groups 2A and 7A.

2.69 In group 1A, sodium (Na) and potassium (K), and in group 5A, antimony (Sb).

Atomic Theory (Sections 2.4 and 2.5)

2.70 The law of mass conservation in terms of Dalton's atomic theory states that chemical reactions only rearrange the way that atoms are combined; the atoms themselves are not changed.
The law of definite proportions in terms of Dalton's atomic theory states that the chemical combination of elements to make different substances occurs when atoms join together in small, whole-number ratios.

2.71 The law of multiple proportions states that if two elements combine in different ways to form different substances, the mass ratios are small, whole-number multiples of each other. This is very similar to Dalton's statement that the chemical combination of elements to make different substances occurs when atoms join together in small, whole-number ratios.

2.72 In any chemical reaction, the combined mass of the final products equals the combined mass of the starting reactants.
mass of reactants = mass of products
mass of reactants = mass of Hg + mass of O_2 = 114.0 g + 12.8 g = 126.8 g
mass of products = mass of HgO + mass of left over O_2 = 123.1 g + mass of left over O_2
126.8 g = 123.1 g + mass of left over O_2
mass of left over O_2 = 126.8 g – 123.1 g = 3.7 g

2.73 In any chemical reaction, the combined mass of the final products equals the combined mass of the starting reactants.
mass of reactants = mass of products
mass of reactants = mass of $CaCO_3$ = 612 g
mass of products = mass of CaO + mass of CO_2 = 343 g + mass of CO_2
612 g = 343 g + mass of CO_2
mass of CO_2 = 612 g – 343 g = 269 g

2.74 For the "other" compound: C:H mass ratio = (32.0 g C) / (8.0 g H) = 4
The "other" compound is not methane because the methane C:H mass ratio is 3.

$$\frac{\text{C:H mass ratio in "other"}}{\text{C:H mass ratio in methane}} = \frac{4}{3}$$

2.75 For the "other" compound: B:H mass ratio = (43.2 g B) / (6.0 g H) = 7.2
The "other" compound is not borane because the borane B:H mass ratio is 3.6.

$$\frac{\text{B:H mass ratio in "other"}}{\text{B:H mass ratio in borane}} = \frac{7.2}{3.6} = \frac{2}{1}$$

2.76 First, find the C:H ratio in each compound.
Benzene: C:H mass ratio = (4.61 g C) / (0.39 g H) = 12
Ethane: C:H mass ratio (4.00 g C) / (1.00 g H) = 4.00
Ethylene: C:H mass ratio = (4.29 g C) / (0.71 g H) = 6.0

$$\frac{\text{C:H mass ratio in benzene}}{\text{C:H mass ratio in ethane}} = \frac{12}{4.00} = \frac{3}{1}$$

$$\frac{\text{C:H mass ratio in benzene}}{\text{C:H mass ratio in ethylene}} = \frac{12}{6.0} = \frac{2}{1}$$

$$\frac{\text{C:H mass ratio in ethylene}}{\text{C:H mass ratio in ethane}} = \frac{6.0}{4.00} = \frac{3}{2}$$

2.77 (a) For benzene:

$$4.61 \text{ g x } \frac{1 \text{ u}}{1.6605 \text{ x } 10^{-24} \text{ g}} \text{ x } \frac{1 \text{ C atom}}{12.011} = 2.31 \text{ x } 10^{23} \text{ C atoms}$$

$$0.39 \text{ g x } \frac{1 \text{ u}}{1.6605 \text{ x } 10^{-24} \text{ g}} \text{ x } \frac{1 \text{ H atom}}{1.008} = 2.3 \text{ x } 10^{23} \text{ H atoms}$$

$$\frac{C}{H} = \frac{2.31 \times 10^{23} \text{ C atoms}}{2.3 \times 10^{23} \text{ H atoms}} = \frac{1 \text{ C}}{1 \text{ H}}$$ A possible formula for benzene is CH.

For ethane:

$$4.00 \text{ g} \times \frac{1 \text{ u}}{1.6605 \times 10^{-24} \text{ g}} \times \frac{1 \text{ C atom}}{12.011} = 2.01 \times 10^{23} \text{ C atoms}$$

$$1.00 \text{ g} \times \frac{1 \text{ u}}{1.6605 \times 10^{-24} \text{ g}} \times \frac{1 \text{ H atom}}{1.008} = 5.97 \times 10^{23} \text{ H atoms}$$

$$\frac{C}{H} = \frac{2.01 \times 10^{23} \text{ C atoms}}{5.97 \times 10^{23} \text{ H atoms}} = \frac{1 \text{ C}}{3 \text{ H}}$$ A possible formula for ethane is CH_3.

For ethylene:

$$4.29 \text{ g} \times \frac{1 \text{ u}}{1.6605 \times 10^{-24} \text{ g}} \times \frac{1 \text{ C atom}}{12.011} = 2.15 \times 10^{23} \text{ C atoms}$$

$$0.71 \text{ g} \times \frac{1 \text{ u}}{1.6605 \times 10^{-24} \text{ g}} \times \frac{1 \text{ H atom}}{1.008} = 4.2 \times 10^{23} \text{ H atoms}$$

$$\frac{C}{H} = \frac{2.15 \times 10^{23} \text{ C atoms}}{4.2 \times 10^{23} \text{ H atoms}} = \frac{1 \text{ C}}{2 \text{ H}}$$ A possible formula for ethylene is CH_2.

(b) The results in part (a) give the smallest whole-number ratio of C to H for benzene ($CH \times 6 = C_6H_6$), ethane ($CH_3 \times 2 = C_2H_6$), and ethylene ($CH_2 \times 2 = C_2H_4$), and these ratios are consistent with their modern formulas.

2.78 Assume a 100.0 g sample for each compound and then find the O:C ratio in each compound.
Compound 1: O:C mass ratio = (57.1 g O)/(42.9 g C) = 1.33
Compound 2: O:C mass ratio = (72.7 g O)/(27.3 g C) = 2.66

$$\frac{\text{O:C mass ratio in compound 2}}{\text{O:C mass ratio in compound 1}} = \frac{2.66}{1.33} = \frac{2}{1}$$

If compound 1 is CO, and the O:C mass ratio is 2 times that of compound 1, then the formula for compound 2 is CO_2.

2.79 First, find the O:C ratio in each compound.
Carbon suboxide: O:C mass ratio = (1.18 g O)/(1.32 g C) = 0.894
Carbon monoxide: O:C mass ratio = (16.00 g O)/(12.00 g C) = 1.33

$$\frac{\text{O:C mass ratio for carbon suboxide}}{\text{O:C mass ratio for CO}} = \frac{0.894}{1.33} = \frac{0.67}{1}$$

The O:C mass ratio for carbon suboxide is 2/3 that for CO, then the formula for carbon suboxide is $CO_{2/3}$. To get integer subscripts, multiply both subscripts by 3. The formula for carbon suboxide is C_3O_2.

2.80 Assume a 1.00 g sample of the binary compound of zinc and sulfur.
0.671 x 1.00 g = 0.671 g Zn; 0.329 x 1.00 g = 0.329 g S

$$0.671 \text{ g} \times \frac{1 \text{ u}}{1.6605 \times 10^{-24} \text{ g}} \times \frac{1 \text{ Zn atom}}{65.39} = 6.18 \times 10^{21} \text{ Zn atoms}$$

$$0.329 \text{ g x } \frac{1}{1.6605 \times 10^{-24} \text{ g}} \text{ x } \frac{1 \text{ S atom}}{32.066} = 6.18 \times 10^{21} \text{ S atoms}$$

$$\frac{\text{Zn}}{\text{S}} = \frac{6.18 \times 10^{21} \text{ Zn atoms}}{6.18 \times 10^{21} \text{ S atoms}} = \frac{1}{1}$$

2.81 Assume a 1.000 g sample of one of the binary compounds.
0.3104 x 1.000 g = 0.3104 g Ti; 0.6896 x 1.000 g = 0.6896 g Cl

$$0.3104 \text{ g x } \frac{1 \text{ u}}{1.6605 \times 10^{-24} \text{ g}} \text{ x } \frac{1 \text{ Ti atom}}{47.88} = 3.90 \times 10^{21} \text{ Ti atoms}$$

$$0.6896 \text{ g x } \frac{1 \text{ u}}{1.6605 \times 10^{-24} \text{ g}} \text{ x } \frac{1 \text{ Cl atom}}{35.453} = 1.17 \times 10^{22} \text{ Cl atoms}$$

$$\frac{\text{Cl}}{\text{Ti}} = \frac{1.17 \times 10^{22}}{3.90 \times 10^{21}} = \frac{3}{1}$$

Assume a 1.000 g sample of the other binary compound.
0.2524 x 1.000 g = 0.2524 g Ti; 0.7476 x 1.000 g = 0.7476 g Cl

$$0.2524 \text{ g x } \frac{1 \text{ u}}{1.6605 \times 10^{-24} \text{ g}} \text{ x } \frac{1 \text{ Ti atom}}{47.88} = 3.17 \times 10^{21} \text{ Ti atoms}$$

$$0.7476 \text{ g x } \frac{1 \text{ u}}{1.6605 \times 10^{-24} \text{ g}} \text{ x } \frac{1 \text{ Cl atom}}{35.453} = 1.27 \times 10^{22} \text{ Cl atoms}$$

$$\frac{\text{Cl}}{\text{Ti}} = \frac{1.27 \times 10^{22}}{3.17 \times 10^{21}} = \frac{4}{1}$$

Elements and Atoms (Sections 2.6 –2.8)

2.82 electron

2.83 The strength of the magnetic or electric field affects the magnitude of the deflection of the cathode ray in Thomson's experiment.

2.84 (d) The mass to charge ratio of the electron.

2.85 (c) The charge of the electron.

2.86 (a) -1.010×10^{-18} C because it is not an integer multiple of the electron charge.

2.87 The results of Rutherford's gold foil experiment showed that atoms contained a very small but massive positively charged nucleus.

2.88 (a) The alpha particles would pass right through the gold foil with little to no deflection.

2.89 Assume the nucleus has a diameter of 1 mm.
1 mm = 1×10^{-3} m

$$1 \times 10^{-3}\ \text{m (nucleus diameter)} \times \frac{1 \times 10^{-10}\ \text{m (atom diameter)}}{1 \times 10^{-15}\ \text{m (nucleus diameter)}} = 100\ \text{m (atom diameter)}$$

2.90 350 pm = 350×10^{-12} m

$$\text{Pb atoms} = 0.25\ \text{in} \times \frac{2.54\ \text{cm}}{1\ \text{in}} \times \frac{1 \times 10^{-2}\ \text{m}}{1\ \text{cm}} \times \frac{1\ \text{Pb atom}}{350 \times 10^{-12}\ \text{m}} = 1.8 \times 10^{7}\ \text{Pb atoms thick}$$

2.91 1 mm = 1×10^{-3} m; 150 pm = 150×10^{-12} m

$$\text{C atoms} = 1 \times 10^{-3}\ \text{m} \times \frac{1\ \text{C atom}}{150 \times 10^{-12}\ \text{m}} = 7 \times 10^{6}\ \text{C atoms}$$

2.92 The atomic number is equal to the number of protons.
The mass number is equal to the sum of the number of protons and the number of neutrons.

2.93 The atomic number is equal to the number of protons.
The atomic mass is the weighted average mass of the various isotopes for a particular element.

2.94 The subscript giving the atomic number of an atom is often left off of an isotope symbol because one can readily look up the atomic number in the periodic table.

2.95 Te has isotopes with more neutrons than the isotopes of I.

2.96 (a) carbon, C (b) argon, Ar (c) vanadium, V

2.97 $^{137}_{55}\text{Cs}$

2.98 (a) $^{220}_{86}\text{Rn}$ (b) $^{210}_{84}\text{Po}$ (c) $^{197}_{79}\text{Au}$

2.99 (a) $^{140}_{58}\text{Ce}$ (b) $^{60}_{27}\text{Co}$

2.100 (a) $^{15}_{7}\text{N}$, 7 protons, 7 electrons, (15 – 7) = 8 neutrons
(b) $^{60}_{27}\text{Co}$, 27 protons, 27 electrons, (60 – 27) = 33 neutrons
(c) $^{131}_{53}\text{I}$, 53 protons, 53 electrons, (131 – 53) = 78 neutrons
(d) $^{142}_{58}\text{Ce}$, 58 protons, 58 electrons, (142 – 58) = 84 neutrons

2.101 (a) ^{27}Al, 13 protons and (27 – 13) = 14 neutrons
(b) ^{32}S, 16 protons and (32 – 16) = 16 neutrons
(c) ^{64}Zn, 30 protons and (64 – 30) = 34 neutrons
(d) ^{207}Pb, 82 protons and (207 – 82) = 125 neutrons

2.102 (a) $^{24}_{12}Mg$, magnesium (b) $^{58}_{28}Ni$, nickel

(c) $^{104}_{46}Pd$, palladium (d) $^{183}_{74}W$, tungsten

2.103 (a) $^{202}_{80}Hg$, mercury (b) $^{195}_{78}Pt$, platinum

(c) $^{184}_{76}Os$, osmium (d) $^{209}_{83}Bi$, bismuth

2.104 $^{12}_{5}C$, the atomic number for carbon is 6, not 5.

$^{33}_{35}Br$, the mass number must be greater than the atomic number.

$^{11}_{5}Bo$, the element symbol for boron is B.

2.105 $^{14}_{7}Ni$, the element symbol for nitrogen is N.

$^{73}_{23}Ge$, the atomic number for germanium is 32, not 23.

$^{1}_{2}He$, the mass number must be greater than the atomic number.

Atomic Weight and Moles (Section 2.9)

2.106 An element's atomic mass is the weighted average of the isotopic masses of the element's naturally occurring isotopes. The atomic mass for Cu (63.546) must fall between the masses of its two isotopes. If one isotope is ^{65}Cu, the other isotope must be ^{63}Cu, and not ^{66}Cu. If the other isotope was ^{66}Cu, the atomic mass for Cu would be greater than 65.

2.107 An element's atomic mass is the weighted average of the isotopic masses of the element's naturally occurring isotopes. The atomic mass for S (32.065) must fall between the masses of its lightest and heaviest isotopes. If three of its isotopes are ^{33}S, ^{34}S, and ^{36}S, the other isotope must be ^{32}S, and not ^{35}S. If the other isotope was ^{35}S, the atomic mass for S would be greater than 33.

2.108 (0.199 x 10.0129) + (0.801 x 11.009 31) = 10.8 for B

2.109 (0.5184 x 106.9051) + (0.4816 x 108.9048) = 107.9 for Ag

2.110 24.305 = (0.7899 x 23.985) + (0.1000 x 24.986) + (0.1101 x Z)
Solve for Z. Z = 25.982 for ^{26}Mg.

2.111 The total abundance of all three isotopes must be 100.00%. The natural abundance of ^{29}Si is 4.68%. The natural abundance of ^{28}Si and ^{30}Si together must be 100.00% – 4.68% = 95.32%. Let Y be the natural abundance of ^{28}Si and [95.32 – Y] the natural abundance of ^{30}Si.
28.0855 = (0.0468 x 28.9765) + (Y x 27.9769) + ([0.9532 – Y] x 29.9738)

Solve for Y. $Y = \frac{-1.842}{-1.997} = 0.9224$

^{28}Si natural abundance = 92.2% ^{30}Si natural abundance = 95.32 – 92.24 = 3.1%

2.112 atomic mass = (0.6915 x 62.93) + (0.3085 x 64.93) = 63.55

2.113 $2.15 \text{ g Cu} \times \frac{1 \text{ mol Cu}}{63.55 \text{ g Cu}} \times \frac{6.022 \times 10^{23} \text{ Cu atoms}}{1 \text{ mol Cu atoms}} = 2.04 \times 10^{22} \text{ Cu atoms}$

2.114 (a) $\text{g Ti} = 1.505 \text{ mol Ti} \times \frac{47.867 \text{ g Ti}}{1 \text{ mol Ti}} = 72.04 \text{ g Ti}$

(b) $\text{g Na} = 0.337 \text{ mol Na} \times \frac{22.989\ 770 \text{ g Na}}{1 \text{ mol Na}} = 7.75 \text{ g Na}$

(c) $\text{g U} = 2.583 \text{ mol U} \times \frac{238.028\ 91 \text{ g U}}{1 \text{ mol U}} = 614.8 \text{ g U}$

2.115 (a) $\text{mol Ti} = 11.51 \text{ g Ti} \times \frac{1 \text{ mol Ti}}{47.867 \text{ g Ti}} = 0.2405 \text{ mol Ti}$

(b) $\text{mol Na} = 29.127 \text{ g Na} \times \frac{1 \text{ mol Na}}{22.989\ 770 \text{ g Na}} = 1.2670 \text{ mol Na}$

(c) $\text{mol U} = 1.477 \text{ kg} \times \frac{1000 \text{ g}}{1 \text{ kg}} \times \frac{1 \text{ mol U}}{238.028\ 91 \text{ g U}} = 6.205 \text{ mol U}$

2.116 The mass of 6.02×10^{23} atoms is its atomic mass expressed in grams. If the atomic mass of an element is X, then 6.02×10^{23} atoms of this element weighs X grams.

2.117 $\text{mass} = \frac{x \text{ g}}{6.02 \times 10^{23} \text{ atoms}} \times 3.17 \times 10^{20} \text{ atoms} = (x) \times 5.27 \times 10^{-4} \text{ g}$

2.118 The mass of 6.02×10^{23} atoms is its atomic mass expressed in grams. If the mass of 6.02×10^{23} atoms of element Y is 83.80 g, then the atomic mass of Y is 83.80. Y is Kr.

2.119 $\text{atomic mass in g} = \frac{0.815 \text{ g}}{4.61 \times 10^{21} \text{ atoms}} \times 6.02 \times 10^{23} \text{ atoms} = 106 \text{ g};\ Z = \text{Pd}$

Chemical Compounds (Sections 2.10 and 2.11)

2.120 A covalent bond results when two atoms share several (usually two) of their electrons. An ionic bond results from a complete transfer of one or more electrons from one atom to another. The C–H bonds in methane (CH_4) are covalent bonds. The bond in NaCl (Na^+Cl^-) is an ionic bond.

2.121 Covalent bonds typically form between nonmetals. (a) B–Br, (c) Br–Cl, and (d) O–Br are covalent bonds.
Ionic bonds typically form between a metal and a nonmetal. (b) Na–Br is an ionic bond.

2.122 Element symbols are composed of one or two letters. If the element symbol is two letters, the first letter is uppercase and the second is lowercase. CO stands for carbon and oxygen in carbon monoxide.

2.123 (a) The formula of ammonia is NH_3.
(b) The ionic solid potassium chloride has the formula KCl.
(c) Cl^- is an anion.
(d) CH_4 is a neutral molecule.

2.124 (a) Be^{2+}, 4 protons and 2 electrons
(b) Rb^+, 37 protons and 36 electrons
(c) Se^{2-}, 34 protons and 36 electrons
(d) Au^{3+}, 79 protons and 76 electrons

2.125 (a) A +2 cation that has 36 electrons must have 38 protons. X = Sr.
(b) A –1 anion that has 36 electrons must have 35 protons. X = Br.

2.126 C_3H_8O

2.127 $C_3H_6O_3$

2.128

```
    H   H   H   H
    |   |   |   |
H—C—C—C—C—H
    |   |   |   |
    H   H   H   H
```

2.129

```
        H  H
     H  |  |  H
      \ C—C /
  H    /    \    H
   \ C        C /
  H /  \    /  \ H
        C—C
     H /  |  | \ H
        H  H
```

2.130

```
              H              H
              |              |
          H—C—H       H—C—H
    H         |      H       |       H
    |         |      |       |       |
H—C———C———C———C———C—H
    |         |      |       |       |
    H         |      H       H       H
          H—C—H
              |
              H
```


2.131

```
          H          H
         /            \
  H—O—C             C—O—H
      / \     O     / \
     H   C  /   \  C    H
    H'    \       /  \O—H
      H—C———C—H
          /      \
         O        O—H
        /
       H
```

Naming Compounds (Section 2.12)

2.132 (a) CsF, cesium fluoride (b) K_2O, potassium oxide (c) CuO, copper(II) oxide

2.133 (a) BaS, barium sulfide (b) $BeBr_2$, beryllium bromide (c) $FeCl_3$, iron(III) chloride

2.134 (a) potassium chloride, KCl (b) tin(II) bromide, $SnBr_2$ (c) calcium oxide, CaO
(d) barium chloride, $BaCl_2$ (e) aluminum hydride, AlH_3

2.135 (a) vanadium(III) chloride, VCl_3 (b) manganese(IV) oxide, MnO_2
(c) copper(II) sulfide, CuS (d) aluminum oxide, Al_2O_3

2.136 (a) calcium acetate, $Ca(CH_3CO_2)_2$ (b) iron(II) cyanide, $Fe(CN)_2$
(c) sodium dichromate, $Na_2Cr_2O_7$ (d) chromium(III) sulfate, $Cr_2(SO_4)_3$
(e) mercury(II) perchlorate, $Hg(ClO_4)_2$

2.137 (a) lithium phosphate, Li_3PO_4 (b) magnesium hydrogen sulfate, $Mg(HSO_4)_2$
(c) manganese(II) nitrate, $Mn(NO_3)_2$ (d) chromium(III) sulfate, $Cr_2(SO_4)_3$

2.138 (a) $Ca(ClO)_2$, calcium hypochlorite
(b) $Ag_2S_2O_3$, silver(I) thiosulfate or silver thiosulfate
(c) NaH_2PO_4, sodium dihydrogen phosphate (d) $Sn(NO_3)_2$, tin(II) nitrate
(e) $Pb(CH_3CO_2)_4$, lead(IV) acetate (f) $(NH_4)_2SO_4$, ammonium sulfate

2.139 (a) Ba^{2+}, barium ion (b) Cs^+, cesium ion (c) V^{3+}, vanadium(III) ion
(d) HCO_3^-, hydrogen carbonate ion (e) NH_4^+, ammonium ion (f) Ni^{2+}, nickel(II) ion
(g) NO_2^-, nitrite ion (h) ClO_2^-, chlorite ion
(i) Mn^{2+}, manganese(II) ion (j) ClO_4^-, perchlorate ion

2.140 (a) $CaBr_2$ (b) $CaSO_4$ (c) $Al_2(SO_4)_3$

2.141 (a) $NaNO_3$ (b) K_2SO_4 (c) $SrCl_2$

2.142 (a) $CaCl_2$ (b) CaO (c) CaS

2.143 (a) RbBr (b) Rb_3N (c) Rb_2Se

2.144 (a) sulfite ion, SO_3^{2-} (b) phosphate ion, PO_4^{3-} (c) zirconium(IV) ion, Zr^{4+}
(d) chromate ion, CrO_4^{2-} (e) acetate ion, $CH_3CO_2^-$ (f) thiosulfate ion, $S_2O_3^{2-}$

2.145 (a) Zn^{2+} (b) Fe^{3+} (c) Ti^{4+} (d) Sn^{2+} (e) Hg_2^{2+} (f) Mn^{4+} (g) K^+ (h) Cu^{2+}

2.146 (a) CCl_4, carbon tetrachloride (b) ClO_2, chlorine dioxide
(c) N_2O, dinitrogen monoxide (d) N_2O_3, dinitrogen trioxide

2.147 (a) NCl_3, nitrogen trichloride (b) P_4O_6, tetraphosphorus hexoxide
(c) S_2F_2, disulfur difluoride

2.148 (a) NO, nitrogen monoxide (b) N_2O, dinitrogen monoxide (c) NO_2, nitrogen dioxide
(d) N_2O_4, dinitrogen tetroxide (e) N_2O_5, dinitrogen pentoxide

2.149 (a) SO, sulfur monoxide (b) SO_2, disulfur dioxide (c) S_5O, pentasulfur monoxide
(d) S_7O_2, heptasulfur dioxide (e) SO_3, sulfur trioxide

2.150 (a) Na^+ and SO_4^{2-}; therefore the formula is Na_2SO_4
(b) Ba^{2+} and PO_4^{3-}; therefore the formula is $Ba_3(PO_4)_2$
(c) Ga^{3+} and SO_4^{2-}; therefore the formula is $Ga_2(SO_4)_3$

2.151 (a) sodium peroxide, Na_2O_2 (b) aluminum bromide, $AlBr_3$
(c) chromium(III) sulfate, $Cr_2(SO_4)_3$

Chapter Problems

2.152 atomic mass = (0.205 x 69.924) + (0.274 x 71.922)
+ (0.078 x 72.923) + (0.365 x 73.921) + (0.078 x 75.921) = 72.6

2.153 Deuterium is ^2H and deuterium fluoride is ^2HF.
^2H has 1 proton, 1 neutron, and 1 electron.
F has 9 protons, 10 neutrons, and 9 electrons.
^2HF has 10 protons, 11 neutrons, and 10 electrons.
Chemically, ^2HF is like HF and is a weak acid.

2.154 For NH_3, $(2.34\text{ g N})\left(\dfrac{3 \times 1.0079\text{ H}}{14.0067\text{ N}}\right) = 0.505\text{ g H}$

For N_2H_4, $(2.34\text{ g N})\left(\dfrac{4 \times 1.0079\text{ H}}{2 \times 14.0067\text{ N}}\right) = 0.337\text{ g H}$

2.155 $\text{g N} = (1.575\text{ g H})\left(\dfrac{3.670\text{ g N}}{0.5275\text{ g H}}\right) = 10.96\text{ g N}$

From Problem 2.154:

for NH_3, $\dfrac{\text{g N}}{\text{g H}} = \dfrac{2.34 \text{ g N}}{0.505 \text{ g H}} = 4.63$

for N_2H_4, $\dfrac{\text{g N}}{\text{g H}} = \dfrac{2.34 \text{ g N}}{0.337 \text{ g H}} = 6.94$

for compound X, $\dfrac{\text{g N}}{\text{g H}} = \dfrac{10.96 \text{ g N}}{1.575 \text{ g H}} = 6.96$; X is N_2H_4

2.156 (a) I (b) Kr

2.157 $^{1}H^{35}Cl$ has 18 protons, 18 neutrons, and 18 electrons.
$^{1}H^{37}Cl$ has 18 protons, 20 neutrons, and 18 electrons.
$^{2}H^{35}Cl$ has 18 protons, 19 neutrons, and 18 electrons.
$^{2}H^{37}Cl$ has 18 protons, 21 neutrons, and 18 electrons.
$^{3}H^{35}Cl$ has 18 protons, 20 neutrons, and 18 electrons.
$^{3}H^{37}Cl$ has 18 protons, 22 neutrons, and 18 electrons.

2.158 $\dfrac{12.0000}{15.9994} = \dfrac{X}{16.0000}$; X = 12.0005 for ^{12}C prior to 1961.

2.159 $\dfrac{39.9626}{15.9994} = \dfrac{X}{16.0000}$; X = 39.9641 for ^{40}Ca prior to 1961.

2.160 molecular mass = (8 x 12.011) + (9 x 1.0079) + (1 x 14.0067) + (2 x 15.9994) = 151.165 g/mol

2.161 mass % C = $\dfrac{8 \text{ x } 12.011}{151.165}$ x 100 = 63.565%

mass % H = $\dfrac{9 \text{ x } 1.0079}{151.165}$ x 100 = 6.0008%

mass % N = $\dfrac{14.0067}{151.165}$ x 100 = 9.2658%

mass % O = $\dfrac{2 \text{ x } 15.9994}{151.165}$ x 100 = 21.168%

2.162

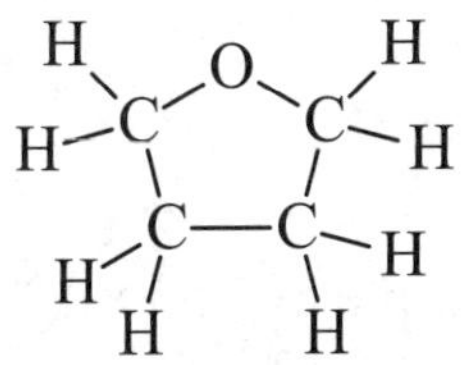

```
 H    O    H
  \ /   \ /
H—C       C—H
   \     /
    C—C—H
H /  \
   H   H
```

2.163 (a) Arrange the droplet charges in increasing order.
2.21×10^{-16} C
4.42×10^{-16} C
4.98×10^{-16} C
6.64×10^{-16} C
7.74×10^{-16} C

Find the smallest charge difference between droplet charges.
$(4.42 - 2.21) \times 10^{-16}$ C = 2.21×10^{-16} C
$(4.98 - 4.42) \times 10^{-16}$ C = 0.56×10^{-16} C
$(6.64 - 4.98) \times 10^{-16}$ C = 1.66×10^{-16} C
$(7.74 - 6.64) \times 10^{-16}$ C = 1.10×10^{-16} C
The smallest difference (0.56×10^{-16} C) is the approximate value for the charge on one blorvek.

To get a more accurate value, divide the droplet charges by the 0.56×10^{-16} C to determine the total number of blorveks.
2.21×10^{-16} C/0.56×10^{-16} C = 4
4.42×10^{-16} C/0.56×10^{-16} C = 8
4.98×10^{-16} C/0.56×10^{-16} C = 9
6.64×10^{-16} C/0.56×10^{-16} C = 12
7.74×10^{-16} C/0.56×10^{-16} C = 14
There are (4 + 8 + 9 + 12 + 14) = 47 blorveks on the 5 droplets with a total charge of $(2.21 + 4.42 + 4.98 + 6.64 + 7.74) \times 10^{-16}$ C = 25.99×10^{-16} C.

The charge on one blorvek is $\dfrac{25.99 \times 10^{-16}\ \text{C}}{47\ \text{blorveks}} = 5.53 \times 10^{-17}$ C/blorvek.

(b) Again arrange the droplet charges in increasing order.
2.21×10^{-16} C
4.42×10^{-16} C
4.98×10^{-16} C
5.81×10^{-16} C
6.64×10^{-16} C
7.74×10^{-16} C

Find the smallest charge difference between droplet charges.
$(4.42 - 2.21) \times 10^{-16}$ C = 2.21×10^{-16} C
$(4.98 - 4.42) \times 10^{-16}$ C = 0.56×10^{-16} C
$(5.81 - 4.98) \times 10^{-16}$ C = 0.83×10^{-16} C
$(6.64 - 5.81) \times 10^{-16}$ C = 0.83×10^{-16} C
$(7.74 - 6.64) \times 10^{-16}$ C = 1.10×10^{-16} C
The new charge difference (0.83×10^{-16} C) is not an integer multiple of 5.53×10^{-17} C, which means the charge on the blorvek must be smaller than 5.53×10^{-17} C.
The difference between 0.83×10^{-16} C and 0.56×10^{-16} C = 0.27×10^{-16} C is the approximate value for the charge on one blorvek.

To get a more accurate value, divide the droplet charges by the 0.27×10^{-16} C to determine the total number of blorveks.
$2.21 \times 10^{-16}\ \text{C}/0.27 \times 10^{-16}\ \text{C} = 8$
$4.42 \times 10^{-16}\ \text{C}/0.27 \times 10^{-16}\ \text{C} = 16$
$4.98 \times 10^{-16}\ \text{C}/0.27 \times 10^{-16}\ \text{C} = 18$
$5.81 \times 10^{-16}\ \text{C}/0.27 \times 10^{-16}\ \text{C} = 21$
$6.64 \times 10^{-16}\ \text{C}/0.27 \times 10^{-16}\ \text{C} = 24$
$7.74 \times 10^{-16}\ \text{C}/0.27 \times 10^{-16}\ \text{C} = 28$
There are $(8 + 16 + 18 + 21 + 24 + 28) = 115$ blorveks on the 6 droplets with a total charge of $(2.21 + 4.42 + 4.98 + 5.81 + 6.64 + 7.74) \times 10^{-16}\ \text{C} = 31.80 \times 10^{-16}$ C.

The charge on one blorvek is $\dfrac{31.80 \times 10^{-16}\ \text{C}}{115\ \text{blorveks}} = 2.77 \times 10^{-17}$ C/blorvek.

3 Mass Relationships in Chemical Reactions

3.1 $3\ A_2 + 2\ B \rightarrow 2\ BA_3$

3.2

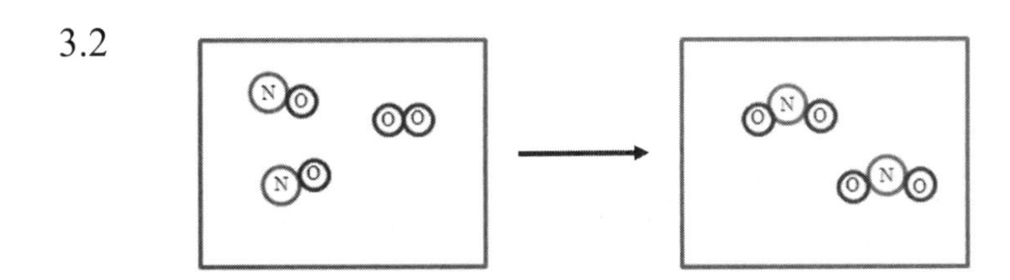

3.3 (a) $C_6H_{12}O_6 \rightarrow 2\ C_2H_6O + 2\ CO_2$
(b) $2\ NaClO_3 \rightarrow 2\ NaCl + O_2$
(c) $4\ NH_3 + Cl_2 \rightarrow N_2H_4 + 2\ NH_4Cl$

3.4 $8\ KClO_3 + C_{12}H_{22}O_{11} \rightarrow 8\ KCl + 12\ CO_2 + 11\ H_2O$

3.5 (a) Fe_2O_3 $2(55.85) + 3(16.00) = 159.7$
(b) H_2SO_4 $2(1.01) + 1(32.07) + 4(16.00) = 98.1$
(c) $C_6H_8O_7$ $6(12.01) + 8(1.01) + 7(16.00) = 192.1$
(d) $C_{16}H_{18}N_2O_4S$ $16(12.01) + 18(1.01) + 2(14.01) + 4(16.00) + 1(32.07) = 334.4$

3.6 sucrose, $C_{12}H_{22}O_{11}$ $12(12.01) + 22(1.01) + 11(16.00) = 342.3$
molar mass = 342.3 g/mol

3.7 $NaHCO_3$, 84.0

$$5.26\text{ g }NaHCO_3 \times \frac{1\text{ mol }NaHCO_3}{84.0\text{ g }NaHCO_3} = 0.0626\text{ mol }NaHCO_3$$

3.8 glucose, $C_6H_{12}O_6$, 180.0

(a) $$0.0833\text{ mol glucose} \times \frac{180.0\text{ g glucose}}{1\text{ mol glucose}} = 15.0\text{ g glucose}$$

(b) $$15.0\text{ g glucose} \times \frac{1\text{ glucose tablet}}{3.75\text{ g glucose}} = 4\text{ glucose tablets}$$

(c) $$0.0833\text{ mol glucose} \times \frac{6.022 \times 10^{23}\text{ glucose molecules}}{1\text{ mol glucose}} = 5.02 \times 10^{22}\text{ glucose molecules}$$

3.9 salicylic acid, $C_7H_6O_3$, 138.1; acetic anhydride, $C_4H_6O_3$, 102.1

(a) $$4.50\text{ g }C_7H_6O_3 \times \frac{1\text{ mol }C_7H_6O_3}{138.1\text{ g }C_7H_6O_3} \times \frac{1\text{ mol }C_4H_6O_3}{1\text{ mol }C_7H_6O_3} \times \frac{102.1\text{ g }C_4H_6O_3}{1\text{ mol }C_4H_6O_3} = 3.33\text{ g }C_4H_6O_3$$

3.10 salicylic acid, $C_7H_6O_3$, 138.1; acetic anhydride, $C_4H_6O_3$, 102.1
aspirin, $C_9H_8O_4$, 180.2; acetic acid, CH_3CO_2H, 60.1

(a) $10.0\text{ g } C_9H_8O_4 \times \dfrac{1\text{ mol } C_9H_8O_4}{180.2\text{ g } C_9H_8O_4} \times \dfrac{1\text{ mol } C_7H_6O_3}{1\text{ mol } C_9H_8O_4} \times \dfrac{138.1\text{ g } C_7H_6O_3}{1\text{ mol } C_7H_6O_3} = 7.66\text{ g } C_7H_6O_3$

(b) $10.0\text{ g } C_9H_8O_4 \times \dfrac{1\text{ mol } C_9H_8O_4}{180.2\text{ g } C_9H_8O_4} \times \dfrac{1\text{ mol } CH_3CO_2H}{1\text{ mol } C_9H_8O_4} \times \dfrac{60.1\text{ g } CH_3CO_2H}{1\text{ mol } CH_3CO_2H} = 3.34\text{ g } CH_3CO_2H$

3.11 C_2H_4, 28.1; C_2H_6O, 46.1

$$4.6\text{ g } C_2H_4 \times \frac{1\text{ mol } C_2H_4}{28.1\text{ g } C_2H_4} \times \frac{1\text{ mol } C_2H_6O}{1\text{ mol } C_2H_4} \times \frac{46.1\text{ g } C_2H_6O}{1\text{ mol } C_2H_6O} = 7.5\text{ g } C_2H_6O$$
(theoretical yield)

$$\text{Percent yield} = \frac{\text{Actual yield}}{\text{Theoretical yield}} \times 100\% = \frac{4.7\text{ g}}{7.5\text{ g}} \times 100\% = 63\%$$

3.12 C_2H_6O, 46.1; $C_4H_{10}O$, 74.1

(a) $40.0\text{ g } C_2H_6O \times \dfrac{1\text{ mol } C_2H_6O}{46.1\text{ g } C_2H_6O} \times \dfrac{1\text{ mol } C_4H_{10}O}{2\text{ mol } C_2H_6O} \times \dfrac{74.1\text{ g } C_4H_{10}O}{1\text{ mol } C_4H_{10}O} = 32.1\text{ g } C_4H_{10}O$ (theoretical yield)

actual yield = (32.1 g)(0.870) = 27.9 g $C_4H_{10}O$

(b) 100.0 g = (theoretical yield)(0.870)
theoretical yield = 100.0 g/0.870 = 115 g $C_4H_{10}O$

$$115\text{ g } C_4H_{10}O \times \frac{1\text{ mol } C_4H_{10}O}{74.1\text{ g } C_4H_{10}O} \times \frac{2\text{ mol } C_2H_6O}{1\text{ mol } C_4H_{10}O} \times \frac{46.1\text{ g } C_2H_6O}{1\text{ mol } C_2H_6O} = 143\text{ g } C_2H_6O$$

3.13 (a) $A + B_2 \rightarrow AB_2$
There is a 1:1 stoichiometry between the two reactants. A is the limiting reactant because there are fewer reactant A's than there are reactant B_2's.
(b) 1.0 mol of AB_2 can be made from 1.0 mol of A and 1.0 mol of B_2.

3.14

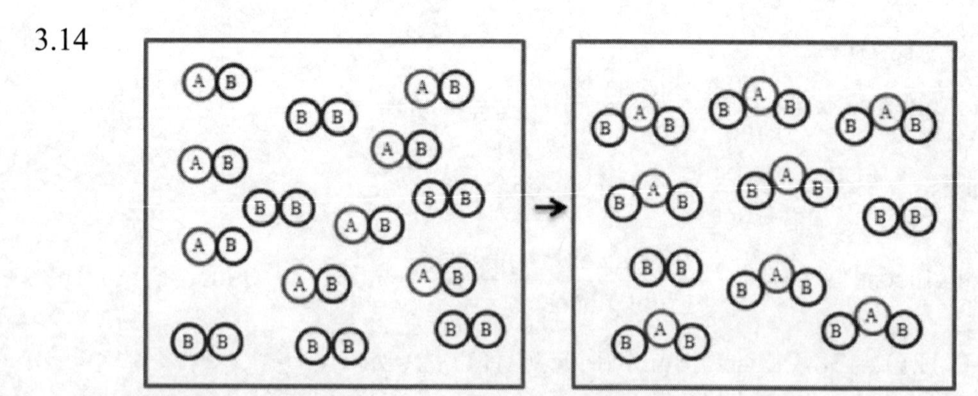

B_2 is in excess, AB is the limiting reactant.

3.15 Li_2O, 29.9: 65 kg = 65,000 g; H_2O, 18.0: 80.0 kg = 80,000 g; LiOH, 23.9

(a) $65{,}000\text{ g }Li_2O \times \dfrac{1\text{ mol }Li_2O}{29.9\text{ g }Li_2O} = 2.17 \times 10^3\text{ mol }Li_2O$

$80{,}000\text{ g }H_2O \times \dfrac{1\text{ mol }H_2O}{18.0\text{ g }H_2O} = 4.44 \times 10^3\text{ mol }H_2O$

The reaction stoichiometry between Li_2O and H_2O is one to one. There are twice as many moles of H_2O as there are moles of Li_2O. Therefore, Li_2O is the limiting reactant.

(b) $(4.44 \times 10^3\text{ mol} - 2.17 \times 10^3\text{ mol}) = 2.27 \times 10^3\text{ mol }H_2O$ remaining

$2.27 \times 10^3\text{ mol }H_2O \times \dfrac{18.0\text{ g }H_2O}{1\text{ mol }H_2O} = 40{,}860\text{ g }H_2O = 40.9\text{ kg} = 41\text{ kg }H_2O$

(c) $2.17 \times 10^3\text{ mol }Li_2O \times \dfrac{2\text{ mol LiOH}}{1\text{ mol }Li_2O} \times \dfrac{23.9\text{ g LiOH}}{1\text{ mol LiOH}} \times \dfrac{1\text{ kg}}{1000\text{ g}} = 104\text{ kg LiOH}$

3.16 $LiHCO_3$, 68.0; LiOH, 23.9

$500.0\text{ g }LiHCO_3 \times \dfrac{1\text{ mol }LiHCO_3}{68.0\text{ g }LiHCO_3} \times \dfrac{1\text{ mol LiOH}}{1\text{ mol }LiHCO_3} \times \dfrac{23.9\text{ g LiOH}}{1\text{ mol LiOH}} = 175.7\text{ g LiOH}$

You start with 400.0 g of LiOH. 500.0 g of $LiHCO_3$ are produced from 175.7 g of LiOH. Over 200 g of LiOH remain. Additional CO_2 can be removed.

3.17 Assume a 100.0 g sample. From the percent composition data, a 100.0 g sample contains 14.25 g C, 56.93 g O, and 28.83 g Mg.

$14.25\text{ g C} \times \dfrac{1\text{ mol C}}{12.011\text{ g C}} = 1.19\text{ mol C}$

$56.93\text{ g O} \times \dfrac{1\text{ mol O}}{15.999\text{ g O}} = 3.56\text{ mol O}$

$28.83\text{ g Mg} \times \dfrac{1\text{ mol Mg}}{24.305\text{ g Mg}} = 1.19\text{ mol Mg}$

$Mg_{1.19}C_{1.19}O_{3.56}$; divide each subscript by the smallest, 1.19.

$Mg_{1.19/1.19}C_{1.19/1.19}O_{3.56/1.19}$

The empirical formula is $MgCO_3$.

3.18 Assume a 100.0 g sample. From the percent composition data, a 100.0 g sample contains 40.0 g C, 13.3 g H, and 46.7 g N.

$40.0\text{ g C} \times \dfrac{1\text{ mol C}}{12.01\text{ g C}} = 3.33\text{ mol C}$

$13.3\text{ g H} \times \dfrac{1\text{ mol H}}{1.01\text{ g H}} = 13.2\text{ mol H}$

$46.7\text{ g N} \times \dfrac{1\text{ mol N}}{14.01\text{ g N}} = 3.33\text{ mol N}$

$C_{3.33}H_{13.2}N_{3.33}$; divide each subscript by the smallest, 3.33.

$C_{3.33/3.33}H_{13.2/3.33}N_{3.33/3.33}$

The empirical formula is CH_4N, 30.0.
60.0 / 30.0 = 2; molecular formula = $C_{(2 \times 1)}H_{(2 \times 4)}N_{(2 \times 1)} = C_2H_8N_2$

3.19 glucose, $C_6H_{12}O_6$, 180.0
Divide each subscript by the smallest, 6, to get the empirical formula.
$C_{6/6}H_{12/6}O_{6/6}$
The empirical formula is CH_2O.
Percent composition:

$$\%C = \frac{6 \times 12.0 \text{ g}}{180.0 \text{ g}} \times 100\% = 40.0\%$$

$$\%H = \frac{12 \times 1.01 \text{ g}}{180.0 \text{ g}} \times 100\% = 6.7\%$$

$$\%O = \frac{6 \times 16.0 \text{ g}}{180.0 \text{ g}} \times 100\% = 53.3\%$$

3.20
$$1.161 \text{ g } H_2O \times \frac{1 \text{ mol } H_2O}{18.0 \text{ g } H_2O} \times \frac{2 \text{ mol H}}{1 \text{ mol } H_2O} = 0.129 \text{ mol H}$$

$$2.818 \text{ g } CO_2 \times \frac{1 \text{ mol } CO_2}{44.0 \text{ g } CO_2} \times \frac{1 \text{ mol C}}{1 \text{ mol } CO_2} = 0.0640 \text{ mol C}$$

$$0.129 \text{ mol H} \times \frac{1.01 \text{ g H}}{1 \text{ mol H}} = 0.130 \text{ g H}$$

$$0.0640 \text{ mol C} \times \frac{12.0 \text{ g C}}{1 \text{ mol C}} = 0.768 \text{ g C}$$

1.00 g total – (0.130 g H + 0.768 g C) = 0.102 g O

$$0.102 \text{ g O} \times \frac{1 \text{ mol O}}{16.0 \text{ g O}} = 0.006\ 38 \text{ mol O}$$

$C_{0.0640}H_{0.129}O_{0.006\ 38}$; divide each subscript by the smallest, 0.006 38.
$C_{0.0640 / 0.006\ 38}H_{0.129 / 0.006\ 38}O_{0.006\ 38 / 0.006\ 38}$
$C_{10.03}H_{20.22}O_1$
The empirical formula is $C_{10}H_{20}O$.

3.21
$$0.697 \text{ g } H_2O \times \frac{1 \text{ mol } H_2O}{18.0 \text{ g } H_2O} \times \frac{2 \text{ mol H}}{1 \text{ mol } H_2O} = 0.0774 \text{ mol H}$$

$$1.55 \text{ g } CO_2 \times \frac{1 \text{ mol } CO_2}{44.0 \text{ g } CO_2} \times \frac{1 \text{ mol C}}{1 \text{ mol } CO_2} = 0.0352 \text{ mol C}$$

$C_{0.0352}H_{0.0774}$; divide each subscript by the smaller, 0.0352.
$C_{0.0352 / 0.0352}H_{0.0774 / 0.0352}$
CH_2
The empirical formula is CH_2, 14.0.
142.0 / 14.0 = 10.1 = 10; molecular formula = $C_{(10 \times 1)}H_{(10 \times 2)} = C_{10}H_{20}$

3.22 (a) $0.138 \text{ g } H_2O \times \frac{1 \text{ mol } H_2O}{18.0 \text{ g } H_2O} \times \frac{2 \text{ mol H}}{1 \text{ mol } H_2O} = 0.0153 \text{ mol H}$

$1.617 \text{ g } CO_2 \times \frac{1 \text{ mol } CO_2}{44.0 \text{ g } CO_2} \times \frac{1 \text{ mol C}}{1 \text{ mol } CO_2} = 0.0368 \text{ mol C}$

$0.0153 \text{ mol H} \times \frac{1.01 \text{ g H}}{1 \text{ mol H}} = 0.016 \text{ g H}$

$0.0368 \text{ mol C} \times \frac{12.0 \text{ g C}}{1 \text{ mol C}} = 0.442 \text{ g C}$

1.0 g total – (0.016 g H + 0.442 g C) = 0.542 g Cl

$0.542 \text{ g Cl} \times \frac{1 \text{ mol Cl}}{35.5 \text{ g Cl}} = 0.0153 \text{ mol Cl}$

$C_{0.0368}H_{0.0153}Cl_{0.0153}$; divide each subscript by the smallest, 0.0153.

$C_{0.0368 / 0.0153}H_{0.0153 / 0.0153}Cl_{0.0153 / 0.0153}$

$C_{2.5}HCl$

Multiply each subscript by 2 to get all integer subscripts.

$C_{(2 \times 2.5)}H_{(2 \times 1)}Cl_{(2 \times 1)}$

The empirical formula is $C_5H_2Cl_2$, 133.0.

(b) 326.26/133.0 = 2.45 ≈ 2.5; molecular formula = $C_{(2.5 \times 5)}H_{(2.5 \times 2)}Cl_{(2.5 \times 2)} = C_{12}H_5Cl_5$

(c) No, because Cl does not form a useful oxide, so both O and Cl would have to be obtained by mass difference but that is impossible. You can only determine one element by mass difference.

3.23 The empirical formula is C_6H_5, 77.0.

The peak with the largest mass in the mass spectrum occurs near a mass of 154, which would be the molecular weight of the compound.

154/77.0 = 2; molecular formula = $C_{(2 \times 6)}H_{(2 \times 5)} = C_{12}H_{10}$

3.24 $0.57 \text{ g } H_2O \times \frac{1 \text{ mol } H_2O}{18.0 \text{ g } H_2O} \times \frac{2 \text{ mol H}}{1 \text{ mol } H_2O} = 0.063 \text{ mol H}$

$2.79 \text{ g } CO_2 \times \frac{1 \text{ mol } CO_2}{44.0 \text{ g } CO_2} \times \frac{1 \text{ mol C}}{1 \text{ mol } CO_2} = 0.063 \text{ mol C}$

$0.063 \text{ mol H} \times \frac{1.01 \text{ g H}}{1 \text{ mol H}} = 0.063 \text{ g H}$

$0.063 \text{ mol C} \times \frac{12.0 \text{ g C}}{1 \text{ mol C}} = 0.76 \text{ g C}$

1.00 g total – (0.063 g H + 0.76 g C) = 0.18 g N

$0.18 \text{ g N} \times \frac{1 \text{ mol N}}{14.0 \text{ g N}} = 0.013 \text{ mol N}$

$C_{0.063}H_{0.063}N_{0.013}$; divide each subscript by the smallest, 0.013.

$C_{0.063 / 0.013}H_{0.063 / 0.013}N_{0.013 / 0.013}$

$C_{4.85}H_{4.85}N$

The empirical formula is C_5H_5N, 79.0.

The peak with the largest mass in the mass spectrum occurs near a mass of 79, which would be the molecular weight of the compound, therefore the empirical formula and molecular formula are the same.

3.25 (a) C_2H_5OH, 46.1; CO_2, 44.0

$C_2H_5OH + 3O_2 \rightarrow 2\,CO_2 + 3\,H_2O$

$$14.7\text{ gal} \times \frac{1\text{ L}}{0.2642\text{ gal}} \times \frac{0.79\text{ kg}}{1\text{ L}} = 44\text{ kg } C_2H_5OH$$

$$44\text{ kg } C_2H_5OH \times \frac{1000\text{ g}}{1\text{ kg}} \times \frac{1\text{ mol } C_2H_5OH}{46.1\text{ g } C_2H_5OH} \times \frac{2\text{ mol } CO_2}{1\text{ mol } C_2H_5OH}$$

$$\times \frac{44.0\text{ g } CO_2}{1\text{ mol } CO_2} \times \frac{1\text{ kg}}{1000\text{ g}} = 84\text{ kg } CO_2$$

(b) C_3H_8, 44.1; CO_2, 44.0

$C_3H_8 + 5O_2 \rightarrow 3\,CO_2 + 4\,H_2O$

$$13.7\text{ gal} \times \frac{1\text{ L}}{0.2642\text{ gal}} \times \frac{0.49\text{ kg}}{1\text{ L}} = 25\text{ kg } C_2H_5OH$$

$$25\text{ kg } C_2H_5OH \times \frac{1000\text{ g}}{1\text{ kg}} \times \frac{1\text{ mol } C_3H_8}{44.1\text{ g } C_3H_8} \times \frac{3\text{ mol } CO_2}{1\text{ mol } C_3H_8}$$

$$\times \frac{44.0\text{ g } CO_2}{1\text{ mol } CO_2} \times \frac{1\text{ kg}}{1000\text{ g}} = 75\text{ kg } CO_2$$

(c) CH_4, 16.0; CO_2, 44.0

$CH_4 + 2O_2 \rightarrow CO_2 + 2\,H_2O$

$$25.7\text{ kg } C_2H_5OH \times \frac{1000\text{ g}}{1\text{ kg}} \times \frac{1\text{ mol } CH_4}{16.0\text{ g } CH_4} \times \frac{1\text{ mol } CO_2}{1\text{ mol } CH_4}$$

$$\times \frac{44.0\text{ g } CO_2}{1\text{ mol } CO_2} \times \frac{1\text{ kg}}{1000\text{ g}} = 70.7\text{ kg } CO_2$$

(d) Electricity from a Coal-Burning Power Plant

Assume an electric car is driven for 250 miles.

$$250\text{ mi} \times \frac{35\text{ kWh}}{100\text{ mi}} \times \frac{0.94\text{ kg } CO_2}{1\text{ kWh}} = 82\text{ kg } CO_2$$

(e) Electricity from a Natural Gas-Burning Power Plant

Assume an electric car is driven for 250 miles.

$$250\text{ mi} \times \frac{35\text{ kWh}}{100\text{ mi}} \times \frac{0.55\text{ kg } CO_2}{1\text{ kWh}} = 48\text{ kg } CO_2$$

(f) Electricity from natural gas produced the least amount of CO_2 when burned to provide energy for a car.

3.26 (a) $6\,CO_2 + 6\,H_2O \rightarrow C_6H_{12}O_6 + 6\,O_2$

(b) CO_2, 44.0; $C_6H_{12}O_6$, 180.0

$$1400 \text{ lb } CO_2 \text{ x } \frac{1 \text{ kg}}{2.2 \text{ lb}} \text{ x } \frac{1000 \text{ g}}{1 \text{ kg}} \text{ x } \frac{1 \text{ mol } CO_2}{44.0 \text{ g } CO_2}$$
$$\text{x } \frac{1 \text{ mol } C_6H_{12}O_6}{6 \text{ mol } CO_2} \text{ x } \frac{180.0 \text{ g } C_6H_{12}O_6}{1 \text{ mol } C_6H_{12}O_6} \text{ x } \frac{1 \text{ kg}}{1000 \text{ g}} = 434 \text{ kg } C_6H_{12}O_6$$

3.27 (a) CO_2, 44.0; $HOCH_2CH_2NH_2$, 61.1

$CO_2 + 2\ HOCH_2CH_2NH_2 \rightarrow HOCH_2CH_2NH_3^+ + HOCH_2CH_2NHCO_2^-$

$$20 \text{ x } 10^6 \text{ tons x } \frac{907.2 \text{ kg}}{1 \text{ ton}} \text{ x } \frac{1000 \text{ g}}{1 \text{ kg}} \text{ x } \frac{1 \text{ mol } CO_2}{44.0 \text{ g } CO_2}$$
$$\text{x } \frac{2 \text{ mol } HOCH_2CH_2NH_2}{1 \text{ mol } CO_2} \text{ x } \frac{61.1 \text{ g } HOCH_2CH_2NH_2}{1 \text{ mol } HOCH_2CH_2NH_2}$$
$$\text{x } \frac{1 \text{ kg}}{1000 \text{ g}} \text{ x } \frac{1 \text{ ton}}{907.2 \text{ kg}} = 55.5 \text{ x } 10^6 \text{ tons } HOCH_2CH_2NH_2$$

(b)
$$25 \text{ x } 10^6 \text{ tons } HOCH_2CH_2NH_2 \text{ x } \frac{907.2 \text{ kg}}{1 \text{ ton}} \text{ x } \frac{1000 \text{ g}}{1 \text{ kg}} \text{ x } \frac{1 \text{ mol } HOCH_2CH_2NH_2}{61.1 \text{ g } HOCH_2CH_2NH_2}$$
$$\text{x } \frac{1 \text{ mol } CO_2}{2 \text{ mol } HOCH_2CH_2NH_2} \text{ x } \frac{44.0 \text{ g } CO_2}{1 \text{ mol } CO_2}$$
$$\text{x } \frac{1 \text{ kg}}{1000 \text{ g}} \text{ x } \frac{1 \text{ ton}}{907.2 \text{ g}} = 9 \text{ x } 10^6 \text{ tons } CO_2$$

($20 \text{ x } 10^6$ tons – $9 \text{ x } 10^6$ tons) = $11 \text{ x } 10^6$ (11 million) tons of CO_2 would escape into the atmosphere if only 25 million tons of $HOCH_2CH_2NH_2$ were used.

3.28 Open response question

Conceptual Problems

3.29 (c) $2\ A + B_2 \rightarrow A_2B_2$

3.30 reactants, box (d), and products, box (c)

3.31 The molecular formula for cytosine is $C_4H_5N_3O$.

$$\text{mol } CO_2 = 0.001 \text{ mol cyt x } \frac{4 \text{ C}}{\text{cyt}} \text{ x } \frac{1 \text{ } CO_2}{\text{C}} = 0.004 \text{ mol } CO_2$$

$$\text{mol } H_2O = 0.001 \text{ mol cyt x } \frac{5 \text{ H}}{\text{cyt}} \text{ x } \frac{1 \text{ } H_2O}{2 \text{ H}} = 0.0025 \text{ mol } H_2O$$

3.32 $C_{17}H_{18}F_3NO$ 17(12.01) + 18(1.01) + 3(19.00) + 1(14.01) + 1(16.00) = 309.36

3.33 $C_3H_7NO_2S$, 121.2

$$\%C = \frac{3 \text{ x } 12.0 \text{ g}}{121.2 \text{ g}} \text{ x } 100\% = 29.7\%$$

$$\%H = \frac{7 \text{ x } 1.01 \text{ g}}{121.2 \text{ g}} \text{ x } 100\% = 5.83\%$$

$$\%N = \frac{1 \text{ x } 14.0 \text{ g}}{121.2 \text{ g}} \text{ x } 100\% = 11.6\%$$

$$\%O = \frac{2 \text{ x } 16.0 \text{ g}}{121.2 \text{ g}} \text{ x } 100\% = 26.4\%$$

$$\%S = \frac{1 \text{ x } 32.1 \text{ g}}{121.2 \text{ g}} \text{ x } 100\% = 26.5\%$$

3.34 (a) $A_2 + 3\ B_2 \rightarrow 2\ AB_3$; B_2 is the limiting reactant because it is completely consumed.
(b) For 1.0 mol of A_2, 3.0 mol of B_2 are required. Because only 1.0 mol of B_2 is available, B_2 is the limiting reactant.

$$1 \text{ mol } B_2 \text{ x } \frac{2 \text{ mol } AB_3}{3 \text{ mol } B_2} = 2/3 \text{ mol } AB_3$$

3.35 $C_xH_y \xrightarrow{O_2} 3\ CO_2 + 4\ H_2O$; x is equal to the coefficient for CO_2 and y is equal to 2 times the coefficient for H_2O. The empirical formula for the hydrocarbon is C_3H_8.

Section Problems
Balancing Equations (Section 3.2)

3.36 Equation (b) is balanced, (a) is not balanced.

3.37 (a) and (c) are not balanced, (b) is balanced.
(a) $2\ Al + Fe_2O_3 \rightarrow Al_2O_3 + 2\ Fe$ (balanced)
(c) $4\ Au + 8\ NaCN + O_2 + 2\ H_2O \rightarrow 4\ NaAu(CN)_2 + 4\ NaOH$ (balanced)

3.38 (a) $Mg + 2\ HNO_3 \rightarrow H_2 + Mg(NO_3)_2$
(b) $CaC_2 + 2\ H_2O \rightarrow Ca(OH)_2 + C_2H_2$
(c) $2\ S + 3\ O_2 \rightarrow 2\ SO_3$
(d) $UO_2 + 4\ HF \rightarrow UF_4 + 2\ H_2O$

3.39 (a) $2\ NH_4NO_3 \rightarrow 2\ N_2 + O_2 + 4\ H_2O$
(b) $C_2H_6O + O_2 \rightarrow C_2H_4O_2 + H_2O$
(c) $C_2H_8N_2 + 2\ N_2O_4 \rightarrow 3\ N_2 + 2\ CO_2 + 4\ H_2O$

3.40 (a) $SiCl_4 + 2\ H_2O \rightarrow SiO_2 + 4\ HCl$
(b) $P_4O_{10} + 6\ H_2O \rightarrow 4\ H_3PO_4$
(c) $CaCN_2 + 3\ H_2O \rightarrow CaCO_3 + 2\ NH_3$
(d) $3\ NO_2 + H_2O \rightarrow 2\ HNO_3 + NO$

3.41 (a) $VCl_3 + 3\ Na + 6\ CO \rightarrow V(CO)_6 + 3\ NaCl$
(b) $RuI_3 + 5\ CO + 3\ Ag \rightarrow Ru(CO)_5 + 3AgI$
(c) $2\ CoS + 8\ CO + 4\ Cu \rightarrow Co_2(CO)_8 + 2\ Cu_2S$

Molecular Weights and Stoichiometry (Section 3.3)

3.42 (a) Hg_2Cl_2: 2(200.59) + 2(35.45) = 472.1
(b) $C_4H_8O_2$: 4(12.01) + 8(1.01) + 2(16.00) = 88.1
(c) CF_2Cl_2: 1(12.01) + 2(19.00) + 2(35.45) = 120.9

3.43 (a) (1 x 30.97) + (Y x 35.45) = 137.3; Solve for Y; Y = 3.
The formula is PCl_3.
(b) (10 x 12.01) + (14 x 1.008) + (Z x 14.01) = 162.2
Solve for Z; Z = 2. The formula is $C_{10}H_{14}N_2$.

3.44 (a) $C_{33}H_{35}FN_2O_5$: 33(12.01) + 35(1.01) + 1(19.00) + 2(14.01) + 5(16.00) = 558.7
(b) $C_{22}H_{27}F_3O_4S$: 22(12.01) + 27(1.01) + 3(19.00) + 4(16.00) + 1(32.06) = 444.5
(c) $C_{16}H_{16}ClNO_2S$: 16(12.01) + 16(1.01) + 1(35.45) + 1(14.01) + 2(16.00) + 1(32.06) = 321.8

3.45 (a) $C_6H_6Cl_2O_3$: 6(12.01) + 6(1.01) + 2(35.45) + 3(16.00) = 197.0
(b) $C_{15}H_{22}ClNO_2$: 15(12.01) + 22(1.01) + 1(35.45) + 1(14.01) + 2(16.00) = 283.8
(c) $C_8H_6Cl_2O_3$: 8(12.01) + 6(1.01) + 2(35.45) + 3(16.00) = 221.0

3.46 One mole equals the atomic weight or molecular weight in grams.
(a) Ti, 47.87 g (b) Br_2, 159.81 g (c) Hg, 200.59 g (d) H_2O, 18.02 g

3.47 (a) $1.00\text{ g Cr} \times \dfrac{1\text{ mol Cr}}{52.0\text{ g Cr}} = 0.0192\text{ mol Cr}$

(b) $1.00\text{ g Cl}_2 \times \dfrac{1\text{ mol Cl}_2}{70.9\text{ g Cl}_2} = 0.0141\text{ mol Cl}_2$

(c) $1.00\text{ g Au} \times \dfrac{1\text{ mol Au}}{197.0\text{ g Au}} = 0.005\ 08\text{ mol Au}$

(d) $1.00\text{ g NH}_3 \times \dfrac{1\text{ mol NH}_3}{17.0\text{ g NH}_3} = 0.0588\text{ mol NH}_3$

3.48 There are 3 ions (one Mg^{2+} and 2 Cl^-) per formula unit of $MgCl_2$.
$MgCl_2$, 95.2

$$27.5\text{ g MgCl}_2 \times \frac{1\text{ mol MgCl}_2}{95.2\text{ g MgCl}_2} \times \frac{3\text{ mol ions}}{1\text{ mol MgCl}_2} = 0.867\text{ mol ions}$$

3.49 There are 3 F^- anions per formula unit of AlF_3.
AlF_3, 84.0

$$35.6\text{ g AlF}_3 \times \frac{1\text{ mol AlF}_3}{84.0\text{ g AlF}_3} \times \frac{3\text{ mol anions}}{1\text{ mol AlF}_3} = 1.27\text{ mol F}^-$$

3.50 Molar mass = $\frac{3.28 \text{ g}}{0.0275 \text{ mol}}$ = 119 g/mol; molecular weight = 119.

3.51 Molar mass = $\frac{221.6 \text{ g}}{0.5731 \text{ mol}}$ = 386.7 g/mol; molecular weight = 386.7.

3.52 $FeSO_4$, 151.9; 300 mg = 0.300 g

$0.300 \text{ g } FeSO_4 \text{ x } \frac{1 \text{ mol } FeSO_4}{151.9 \text{ g } FeSO_4} = 1.97 \text{ x } 10^{-3} \text{ mol } FeSO_4$

$1.97 \text{ x } 10^{-3} \text{ mol } FeSO_4 \text{ x } \frac{6.022 \text{ x } 10^{23} \text{ Fe(II) atoms}}{1 \text{ mol } FeSO_4} = 1.19 \text{ x } 10^{21} \text{ Fe(II) atoms}$

3.53 $0.0001 \text{ g C x } \frac{1 \text{ mol C}}{12.0 \text{ g C}} \text{ x } \frac{6.02 \text{ x } 10^{23} \text{ C atoms}}{1 \text{ mol C}} = 5 \text{ x } 10^{18} \text{ C atoms}$

3.54 $C_8H_{10}N_4O_2$, 194.2; 125 mg = 0.125 g

$0.125 \text{ g caffeine x } \frac{1 \text{ mol caffeine}}{194.2 \text{ g caffeine}} = 6.44 \text{ x } 10^{-4} \text{ mol caffeine}$

$0.125 \text{ g caffeine x } \frac{1 \text{ mol caffeine}}{194.2 \text{ g caffeine}} \text{ x } \frac{6.022 \text{ x } 10^{23} \text{ molecules}}{1 \text{ mol}} = 3.88 \text{ x } 10^{20} \text{ caffeine molecules}$

3.55 (a) $0.0015 \text{ mol Na x } \frac{23.0 \text{ g Na}}{1 \text{ mol Na}} = 0.034 \text{ g Na}$

(b) $0.0015 \text{ mol Pb x } \frac{207.2 \text{ g Pb}}{1 \text{ mol Pb}} = 0.31 \text{ g Pb}$

(c) $C_{16}H_{13}ClN_2O$, 284.7

$0.0015 \text{ mol diazepam x } \frac{284.7 \text{ g diazepam}}{1 \text{ mol diazepam}} = 0.43 \text{ g diazepam}$

3.56 By definition, $6.022 \text{ x } 10^{23}$ particles are 1.000 mole of particles.
mol Ar = 0.2500 mol and mol "other" = 0.7500 mol

mass Ar = $0.2500 \text{ mol Ar x } \frac{39.95 \text{ g Ar}}{1 \text{ mol Ar}} = 9.99 \text{ g Ar}$

mass "other" = total mass – mass Ar = 25.12 g – 9.99 g = 15.13 g "other"

molar mass "other" = $\frac{15.13 \text{ g "other"}}{0.7500 \text{ mol "other"}} = 20.17 \text{ g/mol}$

The "other" element is neon (Ne).

3.57 He, 4.00; Kr, 83.80
By definition, $6.022 \text{ x } 10^{23}$ particles are 1.000 mole of particles.
mass of $6.022 \text{ x } 10^{23}$ particles = (0.30)(4.00 g He) + (0.70)(83.80 g Ar) = 60 g

number of particles in sample = $107.75 \text{ g x } \frac{6.022 \text{ x } 10^{23} \text{ particles}}{60 \text{ g}} = 1.1 \text{ x } 10^{24} \text{ particles}$

3.58 TiO_2, 79.87; $100.0 \text{ kg Ti} \times \dfrac{79.87 \text{ kg } TiO_2}{47.87 \text{ kg Ti}} = 166.8 \text{ kg } TiO_2$

3.59 Fe_2O_3, 159.7; $\% \text{ Fe} = \dfrac{2(55.85 \text{ g}) \text{ Fe}}{159.7 \text{ g } Fe_2O_3} \times 100\% = 69.94\%$

mass Fe = (0.6994)(105 kg) = 73.4 kg

3.60 (a) $2\ Fe_2O_3 + 3\ C \rightarrow 4\ Fe + 3\ CO_2$

(b) Fe_2O_3, 159.7; $525 \text{ g } Fe_2O_3 \times \dfrac{1 \text{ mol } Fe_2O_3}{159.7 \text{ g } Fe_2O_3} \times \dfrac{3 \text{ mol C}}{2 \text{ mol } Fe_2O_3} = 4.93 \text{ mol C}$

(c) $4.93 \text{ mol C} \times \dfrac{12.01 \text{ g C}}{1 \text{ mol C}} = 59.2 \text{ g C}$

3.61 (a) $Fe_2O_3 + 3\ CO \rightarrow 2\ Fe + 3\ CO_2$

(b) Fe_2O_3, 159.7; CO, 28.01

$$3.02 \text{ g } Fe_2O_3 \times \frac{1 \text{ mol } Fe_2O_3}{159.7 \text{ g } Fe_2O_3} \times \frac{3 \text{ mol CO}}{1 \text{ mol } Fe_2O_3} \times \frac{28.01 \text{ g CO}}{1 \text{ mol CO}} = 1.59 \text{ g CO}$$

(c) $1.68 \text{ mol } Fe_2O_3 \times \dfrac{3 \text{ mol CO}}{1 \text{ mol } Fe_2O_3} \times \dfrac{28.01 \text{ g CO}}{1 \text{ mol CO}} = 141 \text{ g CO}$

3.62 (a) $2\ Mg + O_2 \rightarrow 2\ MgO$

(b) Mg, 24.30; O_2, 32.00; MgO, 40.30

$$25.0 \text{ g Mg} \times \frac{1 \text{ mol Mg}}{24.30 \text{ g Mg}} \times \frac{1 \text{ mol } O_2}{2 \text{ mol Mg}} \times \frac{32.00 \text{ g } O_2}{1 \text{ mol } O_2} = 16.5 \text{ g } O_2$$

$$25.0 \text{ g Mg} \times \frac{1 \text{ mol Mg}}{24.30 \text{ g Mg}} \times \frac{2 \text{ mol MgO}}{2 \text{ mol Mg}} \times \frac{40.30 \text{ g MgO}}{1 \text{ mol MgO}} = 41.5 \text{ g MgO}$$

(c) $25.0 \text{ g } O_2 \times \dfrac{1 \text{ mol } O_2}{32.00 \text{ g } O_2} \times \dfrac{2 \text{ mol Mg}}{1 \text{ mol } O_2} \times \dfrac{24.30 \text{ g Mg}}{1 \text{ mol Mg}} = 38.0 \text{ g Mg}$

$$25.0 \text{ g } O_2 \times \frac{1 \text{ mol } O_2}{32.00 \text{ g } O_2} \times \frac{2 \text{ mol MgO}}{1 \text{ mol } O_2} \times \frac{40.30 \text{ g MgO}}{1 \text{ mol MgO}} = 63.0 \text{ g MgO}$$

3.63 $C_2H_4 + H_2O \rightarrow C_2H_6O$; C_2H_4, 28.05; H_2O, 18.02; C_2H_6O, 46.07

(a) $0.133 \text{ mol } H_2O \times \dfrac{1 \text{ mol } C_2H_4}{1 \text{ mol } H_2O} \times \dfrac{28.05 \text{ g } C_2H_4}{1 \text{ mol } C_2H_4} = 3.73 \text{ g } C_2H_4$

$$0.133 \text{ mol } H_2O \times \frac{1 \text{ mol } C_2H_6O}{1 \text{ mol } H_2O} \times \frac{46.07 \text{ g } C_2H_6O}{1 \text{ mol } C_2H_6O} = 6.13 \text{ g } C_2H_6O$$

(b) $0.371 \text{ mol } C_2H_4 \text{ x } \frac{1 \text{ mol } H_2O}{1 \text{ mol } C_2H_4} \text{ x } \frac{18.02 \text{ g } H_2O}{1 \text{ mol } H_2O} = 6.69 \text{ g } H_2O$

$0.371 \text{ mol } C_2H_4 \text{ x } \frac{1 \text{ mol } C_2H_6O}{1 \text{ mol } C_2H_4} \text{ x } \frac{46.07 \text{ g } C_2H_6O}{1 \text{ mol } C_2H_6O} = 17.1 \text{ g } C_2H_6O$

3.64 (a) $2\ HgO \rightarrow 2\ Hg + O_2$

(b) HgO, 216.6; Hg, 200.6; O_2, 32.0

$45.5 \text{ g HgO x } \frac{1 \text{ mol HgO}}{216.6 \text{ g HgO}} \text{ x } \frac{2 \text{ mol Hg}}{2 \text{ mol HgO}} \text{ x } \frac{200.6 \text{ g Hg}}{1 \text{ mol Hg}} = 42.1 \text{ g Hg}$

$45.5 \text{ g HgO x } \frac{1 \text{ mol HgO}}{216.6 \text{ g HgO}} \text{ x } \frac{1 \text{ mol } O_2}{2 \text{ mol HgO}} \text{ x } \frac{32.00 \text{ g } O_2}{1 \text{ mol } O_2} = 3.36 \text{ g } O_2$

(c) $33.3 \text{ g } O_2 \text{ x } \frac{1 \text{ mol } O_2}{32.00 \text{ g } O_2} \text{ x } \frac{2 \text{ mol HgO}}{1 \text{ mol } O_2} \text{ x } \frac{216.6 \text{ g HgO}}{1 \text{ mol HgO}} = 451 \text{ g HgO}$

3.65 5.60 kg = 5600 g; $TiCl_4$, 189.7; TiO_2, 79.87

$5600 \text{ g } TiCl_4 \text{ x } \frac{1 \text{ mol } TiCl_4}{189.7 \text{ g } TiCl_4} \text{ x } \frac{1 \text{ mol } TiO_2}{1 \text{ mol } TiCl_4} \text{ x } \frac{79.87 \text{ g } TiO_2}{1 \text{ mol } TiO_2} = 2358 \text{ g } TiO_2 = 2.36 \text{ kg } TiO_2$

3.66 $2.00 \text{ g Ag x } \frac{1 \text{ mol Ag}}{107.9 \text{ g Ag}} = 0.0185 \text{ mol Ag}; \ 0.657 \text{ g Cl x } \frac{1 \text{ mol Cl}}{35.45 \text{ g Cl}} = 0.0185 \text{ mol Cl}$

$Ag_{0.0185}Cl_{0.0185}$; divide both subscripts by 0.0185. The empirical formula is AgCl.

3.67 $5.0 \text{ g Al x } \frac{1 \text{ mol Al}}{27.0 \text{ g Al}} = 0.19 \text{ mol Al}; \ 4.45 \text{ g O x } \frac{1 \text{ mol O}}{16.0 \text{ g O}} = 0.28 \text{ mol O}$

$Al_{0.19}O_{0.28}$; divide both subscripts by the smaller, 0.19.

$Al_{0.19/0.19}O_{0.28/0.19}$

$Al_1O_{1.5}$; multiply both subscripts by 2 to obtain integers.

The empirical formula is Al_2O_3.

3.68 N_2H_4, 32.05; I_2, 253.8; HI, 127.9

(a) $36.7 \text{ g } N_2H_4 \text{ x } \frac{1 \text{ mol } N_2H_4}{32.05 \text{ g } N_2H_4} \text{ x } \frac{2 \text{ mol } I_2}{1 \text{ mol } N_2H_4} \text{ x } \frac{253.8 \text{ g } I_2}{1 \text{ mol } I_2} = 581 \text{ g } I_2$

(b) $115.7 \text{ g } N_2H_4 \text{ x } \frac{1 \text{ mol } N_2H_4}{32.05 \text{ g } N_2H_4} \text{ x } \frac{4 \text{ mol HI}}{1 \text{ mol } N_2H_4} \text{ x } \frac{127.9 \text{ g HI}}{1 \text{ mol HI}} = 1847 \text{ g HI}$

3.69 H_2S, 34.08; I_2, 253.8; HI, 127.9

(a) $49.2 \text{ g } H_2S \text{ x } \frac{1 \text{ mol } H_2S}{34.08 \text{ g } H_2S} \text{ x } \frac{1 \text{ mol } I_2}{1 \text{ mol } H_2S} \text{ x } \frac{253.8 \text{ g } I_2}{1 \text{ mol } I_2} = 366 \text{ g } I_2$

(b) $95.4 \text{ g } H_2S \times \frac{1 \text{ mol } H_2S}{34.08 \text{ g } H_2S} \times \frac{2 \text{ mol HI}}{1 \text{ mol } H_2S} \times \frac{127.9 \text{ g HI}}{1 \text{ mol HI}} = 716 \text{ g HI}$

Limiting Reactants and Reaction Yield (Sections 3.4 and 3.5)

3.70 $3.44 \text{ mol } N_2 \times \frac{3 \text{ mol } H_2}{1 \text{ mol } N_2} = 10.3 \text{ mol } H_2$ required.

Because there is only 1.39 mol H_2, H_2 is the limiting reactant.

$1.39 \text{ mol } H_2 \times \frac{2 \text{ mol } NH_3}{3 \text{ mol } H_2} \times \frac{17.03 \text{ g } NH_3}{1 \text{ mol } NH_3} = 15.8 \text{ g } NH_3$

$1.39 \text{ mol } H_2 \times \frac{1 \text{ mol } N_2}{3 \text{ mol } H_2} \times \frac{28.01 \text{ g } N_2}{1 \text{ mol } N_2} = 13.0 \text{ g } N_2$ reacted

$3.44 \text{ mol } N_2 \times \frac{28.01 \text{ g } N_2}{1 \text{ mol } N_2} = 96.3 \text{ g } N_2$ initially

$(96.3 \text{ g} - 13.0 \text{ g}) = 83.3 \text{ g } N_2$ left over

3.71 H_2, 2.016; Cl_2, 70.91; HCl 36.46

$3.56 \text{ g } H_2 \times \frac{1 \text{ mol } H_2}{2.016 \text{ g } H_2} = 1.77 \text{ mol } H_2$

$8.94 \text{ g } Cl_2 \times \frac{1 \text{ mol } Cl_2}{70.91 \text{ g } Cl_2} = 0.126 \text{ mol } Cl_2$

Because the reaction stoichiometry between H_2 and Cl_2 is one to one, Cl_2 is the limiting reactant.

$0.126 \text{ mol } Cl_2 \times \frac{2 \text{ mol HCl}}{1 \text{ mol } Cl_2} \times \frac{36.46 \text{ g HCl}}{1 \text{ mol HCl}} = 9.19 \text{ g HCl}$

3.72 C_2H_4, 28.05; Cl_2, 70.91; $C_2H_4Cl_2$, 98.96

$15.4 \text{ g } C_2H_4 \times \frac{1 \text{ mol } C_2H_4}{28.05 \text{ g } C_2H_4} = 0.549 \text{ mol } C_2H_4$

$3.74 \text{ g } Cl_2 \times \frac{1 \text{ mol } Cl_2}{70.91 \text{ g } Cl_2} = 0.0527 \text{ mol } Cl_2$

Because the reaction stoichiometry between C_2H_4 and Cl_2 is one to one, Cl_2 is the limiting reactant.

$0.0527 \text{ mol } Cl_2 \times \frac{1 \text{ mol } C_2H_4Cl_2}{1 \text{ mol } Cl_2} \times \frac{98.96 \text{ g } C_2H_4Cl_2}{1 \text{ mol } C_2H_4Cl_2} = 5.22 \text{ g } C_2H_4Cl_2$

3.73 (a) NaCl, 58.44; $AgNO_3$, 169.9; AgCl, 143.3; $NaNO_3$, 85.00

$NaCl + AgNO_3 \rightarrow AgCl + NaNO_3$

$$1.3\ g\ NaCl \times \frac{1\ mol\ NaCl}{58.44\ g\ NaCl} = 0.022\ mol\ NaCl$$

$$3.5\ g\ AgNO_3 \times \frac{1\ mol\ AgNO_3}{169.9\ g\ AgNO_3} = 0.021\ mol\ AgNO_3$$

Because the reaction stoichiometry between NaCl and $AgNO_3$ is one to one, $AgNO_3$ is the limiting reactant.

$$0.021\ mol\ AgNO_3 \times \frac{1\ mol\ AgCl}{1\ mol\ AgNO_3} \times \frac{143.3\ g\ AgCl}{1\ mol\ AgCl} = 3.0\ g\ AgCl$$

$$0.021\ mol\ AgNO_3 \times \frac{1\ mol\ NaNO_3}{1\ mol\ AgNO_3} \times \frac{85.00\ g\ NaNO_3}{1\ mol\ NaNO_3} = 1.8\ g\ NaNO_3$$

$$0.021\ mol\ AgNO_3 \times \frac{1\ mol\ NaCl}{1\ mol\ AgNO_3} \times \frac{58.44\ g\ NaCl}{1\ mol\ NaCl} = 1.2\ g\ NaCl\ reacted$$

(1.3 g – 1.2 g) = 0.1 g NaCl left over

(b) $BaCl_2$, 208.2; H_2SO_4, 98.08; $BaSO_4$, 233.4; HCl, 36.46

$BaCl_2 + H_2SO_4 \rightarrow BaSO_4 + 2\ HCl$

$$2.65\ g\ BaCl_2 \times \frac{1\ mol\ BaCl_2}{208.2\ g\ BaCl_2} = 0.0127\ mol\ BaCl_2$$

$$6.78\ g\ H_2SO_4 \times \frac{1\ mol\ H_2SO_4}{98.08\ g\ H_2SO_4} = 0.0691\ mol\ H_2SO_4$$

Because the reaction stoichiometry between $BaCl_2$ and H_2SO_4 is one to one, $BaCl_2$ is the limiting reactant.

$$0.0127\ mol\ BaCl_2 \times \frac{1\ mol\ BaSO_4}{1\ mol\ BaCl_2} \times \frac{233.4\ g\ BaSO_4}{1\ mol\ BaSO_4} = 2.96\ g\ BaSO_4$$

$$0.0127\ mol\ BaCl_2 \times \frac{2\ mol\ HCl}{1\ mol\ BaCl_2} \times \frac{36.46\ g\ HCl}{1\ mol\ HCl} = 0.926\ g\ HCl$$

$$0.0127\ mol\ BaCl_2 \times \frac{1\ mol\ H_2SO_4}{1\ mol\ BaCl_2} \times \frac{98.1\ g\ H_2SO_4}{1\ mol\ H_2SO_4} = 1.25\ g\ H_2SO_4\ reacted$$

(6.78 g – 1.25 g) = 5.53 g H_2SO_4 left over

3.74 H_2SO_4, 98.08; $NiCO_3$, 118.7; $NiSO_4$, 154.8

(a) $$14.5\ g\ NiCO_3 \times \frac{1\ mol\ NiCO_3}{118.7\ g\ NiCO_3} \times \frac{1\ mol\ H_2SO_4}{1\ mol\ NiCO_3} \times \frac{98.08\ g\ H_2SO_4}{1\ mol\ H_2SO_4} = 12.0\ g\ H_2SO_4$$

(b) $$14.5\ g\ NiCO_3 \times \frac{1\ mol\ NiCO_3}{118.7\ g\ NiCO_3} \times \frac{1\ mol\ NiSO_4}{1\ mol\ NiCO_3} \times \frac{154.8\ g\ NiSO_4}{1\ mol\ NiSO_4} \times 0.789 = 14.9\ g\ NiSO_4$$

3.75 O_2, 32.00; N_2, 28.01; N_2H_4, 32.05

(a) $$50.0\ g\ N_2H_4 \times \frac{1\ mol\ N_2H_4}{32.05\ g\ N_2H_4} \times \frac{1\ mol\ O_2}{1\ mol\ N_2H_4} \times \frac{32.00\ g\ O_2}{1\ mol\ O_2} = 49.9\ g\ O_2$$

(b) $50.0 \text{ g } N_2H_4 \times \frac{1 \text{ mol } N_2H_4}{32.05 \text{ g } N_2H_4} \times \frac{1 \text{ mol } N_2}{1 \text{ mol } N_2H_4} \times \frac{28.01 \text{ g } N_2}{1 \text{ mol } N_2} \times 0.855 = 37.4 \text{ g } N_2$

3.76 $CaCO_3$, 100.1; HCl, 36.46

$CaCO_3 + 2\,HCl \rightarrow CaCl_2 + H_2O + CO_2$

$$2.35 \text{ g } CaCO_3 \times \frac{1 \text{ mol } CaCO_3}{100.1 \text{ g } CaCO_3} = 0.0235 \text{ mol } CaCO_3$$

$$2.35 \text{ g HCl} \times \frac{1 \text{ mol HCl}}{36.46 \text{ g HCl}} = 0.0645 \text{ mol HCl}$$

The reaction stoichiometry is 1 mole of $CaCO_3$ for every 2 moles of HCl. For 0.0235 mol $CaCO_3$, we only need 2(0.0235 mol) = 0.0470 mol HCl. We have 0.0645 mol HCl, therefore $CaCO_3$ is the limiting reactant.

$$0.0235 \text{ mol } CaCO_3 \times \frac{1 \text{ mol } CO_2}{1 \text{ mol } CaCO_3} \times \frac{22.4 \text{ L}}{1 \text{ mol } CO_2} = 0.526 \text{ L } CO_2$$

3.77 $2\,NaN_3 \rightarrow 3\,N_2 + 2\,Na$; NaN_3, 65.01; N_2, 28.01

$$38.5 \text{ g } NaN_3 \times \frac{1 \text{ mol } NaN_3}{65.01 \text{ g } NaN_3} \times \frac{3 \text{ mol } N_2}{2 \text{ mol } NaN_3} \times \frac{47.0 \text{ L}}{1.00 \text{ mol } N_2} = 41.8 \text{ L}$$

3.78 $CH_3CO_2H + C_5H_{12}O \rightarrow C_7H_{14}O_2 + H_2O$

CH_3CO_2H, 60.05; $C_5H_{12}O$, 88.15; $C_7H_{14}O_2$, 130.19

$$3.58 \text{ g } CH_3CO_2H \times \frac{1 \text{ mol } CH_3CO_2H}{60.05 \text{ g } CH_3CO_2H} = 0.0596 \text{ mol } CH_3CO_2H$$

$$4.75 \text{ g } C_5H_{12}O \times \frac{1 \text{ mol } C_5H_{12}O}{88.15 \text{ g } C_5H_{12}O} = 0.0539 \text{ mol } C_5H_{12}O$$

Because the reaction stoichiometry between CH_3CO_2H and $C_5H_{12}O$ is one to one, isopentyl alcohol ($C_5H_{12}O$) is the limiting reactant.

$$0.0539 \text{ mol } C_5H_{12}O \times \frac{1 \text{ mol } C_7H_{14}O_2}{1 \text{ mol } C_5H_{12}O} \times \frac{130.19 \text{ g } C_7H_{14}O_2}{1 \text{ mol } C_7H_{14}O_2} = 7.02 \text{ g } C_7H_{14}O_2$$

7.02 g $C_7H_{14}O_2$ is the theoretical yield. Actual yield = (7.02 g)(0.45) = 3.2 g.

3.79 $K_2PtCl_4 + 2\,NH_3 \rightarrow 2\,KCl + Pt(NH_3)_2Cl_2$

K_2PtCl_4, 415.1; NH_3, 17.03; $Pt(NH_3)_2Cl_2$, 300.0

$$55.8 \text{ g } K_2PtCl_4 \times \frac{1 \text{ mol } K_2PtCl_4}{415.1 \text{ g } K_2PtCl_4} = 0.134 \text{ mol } K_2PtCl_4$$

$$35.6 \text{ g } NH_3 \times \frac{1 \text{ mol } NH_3}{17.03 \text{ g } NH_3} = 2.09 \text{ mol } NH_3$$

Only 2(0.134) = 0.268 mol NH_3 are needed to react with 0.134 mol K_2PtCl_4. Therefore, the NH_3 is in excess and K_2PtCl_4 is the limiting reactant.

$$0.134 \text{ mol } K_2PtCl_4 \times \frac{1 \text{ mol } Pt(NH_3)_2Cl_2}{1 \text{ mol } K_2PtCl_4} \times \frac{300.0 \text{ g } Pt(NH_3)_2Cl_2}{1 \text{ mol } Pt(NH_3)_2Cl_2} = 40.2 \text{ g } Pt(NH_3)_2Cl_2$$

40.2 g $Pt(NH_3)_2Cl_2$ is the theoretical yield.
Actual yield = (40.2 g)(0.95) = 38 g $Pt(NH_3)_2Cl_2$.

3.80 $CH_3CO_2H + C_5H_{12}O \rightarrow C_7H_{14}O_2 + H_2O$

CH_3CO_2H, 60.05; $C_5H_{12}O$, 88.15; $C_7H_{14}O_2$, 130.19

$$1.87 \text{ g } CH_3CO_2H \times \frac{1 \text{ mol } CH_3CO_2H}{60.05 \text{ g } CH_3CO_2H} = 0.0311 \text{ mol } CH_3CO_2H$$

$$2.31 \text{ g } C_5H_{12}O \times \frac{1 \text{ mol } C_5H_{12}O}{88.15 \text{ g } C_5H_{12}O} = 0.0262 \text{ mol } C_5H_{12}O$$

Because the reaction stoichiometry between CH_3CO_2H and $C_5H_{12}O$ is one to one, isopentyl alcohol ($C_5H_{12}O$) is the limiting reactant.

$$0.0262 \text{ mol } C_5H_{12}O \times \frac{1 \text{ mol } C_7H_{14}O_2}{1 \text{ mol } C_5H_{12}O} \times \frac{130.19 \text{ g } C_7H_{14}O_2}{1 \text{ mol } C_7H_{14}O_2} = 3.41 \text{ g } C_7H_{14}O_2$$

3.41 g $C_7H_{14}O_2$ is the theoretical yield.

$$\% \text{ Yield} = \frac{\text{Actual yield}}{\text{Theoretical yield}} \times 100\% = \frac{2.96 \text{ g}}{3.41 \text{ g}} \times 100\% = 86.8\%$$

3.81 $K_2PtCl_4 + 2\,NH_3 \rightarrow 2\,KCl + Pt(NH_3)_2Cl_2$

K_2PtCl_4, 415.1; NH_3, 17.03; $Pt(NH_3)_2Cl_2$, 300.0

$$3.42 \text{ g } K_2PtCl_4 \times \frac{1 \text{ mol } K_2PtCl_4}{415.1 \text{ g } K_2PtCl_4} = 0.008\ 24 \text{ mol } K_2PtCl_4$$

$$1.61 \text{ g } NH_3 \times \frac{1 \text{ mol } NH_3}{17.03 \text{ g } NH_3} = 0.0945 \text{ mol } NH_3$$

Only 2 x (0.008 24) = 0.0165 mol of NH_3 are needed to react with 0.008 24 mol K_2PtCl_4. Therefore, the NH_3 is in excess and K_2PtCl_4 is the limiting reactant.

$$0.008\ 24 \text{ mol } K_2PtCl_4 \times \frac{1 \text{ mol } Pt(NH_3)_2Cl_2}{1 \text{ mol } K_2PtCl_4} \times \frac{300.0 \text{ g } Pt(NH_3)_2Cl_2}{1 \text{ mol } Pt(NH_3)_2Cl_2} = 2.47 \text{ g } Pt(NH_3)_2Cl_2$$

2.47 g $Pt(NH_3)_2Cl_2$ is the theoretical yield. 2.08 g $Pt(NH_3)_2Cl_2$ is the actual yield.

$$\% \text{ Yield} = \frac{\text{Actual yield}}{\text{Theoretical yield}} \times 100\% = \frac{2.08 \text{ g}}{2.47 \text{ g}} \times 100\% = 84.2\%$$

Formulas and Elemental Analysis (Sections 3.6 and 3.7)

3.82 CH_4N_2O, 60.1

$$\% \text{C} = \frac{12.0 \text{ g C}}{60.1 \text{ g}} \times 100\% = 20.0\%$$

$$\% \text{H} = \frac{4 \times 1.01 \text{ g H}}{60.1 \text{ g}} \times 100\% = 6.72\%$$

$$\% N = \frac{2 \text{ x } 14.0 \text{ g N}}{60.1 \text{ g}} \text{ x } 100\% = 46.6\%$$

$$\% O = \frac{16.0 \text{ g O}}{60.1 \text{ g}} \text{ x } 100\% = 26.6\%$$

3.83 (a) $Cu_2(OH)_2CO_3$, 221.1

$$\% Cu = \frac{2 \text{ x } 63.5 \text{ g Cu}}{221.1 \text{ g}} \text{ x } 100\% = 57.4\%$$

$$\% O = \frac{5 \text{ x } 16.0 \text{ g O}}{221.1 \text{ g}} \text{ x } 100\% = 36.2\%$$

$$\% C = \frac{12.0 \text{ g C}}{221.1 \text{ g}} \text{ x } 100\% = 5.43\%$$

$$\% H = \frac{2 \text{ x } 1.01 \text{ g H}}{221.1 \text{ g}} \text{ x } 100\% = 0.914\%$$

(b) $C_8H_9NO_2$, 151.2

$$\% C = \frac{8 \text{ x } 12.0 \text{ g C}}{151.2 \text{ g}} \text{ x } 100\% = 63.5\%$$

$$\% H = \frac{9 \text{ x } 1.01 \text{ g H}}{151.2 \text{ g}} \text{ x } 100\% = 6.01\%$$

$$\% N = \frac{14.0 \text{ g N}}{151.2 \text{ g}} \text{ x } 100\% = 9.26\%$$

$$\% O = \frac{2 \text{ x } 16.0 \text{ g O}}{151.2 \text{ g}} \text{ x } 100\% = 21.2\%$$

(c) $Fe_4[Fe(CN)_6]_3$, 859.2

$$\% Fe = \frac{7 \text{ x } 55.85 \text{ g Fe}}{859.2 \text{ g}} \text{ x } 100\% = 45.50\%$$

$$\% C = \frac{18 \text{ x } 12.01 \text{ g C}}{859.2 \text{ g}} \text{ x } 100\% = 25.16\%$$

$$\% N = \frac{18 \text{ x } 14.01 \text{ g N}}{859.2 \text{ g}} \text{ x } 100\% = 29.35\%$$

3.84 Assume a 100.0 g sample of liquid. From the percent composition data, a 100.0 g sample of liquid contains 5.57 g H, 28.01 g Cl, and 66.42 g C.

$$66.42 \text{ g C x } \frac{1 \text{ mol C}}{12.01 \text{ g C}} = 5.530 \text{ mol C}$$

$$5.57 \text{ g H x } \frac{1 \text{ mol H}}{1.01 \text{ g H}} = 5.51 \text{ mol H}$$

$$28.01 \text{ g Cl x } \frac{1 \text{ mol Cl}}{35.45 \text{ g Cl}} = 0.7901 \text{ mol Cl}$$

$C_{5.530}H_{5.51}Cl_{0.7901}$; divide each subscript by the smallest, 0.7901.
$C_{5.530 / 0.7901}H_{5.51 / 0.7901}Cl_{0.7901 / 0.7901}$
C_7H_7Cl
The empirical formula is C_7H_7Cl, 126.59.
Because the molecular weight equals the empirical formula weight, the empirical formula is also the molecular formula.

3.85 Assume a 100.0 g sample of liquid. From the percent composition data, a 100.0 g sample of liquid contains 34.31 g C, 5.28 g H, and 60.41 g I.

$$34.31 \text{ g C} \times \frac{1 \text{ mol C}}{12.01 \text{ g C}} = 2.857 \text{ mol C}$$

$$5.28 \text{ g H} \times \frac{1 \text{ mol H}}{1.01 \text{ g H}} = 5.23 \text{ mol H}$$

$$60.41 \text{ g I} \times \frac{1 \text{ mol I}}{126.9 \text{ g I}} = 0.4760 \text{ mol I}$$

$C_{2.857}H_{5.23}I_{0.4760}$; divide each subscript by the smallest, 0.4760.
$C_{2.857 / 0.4760}H_{5.23 / 0.4760}I_{0.4760 / 0.4760}$
$C_6H_{11}I$
The empirical formula is $C_6H_{11}I$, 210.07.
Because the molecular weight equals the empirical formula weight, the empirical formula is also the molecular formula.

3.86 Assume a 100.0 g sample. From the percent composition data, a 100.0 g sample contains 24.25 g F and 75.75 g Sn.

$$24.25 \text{ g F} \times \frac{1 \text{ mol F}}{19.00 \text{ g F}} = 1.276 \text{ mol F}$$

$$75.75 \text{ g Sn} \times \frac{1 \text{ mol Sn}}{118.7 \text{ g Sn}} = 0.6382 \text{ mol Sn}$$

$Sn_{0.6382}F_{1.276}$; divide each subscript by the smaller, 0.6382.
$Sn_{0.6382 / 0.6382}F_{1.276 / 0.6382}$
The empirical formula is SnF_2.

3.87 (a) Assume a 100.0 g sample. From the percent composition data, a 100.0 g sample contains 75.69 g C, 15.51 g O, and 8.80 g H.

$$75.69 \text{ g C} \times \frac{1 \text{ mol C}}{12.01 \text{ g C}} = 6.302 \text{ mol C}$$

$$8.80 \text{ g H} \times \frac{1 \text{ mol H}}{1.01 \text{ g H}} = 8.71 \text{ mol H}$$

$$15.51 \text{ g O} \times \frac{1 \text{ mol O}}{16.00 \text{ g O}} = 0.9694 \text{ mol O}$$

$C_{6.302}H_{8.71}O_{0.9694}$; divide each subscript by the smallest, 0.9694.
$C_{6.302 / 0.9694}H_{8.71 / 0.9694}O_{0.9694 / 0.9694}$
$C_{6.5}H_9O$; multiply each subscript by 2 to obtain integers.
The empirical formula is $C_{13}H_{18}O_2$.

(b) Assume a 100.0 g sample. From the percent composition data, a 100.0 g sample contains 72.36 g Fe, and 27.64 g O.

$$72.36 \text{ g Fe x } \frac{1 \text{ mol Fe}}{55.85 \text{ g Fe}} = 1.296 \text{ mol Fe}$$

$$27.64 \text{ g O x } \frac{1 \text{ mol O}}{16.00 \text{ g O}} = 1.727 \text{ mol O}$$

$Fe_{1.296}O_{1.727}$; divide each subscript by the smaller, 1.296.
$Fe_{1.296 / 1.296}O_{1.727 / 1.296}$
$FeO_{1.333}$; multiply each subscript by 3 to obtain integers.
The empirical formula is Fe_3O_4.
(c) Assume a 100.0 g sample. From the percent composition data, a 100.0 g sample contains 34.91 g O, 15.32 g Si, and 49.77 g Zr.

$$49.77 \text{ g Zr x } \frac{1 \text{ mol Zr}}{91.22 \text{ g Zr}} = 0.5456 \text{ mol Zr}$$

$$15.32 \text{ g Si x } \frac{1 \text{ mol Si}}{28.09 \text{ g Si}} = 0.5454 \text{ mol Si}$$

$$34.91 \text{ g O x } \frac{1 \text{ mol O}}{16.00 \text{ g O}} = 2.182 \text{ mol O}$$

$Zr_{0.5456}Si_{0.5454}O_{2.182}$; divide each subscript by the smallest, 0.5454.
$Zr_{0.5456 / 0.5454}Si_{0.5454/ 0.5454}O_{2.182 / 0.5454}$
The empirical formula is $ZrSiO_4$.

3.88 Mass of toluene sample = 45.62 mg = 0.045 62 g; mass of CO_2 = 152.5 mg = 0.1525 g; mass of H_2O = 35.67 mg = 0.035 67 g

$$0.1525 \text{ g } CO_2 \text{ x } \frac{1 \text{ mol } CO_2}{44.01 \text{ g } CO_2} \text{ x } \frac{1 \text{ mol C}}{1 \text{ mol } CO_2} = 0.003\ 465 \text{ mol C}$$

$$\text{mass C} = 0.003\ 465 \text{ mol C x } \frac{12.011 \text{ g C}}{1 \text{ mol C}} = 0.041\ 62 \text{ g C}$$

$$0.035\ 67 \text{ g } H_2O \text{ x } \frac{1 \text{ mol } H_2O}{18.02 \text{ g } H_2O} \text{ x } \frac{2 \text{ mol H}}{1 \text{ mol } H_2O} = 0.003\ 959 \text{ mol H}$$

$$\text{mass H} = 0.003\ 959 \text{ mol H x } \frac{1.008 \text{ g H}}{1 \text{ mol H}} = 0.003\ 991 \text{ g H}$$

The (mass C + mass H) = 0.041 62 g + 0.003 991 g = 0.045 61 g. The calculated mass of (C + H) essentially equals the mass of the toluene sample, this means that toluene contains only C and H and no other elements.
$C_{0.003\ 465}H_{0.003\ 959}$; divide each subscript by the smaller, 0.003 465.
$C_{0.003\ 465 / 0.003\ 465}H_{0.003\ 959 / 0.003\ 465}$
$CH_{1.14}$; multiply each subscript by 7 to obtain integers.
The empirical formula is C_7H_8.

3.89 5.024 mg = 0.005 024 g; 13.90 mg = 0.013 90 g; 6.048 mg = 0.006 048 g

$$0.013\ 90\ \text{g}\ CO_2 \times \frac{1\ \text{mol}\ CO_2}{44.01\ \text{g}\ CO_2} \times \frac{1\ \text{mol C}}{1\ \text{mol}\ CO_2} = 3.158 \times 10^{-4}\ \text{mol C}$$

$$0.006\ 048\ \text{g}\ H_2O \times \frac{1\ \text{mol}\ H_2O}{18.02\ \text{g}\ H_2O} \times \frac{2\ \text{mol H}}{1\ \text{mol}\ H_2O} = 6.713 \times 10^{-4}\ \text{mol H}$$

$$3.158 \times 10^{-4}\ \text{mol C} \times \frac{12.01\ \text{g C}}{1\ \text{mol C}} = 0.003\ 793\ \text{g C}$$

$$6.713 \times 10^{-4}\ \text{mol H} \times \frac{1.008\ \text{g H}}{1\ \text{mol H}} = 0.000\ 676\ 7\ \text{g H}$$

mass N = 0.005 024 g – (0.003 793 g + 0.000 676 7 g) = 0.000 554 g N

$$0.000\ 554\ \text{g N} \times \frac{1\ \text{mol N}}{14.01\ \text{g N}} = 3.95 \times 10^{-5}\ \text{mol N}$$

Scale each mol quantity to eliminate exponents.

$C_{3.158}H_{6.713}N_{0.395}$; divide each subscript by the smallest, 0.395.

$C_{3.158/0.395}H_{6.713/0.395}N_{0.395/0.395}$

The empirical formula is $C_8H_{17}N$.

3.90 Let X equal the molecular weight of cytochrome c.

$$0.0043 = \frac{55.847\ \text{u}}{X}; \quad X = \frac{55.847\ \text{u}}{0.0043} = 13{,}000\ \text{u}$$

3.91 Let X equal the molecular weight of nitrogenase.

$$0.000\ 872 = \frac{2 \times 95.94\ \text{u}}{X}; \quad X = \frac{2 \times 95.94\ \text{u}}{0.000\ 872} = 220{,}000\ \text{u}$$

3.92 Let X equal the molecular weight of disilane.

$$0.9028 = \frac{2 \times 28.09\ \text{u}}{X}; \quad X = \frac{2 \times 28.09\ \text{u}}{0.9028} = 62.23\ \text{u}$$

62.23 – 2(Si atomic weight) = 62.23 – 2(28.09) = 6.05

6.05 is the total mass of H atoms.

$$6.05\ \text{u} \times \frac{1\ \text{H atom}}{1.01\ \text{u}} = 6\ \text{H atoms}$$; Disilane is Si_2H_6.

3.93 Let X equal the molecular weight of MS_2.

$$0.4006 = \frac{2 \times 32.07\ \text{u}}{X}; \quad X = \frac{2 \times 32.07\ \text{u}}{0.4006} = 160.1\ \text{u}$$

Atomic weight of M = 160.1 – 2(S atomic weight)
= 160.1 – 2(32.07) = 95.96

M is Mo.

3.94 $C_{12}Br_{10}$, 943.2; $C_{12}Br_{10}O$, 959.2; 17.33 mg = 0.017 33 g

For $C_{12}Br_{10}$, $\% C = \frac{12 \text{ x } 12.01 \text{ g C}}{943.2 \text{ g}} \text{ x } 100\% = 15.28\%$

For $C_{12}Br_{10}O$, $\% C = \frac{12 \text{ x } 12.01 \text{ g C}}{959.2 \text{ g}} \text{ x } 100\% = 15.03\%$

Calculate the mass of C in 17.33 mg of CO_2.

$$0.017\ 33 \text{ g } CO_2 \text{ x } \frac{1 \text{ mol } CO_2}{44.01 \text{ g } CO_2} \text{ x } \frac{1 \text{ mol C}}{1 \text{ mol } CO_2} \text{ x } \frac{12.01 \text{ g C}}{1 \text{ mol C}} = 0.004\ 729 \text{ g} = 4.729 \text{ mg C}$$

Calculate the %C in the 31.472 mg sample.

$$\%C = \frac{4.729 \text{ mg C}}{31.472 \text{ mg}} \text{ x } 100\% = 15.03\%$$

Decabrom is $C_{12}Br_{10}O$.

3.95 Mass of amphetamine sample = 42.92 mg = 0.042 92 g;
mass of CO_2 = 125.75 mg = 0.125 75;
mass of H_2O = 37.187 mg = 0.037 187 g

$$0.125\ 75\text{g } CO_2 \text{ x } \frac{1 \text{ mol } CO_2}{44.01 \text{ g } CO_2} \text{ x } \frac{1 \text{ mol C}}{1 \text{ mol } CO_2} = 0.002\ 857 \text{ mol C}$$

$$\text{mass C} = 0.002\ 857 \text{ mol C x } \frac{12.01 \text{ g C}}{1 \text{ mol C}} = 0.034\ 32 \text{ g} = 34.32 \text{ mg C}$$

$$0.037\ 187 \text{ g } H_2O \text{ x } \frac{1 \text{ mol } H_2O}{18.02 \text{ g } H_2O} \text{ x } \frac{2 \text{ mol H}}{1 \text{ mol } H_2O} = 0.004\ 127 \text{ mol H}$$

$$\text{mass H} = 0.004\ 127 \text{ mol H x } \frac{1.008 \text{ g H}}{1 \text{ mol H}} = 0.004\ 160 \text{ g} = 4.160 \text{ mg H}$$

mass N = 42.92 mg – mass H – mass C = 42.92 mg – 4.16 mg – 34.32 mg = 4.44 mg = 0.004 44 g N

$$0.004\ 44 \text{ g N x } \frac{1 \text{ mol N}}{14.01 \text{ g N}} = 0.000\ 317 \text{ mol N}$$

$C_{0.002\ 857}H_{0.004\ 127}N_{0.000\ 317}$; divide each subscript by the smallest, 0.000 317.

$C_{0.002\ 857 / 0.000\ 317}H_{0.004\ 127 / 0.000\ 317}N_{0.000\ 317 / 0.000\ 317}$

$C_9H_{13}N$

The empirical formula is $C_9H_{13}N$, 135.20.

Because the molecular weight is less than 160 g/mol, the empirical formula is also the molecular formula.

Mass Spectrometry (Section 3.8)

3.96 A neutral molecule will travel in a straight, undeflected, path in a mass spectrometer. Ionization is necessary as electric and magnetic fields will only exert a force on a charged species, not a neutral molecule. Ions of different masses are then accelerated by an electric field and passed between the poles of a strong magnet, which deflects them through a curved, evacuated pipe. The radius of deflection of a charged ion, M^+, as it passes between the magnet poles depends on its mass, with lighter ions deflected more strongly than heavier ones.

3.97 High mass accuracy is often needed to make an identification of a compound. For example, two compounds with different molecular formulas can have very similar masses, C_5H_8O = 84.0570 and C_6H_{12} = 84.0934. Highly accurate mass measurements are used to confirm the identity of molecules.

3.98 Mass of the sample is 70.042 11.
For C_5H_{10}, mass = 5(12.000 000) + 10(1.007 825) = 70.078 250
For C_4H_6O, mass = 4(12.000 000) + 6(1.007 825) + 1(15.994 915) = 70.041 865
For $C_3H_6N_2$, mass = 3(12.000 000) + 6(1.007 825) + 2(14.003 074) = 70.053 098
The sample is C_4H_6O.

3.99 Mass of the sample is 58.077 46.
For C_4H_{10}, mass = 4(12.000 000) + 10(1.007 825) = 58.078 250
For C_3H_6O, mass = 3(12.000 000) + 6(1.007 825) + 1(15.994 915) = 58.041 865
For $C_2H_6N_2$, mass = 2(12.000 000) + 6(1.007 825) + 2(14.003 074) = 58.053 098
The sample is C_4H_{10}.

3.100 mass of CO_2 = 169.2 mg = 0.1692 g; mass of H_2O = 34.6 mg = 0.0346 g

$$0.1692 \text{ g } CO_2 \text{ x } \frac{1 \text{ mol } CO_2}{44.01 \text{ g } CO_2} \text{ x } \frac{1 \text{ mol C}}{1 \text{ mol } CO_2} = 0.003\ 845 \text{ mol C}$$

$$0.0346 \text{ g } H_2O \text{ x } \frac{1 \text{ mol } H_2O}{18.02 \text{ g } H_2O} \text{ x } \frac{2 \text{ mol H}}{1 \text{ mol } H_2O} = 0.003\ 84 \text{ mol H}$$

$C_{0.003\ 845}H_{0.003\ 84}$; divide each subscript by the smaller, 0.003 84.
$C_{0.003\ 845 / 0.003\ 84}H_{0.003\ 84 / 0.003\ 84}$
The empirical formula is CH, 13.0.
The peak with the largest mass in the mass spectrum occurs near a mass of 78, which would be the molecular weight of the compound.
78/13.0 = 6; molecular formula = $C_{(6 \text{ x } 1)}H_{(6 \text{ x } 1)} = C_6H_6$

3.101 mass of 1,2,3-benzenetriol = 150.0 mg = 0.1500 g;
mass of CO_2 = 314.2 mg = 0.3142 g; mass of H_2O = 64.3 mg = 0.0643 g

$$0.3142 \text{ g } CO_2 \text{ x } \frac{1 \text{ mol } CO_2}{44.01 \text{ g } CO_2} \text{ x } \frac{1 \text{ mol C}}{1 \text{ mol } CO_2} = 0.007\ 139 \text{ mol C}$$

$$0.0643 \text{ g } H_2O \text{ x } \frac{1 \text{ mol } H_2O}{18.02 \text{ g } H_2O} \text{ x } \frac{2 \text{ mol H}}{1 \text{ mol } H_2O} = 0.007\ 14 \text{ mol H}$$

$$0.007\ 139 \text{ mol C x } \frac{12.01 \text{ g C}}{1 \text{ mol C}} = 0.085\ 74 \text{ g C}$$

$$0.007\ 14 \text{ mol H x } \frac{1.008 \text{ g H}}{1 \text{ mol H}} = 0.007\ 20 \text{ g H}$$

mass O = 0.1500 g – (0.085 74 g + 0.007 20 g) = 0.0571 g O

$$0.0571 \text{ g O x } \frac{1 \text{ mol O}}{16.00 \text{ g O}} = 0.003\ 57 \text{ mol O}$$

$C_{0.007\,139}H_{0.007\,14}O_{0.003\,57}$; divide each subscript by the smallest, 0.003 57.
$C_{0.007\,139\,/\,0.003\,57}\,H_{0.007\,14\,/\,0.003\,57}O_{0.003\,57\,/\,0.003\,57}$
The empirical formula is C_2H_2O, 42.0.
The peak with the largest mass in the mass spectrum occurs near a mass of 125, which would be the molecular weight of the compound.
125/42.0 = 2.97 = 3; molecular formula = $C_{(3\,x\,2)}H_{(3\,x\,2)}O_{(3\,x\,1)} = C_6H_6O_3$

Chapter Problems

3.102 NaCl, 58.4; KCl, 74.6; $CaCl_2$, 111.0; 500 mL = 0.500 L

$$4.30\text{ g NaCl x }\frac{1\text{ mol NaCl}}{58.4\text{ g NaCl}} = 0.0736\text{ mol NaCl}$$

$$0.150\text{ g KCl x }\frac{1\text{ mol KCl}}{74.6\text{ g KCl}} = 0.002\ 01\text{ mol KCl}$$

$$0.165\text{ g CaCl}_2\text{ x }\frac{1\text{ mol CaCl}_2}{111.0\text{ g CaCl}_2} = 0.001\ 49\text{ mol CaCl}_2$$

0.0736 mol + 0.002 01 mol + 2(0.001 49 mol) = 0.0786 mol Cl^-

$$Na^+\text{ molarity} = \frac{0.0736\text{ mol}}{0.500\text{ L}} = 0.147\text{ M}$$

$$Ca^{2+}\text{ molarity} = \frac{0.001\ 49\text{ mol}}{0.500\text{ L}} = 0.002\ 98\text{ M}$$

$$K^+\text{ molarity} = \frac{0.002\ 01\text{ mol}}{0.500\text{ L}} = 0.004\ 02\text{ M}$$

$$Cl^-\text{ molarity} = \frac{0.0786\text{ mol}}{0.500\text{ L}} = 0.157\text{ M}$$

3.103 (a) $4\ C_6H_5NO_2 + 29\ O_2 \rightarrow 24\ CO_2 + 10\ H_2O + 4\ NO_2$
(b) $2\ Au + 6\ H_2SeO_4 \rightarrow Au_2(SeO_4)_3 + 3\ H_2SeO_3 + 3\ H_2O$
(c) $6\ NH_4ClO_4 + 8\ Al \rightarrow 4\ Al_2O_3 + 3\ N_2 + 3\ Cl_2 + 12\ H_2O$

3.104 H_2S, 34.08; O_2, 32.00; Ag_2S, 247.8
(a) $4\ Ag + 2\ H_2S + O_2 \rightarrow 2\ Ag_2S + 2\ H_2O$
(b) Compute the theoretical yield for Ag_2S from each reactant to determine the limiting reactant.

$$496\text{ g Ag x }\frac{1\text{ mol Ag}}{107.9\text{ g Ag}}\text{ x }\frac{2\text{ mol Ag}_2\text{S}}{4\text{ mol Ag}}\text{ x }\frac{247.8\text{ g Ag}_2\text{S}}{1\text{ mol Ag}_2\text{S}} = 570\text{ g Ag}_2\text{S}$$

$$80.0\text{ g H}_2\text{S x }\frac{1\text{ mol H}_2\text{S}}{34.08\text{ g H}_2\text{S}}\text{ x }\frac{2\text{ mol Ag}_2\text{S}}{2\text{ mol H}_2\text{S}}\text{ x }\frac{247.8\text{ g Ag}_2\text{S}}{1\text{ mol Ag}_2\text{S}} = 582\text{ g Ag}_2\text{S}$$

$$40.0\text{ g O}_2\text{ x }\frac{1\text{ mol O}_2}{32.00\text{ g O}_2}\text{ x }\frac{2\text{ mol Ag}_2\text{S}}{1\text{ mol O}_2}\text{ x }\frac{247.8\text{ g Ag}_2\text{S}}{1\text{ mol Ag}_2\text{S}} = 619\text{ g Ag}_2\text{S}$$

The smallest theoretical yield (570 g) means that Ag is the limiting reactant.
With a 95% yield, the mass of Ag_2S produced = (0.95)(570 g) = 541 g Ag_2S

3.105 (a) $C_6H_{12}O_6$, 180.2

$$\% C = \frac{6 \text{ x } 12.01 \text{ g C}}{180.2 \text{ g}} \text{ x } 100\% = 39.99\%$$

$$\% H = \frac{12 \text{ x } 1.008 \text{ g H}}{180.2 \text{ g}} \text{ x } 100\% = 6.713\%$$

$$\% O = \frac{6 \text{ x } 16.00 \text{ g O}}{180.2 \text{ g}} \text{ x } 100\% = 53.27\%$$

(b) H_2SO_4, 98.08

$$\% H = \frac{2 \text{ x } 1.008 \text{ g H}}{98.08 \text{ g}} \text{ x } 100\% = 2.055\%$$

$$\% S = \frac{32.07 \text{ g S}}{98.08 \text{ g}} \text{ x } 100\% = 32.70\%$$

$$\% O = \frac{4 \text{ x } 16.00 \text{ g O}}{98.08 \text{ g}} \text{ x } 100\% = 65.25\%$$

(c) $KMnO_4$, 158.0

$$\% K = \frac{39.10 \text{ g K}}{158.0 \text{ g}} \text{ x } 100\% = 24.75\%$$

$$\% Mn = \frac{54.94 \text{ g Mn}}{158.0 \text{ g}} \text{ x } 100\% = 34.77\%$$

$$\% O = \frac{4 \text{ x } 16.00 \text{ g O}}{158.0 \text{ g}} \text{ x } 100\% = 40.51\%$$

(d) $C_7H_5NO_3S$, 183.2

$$\% C = \frac{7 \text{ x } 12.01 \text{ g C}}{183.2 \text{ g}} \text{ x } 100\% = 45.89\%$$

$$\% H = \frac{5 \text{ x } 1.008 \text{ g H}}{183.2 \text{ g}} \text{ x } 100\% = 2.751\%$$

$$\% N = \frac{14.01 \text{ g N}}{183.2 \text{ g}} \text{ x } 100\% = 7.647\%$$

$$\% O = \frac{3 \text{ x } 16.00 \text{ g O}}{183.2 \text{ g}} \text{ x } 100\% = 26.20\%$$

$$\% S = \frac{32.07 \text{ g S}}{183.2 \text{ g}} \text{ x } 100\% = 17.51\%$$

3.106 (a) Assume a 100.0 g sample of aspirin. From the percent composition data, a 100.0 g sample contains 60.00 g C, 35.52 g O, and 4.48 g H.

$$60.00 \text{ g C x } \frac{1 \text{ mol C}}{12.01 \text{ g C}} = 4.996 \text{ mol C}$$

$$35.52 \text{ g O x } \frac{1 \text{ mol O}}{16.00 \text{ g O}} = 2.220 \text{ mol O}$$

$4.48\ g\ H \times \frac{1\ mol\ H}{1.01\ g\ H} = 4.44\ mol\ H$

$C_{4.996}H_{4.44}O_{2.220}$; divide each subscript by the smallest, 2.220.

$C_{4.996\,/\,2.220}H_{4.44\,/\,2.220}O_{2.220\,/\,2.220}$

$C_{2.25}H_2O_1$; multiply each subscript by 4 to obtain integers.

The empirical formula is $C_9H_8O_4$.

(b) Assume a 100.0 g sample of ilmenite. From the percent composition data, a 100.0 g sample contains 31.63 g O, 31.56 g Ti, and 36.81 g Fe.

$31.63\ g\ O \times \frac{1\ mol\ O}{16.00\ g\ O} = 1.977\ mol\ O$

$31.56\ g\ Ti \times \frac{1\ mol\ Ti}{47.87\ g\ Ti} = 0.6593\ mol\ Ti$

$36.81\ g\ Fe \times \frac{1\ mol\ Fe}{55.85\ g\ Fe} = 0.6591\ mol\ Fe$

$Fe_{0.6591}Ti_{0.6593}O_{1.977}$; divide each subscript by the smallest, 0.6591.

$Fe_{0.6591\,/\,0.6591}Ti_{0.6593\,/\,0.6591}O_{1.977\,/\,0.6591}$

The empirical formula is $FeTiO_3$.

(c) Assume a 100.0 g sample of sodium thiosulfate. From the percent composition data, a 100.0 g sample contains 30.36 g O, 29.08 g Na, and 40.56 g S.

$30.36\ g\ O \times \frac{1\ mol\ O}{16.00\ g\ O} = 1.897\ mol\ O$

$29.08\ g\ Na \times \frac{1\ mol\ Na}{22.99\ g\ Na} = 1.265\ mol\ Na$

$40.56\ g\ S \times \frac{1\ mol\ S}{32.07\ g\ S} = 1.265\ mol\ S$

$Na_{1.265}S_{1.265}O_{1.897}$; divide each subscript by the smallest, 1.265.

$Na_{1.265\,/\,1.265}S_{1.265\,/\,1.265}O_{1.897\,/\,1.265}$

$NaSO_{1.5}$; multiply each subscript by 2 to obtain integers.

The empirical formula is $Na_2S_2O_3$.

3.107 WCl_6, 396.6; W_6Cl_{12}, 1528

(a) $6\ WCl_6 + 8\ Bi \rightarrow W_6Cl_{12} + 8\ BiCl_3$

(b) $150.0\ g\ WCl_6 \times \frac{1\ mol\ WCl_6}{396.6\ g\ WCl_6} \times \frac{8\ mol\ Bi}{6\ mol\ WCl_6} \times \frac{209.0\ g\ Bi}{1\ mol\ Bi} = 105.4\ g\ Bi$

(c) Compute the theoretical yield for W_6Cl_{12} from each reactant to determine the limiting reactant.

$228\ g\ WCl_6 \times \frac{1\ mol\ WCl_6}{396.6\ g\ WCl_6} \times \frac{1\ mol\ W_6Cl_{12}}{6\ mol\ WCl_6} \times \frac{1528\ g\ W_6Cl_{12}}{1\ mol\ W_6Cl_{12}} = 146\ g\ W_6Cl_{12}$

$175\ g\ Bi \times \frac{1\ mol\ Bi}{209.0\ g\ Bi} \times \frac{1\ mol\ W_6Cl_{12}}{8\ mol\ Bi} \times \frac{1528\ g\ W_6Cl_{12}}{1\ mol\ W_6Cl_{12}} = 160\ g\ W_6Cl_{12}$

The smaller theoretical yield (146 g) means that WCl_6 is the limiting reactant and 146 g W_6Cl_{12} are produced.

3.108 NaH, 24.00; B_2H_6, 27.67; $NaBH_4$, 37.83

(a) $2\ NaH + B_2H_6 \rightarrow 2\ NaBH_4$

$$8.55\text{ g NaH} \times \frac{1\text{ mol NaH}}{24.00\text{ g NaH}} = 0.356\text{ mol NaH}$$

$$6.75\text{ g } B_2H_6 \times \frac{1\text{ mol } B_2H_6}{27.67\text{ g } B_2H_6} = 0.244\text{ mol } B_2H_6$$

For 0.244 mol B_2H_6, 2 x (0.244) = 0.488 mol NaH are needed. Because only 0.356 mol of NaH is available, NaH is the limiting reactant.

$$0.356\text{ mol NaH} \times \frac{2\text{ mol } NaBH_4}{2\text{ mol NaH}} \times \frac{37.83\text{ g } NaBH_4}{1\text{ mol } NaBH_4} = 13.5\text{ g } NaBH_4\text{ produced}$$

(b) $$0.356\text{ mol NaH} \times \frac{1\text{ mol } B_2H_6}{2\text{ mol NaH}} \times \frac{27.67\text{ g } B_2H_6}{1\text{ mol } B_2H_6} = 4.93\text{ g } B_2H_6\text{ reacted}$$

B_2H_6 left over = 6.75 g – 4.93 g = 1.82 g B_2H_6

3.109 Assume a 100.0 g sample of ferrocene. From the percent composition data, a 100.0 g sample contains 5.42 g H, 64.56 g C, and 30.02 g Fe.

$$5.42\text{ g H} \times \frac{1\text{ mol H}}{1.01\text{ g H}} = 5.37\text{ mol H}$$

$$64.56\text{ g C} \times \frac{1\text{ mol C}}{12.01\text{ g C}} = 5.376\text{ mol C}$$

$$30.02\text{ g Fe} \times \frac{1\text{ mol Fe}}{55.85\text{ g Fe}} = 0.5375\text{ mol Fe}$$

$C_{5.376}H_{5.37}Fe_{0.5375}$; divide each subscript by the smallest, 0.5375.

$C_{5.376\,/\,0.5375}H_{5.37\,/\,0.5375}Fe_{0.5375\,/\,0.5375}$

The empirical formula is $C_{10}H_{10}Fe$.

3.110 Mass of 1 HCl molecule = $(36.5\ \frac{\text{u}}{\text{molecule}})(1.6605 \times 10^{-24}\ \frac{\text{g}}{\text{u}}) = 6.06 \times 10^{-23}$ g/molecule

$$\text{Avogadro's number} = \left(\frac{36.5\text{ g/mol}}{6.06 \times 10^{-23}\text{ g/molecule}}\right) = 6.02 \times 10^{23}\text{ molecules/mol}$$

3.111 23.46 mg = 0.023 46 g; 20.42 mg = 0.02042 g; 33.27 mg = 0.033 27 g

$$0.033\ 27\text{ g } CO_2 \times \frac{1\text{ mol } CO_2}{44.01\text{ g } CO_2} \times \frac{1\text{ mol C}}{1\text{ mol } CO_2} = 7.560 \times 10^{-4}\text{ mol C}$$

$$0.020\ 42\text{ g } H_2O \times \frac{1\text{ mol } H_2O}{18.02\text{ g } H_2O} \times \frac{2\text{ mol H}}{1\text{ mol } H_2O} = 2.266 \times 10^{-3}\text{ mol H}$$

$$7.560 \times 10^{-4}\text{ mol C} \times \frac{12.01\text{ g C}}{1\text{ mol C}} = 0.009\ 080\text{ g C}$$

$$2.266 \times 10^{-3}\text{ mol H} \times \frac{1.008\text{ g H}}{1\text{ mol H}} = 0.002\ 284\text{ g H}$$

mass O = 0.023 46 g – (0.009 080 g + 0.002 284 g) = 0.012 10 g O

$$0.012\ 10 \text{ g O} \times \frac{1 \text{ mol O}}{16.00 \text{ g O}} = 7.563 \times 10^{-4} \text{ mol O}$$

Scale each mol quantity to eliminate exponents.

$C_{0.7560}H_{2.266}O_{0.7563}$; divide each subscript by the smallest, 0.7560.

$C_{0.7560/0.7560}H_{2.266/0.7560}O_{0.7563/0.7560}$

The empirical formula is CH_3O, 31.0.

62.0 / 31.0 = 2; molecular formula = $C_{(2 \times 1)}H_{(2 \times 3)}O_{(2 \times 1)} = C_2H_6O_2$

3.112 High-resolution mass spectrometry is capable of measuring the mass of molecules with a particular isotopic composition.

3.113 (a) $CO(NH_2)_2(aq) + 6\ HOCl(aq) \rightarrow 2\ NCl_3(aq) + CO_2(aq) + 5\ H_2O(l)$
(b) $2\ Ca_3(PO_4)_2(s) + 6\ SiO_2(s) + 10\ C(s) \rightarrow P_4(g) + 6\ CaSiO_3(l) + 10\ CO(g)$

3.114 The combustion reaction is: $2\ C_8H_{18} + 25\ O_2 \rightarrow 16\ CO_2 + 18\ H_2O$

C_8H_{18}, 114.23; CO_2, 44.01

$$\text{pounds } CO_2 = 1.00 \text{ gal} \times \frac{3.7854 \text{ L}}{1 \text{ gal}} \times \frac{1000 \text{ mL}}{1 \text{ L}} \times \frac{0.703 \text{ g } C_8H_{18}}{1 \text{ mL}} \times \frac{1 \text{ mol } C_8H_{18}}{114.23 \text{ g } C_8H_{18}} \times$$

$$\frac{16 \text{ mol } CO_2}{2 \text{ mol } C_8H_{18}} \times \frac{44.01 \text{ g } CO_2}{1 \text{ mol } CO_2} \times \frac{1 \text{ lb}}{453.59 \text{ g}} = 18.1 \text{ lb } CO_2$$

3.115 AgCl, 143.32; CO_2, 44.01; H_2O, 18.02

$$\text{mol Cl in 1.00 g of X} = 1.95 \text{ g AgCl} \times \frac{1 \text{ mol AgCl}}{143.32 \text{ g AgCl}} \times \frac{1 \text{ mol Cl}}{1 \text{ mol AgCl}} = 0.0136 \text{ mol Cl}$$

$$\text{mass Cl} = 0.0136 \text{ mol Cl} \times \frac{35.453 \text{ g Cl}}{1 \text{ mol Cl}} = 0.482 \text{ g Cl}$$

$$\text{mol C in 1.00 g of X} = 0.900 \text{ g } CO_2 \times \frac{1 \text{ mol } CO_2}{44.01 \text{ g } CO_2} \times \frac{1 \text{ mol C}}{1 \text{ mol } CO_2} = 0.0204 \text{ mol C}$$

$$\text{mass C} = 0.0204 \text{ mol C} \times \frac{12.011 \text{ g C}}{1 \text{ mol C}} = 0.245 \text{ g C}$$

$$\text{mol H in 1.00 g of X} = 0.735 \text{ g } H_2O \times \frac{1 \text{ mol } H_2O}{18.02 \text{ g } H_2O} \times \frac{2 \text{ mol H}}{1 \text{ mol } H_2O} = 0.0816 \text{ mol H}$$

$$\text{mass H} = 0.0816 \text{ mol H} \times \frac{1.008 \text{ g H}}{1 \text{ mol H}} = 0.0823 \text{ g H}$$

mass N = 1.00 g – mass Cl – mass C – mass H = 1.00 – 0.482 g – 0.245 g – 0.0823 g = 0.19 g N

$$\text{mol N in 1.00 g of X} = 0.19 \text{ g N} \times \frac{1 \text{ mol N}}{14.01 \text{ g N}} = 0.014 \text{ mol N}$$

Determine empirical formula.

$C_{0.0204}H_{0.0816}N_{0.014}Cl_{0.0136}$; divide each subscript by the smallest, 0.0136.

$C_{0.0204/0.0136}H_{0.0816/0.0136}N_{0.014/0.0136}Cl_{0.0136/0.0136}$

$C_{1.5}H_6NCl$, multiply each subscript by 2 to get integers.

The empirical formula is $C_3H_{12}N_2Cl_2$.

3.116 $CaCO_3$, 100.09

$$\% \text{ Ca} = \frac{40.08 \text{ g Ca}}{100.09 \text{ g}} \text{ x } 100\% = 40.04\%$$

$$\% \text{ C} = \frac{12.01 \text{ g C}}{100.09 \text{ g}} \text{ x } 100\% = 12.00\%$$

$$\% \text{ O} = \frac{3 \text{ x } 16.00 \text{ g O}}{100.09 \text{ g}} \text{ x } 100\% = 47.96\%$$

Because the mass %'s for the pulverized rock are different from the mass %'s for pure $CaCO_3$ calculated here, the pulverized rock cannot be pure $CaCO_3$.

3.117 C_2H_6O, 46.07; H_2O, 18.02

Let X = mass of H_2O in the 10.00 g sample.

Let Y = mass of ethanol (C_2H_6O) in the 10.00 g sample.

X + Y = 10.00 g and Y = 10.00 g – X

mass of collected H_2O = 11.27 g

$$\text{mass of collected } H_2O = X + \left(Y \text{ x } \frac{1 \text{ mol } C_2H_6O}{46.07 \text{ g } C_2H_6O} \text{ x } \frac{3 \text{ mol } H_2O}{1 \text{ mol } C_2H_6O} \text{ x } \frac{18.02 \text{ g } H_2O}{1 \text{ mol } H_2O} \right)$$

Substitute for Y.

$$11.27 \text{ g} = X + \left((10.00 \text{ g} - X) \text{ x } \frac{1 \text{ mol } C_2H_6O}{46.07 \text{ g } C_2H_6O} \text{ x } \frac{3 \text{ mol } H_2O}{1 \text{ mol } C_2H_6O} \text{ x } \frac{18.02 \text{ g } H_2O}{1 \text{ mol } H_2O} \right)$$

11.27 g = X + (10.00 g – X)(1.173)

11.27 g = X + 11.73 g – 1.173 X

0.173 X = 11.73 g – 11.27 g = 0.46 g

$$X = \frac{0.46 \text{ g}}{0.173} = 2.7 \text{ g } H_2O$$

Y = 10.00 g – X = 10.00 g – 2.7 g = 7.3 g C_2H_6O

3.118 FeO, 71.85; Fe_2O_3, 159.7

Let X equal the mass of FeO and Y the mass of Fe_2O_3 in the 10.0 g mixture. Therefore, X + Y = 10.0 g.

$$\text{mol Fe} = 7.43 \text{ g x } \frac{1 \text{ mol Fe}}{55.85 \text{ g Fe}} = 0.133 \text{ mol Fe}$$

mol FeO + 2 x mol Fe_2O_3 = 0.133 mol Fe

$$X \text{ x } \frac{1 \text{ mol FeO}}{71.85 \text{ g FeO}} + 2 \text{ x } \left(Y \text{ x } \frac{1 \text{ mol } Fe_2O_3}{159.7 \text{ g } Fe_2O_3} \right) = 0.133 \text{ mol Fe}$$

Rearrange to get X = 10.0 g – Y and then substitute it into the equation above to solve for Y.

$$(10.0 \text{ g} - Y) \text{ x } \frac{1 \text{ mol FeO}}{71.85 \text{ g FeO}} + 2 \text{ x } \left(Y \text{ x } \frac{1 \text{ mol } Fe_2O_3}{159.7 \text{ g } Fe_2O_3} \right) = 0.133 \text{ mol Fe}$$

$$\frac{10.0\text{ mol}}{71.85} - \frac{\text{Y mol}}{71.85\text{ g}} + \frac{2\text{ Y mol}}{159.7\text{ g}} = 0.133\text{ mol}$$

$$-\frac{\text{Y mol}}{71.85\text{ g}} + \frac{2\text{ Y mol}}{159.7\text{ g}} = 0.133\text{ mol} - \frac{10.0\text{ mol}}{71.85} = -0.0062\text{ mol}$$

$$\frac{(-\text{Y mol})(159.7\text{ g}) + (2\text{ Y mol})(71.85\text{ g})}{(71.85\text{ g})(159.7\text{ g})} = -0.0062\text{ mol}$$

$$\frac{-16.0\text{ Y mol}}{11474\text{ g}} = -0.0062\text{ mol}; \quad \frac{16.0\text{ Y}}{11474\text{ g}} = 0.0062$$

$Y = (0.0062)(11474\text{ g})/16.0 = 4.44\text{ g} = 4.4\text{ g } Fe_2O_3$

$X = 10.0\text{ g} - Y = 10.0\text{ g} - 4.4\text{ g} = 5.6\text{ g FeO}$

3.119 AgCl, 143.32

Find the mass of Cl in 1.68 g of AgCl.

$$\text{mol Cl in 1.68 g of AgCl} = 1.68\text{ g AgCl x } \frac{1\text{ mol AgCl}}{143.32\text{ g AgCl}} \text{ x } \frac{1\text{ mol Cl}}{1\text{ mol AgCl}} = 0.0117\text{ mol Cl}$$

$$\text{mass Cl} = 0.0117\text{ mol Cl x } \frac{35.453\text{ g Cl}}{1\text{ mol Cl}} = 0.415\text{ g Cl}$$

All of the Cl in AgCl came from XCl_3.

Find the mass of X in 0.634 g of XCl_3.

Mass of X = 0.634 g – 0.415 g = 0.219 g X

$$0.0117\text{ mol Cl x } \frac{1\text{ mol X}}{3\text{ mol Cl}} = 0.00390\text{ mol X}$$

$$\text{molar mass of X} = \frac{0.219\text{ g}}{0.00390\text{ mol}} = 56.2\text{ g/mol; X = Fe}$$

3.120 $C_6H_{12}O_6 + 6\,O_2 \rightarrow 6\,CO_2 + 6\,H_2O$; $C_6H_{12}O_6$, 180.16; CO_2, 44.01

$$66.3\text{ g } C_6H_{12}O_6 \text{ x } \frac{1\text{ mol } C_6H_{12}O_6}{180.16\text{ g } C_6H_{12}O_6} \text{ x } \frac{6\text{ mol } CO_2}{1\text{ mol } C_6H_{12}O_6} \text{ x } \frac{44.01\text{ g } CO_2}{1\text{ mol } CO_2} = 97.2\text{ g } CO_2$$

$$66.3\text{ g } C_6H_{12}O_6 \text{ x } \frac{1\text{ mol } C_6H_{12}O_6}{180.16\text{ g } C_6H_{12}O_6} \text{ x } \frac{6\text{ mol } CO_2}{1\text{ mol } C_6H_{12}O_6} \text{ x } \frac{25.4\text{ L } CO_2}{1\text{ mol } CO_2} = 56.1\text{ L } CO_2$$

3.121 Mass of Cu = 2.196 g; mass of S = 2.748 g – 2.196 g = 0.552 g S

(a) $\%Cu = \frac{2.196\text{ g}}{2.748\text{ g}} \text{ x } 100\% = 79.91\%$

$\%S = \frac{0.552\text{ g}}{2.748\text{ g}} \text{ x } 100\% = 20.1\%$

(b) $2.196\text{ g Cu x } \frac{1\text{ mol Cu}}{63.55\text{ g Cu}} = 0.034\,55\text{ mol Cu}$

$0.552\text{ g S x } \frac{1\text{ mol S}}{32.07\text{ g S}} = 0.0172\text{ mol S}$

$Cu_{0.03455}S_{0.0172}$; divide each subscript by the smaller, 0.0172.
$Cu_{0.03455/0.0172}S_{0.0172/0.0172}$
The empirical formula is Cu_2S.
(c) Cu_2S, 159.16

$$\frac{5.6\ g\ Cu_2S}{1\ cm^3} \times \frac{1\ mol\ Cu_2S}{159.16\ g\ Cu_2S} \times \frac{2\ mol\ Cu^+\ ions}{1\ mol\ Cu_2S} \times \frac{6.022 \times 10^{23}\ Cu^+\ ions}{1\ mol\ Cu^+\ ions}$$
$$= 4.2 \times 10^{22}\ Cu^+\ ions/cm^3$$

3.122 Mass of added Cl = mass of XCl_5 – mass of XCl_3 = 13.233 g – 8.729 g = 4.504 g

$$\text{mass of Cl in } XCl_5 = 5\ Cl\text{'s} \times \frac{4.504\ g}{2\ Cl\text{'s}} = 11.26\ g\ Cl$$

mass of X in XCl_5 = 13.233 g – 11.26 g = 1.973 g X

$$11.26\ g\ Cl \times \frac{1\ mol\ Cl}{35.45\ g\ Cl} = 0.3176\ mol\ Cl$$

$$0.3176\ mol\ Cl \times \frac{1\ mol\ X}{5\ mol\ Cl} = 0.063\ 52\ mol\ X$$

$$\text{molar mass of X} = \frac{1.973\ g\ X}{0.063\ 52\ mol\ X} = 31.1\ g/mol;\ \text{atomic weight} = 31.1,\ X = P$$

3.123 PCl_3, 137.33; PCl_5, 208.24
Let Y = mass of PCl_3 in the mixture, and (10.00 – Y) = mass of PCl_5 in the mixture.

$$\text{fraction Cl in } PCl_3 = \frac{(3)(35.453\ g/mol)}{137.33\ g/mol} = 0.774\ 48$$

$$\text{fraction Cl in } PCl_5 = \frac{(5)(35.453\ g/mol)}{208.24\ g/mol} = 0.851\ 25$$

(mass of Cl in PCl_3) + (mass of Cl in PCl_5) = mass of Cl in the mixture
0.774 48Y + 0.851 25(10.00 g – Y) = (0.8104)(10.00 g)
Y = 5.32 g PCl_3 and 10.00 – Y = 4.68 g PCl_5

3.124 NH_4NO_3, 80.04; $(NH_4)_2HPO_4$, 132.06
Assume you have a 100.0 g sample of the mixture.
Let X = grams of NH_4NO_3 and (100.0 – X) = grams of $(NH_4)_2HPO_4$.
Both compounds contain 2 nitrogen atoms per formula unit.
Because the mass % N in the sample is 30.43%, the 100.0 g sample contains 30.43 g N.

$$\text{mol } NH_4NO_3 = (X) \times \frac{1\ mol\ NH_4NO_3}{80.04\ g}$$

$$\text{mol } (NH_4)_2HPO_4 = (100.0 - X) \times \frac{1\ mol\ (NH_4)_2HPO_4}{132.06\ g}$$

$$\text{mass N} = \left(\left((X) \times \frac{1\ mol\ NH_4NO_3}{80.04\ g}\right) + \left((100.0 - X) \times \frac{1\ mol\ (NH_4)_2HPO_4}{132.06\ g}\right)\right) \times \left(\frac{2\ mol\ N}{1\ mol\ ammonium\ cmpds}\right) \times \left(\frac{14.0067\ g\ N}{1\ mol\ N}\right) = 30.43\ g$$

Solve for X.

$$\left(\frac{X}{80.04} + \frac{100.0 - X}{132.06}\right)(2)(14.0067) = 30.43$$

$$\left(\frac{X}{80.04} + \frac{100.0 - X}{132.06}\right) = 1.08627$$

$$\frac{(132.06)(X) + (100.0 - X)(80.04)}{(80.04)(132.06)} = 1.08627$$

$(132.06)(X) + (100.0 - X)(80.04) = (1.08627)(80.04)(132.06)$

$132.06X + 8004 - 80.04X = 11481.96$

$132.06X - 80.04X = 11481.96 - 8004$

$52.02X = 3477.96$

$$X = \frac{3477.96}{52.02} = 66.86 \text{ g } NH_4NO_3$$

$(100.0 - X) = (100.0 - 66.86) = 33.14 \text{ g } (NH_4)_2HPO_4$

$$\frac{\text{mass}_{NH_4NO_3}}{\text{mass}_{(NH_4)_2HPO_4}} = \frac{66.86 \text{ g}}{33.14 \text{ g}} = 2.018$$

The mass ratio of NH_4NO_3 to $(NH_4)_2HPO_4$ in the mixture is 2 to 1.

3.125 $Na_2CO_3 \rightarrow Na_2O + CO_2$; Na_2CO_3, 106; Na_2O, 62

$CaCO_3 \rightarrow CaO + CO_2$; $CaCO_3$, 100; CaO, 56

In a 0.35 kg sample of glass there would be:

$0.12 \text{ x } 0.35 \text{ kg} = 0.042 \text{ kg} = 42 \text{ g of } Na_2O$

$0.13 \text{ x } 0.35 \text{ kg} = 0.045 \text{ kg} = 45 \text{ g of CaO}$

$350 \text{ g} - 42 \text{ g} - 45 \text{ g} = 263 \text{ g of } SiO_2$

$$\text{mass } Na_2CO_3 = 42 \text{ g } Na_2O \text{ x } \frac{1 \text{ mol } Na_2O}{62 \text{ g } Na_2O} \text{ x } \frac{1 \text{ mol } Na_2CO_3}{1 \text{ mol } Na_2O} \text{ x } \frac{106 \text{ g } Na_2CO_3}{1 \text{ mol } Na_2CO_3} = 72 \text{ g } Na_2CO_3$$

$$\text{mass } CaCO_3 = 45 \text{ g CaO x } \frac{1 \text{ mol CaO}}{56 \text{ g CaO}} \text{ x } \frac{1 \text{ mol } CaCO_3}{1 \text{ mol CaO}} \text{ x } \frac{100 \text{ g } CaCO_3}{1 \text{ mol } CaCO_3} = 80 \text{ g } CaCO_3$$

To make 0.35 kg of glass, start with 72 g Na_2CO_3, 80 g $CaCO_3$, and 263 g SiO_2.

3.126 (a) 56.0 mL = 0.0560 L

$$\text{mol } X_2 = (0.0560 \text{ L } X_2)\left(\frac{1 \text{ mol}}{22.41 \text{ L}}\right) = 0.00250 \text{ mol } X_2$$

$\text{mass } X_2 = 1.12 \text{ g } MX_2 - 0.720 \text{ g MX} = 0.40 \text{ g } X_2$

$$\text{molar mass } X_2 = \frac{0.40 \text{ g}}{0.00250 \text{ mol}} = 160 \text{ g/mol}$$

atomic weight of X = 160/2 = 80; X is Br.

(b) mol MX = 0.00250 mol X_2 x $\frac{2 \text{ mol MX}}{1 \text{ mol } X_2}$ = 0.00500 mol MX

mass of X in MX = 0.00500 mol MX x $\frac{1 \text{ mol X}}{1 \text{ mol MX}}$ x $\frac{80 \text{ g X}}{1 \text{ mol X}}$ = 0.40 g X

mass of M in MX = 0.720 g MX – 0.40 g X = 0.32 g M

molar mass M = $\frac{0.32 \text{ g}}{0.00500 \text{ mol}}$ = 64 g/mol

atomic weight of X = 64; M is Cu.

4 Reactions in Aqueous Solution

4.1 $C_{12}H_{22}O_{11}$, 342.3; 355 mL = 0.355 L

$$43.0 \text{ g } C_{12}H_{22}O_{11} \text{ x } \frac{1 \text{ mol } C_{12}H_{22}O_{11}}{342.3 \text{ g } C_{12}H_{22}O_{11}} = 0.126 \text{ mol } C_6H_{12}O_6$$

$$\text{molarity} = \frac{0.126 \text{ mol}}{0.355 \text{ L}} = 0.355 \text{ M}$$

4.2 The solution is not 1.00 M. The solution was incorrectly prepared by combining 0.500 mol of solute with 500 mL of water; resulting in a volume larger than 500 mL. Instead the solute should be added to a 500 mL volumetric flask and solvent added to bring the final volume to 500 mL.

4.3 (a) 125 mL = 0.125 L; (0.20 mol/L)(0.125 L) = 0.025 mol $NaHCO_3$
(b) 650.0 mL = 0.6500 L; (2.50 mol/L)(0.6500 L) = 1.62 mol H_2SO_4

4.4 (a) NaOH, 40.0; 500.0 mL = 0.5000 L

$$\frac{1.25 \text{ mol NaOH}}{\text{L}} \text{ x } 0.500 \text{ L x } \frac{40.0 \text{ g NaOH}}{1 \text{ mol NaOH}} = 25.0 \text{ g NaOH}$$

(b) $C_6H_{12}O_6$, 180.2

$$\frac{0.250 \text{ mol } C_6H_{12}O_6}{\text{L}} \text{ x } 1.50 \text{ L x } \frac{180.2 \text{ g } C_6H_{12}O_6}{1 \text{ mol } C_6H_{12}O_6} = 67.6 \text{ g } C_6H_{12}O_6$$

4.5 (a) $C_{27}H_{46}O$, 386.7; 750 mL = 0.750 L

$$\frac{0.005 \text{ mol } C_{27}H_{46}O}{\text{L}} \text{ x } 0.750 \text{ L x } \frac{386.7 \text{ g } C_{27}H_{46}O}{1 \text{ mol } C_{27}H_{46}O} = 1 \text{ g } C_{27}H_{46}O$$

(b)

$$25 \text{ mg } C_{27}H_{46}O \text{ x } \frac{1 \text{ x } 10^{-3} \text{ g}}{1 \text{ mg}} \text{ x } \frac{1 \text{ mol } C_{27}H_{46}O}{386.7 \text{ g } C_{27}H_{46}O}$$

$$\text{x } \frac{1 \text{ L}}{0.005 \text{ mol } C_{27}H_{46}O} \text{ x } \frac{1 \text{ mL}}{1 \text{ x } 10^{-3} \text{ L}} = 13 \text{ mL blood}$$

4.6 $M_i \text{ x } V_i = M_f \text{ x } V_f$; $M_f = \frac{M_i \text{ x } V_i}{V_f} = \frac{3.50 \text{ M x } 75.0 \text{ mL}}{400.0 \text{ mL}} = 0.656 \text{ M}$

4.7 $M_i \text{ x } V_i = M_f \text{ x } V_f$; $V_i = \frac{M_f \text{ x } V_f}{M_i} = \frac{0.500 \text{ M x } 250.0 \text{ mL}}{18.0 \text{ M}} = 6.94 \text{ mL}$

4.8 $FeBr_3$ contains 3 Br^- ions. The molar concentration of Br^- ions = 3 x 0.225 M = 0.675 M.

4.9 (a) A_2Y is the strongest electrolyte because all the molecules have dissociated into ions. A_2X is the weakest because only 1 out of 5 molecules have dissociated into ions.
(b) In a 0.350 M solution of A_2Y, the molar concentration of A ions = 2 x 0.350 = 0.700 M. In a 0.350 M solution of A_2Y, the molar concentration of Y ions = 1 x 0.350 = 0.350 M.
(c) There are 5 A_2X molecules initially in the solution. Only one of them is ionized. The percent ionization = (1/5) x 100% = 20%

4.10 (a) Ionic equation:
$2\ Ag^+(aq) + 2\ NO_3^-(aq) + 2\ Na^+(aq) + CrO_4^{2-}(aq) \rightarrow Ag_2CrO_4(s) + 2\ Na^+(aq) + 2\ NO_3^-(aq)$
Delete spectator ions from the ionic equation to get the net ionic equation.
Net ionic equation: $2\ Ag^+(aq) + CrO_4^{2-}(aq) \rightarrow Ag_2CrO_4(s)$
(b) Ionic equation:
$2\ H^+(aq) + SO_4^{2-}(aq) + MgCO_3(s) \rightarrow H_2O(l) + CO_2(g) + Mg^{2+}(aq) + SO_4^{2-}(aq)$
Delete spectator ions from the ionic equation to get the net ionic equation.
Net ionic equation: $2\ H^+(aq) + MgCO_3(s) \rightarrow H_2O(l) + CO_2(g) + Mg^{2+}(aq)$
(c) Ionic equation:
$Hg^{2+}(aq) + 2\ NO_3^-(aq) + 2\ NH_4^+(aq) + 2\ I^-(aq) \rightarrow HgI_2(s) + 2\ NH_4^+(aq) + 2\ NO_3^-(aq)$
Delete spectator ions from the ionic equation to get the net ionic equation.
Net ionic equation: $Hg^{2+}(aq) + 2\ I^-(aq) \rightarrow HgI_2(s)$

4.11 Spectator ions: Ca^{2+} and NO_3^-
Net ionic equation: $Ag^+(aq) + Cl^-(aq) \rightarrow AgCl(s)$
Molecular equation: $2\ AgNO_3(aq) + CaCl_2(aq) \rightarrow 2\ AgCl(s) + Ca(NO_3)_2(aq)$

4.12 (a) Ionic equation:
$Ni^{2+}(aq) + 2\ Cl^-(aq) + 2\ NH_4^+(aq) + S^{2-}(aq) \rightarrow NiS(s) + 2\ NH_4^+(aq) + 2\ Cl^-(aq)$
Delete spectator ions from the ionic equation to get the net ionic equation.
Net ionic equation: $Ni^{2+}(aq) + S^{2-}(aq) \rightarrow NiS(s)$
(b) Ionic equation:
$2\ Na^+(aq) + CrO_4^{2-}(aq) + Pb^{2+}(aq) + 2\ NO_3^-(aq) \rightarrow PbCrO_4(s) + 2\ Na^+(aq) + 2\ NO_3^-(aq)$
Delete spectator ions from the ionic equation to get the net ionic equation.
Net ionic equation: $Pb^{2+}(aq) + CrO_4^{2-}(aq) \rightarrow PbCrO_4(s)$
(c) Ionic equation:
$2\ Ag^+(aq) + 2\ ClO_4^-(aq) + Ca^{2+}(aq) + 2\ Br^-(aq) \rightarrow 2\ AgBr(s) + Ca^{2+}(aq) + 2\ ClO_4^-(aq)$
Delete spectator ions from the ionic equation and reduce coefficients to get the net ionic equation.
Net ionic equation: $Ag^+(aq) + Br^-(aq) \rightarrow AgBr(s)$
(d) Ionic equation:
$Zn^{2+}(aq) + 2\ Cl^-(aq) + 2\ K^+(aq) + CO_3^{2-}(aq) \rightarrow ZnCO_3(s) + 2\ K^+(aq) + 2\ Cl^-(aq)$
Delete spectator ions from the ionic equation to get the net ionic equation.
Net ionic equation: $Zn^{2+}(aq) + CO_3^{2-}(aq) \rightarrow ZnCO_3(s)$

4.13 $3\ CaCl_2(aq) + 2\ Na_3PO_4(aq) \rightarrow Ca_3(PO_4)_2(s) + 6\ NaCl(aq)$
Ionic equation:
$3\ Ca^{2+}(aq) + 6\ Cl^-(aq) + 6\ Na^+(aq) + 2\ PO_4^{3-}(aq) \rightarrow Ca_3(PO_4)_2(s) + 6\ Na^+(aq) + 6\ Cl^-(aq)$
Delete spectator ions from the ionic equation to get the net ionic equation.
Net ionic equation: $3\ Ca^{2+}(aq) + 2\ PO_4^{3-}(aq) \rightarrow Ca_3(PO_4)_2(s)$

4.14 A precipitate results from the reaction. The precipitate contains cations and anions in a 3:2 ratio. The precipitate is either $Mg_3(PO_4)_2$ or $Zn_3(PO_4)_2$.

4.15 The precipitate is $PbBr_2$.

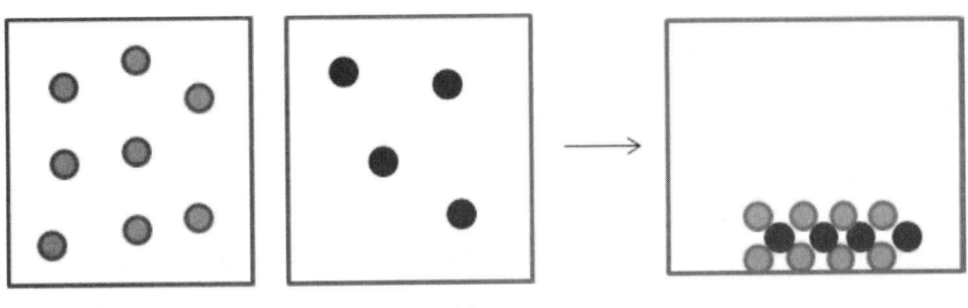

4.16 (a) HI, hydroiodic acid (b) $HBrO_2$, bromous acid (c) H_2CrO_4, chromic acid

4.17 (a) H_3PO_3 (b) H_2Se

4.18 (a) Ionic equation:
$2\ Cs^+(aq) + 2\ OH^-(aq) + 2\ H^+(aq) + SO_4^{2-}(aq) \rightarrow 2\ Cs^+(aq) + SO_4^{2-}(aq) + 2\ H_2O(l)$
Delete spectator ions from the ionic equation and reduce coefficients to get the net ionic equation.
Net ionic equation: $H^+(aq) + OH^-(aq) \rightarrow H_2O(l)$
(b) Ionic equation:
$Ca^{2+}(aq) + 2\ OH^-(aq) + 2\ CH_3CO_2H(aq) \rightarrow Ca^{2+}(aq) + 2\ CH_3CO_2^-(aq) + 2\ H_2O(l)$
Delete spectator ions from the ionic equation and reduce coefficients to get the net ionic equation.
Net ionic equation: $CH_3CO_2H(aq) + OH^-(aq) \rightarrow CH_3CO_2^-(aq) + H_2O(l)$

4.19 $Mg^{2+}(aq) + 2\ OH^-(aq) + 2\ H^+(aq) + 2\ Cl^-(aq) \rightarrow Mg^{2+}(aq) + 2\ Cl^-(aq) + H_2O(l)$
$H^+(aq) + OH^-(aq) \rightarrow H_2O(l)$

4.20 50.0 mL = 0.0500 L
(0.100 mol/L)(0.0500 L) = 5.00×10^{-3} mol NaOH

$$5.00 \times 10^{-3}\ \text{mol NaOH} \times \frac{1\ \text{mol}\ H_2SO_4}{2\ \text{mol NaOH}} = 2.50 \times 10^{-3}\ \text{mol}\ H_2SO_4$$

$$2.50 \times 10^{-3}\ \text{mol}\ H_2SO_4 \times \frac{1\ \text{L}}{0.250\ \text{mol}\ H_2SO_4} \times \frac{1000\ \text{mL}}{1\ \text{L}} = 10.0\ \text{mL}$$

4.21 25.0 mL = 0.0250 L; 68.5 mL = 0.0685 L
From the reaction stoichiometry, moles KOH = moles HNO_3
(0.150 mol/L)(0.0250 L) = 3.75×10^{-3} mol KOH = 3.75×10^{-3} mol HNO_3

$$\text{molarity} = \frac{3.75 \times 10^{-3}\ \text{mol}}{0.0685\ \text{L}} = 0.0547\ \text{M}$$

4.22 25.0 mL = 0.0250 L; 94.7 mL = 0.0947 L
From the reaction stoichiometry, moles NaOH = moles CH_3CO_2H
(0.200 mol/L)(0.0947 L) = 0.018 94 mol NaOH = 0.018 94 mol CH_3CO_2H

$$\text{molarity} = \frac{0.018\ 94\ \text{mol}}{0.0250\ \text{L}} = 0.758\ \text{M}$$

4.23 $$\text{vol } H^+ \text{ solution} = 8\ OH^- \times \frac{1\ H^+}{1\ OH^-} \times \frac{100\ \text{mL}}{12\ H^+} = 66.7\ \text{mL}$$

4.24 (a) $SnCl_4$: Cl −1, Sn +4
(b) CrO_3: O −2, Cr +6
(c) $VOCl_3$: O −2, Cl −1, V +5
(d) V_2O_3: O −2, V +3
(e) HNO_3: O −2, H +1, N +5
(f) $FeSO_4$: O −2, S +6, Fe +2

4.25 (a) +2, ClO, chlorine monoxide; +3, Cl_2O_3, dichlorine trioxide; +6, ClO_3, chlorine trioxide; +7, Cl_2O_7, dichlorine heptoxide
(b) Cl_2O_7 cannot react with O_2 because Cl has its maximum oxidation number, +7, and cannot be further oxidized.

4.26 (a) $SnO_2(s) + 2\ C(s) \rightarrow Sn(s) + 2\ CO(g)$
C is oxidized (its oxidation number increases from 0 to +2). C is the reducing agent.
The Sn in SnO_2 is reduced (its oxidation number decreases from +4 to 0). SnO_2 is the oxidizing agent.
(b) $Sn^{2+}(aq) + 2\ Fe^{3+}(aq) \rightarrow Sn^{4+}(aq) + 2\ Fe^{2+}(aq)$
Sn^{2+} is oxidized (its oxidation number increases from +2 to +4). Sn^{2+} is the reducing agent.
Fe^{3+} is reduced (its oxidation number decreases from +3 to +2). Fe^{3+} is the oxidizing agent.
(c) $4\ NH_3(g) + 5\ O_2(g) \rightarrow 4\ NO(g) + 6\ H_2O(l)$
The N in NH_3 is oxidized (its oxidation number increases from −3 to +2). NH_3 is the reducing agent.
Each O in O_2 is reduced (its oxidation number decreases from 0 to −2). O_2 is the oxidizing agent.

4.27 (a) C in C_2H_5OH oxidized and Cr in $K_2Cr_2O_7$ gets reduced.
(b) $K_2Cr_2O_7$ is the oxidizing agent; C_2H_5OH is the reducing agent.

4.28 (a) Pt is below H in the activity series; therefore NO REACTION.
(b) Mg is below Ca in the activity series; therefore NO REACTION.

4.29 "Any element higher in the activity series will react with the ion of any element lower in the activity series."

$A + D^+ \rightarrow A^+ + D$; therefore A is higher than D.
$B^+ + D \rightarrow B + D^+$; therefore D is higher than B.
$C^+ + D \rightarrow C + D^+$; therefore D is higher than C.
$B + C^+ \rightarrow B^+ + C$; therefore B is higher than C.
The net result is A > D > B > C.

4.30 31.50 mL = 0.031 50 L; 10.00 mL = 0.010 00 L

$$0.031\ 50\ \text{L} \times \frac{0.105\ \text{mol BrO}_3^-}{1\ \text{L}} \times \frac{6\ \text{mol Fe}^{2+}}{1\ \text{mol BrO}_3^-} = 1.98 \times 10^{-2}\ \text{mol Fe}^{2+}$$

$$\text{molarity} = \frac{1.98 \times 10^{-2}\ \text{mol Fe}^{2+}}{0.010\ 00\ \text{L}} = 1.98\ \text{M Fe}^{2+}\ \text{solution}$$

4.31 14.92 mL = 0.014 92 L

$$0.014\ 92\ \text{L} \times \frac{0.100\ \text{mol Cr}_2\text{O}_7^{-2}}{1\ \text{L}} \times \frac{6\ \text{mol Fe}^{2+}}{1\ \text{mol Cr}_2\text{O}_7^{-2}} \times \frac{55.85\ \text{g Fe}^{2+}}{1\ \text{mol Fe}^{2+}} \times \frac{1000\ \text{mg}}{1\ \text{g}} = 50.0\ \text{mg Fe}^{2+}$$

4.32 Sodium chloride, sodium citrate, and potassium dihydrogen phosphate are strong electrolytes since they are soluble ionic compounds; citric acid and vitamin B3 are weak electrolytes due the presence of the carboxylic acid (CO_2H) group; and fructose is a nonelectrolyte. (b) Sodium chloride, potassium dihydrogen phosphate, and sodium citrate replenish electrolytes with important biological functions.

4.33 150 mg = 0.15 g; 35 mg = 0.035 g; 360 mL = 0.36 L

$$\text{molarity} = \frac{\left(0.15\ \text{g Na}^+ \times \frac{1\ \text{mol Na}^+}{23\ \text{g Na}^+}\right)}{0.36\ \text{L}} = 0.018\ \text{M}$$

$$\text{molarity} = \frac{\left(0.35\ \text{g K}^+ \times \frac{1\ \text{mol K}^+}{39\ \text{g K}^+}\right)}{0.36\ \text{L}} = 0.0025\ \text{M}$$

4.34 NaCl, 58.4; 0.500 L = 500 mL

$$0.416\ \text{mg Na}^+/\text{mL} \times \frac{1 \times 10^{-3}\ \text{g}}{1\ \text{mg}} \times 500\ \text{mL} \times \frac{1\ \text{mol Na}^+}{23.0\ \text{g Na}^+}$$

$$\times \frac{1\ \text{mol NaCl}}{1\ \text{mol Na}^+} \times \frac{58.4\ \text{g NaCl}}{1\ \text{mol NaCl}} = 0.528\ \text{g NaCl}$$

4.35 (a) Ba^{2+} (b) $3\ Ba^{2+}(aq) + 2\ PO_4^{3-}(aq) \rightarrow Ba_3(PO_4)_2(s)$

4.36 AgCl, 143.3; 172 mg = 0.172 g; 100.0 mL = 0.1000 L

$$0.172 \text{ g AgCl} \times \frac{1 \text{ mol AgCl}}{143.3 \text{ g AgCl}} \times \frac{1 \text{ mol Cl}^-}{1 \text{ mol AgCl}} = 0.001\ 20 \text{ mol Cl}^-$$

$$\text{Cl}^- \text{ molarity} = \frac{0.001\ 20 \text{ mol Cl}^-}{0.1000 \text{ L}} = 0.0120 \text{ M}$$

4.37 35.6 mL = 0.0356 L; 25.0 mL = 0.0250 L
(0.0356 L)(0.0400 mol/L) = 0.001 42 mol NaOH

$$0.001\ 42 \text{ mol NaOH} \times \frac{1 \text{ mol } H_3C_6H_5O_7}{3 \text{ mol NaOH}} = 4.75 \times 10^{-4} \text{ mol } H_3C_6H_5O_7$$

$$H_3C_6H_5O_7 \text{ molarity} = \frac{4.75 \times 10^{-4} \text{ mol}}{0.0250 \text{ L}} = 0.0190 \text{ M}$$

Conceptual Problems

4.38 The concentration of a solution is cut in half when the volume is doubled. This is best represented by box (b).

4.39 (a) AX_3 is the strongest electrolyte because it is completely dissociated into ions.
AY_3 is the weakest electrolyte because it is the least dissociated of the three substances.
(b) In a 0.500 M solution of AX_3, the molar concentration of A ions = 1 x 0.500= 0.500 M.
In a 0.500 M solution of AX_3, the molar concentration of X ions = 3 x 0.500 = 1.50 M.
(c) There are 5 AZ_3 molecules initially in the solution. Four of them are ionized. The percent ionization = (4/5) x 100% = 80%

4.40 (a) 2 $Na^+(aq)$ + $CO_3^{2-}(aq)$ does not form a precipitate. This is represented by box (1).
(b) $Ba^{2+}(aq)$ + $CrO_4^{2-}(aq)$ → $BaCrO_4(s)$. This is represented by box (2).
(c) 2 $Ag^+(aq)$ + $SO_4^{2-}(aq)$ → $Ag_2SO_4(s)$. This is represented by box (3).

4.41 In the precipitate there are two cations (blue) for each anion (red). Looking at the ions in the list, the anion must have a –2 charge and the cation a +1 charge for charge neutrality of the precipitate. The cation must be Ag^+ because all Na^+ salts are soluble. Ag_2CrO_4 and Ag_2CO_3 are insoluble and consistent with the observed result.

4.42 HY is the strongest acid because it is completely dissociated.
HX is the weakest acid because it is the least dissociated.

4.43 One OH^- will react with each available H^+ on the acid forming H_2O. The acid is identified by how many of the 12 OH^- react with three molecules of each acid.
(a) Three HF's react with three OH^-, leaving nine OH^- unreacted (box 2).
(b) Three H_2SO_3's react with six OH^-, leaving six OH^- unreacted (box 3).
(c) Three H_3PO_4's react with nine OH^-, leaving three OH^- unreacted (box 1).

4.44 The concentration in the buret is three times that in the flask. The NaOCl concentration is 0.040 M. Because the I^- concentration in the buret is three times the OCl^- concentration in the flask and the reaction requires 2 I^- ions per OCl^- ion, 2/3 or 67% of the I^- solution from the buret must be added to the flask to react with all of the OCl^-.

4.45 (a) Ionic equation:
$K^+(aq) + Cl^-(aq) + Ag^+(aq) + NO_3^-(aq) \rightarrow AgCl(s) + K^+(aq) + NO_3^-(aq)$
(b) Ionic equation:
$HF(aq) + K^+(aq) + OH^-(aq) \rightarrow K^+(aq) + F^-(aq) + H_2O(l)$
(c) Ionic equation:
$Ba^{2+}(aq) + 2\ Cl^-(aq) + 2\ Na^+(aq) + SO_4^{2-}(aq) \rightarrow BaSO_4(s) + 2\ Na^+(aq) + 2\ Cl^-(aq)$
Reaction (c) would have the highest initial conductivity because of the 3 net ions for each $BaCl_2$ (a strong electrolyte).
Reaction (b) would have the lowest (almost zero) initial conductivity because HF is a very weak acid/electrolyte.
Reaction (a) would have an intermediate initial conductivity between that for reactions (b) and (c).
Figure (1) is for reaction (a); figure (2) is for reaction (b); and figure (3) is for reaction (c).

4.46 (a) $Sr^+ + At \rightarrow Sr + At^+$ No reaction.
(b) $Si + At^+ \rightarrow Si^+ + At$ Reaction would occur.
(c) $Sr + Si^+ \rightarrow Sr^+ + Si$ Reaction would occur.

4.47 "Any element higher in the activity series will react with the ion of any element lower in the activity series."
$BLUE + RED^+ \rightarrow BLUE^+ + RED$; therefore BLUE is higher than RED.
$RED + GREEN^+ \rightarrow RED^+ + GREEN$; therefore RED is higher than GREEN.
The net result is BLUE > RED > GREEN. Because BLUE is above GREEN, the following reaction will occur: $BLUE + GREEN^+ \rightarrow BLUE^+ + GREEN$

Section Problems
Molarity and Dilution (Sections 4.1 and 4.2)

4.48 (a) 35.0 mL = 0.0350 L; $\dfrac{1.200\text{ mol } HNO_3}{L} \times 0.0350\text{ L} = 0.0420\text{ mol } HNO_3$

(b) 175 mL = 0.175 L; $\dfrac{0.67\text{ mol } C_6H_{12}O_6}{L} \times 0.175\text{ L} = 0.12\text{ mol } C_6H_{12}O_6$

4.49 (a) C_2H_6O, 46.07; 250.0 mL = 0.2500 L
$\dfrac{0.600\text{ mol } C_2H_6O}{L} \times 0.2500\text{ L} = 0.150\text{ mol } C_2H_6O$
(0.150 mol)(46.07 g/mol) = 6.91 g C_2H_6O
(b) H_3BO_3, 61.83; 167 mL = 0.167 L
$\dfrac{0.200\text{ mol } H_3BO_3}{L} \times 0.167\text{ L} = 0.0334\text{ mol } H_3BO_3$
(0.0334 mol)(61.83 g/mol) = 2.07 g H_3BO_3

4.50 $BaCl_2$, 208.2

$$15.0 \text{ g } BaCl_2 \text{ x } \frac{1 \text{ mol } BaCl_2}{208.2 \text{ g } BaCl_2} = 0.0720 \text{ mol } BaCl_2$$

$$0.0720 \text{ mol x } \frac{1.0 \text{ L}}{0.45 \text{ mol}} = 0.16 \text{ L}; \quad 0.16 \text{ L} = 160 \text{ mL}$$

4.51 $$0.0171 \text{ mol KOH x } \frac{1.00 \text{ L}}{0.350 \text{ mol KOH}} = 0.0489 \text{ L}; \ 0.0489 \text{ L} = 48.9 \text{ mL}$$

4.52 NaCl, 58.4; 400 mg = 0.400 g; 100 mL = 0.100 L

$$0.400 \text{ g NaCl x } \frac{1 \text{ mol NaCl}}{58.4 \text{ g NaCl}} = 0.006\ 85 \text{ mol NaCl}$$

$$\text{molarity} = \frac{0.006\ 85 \text{ mol}}{0.100 \text{ L}} = 0.0685 \text{ M}$$

4.53 $C_6H_{12}O_6$, 180.2; 90 mg = 0.090 g; 100 mL = 0.100 L

$$0.090 \text{ g } C_6H_{12}O_6 \text{ x } \frac{1 \text{ mol } C_6H_{12}O_6}{180.2 \text{ g } C_6H_{12}O_6} = 0.000\ 50 \text{ mol } C_6H_{12}O_6$$

$$\text{molarity} = \frac{0.000\ 50 \text{ mol}}{0.100 \text{ L}} = 0.0050 \text{ M} = 5.0 \text{ x } 10^{-3} \text{ M}$$

4.54 $$3.045 \text{ g Cu x } \frac{1 \text{ mol Cu}}{63.546 \text{ g Cu}} = 0.047\ 92 \text{ mol Cu}; \ 50.0 \text{ mL} = 0.0500 \text{ L}$$

$$Cu(NO_3)_2 \text{ molarity} = \frac{0.047\ 92 \text{ mol}}{0.0500 \text{ L}} = 0.958 \text{ M}$$

4.55 250.0 mL = 0.2500 L

mass of Zn in penny = (2.482 g)(0.975) = 2.42 g Zn

$$2.42 \text{ g Zn x } \frac{1 \text{ mol Zn}}{65.38 \text{ g Zn}} = 0.0370 \text{ mol Zn}$$

mol $Zn(NO_3)_2$ = mol Zn = 0.0370 mol

$$Zn(NO_3)_2 \text{ molarity} = \frac{0.0370 \text{ mol}}{0.2500 \text{ L}} = 0.148 \text{ M}$$

4.56 $M_f \text{ x } V_f = M_i \text{ x } V_i$; $$M_f = \frac{M_i \text{ x } V_i}{V_f} = \frac{12.0 \text{ M x } 35.7 \text{ mL}}{250.0 \text{ mL}} = 1.71 \text{ M HCl}$$

4.57 $M_f \text{ x } V_f = M_i \text{ x } V_i$; $$V_f = \frac{M_i \text{ x } V_i}{M_f} = \frac{0.0913 \text{ M x } 70.00 \text{ mL}}{0.0150 \text{ M}} = 426 \text{ mL}$$

4.58 $CaCl_2$, 111.0; 500 mL = 0.500 L
(0.500 L)(0.33 mol/L) = 0.165 mol $CaCl_2$

$$0.165 \text{ mol } CaCl_2 \times \frac{111.0 \text{ g } CaCl_2}{1 \text{ mol } CaCl_2} = 18.3 \text{ g } CaCl_2$$

Place 18.3 g of $CaCl_2$ in a 500 mL volumetric flask and fill it to the mark with water.

4.59 250 mL = 0.250 L

$$M_f \times V_f = M_i \times V_i; \quad V_i = \frac{M_f \times V_f}{M_i} = \frac{0.150 \text{ M} \times 250 \text{ mL}}{3.00 \text{ M}} = 12.5 \text{ mL}$$

Using a 25 mL graduated cylinder, measure out 12.5 mL of 3.00 M $CaCl_2$ solution. Pour the 12.5 mL into a 250 mL volumetric flask and fill it to the mark with water.

Electrolytes, Net Ionic Equations, and Aqueous Reactions (Sections 4.3–4.5)

4.60 (a) strong electrolyte, bright (b) nonelectrolyte, dark (c) weak electrolyte, dim

4.61 (a) weak electrolyte, dim (b) strong electrolyte, bright (c) nonelectrolyte, dark

4.62 $Ba(OH)_2$ is soluble in aqueous solution, dissociates into $Ba^{2+}(aq)$ and 2 $OH^-(aq)$, and conducts electricity. In aqueous solution, H_2SO_4 dissociates into $H^+(aq)$ and $HSO_4^-(aq)$. H_2SO_4 solutions conduct electricity. When equal molar solutions of $Ba(OH)_2$ and H_2SO_4 are mixed, the insoluble $BaSO_4$ is formed along with two H_2O. In water, $BaSO_4$ does not produce any appreciable amount of ions and the mixture does not conduct electricity.

4.63 H_2O is polar and a good H^+ acceptor. It allows the polar HCl to dissociate into ions in aqueous solution: $HCl + H_2O \rightarrow H_3O^+ + Cl^-$.
$CHCl_3$ is not very polar and not a H^+ acceptor and so does not allow the polar HCl to dissociate into ions.

4.64 (a) HBr, strong electrolyte
(b) HF, weak electrolyte
(c) $NaClO_4$, strong electrolyte
(d) $(NH_4)_2CO_3$, strong electrolyte
(e) NH_3, weak electrolyte
(f) C_2H_5OH, nonelectrolyte

4.65 It is possible for a molecular compound to be a strong electrolyte. For example, HCl is a molecular compound when pure, but dissociates completely to give H^+ and Cl^- ions when it dissolves in water.

4.66 (a) K_2CO_3 contains 3 ions (2 K^+ and 1 CO_3^{2-}).
The molar concentration of ions = 3 x 0.750 M = 2.25 M.
(b) $AlCl_3$ contains 4 ions (1 Al^{3+} and 3 Cl^-).
The molar concentration of ions = 4 x 0.355 M = 1.42 M.

4.67 (a) CH_3OH is a nonelectrolyte. The ion concentration from CH_3OH is zero.
(b) $HClO_4$ is a strong acid.
$HClO_4(aq) \rightarrow H^+(aq) + ClO_4^-(aq)$

In solution, there are 2 moles of ions per mole of $HClO_4$.
The molar concentration of ions = 2 × 0.225 M = 0.450 M.

4.68 (a) precipitation (b) redox (c) acid-base neutralization

4.69 (a) redox (b) precipitation (c) acid-base neutralization

4.70 (a) Ionic equation:
$Hg^{2+}(aq) + 2\ NO_3^-(aq) + 2\ Na^+(aq) + 2\ I^-(aq) \rightarrow 2\ Na^+(aq) + 2\ NO_3^-(aq) + HgI_2(s)$
Delete spectator ions from the ionic equation to get the net ionic equation.
Net ionic equation: $Hg^{2+}(aq) + 2\ I^-(aq) \rightarrow HgI_2(s)$

(b) $2\ HgO(s) \xrightarrow{\text{Heat}} 2\ Hg(l) + O_2(g)$
(c) Ionic equation:
$H_3PO_4(aq) + 3\ K^+(aq) + 3\ OH^-(aq) \rightarrow 3\ K^+(aq) + PO_4^{3-}(aq) + 3\ H_2O(l)$
Delete spectator ions from the ionic equation to get the net ionic equation.
Net ionic equation: $H_3PO_4(aq) + 3\ OH^-(aq) \rightarrow PO_4^{3-}(aq) + 3\ H_2O(l)$

4.71 (a) $S_8(s) + 8\ O_2(g) \rightarrow 8\ SO_2(g)$
(b) Ionic equation:
$Ni^{2+}(aq) + 2\ Cl^-(aq) + 2\ Na^+(aq) + S^{2-}(aq) \rightarrow NiS(s) + 2\ Na^+(aq) + 2\ Cl^-(aq)$
Delete spectator ions from the ionic equation to get the net ionic equation.
Net ionic equation: $Ni^{2+}(aq) + S^{2-}(aq) \rightarrow NiS(s)$
(c) Ionic equation:
$2\ CH_3CO_2H(aq) + Ba^{2+}(aq) + 2\ OH^-(aq) \rightarrow 2\ CH_3CO_2^-(aq) + Ba^{2+}(aq) + 2\ H_2O(l)$
Delete spectator ions from the ionic equation to get the net ionic equation.
Net ionic equation: $CH_3CO_2H(aq) + OH^-(aq) \rightarrow CH_3CO_2^-(aq) + H_2O(l)$

Precipitation Reactions and Solubility Guidelines (Section 4.6)

4.72 (a) $PbSO_4$, insoluble (b) $Ba(NO_3)_2$, soluble
(c) $SnCO_3$, insoluble (d) $(NH_4)_3PO_4$, soluble

4.73 (a) ZnS, insoluble (b) $Au_2(CO_3)_3$, insoluble
(c) $PbCl_2$, insoluble (soluble in hot water) (d) Na_2S, soluble

4.74 (a) No precipitate will form. (b) $Fe^{2+}(aq) + 2\ OH^-(aq) \rightarrow Fe(OH)_2(s)$
(c) No precipitate will form. (d) No precipitate will form.

4.75 (a) $Mn^{2+}(aq) + S^{2-}(aq) \rightarrow MnS(s)$
(b) No precipitate will form.
(c) $3\ Hg^{2+}(aq) + 2\ PO_4^{3-}(aq) \rightarrow Hg_3(PO_4)_2(s)$
(d) $Ba^{2+}(aq) + 2\ OH^-(aq) \rightarrow Ba(OH)_2(s)$

4.76 (a) 0.10 M $LiNO_3$ will not form a precipitate.
(b) $BaSO_4(s)$ will precipitate. (c) AgCl(s) will precipitate.

4.77 (a) $Mg(OH)_2(s)$ will precipitate. (b) 0.10 M NH_4Br will not form a precipitate.
(c) $Fe(OH)_2(s)$ will precipitate.

4.78 (a) $Pb(NO_3)_2(aq) + Na_2SO_4(aq) \rightarrow PbSO_4(s) + 2\ NaNO_3(aq)$
(b) $3\ MgCl_2(aq) + 2\ K_3PO_4(aq) \rightarrow Mg_3(PO_4)_2(s) + 6\ KCl(aq)$
(c) $ZnSO_4(aq) + Na_2CrO_4(aq) \rightarrow ZnCrO_4(s) + Na_2SO_4(aq)$

4.79 (a) $AlCl_3(aq) + 3\ NaOH(aq) \rightarrow Al(OH)_3(s) + 3\ NaCl(aq)$
(b) $Fe(NO_3)_2(aq) + Na_2S(aq) \rightarrow FeS(s) + 2\ NaNO_3(aq)$
(c) $CoSO_4(aq) + K_2CO_3(aq) \rightarrow CoCO_3(s) + K_2SO_4(aq)$

4.80 $Ag^+(aq) + NO_3^-(aq) + H^+(aq) + Cl^-(aq) \rightarrow AgCl(s) + H^+(aq) + NO_3^-(aq)$

$$\text{mol } Cl^- = 30.0 \text{ mL} \times \frac{1 \times 10^{-3}\text{ L}}{1 \text{ mL}} \times 0.150 \text{ mol/L} = 0.00450 \text{ mol}$$

$$\text{mol } Ag^+ = 25.0 \text{ mL} \times \frac{1 \times 10^{-3}\text{ L}}{1 \text{ mL}} \times 0.200 \text{ mol/L} = 0.00500 \text{ mol}$$

Because the mole ratio between Ag^+ and Cl^- is 1 to 1, Cl^- is the limiting reactant because it is the smaller mol quantity. 0.00450 mol of AgCl is produced.

AgCl, 143.3; $\text{mass AgCl} = 0.00450 \text{ mol AgCl} \times \frac{143.3 \text{ g AgCl}}{1 \text{ mol AgCl}} = 0.645 \text{ g AgCl}$

4.81 $Ba^{2+}(aq) + 2\ Cl^-(aq) + 2\ Na^+(aq) + CO_3^{2-}(aq) \rightarrow BaCO_3(s) + 2\ Na^+(aq) + 2\ Cl^-(aq)$

$$\text{mol } CO_3^{2-} = 40.0 \text{ mL} \times \frac{1 \times 10^{-3}\text{ L}}{1 \text{ mL}} \times 0.150 \text{ mol/L} = 0.00600 \text{ mol}$$

$$\text{mol } Ba^{2+} = 55.0 \text{ mL} \times \frac{1 \times 10^{-3}\text{ L}}{1 \text{ mL}} \times 0.100 \text{ mol/L} = 0.00550 \text{ mol}$$

Because the mole ratio between Ba^{2+} and CO_3^{2-} is 1 to 1, Ba^{2+} is the limiting reactant because it is the smaller mol quantity. 0.00550 mol of $BaCO_3$ is produced.
$BaCO_3$, 197.3

$$\text{mass } BaCO_3 = 0.00550 \text{ mol } BaCO_3 \times \frac{197.3 \text{ g } BaCO_3}{1 \text{ mol } BaCO_3} = 1.09 \text{ g } BaCO_3$$

4.82 Add HCl(aq); it will selectively precipitate AgCl(s).

4.83 Add $Na_2SO_4(aq)$; it will selectively precipitate $BaSO_4(s)$.

4.84 Ag^+ is eliminated because it would have precipitated as AgCl(s); Ba^{2+} is eliminated because it would have precipitated as $BaSO_4(s)$. The solution might contain Cs^+ and/or NH_4^+. Neither of these will precipitate with OH^-, SO_4^{2-}, or Cl^-.

4.85 Cl^- is eliminated because it would have precipitated as AgCl(s). OH^- is eliminated because it would have precipitated as either AgOH(s) or $Cu(OH)_2(s)$. SO_4^{2-} is eliminated because it would have precipitated as $BaSO_4(s)$. The solution might contain NO_3^- because all nitrates are soluble.

Acids, Bases, and Neutralization Reactions (Section 4.7)

4.86 Add the solution to an active metal, such as magnesium. Bubbles of H_2 gas indicate the presence of an acid.

4.87 We use a double arrow to show the dissociation of a weak acid or weak base in aqueous solution to indicate the equilibrium between reactants and products.

4.88 (a) $2\ H^+(aq) + 2\ ClO_4^-(aq) + Ca^{2+}(aq) + 2\ OH^-(aq) \rightarrow Ca^{2+}(aq) + 2\ ClO_4^-(aq) + 2\ H_2O(l)$
(b) $CH_3CO_2H(aq) + Na^+(aq) + OH^-(aq) \rightarrow CH_3CO_2^-(aq) + Na^+(aq) + H_2O(l)$

4.89 (a) $2\ H^+(aq) + 2\ Br^-(aq) + Ca^{2+}(aq) + 2\ OH^-(aq) \rightarrow Ca^{2+}(aq) + 2\ Br^-(aq) + 2\ H_2O(l)$
(b) $Ba^{2+}(aq) + 2\ OH^-(aq) + 2\ H^+(aq) + 2\ NO_3^-(aq) \rightarrow Ba^{2+}(aq) + 2\ NO_3^-(aq) + 2\ H_2O(l)$

4.90 (a) $LiOH(aq) + HI(aq) \rightarrow LiI(aq) + H_2O(l)$
Ionic equation: $Li^+(aq) + OH^-(aq) + H^+(aq) + I^-(aq) \rightarrow Li^+(aq) + I^-(aq) + H_2O(l)$
Delete spectator ions from the ionic equation to get the net ionic equation.
Net ionic equation: $H^+(aq) + OH^-(aq) \rightarrow H_2O(l)$
(b) $2\ HBr(aq) + Ca(OH)_2(aq) \rightarrow CaBr_2(aq) + 2\ H_2O(l)$
Ionic equation:
$2\ H^+(aq) + 2\ Br^-(aq) + Ca^{2+}(aq) + 2\ OH^-(aq) \rightarrow Ca^{2+}(aq) + 2\ Br^-(aq) + 2\ H_2O(l)$
Delete spectator ions from the ionic equation to get the net ionic equation.
Net ionic equation: $H^+(aq) + OH^-(aq) \rightarrow H_2O(l)$

4.91 (a) $2\ Fe(OH)_3(s) + 3\ H_2SO_4(aq) \rightarrow Fe_2(SO_4)_3(aq) + 6\ H_2O(l)$
Ionic equation and net ionic equation are the same.
$2\ Fe(OH)_3(s) + 3\ H^+(aq) + 3\ HSO_4^-(aq) \rightarrow 2\ Fe^{3+}(aq) + 3\ SO_4^{2-}(aq) + 6\ H_2O(l)$
(b) $HClO_3(aq) + NaOH(aq) \rightarrow NaClO_3(aq) + H_2O(l)$
Ionic equation $H^+(aq) + ClO_3^-(aq) + Na^+(aq) + OH^-(aq) \rightarrow Na^+(aq) + ClO_3^-(aq) + H_2O(l)$
Delete spectator ions from the ionic equation to get the net ionic equation.
Net ionic equation: $H^+(aq) + OH^-(aq) \rightarrow H_2O(l)$

Solution Stoichiometry and Titration (Sections 4.8 and 4.9)

4.92 $2\ HBr(aq) + K_2CO_3(aq) \rightarrow 2\ KBr(aq) + CO_2(g) + H_2O(l)$
K_2CO_3, 138.2; 450 mL = 0.450 L

$$\frac{0.500\ \text{mol HBr}}{\text{L}} \times 0.450\ \text{L} = 0.225\ \text{mol HBr}$$

$$0.225\ \text{mol HBr} \times \frac{1\ \text{mol}\ K_2CO_3}{2\ \text{mol HBr}} \times \frac{138.2\ \text{g}\ K_2CO_3}{1\ \text{mol}\ K_2CO_3} = 15.5\ \text{g}\ K_2CO_3$$

4.93 $2\ C_4H_{10}S + NaOCl \rightarrow C_8H_{18}S_2 + NaCl + H_2O$
$C_4H_{10}S$, 90.19; 5.00 mL = 0.005 00 L

$$\frac{0.0985\ \text{mol NaOCl}}{\text{L}} \times 0.005\ 00\ \text{L} = 4.925 \times 10^{-4}\ \text{mol NaOCl}$$

$$4.925 \times 10^{-4} \text{ mol NaOCl} \times \frac{2 \text{ mol } C_4H_{10}S}{1 \text{ mol NaOCl}} \times \frac{90.19 \text{ g } C_4H_{10}S}{1 \text{ mol } C_4H_{10}S} = 0.0888 \text{ g } C_4H_{10}S$$

4.94 $H_2C_2O_4$, 90.0

$$3.225 \text{ g } H_2C_2O_4 \times \frac{1 \text{ mol } H_2C_2O_4}{90.0 \text{ g } H_2C_2O_4} \times \frac{2 \text{ mol } KMnO_4}{5 \text{ mol } H_2C_2O_4} = 0.0143 \text{ mol } KMnO_4$$

$$0.0143 \text{ mol} \times \frac{1 \text{ L}}{0.250 \text{ mol}} = 0.0572 \text{ L} = 57.2 \text{ mL}$$

4.95 $H_2C_2O_4$, 90.0; 400.0 mL = 0.4000 L; 25.0 mL = 0.0250 L

$$12.0 \text{ g } H_2C_2O_4 \times \frac{1 \text{ mol } H_2C_2O_4}{90.0 \text{ g } H_2C_2O_4} = 0.133 \text{ mol } H_2C_2O_4$$

$$\text{molarity} = \frac{0.133 \text{ mol}}{0.4000 \text{ L}} = 0.333 \text{ M } H_2C_2O_4$$

$H_2C_2O_4(aq) + 2\ KOH(aq) \rightarrow K_2C_2O_4(aq) + 2\ H_2O(l)$

$$\frac{0.333 \text{ mol } C_2H_2O_4}{\text{L}} \times 0.0250 \text{ L} = 0.008\ 32 \text{ mol } H_2C_2O_4$$

$$0.008\ 32 \text{ mol } H_2C_2O_4 \times \frac{2 \text{ mol KOH}}{1 \text{ mol } H_2C_2O_4} = 0.0166 \text{ mol KOH}$$

$$0.0166 \text{ mol} \times \frac{1 \text{ L}}{0.100 \text{ mol}} = 0.166 \text{ L};\ 0.166 \text{ L} = 166 \text{ mL}$$

4.96 (a) $LiOH(aq) + HBr(aq) \rightarrow LiBr(aq) + H_2O(l)$
25.0 mL = 0.0250 L and 75.0 mL = 0.0750 L
(0.240 mol/L LiOH)(0.0250 L) = 0.006 00 mol LiOH
(0.200 mol/L HBr)(0.0750 L) = 0.0150 mol HBr
HBr and LiOH react in a one to one mole ratio.
mol HBr left over = 0.0150 mol HBr – 0.006 00 mol LiOH = 0.0090 mol HBr
$KOH(aq) + HBr(aq) \rightarrow KBr(aq) + H_2O(l)$
HBr and KOH react in a one to one mole ratio.
0.0090 mol KOH will neutralize the solution.

$$\text{volume} = \frac{\text{mol}}{\text{M}} = \frac{0.0090 \text{ mol KOH}}{1.00 \text{ mol/L}} = 0.0090 \text{ L} = 9.0 \text{ mL KOH}$$

(b) $HCl(aq) + NaOH(aq) \rightarrow NaCl(aq) + H_2O(l)$
45.0 mL = 0.0450 L and 10.0 mL = 0.0100 L
(0.300 mol/L HCl)(0.0450 L) = 0.0135 mol HCl
(0.250 mol/L NaOH)(0.0100 L) = 0.002 50 mol NaOH
HCl and NaOH react in a one to one mole ratio.
mol HCl left over = 0.0135 mol HCl – 0.002 50 mol NaOH = 0.0110 mol HCl
$KOH(aq) + HCl(aq) \rightarrow KCl(aq) + H_2O(l)$
HCl and KOH react in a one to one mole ratio.

0.0110 mol KOH will neutralize the solution.

$$\text{volume} = \frac{\text{mol}}{\text{M}} = \frac{0.0110 \text{ mol KOH}}{1.00 \text{ mol/L}} = 0.0110 \text{ L} = 11.0 \text{ mL KOH}$$

4.97 (a) $HNO_3(aq) + KOH(aq) \rightarrow KNO_3(aq) + H_2O(l)$
100.0 mL = 0.1000 L and 400.0 mL = 0.4000 L
(0.160 mol/L HNO_3)(0.1000 L) = 0.0160 mol HNO_3
(0.100 mol/L KOH)(0.4000 L) = 0.0400 mol KOH
HNO_3 and KOH react in a one to one mole ratio.
mol KOH left over = 0.0400 mol KOH – 0.0160 mol HNO_3 = 0.0240 mol KOH
$HCl(aq) + KOH(aq) \rightarrow KCl(aq) + H_2O(l)$
HCl and KOH react in a one to one mole ratio.
0.0240 mol HCl will neutralize the solution.

$$\text{volume} = \frac{\text{mol}}{\text{M}} = \frac{0.0240 \text{ mol HCl}}{2.00 \text{ mol/L}} = 0.0120 \text{ L} = 12.0 \text{ mL HCl}$$

(b) $NaOH(aq) + HBr(aq) \rightarrow NaBr(aq) + H_2O(l)$
350.0 mL = 0.3500 L and 150.0 mL = 0.1500 L
(0.120 mol/L NaOH)(0.3500 L) = 0.0420 mol NaOH
(0.190 mol/L HBr)(0.1500 L) = 0.0285 mol HBr
NaOH and HBr react in a one to one mole ratio.
mol NaOH left over = 0.0420 mol NaOH – 0.0285 mol HBr = 0.0135 mol NaOH
$HCl(aq) + NaOH(aq) \rightarrow NaCl(aq) + H_2O(l)$
HCl and NaOH react in a one to one mole ratio.
0.0135 mol HCl will neutralize the solution.

$$\text{volume} = \frac{\text{mol}}{\text{M}} = \frac{0.0135 \text{ mol HCl}}{2.00 \text{ mol/L}} = 0.006\ 75 \text{ L} = 6.75 \text{ mL HCl}$$

4.98 (a) $HBr(aq) + KOH(aq) \rightarrow KBr(aq) + H_2O(l)$
50.0 mL = 0.0500 L and 30.0 mL = 0.0300 L
(0.100 mol/L HBr)(0.0500 L) = 0.005 00 mol HBr
(0.200 mol/L KOH)(0.0300 L) = 0.006 00 mol KOH
HBr and KOH react in a one to one mole ratio. The resulting solution is basic because there is an excess of KOH.

(b) $2\ HCl(aq) + Ba(OH)_2(aq) \rightarrow BaCl_2(aq) + 2\ H_2O(l))$
100.0 mL = 0.1000 L and 75 mL = 0.0750 L
(0.0750 mol/L HCl)(0.1000 L) = 0.007 50 mol HCl
(0.100 mol/L $Ba(OH)_2$)(0.0750 L) = 0.007 50 mol $Ba(OH)_2$

$$0.007\ 50 \text{ mol HCl} \times \frac{1 \text{ mol } Ba(OH)_2}{2 \text{ mol HCl}} = 0.003\ 75 \text{ mol } Ba(OH)_2 \text{ needed}$$

The resulting solution is basic because there is an excess of $Ba(OH)_2$.

4.99 (a) $HClO_4(aq) + NaOH(aq) \rightarrow NaClO_4(aq) + H_2O(l)$
65.0 mL = 0.0650 L and 40.0 mL = 0.0400 L

$(0.0500 \text{ mol/L } HClO_4)(0.0650 \text{ L}) = 0.003\ 25 \text{ mol } HClO_4$
$(0.0750 \text{ mol/L NaOH})(0.0400 \text{ L}) = 0.003\ 00 \text{ mol NaOH}$
$HClO_4$ and NaOH react in a one to one mole ratio. The resulting solution is acidic because there is an excess of $HClO_4$.
(b) $2\ HNO_3(aq) + Ca(OH)_2(aq) \rightarrow Ca(NO_3)_2(aq) + 2\ H_2O(l)$
125.0 mL = 0.1250 L and 90.0 mL = 0.0900 L
$(0.100 \text{ mol/L } HNO_3)(0.1250 \text{ L}) = 0.0125 \text{ mol } HNO_3$
$(0.0750 \text{ mol/L } Ca(OH)_2)(0.0900 \text{ L}) = 0.006\ 75 \text{ mol } Ca(OH)_2$

$$0.0125 \text{ mol } HNO_3 \times \frac{1 \text{ mol } Ca(OH)_2}{2 \text{ mol } HNO_3} = 0.006\ 25 \text{ mol } Ca(OH)_2 \text{ needed}$$

The resulting solution is basic because there is an excess of $Ca(OH)_2$.

Redox Reactions, Oxidation Numbers, and Activity Series (Sections 4.10–4.12)

4.100 The best reducing agents are at the bottom left of the periodic table. The best oxidizing agents are at the top right of the periodic table (excluding the noble gases).

4.101 The most easily reduced elements in the periodic table are in the top-right corner, excluding group 8A.
The most easily oxidized elements in the periodic table are in the bottom-left corner.

4.102 (a) An oxidizing agent gains electrons.
(b) A reducing agent loses electrons.
(c) A substance undergoing oxidation loses electrons.
(d) A substance undergoing reduction gains electrons.

4.103 (a) In a redox reaction, the oxidation number decreases for an oxidizing agent.
(b) In a redox reaction, the oxidation number increases for a reducing agent.
(c) In a redox reaction, the oxidation number increases for a substance undergoing oxidation.
(d) In a redox reaction, the oxidation number decreases for a substance undergoing reduction.

4.104 (a) NO_2 O −2, N +4; (b) SO_3 O −2, S +6
(c) $COCl_2$ O −2, Cl −1, C +4; (d) CH_2Cl_2 Cl −1, H +1, C 0
(e) $KClO_3$ O −2, K +1, Cl +5; (f) HNO_3 O −2, H +1, N +5

4.105 (a) $VOCl_3$ O −2, Cl −1, V +5
(b) $CuSO_4$ O −2, S +6, Cu +2
(c) CH_2O O −2, H +1, C 0
(d) Mn_2O_7 O −2, Mn +7
(e) OsO_4 O −2, Os +8
(f) H_2PtCl_6 Cl −1, H +1, Pt +4

4.106 (a) ClO_3^- O −2, Cl +5; (b) SO_3^{2-} O −2, S +4
(c) $C_2O_4^{2-}$ O −2, C +3; (d) NO_2^- O −2, N +3
(e) BrO^- O −2, Br +1; (f) AsO_4^{3-} O −2, As +5

4.107 (a) $Cr(OH)_4^-$ O −2, H +1, Cr +3
(b) $S_2O_3^{2-}$ O −2, S +2
(c) NO_3^- O −2, N +5
(d) MnO_4^{2-} O −2, Mn +6
(e) HPO_4^{2-} O −2, H +1, P +5
(f) $V_2O_7^{4-}$ O −2, V +5

4.108 (a) +1, N_2O, nitrous oxide; +2, NO, nitric oxide; +4, N_2O_4, dinitrogen tetroxide; +5, N_2O_5, dinitrogen pentoxide
(b) N_2O_5 cannot react with O_2 because N has its maximum oxidation number, +5, and cannot be further oxidized.

4.109 (a) tetraphosphorus hexoxide, P_4O_6, P oxidation number = +12/4 = +3
tetraphosphorus decoxide, P_4O_{10}, P oxidation number = +20/4 = +5
(b) P_4O_6, because it has the lower P oxidation number.

4.110 (a) $Ca(s) + Sn^{2+}(aq) \rightarrow Ca^{2+}(aq) + Sn(s)$
Ca(s) is oxidized (oxidation number increases from 0 to +2).
$Sn^{2+}(aq)$ is reduced (oxidation number decreases from +2 to 0).
(b) $ICl(s) + H_2O(l) \rightarrow HCl(aq) + HOI(aq)$
No oxidation numbers change. The reaction is not a redox reaction.

4.111 (a) $Si(s) + 2\ Cl_2(g) \rightarrow SiCl_4(l)$
Si(s) is oxidized (oxidation number increases from 0 to +4).
$Cl_2(g)$ is reduced (oxidation number decreases from 0 to −1).
(b) $Cl_2(g) + 2\ NaBr(aq) \rightarrow Br_2(aq) + 2\ NaCl(aq)$
$Br^-(aq)$ is oxidized (oxidation number increases from −1 to 0).
$Cl_2(g)$ is reduced (oxidation number decreases from 0 to −1).

4.112 (a) Zn is below Na^+; therefore no reaction.
(b) Pt is below H^+; therefore no reaction.
(c) Au is below Ag^+; therefore no reaction.
(d) Ag is above Au^{3+}; the reaction is $Au^{3+}(aq) + 3\ Ag(s) \rightarrow 3\ Ag^+(aq) + Au(s)$.

4.113 Sr is more metallic than Sb because it is in the same period and to the left of Sb on the periodic table. Sr is the better reducing agent.
$2\ Sb^{3+}(aq) + 3\ Sr(s) \rightarrow 2\ Sb(s) + 3\ Sr^{2+}(aq)$ will occur, the reverse will not.

4.114 (a) "Any element higher in the activity series will react with the ion of any element lower in the activity series."
$A + B^+ \rightarrow A^+ + B$; therefore A is higher than B.
$C^+ + D \rightarrow$ no reaction; therefore C is higher than D.
$B + D^+ \rightarrow B^+ + D$; therefore B is higher than D.
$B + C^+ \rightarrow B^+ + C$; therefore B is higher than C.
The net result is $A > B > C > D$.

(b) (1) C is below A^+; therefore no reaction.
(2) D is below A^+; therefore no reaction.

4.115 (a) "Any element higher in the activity series will react with the ion of any element lower in the activity series."
$2\ A + B^{2+} \rightarrow 2\ A^+ + B$; therefore A is higher than B.
$B + D^{2+} \rightarrow B^{2+} + D$; therefore B is higher than D.
$A^+ + C \rightarrow$ no reaction; therefore A is higher than C.
$2\ C + B^{2+} \rightarrow 2\ C^+ + B$; therefore C is higher than B.
The net result is A > C > B > D.
(b) (1) D is below A^+; therefore no reaction.
(2) C is above D^{2+}; therefore the reaction will occur.

Redox Titrations (Section 4.13)

4.116 $I_2(aq) + 2\ S_2O_3^{2-}(aq) \rightarrow S_4O_6^{2-}(aq) + 2\ I^-(aq)$; 35.20 mL = 0.032 50 L

$$0.035\ 20\ \text{L} \times \frac{0.150\ \text{mol}\ S_2O_3^{2-}}{\text{L}} \times \frac{1\ \text{mol}\ I_2}{2\ \text{mol}\ S_2O_3^{2-}} \times \frac{253.8\ \text{g}\ I_2}{1\ \text{mol}\ I_2} = 0.670\ \text{g}\ I_2$$

4.117 $$2.486\ \text{g}\ I_2 \times \frac{1\ \text{mol}\ I_2}{253.8\ \text{g}\ I_2} \times \frac{2\ \text{mol}\ S_2O_3^{2-}}{1\ \text{mol}\ I_2} = 1.959 \times 10^{-2}\ \text{mol}\ S_2O_3^{2-}$$

$$1.959 \times 10^{-2}\ \text{mol} \times \frac{1\ \text{L}}{0.250\ \text{mol}} \times \frac{1\ \text{mL}}{1 \times 10^{-3}\ \text{L}} = 78.4\ \text{mL}$$

4.118 46.99 mL = 0.046 99 L; 50.00 mL = 0.050 00 L
$(0.2004\ \text{mol/L}\ Cr_2O_7^{2-})(0.046\ 99\ \text{L}) = 0.009\ 417\ \text{mol}\ Cr_2O_7^{2-}$

$$0.009\ 417\ \text{mol}\ Cr_2O_7^{2-} \times \frac{6\ \text{mol}\ Fe^{2+}}{1\ \text{mol}\ Cr_2O_7^{2-}} = 0.056\ 50\ \text{mol}\ Fe^{2+}$$

$$[Fe^{2+}] = \frac{0.056\ 50\ \text{mol}}{0.050\ 00\ \text{L}} = 1.130\ \text{M}$$

4.119 $FeSO_4$, 151.9; 18.72 mL = 0.018 72 L
$(0.1500\ \text{mol/L}\ Cr_2O_7^{2-})(0.018\ 72\ \text{L}) = 0.002\ 808\ \text{mol}\ Cr_2O_7^{2-}$

$$0.002\ 808\ \text{mol}\ Cr_2O_7^{2-} \times \frac{6\ \text{mol}\ Fe^{2+}}{1\ \text{mol}\ Cr_2O_7^{2-}} \times \frac{1\ \text{mol}\ FeSO_4}{1\ \text{mol}\ Fe^{2+}} \times \frac{151.9\ \text{g}\ FeSO_4}{1\ \text{mol}\ FeSO_4} = 2.559\ \text{g}\ FeSO_4$$

4.120 $3\ H_3AsO_3(aq) + BrO_3^-(aq) \rightarrow Br^-(aq) + 3\ H_3AsO_4(aq)$
22.35 mL = 0.022 35 L and 50.00 mL = 0.050 00 L

$$0.022\ 35\ \text{L} \times \frac{0.100\ \text{mol}\ BrO_3^-}{\text{L}} \times \frac{3\ \text{mol}\ H_3AsO_3}{1\ \text{mol}\ BrO_3^-} = 6.70 \times 10^{-3}\ \text{mol}\ H_3AsO_3$$

$$\text{molarity} = \frac{6.70 \times 10^{-3}\ \text{mol}}{0.050\ 00\ \text{L}} = 0.134\ \text{M As(III)}$$

4.121 As_2O_3, 197.8; 28.55 mL = 0.028 55 L

$$1.550 \text{ g } As_2O_3 \times \frac{1 \text{ mol } As_2O_3}{197.84 \text{ g } As_2O_3} \times \frac{2 \text{ mol } H_3AsO_3}{1 \text{ mol } As_2O_3} \times \frac{1 \text{ mol } BrO_3^-}{3 \text{ mol } H_3AsO_3}$$

$$= 5.223 \times 10^{-3} \text{ mol } BrO_3^-; \quad KBrO_3 \text{ molarity} = \frac{5.223 \times 10^{-3} \text{ mol}}{0.028\ 55 \text{ L}} = 0.1829 \text{ M}$$

4.122 $2\ Fe^{3+}(aq) + Sn^{2+}(aq) \rightarrow 2\ Fe^{2+}(aq) + Sn^{4+}(aq)$; 13.28 mL = 0.013 28 L

$$0.013\ 28 \text{ L} \times \frac{0.1015 \text{ mol } Sn^{2+}}{\text{L}} \times \frac{2 \text{ mol } Fe^{3+}}{1 \text{ mol } Sn^{2+}} \times \frac{55.845 \text{ g } Fe^{3+}}{1 \text{ mol } Fe^{3+}} = 0.1506 \text{ g } Fe^{3+}$$

$$\text{mass \% Fe} = \frac{0.1506 \text{ g}}{0.1875 \text{ g}} \times 100\% = 80.32\%$$

4.123 Fe_2O_3, 159.69; 23.84 mL = 0.023 84 L

$$1.4855 \text{ g } Fe_2O_3 \times \frac{1 \text{ mol } Fe_2O_3}{159.69 \text{ g } Fe_2O_3} \times \frac{2 \text{ mol } Fe^{3+}}{1 \text{ mol } Fe_2O_3} \times \frac{1 \text{ mol } Sn^{2+}}{2 \text{ mol } Fe^{3+}} = 0.009\ 302 \text{ mol } Sn^{2+}$$

$$Sn^{2+} \text{ molarity} = \frac{0.009\ 302 \text{ mol}}{0.023\ 84 \text{ L}} = 0.3902 \text{ M}$$

4.124 $C_2H_5OH(aq) + 2\ Cr_2O_7^{2-}(aq) + 16\ H^+(aq) \rightarrow 2\ CO_2(g) + 4\ Cr^{3+}(aq) + 11\ H_2O(l)$
C_2H_5OH, 46.1; 8.76 mL = 0.008 76 L

$$0.008\ 76 \text{ L} \times \frac{0.049\ 88 \text{ mol } Cr_2O_7^{2-}}{\text{L}} \times \frac{1 \text{ mol } C_2H_5OH}{2 \text{ mol } Cr_2O_7^{2-}} \times \frac{46.1 \text{ g } C_2H_5OH}{1 \text{ mol } C_2H_5OH}$$

$$= 0.0101 \text{ g } C_2H_5OH$$

$$\text{mass \% } C_2H_5OH = \frac{0.0101 \text{ g}}{10.002 \text{ g}} \times 100\% = 0.101\%$$

4.125 21.08 mL = 0.021 08 L

$$0.021\ 08 \text{ L} \times \frac{9.88 \times 10^{-4} \text{ mol } MnO_4^-}{\text{L}} \times \frac{5 \text{ mol } H_2C_2O_4}{2 \text{ mol } MnO_4^-} \times \frac{1 \text{ mol } Ca^{2+}}{1 \text{ mol } H_2C_2O_4} \times \frac{40.08 \text{ g } Ca^{2+}}{1 \text{ mol } Ca^{2+}}$$

$$= 0.002\ 09 \text{ g} = 2.09 \text{ mg}$$

Chapter Problems

4.126 NaCl, 58.4; KCl, 74.6; $CaCl_2$, 111.0; 500 mL = 0.500 L

$$4.30 \text{ g NaCl} \times \frac{1 \text{ mol NaCl}}{58.4 \text{ g NaCl}} = 0.0736 \text{ mol NaCl}$$

$$0.150 \text{ g KCl} \times \frac{1 \text{ mol KCl}}{74.6 \text{ g KCl}} = 0.002\ 01 \text{ mol KCl}$$

$$0.165 \text{ g } CaCl_2 \times \frac{1 \text{ mol } CaCl_2}{111.0 \text{ g } CaCl_2} = 0.001\ 49 \text{ mol } CaCl_2$$

$$0.0736 \text{ mol} + 0.002\ 01 \text{ mol} + 2(0.001\ 49 \text{ mol}) = 0.0786 \text{ mol } Cl^-$$

$$Na^+ \text{ molarity} = \frac{0.0736 \text{ mol}}{0.500 \text{ L}} = 0.147 \text{ M}$$

$$Ca^{2+} \text{ molarity} = \frac{0.001\ 49 \text{ mol}}{0.500 \text{ L}} = 0.002\ 98 \text{ M}$$

$$K^+ \text{ molarity} = \frac{0.002\ 01 \text{ mol}}{0.500 \text{ L}} = 0.004\ 02 \text{ M}$$

$$Cl^- \text{ molarity} = \frac{0.0786 \text{ mol}}{0.500 \text{ L}} = 0.157 \text{ M}$$

4.127 (a) $1.0 \times 10^{-11} \text{ g/mL} \times \frac{1 \text{ mol Au}}{197.0 \text{ g Au}} \times \frac{1 \text{ mL}}{1 \times 10^{-3} \text{ L}} = 5.1 \times 10^{-11} \text{ mol/L}$

(b) $1.0 \times 10^{-11} \text{ g/mL} \times \frac{1 \text{ mL}}{1 \times 10^{-3} \text{ L}} \times 1.3 \times 10^{21} \text{ L} = 1.3 \times 10^{13} \text{ g Au}$

4.128 Na_2SO_4, 142.04; Na_3PO_4, 163.94; Li_2SO_4, 109.95; 100.00 mL = 0.100 00 L

$$0.550 \text{ g } Na_2SO_4 \times \frac{1 \text{ mol } Na_2SO_4}{142.04 \text{ g } Na_2SO_4} = 0.003\ 872 \text{ mol } Na_2SO_4$$

$$1.188 \text{ g } Na_3PO_4 \times \frac{1 \text{ mol } Na_3PO_4}{163.94 \text{ g } Na_3PO_4} = 0.007\ 247 \text{ mol } Na_3PO_4$$

$$0.223 \text{ g } Li_2SO_4 \times \frac{1 \text{ mol } Li_2SO_4}{109.95 \text{ g } Li_2SO_4} = 0.002\ 028 \text{ mol } Li_2SO_4$$

$$Na^+ \text{ molarity} = \frac{(2 \times 0.003\ 872 \text{ mol}) + (3 \times 0.007\ 247 \text{ mol})}{0.100\ 00 \text{ L}} = 0.295 \text{ M}$$

$$Li^+ \text{ molarity} = \frac{2 \times 0.002\ 028 \text{ mol}}{0.100\ 00 \text{ L}} = 0.0406 \text{ M}$$

$$SO_4^{2-} \text{ molarity} = \frac{(1 \times 0.003\ 872 \text{ mol}) + (1 \times 0.002\ 028 \text{ mol})}{0.100\ 00 \text{ L}} = 0.0590 \text{ M}$$

$$PO_4^{3-} \text{ molarity} = \frac{1 \times 0.007\ 247 \text{ mol}}{0.100\ 00 \text{ L}} = 0.0725 \text{ M}$$

4.129 Let X equal the mass of benzoic acid and Y the mass of gallic acid in the 1.00 g mixture. Therefore, X + Y = 1.00 g.
Because both acids contain only one acidic hydrogen, there is a 1 to 1 mol ratio between each acid and NaOH in the acid-base titration.
In the titration, mol benzoic acid + mol gallic acid = mol NaOH.

Therefore, $X \times \frac{1 \text{ mol BA}}{122 \text{ g BA}} + Y \times \frac{1 \text{ mol GA}}{170 \text{ g GA}} = \text{mol NaOH}$

$$\text{mol NaOH} = 14.7\text{ mL} \times \frac{1 \times 10^{-3}\text{ L}}{1\text{ mL}} \times \frac{0.500\text{ mol NaOH}}{1\text{ L}} = 0.00735\text{ mol NaOH}$$

We have two unknowns, X and Y, and two equations.

$X + Y = 1.00\text{ g}$

$$X \times \frac{1\text{ mol BA}}{122\text{ g BA}} + Y \times \frac{1\text{ mol GA}}{170\text{ g GA}} = 0.00735\text{ mol NaOH}$$

Rearrange to get $X = 1.00\text{ g} - Y$ and then substitute it into the equation above to solve for Y.

$$(1.00\text{ g} - Y) \times \frac{1\text{ mol BA}}{122\text{ g BA}} + Y \times \frac{1\text{ mol GA}}{170\text{ g GA}} = 0.00735\text{ mol NaOH}$$

$$\frac{1\text{ mol}}{122} - \frac{Y\text{ mol}}{122\text{ g}} + \frac{Y\text{ mol}}{170\text{ g}} = 0.00735\text{ mol}$$

$$-\frac{Y\text{ mol}}{122\text{ g}} + \frac{Y\text{ mol}}{170\text{ g}} = 0.00735\text{ mol} - \frac{1\text{ mol}}{122} = -8.47 \times 10^{-4}\text{ mol}$$

$$\frac{(-Y\text{ mol})(170\text{ g}) + (Y\text{ mol})(122\text{ g})}{(170\text{ g})(122\text{ g})} = -8.47 \times 10^{-4}\text{ mol}$$

$$\frac{-48\text{ Y mol}}{20740\text{ g}} = -8.47 \times 10^{-4}\text{ mol}; \quad \frac{48\text{ Y}}{20740\text{ g}} = 8.47 \times 10^{-4}$$

$$Y = \frac{(20740\text{ g})(8.47 \times 10^{-4})}{48} = 0.366\text{ g}$$

$X = 1.00\text{ g} - 0.366\text{ g} = 0.634\text{ g}$

In the 1.00 g mixture there is 0.63 g of benzoic acid and 0.37 g of gallic acid.

4.130 $H_2C_2O_4$, 90.03; 22.35 mL = 0.022 35 L

$$0.5170\text{ g } H_2C_2O_4 \times \frac{1\text{ mol } H_2C_2O_4}{90.03\text{ g } H_2C_2O_4} \times \frac{2\text{ mol } KMnO_4}{5\text{ mol } H_2C_2O_4} = 2.297 \times 10^{-3}\text{ mol } KMnO_4$$

$$KMnO_4\text{ molarity} = \frac{2.297 \times 10^{-3}\text{ mol}}{0.022\ 35\text{ L}} = 0.1028\text{ M}$$

4.131 (a) $XOCl_2 + 2\,H_2O \rightarrow 2\,HCl + H_2XO_3$

(b) 96.1 mL = 0.0961 L

mol NaOH = (0.1225 mol/L)(0.0961 L) = 0.01177 mol NaOH

$$\text{mol } H^+ = 0.01177\text{ mol NaOH} \times \frac{1\text{ mol } H^+}{1\text{ mol NaOH}} = 0.01177\text{ mol } H^+$$

Of the total H^+ concentration, half comes from HCl and half comes from H_2XO_3.

$$\text{mol } H_2XO_3 = \frac{0.01177\text{ mol } H^+}{2} \times \frac{1\text{ mol } H_2XO_3}{2\text{ mol } H^+} = 2.943 \times 10^{-3}\text{ mol } H_2XO_3$$

$$\text{mol } XOCl_2 = 2.943 \times 10^{-3}\text{ mol } H_2XO_3 \times \frac{1\text{ mol } XOCl_2}{1\text{ mol } H_2XO_3} = 2.943 \times 10^{-3}\text{ mol } XOCl_2$$

$$\text{molar mass } XOCl_2 = \frac{0.350\text{ g } XOCl_2}{2.943 \times 10^{-3}\text{ mol } XOCl_2} = 118.9\text{ g/mol}$$

molecular mass of $XOCl_2$ = 118.9

atomic weight of X = 118.9 – 16.0 – 2(35.45) = 32.0: X = S

4.132 57.91 mL = 0.057 91 L

$$0.057\ 91\ \text{L} \times \frac{0.1018\ \text{mol Ce}^{4+}}{\text{L}} \times \frac{1\ \text{mol Fe}^{2+}}{1\ \text{mol Ce}^{4+}} \times \frac{55.85\ \text{g Fe}^{2+}}{1\ \text{mol Fe}^{2+}} = 0.3292\ \text{g Fe}^{2+}$$

$$\text{mass \% Fe} = \frac{0.3292\ \text{g}}{1.2284\ \text{g}} \times 100\% = 26.80\%$$

4.133 (a) C_2H_6 H +1, C –3

(b) $Na_2B_4O_7$ O –2, Na +1, B +3

(c) Mg_2SiO_4 O –2, Mg +2, Si +4

4.134 (a) "Any element higher in the activity series will react with the ion of any element lower in the activity series."

$C + B^+ \rightarrow C^+ + B$; therefore C is higher than B.

$A^+ + D \rightarrow$ no reaction; therefore A is higher than D.

$C^+ + A \rightarrow$ no reaction; therefore C is higher than A.

$D + B^+ \rightarrow D^+ + B$; therefore D is higher than B.

The net result is C > A > D > B.

(b) (1) The reaction, $A^+ + C \rightarrow A + C^+$, will occur because C is above A in the activity series.

(2) The reaction, $A^+ + B \rightarrow A + B^+$, will not occur because B is below A in the activity series.

4.135 (a) platinum (b) silver (c) chromium

4.136 10.49 mL = 0.010 49 L

$(0.100\ \text{mol/L}\ S_2O_3^{2-})(0.010\ 49\ \text{L}) = 0.001\ 049\ \text{mol}\ S_2O_3^{2-}$

$$0.001\ 049\ \text{mol}\ S_2O_3^{2-} \times \frac{1\ \text{mol}\ I_3^-}{2\ \text{mol}\ S_2O_3^{2-}} \times \frac{2\ \text{mol}\ Cu^{2+}}{1\ \text{mol}\ I_3^-} = 0.001\ 049\ \text{mol}\ Cu^{2-}$$

$$0.001\ 049\ \text{mol}\ Cu^{2-} \times \frac{63.55\ \text{g Cu}}{1\ \text{mol}\ Cu^{2+}} = 0.066\ 66\ \text{g Cu}$$

$$\text{mass \% Cu} = \frac{0.066\ 66\ \text{g Cu}}{14.98\ \text{g sample}} \times 100\% = 0.4450\ \%\ \text{Cu}$$

4.137 (a) $K_{sp} = [Ag^+]^2[CrO_4^{2-}]$

(b) $Ag_2CrO_4(s) \rightleftarrows 2\ Ag^+(aq) + CrO_4^{2-}(aq)$

                2x   x

In a saturated solution 2x = $[Ag^+]$ and x = $[CrO_4^{2-}]$.

$K_{sp} = [Ag^+]^2[CrO_4^{2-}] = 1.1 \times 10^{-12} = (2x)^2(x) = 4x^3$; Solve for x; $x = 6.5 \times 10^{-5}$ M

$[Ag^+] = 2x = 2(6.5 \times 10^{-5}\ M) = 1.3 \times 10^{-4}$ M; $[CrO_4^{2-}] = x = 6.5 \times 10^{-5}$ M

4.138 $MgF_2(s) \rightleftarrows Mg^{2+}(aq) + 2\,F^-(aq)$

$\qquad\qquad\qquad\quad x \qquad\quad 2x$

$[Mg^{2+}] = x = 2.6 \times 10^{-4}$ M and $[F^-] = 2x = 2(2.6 \times 10^{-4}\ M) = 5.2 \times 10^{-4}$ M in a saturated solution.

$K_{sp} = [Mg^{2+}][F^-]^2 = (2.6 \times 10^{-4}\ M)(5.2 \times 10^{-4}\ M)^2 = 7.0 \times 10^{-11}$

4.139 65.20 mL = 0.065 20 L

$$1.926 \text{ g succinic acid} \times \frac{1 \text{ mol succinic acid}}{118.1 \text{ g succinic acid}} = 0.016\,31 \text{ mol succinic acid}$$

$$0.5000\ \frac{\text{mol NaOH}}{1\text{ L}} \times 0.065\,20\text{ L} = 0.032\,60 \text{ mol NaOH}$$

$$\frac{0.032\,60 \text{ mol NaOH}}{0.016\,31 \text{ mol succinic acid}} = 2;$$ therefore succinic acid has two acidic hydrogens.

4.140 (a) Add HCl to precipitate Hg_2Cl_2. $\quad Hg_2^{2+}(aq) + 2Cl^-(aq) \rightarrow Hg_2Cl_2(s)$
(b) Add H_2SO_4 to precipitate $PbSO_4$. $\quad Pb^{2+}(aq) + SO_4^{2-}(aq) \rightarrow PbSO_4(s)$
(c) Add Na_2CO_3 to precipitate $CaCO_3$. $\quad Ca^{2+}(aq) + CO_3^{2-}(aq) \rightarrow CaCO_3(s)$
(d) Add Na_2SO_4 to precipitate $BaSO_4$. $\quad Ba^{2+}(aq) + SO_4^{2-}(aq) \rightarrow BaSO_4(s)$

4.141 (a) Add $AgNO_3$ to precipitate AgCl. $\quad Ag^+(aq) + Cl^-(aq) \rightarrow AgCl(s)$
(b) Add $NiCl_2$ to precipitate NiS. $\quad Ni^{2+}(aq) + S^{2-}(aq) \rightarrow NiS(s)$
(c) Add $CaCl_2$ to precipitate $CaCO_3$. $\quad Ca^{2+}(aq) + CO_3^{2-}(aq) \rightarrow CaCO_3(s)$
(d) Add $MgCl_2$ to precipitate $Mg(OH)_2$. $\quad Mg^{2+}(aq) + 2\,OH^-(aq) \rightarrow Mg(OH)_2(s)$

4.142 100.0 mL = 0.1000 L; 47.14 mL = 0.047 14 L

$$\text{mol HCl and HBr} = \text{mol H}^+ = 0.1235\ \frac{\text{mol NaOH}}{1\text{ L}} \times 0.047\,14\text{ L} = 5.8218 \times 10^{-3}\text{ mol}$$

mass of AgCl and AgBr = 0.9974 g; mol Ag = mol H^+ = 5.8218×10^{-3} mol

$$\text{mass of Ag} = 5.8218 \times 10^{-3}\text{ mol Ag} \times \frac{107.87\text{ g Ag}}{1\text{ mol Ag}} = 0.6280\text{ g Ag}$$

mass of Cl and Br = 0.9974 g – 0.6280 g = 0.3694 g of Cl and Br
Let Y = moles Cl and Z = moles Br in 0.3694 g of Cl and Br.
Let (Y + Z) = moles Ag in 0.6280 g Ag.
For Ag: $\quad 0.6280\text{ g} = (Y + Z) \times 107.87\text{ g}$
For Cl and Br: $\quad 0.3694\text{ g} = (Y \times 35.453\text{ g}) + (Z \times 79.904\text{ g})$
Solve the simultaneous equations for Y and Z.

Rearrange the Ag equation: $$\left(\frac{0.6280\text{ g}}{107.87\text{ g}} - Z\right) = Y$$

Substitute for Y in the Cl and Br equation above and solve for Z.

$$0.3694\text{ g} = \left[\left(\frac{0.6280\text{ g}}{107.87\text{ g}} - Z\right) \times 35.453\text{ g}\right] + (Z \times 79.904\text{ g}); \qquad Z = \frac{0.1630}{44.451} = 3.667 \times 10^{-3}$$

$$Y = \left(\frac{0.6280\text{ g}}{107.87\text{ g}} - Z\right) = \left(\frac{0.6280\text{ g}}{107.87\text{ g}} - 3.667 \times 10^{-3}\right) = 2.155 \times 10^{-3}$$

$$\text{HCl molarity} = \frac{2.155 \times 10^{-3}\ \text{mol}}{0.1000\ \text{L}} = 0.021\ 55\ \text{M}$$

$$\text{HBr molarity} = \frac{3.667 \times 10^{-3}\ \text{mol}}{0.1000\ \text{L}} = 0.036\ 67\ \text{M}$$

4.143 CuO, 79.55; Cu_2O, 143.09

Let X equal the mass of CuO and Y the mass of Cu_2O in the 10.50 g mixture. Therefore, X + Y = 10.50 g.

$$\text{mol Cu} = 8.66\ \text{g} \times \frac{1\ \text{mol Cu}}{63.546\ \text{g Cu}} = 0.1363\ \text{mol Cu}$$

$$\text{mol CuO} + 2 \times \text{mol Cu}_2\text{O} = 0.1363\ \text{mol Cu}$$

$$X \times \frac{1\ \text{mol CuO}}{79.55\ \text{g CuO}} + 2 \times \left(Y \times \frac{1\ \text{mol Cu}_2\text{O}}{143.09\ \text{g Cu}_2\text{O}}\right) = 0.1363\ \text{mol Cu}$$

Rearrange to get X = 10.50 g – Y and then substitute it into the equation above to solve for Y.

$$(10.50\ \text{g} - Y) \times \frac{1\ \text{mol CuO}}{79.55\ \text{g CuO}} + 2 \times \left(Y \times \frac{1\ \text{mol Cu}_2\text{O}}{143.09\ \text{g Cu}_2\text{O}}\right) = 0.1363\ \text{mol Cu}$$

$$\frac{10.50\ \text{mol}}{79.55} - \frac{Y\ \text{mol}}{79.55\ \text{g}} + \frac{2\ Y\ \text{mol}}{143.09\ \text{g}} = 0.1363\ \text{mol}$$

$$-\frac{Y\ \text{mol}}{79.55\ \text{g}} + \frac{2\ Y\ \text{mol}}{143.09\ \text{g}} = 0.1363\ \text{mol} - \frac{10.50\ \text{mol}}{79.55} = 0.0043\ \text{mol}$$

$$\frac{(-Y\ \text{mol})(143.09\ \text{g}) + (2\ Y\ \text{mol})(79.55\ \text{g})}{(79.55\ \text{g})(143.09\ \text{g})} = 0.0043\ \text{mol}$$

$$\frac{16.01\ Y\ \text{mol}}{11383\ \text{g}} = 0.0043\ \text{mol}; \quad \frac{16.01\ Y}{11383\ \text{g}} = 0.0043$$

Y = (0.0043)(11383 g)/16.01 = 3.06 g Cu_2O

X = 10.50 g – Y = 10.50 g – 3.06 g = 7.44 g CuO

4.144 (a) PbI_2, 461.01

$Pb(NO_3)_2(aq) + 2\ KI(aq) \rightarrow PbI_2(s) + 2\ KNO_3(aq)$

75.0 mL = 0.0750 L and 100.0 mL = 0.1000 L

mol $Pb(NO_3)_2$ = (0.0750 L)(0.100 mol/L) = 7.50×10^{-3} mol $Pb(NO_3)_2$

mol KI = (0.1000 L)(0.190 mol/L) = 1.90×10^{-2} mol KI

$$\text{mols KI needed} = 7.50 \times 10^{-3}\ \text{mol Pb(NO}_3)_2 \times \frac{2\ \text{mol KI}}{1\ \text{mol Pb(NO}_3)_2} = 1.50 \times 10^{-2}\ \text{mol KI}$$

There is an excess of KI, so $Pb(NO_3)_2$ is the limiting reactant.

$$\text{mass PbI}_2 = 7.50 \times 10^{-3}\ \text{mol Pb(NO}_3)_2 \times \frac{1\ \text{mol PbI}_2}{1\ \text{mol Pb(NO}_3)_2} \times \frac{461.01\ \text{g PbI}_2}{1\ \text{mol PbI}_2} = 3.46\ \text{g PbI}_2$$

(b) Because $Pb(NO_3)_2$ is the limiting reactant, Pb^{2+} is totally consumed and $[Pb^{2+}] = 0$.

mol K^+ = mol KI = 1.90×10^{-2} mol

$\text{mol } NO_3^- = 7.50 \times 10^{-3} \text{ mol } Pb(NO_3)_2 \times \dfrac{2 \text{ mol } NO_3^-}{1 \text{ mol } Pb(NO_3)_2} = 0.0150 \text{ mol } NO_3^-$

$\text{mol } I^- = (\text{initial mol KI}) - (\text{mol KI needed}) = 0.0190 \text{ mol} - 0.0150 \text{ mol} = 0.0040 \text{ mol } I^-$
total volume = 0.0750 L + 0.1000 L = 0.1750 L

$[K^+] = \dfrac{0.0190 \text{ mol}}{0.1750 \text{ L}} = 0.109 \text{ M}$

$[NO_3^-] = \dfrac{0.0150 \text{ mol}}{0.1750 \text{ L}} = 0.0857 \text{ M}$

$[I^-] = \dfrac{0.0040 \text{ mol}}{0.1750 \text{ L}} = 0.023 \text{ M}$

4.145 Mass O in M_2O_3 = 1.890 g M_2O_3 – 1.000 g M = 0.890 g O

$0.890 \text{ g O} \times \dfrac{1 \text{ mol O}}{16.00 \text{ g O}} = 0.0556 \text{ mol O}$

$0.0556 \text{ mol O} \times \dfrac{2 \text{ mol M}}{3 \text{ mol O}} = 0.0371 \text{ mol M}$

$\text{molar mass M} = \dfrac{1.000 \text{ g M}}{0.0371 \text{ mol M}} = 26.97 \text{ g/mol}$; M is Al

4.146 Mass S in MS = 1.504 g MS – 1.000 g M = 0.504 g S

$0.504 \text{ g S} \times \dfrac{1 \text{ mol S}}{32.06 \text{ g S}} = 0.0157 \text{ mol S}$

$0.0157 \text{ mol S} \times \dfrac{1 \text{ mol M}}{1 \text{ mol S}} = 0.0157 \text{ mol M}$

$\text{molar mass M} = \dfrac{1.000 \text{ g M}}{0.0157 \text{ mol M}} = 63.69 \text{ g/mol}$; M is Cu

4.147 CH_3CO_2H, 60.05; $H_2C_2O_4$, 90.03
27.15 mL = 0.027 15 L and 15.05 mL = 0.015 05 L
(0.0247 mol/L $KMnO_4$)(0.015 05 L) = 3.72×10^{-4} mol $KMnO_4$

$3.72 \times 10^{-4} \text{ mol } KMnO_4 \times \dfrac{5 \text{ mol } H_2C_2O_4}{2 \text{ mol } K_2MnO_4} = 9.30 \times 10^{-4} \text{ mol } H_2C_2O_4$

$2\,NaOH(aq) + H_2C_2O_4(aq) \rightarrow Na_2C_2O_4(aq) + 2\,H_2O(l)$

$9.30 \times 10^{-4} \text{ mol } H_2C_2O_4 \times \dfrac{2 \text{ mol NaOH}}{1 \text{ mol } H_2C_2O_4} = 1.86 \times 10^{-3} \text{ mol NaOH}$

$\text{volume} = \dfrac{\text{mol}}{\text{M}} = \dfrac{1.86 \times 10^{-3} \text{ mol NaOH}}{0.100 \text{ mol/L}} = 0.0186 \text{ L} = 18.6 \text{ mL NaOH reacted with } H_2C_2O_4$

remaining NaOH = 27.15 mL – 18.6 mL = 8.5 mL = 0.0085 L
(0.100 mol/L NaOH)(0.0085 L) = 8.5×10^{-4} mol NaOH reacted with CH_3CO_2H
$NaOH(aq) + CH_3CO_2H(aq) \rightarrow NaCH_3CO_2(aq) + H_2O(l)$

$8.5 \times 10^{-4} \text{ mol NaOH} \times \dfrac{1 \text{ mol } CH_3CO_2H}{1 \text{ mol NaOH}} = 8.5 \times 10^{-4} \text{ mol } CH_3CO_2H$

$8.5 \times 10^{-4} \text{ mol } CH_3CO_2H \times \dfrac{60.05 \text{ g } CH_3CO_2H}{1 \text{ mol } CH_3CO_2H} = 0.051 \text{ g } CH_3CO_2H$

$9.30 \times 10^{-4} \text{ mol } H_2C_2O_4 \times \dfrac{90.03 \text{ g } H_2C_2O_4}{1 \text{ mol } H_2C_2O_4} = 0.0837 \text{ g } H_2C_2O_4$

mass of sample = 0.051 g CH_3CO_2H + 0.0837 g $H_2C_2O_4$ = 0.135 g

mass % $CH_3CO_2H = \dfrac{0.051 \text{ g } CH_3CO_2H}{0.135 \text{ g}} \times 100\% = 38\%$

mass % $H_2C_2O_4 = \dfrac{0.0837 \text{ g } H_2C_2O_4}{0.135 \text{ g}} \times 100\% = 62.0\%$

4.148 48.39 mL = 0.048 39 L

$(0.1116 \text{ mol/L } MnO_4^-)(0.048\ 39 \text{ L}) = 5.400 \times 10^{-3} \text{ mol } MnO_4^-$

$5.400 \times 10^{-3} \text{ mol } MnO_4^- \times \dfrac{5 \text{ mol } Fe^{2+}}{1 \text{ mol } MnO_4^-} \times \dfrac{55.84 \text{ g } Fe^{2+}}{1 \text{ mol } Fe^{2+}} = 1.508 \text{ g } Fe^{2+}$

mass % Fe = $\dfrac{1.508 \text{ g } Fe^{2+}}{2.368 \text{ g}} \times 100\% = 63.68\%$

4.149 $FeCl_2$, 126.8; NaCl, 58.44; AgCl, 143.3; 14.28 mL = 0.014 28 L

$(0.198 \text{ mol/L } MnO_4^-)(0.014\ 28 \text{ L}) = 0.002\ 83 \text{ mol } MnO_4^-$

$0.002\ 83 \text{ mol } MnO_4^- \times \dfrac{5 \text{ mol } Fe^{2+}}{1 \text{ mol } MnO_4^-} = 0.0141 \text{ mol } Fe^{2+}$

$AgNO_3$(aq) reacts with a solution of $FeCl_2$(aq) and NaCl(aq) to precipitate AgCl(s).

$7.0149 \text{ g AgCl} \times \dfrac{1 \text{ mol AgCl}}{143.3 \text{ g AgCl}} = 0.048\ 95 \text{ mol AgCl}$

$0.048\ 95 \text{ mol AgCl} \times \dfrac{1 \text{ mol } Cl^-}{1 \text{ mol AgCl}} = 0.048\ 95 \text{ mol } Cl^-$

$0.0141 \text{ mol } Fe^{2+} \times \dfrac{2 \text{ mol } Cl^-}{1 \text{ mol } Fe^{2+}} = 0.0282 \text{ mol } Cl^- \text{ from } FeCl_2$

mol Cl^- from NaCl = 0.048 95 mol Cl^- – 0.0282 mol Cl^- from $FeCl_2$ = 0.0207 mol Cl^-

mol NaCl = mol Cl^- from NaCl = 0.0207 mol NaCl

mol $FeCl_2$ = mol Fe^{2+} = 0.0141 mol $FeCl_2$

$0.0207 \text{ mol NaCl} \times \dfrac{58.44 \text{ g NaCl}}{1 \text{ mol NaCl}} = 1.21 \text{ g NaCl}$

$0.0141 \text{ mol } FeCl_2 \times \dfrac{126.8 \text{ g } FeCl_2}{1 \text{ mol } FeCl_2} = 1.79 \text{ g } FeCl_2$

total mass = 1.21 g + 1.79 g = 3.00 g

$$\text{mass \% NaCl} = \frac{1.21\text{ g NaCl}}{3.00\text{ g}} \times 100\% = 40.3\%$$

$$\text{mass \% FeCl}_2 = \frac{1.79\text{ g FeCl}_2}{3.00\text{ g}} \times 100\% = 59.7\%$$

Multiconcept Problems

4.150 Let SA stand for salicylic acid.

$$\text{mol C in 1.00 g of SA} = 2.23\text{ g CO}_2 \times \frac{1\text{ mol CO}_2}{44.01\text{ g CO}_2} \times \frac{1\text{ mol C}}{1\text{ mol CO}_2} = 0.0507\text{ mol C}$$

$$\text{mass C} = 0.0507\text{ mol C} \times \frac{12.011\text{ g C}}{1\text{ mol C}} = 0.609\text{ g C}$$

$$\text{mol H in 1.00 g of SA} = 0.39\text{ g H}_2\text{O} \times \frac{1\text{ mol H}_2\text{O}}{18.02\text{ g H}_2\text{O}} \times \frac{2\text{ mol H}}{1\text{ mol H}_2\text{O}} = 0.043\text{ mol H}$$

$$\text{mass H} = 0.043\text{ mol H} \times \frac{1.008\text{ g H}}{1\text{ mol H}} = 0.043\text{ g H}$$

mass O = 1.00 g – mass C – mass H = 1.00 – 0.609 g – 0.043 g = 0.35 g O

$$\text{mol O in 1.00 g of} = 0.35\text{ g N} \times \frac{1\text{ mol O}}{16.00\text{ g O}} = 0.022\text{ mol O}$$

Determine empirical formula.
$C_{0.0507}H_{0.043}O_{0.022}$; divide each subscript by the smallest, 0.022.
$C_{0.0507 / 0.022}H_{0.043 / 0.022}O_{0.022 / 0.022}$
$C_{2.3}H_2O$, multiply each subscript by 3 to get integers.
The empirical formula is $C_7H_6O_3$. The empirical formula mass = 138.12 g/mol.

Because salicylic acid has only one acidic hydrogen, there is a 1 to 1 mol ratio between salicylic acid and NaOH in the acid-base titration.

$$\text{mol SA in 1.00 g SA} = 72.4\text{ mL} \times \frac{1 \times 10^{-3}\text{ L}}{1\text{ mL}} \times \frac{0.100\text{ mol NaOH}}{1\text{ L}} \times \frac{1\text{ mol SA}}{1\text{ mol NaOH}} = 0.00724\text{ mol SA}$$

$$\text{SA molar mass} = \frac{1.00\text{ g}}{0.00724\text{ mol}} = 138\text{ g/mol}$$

Because the empirical formula mass and the molar mass are the same, the empirical formula is the molecular formula for salicylic acid.

4.151 (a) $\text{mol C} = 4.83\text{ g CO}_2 \times \frac{1\text{ mol CO}_2}{44.01\text{ g CO}_2} \times \frac{1\text{ mol C}}{1\text{ mol CO}_2} = 0.110\text{ mol C}$

$$\text{mass C} = 0.110\text{ mol C} \times \frac{12.011\text{ g C}}{1\text{ mol C}} = 1.32\text{ g C}$$

$$\text{mol H} = 1.48\text{ g H}_2\text{O} \times \frac{1\text{ mol H}_2\text{O}}{18.02\text{ g H}_2\text{O}} \times \frac{2\text{ mol H}}{1\text{ mol H}_2\text{O}} = 0.164\text{ mol H}$$

mass H = 0.164 mol H x $\frac{1.008 \text{ g H}}{1 \text{ mol H}}$ = 0.165 g H

109.8 mL = 0.1098 L

mol NaOH = (0.1098 L)(1.00 mol/L) = 0.110 mol NaOH

$H_2SO_4(aq) + 2\ NaOH(aq) \rightarrow Na_2SO_4(aq) + 2\ H_2O(l)$

mol H_2SO_4 = 0.110 mol NaOH x $\frac{1 \text{ mol } H_2SO_4}{2 \text{ mol NaOH}}$ = 0.0550 mol H_2SO_4

mol S = 0.0550 mol H_2SO_4 x $\frac{1 \text{ mol S}}{1 \text{ mol } H_2SO_4}$ = 0.0550 mol S

mass S = 0.0550 mol S x $\frac{32.06 \text{ g S}}{1 \text{ mol S}}$ = 1.76 g S

mass O = 5.00 g – mass C – mass H – mass S = 5.00 g –1.32 g – 0.165 g – 1.76 g = 1.75 g O

mol O = 1.75 g O x $\frac{1 \text{ mol O}}{16.00 \text{ g O}}$ = 0.109 mol O

$C_{0.110}H_{0.164}O_{0.109}S_{0.0550}$; divide each subscript by the smallest, 0.0550.

$C_{0.110/0.0550}H_{0.164/0.0550}O_{0.109/0.0550}S_{0.0550/0.0550}$

The empirical formula is $C_2H_3O_2S$.

The empirical formula mass = 91.1 g/mol.

(b) 54.9 mL = 0.0549 L

mol NaOH = (0.0549 L)(1.00 mol/L) = 0.0549 mol NaOH

Because X has two acidic hydrogens, two mol of NaOH are required to titrate 1 mol of X.

mol X = 0.0549 mol NaOH x $\frac{1 \text{ mol X}}{2 \text{ mol NaOH}}$ = 0.0274 mol X

X molar mass = $\frac{5.00 \text{ g}}{0.0274 \text{ mol}}$ = 182 g/mol

Because the molar mass is twice the empirical formula mass, the molecular formula is twice the empirical formula.

The molecular formula is $C_{(2 \times 2)}H_{(2 \times 3)}O_{(2 \times 2)}S_{(2 \times 1)} = C_4H_6O_4S_2$.

4.152 100.00 mL = 0.100 00 L; 71.02 mL = 0.071 02 L

mol H_2SO_4 = $\frac{0.1083 \text{ mol } H_2SO_4}{L}$ x 0.100 00 L = 0.010 83 mol H_2SO_4

mol NaOH = $\frac{0.1241 \text{ mol NaOH}}{L}$ x 0.071 02 L = 0.008 814 mol NaOH

$H_2SO_4 + 2\ NaOH \rightarrow Na_2SO_4 + 2\ H_2O$

mol H_2SO_4 reacted with NaOH = 0.008 814 mol NaOH x $\frac{1 \text{ mol } H_2SO_4}{2 \text{ mol NaOH}}$ = 0.004 407 mol H_2SO_4

mol H_2SO_4 reacted with MCO_3 = 0.010 83 mol – 0.004 407 mol = 0.006 423 mol H_2SO_4

mol H_2SO_4 reacted with MCO_3 = mol CO_3^{2-} in MCO_3 = mol CO_2 produced = 0.006 423 mol CO_2

(a) CO_3^{2-}, 60.01; $0.006\,423 \text{ mol } CO_3^{2-} \times \dfrac{60.01 \text{ g } CO_3^{2-}}{1 \text{ mol } CO_3^{2-}} = 0.3854 \text{ g } CO_3^{2-}$

mass of M = 1.268 g – 0.3854 g = 0.8826 g M

molar mass of M = $\dfrac{0.8826 \text{ g}}{0.006\,423 \text{ mol}} = 137.4 \text{ g/mol}$; M is Ba

(b) $0.006\,423 \text{ mol } CO_2 \times \dfrac{44.01 \text{ g } CO_2}{1 \text{ mol } CO_2} \times \dfrac{1 \text{ L}}{1.799 \text{ g}} = 0.1571 \text{ L } CO_2$

4.153 (a) (i) $M_2O_3(s) + 3\,C(s) + 3\,Cl_2(g) \rightarrow 2\,MCl_3(l) + 3\,CO(g)$

(ii) $2\,MCl_3(l) + 3\,H_2(g) \rightarrow 2\,M(s) + 6\,HCl(g)$

(b) $HCl(aq) + NaOH(aq) \rightarrow H_2O(l) + NaCl(aq)$

144.2 mL = 0.1442 L

mol NaOH = (0.511 mol/L)(0.1442 L) = 0.07369 mol NaOH

mol HCl = $0.07369 \text{ mol NaOH} \times \dfrac{1 \text{ mol HCl}}{1 \text{ mol NaOH}} = 0.07369 \text{ mol HCl}$

mol M = $0.07369 \text{ mol HCl} \times \dfrac{2 \text{ mol M}}{6 \text{ mol HCl}} = 0.02456 \text{ mol M}$

mol M_2O_3 = $0.02456 \text{ mol M} \times \dfrac{2 \text{ mol } MCl_3}{2 \text{ mol M}} \times \dfrac{1 \text{ mol } M_2O_3}{2 \text{ mol } MCl_3} = 0.01228 \text{ mol } M_2O_3$

molar mass M_2O_3 = $\dfrac{0.855 \text{ g}}{0.01228 \text{ mol}} = 69.6 \text{ g/mol}$; molecular mass M_2O_3 = 69.6

atomic weight of M = $\dfrac{69.6 \text{ amu} - (3 \times 16.0 \text{ amu})}{2} = 10.8$; M = B

(c) mass of M = $0.02456 \text{ mol M} \times \dfrac{10.81 \text{ g M}}{1 \text{ mol M}} = 0.265 \text{ g M}$

4.154 NaOH, 40.00; $Ba(OH)_2$, 171.34

Let X equal the mass of NaOH and Y the mass of $Ba(OH)_2$ in the 10.0 g mixture.
Therefore, X + Y = 10.0 g.

mol HCl = $108.9 \text{ mL} \times \dfrac{1 \times 10^{-3} \text{ L}}{1 \text{ mL}} \times \dfrac{1.50 \text{ mol HCl}}{1 \text{ L}} = 0.163 \text{ mol HCl}$

mol NaOH + 2 x mol $Ba(OH)_2$ = 0.163 mol HCl

$$X \times \frac{1 \text{ mol NaOH}}{40.00 \text{ g NaOH}} + 2 \times \left(Y \times \frac{1 \text{ mol } Ba(OH)_2}{171.34 \text{ g } Ba(OH)_2} \right) = 0.163 \text{ mol HCl}$$

Rearrange to get X = 10.0 g – Y and then substitute it into the equation above to solve for Y.

$$(10.0 \text{ g} - Y) \times \frac{1 \text{ mol NaOH}}{40.00 \text{ g NaOH}} + 2 \times \left(Y \times \frac{1 \text{ mol } Ba(OH)_2}{171.34 \text{ g } Ba(OH)_2} \right) = 0.163 \text{ mol HCl}$$

$$\frac{10.00 \text{ mol}}{40.00} - \frac{Y \text{ mol}}{40.00 \text{ g}} + \frac{2\,Y \text{ mol}}{171.34 \text{ g}} = 0.163 \text{ mol}$$

$$-\frac{\text{Y mol}}{40.00 \text{ g}} + \frac{2 \text{ Y mol}}{171.34 \text{ g}} = 0.163 \text{ mol} - \frac{10.00 \text{ mol}}{40.00} = -0.087 \text{ mol}$$

$$\frac{(-\text{Y mol})(171.34 \text{ g}) + (2 \text{ Y mol})(40.00 \text{ g})}{(40.00 \text{ g})(171.34 \text{ g})} = -0.087 \text{ mol}$$

$$\frac{-91.34 \text{ Y mol}}{6853.6 \text{ g}} = -0.087 \text{ mol}; \quad \frac{91.34 \text{ Y}}{6853.6 \text{ g}} = 0.087$$

Y = (0.087)(6853.6 g)/91.34 = 6.5 g $Ba(OH)_2$
X = 10.0 g – Y = 10.0 g – 6.5 g = 3.5 g NaOH

4.155 100.0 mL = 0.1000 L and 50.0 mL = 0.0500 L
mol Na_2SO_4 = (0.1000 L)(0.100 mol/L) = 0.0100 mol Na_2SO_4
mol SO_4^{2-} = mol Na_2SO_4 = 0.0100 mol SO_4^{2-}

$$\text{mol } Na^+ = 0.0100 \text{ mol } Na_2SO_4 \times \frac{2 \text{ mol } Na^+}{1 \text{ mol } Na_2SO_4} = 0.0200 \text{ mol } Na^+$$

mol $ZnCl_2$ = (0.0500 L)(0.300 mol/L) = 0.0150 mol $ZnCl_2$
mol Zn^{2+} = mol $ZnCl_2$ = 0.0150 mol Zn^{2+}

$$\text{mol } Cl^- = 0.0150 \text{ mol } ZnCl_2 \times \frac{2 \text{ mol } Cl^-}{1 \text{ mol } ZnCl_2} = 0.0300 \text{ mol } Cl^-$$

mol $Ba(CN)_2$ = (0.1000 L)(0.200 mol/L) = 0.0200 mol $Ba(CN)_2$
mol Ba^{2+} = mol $Ba(CN)_2$ = 0.0200 mol Ba^{2+}

$$\text{mol } CN^- = 0.0200 \text{ mol } Ba(CN)_2 \times \frac{2 \text{ mol } CN^-}{1 \text{ mol } Ba(CN)_2} = 0.0400 \text{ mol } CN^-$$

The following two reactions will take place to form precipitates.
$Zn^{2+}(aq) + 2\ CN^-(aq) \rightarrow Zn(CN)_2(s)$
$Ba^{2+}(aq) + SO_4^{2-}(aq) \rightarrow BaSO_4(s)$

$$\text{For } Zn^{2+}, \text{ mol } CN^- \text{ needed} = 0.0150 \text{ mol } Zn^{2+} \times \frac{2 \text{ mol } CN^-}{1 \text{ mol } Zn^{2+}} = 0.0300 \text{ mol } CN^- \text{ needed}$$

CN^- is in excess, so Zn^{2+} is the limiting reactant and is totally consumed.
mol CN^- remaining after reaction= 0.0400 mol – 0.0300 mol = 0.0100 mol CN^-
For Ba^{2+}, mol SO_4^{2-} needed = mol Ba^{2+} = 0.0200 mol SO_4^{2-} needed
Ba^{2+} is in excess, so SO_4^{2-} is the limiting reactant and is totally consumed.
mol Ba^{2+} remaining after reaction = 0.0200 mol – 0.0100 mol = 0.0100 mol Ba^{2+}
total volume = 0.1000 L + 0.0500 L + 0.1000 L = 0.2500 L
$[Zn^{2+}] = [SO_4^{2-}] = 0$

$$[Na^+] = \frac{0.0200 \text{ mol}}{0.2500 \text{ L}} = 0.0800 \text{ M}$$

$$[Cl^-] = \frac{0.0300 \text{ mol}}{0.2500 \text{ L}} = 0.120 \text{ M}$$

$$[CN^-] = \frac{0.0100 \text{ mol}}{0.2500 \text{ L}} = 0.0400 \text{ M}$$

$$[Ba^{2+}] = \frac{0.0100 \text{ mol}}{0.2500 \text{ L}} = 0.0400 \text{ M}$$

4.156 KNO_3, 101.10; $BaCl_2$, 208.24; NaCl, 58.44; $BaSO_4$, 233.40; AgCl, 143.32

(a) The two precipitates are $BaSO_4(s)$ and AgCl(s).

(b) H_2SO_4 only reacts with $BaCl_2$.

$H_2SO_4(aq) + BaCl_2(aq) \rightarrow BaSO_4(s) + 2\ HCl(aq)$

Calculate the number of moles of $BaCl_2$ in 100.0 g of the mixture.

$$\text{mol } BaCl_2 = 67.3 \text{ g } BaSO_4 \times \frac{1 \text{ mol } BaSO_4}{233.40 \text{ g } BaSO_4} \times \frac{1 \text{ mol } BaCl_2}{1 \text{ mol } BaSO_4} = 0.288 \text{ mol } BaCl_2$$

Calculate mass and moles of $BaCl_2$ in 250.0 g sample.

$$\text{mass } BaCl_2 = 0.288 \text{ mol } BaCl_2 \times \frac{208.24 \text{ g } BaCl_2}{1 \text{ mol } BaCl_2} \times \frac{250.0 \text{ g}}{100.0 \text{ g}} = 150. \text{ g } BaCl_2$$

$$\text{mol } BaCl_2 = 150. \text{ g } BaCl_2 \times \frac{1 \text{ mol } BaCl_2}{208.24 \text{ g } BaCl_2} = 0.720 \text{ mol } BaCl_2$$

$AgNO_3$ reacts with both NaCl and $BaCl_2$ in the remaining 150.0 g of the mixture.

$3\ AgNO_3(aq) + NaCl(aq) + BaCl_2(aq) \rightarrow 3\ AgCl(s) + NaNO_3(aq) + Ba(NO_3)_2(aq)$

Calculate the moles of AgCl that would have been produced from the 250.0 g mixture.

$$\text{mol AgCl} = 197.6 \text{ g AgCl} \times \frac{1 \text{ mol AgCl}}{143.32 \text{ g AgCl}} \times \frac{250.0 \text{ g}}{150.0 \text{ g}} = 2.30 \text{ mol AgCl}$$

mol AgCl = 2 x (mol $BaCl_2$) + mol NaCl

Calculate the moles and mass of NaCl in the 250.0 g mixture.

2.30 mol AgCl = 2 x 0.720 mol $BaCl_2$ + mol NaCl

mol NaCl = 2.30 mol – 2(0.720 mol) = 0.86 mol NaCl

$$\text{mass NaCl} = 0.86 \text{ mol NaCl} \times \frac{58.44 \text{ g NaCl}}{1 \text{ mol NaCl}} = 50. \text{ g NaCl}$$

Calculate the mass of KNO_3 in the 250.0 g mixture.

total mass = mass $BaCl_2$ + mass NaCl + mass KNO_3

250.0 g = 150. g $BaCl_2$ + 50. g NaCl + mass KNO_3

mass KNO_3 = 250.0 g – 150. g $BaCl_2$ – 50. g NaCl = 50. g KNO_3

4.157 100.0 mL = 0.1000 L; 50.0 mL = 0.0500 L; 250.0 mL = 0.2500 L

After step (2): $BaCl_2(aq) + 2\ AgNO_3(aq) \rightarrow AgCl(s) + Ba(NO_3)_2(aq)$

mol $BaCl_2$ = (0.1000 L)(0.100 mol/L) = 0.0100 mol $BaCl_2$

mol Ba^{2+} = mol $BaCl_2$ = 0.0100 mol Ba^{2+}

$$\text{mol } Cl^- = 0.0100 \text{ mol } BaCl_2 \times \frac{2 \text{ mol } Cl^-}{1 \text{ mol } BaCl_2} = 0.0200 \text{ mol } Cl^-$$

mol $AgNO_3$ = (0.0500 L)(0.100 mol/L) = 0.00500 mol $AgNO_3$

mol Ag^+ = mol $AgNO_3$ = 0.00500 mol Ag^+

mol NO_3^- = mol $AgNO_3$ = 0.00500 mol NO_3^-

0.00500 mol Ag^+ requires only 0.00500 mol Cl^-, so Ag^+ is the limiting reactant and totally consumed.

mol Cl^- remaining after reaction = 0.0200 mol – 0.00500 mol = 0.0150 mol Cl^-

After step (3): $Ba^{2+}(aq) + H_2SO_4(aq) \rightarrow BaSO_4(s) + 2\ H^+(aq)$

mol H_2SO_4 = (0.0500 L)(0.100 mol/L) = 0.00500 mol H_2SO_4

mol SO_4^{2-} = mol H_2SO_4 = 0.00500 mol SO_4^{2-}

mol H^+ = 0.00500 mol H_2SO_4 x $\frac{2 \text{ mol } H^+}{1 \text{ mol } H_2SO_4}$ = 0.0100 mol H^+

0.0100 mol Ba^{2+} requires 0.0100 mol SO_4^{2-}, so SO_4^{2-} is the limiting reactant and is totally consumed.
mol Ba^{2+} remaining after reaction = 0.0100 mol – 0.00500 mol = 0.00500 mol Ba^{2+}

After step (4): $NH_3(aq) + H^+(aq) \rightarrow NH_4^+(aq)$
mol NH_3 = (0.2500 L)(0.100 mol/L) = 0.0250 mol NH_3
0.0250 mol NH_3 requires 0.0250 mol H^+, so H^+ is the limiting reactant and is totally consumed.
mol NH_3 remaining after reaction = 0.0250 mol – 0.0100 mol = 0.0150 mol NH_3
mol NH_4^+ = mol H^+ before reaction = 0.0100 mol NH_4^+
total volume = 0.1000 L + 0.0500 L + 0.0500 L + 0.2500 L = 0.4500 L

$[Ba^{2+}] = \frac{0.00500 \text{ mol}}{0.4500 \text{ L}} = 0.0111 \text{ M}$

$[Cl^-] = \frac{0.0150 \text{ mol}}{0.4500 \text{ L}} = 0.0333 \text{ M}$

$[NO_3^-] = \frac{0.00500 \text{ mol}}{0.4500 \text{ L}} = 0.0111 \text{ M}$

$[NH_3] = \frac{0.0150 \text{ mol}}{0.4500 \text{ L}} = 0.0333 \text{ M}$

$[NH_4^+] = \frac{0.0100 \text{ mol}}{0.4500 \text{ L}} = 0.0222 \text{ M}$

4.158 (a) $Cr^{2+}(aq) + Cr_2O_7^{2-}(aq) \rightarrow Cr^{3+}(aq)$
$[Cr^{2+}(aq) \rightarrow Cr^{3+}(aq) + e^-] \times 6$ (oxidation half reaction)

$Cr_2O_7^{2-}(aq) \rightarrow Cr^{3+}(aq)$
$Cr_2O_7^{2-}(aq) \rightarrow 2\ Cr^{3+}(aq)$
$Cr_2O_7^{2-}(aq) \rightarrow 2\ Cr^{3+}(aq) + 7\ H_2O(l)$
$14\ H^+(aq) + Cr_2O_7^{2-}(aq) \rightarrow 2\ Cr^{3+}(aq) + 7\ H_2O(l)$
$6\ e^- + 14\ H^+(aq) + Cr_2O_7^{2-}(aq) \rightarrow 2\ Cr^{3+}(aq) + 7\ H_2O(l)$ (reduction half reaction)

Combine the two half reactions.
$14\ H^+(aq) + Cr_2O_7^{2-}(aq) + 6\ Cr^{2+}(aq) \rightarrow 8\ Cr^{3+}(aq) + 7\ H_2O(l)$

(b) total volume = 100.0 ml + 20.0 mL = 120.0 mL = 0.1200 L
Initial moles:

$0.120 \frac{\text{mol } Cr(NO_3)_2}{1 \text{ L}} \times 0.1000 \text{ L} = 0.0120 \text{ mol } Cr(NO_3)_2$

$0.500 \frac{\text{mol } HNO_3}{1 \text{ L}} \times 0.1000 \text{ L} = 0.0500 \text{ mol } HNO_3$

$$0.250\ \frac{\text{mol } K_2Cr_2O_7}{1\text{ L}} \times 0.0200\text{ L} = 0.005\ 00\text{ mol } K_2Cr_2O_7$$

Check for the limiting reactant. 0.0120 mol of Cr^{2+} requires (0.0120)/6 = 0.00200 mol $Cr_2O_7^{2-}$ and (14/6)(0.0120) = 0.0280 mol H^+. Both are in excess of the required amounts, so Cr^{2+} is the limiting reactant.

	$14\ H^+(aq)$ +	$Cr_2O_7^{2-}(aq)$ +	$6\ Cr^{2+}(aq)$ →	$8\ Cr^{3+}(aq)$ + $7\ H_2O(l)$
Initial moles	0.0500	0.00500	0.0120	0
Change	−14x	−x	−6x	+8x

Because Cr^{2+} is the limiting reactant, 6x = 0.0120 and x = 0.00200

Final moles	0.0220	0.00300	0	0.0160

$$\text{mol } K^+ = 0.00500\text{ mol } K_2Cr_2O_7 \times \frac{2\text{ mol } K^+}{1\text{ mol } K_2Cr_2O_7} = 0.0100\text{ mol } K^+$$

$$\text{mol } NO_3^- = 0.0120\text{ mol } Cr(NO_3)_2 \times \frac{2\text{ mol } NO_3^-}{1\text{ mol } Cr(NO_3)_2}$$

$$+ 0.0500\text{ mol } HNO_3 \times \frac{1\text{ mol } NO_3^-}{1\text{ mol } HNO_3} = 0.0740\text{ mol } NO_3^-$$

mol H^+ = 0.0220 mol; mol $Cr_2O_7^{2-}$ = 0.00300 mol; mol Cr^{3+} = 0.01600 mol

Check for charge neutrality.
Total moles of +charge = 0.0100 + 0.0220 + 3 x (0.01600) = 0.0800 mol +charge
Total moles of −charge = 0.0740 + 2 x (0.00300) = 0.0800 mol −charge
The charges balance and there is electrical neutrality in the solution after the reaction.

$$K^+\text{ molarity} = \frac{0.0100\text{ mol } K^+}{0.1200\text{ L}} = 0.0833\text{ M}$$

$$NO_3^-\text{ molarity} = \frac{0.0740\text{ mol } NO_3^-}{0.1200\text{ L}} = 0.617\text{ M}$$

$$H^+\text{ molarity} = \frac{0.0220\text{ mol } H^+}{0.1200\text{ L}} = 0.183\text{ M}$$

$$Cr_2O_7^{2-}\text{ molarity} = \frac{0.00300\text{ mol } Cr_2O_7^{2-}}{0.1200\text{ L}} = 0.0250\text{ M}$$

$$Cr^{3+}\text{ molarity} = \frac{0.0160\text{ mol } Cr^{3+}}{0.1200\text{ L}} = 0.133\text{ M}$$

4.159 (a) (1) $I^-(aq) \rightarrow I_3^-(aq)$
$3\ I^-(aq) \rightarrow I_3^-(aq)$
$3\ I^-(aq) \rightarrow I_3^-(aq) + 2\ e^-$ (oxidation half reaction)

$HNO_2(aq) \rightarrow NO(g)$
$HNO_2(aq) \rightarrow NO(g) + H_2O(l)$
$H^+(aq) + HNO_2(aq) \rightarrow NO(g) + H_2O(l)$

$[e^- + H^+(aq) + HNO_2(aq) \rightarrow NO(g) + H_2O(l)] \times 2$ (reduction half reaction)

Combine the two half reactions.
$3\ I^-(aq) + 2\ H^+(aq) + 2\ HNO_2(aq) \rightarrow I_3^-(aq) + 2\ NO(g) + 2\ H_2O(l)$

(2) $S_2O_3^{2-}(aq) \rightarrow S_4O_6^{2-}(aq)$
$2\ S_2O_3^{2-}(aq) \rightarrow S_4O_6^{2-}(aq)$
$2\ S_2O_3^{2-}(aq) \rightarrow S_4O_6^{2-}(aq) + 2\ e^-$ (oxidation half reaction)

$I_3^-(aq) \rightarrow I^-(aq)$
$I_3^-(aq) \rightarrow 3\ I^-(aq)$
$2\ e^- + I_3^-(aq) \rightarrow 3\ I^-(aq)$ (reduction half reaction)

Combine the two half reactions.
$2\ S_2O_3^{2-}(aq) + I_3^-(aq) \rightarrow S_4O_6^{2-}(aq) + 3\ I^-(aq)$

(b) 18.77 mL = 0.018 77 L; NO_2^-, 46.01

$$0.1500\ \frac{\text{mol } S_2O_3^{2-}}{1\text{ L}} \times 0.018\ 77\text{ L} = 0.002\ 815\ 5\text{ mol } S_2O_3^{2-}$$

$$\text{mass } NO_2^- = 0.002\ 815\ 5\text{ mol } S_2O_3^{2-} \times \frac{1\text{ mol } I_3^-}{2\text{ mol } S_2O_3^{2-}} \times \frac{2\text{ mol } NO_2^-}{1\text{ mol } I_3^-} \times \frac{46.01\text{ g } NO_2^-}{1\text{ mol } NO_2^-} = 0.1295\text{ g } NO_2^-$$

$$\text{mass \% } NO_2^- = \frac{0.1295\text{ g}}{2.935\text{ g}} \times 100\% = 4.412\%$$

4.160 (a) (1) $Cu(s) \rightarrow Cu^{2+}(aq)$
$[Cu(s) \rightarrow Cu^{2+}(aq) + 2\ e^-] \times 3$ (oxidation half reaction)

$NO_3^-(aq) \rightarrow NO(g)$
$NO_3^-(aq) \rightarrow NO(g) + 2\ H_2O(l)$
$4\ H^+(aq) + NO_3^-(aq) \rightarrow NO(g) + 2\ H_2O(l)$
$[3\ e^- + 4\ H^+(aq) + NO_3^-(aq) \rightarrow NO(g) + 2\ H_2O(l)] \times 2$ (reduction half reaction)

Combine the two half reactions.
$3\ Cu(s) + 8\ H^+(aq) + 2\ NO_3^-(aq) \rightarrow 3\ Cu^{2+}(aq) + 2\ NO(g) + 4\ H_2O(l)$

(2) $Cu^{2+}(aq) + SCN^-(aq) \rightarrow CuSCN(s)$
$[e^- + Cu^{2+}(aq) + SCN^-(aq) \rightarrow CuSCN(s)] \times 2$ (reduction half reaction)

$HSO_3^-(aq) \rightarrow HSO_4^-(aq)$
$H_2O(l) + HSO_3^-(aq) \rightarrow HSO_4^-(aq)$
$H_2O(l) + HSO_3^-(aq) \rightarrow HSO_4^-(aq) + 2\ H^+(aq)$

$H_2O(l) + HSO_3^-(aq) \rightarrow HSO_4^-(aq) + 2\ H^+(aq) + 2\ e^-$
(oxidation half reaction)

Combine the two half reactions.

$2\ Cu^{2+}(aq) + 2\ SCN^-(aq) + H_2O(l) + HSO_3^-(aq) \rightarrow$
$2\ CuSCN(s) + HSO_4^-(aq) + 2\ H^+(aq)$

(3) $Cu^+(aq) \rightarrow Cu^{2+}(aq)$
$[Cu^+(aq) \rightarrow Cu^{2+}(aq) + e^-] \times 10$ (oxidation half reaction)

$IO_3^-(aq) \rightarrow I_2(aq)$
$2\ IO_3^-(aq) \rightarrow I_2(aq)$
$2\ IO_3^-(aq) \rightarrow I_2(aq) + 6\ H_2O(l)$
$12\ H^+(aq) + 2\ IO_3^-(aq) \rightarrow I_2(aq) + 6\ H_2O(l)$
$10\ e^- + 12\ H^+(aq) + 2\ IO_3^-(aq) \rightarrow I_2(aq) + 6\ H_2O(l)$ (reduction half reaction)
Combine the two half reactions.
$10\ Cu^+(aq) + 12\ H^+(aq) + 2\ IO_3^-(aq) \rightarrow 10\ Cu^{2+}(aq) + I_2(aq) + 6\ H_2O(l)$

(4) $I_2(aq) \rightarrow I^-(aq)$
$I_2(aq) \rightarrow 2\ I^-(aq)$
$2\ e^- + I_2(aq) \rightarrow 2\ I^-(aq)$ (reduction half reaction)

$S_2O_3^{2-}(aq) \rightarrow S_4O_6^{2-}(aq)$
$2\ S_2O_3^{2-}(aq) \rightarrow S_4O_6^{2-}(aq)$
$2\ S_2O_3^{2-}(aq) \rightarrow S_4O_6^{2-}(aq) + 2\ e^-$ (oxidation half reaction)

Combine the two half reactions.
$I_2(aq) + 2\ S_2O_3^{2-}(aq) \rightarrow 2\ I^-(aq) + S_4O_6^{2-}(aq)$

(5) $2\ ZnNH_4PO_4 \rightarrow Zn_2P_2O_7 + H_2O + 2\ NH_3$

(b) 10.82 mL = 0.01082 L
mol $S_2O_3^{2-}$ = (0.1220 mol/L)(0.01082 L) = 0.00132 mol $S_2O_3^{2-}$

$$\text{mol } I_2 = 0.00132 \text{ mol } S_2O_3^{2-} \times \frac{1 \text{ mol } I_2}{2 \text{ mol } S_2O_3^{2-}} = 6.60 \times 10^{-4} \text{ mol } I_2$$

$$\text{mol } Cu^+ = 6.60 \times 10^{-4} \text{ mol } I_2 \times \frac{10 \text{ mol } Cu^+}{1 \text{ mol } I_2} = 6.60 \times 10^{-3} \text{ mol } Cu^+ \text{ (Cu)}$$

g Cu = (6.60×10^{-3} mol)(63.546 g/mol) = 0.419 g Cu

$$\text{mass \% Cu in brass} = \frac{0.419 \text{ g Cu}}{0.544 \text{ g brass}} \times 100\% = 77.1\% \text{ Cu}$$

(c) $Zn_2P_2O_7$, 304.72

$$\text{mass \% Zn in } Zn_2P_2O_7 = \frac{2 \times 65.39 \text{ g}}{304.72 \text{ g}} \times 100\% = 42.92\%$$

mass of Zn in $Zn_2P_2O_7$ = (0.4292)(0.246 g) = 0.106 g Zn

$$\text{mass \% Zn in brass} = \frac{0.106 \text{ g Zn}}{0.544 \text{ g brass}} \times 100\% = 19.5\% \text{ Zn}$$

4.161 (a) $BaSO_4$, 233.38

$$\text{mol S} = 7.19\text{ g BaSO}_4 \text{ x } \frac{1\text{ mol BaSO}_4}{233.38\text{ g BaSO}_4} \text{ x } \frac{1\text{ mol S}}{1\text{ mol BaSO}_4} = 0.0308\text{ mol S}$$

$$\text{theoretical mol S} = \frac{0.0308\text{ mol S}}{0.913} = 0.0337\text{ mol S}$$

(b) Assume n = 1:

$$\text{mol Cl in MCl}_5 = 0.0337\text{ mol S x } \frac{5\text{ mol Cl}}{1\text{ mol S}} = 0.168\text{ mol Cl}$$

$$\text{mass Cl} = 0.168\text{ mol Cl x } \frac{35.453\text{ g Cl}}{1\text{ mol Cl}} = 5.97\text{ g Cl}$$

This is impossible because the initial mass of MCl_5 was only 4.61 g.

Assume n = 2:

$$\text{mol Cl in MCl}_5 = 0.0337\text{ mol S x } \frac{5\text{ mol Cl}}{2\text{ mol S}} = 0.0842\text{ mol Cl}$$

$$\text{mass Cl} = 0.0842\text{ mol Cl x } \frac{35.453\text{ g Cl}}{1\text{ mol Cl}} = 2.99\text{ g Cl}$$

mass M = 4.61g – 2.99 g = 1.62 g M

$$\text{mol M} = 0.0337\text{ mol S x } \frac{1\text{ mol M}}{2\text{ mol S}} = 0.0168\text{ mol}$$

$$\text{M molar mass} = \frac{1.62\text{ g}}{0.0168\text{ mol}} = 96.4\text{ g/mol}; \quad \text{M atomic weight} = 96.4$$

96.4 is reasonable and suggests that M is Mo.

Assume n = 3:

$$\text{mol Cl in MCl}_5 = 0.0337\text{ mol S x } \frac{5\text{ mol Cl}}{3\text{ mol S}} = 0.0562\text{ mol Cl}$$

$$\text{mass Cl} = 0.0562\text{ mol Cl x } \frac{35.453\text{ g Cl}}{1\text{ mol Cl}} = 1.99\text{ g Cl}$$

mass M = 4.61g – 1.99 g = 2.62 g M

$$\text{mol M} = 0.0337\text{ mol S x } \frac{1\text{ mol M}}{3\text{ mol S}} = 0.0112\text{ mol}$$

$$\text{M molar mass} = \frac{2.62\text{ g}}{0.0112\text{ mol}} = 234\text{ g/mol}; \quad \text{M atomic weight} = 234$$

234 is between Pa and U, which is highly unlikely for a lubricant.

Assume n = 4:

$$\text{mol Cl in MCl}_5 = 0.0337\text{ mol S x } \frac{5\text{ mol Cl}}{4\text{ mol S}} = 0.0421\text{ mol Cl}$$

$$\text{mass Cl} = 0.0421\text{ mol Cl x } \frac{35.453\text{ g Cl}}{1\text{ mol Cl}} = 1.49\text{ g Cl}$$

mass M = 4.61g – 1.49 g = 3.12 g M

mol M = 0.0337 mol S x $\frac{1 \text{ mol M}}{4 \text{ mol S}}$ = 0.00842 mol

M molar mass = $\frac{3.12 \text{ g}}{0.008\,42 \text{ mol}}$ = 371 g/mol; M atomic weight = 371

No known elements have a mass as great as 371.

(c) M is most likely Mo and the metal sulfide is MoS_2.

(d) (1) $2\,MoCl_5(s) + 5\,Na_2S(s) \rightarrow 2\,MoS_2(s) + S(l) + 10\,NaCl(s)$

(2) $2\,MoS_2(s) + 7\,O_2(g) \rightarrow 2\,MoO_3(s) + 4\,SO_2(g)$

(3) $SO_2(g) + 2\,Fe^{3+}(aq) + 2\,H_2O(l) \rightarrow 2\,Fe^{2+}(aq) + SO_4^{2-}(aq) + 4\,H^+(aq)$

(4) $SO_4^{2-}(aq) + Ba^{2+}(aq) \rightarrow BaSO_4(s)$

4.162 (a) $H_3MO_3(aq) \rightarrow H_3MO_4(aq)$

$H_3MO_3(aq) + H_2O(l) \rightarrow H_3MO_4(aq)$

$H_3MO_3(aq) + H_2O(l) \rightarrow H_3MO_4(aq) + 2\,H^+(aq)$

$[H_3MO_3(aq) + H_2O(l) \rightarrow H_3MO_4(aq) + 2\,H^+(aq) + 2\,e^-]$ x 5 (oxidation half reaction)

$MnO_4^-(aq) \rightarrow Mn^{2+}(aq)$

$MnO_4^-(aq) \rightarrow Mn^{2+}(aq) + 4\,H_2O(l)$

$MnO_4^-(aq) + 8\,H^+(aq) \rightarrow Mn^{2+}(aq) + 4\,H_2O(l)$

$[MnO_4^-(aq) + 8\,H^+(aq) + 5\,e^- \rightarrow Mn^{2+}(aq) + 4\,H_2O(l)]$ x 2 (reduction half reaction)

Combine the two half reactions.

$5\,H_3MO_3(aq) + 5\,H_2O(l) + 2\,MnO_4^-(aq) + 16\,H^+(aq) \rightarrow$
$5\,H_3MO_4(aq) + 10\,H^+(aq) + 2\,Mn^{2+}(aq) + 8\,H_2O(l)$

$5\,H_3MO_3(aq) + 2\,MnO_4^-(aq) + 6\,H^+(aq) \rightarrow 5\,H_3MO_4(aq) + 2\,Mn^{2+}(aq) + 3\,H_2O(l)$

(b) 10.7 mL = 0.0107 L

mol MnO_4^- = (0.0107 L)(0.100 mol/L) = 1.07 x 10^{-3} mol MnO_4^-

mol H_3MO_3 = 1.07 x 10^{-3} mol MnO_4^- x $\frac{5 \text{ mol } H_3MO_3}{2 \text{ mol } MnO_4^-}$ = 2.67 x 10^{-3} mol H_3MO_3

mol M_2O_3 = 2.67 x 10^{-3} mol H_3MO_3 x $\frac{1 \text{ mol } M_2O_3}{2 \text{ mol } H_3MO_3}$ = 1.34 x 10^{-3} mol M_2O_3

mol M in M_2O_3 = 1.34 x 10^{-3} mol M_2O_3 x $\frac{2 \text{ mol M}}{1 \text{ mol } M_2O_3}$ = 2.68 x 10^{-3} mol M

(c) M molar mass = $\frac{0.200 \text{ g}}{2.68 \times 10^{-3} \text{ mol}}$ = 74.6 g/mol; M atomic weight = 74.6

M is As.

5 Periodicity and the Electronic Structure of Atoms

5.1 $\nu = 102.5 \text{ MHz} = 102.5 \times 10^6 \text{ Hz} = 102.5 \times 10^6 \text{ s}^{-1}$

$$\lambda = \frac{c}{\nu} = \frac{3.00 \times 10^8 \text{ m/s}}{102.5 \times 10^6 \text{ s}^{-1}} = 2.93 \text{ m}$$

$\nu = 9.55 \times 10^{17} \text{ Hz} = 9.55 \times 10^{17} \text{ s}^{-1}$

$$\lambda = \frac{c}{\nu} = \frac{3.00 \times 10^8 \text{ m/s}}{9.55 \times 10^{17} \text{ s}^{-1}} = 3.14 \times 10^{-10} \text{ m}$$

5.2 The wave with the shorter wavelength (b) has the higher frequency. The wave with the larger amplitude (b) represents the more intense beam of light. The wave with the shorter wavelength (b) represents blue light. The wave with the longer wavelength (a) represents red light.

5.3 IR, $\lambda = 1.55 \times 10^{-6}$ m

$$E = \frac{hc}{\lambda} = (6.626 \times 10^{-34} \text{ J}\cdot\text{s})\left(\frac{3.00 \times 10^8 \text{ m/s}}{1.55 \times 10^{-6} \text{ m}}\right)(6.022 \times 10^{23}\text{/mol})$$

$E = 7.72 \times 10^4$ J/mol $= 77.2$ kJ/mol

UV, $\lambda = 250 \text{ nm} = 250 \times 10^{-9}$ m

$$E = \frac{hc}{\lambda} = (6.626 \times 10^{-34} \text{ J}\cdot\text{s})\left(\frac{3.00 \times 10^8 \text{ m/s}}{250 \times 10^{-9} \text{ m}}\right)(6.022 \times 10^{23}\text{/mol})$$

$E = 4.79 \times 10^5$ J/mol $= 479$ kJ/mol

X ray, $\lambda = 5.49 \text{ nm} = 5.49 \times 10^{-9}$ m

$$E = \frac{hc}{\lambda} = (6.626 \times 10^{-34} \text{ J}\cdot\text{s})\left(\frac{3.00 \times 10^8 \text{ m/s}}{5.49 \times 10^{-9} \text{ m}}\right)(6.022 \times 10^{23}\text{/mol})$$

$E = 2.18 \times 10^7$ J/mol $= 2.18 \times 10^4$ kJ/mol

5.4 $$E = \frac{hc}{\lambda} = (6.626 \times 10^{-34} \text{ J}\cdot\text{s})\left(\frac{3.00 \times 10^8 \text{ m/s}}{2.3 \times 10^{-3} \text{ m}}\right)(6.022 \times 10^{23}\text{/mol})$$

$E = 52$ J/mol $= 0.052$ kJ/mol

$$74 \text{ kJ} \times \frac{1 \text{ mol photons}}{0.052 \text{ kJ}} = 1.4 \times 10^3 \text{ mol photons}$$

5.5 $$E = \frac{hc}{\lambda} = (6.626 \times 10^{-34} \text{ J}\cdot\text{s})\left(\frac{3.00 \times 10^8 \text{ m/s}}{390 \times 10^{-9} \text{ m}}\right)(6.022 \times 10^{23}\text{/mol})$$

$E = 3.07 \times 10^5$ J/mol = 307 kJ/mol
The energy is less than the work function. Electrons will not be ejected.

5.6 (a) Ag is predicted to have the higher work function because Rb is further left on the periodic table and holds its electrons less tightly.
(b) Rb because lower energies correspond to longer wavelength.

5.7 $m = 2$; $R_\infty = 1.097 \times 10^{-2}\ \text{nm}^{-1}$

$$\frac{1}{\lambda} = R_\infty\left[\frac{1}{m^2} - \frac{1}{n^2}\right];\quad \frac{1}{\lambda} = R_\infty\left[\frac{1}{2^2} - \frac{1}{7^2}\right];\quad \frac{1}{\lambda} = 2.519 \times 10^{-3}\ \text{nm}^{-1};\quad \lambda = 397.0\ \text{nm}$$

5.8 $m = 3$; $R_\infty = 1.097 \times 10^{-2}\ \text{nm}^{-1}$

(a) $$\frac{1}{\lambda} = R_\infty\left[\frac{1}{m^2} - \frac{1}{n^2}\right];\quad \frac{1}{\lambda} = R_\infty\left[\frac{1}{3^2} - \frac{1}{4^2}\right];\quad \frac{1}{\lambda} = 5.333 \times 10^{-4}\ \text{nm}^{-1};\quad \lambda = 1875\ \text{nm}$$

(b) $$\frac{1}{\lambda} = R_\infty\left[\frac{1}{m^2} - \frac{1}{n^2}\right];\quad \frac{1}{\lambda} = R_\infty\left[\frac{1}{3^2} - \frac{1}{\infty^2}\right];\quad \frac{1}{\lambda} = 1.219 \times 10^{-3}\ \text{nm}^{-1};\quad \lambda = 820.4\ \text{nm}$$

5.9 $$\lambda = \frac{h}{mv} = \frac{6.626 \times 10^{-34}\ \text{kg m}^2\ \text{s}^{-1}}{(1150\ \text{kg})(24.6\ \text{m/s})} = 2.34 \times 10^{-38}\ \text{m}$$
This wavelength is shorter than the diameter of an atom.

5.10 $\lambda = 1\text{nm}/10 = 1 \times 10^{-10}$ m

$$\lambda = \frac{h}{mv};\quad v = \frac{h}{m\lambda} = \frac{6.626 \times 10^{-34}\ \text{kg m}^2\ \text{s}^{-1}}{(9.1 \times 10^{-31}\ \text{kg})(1 \times 10^{-10}\ \text{m})} = 7 \times 10^6\ \text{m/s}$$

5.11 (a) 2p (b) 4f (c) 3d

5.12

n	l	m_l	Orbital	No. of Orbitals
5	0	0	5s	1
	1	−1, 0, +1	5p	3
	2	−2, −1, 0, +1, +2	5d	5
	3	−3, −2, −1, 0, +1, +2, +3	5f	7
	4	−4, −3, −2, −1, 0, +1, +2, +3, +4	5g	9

There are 25 possible orbitals in the fifth shell.

5.13 $n = 4$, $l = 0$, 4s

5.14 The g orbitals have four nodal planes.

5.15 (a) Ti, $1s^2 2s^2 2p^6 3s^2 3p^6 4s^2 3d^2$ or $[Ar]\ 4s^2 3d^2$

[Ar] ↑↓ (4s) ↑ ↑ _ _ _ (3d)

(b) Zn, $1s^2 2s^2 2p^6 3s^2 3p^6 4s^2 3d^{10}$ or $[Ar]\ 4s^2 3d^{10}$

[Ar] ↑↓ (4s) ↑↓ ↑↓ ↑↓ ↑↓ ↑↓ (3d)

(c) Sn, $1s^2 2s^2 2p^6 3s^2 3p^6 4s^2 3d^{10} 4p^6 5s^2 4d^{10} 5p^2$ or $[Kr]\ 5s^2 4d^{10} 5p^2$

[Kr] ↑↓ (5s) ↑↓ ↑↓ ↑↓ ↑↓ ↑↓ (4d) ↑ ↑ _ (5p)

(d) Pb, $[Xe]\ 6s^2 4f^{14} 5d^{10} 6p^2$

[Xe] ↑↓ (6s) ↑↓ ↑↓ ↑↓ ↑↓ ↑↓ ↑↓ ↑↓ (4f) ↑↓ ↑↓ ↑↓ ↑↓ ↑↓ (5d) ↑ ↑ _ (6p)

5.16 (a) 43 electrons = Tc (b) 28 electrons = Ni

5.17 (a) Sn; atoms get larger as you go down a group.
(b) Lu; atoms get smaller as you go across a period.

5.18 Iodine has the largest atomic radius of the three halogens. C–I would be the longest bond length.

5.19 (a) $[Xe]\ 6s^2 4f^{14} 5d^{10}$

(b) [Xe] ↑↓ (6s) ↑↓ ↑↓ ↑↓ ↑↓ ↑↓ ↑↓ ↑↓ (4f) ↑↓ ↑↓ ↑↓ ↑↓ ↑↓ (5d)

(c) There are no unpaired electrons.

5.20 (a) $[Xe]\ 6s^1 4f^{14} 5d^{10} 6p^1$

(b) [Xe] ↑ (6s) ↑↓ ↑↓ ↑↓ ↑↓ ↑↓ ↑↓ ↑↓ (4f) ↑↓ ↑↓ ↑↓ ↑↓ ↑↓ (5d) ↑ _ _ (6p)

(c) There are 2 unpaired electrons.

5.21 (a) 7d, $n = 7$, $l = 2$, $m_l = -2, -1, 0, 1, 2$
(b) 6p, $n = 6$, $l = 1$, $m_l = -1, 0, 1$

(c) $$E = \frac{hc}{\lambda} = (6.626 \times 10^{-34}\ J \cdot s)\left(\frac{3.00 \times 10^8\ m/s}{434.7 \times 10^{-9}\ m}\right)(6.022 \times 10^{23}\ /mol)$$

$E = 2.75 \times 10^5$ J/mol = 275 kJ/mol

5.22 The shortest wavelength corresponds to the highest energy, therefore, 126.8 nm corresponds to 8p → 6s; 140.2 nm corresponds to 7p → 6s; and 185.0 nm corresponds to 6p → 6s.

5.23 (a) The fluorescent bulb does not emit all the wavelengths of light that would be emitted from a white light source. Notice that there are dark regions between the colored peaks.
(b) Fluorescent light does appear as "white light" because its line spectrum has contributions from all the colors (blue, green, yellow, orange, and red).

Conceptual Problems

5.24 The wave with the larger amplitude (a) has the greater intensity. The wave with the shorter wavelength (a) has the higher energy radiation. The wave with the shorter wavelength (a) represents yellow light. The wave with the longer wavelength (b) represents infrared radiation.

5.25 (a) Transitions (a) and (b) are absorptions. (a) is of lower energy and longer wavelength.
(b) Transitions (c) and (d) are emissions. (d) is of higher energy and shorter wavelength.

5.26 (a) $3p_y$ $n = 3, l = 1$ (b) $4d_{z^2}$ $n = 4, l = 2$

5.27
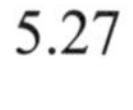

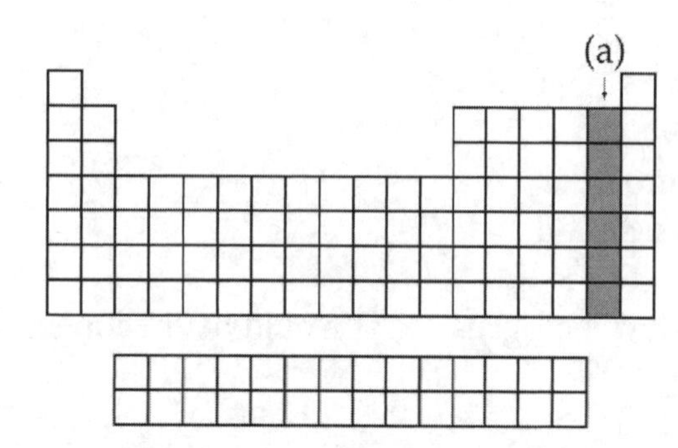

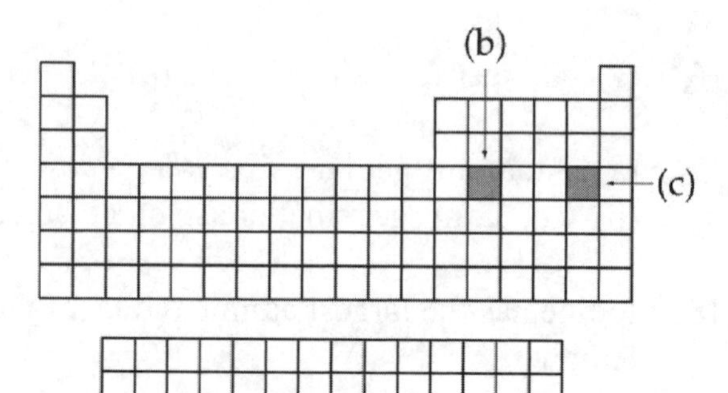

5.28 The green element, molybdenum, has an anomalous electron configuration. Its predicted electron configuration is [Ar] $5s^2\ 4d^4$. Its anomalous electron configuration is [Ar] $5s^1\ 4d^5$ because of the resulting half-filled d-orbitals.

5.29 [Ar] $4s^2\ 3d^{10}\ 4p^1$ is Ga.

5.30 There are 34 total electrons in the atom, so there are also 34 protons in the nucleus. The atom is selenium (Se)

Se, [Ar] $\underline{\uparrow\downarrow}$ $\underline{\uparrow\downarrow}$ $\underline{\uparrow\downarrow}$ $\underline{\uparrow\downarrow}$ $\underline{\uparrow\downarrow}$ $\underline{\uparrow\downarrow}$ $\underline{\uparrow\downarrow}$ $\underline{\uparrow}$ $\underline{\uparrow}$

4s 3d 4p

5.31 Ca and Br are in the same period, with Br to the far right of Ca. Ca is larger than Br. Sr is directly below Ca in the same group, and is larger than Ca. The result is Sr (215 pm) > Ca (197 pm) > Br (114 pm)

Section Problems

Electromagnetic Energy and Atomic Spectra (Sections 5.1–5.4)

5.32 Violet has the higher frequency and energy. Red has the higher wavelength.

5.33 Ultraviolet light has the higher frequency and the greater energy. Infrared light has the longer wavelength.

5.34 1.15×10^{-7} m = 115×10^{-9} m = 115 nm = UV
2.0×10^{-6} m = 2000×10^{-9} m = 2000 nm = IR
The visible region is (380 to 780 nm) is completely within this range. The ultraviolet and infrared regions are partially in this range.

5.35 290 MHz = 290×10^6 Hz = 2.9×10^8 Hz = radio waves
90 GHz = 90×10^9 Hz = 9.0×10^{10} Hz = microwaves
Radio waves and microwaves are in this region.

5.36 $\lambda = \frac{c}{\nu} = \frac{3.00 \times 10^8 \text{ m/s}}{5.5 \times 10^{15} \text{ s}^{-1}} = 5.5 \times 10^{-8} \text{ m}$

5.37 $\lambda = \frac{c}{\nu} = \frac{3.00 \times 10^8 \text{ m/s}}{4.33 \times 10^{-3} \text{ m}} = 6.93 \times 10^{10} \text{ s}^{-1} = 6.93 \times 10^{10} \text{ Hz}$

5.38 (a) ν = 99.5 MHz = $99.5 \times 10^6 \text{ s}^{-1}$
$E = h\nu = (6.626 \times 10^{-34} \text{ J·s})(99.5 \times 10^6 \text{ s}^{-1})(6.022 \times 10^{23} \text{ /mol})$
$E = 3.97 \times 10^{-2}$ J/mol = 3.97×10^{-5} kJ/mol
ν = 1150 kHz = $1150 \times 10^3 \text{ s}^{-1}$
$E = h\nu = (6.626 \times 10^{-34} \text{ J·s})(1150 \times 10^3 \text{ s}^{-1})(6.022 \times 10^{23} \text{ /mol})$
$E = 4.589 \times 10^{-4}$ J/mol = 4.589×10^{-7} kJ/mol
The FM radio wave (99.5 MHz) has the higher energy.
(b) $\lambda = 3.44 \times 10^{-9}$ m

$$E = \frac{hc}{\lambda} = (6.626 \times 10^{-34} \text{ J·s})\left(\frac{3.00 \times 10^8 \text{ m/s}}{3.44 \times 10^{-9} \text{ m}}\right)(6.022 \times 10^{23}/\text{mol})$$

$E = 3.48 \times 10^7$ J/mol = 3.48×10^4 kJ/mol
$\lambda = 6.71 \times 10^{-2}$ m

$$E = \frac{hc}{\lambda} = (6.626 \times 10^{-34} \text{ J·s})\left(\frac{3.00 \times 10^8 \text{ m/s}}{6.71 \times 10^{-2} \text{ m}}\right)(6.022 \times 10^{23}/\text{mol})$$

E = 1.78 J/mol = 1.78×10^{-3} kJ/mol
The X ray ($\lambda = 3.44 \times 10^{-9}$ m) has the higher energy.

5.39 ν = 400 MHz = $400 \times 10^6 \text{ s}^{-1}$
$E = (6.626 \times 10^{-34} \text{ J·s})(400 \times 10^6 \text{ s}^{-1})(6.02 \times 10^{23}/\text{mol}) = 0.160 \text{ J/mol} = 1.60 \times 10^{-4} \text{ kJ/mol}$

5.40 (a) $\lambda = \frac{c}{\nu} = \frac{3.00 \times 10^8 \text{ m/s}}{825 \times 10^6 \text{ s}^{-1}} \times \frac{1 \text{ cm}}{1 \times 10^{-2} \text{ m}} = 36.4 \text{ cm}$

(b) $\lambda = \frac{c}{\nu} = \frac{3.00 \times 10^8 \text{ m/s}}{875 \times 10^6 \text{ s}^{-1}} \times \frac{1 \text{ cm}}{1 \times 10^{-2} \text{ m}} = 34.3 \text{ cm}$

5.41 (a) $\nu = \frac{c}{\lambda} = \frac{3.00 \times 10^8 \text{ m/s}}{1300 \times 10^{-9} \text{ m}} = 2.3 \times 10^{14} \text{ s}^{-1}$

(b) $\text{time} = \dfrac{\text{distance}}{\text{speed}} = \dfrac{12 \text{ x } 10^3 \text{ m}}{3.00 \text{ x } 10^8 \text{ m/s}} = 4.0 \text{ x } 10^{-5} \text{ s}$

5.42 (a) $E = 90.5 \text{ kJ/mol x } \dfrac{1000 \text{ J}}{1 \text{ kJ}} \text{ x } \dfrac{1 \text{ mol}}{6.02 \text{ x } 10^{23}} = 1.50 \text{ x } 10^{-19} \text{ J}$

$\nu = \dfrac{E}{h} = \dfrac{1.50 \text{ x } 10^{-19} \text{ J}}{6.626 \text{ x } 10^{-34} \text{ J·s}} = 2.27 \text{ x } 10^{14} \text{ s}^{-1}$

$\lambda = \dfrac{c}{\nu} = \dfrac{3.00 \text{ x } 10^8 \text{ m/s}}{2.27 \text{ x } 10^{14} \text{ s}^{-1}} = 1.32 \text{ x } 10^{-6} \text{ m} = 1320 \text{ x } 10^{-9} \text{ m} = 1320 \text{ nm, near IR}$

(b) $E = 8.05 \text{ x } 10^{-4} \text{ kJ/mol x } \dfrac{1000 \text{ J}}{1 \text{ kJ}} \text{ x } \dfrac{1 \text{ mol}}{6.02 \text{ x } 10^{23}} = 1.34 \text{ x } 10^{-24} \text{ J}$

$\nu = \dfrac{E}{h} = \dfrac{1.34 \text{ x } 10^{-24} \text{ J}}{6.626 \text{ x } 10^{-34} \text{ J·s}} = 2.02 \text{ x } 10^{9} \text{ s}^{-1}$

$\lambda = \dfrac{c}{\nu} = \dfrac{3.00 \text{ x } 10^8 \text{ m/s}}{2.02 \text{ x } 10^{9} \text{ s}^{-1}} = 0.149 \text{ m, radio wave}$

(c) $E = 1.83 \text{ x } 10^3 \text{ kJ/mol x } \dfrac{1000 \text{ J}}{1 \text{ kJ}} \text{ x } \dfrac{1 \text{ mol}}{6.02 \text{ x } 10^{23}} = 3.04 \text{ x } 10^{-18} \text{ J}$

$\nu = \dfrac{E}{h} = \dfrac{3.04 \text{ x } 10^{-18} \text{ J}}{6.626 \text{ x } 10^{-34} \text{ J·s}} = 4.59 \text{ x } 10^{15} \text{ s}^{-1}$

$\lambda = \dfrac{c}{\nu} = \dfrac{3.00 \text{ x } 10^8 \text{ m/s}}{4.59 \text{ x } 10^{15} \text{ s}^{-1}} = 6.54 \text{ x } 10^{-8} \text{ m} = 65.4 \text{ x } 10^{-9} \text{ m} = 65.4 \text{ nm, UV}$

5.43 (a) $E = h\nu = (6.626 \text{ x } 10^{-34} \text{ J·s})(5.97 \text{ x } 10^{19} \text{ s}^{-1})\left(\dfrac{1 \text{ kJ}}{1000 \text{ J}}\right)(6.022 \text{ x } 10^{23}\text{/mol})$

$E = 2.38 \text{ x } 10^7 \text{ kJ/mol}$

(b) $E = h\nu = (6.626 \text{ x } 10^{-34} \text{ J·s})(1.26 \text{ x } 10^{6} \text{ s}^{-1})\left(\dfrac{1 \text{ kJ}}{1000 \text{ J}}\right)(6.022 \text{ x } 10^{23}\text{/mol})$

$E = 5.03 \text{ x } 10^{-7} \text{ kJ/mol}$

(c) $E = h\nu = (6.626 \text{ x } 10^{-34} \text{ J·s})\left(\dfrac{3.00 \text{ x } 10^8 \text{ m/s}}{2.57 \text{ x } 10^2 \text{ m}}\right)\left(\dfrac{1 \text{ kJ}}{1000 \text{ J}}\right)(6.022 \text{ x } 10^{23}\text{/mol})$

$E = 4.66 \text{ x } 10^{-7} \text{ kJ/mol}$

5.44 (a) $\lambda = \dfrac{c}{\nu} = \dfrac{3.00 \text{ x } 10^8 \text{ m/s}}{3.85 \text{ x } 10^{14} \text{ s}^{-1}} = 7.79 \text{ x } 10^{-7} \text{ m} = 779 \text{ x } 10^{-9} \text{ m} = 779 \text{ nm}$

$E = h\nu = (6.626 \text{ x } 10^{-34} \text{ J·s})(3.85 \text{ x } 10^{14} \text{ s}^{-1}) = 2.55 \text{ x } 10^{-19} \text{ J}$

(b) $\lambda = \dfrac{c}{\nu} = \dfrac{3.00 \text{ x } 10^8 \text{ m/s}}{4.62 \text{ x } 10^{14} \text{ s}^{-1}} = 6.49 \text{ x } 10^{-7} \text{ m} = 649 \text{ x } 10^{-9} \text{ m} = 649 \text{ nm}$

$E = h\nu = (6.626 \text{ x } 10^{-34} \text{ J·s})(4.62 \text{ x } 10^{14} \text{ s}^{-1}) = 3.06 \text{ x } 10^{-19} \text{ J}$

(c) $\lambda = \frac{c}{\nu} = \frac{3.00 \times 10^8 \text{ m/s}}{7.41 \times 10^{14} \text{ s}^{-1}} = 4.05 \times 10^{-7} \text{ m} = 405 \times 10^{-9} \text{ m} = 405 \text{ nm}$

$E = h\nu = (6.626 \times 10^{-34} \text{ J}\cdot\text{s})(7.41 \times 10^{14} \text{ s}^{-1}) = 4.91 \times 10^{-19} \text{ J}$

5.45 (a) $\lambda = \frac{c}{\nu} = \frac{3.00 \times 10^8 \text{ m/s}}{4.35 \times 10^{14} \text{ s}^{-1}} = 6.90 \times 10^{-7} \text{ m} = 690 \times 10^{-9} \text{ m} = 690 \text{ nm}$, yes.

(b) $E = 43 \text{ kJ/mol} \times \frac{1000 \text{ J}}{1 \text{ kJ}} \times \frac{1 \text{ mol}}{6.02 \times 10^{23}} = 7.1 \times 10^{-20} \text{ J}$

$\nu = \frac{E}{h} = \frac{7.1 \times 10^{-20} \text{ J}}{6.626 \times 10^{-34} \text{ J}\cdot\text{s}} = 1.1 \times 10^{14} \text{ s}^{-1}$

$\lambda = \frac{c}{\nu} = \frac{3.00 \times 10^8 \text{ m/s}}{1.1 \times 10^{14} \text{ s}^{-1}} = 2.8 \times 10^{-6} \text{ m} = 2800 \times 10^{-9} \text{ m} = 2800 \text{ nm}$, no.

(c) $\lambda = \frac{c}{\nu} = \frac{3.00 \times 10^8 \text{ m/s}}{706 \times 10^{12} \text{ s}^{-1}} = 4.25 \times 10^{-7} \text{ m} = 425 \times 10^{-9} \text{ m} = 425 \text{ nm}$, yes.

5.46 For n = 3; $\lambda = 656.3 \text{ nm} = 656.3 \times 10^{-9} \text{ m}$

$$E = \frac{hc}{\lambda} = (6.626 \times 10^{-34} \text{ J}\cdot\text{s})\left(\frac{2.998 \times 10^8 \text{ m/s}}{656.3 \times 10^{-9} \text{ m}}\right)\left(\frac{1 \text{ kJ}}{1000 \text{ J}}\right)(6.022 \times 10^{23}/\text{mol})$$

$E = 182.3 \text{ kJ/mol}$

For n = 4; $\lambda = 486.1 \text{ nm} = 486.1 \times 10^{-9} \text{ m}$

$$E = \frac{hc}{\lambda} = (6.626 \times 10^{-34} \text{ J}\cdot\text{s})\left(\frac{2.998 \times 10^8 \text{ m/s}}{486.1 \times 10^{-9} \text{ m}}\right)\left(\frac{1 \text{ kJ}}{1000 \text{ J}}\right)(6.022 \times 10^{23}/\text{mol})$$

$E = 246.1 \text{ kJ/mol}$

For n = 5; $\lambda = 434.0 \text{ nm} = 434.0 \times 10^{-9} \text{ m}$

$$E = \frac{hc}{\lambda} = (6.626 \times 10^{-34} \text{ J}\cdot\text{s})\left(\frac{2.998 \times 10^8 \text{ m/s}}{434.0 \times 10^{-9} \text{ m}}\right)\left(\frac{1 \text{ kJ}}{1000 \text{ J}}\right)(6.022 \times 10^{23}/\text{mol})$$

$E = 275.6 \text{ kJ/mol}$

5.47 $486.1 \text{ nm} = 486.1 \times 10^{-9} \text{ m}$

$$E = \frac{hc}{\lambda} = (6.626 \times 10^{-34} \text{ J}\cdot\text{s})\left(\frac{2.998 \times 10^8 \text{ m/s}}{486.1 \times 10^{-9} \text{ m}}\right)\left(\frac{1 \text{ kJ}}{1000 \text{ J}}\right)(6.022 \times 10^{23}/\text{mol})$$

$E = 246.1 \text{ kJ/mol}$

5.48 $m = 1, n = \infty;\ R_\infty = 1.097 \times 10^{-2} \text{ nm}^{-1}$

$$\frac{1}{\lambda} = R_\infty\left[\frac{1}{m^2} - \frac{1}{n^2}\right];\ \frac{1}{\lambda} = R_\infty\left[\frac{1}{1^2} - \frac{1}{\infty^2}\right] = R_\infty = 1.097 \times 10^{-2} \text{ nm}^{-1};\ \lambda = 91.16 \text{ nm}$$

$$E = \frac{hc}{\lambda} = (6.626 \times 10^{-34}\ J \cdot s)\left(\frac{2.998 \times 10^{8}\ m/s}{91.16 \times 10^{-9}\ m}\right)\left(\frac{1\ kJ}{1000\ J}\right)(6.022 \times 10^{23}/mol)$$

$E = 1312\ kJ/mol$

5.49 $m = 2, n = \infty;\ R_\infty = 1.097 \times 10^{-2}\ nm^{-1}$

$$\frac{1}{\lambda} = R_\infty\left[\frac{1}{m^2} - \frac{1}{n^2}\right];\ \frac{1}{\lambda} = R_\infty\left[\frac{1}{2^2} - \frac{1}{\infty^2}\right] = \frac{R_\infty}{4} = 2.743 \times 10^{-3}\ nm^{-1};\ \lambda = 364.6\ nm$$

$$E = \frac{hc}{\lambda} = (6.626 \times 10^{-34}\ J \cdot s)\left(\frac{2.998 \times 10^{8}\ m/s}{364.6 \times 10^{-9}\ m}\right)\left(\frac{1\ kJ}{1000\ J}\right)(6.022 \times 10^{23}/mol)$$

$E = 328.1\ kJ/mol$

5.50 $m = 4, n = 5;\ R_\infty = 1.097 \times 10^{-2}\ nm^{-1}$

$$\frac{1}{\lambda} = R_\infty\left[\frac{1}{m^2} - \frac{1}{n^2}\right];\ \frac{1}{\lambda} = R_\infty\left[\frac{1}{4^2} - \frac{1}{5^2}\right] = 2.468 \times 10^{-4}\ nm^{-1};\ \lambda = 4051\ nm$$

$$E = \frac{hc}{\lambda} = (6.626 \times 10^{-34}\ J \cdot s)\left(\frac{2.998 \times 10^{8}\ m/s}{4051 \times 10^{-9}\ m}\right)\left(\frac{1\ kJ}{1000\ J}\right)(6.022 \times 10^{23}/mol)$$

$E = 29.55\ kJ/mol$, IR

$m = 4, n = 6;\ R_\infty = 1.097 \times 10^{-2}\ nm^{-1}$

$$\frac{1}{\lambda} = R_\infty\left[\frac{1}{m^2} - \frac{1}{n^2}\right];\ \frac{1}{\lambda} = R_\infty\left[\frac{1}{4^2} - \frac{1}{6^2}\right] = 3.809 \times 10^{-4}\ nm^{-1};\ \lambda = 2625\ nm$$

$$E = \frac{hc}{\lambda} = (6.626 \times 10^{-34}\ J \cdot s)\left(\frac{2.998 \times 10^{8}\ m/s}{2625 \times 10^{-9}\ m}\right)\left(\frac{1\ kJ}{1000\ J}\right)(6.022 \times 10^{23}/mol)$$

$E = 45.60\ kJ/mol$, IR

5.51 $m = 3, n = 4;\ R_\infty = 1.097 \times 10^{-2}\ nm^{-1}$

$$\frac{1}{\lambda} = R_\infty\left[\frac{1}{m^2} - \frac{1}{n^2}\right];\ \frac{1}{\lambda} = R_\infty\left[\frac{1}{3^2} - \frac{1}{4^2}\right] = 5.333 \times 10^{-4}\ nm^{-1};\ \lambda = 1875\ nm$$

$$E = \frac{hc}{\lambda} = (6.626 \times 10^{-34}\ J \cdot s)\left(\frac{2.998 \times 10^{8}\ m/s}{1875 \times 10^{-9}\ m}\right)\left(\frac{1\ kJ}{1000\ J}\right)(6.022 \times 10^{23}/mol)$$

$E = 63.80\ kJ/mol$, IR

$m = 3, n = 5;\ R_\infty = 1.097 \times 10^{-2}\ nm^{-1}$

$$\frac{1}{\lambda} = R_\infty\left[\frac{1}{m^2} - \frac{1}{n^2}\right];\ \frac{1}{\lambda} = R_\infty\left[\frac{1}{3^2} - \frac{1}{5^2}\right] = 7.801 \times 10^{-4}\ nm^{-1};\ \lambda = 1282\ nm$$

$$E = \frac{hc}{\lambda} = (6.626 \times 10^{-34}\ J \cdot s)\left(\frac{2.998 \times 10^{8}\ m/s}{1282 \times 10^{-9}\ m}\right)\left(\frac{1\ kJ}{1000\ J}\right)(6.022 \times 10^{23}/mol)$$

$E = 93.00\ kJ/mol$, IR

5.52 Both (c) & (d) are below the threshold energy and no electrons would be ejected. (b) would eject the least number of electrons.

5.53 (a) is above the threshold energy and has a high amplitude. It will cause the largest number of electrons to be ejected.

5.54 $$E = (436 \text{ kJ/mol})\left(\frac{1000 \text{ J}}{1 \text{ kJ}}\right)\left(\frac{1 \text{ mol}}{6.022 \times 10^{23} \text{ photon}}\right) = 7.24 \times 10^{-19} \text{ J/photon}$$

$$\nu = \frac{E}{h} = \frac{7.24 \times 10^{-19} \text{ J}}{6.626 \times 10^{-34} \text{ J·s}} = 1.09 \times 10^{15} \text{ s}^{-1} = 1.09 \times 10^{15} \text{ Hz}$$

5.55 $\lambda = 234 \text{ nm} = 234 \times 10^{-9} \text{ m} = 2.34 \times 10^{-7} \text{ m}$

$$E = \frac{hc}{\lambda} = (6.626 \times 10^{-34} \text{ J·s})\left(\frac{3.00 \times 10^{8} \text{ m/s}}{2.34 \times 10^{-7} \text{ m}}\right)(6.022 \times 10^{23}/\text{mol})$$

$E = 5.12 \times 10^{5} \text{ J/mol} = 5.12 \times 10^{2} \text{ kJ/mol} = 512 \text{ kJ/mol}$

5.56 The deuterium lamp produces a continuous emission spectrum.

5.57 The sodium-vapor lamp produces a line emission spectrum.

Particles and Waves (Section 5.5)

5.58 $$\lambda = \frac{h}{mv} = \frac{6.626 \times 10^{-34} \text{ kg m}^2 \text{ s}^{-1}}{(9.11 \times 10^{-31} \text{ kg})(0.99 \times 3.00 \times 10^{8} \text{ m/s})} = 2.45 \times 10^{-12} \text{ m}, \gamma \text{ ray}$$

5.59 $$\lambda = \frac{h}{mv} = \frac{6.626 \times 10^{-34} \text{ kg m}^2 \text{ s}^{-1}}{(1.673 \times 10^{-27} \text{ kg})(0.25 \times 3.00 \times 10^{8} \text{ m/s})} = 5.28 \times 10^{-15} \text{ m}, \gamma \text{ ray}$$

5.60 $156 \text{ km/h} = 156 \times 10^{3} \text{ m}/3600 \text{ s} = 43.3 \text{ m/s}$; $145 \text{ g} = 0.145 \text{ kg}$

$$\lambda = \frac{h}{mv} = \frac{6.626 \times 10^{-34} \text{ kg m}^2 \text{ s}^{-1}}{(0.145 \text{ kg})(43.3 \text{ m/s})} = 1.06 \times 10^{-34} \text{ m}$$

The wavelength is too small, compared to the object, to observe.

5.61 $1.55 \text{ mg} = 1.55 \times 10^{-3} \text{ g} = 1.55 \times 10^{-6} \text{ kg}$

$$\lambda = \frac{h}{mv} = \frac{6.626 \times 10^{-34} \text{ kg m}^2 \text{ s}^{-1}}{(1.55 \times 10^{-6} \text{ kg})(1.38 \text{ m/s})} = 3.10 \times 10^{-28} \text{ m}$$

The wavelength is too small, compared to the object, to observe.

5.62 $145 \text{ g} = 0.145 \text{ kg}$; $0.500 \text{ nm} = 0.500 \times 10^{-9} \text{ m}$

$$v = \frac{h}{m\lambda} = \frac{6.626 \times 10^{-34} \text{ kg m}^2 \text{ s}^{-1}}{(0.145 \text{ kg})(0.500 \times 10^{-9} \text{ m})} = 9.14 \times 10^{-24} \text{ m/s}$$

5.63 750 nm = 750 x 10^{-9} m

$$v = \frac{h}{m\lambda} = \frac{6.626 \times 10^{-34}\ \text{kg m}^2\ \text{s}^{-1}}{(9.11 \times 10^{-31}\ \text{kg})(750 \times 10^{-9}\ \text{m})} = 970\ \text{m/s}$$

5.64 0.68 g = 0.68 x 10^{-3} kg

$$(\Delta x)(\Delta mv) \geq \frac{h}{4\pi};\quad \Delta x \geq \frac{h}{4\pi(\Delta mv)} = \frac{6.626 \times 10^{-34}\ \text{kg m}^2\ \text{s}^{-1}}{4\pi(0.68 \times 10^{-3}\ \text{kg})(0.1\ \text{m/s})} = 8 \times 10^{-31}\ \text{m}$$

5.65 $$4.0026\ \text{amu} \times \frac{1.660\,540 \times 10^{-27}\ \text{kg}}{1\ \text{amu}} = 6.6465 \times 10^{-27}\ \text{kg};\quad (\Delta x)(\Delta mv) \geq \frac{h}{4\pi}$$

$$\Delta x \geq \frac{h}{4\pi(\Delta mv)} = \frac{6.626 \times 10^{-34}\ \text{kg m}^2\ \text{s}^{-1}}{4\pi(6.6465 \times 10^{-27}\ \text{kg})(0.01 \times 1.36 \times 10^{3}\ \text{m/s})} = 5.833 \times 10^{-10}\ \text{m}$$

Orbitals and Quantum Mechanics (Sections 5.6–5.9)

5.66 The Heisenberg uncertainty principle states that one can never know both the position and the velocity of an electron beyond a certain level of precision. This means we cannot think of electrons circling the nucleus in specific orbital paths, but we can think of electrons as being found in certain three-dimensional regions of space around the nucleus, called orbitals.

5.67 The probability of finding the electron drops off rapidly as distance from the nucleus increases, although it never drops to zero, even at large distances. As a result, there is no definite boundary or size for an orbital. However, we usually imagine the boundary surface of an orbital enclosing the volume where an electron spends 95% of its time.

5.68 n is the principal quantum number. The size and energy level of an orbital depends on n. l is the angular-momentum quantum number. l defines the three-dimensional shape of an orbital. m_l is the magnetic quantum number. m_l defines the spatial orientation of an orbital. m_s is the spin quantum number. m_s indicates the spin of the electron and can have either of two values, +½ or –½.

5.69 (a) is not allowed because for $l = 0$, $m_l = 0$ only.
(b) is allowed.
(c) is not allowed because for n = 4, l = 0, 1, 2, or 3 only.

5.70 (a) 4s n = 4; $l = 0$; $m_l = 0$; $m_s = \pm½$
(b) 3p n = 3; $l = 1$; $m_l = -1, 0, +1$; $m_s = \pm½$
(c) 5f n = 5; $l = 3$; $m_l = -3, -2, -1, 0, +1, +2, +3$; $m_s = \pm½$
(d) 5d n = 5; $l = 2$; $m_l = -2, -1, 0, +1, +2$; $m_s = \pm½$

5.71 (a) 3s (b) 2p (c) 4f (d) 4d

5.72 A 4s orbital has three nodal surfaces.

4s orbital

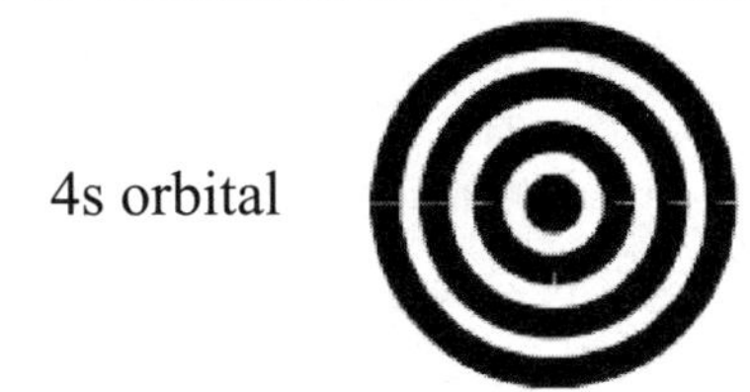

nodes are white

regions of maximum electron probability are black

5.73 The number of nodal planes in a subshell equals the value of the l quantum number.

5.74 Co $1s^2\ 2s^2\ 2p^6\ 3s^2\ 3p^6\ 4s^2\ 3d^7$
(a) is not allowed because for $l = 0$, $m_l = 0$ only.
(b) is not allowed because n = 4 and $l = 2$ is for a 4d orbital.
(c) is allowed because n = 3 and $l = 1$ is for a 3p orbital.

5.75 Se $1s^2\ 2s^2\ 2p^6\ 3s^2\ 3p^6\ 4s^2\ 3d^{10}\ 4p^4$
(a) is not allowed because for n = 3, $l = 0$, 1 or 2.
(b) is not allowed because n = 4 and $l = 2$ is for a 4d orbital.
(c) is allowed because n = 4 and $l = 1$ is for a 4p orbital.

5.76 For n = 5, the maximum number of electrons will occur when the 5g orbital is filled:
$[Rn]\ 7s^2\ 5f^{14}\ 6d^{10}\ 7p^6\ 8s^2\ 5g^{18}$ = 138 electrons

5.77 n = 4, $l = 0$ is a 4s orbital. The electron configuration is $1s^2\ 2s^2\ 2p^6\ 3s^2\ 3p^6\ 4s^2$. The number of electrons is 20.

5.78 $\lambda = 330 \text{ nm} = 330 \times 10^{-9} \text{ m}$

$$E = \frac{hc}{\lambda} = (6.626 \times 10^{-34} \text{ J}\cdot\text{s})\left(\frac{3.00 \times 10^8 \text{ m/s}}{330 \times 10^{-9} \text{ m}}\right)\left(\frac{1 \text{ kJ}}{1000 \text{ J}}\right)(6.022 \times 10^{23}/\text{mol})$$

$$E = 363 \text{ kJ/mol}$$

5.79 $795 \text{ nm} = 795 \times 10^{-9} \text{ m}$

$$E = \frac{hc}{\lambda} = (6.626 \times 10^{-34} \text{ J}\cdot\text{s})\left(\frac{3.00 \times 10^8 \text{ m/s}}{795 \times 10^{-9} \text{ m}}\right)\left(\frac{1 \text{ kJ}}{1000 \text{ J}}\right)(6.022 \times 10^{23}/\text{mol})$$

$$E = 151 \text{ kJ/mol}$$

5.80 C $1s^2\ 2s^2\ 2p^2$

n	l	m_l	m_s
1	0	0	+½
1	0	0	–½
2	0	0	+½
2	0	0	–½
2	1	–1	+½
2	1	0	+½

5.81 O $1s^2 2s^2 2p^4$

n	l	m_l	m_s
1	0	0	+½
1	0	0	–½
2	0	0	+½
2	0	0	–½
2	1	–1	+½
2	1	0	+½
2	1	+1	+½
2	1	–1	–½

5.82 Sr [Kr] $5s^2$

n	l	m_l	m_s
5	0	0	+½
5	0	0	–½

5.83 Mo [Kr] $5s^2 4d^4$

(c) are valid quantum numbers for a 4d electron.

Electron Configurations (Sections 5.10–5.13)

5.84 Part of the electron-nucleus attraction is canceled by the electron-electron repulsion, an effect we describe by saying that the electrons are shielded from the nucleus by the other electrons. The net nuclear charge actually felt by an electron is called the effective nuclear charge, Z_{eff}, and is often substantially lower than the actual nuclear charge, Z_{actual}.

$$Z_{eff} = Z_{actual} - \text{electron shielding}$$

5.85 Electron shielding gives rise to energy differences among 3s, 3p, and 3d orbitals in multielectron atoms because of the differences in orbital shape. For example, the 3s orbital is spherical and has a large probability density near the nucleus, while the 3p orbital is dumbbell shaped with a node at the nucleus. An electron in a 3s orbital can penetrate closer to the nucleus than an electron in a 3p orbital can and feels less of a shielding effect from other electrons. Generally, for any given value of the principal quantum number n, a lower value of l corresponds to a higher value of Z_{eff} and to a lower energy for the orbital.

5.86 4s > 4d > 4f

5.87 K < Ca < Se < Kr

5.88 The number of elements in successive periods of the periodic table increases by the progression 2, 8, 18, 32 because the principal quantum number n increases by 1 from one period to the next. As the principal quantum number increases, the number of orbitals in a shell increases. The progression of elements parallels the number of electrons in a particular shell.

5.89 The n and *l* quantum numbers determine the energy level of an orbital in a multielectron atom.

5.90 (a) 5d (b) 4s (c) 6s

5.91 (a) 2p < 3p < 5s < 4d (b) 2s < 4s < 3d < 4p (c) 3d < 4p < 5p < 6s

5.92 (a) 3d after 4s (b) 4p after 3d(c) 6d after 5f (d) 6s after 5p

5.93 (a) 3s before 3p (b) 3d before 4p (c) 6s before 4f (d) 4f before 5d

5.94
(a) Ti, Z = 22 $1s^2 2s^2 2p^6 3s^2 3p^6 4s^2 3d^2$
(b) Ru, Z = 44 $1s^2 2s^2 2p^6 3s^2 3p^6 4s^2 3d^{10} 4p^6 5s^2 4d^6$
(c) Sn, Z = 50 $1s^2 2s^2 2p^6 3s^2 3p^6 4s^2 3d^{10} 4p^6 5s^2 4d^{10} 5p^2$
(d) Sr, Z = 38 $1s^2 2s^2 2p^6 3s^2 3p^6 4s^2 3d^{10} 4p^6 5s^2$
(e) Se, Z = 34 $1s^2 2s^2 2p^6 3s^2 3p^6 4s^2 3d^{10} 4p^4$

5.95
(a) Z = 55, Cs [Xe] $6s^1$ (b) Z = 40, Zr [Kr] $5s^2 4d^2$
(c) Z = 80, Hg[Xe] $6s^2 4f^{14} 5d^{10}$ (d) Z = 62, Sm [Xe] $6s^2 4f^6$

5.96
(a) Rb, Z = 37 [Kr] ↑ (5s)
(b) W, Z = 74 [Xe] ↑↓ (6s) ↑↓ ↑↓ ↑↓ ↑↓ ↑↓ ↑↓ ↑↓ (4f) ↑ ↑ ↑ ↑ _ (5d)
(c) Ge, Z = 32 [Ar] ↑↓ (4s) ↑↓ ↑↓ ↑↓ ↑↓ ↑↓ (3d) ↑ ↑ _ (4p)
(d) Zr, Z = 40 [Kr] ↑↓ (5s) ↑ ↑ _ _ _ (4d)

5.97
(a) Z = 25, Mn [Ar] ↑↓ (4s) ↑ ↑ ↑ ↑ ↑ (3d)
(b) Z = 56, Ba [Xe] ↑↓ (6s)
(c) Z = 28, Ni [Ar] ↑↓ (4s) ↑↓ ↑↓ ↑↓ ↑ ↑ (3d)
(d) Z = 47, Ag [Kr] ↑ (5s) ↑↓ ↑↓ ↑↓ ↑↓ ↑↓ (4d)

5.98
(a) O $1s^2 2s^2 2p^4$ ↑↓ ↑ ↑ (2p) 2 unpaired e^-
(b) Si $1s^2 2s^2 2p^6 3s^2 3p^2$ ↑ ↑ _ (3p) 2 unpaired e^-

(c) K [Ar] $4s^1$ 1 unpaired e^-

(d) As [Ar] $4s^2 3d^{10} 4p^3$ ↑ ↑ ↑ (4p) 3 unpaired e^-

5.99 (a) Z = 31, Ga (b) Z = 46, Pd

5.100 Order of orbital filling:
1s→2s→2p→3s→3p→4s→3d→4p→5s→4d→5p→6s→4f→5d→6p→7s→5f→6d→7p→8s→5g
Z = 121

5.101 A g orbital would begin filling at atomic number = 121 (see 5.82). There are nine g orbitals that can each hold two electrons. The first element to have a filled g orbital would be atomic number = 138.

5.102 Na^+ $1s^2 2s^2 2p^6$

5.103 Cl^- $1s^2 2s^2 2p^6 3s^2 3p^6$

Electron Configurations and Periodic Properties (Section 5.14)

5.104 Atomic radii increase down a group because the electron shells are farther away from the nucleus.

5.105 Across a period, the effective nuclear charge increases, causing a decrease in atomic radii.

5.106 F < O < S

5.107 Cl < As < K < Rb

5.108 (a) K, lower in group 1A
(b) Ta, lower in group 5B
(c) V, farther to the left in same period
(d) Ba, four periods lower and only one group to the right

5.109 (a) Ge, lower in group 4A
(b) Pt, lower in group 8B
(c) Sn, farther to the left in same period
(d) Rb, lower in group 1A

5.110 Z = 116 [Rn] $7s^2 5f^{14} 6d^{10} 7p^4$

5.111 Z = 119 [Rn] $7s^2 5f^{14} 6d^{10} 7p^6 8s^1$

Chapter Problems

5.112 m = 2; n = 3; R = 1.097 x 10^{-2} nm^{-1}

$$\frac{1}{\lambda} = Z^2R\left[\frac{1}{m^2} - \frac{1}{n^2}\right]; \ \frac{1}{\lambda} = (2^2)R\left[\frac{1}{2^2} - \frac{1}{3^2}\right] = 6.094 \times 10^{-3}\ nm^{-1}$$

λ = 164 nm

5.113 m = 1; n = 4; R = 1.097 x 10^{-2} nm^{-1}

$$\frac{1}{\lambda} = Z^2R\left[\frac{1}{m^2} - \frac{1}{n^2}\right]; \ \frac{1}{\lambda} = (3^2)R\left[\frac{1}{1^2} - \frac{1}{4^2}\right] = 9.256 \times 10^{-2}\ nm^{-1}$$

λ = 10.8 nm

5.114 m = 2; R_∞ = 1.097 x 10^{-2} nm^{-1}

$$\frac{1}{\lambda} = R_\infty\left[\frac{1}{m^2} - \frac{1}{n^2}\right]; \ \frac{1}{\lambda} = R_\infty\left[\frac{1}{2^2} - \frac{1}{6^2}\right] = 2.438 \times 10^{-3}\ nm^{-1}$$

λ = 410.2 nm = 410.2 x 10^{-9} m

$$E = \frac{hc}{\lambda} = (6.626 \times 10^{-34}\ J\cdot s)\left(\frac{2.998 \times 10^8\ m/s}{410.2 \times 10^{-9}\ m}\right)\left(\frac{1\ kJ}{1000\ J}\right)(6.022 \times 10^{23}/mol)$$

E = 291.6 kJ/mol

5.115 m = 5; R_∞ = 1.097 x 10^{-2} nm^{-1}

$$\frac{1}{\lambda} = R_\infty\left[\frac{1}{m^2} - \frac{1}{n^2}\right]$$

$$n = 6, \ \frac{1}{\lambda} = R_\infty\left[\frac{1}{5^2} - \frac{1}{6^2}\right] = 1.341 \times 10^{-4}\ nm^{-1}; \ \lambda = 7458\ nm = 7458 \times 10^{-9}\ m$$

$$E = \frac{hc}{\lambda} = (6.626 \times 10^{-34}\ J\cdot s)\left(\frac{2.998 \times 10^8\ m/s}{7458 \times 10^{-9}\ m}\right)\left(\frac{1\ kJ}{1000\ J}\right)(6.022 \times 10^{23}/mol)$$

E = 16.04 kJ/mol

$$n = 7, \ \frac{1}{\lambda} = R_\infty\left[\frac{1}{5^2} - \frac{1}{7^2}\right] = 2.149 \times 10^{-4}\ nm^{-1}; \ \lambda = 4653\ nm = 4653 \times 10^{-9}\ m$$

$$E = \frac{hc}{\lambda} = (6.626 \times 10^{-34}\ J\cdot s)\left(\frac{2.998 \times 10^8\ m/s}{4653 \times 10^{-9}\ m}\right)\left(\frac{1\ kJ}{1000\ J}\right)(6.022 \times 10^{23}/mol)$$

E = 25.71 kJ/mol

The lines in this series are in the infrared region of the electromagnetic spectrum.

5.116 m = 5, n = ∞; R_∞ = 1.097 x 10^{-2} nm^{-1}

$$\frac{1}{\lambda} = R_\infty\left[\frac{1}{5^2} - \frac{1}{\infty^2}\right] = R_\infty\left[\frac{1}{25}\right] = 4.388 \times 10^{-4}\ nm^{-1}; \ \lambda = 2279\ nm$$

5.117 (a) $E = \left(142\ \frac{\text{kJ}}{\text{mol}}\right)\left(\frac{1000\text{ J}}{1\text{ kJ}}\right)\left(\frac{1\text{ mol}}{6.02 \times 10^{23}}\right) = 2.36 \times 10^{-19}\text{ J}$

$E = \frac{hc}{\lambda},\quad \lambda = \frac{hc}{E} = \frac{(6.626 \times 10^{-34}\text{ J}\cdot\text{s})(3.00 \times 10^{8}\text{ m/s})}{2.36 \times 10^{-19}\text{ J}}$

$\lambda = 8.42 \times 10^{-7}$ m (infrared)

(b) $E = \left(4.55 \times 10^{-2}\ \frac{\text{kJ}}{\text{mol}}\right)\left(\frac{1000\text{ J}}{1\text{ kJ}}\right)\left(\frac{1\text{ mol}}{6.02 \times 10^{23}}\right) = 7.56 \times 10^{-23}\text{ J}$

$E = \frac{hc}{\lambda},\quad \lambda = \frac{hc}{E} = \frac{(6.626 \times 10^{-34}\text{ J}\cdot\text{s})(3.00 \times 10^{8}\text{ m/s})}{7.56 \times 10^{-23}\text{ J}}$

$\lambda = 2.63 \times 10^{-3}$ m (microwave)

(c) $E = \left(4.81 \times 10^{4}\ \frac{\text{kJ}}{\text{mol}}\right)\left(\frac{1000\text{ J}}{1\text{ kJ}}\right)\left(\frac{1\text{ mol}}{6.02 \times 10^{23}}\right) = 7.99 \times 10^{-17}\text{ J}$

$E = \frac{hc}{\lambda},\quad \lambda = \frac{hc}{E} = \frac{(6.626 \times 10^{-34}\text{ J}\cdot\text{s})(3.00 \times 10^{8}\text{ m/s})}{7.99 \times 10^{-17}\text{ J}}$

$\lambda = 2.49 \times 10^{-9}$ m (X ray)

5.118 (a) $E = h\nu = (6.626 \times 10^{-34}\text{ J}\cdot\text{s})(3.79 \times 10^{11}\text{ s}^{-1})\left(\frac{1\text{ kJ}}{1000\text{ J}}\right)(6.022 \times 10^{23}/\text{mol}) = 0.151\text{ kJ/mol}$

(b) $E = h\nu = (6.626 \times 10^{-34}\text{ J}\cdot\text{s})(5.45 \times 10^{4}\text{ s}^{-1})\left(\frac{1\text{ kJ}}{1000\text{ J}}\right)(6.022 \times 10^{23}/\text{mol}) = 2.17 \times 10^{-8}\text{ kJ/mol}$

(c) $E = h\nu = (6.626 \times 10^{-34}\text{ J}\cdot\text{s})\left(\frac{3.00 \times 10^{8}\text{ m/s}}{4.11 \times 10^{-5}\text{ m}}\right)\left(\frac{1\text{ kJ}}{1000\text{ J}}\right)(6.022 \times 10^{23}/\text{mol}) = 2.91\text{ kJ/mol}$

5.119 $\nu = 9{,}192{,}631{,}770\text{ s}^{-1} = 9.19263 \times 10^{9}\text{ s}^{-1}$

$E = h\nu = (6.626 \times 10^{-34}\text{ J}\cdot\text{s})(9.19263 \times 10^{9}\text{ s}^{-1})\left(\frac{1\text{ kJ}}{1000\text{ J}}\right)(6.022 \times 10^{23}/\text{mol}) = 3.668 \times 10^{-3}\text{ kJ/mol}$

5.120 (a) Ra [Rn] $7s^2$ [Rn] ↑↓ (7s)

(b) Sc [Ar] $4s^2\,3d^1$ [Ar] ↑↓ (4s) ↑ _ _ _ _ (3d)

(c) Lr [Rn] $7s^2\,5f^{14}\,6d^1$ [Rn] ↑↓ (7s) ↑↓ ↑↓ ↑↓ ↑↓ ↑↓ ↑↓ ↑↓ (5f) ↑ _ _ _ _ (6d)

(d) B [He] $2s^2\,2p^1$ [He] ↑↓ (2s) ↑ _ _ (2p)

(e) Te [Kr] $5s^2\,4d^{10}\,5p^4$ [Kr] ↑↓ (5s) ↑↓ ↑↓ ↑↓ ↑↓ ↑↓ (4d) ↑↓ ↑ ↑ (5p)

5.121 (a) row 1   n = 1, $l = 0$ 1s   2 elements
$l = 1$ 1p   6 elements
$l = 2$ 1d   10 elements
row 2   n = 2, $l = 0$ 2s   2 elements
$l = 1$ 2p   6 elements
$l = 2$ 2d   10 elements
$l = 3$ 2f   14 elements

There would be 50 elements in the first two rows.
(b) There would be 18 elements in the first row [see (a) above]. The fifth element in the second row would have atomic number = 23.
(c) Z = 12 $\underline{\uparrow\downarrow}$ (1s) $\underline{\uparrow\downarrow}$ $\underline{\uparrow\downarrow}$ $\underline{\uparrow\downarrow}$ (1p) $\underline{\uparrow}$ $\underline{\uparrow}$ $\underline{\uparrow}$ $\underline{\uparrow}$ $\underline{\ \ }$ (1d)

5.122 206.5 kJ = 206.5 x 10^3 J; $E = \frac{206.5 \times 10^3\ J}{1\ mol} \times \frac{1\ mol}{6.022 \times 10^{23}} = 3.429 \times 10^{-19}\ J$

$$E = \frac{hc}{\lambda}, \quad \lambda = \frac{hc}{E} = \frac{(6.626 \times 10^{-34}\ J \cdot s)(3.00 \times 10^8\ m/s)}{3.429 \times 10^{-19}\ J} = 5.797 \times 10^{-7}\ m = 580.\ nm$$

5.123 780 nm is at the red end of the visible region of the electromagnetic spectrum.
780 nm = 780 x 10^{-9} m

$$E = \frac{hc}{\lambda} = (6.626 \times 10^{-34}\ J \cdot s)\left(\frac{3.00 \times 10^8\ m/s}{780 \times 10^{-9}\ m}\right)\left(\frac{1\ kJ}{1000\ J}\right)(6.022 \times 10^{23}/mol) = 153\ kJ/mol$$

5.124 (a) Sr, Z = 38   [Kr] $\underline{\uparrow\downarrow}$ (5s)

(b) Cd, Z = 48   [Kr] $\underline{\uparrow\downarrow}$ (5s) $\underline{\uparrow\downarrow}$ $\underline{\uparrow\downarrow}$ $\underline{\uparrow\downarrow}$ $\underline{\uparrow\downarrow}$ $\underline{\uparrow\downarrow}$ (4d)

(c) Z = 22, Ti   [Ar] $\underline{\uparrow\downarrow}$ (4s) $\underline{\uparrow}$ $\underline{\uparrow}$ $\underline{\ \ }$ $\underline{\ \ }$ $\underline{\ \ }$ (3d)

(d) Z = 34, Se   [Ar] $\underline{\uparrow\downarrow}$ (4s) $\underline{\uparrow\downarrow}$ $\underline{\uparrow\downarrow}$ $\underline{\uparrow\downarrow}$ $\underline{\uparrow\downarrow}$ $\underline{\uparrow\downarrow}$ (3d) $\underline{\uparrow\downarrow}$ $\underline{\uparrow}$ $\underline{\uparrow}$ (4p)

5.125 La ([Xe] $6s^2$ $5d^1$) is directly below Y ([Kr] $5s^2$ $4d^1$) in the periodic table. Both have similar valence electron configurations, but for La the valence electrons are one shell farther out leading to its larger radius.

Although Hf ([Xe] $6s^2$ $4f^{14}$ $5d^2$) is directly below Zr ([Kr] $5s^2$ $4d^2$) in the periodic table, Zr and Hf have almost identical atomic radii because the 4f electrons in Hf are not effective in shielding the valence electrons. The valence electrons in Hf are drawn in closer to the nucleus by the higher Z_{eff}.

5.126 (a) $\lambda = \frac{c}{\nu} = \frac{3.00 \times 10^8 \text{ m/s}}{2.9 \times 10^{18} \text{ s}^{-1}} = 1.0 \times 10^{-10} \text{ m}$

(b) $E = h\nu = (6.626 \times 10^{-34} \text{ J}\cdot\text{s})(2.9 \times 10^{18} \text{ s}^{-1})\left(\frac{1 \text{ kJ}}{1000 \text{ J}}\right)(6.022 \times 10^{23}/\text{mol})$

$E = 1.2 \times 10^6 \text{ kJ/mol}$

(c) X rays

5.127 (a) $\nu = \frac{c}{\lambda} = \frac{3.00 \times 10^8 \text{ m/s}}{450 \times 10^{-9} \text{ m}} = 6.67 \times 10^{14} \text{ s}^{-1}$

(b) $E = h\nu = (6.626 \times 10^{-34} \text{ J}\cdot\text{s})(6.67 \times 10^{14} \text{ s}^{-1})\left(\frac{1 \text{ kJ}}{1000 \text{ J}}\right)(6.022 \times 10^{23}/\text{mol})$

$E = 266 \text{ kJ/mol}$

(c) Blue

5.128 For K, $Z_{eff} = \sqrt{\frac{(418.8 \text{ kJ/mol})(4^2)}{1312 \text{ kJ/mol}}} = 2.26$

For Kr, $Z_{eff} = \sqrt{\frac{(1350.7 \text{ kJ/mol})(4^2)}{1312 \text{ kJ/mol}}} = 4.06$

5.129 75 W = 75 J/s; 550 nm = 550×10^{-9} m; (0.05)(75 J/s) = 3.75 J/s

$E = \frac{hc}{\lambda} = (6.626 \times 10^{-34} \text{ J}\cdot\text{s})\left(\frac{3.00 \times 10^8 \text{ m/s}}{550 \times 10^{-9} \text{ m}}\right) = 3.61 \times 10^{-19} \text{ J/photon}$

$\text{number of photons} = \frac{3.75 \text{ J/s}}{3.61 \times 10^{-19} \text{ J/photon}} = 1.0 \times 10^{19} \text{ photons/s}$

5.130 $q = (350 \text{ g})(4.184 \text{ J/g}\cdot{}^\circ\text{C})(95\ {}^\circ\text{C} - 20\ {}^\circ\text{C}) = 109{,}830 \text{ J}$

$\lambda = 15.0 \text{ cm} = 15.0 \times 10^{-2} \text{ m}$

$E = (6.626 \times 10^{-34} \text{ J}\cdot\text{s})\left(\frac{3.00 \times 10^8 \text{ m/s}}{15.0 \times 10^{-2} \text{ m}}\right) = 1.33 \times 10^{-24} \text{ J/photon}$

$\text{number of photons} = \frac{109{,}830 \text{ J}}{1.33 \times 10^{-24} \text{ J/photon}} = 8.3 \times 10^{28} \text{ photons}$

5.131 $E = \left(310 \frac{\text{kJ}}{\text{mol}}\right)\left(\frac{1000 \text{ J}}{1 \text{ kJ}}\right)\left(\frac{1 \text{ mol}}{6.022 \times 10^{23}}\right) = 5.15 \times 10^{-19} \text{ J}$

$E = \frac{hc}{\lambda}, \quad \lambda = \frac{hc}{E} = \frac{(6.626 \times 10^{-34} \text{ J}\cdot\text{s})(3.00 \times 10^8 \text{ m/s})}{5.15 \times 10^{-19} \text{ J}} = 3.86 \times 10^{-7} \text{ m} = 386 \text{ nm}$

5.132 48.2 nm = 48.2×10^{-9} m

$$E(\text{photon}) = 6.626 \times 10^{-34}\ \text{J}\cdot\text{s} \times \frac{3.00 \times 10^{8}\ \text{m/s}}{48.2 \times 10^{-9}\ \text{m}} \times \frac{1\ \text{kJ}}{1000\ \text{J}} \times \frac{6.022 \times 10^{23}}{\text{mol}} = 2.48 \times 10^{3}\ \text{kJ/mol}$$

$$E_K = E(\text{electron}) = \tfrac{1}{2}(9.109 \times 10^{-31}\ \text{kg})(2.371 \times 10^{6}\ \text{m/s})^2\left(\frac{1\ \text{kJ}}{1000\ \text{J}}\right)\left(\frac{6.022 \times 10^{23}}{\text{mol}}\right)$$

$$E_K = 1.54 \times 10^{3}\ \text{kJ/mol}$$

$E(\text{photon}) = E_i + E_K$; $E_i = E(\text{photon}) - E_K = (2.48 \times 10^3) - (1.54 \times 10^3) = 940$ kJ/mol

5.133 Charge on electron = 1.602×10^{-19} C; $1\ \text{V}\cdot\text{C} = 1\ \text{J} = 1\ \text{kg m}^2/\text{s}^2$

(a) $E_K = (30{,}000\ \text{V})(1.602 \times 10^{-19}\ \text{C}) = 4.806 \times 10^{-15}$ J

$$E_K = \tfrac{1}{2}mv^2;\quad v = \sqrt{\frac{2E_K}{m}} = \sqrt{\frac{2 \times 4.806 \times 10^{-15}\ \text{kg m}^2/\text{s}^2}{9.109 \times 10^{-31}\ \text{kg}}} = 1.03 \times 10^{8}\ \text{m/s}$$

$$\lambda = \frac{h}{mv} = \frac{6.626 \times 10^{-34}\ \text{kg m}^2/\text{s}}{(9.109 \times 10^{-31}\ \text{kg})(1.03 \times 10^{8}\ \text{m/s})} = 7.06 \times 10^{-12}\ \text{m}$$

$$\text{(b)}\ E = \frac{hc}{\lambda} = (6.626 \times 10^{-34}\ \text{J}\cdot\text{s})\left(\frac{3.00 \times 10^{8}\ \text{m/s}}{1.54 \times 10^{-10}\ \text{m}}\right) = 1.29 \times 10^{-15}\ \text{J/photon}$$

5.134 Substitute the equation for the orbit radius, r, into the equation for the energy level, E, to get

$$E = \frac{-Ze^2}{2\left(\frac{n^2 a_o}{Z}\right)} = \frac{-Z^2e^2}{2a_o n^2}$$

Let E_1 be the energy of an electron in a lower orbit and E_2 the energy of an electron in a higher orbit. The difference between the two energy levels is

$$\Delta E = E_2 - E_1 = \frac{-Z^2e^2}{2a_o n_2^2} - \frac{-Z^2e^2}{2a_o n_1^2} = \frac{-Z^2e^2}{2a_o n_2^2} + \frac{Z^2e^2}{2a_o n_1^2} = \frac{Z^2e^2}{2a_o n_1^2} - \frac{Z^2e^2}{2a_o n_2^2}$$

$$\Delta E = \frac{Z^2e^2}{2a_o}\left[\frac{1}{n_1^2} - \frac{1}{n_2^2}\right]$$

Because Z, e, and a_o are constants, this equation shows that ΔE is proportional to $\left[\frac{1}{n_1^2} - \frac{1}{n_2^2}\right]$ where n_1 and n_2 are integers with $n_2 > n_1$.

This is similar to the Balmer-Rydberg equation where $1/\lambda$ or ν for the emission spectra of atoms is proportional to $\left[\frac{1}{m^2} - \frac{1}{n^2}\right]$ where m and n are integers with $n > m$.

5.135 (a) $\underline{\uparrow\downarrow}$ $\underline{\uparrow\downarrow}$ $\underline{\uparrow\downarrow}$ $\underline{\uparrow\downarrow}$ $\underline{\uparrow\downarrow}$ $\underline{\uparrow\downarrow}$ $\underline{\uparrow\downarrow}$ $\underline{\uparrow}$ $\underline{\uparrow}$ $\underline{}$

1s 2s 2p 3s 3p 4s

Two partially filled orbitals.

(b) The element in the 3rd column and 4th row under these new rules would have an atomic number of 30 and be in the s-block.

5.136 (a) 3d, $n = 3, l = 2$

(b) 2p, $n = 2, l = 1, m_l = -1, 0, +1$

3p, $n = 3, l = 1, m_l = -1, 0, +1$

3d, $n = 3, l = 2, m_l = -2, -1, 0, +1, +2$

(c) N, $1s^2 2s^2 2p^3$ so the 3s, 3p, and 3d orbitals are empty.

(d) C, $1s^2 2s^2 2p^2$ so the 1s and 2s orbitals are filled.

(e) Be, $1s^2 2s^2$ so the 2s orbital contains the outermost electrons.

(f) 2p and 3p ($\underline{\uparrow}$ $\underline{\uparrow}$ __) and 3d ($\underline{\uparrow}$ $\underline{\uparrow}$ __ __ __).

5.137 $\lambda = 1.03 \times 10^{-7}$ m $= 103 \times 10^{-9}$ m $= 103$ nm

$$\frac{1}{\lambda} = R_\infty\left[\frac{1}{m^2} - \frac{1}{n^2}\right],\ R_\infty = 1.097 \times 10^{-2}\ \text{nm}^{-1}$$

$$\frac{1}{103\ \text{nm}} = (1.097 \times 10^{-2}\ \text{nm}^{-1})\left[\frac{1}{1^2} - \frac{1}{n^2}\right],\ \text{solve for n.}$$

$$\frac{(1/103\ \text{nm})}{(1.097 \times 10^{-2}\ \text{nm}^{-1})} - 1 = -\frac{1}{n^2}$$

$$\frac{1}{n^2} = 0.115;\ n^2 = \frac{1}{0.115};\ n = \sqrt{\frac{1}{0.115}} = 2.95$$

The electron jumps to the third shell.

5.138 (a) $E = h\nu$; $\nu = \frac{E}{h} = \frac{7.21 \times 10^{-19}\ \text{J}}{6.626 \times 10^{-34}\ \text{J}\cdot\text{s}} = 1.09 \times 10^{15}\ \text{s}^{-1}$

(b) $E(\text{photon}) = E_i + E_K$; from (a), $E_i = 7.21 \times 10^{-19}$ J

$$E(\text{photon}) = \frac{hc}{\lambda} = (6.626 \times 10^{-34}\ \text{J}\cdot\text{s})\left(\frac{3.00 \times 10^8\ \text{m/s}}{2.50 \times 10^{-7}\ \text{m}}\right) = 7.95 \times 10^{-19}\ \text{J}$$

$E_K = E(\text{photon}) - E_i = (7.95 \times 10^{-19}\ \text{J}) - (7.21 \times 10^{-19}\ \text{J}) = 7.4 \times 10^{-20}$ J

Calculate the electron velocity from the kinetic energy, E_K.

$E_K = 7.4 \times 10^{-20}\ \text{J} = 7.4 \times 10^{-20}\ \text{kg}\cdot\text{m}^2/\text{s}^2 = \frac{1}{2}mv^2 = \frac{1}{2}(9.109 \times 10^{-31}\ \text{kg})v^2$

$$v = \sqrt{\frac{2 \times (7.4 \times 10^{-20}\ \text{kg}\cdot\text{m}^2/\text{s}^2)}{9.109 \times 10^{-31}\ \text{kg}}} = 4.0 \times 10^5\ \text{m/s}$$

$$\text{de Broglie wavelength} = \frac{h}{mv} = \frac{6.626 \times 10^{-34}\ \text{kg}\cdot\text{m}^2/\text{s}}{(9.109 \times 10^{-31}\ \text{kg})(4.0 \times 10^5\ \text{m/s})} = 1.8 \times 10^{-9}\ \text{m} = 1.8\ \text{nm}$$

Multiconcept Problems

5.139 (a) $E = h\nu;\ \nu = \frac{E}{h} = \frac{4.70 \times 10^{-16}\ J}{6.626 \times 10^{-34}\ J \cdot s} = 7.09 \times 10^{17}\ s^{-1}$

(b) $\lambda = \frac{c}{\nu} = \frac{3.00 \times 10^{8}\ m/s}{7.09 \times 10^{17}\ s^{-1}} = 4.23 \times 10^{-10}\ m = 0.423 \times 10^{-9}\ m = 0.423\ nm$

(c) $\lambda = \frac{h}{mv};\ v = \frac{h}{m\lambda} = \frac{6.626 \times 10^{-34}\ kg \cdot m^2/s}{(9.11 \times 10^{-31}\ kg)(4.23 \times 10^{-10}\ m)} = 1.72 \times 10^{6}\ m/s$

(d) $E_K = \frac{mv^2}{2} = \frac{(9.11 \times 10^{-31}\ kg)(1.72 \times 10^{6}\ m/s)^2}{2} = 1.35 \times 10^{-18}\ kg \cdot m^2/s^2 = 1.35 \times 10^{-18}\ J$

5.140 (a) 5f subshell: $n = 5,\ l = 3,\ m_l = -3, -2, -1, 0, +1, +2, +3$
3d subshell: $n = 3,\ l = 2,\ m_l = -2, -1, 0, +1, +2$

(b) In the H atom the subshells in a particular energy level are all degenerate, i.e., all have the same energy. Therefore, you only need to consider the principal quantum number, n, to calculate the wavelength emitted for an electron that drops from the 5f to the 3d subshell.

$m = 3,\ n = 5;\ R_\infty = 1.097 \times 10^{-2}\ nm^{-1}$

$\frac{1}{\lambda} = R_\infty\left[\frac{1}{m^2} - \frac{1}{n^2}\right];\ \frac{1}{\lambda} = R_\infty\left[\frac{1}{3^2} - \frac{1}{5^2}\right];\ \frac{1}{\lambda} = 7.801 \times 10^{-4}\ nm^{-1};\ \lambda = 1282\ nm$

(c) $m = 3,\ n = \infty;\ R_\infty = 1.097 \times 10^{-2}\ nm^{-1}$

$\frac{1}{\lambda} = R_\infty\left[\frac{1}{m^2} - \frac{1}{n^2}\right];\ \frac{1}{\lambda} = R_\infty\left[\frac{1}{3^2} - \frac{1}{\infty^2}\right];\ \frac{1}{\lambda} = R_\infty\left[\frac{1}{3^2}\right] = 1.219 \times 10^{-3}\ nm^{-1};\ \lambda = 820.4\ nm$

$E = (6.626 \times 10^{-34}\ J \cdot s)\left(\frac{3.00 \times 10^{8}\ m/s}{820.4 \times 10^{-9}\ m}\right)(6.022 \times 10^{23}/mol) = 1.46 \times 10^{5}\ J/mol = 146\ kJ/mol$

5.141 (a) [Kr] $5s^2\ 4d^{10}\ 5p^6$ (b) [Kr] $5s^2\ 4d^{10}\ 5p^5\ 6s^1$

(c) Both Xe^* and Cs have a single electron in the 6s orbital with similar effective nuclear charges. Therefore the 6s electrons in both cases are held with similar strengths and require almost the same energy to remove.

6 Ionic Compounds: Periodic Trends and Bonding Theory

6.1 (a) Ra^{2+} [Rn] (b) Ni^{2+} [Ar] $3d^8$ (c) N^{3-} [Ne]

6.2 (b) Ti^{4+}, Ca^{2+}, and Cl^- are isoelectronic. They all have the electron configuration of Ar.
(c) Na^+, Mg^{2+}, and Al^{3+} are isoelectronic. They all have the electron configuration of Ne.

6.3 (a) O^{2-}; decrease in effective nuclear charge and an increase in electron-electron repulsions leads to the anion being larger.
(b) Fe; in Fe^{3+} electrons are removed from a larger valence shell and there is an increase in effective nuclear charge leading to the smaller cation.
(c) H^-; decrease in effective nuclear charge and an increase in electron-electron repulsions leads to the anion being larger.

6.4 K^+, Ca^{2+}, and Cl^- are isoelectronic. The Z_{eff} for $Ca^{2+} > K^+ > Cl^-$.
In terms of size, $Cl^- > K^+ > Ca^{2+}$. Cl^- is yellow, K^+ is green, and Ca^{2+} is red.

6.5 Ionization energy generally increases from left to right across a row of the periodic table and decreases from top to bottom down a group.
(a) Br (b) S (c) Se (d) Ne

6.6 (c) < (b) < (a) < (d)

6.7 (a) Be $1s^2 2s^2$; N $1s^2 2s^2 2p^3$
Be would have the larger third ionization energy because this electron would come from the 1s orbital.
(b) Ga [Ar] $4s^2 3d^{10} 4p^1$; Ge [Ar] $4s^2 3d^{10} 4p^2$
Ga would have the larger fourth ionization energy because this electron would come from the 3d orbitals. Ge would be losing a 4s electron.

6.8 The first 3 electrons are relatively easy to remove compared to the fourth and fifth electrons. The atom is Al.

6.9 Cr [Ar] $4s^1 3d^5$; Mn [Ar] $4s^2 3d^5$; Fe [Ar] $4s^2 3d^6$
Cr can accept an electron into a 4s orbital. The 4s orbital is lower in energy than a 3d orbital. Both Mn and Fe accept the added electron into a 3d orbital that contains an electron, but Mn has a lower value of Z_{eff}. Therefore, Mn has a less negative E_{ea} than either Cr or Fe.

6.10 The least favorable E_{ea} is for Kr (red) because it is a noble gas with a filled set of 4p orbitals. The most favorable E_{ea} is for Ge (blue) because the 4p orbitals would become half filled. In addition, Z_{eff} is larger for Ge than it is for K (green).

6.11 (a) Rb would lose one electron and adopt the Kr noble gas configuration.
(b) Ba would lose two electrons and adopt the Xe noble gas configuration.
(c) Ga would lose three electrons and adopt an Ar-like noble gas configuration (note that Ga^{3+} has ten 3d electrons in addition to the two 3s and six 3p electrons).
(d) F would gain one electron and adopt the Ne noble gas configuration.

6.12 Group 6A elements will gain 2 electrons. The ion charge will be 2–.

6.13

$Mg(s) \rightarrow Mg(g)$	+147.7	kJ/mol
$Mg(g) \rightarrow Mg^{+}(g) + e^{-}$	+737.7	kJ/mol
$Mg^{+}(g) \rightarrow Mg^{2+}(g) + e^{-}$	+1450.7	kJ/mol
$F_2(g) \rightarrow 2\,F(g)$	+158	kJ/mol
$2[F(g) + e^{-} \rightarrow F^{-}(g)]$	2(–328)	kJ/mol
$Mg^{2+}(g) + 2\,F^{-}(g) \rightarrow MgF_2(s)$	–2957	kJ/mol
Sum =	–1119	kJ/mol for $Mg(s) + F_2(g) \rightarrow MgF_2(s)$

6.14

$Li(s) \rightarrow Li(g)$	+161	kJ/mol
$Li(s) \rightarrow Li(g) + e^{-}$	+520	kJ/mol
$\frac{1}{2}\,[Cl_2(g) \rightarrow 2\,Cl(g)]$	+243/2	kJ/mol
$Cl(g) + e^{-} \rightarrow Cl^{-}(g)$	–349	kJ/mol
$Li^{+}(g) + Cl^{-}(g) \rightarrow LiCl(s)$	–U	kJ/mol
Sum =	–409	kJ/mol for $Li(s) + \frac{1}{2}\,Cl_2(g) \rightarrow LiCl(s)$

electrostatic attraction = –U = – 409 – 161 – 520 – 243/2 + 349 = –863 kJ/mol

6.15 (a) KCl has the higher lattice energy because of the smaller K^{+}.
(b) CaF_2 has the higher lattice energy because of the smaller Ca^{2+}.
(c) CaO has the higher lattice energy because of the higher charge on both the cation and anion.

6.16 The anions are larger than the cations. Cl^{-} is larger than O^{2-} because it is below it in the periodic table. Therefore, (a) is NaCl and (b) is MgO. Because of the higher ion charge and shorter cation–anion distance, MgO has the larger lattice energy.

6.17 In ionic liquids the cation has an irregular shape and one or both of the ions are large and bulky to disperse charges over a large volume. Both factors minimize the crystal lattice energy, making the solid less stable and favoring the liquid.

6.18 (a) Iodide ions are larger than bromide ions. Tetraheptylammonium bromide corresponds to picture (ii) and tetraheptylammonium iodide corresponds to picture (i).
(b) Tetraheptylammonium bromide has the larger lattice energy because bromide ions are smaller than iodide ions.
(c) Tetraheptylammonium bromide has melting point of 88 °C and tetraheptylammonium iodide has a melting point of 39 °C.

6.19 (a) F^{-}, $1s^2 2s^2 2p^6$
Se^{2-}, $1s^2 2s^2 2p^6 3s^2 3p^6 4s^2 3d^{10} 4p^6$

O^{2-}, $1s^2 2s^2 2p^6$
Br^-, $1s^2 2s^2 2p^6 3s^2 3p^6 4s^2 3d^{10} 4p^6$
(b) F^- and O^{2-} are isoelectronic; Se^{2-} and Br^- are isoelectronic.
(c) The larger Br^-.

Conceptual Problems

6.20

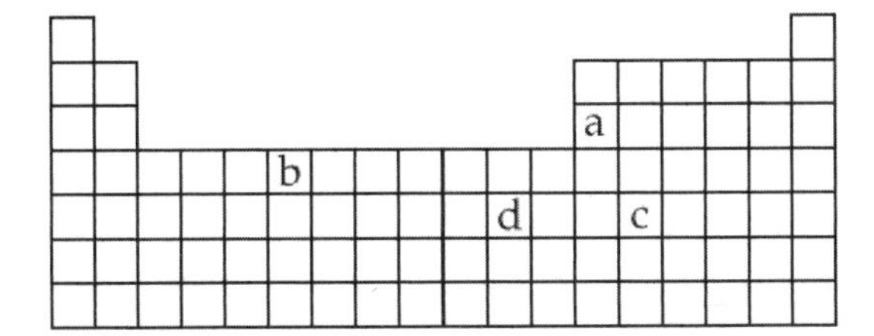

(a) Al^{3+}
(b) Cr^{3+}
(c) Sn^{2+}
(d) Ag^+

6.21 The first sphere gets larger on going from reactant to product. This is consistent with it being a nonmetal gaining an electron and becoming an anion. The second sphere gets smaller on going from reactant to product. This is consistent with it being a metal losing an electron and becoming a cation.

6.22

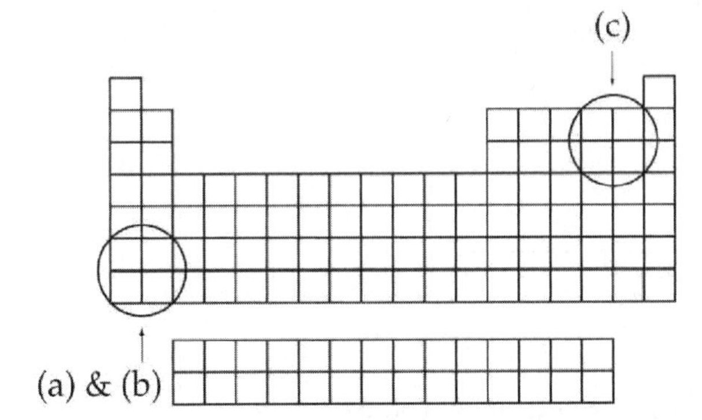

6.23 Ca (red) would have the largest third ionization energy of the three because the electron being removed is from a filled valence shell. For Al (green) and Kr (blue), the electron being removed is from a partially filled valence shell. The third ionization energy for Kr would be larger than that for Al because the electron being removed from Kr is coming out of a 4p orbital while the electron being removed from Al makes Al isoelectronic with Ne. In addition, Z_{eff} is larger for Kr than for Al. The ease of losing its third electron is Al < Kr < Ca.

6.24 The first 2 electrons are relatively easy to remove compared to the third and fourth electron. The atom is Mg.

6.25 (a) $MgSO_4$ (b) Li_2CO_3 (c) $FeCl_2$ (d) $Ca_3(PO_4)_2$

6.26 (a) shows an extended array, which represents an ionic compound.
(b) shows discrete units, which represent a covalent compound.

6.27 (a) I_2 (b) Na (c) NaCl (d) Cl_2

6.28 (c) has the largest lattice energy because the charges are closest together.
(a) has the smallest lattice energy because the charges are farthest apart.

6.29 All the ions in both drawings are singly charged, so only the size of the ions is important. The ions in drawing (b) are smaller and closer together, so (b) has the larger lattice energy.

6.30 Green, CBr_4; Blue, SrF_2; Red, PbS or PbS_2

6.31

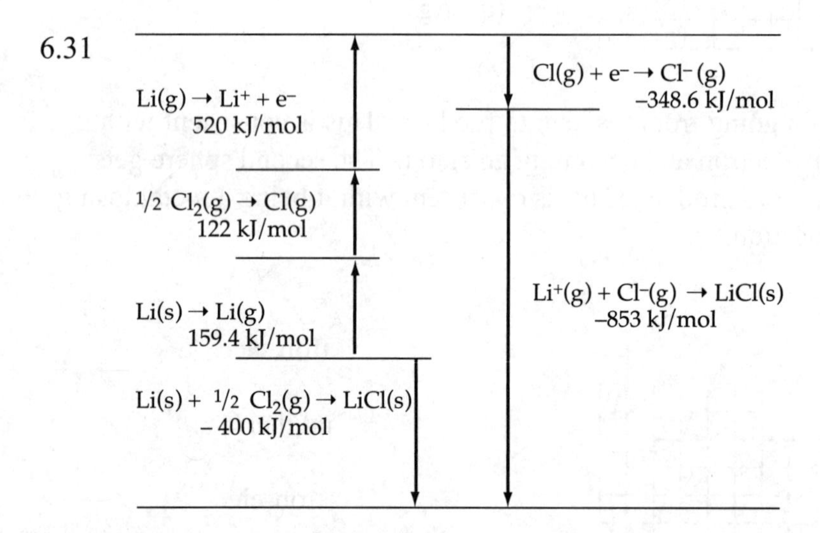

Section Problems

Ions, Ionization Energy, and Electron Affinity (Sections 6.1–6.5)

6.32 A covalent bond results when two atoms share several (usually two) of their electrons. An ionic bond results from a complete transfer of one or more electrons from one atom to another.

6.33 (a) and (d), covalent; (b) and (c), ionic

6.34 A molecule is the unit of matter that results when two or more atoms are joined by covalent bonds. An ion results when an atom gains or loses electrons.

6.35 (a), (b), and (d), ions; (c), molecule

6.36 (a) Be^{2+}, 4 protons and 2 electrons (b) Rb^{+}, 37 protons and 36 electrons
(c) Se^{2-}, 34 protons and 36 electrons (d) Au^{3+}, 79 protons and 76 electrons

6.37 (a) A 2+ cation that has 36 electrons must have 38 protons. X = Sr.
(b) A 1– anion that has 36 electrons must have 35 protons. X = Br.

6.38 (a) La^{3+}, [Xe] (b) Ag^{+}, [Kr] $4d^{10}$ (c) Sn^{2+}, [Kr] $5s^2 4d^{10}$

6.39 (a) Se^{2-}, [Kr] (b) N^{3-}, [Ne]

6.40 Ca^{2+}, [Ar]; Ti^{2+}, [Ar] $3d^2$

6.41 Z = 30, Zn

6.42 The neutral atom contains 12 e^- and is Mg. The ion is Mg^{2+}.

6.43 [Kr] $4d^3$, Mo^{3+}; [Kr] $5s^2 4d^2$, Zr

6.44 Cr^{2+} [Ar] $3d^4$ ↑ ↑ ↑ ↑ __ (3d)

Fe^{2+} [Ar] $3d^6$ ↑↓ ↑ ↑ ↑ ↑ (3d)

6.45 Fe^{3+} [Ar] $3d^5$ ↑ ↑ ↑ ↑ ↑ (3d)

6.46 (a) S^{2-}; decrease in effective nuclear charge and an increase in electron-electron repulsions leads to the anion being larger.
(b) Ca; in Ca^{2+} electrons are removed from a larger valence shell and there is an increase in effective nuclear charge leading to the smaller cation.
(c) O^{2-}; decrease in effective nuclear charge and an increase in electron-electron repulsions leads to O^{2-} being larger.

6.47 (a) Rb; in Rb^+ the electron is removed from a larger valence shell and there is an increase in effective nuclear charge leading to the smaller cation.
(b) N^{3-}; decrease in effective nuclear charge and an increase in electron-electron repulsions leads to the anion being larger.
(c) Cr^{3+}; in Cr^{6+} electrons are removed from a larger valence shell and there is an increase in effective nuclear charge leading to Cr^{6+} being the smaller cation.

6.48 Sr^{2+}, Se^{2-}, Br^-, and Rb^+ are isoelectronic. The Z_{eff} for $Sr^{2+} > Rb^+ > Br^- > Se^{2-}$. The smallest ion has the largest Z_{eff}.
Ions arranged from smallest to largest are $Sr^{2+} < Rb^+ < Br^- < Se^{2-}$.

6.49 Mg^{2+}, O^{2-}, F^-, and Na^+ are isoelectronic. The Z_{eff} for $Mg^{2+} > Na^+ > F^- > O^{2-}$. The smallest ion has the largest Z_{eff}.
Ions arranged from smallest to largest are $Mg^{2+} < Na^+ < F^- < O^{2-}$.

6.50 The largest E_{i1} are found in Group 8A because of the largest values of Z_{eff}.
The smallest E_{i1} are found in Group 1A because of the smallest values of Z_{eff}.

6.51 Fr would have the smallest ionization energy, and He would have the largest.

6.52 Using Figure 6.3 as a reference:

	Lowest E_{i1}	Highest E_{i1}
(a)	K	Li
(b)	B	Cl
(c)	Ca	Cl

6.53 Using Figure 6.3 as a reference:
(a) Na < I < P
(b) Sr < Mg < P
(c) Cs < Ca < Se

6.54 (a) K [Ar] $4s^1$ Ca [Ar] $4s^2$
Ca has the smaller second ionization energy because it is easier to remove the second 4s valence electron in Ca than it is to remove the second electron in K from the filled 3p orbitals.
(b) Ca [Ar] $4s^2$ Ga [Ar] $4s^2 3d^{10} 4p^1$
Ca has the larger third ionization energy because it is more difficult to remove the third electron in Ca from the filled 3p orbitals than it is to remove the third electron (second 4s valence electron) from Ga.

6.55 Sn has a smaller fourth ionization energy than Sb because of a smaller Z_{eff}.
Br has a larger sixth ionization energy than Se because of a larger Z_{eff}.

6.56 (a) $1s^2 2s^2 2p^6 3s^2 3p^3$ is P (b) $1s^2 2s^2 2p^6 3s^2 3p^6$ is Ar (c) $1s^2 2s^2 2p^6 3s^2 3p^6 4s^2$ is Ca
Ar has the highest E_{i2}. Ar has a higher Z_{eff} than P. The 4s electrons in Ca are easier to remove than any 3p electrons.
Ar has the lowest E_{i7}. It is difficult to remove 3p electrons from Ca, and it is difficult to remove 2p electrons from P.

6.57 (b) Cl has the highest E_{i1} and smallest E_{i4}.

6.58 The likely second row element is boron because it has three valence electrons. The large fourth ionization energy is from an inner shell 1s electron.

6.59 The likely second row element is beryllium because it has two valence electrons. The large third and fourth ionization energies are from inner shell 1s electrons.

6.60 The relationship between the electron affinity of a univalent cation and the ionization energy of the neutral atom is that they have the same magnitude but opposite signs.

6.61 The relationship between the ionization energy of a univalent anion and the electron affinity of the neutral atom is that they have the same magnitude but opposite signs.

6.62 Na^+ has a more negative electron affinity than either Na or Cl because of its positive charge.

6.63 Br would have a more negative electron affinity than Br^- because Br^- has no room in its valence shell for an additional electron.

6.64 Energy is usually released when an electron is added to a neutral atom but absorbed when an electron is removed from a neutral atom because of the positive Z_{eff}.

6.65 E_{i1} increases steadily across the periodic table from Group 1A to Group 8A because electrons are being removed from the same shell and Z_{eff} is increasing. The electron affinity increases irregularly from 1A to 7A and then falls dramatically for Group 8A because the additional electron goes into the next higher shell.

6.66 The electron-electron repulsion is large and Z_{eff} is low.

6.67 The 3p orbitals in P are half-filled. The electron affinity for Si is more negative because the added electron is going into an empty 3p orbital. The electron affinity for S is more negative because of a higher Z_{eff}.

Octet Rule, Ionic Bonds, and Lattice Energy (Sections 6.6–6.8)

6.68 (a) [Ne], N^{3-} (b) [Ar], Ca^{2+} (c) [Ar], S^{2-} (d) [Kr], Br^-

6.69 (a) Mg^{2+} and Cl^-, $MgCl_2$, magnesium chloride
(b) Ca^{2+} and O^{2-}, CaO, calcium oxide
(c) Li^+ and N^{3-}, Li_3N, lithium nitride
(d) Al^{3+} and O^{2-}, Al_2O_3, aluminum oxide

6.70 (a) Because X reacts by losing electrons, it is likely to be a metal.
(b) Because Y reacts by gaining electrons, it is likely to be a nonmetal.
(c) X_2Y_3
(d) X is likely to be in group 3A and Y is likely to be in group 6A.

6.71 (a) Because X reacts by losing electrons, it is likely to be a metal.
(b) Because Y reacts by gaining electrons, it is likely to be a nonmetal.
(c) XY_2
(d) X is likely to be in group 2A and Y is likely to be in group 7A.

6.72 $MgCl_2$ > LiCl > KCl > KBr

6.73 $AlBr_3$ > CaO > $MgBr_2$ > LiBr

6.74 $Li \rightarrow Li^+ + e^-$ +520 kJ/mol
$Br + e^- \rightarrow Br^-$ $\underline{-325 \text{ kJ/mol}}$
+195 kJ/mol

6.75 The total energy = (376 kJ/mol) + (–349 kJ/mol) = +27 kJ/mol, which is unfavorable because it is positive.

6.76 $Li(s) \rightarrow Li(g)$ +159.4 kJ/mol

$Li(s) \rightarrow Li(g) + e^-$ +520 kJ/mol

$\frac{1}{2}[Br_2(l) \rightarrow Br_2(g)]$ +15.4 kJ/mol

$\frac{1}{2}[Br_2(g) \rightarrow 2\,Br(g)]$ +112 kJ/mol

$Br(g) + e^- \rightarrow Br^-(g)$ −325 kJ/mol

$Li^+(g) + Br^-(g) \rightarrow LiBr(s)$ −807 kJ/mol

Sum = −325 kJ/mol for $Li(s) + \frac{1}{2}Br_2(l) \rightarrow LiBr(s)$

6.77 (a) $Li(s) \rightarrow Li(g)$ +159.4 kJ/mol

$Li(g) \rightarrow Li^+(g) + e^-$ +520 kJ/mol

$\frac{1}{2}[F_2(g) \rightarrow 2\,F(g)]$ +79 kJ/mol

$F(g) + e^- \rightarrow F^-(g)$ −328 kJ/mol

$Li^+(g) + F^-(g) \rightarrow LiF(s)$ −1036 kJ/mol

Sum = −606 kJ/mol for $Li(s) + \frac{1}{2}F_2(g) \rightarrow LiF(s)$

(b) $Ca(s) \rightarrow Ca(g)$ +178.2 kJ/mol

$Ca(g) \rightarrow Ca^+(g) + e^-$ +589.8 kJ/mol

$Ca^+(g) \rightarrow Ca^{2+}(g) + e^-$ +1145 kJ/mol

$F_2(g) \rightarrow 2\,F(g)$ +158 kJ/mol

$2[F(g) + e^- \rightarrow F^-(g)]$ 2(−328) kJ/mol

$Ca^{2+}(g) + 2\,F^- \rightarrow CaF_2(s)$ −2630 kJ/mol

Sum = −1215 kJ/mol for $Ca(s) + F_2(g) \rightarrow CaF_2(s)$

6.78 $Na(s) \rightarrow Na(g)$ +107.3 kJ/mol

$Na(g) \rightarrow Na^+(g) + e^-$ +495.8 kJ/mol

$\frac{1}{2}[H_2(g) \rightarrow 2\,H(g)]$ ½(+435.9) kJ/mol

$H(g) + e^- \rightarrow H^-(g)$ −72.8 kJ/mol

$Na^+(g) + H^-(g) \rightarrow NaH(s)$ −U

Sum = −60 kJ/mol for $Na(s) + \frac{1}{2}H_2(g) \rightarrow NaH(s)$

$-U = -60 - 107.3 - 495.8 - 435.9/2 + 72.8 = -808$ kJ/mol; $U = 808$ kJ/mol

6.79 $Ca(s) \rightarrow Ca(g)$ +178.2 kJ/mol

$Ca(g) \rightarrow Ca^+(g) + e^-$ +589.8 kJ/mol

$Ca^+(g) \rightarrow Ca^{2+}(g) + e^-$ +1145 kJ/mol

$H_2(g) \rightarrow 2\,H(g)$ +435.9 kJ/mol

$2[H(g) + e^- \rightarrow H^-(g)]$ 2(−72.8) kJ/mol

$Ca^{2+}(g) + 2\,H^-(g) \rightarrow CaH_2(s)$ −U

Sum = −186.2 kJ/mol for $Ca(s) + H_2(g) \rightarrow CaH_2(s)$

$-U = -186.2 - 178.2 - 589.8 - 1145 - 435.9 + 2(72.8) = -2390$ kJ/mol; $U = 2390$ kJ/mol

6.80 $Cs(s) \rightarrow Cs(g)$ +76.1 kJ/mol

$Cs(g) \rightarrow Cs^+(g) + e^-$ +375.7 kJ/mol

$\frac{1}{2}[F_2(g) \rightarrow 2\,F(g)]$ +79 kJ/mol

$F(g) + e^- \rightarrow F^-(g)$ −328 kJ/mol

$Cs^+(g) + F^-(g) \rightarrow CsF(s)$ −740 kJ/mol

Sum = −537 kJ/mol for $Cs(s) + \frac{1}{2}F_2(g) \rightarrow CsF(s)$

6.81

$Cs(s) \rightarrow Cs(g)$	+76.1 kJ/mol
$Cs(g) \rightarrow Cs^{+}(g) + e^{-}$	+375.7 kJ/mol
$Cs^{+}(g) \rightarrow Cs^{2+}(g) + e^{-}$	+2422 kJ/mol
$F_2(g) \rightarrow 2\ F(g)$	+158 kJ/mol
$2[F(g) + e^{-} \rightarrow F^{-}(g)]$	2(−328) kJ/mol
$Cs^{2+}(g) + 2\ F^{-}(g) \rightarrow CsF_2(s)$	−2347 kJ/mol

Sum = +29 kJ/mol for $Cs(s) + F_2(g) \rightarrow CsF_2(s)$

The overall reaction absorbs 29 kJ/mol.
In the reaction of cesium with fluorine, CsF will form because the overall energy for the formation of CsF is negative, whereas it is positive for CsF_2.

6.82

$Ca(s) \rightarrow Ca(g)$	+178.2 kJ/mol
$Ca(g) \rightarrow Ca^{+}(g) + e^{-}$	+589.8 kJ/mol
$\frac{1}{2}[Cl_2(g) \rightarrow 2\ Cl(g)]$	+121.5 kJ/mol
$Cl(g) + e^{-} \rightarrow Cl^{-}(g)$	−348.6 kJ/mol
$Ca^{+}(g) + Cl^{-}(g) \rightarrow CaCl(s)$	−717 kJ/mol

Sum = −176 kJ/mol for $Ca(s) + \frac{1}{2}\ Cl_2(g) \rightarrow CaCl(s)$

6.83

$Ca(s) \rightarrow Ca(g)$	+178.2 kJ/mol
$Ca(g) \rightarrow Ca^{+}(g) + e^{-}$	+589.8 kJ/mol
$Ca^{+}(g) \rightarrow Ca^{2+}(g) + e^{-}$	+1145 kJ/mol
$Cl_2(g) \rightarrow 2\ Cl(g)$	+243 kJ/mol
$2[Cl(g) + e^{-} \rightarrow Cl^{-}(g)]$	2(−348.6) kJ/mol
$Ca^{2+}(g) + 2\ Cl^{-}(g) \rightarrow CaCl_2(s)$	−2258 kJ/mol

Sum = −799 kJ/mol for $Ca(s) + Cl_2(g) \rightarrow CaCl_2(s)$

In the reaction of calcium with chlorine, $CaCl_2$ will form because the overall energy for the formation of $CaCl_2$ is much more negative than for the formation of CaCl.

6.84

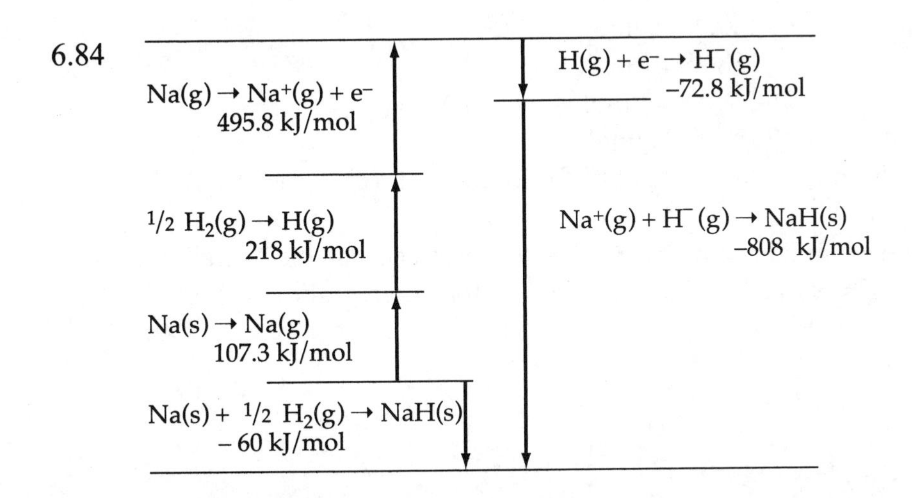

6.85

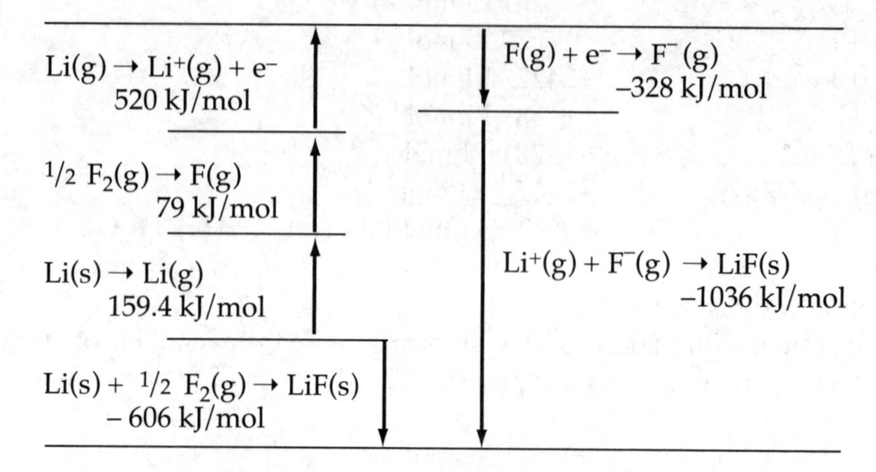

Chapter Problems

6.86 Cu^{2+} has fewer electrons and a larger effective nuclear charge; therefore it has the smaller ionic radius.

6.87 $S^{2-} > Ca^{2+} > Sc^{3+} > Ti^{4+}$, Z_{eff} increases on going from S^{2-} to Ti^{4+}.

6.88

$Mg(s) \rightarrow Mg(g)$	+147.7	kJ/mol
$Mg(g) \rightarrow Mg^{+}(g) + e^{-}$	+737.7	kJ/mol
$\frac{1}{2} F_2(g) \rightarrow F(g)$	+79	kJ/mol
$F(g) + e^{-} \rightarrow F^{-}(g)$	−328	kJ/mol
$Mg^{+}(g) + F^{-}(g) \rightarrow MgF(s)$	−930	kJ/mol
Sum =	−294	kJ/mol for $Mg(s) + \frac{1}{2} F_2(g) \rightarrow MgF(s)$

$Mg(s) \rightarrow Mg(g)$	+147.7	kJ/mol
$Mg(g) \rightarrow Mg^{+}(g) + e^{-}$	+737.7	kJ/mol
$Mg^{+}(g) \rightarrow Mg^{2+}(g) + e^{-}$	+1450.7	kJ/mol
$F_2(g) \rightarrow 2 F(g)$	+158	kJ/mol
$2[F(g) + e^{-} \rightarrow F^{-}(g)]$	2(−328)	kJ/mol
$Mg^{2+}(g) + 2 F^{-}(g) \rightarrow MgF_2(s)$	−2952	kJ/mol
Sum =	−1114	kJ/mol for $Mg(s) + F_2(g) \rightarrow MgF_2(s)$

In the reaction of magnesium with fluorine, MgF_2 will form because the overall energy for the formation of MgF_2 is much more negative than for the formation of MgF.

6.89

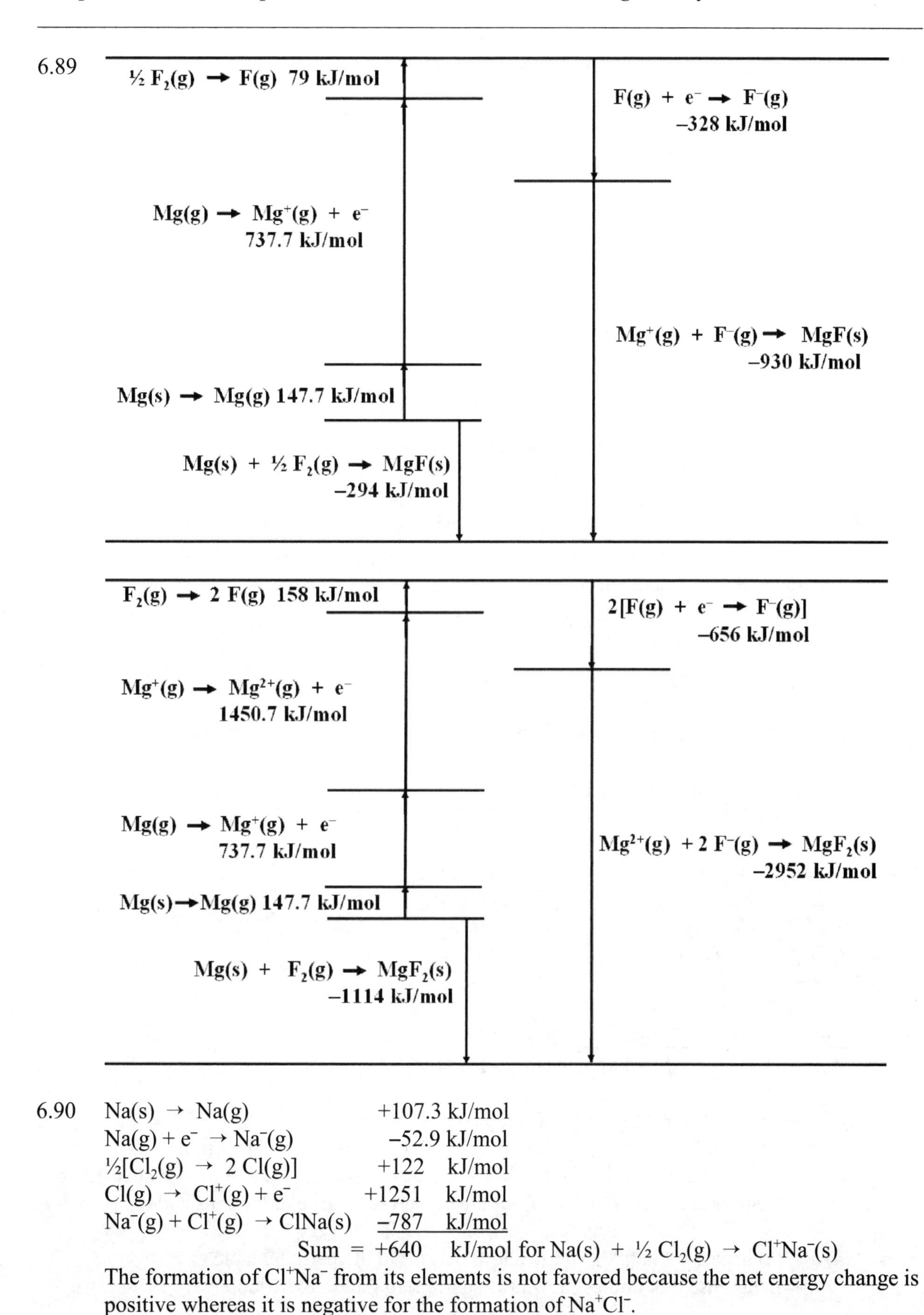

6.90

Na(s) → Na(g)	+107.3 kJ/mol
Na(g) + e^- → Na^-(g)	–52.9 kJ/mol
½[Cl_2(g) → 2 Cl(g)]	+122 kJ/mol
Cl(g) → Cl^+(g) + e^-	+1251 kJ/mol
Na^-(g) + Cl^+(g) → ClNa(s)	–787 kJ/mol
Sum =	+640 kJ/mol for Na(s) + ½ Cl_2(g) → Cl^+Na^-(s)

The formation of Cl^+Na^- from its elements is not favored because the net energy change is positive whereas it is negative for the formation of Na^+Cl^-.

6.91

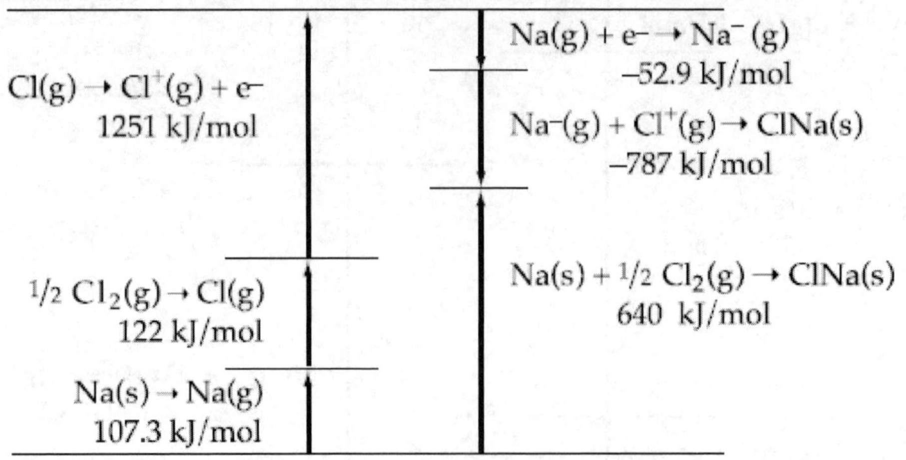

$Na(g) + e^- \rightarrow Na^-(g)$
–52.9 kJ/mol

$Cl(g) \rightarrow Cl^+(g) + e^-$
1251 kJ/mol

$Na^-(g) + Cl^+(g) \rightarrow ClNa(s)$
–787 kJ/mol

$Na(s) + 1/2\ Cl_2(g) \rightarrow ClNa(s)$
640 kJ/mol

$1/2\ Cl_2(g) \rightarrow Cl(g)$
122 kJ/mol

$Na(s) \rightarrow Na(g)$
107.3 kJ/mol

6.92 When moving diagonally down and right on the periodic table, the increase in atomic radius caused by going to a larger shell is offset by a decrease caused by a higher Z_{eff}. Thus, there is little net change in the charge density.

6.93 (a) Assume a 100.0 g sample. From the percent composition data, a 100.0 g sample contains 57.67 g Cs and 42.33 g Pt.

$$57.67 \text{ g Cs x } \frac{1 \text{ mol Cs}}{132.9 \text{ g Cs}} = 0.4339 \text{ mol Cs}$$

$$42.33 \text{ g Pt x } \frac{1 \text{ mol Pt}}{195.1 \text{ g Pt}} = 0.2170 \text{ mol Pt}$$

$Cs_{0.4339}Pt_{0.2170}$; divide each subscript by the smaller, 0.2170.
$Cs_{0.4339\,/\,0.2170}Pt_{0.2170\,/\,0.2170}$
The empirical formula is Cs_2Pt.
(b) Cs^+, [Xe]   (c) Pt^{-2}, [Xe] $6s^2\ 4f^{14}\ 5d^{10}$

6.94

$Mg(s) \rightarrow Mg(g)$	+147.7 kJ/mol
$Mg(g) \rightarrow Mg^+(g) + e^-$	+738 kJ/mol
$Mg^+(g) \rightarrow Mg^{2+}(g) + e^-$	+1451 kJ/mol
$½[O_2(g) \rightarrow 2\ O(g)]$	+249.2 kJ/mol
$O(g) + e^- \rightarrow O^-(g)$	–141.0 kJ/mol
$O^-(g) + e^- \rightarrow O^{2-}(g)$	E_{ea2}
$Mg^{2+}(g) + O^{2-}(g) \rightarrow MgO(s)$	–3791 kJ/mol
$Mg(s) + ½O_2(g) \rightarrow MgO(s)$	–601.7 kJ/mol

$147.7 + 738 + 1451 + 249.2 - 141.0 + E_{ea2} - 3791 = -601.7$
$E_{ea2} = -147.7 - 738 - 1451 - 249.2 + 141.0 + 3791 - 601.7 = +744$ kJ/mol
Because E_{ea2} is positive, O^{2-} is not stable in the gas phase. It is stable in MgO because of the large lattice energy that results from the +2 and –2 charge of the ions and their small size.

6.95 (a) (i) Ra because it is farthest down (7th period) in the periodic table.
(ii) In because it is farthest down (5th period) in the periodic table.

(b) (i) Tl and Po are farthest down (6th period) but Tl is larger because it is to the left of Po and thus has the smaller ionization energy.
(ii) Cs and Bi are farthest down (6th period) but Cs is larger because it is to the left of Bi and thus has the smaller ionization energy.

6.96 (a) The more negative the E_{ea}, the greater the tendency of the atom to accept an electron, and the more stable the anion that results. Be, N, O, and F are all second row elements. F has the most negative E_{ea} of the group because the anion that forms, F^-, has a complete octet of electrons and its nucleus has the highest effective nuclear charge.
(b) Se^{2-} and Rb^+ are below O^{2-} and F^- in the periodic table and are the larger of the four. Se^{2-} and Rb^+ are isoelectronic, but Rb^+ has the higher effective nuclear charge so it is smaller. Therefore Se^{2-} is the largest of the four ions.

6.97

$Ca(s) \rightarrow Ca(g)$	+178 kJ/mol
$Ca(g) \rightarrow Ca^+(g)$	+590 kJ/mol
$Ca^+(g) \rightarrow Ca^{2+}(g)$	+1145 kJ/mol
$2\ C(s) \rightarrow 2\ C(g)$	2(+717 kJ/mol)
$2\ C(g) \rightarrow C_2(g)$	–614 kJ/mol
$C_2(g) \rightarrow C_2^-(g)$	–315 kJ/mol
$C_2^-(g) \rightarrow C_2^{2-}(g)$	+410 kJ/mol
$Ca^{2+}(g) + C_2^{2-}(g) \rightarrow CaC_2(s)$	–U
$Ca(s) + 2\ C(s) \rightarrow CaC_2(s)$	–60 kJ/mol

$-U = -60 - 178 - 590 - 1145 - 2(717) + 614 + 315 - 410 = -2888$ kJ/mol
$U = 2888$ kJ/mol

6.98

$Cr(s) \rightarrow Cr(g)$	+397 kJ/mol
$Cr(g) \rightarrow Cr^+(g)$	+652 kJ/mol
$Cr^+(g) \rightarrow Cr^{2+}(g)$	+1588 kJ/mol
$Cr^{2+}(g) \rightarrow Cr^{3+}(g)$	+2882 kJ/mol
$\frac{1}{2}(I_2(s) \rightarrow I_2(g))$	+62/2 kJ/mol
$\frac{1}{2}(I_2(g) \rightarrow 2\ I(g))$	+151/2 kJ/mol
$I(g) + e^- \rightarrow I^-(g)$	–295 kJ/mol
$Cl_2(g) \rightarrow 2\ Cl(g)$	+243 kJ/mol
$2(Cl(g) + e^- \rightarrow Cl^-(g))$	2(–349) kJ/mol
$Cr^{3+}(g) + 2\ Cl^-(g) + I^-(g) \rightarrow CrCl_2I(s)$	–U
$Cr(s) + Cl_2(g) + \frac{1}{2} I_2(g) \rightarrow CrCl_2I(s)$	– 420 kJ/mol

$-U = -420 - 397 - 652 - 1588 - 2882 - 62/2 - 151/2 + 295 - 243 + 2(349) = -5295.5$ kJ/mol
$U = 5295$ kJ/mol

Multiconcept Problems

6.99 (a) $E = (703\ kJ/mol)(1000\ J/1\ kJ)/(6.022 \times 10^{23}\ photons/mol) = 1.17 \times 10^{-18}$ J/photon

$$E = \frac{hc}{\lambda}$$

$$\lambda = \frac{hc}{E} = \frac{(6.626 \times 10^{-34}\ J\cdot s)(3.00 \times 10^{8}\ m/s)}{1.17 \times 10^{-18}\ J} = 1.70 \times 10^{-7}\ m = 170 \times 10^{-9}\ m = 170\ nm$$

(b) Bi [Xe] $6s^2\,4f^{14}\,5d^{10}\,6p^3$
Bi^+ [Xe] $6s^2\,4f^{14}\,5d^{10}\,6p^2$

(c) n = 6, l = 1

(d) Element 115 would be directly below Bi in the periodic table. The valence electron is farther from the nucleus and less strongly held than in Bi. The ionization energy for element 115 would be less than that for Bi.

6.100 (a) Fe [Ar] $4s^2\,3d^6$
Fe^{2+} [Ar] $3d^6$
Fe^{3+} [Ar] $3d^5$

(b) A 3d electron is removed on going from Fe^{2+} to Fe^{3+}. For the 3d electron, n = 3 and l = 2.

(c) $E(J/photon) = 2952\ kJ/mol \times \frac{1\ mol\ photons}{6.022 \times 10^{23}\ photons} \times \frac{1000\ J}{1\ kJ} = 4.90 \times 10^{-18}\ J/photon$

$$E = \frac{hc}{\lambda}$$

$$\lambda = \frac{hc}{E} = \frac{(6.626 \times 10^{-34}\ J\cdot s)(3.00 \times 10^{8}\ m/s)}{4.90 \times 10^{-18}\ J} = 4.06 \times 10^{-8}\ m = 40.6 \times 10^{-9}\ m = 40.6\ nm$$

(d) Ru is directly below Fe in the periodic table and the two metals have similar electron configurations. The electron removed from Ru to go from Ru^{2+} to Ru^{3+} is a 4d electron. The electron with the higher principal quantum number, n = 4, is farther from the nucleus, less tightly held, and requires less energy to remove.

6.101 (a) 58.4 nm = 58.4×10^{-9} m

$$E(photon) = 6.626 \times 10^{-34}\ J{\cdot}s \times \frac{3.00 \times 10^{8}\ m/s}{58.4 \times 10^{-9}\ m} \times \frac{1\ kJ}{1000\ J} \times \frac{6.022 \times 10^{23}}{mol} = 2050\ kJ/mol$$

$$E_K = E(electron) = \tfrac{1}{2}(9.109 \times 10^{-31}\ kg)(2.450 \times 10^{6}\ m/s)^2\left(\frac{1\ kJ}{1000\ J}\right)\left(\frac{6.022 \times 10^{23}}{mol}\right)$$

E_K = 1646 kJ/mol

$E(photon) = E_i + E_K$; $E_i = E(photon) - E_K = 2050 - 1646 = 404\ kJ/mol$

(b) 142 nm = 142×10^{-9} m

$$E(photon) = 6.626 \times 10^{-34}\ J\cdot s \times \frac{3.00 \times 10^{8}\ m/s}{142 \times 10^{-9}\ m} \times \frac{1\ kJ}{1000\ J} \times \frac{6.022 \times 10^{23}}{mol} = 843\ kJ/mol$$

$$E_K = E(electron) = \tfrac{1}{2}(9.109 \times 10^{-31}\ kg)(1.240 \times 10^{6}\ m/s)^2\left(\frac{1\ kJ}{1000\ J}\right)\left(\frac{6.022 \times 10^{23}}{mol}\right)$$

E_K = 422 kJ/mol

$E(photon) = E_i + E_K$; $E_i = E(photon) - E_K = 843 - 422 = 421\ kJ/mol$

7 Covalent Bonding and Electron-Dot Structures

7.1

(a) $SiCl_4$	chlorine	EN = 3.0	
	silicon	EN = 1.8	
		ΔEN = 1.2	The Si–Cl bond is polar covalent.
(b) CsBr	bromine	EN = 2.8	
	cesium	EN = 0.7	
		ΔEN = 2.1	The Cs^+Br^- bond is ionic.
(c) $FeBr_3$	bromine	EN = 2.8	
	iron	EN = 1.8	
		ΔEN = 1.0	The Fe–Br bond is polar covalent.
(d) CH_4	carbon	EN = 2.5	
	hydrogen	EN = 2.1	
		ΔEN = 0.4	The C–H bond is polar covalent.

7.2 H is positively polarized (blue). O is negatively polarized (red). This is consistent with the electronegativity values for O (3.5) and H (2.1). The more negatively polarized atom should be the one with the larger electronegativity.

7.3 (a) H:S:H (b) H / :Cl:C:Cl: / :Cl:

7.4 (a) OF_2 (b) $SiCl_4$

7.5 (a) H–C(H)(F)–F (b) :Cl–Al(Cl)–Cl: (c) $[O–Cl(O)(O)–O]^-$

(d) PCl_5: :Cl–P(Cl)(Cl)(Cl)–Cl: (e) $XeOF_4$: O=Xe(F)(F)(F)(F) (e) $[H–N(H)(H)–H]^+$

7.6 (a) is incorrect, the least electronegative atom (P) should be placed in the center.
(b) correct
(c) is incorrect, oxygen does not have a complete octet.

7.7 (a) :C≡O: (b) H—C≡N: (c) 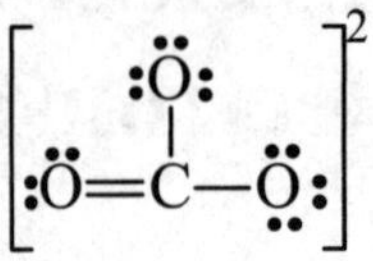 [:Ö: / | / :Ö=C—Ö:]$^{2-}$ (d) 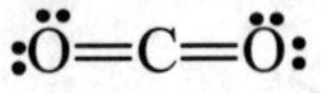 :Ö=C=Ö:

7.8 (a) correct
(b) is incorrect, the electron-dot structure has 34 electrons, but there are only 32 valence electrons in phosphate.
(c) correct
(d) is incorrect, the oxygen with a double bond does not have a complete octet and there are only 30 electrons in the electron-dot structure while there are 32 valence electrons.

Electron-dot structures are a model for bonding and it is possible to draw several structures that meet the criteria for a valid structure. If more than one electron-dot structure can be drawn, the molecule exhibits resonance.

7.9 (a) 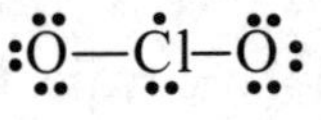 :Ö—Ċl—Ö: (b) 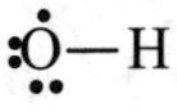 :Ȯ—H (c)

```
   H  H
   |  |
H—C—C•
   |  |
   H  H
```

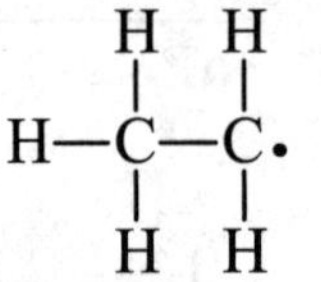

7.10 O_2^- is a radical and the presence of an unpaired electron leads to a highly reactive species.

7.11 (a)

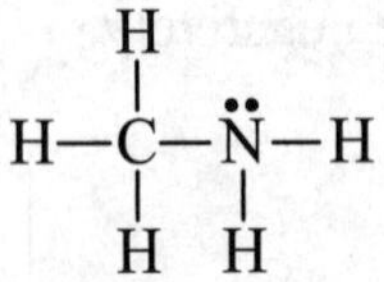

```
   H
   |  ..
H—C—N—H
   |  |
   H  H
```

(b)

```
   H  H
   |  |
H—C=C—H
```

(c)

```
   H  H  H
   |  |  |
H—C—C—C—H
   |  |  |
   H  H  H
```

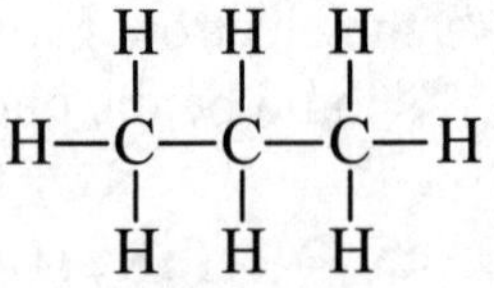

(d) 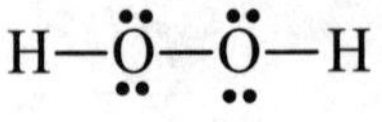 H—Ö—Ö—H

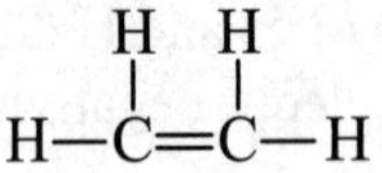

(e)

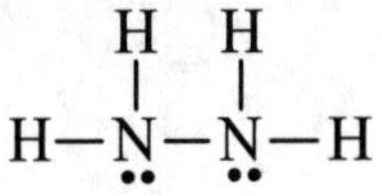

```
   H  H
   |  |
H—N—N—H
   ..  ..
```

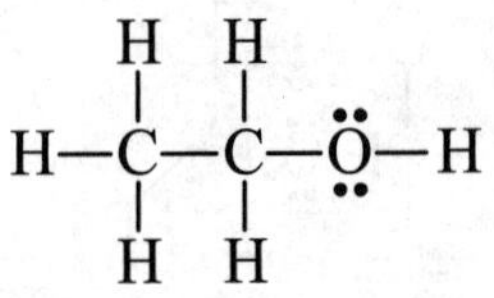

7.12

```
   H  H                   H      H
   |  |   ..              |  ..  |
H—C—C—O—H   and   H—C—O—C—H
   |  |   ..              |  ..  |
   H  H                   H      H
```

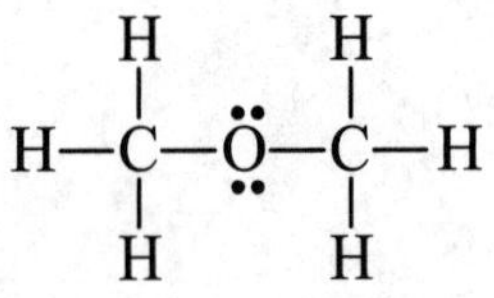

7.13 Molecular formula: $C_4H_5N_3O$

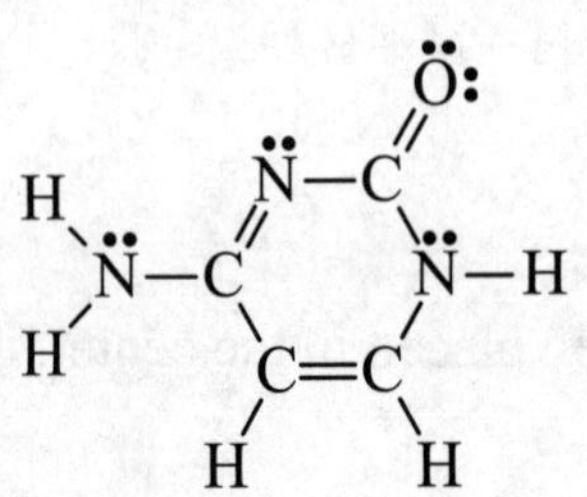

```
              ..
              O:
              //
         ..  /
         N—C
H       //    \ ..
 \ ..  /       N—H
  N—C         /
 /     \     /
H       C=C
       /    \
      H      H
```

7.14 (a)

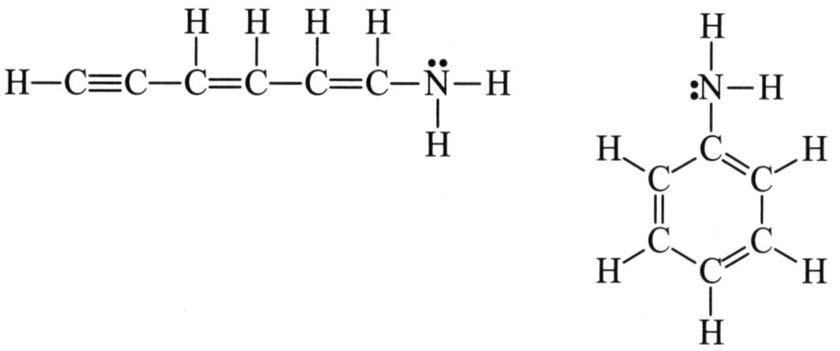

```
        H   H   H   H
        |   |   |   |   ..
H—C≡C—C═C—C═C—N—H
                  |
                  H
```

```
        H
        |
       :N—H
        |
  H   C    H
    C    C
    ‖    |
    C    C
  H   C    H
      |
      H
```

(b)

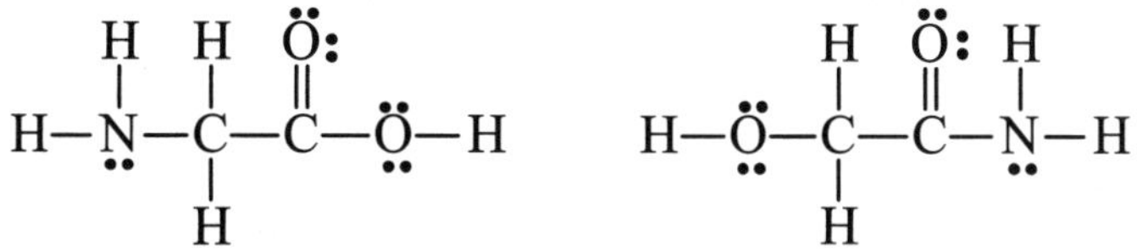

```
    H   H   Ö:                H   Ö:  H
    |   |   ‖                 |   ‖   |
H—N—C—C—Ö—H          H—Ö—C—C—N—H
  ..  |     ..              ..  |     ..
      H                         H
```

7.15 (a)

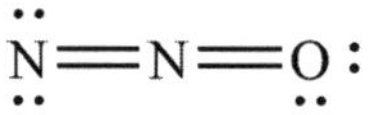

```
..
N═N═O:
..    ..
```

(b)

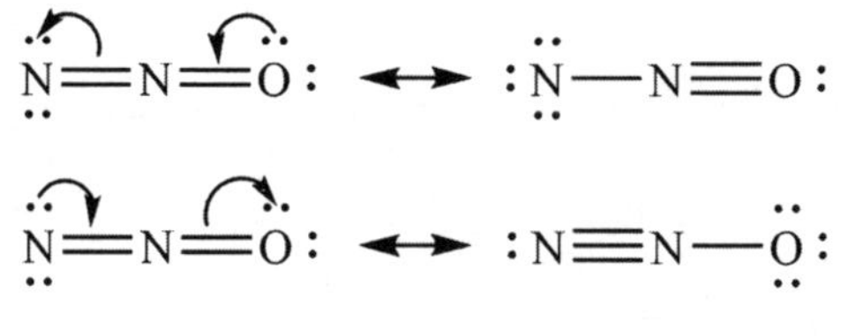

```
 ..           ..
 N═N═O:  ←→  :N—N≡O:
 ..           ..

 ..                  ..
 N═N═O:  ←→  :N≡N—O:
 ..                  ..
```

7.16 (a)

```
     ..                 ..
     S                  S
:O       O:   ←→   :O      O:
 ..      ..          ..     ..
```

(b)

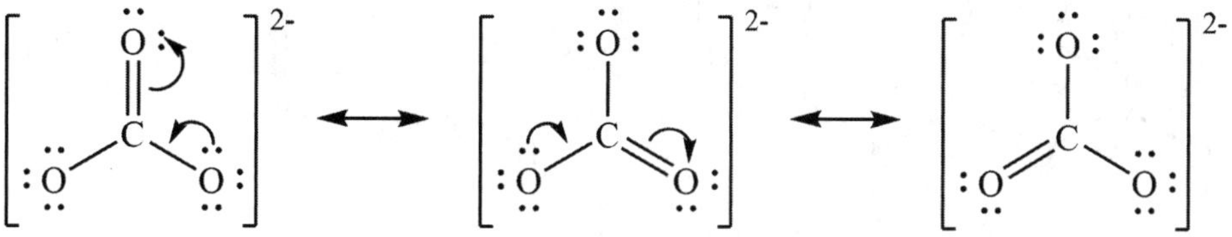

```
[    ..       ]2-      [   :O:      ]2-      [   :O:      ]2-
[    O:       ]        [    |       ]        [    |       ]
[    ‖        ]  ←→    [    C       ]  ←→    [    C       ]
[    C        ]        [ :O    O:   ]        [ :O    O:   ]
[ :O    O:    ]        [  ..   ..   ]        [  ..   ..   ]
[  ..   ..    ]
```

(c)

```
  ..            ..            ..            ..
 :F:           :F:           :F:           :F
  |             |             |             ‖
  B     ←→      B     ←→      B     ←→      B
:F:  :F:      :F:   F:      :F   :F:      :F:  :F:
 ..   ..       ..   ..       ..   ..       ..   ..
```

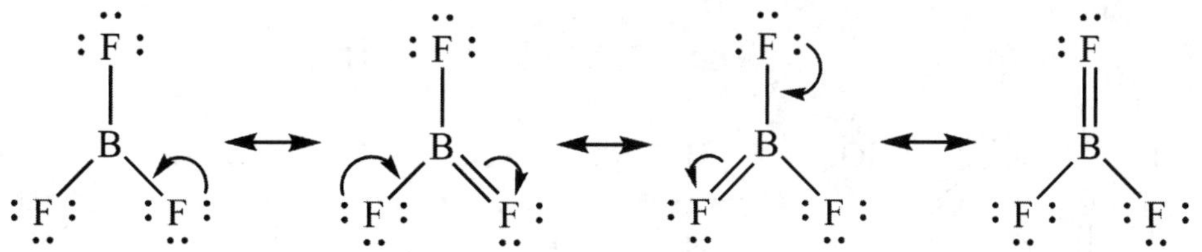

7.17 (a)

```
        H                     ..
        |       H   H         O:
        N                     ‖
H—C          ..     C         C     ..   H
    ‖        C          C          O
    N═C            :N       H      ..
    ..  \          /  \
          H       H    H
```

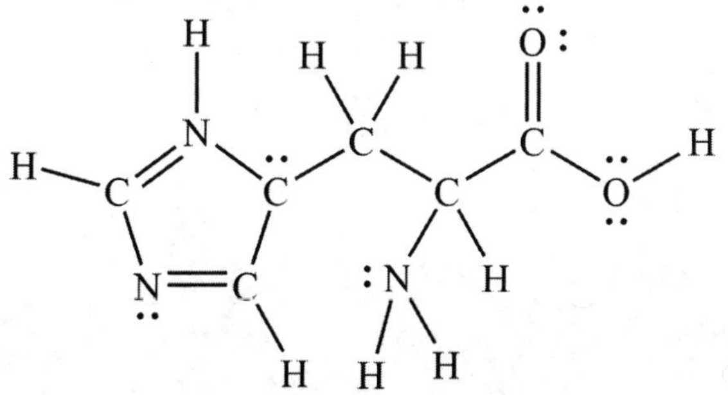

This is a valid resonance structure.

(b)

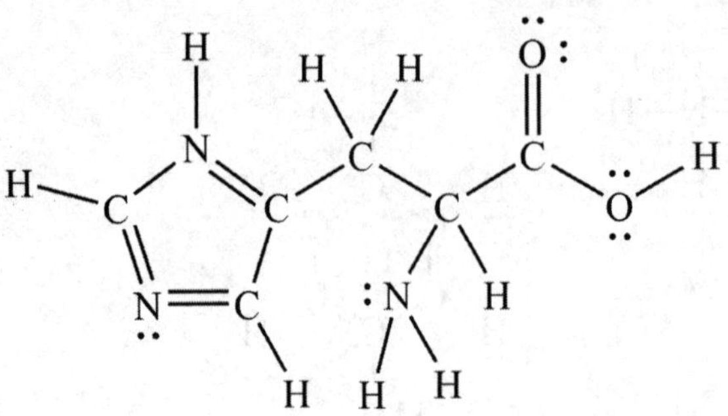

This is not a valid resonance structure. The N with two double bonds has 10 electrons, not 8.

7.18 (a)

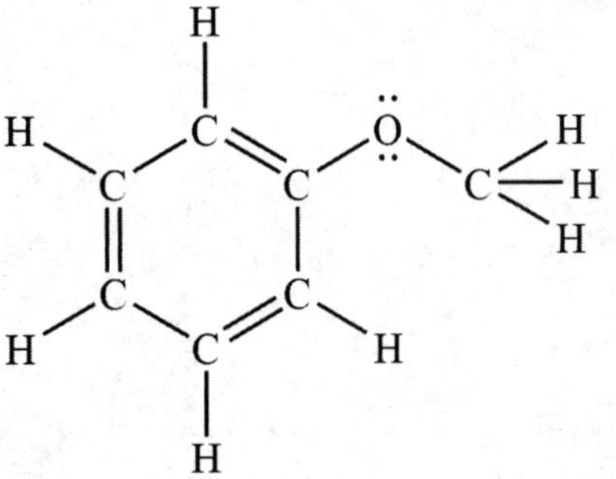

(b)

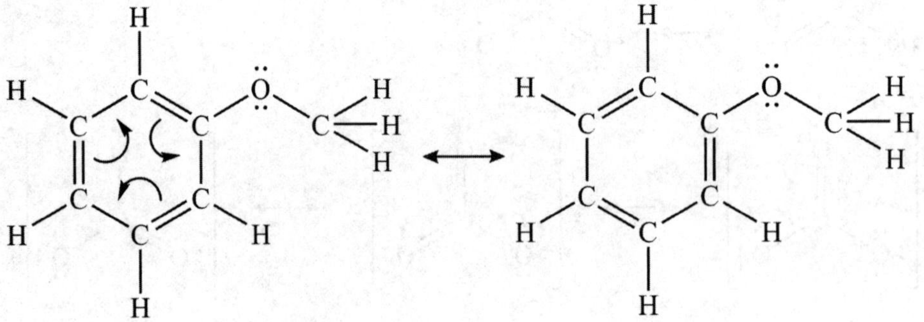

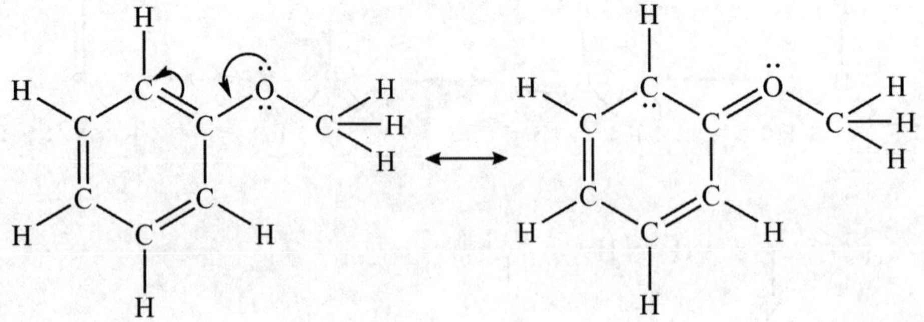

7.19 (a) $\left[:\ddot{N}=C=\ddot{O}:\right]^-$

For nitrogen:	Isolated nitrogen valence electrons	5
	Bound nitrogen bonding electrons	4
	Bound nitrogen nonbonding electrons	4
	Formal charge = 5 – ½(4) – 4 = –1	
For carbon:	Isolated carbon valence electrons	4
	Bound carbon bonding electrons	8
	Bound carbon nonbonding electrons	0
	Formal charge = 4 – ½ (8) – 0 = 0	
For oxygen:	Isolated oxygen valence electrons	6
	Bound oxygen bonding electrons	4
	Bound oxygen nonbonding electrons	4
	Formal charge = 6 – ½(4) – 4 = 0	

(b) $:\ddot{\underset{\cdot\cdot}{O}}-\ddot{O}=\ddot{O}:$

For left oxygen:	Isolated oxygen valence electrons	6
	Bound oxygen bonding electrons	2
	Bound oxygen nonbonding electrons	6
	Formal charge = 6 – ½(2) – 6 = –1	
For central oxygen:	Isolated oxygen valence electrons	6
	Bound oxygen bonding electrons	6
	Bound oxygen nonbonding electrons	2
	Formal charge = 6 – ½(6) – 2 = +1	
For right oxygen:	Isolated oxygen valence electrons	6
	Bound oxygen bonding electrons	4
	Bound oxygen nonbonding electrons	4
	Formal charge = 6 – ½(4) – 4 = 0	

7.20 (a) $\left[\ddot{\underset{\cdot\cdot}{N}}=C=\ddot{\underset{\cdot\cdot}{O}}\right]^- \longleftrightarrow \left[:N\equiv C-\ddot{\underset{\cdot\cdot}{O}}:\right]^-$

$\left[\ddot{\underset{\cdot\cdot}{N}}=C=\ddot{\underset{\cdot\cdot}{O}}\right]^- \longleftrightarrow \left[:\ddot{\underset{\cdot\cdot}{N}}-C\equiv O:\right]^-$

(b) $\left[\ddot{\underset{\cdot\cdot}{N}}=C=\ddot{\underset{\cdot\cdot}{O}}\right]^-$

For nitrogen:	Isolated nitrogen valence electrons	5
	Bound nitrogen bonding electrons	4
	Bound nitrogen nonbonding electrons	4
	Formal charge = 5 – ½(4) – 4 = –1	

For carbon:	Isolated carbon valence electrons	4
	Bound carbon bonding electrons	8
	Bound carbon nonbonding electrons	0
	Formal charge = 4 – ½(8) – 0 = 0	
For oxygen:	Isolated oxygen valence electrons	6
	Bound oxygen bonding electrons	4
	Bound oxygen nonbonding electrons	4
	Formal charge = 6 – ½(4) – 4 = 0	

$$\left[:\mathrm{N}\equiv\mathrm{C}-\underset{\cdot\cdot}{\ddot{\mathrm{O}}}:\right]^{-}$$

For nitrogen:	Isolated nitrogen valence electrons	5
	Bound nitrogen bonding electrons	6
	Bound nitrogen nonbonding electrons	2
	Formal charge = 5 – ½(6) – 2 = 0	
For carbon:	Isolated carbon valence electrons	4
	Bound carbon bonding electrons	8
	Bound carbon nonbonding electrons	0
	Formal charge = 4 – ½(8) – 0 = 0	
For oxygen:	Isolated oxygen valence electrons	6
	Bound oxygen bonding electrons	2
	Bound oxygen nonbonding electrons	6
	Formal charge = 6 – ½(2) – 6 = –1	

$$\left[:\underset{\cdot\cdot}{\ddot{\mathrm{N}}}-\mathrm{C}\equiv\mathrm{O}:\right]^{-}$$

For nitrogen:	Isolated nitrogen valence electrons	5
	Bound nitrogen bonding electrons	2
	Bound nitrogen nonbonding electrons	6
	Formal charge = 5 – ½(2) – 6 = –2	
For carbon:	Isolated carbon valence electrons	4
	Bound carbon bonding electrons	8
	Bound carbon nonbonding electrons	0
	Formal charge = 4 – ½(8) – 0 = 0	
For oxygen:	Isolated oxygen valence electrons	6
	Bound oxygen bonding electrons	6
	Bound oxygen nonbonding electrons	2
	Formal charge = 6 – ½(6) – 2 = +1	

$$[\text{N}{=}\text{C}{=}\text{O}]^- \longleftrightarrow [\text{:N}{\equiv}\text{C}{-}\text{O:}]^- \longleftrightarrow [\text{:}\overset{-2}{\text{N}}{-}\text{C}{\equiv}\overset{+}{\text{O}}\text{:}]^-$$

The first two structures make the largest contribution to the resonance hybrid because the –1 formal charge is on either of the electronegative N or O. The third structure has a +1 formal charge on O and does not significantly contribute to the resonance hybrid.
(c) Carbon–nitrogen because of the triple bond contribution to the resonance hybrid.

7.21

$$\text{H}_3\text{C}{-}\text{C}({=}\text{O}){-}\text{O}{-}\text{H}$$

All atoms in this structure have 0 formal charge.

$$\text{H}_3\text{C}{-}\text{C}({-}\overset{-1}{\text{O}}){=}\overset{+1}{\text{O}}{-}\text{H}$$

For top oxygen:	Isolated oxygen valence electrons	6
	Bound oxygen bonding electrons	2
	Bound oxygen nonbonding electrons	6
	Formal charge = 6 – ½(2) – 6 = –1	

For right oxygen:	Isolated oxygen valence electrons	6
	Bound oxygen bonding electrons	6
	Bound oxygen nonbonding electrons	2
	Formal charge = 6 – ½(6) – 2 = +1	

All the other atoms in this structure have 0 formal charge.

The structure without formal charges makes a larger contribution to the resonance hybrid because energy is required to separate + and – charges. Thus, the actual electronic structure of acetic acid is closer to that of the more favorable, lower energy structure.

7.22

All atoms in structures 1 and 2 have 0 formal charge.

In structure 3:

For carbon:	Isolated carbon valence electrons	4
	Bound carbon bonding electrons	6
	Bound carbon nonbonding electrons	2
	Formal charge = $4 - \frac{1}{2}(6) - 2 = -1$	
For oxygen:	Isolated oxygen valence electrons	6
	Bound oxygen bonding electrons	6
	Bound oxygen nonbonding electrons	2
	Formal charge = $6 - \frac{1}{2}(6) - 2 = +1$	

All the other atoms in structure 3 have 0 formal charge.

The structures 1 and 2 without formal charges make the larger contribution to the resonance hybrid because energy is required to separate + and – charges.

7.23 $\Delta EN(P{=}O) = 1.4$ and $\Delta EN(P{=}S) = 0.4$. Both bonds are polar covalent, but the phosphorus-oxygen bond is more polar.

7.24 For the reaction between an organophosphate insecticide and the enzyme to occur, the phosphorus atom must bear a positive charge. Greater positive charge leads to increased rate of reaction and increased toxicity of the insecticide. The more electronegative the X group, the more positive the phosphorus.
(a) Cl (b) CF_3

7.25 Phosphorus is in the third row of the periodic table and can utilize d orbitals to hold extra electrons, and therefore form more than four bonds.

7.26 $C_{11}H_{19}N_2PSO_3$

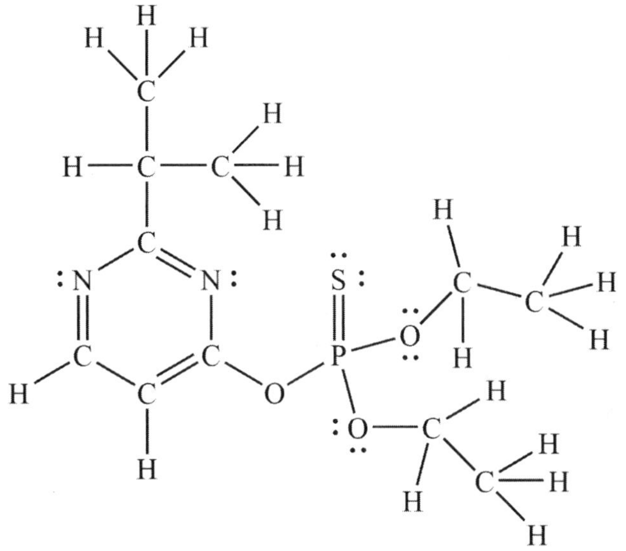

7.27

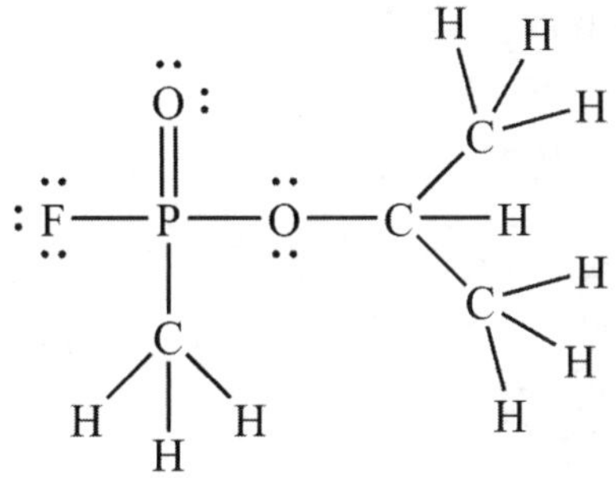

For phosphorus:	Isolated phosphorus valence electrons	5
	Bound phosphorus bonding electrons	10
	Bound phosphorus nonbonding electrons	0
	Formal charge = 5 – ½(10) – 0 = 0	
For top oxygen:	Isolated oxygen valence electrons	6
	Bound oxygen bonding electrons	4
	Bound oxygen nonbonding electrons	4
	Formal charge = 6 – ½(4) – 4 = 0	
For right oxygen:	Isolated oxygen valence electrons	6
	Bound oxygen bonding electrons	4
	Bound oxygen nonbonding electrons	4
	Formal charge = 6 – ½(4) – 4 = 0	
For fluorine:	Isolated fluorine valence electrons	7
	Bound fluorine bonding electrons	2
	Bound fluorine nonbonding electrons	6
	Formal charge = 7 – ½(2) – 6 = 0	

For carbon:	Isolated carbon valence electrons	4
	Bound carbon bonding electrons	8
	Bound carbon nonbonding electrons	0
	Formal charge = 4 – ½(8) – 0 = 0	

7.28 (a)

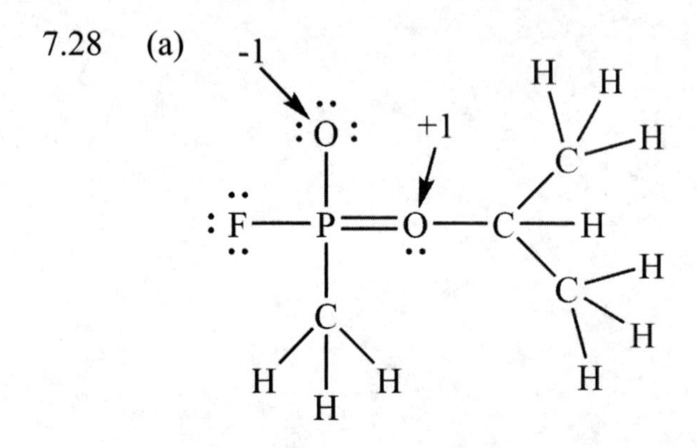

For fluorine:	Isolated fluorine valence electrons	7
	Bound fluorine bonding electrons	2
	Bound fluorine nonbonding electrons	6
	Formal charge = 7 – ½(2) – 6 = 0	
For phosphorus:	Isolated phosphorus valence electrons	5
	Bound phosphorus bonding electrons	10
	Bound phosphorus nonbonding electrons	0
	Formal charge = 5 – ½(10) – 0 = 0	
For top oxygen:	Isolated oxygen valence electrons	6
	Bound oxygen bonding electrons	2
	Bound oxygen nonbonding electrons	6
	Formal charge = 6 – ½(2) – 6 = –1	
For right oxygen:	Isolated oxygen valence electrons	6
	Bound oxygen bonding electrons	6
	Bound oxygen nonbonding electrons	2
	Formal charge = 6 – ½(6) – 2 = +1	

(b)

For fluorine:	Isolated fluorine valence electrons	7
	Bound fluorine bonding electrons	4
	Bound fluorine nonbonding electrons	4
	Formal charge = 7 – ½(4) – 4 = +1	
For phosphorus:	Isolated phosphorus valence electrons	5
	Bound phosphorus bonding electrons	12
	Bound phosphorus nonbonding electrons	0
	Formal charge = 5 – ½(12) – 0 = –1	
For top oxygen:	Isolated oxygen valence electrons	6
	Bound oxygen bonding electrons	4
	Bound oxygen nonbonding electrons	4
	Formal charge = 6 – ½(4) – 4 = 0	
For right oxygen:	Isolated oxygen valence electrons	6
	Bound oxygen bonding electrons	4
	Bound oxygen nonbonding electrons	4
	Formal charge = 6 – ½(4) – 4 = 0	

(c)

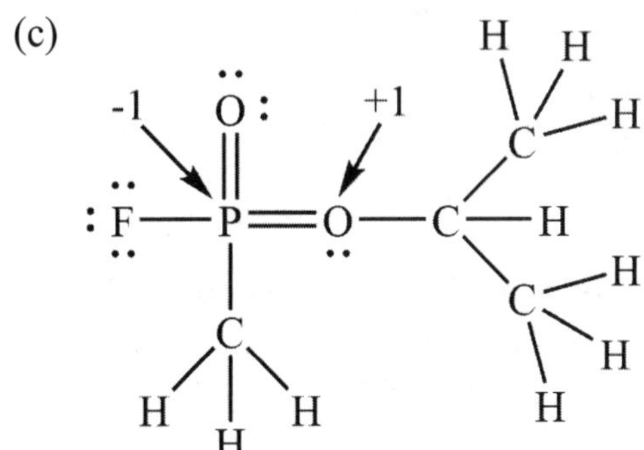

For fluorine:	Isolated fluorine valence electrons	7
	Bound fluorine bonding electrons	2
	Bound fluorine nonbonding electrons	6
	Formal charge = 7 – ½(2) – 6 = 0	
For phosphorus:	Isolated phosphorus valence electrons	5
	Bound phosphorus bonding electrons	12
	Bound phosphorus nonbonding electrons	0
	Formal charge = 5 – ½(12) – 0 = –1	
For top oxygen:	Isolated oxygen valence electrons	6
	Bound oxygen bonding electrons	4
	Bound oxygen nonbonding electrons	4
	Formal charge = 6 – ½(4) – 4 = 0	

For right oxygen:	Isolated oxygen valence electrons	6
	Bound oxygen bonding electrons	6
	Bound oxygen nonbonding electrons	2
	Formal charge = 6 – ½(6) – 2 = +1	

(d)

This structure is not valid because one carbon has five bonds.

The resonance structure in problem 7.27 is the major contributor to the resonance hybrid because formal charges on all atoms are zero.

Conceptual Problems

7.29 (a) A (b) D (c) B (d) C

7.30 C–D is the stronger bond. A–B is the longer bond.

7.31 As the electrostatic potential maps are drawn, the Li and Cl are at the tops of each map. The red area is for a negatively polarized region (associated with Cl). The blue area is for a positively polarized region (associated with Li). Map (a) is for CH_3Cl and Map (b) is for CH_3Li.

7.32 (a) fluoroethane (b) ethane (c) ethanol (d) acetaldehyde

7.33 (a) is ionic; (b) and (c) are covalent.

7.34 (a) $C_8H_9NO_2$

7.35 (a) $C_{13}H_{10}N_2O_4$

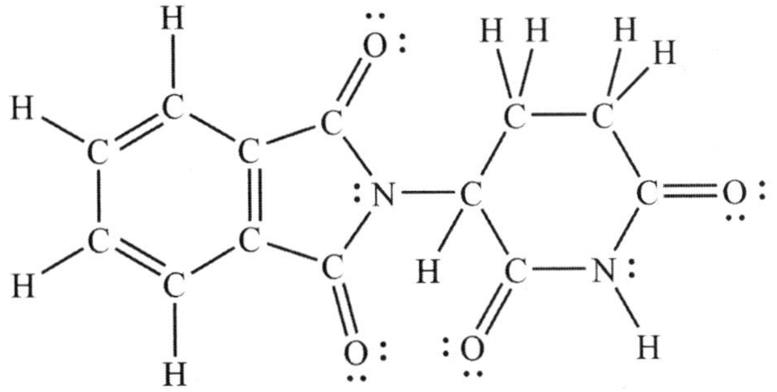

7.36

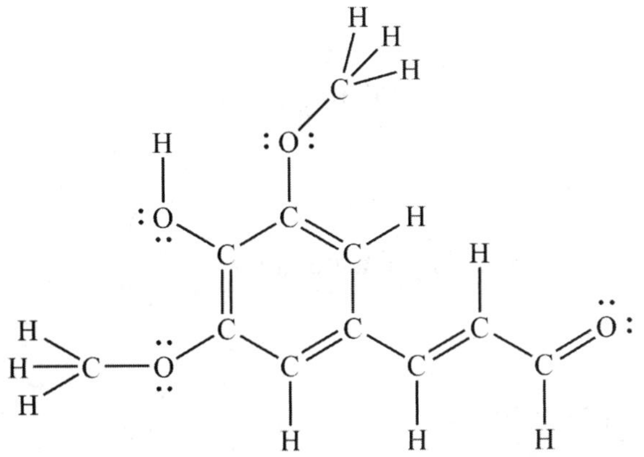

7.37

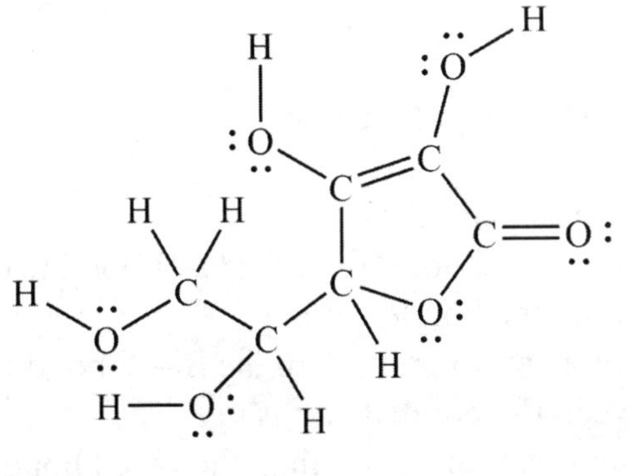

Section Problems

Strengths of Covalent Bonds and Electronegativity (Sections 7.1–7.4)

7.38 (a) ionic (b) nonpolar covalent (c) covalent

7.39 (a) Attractive forces between the positively charged nuclei and the electrons in both atoms occur when the atoms are close together and a covalent bond forms.

7.40 Electronegativity increases from left to right across a period and decreases down a group.

7.41 Z = 119 would be below francium and have a very low electronegativity.

7.42 K < Li < Mg < Pb < C < Br

7.43 Cl > C > Cu > Ca > Cs

7.44 (a) HF

fluorine	EN = 4.0	
hydrogen	EN = 2.1	
	ΔEN = 1.9	HF is polar covalent.

(b) HI

iodine	EN = 2.5	
hydrogen	EN = 2.1	
	ΔEN = 0.4	HI is polar covalent.

(c) $PdCl_2$

chlorine	EN = 3.0	
palladium	EN = 2.2	
	ΔEN = 0.8	$PdCl_2$ is polar covalent.

(d) BBr_3

bromine	EN = 2.8	
boron	EN = 2.0	
	ΔEN = 0.8	BBr_3 is polar covalent.

(e) NaOH — $Na^+ – OH^-$ is ionic

OH^-

oxygen	EN = 3.5	
hydrogen	EN = 2.1	
	ΔEN = 1.4	OH^- is polar covalent.

(f) CH_3Li

lithium	EN = 1.0	
carbon	EN = 2.5	
	ΔEN = 1.5	CH_3Li is polar covalent.

7.45 The electronegativity for each element is shown in parentheses.
(a) C (2.5), H (2.1), Cl (3.0): The C–Cl bond is more polar than the C–H bond because of the larger electronegativity difference between the bonded atoms.
(b) Si (1.8), Li (1.0), Cl (3.0): The Si–Cl bond is more polar than the Si–Li bond because of the larger electronegativity difference between the bonded atoms.
(c) N (3.0), Cl (3.0), Mg (1.2): The N–Mg bond is more polar than the N–Cl bond because of the larger electronegativity difference between the bonded atoms.

7.46 (a) $\overset{\delta-}{C} - \overset{\delta+}{H}$ $\overset{\delta+}{C} - \overset{\delta-}{Cl}$ (b) $\overset{\delta-}{Si} - \overset{\delta+}{Li}$ $\overset{\delta+}{Si} - \overset{\delta-}{Cl}$

(c) N – Cl $\overset{\delta-}{N} - \overset{\delta+}{Mg}$

7.47 (a) $\overset{\delta-}{F} - \overset{\delta+}{H}$ (b) $\overset{\delta-}{I} - \overset{\delta+}{H}$ (c) $\overset{\delta-}{Cl} - \overset{\delta+}{Pd}$

(d) $\overset{\delta-}{Br} - \overset{\delta+}{B}$ (e) $\overset{\delta-}{O} - \overset{\delta+}{H}$

7.48 (a) MgO, $BaCl_2$, $LiBr$ (b) P_4 (c) $CdBr_2$, BrF_3, NF_3, $POCl_3$

7.49 (a) $CaCl_2$, NaF, LiF (c) S_8 (c) $SOCl_2$, CBr_4, $BrCl$, AsH_3

7.50 (a) CCl_4

chlorine	EN = 3.0
carbon	EN = 2.5
	ΔEN = 0.5

(b) $BaCl_2$

chlorine	EN = 3.0
barium	EN = 0.9
	ΔEN = 2.1

(c) $TiCl_3$

chlorine	EN = 3.0
titanium	EN = 1.5
	ΔEN = 1.5

(d) Cl_2O

oxygen	EN = 3.5
chlorine	EN = 3.0
	ΔEN = 0.5

Increasing ionic character: $CCl_4 \sim ClO_2 < TiCl_3 < BaCl_2$

7.51 (a) NH_3

nitrogen	EN = 3.0
hydrogen	EN = 2.1
	ΔEN = 0.9

(b) NCl_3

nitrogen	EN = 3.0
chlorine	EN = 3.0
	ΔEN = 0.0

(c) Na_3N

nitrogen	EN = 3.0
sodium	EN = 0.9
	ΔEN = 2.1

(d) NO_2

oxygen	EN = 3.5
nitrogen	EN = 3.0
	ΔEN = 0.5

Increasing ionic character: $NCl_3 < NO_2 < NH_3 < Na_3N$

7.52 (a) $MgBr_2$ (b) PBr_3

7.53 (a) $CaCl_2$ (b) $SiCl_4$

7.54 $\text{H}-\ddot{\text{N}}=\ddot{\text{N}}-\text{H}$ $\quad$ $\text{H}_2\ddot{\text{N}}-\ddot{\text{N}}\text{H}_2$

N_2H_2 has the stronger N–N bond because of the higher bond order.

7.55 $\cdot\ddot{\text{N}}=\ddot{\text{O}}:$ $\quad$ $:\ddot{\text{O}}=\dot{\text{N}}-\ddot{\underset{\cdot\cdot}{\text{O}}}: \longleftrightarrow :\ddot{\underset{\cdot\cdot}{\text{O}}}-\dot{\text{N}}=\ddot{\text{O}}:$

NO has the stronger N–O bond because of the higher bond order.

7.56 C–F 450 kJ/mol $\Delta EN = EN(F) - EN(C) = 4.0 - 2.5 = 1.5$
N–F 270 kJ/mol $\Delta EN = EN(F) - EN(N) = 4.0 - 3.0 = 1.0$
O–F 180 kJ/mol $\Delta EN = EN(F) - EN(O) = 4.0 - 3.5 = 0.5$
F–F 159 kJ/mol $\Delta EN = EN(F) - EN(F) = 4.0 - 4.0 = 0$
In general, increased bond polarity leads to increased bond strength.

7.57 C–F 450 kJ/mol $\Delta EN = EN(F) - EN(C) = 4.0 - 2.5 = 1.5$
C–Cl 330 kJ/mol $\Delta EN = EN(Cl) - EN(C) = 3.0 - 2.5 = 0.5$
C–Br 270 kJ/mol $\Delta EN = EN(Br) - EN(C) = 2.8 - 2.5 = 0.3$
C–I 240 kJ/mol $\Delta EN = EN(I) - EN(C) = 2.5 - 2.5 = 0$
In general, increased bond polarity leads to increased bond strength.

7.58 N–H $\Delta EN = EN(N) - EN(H) = 3.0 - 2.1 = 0.9$
O–H $\Delta EN = EN(O) - EN(H) = 3.5 - 2.1 = 1.4$
S–H $\Delta EN = EN(S) - EN(H) = 2.5 - 2.1 = 0.4$
In general, increased bond polarity leads to increased bond strength. The most polar bond is the O–H bond and should be the strongest of the three.

7.59 All three bonds are nonpolar. In general, the longer the bond length, the weaker the bond. I > Br > Cl. The I–I bond is the longest and should be the weakest of the three.

7.60 (a) Phosphorus trichloride (b) Dinitrogen trioxide (c) Tetraphosphorus heptoxide
(d) Bromine trifluoride (e) Nitrogen trichloride (f) Tetraphosphorus hexoxide;
(g) Disulfur difluoride (h) Selenium dioxide

7.61 (a) S_2Cl_2 (b) ICl (c) NI_3 (d) Cl_2O (e) ClO_3 (f) S_4N_4

7.62 (b) and (c) are ionic compounds; (a) is a covalent compound and is most likely a gas at room temperature.

7.63 (a) and (b) are covalent compounds; (c) is an ionic compound and is most likely a solid at room temperature.

Electron-Dot Structures and Resonance (Sections 7.5–7.9)

7.64 The octet rule states that main-group elements tend to react so that they attain a noble gas electron configuration with filled s and p sublevels (8 electrons) in their valence electron shells. The transition metals are characterized by partially filled d orbitals that can be used to expand their valence shell beyond the normal octet of electrons.

7.65 (a) $AlCl_3$ Al has only 6 electrons around it. (c) PCl_5 P has 10 electrons around it.

7.66 (a) :Br: / :Br—C—Br: / :Br: (b) :Cl—N—Cl: / :Cl: (c) H H / H—C—C—Cl: / H H (d) $[$:F: / :F—B—F: / :F:$]^{-}$

(e) $[$:O—O:$]^{2-}$ (f) $[$:N≡O:$]^{+}$

7.67 (a) :Cl—Sb—Cl: / :Cl: (b) :F—Kr—F: (c) :O—Cl—O:

(d) :F: / :F: / P / :F— —F: / :F: (e) :O: / H—O—P—O—H / :O: / H (f) :Cl—Se—Cl:

7.68 (c) is the correct electron-dot for XeF_5^+ because it accounts for the required number of bonding (10) and lone pair (32) electrons.

7.69

H:O:H + H^+ ⟶ $[$H / H:O:H$]^+$

hydronium ion

7.70

:O O: / ‖ ‖ / H—O—C—C—O—H

7.71 :S=C=S: ; CS_2 has two double bonds.

7.72 (a) The anion has 32 valence electrons. Each Cl has seven valence electrons (28 total). The minus one charge on the anion accounts for one valence electron. This leaves three valence electrons for X. X is Al.

(b) The cation has eight valence electrons. Each H has one valence electron (4 total). X is left with four valence electrons. Since this is a cation, one valence electron was removed from X. X has five valence electrons. X is P.

7.73 (a) This fourth-row element has six valence electrons. It is Se.
(b) This fourth-row element has eight valence electrons. It is Kr.

7.74 (a)

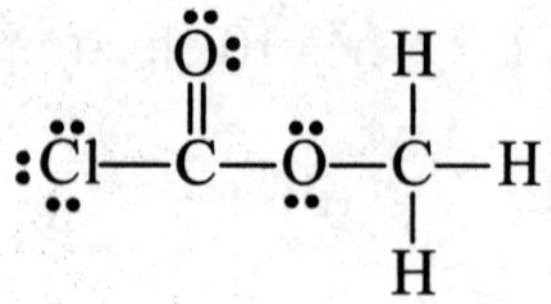

:Ö: H
:C̈l—C—Ö—C—H
H

(b)

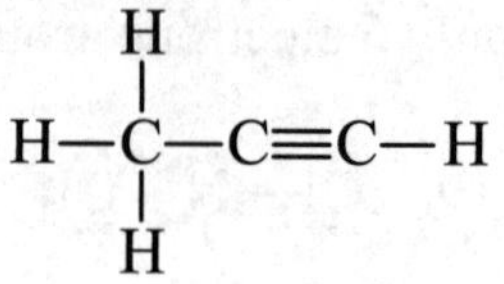

H
H—C—C≡C—H
H

7.75 (a)

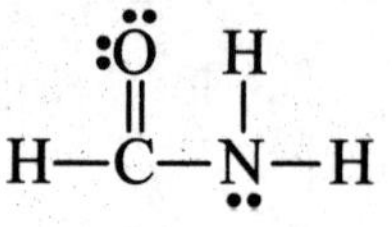

:Ö H
H—C—N̈—H

(b)

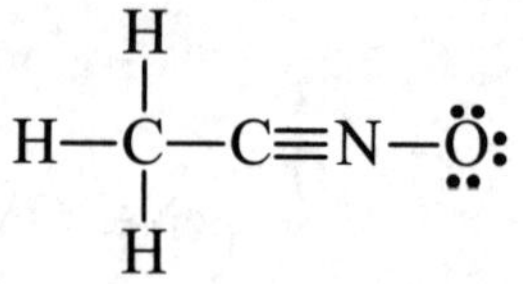

H
H—C—C≡N—Ö:
H

7.76

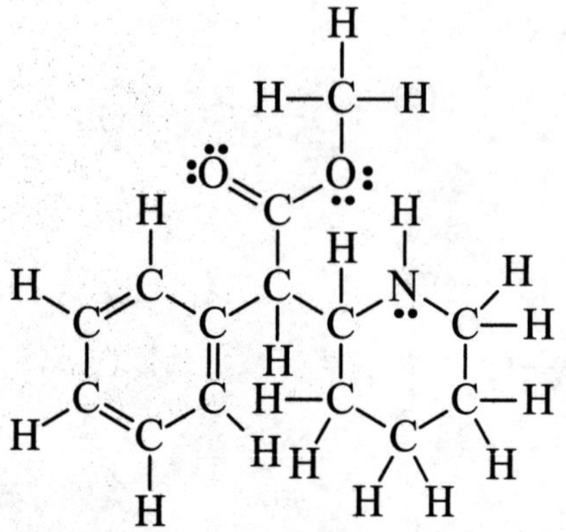

H
H—C—H
:Ö═C—Ö:
H C H H
H C C N H
H C C C C—H
H C C H—C C—H
H C H H C H
H H H

7.77

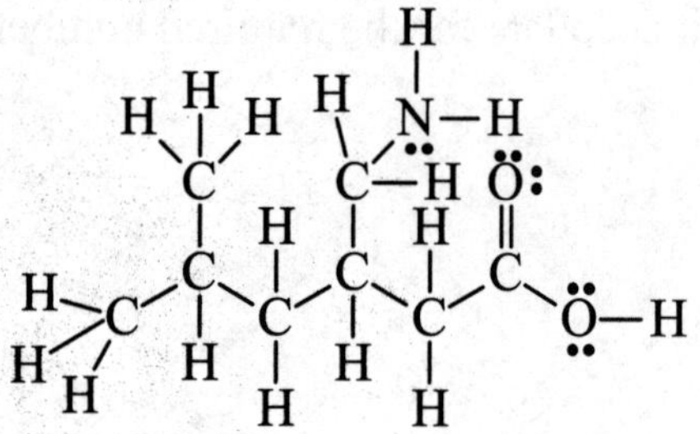

H
H H H H N—H
C C—H Ö:
H H
H—C C C C C C Ö—H
H H H H H
H H H

7.78 (a)

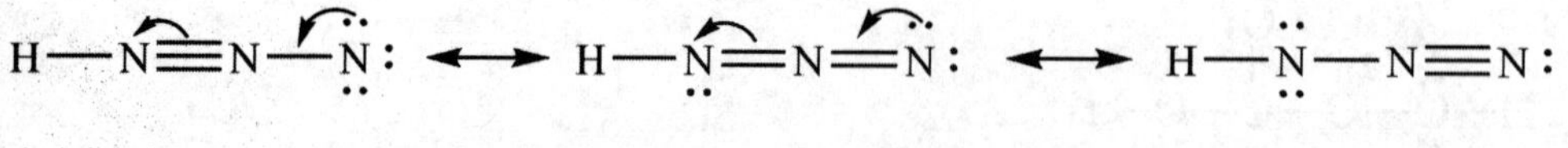

H—N≡N—N̈: ⟷ H—N̈═N═N̈: ⟷ H—N̈—N≡N:

(b)

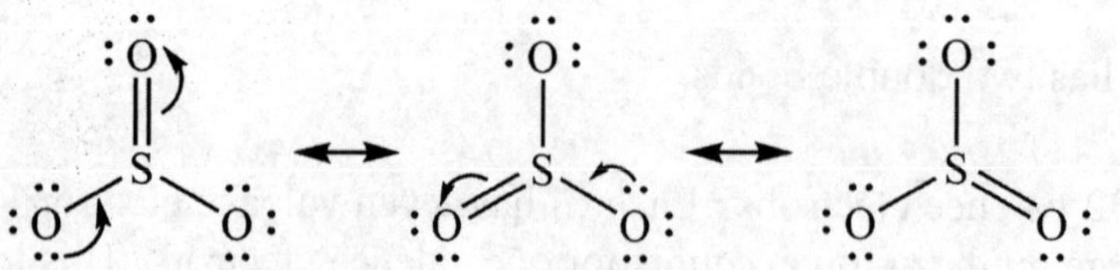

:Ö: :Ö: :Ö:
S ⟷ S ⟷ S
:Ö: :Ö: :Ö: :Ö: :Ö: :Ö:

(c)

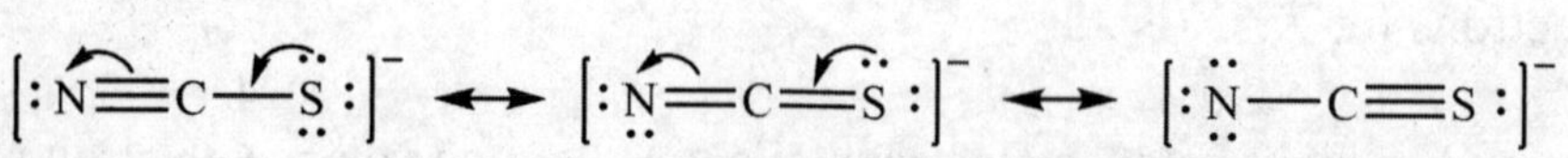

[:N≡C—S̈:]⁻ ⟷ [:N̈═C═S̈:]⁻ ⟷ [:N̈—C≡S:]⁻

7.79 (a) 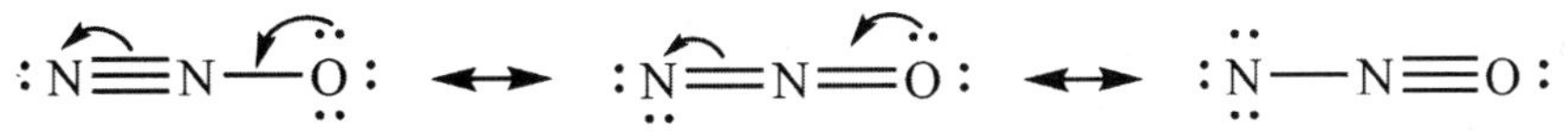 :N≡N—Ö: ⟷ :N̈=N=Ö: ⟷ :N̈—N≡O:

(b)  ·N̈=Ö: ⟷ :N̈=Ö·

(c)

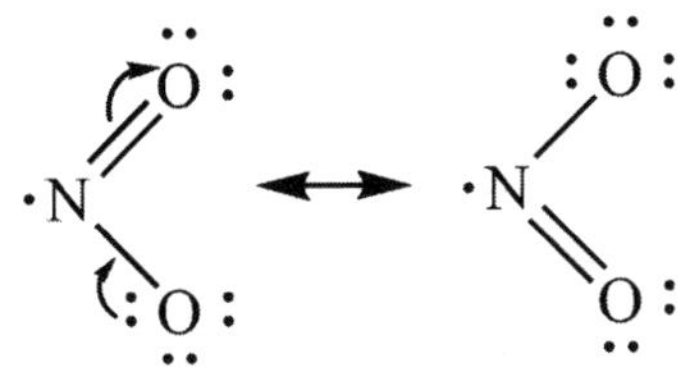

```
      ..                 ..
     O:                 :O:
    //                  /
 ·N        ⟷        ·N
    \                   \\
    :O:                  O:
     ..                  ..
```

(d)

```
                 ..              ..                    ..
                 O:             :O:                   :O:
   ..   ..      //         ..  ..  /           ..        /
 :O=N—N       ⟷   :O=N—N        ⟷    :O=N=N
                \                    \\                    \
                :O:                   O:                  :O:
                 ..                   ..                   ..
```

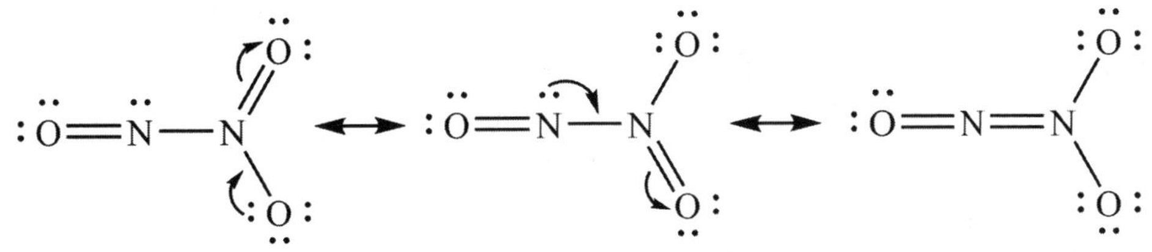

7.80 (a) yes (b) no (c) yes (d) yes

7.81 (a) yes (b) no (c) yes

7.82

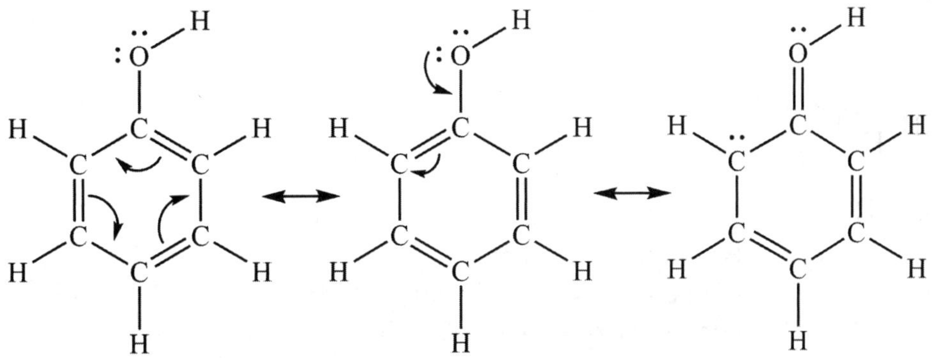

```
        H                 H                  H
   ..  /             ..  /              ..  /
  :O                :O                  O
   |                 |                  ||
H  C    H        H   C    H       H ..  C    H
 \/ \\ /          \ // \ /         \/  / \  /
  C    C     ⟷    C     C    ⟷    C      C
  ||   |          |     ||         |      ||
  C    C          C     C          C      C
 / \\ // \       / \\  / \        /  \\  / \
H    C    H     H    C     H     H     C    H
     |               |                 |
     H               H                 H
```

7.83

```
H          H    H          H
 \    ..  /      \  ..     /
  C=C—N     ⟷    C—C=N
 /  |    \       /   |   \
H   H     H     H    H    H
```

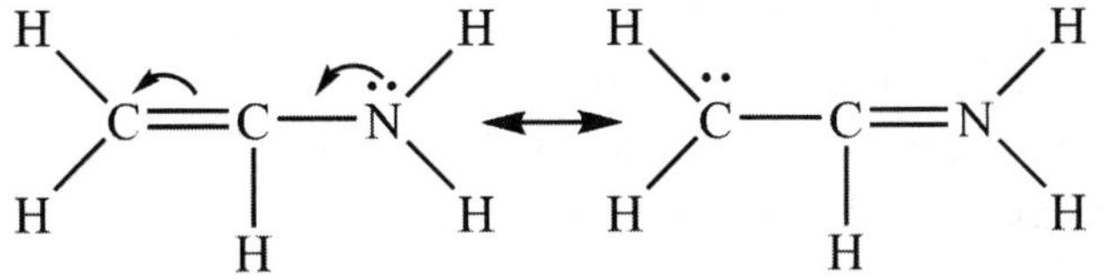

Formal Charges (Section 7.10)

7.84 :C≡O:

For carbon:	Isolated carbon valence electrons	4
	Bound carbon bonding electrons	6
	Bound carbon nonbonding electrons	2
	Formal charge = 4 – ½(6) – 2 = –1	

For oxygen:	Isolated oxygen valence electrons	6
	Bound oxygen bonding electrons	6
	Bound oxygen nonbonding electrons	2
	Formal charge = 6 – ½(6) – 2 = +1	

7.85 (a)

```
      H
      |
  H—N̤—Ö̤—H
```

For hydrogen:	Isolated hydrogen valence electrons	1
	Bound hydrogen bonding electrons	2
	Bound hydrogen nonbonding electrons	0
	Formal charge = 1 – ½(2) – 0 = 0	
For nitrogen:	Isolated nitrogen valence electrons	5
	Bound nitrogen bonding electrons	6
	Bound nitrogen nonbonding electrons	2
	Formal charge = 5 – ½(6) – 2 = 0	
For oxygen:	Isolated oxygen valence electrons	6
	Bound oxygen bonding electrons	4
	Bound oxygen nonbonding electrons	4
	Formal charge = 6 – ½(4) – 4 = 0	

(b)

```
 ┌           H   ┐ –
 │           |   │
 │   H—N̤̈—C—H   │
 │           |   │
 └           H   ┘
```

For hydrogen:	Isolated hydrogen valence electrons	1
	Bound hydrogen bonding electrons	2
	Bound hydrogen nonbonding electrons	0
	Formal charge = 1 – ½(2) – 0 = 0	
For nitrogen:	Isolated nitrogen valence electrons	5
	Bound nitrogen bonding electrons	4
	Bound nitrogen nonbonding electrons	4
	Formal charge = 5 – ½(4) – 4 = –1	
For carbon:	Isolated carbon valence electrons	4
	Bound carbon bonding electrons	8
	Bound carbon nonbonding electrons	0
	Formal charge = 4 – ½(8) – 0 = 0	

(c) 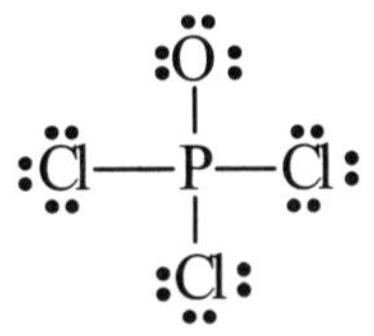 :Ö:
|
:C̤̈l—P—C̤̈l:
|
:C̤l:

For chlorine:
Isolated chlorine valence electrons 7
Bound chlorine bonding electrons 2
Bound chlorine nonbonding electrons 6
Formal charge = 7 – ½(2) – 6 = 0

For oxygen:
Isolated oxygen valence electrons 6
Bound oxygen bonding electrons 2
Bound oxygen nonbonding electrons 6
Formal charge = 6 – ½(2) – 6 = –1

For phosphorus:
Isolated phosphorus valence electrons 5
Bound phosphorus bonding electrons 8
Bound phosphorus nonbonding electrons 0
Formal charge = 5 – ½(8) – 0 = +1

7.86 [:Ö—C̤̈l—Ö:]$^-$

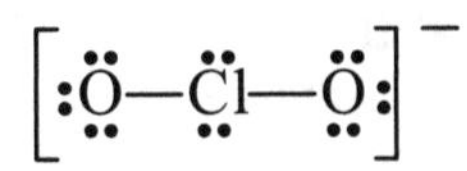

For both oxygens:
Isolated oxygen valence electrons 6
Bound oxygen bonding electrons 2
Bound oxygen nonbonding electrons 6
Formal charge = 6 – ½(2) – 6 = –1

For chlorine:
Isolated chlorine valence electrons 7
Bound chlorine bonding electrons 4
Bound chlorine nonbonding electrons 4
Formal charge = 7 – ½(4) – 4 = +1

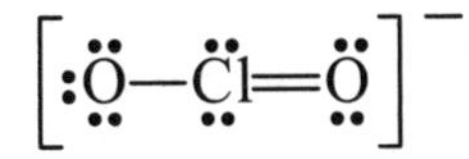

[:Ö—C̤̈l=Ö]$^-$

For left oxygen:
Isolated oxygen valence electrons 6
Bound oxygen bonding electrons 2
Bound oxygen nonbonding electrons 6
Formal charge = 6 – ½(2) – 6 = –1

For right oxygen:
Isolated oxygen valence electrons 6
Bound oxygen bonding electrons 4
Bound oxygen nonbonding electrons 4
Formal charge = 6 – ½(4) – 4 = 0

For chlorine:	Isolated chlorine valence electrons	7
	Bound chlorine bonding electrons	6
	Bound chlorine nonbonding electrons	4
	Formal charge = 7 – ½(6) – 4 = 0	

7.87

```
         :O:
          ||
   ..     ..
HO—S—OH
  ..      ..     ..
```

For sulfur:	Isolated sulfur valence electrons	6
	Bound sulfur bonding electrons	8
	Bound sulfur nonbonding electrons	2
	Formal charge = 6 – ½(8) – 2 = 0	
For doubly bound oxygen:	Isolated oxygen valence electrons	6
	Bound oxygen bonding electrons	4
	Bound oxygen nonbonding electrons	4
	Formal charge = 6 – ½(4) – 4 = 0	
For oxygen bound to hydrogen:	Isolated oxygen valence electrons	6
	Bound oxygen bonding electrons	4
	Bound oxygen nonbonding electrons	4
	Formal charge = 6 – ½(4) – 4 = 0	
For hydrogen:	Isolated hydrogen valence electrons	1
	Bound hydrogen bonding electrons	2
	Bound hydrogen nonbonding electrons	0
	Formal charge = 1 – ½(2) – 0 = 0	

```
          ..
         :O:
          |
   ..     ..
HO—S—OH
  ..      ..     ..
```

For sulfur:	Isolated sulfur valence electrons	6
	Bound sulfur bonding electrons	6
	Bound sulfur nonbonding electrons	2
	Formal charge = 6 – ½(6) – 2 = +1	
For oxygen not bound to hydrogen:	Isolated oxygen valence electrons	6
	Bound oxygen bonding electrons	2
	Bound oxygen nonbonding electrons	6
	Formal charge = 6 – ½(2) – 6 = – 1	
For oxygen bound to hydrogen:	Isolated oxygen valence electrons	6
	Bound oxygen bonding electrons	4
	Bound oxygen nonbonding electrons	4
	Formal charge = 6 – ½(4) – 4 = 0	

For hydrogen:	Isolated hydrogen valence electrons	1
	Bound hydrogen bonding electrons	2
	Bound hydrogen nonbonding electrons	0
	Formal charge = 1 – ½(2) – 0 = 0	

7.88 (a) $H_2C{=}N{=}\ddot{N}{:}$

For hydrogen:	Isolated hydrogen valence electrons	1
	Bound hydrogen bonding electrons	2
	Bound hydrogen nonbonding electrons	0
	Formal charge = 1 – ½(2) – 0 = 0	
For nitrogen: (central)	Isolated nitrogen valence electrons	5
	Bound nitrogen bonding electrons	8
	Bound nitrogen nonbonding electrons	0
	Formal charge = 5 – ½(8) – 0 = +1	
For nitrogen: (terminal)	Isolated nitrogen valence electrons	5
	Bound nitrogen bonding electrons	4
	Bound nitrogen nonbonding electrons	4
	Formal charge = 5 – ½(4) – 4 = –1	
For carbon:	Isolated carbon valence electrons	4
	Bound carbon bonding electrons	8
	Bound carbon nonbonding electrons	0
	Formal charge = 4 – ½(8) – 0 = 0	

(b) $H_2C{-}\ddot{N}{=}\ddot{N}{:}$

For hydrogen:	Isolated hydrogen valence electrons	1
	Bound hydrogen bonding electrons	2
	Bound hydrogen nonbonding electrons	0
	Formal charge = 1 – ½(2) – 0 = 0	
For nitrogen: (central)	Isolated nitrogen valence electrons	5
	Bound nitrogen bonding electrons	6
	Bound nitrogen nonbonding electrons	2
	Formal charge = 5 – ½(6) – 2 = 0	
For nitrogen: (terminal)	Isolated nitrogen valence electrons	5
	Bound nitrogen bonding electrons	4
	Bound nitrogen nonbonding electrons	4
	Formal charge = 5 – ½(4) – 4 = –1	

For carbon:	Isolated carbon valence electrons	4
	Bound carbon bonding electrons	6
	Bound carbon nonbonding electrons	0
	Formal charge = 4 – ½(6) – 0 = +1	

Structure (a) is more important because of the octet of electrons around carbon.

7.89

$$\left[\begin{array}{c} :\ddot{O}: \\ \| \\ H-C-\ddot{C}-H \\ | \\ H \end{array}\right]^{-}$$

For oxygen:	Isolated oxygen valence electrons	6
	Bound oxygen bonding electrons	4
	Bound oxygen nonbonding electrons	4
	Formal charge = 6 – ½(4) – 4 = 0	
For left carbon:	Isolated carbon valence electrons	4
	Bound carbon bonding electrons	8
	Bound carbon nonbonding electrons	0
	Formal charge = 4 – ½(8) – 0 = 0	
For right carbon:	Isolated carbon valence electrons	4
	Bound carbon bonding electrons	6
	Bound carbon nonbonding electrons	2
	Formal charge = 4 – ½(6) – 2 = –1	

$$\left[\begin{array}{c} :\ddot{\underset{..}{O}}: \\ | \\ H-C=C-H \\ | \\ H \end{array}\right]^{-}$$

For oxygen:	Isolated oxygen valence electrons	6
	Bound oxygen bonding electrons	2
	Bound oxygen nonbonding electrons	6
	Formal charge = 6 – ½(2) – 6 = –1	
For left carbon:	Isolated carbon valence electrons	4
	Bound carbon bonding electrons	8
	Bound carbon nonbonding electrons	0
	Formal charge = 4 – ½(8) – 0 = 0	
For right carbon:	Isolated carbon valence electrons	4
	Bound carbon bonding electrons	8
	Bound carbon nonbonding electrons	0
	Formal charge = 4 – ½(8) – 0 = 0	

The second structure is more important because of the –1 formal charge on the more electronegative oxygen.

7.90 [Electron-dot resonance structures: $Cl_3C{-}N^{+}(=O){-}O^{-}$ ⟷ $Cl_3C{-}N^{+}(-O^{-}){=}O$]

For chlorine:	Isolated chlorine valence electrons	7
	Bound chlorine bonding electrons	2
	Bound chlorine nonbonding electrons	6
	Formal charge = 7 – ½(2) – 6 = 0	
For carbon:	Isolated carbon valence electrons	4
	Bound carbon bonding electrons	8
	Bound carbon nonbonding electrons	0
	Formal charge = 4 – ½(8) – 0 = 0	
For nitrogen:	Isolated nitrogen valence electrons	5
	Bound nitrogen bonding electrons	8
	Bound nitrogen nonbonding electrons	0
	Formal charge = 5 – ½(8) – 0 = +1	
For oxygen: (double bonded)	Isolated oxygen valence electrons	6
	Bound oxygen bonding electrons	4
	Bound oxygen nonbonding electrons	4
	Formal charge = 6 – ½(4) – 4 = 0	
For oxygen: (single bonded)	Isolated oxygen valence electrons	6
	Bound oxygen bonding electrons	2
	Bound oxygen nonbonding electrons	6
	Formal charge = 6 – ½(2) – 6 = –1	

7.91 [Electron-dot resonance structures: $O{=}N{-}Cl$ ⟷ $^{-}O{-}N{=}Cl^{+}$]

For chlorine: (single bonded)	Isolated chlorine valence electrons	7
	Bound chlorine bonding electrons	2
	Bound chlorine nonbonding electrons	6
	Formal charge = 7 – ½(2) – 6 = 0	
For chlorine: (double bonded)	Isolated chlorine valence electrons	7
	Bound chlorine bonding electrons	4
	Bound chlorine nonbonding electrons	4
	Formal charge = 7 – ½(4) – 4 = +1	

For nitrogen:	Isolated nitrogen valence electrons	5
	Bound nitrogen bonding electrons	6
	Bound nitrogen nonbonding electrons	2
	Formal charge = $5 - \frac{1}{2}(6) - 2 = 0$	

For oxygen: (double bonded)	Isolated oxygen valence electrons	6
	Bound oxygen bonding electrons	4
	Bound oxygen nonbonding electrons	4
	Formal charge = $6 - \frac{1}{2}(4) - 4 = 0$	

For oxygen: (single bonded)	Isolated oxygen valence electrons	6
	Bound oxygen bonding electrons	2
	Bound oxygen nonbonding electrons	6
	Formal charge = $6 - \frac{1}{2}(2) - 6 = -1$	

:Ö=N̈—C̈l: is the larger contributor to the resonance hybrid because it has no formal charges.

7.92

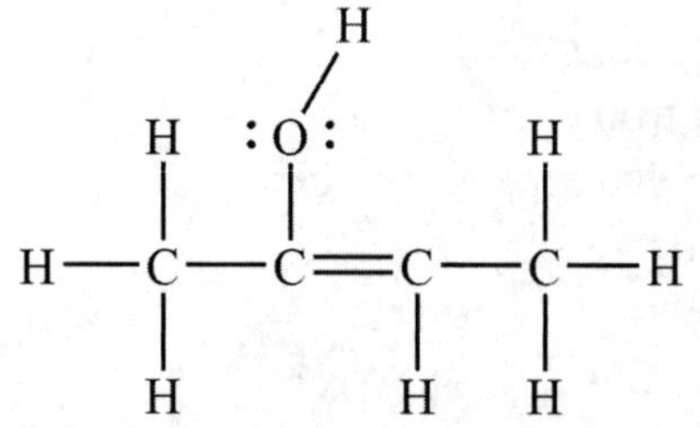

```
               H
              /
     H   :O:        H
     |    |         |
H ── C ── C ══ C ── C ── H
     |         |    |
     H         H    H
```

All atoms have 0 formal charge.

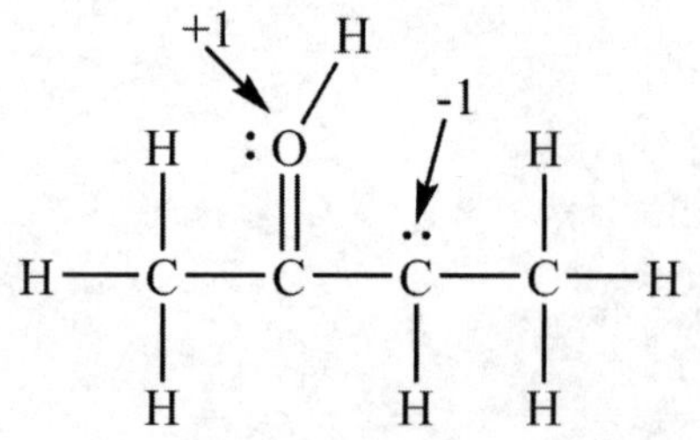

```
   +1     H
     ↘   /     -1
     H  :O    ↙    H
     |   ‖   ..    |
H ── C ── C ── C ── C ── H
     |         |    |
     H         H    H
```

For oxygen:	Isolated oxygen valence electrons	6
	Bound oxygen bonding electrons	6
	Bound oxygen nonbonding electrons	2
	Formal charge = $6 - \frac{1}{2}(6) - 2 = +1$	

For carbon:	Isolated carbon valence electrons	4
	Bound carbon bonding electrons	6
	Bound carbon nonbonding electrons	2
	Formal charge = $4 - \frac{1}{2}(6) - 2 = -1$	

All other atoms have 0 formal charge.

The original structure is the larger contributor.

7.93

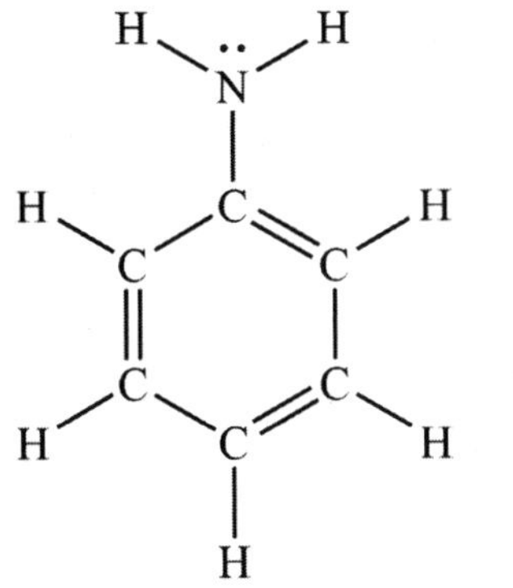

All atoms have 0 formal charge.

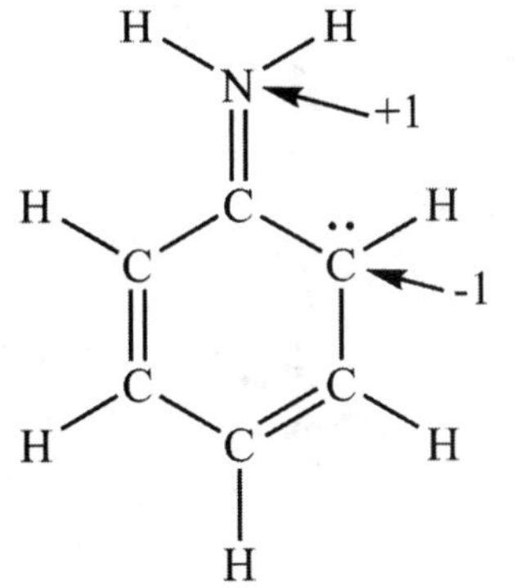

For nitrogen:	Isolated nitrogen valence electrons	5
	Bound nitrogen bonding electrons	8
	Bound nitrogen nonbonding electrons	0
	Formal charge = 5 – ½(8) – 0 = +1	
For carbon:	Isolated carbon valence electrons	4
	Bound carbon bonding electrons	6
	Bound carbon nonbonding electrons	2
	Formal charge = 4 – ½(6) – 2 = –1	

All other atoms have 0 formal charge.

The original structure is the larger contributor.

Chapter Problems

7.94 (a)

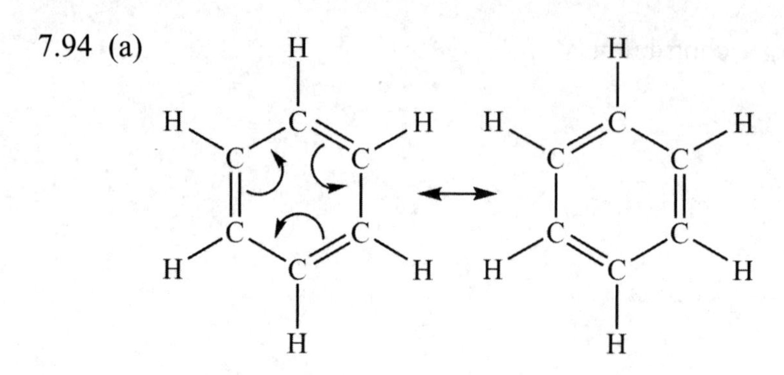

(b) ii)

7.95 (a) reactants

F—B(—F)—F $\qquad$ $H_3C—\ddot{\underset{\cdot\cdot}{O}}—CH_3$

For boron:		
	Isolated boron valence electrons	3
	Bound boron bonding electrons	6
	Bound boron nonbonding electrons	0
	Formal charge $= 3 – ½(6) – 0 = 0$	

For oxygen:		
	Isolated oxygen valence electrons	6
	Bound oxygen bonding electrons	4
	Bound oxygen nonbonding electrons	4
	Formal charge $= 6 – ½(4) – 4 = 0$	

product

F—B(—F)(—F)—O:(CH_3)(CH_3)

For boron:		
	Isolated boron valence electrons	3
	Bound boron bonding electrons	8
	Bound boron nonbonding electrons	0
	Formal charge $= 3 – ½(8) – 0 = –1$	

For oxygen:		
	Isolated oxygen valence electrons	6
	Bound oxygen bonding electrons	6
	Bound oxygen nonbonding electrons	2
	Formal charge $= 6 – ½(6) – 2 = +1$	

7.96 (a) SF_3–SF structure: :F: above S; :F–S–S–F: ; :F: below S

(b) $H-\underset{H}{\overset{H}{C}}-C\equiv C-C(=\ddot{O}:)-\ddot{O}:^-$

7.97 (a) $H-C\equiv\overset{+}{N}-\ddot{S}:^-$

For hydrogen:	Isolated hydrogen valence electrons	1
	Bound hydrogen bonding electrons	2
	Bound hydrogen nonbonding electrons	0
	Formal charge = 1 – ½(2) – 0 = 0	
For carbon:	Isolated carbon valence electrons	4
	Bound carbon bonding electrons	8
	Bound carbon nonbonding electrons	0
	Formal charge = 4 – ½(8) – 0 = 0	
For nitrogen:	Isolated nitrogen valence electrons	5
	Bound nitrogen bonding electrons	8
	Bound nitrogen nonbonding electrons	0
	Formal charge = 5 – ½(8) – 0 = +1	
For sulfur:	Isolated sulfur valence electrons	6
	Bound sulfur bonding electrons	2
	Bound sulfur nonbonding electrons	6
	Formal charge = 6 – ½(2) – 6 = –1	

(b) $H-\ddot{N}=C=\ddot{S}:$

For hydrogen:	Isolated hydrogen valence electrons	1
	Bound hydrogen bonding electrons	2
	Bound hydrogen nonbonding electrons	0
	Formal charge = 1 – ½(2) – 0 = 0	
For nitrogen:	Isolated nitrogen valence electrons	5
	Bound nitrogen bonding electrons	6
	Bound nitrogen nonbonding electrons	2
	Formal charge = 5 – ½(6) – 2 = 0	
For carbon:	Isolated carbon valence electrons	4
	Bound carbon bonding electrons	8
	Bound carbon nonbonding electrons	0
	Formal charge = 4 – ½(8) – 0 = 0	

For sulfur:

Isolated sulfur valence electrons 6
Bound sulfur bonding electrons 4
Bound sulfur nonbonding electrons 4
Formal charge = 6 – ½(4) – 4 = 0

(c) H—N̈=C=S̈: is more stable because it has no formal charges.

7.98

```
   :Ö:⁻                      :Ö:⁻                        :Ö:⁻
    |                         |                           |
   :O:        <——>           :O:         <——>            :O +
    |                         |                           ||
:Ö=S—Ö:⁻               ⁻:Ö—S=Ö:                    ⁻:Ö—S—Ö:⁻
   +2                        +2                          +2
```

For

```
   :Ö:
    |
   :O:
    |
:Ö=S—Ö:
```

For oxygen: (top)

Isolated oxygen valence electrons 6
Bound oxygen bonding electrons 2
Bound oxygen nonbonding electrons 6
Formal charge = 6 – ½(2) – 6 = –1

For oxygen: (middle)

Isolated oxygen valence electrons 6
Bound oxygen bonding electrons 4
Bound oxygen nonbonding electrons 4
Formal charge = 6 – ½(4) – 4 = 0

For oxygen: (left)

Isolated oxygen valence electrons 6
Bound oxygen bonding electrons 4
Bound oxygen nonbonding electrons 4
Formal charge = 6 – ½(4) – 4 = 0

For oxygen: (right)

Isolated oxygen valence electrons 6
Bound oxygen bonding electrons 2
Bound oxygen nonbonding electrons 6
Formal charge = 6 – ½(2) – 6 = –1

For sulfur:

Isolated sulfur valence electrons 6
Bound sulfur bonding electrons 8
Bound sulfur nonbonding electrons 0
Formal charge = 6 – ½(8) – 0 = +2

For
```
       ..
      :O:
       |
      :O:
  ..   |   ..
 :O—S=O:
  ..
```

For oxygen: (top)	Isolated oxygen valence electrons	6
	Bound oxygen bonding electrons	2
	Bound oxygen nonbonding electrons	6
	Formal charge = 6 – ½(2) – 6 = –1	

For oxygen: (middle)	Isolated oxygen valence electrons	6
	Bound oxygen bonding electrons	4
	Bound oxygen nonbonding electrons	4
	Formal charge = 6 – ½(4) – 4 = 0	

For oxygen: (left)	Isolated oxygen valence electrons	6
	Bound oxygen bonding electrons	2
	Bound oxygen nonbonding electrons	6
	Formal charge = 6 – ½(2) – 6 = –1	

For oxygen: (right)	Isolated oxygen valence electrons	6
	Bound oxygen bonding electrons	4
	Bound oxygen nonbonding electrons	4
	Formal charge = 6 – ½(4) – 4 = 0	

For sulfur:	Isolated sulfur valence electrons	6
	Bound sulfur bonding electrons	8
	Bound sulfur nonbonding electrons	0
	Formal charge = 6 – ½(8) – 0 = +2	

For
```
       ..
      :O:
       |
      :O
       ||
  ..       ..
 :O—S—O:
  ..       ..
```

For oxygen: (top)	Isolated oxygen valence electrons	6
	Bound oxygen bonding electrons	2
	Bound oxygen nonbonding electrons	6
	Formal charge = 6 – ½(2) – 6 = –1	

For oxygen: (middle)	Isolated oxygen valence electrons	6
	Bound oxygen bonding electrons	6
	Bound oxygen nonbonding electrons	2
	Formal charge = 6 – ½(6) – 2 = +1	

For oxygen: (left)
Isolated oxygen valence electrons 6
Bound oxygen bonding electrons 2
Bound oxygen nonbonding electrons 6
Formal charge = 6 – ½(2) – 6 = –1

For oxygen: (right)
Isolated oxygen valence electrons 6
Bound oxygen bonding electrons 2
Bound oxygen nonbonding electrons 6
Formal charge = 6 – ½(2) – 6 = –1

For sulfur:
Isolated sulfur valence electrons 6
Bound sulfur bonding electrons 8
Bound sulfur nonbonding electrons 0
Formal charge = 6 – ½(8) – 0 = +2

7.99

```
    H                                 H   +          —
    |     ..     ..                   |               ..
H—C—N=C=O:       <——>        H—C—N≡C—O:
    |                                 |               ..
    H                                 H
```

```
        H
        |   ..     ..
For H—C—N=C=O:
        |
        H
```

For hydrogen:
Isolated hydrogen valence electrons 1
Bound hydrogen bonding electrons 2
Bound hydrogen nonbonding electrons 0
Formal charge = 1 – ½(2) – 0 = 0

For carbon: (left)
Isolated carbon valence electrons 4
Bound carbon bonding electrons 8
Bound carbon nonbonding electrons 0
Formal charge = 4 – ½(8) – 0 = 0

For nitrogen:
Isolated nitrogen valence electrons 5
Bound nitrogen bonding electrons 6
Bound nitrogen nonbonding electrons 2
Formal charge = 5 – ½(6) – 2 = 0

For carbon: (right)
Isolated carbon valence electrons 4
Bound carbon bonding electrons 8
Bound carbon nonbonding electrons 0
Formal charge = 4 – ½(8) – 0 = 0

For oxygen:
Isolated oxygen valence electrons 6
Bound oxygen bonding electrons 4
Bound oxygen nonbonding electrons 4
Formal charge = 6 – ½(4) – 4 = 0

For $H{-}C(H)(H){-}\overset{+}{N}{\equiv}C{-}\overset{-}{\ddot{\underset{..}{O}}}{:}$

For hydrogen:	Isolated hydrogen valence electrons	1
	Bound hydrogen bonding electrons	2
	Bound hydrogen nonbonding electrons	0
	Formal charge = 1 – ½(2) – 0 = 0	
For carbon: (left)	Isolated carbon valence electrons	4
	Bound carbon bonding electrons	8
	Bound carbon nonbonding electrons	0
	Formal charge = 4 – ½(8) – 0 = 0	
For nitrogen:	Isolated nitrogen valence electrons	5
	Bound nitrogen bonding electrons	8
	Bound nitrogen nonbonding electrons	0
	Formal charge = 5 – ½(8) – 0 = +1	
For carbon: (right)	Isolated carbon valence electrons	4
	Bound carbon bonding electrons	8
	Bound carbon nonbonding electrons	0
	Formal charge = 4 – ½(8) – 0 = 0	
For oxygen:	Isolated oxygen valence electrons	6
	Bound oxygen bonding electrons	2
	Bound oxygen nonbonding electrons	6
	Formal charge = 6 – ½(2) – 6 = –1	

7.100

```
      :Cl: :Ö—H
        |   |
  :Cl—C—C—Ö—H
        |   |
      :Cl:  H
```

7.101 $CH_4(g) + Cl_2(g) \rightarrow CH_3Cl(g) + HCl(g)$

Energy change = D (Reactant bonds) – D (Product bonds)

Energy change = $[4\ D_{C\text{–}H} + D_{Cl\text{–}Cl}] - [3\ D_{C\text{–}H} + D_{C\text{–}Cl} + D_{H\text{–}Cl}]$

Energy change = [(4 mol)(410 kJ/mol) + (1 mol)(243 kJ/mol)]
– [(3 mol)(410 kJ/mol) + (1 mol)(330 kJ/mol) + (1 mol)(432 kJ/mol)] = –109 kJ

7.102

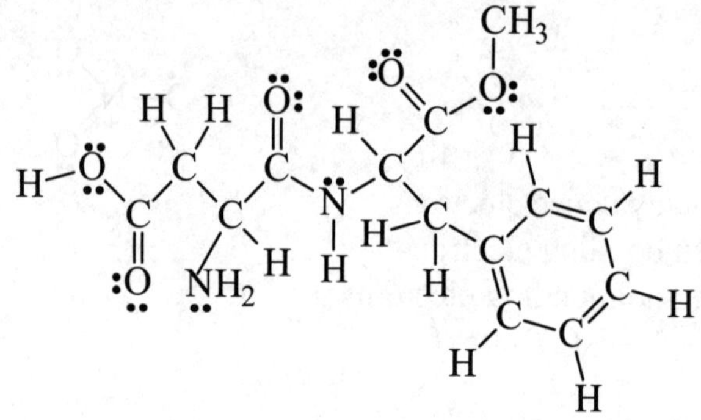

CH3
:Ö O:
H H Ö: H C H
H—Ö C C N C H
C C H—C C=C
:O NH2 H H H C—H
H C—C
H

7.103

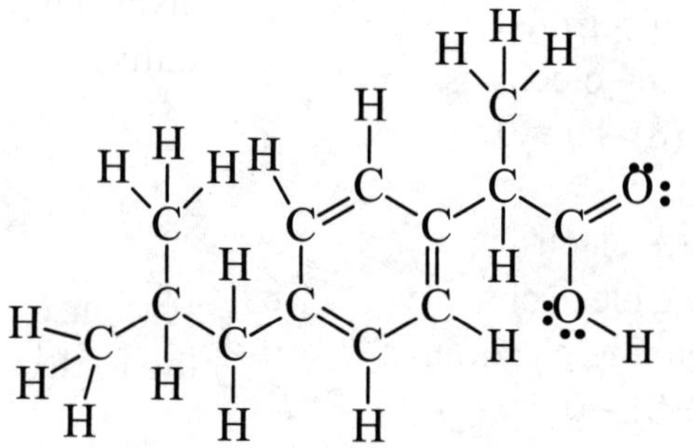

H H H
H C
H H H H C Ö:
C C=C C C
H C—H
H—C C C C :O H
H H H H H
H H H

7.104 (a)

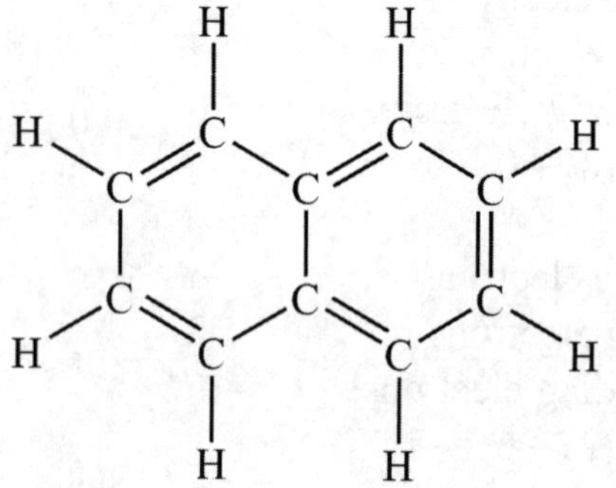

H H
H C C H
C C C
C C C
H C C H
H H

(b)

H H H H
H C C H H C C H
C C C ↔ C C C
C C C C C C
H C C H H C C H
H H H H

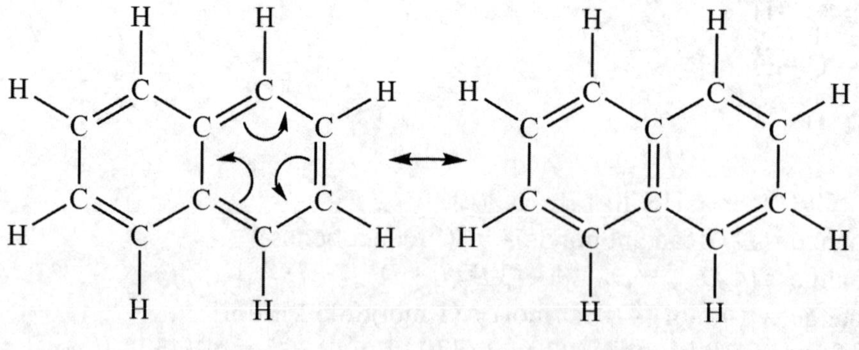

7.105 Structures (a) and (b) are resonance hybrids of the same molecule.

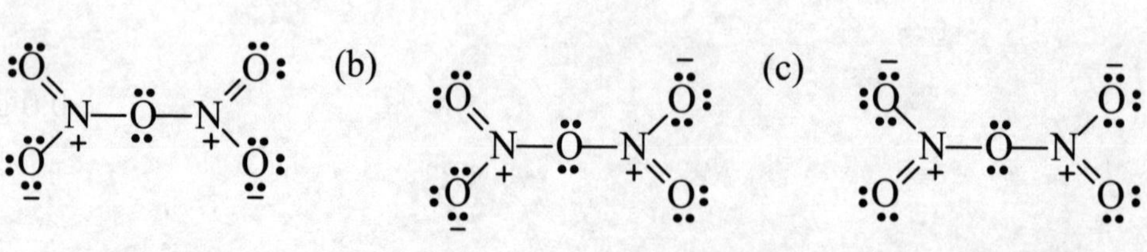

7.106 (a) :Ö Ö: (b) :Ö Ö: (c) :Ö Ö:

N—Ö—N N—Ö—N N—Ö—N

:O + + O: :O + + O: :O + + O:

(d) 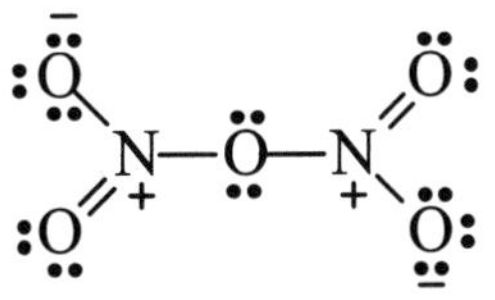 (e) 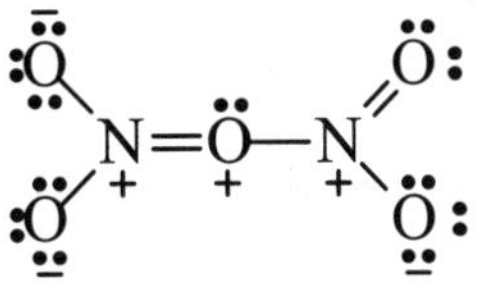 (f)

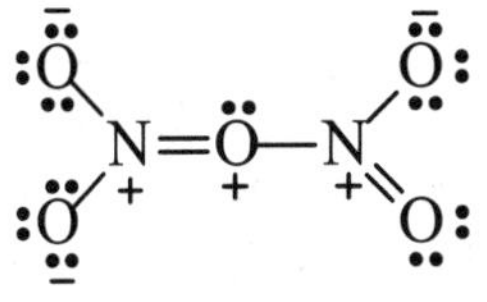

(g) 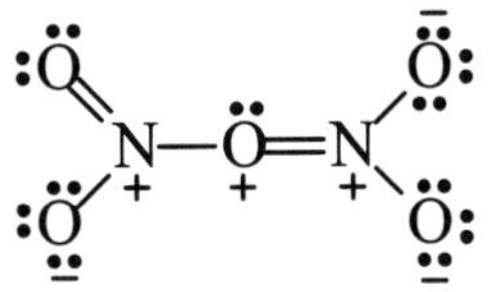 (h)

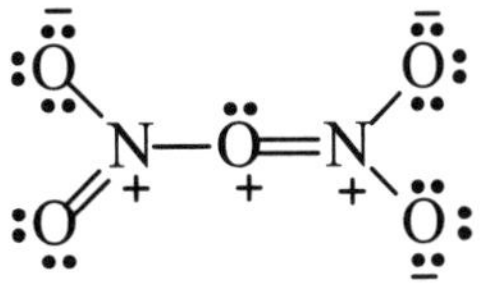

Structures (a) – (d) make more important contributions to the resonance hybrid because of only −1 and 0 formal charges on the oxygens. A +1 formal charge is unlikely.

7.107 (a) (1) 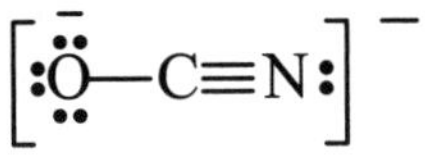 $[:\ddot{O}-C\equiv N:]^-$ (2) 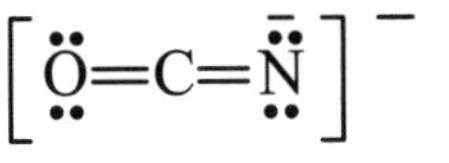 $[\ddot{O}=C=\ddot{N}]^-$ (3) 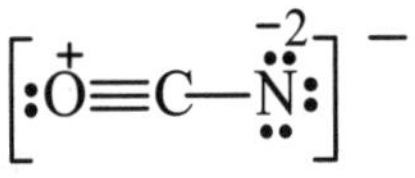 $[:O\equiv C-\ddot{N}:]^-$

(b) Structure (1) makes the greatest contribution to the resonance hybrid because of the −1 formal charge on the oxygen. Structure (3) makes the least contribution to the resonance hybrid because of the +1 formal charge on the oxygen.

Multiconcept Problems

7.108 (a) Assume a 100.0 g sample. From the percent composition data, a 100.0 g sample contains 47.5 g S and 52.5 g Cl.

$$47.5\ \text{g S} \times \frac{1\ \text{mol S}}{32.065\ \text{g S}} = 1.48\ \text{mol S}$$

$$52.5\ \text{g Cl} \times \frac{1\ \text{mol Cl}}{35.453\ \text{g Cl}} = 1.48\ \text{mol Cl}$$

Because the two mole quantities are the same, the empirical formula is SCl.

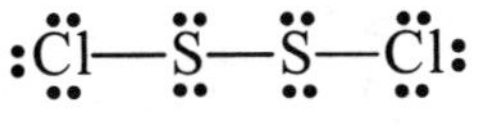

$:\ddot{Cl}-\ddot{S}-\ddot{S}-\ddot{Cl}:$

For sulfur:	Isolated sulfur valence electrons	6
	Bound sulfur bonding electrons	4
	Bound sulfur nonbonding electrons	4
	Formal charge = 6 – ½(4) – 4 = 0	
For chlorine:	Isolated chlorine valence electrons	7
	Bound chlorine bonding electrons	2
	Bound chlorine nonbonding electrons	6
	Formal charge = 7 – ½(2) – 6 = 0	

7.109 (a) Assume a 100.0 g sample. From the percent composition data, a 100.0 g sample contains 47.5 g S and 52.5 g Cl.

$$69.6\ \text{g S} \times \frac{1\ \text{mol S}}{32.065\ \text{g S}} = 2.17\ \text{mol S}$$

$30.4\ \text{g N} \times \dfrac{1\ \text{mol N}}{14.007\ \text{g N}} = 2.17\ \text{mol N}$

Because the two mole quantities are the same, the empirical formula is SN.
The empirical formula weight = 47.07

$\text{Multiple} = \dfrac{\text{molecular weight}}{\text{empirical formula weight}} = \dfrac{184.3}{47.07} = 4.00$

The molecular formula is S_4N_4.

(b)

```
     ..   ..
     S — N
   //  ..  \
 :N        :S:
  |          |
 :S:        N:
   \ ..   //
     N — S
     ..   ..
```

For sulfur:	Isolated sulfur valence electrons	6
(1 lone pair)	Bound sulfur bonding electrons	6
	Bound sulfur nonbonding electrons	2
	Formal charge = 6 – ½(6) – 2 = +1	

For sulfur:	Isolated sulfur valence electrons	6
(2 lone pairs)	Bound sulfur bonding electrons	4
	Bound sulfur nonbonding electrons	4
	Formal charge = 6 – ½(4) – 4 = 0	

For nitrogen:	Isolated nitrogen valence electrons	5
(1 lone pair)	Bound nitrogen bonding electrons	6
	Bound nitrogen nonbonding electrons	2
	Formal charge = 5 – ½(6) – 2 = 0	

For nitrogen:	Isolated nitrogen valence electrons	5
(2 lone pairs)	Bound nitrogen bonding electrons	4
	Bound nitrogen nonbonding electrons	4
	Formal charge = 5 – ½(4) – 4 = –1	

```
     ..   ..
     S — N
   //      \\
 :N        :S:
  |          |
 :S:        N:
   \\     //
     N — S
     ..   ..
```

For sulfur:	Isolated sulfur valence electrons	6
(1 lone pair)	Bound sulfur bonding electrons	6
	Bound sulfur nonbonding electrons	2
	Formal charge = 6 – ½(6) – 2 = +1	

For sulfur:	Isolated sulfur valence electrons	6
(2 lone pairs)	Bound sulfur bonding electrons	6
	Bound sulfur nonbonding electrons	4
	Formal charge = 6 – ½(6) – 4 = –1	

For nitrogen:	Isolated nitrogen valence electrons	5
	Bound nitrogen bonding electrons	6
	Bound nitrogen nonbonding electrons	2
	Formal charge = 5 – ½(6) – 2 = 0	

(c) $\text{Multiple} = \dfrac{\text{molecular weight}}{\text{empirical formula weight}} = \dfrac{92.2}{47.07} = 1.96 = 2$

The molecular formula is S_2N_2. Two possible structures are shown here.

N=S / S—N (ring with single bonds) and N—N / S—S (ring with double bonds)

7.110 (a) $[:\ddot{O}—H]^-$ $:\ddot{O}—H$

(b) The oxygen in OH has a half-filled 2p orbital that can accept the additional electron. For a 2p orbital, n = 2 and l = 1.

(c) The electron affinity for OH is slightly more negative than for an O atom because when OH gains an additional electron, it achieves an octet configuration.

7.111 (a) (4 orbitals)(3 electrons) = 12 outer-shell electrons

(b) 3 electrons

(c) $1s^3\,2s^3\,2p^6$; :X:

(d) :X::X:

7.112 (a) $[O_3Cr—O—CrO_3]^{2-}$ (each Cr: two Cr=O double bonds, one Cr—O single bond to a terminal oxygen, and one bridging O)

All formal charges are zero except for the two single bonded oxygens which each have a formal charge of –1.

(b) Each Cr atom has 6 pairs of electrons around it.

Covalent Compounds: Bonding Theories and Molecular Structure

8.1

	Number of Bonded Atoms	Number of Lone Pairs	Shape
(a) O_3	2	1	bent
(b) H_3O^+	3	1	trigonal pyramidal
(c) XeF_2	2	3	linear
(d) PF_6^-	6	0	octahedral
(e) $XeOF_4$	5	1	square pyramidal
(f) AlH_4^-	4	0	tetrahedral
(g) BF_4^-	4	0	tetrahedral
(h) $SiCl_4$	4	0	tetrahedral
(i) ICl_4^-	4	2	square planar
(j) $AlCl_3$	3	0	trigonal planar

8.2 (a) 4 charge clouds, tetrahedral charge cloud arrangement, tetrahedral molecular geometry
(b) 5 charge clouds, trigonal bipyramidal charge cloud arrangement, seesaw molecular geometry.

8.3

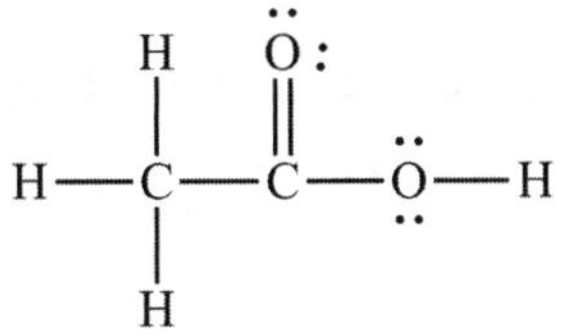

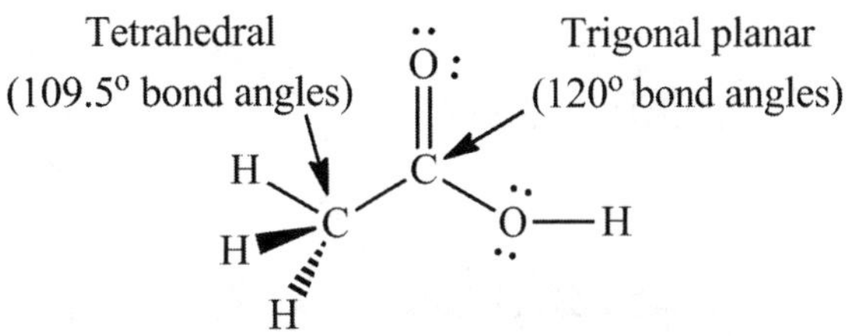

Tetrahedral (109.5° bond angles)

Trigonal planar (120° bond angles)

8.4 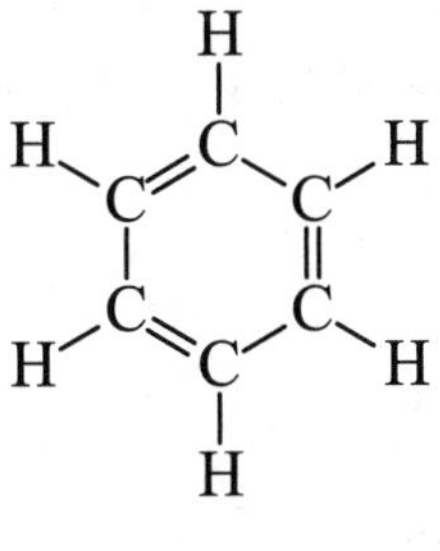 The bond angle around every carbon is 120° (trigonal planar). Benzene is a planar hexagon.

8.5 CH_2Cl_2; The C is sp^3 hybridized. The C–H bonds are formed by the overlap of one singly occupied sp^3 orbital on C with a singly occupied H 1s orbital. The C–Cl bonds are formed by the overlap of one singly occupied sp^3 orbital on C with a singly occupied Cl 2p orbital.

8.6

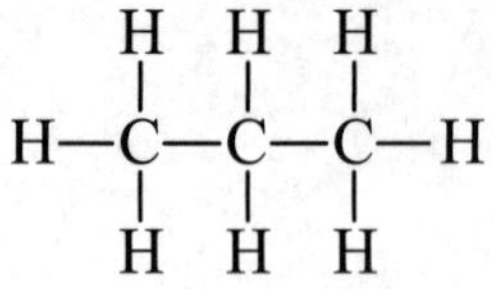

```
    H   H   H
    |   |   |
H—C—C—C—H
    |   |   |
    H   H   H
```

Each C is sp^3 hybridized. The C–C bonds are formed by the overlap of one singly occupied sp^3 hybrid orbital from each C. The C–H bonds are formed by the overlap of one singly occupied sp^3 orbital on C with a singly occupied H 1s orbital.

8.7

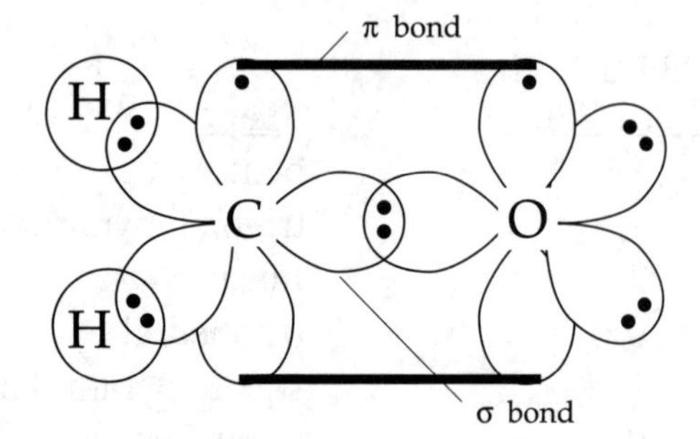

The carbon in formaldehyde is sp^2 hybridized.

8.8

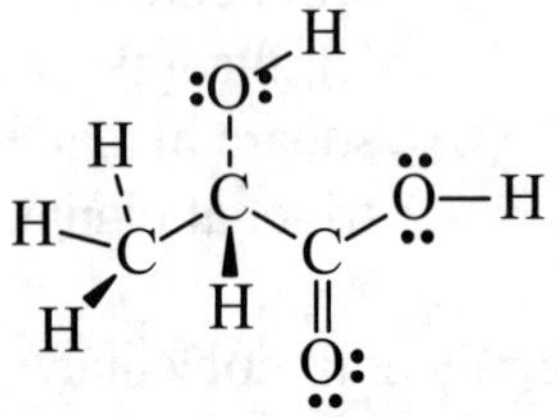

The two Cs with four single bonds are sp^3 hybridized. The C with a double bond is sp^2 hybridized. The C–C bonds are formed by the overlap of one singly occupied sp^3 or sp^2 hybrid orbital from each C. The C–H bonds are formed by the overlap of one singly occupied sp^3 orbital on C with a singly occupied H 1s orbital.

8.9

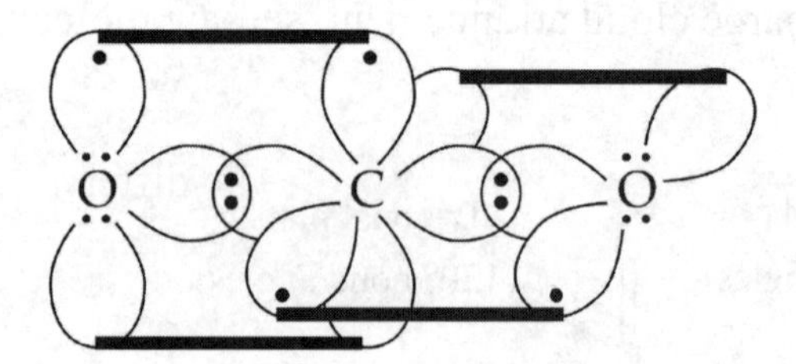

In CO_2 the carbon is sp hybridized.

8.10

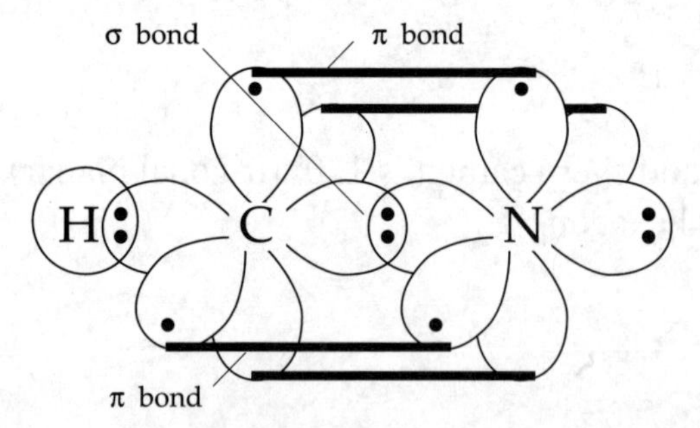

In HCN the carbon is sp hybridized.

8.11 (a)

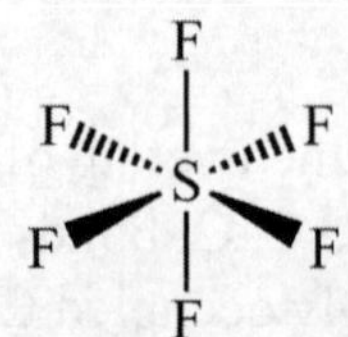

SF_6 has polar covalent bonds but the molecule is symmetrical (octahedral). The individual bond polarities cancel, and the molecule has no dipole moment.

(b) H, F, C=C, H, F

The C–F bonds in CH_2CF_2 are polar covalent bonds, and the molecule is polar.

(c)

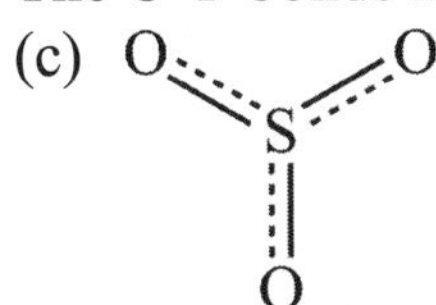

SO_3 has polar covalent bonds but the molecule is symmetrical (trigonal planar). The individual bond polarities cancel, and the molecule has no dipole moment.

(d)

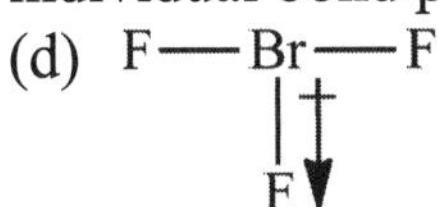

The Br–F bonds in BrF_3 are polar covalent bonds, and the molecule is polar.

8.12 (a) CF_4 (b) CH_2F_2 (c) CHF_3 (d) CH_3F

8.13 $\mu = Q \times r = (1.60 \times 10^{-19}\ C)(92 \times 10^{-12}\ m)\left(\dfrac{1\ D}{3.336 \times 10^{-30}\ C\cdot m}\right) = 4.41\ D$

$\%\text{ ionic character for HF} = \dfrac{1.83\ D}{4.41\ D} \times 100\% = 41\%$

HF has more ionic character than HCl. HCl has only 18% ionic character.

8.14 HBr is predicted to have greater percent ionic character than HI because the difference in electronegativity between hydrogen and bromine (0.7) is greater than between hydrogen and iodine (0.4).

HBr, $\mu = Q \times r = (1.60 \times 10^{-19}\ C)(142 \times 10^{-12}\ m)\left(\dfrac{1\ D}{3.336 \times 10^{-30}\ C\cdot m}\right) = 6.81\ D$

$\%\text{ ionic character for HF} = \dfrac{0.82\ D}{6.81\ D} \times 100\% = 12\%$

HI, $\mu = Q \times r = (1.60 \times 10^{-19}\ C)(161 \times 10^{-12}\ m)\left(\dfrac{1\ D}{3.336 \times 10^{-30}\ C\cdot m}\right) = 7.72\ D$

$\%\text{ ionic character for HF} = \dfrac{0.38\ D}{7.72\ D} \times 100\% = 4.9\%$

The calculation of percent ionic character supports the prediction.

8.15 (b) and (c) are correct depictions of hydrogen bonding. (a) and (d) are incorrect because hydrogen is covalently bonded to C, which is not one of the highly electronegative elements F, O, or N.

8.16 Hydrogen bonds are shown as dashed lines.

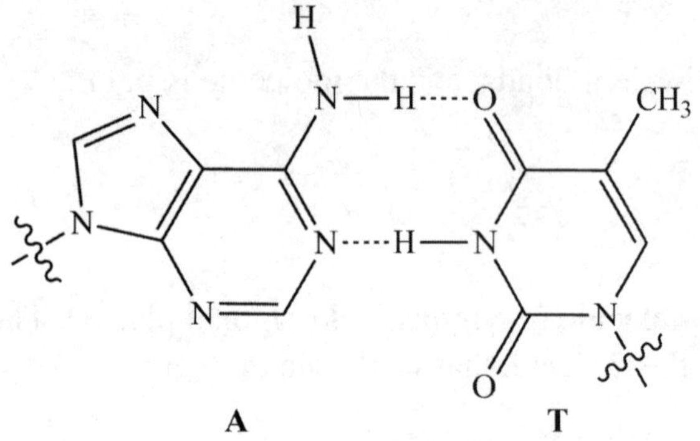

A T

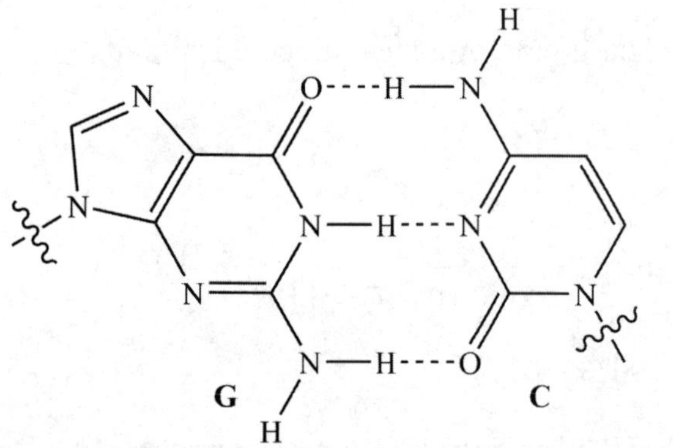

G C

Because of the three hydrogen bonds, DNA regions that are high in G–C pairs would have the higher melting point.

8.17 (a) Both CH_3F and HNO_3 have net dipole moments and dipole–dipole forces.
(b) Only HNO_3 can hydrogen bond.
(c) Ar has fewer electrons than Cl_2 and CCl_4, and has the smallest dispersion forces.
(d) CCl_4 is larger than Ar and Cl_2, has more electrons and the largest dispersion forces.

8.18

H_2S	dipole-dipole, dispersion
CH_3OH	hydrogen bonding, dipole-dipole, dispersion
C_2H_6	dispersion
Ar	dispersion

$Ar < C_2H_6 < H_2S < CH_3OH$

8.19 For He_2^+ σ^*_{1s} $\underline{\uparrow}$
σ_{1s} $\underline{\uparrow\downarrow}$

$$He_2^+ \text{ Bond order} = \frac{\left(\begin{array}{c}\text{number of}\\ \text{bonding electrons}\end{array}\right) - \left(\begin{array}{c}\text{number of}\\ \text{antibonding electrons}\end{array}\right)}{2} = \frac{2-1}{2} = 1/2$$

He_2^+ should be stable with a bond order of 1/2.

8.20 The bond order in He_2^{2+} is 1, which is greater than the bond order of 1/2 in He_2^+; therefore He_2^{2+} is predicted to have a stronger bond and be a more stable species

8.21 For B_2

σ^*_{2p}	—
π^*_{2p}	— —
σ_{2p}	—
π_{2p}	↑ ↑
σ^*_{2s}	↑↓
σ_{2s}	↑↓

$$B_2 \text{ Bond order} = \frac{\left(\begin{array}{c}\text{number of}\\ \text{bonding electrons}\end{array}\right) - \left(\begin{array}{c}\text{number of}\\ \text{antibonding electrons}\end{array}\right)}{2} = \frac{4-2}{2} = 1$$

B_2 is paramagnetic because it has two unpaired electrons in the π_{2p} molecular orbitals.

For C_2

σ^*_{2p}	—
π^*_{2p}	— —
σ_{2p}	—
π_{2p}	↑↓ ↑↓
σ^*_{2s}	↑↓
σ_{2s}	↑↓

C_2 Bond order $= \frac{6-2}{2} = 2$; C_2 is diamagnetic because all electrons are paired.

8.22 The bond orders are: $O_2^{2-} = 1$, $O_2^- = 1.5$, $O_2 = 2$, $O_2^+ = 2.5$, $O_2^{2+} = 3$.
The order from weakest to strongest bond is: $O_2^{2-} < O_2^- < O_2 < O_2^+ < O_2^{2+}$.
The order from shortest to longest bond is: $O_2^{2+} < O_2^+ < O_2 < O_2^- < O_2^{2-}$.

8.23

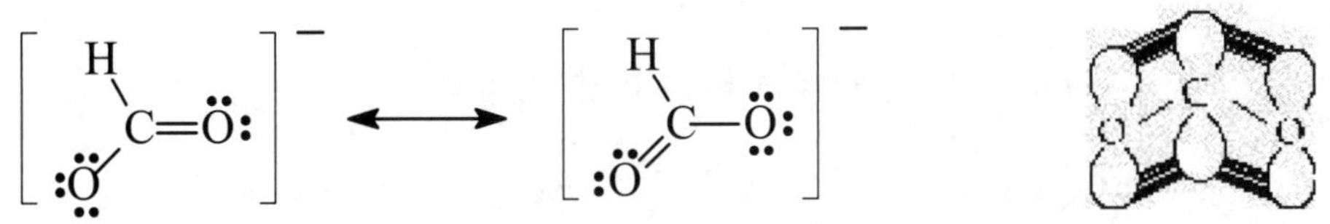

8.24

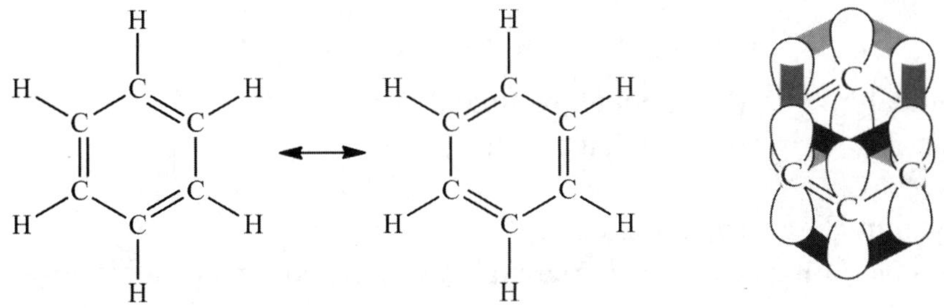

8.25 (b), (c), and (e) are chiral.

8.26 The mirror image of molecule (a) has the same shape as (a) and is identical to it in all respects, so there is no handedness associated with it. The mirror image of molecule (b) is different than (b) so there is a handedness to this molecule.

8.27 Only (a), lactic acid, is chiral.

8.28

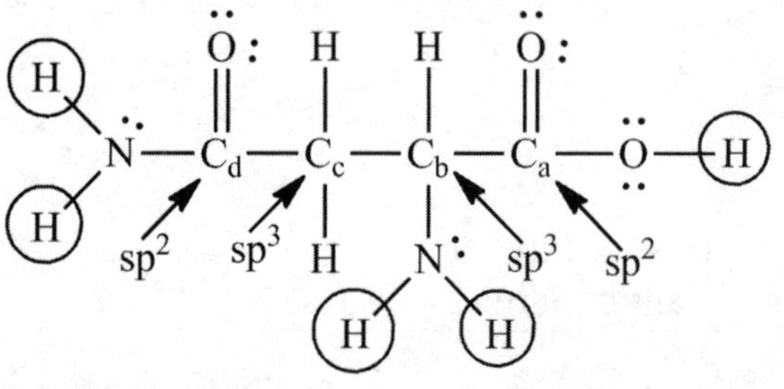

O: H H O:
H N—C_d—C_c—C_b—C_a—O—H
H sp² sp³ H N sp³ sp²
H H

(a) C_b is the chiral center. There are four different groups attached to it.
(b) 16 σ bonds and 2 π bonds.
(c) and (d) see figure

8.29 (a)

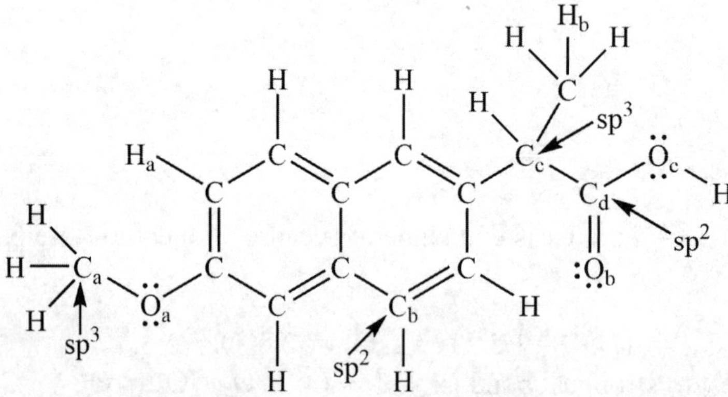

H_b
H H
C
H H H
sp³
H_a C C C_c O_c
C C C C_d H_c
H sp²
H—C_a C C C :O_b
H O_a C C_b H
sp³ sp²
H H

(b) C_c is the chiral center. There are four different groups attached to it.
(c) 32 σ bonds and 6 π bonds.
(d) see figure
(e) All three oxygens and H_c can participate in hydrogen bonding.

8.30 (a) C_c (b) C_a, sp^3; C_b, sp^2; C_c, sp^3; C_d, sp^2 (c) yes, because of the O.

Conceptual Problems

8.31 (a) square pyramidal (b) trigonal pyramidal
(c) square planar (d) trigonal planar

8.32 (a) trigonal bipyramidal (b) tetrahedral
(c) square pyramidal (4 ligands in the horizontal plane, including one hidden)

8.33 Molecular model (c) does not have a tetrahedral central atom. It is square planar.

8.34 (a) sp^2 (b) sp (c) sp^3

8.35 The expected hybridizations of C and N in urea are sp^2 and sp^3, respectively. The expected bond angles are (i) N–C–O and N–C–N, ~120°, and (ii) C–N–H and H–N–H, ~109°. Based on the molecular model, the C and N are both sp^2 hybridized and all bond angles are ~120°.

8.36 (a) $C_8H_9NO_2$
(b), (c), and (d)

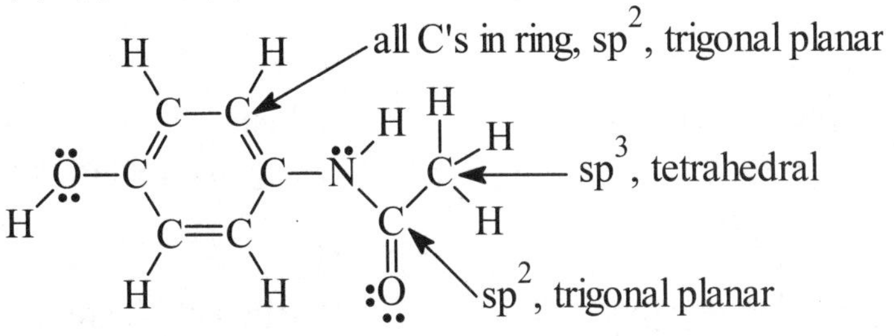

all C's in ring, sp^2, trigonal planar

sp^3, tetrahedral

sp^2, trigonal planar

8.37 (a) $C_{13}H_{10}N_2O_4$
(b), (c), and (d)

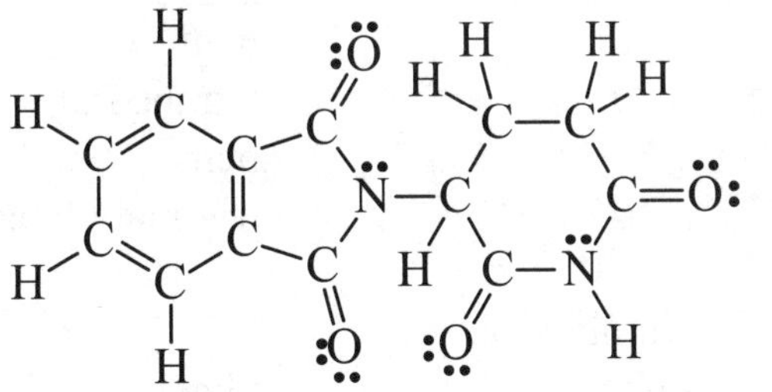

All carbons that have only single bonds are sp^3 hybridized and have tetrahedral geometries. All carbons that have double bonds are sp^2 hybridized and have trigonal planar geometries.

8.38 The electronegative O atoms are electron rich (red), while the rest of the molecule is electron poor (blue).

8.39 net

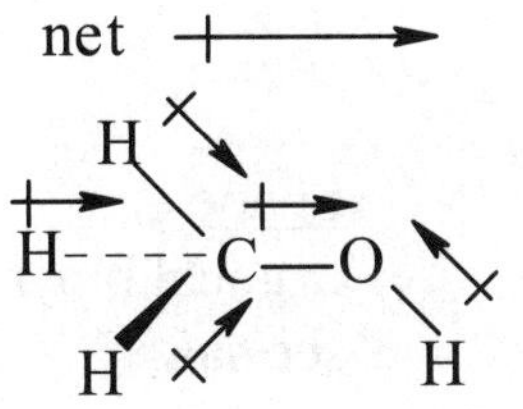

8.40 The N atom is electron rich (red) because of its high electronegativity. The C and H atoms are electron poor (blue) because they are less electronegative.

8.41 (a) (i) is *trans* 1,2 dichloroethylene and (ii) is *cis* 1,2 dichloroethylene
(b) *cis* 1,2 dichloroethylene has a dipole moment of 2.39 D and *trans* 1,2 dichloroethylene has a dipole moment of 0.00 D.
(c) The polar molecule, *cis* 1,2 dichloroethylene, has the higher boiling point.

Section Problems
The VSEPR Model (Section 8.1)

8.42 From data in Table 8.1:
(a) trigonal planar (b) trigonal bipyramidal (c) linear (d) octahedral

8.43 From data in Table 8.1: (a) T shaped (b) bent (c) square planar

8.44 From data in Table 8.1:
(a) tetrahedral, 4 (b) octahedral, 6 (c) bent, 3 or 4
(d) linear, 2 or 5 (e) square pyramidal, 6 (f) trigonal pyramidal, 4

8.45 From data in Table 8.1:
(a) seesaw, 5 (b) square planar, 6 (c) trigonal bipyramidal, 5
(d) T shaped, 5 (e) trigonal planar, 3 (f) linear, 2 or 5

8.46

	Number of Bonded Atoms	Number of Lone Pairs	Shape
(a) H_2Se	2	2	bent
(b) $TiCl_4$	4	0	tetrahedral
(c) O_3	2	1	bent
(d) GaH_3	3	0	trigonal planar

8.47

	Number of Bonded Atoms	Number of Lone Pairs	Shape
(a) XeO_4	4	0	tetrahedral
(b) SO_2Cl_2	4	0	tetrahedral
(c) OsO_4	4	0	tetrahedral
(d) SeO_2	2	1	bent

8.48

	Number of Bonded Atoms	Number of Lone Pairs	Shape
(a) SbF_5	5	0	trigonal bipyramidal
(b) IF_4^+	4	1	see saw
(c) SeO_3^{2-}	3	1	trigonal pyramidal
(d) CrO_4^{2-}	4	0	tetrahedral

8.49

	Number of Bonded Atoms	Number of Lone Pairs	Shape
(a) NO_3^-	3	0	trigonal planar
(b) NO_2^+	2	0	linear
(c) NO_2^-	2	1	bent

8.50

	Number of Bonded Atoms	Number of Lone Pairs	Shape
(a) PO_4^{3-}	4	0	tetrahedral
(b) MnO_4^-	4	0	tetrahedral
(c) SO_4^{2-}	4	0	tetrahedral
(d) SO_3^{2-}	3	1	trigonal pyramidal
(e) ClO_4^-	4	0	tetrahedral
(f) SCN^- (C is the central atom)	2	0	linear

8.51

	Number of Bonded Atoms	Number of Lone Pairs	Shape
(a) XeF_3^+	3	2	T shaped
(b) SF_3^+	3	1	trigonal pyramidal
(c) ClF_2^+	2	2	bent
(d) CH_3^+	3	0	trigonal planar

8.52 (a) In SF_2 the sulfur is bound to two fluorines and contains two lone pairs of electrons. SF_2 is bent and the F–S–F bond angle is approximately 109°.
(b) In N_2H_2 each nitrogen is bound to the other nitrogen and one hydrogen. Each nitrogen has one lone pair of electrons. The H–N–N bond angle is approximately 120°.
(c) In KrF_4 the krypton is bound to four fluorines and contains two lone pairs of electrons. KrF_4 is square planar, and the F–Kr–F bond angle is 90°.
(d) In NOCl the nitrogen is bound to one oxygen and one chlorine and contains one lone pair of electrons. NOCl is bent, and the Cl–N–O bond angle is approximately 120°.

8.53 (a) In PCl_6^- the phosphorus is bound to six chlorines. There are no lone pairs of electrons on the phosphorus. PCl_6^- is octahedral, and the Cl–P–Cl bond angle is 90°.
(b) In ICl_2^- the iodine is bound to two chlorines and contains three lone pairs of electrons. ICl_2^- is linear, and the Cl–I–Cl bond angle is 180°.
(c) In SO_4^{2-} the sulfur is bound to four oxygens. There are no lone pairs of electrons on the sulfur. SO_4^{2-} is tetrahedral, and the O–S–O bond angle is 109.5°.
(d) In BO_3^{3-} the boron is bound to three oxygens. There are no lone pairs of electrons on the boron. BO_3^{3-} is trigonal planar, and the O–B–O bond angle is 120°.

8.54

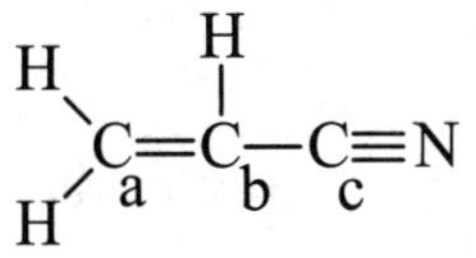

$H_2C_a{=}C_b(H){-}C_c{\equiv}N$

$H-C_a-H$	~ 120°	C_b-C_c-N	180°
$H-C_a-C_b$	~ 120°	C_a-C_b-H	~ 120°
$C_a-C_b-C_c$	~ 120°	$H-C_b-C_c$	~ 120°

8.55

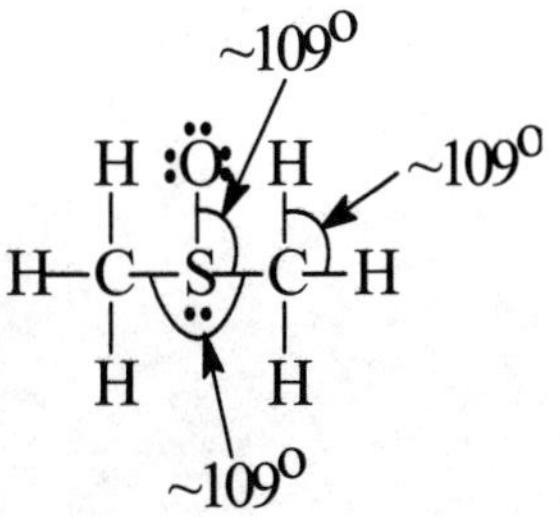

$H_3C-\ddot{S}(:\ddot{O}:)-CH_3$

~109° (C–S–O), ~109° (H–C–S), ~109° (C–S–C)

8.56

```
        :F:
         |
    :F—C—F:
         |
 :F,,,,  |  ,,,,F:
        S
 :F ◄   |   ► F:
         |
        :F:
```

The bond angles are 109.5° around carbon and 90° around S.

8.57

```
    H       :F:
    |        |
H—C—Ö—C—F:
    |        |
    H       :F:
```

The geometry is tetrahedral around both carbons and bent around the oxygen.

8.58 All six carbons in cyclohexane are bonded to two other carbons and two hydrogens (i.e., four charge clouds). The geometry about each carbon is tetrahedral with a C–C–C bond angle of approximately 109°. Because the geometry about each carbon is tetrahedral, the cyclohexane ring cannot be flat.

8.59 All six carbon atoms are sp^2 hybridized and the bond angles are ~120°. The geometry about each carbon is trigonal planar.

Valence Bond Theory and Hybridization (Sections 8.2–8.4)

8.60 In a π bond, the shared electrons occupy a region above and below a line connecting the two nuclei. A σ bond has its shared electrons located along the axis between the two nuclei.

8.61 Using the data here, the difference in energy between a carbon-carbon double bond and a carbon-carbon single bond is (728 kJ/mol – 350 kJ/mol) = 378 kJ/mol. This represents the energy of a π bond, which from this data would indicate that a π bond is stronger than a σ bond. However, numerous other sources list the average carbon-carbon double bond energy as ~615 kJ/mol, this would lead to a π bond energy of (615 kJ/mol – 350 kJ/mol) = 265 kJ/mol. This data would indicate that a σ bond is stronger than a π bond.

8.62 See Table 8.2. (a) sp (b) sp^2 (c) sp^3

8.63 See Table 8.2. (a) tetrahedral (b) trigonal planar (c) linear

8.64 (a) sp^2 (b) sp^2 (c) sp^3 (d) sp^2

8.65 (a) sp^3 (b) sp^2 (c) sp^2 (d) sp^3

8.66

```
        O   O   H   O
        ‖   ‖   |   ‖
H—O—C—C—C—C—O—H
       a    b  | c   d
                H
```

Carbons a, b, and d are sp^2 hybridized and carbon c is sp^3 hybridized.

The bond angles around carbons a, b, and d are ~120°. The bond angles around carbon c are ~109°. The terminal H–O–C bond angles are ~109°.

8.67 (a)

```
     H   H   :Ö:
     |   |    ‖
H—N—C—C—Ö—H
    ..  |       ..
        H
```

(b) H–C–H, ~109°; O–C–O, ~120°; H–N–H, ~107°

(c) N, sp^3; left C, sp^3; right C, sp^2

8.68

```
            :Cl:
O    C
            :Cl:
```

In Cl_2CO the carbon is sp2 hybridized.

8.69

```
                    H    sp³
                    |  ↙
H—C≡C—C—H
     ↖ ↗          |
      sp           H
```

Dipole Moments and Intermolecular Forces (Sections 8.5 and 8.6)

8.70 If a molecule has polar covalent bonds, the molecular shape (and location of lone pairs of electrons) determines whether the bond dipoles cancel and thus whether the molecule has a dipole moment.

8.71 Dipole-dipole forces arise between molecules that have permanent dipole moments. London dispersion forces arise between molecules as a result of induced temporary dipoles.

8.72 (a) $CHCl_3$ has a permanent dipole moment. Dipole-dipole intermolecular forces are important. London dispersion forces are also present.

(b) O_2 has no dipole moment. London dispersion intermolecular forces are important.

(c) Polyethylene, C_nH_{2n+2}. London dispersion intermolecular forces are important.

(d) CH_3OH has a permanent dipole moment. Dipole-dipole intermolecular forces and hydrogen bonding are important. London dispersion forces are also present.

8.73 (a) Xe has no dipole-dipole forces
(b) HF has the largest hydrogen bond forces
(c) Xe has the largest dispersion forces

8.74 For CH_3OH and CH_4, dispersion forces are small. CH_3OH can hydrogen bond; CH_4 cannot. This accounts for the large difference in boiling points.
For 1-decanol and decane, dispersion forces are comparable and relatively large along the C–H chain. 1-decanol can hydrogen bond; decane cannot. This accounts for the 57 °C higher boiling point for 1-decanol.

8.75 (a) C_8H_{18} has the larger dispersion forces because of its longer hydrocarbon chain.
(b) HI has the larger dispersion forces because of the larger, more polarizable iodine.
(c) H_2Se has the larger dispersion forces because of the more polarizable and less electronegative Se.

8.76 (a)

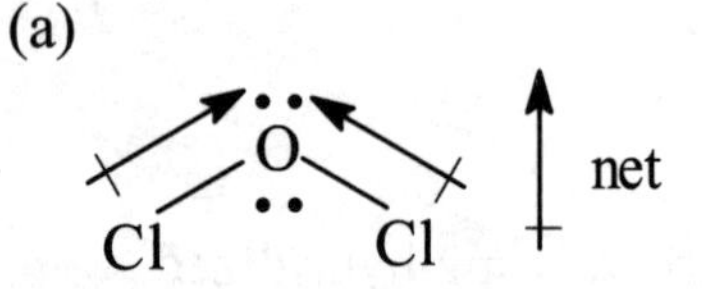

(b) F F Xe F F net dipole moment = 0

(c) CH_3 — C - - - H, Cl, H net

(d) F, B, F, F net dipole moment = 0

8.77 (a) N, F, F, F net

(b) H_3C - - N, H, H net

(c) F—Xe—F
net dipole moment = 0

(d) Cl, Cl, P, Cl, Cl, Cl net dipole moment = 0

8.78 $\mu = Q \times r = (1.60 \times 10^{-19}\ C)(213.9 \times 10^{-12}\ m)\left(\dfrac{1\ D}{3.336 \times 10^{-30}\ C \cdot m}\right) = 10.26\ D$

% ionic character for BrCl = $\dfrac{0.518\ D}{10.26\ D} \times 100\% = 5.05\%$

8.79 $\mu = Q \times r = (1.60 \times 10^{-19}\ C)(162.8 \times 10^{-12}\ m)\left(\dfrac{1\ D}{3.336 \times 10^{-30}\ C \cdot m}\right) = 7.81\ D$

% ionic character for ClF = $\dfrac{0.887\ D}{7.81\ D} \times 100\% = 11.4\%$

8.80

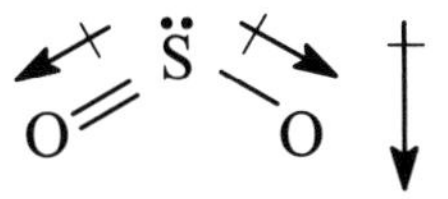

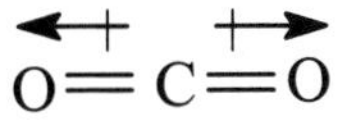

O=S(:)–O   net (dipole arrow)

O=C=O
net dipole moment = 0

SO_2 is bent and the individual bond dipole moments add to give the molecule a net dipole moment.
CO_2 is linear and the individual bond dipole moments point in opposite directions to cancel each other out. CO_2 has no net dipole moment.

8.81

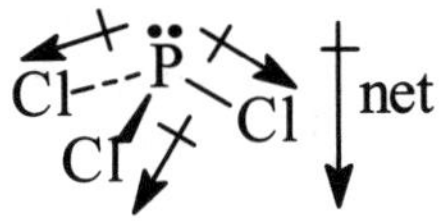

Cl–P(:)(Cl)–Cl   net

Cl, Cl–P–Cl, Cl, Cl   net dipole moment = 0

In both PCl_3 and PCl_5 the P–Cl bond is polar covalent. PCl_3 is trigonal pyramidal and the bond dipoles add to give the molecule a net dipole moment. PCl_5 is trigonal bipyramidal and the bond dipoles cancel. PCl_5 has no dipole moment.

8.82 (a) 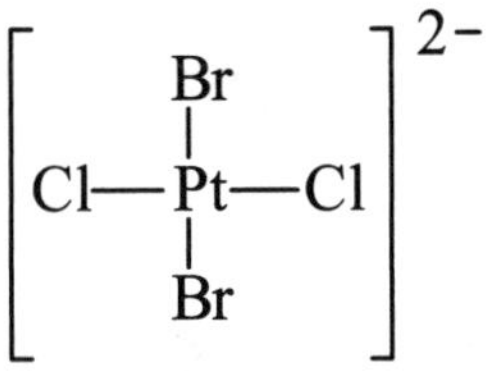 $[PtBr_2Cl_2]^{2-}$: Br above Pt, Cl—Pt—Cl, Br below Pt

(b) 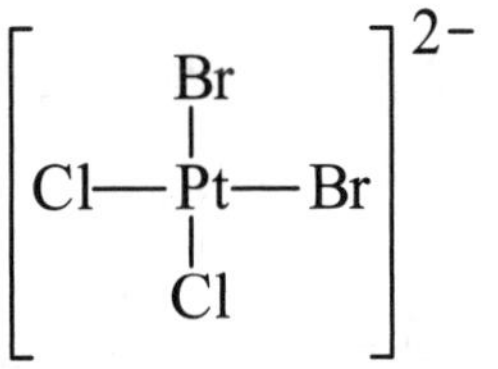 $[PtBr_2Cl_2]^{2-}$: Br above Pt, Cl—Pt—Br, Cl below Pt

8.83 SiF_4 is tetrahedral and nonpolar. SF_4 has one lone pair of electrons, adopts a see saw geometry, and is polar.

8.84

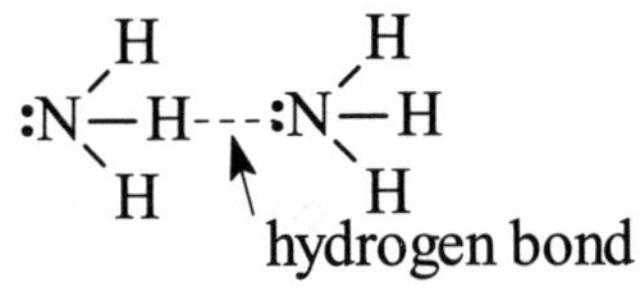

:N(H)(H)—H- - -:N(H)(H)—H

hydrogen bond

8.85

hydrogen bond (intramolecular O–H- - -O in HO–CH$_2$–CH$_2$–CH$_2$–OH ring arrangement: H–O, CH$_2$, CH$_2$, O–H, C H$_2$)

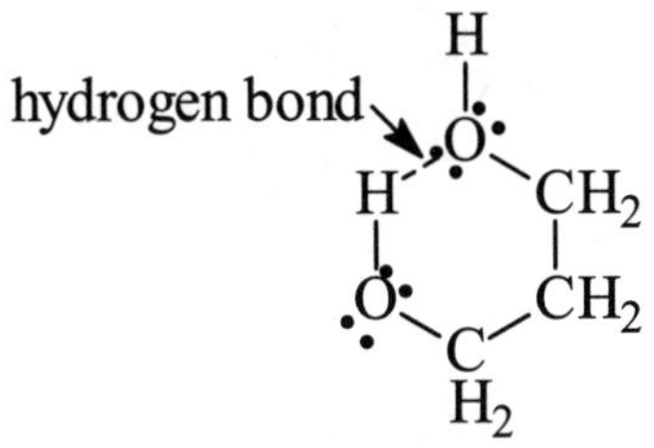

8.86 Illustrations (ii) and (iii) depict the hydrogen bonding that occurs between methylamine and water.

8.87 (a) No.
(b) Illustration (ii) depicts the hydrogen bonding that occurs between dimethyl ether and water.

Molecular Orbital Theory (Sections 8.7– 8.9)

8.88 Electrons in a bonding molecular orbital spend most of their time in the region between the two nuclei, helping to bond the atoms together. Electrons in an antibonding molecular orbital cannot occupy the central region between the nuclei and cannot contribute to bonding.

8.89 The additive combination of two 2s orbitals is lower in energy than the two isolated 2s orbitals and is called a bonding molecular orbital. The subtractive combination of two 2s orbitals is higher in energy than the two isolated 2s orbitals and is called an antibonding molecular orbital.

8.90

	O_2^+	O_2	O_2^-
σ^*_{2p}	—	—	—
π^*_{2p}	↑ —	↑ ↑	↑↓ ↑
π_{2p}	↑↓ ↑↓	↑↓ ↑↓	↑↓ ↑↓
σ_{2p}	↑↓	↑↓	↑↓
σ^*_{2s}	↑↓	↑↓	↑↓
σ_{2s}	↑↓	↑↓	↑↓

$$\text{Bond order} = \frac{\left(\begin{array}{c}\text{number of}\\ \text{bonding electrons}\end{array}\right) - \left(\begin{array}{c}\text{number of}\\ \text{antibonding electrons}\end{array}\right)}{2}$$

O_2^+ bond order $= \dfrac{8 - 3}{2} = 2.5$ O_2 bond order $= \dfrac{8 - 4}{2} = 2$

O_2^- bond order $= \dfrac{8 - 5}{2} = 1.5$

All are stable with bond orders between 1.5 and 2.5. All have unpaired electrons.

8.91

	N_2^+	N_2	N_2^-
σ^*_{2p}	—	—	—
π^*_{2p}	— —	— —	↑ —
σ_{2p}	↑	↑↓	↑↓
π_{2p}	↑↓ ↑↓	↑↓ ↑↓	↑↓ ↑↓
σ^*_{2s}	↑↓	↑↓	↑↓
σ_{2s}	↑↓	↑↓	↑↓

$$\text{Bond order} = \frac{\left(\begin{array}{c}\text{number of}\\ \text{bonding electrons}\end{array}\right) - \left(\begin{array}{c}\text{number of}\\ \text{antibonding electrons}\end{array}\right)}{2}$$

N_2^+ bond order $= \dfrac{7 - 2}{2} = 2.5$ N_2 bond order $= \dfrac{8 - 2}{2} = 3$

N_2^- bond order $= \dfrac{8 - 3}{2} = 2.5$

All are stable with bond orders of either 3 or 2.5. N_2^+ and N_2^- contain unpaired electrons.

8.92

	C_2	C_2^-
σ^*_{2p}	__	__
π^*_{2p}	__ __	__ __
σ_{2p}	__	↑
π_{2p}	↑↓ ↑↓	↑↓ ↑↓
σ^*_{2s}	↑↓	↑↓
σ_{2s}	↑↓	↑↓

$$\text{Bond order} = \frac{\begin{pmatrix}\text{number of}\\ \text{bonding electrons}\end{pmatrix} - \begin{pmatrix}\text{number of}\\ \text{antibonding electrons}\end{pmatrix}}{2}$$

(a) C_2 bond order $= \dfrac{6 - 2}{2} = 2$

(b) Add one electron because it will go into a bonding molecular orbital.

(c) C_2^- bond order $= \dfrac{7 - 2}{2} = 2.5$

8.93

	O_2	O_2^+
σ^*_{2p}	__	__
π^*_{2p}	↑ ↑	↑ __
π_{2p}	↑↓ ↑↓	↑↓ ↑↓
σ_{2p}	↑↓	↑↓
σ^*_{2s}	↑↓	↑↓
σ_{2s}	↑↓	↑↓

$$\text{Bond order} = \frac{\begin{pmatrix}\text{number of}\\ \text{bonding electrons}\end{pmatrix} - \begin{pmatrix}\text{number of}\\ \text{antibonding electrons}\end{pmatrix}}{2}$$

(a) O_2 bond order $= \dfrac{8 - 4}{2} = 2$

(b) Remove one electron because it will come out of an anti bonding molecular orbital.

(c) O_2^+ bond order $= \dfrac{8 - 3}{2} = 2.5$

8.94

(a) C_2^{2-} diamagnetic

σ^*_{2p}	__
π^*_{2p}	__ __
σ_{2p}	↑↓
π_{2p}	↑↓ ↑↓
σ^*_{2s}	↑↓
σ_{2s}	↑↓

(b) C_2^{2+} paramagnetic

σ^*_{2p}	__
π^*_{2p}	__ __
σ_{2p}	__
π_{2p}	↑ ↑
σ^*_{2s}	↑↓
σ_{2s}	↑↓

(c) F_2^- paramagnetic

σ^*_{2p}	↑
π^*_{2p}	↑↓ ↑↓
π_{2p}	↑↓ ↑↓
σ_{2p}	↑↓
σ^*_{2s}	↑↓
σ_{2s}	↑↓

(d) Cl_2 diamagnetic

σ^*_{2p}	__
π^*_{2p}	↑↓ ↑↓
π_{2p}	↑↓ ↑↓
σ_{2p}	↑↓
σ^*_{2s}	↑↓
σ_{2s}	↑↓

(e) Li_2^+ paramagnetic

σ^*_{2p}	__
π^*_{2p}	__ __
σ_{2p}	__
π_{2p}	↑ __
σ^*_{2s}	↑↓
σ_{2s}	↑↓

8.95

(a) O_2^{2+}	(b) N_2^{2+}	(c) C_2^{+}	(d) F_2^{2+}	(e) Cl_2^{+}
— σ^*_{2p}	— σ^*_{2p}	— σ^*_{2p}	— σ^*_{2p}	— σ^*_{2p}
— — π^*_{2p}	— — π^*_{2p}	— — π^*_{2p}	↑ ↑ π^*_{2p}	↑↓ ↑ π^*_{2p}
↑↓ ↑↓ π_{2p}	— σ_{2p}	— σ_{2p}	↑↓ ↑↓ π_{2p}	↑↓ ↑↓ π_{2p}
↑↓ σ_{2p}	↑↓ ↑↓ π_{2p}	↑↓ ↑ π_{2p}	↑↓ σ_{2p}	↑↓ σ_{2p}
↑↓ σ^*_{2s}	↑↓ σ^*_{2s}	↑↓ σ^*_{2s}	↑↓ σ^*_{2s}	↑↓ σ^*_{2s}
↑↓ σ_{2s}	↑↓ σ_{2s}	↑↓ σ_{2s}	↑↓ σ_{2s}	↑↓ σ_{2s}
diamagnetic	diamagnetic	paramagnetic	paramagnetic	paramagnetic

8.96

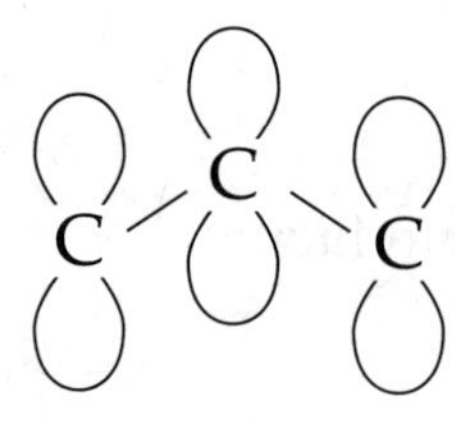

p orbitals in allyl cation

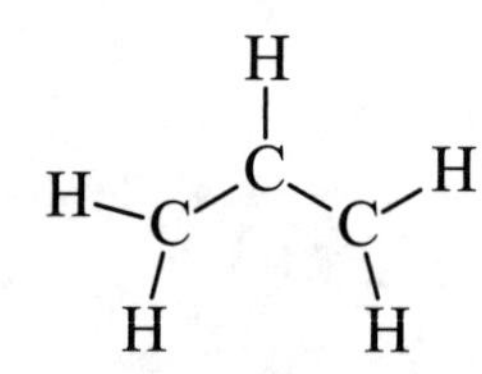

allyl cation showing only the σ bonds (each C is sp^2 hybridized)

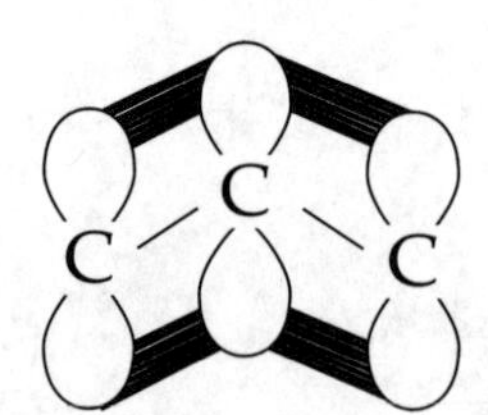

delocalized MO model for π bonding in the allyl cation

8.97

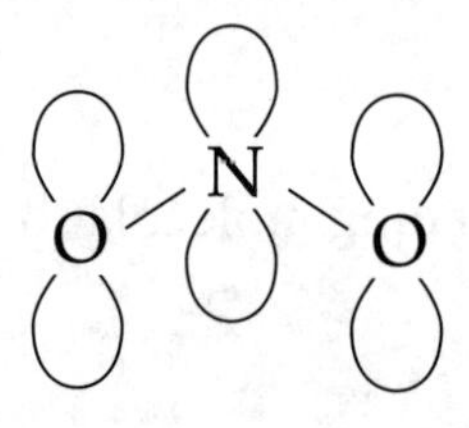

p orbitals in NO_2^-

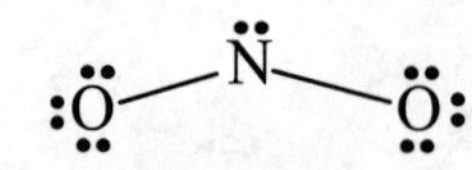

NO_2^- showing only the σ bonds (N is sp^2 hybridized)

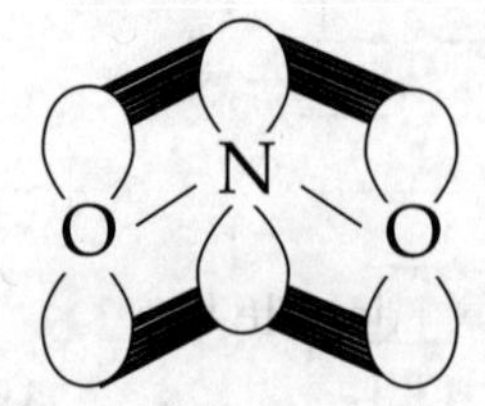

delocalized MO model for π bonding in NO_2^-

Chapter Problems

8.98 Because chlorine is larger than fluorine, the charge separation is larger in CH_3Cl compared to CH_3F, resulting in CH_3Cl having a slightly larger dipole moment.

8.99 Al_2O_3, ionic (greater lattice energy than NaCl because of higher ion charges); F_2, dispersion; H_2O, H–bonding, dipole-dipole; Br_2, dispersion (larger and more polarizable than F_2), ICl, dipole-dipole, NaCl, ionic
rank according to normal boiling points: $F_2 < Br_2 < ICl < H_2O < NaCl < Al_2O_3$

8.100

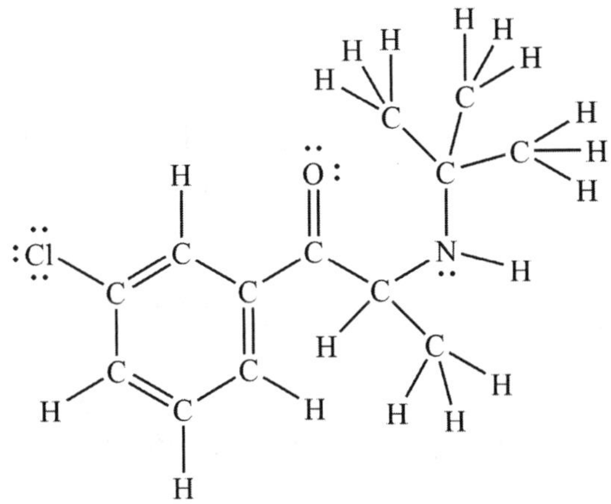

(a) There are 34 σ bonds and 4 π bonds.
(b) and (c) Each C with four single bonds is sp^3 hybridized with bond angles of 109.5°. Each C with a double bond is sp^2 hybridized with bond angles of 120°.
(d) The nitrogen is sp^3 hybridized.

8.101

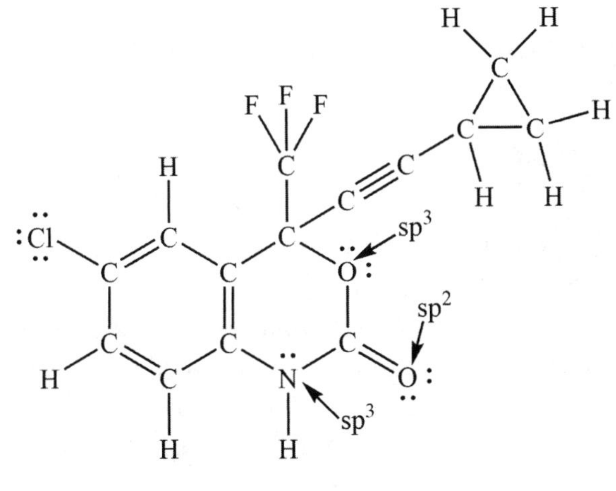

(a) There are 32 σ bonds and 6 π bonds.
(b) Each C with four single bonds is sp^3 hybridized. Each C with a double bond is sp^2 hybridized. Each C with a triple bond is sp hybridized.
(c) and (d) see figure

8.102 Both the B and N are sp^2 hybridized. All bond angles are ~120°. The overall geometry of the molecule is planar.

8.103 The triply bonded carbon atoms are sp hybridized. The theoretical bond angle for C–C≡C is 180°. Benzyne is so reactive because the C–C≡C bond angle is closer to 120° and is very strained.

8.104 (a) H—C≡C—H (b) H—N̈=N̈—H (c) :Ö: bonded to S; :C̈l—S̈—C̈l:

8.105

	Number of Bonded Atoms	Number of Lone Pairs	Shape
(a) BF_3	3	0	trigonal planar (120°)
PF_3	3	1 .	trigonal pyramidal (~107°)
PF_3 has the smaller F–X–F angles.			
(b) PCl_4^+	4	0	tetrahedral (109.5°)
ICl_2^-	2	3	linear (180°)
PCl_4^+ has the smaller Cl–X–Cl angles.			
(c) CCl_3^-	3	1	trigonal pyramidal (~107°)
PCl_6^-	6	0	octahedral (90°)
PCl_6^- has the smaller Cl–X–Cl angles.			

8.106 (a) (1) $[:\ddot{O}—C≡N:]^-$ (2) $[\ddot{O}=C=\ddot{N}]^-$ (3) $[:O≡C—\ddot{N}:]^-$

(b) Structure (1) makes the greatest contribution to the resonance hybrid because of the −1 formal charge on the oxygen. Structure (3) makes the least contribution to the resonance hybrid because of the +1 formal charge on the oxygen.
(c) and (d) OCN^- is linear because the C has 2 charge clouds. It is sp hybridized in all three resonance structures. It forms two π bonds.

8.107

21 σ bonds
5 π bonds
Each C with a double bond is sp^2 hybridized.
The $–CH_3$ carbon is sp^3 hybridized.

8.108 $[H—C≡N—\ddot{Xe}—\ddot{F}:]^+$ Both the carbon and nitrogen are sp hybridized.

8.109 (a)

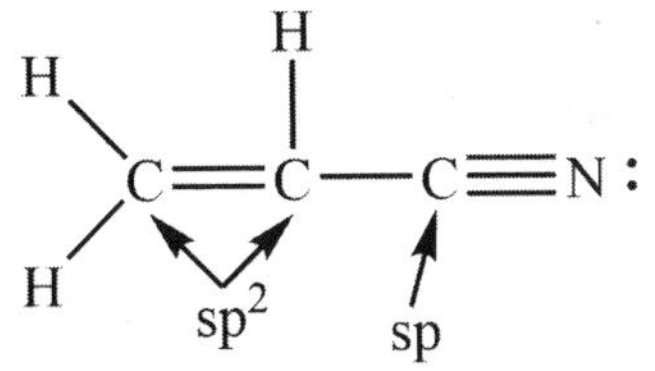

(b) There are 6 σ bonds and 3 π bonds.
(c) see figure
(d) The shortest bond is the C–N triple bond.

8.110 Every carbon is sp^2 hybridized. There are 18 σ bonds and 5 π bonds.

8.111

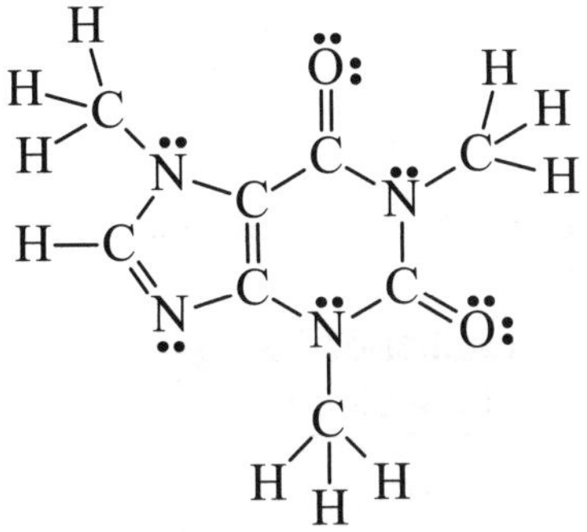

Each C with four single bonds is sp^3 hybridized.
Each C with a double bond is sp^2 hybridized.

8.112

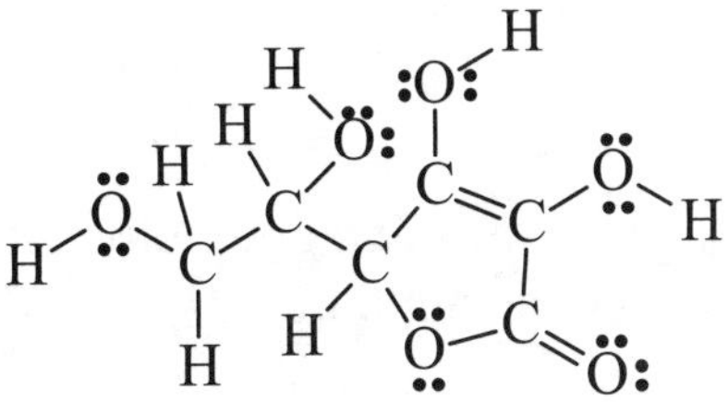

Each C with four single bonds is sp^3 hybridized.
Each C with a double bond is sp^2 hybridized.

8.113 Li_2

σ^*_{2s} —
σ_{2s} ↑↓
σ^*_{1s} ↑↓
σ_{1s} ↑↓

$$Li_2 \text{ Bond order} = \frac{\left(\begin{array}{c}\text{number of}\\ \text{bonding electrons}\end{array}\right) - \left(\begin{array}{c}\text{number of}\\ \text{antibonding electrons}\end{array}\right)}{2} = \frac{4-2}{2} = 1$$

The bond order for Li_2 is 1, and the molecule is likely to be stable.

8.114 C_2^{2-}

σ^*_{2p} —
π^*_{2p} — —
σ_{2p} ↑↓
π_{2p} ↑↓ ↑↓
σ^*_{2s} ↑↓
σ_{2s} ↑↓

$$\text{Bond order} = \frac{\begin{pmatrix}\text{number of}\\ \text{bonding electrons}\end{pmatrix} - \begin{pmatrix}\text{number of}\\ \text{antibonding electrons}\end{pmatrix}}{2}$$

C_2^{2-} bond order $= \dfrac{8 - 2}{2} = 3$; there is a triple bond between the two carbons.

8.115 (a)

	S_2	S_2^{2-}
σ^*_{3p}	—	—
π^*_{3p}	↑ ↑	↑↓ ↑↓
π_{3p}	↑↓ ↑↓	↑↓ ↑↓
σ_{3p}	↑↓	↑↓
σ^*_{3s}	↑↓	↑↓
σ_{3s}	↑↓	↑↓

(b) S_2 would be paramagnetic with two unpaired electrons in the π^*_{3p} MOs.

(c) $$\text{Bond order} = \frac{\begin{pmatrix}\text{number of}\\ \text{bonding electrons}\end{pmatrix} - \begin{pmatrix}\text{number of}\\ \text{antibonding electrons}\end{pmatrix}}{2}$$

S_2 bond order $= \dfrac{8 - 4}{2} = 2$

(d) S_2^{2-} bond order $= \dfrac{8 - 6}{2} = 1$

The two added electrons go into the antibonding π^*_{3p} MOs, the bond order drops from 2 to 1, and the bond length in S_2^{2-} should be longer than the bond length in S_2.

8.116 (a) CO

	CO
σ^*_{2p}	—
π^*_{2p}	— —
σ_{2p}	↑↓
π_{2p}	↑↓ ↑↓
σ^*_{2s}	↑↓
σ_{2s}	↑↓

(b) All electrons are paired, CO is diamagnetic.

(c)

$$\text{Bond order} = \frac{\begin{pmatrix}\text{number of}\\ \text{bonding electrons}\end{pmatrix} - \begin{pmatrix}\text{number of}\\ \text{antibonding electrons}\end{pmatrix}}{2}$$

CO bond order $= \dfrac{8 - 2}{2} = 3$

The bond order here matches that predicted by the electron-dot structure (:C≡O:).

(d)

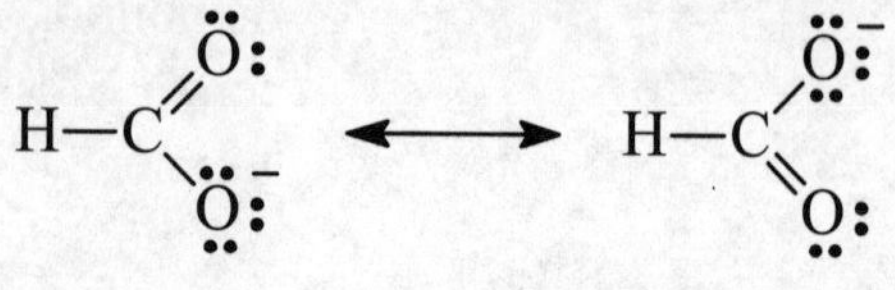

H—C(=Ö:)—Ö:⁻ ⟷ H—C(—Ö:⁻)=Ö:

8.117 (a)

The left S has 5 electron clouds (4 bonding, 1 lone pair). The geometry about this S is seesaw. The right S has 4 electron clouds (2 bonding, 2 lone pairs). The geometry about this S is bent.

(b)

The left C has 4 electron clouds (4 bonding, 0 lone pairs). The geometry about this C is tetrahedral. The right C has 3 electron clouds (3 bonding, 0 lone pairs). The geometry about this C is trigonal planar. The central two C's have 2 electron clouds (2 bonding, 0 lone pairs). The geometry about these two C's is linear.

8.118

Multiconcept Problems

8.119 (a)

(b) Each Cr atom has 6 pairs of electrons around it. The likely geometry about each Cr atom is tetrahedral because each Cr has 4 charge clouds.

8.120 (a)

polar polar nonpolar

(b) All three molecules are planar. The first two structures are polar because they both have an unsymmetrical distribution of atoms about the center of the molecule (the middle of the double bond), and bond polarities do not cancel. Structure 3 is nonpolar because the H's and Cl's, respectively, are symmetrically distributed about the center of the molecule, both being opposite each other. In this arrangement, bond polarities cancel.

(c) 200 nm = 200 x 10^{-9} m

$$E = \frac{hc}{\lambda} = \frac{(6.626 \times 10^{-34}\ J\cdot s)(3.00 \times 10^{8}\ m/s)}{200 \times 10^{-9}\ m}(6.022 \times 10^{23}/mol)$$

E = 5.99 x 10^5 J/mol = 599 kJ/mol

(d)

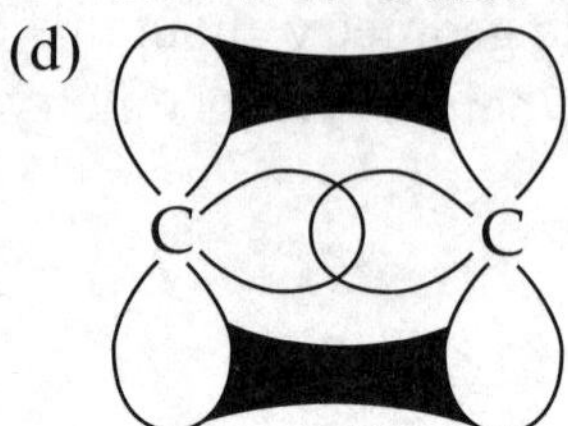

The π bond must be broken before rotation can occur.

8.121 (a) Each carbon is sp^2 hybridized.

(b) and (c)

antibonding	—
antibonding	— —
nonbonding	↑↓ ↑↓
bonding	↑↓ ↑↓
bonding	↑↓

(d) The cyclooctatetraene dianion has only paired electrons and is diamagnetic.

9 Thermochemistry: Chemical Energy

9.1 $\Delta V = (4.3\ L - 8.6\ L) = -4.3\ L$

$w = -P\Delta V = -(44\ atm)(-4.3\ L) = +189.2\ L\cdot atm$

$w = (189.2\ L\cdot atm)\left(101\ \frac{J}{L\cdot atm}\right) = +1.9 \times 10^4\ J$

The positive sign for the work indicates that the surroundings do work on the system. Energy flows into the system.

9.2 $w = -P\Delta V = -(2.5\ atm)(3\ L - 2\ L) = -2.5\ L\cdot atm$

$w = (-2.5\ L\cdot atm)\left(101\ \frac{J}{L\cdot atm}\right) = -252.5\ J = -250\ J = -0.25\ kJ$

The negative sign indicates that the expanding system loses work energy and does work on the surroundings.

9.3 $\Delta H^\circ = \frac{-484\ kJ}{2\ mol\ H_2}$

$P\Delta V = (1.00\ atm)(-24.4\ L) = -24.4\ L\cdot atm$

$P\Delta V = (-24.4\ L\cdot atm)\left(101\ \frac{J}{L\cdot atm}\right) = -2464\ J = -2.46\ kJ$

$w = -P\Delta V = +2.46\ kJ$

$\Delta H^\circ = \frac{-484\ kJ}{2\ mol\ H_2}$

$\Delta E = \Delta H - P\Delta V = -484\ kJ - (-2.46\ kJ) = -481.5\ kJ = -482\ kJ$

9.4 (a) $w = -P\Delta V$ is positive and $P\Delta V$ is negative for this reaction because the system volume is decreased at constant pressure.

(b) $P\Delta V$ is small compared to ΔE.

$\Delta H = \Delta E + P\Delta V$; ΔH is negative. Its value is slightly more negative than ΔE.

9.5 (a) H_2, 2.016

$10.00\ g\ H_2 \times \frac{1\ mol\ H_2}{2.016\ g\ H_2} \times \frac{-571.6\ kJ}{2\ mol\ H_2} = -1418\ kJ$

1418 kJ of heat are evolved.

(b) $5.500\ mol\ H_2O \times \frac{571.6\ kJ}{2\ mol\ H_2O} = 1572\ kJ$

1572 kJ of heat are absorbed.

9.6 C_3H_8, 44.09

$$1.8 \times 10^6 \text{ kJ} \times \frac{1 \text{ mol } C_3H_8}{2044 \text{ kJ}} \times \frac{44.09 \text{ g } C_3H_8}{1 \text{ mol } C_3H_8} \times \frac{1 \text{ kg}}{1000 \text{ g}} = 39 \text{ kg } C_3H_8$$

9.7 (a) heat transfers from system to surroundings, exothermic, ΔH° is negative
(b) heat transfers from system to surroundings, exothermic, ΔH° is negative
(c) heat transfers from surroundings to system, endothermic, ΔH° is positive

9.8 LiF, 25.9; $BaSO_4$, 233.4
Only LiF and $BaSO_4$ will produce endothermic reactions and lower the water temperature.

$$10.00 \text{ g LiF} \times \frac{1 \text{ mol LiF}}{25.9 \text{ g LiF}} \times \frac{5.5 \text{ kJ}}{1 \text{ mol LiF}} = 2.12 \text{ kJ}$$

$$10.00 \text{ g } BaSO_4 \times \frac{1 \text{ mol } BaSO_4}{233.4 \text{ g } BaSO_4} \times \frac{26.3 \text{ kJ}}{1 \text{ mol } BaSO_4} = 1.13 \text{ kJ}$$

10.00 g of LiF would result in the largest temperature decrease.

9.9 q = (specific heat capacity) x m x ΔT

$$\text{specific heat capacity} = \frac{q}{m \times \Delta T} = \frac{97.2 \text{ J}}{(75.0 \text{ g})(10.0\,^\circ\text{C})} = 0.130 \text{ J/(g} \cdot {}^\circ\text{C)}$$

$$\text{molar heat capacity} = 0.130 \text{ J/(g} \cdot {}^\circ\text{C)} \times \frac{207.2 \text{ g Pb}}{1 \text{ mol Pb}} = 26.9 \text{ J/(mol} \cdot {}^\circ\text{C)}$$

9.10 $$\text{heat capacity} = \frac{1650 \text{ J}}{2.00\,^\circ\text{C}} = 825 \text{ J/}^\circ\text{C}$$

9.11 25.0 mL = 0.0250 L and 50.0 mL = 0.0500 L
mol H_2SO_4 = (1.00 mol/L)(0.0250 L) = 0.0250 mol H_2SO_4
mol NaOH = (1.00 mol/L)(0.0500 L) = 0.0500 mol NaOH
NaOH and H_2SO_4 are present in a 2:1 mol ratio. This matches the stoichiometric ratio in the balanced equation.
q = (specific heat) x m x ΔT
m = (25.0 mL + 50.0 mL)(1.00 g/mL) = 75.0 g

$$q = \left(4.18 \frac{\text{J}}{\text{g} \cdot {}^\circ\text{C}}\right)(75.0 \text{ g})(33.9\,^\circ\text{C} - 25.0\,^\circ\text{C}) = 2790 \text{ J}$$

$$\text{mol } H_2SO_4 = 0.0250 \text{ L} \times 1.00 \frac{\text{mol}}{\text{L}} H_2SO_4 = 0.0250 \text{ mol } H_2SO_4$$

$$\text{Heat evolved per mole of } H_2SO_4 = \frac{2.79 \times 10^3 \text{ J}}{0.0250 \text{ mol } H_2SO_4} = 1.1 \times 10^5 \text{ J/mol } H_2SO_4$$

Because the reaction evolves heat, the sign for ΔH is negative.

$$\Delta H = -1.1 \times 10^5 \text{ J/mol} \times \frac{1 \text{ kJ}}{1000 \text{ J}} = -1.1 \times 10^2 \text{ kJ/mol}$$

9.12 $q = (4.18 \frac{J}{g \cdot {}^{o}C})(250.0\ g)(18.6\ {}^{o}C) + (623\ J/{}^{o}C)(18.6\ {}^{o}C) = 31{,}025\ J$

$\Delta H = -q = \Delta E = -31{,}025\ J/2\ g \times \frac{1\ kJ}{1000\ J} = -15.5\ kJ/g$

$\Delta H = -q = \Delta E = -15.5\ kJ/g \times \frac{1\ Cal}{4.184\ kJ} = -3.71\ Cal/g$

9.13

	$C(s) + O_2(g) \rightarrow CO_2(g)$	$\Delta H^o = -393.5\ kJ$
	$H_2O(g) \rightarrow 1/2\ O_2(g) + H_2(g)$	$\Delta H^o = +483.6/2\ kJ$
	$CO2(g) \rightarrow 1/2\ O_2(g) + CO(g)$	$\Delta H^o = +566.0/2\ kJ$
sum	$C(s) + H_2O(g) \rightarrow CO(g) + H_2(g)$	$\Delta H^o = +131.3\ kJ$

9.14 (a) $A + 2\ B \rightarrow D$; $\Delta H^o = -100\ kJ + (-50\ kJ) = -150\ kJ$

(b) The red arrow corresponds to step 1: $A + B \rightarrow C$
The green arrow corresponds to step 2: $C + B \rightarrow D$
The blue arrow corresponds to the overall reaction.

(c) The top energy level represents A + 2 B.
The middle energy level represents C + B.
The bottom energy level represents D.

9.15 $4\ NH_3(g) + 5\ O_2(g) \rightarrow 4\ NO(g) + 6\ H_2O(g)$

$\Delta H^o_{rxn} = [4\ \Delta H^o_f\ (NO) + 6\ \Delta H^o_f\ (H_2O)] - [4\ \Delta H^o_f\ (NH_3)]$

$\Delta H^o_{rxn} = [(4\ mol)(91.3\ kJ/mol) + (6\ mol)(-241.8\ kJ/mol)] - [(4\ mol)(-46.1\ kJ/mol)]$

$\Delta H^o_{rxn} = -901.2\ kJ$

9.16 $2\ C_8H_{18}(l) + 25\ O_2(g) \rightarrow 8\ CO_2(g) + 9\ H_2O(l)$

$\Delta H^o_{rxn} = \Delta H^o_c = -5220\ kJ$

$\Delta H^o_{rxn} = [8\ \Delta H^o_f(CO_2) + 9\ \Delta H^o_f(H_2O)] - [2\ \Delta H^o_f(C_8H_{18})]$

$-5220\ kJ = [(8\ mol)(-393.5\ kJ/mol) + (9\ mol)(-285.8\ kJ/mol)] - [(2\ mol)(\Delta H^o_f(C_8H_{18}))]$

Solve for $\Delta H^o_f(C_8H_{18})$

$-5220\ kJ = -5720.2\ kJ - (2\ mol)(\Delta H^o_f(C_8H_{18}))$

$+500\ kJ = -(2\ mol)(\Delta H^o_f(C_8H_{18}))$

$\Delta H^o_f(C_8H_{18}) = \frac{500\ kJ}{-2\ mol} = -250\ kJ/mol$

9.17 $H_2C{=}CH_2(g) + H_2O(g) \rightarrow C_2H_5OH(g)$

$\Delta H^o_{rxn} = D\ \text{(Reactant bonds)} - D\ \text{(Product bonds)}$

$\Delta H^o_{rxn} = (D_{C=C} + 4\ D_{C-H} + 2\ D_{O-H}) - (D_{C-C} + D_{C-O} + 5\ D_{C-H} + D_{O-H})$

$\Delta H^o_{rxn} = [(1\ mol)(728\ kJ/mol) + (4\ mol)(410\ kJ/mol) + (2\ mol)(460\ kJ/mol)]$
$- [(1\ mol)(350\ kJ/mol) + (1\ mol)(350\ kJ/mol) + (5\ mol)(410\ kJ/mol) + (1\ mol)(460\ kJ/mol)]$

$\Delta H^o_{rxn} = +78\ kJ$

9.18 $\Delta H^o_{rxn} = D\ \text{(Reactant bonds)} - D\ \text{(Product bonds)}$

(a) $2\ C_6H_6(g) + 15\ O_2(g) \rightarrow 12\ CO_2(g) + 6\ H_2O(g)$

$\Delta H^o_{rxn} = (12\ D_{C-C} + 12\ D_{C-H} + 15\ D_{O=O}) - (24\ D_{C=O} + 12\ D_{O-H}) = -6339\ kJ$

$\Delta H^\circ_{rxn} = [(12\ mol)(D_{C-C}) + (12\ mol)(410\ kJ/mol) + (15\ mol)(498\ kJ/mol)]$
$\quad - [(24\ mol)(804\ kJ/mol) + (12\ mol)(460\ kJ/mol)] = -6339\ kJ$

$(12\ mol)(D_{C-C}) = -6339\ kJ + [(24\ mol)(804\ kJ/mol) + (12\ mol)(460\ kJ/mol)]$
$\quad - [(12\ mol)(410\ kJ/mol) + (15\ mol)(498\ kJ/mol)]$

$(12\ mol)(D_{C-C}) = 6087\ kJ$

$D_{C-C} = \dfrac{6087\ kJ}{12\ mol} = 507\ kJ/mol$ for C– C bond in benzene

9.19 ΔS° is negative because the reaction decreases the number of moles of gaseous molecules.

9.20 The reaction proceeds from a solid and a gas (reactants) to all gas (product). Randomness increases, so ΔS° is positive.

9.21 The reaction involves only bond making, so it is exothermic and ΔH is negative. There are more reactant atoms than product molecules. The randomness of the system decreases on going from reactant to product, therefore ΔS is negative. Because the reaction is spontaneous, ΔG is negative.

9.22 (a) $2\ A_2 + B_2 \rightarrow 2\ A_2B$
(b) Because the reaction is exothermic, ΔH is negative. There are more reactant molecules than product molecules. The randomness of the system decreases on going from reactant to product, therefore ΔS is negative.
(c) Because $\Delta G = \Delta H - T\Delta S$, a reaction with both ΔH and ΔS negative is favored at low temperatures where the negative ΔH term is larger than the positive $- T\Delta S$, and ΔG is negative.

9.23 $\Delta G^\circ = \Delta H^\circ - T\Delta S^\circ = (-92.2\ kJ) - (298\ K)(-0.199\ kJ/K) = -32.9\ kJ$
Because ΔG° is negative, the reaction is spontaneous.
Set $\Delta G^\circ = 0$ and solve for T.

$$\Delta G^\circ = 0 = \Delta H^\circ - T\Delta S^\circ; \quad T = \frac{\Delta H^\circ}{\Delta S^\circ} = \frac{-92.2\ kJ}{-0.199\ kJ/K} = 463\ K = 190\ ^\circ C$$

9.24 $\Delta G^\circ = \Delta H^\circ - T\Delta S^\circ$
Set $\Delta G^\circ = 0$ and solve for T.

$$\Delta G^\circ = 0 = \Delta H^\circ - T\Delta S^\circ; \quad T = \frac{\Delta H^\circ}{\Delta S^\circ} = \frac{+75.0\ kJ}{+0.231\ kJ/K} = 325\ K = 52\ ^\circ C$$

The reaction is at equilibrium at 325 K (52 °C). The temperature should be increased to make the reaction spontaneous.

9.25 (a) $C_2H_6O(l) + 3\ O_2(g) \rightarrow 2\ CO_2(g) + 3\ H_2O(l)$
(b) $\Delta H^\circ_c = [2\ \Delta H^\circ_f(CO_2) + 3\ \Delta H^\circ_f(H_2O)] - \Delta H^\circ_f(C_2H_6O)$
$\Delta H^\circ_c = [(2\ mol)(-393.5\ kJ/mol) + (3\ mol)(-285.8\ kJ/mol)] - [(1\ mol)(-277.7\ kJ/mol)]$
$\Delta H^\circ_c = -1367\ kJ$

(c) C_2H_6O, 46.1; $\Delta H^\circ_c = -1367\ kJ/mol \times \dfrac{1\ mol\ C_2H_6O}{46.1\ g\ C_2H_6O} = -29.7\ kJ/g$

(d) $\Delta H^\circ_c = (-29.7\ kJ/g)(0.789\ g/mL) = -23.4\ kJ/mL$

9.26 $\Delta H^{o}_{rxn} = D\text{ (Reactant bonds)} - D\text{ (Product bonds)}$

$C_2H_6O(l) + 3\ O_2(g) \rightarrow 2\ CO_2(g) + 3\ H_2O(l)$

$\Delta H^{o}_{c} = (D_{C\text{–}C} + 5\ D_{C\text{–}H} + D_{C\text{–}O} + D_{O\text{–}H} + 3\ D_{O=O}) - (4\ D_{C=O} + 6\ D_{O\text{–}H})$

$\Delta H^{o}_{c} = [(1\text{ mol})(350\text{ kJ/mol}) + (5\text{ mol})(410\text{ kJ/mol}) + (1\text{ mol})(350\text{ kJ/mol})$
$+ (1\text{ mol})(460\text{ kJ/mol}) + (3\text{ mol})(498\text{ kJ/mol})]$
$- [(4\text{ mol})(804\text{ kJ/mol}) + (6\text{ mol})(460\text{ kJ/mol})] = -1272\text{ kJ/mol}$

9.27 (a) $\Delta V = (49.0\text{ L} - 73.5\text{ L}) = -24.5\text{ L}$

$P\Delta V = (1.00\text{ atm})(-24.5\text{ L}) = -24.5\text{ L}\cdot\text{atm}$

$P\Delta V = (-24.5\text{ L}\cdot\text{atm})(101\ \frac{\text{J}}{\text{L}\cdot\text{atm}}) = -2475\text{ J} = -2.47\text{ kJ}$

$w = -P\Delta V = +2.47\text{ kJ}$

Work energy is transferred to the system.

(b) $\Delta E = \Delta H - P\Delta V = -1367\text{ kJ} - (-2.47\text{ kJ}) = -1365\text{ kJ/mol}$

9.28 (a) $q = (4.18\ \frac{\text{J}}{\text{g}\cdot{}^{o}\text{C}})(300.0\text{ g})(6.85\,^{o}\text{C}) + (675\text{ J/}^{o}\text{C})(6.85\,^{o}\text{C}) = 13{,}214\text{ J}$

$\Delta H = -q = \Delta E = -13{,}214\text{ J/0.350 g} \times \frac{1\text{ kJ}}{1000\text{ J}} = -37.8\text{ kJ/g}$

(b) $C_{19}H_{38}O_2$, 298.5

$\Delta E = -37.8\text{ kJ/g} \times \frac{298.5\text{ g } C_{19}H_{38}O_2}{1\text{ mol } C_{19}H_{38}O_2} = -1.13 \times 10^{4}\text{ kJ/mol}$

(c) $\Delta E = (-37.8\text{ kJ/g})(0.880\text{ g/mL}) = -33.3\text{ kJ/mL}$

9.29 C_2H_6O, 46.1

$\text{heat} = 35.8\text{ kg } C_2H_6O \times \frac{1000\text{ g}}{1\text{ kg}} \times \frac{1\text{ mol } C_2H_6O}{46.1\text{ g } C_2H_6O} \times \frac{-1367\text{ kJ}}{1\text{ mol } C_2H_6O} \times \frac{1\text{ MJ}}{1000\text{ kJ}} = -1060\text{ MJ}$

$C_{19}H_{38}O_2$, 298.5

$\text{heat} = 39.9\text{ kg } C_{19}H_{38}O_2 \times \frac{1000\text{ g}}{1\text{ kg}} \times \frac{1\text{ mol } C_{19}H_{38}O_2}{298.5\text{ g } C_{19}H_{38}O_2} \times \frac{-11{,}236\text{ kJ}}{1\text{ mol } C_{19}H_{38}O_2} \times \frac{1\text{ MJ}}{1000\text{ kJ}} = -1500\text{ MJ}$

heat (octane) = 1530 MJ

Conceptual Problems

9.30 $\Delta H > 0$ and $w < 0$

9.31 (a) $w = -P\Delta V$, $\Delta V > 0$; therefore $w < 0$ and the system is doing work on the surroundings.

(b) Since the temperature has increased there has been an enthalpy change. The system evolved heat, the reaction is exothermic, and $\Delta H < 0$.

9.32 (a) (b)

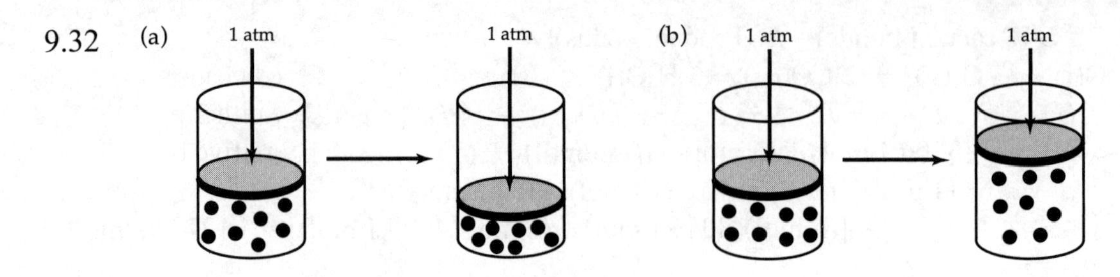

9.33 $\Delta H = \Delta E + P\Delta V$

$\Delta H - \Delta E = P\Delta V$

$$\Delta V = \frac{\Delta H - \Delta E}{P} = \frac{[-35.0\text{ kJ} - (-34.8\text{ kJ})]}{1\text{ atm}} \times \frac{1\text{ L}\cdot\text{atm}}{101 \times 10^{-3}\text{ kJ}} = -2\text{ L}$$

$\Delta V = -2\text{ L} = V_{final} - V_{initial} = V_{final} - 5\text{ L}$; $V_{final} = -2\text{ L} - (-5\text{ L}) = 3\text{ L}$

The volume decreases from 5 L to 3 L.

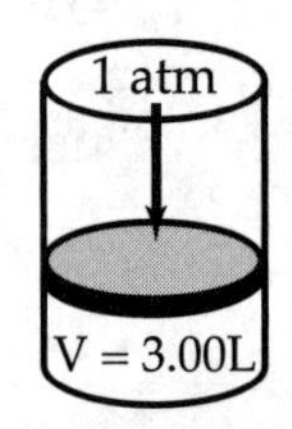

9.34 (a) Diagram 1 (b) $\Delta H° = +30$ kJ

9.35

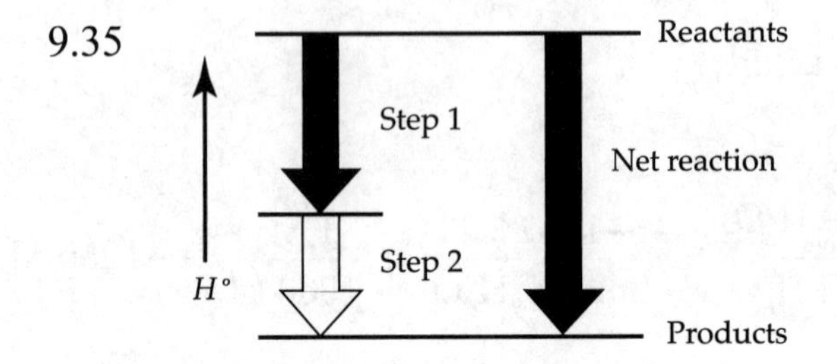

9.36 (a) $2\ AB_2 \rightarrow A_2 + 2\ B_2$

(b) Because the reaction is exothermic, ΔH is negative. Because the number of molecules increases going from reactants to products, ΔS is positive.

(c) When ΔH is negative and ΔS is positive, the reaction is likely to be spontaneous at all temperatures.

9.37 $\Delta H° = +55$ kJ

$\Delta S°$ is positive because randomness increases when a solid is converted to a gas.

$\Delta G° = \Delta H° - T\Delta S°$; For the reaction to be spontaneous, $\Delta G°$ must be negative.

Because $\Delta H°$ and $\Delta S°$ are both positive, the reaction is spontaneous at some higher temperatures, but nonspontaneous at some lower temperatures.

9.38 The change is the spontaneous conversion of a liquid to a gas. ΔG is negative because the change is spontaneous. The conversion of a liquid to a gas is endothermic, therefore ΔH is positive. ΔS is positive because randomness increases when a liquid is converted to a gas.

9.39 (a) $2\ A_3 \rightarrow 3\ A_2$
(b) Because the reaction is spontaneous, ΔG is negative. ΔS is positive because the number of molecules increases in going from reactant to products. ΔH could be either positive or negative and the reaction would still be spontaneous. ΔH is probably positive because there is more bond breaking than bond making.

Section Problems
Heat, Work, and Energy (Sections 9.1–9.3)

9.40 Heat is the energy transferred from one object to another as the result of a temperature difference between them. Temperature is a measure of the kinetic energy of molecular motion.
Energy is the capacity to do work or supply heat. Work is defined as the distance moved times the force that opposes the motion (w = d x F).
Kinetic energy is the energy of motion. Potential energy is stored energy.

9.41 Internal energy is the sum of kinetic and potential energies for each particle in the system.

9.42 Car: $E_K = \frac{1}{2}(1400\ \text{kg})\left(\frac{115 \times 10^3\ \text{m}}{3600\ \text{s}}\right)^2 = 7.1 \times 10^5\ \text{J}$

Truck: $E_K = \frac{1}{2}(12{,}000\ \text{kg})\left(\frac{38 \times 10^3\ \text{m}}{3600\ \text{s}}\right)^2 = 6.7 \times 10^5\ \text{J}$

The car has more kinetic energy.

9.43 Heat = q = 7.1×10^5 J (from Problem 9.42)
q = (specific heat) x m x ΔT

$$m = \frac{q}{(\text{specific heat}) \times \Delta T} = \frac{7.1 \times 10^5\ \text{J}}{\left(4.18\ \frac{\text{J}}{\text{g}\cdot{}^\circ\text{C}}\right)(50\ {}^\circ\text{C} - 20\ {}^\circ\text{C})} = 5.7 \times 10^3\ \text{g of water}$$

9.44 (a) and (b) are state functions; (c) is not.

9.45 (a), (b) and (c) are state functions.

9.46 $w = -P\Delta V = -(3.6\ \text{atm})(3.4\ \text{L} - 3.2\ \text{L}) = -0.72\ \text{L}\cdot\text{atm}$

$w = (-0.72\ \text{L}\cdot\text{atm})\left(\frac{101\ \text{J}}{1\ \text{L}\cdot\text{atm}}\right) = -72.7\ \text{J} = -70\ \text{J}$; The energy change is negative.

9.47 $V_{initial} = 50.0\ \text{mL} + 50.0\ \text{mL} = 100.0\ \text{mL} = 0.1000\ \text{L}$
$V_{final} = 50.0\ \text{mL} = 0.0500\ \text{L}$
$\Delta V = V_{final} - V_{initial} = (0.0500\ \text{L} - 0.1000\ \text{L}) = -0.0500\ \text{L}$
$w = -P\Delta V = -(1.5\ \text{atm})(-0.0500\ \text{L}) = +0.075\ \text{L}\cdot\text{atm}$

$$w = (+0.075\ \text{L}\cdot\text{atm})\left(\frac{101\ \text{J}}{1\ \text{L}\cdot\text{atm}}\right) = +7.6\ \text{J}$$

The positive sign for the work indicates that the surroundings do work on the system. Energy flows into the system.

9.48 (a) 100 W = 100 J/s

$$50\ \text{Cal} \times \frac{1000\ \text{cal}}{1\ \text{Cal}} \times \frac{4.184\ \text{J}}{1\ \text{cal}} \times \frac{1\ \text{s}}{100\ \text{J}} \times \frac{1\ \text{min}}{60\ \text{s}} = 35\ \text{min}$$

(b) 23 W = 23 J/s

$$50\ \text{Cal} \times \frac{1000\ \text{cal}}{1\ \text{Cal}} \times \frac{4.184\ \text{J}}{1\ \text{cal}} \times \frac{1\ \text{s}}{23\ \text{J}} \times \frac{1\ \text{min}}{60\ \text{s}} = 150\ \text{min}$$

9.49 (a) 275 W = 275 J/s

$$440\ \text{Cal} \times \frac{1000\ \text{cal}}{1\ \text{Cal}} \times \frac{4.184\ \text{J}}{1\ \text{cal}} \times \frac{1\ \text{s}}{275\ \text{J}} \times \frac{1\ \text{min}}{60\ \text{s}} \times \frac{1\ \text{h}}{60\ \text{min}} = 1.86\ \text{h}$$

(b) 175 W = 175 J/s

$$440\ \text{Cal} \times \frac{1000\ \text{cal}}{1\ \text{Cal}} \times \frac{4.184\ \text{J}}{1\ \text{cal}} \times \frac{1\ \text{s}}{175\ \text{J}} \times \frac{1\ \text{min}}{60\ \text{s}} \times \frac{1\ \text{h}}{60\ \text{min}} = 2.92\ \text{h}$$

9.50 (a) $w = -P\Delta V = -(0.975\ \text{atm})(18.0\ \text{L} - 12.0\ \text{L}) = -5.85\ \text{L}\cdot\text{atm}$

$$w = (-5.85\ \text{L}\cdot\text{atm})\left(\frac{101.325\ \text{J}}{1\ \text{L}\cdot\text{atm}}\right) = -593\ \text{J}$$

(b) $\Delta V = 6.0\ \text{L} = 6000\ \text{mL} = 6000\ \text{cm}^3$; $r = \frac{17.0\ \text{cm}}{2} = 8.50\ \text{cm}$

$$V = \pi r^2 h;\ h = \frac{V}{\pi r^2} = \frac{6000\ \text{cm}^3}{\pi(8.50\ \text{cm})^2} = 26.4\ \text{cm}$$

9.51 (a) $r = \frac{40.0\ \text{cm}}{2} = 20.0\ \text{cm}$

$V = \pi r^2 h = \pi(20.0\ \text{cm})^2(-65.0\ \text{cm}) = -81{,}700\ \text{cm}^3 = -81{,}700\ \text{mL} = -81.7\ \text{L}$

(b) $w = -P\Delta V = -(0.905\ \text{atm})(-81.7\ \text{L}) = 73.9\ \text{L}\cdot\text{atm}$

$$w = (73.9\ \text{L}\cdot\text{atm})\left(\frac{101.325\ \text{J}}{1\ \text{L}\cdot\text{atm}}\right) = 7.49 \times 10^3\ \text{J}$$

Energy and Enthalpy (Section 9.4)

9.52 $\Delta E = q_v$ is the heat change associated with a reaction at constant volume. Since $\Delta V = 0$, no PV work is done.
$\Delta H = q_p$ is the heat change associated with a reaction at constant pressure. Since $\Delta V \neq 0$, PV work can also be done.

9.53 ΔH is negative for an exothermic reaction. ΔH is positive for an endothermic reaction.

9.54 $\Delta H = \Delta E + P\Delta V$; ΔH and ΔE are nearly equal when there are no gases involved in a chemical reaction, or, if gases are involved, $\Delta V = 0$ (i.e., there are the same number of reactant and product gas molecules).

9.55 Heat is lost on going from $H_2O(g) \rightarrow H_2O(l) \rightarrow H_2O(s)$.
$H_2O(g)$ has the highest enthalpy content. $H_2O(s)$ has the lowest enthalpy content.

9.56 $\Delta V = 448$ L and assume $P = 1.00$ atm

$w = -P\Delta V = -(1.00 \text{ atm})(448 \text{ L}) = -448 \text{ L}\cdot\text{atm}$

$$w = -(448 \text{ L}\cdot\text{atm})\left(101 \frac{\text{J}}{\text{L}\cdot\text{atm}}\right) = -4.52 \times 10^4 \text{ J}$$

$$w = -4.52 \times 10^4 \text{ J} \times \frac{1 \text{ kJ}}{1000 \text{ J}} = -45.2 \text{ kJ}$$

9.57 $$\Delta H^\circ = -484 \frac{\text{kJ}}{2 \text{ mol } H_2}$$

$P\Delta V = (1.00 \text{ atm})(-5.6 \text{ L}) = -5.6 \text{ L}\cdot\text{atm}$

$$P\Delta V = (-5.6 \text{ L}\cdot\text{atm})\left(101 \frac{\text{J}}{\text{L}\cdot\text{atm}}\right) = -565.6 \text{ J} = -570 \text{ J} = -0.57 \text{ kJ}$$

$w = -P\Delta V = 570 \text{ J} = 0.57 \text{ kJ}$

$$\Delta H = \frac{-121 \text{ kJ}}{0.50 \text{ mol } H_2}$$

$\Delta E = \Delta H - P\Delta V = -121 \text{ kJ} - (-0.57 \text{ kJ}) = -120.43 \text{ kJ} = -120 \text{ kJ}$

9.58 $P\Delta V = -7.6$ J (from Problem 9.47)
$\Delta H = \Delta E + P\Delta V$
$\Delta E = \Delta H - P\Delta V = -0.31 \text{ kJ} - (-7.6 \times 10^{-3} \text{ kJ}) = -0.30 \text{ kJ}$

9.59 $\Delta H = -244$ kJ and $w = -P\Delta V = 35$ kJ; therefore $P\Delta V = -35$ kJ
$\Delta E = \Delta H - P\Delta V = -244 \text{ kJ} - (-35 \text{ kJ}) = -209 \text{ kJ}$
For the system: $\Delta H = -244$ kJ and $\Delta E = -209$ kJ
ΔH and ΔE for the surroundings are just the opposite of what they are for the system.
For the surroundings: $\Delta H = 244$ kJ and $\Delta E = 209$ kJ

9.60 $$\Delta H = \Delta E + P\Delta V = 44.0 \text{ kJ} + (1.00 \text{ atm})(14.0 \text{ L})\left(\frac{101.325 \text{ J}}{1 \text{ L}\cdot\text{atm}} \times \frac{1 \text{ kJ}}{1000 \text{ J}}\right) = 45.4 \text{ kJ}$$

9.61 $$\Delta H = \Delta E + P\Delta V = 71.5 \text{ kJ} + (1.10 \text{ atm})(-13.6 \text{ L})\left(\frac{101.325 \text{ J}}{1 \text{ L}\cdot\text{atm}} \times \frac{1 \text{ kJ}}{1000 \text{ J}}\right) = 70.0 \text{ kJ}$$

Thermochemical Equations for Chemical and Physical Changes (Sections 9.5 and 9.6)

9.62 (a) heat is transferred to the system, endothermic, $\Delta H^\circ > 0$
(b) heat is transferred to the surroundings, exothermic, $\Delta H^\circ < 0$
(c) heat is transferred to the system, endothermic, $\Delta H^\circ > 0$

9.63 (a) heat is transferred to the system, endothermic, $\Delta H^\circ > 0$
(b) heat is transferred to the system, endothermic, $\Delta H^\circ > 0$
(c) heat is transferred to the surroundings, exothermic, $\Delta H^\circ < 0$

9.64 $C_4H_{10}O$, 74.12; mass of $C_4H_{10}O$ = (0.7138 g/mL)(100 mL) = 71.38 g

$$\text{mol } C_4H_{10}O = 71.38 \text{ g} \times \frac{1 \text{ mol}}{74.12 \text{ g}} = 0.9630 \text{ mol}$$

$$q = n \times \Delta H_{vap} = 0.9630 \text{ mol} \times 26.5 \text{ kJ/mol} = 25.5 \text{ kJ}$$

9.65 Assume 100 mL of H_2O = 100 g; H_2O, 18.02

$$100 \text{ g} \times \frac{1 \text{ mol } H_2O}{18.02 \text{ g } H_2O} \times \frac{40.7 \text{ kJ}}{1 \text{ mol } H_2O} = 226 \text{ kJ}$$

The heat to vaporize 100 mL of H_2O is much greater than that to vaporize 100 mL of diethyl ether.

9.66 Al, 26.98

$$\text{mol Al} = 5.00 \text{ g} \times \frac{1 \text{ mol}}{26.98 \text{ g}} = 0.1853 \text{ mol}$$

$$q = n \times \Delta H^\circ = 0.1853 \text{ mol Al} \times \frac{-1408.4 \text{ kJ}}{2 \text{ mol Al}} = -131 \text{ kJ}$$; 131 kJ is released.

9.67 Na, 22.99; $\Delta H^\circ = -368.4$ kJ/2 mol Na $= -184.2$ kJ/mol Na

$$1.00 \text{ g Na} \times \frac{1 \text{ mol Na}}{22.99 \text{ g Na}} \times \frac{-184.2 \text{ kJ}}{1 \text{ mol Na}} = -8.01 \text{ kJ}$$

8.01 kJ of heat is evolved. The reaction is exothermic.

9.68 (a) C_3H_8, 44.10; $\Delta H^\circ = -2220$ kJ/mol C_3H_8

$$15.5 \text{ g} \times \frac{1 \text{ mol } C_3H_8}{44.10 \text{ g } C_3H_8} \times \frac{-2220 \text{ kJ}}{1 \text{ mol } C_3H_8} = -780. \text{ kJ}$$

780. kJ of heat is evolved.

(b) $Ba(OH)_2 \cdot 8\,H_2O$, 315.5; $\Delta H^\circ = +80.3$ kJ/mol $Ba(OH)_2 \cdot 8\,H_2O$

$$4.88 \text{ g} \times \frac{1 \text{ mol } Ba(OH)_2 \cdot 8\,H_2O}{315.5 \text{ g } Ba(OH)_2 \cdot 8\,H_2O} \times \frac{80.3 \text{ kJ}}{1 \text{ mol } Ba(OH)_2 \cdot 8\,H_2O} = +1.24 \text{ kJ}$$

1.24 kJ of heat is absorbed.

9.69 CH_3NO_2, 61.04

$$q = 100.0 \text{ g } CH_3NO_2 \text{ x } \frac{1 \text{ mol } CH_3NO_2}{61.04 \text{ g } CH_3NO_2} \text{ x } \frac{2441.6 \text{ kJ}}{4 \text{ mol } CH_3NO_2} = 1.000 \text{ x } 10^3 \text{ kJ}$$

9.70 Fe_2O_3, 159.7; mol $Fe_2O_3 = 2.50 \text{ g x } \frac{1 \text{ mol}}{159.7 \text{ g}} = 0.015\ 65 \text{ mol}$

$$q = n \text{ x } \Delta H^\circ = 0.015\ 65 \text{ mol } Fe_2O_3 \text{ x } \frac{-24.8 \text{ kJ}}{1 \text{ mol } Fe_2O_3} = -0.388 \text{ kJ};\ 0.388 \text{ kJ is evolved.}$$

Because ΔH is negative, the reaction is exothermic.

9.71 CaO, 56.08

$$\text{mol CaO} = 233.0 \text{ g x } \frac{1 \text{ mol}}{56.08 \text{ g}} = 4.155 \text{ mol}$$

$$q = n \text{ x } \Delta H^\circ = 4.155 \text{ mol CaO x } \frac{464.6 \text{ kJ}}{1 \text{ mol CaO}} = 1930 \text{ kJ};\ 1930 \text{ kJ is absorbed.}$$

Because ΔH is positive, the reaction is endothermic.

Calorimetry and Heat Capacity (Section 9.7)

9.72 Heat capacity is the amount of heat required to raise the temperature of a substance a given amount. Specific heat is the amount of heat necessary to raise the temperature of exactly 1 g of a substance by exactly 1 °C.

9.73 A measurement carried out in a bomb calorimeter is done at constant volume and therefore ΔE is obtained.

9.74 Na, 22.99

$$\text{specific heat} = 28.2 \frac{\text{J}}{\text{mol}\cdot{}^\circ\text{C}} \text{ x } \frac{1 \text{ mol}}{22.99 \text{ g}} = 1.23 \text{ J/(g}\cdot{}^\circ\text{C)}$$

9.75 $q = (\text{specific heat}) \text{ x m x } \Delta T$

$$\text{specific heat} = \frac{q}{\text{m x } \Delta T} = \frac{89.7 \text{ J}}{(33.0 \text{ g})(5.20\ {}^\circ\text{C})} = 0.523 \text{ J/(g}\cdot{}^\circ\text{C)}$$

$$C_m = [0.523 \text{ J/(g}\cdot{}^\circ\text{C)}](47.88 \text{ g/mol}) = 25.0 \text{ J/(mol}\cdot{}^\circ\text{C)}$$

9.76 $q = (\text{specific heat}) \text{ x m x } \Delta T = (4.18 \frac{\text{J}}{\text{g}\cdot{}^\circ\text{C}})(350 \text{ g})(3\ {}^\circ\text{C} - 25\ {}^\circ\text{C}) = -3.2 \text{ x } 10^4 \text{ J}$

$$q = -3.2 \text{ x } 10^4 \text{ J x } \frac{1 \text{ kJ}}{1000 \text{ J}} = -32 \text{ kJ}$$

9.77 $q = (\text{specific heat}) \times m \times \Delta T = (4.18 \frac{J}{g \cdot ^\circ C})(250.0\ g)(80.0\ ^\circ C - 25.0\ ^\circ C) = 5.75 \times 10^4\ J$

$q = 5.75 \times 10^4\ J \times \frac{1\ kJ}{1000\ J} = 57.5\ kJ$

9.78 NH_4NO_3, 80.04; assume 125 mL = 125 g H_2O

$50.0\ g\ NH_4NO_3 \times \frac{1\ mol\ NH_4NO_3}{80.04\ g\ NH_4NO_3} = 0.625\ mol\ NH_4NO_3$

$q_p = \Delta H \times n = (+25.7\ kJ/mol)(0.625\ mol) = 16.1\ kJ = 16{,}100\ J$

$q_{soln} = -q_p = -16{,}100\ J$

$q_{soln} = (\text{specific heat}) \times m \times \Delta T$

$$\Delta T = \frac{q_{soln}}{(\text{specific heat}) \times m} = \frac{-16{,}100\ J}{\left(4.18 \frac{J}{g \cdot ^\circ C}\right)(50\ g + 125\ g)} = -22.0\ ^\circ C$$

$\Delta T = -22.0\ ^\circ C = T_{final} - T_{initial} = T_{final} - 25.0\ ^\circ C$

$T_{final} = -22.0\ ^\circ C + 25.0\ ^\circ C = 3.0\ ^\circ C$

9.79 LiCl, 42.4; assume 125 mL = 125 g H_2O

$25.0\ g\ LiCl \times \frac{1\ mol\ LiCl}{42.4\ g\ LiCl} = 0.590\ mol\ LiCl$

$q_p = \Delta H \times n = (-36.9\ kJ/mol)(0.590\ mol) = -21.8\ kJ = -21{,}800\ J$

$q_{soln} = -q_p = +21{,}800\ J$

$q_{soln} = (\text{specific heat}) \times m \times \Delta T$

$$\Delta T = \frac{q_{soln}}{(\text{specific heat}) \times m} = \frac{21{,}800\ J}{\left(4.18 \frac{J}{g \cdot ^\circ C}\right)(25\ g + 125\ g)} = 34.8\ ^\circ C$$

$\Delta T = 34.8\ ^\circ C = T_{final} - T_{initial} = T_{final} - 25.0\ ^\circ C$

$T_{final} = 34.8\ ^\circ C + 25.0\ ^\circ C = 59.8\ ^\circ C$

9.80 Mass of solution = 50.0 g + 1.045 g = 51.0 g

$q = (\text{specific heat}) \times m \times \Delta T$

$$q = \left(4.18 \frac{J}{g \cdot ^\circ C}\right)(51.0\ g)(32.3\ ^\circ C - 25.0\ ^\circ C) = 1.56 \times 10^3\ J = 1.56\ kJ$$

CaO, 56.08; $\text{mol CaO} = 1.045\ g \times \frac{1\ mol}{56.08\ g} = 0.018\ 63\ mol$

$\text{Heat evolved per mole of CaO} = \frac{1.56\ kJ}{0.018\ 63\ mol} = 83.7\ kJ/mol\ CaO$

Because the reaction evolves heat, the sign for ΔH is negative. $\Delta H = -83.7\ kJ$

9.81 NaOH, 40.00; HCl, 36.46

$$8.00 \text{ g NaOH x } \frac{1 \text{ mol NaOH}}{40.00 \text{ g NaOH}} = 0.200 \text{ mol NaOH}$$

$$8.00 \text{ g HCl x } \frac{1 \text{ mol HCl}}{36.46 \text{ g HCl}} = 0.219 \text{ mol HCl}$$

Because the reaction stoichiometry between NaOH and HCl is 1:1, the NaOH is the limiting reactant.

$q_P = -q_{soln} = -(\text{specific heat}) \text{ x m x } \Delta T$

$$= -\left(4.18 \frac{\text{J}}{\text{g} \cdot {}^\circ\text{C}}\right)\left(\frac{1 \text{ kJ}}{1000 \text{ J}}\right)(316 \text{ g})(33.5\ {}^\circ\text{C} - 25.0\ {}^\circ\text{C}) = -11.2 \text{ kJ}$$

$\Delta H = q_p/n = (-11.2 \text{ kJ})/(0.200 \text{ mol}) = -56 \text{ kJ/mol}$

When 10.00 g of HCl in 248.0 g of water is added, the same temperature increase is observed because the mass of NaOH is the same and it is still the limiting reactant. The mass of the solution is also the same.

9.82 C_6H_6, 78.11

$$\Delta E = q_V = -q_{H_2O} = -\left(4.18 \frac{\text{J}}{\text{g} \cdot {}^\circ\text{C}}\right)(250.0 \text{ g})(4.53\ {}^\circ\text{C}) - (525 \text{ J/}{}^\circ\text{C})(4.53\ {}^\circ\text{C})$$

$\Delta E = -7112 \text{ J} = -7.11 \text{ kJ}$

$$0.187 \text{ g } C_6H_6 \text{ x } \frac{1 \text{ mol } C_6H_6}{78.11 \text{ g } C_6H_6} = 0.002\ 39 \text{ mol } C_6H_6$$

$\Delta E(\text{per mole}) = (-7.11 \text{ kJ})/(0.002\ 39 \text{ mol}) = -2.97 \text{ x } 10^3 \text{ kJ/mol}$

$\Delta E \text{ (per gram } C_6H_6) = (-2.97 \text{ x } 10^3 \text{ kJ/mol})/(78.11 \text{ g/mol}) = -38.1 \text{ kJ/g}$

9.83 C_2H_6O, 46.07

$$\Delta E = q_V = -q_{H_2O} = -\left(4.18 \frac{\text{J}}{\text{g} \cdot {}^\circ\text{C}}\right)(250.0 \text{ g})(9.15\ {}^\circ\text{C}) - (575 \text{ J/}{}^\circ\text{C})(9.15\ {}^\circ\text{C})$$

$\Delta E = -14{,}823 \text{ J} = -14.8 \text{ kJ}$

$$0.500 \text{ g } C_2H_6O \text{ x } \frac{1 \text{ mol } C_2H_6O}{46.07 \text{ g } C_2H_6O} = 0.0109 \text{ mol } C_2H_6O$$

$\Delta E(\text{per mole}) = (-14.8 \text{ kJ})/(0.0109 \text{ mol}) = -1.36 \text{ x } 10^3 \text{ kJ/mol}$

$\Delta E \text{ (per gram } C_2H_6O) = (-1.36 \text{ x } 10^3 \text{ kJ/mol})/(46.07 \text{ g/mol}) = -29.5 \text{ kJ/g}$

Hess's Law and Heats of Formation (Sections 9.8 and 9.9)

9.84 The standard state of an element is its most stable form at 1 atm and the specified temperature, usually 25 °C.

Because enthalpy is a state function, ΔH is the same regardless of the path taken between reactants and products. Thus, the sum of the enthalpy changes for the individual steps in a reaction is equal to the overall enthalpy change for the entire reaction, a relationship known as Hess's law. Hess's law works because of the law of the conservation of energy.

9.85 $\Delta H^\circ_{net} = \Delta H^\circ_1 + \Delta H^\circ_2 = -174.2\text{ kJ} + (-318.4\text{ kJ}) = -492.6\text{ kJ}$

9.86

$CH_4(g) + Cl_2(g) \rightarrow CH_3Cl(g) + HCl(g)$ $\quad \Delta H^\circ_1 = -98.3\text{ kJ}$

$\underline{CH_3Cl(g) + Cl_2(g) \rightarrow CH_2Cl_2(g) + HCl(g)}$ $\quad \Delta H^\circ_2 = -104\text{ kJ}$

Sum $CH_4(g) + 2\ Cl_2(g) \rightarrow CH_2Cl_2(g) + 2\ HCl(g)$

$\Delta H^\circ = \Delta H^\circ_1 + \Delta H^\circ_2 = -202\text{ kJ}$

9.87

$2\ CH_4(g) + 4\ O_2(g) \rightarrow 2\ CO_2(g) + 4\ H_2O(l)$ $\quad \Delta H^\circ_1 = 2(-890.3\text{ kJ})$

$C_2H_6(g) \rightarrow C_2H_4(g) + H_2(g)$ $\quad \Delta H^\circ_2 = +136.3\text{ kJ}$

$2\ CO_2(g) + 3\ H_2O(l) \rightarrow C_2H_6(g) + 7/2\ O_2(g)$ $\quad \Delta H^\circ_3 = \dfrac{3120.8\text{ kJ}}{2}$

$\underline{H_2O(l) \rightarrow H_2(g) + 1/2\ O_2(g)}$ $\quad \underline{\Delta H^\circ_4 = +285.8\text{ kJ}}$

Sum $2\ CH_4(g) \rightarrow C_2H_4(g) + 2\ H_2(g)$ $\quad \Delta H^\circ = +201.9\text{ kJ}$

9.88 A compound's standard heat of formation is the amount of heat associated with the formation of 1 mole of a compound from its elements (in their standard states).

9.89 The standard state of an element is, by definition, the most stable form of the element at 1 atm and 25 °C. Elements always have $\Delta H^\circ_f = 0$ because the enthalpy change for the formation of an element from itself is zero and the standard state of elements is the reference point from which all enthalpy changes are measured.

9.90 (a) Cl_2, gas (b) Hg, liquid (c) CO_2, gas (d) Ga, solid

9.95 (a) NH_3, gas (b) Fe, solid (c) N_2, gas (d) Br_2, liquid

9.92 (a) $2\ Fe(s) + 3/2\ O_2(g) \rightarrow Fe_2O_3(s)$

(b) $12\ C(s) + 11\ H_2(g) + 11/2\ O_2(g) \rightarrow C_{12}H_{22}O_{11}(s)$

(c) $U(s) + 3\ F_2(g) \rightarrow UF_6(s)$

9.93 (a) $2\ C(s) + 3\ H_2(g) + 1/2\ O_2(g) \rightarrow C_2H_6O(l)$

(b) $2\ Na(s) + S(s) + 2\ O_2(g) \rightarrow Na_2SO_4(s)$

(c) $C(s) + H_2(g) + Cl_2(g) \rightarrow CH_2Cl_2(l)$

9.94

$S(s) + O_2(g) \rightarrow SO_2(g)$ $\quad \Delta H^\circ_1 = -296.8\text{ kJ}$

$\underline{SO_2 + ½\ O_2(g) \rightarrow SO_3(g)}$ $\quad \Delta H^\circ_2 = -98.9\text{ kJ}$

Sum $S(s) + 3/2\ O_2(g) \rightarrow SO_3(g)$ $\quad \Delta H^\circ_3 = \Delta H^\circ_1 + \Delta H^\circ_2$

$\Delta H^\circ_f = \Delta H^\circ_3 = -296.8\text{ kJ} + (-98.9\text{ kJ}) = -395.7\text{ kJ/mol}$

9.95 $\Delta H^\circ_{rxn} = [12\ \Delta H^\circ_f(CO_2) + 6\ \Delta H^\circ_f(H_2O)] - [2\ \Delta H^\circ_f(C_6H_6)]$

$-6534\text{ kJ} = [(12\text{ mol})(-393.5\text{ kJ/mol}) + (6\text{ mol})(-285.8\text{ kJ/mol})] - [(2\text{ mol})(\Delta H^\circ_f(C_6H_6))]$

Solve for $\Delta H^\circ_f(C_6H_6)$.

$-6534\text{ kJ} = -6436.8\text{ kJ} - [(2\text{ mol})(\Delta H^\circ_f(C_6H_6))]$; $\quad 97.2\text{ kJ} = (2\text{ mol})(\Delta H^\circ_f(C_6H_6))$

$\Delta H^\circ_f(C_6H_6) = +48.6\text{ kJ/mol}$

9.96

	$SO_3(g) + H_2O(l) \rightarrow H_2SO_4(aq)$	$\Delta H°_1 = -227.8$ kJ
	$H_2(g) + ½\ O_2(g) \rightarrow H_2O(l)$	$\Delta H°_2 = \Delta H°_f = -285.8$ kJ
	$S(s) + 3/2\ O_2(g) \rightarrow SO_3(g)$	$\Delta H°_3 = \Delta H°_f = -395.7$
Sum	$S(s) + H_2(g) + 2\ O_2(g) \rightarrow H_2SO_4(aq)$	$\Delta H°_f\ (H_2SO_4) = ?$

$\Delta H°_f\ (H_2SO_4) = \Delta H°_1 + \Delta H°_2 + \Delta H°_3 = -909.3$ kJ

9.97 $\Delta H°_{rxn} = [\Delta H°_f(CH_3CO_2H) + \Delta H°_f(H_2O)] - \Delta H°_f(CH_3CH_2OH)$
$\Delta H°_{rxn} = [(1\text{ mol})(-484.5\text{ kJ/mol}) + (1\text{mol})(-285.8\text{ kJ/mol})] - [(1\text{ mol})(-277.7\text{ kJ/mol})]$
$\Delta H°_{rxn} = -492.6$ kJ

9.98 $C_8H_8(l) + 10\ O_2(g) \rightarrow 8\ CO_2(g) + 4\ H_2O(l)$
$\Delta H°_{rxn} = \Delta H°_c = -4395$ kJ
$\Delta H°_{rxn} = [8\ \Delta H°_f(CO_2) + 4\ \Delta H°_f(H_2O)] - \Delta H°_f(C_8H_8)$
$-4395\text{ kJ} = [(8\text{ mol})(-393.5\text{ kJ/mol}) + (4\text{ mol})(-285.8\text{ kJ/mol})] - [(1\text{ mol})(\Delta H°_f(C_8H_8))]$
Solve for $\Delta H°_f(C_8H_8)$
$-4395\text{ kJ} = -4291.2\text{ kJ} - (1\text{ mol})(\Delta H°_f(C_8H_8))$
$-103.8\text{ kJ} = -(1\text{ mol})(\Delta H°_f(C_8H_8))$

$$\Delta H°_f(C_8H_8) = \frac{-103.8\text{ kJ}}{-1\text{ mol}} = +103.8\text{ kJ/mol} = +104\text{ kJ/mol}$$

9.99 $C_5H_{12}O(l) + 15/2\ O_2(g) \rightarrow 5\ CO_2(g) + 6\ H_2O(l)$
$\Delta H°_{rxn} = [5\ \Delta H°_f(CO_2) + 6\ \Delta H°_f(H_2O)] - \Delta H°_f(C_5H_{12}O)$
$\Delta H°_{rxn} = [(5\text{ mol})(-393.5\text{ kJ/mol}) + (6\text{ mol})(-285.8\text{ kJ/mol})] - [(1\text{ mol})(-313.6\text{ kJ/mol})]$
$\Delta H°_{rxn} = -3369$ kJ

9.100 $\Delta H°_{rxn} = \Delta H°_f(\text{MTBE}) - [\Delta H°_f(\text{2-Methylpropene}) + \Delta H°_f(CH_3OH)]$
$-57.5\text{ kJ} = -313.6\text{ kJ} - [(1\text{ mol})(\Delta H°_f(\text{2-Methylpropene})) + (-239.2\text{ kJ})]$
Solve for $\Delta H°_f$(2-Methylpropene).
$-16.9\text{ kJ} = (1\text{ mol})(\Delta H°_f(\text{2-Methylpropene}))$
$\Delta H°_f(\text{2-Methylpropene}) = -16.9$ kJ/mol

9.101 $C_{51}H_{88}O_6(l) + 70\ O_2(g) \rightarrow 51\ CO_2(g) + 44\ H_2O(l)$
$\Delta H°_{rxn} = [51\ \Delta H°_f(CO_2) + 44\ \Delta H°_f(H_2O)] - \Delta H°_f(C_{51}H_{88}O_6)$
$\Delta H°_{rxn} = [(51\text{ mol})(-393.5\text{ kJ/mol}) + (44\text{ mol})(-285.8\text{ kJ/mol})] - [(1\text{ mol})(-1310\text{ kJ/mol})]$
$\Delta H°_{rxn} = -3.133 \times 10^4$ kJ/mol $C_{51}H_{88}O_6$
$C_{51}H_{88}O_6$, 797.25

$$q = -3.133 \times 10^4\ \frac{\text{kJ}}{\text{mol}} \times \frac{1\text{ mol}}{797.25\text{ g}} \times 0.94\ \frac{\text{g}}{\text{mL}} = -37\text{ kJ/mL};\ 37\text{ kJ released per mL}$$

9.102 $CaCO_3(s) \rightarrow CaO(s) + CO_2(g)$
$\Delta H°_{rxn} = [\Delta H°_f(CaO) + \Delta H°_f(CO_2)] - \Delta H°_f(CaCO_3)$
$\Delta H°_{rxn} = [(1\text{ mol})(-634.9\text{ kJ/mol}) + (1\text{ mol})(-393.5\text{ kJ/mol})] - [(1\text{ mol})(-1207.6\text{ kJ/mol})]$
$\Delta H°_{rxn} = +179.2$ kJ

9.103 $3\ N_2O_4(g) + 2\ H_2O(l) \rightarrow 4\ HNO_3(aq) + 2\ NO(g)$

$\Delta H^\circ_{rxn} = [4\ \Delta H^\circ_f(HNO_3) + 2\ \Delta H^\circ_f(NO)] - [3\ \Delta H^\circ_f(N_2O_4) + 2\ \Delta H^\circ_f(H_2O)]$

$\Delta H^\circ_{rxn} = [(4\ mol)(-207.4\ kJ/mol) + (2\ mol)(91.3\ kJ/mol)] - [(3\ mol)(11.1\ kJ/mol) + (2\ mol)(-285.8\ kJ/mol)] = -108.7\ kJ$

9.104 $CaCO_3(s) \rightarrow CaO(s) + CO_2(g)$

$\Delta H^\circ_{rxn} = [\Delta H^\circ_f(CaO) + \Delta H^\circ_f(CO_2)] - \Delta H^\circ_f(CaCO_3)$

$\Delta H^\circ_{rxn} = [(1\ mol)(-634.9\ kJ/mol) + (1\ mol)(-393.5\ kJ/mol)] - [(1\ mol)(-1207.6\ kJ/mol)]$

$\Delta H^\circ_{rxn} = +179.2\ kJ$

9.105 $6\ CO_2(g) + 6\ H_2O(l) \rightarrow C_6H_{12}O_6(s) + 6\ O_2(g)$

$\Delta H^\circ_{rxn} = \Delta H^\circ_f(C_6H_{12}O_6) - [6\ \Delta H^\circ_f(CO_2) + 6\ \Delta H^\circ_f(H_2O(l))]$

$\Delta H^\circ_{rxn} = [(1\ mol)(-1273.3\ kJ/mol)] - [(6\ mol)(-393.5\ kJ/mol) + (6\ mol)(-285.8\ kJ/mol)]$

$\Delta H^\circ_{rxn} = +2802.5\ kJ = +2803\ kJ$

Bond Dissociation Energies (Section 9.10)

9.106 $H_2C{=}CH_2(g) + H_2(g) \rightarrow CH_3CH_3(g)$

$\Delta H^\circ_{rxn} = D\ \text{(Reactant bonds)} - D\ \text{(Product bonds)}$

$\Delta H^\circ_{rxn} = (D_{C=C} + 4\ D_{C-H} + D_{H-H}) - (6\ D_{C-H} + D_{C-C})$

$\Delta H^\circ_{rxn} = [(1\ mol)(728\ kJ/mol) + (4\ mol)(410\ kJ/mol) + (1\ mol)(436\ kJ/mol)] - [(6\ mol)(410\ kJ/mol) + (1\ mol)(350\ kJ/mol)] = -6\ kJ$

9.107 $2\ NH_3(g) + Cl_2(g) \rightarrow N_2H_4(g) + 2\ HCl(g)$

$\Delta H^\circ_{rxn} = D\ \text{(Reactant bonds)} - D\ \text{(Product bonds)}$

$\Delta H^\circ_{rxn} = (6\ D_{N-H} + D_{Cl-Cl}) - (D_{N-N} + 4\ D_{N-H} + 2\ D_{H-Cl})$

$\Delta H^\circ_{rxn} = [(6\ mol)(390\ kJ/mol) + (1\ mol)(243\ kJ/mol)] - [(1\ mol)(240\ kJ/mol) + (4\ mol)(390\ kJ/mol) + (2\ mol)(432\ kJ/mol)] = -81\ kJ$

9.108 $\Delta H^\circ_{rxn} = D\ \text{(Reactant bonds)} - D\ \text{(Product bonds)}$

(a) $2\ CH_4(g) \rightarrow C_2H_6(g) + H_2(g)$

$\Delta H^\circ_{rxn} = (8\ D_{C-H}) - (D_{C-C} + 6\ D_{C-H} + D_{H-H})$

$\Delta H^\circ_{rxn} = [(8\ mol)(410\ kJ/mol)] - [(1\ mol)(350\ kJ/mol) + (6\ mol)(410\ kJ/mol) + (1\ mol)(436\ kJ/mol)] = +34\ kJ$

(b) $C_2H_6(g) + F_2(g) \rightarrow C_2H_5F(g) + HF(g)$

$\Delta H^\circ_{rxn} = (6\ D_{C-H} + D_{C-C} + D_{F-F}) - (5\ D_{C-H} + D_{C-C} + D_{C-F} + D_{H-F})$

$\Delta H^\circ_{rxn} = [(6\ mol)(410\ kJ/mol) + (1\ mol)(350\ kJ/mol) + (1\ mol)(159\ kJ/mol)] - [(5\ mol)(410\ kJ/mol) + (1\ mol)(350\ kJ/mol) + (1\ mol)(450\ kJ/mol) + (1\ mol)(570\ kJ/mol)] = -451\ kJ$

(c) $N_2(g) + 3\ H_2(g) \rightarrow 2\ NH_3(g)$

The bond dissociation energy for N_2 is 945 kJ/mol.

$\Delta H^\circ_{rxn} = (D_{N_2} + 3\ D_{H-H}) - (6\ D_{N-H})$

$\Delta H^\circ_{rxn} = [(1\ mol)(945\ kJ/mol) + (3\ mol)(436\ kJ/mol)] - [(6\ mol)(390\ kJ/mol)] = -87\ kJ$

9.109 $CH_3CH{=}CH_2 + H_2O \rightarrow CH_3CH(OH)CH_3$

$\Delta H^\circ_{rxn} = D$ (Reactant bonds) – D (Product bonds)

$\Delta H^\circ_{rxn} = (D_{C=C} + D_{C\text{–}C} + 6\,D_{C\text{–}H} + 2\,D_{O\text{–}H}) - (2\,D_{C\text{–}C} + 7\,D_{C\text{–}H} + D_{C\text{–}O} + D_{O\text{–}H})$

ΔH°_{rxn} = [(1 mol)(728 kJ/mol) + (1 mol)(350 kJ/mol) + (6 mol)(410 kJ/mol) + (2 mol)(460 kJ/mol)] – [(2 mol)(350 kJ/mol) + (7 mol)(410 kJ/mol) + (1 mol)(350 kJ/mol) + (1 mol)(460 kJ/mol)] = +78 kJ

Fossil Fuels and Heats of Combustion (Section 9.11)

9.110 $C_4H_{10}(l) + \frac{13}{2}\,O_2(g) \rightarrow 4\,CO_2(g) + 5\,H_2O(g)$

$\Delta H^\circ_{rxn} = [4\,\Delta H^\circ_f(CO_2) + 5\,\Delta H^\circ_f(H_2O)] - \Delta H^\circ_f(C_4H_{10})$

ΔH°_{rxn} = [(4 mol)(–393.5 kJ/mol) + (5 mol)(–241.8 kJ/mol)] – [(1 mol)(–147.5 kJ/mol)]

ΔH°_{rxn} = –2635.5 kJ; ΔH°_C = –2635.5 kJ/mol

C_4H_{10}, 58.12; $\Delta H^\circ_C = \left(-2635.5\,\frac{kJ}{mol}\right)\left(\frac{1\ mol}{58.12\ g}\right) = -45.35\ kJ/g$

$$\Delta H^\circ_C = \left(-45.35\,\frac{kJ}{g}\right)\left(0.579\,\frac{g}{mL}\right) = -26.3\ kJ/mL$$

9.111 $\Delta H^\circ_{rxn} = D$ (Reactant bonds) – D (Product bonds)

$C_2H_6(g) + 7/2\,O_2(g) \rightarrow 2\,CO_2(g) + 3\,H_2O(g)$

$\Delta H^\circ_{rxn} = (1\,D_{C\text{–}C} + 6\,D_{C\text{–}H} + 7/2\,D_{O=O}) - (4\,D_{C=O} + 6\,D_{O\text{–}H})$

ΔH°_{rxn} = [(1 mol)(350 kJ/mol) + (6 mol)(410 kJ/mol) + (7/2 mol)(498 kJ/mol)] – [(4 mol)(804 kJ/mol) + (6 mol)(460 kJ/mol)] = –1423 kJ

Free Energy and Entropy (Sections 9.12 and 9.13)

9.112 Entropy is a measure of molecular randomness.

9.113 $\Delta G = \Delta H - T\Delta S$; ΔH is usually more important because it is usually much larger than $T\Delta S$.

9.114 A reaction can be spontaneous yet endothermic if ΔS is positive (more randomness) and the $T\Delta S$ term is larger than ΔH.

9.115 A reaction can be nonspontaneous yet exothermic if ΔS is negative (less randomness) and the temperature is high enough so that the $T\Delta S$ term is more negative than ΔH.

9.116 (a) positive (more randomness) (b) negative (less randomness)

9.117 (a) positive (more randomness) (b) negative (less randomness)
(c) positive (more randomness)

9.118 (a) zero (equilibrium) (b) zero (equilibrium)
(c) negative (spontaneous)

9.119 Because the mixing of gas molecules is spontaneous, ΔG is negative. The mixture of gas molecules is more random so ΔS is positive. For the diffusion of gases, ΔH is approximately zero.

9.120 ΔS is positive. The reaction increases the total number of molecules.

9.121 $\Delta S < 0$. The reaction decreases the number of gas molecules.

9.122 $\Delta G = \Delta H - T\Delta S$

(a) $\Delta G = -48\ \text{kJ} - (400\ \text{K})(135 \times 10^{-3}\ \text{kJ/K}) = -102\ \text{kJ}$
$\Delta G < 0$, spontaneous; $\Delta H < 0$, exothermic.
(b) $\Delta G = -48\ \text{kJ} - (400\ \text{K})(-135 \times 10^{-3}\ \text{kJ/K}) = +6\ \text{kJ}$
$\Delta G > 0$, nonspontaneous; $\Delta H < 0$, exothermic.
(c) $\Delta G = +48\ \text{kJ} - (400\ \text{K})(135 \times 10^{-3}\ \text{kJ/K}) = -6\ \text{kJ}$
$\Delta G < 0$, spontaneous; $\Delta H > 0$, endothermic.
(d) $\Delta G = +48\ \text{kJ} - (400\ \text{K})(-135 \times 10^{-3}\ \text{kJ/K}) = +102\ \text{kJ}$
$\Delta G > 0$, nonspontaneous; $\Delta H > 0$, endothermic.

9.123 $\Delta G = \Delta H - T\Delta S$

(a) $\Delta G = -128\ \text{kJ} - (500\ \text{K})(35 \times 10^{-3}\ \text{kJ/K}) = -146\ \text{kJ}$
$\Delta G < 0$, spontaneous; $\Delta H < 0$, exothermic
(b) $\Delta G = +67\ \text{kJ} - (250\ \text{K})(-140 \times 10^{-3}\ \text{kJ/K}) = +102\ \text{kJ}$
$\Delta G > 0$, nonspontaneous; $\Delta H > 0$, endothermic
(c) $\Delta G = +75\ \text{kJ} - (800\ \text{K})(95 \times 10^{-3}\ \text{kJ/K}) = -1\ \text{kJ}$
$\Delta G < 0$, spontaneous; $\Delta H > 0$, endothermic

9.124 $\Delta G = \Delta H - T\Delta S$; Set $\Delta G = 0$ and solve for T (the crossover temperature).

$$T = \frac{\Delta H}{\Delta S} = \frac{-33\ \text{kJ}}{-0.058\ \text{kJ/K}} = 570\ \text{K}$$

9.125 Because $\Delta H > 0$ and $\Delta S < 0$, the reaction is nonspontaneous at all temperatures. There is no crossover temperature.

9.126 (a) $\Delta H < 0$ and $\Delta S > 0$; reaction is spontaneous at all temperatures.
(b) $\Delta H < 0$ and $\Delta S < 0$; reaction has a crossover temperature.
(c) $\Delta H > 0$ and $\Delta S > 0$; reaction has a crossover temperature.
(d) $\Delta H > 0$ and $\Delta S < 0$; reaction is nonspontaneous at all temperatures.

9.127 (a) $\Delta H < 0$ and $\Delta S < 0$. The reaction is favored by enthalpy but not by entropy.
$\Delta G^\circ = \Delta H^\circ - T\Delta S^\circ = -217.5\ \text{kJ/mol} - (298\ \text{K})[-233.9 \times 10^{-3}\ \text{kJ/(K} \cdot \text{mol)}] = -147.8\ \text{kJ}$
(b) The reaction has a crossover temperature. Set $\Delta G = 0$ and solve for T (the crossover temperature).
$\Delta G^\circ = 0 = \Delta H^\circ - T\Delta S^\circ$

$$T = \frac{\Delta H^\circ}{\Delta S^\circ} = \frac{217.5\ \text{kJ/mol}}{233.9 \times 10^{-3}\ \text{kJ/(K} \cdot \text{mol)}} = 929.9\ \text{K}$$

9.128 $T = -114.1\ ^\circ C = 273.15 + (-114.1) = 159.0\ K$

$\Delta G_{fus} = \Delta H_{fus} - T\Delta S_{fus}$; $\Delta G = 0$ at the melting point temperature.

Set $\Delta G = 0$ and solve for ΔS_{fus}.

$\Delta G = 0 = \Delta H_{fus} - T\Delta S_{fus}$

$$\Delta S_{fus} = \frac{\Delta H_{fus}}{T} = \frac{5.02\ kJ/mol}{159.0\ K} = 0.0316\ kJ/(K \cdot mol) = 31.6\ J/(K \cdot mol)$$

9.129 $T = 61.2\ ^\circ C = 273.15 + (61.2) = 334.4\ K$

$\Delta G_{vap} = \Delta H_{vap} - T\Delta S_{vap}$; $\Delta G = 0$ at the boiling point temperature.

Set $\Delta G = 0$ and solve for ΔS_{vap}.

$\Delta G = 0 = \Delta H_{vap} - T\Delta S_{vap}$

$$\Delta S_{vap} = \frac{\Delta H_{vap}}{T} = \frac{29.2\ kJ/mol}{334.4\ K} = 0.0873\ kJ/(K \cdot mol) = 87.3\ J/(K \cdot mol)$$

Chapter Problems

9.130 $\Delta H = \Delta E + P\Delta V$

$\Delta H = -5.67\ kJ = -7.20\ kJ + P\Delta V$

$P\Delta V = +1.53\ kJ$

$w = -P\Delta V = -1.53\ kJ$

$$w = -P\Delta V = -1.53\ kJ \times \frac{1000\ J}{1\ kJ} \times \frac{1}{\left(\frac{101\ J}{L \cdot atm}\right)} = -15.1\ L \cdot atm$$

$P\Delta V = 15.1\ L \cdot atm = (1\ atm)(\Delta V)$

$\Delta V = 15.1\ L$

9.131 $\Delta H = -1256.2\ kJ/mol\ C_2H_2$; C_2H_2, 26.04

$w = -P\Delta V = -(1.00\ atm)(-2.80\ L) = 2.80\ L \cdot atm$

$$w = (2.80\ L \cdot atm)\left(\frac{101\ J}{1\ L \cdot atm}\right) = 283\ J = 0.283\ kJ$$

$$6.50\ g \times \frac{1\ mol\ C_2H_2}{26.04\ g\ C_2H_2} = 0.250\ mol\ C_2H_2$$

$q = (-1256.2\ kJ/mol)(0.250\ mol) = -314\ kJ$

$\Delta E = \Delta H - P\Delta V = -314\ kJ - (-0.283\ kJ) = -314\ kJ$

9.132 C_2H_4, 28.05; HCl, 36.46

$w = -P\Delta V = -(1.00\ atm)(-71.5\ L) = 71.5\ L \cdot atm$

$$w = (71.5\ L \cdot atm)\left(\frac{101\ J}{1\ L \cdot atm}\right) = 7222\ J = 7.22\ kJ$$

$$89.5\ g\ C_2H_4 \times \frac{1\ mol\ C_2H_4}{28.05\ g\ C_2H_4} = 3.19\ mol\ C_2H_4; \quad 125\ g\ HCl \times \frac{1\ mol\ HCl}{36.46\ g\ HCl} = 3.43\ mol\ HCl$$

Because the reaction stoichiometry between C_2H_4 and HCl is 1:1, C_2H_4 is the limiting reactant.

$\Delta H° = -72.3$ kJ/mol C_2H_4
$q = (-72.3 \text{ kJ/mol})(3.19 \text{ mol}) = -231$ kJ
$\Delta E = \Delta H - P\Delta V = -231 \text{ kJ} - (-7.22 \text{ kJ}) = -224$ kJ

9.133 $Mg(s) + 2\,HCl(aq) \rightarrow MgCl_2(aq) + H_2(g)$

$$\text{mol Mg} = 1.50 \text{ g} \times \frac{1 \text{ mol}}{24.3 \text{ g}} = 0.0617 \text{ mol Mg}$$

$$\text{mol HCl} = 0.200 \text{ L} \times 6.00 \frac{\text{mol}}{\text{L}} = 1.20 \text{ mol HCl}$$

There is an excess of HCl. Mg is the limiting reactant.

$$q = \left(4.18 \frac{\text{J}}{\text{g} \cdot °\text{C}}\right)(200 \text{ g})(42.9\,°\text{C} - 25.0\,°\text{C}) + \left(776 \frac{\text{J}}{°\text{C}}\right)(42.9\,°\text{C} - 25.0\,°\text{C}) = 2.89 \times 10^4 \text{ J}$$

$$q = 2.89 \times 10^4 \text{ J} \times \frac{1 \text{ kJ}}{1000 \text{ J}} = 28.9 \text{ kJ}$$

$$\text{Heat evolved per mole of Mg} = \frac{28.9 \text{ kJ}}{0.0617 \text{ mol}} = 468 \text{ kJ/mol}$$

Because the reaction evolves heat, the sign for ΔH is negative. $\Delta H = -468$ kJ

9.134 (a) $C(s) + CO_2(g) \rightarrow 2\,CO(g)$
$\Delta H°_{rxn} = [2\,\Delta H°_f(CO)] - \Delta H°_f(CO_2)$
$\Delta H°_{rxn} = [(2 \text{ mol})(-110.5 \text{ kJ/mol})] - [(1 \text{ mol})(-393.5 \text{ kJ/mol})] = +172.5$ kJ
(b) $2\,H_2O_2(aq) \rightarrow 2\,H_2O(l) + O_2(g)$
$\Delta H°_{rxn} = [2\,\Delta H°_f(H_2O)] - [2\,\Delta H°_f(H_2O_2)]$
$\Delta H°_{rxn} = [(2 \text{ mol})(-285.8 \text{ kJ/mol})] - [(2 \text{ mol})(-191.2 \text{ kJ/mol})] = -189.2$ kJ
(c) $Fe_2O_3(s) + 3\,CO(g) \rightarrow 2\,Fe(s) + 3\,CO_2(g)$
$\Delta H°_{rxn} = [3\,\Delta H°_f(CO_2)] - [\Delta H°_f(Fe_2O_3) + 3\,\Delta H°_f(CO)]$
$\Delta H°_{rxn} = [(3 \text{ mol})(-393.5 \text{ kJ/mol})]$
$- [(1 \text{ mol})(-824.2 \text{ kJ/mol}) + (3 \text{ mol})(-110.5 \text{ kJ/mol})] = -24.8$ kJ

9.135

	Reaction	
	$2\,NO(g) + O_2(g) \rightarrow 2\,NO_2(g)$	$\Delta H°_1 = 2(-58.1 \text{ kJ})$
	$2\,NO_2(g) \rightarrow N_2O_4(g)$	$\Delta H°_2 = -55.3$ kJ
Sum	$2\,NO(g) + O_2(g) \rightarrow N_2O_4(g)$	
	$\Delta H° = \Delta H°_1 + \Delta H°_2 = -171.5$ kJ	

9.136 $\Delta G = \Delta H - T\Delta S$; at equilibrium $\Delta G = 0$. Set $\Delta G = 0$ and solve for T.

$$\Delta G = 0 = \Delta H - T\Delta S; \quad T = \frac{\Delta H}{\Delta S} = \frac{30.91 \text{ kJ/mol}}{93.2 \times 10^{-3} \text{ kJ/(K} \cdot \text{mol)}} = 332 \text{ K} = 59\,°\text{C}$$

9.137 $G_{fus} = \Delta H_{fus} - T\Delta S_{fus}$; at the melting point $\Delta G = 0$. Set $\Delta G = 0$ and solve for T (the melting point).

$$\Delta G = 0 = \Delta H_{fus} - T\Delta S_{fus}; \quad T = \frac{\Delta H_{fus}}{\Delta S_{fus}} = \frac{9.95 \text{ kJ}}{0.0357 \text{ kJ/K}} = 279 \text{ K}$$

9.138 $HgS(s) + O_2(g) \rightarrow Hg(l) + SO_2(g)$

(a) $\Delta H^\circ_{rxn} = \Delta H^\circ_f(SO_2) - \Delta H^\circ_f(HgS)$

$\Delta H^\circ_{rxn} = [(1\ mol)(-296.8\ kJ/mol)] - [(1\ mol)(-58.2\ kJ/mol)] = -238.6\ kJ$

(b) and (c) Because $\Delta H < 0$ and $\Delta S > 0$, the reaction is spontaneous at all temperatures.

9.139 (a) $\Delta H^\circ_{rxn} = \Delta H^\circ_f(CH_3OH) - \Delta H^\circ_f(CO)$

$\Delta H^\circ_{rxn} = [(1\ mol)(-238.7\ kJ/mol)] - [(1\ mol)(-110.5\ kJ/mol)] = -128.2\ kJ$

(b) $\Delta G^\circ = \Delta H^\circ - T\Delta S^\circ = -128.2\ kJ - (298\ K)(-332 \times 10^{-3}\ kJ/K) = -29.3\ kJ$

(c) Step 1 is spontaneous since $\Delta G^\circ < 0$.

(d) ΔH°, because it is larger than $T\Delta S^\circ$.

(e) Set $\Delta G = 0$ and solve for T.

$\Delta G = 0 = \Delta H - T\Delta S$

$T = \dfrac{\Delta H}{\Delta S} = \dfrac{128.2\ kJ}{332 \times 10^{-3}\ kJ/K} = 386\ K$; The reaction is spontaneous below 386 K.

(f) $\Delta H^\circ_{rxn} = \Delta H^\circ_f(CH_4) - \Delta H^\circ_f(CH_3OH)$

$\Delta H^\circ_{rxn} = [(1\ mol)(-74.8\ kJ/mol)] - [(1\ mol)(-238.7\ kJ/mol)] = +163.9\ kJ$

(g) $\Delta G^\circ = \Delta H^\circ - T\Delta S^\circ = +163.9\ kJ - (298\ K)(162 \times 10^{-3}\ kJ/K) = +115.6\ kJ$

(h) Step 2 is nonspontaneous since $\Delta G^\circ > 0$.

(i) ΔH°, because it is larger than $T\Delta S^\circ$.

(j) Set $\Delta G = 0$ and solve for T.

$\Delta G = 0 = \Delta H - T\Delta S$

$T = \dfrac{\Delta H}{\Delta S} = \dfrac{163.9\ kJ}{162 \times 10^{-3}\ kJ/K} = 1012\ K$; The reaction is spontaneous above 1012 K.

(k) $\Delta G^\circ_{overall} = \Delta G^\circ_1 + \Delta G^\circ_2 = -29.3\ kJ + 115.6\ kJ = +86.3\ kJ$

$\Delta H^\circ_{overall} = \Delta H^\circ_1 + \Delta H^\circ_2 = -128.2\ kJ + 163.9\ kJ = +35.7\ kJ$

$\Delta S^\circ_{overall} = \Delta S^\circ_1 + \Delta S^\circ_2 = -332\ J/K + 162\ J/K = -170\ J/K$

(l) The overall reaction is nonspontaneous since $\Delta G^\circ_{overall} > 0$ at 298 K.

(m) The two reactions should be run separately. Run step 1 below 386 K and run step 2 above 1012 K.

9.140 (a) $2\ C_8H_{18}(l) + 25\ O_2(g) \rightarrow 16\ CO_2(g) + 18\ H_2O(l)$

(b) $C_8H_{18}(l) + 25/2\ O_2(g) \rightarrow 8\ CO_2(g) + 9\ H_2O(l)$

$\Delta H^\circ_{rxn} = \Delta H^\circ_c = -5461\ kJ$

$\Delta H^\circ_{rxn} = [8\ \Delta H^\circ_f(CO_2) + 9\ \Delta H^\circ_f(H_2O)] - \Delta H^\circ_f(C_8H_{18})$

$-5461\ kJ = [(8\ mol)(-393.5\ kJ/mol) + (9\ mol)(-285.8\ kJ/mol)] - [(1\ mol)(\Delta H^\circ_f(C_8H_{18}))]$

Solve for $\Delta H^\circ_f(C_8H_{18})$.

$-5461\ kJ = -5720.2\ kJ - [(1\ mol)(\Delta H^\circ_f(C_8H_{18}))]$

$259\ kJ = -(1\ mol)(\Delta H^\circ_f(C_8H_{18}))$

$\Delta H^\circ_f(C_8H_{18}) = -259\ kJ/mol$

9.141 Assume 1.00 kg of H_2O; $1\ kg{\cdot}m^2/s^2 = 1\ J$

$E_p = (1.00\ kg)(9.81\ m/s^2)(739\ m) = 7250\ kg{\cdot}m^2/s^2 = 7250\ J$

$q = \text{specific heat} \times m \times \Delta T$

$\Delta T = \dfrac{q}{m \times \text{specific heat}} = \dfrac{7250\ J}{(1000\ g)(4.18\ J/(g{\cdot}^\circ C))} = 1.73\ ^\circ C$ (temperature rise)

9.142 (a) $\Delta S_{total} = \Delta S_{system} + \Delta S_{surr}$ and $\Delta S_{surr} = -\Delta H/T$

$\Delta S_{total} = \Delta S_{system} + (-\Delta H/T) = \Delta S_{system} - \Delta H/T$

$\Delta S_{system} = \Delta S_{total} + \Delta H/T$

$\Delta G = \Delta H - T\Delta S$ (substitute ΔS_{system} for ΔS in this equation)

$\Delta G = \Delta H - T(\Delta S_{total} + \Delta H/T) = -T\Delta S_{total}$

$\Delta G = -T\Delta S_{total}$ For a spontaneous reaction, if $\Delta S_{total} > 0$ then $\Delta G < 0$.

(b) $\Delta G^\circ = \Delta H^\circ - T\Delta S^\circ$

$\Delta H^\circ = \Delta G^\circ + T\Delta S^\circ$

$$\Delta S_{surr} = -\frac{\Delta H^\circ}{T} = -\frac{[\Delta G^\circ + T\Delta S^\circ]}{T} = -\frac{[2879 \times 10^3 \text{ J/mol} + (298 \text{ K})(-262 \text{ J/(K·mol)})]}{298 \text{ K}}$$

$\Delta S_{surr} = -9399$ J/(K· mol)

9.143

$3/2\ NO_2(g) + 1/2\ H_2O(l) \rightarrow HNO_3(aq) + 1/2\ NO(g)$ $\quad \Delta H^\circ_1 = \frac{-137.3 \text{ kJ}}{2}$

$3/2\ NO(g) + 3/4\ O_2(g) \rightarrow 3/2\ NO_2(g)$ $\quad \Delta H^\circ_2 = \frac{(3)(-116.2 \text{ kJ})}{4}$

$1/2\ N_2(g) + 3/2\ H_2(g) \rightarrow NH_3(g)$ $\quad \Delta H^\circ_3 = -46.1$ kJ

$NH_3(g) + 5/4\ O_2(g) \rightarrow NO(g) + 6/4\ H_2O(l)$ $\quad \Delta H^\circ_4 = \frac{-1165.2 \text{ kJ}}{4}$

$H_2O(l) \rightarrow 1/2\ O_2(g) + H_2(g)$ $\quad \Delta H^\circ_5 = +285.8$ kJ

Sum $1/2\ H_2(g) + 1/2\ N_2(g) + 3/2\ O_2(g) \rightarrow HNO_3(aq)$ $\quad \Delta H^\circ = -207.4$ kJ

9.144 $q_{Mo} = (110.0 \text{ g})(\text{specific heat Mo})(28.0\ ^\circ\text{C} - 100.0\ ^\circ\text{C})$

$q_{H_2O} = (150.0 \text{ g})[4.184 \text{ J/(g} \cdot {}^\circ\text{C)}](28.0\ ^\circ\text{C} - 24.6\ ^\circ\text{C})$

$q_{Mo} = -q_{H_2O}$

$(110.0 \text{ g})(\text{specific heat Mo})(28.0\ ^\circ\text{C} - 100.0\ ^\circ\text{C}) = -(150.0 \text{ g})[4.184 \text{ J/(g} \cdot {}^\circ\text{C)}](28.0\ ^\circ\text{C} - 24.6\ ^\circ\text{C})$

$$\text{specific heat Mo} = \frac{-(150.0 \text{ g})[4.184 \text{ J/(g·}^\circ\text{C)}](28.0\ ^\circ\text{C} - 24.6\ ^\circ\text{C})}{(110.0 \text{ g})(28.0\ ^\circ\text{C} - 100.0\ ^\circ\text{C})} = 0.27 \text{ J/(g} \cdot {}^\circ\text{C)}$$

9.145 $q_{ice\ tea} = -q_{ice}$

$$q_{ice\ tea} = (4.18 \frac{\text{J}}{\text{g·}^\circ\text{C}})(400.0 \text{ g})(10.0\ ^\circ\text{C} - 80.0\ ^\circ\text{C}) = -1.17 \times 10^5 \text{ J}$$

H_2O, 18.02

$$q_{ice} = 1.17 \times 10^5 \text{ J} = (6.01 \text{ kJ/mol})\left(\frac{1000 \text{ J}}{1 \text{ kJ}}\right)\left(m_{ice} \times \frac{1 \text{ mol } H_2O}{18.02 \text{ g } H_2O}\right) + \left(4.18 \frac{\text{J}}{(\text{g·}^\circ\text{C})}\right)(m_{ice})(10.0\ ^\circ\text{C} - 0.0\ ^\circ\text{C})$$

Solve for the mass of ice, m_{ice}.

$1.17 \times 10^5 \text{ J} = (3.34 \times 10^2 \text{ J/g})(m_{ice}) + (41.8 \text{ J/g})(m_{ice}) = (3.76 \times 10^2 \text{ J/g})(m_{ice})$

$$m_{ice} = \frac{1.17 \times 10^5 \text{ J}}{3.76 \times 10^2 \text{ J/g}} = 311 \text{ g of ice}$$

9.146 There is a large excess of NaOH. 5.00 mL = 0.005 00 L

mol citric acid = (0.005 00 L)(0.64 mol/L) = 0.0032 mol citric acid

$q_{H_2O} = (51.6\ g)[4.0\ J/(g \cdot {}^\circ C)](27.9\ {}^\circ C - 26.0\ {}^\circ C) = 392\ J$

$q_{rxn} = -q_{H_2O} = -392\ J$

$$\Delta H = -\frac{392\ J}{0.0032\ mol} \times \frac{1\ kJ}{1000\ J} = -123\ kJ/mol = -120\ kJ/mol\ citric\ acid$$

9.147 $CsOH(aq) + HCl(aq) \rightarrow CsCl(aq) + H_2O(l)$

$$mol\ CsOH = 0.100\ L \times \frac{0.200\ mol\ CsOH}{1.00\ L} = 0.0200\ mol\ CsOH$$

$$mol\ HCl = 0.050\ L \times \frac{0.400\ mol\ HCl}{1.00\ L} = 0.0200\ mol\ HCl$$

The reactants were mixed in equal mole amounts. volume = 150 mL; mass = 150 g

$$q_{solution} = \left(4.2 \frac{J}{(g \cdot {}^\circ C)}\right)(150\ g)(24.28\ {}^\circ C - 22.50\ {}^\circ C) = 1121\ J$$

$q_{reaction} = -q_{solution} = -1121\ J$

$$\Delta H = \frac{q_{reaction}}{mol\ CsOH} = \frac{-1121\ J}{0.0200\ mol\ CsOH} \times \frac{1\ kJ}{1000\ J} = -56\ kJ/mol\ CsOH$$

9.148 $NaNO_3$, 84.99; KF, 58.10

For $NaNO_3(s) \rightarrow NaNO_3(aq)$, $q = 20.4\ kJ/mol \times \dfrac{1\ mol\ NaNO_3}{84.99\ g\ NaNO_3} = 0.240\ kJ/g$

For $KF(s) \rightarrow KF(aq)$, $q = -17.7\ kJ/mol \times \dfrac{1\ mol\ KF}{58.10\ g\ KF} = -0.305\ kJ/g$

$q_{soln} = (110.0\ g)[4.18\ J/(g \cdot {}^\circ C)](2.22\ {}^\circ C) = 1021\ J = 1.02\ kJ$

$q_{rxn} = -q_{soln} = -1.02\ kJ$

Let X = mass of $NaNO_3$ and Y = mass of KF

X + Y = 10.0 g, so Y = 10.0 g – X

$q_{rxn} = -1.02\ kJ = X(0.240\ kJ/g) + Y(-0.305\ kJ/g)$ (substitute for Y and solve for X)

$-1.02\ kJ = X(0.240\ kJ/g) + (10.0\ g - X)(-0.305\ kJ/g)$

$-1.02\ kJ = (0.240\ kJ)X - 3.05\ kJ + (0.305\ kJ)X$

$2.03\ kJ = (0.545\ kJ)X$

$$X = \frac{2.03\ kJ}{0.545\ kJ} = 3.72\ g\ NaNO_3$$

Y = 10.0 g – X = 10.0 g – 3.72 g = 6.28 g KF = 6.3 g KF

9.149

	ΔH
$4\ CO(g) + 2\ O_2(g) \rightarrow 4\ CO_2(g)$	2(–566.0 kJ)
$2\ NO_2(g) \rightarrow 2\ NO(g) + O_2(g)$	+116.2 kJ
$2\ NO(g) \rightarrow O_2(g) + N_2(g)$	2(– 91.3 kJ)
$4\ CO(g) + 2\ NO_2(g) \rightarrow 4\ CO_2(g) + N_2(g)$	– 1198.4 kJ

Multiconcept Problems

9.150 (a) Each S has 2 bonding pairs and 2 lone pairs of electrons. Each S is sp^3 hybridized and the geometry around each S is bent.

(b) $\Delta H = D$ (reactant bonds) – D (product bonds) = (8 $D_{S\text{-}S}$) – (4 $D_{S=S}$) = +237 kJ

ΔH = [(8 mol)(225 kJ/mol) – [(4 mol)($D_{S=S}$)] = +237 kJ

– (4 mol)($D_{S=S}$) = 237 kJ – 1800 kJ = –1563 kJ

$D_{S=S}$ = (1563 kJ)/(4 mol) = 391 kJ/mol

(c)

	S_2
σ^*_{3p}	__
π^*_{3p}	↑ ↑
π_{3p}	↑↓ ↑↓
σ_{3p}	↑↓
σ^*_{3s}	↑↓
σ_{3s}	↑↓

S_2 should be paramagnetic with two unpaired electrons in the π^*_{3p} MOs.

9.151 (a)

```
      :Ö
       ||
       C
     /   \
  :Cl:   :Cl:
```

(b) $C(g) + \frac{1}{2} O_2(g) + Cl_2(g) \rightarrow COCl_2(g)$

$\Delta H^\circ_f = \Delta H^\circ_f(C(g)) + (\frac{1}{2} D_{O=O} + D_{Cl\text{-}Cl}) - (D_{C=O} + 2\, D_{C\text{-}Cl})$

ΔH°_f = (716.7 kJ) + [(½ mol)(498 kJ/mol) + (1 mol)(243 kJ/mol)]

– [(1 mol)(732 kJ/mol) + (2 mol)(330 kJ/mol)]

ΔH°_{rxn} = – 183 kJ per mol $COCl_2$; From Appendix B, $\Delta H^\circ_f(COCl_2)$ = –219.1 kJ/mol

The calculation of ΔH°_f from bond energies is only an estimate because the bond energies are average values derived from many different compounds.

9.152 (a) (1) $2\, CH_3CO_2H(l) + Na_2CO_3(s) \rightarrow 2\, CH_3CO_2Na(aq) + CO_2(g) + H_2O(l)$

(2) $CH_3CO_2H(l) + NaHCO_3(s) \rightarrow CH_3CO_2Na(aq) + CO_2(g) + H_2O(l)$

(b) CH_3CO_2H, 60.05; Na_2CO_3, 105.99; $NaHCO_3$, 84.01

$$1 \text{ gal} \times \frac{3.7854 \text{ L}}{1 \text{ gal}} \times \frac{1 \text{ mL}}{1 \times 10^{-3} \text{ L}} \times \frac{1.049 \text{ g } CH_3CO_2H}{1 \text{ mL}} = 3971 \text{ g } CH_3CO_2H$$

$$3971 \text{ g } CH_3CO_2H \times \frac{1 \text{ mol } CH_3CO_2H}{60.05 \text{ g } CH_3CO_2H} = 66.13 \text{ mol } CH_3CO_2H$$

For reaction (1)

$$66.13 \text{ mol } CH_3CO_2H \times \frac{1 \text{ mol } Na_2CO_3}{2 \text{ mol } CH_3CO_2H} \times \frac{105.99 \text{ g } Na_2CO_3}{1 \text{ mol } Na_2CO_3} \times \frac{1 \text{ kg}}{1000 \text{ g}} = 3.505 \text{ kg } Na_2CO_3$$

For reaction (2)

$$66.13 \text{ mol } CH_3CO_2H \times \frac{1 \text{ mol } NaHCO_3}{1 \text{ mol } CH_3CO_2H} \times \frac{84.01 \text{ g } NaHCO_3}{1 \text{ mol } NaHCO_3} \times \frac{1 \text{ kg}}{1000 \text{ g}} = 5.556 \text{ kg } NaHCO_3$$

(c) $2\ CH_3CO_2H(l) + Na_2CO_3(s) \rightarrow 2\ CH_3CO_2Na(aq) + CO_2(g) + H_2O(l)$

$\Delta H^\circ_{rxn} = [2\ \Delta H^\circ_f(CH_3CO_2Na) + \Delta H^\circ_f(CO_2) + \Delta H^\circ_f(H_2O)] - [2\ \Delta H^\circ_f(CH_3CO_2H) + \Delta H^\circ_f(Na_2CO_3)]$

$\Delta H^\circ_{rxn} = [(2\text{ mol})(-726.1\text{ kJ/mol}) + (1\text{ mol})(-393.5\text{ kJ/mol}) + (1\text{ mol})(-285.8\text{ kJ/mol})] - [(2\text{ mol})(-484.5\text{ kJ/mol}) + (1\text{ mol})(-1130.7\text{ kJ/mol})]$

$\Delta H^\circ_{rxn} = -31.8$ kJ for 2 mol CH_3CO_2H

$$\text{Heat} = -\frac{31.8\text{ kJ}}{2\text{ mol } CH_3CO_2H} \times 66.13\text{ mol } CH_3CO_2H = -1050\text{ kJ (liberated)}$$

$CH_3CO_2H(l) + NaHCO_3(s) \rightarrow CH_3CO_2Na(aq) + CO_2(g) + H_2O(l)$

$\Delta H^\circ_{rxn} = [\Delta H^\circ_f(CH_3CO_2Na) + \Delta H^\circ_f(CO_2) + \Delta H^\circ_f(H_2O)] - [\Delta H^\circ_f(CH_3CO_2H) + \Delta H^\circ_f(NaHCO_3)]$

$\Delta H^\circ_{rxn} = [(1\text{ mol})(-726.1\text{ kJ/mol}) + (1\text{ mol})(-393.5\text{ kJ/mol}) + (1\text{ mol})(-285.8\text{ kJ/mol})] - [(1\text{ mol})(-484.5\text{ kJ/mol}) + (1\text{ mol})(-950.8\text{ kJ/mol})]$

$\Delta H^\circ_{rxn} = +29.9$ kJ for 1 mol CH_3CO_2H

$$q = \frac{29.9\text{ kJ}}{1\text{ mol } CH_3CO_2H} \times 66.13\text{ mol } CH_3CO_2H = +1980\text{ kJ (absorbed)}$$

9.153 (a) $2\ K(s) + 2\ H_2O(l) \rightarrow 2\ KOH(aq) + H_2(g)$

(b) $\Delta H^\circ_{rxn} = [2\ \Delta H^\circ_f(KOH)] - [2\ \Delta H^\circ_f(H_2O)]$

$\Delta H^\circ_{rxn} = [(2\text{ mol})(-482.4\text{ kJ/mol})] - [(2\text{ mol})(-285.8\text{ kJ/mol})] = -393.2\text{ kJ}$

(c) The reaction produces 393.2 kJ/ 2 mol K = 196.6 kJ/ mol K. Assume that the mass of the water does not change and that the specific heat = 4.18 J/(g·°C) for the solution that is produced.

$$q = 7.55\text{ g K} \times \frac{1\text{ mol K}}{39.10\text{ g K}} \times \frac{196.6\text{ kJ}}{1\text{ mol K}} \times \frac{1000\text{ J}}{1\text{ kJ}} = 3.80 \times 10^4\text{ J}$$

$q = (\text{specific heat}) \times m \times \Delta T$

$$\Delta T = \frac{q}{(\text{specific heat}) \times m} = \frac{3.80 \times 10^4\text{ J}}{[4.18\text{ J/(g}\cdot{}^\circ\text{C)}](400.0\text{ g})} = 22.7\ {}^\circ\text{C}$$

$\Delta T = T_{final} - T_{initial}$

$T_{final} = \Delta T + T_{initial} = 22.7\ {}^\circ\text{C} + 25.0\ {}^\circ\text{C} = 47.7\ {}^\circ\text{C}$

(d) $$7.55\text{ g K} \times \frac{1\text{ mol K}}{39.10\text{ g K}} \times \frac{2\text{ mol KOH}}{2\text{ mol K}} = 0.193\text{ mol KOH}$$

Assume that the mass of the solution does not change during the reaction and that the solution has a density of 1.00 g/mL.

$$\text{solution volume} = 400.0\text{ g} \times \frac{1.00\text{ mL}}{1\text{ g}} \times \frac{1 \times 10^{-3}\text{ L}}{1\text{ mL}} = 0.400\text{ L}$$

$$\text{molarity} = \frac{0.193\text{ mol KOH}}{0.400\text{ L}} = 0.483\text{ M}$$

$2\ KOH(aq) + H_2SO_4(aq) \rightarrow K_2SO_4(aq) + 2\ H_2O(l)$

$$0.193\text{ mol KOH} \times \frac{1\text{ mol } H_2SO_4}{2\text{ mol KOH}} \times \frac{1000\text{ mL}}{0.554\text{ mol } H_2SO_4} = 174\text{ mL of } 0.554\text{ M } H_2SO_4$$

9.154 (a)

```
   H   H
   |   |
H—N—N—H
  ••  ••
```

Each N is sp^3 hybridized and the geometry about each N is trigonal pyramidal.

(b)

$2/4\ NH_3(g) + 3/4\ N_2O(g) \rightarrow N_2(g) + 3/4\ H_2O(l)$ $\quad \Delta H°_1 = \frac{-1011.2\ kJ}{4}$

$1/4\ H_2O(l) + 1/4\ N_2H_4(l) \rightarrow 1/8\ O_2(g) + 1/2\ NH_3(g)$ $\quad \Delta H°_2 = \frac{+286\ kJ}{8}$

$3/4\ N_2H_4(l) + 3/4\ H_2O(l) \rightarrow 3/4\ N_2O(g) + 9/4\ H_2(g)$ $\quad \Delta H°_3 = \frac{(3)(+317\ kJ)}{4}$

$9/4\ H_2(g) + 9/8\ O_2(g) \rightarrow 9/4\ H_2O(l)$ $\quad \Delta H°_4 = \frac{(9)(-285.8\ kJ)}{4}$

Sum $\quad N_2H_4(l) + O_2(g) \rightarrow N_2(g) + 2\ H_2O(l)$ $\quad \Delta H° = -622\ kJ$

(c) N_2H_4, 32.045

$$\text{mol } N_2H_4 = 100.0\ g\ N_2H_4 \times \frac{1\ mol\ N_2H_4}{32.045\ g\ N_2H_4} = 3.12\ mol\ N_2H_4$$

$$q = (3.12\ mol\ N_2H_4)(622\ kJ/mol) = 1940\ kJ$$

9.155 Assume 100.0 g of Y.

$$\text{mol F} = 61.7\ g\ F \times \frac{1\ mol\ F}{19.00\ g\ F} = 3.25\ mol\ F$$

$$\text{mol Cl} = 38.3\ g\ Cl \times \frac{1\ mol\ Cl}{35.45\ g\ Cl} = 1.08\ mol\ Cl$$

$Cl_{1.08}F_{3.25}$; divide each subscript by the smaller of the two, 1.08.

$Cl_{1.08/1.08}F_{3.25/1.08}$

ClF_3

(a) Y is ClF_3 and X is ClF

(b)

```
 ••   ••   ••
:F — Cl — F:
 ••   |    ••
     :F:
      ••
```

There are five electron clouds around the Cl (3 bonding and 2 lone pairs). The geometry is T-shaped.

(c)

	ΔH
$Cl_2O(g) + 3\ OF_2(g) \rightarrow 2\ O_2(g) + 2\ ClF_3(g)$	−532.8 kJ
$2\ ClF(g) + O_2(g) \rightarrow Cl_2O(g) + OF_2(g)$	+205.4 kJ
$O_2(g) + 2\ F_2(g) \rightarrow 2\ OF_2(g)$	2(+24.5 kJ)
$2\ ClF(g) + 2\ F_2(g) \rightarrow 2\ ClF_3(g)$	−278.4 kJ

Divide reaction coefficients and ΔH by 2.

$ClF(g) + F_2(g) \rightarrow ClF_3(g)$ $\quad \Delta H = -278.4\ kJ/2 = -139.2\ kJ/mol\ ClF_3$

(d) ClF, 54.45

$$q = 25.0\ g\ ClF \times \frac{1\ mol\ ClF}{54.45\ g\ ClF} \times \frac{-139.2\ kJ}{1\ mol\ ClF} \times 0.875 = -55.9\ kJ$$

55.9 kJ is released in this reaction.

10 Gases: Their Properties and Behavior

10.1 $P = 28.48 \text{ in Hg} \times \dfrac{2.54 \text{ cm}}{1 \text{ in}} \times \dfrac{10 \text{ mm}}{1 \text{ cm}} \times \dfrac{1 \text{ atm}}{760 \text{ mm Hg}} = 0.952 \text{ atm}$

$P = 28.48 \text{ in Hg} \times \dfrac{2.54 \text{ cm}}{1 \text{ in}} \times \dfrac{10 \text{ mm}}{1 \text{ cm}} \times \dfrac{101{,}325 \text{ Pa}}{760 \text{ mm Hg}} = 9.64 \times 10^4 \text{ Pa}$

$P = 9.64 \times 10^4 \text{ Pa} \times \dfrac{1 \text{ bar}}{10^5 \text{ Pa}} = 0.964 \text{ bar}$

10.2 (a) $P = 760 \text{ mm Hg} \times \dfrac{13.6 \text{ mm H}_2\text{O}}{1 \text{ mm Hg}} \times \dfrac{1 \times 10^{-3} \text{ m}}{1 \text{ mm}} = 10.3 \text{ m H}_2\text{O}$

(b) A barometer filled with water would be too tall to be practical.

10.3 The pressure in the flask is less than 0.975 atm because the liquid level is higher on the side connected to the flask. The 24.7 cm of Hg is the difference between the two pressures.

$\text{Pressure difference} = 24.7 \text{ cm Hg} \times \dfrac{1.00 \text{ atm}}{76.0 \text{ cm Hg}} = 0.325 \text{ atm}$

Pressure in flask = 0.975 atm – 0.325 atm = 0.650 atm

10.4 (a)

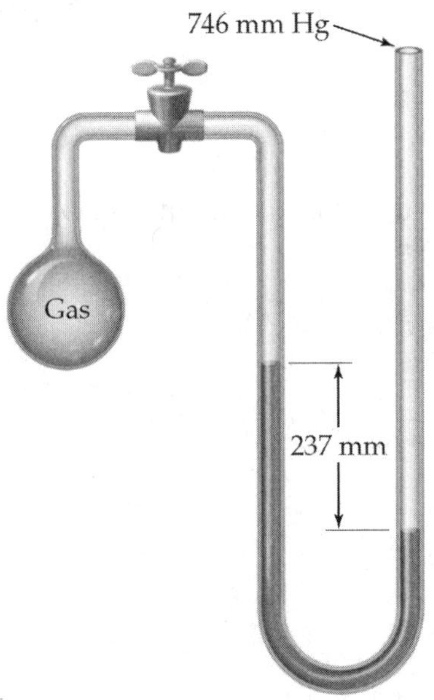

746 mm Hg

Gas

237 mm

(b) (13.6 g/mL)/(0.822 g/mL) = 16.5

$P_{bulb} = 746 \text{ mm Hg} - \left(237 \text{ mm min oil} \times \dfrac{1 \text{ mm Hg}}{16.5 \text{ mm min oil}}\right) = 732 \text{ mm Hg}$

10.5 (a) Assume an initial volume of 1.00 L.
First consider the volume change resulting from a change in the number of moles with the pressure and temperature constant.

$$\frac{V_i}{n_i} = \frac{V_f}{n_f};\ V_f = \frac{V_i n_f}{n_i} = \frac{(1.00\ \text{L})(0.225\ \text{mol})}{0.3\ \text{mol}} = 0.75\ \text{L}$$

Now consider the volume change from 0.75 L as a result of a change in temperature with the number of moles and the pressure constant.

$$\frac{V_i}{T_i} = \frac{V_f}{T_f};\ V_f = \frac{V_i T_f}{T_i} = \frac{(0.75\ \text{L})(400\ \text{K})}{300\ \text{K}} = 1.0\ \text{L}$$

There is no net change in the volume as a result of the decrease in the number of moles of gas and a temperature increase.

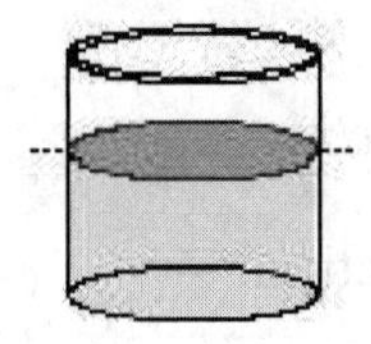

(b) Assume an initial volume of 1.00 L.
First consider the volume change resulting from a change in the number of moles with the pressure and temperature constant.

$$\frac{V_i}{n_i} = \frac{V_f}{n_f};\ V_f = \frac{V_i n_f}{n_i} = \frac{(1.00\ \text{L})(0.225\ \text{mol})}{0.3\ \text{mol}} = 0.75\ \text{L}$$

Now consider the volume change from 0.75 L as a result of a change in temperature with the number of moles and the pressure constant.

$$\frac{V_i}{T_i} = \frac{V_f}{T_f};\ V_f = \frac{V_i T_f}{T_i} = \frac{(0.75\ \text{L})(200\ \text{K})}{300\ \text{K}} = 0.5\ \text{L}$$

The volume would be cut in half as a result of the decrease in the number of moles of gas and a temperature decrease.

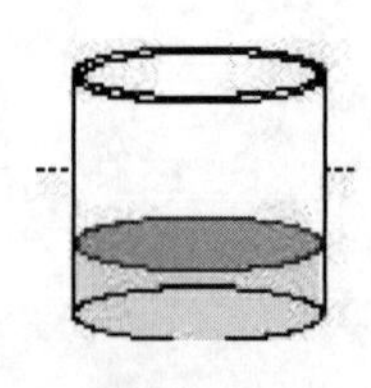

10.6

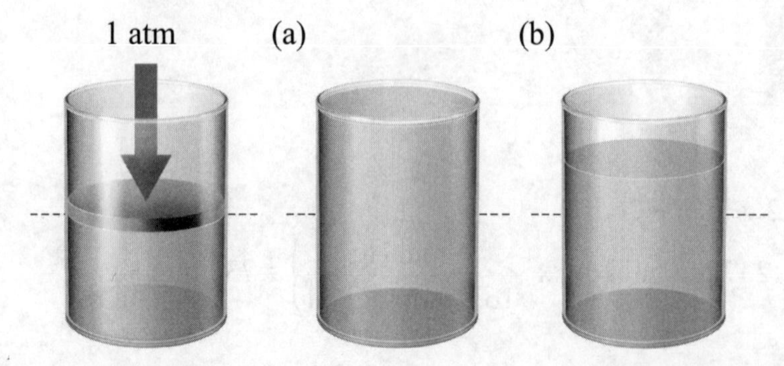

10.7 $n = \frac{PV}{RT} = \frac{(1.000\text{ atm})(1.000 \times 10^5\text{ L})}{\left(0.082\ 06\ \frac{\text{L}\cdot\text{atm}}{\text{K}\cdot\text{mol}}\right)(273.15\text{ K})} = 4.461 \times 10^3\text{ mol } CH_4$

CH_4, 16.04; mass $CH_4 = (4.461 \times 10^3\text{ mol})\left(\frac{16.04\text{ g}}{1\text{ mol}}\right) = 7.155 \times 10^4\text{ g } CH_4$

10.8 C_3H_8, 44.10; V = 350 mL = 0.350 L; T = 20 °C = 293 K

$n = 3.2\text{ g} \times \frac{1\text{ mol } C_3H_8}{44.10\text{ g } C_3H_8} = 0.073\text{ mol } C_3H_8$

$P = \frac{nRT}{V} = \frac{(0.073\text{ mol})\left(0.082\ 06\ \frac{\text{L}\cdot\text{atm}}{\text{K}\cdot\text{mol}}\right)(293\text{ K})}{0.350\text{ L}} = 5.0\text{ atm}$

10.9 The volume and number of moles of gas remain constant.

$\frac{nR}{V} = \frac{P_i}{T_i} = \frac{P_f}{T_f}; \quad T_f = \frac{P_f T_i}{P_i} = \frac{(2.37\text{ atm})(273\text{ K})}{2.15\text{ atm}} = 301\text{ K} = 28\text{ °C}$

10.10 $\frac{P_i V_i}{T_i} = \frac{P_f V_f}{T_f}$

25.0 °C = 298 K; $V_f = \frac{P_i V_i T_f}{T_i P_f} = \frac{(745\text{ mm Hg})(45.0\text{ L})(225\text{ K})}{(298\text{ K})(178\text{ mm Hg})} = 142\text{ L}$

10.11 $CaCO_3(s) + 2\ HCl(aq) \rightarrow CaCl_2(aq) + CO_2(g) + H_2O(l)$

$CaCO_3$, 100.1; CO_2, 44.01

mole $CO_2 = 33.7\text{ g } CaCO_3 \times \frac{1\text{ mol } CaCO_3}{100.1\text{ g } CaCO_3} \times \frac{1\text{ mol } CO_2}{1\text{ mol } CaCO_3} = 0.337\text{ mol } CO_2$

mass $CO_2 = 0.337\text{ mol } CO_2 \times \frac{44.01\text{ g } CO_2}{1\text{ mol } CO_2} = 14.8\text{ g } CO_2$

$V = \frac{nRT}{P} = \frac{(0.337\text{ mol})\left(0.082\ 06\ \frac{\text{L}\cdot\text{atm}}{\text{K}\cdot\text{mol}}\right)(273\text{ K})}{1.00\text{ atm}} = 7.55\text{ L}$

10.12 $3\ H_2(g) + N_2(g) \rightarrow 2\ NH_3(g)$

$V_{H_2} = 500.0\text{ L } NH_3 \times \frac{3\text{ mol } H_2}{2\text{ mol } NH_3} = 750.0\text{ L } H_2$

$V_{N_2} = 500.0\text{ L } NH_3 \times \frac{1\text{ mol } N_2}{2\text{ mol } NH_3} = 250.0\text{ L } N_2$

10.13 $n = \frac{PV}{RT} = \frac{(1.00 \text{ atm})(1.00 \text{ L})}{\left(0.082\ 06 \frac{\text{L} \cdot \text{atm}}{\text{K} \cdot \text{mol}}\right)(273 \text{ K})} = 0.0446 \text{ mol}$

molar mass $= \frac{1.52 \text{ g}}{0.0446 \text{ mol}} = 34.1$ g/mol; molecular mass = 34.1

$Na_2S(aq) + 2\ HCl(aq) \rightarrow H_2S(g) + 2\ NaCl(aq)$

The foul-smelling gas is H_2S, hydrogen sulfide.

10.14 (a) CO_2, 44.0

$d = \frac{PM}{RT} = \frac{(1 \text{ atm})(44.0 \text{ g/mol})}{\left(0.082\ 06 \frac{\text{L} \cdot \text{atm}}{\text{K} \cdot \text{mol}}\right)(298 \text{ K})} = 1.80 \text{ g/L}$

(b) Carbon dioxide has a higher molar mass (44.0 g/mol) than the major components of air N_2 (28.0 g/mol) and O_2 (32.0 g/mol).

10.15 $X_{O_2} = 0.36$; $X_{N_2} = 0.64$; 25.0 °C = 298 K

(a) $P_{O_2} = P_{tot} \cdot X_{O_2} = (50 \text{ atm})(0.36) = 18 \text{ atm}$

$P_{N_2} = P_{tot} \cdot X_{N_2} = (50 \text{ atm})(0.64) = 32 \text{ atm}$

(b) $n_{O_2} = \frac{PV}{RT} = \frac{(18 \text{ atm})(10.0 \text{ L})}{\left(0.082\ 06 \frac{\text{L} \cdot \text{atm}}{\text{K} \cdot \text{mol}}\right)(298 \text{ K})} = 7.36 \text{ mol } O_2$

$n_{N_2} = \frac{PV}{RT} = \frac{(32 \text{ atm})(10.0 \text{ L})}{\left(0.082\ 06 \frac{\text{L} \cdot \text{atm}}{\text{K} \cdot \text{mol}}\right)(298 \text{ K})} = 13.1 \text{mol } N_2$

10.16 $P_{O_2} = P_{tot} \cdot X_{O_2}$

$0.21 \text{ atm} = (8.38 \text{ atm}) \cdot X_{O_2}$

$X_{O_2} = \frac{0.21 \text{ atm}}{8.38 \text{ atm}} = 0.025$

10.17 (a) $\frac{\text{rate } O_2}{\text{rate Kr}} = \sqrt{\frac{M_{Kr}}{M_{O_2}}} = \sqrt{\frac{83.8}{32.0}}$; $\frac{\text{rate } O_2}{\text{rate Kr}} = 1.62$

O_2 diffuses 1.62 times faster than Kr.

(b) $\frac{\text{rate } C_2H_2}{\text{rate } N_2} = \sqrt{\frac{M_{N_2}}{M_{C_2H_2}}} = \sqrt{\frac{28.0}{26.0}}$; $\frac{\text{rate } C_2H_2}{\text{rate } N_2} = 1.04$

C_2H_2 diffuses 1.04 times faster than N_2.

10.18 SO_2, 64.06

$$\frac{\text{rate}_1}{\text{rate}_2} = 1.414 = \sqrt{\frac{M_{SO_2}}{M_x}}; \quad \sqrt{M_x} = \frac{\sqrt{M_x}}{1.414} = \frac{\sqrt{64.06}}{1.414} = 5.660$$

$M_x = (5.660)^2 = 32.00$; The unknown gas could be O_2.

10.19 110 ppb = X_{O_3} x 10^9 ppb; $\quad X_{O_3} = \dfrac{110}{10^9} = 110 \times 10^{-9}$

volume % = X_{O_3} x 100 = 1.10×10^{-5}%

ppm = X_{O_3} x 10^6 = 0.110 ppm

O_3 molecules = $(X_{O_3})(3.3 \times 10^{-2}\text{ mol})(6.022 \times 10^{23}\text{ molecules/mol}) = 2.2 \times 10^{15}$ O_3 molecules

10.20 95 ppb = X_{O_3} x 10^9 ppb; $\quad X_{O_3} = \dfrac{95}{10^9} = 95 \times 10^{-9}$

$P_{O_3} = P_{tot} \cdot X_{O_3} = (0.79\text{ atm})(95 \times 10^{-9}) = 7.5 \times 10^{-8}$ atm

10.21 Nitrogen and oxygen are diatomic molecules that do not have a dipole moment. As the bond stretches in a vibration, the molecule still does not have a dipole moment. Since no change in dipole moment occurs, IR radiation will not be absorbed.

10.22 The water molecule has a bent geometry and thus all vibrations will result in a change in dipole moment and the absorption of IR radiation.

10.23 The symmetric stretch does not absorb IR radiation; the asymmetric stretch absorbs IR radiation.

10.24 1.82/0.48 = 3.8; The contribution from CO_2 is 3.8 times as large as the contribution from CH_4.

10.25 Although N_2O has a greater potential for warming on a per mass basis, the atmosphere has a much higher concentration of CO_2.

10.26 Water vapor is a potent greenhouse gas and when considering the greenhouse effect, increased levels would cause warming. However, increased levels of water vapor may also cause changes in some types of clouds that have a cooling effect.

Conceptual Problems

10.27

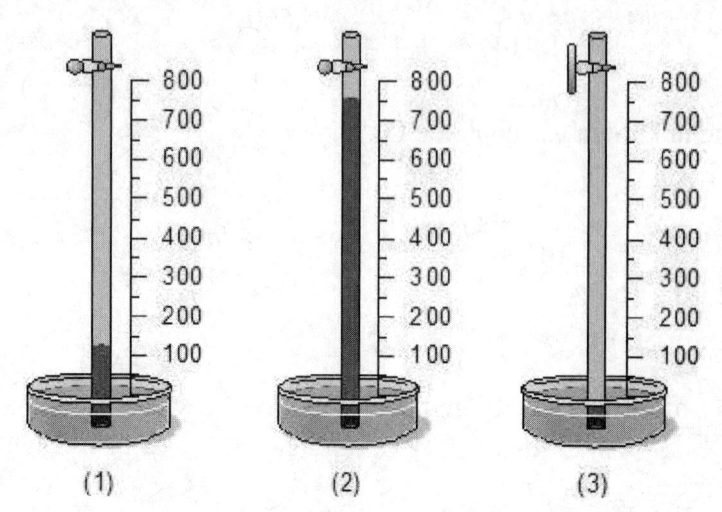

(a) $\frac{P_2}{T_2} = \frac{P_1}{T_1}$; $P_2 = \frac{(T_2)(P_1)}{T_1} = \frac{(248\text{ K})(760\text{ mm Hg})}{298\text{ K}} = 632\text{ mm Hg}$

The column of Hg will rise to ~130 mm Hg inside the tube (drawing 1). The pressure inside the tube (632 mm Hg) plus the pressure of 130 mm Hg equals the external pressure, ~760 mm Hg.

(b) The column of Hg will rise to ~760 mm (drawing 2), which is equal to the external pressure, ~760 mm Hg.

(c) The pressure inside the tube is equal to the external pressure and the Hg level inside the tube will be the same as in the dish (drawing 3).

10.28 The gas pressure in the bulb in mm Hg is equal to the difference in the height of the Hg in the two arms of the manometer.

10.29 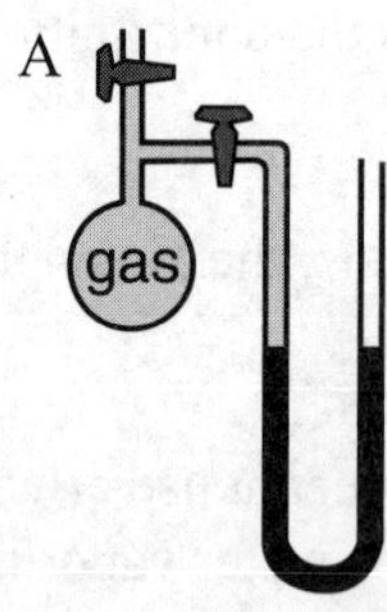

When stopcock A is opened, the pressure in the flask will equal the external pressure, and the level of mercury will be the same in both arms of the manometer.

10.30 The picture on the right will be the same as that on the left, apart from random scrambling of the He and Ar atoms.

10.31 (a)

The volume of a gas is proportional to the kelvin temperature at constant pressure. As the temperature increases from 300 K to 450 K, the volume will increase by a factor of 1.5.

(b)

The volume of a gas is inversely proportional to pressure at constant temperature. As the pressure increases from 1 atm to 2 atm, the volume will decrease by a factor of 2.

(c)

PV = nRT; The amount of gas (n) is constant.

Therefore $nR = \frac{P_iV_i}{T_i} = \frac{P_fV_f}{T_f}$.

Assume V_i = 1 L and solve for V_f.

$$\frac{P_iV_iT_f}{T_iP_f} = \frac{(3\text{ atm})(1\text{ L})(200\text{ K})}{(300\text{ K})(2\text{ atm})} = V_f = 1\text{ L}$$

There is no change in volume.

10.32 The sample remains a gas at 150 K. Drawing (c) represents the gas at this temperature. The gas molecules still fill the container.

10.33 The two gases should mix randomly and homogeneously and this is best represented by drawing (c).

10.34 n_{Total} = (1 black) + (3 blue) + (4 red) + (6 green) = 14

P_{Total} = 420 mm Hg

$$P_x = P_{Total} \times \frac{n_x}{n_{Total}}$$

$$P_{black} = (420\text{ mm Hg}) \times \left(\frac{1}{14}\right) = 30.0\text{ mm Hg};\quad P_{blue} = (420\text{ mm Hg}) \times \left(\frac{3}{14}\right) = 90.0\text{ mm Hg}$$

$$P_{red} = (420\text{ mm Hg}) \times \left(\frac{4}{14}\right) = 120\text{ mm Hg};\quad P_{green} = (420\text{ mm Hg}) \times \left(\frac{6}{14}\right) = 180\text{ mm Hg}$$

10.35 The two gases will be equally distributed among the three flasks.

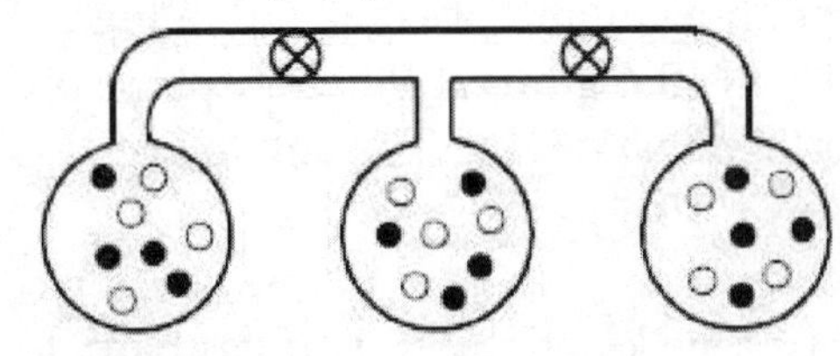

10.36 (a) The temperature has increased by about 10% (from 300 K to 325 K) while the amount and the pressure are unchanged. Thus, the volume should increase by about 10%.

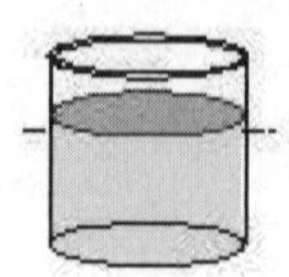

(b) The temperature has increased by a factor of 1.5 (from 300 K to 450 K) and the pressure has increased by a factor of 3 (from 0.9 atm to 2.7 atm) while the amount is unchanged. Thus, the volume should decrease by half (1.5/3 = 0.5).

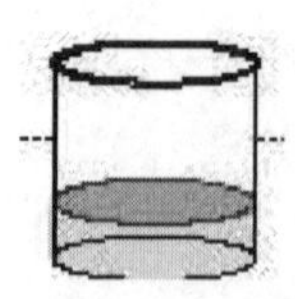

(c) Both the amount and the pressure have increased by a factor of 3 (from 0.075 mol to 0.22 mol and from 0.9 atm to 2.7 atm) while the temperature is unchanged. Thus, the volume is unchanged.

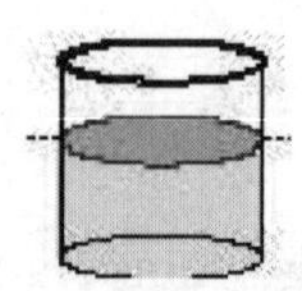

10.37 (a) Because there are more yellow gas molecules than there are blue, the yellow gas molecules have the higher average speed.
(b) Each rate is proportional to the number of effused gas molecules of each type.
$M_{yellow} = 25$

$$\frac{\text{rate}_{blue}}{\text{rate}_{yellow}} = \sqrt{\frac{M_{yellow}}{M_{blue}}};\quad \frac{5}{6} = \sqrt{\frac{25}{M_{blue}}};\quad \left(\frac{5}{6}\right)^2 = \frac{25}{M_{blue}};\quad M_{blue} = \frac{25}{\left(\frac{5}{6}\right)^2} = 36$$

Section Problems
Gases and Gas Pressure (Section 10.1)

10.38 1.00 atm = 14.7 psi

$$1.00 \text{ mm Hg} \times \frac{1 \text{ atm}}{760 \text{ mm Hg}} \times \frac{14.7 \text{ psi}}{1 \text{ atm}} = 1.93 \times 10^{-2} \text{ psi}$$

10.39 1.00 atmosphere pressure can support a column of Hg 0.760 m high. Because the density of H_2O is 1.00 g/mL and that of Hg is 13.6 g/mL, 1.00 atmosphere pressure can support a column of H_2O 13.6 times higher than that of Hg. The column of H_2O supported by 1.00 atmosphere will be (0.760 m)(13.6) = 10.3 m.

10.40 Temperature is a measure of the average kinetic energy of gas particles.

10.41 Gases are much more compressible than solids or liquids because there is a large amount of empty space between individual gas molecules.

10.42 $P = 480 \text{ mm Hg} \times \dfrac{1.00 \text{ atm}}{760 \text{ mm Hg}} = 0.632 \text{ atm}$; $P = 480 \text{ mm Hg} \times \dfrac{101{,}325 \text{ Pa}}{760 \text{ mm Hg}} = 6.40 \times 10^4 \text{ Pa}$

10.43 $P = 352 \text{ torr} \times \dfrac{101{,}325 \text{ Pa}}{760 \text{ torr}} \times \dfrac{1 \text{ kPa}}{1000 \text{ Pa}} = 46.9 \text{ kPa}$

$P = 0.255 \text{ atm} \times \dfrac{760 \text{ mm Hg}}{1.00 \text{ atm}} = 194 \text{ mm Hg}$

$P = 0.0382 \text{ mm Hg} \times \dfrac{101{,}325 \text{ Pa}}{760 \text{ mm Hg}} = 5.09 \text{ Pa}$

10.44 $P_{flask} > 754.3 \text{ mm Hg}$; $P_{flask} = 754.3 \text{ mm Hg} + 176 \text{ mm Hg} = 930 \text{ mm Hg}$

10.45 $P_{flask} < 1.021 \text{ atm}$ (see Figure 10.5)

$P_{difference} = 28.3 \text{ cm Hg} \times \dfrac{1.00 \text{ atm}}{76.0 \text{ cm Hg}} = 0.372 \text{ atm}$

$P_{flask} = 1.021 \text{ atm} - P_{difference} = 1.021 \text{ atm} - 0.372 \text{ atm} = 0.649 \text{ atm}$

10.46 $P_{flask} > 752.3 \text{ mm Hg}$ (see Figure 10.5); If the pressure in the flask can support a column of ethyl alcohol (d = 0.7893 g/mL) 55.1 cm high, then it can only support a column of Hg that is much shorter because of the higher density of Hg.

$55.1 \text{ cm} \times \dfrac{0.7893 \text{ g/mL}}{13.546 \text{ g/mL}} = 3.21 \text{ cm Hg} = 32.1 \text{ mm Hg}$

$P_{flask} = 752.3 \text{ mm Hg} + 32.1 \text{ mm Hg} = 784.4 \text{ mm Hg}$

$P_{flask} = 784.4 \text{ mm Hg} \times \dfrac{101{,}325 \text{ Pa}}{760 \text{ mm Hg}} = 1.046 \times 10^5 \text{ Pa}$

10.47 Compute the height of a column of $CHCl_3$ that 1.00 atm can support.

$760 \text{ mm Hg} \times \dfrac{13.546 \text{ g/mL}}{1.4832 \text{ g/mL}} = 6941 \text{ mm } CHCl_3$; therefore $1.00 \text{ atm} = 6941 \text{ mm } CHCl_3$

The pressure in the flask is less than atmospheric pressure.

$P_{atm} - P_{flask} = 0.849 \text{ atm} - 0.788 \text{ atm} = 0.061 \text{ atm}$

$0.061 \text{ atm} \times \dfrac{6941 \text{ mm } CHCl_3}{1.00 \text{ atm}} = 423 \text{ mm } CHCl_3$

The chloroform will be 423 mm higher in the manometer arm connected to the flask.

10.48

	% Volume
N_2	78.08
O_2	20.95
Ar	0.93
CO_2	0.037

The % volume for a particular gas is proportional to the number of molecules of that gas in a mixture of gases.
Average molecular mass of air
= (0.7808)(mol. mass N_2) + (0.2095)(mol. mass O_2)
+ (0.0093)(at. mass Ar) + (0.000 37)(mol. mass CO_2)
= (0.7808)(28.01) + (0.2095)(32.00)
+ (0.0093)(39.95) + (0.000 37)(44.01) = 28.96

10.49 The % volume for a particular gas is proportional to the number of molecules of that gas in a mixture of gases.
Average molecular weight of a diving-gas
= (0.020)(molecular weight O_2) + (0.980)(atomic weight He)
= (0.020)(32.00) + (0.980)(4.00) = 4.56

The Gas Laws (Sections 10.2 and 10.3)

10.50 (a) $\frac{nR}{V} = \frac{P_i}{T_i} = \frac{P_f}{T_f}$; $\frac{P_i T_f}{T_i} = P_f$

Let P_i = 1 atm, T_i = 100 K, T_f = 300 K

$$P_f = \frac{P_i T_f}{T_i} = \frac{(1\text{ atm})(300\text{ K})}{(100\text{ K})} = 3\text{ atm}$$

The pressure would triple.

(b) $\frac{RT}{V} = \frac{P_i}{n_i} = \frac{P_f}{n_f}$; $\frac{P_i n_f}{n_i} = P_f$

Let P_i = 1 atm, n_i = 3 mol, n_f = 1 mol

$$P_f = \frac{P_i n_f}{n_i} = \frac{(1\text{ atm})(1\text{ mol})}{(3\text{ mol})} = \frac{1}{3}\text{ atm}$$

The pressure would be $\frac{1}{3}$ the initial pressure.

(c) $nRT = P_iV_i = P_fV_f$; $\frac{P_i V_i}{V_f} = P_f$

Let P_i = 1 atm, V_i = 1 L, V_f = 1 – 0.45 L = 0.55 L

$$P_f = \frac{P_i V_i}{V_f} = \frac{(1\text{ atm})(1\text{ L})}{(0.55\text{ L})} = 1.8\text{ atm}$$

The pressure would increase by 1.8 times.

(d) $nR = \frac{P_i V_i}{T_i} = \frac{P_f V_f}{T_f}$; $\frac{P_i V_i T_f}{T_i V_f} = P_f$

Let P_i = 1 atm, V_i = 1 L, T_i = 200 K, V_f = 3 L, T_i = 100 K

$$P_f = \frac{P_i V_i T_f}{T_i V_f} = \frac{(1\text{ atm})(1\text{ L})(100\text{ K})}{(200\text{ K})(3\text{ L})} = 0.17\text{ atm}$$

The pressure would be 0.17 times the initial pressure.

10.51 (a) $\frac{nR}{P} = \frac{V_i}{T_i} = \frac{V_f}{T_f}$; $\frac{V_i T_f}{T_i} = V_f$

Let V_i = 1 L, T_i = 400 K, T_f = 200 K

$$V_f = \frac{V_i T_f}{T_i} = \frac{(1\text{ L})(200\text{ K})}{(400\text{ K})} = 0.5\text{ L}$$

The volume would be halved.

(b) $\frac{RT}{P} = \frac{V_i}{n_i} = \frac{V_f}{n_f}$; $\frac{V_i n_f}{n_i} = V_f$

Let V_i = 1 L, n_i = 4 mol, n_f = 5 mol

$$V_f = \frac{V_i n_f}{n_i} = \frac{(1\text{ L})(5\text{ mol})}{(4\text{ mol})} = 1.25\text{ L}$$

The volume would increase by 1/4.

(c) $nRT = P_i V_i = P_f V_f$; $\frac{P_i V_i}{P_f} = V_f$

Let V_i = 1 L, P_i = 4 atm, P_f = 1 atm

$$V_f = \frac{P_i V_i}{P_f} = \frac{(4\text{ atm})(1\text{ L})}{(1\text{ atm})} = 4\text{ L}$$

The volume would increase by a factor of 4.

(d) $nR = \frac{P_i V_i}{T_i} = \frac{P_f V_f}{T_f}$; $\frac{P_i V_i T_f}{P_f T_i} = V_f$

Let V_i = 1 L, T_i = 200 K, T_f = 400 K, P_i = 1 atm, P_f = 2 atm

$$V_f = \frac{P_i V_i T_f}{P_f T_i} = \frac{(1\text{ atm})(1\text{ L})(400\text{ K})}{(2\text{ atm})(200\text{ K})} = 1\text{ L}$$

There is no volume change.

10.52 They all contain the same number of gas molecules.

10.53 For air, T = 50 °C = 323 K.

$$n = \frac{PV}{RT} = \frac{\left(750\text{ mm Hg x }\frac{1.00\text{ atm}}{760\text{ mm Hg}}\right)(2.50\text{ L})}{\left(0.082\ 06\ \frac{\text{L}\cdot\text{atm}}{\text{K}\cdot\text{mol}}\right)(323\text{ K})} = 0.0931\text{ mol air}$$

For CO_2, T = –10 °C = 263 K

$$n = \frac{PV}{RT} = \frac{\left(765 \text{ mm Hg x } \frac{1.00 \text{ atm}}{760 \text{ mm Hg}}\right)(2.16 \text{ L})}{\left(0.082\ 06 \frac{\text{L} \cdot \text{atm}}{\text{K} \cdot \text{mol}}\right)(263 \text{ K})} = 0.101 \text{ mol } CO_2$$

Because the number of moles of CO_2 is larger than the number of moles of air, the CO_2 sample contains more molecules.

10.54 n and T are constant; therefore $nRT = P_iV_i = P_fV_f$

$$V_f = \frac{P_iV_i}{P_f} = \frac{(150 \text{ atm})(49.0 \text{ L})}{(1.02 \text{ atm})} = 7210 \text{ L}$$

n and P are constant; therefore $\frac{nR}{P} = \frac{V_i}{T_i} = \frac{V_f}{T_f}$

$$V_f = \frac{V_iT_f}{T_i} = \frac{(49.0 \text{ L})(308 \text{ K})}{(293 \text{ K})} = 51.5 \text{ L}$$

10.55 $T_i = 20\ °C = 293\ K$; $nR = \frac{P_iV_i}{T_i} = \frac{P_fV_f}{T_f}$

$$V_f = \frac{P_iV_iT_f}{T_iP_f} = \frac{(140 \text{ atm})(8.0 \text{ L})(273 \text{ K})}{(293 \text{ K})(1.00 \text{ atm})} = 1.0 \text{ x } 10^3 \text{ L}$$

10.56 $15.0 \text{ g } CO_2 \text{ x } \frac{1 \text{ mol } CO_2}{44.0 \text{ g } CO_2} = 0.341 \text{ mol } CO_2$

$$P = \frac{nRT}{V} = \frac{(0.341 \text{ mol})\left(0.082\ 06 \frac{\text{L} \cdot \text{atm}}{\text{K} \cdot \text{mol}}\right)(300 \text{ K})}{(0.30 \text{ L})} = 27.98 \text{ atm}$$

$$27.98 \text{ atm x } \frac{760 \text{ mm Hg}}{1 \text{ atm}} = 2.1 \text{ x } 10^4 \text{ mm Hg}$$

10.57 $2.0 \text{ g } N_2 \text{ x } \frac{1 \text{ mol } N_2}{28.0 \text{ g } N_2} = 0.0714 \text{ mol } N_2$

$$T = \frac{PV}{nR} = \frac{(6.0 \text{ atm})(0.40 \text{ L})}{(0.0714 \text{ mol})\left(0.082\ 06 \frac{\text{L} \cdot \text{atm}}{\text{K} \cdot \text{mol}}\right)} = 410 \text{ K}$$

10.58 $\frac{1 \text{ H atom}}{\text{cm}^3} \text{ x } \frac{1 \text{ mol H}}{6.02 \text{ x } 10^{23} \text{ atoms}} \text{ x } \frac{1000 \text{ cm}^3}{1 \text{ L}} = 1.7 \text{ x } 10^{-21} \text{ mol H/L}$

$$P = \frac{nRT}{V} = \frac{(1.7 \times 10^{-21}\ \text{mol})\left(0.082\ 06\ \frac{\text{L} \cdot \text{atm}}{\text{K} \cdot \text{mol}}\right)(100\ \text{K})}{(1\ \text{L})} = 1.4 \times 10^{-20}\ \text{atm}$$

$$P = 1.4 \times 10^{-20}\ \text{atm} \times \frac{760\ \text{mm Hg}}{1.0\ \text{atm}} = 1 \times 10^{-17}\ \text{mm Hg}$$

10.59 CH_4, 16.04; 5.54 kg = 5.54×10^3 g; T = 20 °C = 293 K

$$P = \frac{nRT}{V} = \frac{\left(5.54 \times 10^3\ \text{g} \times \frac{1\ \text{mol}}{16.04\ \text{g}}\right)\left(0.082\ 06\ \frac{\text{L} \cdot \text{atm}}{\text{K} \cdot \text{mol}}\right)(293\ \text{K})}{(43.8\ \text{L})} = 189.6\ \text{atm}$$

$$P = 189.6\ \text{atm} \times \frac{101{,}325\ \text{Pa}}{1\ \text{atm}} \times \frac{1\ \text{k Pa}}{1000\ \text{Pa}} = 1.92 \times 10^4\ \text{kPa}$$

10.60
$$n = \frac{PV}{RT} = \frac{\left(17{,}180\ \text{kPa} \times \frac{1000\ \text{Pa}}{1\ \text{k Pa}} \times \frac{1\ \text{atm}}{101{,}325\ \text{Pa}}\right)(43.8\ \text{L})}{\left(0.082\ 06\ \frac{\text{L} \cdot \text{atm}}{\text{K} \cdot \text{mol}}\right)(293\text{K})} = 308.9\ \text{mol}$$

$$\text{mass Ar} = 308.9\ \text{mol} \times \frac{39.948\ \text{g}}{1\ \text{mol}} = 12340\ \text{g} = 1.23 \times 10^4\ \text{g}$$

10.61
$$P = 13{,}800\ \text{kPa} \times \frac{1000\ \text{Pa}}{1\ \text{kPa}} \times \frac{1\ \text{atm}}{101{,}325\ \text{Pa}} = 136.2\ \text{atm}$$

n and T are constant; therefore $nRT = P_iV_i = P_fV_f$

$$V_f = \frac{P_iV_i}{P_f} = \frac{(136.2\ \text{atm})(2.30\ \text{L})}{(1.25\ \text{atm})} = 250.6\ \text{L}$$

$$250.6\ \text{L} \times \frac{1\ \text{balloon}}{1.5\ \text{L}} = 167\ \text{balloons}$$

Gas Stoichiometry (Section 10.4)

10.62 For steam, T = 123.0 °C = 396 K

$$n = \frac{PV}{RT} = \frac{(0.93\ \text{atm})(15.0\ \text{L})}{\left(0.082\ 06\ \frac{\text{L} \cdot \text{atm}}{\text{K} \cdot \text{mol}}\right)(396\ \text{K})} = 0.43\ \text{mol steam}$$

For ice, H_2O, 18.02; $n = 10.5\ \text{g} \times \frac{1\ \text{mol}}{18.02\ \text{g}} = 0.583\ \text{mol ice}$

Because the number of moles of ice is larger than the number of moles of steam, the ice contains more H_2O molecules.

10.63 T = 85.0 °C = 358 K

$$n_{Ar} = \frac{PV}{RT} = \frac{\left(1111 \text{ mm Hg x } \dfrac{1.00 \text{ atm}}{760 \text{ mm Hg}}\right)(3.14 \text{ L})}{\left(0.082\ 06\ \dfrac{\text{L} \cdot \text{atm}}{\text{K} \cdot \text{mol}}\right)(358 \text{ K})} = 0.156 \text{ mol Ar}$$

Cl_2, 70.91; $n_{Cl_2} = 11.07 \text{ g } Cl_2 \text{ x } \dfrac{1 \text{ mol } Cl_2}{70.91 \text{ g } Cl_2} = 0.156 \text{ mol } Cl_2$

There are equal numbers of Ar atoms and Cl_2 molecules in their respective samples.

10.64 The containers are identical. Both containers contain the same number of gas molecules. Weigh the containers. Because the molecular mass for O_2 is greater than the molecular mass for H_2, the heavier container contains O_2.

10.65 Assuming that you can see through the flask, Cl_2 gas is greenish and Ar is colorless.

10.66 room volume = 4.0 m x 5.0 m x 2.5 m x $\dfrac{1 \text{ L}}{10^{-3} \text{ m}^3} = 5.0 \text{ x } 10^4 \text{ L}$

$$n_{total} = \frac{PV}{RT} = \frac{(1.0 \text{ atm})(5.0 \text{ x } 10^4 \text{ L})}{\left(0.082\ 06\ \dfrac{\text{L} \cdot \text{atm}}{\text{K} \cdot \text{mol}}\right)(273 \text{ K})} = 2.23 \text{ x } 10^3 \text{ mol}$$

$n_{O_2} = (0.2095)n_{total} = (0.2095)(2.23 \text{ x } 10^3 \text{ mol}) = 467 \text{ mol } O_2$

mass O_2 = 467 mol x $\dfrac{32.0 \text{ g}}{1 \text{ mol}} = 1.5 \text{ x } 10^4 \text{ g } O_2$

10.67 $0.25 \text{ g } O_2 \text{ x } \dfrac{1 \text{ mol } O_2}{32.0 \text{ g } O_2} = 7.8 \text{ x } 10^{-3} \text{ mol } O_2$

$$V = \frac{nRT}{P} = \frac{(7.8 \text{ x } 10^{-3} \text{ mol})\left(0.082\ 06\ \dfrac{\text{L} \cdot \text{atm}}{\text{K} \cdot \text{mol}}\right)(310 \text{ K})}{1.0 \text{ atm}} = 0.198 \text{ L} = 0.200 \text{ L} = 200 \text{ mL } O_2$$

10.68 (a) CH_4, 16.04; $d = \dfrac{16.04 \text{ g}}{22.4 \text{ L}} = 0.716 \text{ g/L}$

(b) CO_2, 44.01; $d = \dfrac{44.01 \text{ g}}{22.4 \text{ L}} = 1.96 \text{ g/L}$

(c) O_2, 32.00; $d = \dfrac{32.00 \text{ g}}{22.4 \text{ L}} = 1.43 \text{ g/L}$

10.69 Average molar mass = (0.270)(molar mass F_2) + (0.730)(molar mass He)
= (0.270)(38.00 g/mol) + (0.730)(4.003 g/mol) = 13.18 g/mol

Assume 1.00 mole of the gas mixture. T = 27.5 °C = 300.6 K

$$V = \frac{nRT}{P} = \frac{(1.00\text{ mol})\left(0.082\ 06\ \frac{\text{L}\cdot\text{atm}}{\text{K}\cdot\text{mol}}\right)(300.6\text{ K})}{\left(714\text{ mm Hg x }\frac{1.00\text{ atm}}{760\text{ mm Hg}}\right)} = 26.3\text{ L}$$

$$d = \frac{13.18\text{ g}}{26.3\text{ L}} = 0.501\text{ g/L}$$

10.70 $$n = \frac{PV}{RT} = \frac{\left(356\text{ mm Hg x }\frac{1.00\text{ atm}}{760\text{ mm Hg}}\right)(1.500\text{ L})}{\left(0.082\ 06\ \frac{\text{L}\cdot\text{atm}}{\text{K}\cdot\text{mol}}\right)(295.5\text{ K})} = 0.0290\text{ mol}$$

$$\text{molar mass} = \frac{0.9847\text{ g}}{0.0290\text{ mol}} = 34.0\text{ g/mol}; \quad \text{molecular weight} = 34.0$$

10.71 (a) Assume 1.000 L gas sample

$$n = \frac{PV}{RT} = \frac{(1.00\text{ atm})(1.000\text{ L})}{\left(0.082\ 06\ \frac{\text{L}\cdot\text{atm}}{\text{K}\cdot\text{mol}}\right)(273\text{ K})} = 0.0446\text{ mol}$$

$$\text{molar mass} = \frac{1.342\text{ g}}{0.0446\text{ mol}} = 30.1\text{ g/mol}; \quad \text{molecular weight} = 30.1$$

(b) Assume 1.000 L gas sample

$$n = \frac{PV}{RT} = \frac{\left(752\text{ mm Hg x }\frac{1.00\text{ atm}}{760\text{ mm Hg}}\right)(1.000\text{ L})}{\left(0.082\ 06\ \frac{\text{L}\cdot\text{atm}}{\text{K}\cdot\text{mol}}\right)(298\text{ K})} = 0.0405\text{ mol}$$

$$\text{molar mass} = \frac{1.053\text{ g}}{0.0405\text{ mol}} = 26.0\text{ g/mol}; \quad \text{molecular weight} = 26.0$$

10.72 $2\ HgO(s) \rightarrow 2\ Hg(l) + O_2(g)$; HgO, 216.59

$$10.57\text{ g HgO x }\frac{1\text{ mol HgO}}{216.59\text{ g HgO}}\text{ x }\frac{1\text{ mol }O_2}{2\text{ mol HgO}} = 0.024\ 40\text{ mol }O_2$$

$$V = \frac{nRT}{P} = \frac{(0.024\ 40\text{ mol})\left(0.082\ 06\ \frac{\text{L}\cdot\text{atm}}{\text{K}\cdot\text{mol}}\right)(273.15\text{ K})}{1.000\text{ atm}} = 0.5469\text{ L}$$

10.73 $2\ HgO(s) \rightarrow 2\ Hg(l) + O_2(g)$; HgO, 216.59

$$\text{mass HgO} = 0.0155\text{ mol }O_2\text{ x }\frac{2\text{ mol HgO}}{1\text{ mol }O_2}\text{ x }\frac{216.59\text{ g HgO}}{1\text{ mol HgO}} = 6.71\text{ g HgO}$$

10.74 $Zn(s) + 2\,HCl(aq) \rightarrow ZnCl_2(aq) + H_2(g)$

(a) $25.5\text{ g Zn} \times \frac{1\text{ mol Zn}}{65.39\text{ g Zn}} \times \frac{1\text{ mol }H_2}{1\text{ mol Zn}} = 0.390\text{ mol }H_2$

$$V = \frac{nRT}{P} = \frac{(0.390\text{ mol})\left(0.082\ 06\ \frac{L\cdot atm}{K\cdot mol}\right)(288\text{ K})}{\left(742\text{ mm Hg} \times \frac{1.00\text{ atm}}{760\text{ mm Hg}}\right)} = 9.44\text{ L}$$

(b) $$n = \frac{PV}{RT} = \frac{\left(350\text{ mm Hg} \times \frac{1.00\text{ atm}}{760\text{ mm Hg}}\right)(5.00\text{ L})}{\left(0.082\ 06\ \frac{L\cdot atm}{K\cdot mol}\right)(303.15\text{ K})} = 0.092\ 56\text{ mol }H_2$$

$$0.092\ 56\text{ mol }H_2 \times \frac{1\text{ mol Zn}}{1\text{ mol }H_2} \times \frac{65.39\text{ g Zn}}{1\text{ mol Zn}} = 6.05\text{ g Zn}$$

10.75 $2\,NH_4NO_3(s) \rightarrow 2\,N_2(g) + 4\,H_2O(g) + O_2(g)$; NH_4NO_3, 80.04

Total moles of gas $= 450\text{ g }NH_4NO_3 \times \frac{1\text{ mol }NH_4NO_3}{80.04\text{ g }NH_4NO_3} \times \frac{7\text{ mol gas}}{2\text{ mol }NH_4NO_3} = 19.68\text{ mol}$

$T = 450\ ^\circ C = 723\text{ K}$

$$V = \frac{nRT}{P} = \frac{(19.68\text{ mol})\left(0.082\ 06\ \frac{L\cdot atm}{K\cdot mol}\right)(723\text{ K})}{(1.00\text{ atm})} = 1.17 \times 10^3\text{ L}$$

10.76 (a) $V_{24h} = (4.50\text{ L/min})(60\text{ min/h})(24\text{ h/day}) = 6480\text{ L}$

$V_{CO_2} = (0.034)V_{24h} = (0.034)(6480\text{ L}) = 220\text{ L}$

$$n = \frac{PV}{RT} = \frac{\left(735\text{ mm Hg} \times \frac{1.00\text{ atm}}{760\text{ mm Hg}}\right)(220\text{ L})}{\left(0.082\ 06\ \frac{L\cdot atm}{K\cdot mol}\right)(298\text{ K})} = 8.70\text{ mol }CO_2$$

$$8.70\text{ mol }CO_2 \times \frac{44.01\text{ g }CO_2}{1\text{ mol }CO_2} = 383\text{ g} = 380\text{ g }CO_2$$

(b) $2\,Na_2O_2(s) + 2\,CO_2(g) \rightarrow 2\,Na_2CO_3(s) + O_2(g)$; Na_2O_2, 77.98

$3.65\text{ kg} = 3650\text{ g}$

$$3650\text{ g }Na_2O_2 \times \frac{1\text{ mol }Na_2O_2}{77.98\text{ g }Na_2O_2} \times \frac{2\text{ mol }CO_2}{2\text{ mol }Na_2O_2} \times \frac{1\text{ day}}{8.70\text{ mol }CO_2} = 5.4\text{ days}$$

10.77 $2\ TiCl_4(g) + H_2(g) \rightarrow 2\ TiCl_3(s) + 2\ HCl(g)$; $TiCl_4$, 189.69

(a) T = 435 °C = 708 K

$$n_{H_2} = \frac{PV}{RT} = \frac{\left(795 \text{ mm Hg} \times \frac{1.00 \text{ atm}}{760 \text{ mm Hg}}\right)(155 \text{ L})}{\left(0.082\ 06 \frac{\text{L} \cdot \text{atm}}{\text{K} \cdot \text{mol}}\right)(708 \text{ K})} = 2.79 \text{ mol } H_2$$

$$2.79 \text{ mol } H_2 \times \frac{2 \text{ mol } TiCl_4}{1 \text{ mol } H_2} \times \frac{189.69 \text{ g } TiCl_4}{1 \text{ mol } TiCl_4} = 1058 \text{ g} = 1060 \text{ g } TiCl_4$$

(b) $n_{HCl} = 2.79 \text{ mol } H_2 \times \frac{2 \text{ mol HCl}}{1 \text{ mol } H_2} = 5.58 \text{ mol HCl}$

$$V = \frac{n_{HCl}RT}{P} = \frac{(5.58 \text{ mol})\left(0.082\ 06 \frac{\text{L} \cdot \text{atm}}{\text{K} \cdot \text{mol}}\right)(273 \text{ K})}{(1.00 \text{ atm})} = 125 \text{ L HCl}$$

10.78 He, 4.00; 365 mL = 0.365 L; 25 °C = 298 K

$$n_{He} = \frac{PV}{RT} = \frac{(7.8 \text{ atm})(0.365 \text{ L})}{\left(0.082\ 06 \frac{\text{L} \cdot \text{atm}}{\text{K} \cdot \text{mol}}\right)(298 \text{ K})} = 0.116 \text{ mol He}$$

$$\text{mass He} = 0.116 \text{ mol He} \times \frac{4.00 \text{ g He}}{1 \text{ mol He}} = 0.464 \text{ g He}$$

10.79 $C_3H_8(g) + 5\ O_2(g) \rightarrow 3\ CO_2(g) + 4\ H_2O(l)$

$$n_{propane} = \frac{PV}{RT} = \frac{(4.5 \text{ atm})(15.0 \text{ L})}{\left(0.082\ 06 \frac{\text{L} \cdot \text{atm}}{\text{K} \cdot \text{mol}}\right)(298 \text{ K})} = 2.76 \text{ mol } C_3H_8$$

$$2.76 \text{ mol } C_3H_8 \times \frac{3 \text{ mol } CO_2}{1 \text{ mol } C_3H_8} = 8.28 \text{ mol } CO_2$$

$$V = \frac{nRT}{P} = \frac{(8.28 \text{ mol})\left(0.082\ 06 \frac{\text{L} \cdot \text{atm}}{\text{K} \cdot \text{mol}}\right)(273 \text{ K})}{1.00 \text{ atm}} = 186 \text{ L} = 190 \text{ L}$$

Dalton's Law and Mole Fraction (Section 10.5)

10.80 Because of Avogadro's Law ($V \propto n$), the % volumes are also % moles.

	% mole
N_2	78.08
O_2	20.95
Ar	0.93
CO_2	0.038

In decimal form, % mole = mole fraction.

$P_{N_2} = X_{N_2} \cdot P_{total} = (0.7808)(1.000 \text{ atm}) = 0.7808 \text{ atm}$

$P_{O_2} = X_{O_2} \cdot P_{total} = (0.2095)(1.000 \text{ atm}) = 0.2095 \text{ atm}$

$P_{Ar} = X_{Ar} \cdot P_{total} = (0.0093)(1.000 \text{ atm}) = 0.0093 \text{ atm}$

$P_{CO_2} = X_{CO_2} \cdot P_{total} = (0.000\ 38)(1.000 \text{ atm}) = 0.000\ 38 \text{ atm}$

Pressures of the rest are negligible.

10.81 $X_{CH_4} = \dfrac{94 \text{ mol}}{100 \text{ mol}} = 0.94$; $P_{CH_4} = X_{CH_4} \cdot P_{total} = (0.94)(1.48 \text{ atm}) = 1.4 \text{ atm}$

$X_{C_2H_6} = \dfrac{4 \text{ mol}}{100 \text{ mol}} = 0.040$; $P_{C_2H_6} = X_{C_2H_6} \cdot P_{total} = (0.040)(1.48 \text{ atm}) = 0.059 \text{ atm}$

$X_{C_3H_8} = \dfrac{1.5 \text{ mol}}{100 \text{ mol}} = 0.015$; $P_{C_3H_8} = X_{C_3H_8} \cdot P_{total} = (0.015)(1.48 \text{ atm}) = 0.022 \text{ atm}$

$X_{C_4H_{10}} = \dfrac{0.5 \text{ mol}}{100 \text{ mol}} = 0.0050$; $P_{C_4H_{10}} = X_{C_4H_{10}} \cdot P_{total} = (0.0050)(1.48 \text{ atm}) = 0.0074 \text{ atm}$

10.82 Assume a 100.0 g sample. $\text{g } CO_2 = 1.00 \text{ g}$ and $\text{g } O_2 = 99.0 \text{ g}$

$\text{mol } CO_2 = 1.00 \text{ g } CO_2 \times \dfrac{1 \text{ mol } CO_2}{44.01 \text{ g } CO_2} = 0.0227 \text{ mol } CO_2$

$\text{mol } O_2 = 99.0 \text{ g } O_2 \times \dfrac{1 \text{ mol } O_2}{32.00 \text{ g } O_2} = 3.094 \text{ mol } O_2$

$n_{total} = 3.094 \text{ mol} + 0.0227 \text{ mol} = 3.117 \text{ mol}$

$X_{O_2} = \dfrac{3.094 \text{ mol}}{3.117 \text{ mol}} = 0.993$; $X_{CO_2} = \dfrac{0.0227 \text{ mol}}{3.117 \text{ mol}} = 0.007\ 28$

$P_{O_2} = X_{O_2} \cdot P_{total} = (0.993)(0.977 \text{ atm}) = 0.970 \text{ atm}$

$P_{CO_2} = X_{CO_2} \cdot P_{total} = (0.007\ 28)(0.977 \text{ atm}) = 0.007\ 11 \text{ atm}$

10.83 From Problem 10.84: $X_{HCl} = 0.026$, $X_{H_2} = 0.094$, $X_{Ne} = 0.88$

$P_{HCl} = X_{HCl} \cdot P_{total} = (0.026)(13{,}800 \text{ kPa}) = 3.6 \times 10^2 \text{ kPa}$

$P_{H_2} = X_{H_2} \cdot P_{total} = (0.094)(13{,}800 \text{ kPa}) = 1.3 \times 10^3 \text{ kPa}$

$P_{Ne} = X_{Ne} \cdot P_{total} = (0.88)(13{,}800 \text{ kPa}) = 1.2 \times 10^4 \text{ kPa}$

10.84 Assume a 100.0 g sample.

$\text{g HCl} = (0.0500)(100.0 \text{ g}) = 5.00 \text{ g}$; $5.00 \text{ g HCl} \times \dfrac{1 \text{ mol HCl}}{36.5 \text{ g HCl}} = 0.137 \text{ mol HCl}$

$\text{g } H_2 = (0.0100)(100.0 \text{ g}) = 1.00 \text{ g}$; $1.00 \text{ g } H_2 \times \dfrac{1 \text{ mol } H_2}{2.016 \text{ g } H_2} = 0.496 \text{ mol } H_2$

$\text{g Ne} = (0.94)(100.0 \text{ g}) = 94 \text{ g}$; $94 \text{ g Ne} \times \dfrac{1 \text{ mol Ne}}{20.18 \text{ g Ne}} = 4.66 \text{ mol Ne}$

$n_{total} = 0.137 + 0.496 + 4.66 = 5.3 \text{ mol}$

$X_{HCl} = \dfrac{0.137 \text{ mol}}{5.3 \text{ mol}} = 0.026; \quad X_{H_2} = \dfrac{0.496 \text{ mol}}{5.3 \text{ mol}} = 0.094; \quad X_{Ne} = \dfrac{4.66 \text{ mol}}{5.3 \text{ mol}} = 0.88$

10.85 Assume a 1.000 L gas sample.

$$n = \frac{PV}{RT} = \frac{(1.000 \text{ atm})(1.000 \text{ L})}{\left(0.082\ 06 \dfrac{\text{L}\cdot\text{atm}}{\text{K}\cdot\text{mol}}\right)(273.15 \text{ K})} = 0.044\ 61 \text{ mol}$$

$$\text{average molar mass} = \frac{1.413 \text{ g}}{0.044\ 61 \text{ mol}} = 31.67 \text{ g/mol}$$

$31.67 = x \cdot M_{Ar} + (1 - x) \cdot M_{N_2}$

$31.67 = (x)(39.948) + (1 - x)(28.013)$

Solve for x: $x = 0.3064, \quad 1 - x = 0.6936$

The mixture contains 30.64% Ar and 69.36% N_2.

Assume 100 moles of gas.

$X_{Ar} = \dfrac{30.64 \text{ mol}}{100 \text{ mol}} = 0.3064; \quad X_{N_2} = \dfrac{69.36 \text{ mol}}{100 \text{ mol}} = 0.6936$

10.86 (a) H_2, 2.016

$$P = \frac{nRT}{V} = \frac{\left(14.2 \text{ g x } \dfrac{1 \text{ mol}}{2.016 \text{ g}}\right)\left(0.082\ 06 \dfrac{\text{L}\cdot\text{atm}}{\text{K}\cdot\text{mol}}\right)(290 \text{ K})}{(100.0 \text{ L})} = 1.68 \text{ atm}$$

(b) $$P = \frac{nRT}{V} = \frac{\left(36.7 \text{ g x } \dfrac{1 \text{ mol}}{39.95 \text{ g}}\right)\left(0.082\ 06 \dfrac{\text{L}\cdot\text{atm}}{\text{K}\cdot\text{mol}}\right)(290 \text{ K})}{(100.0 \text{ L})} = 0.219 \text{ atm}$$

10.87 (a) $$P = \frac{nRT}{V} = \frac{\left(0.776 \text{ g x } \dfrac{1 \text{ mol}}{4.003 \text{ g}}\right)\left(0.082\ 06 \dfrac{\text{L}\cdot\text{atm}}{\text{K}\cdot\text{mol}}\right)(300 \text{ K})}{(20.0 \text{ L})} = 0.238 \text{ atm}$$

$0.238 \text{ atm x } \dfrac{760 \text{ mm Hg}}{1 \text{ atm}} = 181 \text{ mm Hg}$

(b) CO_2, 44.01

$$P = \frac{nRT}{V} = \frac{\left(3.61 \text{ g x } \dfrac{1 \text{ mol}}{44.01 \text{ g}}\right)\left(0.082\ 06 \dfrac{\text{L}\cdot\text{atm}}{\text{K}\cdot\text{mol}}\right)(300 \text{ K})}{(20.0 \text{ L})} = 0.101 \text{ atm}$$

$0.101 \text{ atm x } \dfrac{760 \text{ mm Hg}}{1 \text{ atm}} = 76.8 \text{ mm Hg}$

10.88 $P_{total} = P_{H_2} + P_{H_2O}$; $P_{H_2} = P_{total} - P_{H_2O} = 747 \text{ mm Hg} - 23.8 \text{ mm Hg} = 723 \text{ mm Hg}$

$$n = \frac{PV}{RT} = \frac{\left(723 \text{ mm Hg x } \frac{1.00 \text{ atm}}{760 \text{ mm Hg}}\right)(3.557 \text{ L})}{\left(0.082\ 06 \frac{\text{L} \cdot \text{atm}}{\text{K} \cdot \text{mol}}\right)(298 \text{ K})} = 0.1384 \text{ mol } H_2$$

$$0.1384 \text{ mol } H_2 \text{ x } \frac{1 \text{ mol Mg}}{1 \text{ mol } H_2} \text{ x } \frac{24.3 \text{ g Mg}}{1 \text{ mol Mg}} = 3.36 \text{ g Mg}$$

10.89 $P_{total} = P_{Cl_2} + P_{H_2O} = 755 \text{ mm Hg}$

$P_{Cl_2} = P_{total} - P_{H_2O} = 755 \text{ mm Hg} - 28.7 \text{ mm Hg} = 726.3 \text{ mm Hg}$

(a) $X_{Cl_2} = \frac{P_{Cl_2}}{P_{total}} = \frac{726.3 \text{ mm Hg}}{755 \text{ mm Hg}} = 0.962$

(b) NaCl, 58.44

$$n_{Cl_2} = \frac{PV}{RT} = \frac{\left(726.3 \text{ mm Hg x } \frac{1.00 \text{ atm}}{760 \text{ mm Hg}}\right)(0.597 \text{ L})}{\left(0.082\ 06 \frac{\text{L} \cdot \text{atm}}{\text{K} \cdot \text{mol}}\right)(300 \text{ K})} = 0.0232 \text{ mol } Cl_2$$

$$0.0232 \text{ mol } Cl_2 \text{ x } \frac{2 \text{ mol NaCl}}{1 \text{ mol } Cl_2} \text{ x } \frac{58.44 \text{ g NaCl}}{1 \text{ mol NaCl}} = 2.71 \text{ g NaCl}$$

Kinetic–Molecular Theory and Graham's Law (Sections 10.6 and 10.7)

10.90 The kinetic-molecular theory is based on the following assumptions:
1. A gas consists of tiny particles, either atoms or molecules, moving about at random.
2. The volume of the particles themselves is negligible compared with the total volume of the gas; most of the volume of a gas is empty space.
3. The gas particles act independently; there are no attractive or repulsive forces between particles.
4. Collisions of the gas particles, either with other particles or with the walls of the container, are elastic; that is, the total kinetic energy of the gas particles is constant at constant T.
5. The average kinetic energy of the gas particles is proportional to the Kelvin temperature of the sample.

10.91 Diffusion – The mixing of different gases by random molecular motion and with frequent collisions.
Effusion – The process in which gas molecules escape through a tiny hole in a membrane without collisions.

10.92 Heat is the energy transferred from one object to another as the result of a temperature difference between them. Temperature is a measure of the kinetic energy of molecular motion.

10.93 The atomic mass of He is much less than the molecular mass of the major components of air (N_2 and O_2). The rate of effusion of He through the balloon skin is much faster.

10.94 $u = \sqrt{\frac{3RT}{M}} = \sqrt{\frac{3 \times 8.314 \text{ kg m}^2/(\text{s}^2 \text{ K mol}) \times 220 \text{ K}}{28.0 \times 10^{-3} \text{ kg/mol}}} = 443 \text{ m/s}$

10.95 $u = \sqrt{\frac{3RT}{M}}$, M = molar mass, R = 8.314 J/(K · mol), 1 J = 1 kg · m^2/s^2

at 37 °C = 310 K, $u = \sqrt{\frac{3 \times 8.314 \text{ kg m}^2/(\text{s}^2 \text{ K mol}) \times 310 \text{ K}}{28.01 \times 10^{-3} \text{ kg/mol}}} = 525 \text{ m/s}$

at –25 °C = 248 K, $u = \sqrt{\frac{3 \times 8.314 \text{ kg m}^2/(\text{s}^2 \text{ K mol}) \times 248 \text{ K}}{28.01 \times 10^{-3} \text{ kg/mol}}} = 470 \text{ m/s}$

10.96 For Br_2: $u = \sqrt{\frac{3RT}{M}} = \sqrt{\frac{3 \times 8.314 \text{ kg m}^2/(\text{s}^2 \text{ K mol}) \times 293 \text{ K}}{159.8 \times 10^{-3} \text{ kg/mol}}} = 214 \text{ m/s}$

For Xe: $u = 214 \text{ m/s} = \sqrt{\frac{3 \times 8.314 \text{ kg m}^2/(\text{s}^2 \text{ K mol}) \times T}{131.3 \times 10^{-3} \text{ kg/mol}}}$

Square both sides of the equation and solve for T.

$45796 \text{ m}^2/\text{s}^2 = \frac{3 \times 8.314 \text{ kg m}^2/(\text{s}^2 \text{ K mol}) \times T}{131.3 \times 10^{-3} \text{ kg/mol}}$

T = 241 K = –32 °C

10.97 $u = \sqrt{\frac{3RT}{M}}$, M = molar mass, R = 8.314 J/(K · mol), 1 J = 1 kg · m^2/s^2

O_2, 32.00, 32.00 x 10^{-3} kg/mol

$u = 580 \text{ mi/h} \times \frac{1.6093 \text{ km}}{1 \text{ mi}} \times \frac{1000 \text{ m}}{1 \text{ km}} \times \frac{1 \text{ hr}}{60 \text{ min}} \times \frac{1 \text{ min}}{60 \text{ s}} = 259 \text{ m/s}$

$u = \sqrt{\frac{3RT}{M}}; \quad u^2 = \frac{3RT}{M}$

$T = \frac{u^2 M}{3R} = \frac{(259 \text{ m/s})^2 (32.00 \times 10^{-3} \text{ kg/mol})}{(3)(8.314 \text{ kg} \cdot \text{m}^2/\text{s}^2 \cdot \text{K} \cdot \text{mol})} = 86.1 \text{ K}$

T = 86.1 – 273.15 = –187.0 °C

10.98 For H_2, $u = \sqrt{\frac{3\,RT}{M}} = \sqrt{\frac{3 \times 8.314\ kg\ m^2/(s^2\ K\ mol) \times 150\ K}{2.02 \times 10^{-3}\ kg/mol}} = 1360\ m/s$

For He, $u = \sqrt{\frac{3 \times 8.314\ kg\ m^2/(s^2\ K\ mol) \times 648\ K}{4.00 \times 10^{-3}\ kg/mol}} = 2010\ m/s$

He at 375 °C has the higher average speed.

10.99 UF_6, 352.02; T = 145 °C = 418 K

$u = \sqrt{\frac{3\,RT}{M}} = \sqrt{\frac{3 \times 8.314\ kg\ m^2/(s^2\ K\ mol) \times 418\ K}{352.02 \times 10^{-3}\ kg/mol}} = 172\ m/s$

Ferrari

$145\ \frac{mi}{hr} \times \frac{1.6093\ km}{1\ mi} \times \frac{1000\ m}{1\ km} \times \frac{1\ hr}{3600\ s} = 64.8\ m/s$

The UF_6 molecule has the higher average speed.

10.100 $\frac{rate_{H_2}}{rate_X} = \sqrt{\frac{M_X}{M_{H_2}}}$; $\frac{2.92}{1} = \frac{\sqrt{M_X}}{\sqrt{2.02}}$; $2.92\sqrt{2.02} = \sqrt{M_X}$

$M_X = (2.92\sqrt{2.02})^2 = 17.2$ g/mol; molecular weight = 17.2

10.101 $\frac{rate\ Xe}{rate\ Z} = \sqrt{\frac{M_Z}{M_{Xe}}}$; $\frac{1}{1.86} = \frac{\sqrt{M_Z}}{\sqrt{131.29}}$; $\frac{\sqrt{131.29}}{1.86} = \sqrt{M_Z}$; $\frac{131.29}{(1.86)^2} = M_Z$

$M_Z = 37.9$ g/mol; molecular weight = 37.9; The gas could be F_2.

10.102 HCl, 36.5; F_2, 38.0; Ar, 39.9

$\frac{rate\ HCl}{rate\ Ar} = \sqrt{\frac{M_{Ar}}{M_{HCl}}} = \sqrt{\frac{39.9}{36.5}} = 1.05$; $\frac{rate\ F_2}{rate\ Ar} = \sqrt{\frac{M_{Ar}}{M_{F_2}}} = \sqrt{\frac{39.9}{38.0}} = 1.02$

The relative rates of diffusion are HCl(1.05) > F_2(1.02) > Ar(1.00).

10.103 Because CO and N_2 have the same mass, they will have the same diffusion rates.

10.104 $u = 45\ m/s = \sqrt{\frac{3 \times 8.314\ kg\ m^2/(s^2\ K\ mol) \times T}{4.00 \times 10^{-3}\ kg/mol}}$

Square both sides of the equation and solve for T.

$2025\ m^2/s^2 = \frac{3 \times 8.314\ kg\ m^2/(s^2\ K\ mol) \times T}{4.00 \times 10^{-3}\ kg/mol}$

T = 0.325 K = –272.83 °C (near absolute zero)

10.105 $230\text{ km/h} \times \frac{1000\text{ m}}{1\text{ km}} \times \frac{1\text{ h}}{3600\text{ s}} = 63.9\text{ m/s}$

$$u = 63.9\text{ m/s} = \sqrt{\frac{3 \times 8.314\text{ kg m}^2/(\text{s}^2\text{ K mol}) \times T}{32.0 \times 10^{-3}\text{ kg/mol}}}$$

Square both sides of the equation and solve for T.

$$4083\text{ m}^2/\text{s}^2 = \frac{3 \times 8.314\text{ kg m}^2/(\text{s}^2\text{ K mol}) \times T}{32.0 \times 10^{-3}\text{ kg/mol}}$$

$T = 5.24\text{ K} = 5.24\text{ K} - 273.15 = -267.91\ ^\circ\text{C}$

Real Gases (Section 10.8)

10.106 (a) high (b) larger

10.107 (a) high (b) smaller

10.108 $$P = \frac{nRT}{V} = \frac{(0.500\text{ mol})\left(0.082\ 06\ \frac{\text{L}\cdot\text{atm}}{\text{K}\cdot\text{mol}}\right)(300\text{ K})}{(0.600\text{ L})} = 20.5\text{ atm}$$

$$P = \frac{nRT}{V - nb} - \frac{an^2}{V^2}$$

$$P = \frac{(0.500\text{ mol})\left(0.082\ 06\ \frac{\text{L}\cdot\text{atm}}{\text{K}\cdot\text{mol}}\right)(300\text{ K})}{[(0.600\text{ L}) - (0.500\text{ mol})(0.0387\text{ L/mol})]} - \frac{\left(1.35\ \frac{\text{L}^2\cdot\text{atm}}{\text{mol}^2}\right)(0.500\text{ mol})^2}{(0.600\text{ L})^2} = 20.3\text{ atm}$$

10.109 $$P = \frac{nRT}{V} = \frac{(15.00\text{ mol})\left(0.082\ 06\ \frac{\text{L}\cdot\text{atm}}{\text{K}\cdot\text{mol}}\right)(300\text{ K})}{(0.600\text{ L})} = 615.5\text{ atm}$$

$$P = \frac{nRT}{V - nb} - \frac{an^2}{V^2}$$

$$P = \frac{(15.00\text{ mol})\left(0.082\ 06\ \frac{\text{L}\cdot\text{atm}}{\text{K}\cdot\text{mol}}\right)(300\text{ K})}{[(0.600\text{ L}) - (15.00\text{ mol})(0.0387\text{ L/mol})]} - \frac{\left(1.35\ \frac{\text{L}^2\cdot\text{atm}}{\text{mol}^2}\right)(15.00\text{ mol})^2}{(0.600\text{ L})^2} = 18{,}090\text{ atm}$$

The Earth's Atmosphere and Pollution (Section 10.9)

10.110 Troposphere, stratosphere, mesosphere, and thermosphere. Temperature changes are used to distinguish between different regions of the atmosphere.

10.111 $n_{air} = 5.15 \times 10^{15}\text{ kg air} \times \frac{1000\text{ g}}{1\text{ kg}} \times \frac{1\text{ mol air}}{28.8\text{ g air}} = 1.79 \times 10^{17}\text{ mol air}$

$$V_{air} = \frac{nRT}{P} = \frac{(1.79 \times 10^{17}\ mol)\left(0.082\ 06\ \frac{L \cdot atm}{K \cdot mol}\right)(273\ K)}{1.00\ atm} = 4.00 \times 10^{18}\ L$$

10.112 The force of gravity is strongest at the earth's surface and becomes weaker at higher altitude. Because of this, the troposphere contains about 75% of the mass of the entire atmosphere.

10.113 (a) $X_{O_2} = 0.2095$

(b) $P_{O_2} = P_{tot} \cdot X_{O_2} = (1.0\ atm)(0.2095) = 0.21\ atm$

(c) $P_{O_2} = P_{tot} \cdot X_{O_2} = (0.20\ atm)(0.2095) = 0.042\ atm$

10.114 The more toxic pollutant is the one with the lower exposure limit, in this case $PM_{2.5}$.

10.115 The more toxic pollutant is the one with the lower exposure limit, in this case SO_2.

10.116 $\%\ CO_2 = 0.040\% = X_{CO_2} \times 100$; $\frac{0.040}{100} = X_{CO_2} = 4.0 \times 10^{-4}$

$CO_2\ ppm = X_{CO_2} \times 10^6 = (4.0 \times 10^{-4})(10^6) = 400\ ppm$

10.117 $\%\ N_2O = 5 \times 10^{-5}\% = X_{N_2O} \times 100$; $\frac{5 \times 10^{-5}}{100} = X_{N_2O} = 5 \times 10^{-7}$

$N_2O\ ppm = X_{N_2O} \times 10^6 = (5 \times 10^{-7})(10^6) = 0.5\ ppm$

$N_2O\ ppb = X_{N_2O} \times 10^9 = (5 \times 10^{-7})(10^9) = 500\ ppb$

10.118 SO_2, 64.1

$$0.26 \times 10^{-6}\ g\ SO_2 \times \frac{1\ mol\ SO_2}{64.1\ g\ SO_2} \times \frac{6.022 \times 10^{23}\ SO_2\ molecules}{1\ mol\ SO_2} = 2.44 \times 10^{15}\ SO_2\ molecules/L$$

$$mol\ air/L = n = \frac{PV}{RT} = \frac{(1.00\ atm)(1.00\ L)}{\left(0.082\ 06\ \frac{L \cdot atm}{K \cdot mol}\right)(298\ K)} = 0.0409\ mol\ air/L$$

$$0.0409\ mol\ air/L \times \frac{6.022 \times 10^{23}\ air\ molecules}{1\ mol\ air} = 2.46 \times 10^{22}\ air\ molecules/L$$

$$SO_2\ ppb = \frac{2.44 \times 10^{15}\ SO_2\ molecules}{2.46 \times 10^{22}\ air\ molecules} \times 10^9 = 99\ ppb$$

This exceeds the 1-hour limit of 75 ppb.

10.119 From problem 10.118, 1 L of air at 1.00 atm and 25 °C contains 2.46×10^{22} air molecules.

NO_2 concentration = 100 ppb

Let z = the number of NO_2 molecules in 1.00 L of air.

$$\frac{z}{2.46 \times 10^{22} \text{ air molecules}} = \frac{100}{1 \times 10^{9}}$$

$z = (2.46 \times 10^{22} \text{ molecules})(100)/(1 \times 10^{9}) = 2.46 \times 10^{15}\ NO_2$ molecules/L

$(2.46 \times 10^{15}\ NO_2 \text{ molecules/L})(15{,}000 \text{ L/day}) = 3.69 \times 10^{19}\ NO_2$ molecules/day

NO_2, 46.0

$$\text{mass } NO_2 = 3.69 \times 10^{19}\ NO_2 \text{ molecules} \times \frac{1 \text{ mol } NO_2}{6.022 \times 10^{23}\ NO_2 \text{ molecules}} \times \frac{46.0 \text{ g } NO_2}{1 \text{ mol } NO_2} = 0.0028 \text{ g } NO_2$$

10.120 S, 32.1

$$\text{mass S} = 1 \text{ kg} \times \frac{1000 \text{ g}}{1 \text{ kg}} \times 0.02 = 20 \text{ g S}$$

$$\text{mol S} = 20 \text{ g S} \times \frac{1 \text{ mol S}}{32.1 \text{ g S}} = 0.62 \text{ mol S} = 0.62 \text{ mol } SO_2$$

$$V_{SO_2} = \frac{nRT}{P} = \frac{(0.62 \text{ mol})\left(0.082\ 06 \frac{\text{L} \cdot \text{atm}}{\text{K} \cdot \text{mol}}\right)(298 \text{ K})}{1.0 \text{ atm}} = 15 \text{ L}$$

10.121 (a) $PbSO_3(s) \rightarrow PbO(s) + SO_2(g)$

(b) $PbSO_3$, 287.3; mol SO_2 = mol $PbSO_3$; 300 °C = 573 K

$$\text{mol } PbSO_3 = 250 \text{ g } PbSO_3 \times \frac{1 \text{ mol } PbSO_3}{287.3 \text{ g } PbSO_3} = 0.870 \text{ mol } PbSO_3 = 0.870 \text{ mol } SO_2$$

$$V_{SO_2} = \frac{nRT}{P} = \frac{(0.870 \text{ mol})\left(0.082\ 06 \frac{\text{L} \cdot \text{atm}}{\text{K} \cdot \text{mol}}\right)(573 \text{ K})}{1.00 \text{ atm}} = 40.9 \text{ L}$$

10.122 (a) CO and NO_2 (b) CO, NO_2, and SO_2

10.123 Primary pollutants are those that enter the environment directly from a source such as vehicles or industrial emissions. Primary pollutants are CO, Pb, particulate matter, and SO_2.

10.124 Secondary pollutants are formed by the chemical reaction of a primary pollutant and are not directly emitted from a source. Secondary pollutants are NO_2, O_3, and photochemical smog.

10.125 NO_2 and VOCs

10.126 400 nm = 400 x 10^{-9} m

$$E = \frac{hc}{\lambda} = (6.626 \times 10^{-34}\ J\cdot s)\left(\frac{3.00 \times 10^{8}\ m/s}{400 \times 10^{-9}\ m}\right)\left(\frac{1\ kJ}{1000\ J}\right)(6.022 \times 10^{23}/mol)$$

$E = 299\ kJ/mol$

The N–O bond energy is 299 kJ/mol.

10.127 NO, NO_2, then O_3

NO peaks during the morning rush hour. NO_2 forms as NO reacts with O_2. O_3 forms as sunlight breaks the N–O bond in NO_2 and O reacts with O_2.

10.128 The drastic difference in air quality as measured by $PM_{2.5}$ and O_3 could be from a dramatic change in the weather. June 17 could have been a cool, cloudy, late Spring day with little or no photochemical smog whereas June 28 could have been a clear, bright, hot sunny Summer day that produced a significant amount of photochemical smog.

10.129 O_3 is more prevalent in the Summer (August 8) as a result of photochemical smog compared to late Fall (December 5) and Winter.

The Greenhouse Effect and Climate Change (Sections 10.10 and 10.11)

10.130 (a) visible and UV (b) UV (c) infrared (d) infrared

10.131 (a) the Sun; 483 nm = 483 x 10^{-9} m

$$E = \frac{hc}{\lambda} = (6.626 \times 10^{-34}\ J\cdot s)\left(\frac{3.00 \times 10^{8}\ m/s}{483 \times 10^{-9}\ m}\right)\left(\frac{1\ kJ}{1000\ J}\right)(6.022 \times 10^{23}/mol)$$

$E = 248\ kJ/mol$

(b) the Earth; 10,000 nm = 10,000 x 10^{-9} m

$$E = \frac{hc}{\lambda} = (6.626 \times 10^{-34}\ J\cdot s)\left(\frac{3.00 \times 10^{8}\ m/s}{10{,}000 \times 10^{-9}\ m}\right)\left(\frac{1\ kJ}{1000\ J}\right)(6.022 \times 10^{23}/mol)$$

$E = 12.0\ kJ/mol$

(c) The Sun emits the higher energy radiation.

10.132 Because of nuclear fusion, the Sun emits all forms of electromagnetic radiation. The Earth absorbs visible radiation, warms up, and radiates infrared radiation back to space.

10.133

$$:\ddot{O}{=}\ddot{O}{-}\ddot{\underset{\cdot\cdot}{O}}: \longleftrightarrow :\ddot{\underset{\cdot\cdot}{O}}{-}\ddot{O}{=}\ddot{O}:$$

Ozone is a bent molecule like water. All stretches and bends absorb IR radiation. Ozone is a greenhouse gas.

10.134 CO_2, N_2O, and CH_4

10.135 CO_2 and CH_4 concentrations have been steadily increasing over the past 150 years. Over the prior several hundred thousand years, CO_2 and CH_4 concentrations remained relatively constant.

10.136 The Earth's average temperature has risen about 1 °C since 1900.

10.137 Some climate change indicators are sea and land ice changes, sea level changes, biodiversity, and ocean acidity.

Chapter Problems

10.138 (a), (b), and (e) affect air quality; (c) and (d) affect climate

10.139 $$\frac{\text{rate } ^{35}Cl_2}{\text{rate } ^{37}Cl_2} = \sqrt{\frac{M\ ^{37}Cl_2}{M\ ^{35}Cl_2}} = \sqrt{\frac{74.0}{70.0}} = 1.03$$

$$\frac{\text{rate } ^{35}Cl\,^{37}Cl}{\text{rate } ^{37}Cl_2} = \sqrt{\frac{M\ ^{37}Cl_2}{M\ ^{35}Cl\,^{37}Cl}} = \sqrt{\frac{74.0}{72.0}} = 1.01$$

The relative rates of diffusion are $^{35}Cl_2(1.03) > {}^{35}Cl\,^{37}Cl(1.01) > {}^{37}Cl_2(1.00)$.

10.140 Average molecular mass of air = 28.96; CO_2, 44.01

$$P = 760 \text{ mm Hg} \times \frac{44.01 \text{ g/mol}}{28.96 \text{ g/mol}} = 1155 \text{ mm Hg}$$

10.141 $$V = \frac{nRT}{P} = \frac{(1.00 \text{ mol})\left(0.082\ 06\ \frac{L \cdot atm}{K \cdot mol}\right)(1050 \text{ K})}{(75 \text{ atm})} = 1.1 \text{ L}$$

10.142 $$1 \text{ atm} = \left(\frac{14.70 \text{ lb}}{\text{in.}^2}\right)\left(\frac{1.00 \text{ in.}}{2.54 \text{ cm}}\right)^2\left(\frac{453.59 \text{ g}}{1 \text{ lb}}\right) = 1033.5 \text{ g/cm}^2$$

column height = $(1033.5 \text{ g/cm}^2)(1 \text{ cm}^3/0.89\text{g}) = 1161 \text{ cm} = 1200 \text{ cm} = 12 \text{ m}$

10.143 UF_6, 352.0; 70 °C = 70 + 273 = 343 K

(a) $$P = \frac{nRT}{V} = \frac{\left(512.9 \text{ g} \times \frac{1 \text{ mol}}{352.0 \text{ g}}\right)\left(0.082\ 06\ \frac{L \cdot atm}{K \cdot mol}\right)(343 \text{ K})}{(22.9 \text{ L})} = 1.79 \text{ atm}$$

(b) $$P = \frac{nRT}{V - nb} - \frac{an^2}{V^2}$$

$$n = 512.9 \text{ g} \times \frac{1 \text{ mol } UF_6}{352.0 \text{ g } UF_6} = 1.457 \text{ mol } UF_6$$

$$P = \frac{(1.457\text{ mol})\left(0.082\ 06\ \frac{\text{L}\cdot\text{atm}}{\text{K}\cdot\text{mol}}\right)(343\text{ K})}{[(22.9\text{ L}) - (1.457\text{ mol})(0.1128\text{ L/mol})]} - \frac{\left(15.80\ \frac{\text{L}^2\cdot\text{atm}}{\text{mol}^2}\right)(1.457\text{ mol})^2}{(22.9\text{ L})^2} = 1.74\text{ atm}$$

10.144 $15.0\text{ gal} \times \frac{3.7854\text{ L}}{1\text{ gal}} = 56.8\text{ L};\ \ 25\ ^\circ\text{C} = 25 + 273 = 298\text{ K}$

$$n = \frac{PV}{RT} = \frac{\left(743\text{ mm Hg} \times \frac{1.00\text{ atm}}{760\text{ mm Hg}}\right)(56.8\text{ L})}{\left(0.082\ 06\ \frac{\text{L}\cdot\text{atm}}{\text{K}\cdot\text{mol}}\right)(298\text{ K})} = 2.27\text{ mol gasoline}$$

$$2.27\text{ mol gasoline} \times \frac{105\text{ g gasoline}}{1\text{ mol gasoline}} = 238\text{ g gasoline}$$

$$238\text{ g} \times \frac{1\text{ mL}}{0.75\text{ g}} \times \frac{1\text{ L}}{1000\text{ mL}} \times \frac{1\text{ gal}}{3.7854\text{ L}} \times \frac{20.0\text{ mi}}{1\text{ gal}} = 1.68\text{ mi}$$

10.145 Both tanks contain the same number of gas particles (atoms or molecules) at the same temperature with the same average kinetic energy. Because O_2 is lighter than Kr, the average O_2 velocity is greater than the average Kr velocity.
(a) $Kr < O_2$ (b) $O_2 < Kr$ (c) $Kr < O_2$ (d) Both are the same.

10.146 Both tanks contain the same mass of gas at the same temperature. There are more Ne atoms than N_2 molecules. Because Ne is lighter than N_2, the average Ne velocity is greater than the average N_2 velocity.
(a) $N_2 < Ne$ (b) $N_2 < Ne$ (c) $N_2 < Ne$ (d) Both are the same.

10.147 (a) $16{,}400\text{ ft} \times \frac{12\text{ in}}{1\text{ ft}} \times \frac{2.54\text{ cm}}{1\text{ in}} \times \frac{1\text{ m}}{100\text{ cm}} = 5000\text{ m}$

$P = e^{-h/7000} = e^{-(5000/7000)} = 0.490\text{ atm}$

$0.490\text{ atm} \times \frac{760\text{ mm Hg}}{1\text{ atm}} = 372\text{ mm Hg}$

(b) $28{,}251\text{ ft} \times \frac{12\text{ in}}{1\text{ ft}} \times \frac{2.54\text{ cm}}{1\text{ in}} \times \frac{1\text{ m}}{100\text{ cm}} = 8611\text{ m}$

$P = e^{-h/7000} = e^{-(8611/7000)} = 0.292\text{ atm}$

$0.292\text{ atm} \times \frac{760\text{ mm Hg}}{1\text{ atm}} = 222\text{ mm Hg}$

(c) $P_{O_2} = X_{O_2} \cdot P = (0.2095)(222\text{ mm Hg}) = 46.5\text{ mm Hg}$

10.148 I_2, 253.8; Calculate the I_2 pressure before any I_2 dissociation.

$$P = \frac{nRT}{V} = \frac{\left(42.189\text{ g} \times \frac{1\text{ mol}}{253.8\text{ g}}\right)\left(0.082\ 06\ \frac{\text{L}\cdot\text{atm}}{\text{K}\cdot\text{mol}}\right)(1173\text{ K})}{(10.00\text{ L})} = 1.600\text{ atm}$$

	$I_2(g)$	$\rightarrow$	$2\ I(g)$
before reaction (atm)	1.600		0
change (atm)	–x		+2x
after reaction (atm)	1.600 – x		2x

$1.733 = 1.600 - x + 2x$

$1.733 = 1.600 + x$

$x = 1.733 - 1.600 = 0.133$ atm

$P_{I_2} = 1.600 - x = 1.600 - 0.133 = 1.467$ atm

$P_I = 2x = 2(0.133) = 0.266$ atm

$$X_{I_2} = \frac{1.467\ \text{atm}}{1.733\ \text{atm}} = 0.8465 \text{ and } X_I = \frac{0.266\ \text{atm}}{1.733\ \text{atm}} = 0.153$$

10.149 $$n = \frac{PV}{RT} = \frac{(2.15\ \text{atm})(7.35\ \text{L})}{\left(0.082\ 06\ \frac{\text{L}\cdot\text{atm}}{\text{K}\cdot\text{mol}}\right)(293\ \text{K})} = 0.657\ \text{mol Ar}$$

$$0.657\ \text{mol Ar} \times \frac{39.948\ \text{g Ar}}{1\ \text{mol Ar}} = 26.2\ \text{g Ar}$$

$m_{total} = 478.1\ \text{g} + 26.2\ \text{g} = 504.3\ \text{g}$

10.150 This is initially a Boyle's Law problem, because only P and V are changing while n and T remain fixed. The initial volume for each gas is the volume of their individual bulbs. The final volume for each gas is the total volume of the three bulbs.

$nRT = P_iV_i = P_fV_f$; $V_f = 1.50 + 1.00 + 2.00 = 4.50$ L

For CO_2: $$P_f = \frac{P_iV_i}{V_f} = \frac{(2.13\ \text{atm})(1.50\ \text{L})}{(4.50\ \text{L})} = 0.710\ \text{atm}$$

For H_2: $$P_f = \frac{P_iV_i}{V_f} = \frac{(0.861\ \text{atm})(1.00\ \text{L})}{(4.50\ \text{L})} = 0.191\ \text{atm}$$

For Ar: $$P_f = \frac{P_iV_i}{V_f} = \frac{(1.15\ \text{atm})(2.00\ \text{L})}{(4.50\ \text{L})} = 0.511\ \text{atm}$$

From Dalton's Law, $P_{total} = P_{CO_2} + P_{H_2} + P_{Ar}$

$P_{total} = 0.710\ \text{atm} + 0.191\ \text{atm} + 0.511\ \text{atm} = 1.412\ \text{atm}$

10.151 (a) Bulb A contains $CO_2(g)$ and $N_2(g)$; Bulb B contains $CO_2(g)$, $N_2(g)$, and $H_2O(s)$.

(b) Initial moles of gas = $$n = \frac{PV}{RT} = \frac{\left(564\ \text{mm Hg} \times \frac{1.00\ \text{atm}}{760\ \text{mm Hg}}\right)(1.000\ \text{L})}{\left(0.082\ 06\ \frac{\text{L}\cdot\text{atm}}{\text{K}\cdot\text{mol}}\right)(298\ \text{K})}$$

Initial moles of gas = 0.030 35 mol

$$\text{mol gas in Bulb A} = n = \frac{PV}{RT} = \frac{\left(219 \text{ mm Hg x } \dfrac{1.00 \text{ atm}}{760 \text{ mm Hg}}\right)(1.000 \text{ L})}{\left(0.082\ 06 \dfrac{\text{L}\cdot\text{atm}}{\text{K}\cdot\text{mol}}\right)(298 \text{ K})} = 0.011\ 78 \text{ mol}$$

$$\text{mol gas in Bulb B} = n = \frac{PV}{RT} = \frac{\left(219 \text{ mm Hg x } \dfrac{1.00 \text{ atm}}{760 \text{ mm Hg}}\right)(1.000 \text{ L})}{\left(0.082\ 06 \dfrac{\text{L}\cdot\text{atm}}{\text{K}\cdot\text{mol}}\right)(203 \text{ K})} = 0.017\ 30 \text{ mol}$$

$$n_{H_2O} = n_{\text{initial}} - n_A - n_B = 0.030\ 35 - 0.011\ 78 - 0.017\ 30 = 0.001\ 27 \text{ mol} = 0.0013 \text{ mol } H_2O$$

(c) Bulb A contains $N_2(g)$.
Bulb B contains $N_2(g)$ and $H_2O(s)$.
Bulb C contains $N_2(g)$ and $CO_2(s)$.

$$\text{(d) } n_A = \frac{PV}{RT} = \frac{\left(33.5 \text{ mm Hg x } \dfrac{1.00 \text{ atm}}{760 \text{ mm Hg}}\right)(1.000 \text{ L})}{\left(0.082\ 06 \dfrac{\text{L}\cdot\text{atm}}{\text{K}\cdot\text{mol}}\right)(298 \text{ K})} = 0.001\ 803 \text{ mol}$$

$$n_B = \frac{PV}{RT} = \frac{\left(33.5 \text{ mm Hg x } \dfrac{1.00 \text{ atm}}{760 \text{ mm Hg}}\right)(1.000 \text{ L})}{\left(0.082\ 06 \dfrac{\text{L}\cdot\text{atm}}{\text{K}\cdot\text{mol}}\right)(203 \text{ K})} = 0.002\ 646 \text{ mol}$$

$$n_C = \frac{PV}{RT} = \frac{\left(33.5 \text{ mm Hg x } \dfrac{1.00 \text{ atm}}{760 \text{ mm Hg}}\right)(1.000 \text{ L})}{\left(0.082\ 06 \dfrac{\text{L}\cdot\text{atm}}{\text{K}\cdot\text{mol}}\right)(83 \text{ K})} = 0.006\ 472 \text{ mol}$$

$$n_{N_2} = n_A + n_B + n_C = 0.001\ 803 + 0.002\ 646 + 0.006\ 472 = 0.010\ 92 \text{ mol } N_2$$

$$\text{(e) } n_{CO_2} = n_{\text{initial}} - n_{H_2O} - n_{N_2} = 0.030\ 35 - 0.0013 - 0.010\ 92 = 0.0181 \text{ mol } CO_2$$

10.152 $C_3H_5N_3O_9$, 227.1

$$\text{(a) moles } C_3H_5N_3O_9 = 1.00 \text{ g x } \frac{1 \text{ mol}}{227.1 \text{ g}} = 0.004\ 40 \text{ mol}$$

$$n_{\text{air}} = \frac{PV}{RT} = \frac{(1.00 \text{ atm})(0.500 \text{ L})}{\left(0.082\ 06 \dfrac{\text{L}\cdot\text{atm}}{\text{K}\cdot\text{mol}}\right)(293 \text{ K})} = 0.0208 \text{ mol air}$$

$$\text{(b) moles gas from } C_3H_5N_3O_9 = 0.004\ 40 \text{ mol x } \frac{29 \text{ mol gas}}{4 \text{ mol nitro}}$$

moles gas from $C_3H_5N_3O_9$ = 0.0319 mol gas from $C_3H_5N_3O_9$

n_{total} = 0.0319 mol + 0.0208 mol = 0.0527 mol

(c) $P = \frac{nRT}{V} = \frac{(0.0527 \text{ mol})\left(0.082\ 06 \frac{\text{L} \cdot \text{atm}}{\text{K} \cdot \text{mol}}\right)(698 \text{ K})}{(0.500 \text{ L})} = 6.04 \text{ atm}$

10.153 NH_3, 17.03; mol NH_3 = 45.0 g x $\frac{1 \text{ mol}}{17.03 \text{ g}}$ = 2.64 mol

$$P = \frac{nRT}{V} \quad \text{or} \quad P = \frac{nRT}{(V - nb)} - \frac{an^2}{V^2}$$

(a) At T = 0 °C = 273 K

$$P = \frac{(2.64 \text{ mol})\left(0.082\ 06 \frac{\text{L} \cdot \text{atm}}{\text{K} \cdot \text{mol}}\right)(273 \text{ K})}{(1.000 \text{ L})} = 59.1 \text{ atm}$$

$$P = \frac{(2.64 \text{ mol})\left(0.082\ 06 \frac{\text{L} \cdot \text{atm}}{\text{K} \cdot \text{mol}}\right)(273 \text{ K})}{[(1.000 \text{ L}) - (2.64 \text{ mol})(0.0371 \text{ L/mol})]} - \frac{\left(4.17 \frac{\text{L}^2 \cdot \text{atm}}{\text{mol}^2}\right)(2.64 \text{ mol})^2}{(1.000 \text{ L})^2}$$

P = 65.6 atm – 29.1 atm = 36.5 atm

(b) At T = 50 °C = 323 K

$$P = \frac{(2.64 \text{ mol})\left(0.082\ 06 \frac{\text{L} \cdot \text{atm}}{\text{K} \cdot \text{mol}}\right)(323 \text{ K})}{(1.000 \text{ L})} = 70.0 \text{ atm}$$

$$P = \frac{(2.64 \text{ mol})\left(0.082\ 06 \frac{\text{L} \cdot \text{atm}}{\text{K} \cdot \text{mol}}\right)(323 \text{ K})}{[(1.000 \text{ L}) - (2.64 \text{ mol})(0.0371 \text{ L/mol})]} - \frac{\left(4.17 \frac{\text{L}^2 \cdot \text{atm}}{\text{mol}^2}\right)(2.64 \text{ mol})^2}{(1.000 \text{ L})^2}$$

P = 77.6 atm – 29.1 atm = 48.5 atm

(c) At T = 100 °C = 373 K

$$P = \frac{(2.64 \text{ mol})\left(0.082\ 06 \frac{\text{L} \cdot \text{atm}}{\text{K} \cdot \text{mol}}\right)(373 \text{ K})}{(1.000 \text{ L})} = 80.8 \text{ atm}$$

$$P = \frac{(2.64 \text{ mol})\left(0.082\ 06 \frac{\text{L} \cdot \text{atm}}{\text{K} \cdot \text{mol}}\right)(373 \text{ K})}{[(1.000 \text{ L}) - (2.64 \text{ mol})(0.0371 \text{ L/mol})]} - \frac{\left(4.17 \frac{\text{L}^2 \cdot \text{atm}}{\text{mol}^2}\right)(2.64 \text{ mol})^2}{(1.000 \text{ L})^2}$$

P = 89.6 atm – 29.1 atm = 60.5 atm

At the three temperatures, the van der Waals equation predicts a much lower pressure than does the ideal gas law. This is likely due to the fact that NH_3 can hydrogen bond leading to strong intermolecular forces.

10.154 (a) $n_{total} = \frac{PV}{RT} = \frac{\left(258 \text{ mm Hg} \times \frac{1.00 \text{ atm}}{760 \text{ mm Hg}}\right)(0.500 \text{ L})}{\left(0.082\ 06 \frac{\text{L}\cdot\text{atm}}{\text{K}\cdot\text{mol}}\right)(293 \text{ K})} = 0.007\ 06 \text{ mol}$

(b) $n_B = \frac{PV}{RT} = \frac{\left(344 \text{ mm Hg} \times \frac{1 \text{ atm}}{760 \text{ mm Hg}}\right)(0.250 \text{ L})}{\left(0.082\ 06 \frac{\text{L}\cdot\text{atm}}{\text{K}\cdot\text{mol}}\right)(293 \text{ K})} = 0.004\ 71 \text{ moles}$

(c) $d = \frac{0.218 \text{ g}}{0.250 \text{ L}} = 0.872 \text{ g/L}$

(d) molar mass $= \frac{0.218 \text{ g}}{0.004\ 71 \text{ mol}} = 46.3$ g/mol, NO_2; molecular weight = 46.3

(e) $Hg_2CO_3(s) + 6\ HNO_3(aq) \rightarrow 2\ Hg(NO_3)_2(aq) + 3\ H_2O(l) + CO_2(g) + 2\ NO_2(g)$

10.155 CO_2, 44.01

mol CO_2 = 500.0 g CO_2 x $\frac{1 \text{ mol } CO_2}{44.01 \text{ g } CO_2} = 11.36$ mol CO_2

PV = nRT

$P = \frac{nRT}{V} = \frac{(11.36 \text{ mol})\left(0.082\ 06 \frac{\text{L}\cdot\text{atm}}{\text{K}\cdot\text{mol}}\right)(700 \text{ K})}{(0.800 \text{ L})} = 816 \text{ atm}$

10.156 (a) Let x = mol C_nH_{2n+2} in the reaction mixture.

Combustion of $C_nH_{2n+2} \rightarrow nCO_2 + (n+1)H_2O$ needs $n + \left(\frac{n+1}{2}\right) = \frac{3n+1}{2}$ mol O_2

Balanced equation is: $C_nH_{2n+2}(g) + \left(\frac{3n+1}{2}\right) O_2(g) \rightarrow nCO_2(g) + (n+1)H_2O(g)$

In going from reactants to products, the increase in the number of moles is

$[n + (n+1)] - \left[1 + \frac{3n+1}{2}\right] = \frac{n-1}{2}$ per mol of C_nH_{2n+2} reacted.

Before reaction: total mol $= \frac{PV}{RT} = \frac{(2.000 \text{ atm})(0.4000 \text{ L})}{\left(0.082\ 06 \frac{\text{L}\cdot\text{atm}}{\text{K}\cdot\text{mol}}\right)(298.15 \text{ K})} = 0.032\ 70 \text{ mol}$

After reaction: total mol $= \frac{PV}{RT} = \frac{(2.983 \text{ atm})(0.4000 \text{ L})}{\left(0.082\ 06 \frac{\text{L}\cdot\text{atm}}{\text{K}\cdot\text{mol}}\right)(398.15 \text{ K})} = 0.036\ 52 \text{ mol}$

Difference = 0.032 70 mol – 0.036 52 mol = 0.003 82 mol

Increase in number of mol = $\left(\frac{n-1}{2}\right)$ x = 0.003 82 mol; x = $\frac{2(0.003\ 82)}{n-1}$

Also x = $\frac{g\ C_nH_{2n+2}}{\text{molar mass}} = \frac{0.148\ g}{[12.01n + 1.008(2n+2)]\ g/mol}$

So $\frac{2(0.003\ 82)}{n-1} = \frac{0.148}{14.026n + 2.016}$; 0.148 n – 0.148 = 0.107 n + 0.0154

0.041 n = 0.163; n = $\frac{0.163}{0.041}$ = 4.0

C_nH_{2n+2} is C_4H_{10} (butane); molar mass = (4)(12.01) + (10)(1.008) = 58.12 g/mol

(b) 0.148 g C_4H_{10} x $\frac{1\ mol\ C_4H_{10}}{58.12\ g\ C_4H_{10}}$ = 0.002 55 mol C_4H_{10}

mol O_2 initially = total mol – mol C_4H_{10} = 0.032 70 mol – 0.002 55 mol = 0.030 15 mol O_2

$$P_{C_4H_{10}} = \left(\frac{n_{C_4H_{10}}}{n_{total}}\right) P_{initial} = \left(\frac{0.002\ 55\ mol}{0.032\ 70\ mol}\right)(2.000\ atm) = 0.156\ atm$$

$$P_{O_2} = \left(\frac{n_{O_2}}{n_{total}}\right) P_{initial} = \left(\frac{0.030\ 15\ mol}{0.032\ 70\ mol}\right)(2.000\ atm) = 1.844\ atm$$

(c) $C_4H_{10}(g) + \frac{13}{2} O_2 \rightarrow 4\ CO_2(g) + 5\ H_2O(g)$

0.002 55 mol C_4H_{10} x $\frac{4\ mol\ CO_2}{1\ mol\ C_4H_{10}}$ = 0.0102 mol CO_2

0.002 55 mol C_4H_{10} x $\frac{5\ mol\ H_2O}{1\ mol\ C_4H_{10}}$ = 0.012 75 mol H_2O

mol O_2 unreacted = total mol after reaction – mol CO_2 – mol H_2O
= 0.03652 mol – 0.0102 mol – 0.01275 = 0.01357 mol O_2

$$P_{CO_2} = \left(\frac{n_{CO_2}}{n_{total}}\right) P_{final} = \left(\frac{0.0102\ mol}{0.036\ 52\ mol}\right)(2.983\ atm) = 0.833\ atm$$

$$P_{H_2O} = \left(\frac{n_{H_2O}}{n_{total}}\right) P_{final} = \left(\frac{0.012\ 75\ mol}{0.036\ 52\ mol}\right)(2.983\ atm) = 1.041\ atm$$

$$P_{O_2} = \left(\frac{n_{O_2}}{n_{total}}\right) P_{final} = \left(\frac{0.013\ 57\ mol}{0.036\ 52\ mol}\right)(2.983\ atm) = 1.108\ atm$$

10.157 (a) average molecular mass for natural gas

$$= (0.915)(16.04) + (0.085)(30.07) = 17.2$$

$$\text{total moles of gas} = 15.50\text{ g} \times \frac{1\text{ mol gas}}{17.2\text{ g gas}} = 0.901\text{ mol gas}$$

(b) $$P = \frac{(0.901\text{ mol})\left(0.082\ 06\ \frac{\text{L}\cdot\text{atm}}{\text{K}\cdot\text{mol}}\right)(293\text{ K})}{(15.00\text{ L})} = 1.44\text{ atm}$$

(c) $P_{CH_4} = X_{CH_4} \cdot P_{total} = (1.44\text{ atm})(0.915) = 1.32\text{ atm}$

$P_{C_2H_6} = X_{C_2H_6} \cdot P_{total} = (1.44\text{ atm})(0.085) = 0.12\text{ atm}$

(d) $\Delta H_{combustion}(CH_4) = -890.3\text{ kJ/mol}$ and $\Delta H_{combustion}(C_2H_6) = -1427.7\text{ kJ/mol}$

Heat liberated = (0.915)(0.901 mol)(–890.3 kJ/mol)
+ (0.085)(0.901)(–1427.7 kJ/mol) = –843 kJ

10.158 PV = nRT

$$n_{total\,(initial)} = \frac{PV}{RT} = \frac{(3.00\text{ atm})(10.0\text{ L})}{\left(0.082\ 06\ \frac{\text{L}\cdot\text{atm}}{\text{K}\cdot\text{mol}}\right)(373.1\text{ K})} = 0.980\text{ mol}$$

$$n_{total\,(final)} = \frac{PV}{RT} = \frac{(2.40\text{ atm})(10.0\text{ L})}{\left(0.082\ 06\ \frac{\text{L}\cdot\text{atm}}{\text{K}\cdot\text{mol}}\right)(373.1\text{ K})} = 0.784\text{ mol}$$

	$CS_2(g)$	+	$3\ O_2(g)$	→	$CO_2(g)$	+	$2\ SO_2(g)$
before reaction (mol)	y		0.980 – y		0		0
change (mol)	–x		–3x		+x		+2x
after reaction (mol)	y – x = 0		0.980 – y – 3x		x		2x

$n_{total\,(final)} = (y - x) + (0.980 - y - 3x) + x + 2x = 0.784\text{ mol}$

0.980 mol – 4x + 3x = 0.784 mol

x = 0.980 mol – 0.784 mol = 0.196 mol

mol CO_2 = x = 0.196 mol

$$P_{CO_2} = \frac{nRT}{V} = \frac{(0.196\text{ mol})\left(0.082\ 06\ \frac{\text{L}\cdot\text{atm}}{\text{K}\cdot\text{mol}}\right)(373.1\text{ K})}{(10.0\text{ L})} = 0.600\text{ atm}$$

mol SO_2 = 2x = 2(0.196 mol) = 0.392 mol

$$P_{SO_2} = \frac{nRT}{V} = \frac{(0.392\text{ mol})\left(0.082\ 06\ \frac{\text{L}\cdot\text{atm}}{\text{K}\cdot\text{mol}}\right)(373.1\text{ K})}{(10.0\text{ L})} = 1.20\text{ atm}$$

mol O_2 = 0.980 mol – y – 3x = 0.980 mol – x – 3x = 0.980 – 4(0.196 mol) = 0.196 mol

$P_{O_2} = P_{CO_2} = 0.600\text{ atm}$

10.159 (a) T = 0 °C = 273 K; PV = nRT

$$n_Q = \frac{PV}{RT} = \frac{(0.229\ \text{atm})(0.0500\ \text{L})}{\left(0.082\ 06\ \frac{\text{L}\cdot\text{atm}}{\text{K}\cdot\text{mol}}\right)(273\ \text{K})} = 5.11 \times 10^{-4}\ \text{mol Q}$$

$$\text{Q molar mass} = \frac{0.100\ \text{g Q}}{5.11 \times 10^{-4}\ \text{mol Q}} = 196\ \text{g/mol}$$

Xe molar mass = 131.3 g/mol

O_n molar mass = 196 g/mol – 131.3 g/mol = 65 g/mol; 65/16 ≈ 4

So, n = 4 and XeO_4 is the likely formula for Q.

(b) The decomposition reaction is $XeO_4(g) \rightarrow Xe(g) + 2\ O_2(g)$.

After decomposition $n_{Xe} = n_{XeO_4} = 5.11 \times 10^{-4}$ mol and $n_{O_2} = 2 \times n_{XeO_4} = 1.02 \times 10^{-3}$ mol

T = 100 °C = 373 K

$$P_{Xe} = \frac{nRT}{V} = \frac{(5.11 \times 10^{-4}\ \text{mol})\left(0.082\ 06\ \frac{\text{L}\cdot\text{atm}}{\text{K}\cdot\text{mol}}\right)(373\ \text{K})}{(0.0500\ \text{L})} = 0.313\ \text{atm}$$

$P_{O_2} = 2 \times P_{Xe} = 2 \times 0.313\ \text{atm} = 0.626\ \text{atm}$

$P_{total} = P_{Xe} + P_{O_2} = 0.313\ \text{atm} + 0.626\ \text{atm} = 0.939\ \text{atm}$

10.160 $Ca(ClO_3)_2$, 206.98; $Ca(ClO)_2$, 142.98

(a) $Ca(ClO_3)_2(s) \rightarrow CaCl_2(s) + 3\ O_2(g)$

$Ca(ClO)_2(s) \rightarrow CaCl_2(s) + O_2(g)$

(b) T = 700 °C = 700 + 273 = 973 K

PV = nRT

$$n_{O_2} = \frac{PV}{RT} = \frac{(1.00\ \text{atm})(10.0\ \text{L})}{\left(0.082\ 06\ \frac{\text{L}\cdot\text{atm}}{\text{K}\cdot\text{mol}}\right)(973\ \text{K})} = 0.125\ \text{mol } O_2$$

Let X = mol $Ca(ClO_3)_2$ and let Y = mol $Ca(ClO)_2$

X(206.98 g/mol) + Y(142.98 g/mol) = 10.0 g

3X + Y = 0.125 mol, so Y = 0.125 mol – 3X (substitute for Y and solve for X)

X(206.98 g/mol) + (0.125 mol – 3X)(142.98 g/mol) = 10.0 g

X(206.98 g/mol) + 17.9 g – X(428.94 g/mol) = 10.0 g

X(206.98 g/mol) – X(428.94 g/mol) = 10.0 g – 17.9 g = –7.9 g

X(–221.96 g/mol) = –7.9 g

X = (–7.9 g)/(–221.96 g/mol) = 0.0356 mol $Ca(ClO_3)_2$

Y = 0.125 mol – 3X; Y = 0.125 mol – 3(0.0356 mol) = 0.0182 mol $Ca(ClO)_2$

$$\text{mass } Ca(ClO_3)_2 = 0.0356\ \text{mol } Ca(ClO_3)_2 \times \frac{206.98\ \text{g } Ca(ClO_3)_2}{1\ \text{mol } Ca(ClO_3)_2} = 7.4\ \text{g } Ca(ClO_3)_2$$

mass $Ca(ClO)_2$ = 10.0 g – 7.4 g = 2.6 g $Ca(ClO)_2$

10.161 PCl_3, 137.3; O_2, 32.00; $POCl_3$, 153.3

$2\ PCl_3(g) + O_2(g) \rightarrow 2\ POCl_3(g)$

$$\text{mol PCl}_3 = 25.0 \text{ g x } \frac{1 \text{ mol PCl}_3}{137.3 \text{ g PCl}_3} = 0.182 \text{ mol PCl}_3$$

$$\text{mol O}_2 = 3.00 \text{ g x } \frac{1 \text{ mol O}_2}{32.00 \text{ g O}_2} = 0.0937 \text{ mol O}_2$$

Check for limiting reactant.

$$\text{mol O}_2 \text{ needed } = 0.182 \text{ mol PCl}_3 \text{ x } \frac{1 \text{ mol O}_2}{2 \text{ mol PCl}_3} = 0.0910 \text{ mol O}_2 \text{ needed}$$

There is a slight excess of O_2. PCl_3 is the limiting reactant.

$$\text{mol POCl}_3 = 0.182 \text{ mol PCl}_3 \text{ x } \frac{2 \text{ mol POCl}_3}{2 \text{ mol PCl}_3} = 0.182 \text{ mol POCl}_3$$

mol O_2 left over = 0.0937 mol – 0.0910 mol = 0.0027 mol O_2 left over

T = 200.0 °C = 200.0 + 273.15 = 473.1 K; PV = nRT

$$P = \frac{nRT}{V} = \frac{(0.182 \text{ mol} + 0.0027 \text{ mol})\left(0.082\ 06 \dfrac{\text{L}\cdot\text{atm}}{\text{K}\cdot\text{mol}}\right)(473.1 \text{ K})}{(5.00 \text{ L})} = 1.43 \text{ atm}$$

10.162 (a) T = 225 °C = 225 + 273 = 498 K

$$PV = nRT \qquad P^o_{NOCl} = \frac{nRT}{V} = \frac{(2.00 \text{ mol})\left(0.082\ 06 \dfrac{\text{L}\cdot\text{atm}}{\text{K}\cdot\text{mol}}\right)(498 \text{ K})}{(400.0 \text{ L})} = 0.204 \text{ atm}$$

	2 NOCl(g) →	2 NO(g) +	Cl_2(g)
before reaction (atm)	0.204	0	0
change (atm)	–2x	+2x	+x
after reaction (atm)	0.204 – 2x	2x	x

P_{total} (after rxn) = (0.204 atm – 2x) + 2x + x = 0.246 atm

x = 0.246 atm – 0.204 atm = 0.042 atm

P_{NO} = 2x = 2(0.042) = 0.084 atm; P_{Cl_2} = x = 0.042 atm

P_{NOCl} = 0.204 – 2x = 0.204 – 2(0.042) = 0.120 atm

(b) $\% \text{ NOCl decomposed} = \frac{2x}{P^o_{NOCl}} \text{ x } 100\% = \frac{0.084 \text{ atm}}{0.204 \text{ atm}} \text{ x } 100\% = 41\%$

10.163 O_2, 32.00; O_3, 48.00

	3 O_2(g) →	2 O_3(g)
initial (atm)	32.00	0
change (atm)	–3x	+2x
after rxn (atm)	32.00 – 3x	2x

$P_{Total} = P_{O_2} + P_{O_3}$ = 30.64 atm = 32.00 atm – 3x + 2x = 32.00 atm – x

x = 32.00 atm – 30.64 atm = 1.36 atm

$P_{O_2} = 32.00 - 3x = 32.00 - 3(1.36 \text{ atm}) = 27.92 \text{ atm}$

$P_{O_3} = 2x = 2(1.36 \text{ atm}) = 2.72 \text{ atm}$

$T = 25\ °C = 25 + 273 = 298\ K;\ PV = nRT$

$$n_{O_2} = \frac{PV}{RT} = \frac{(27.92 \text{ atm})(10.00 \text{ L})}{\left(0.082\ 06 \frac{L \cdot atm}{K \cdot mol}\right)(298 \text{ K})} = 11.42 \text{ mol } O_2$$

$$n_{O_3} = \frac{PV}{RT} = \frac{(2.72 \text{ atm})(10.00 \text{ L})}{\left(0.082\ 06 \frac{L \cdot atm}{K \cdot mol}\right)(298 \text{ K})} = 1.11 \text{ mol } O_3$$

$$\text{mass } O_2 = 11.42 \text{ mol } O_2 \times \frac{32.00 \text{ g } O_2}{1 \text{ mol } O_2} = 365.4 \text{ g } O_2$$

$$\text{mass } O_3 = 1.11 \text{ mol } O_3 \times \frac{48.00 \text{ g } O_3}{1 \text{ mol } O_3} = 53.3 \text{ g } O_3$$

$\text{total mass} = 365.4 \text{ g} + 53.3 \text{ g} = 418.7 \text{ g}$

$$\text{mass \% } O_3 = \frac{\text{mass } O_3}{\text{total mass}} = \frac{53.3 \text{ g}}{418.7 \text{ g}} \times 100\% = 12.7\%$$

10.164 $CaCO_3$, 100.09; CaO, 56.08

$$\text{mol CaO (or } CO_2) = 25.0 \text{ g } CaCO_3 \times \frac{1 \text{ mol } CaCO_3}{100.09 \text{ g } CaCO_3} \times \frac{1 \text{ mol CaO or } CO_2}{1 \text{ mol } CaCO_3} = 0.250 \text{ mol}$$

$$\text{mass CaO} = 0.250 \text{ mol CaO} \times \frac{56.08 \text{ g CaO}}{1 \text{ mol CaO}} = 14.02 \text{ g CaO}$$

(a) 500.0 mL = 0.5000 L

$PV = nRT;\quad n_{CO_2} = 0.250 \text{ mol}$

$$P_{CO_2} = \frac{nRT}{V} = \frac{(0.250 \text{ mol})\left(0.082\ 06 \frac{L \cdot atm}{K \cdot mol}\right)(1500 \text{ K})}{(0.5000 \text{ L})} = 61.5 \text{ atm}$$

(b) $V_{CaO} = (14.02 \text{ g})/(3.34 \text{ g/mL}) = 4.20 \text{ mL}$

$V = 500.0 \text{ mL} - 4.20 \text{ mL} = 495.8 \text{ mL} = 0.4958 \text{ L}$

$$P = \frac{nRT}{V - nb} - \frac{an^2}{V^2}$$

$$P = \frac{(0.250 \text{ mol})\left(0.082\ 06 \frac{L \cdot atm}{K \cdot mol}\right)(1500 \text{ K})}{[(0.4958 \text{ L}) - (0.250 \text{ mol})(0.0427 \text{ L/mol})]} - \frac{\left(3.59 \frac{L^2 \cdot atm}{mol^2}\right)(0.250 \text{ mol})^2}{(0.4958 \text{ L})^2} = 62.5 \text{ atm}$$

10.165 NO_2, 46.01

Calculate the NO_2 pressure before any NO_2 dimerization.

$$P = \frac{nRT}{V} = \frac{\left(9.66\text{ g x }\frac{1\text{ mol}}{46.01\text{ g}}\right)\left(0.082\ 06\ \frac{\text{L}\cdot\text{atm}}{\text{K}\cdot\text{mol}}\right)(298\text{ K})}{(6.51\text{ L})} = 0.789\text{ atm}$$

	$2\ NO_2(g)$	→	$N_2O_4(g)$
before reaction (atm)	0.789		0
change (atm)	–2x		+x
after reaction (atm)	0.789 – 2x		x

$0.487 = 0.789 - 2x + x$

$0.487 = 0.789 - x$

$x = 0.789 - 0.487 = 0.302$ atm

$P_{NO_2} = 0.789 - 2x = 0.789 - 2(0.302) = 0.185$ atm

$P_{N_2O_4} = x = = 0.302$ atm

$$X_{NO_2} = \frac{0.185\text{ atm}}{0.487\text{ atm}} = 0.380 \text{ and } X_{N_2O_4} = \frac{0.302\text{ atm}}{0.487\text{ atm}} = 0.620$$

10.166 Assume a container volume of 1.00 L with 3.309 g of gas.

25 ºC = 25 + 273 = 298 K

$$n = \frac{PV}{RT} = \frac{(1.00\text{ atm})(1.00\text{ L})}{\left(0.082\ 06\ \frac{\text{L}\cdot\text{atm}}{\text{K}\cdot\text{mol}}\right)(298\text{ K})} = 0.0409\text{ mol}$$

$$\text{molar mass} = \frac{3.309\text{ g}}{0.0409\text{ mol}} = 80.9\text{ g/mol}$$

The gas could be H_2Se, 81.0.

Multiconcept Problems

10.167 CO_2, 44.01

$CH_4(g) + 2\ O_2(g) \rightarrow CO_2(g) + 2\ H_2O(g)$ $\Delta H° = -802$ kJ

(a) 1.00 atm of CH_4 only requires 2.00 atm O_2, therefore O_2 is in excess.

T = 300 ºC = 300 + 273 = 573 K; PV = nRT

$$n_{CH_4} = \frac{PV}{RT} = \frac{(1.00\text{ atm})(4.00\text{ L})}{\left(0.082\ 06\ \frac{\text{L}\cdot\text{atm}}{\text{K}\cdot\text{mol}}\right)(573\text{ K})} = 0.0851\text{ mol } CH_4$$

$$n_{O_2} = \frac{PV}{RT} = \frac{(4.00\text{ atm})(4.00\text{ L})}{\left(0.082\ 06\ \frac{\text{L}\cdot\text{atm}}{\text{K}\cdot\text{mol}}\right)(573\text{ K})} = 0.340\text{ mol } O_2$$

$$\text{mass } CO_2 = 0.0851\text{ mol } CH_4 \text{ x } \frac{1\text{ mol } CO_2}{1\text{ mol } CH_4} \text{ x } \frac{44.01\text{ g } CO_2}{1\text{ mol } CO_2} = 3.75\text{ g } CO_2$$

(b) $q_{rxn} = 0.0851 \text{ mol } CH_4 \times \dfrac{-802 \text{ kJ}}{1 \text{ mol } CH_4} = -68.3 \text{ kJ}$

	$CH_4(g)$	+	$2\ O_2(g)$	→	$CO_2(g)$	+	$2\ H_2O(g)$
initial (mol)	0.0851		0.340		0		0
change (mol)	–0.0851		–2(0.0851)		+0.0851		+2(0.0851)
after rxn (mol)	0		0.340 – 2(0.0851)		0.0851		0.170

total moles of gas = 0.340 mol – 2(0.0851) mol + 0.0851 mol + 0.170 mol = 0.425 mol gas

$q_{rxn} = -68.3 \text{ kJ} \times \dfrac{1000 \text{ J}}{1 \text{ kJ}} = -68{,}300 \text{ J}$

$q_{vessel} = -q_{rxn} = 68{,}300 \text{ J} = (0.425 \text{ mol})(21 \text{ J/(mol} \cdot {}^\circ\text{C}))(t_f - 300\ {}^\circ\text{C}) +$

$(14.500 \text{ kg})\left(\dfrac{1000 \text{ g}}{1 \text{ kg}}\right)(0.449 \text{ J/(g} \cdot {}^\circ\text{C}))(t_f - 300\ {}^\circ\text{C})$

Solve for t_f.

$68{,}300 \text{ J} = (8.925 \text{ J/}^\circ\text{C} + 6510 \text{ J/}^\circ\text{C})(t_f - 300\ {}^\circ\text{C}) = (6519 \text{ J/}^\circ\text{C})(t_f - 300\ {}^\circ\text{C})$

$\dfrac{68{,}300 \text{ J}}{6519 \text{ J/}^\circ\text{C}} = 10.5\ {}^\circ\text{C} = (t_f - 300\ {}^\circ\text{C})$

$t_f = 300\ {}^\circ\text{C} + 10.5\ {}^\circ\text{C} = 310\ {}^\circ\text{C}$

(c) $T = 310\ {}^\circ\text{C} = 310 + 273 = 583 \text{ K}$

$$P_{CO_2} = \frac{nRT}{V} = \frac{(0.0851 \text{ mol})\left(0.082\ 06\ \dfrac{\text{L} \cdot \text{atm}}{\text{K} \cdot \text{mol}}\right)(583 \text{ K})}{(4.00 \text{ L})} = 1.02 \text{ atm}$$

10.168 $X + 3\ O_2 \rightarrow 2\ CO_2 + 3\ H_2O$

(a) $X = C_2H_6O$

$C_2H_6O + 3\ O_2 \rightarrow 2\ CO_2 + 3\ H_2O$

(b) It is an empirical formula because it is the smallest whole number ratio of atoms. It is also a molecular formula because any higher multiple such as $C_4H_{12}O_2$ does not correspond to a stable electron-dot structure.

(c)

```
     H       H              H   H
     |  ..   |              |   |  ..
  H—C—O—C—H          H—C—C—O—H
     |  ..   |              |   |  ..
     H       H              H   H
```

(d) C_2H_6O, 46.07

$\text{mol } C_2H_6O = 5.000 \text{ g } C_2H_6O \times \dfrac{1 \text{ mol } C_2H_6O}{46.07 \text{ g } C_2H_6O} = 0.1085 \text{ mol } C_2H_6O$

$\Delta H_{combustion} = \dfrac{-144.2 \text{ kJ}}{0.1085 \text{ mol}} = -1328.6 \text{ kJ/mol}$

$\Delta H_{combustion} = [2\ \Delta H^\circ_f(CO_2) + 3\ \Delta H^\circ_f(H_2O)] - \Delta H^\circ_f(C_2H_6O)$

$\Delta H^\circ_f(C_2H_6O) = [2\ \Delta H^\circ_f(CO_2) + 3\ \Delta H^\circ_f(H_2O)] - \Delta H_{combustion}$

$= [(2\text{mol})(-393.5 \text{ kJ/mol}) + (3 \text{ mol})(-241.8 \text{ kJ/mol})] - (-1328.6 \text{ kJ})$

$= -183.8 \text{ kJ/mol}$

10.169 (a) $2\,C_8H_{18}(l) + 25\,O_2(g) \rightarrow 16\,CO_2(g) + 18\,H_2O(g)$

(b) $4.6 \times 10^{10}\text{ L } C_8H_{18} \times \dfrac{1000\text{ mL}}{1\text{ L}} \times \dfrac{0.792\text{ g}}{1\text{ mL}} = 3.64 \times 10^{13}\text{ g } C_8H_{18}$

$$3.64 \times 10^{13}\text{ g } C_8H_{18} \times \frac{1\text{ mol } C_8H_{18}}{114.2\text{ g } C_8H_{18}} \times \frac{16\text{ mol } CO_2}{2\text{ mol } C_8H_{18}} = 2.55 \times 10^{12}\text{ mol } CO_2$$

$$2.55 \times 10^{12}\text{ mol } CO_2 \times \frac{44.0\text{ g } CO_2}{1\text{ mol } CO_2} \times \frac{1\text{ kg}}{1000\text{ g}} = 1.1 \times 10^{11}\text{ kg } CO_2$$

(c) $$V = \frac{nRT}{P} = \frac{(2.55 \times 10^{12}\text{ mol})\left(0.082\ 06\ \frac{\text{L}\cdot\text{atm}}{\text{K}\cdot\text{mol}}\right)(273\text{ K})}{(1.00\text{ atm})} = 5.7 \times 10^{13}\text{ L of } CO_2$$

(d) 12.5 moles of O_2 are needed for each mole of isooctane (from part a).

$12.5\text{ mol } O_2 = (0.210)(n_{air})$; $\quad n_{air} = \dfrac{12.5\text{ mol}}{0.210} = 59.5\text{ mol air}$

$$V = \frac{nRT}{P} = \frac{(59.5\text{ mol})\left(0.082\ 06\ \frac{\text{L}\cdot\text{atm}}{\text{K}\cdot\text{mol}}\right)(273\text{ K})}{(1.00\text{ atm})} = 1.33 \times 10^3\text{ L}$$

10.170 (a) Freezing point of H_2O on the Rankine scale is $(9/5)(273.15) = 492\ ^\circ\text{R}$.

(b) $$R = \frac{PV}{nT} = \frac{(1.00\text{ atm})(22.414\text{ L})}{(1.00\text{ mol})(492\,^\circ\text{R})} = 0.0456\ \frac{\text{L}\cdot\text{atm}}{^\circ\text{R}\cdot\text{mol}}$$

(c) $$P = \frac{(2.50\text{ mol})\left(0.0456\ \frac{\text{L}\cdot\text{atm}}{^\circ\text{R}\cdot\text{mol}}\right)(525\,^\circ\text{R})}{[(0.4000\text{ L}) - (2.50\text{ mol})(0.04278\text{ L/mol})]} - \frac{\left(2.253\ \frac{\text{L}^2\cdot\text{atm}}{\text{mol}^2}\right)(2.50\text{ mol})^2}{(0.4000\text{ L})^2}$$

$P = 204.2\text{ atm} - 88.0\text{ atm} = 116\text{ atm}$

10.171 $$n = \frac{PV}{RT} = \frac{(1\text{ atm})(1323\text{ L})}{\left(0.082\ 06\ \frac{\text{L}\cdot\text{atm}}{\text{K}\cdot\text{mol}}\right)(2223\text{ K})} = 7.25\text{ mol of all gases}$$

(a) $0.004\ 00\text{ mol "nitro"} \times \dfrac{7.25\text{ mol gases}}{1\text{ mol "nitro"}} = 0.0290\text{ mol hot gases}$

(b) $$n = \frac{PV}{RT} = \frac{\left(623\text{ mm Hg} \times \frac{1.00\text{ atm}}{760\text{ mm Hg}}\right)(0.500\text{ L})}{\left(0.082\ 06\ \frac{\text{L}\cdot\text{atm}}{\text{K}\cdot\text{mol}}\right)(263\text{ K})} = 0.0190\text{ mol B + C + D}$$

$n_A = n_{total} - n_{(B+C+D)} = 0.0290 - 0.0190 = 0.0100\text{ mol A}$; $A = H_2O$

(c) $n = \dfrac{PV}{RT} = \dfrac{\left(260 \text{ mm Hg} \times \dfrac{1.00 \text{ atm}}{760 \text{ mm Hg}}\right)(0.500 \text{ L})}{\left(0.082\ 06 \dfrac{\text{L}\cdot\text{atm}}{\text{K}\cdot\text{mol}}\right)(298 \text{ K})} = 0.007\ 00 \text{ mol C} + \text{D}$

$n_B = n_{(B+C+D)} - n_{(C+D)} = 0.0190 - 0.007\ 00 = 0.0120 \text{ mol B};\ B = CO_2$

(d) $n = \dfrac{PV}{RT} = \dfrac{\left(223 \text{ mm Hg} \times \dfrac{1.00 \text{ atm}}{760 \text{ mm Hg}}\right)(0.500 \text{ L})}{\left(0.082\ 06 \dfrac{\text{L}\cdot\text{atm}}{\text{K}\cdot\text{mol}}\right)(298 \text{ K})} = 0.006\ 00 \text{ mol D}$

$n_C = n_{(C+D)} - n_D = 0.007\ 00 - 0.006\ 00 = 0.001\ 00 \text{ mol C};\ C = O_2$

$\text{molar mass D} = \dfrac{0.168 \text{ g}}{0.006\ 00 \text{ mol}} = 28.0 \text{ g/mol};\ D = N_2$

(e) $0.004\ C_3H_5N_3O_9(l) \rightarrow 0.0100\ H_2O(g) + 0.012\ CO_2(g) + 0.001\ O_2(g) + 0.006\ N_2(g)$
Multiply each coefficient by 1000 to obtain integers.
$4\ C_3H_5N_3O_9(l) \rightarrow 10\ H_2O(g) + 12\ CO_2(g) + O_2(g) + 6\ N_2(g)$

10.172 CO_2, 44.01; H_2O, 18.02

(a) $\text{mol C} = 0.3744 \text{ g } CO_2 \times \dfrac{1 \text{ mol } CO_2}{44.01 \text{ g } CO_2} \times \dfrac{1 \text{ mol C}}{1 \text{ mol } CO_2} = 0.008\ 507 \text{ mol C}$

$\text{mass C} = 0.008\ 507 \text{ mol C} \times \dfrac{12.011 \text{ g C}}{1 \text{ mol C}} = 0.1022 \text{ g C}$

$\text{mol H} = 0.1838 \text{ g } H_2O \times \dfrac{1 \text{ mol } H_2O}{18.02 \text{ g } H_2O} \times \dfrac{2 \text{ mol H}}{1 \text{ mol } H_2O} = 0.020\ 400 \text{ mol H}$

$\text{mass H} = 0.020\ 400 \text{ mol H} \times \dfrac{1.008 \text{ g H}}{1 \text{ mol H}} = 0.02056 \text{ g H}$

$\text{mass O} = 0.1500 \text{ g} - 0.1022 \text{ g} - 0.02056 \text{ g} = 0.0272 \text{ g O}$

$\text{mol O} = 0.0272 \text{ g O} \times \dfrac{1 \text{ mol O}}{16.00 \text{ g O}} = 0.001\ 70 \text{ mol O}$

$C_{0.008\ 507}H_{0.020\ 400}O_{0.001\ 70}$; divide each subscript by the smallest, 0.001 70.
$C_{0.008\ 507\ /\ 0.001\ 70}H_{0.020\ 400\ /\ 0.001\ 70}O_{0.001\ 70\ /\ 0.001\ 70}$
The empirical formula is $C_5H_{12}O$.
The empirical formula mass is 88 g/mol.

(b) 1 atm = 101,325 Pa; T = 54.8 °C = 54.8 + 273.15 = 327.9 K
PV = nRT

$$n = \frac{PV}{RT} = \frac{\left(100.0\text{ kPa x }\frac{1.00\text{ atm}}{101.325\text{ kPa}}\right)(1.00\text{ L})}{\left(0.082\ 06\ \frac{\text{L}\cdot\text{atm}}{\text{K}\cdot\text{mol}}\right)(327.9\text{ K})} = 0.0367\text{ mol methyl } \textit{tert}\text{-butyl ether}$$

$$\text{methyl } \textit{tert}\text{-butyl ether molar mass} = \frac{3.233\text{ g}}{0.0367\text{ mol}} = 88.1\text{ g/mol}$$

The empirical formula weight and the molar mass are the same, so the molecular formula and empirical formula are the same. $C_5H_{12}O$ is the molecular formula and 88.15 is the molecular weight for methyl *tert*-butyl ether.

(c) $C_5H_{12}O(l) + 15/2\ O_2(g) \rightarrow 5\ CO_2(g) + 6\ H_2O(l)$

(d) $\Delta H^\circ_{combustion} = [5\ \Delta H^\circ_f(CO_2) + 6\ \Delta H^\circ_f(H_2O(l))] - \Delta H^\circ_f(C_5H_{12}O) = -3368.7\text{ kJ}$

$-3368.7\text{ kJ} = [(5\text{ mol})(-393.5\text{ kJ/mol}) + (6\text{ mol})(-285.8\text{ kJ/mol})] - (1\text{ mol})\Delta H^\circ_f(C_5H_{12}O)$

$(1\text{ mol})\Delta H^\circ_f(C_5H_{12}O) = [(5\text{ mol})(-393.5\text{ kJ/mol}) + (6\text{ mol})(-285.8\text{ kJ/mol})] + 3368.7\text{ kJ}$

$\Delta H^\circ_f(C_5H_{12}O) = -313.6\text{ kJ/mol}$

11 Liquids, Solids, and Phase Changes

11.1 boiling point = 78.4 °C = 351.6 K

$\Delta G = \Delta H_{vap} - T\Delta S_{vap}$; At the boiling point (phase change), $\Delta G = 0$

$\Delta H_{vap} = T\Delta S_{vap}$

$$\Delta S_{vap} = \frac{\Delta H_{vap}}{T} = \frac{38.56 \text{ kJ/mol}}{351.6 \text{ K}} = 0.1097 \text{ kJ/(K}\cdot\text{mol)} = 109.7 \text{ J/(K}\cdot\text{mol)}$$

11.2 $\Delta G = \Delta H - T\Delta S$; at the boiling point (phase change), $\Delta G = 0$.

$$\Delta H = T\Delta S; \quad T = \frac{\Delta H_{vap}}{\Delta S_{vap}} = \frac{29.2 \text{ kJ/mol}}{87.5 \text{ x } 10^{-3} \text{ kJ/(K}\cdot\text{mol)}} = 334 \text{ K}$$

11.3 C_6H_6, 78.1; $\quad 15.0 \text{ g } C_6H_6 \text{ x } \frac{1 \text{ mol } C_6H_6}{78.1 \text{ g } C_6H_6} = 0.192 \text{ mol } C_6H_6$

$q_1 = (0.192 \text{ mol})[136.0 \text{ x } 10^{-3} \text{ kJ/(K}\cdot\text{mol)}](80.1 \text{ °C} - 50 \text{ °C}) = 0.786 \text{ kJ}$

$q_2 = (0.192 \text{ mol})(30.72 \text{ kJ/mol}) = 5.90 \text{ kJ}$

$q_3 = (0.192 \text{ mol})(82.4 \text{ x } 10^{-3} \text{ kJ/(K}\cdot\text{mol)}](100 \text{ °C} - 80.1 \text{ °C}) = 0.315 \text{ kJ}$

$q_{total} = q_1 + q_2 + q_3 = 7.00 \text{ kJ}$; 7.00 kJ of heat is required.

11.4 H_2O, 18.0; $\quad 10.0 \text{ g } H_2O \text{ x } \frac{1 \text{ mol } H_2O}{18.0 \text{ g } H_2O} = 0.556 \text{ mol } H_2O$

$q_1 = (0.556 \text{ mol})[75.4 \text{ x } 10^{-3} \text{ kJ/(K}\cdot\text{mol)}](0 \text{ °C} - 25 \text{ °C}) = -1.05 \text{ kJ}$

$q_2 = (0.556 \text{ mol})(-6.01 \text{ kJ/mol}) = -3.34 \text{ kJ}$

$q_3 = (0.556 \text{ mol})(36.6 \text{ x } 10^{-3} \text{ kJ/(K}\cdot\text{mol)}](-10 \text{ °C} - 0 \text{ °C}) = -0.203 \text{ kJ}$

$q_{total} = q_1 + q_2 + q_3 = -4.59 \text{ kJ}$; 4.59 kJ of heat is removed.

11.5
$$\Delta H_{vap} = \frac{(\ln P_2 - \ln P_1)(R)}{\left(\frac{1}{T_1} - \frac{1}{T_2}\right)}$$

$P_1 = 400 \text{ mm Hg}$; $T_1 = 41.0 \text{ °C} = 314.2 \text{ K}$

$P_2 = 760 \text{ mm Hg}$; $T_2 = 331.9 \text{ K}$

$$\Delta H_{vap} = \frac{[\ln(760) - \ln(400)]\left(8.3145 \frac{\text{J}}{\text{K}\cdot\text{mol}}\right)}{\left(\frac{1}{314.2 \text{ K}} - \frac{1}{331.9 \text{ K}}\right)} = 31{,}442 \text{ J/mol} = 31.4 \text{ kJ/mol}$$

11.6 $\ln\left(\frac{P_2}{P_1}\right) = \ln P_2 - \ln P_1 = \frac{\Delta H_{vap}}{R}\left(\frac{1}{T_1} - \frac{1}{T_2}\right)$

$$(\ln P_2 - \ln P_1)\left(\frac{R}{\Delta H_{vap}}\right) = \frac{1}{T_1} - \frac{1}{T_2}$$

$$\frac{1}{T_1} - (\ln P_2 - \ln P_1)\left(\frac{R}{\Delta H_{vap}}\right) = \frac{1}{T_2}$$

$P_1 = 760$ mm Hg; $T_1 = 100.0$ °C = 373.1 K
$P_2 = 407$ mm Hg;
Solve for T_2, the boiling point for H_2O on top of Pikes Peak

$$\frac{1}{373.1\ \text{K}} - [\ln(407) - \ln(760)]\left(\frac{8.3145 \times 10^{-3}\frac{\text{kJ}}{\text{K}\cdot\text{mol}}}{40.7\ \text{kJ/mol}}\right) = \frac{1}{T_2}$$

$\frac{1}{T_2} = 0.002\ 808$; $T_2 = 356.1$ K = 83.0 °C

11.7 There are several possibilities. Here's one:

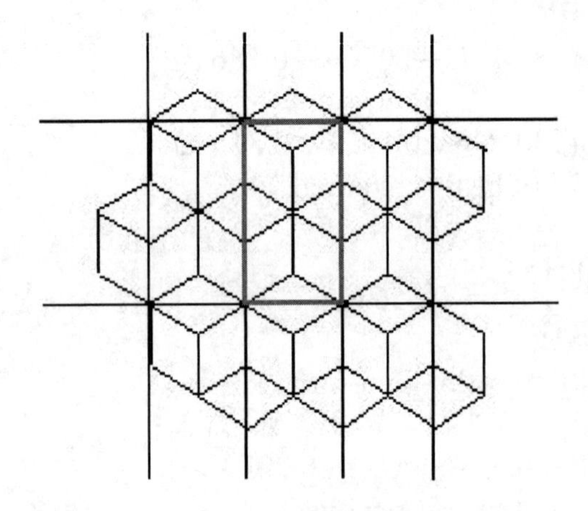

11.8 Here are two possibilities:

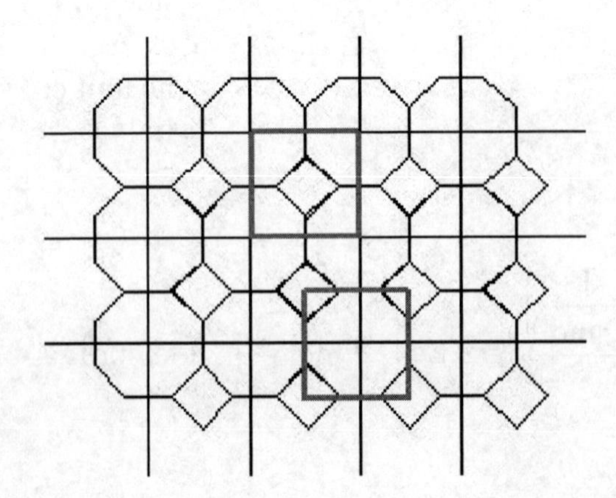

11.9 The face-centered cube is the unit cell for cubic closest packing.

$$r = \sqrt{\frac{d^2}{8}}; \quad d = r \cdot 2 \cdot \sqrt{2} = 197 \text{ pm} \cdot 2 \cdot \sqrt{2} = 557 \text{ pm}$$

11.10 For a simple cube, d = 2r; $r = \frac{d}{2} = \frac{334 \text{ pm}}{2} = 167 \text{ pm}$

11.11 For a simple cube, there is one atom per unit cell.

mass of one Po atom = $209 \text{ g/mol} \times \frac{1 \text{ mol}}{6.022 \times 10^{23} \text{ atoms}} = 3.4706 \times 10^{-22}$ g/atom

unit cell edge = d = 334 pm = 334×10^{-12} m = 3.34×10^{-8} cm

unit cell volume = $d^3 = (3.34 \times 10^{-8} \text{ cm})^3 = 3.7260 \times 10^{-23} \text{ cm}^3$

$$\text{density} = \frac{\text{mass}}{\text{volume}} = \frac{3.4706 \times 10^{-22} \text{ g}}{3.7260 \times 10^{-23} \text{ cm}^3} = 9.31 \text{ g/cm}^3$$

11.12 The unit cell is a face-centered cube. There are four atoms in the unit cell.

$$\text{unit cell volume} = \left(383.3 \times 10^{-12} \text{ m} \times \frac{1 \text{ cm}}{1 \times 10^{-2} \text{ m}}\right)^3 = 5.631 \times 10^{-23} \text{ cm}^3$$

unit cell mass = $(5.631 \times 10^{-23} \text{ cm}^3)(22.67 \text{ g/cm}^3) = 1.277 \times 10^{-21}$ g

$$\text{atom mass} = \frac{1.227 \times 10^{-21} \text{ g}}{4 \text{ atoms}} = 3.192 \times 10^{-22} \text{ g/atom}$$

atomic mass = $(3.192 \times 10^{-22} \text{ g/atom})(6.022 \times 10^{23} \text{ atoms/mol}) = 192.2$ g/mol

The metal is Ir.

11.13 (a) 1/8 S^{2-} at 8 corners = 1 S^{2-}; 1/2 S^{2-} at 6 faces = 3 S^{2-}; 4 Zn^{2+} inside
(b) The formula for zinc sulfide is ZnS.
(b) The oxidation state of Zn is 2+.
(c) The geometry around each zinc is tetrahedral.

11.14 (a) 1/8 Ca^{2+} at 8 corners = 1 Ca^{2+}; 1/2 O^{2-} at 6 faces = 3 O^{2-}; 1 Ti^{4+} inside
The formula for perovskite is $CaTiO_3$.
(b) The oxidation number of Ti is +4 to maintain charge neutrality in the unit cell.
(c) The geometry around titanium, oxygen, and calcium are all octahedral.

11.15 (a) solid → liquid
(b) liquid → gas
(c) gas → liquid → gas

11.16 (a)

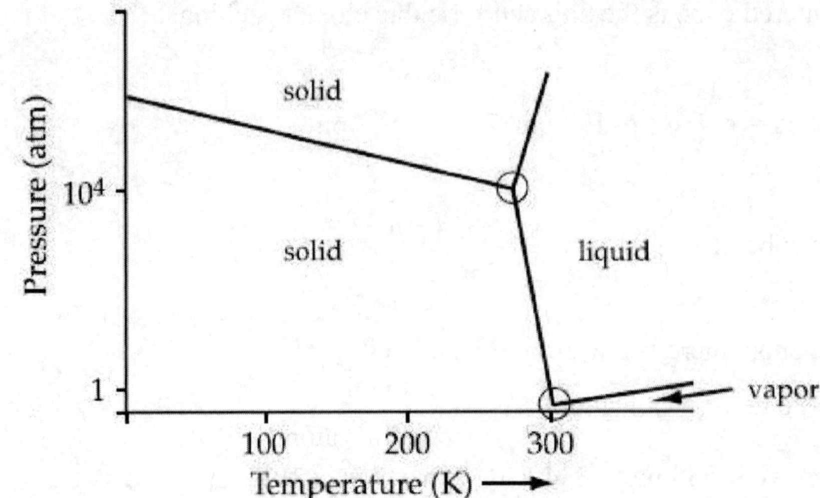

(b) Gallium has two triple points. The one below 1 atm is a solid, liquid, vapor triple point. The one at 10^4 atm is a solid(1), solid(2), liquid triple point.
(c) Increasing the pressure favors the liquid phase, giving the solid/liquid boundary a negative slope. At 1 atm pressure the liquid phase is more dense than the solid phase.

11.17 75 °F = 24 °C; At 70 atm and 24 °C, CO_2 is a liquid.

11.18 (a) $CO_2(s) \rightarrow CO_2(g)$
(b) $CO_2(l) \rightarrow CO_2(g)$
(c) $CO_2(g) \rightarrow CO_2(l) \rightarrow$ supercritical CO_2

11.19 (a) 5.11 atm (b) −56.4 °C (c) 31.1 °C

11.20 (a) For the phase transition, $CO_2(s) \rightarrow CO_2(g)$, the system is becoming more disordered, therefore $\Delta S > 0$.
(b) −78.5 °C
(c) $\Delta G = \Delta H - T\Delta S$; $\Delta G = 0$ at the sublimation temperature (−78.5 °C = 194.6 K). Set $\Delta G = 0$ and solve for ΔS.

$\Delta G = 0 = \Delta H - T\Delta S$

$$\Delta S = \frac{\Delta H}{T} = \frac{26.1 \text{ kJ/mol}}{194.6 \text{ K}} = 0.1341 \text{ kJ/(K·mol)} = 134.1 \text{ J/(K·mol)}$$

11.21 CO_2, 44.01

$$100.0 \text{ g} \times \frac{1 \text{ mol } CO_2}{44.01 \text{ g } CO_2} = 2.272 \text{ mol } CO_2$$

$q_1 = (2.272 \text{ mol})(26.1 \text{ kJ/mol}) = 59.3 \text{ kJ}$

$q_2 = (2.272 \text{ mol})[35.0 \times 10^{-3} \text{ kJ/(mol} \cdot {}^\circ\text{C)}][33\ {}^\circ\text{C} - (-78.5\ {}^\circ\text{C})] = 8.86 \text{ kJ}$

$q_{total} = q_1 + q_2 = 68.2 \text{ kJ}$

Conceptual Problems

11.22 After the volume is decreased, equilibrium is reestablished and the equilibrium vapor pressure will still be 28.0 mm Hg. Vapor pressure depends only on temperature, not on volume.

11.23 (a) cubic closest-packed (b) simple cubic
(c) hexagonal closest-packed (d) body-centered cubic

11.24 (a) cubic closest-packed
(b) 1/8 S^{2-} at 8 corners and 1/2 S^{2-} at 6 faces = 4 S^{2-}; 4 Zn^{2+} inside

11.25 (a) 1/8 Ti^{4+} at 8 corners = 1 Ti^{4+}; 1 Ti^{4+} inside; 1/2 O^{2-} at 2 faces (2/face) = 2 O^{2-}; 2 O^{2-} inside.
(b) TiO_2
(c) The oxidation state of titanium is +4.

11.26 (a) normal boiling point ≈ 300 K; normal melting point ≈ 180 K
(b) (i) solid (ii) gas (iii) supercritical fluid

11.27 (a), (c), (d)

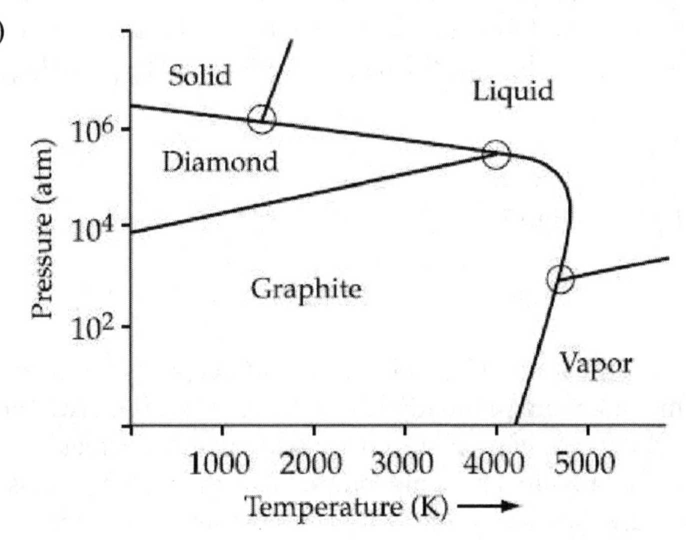

(b) There are three triple points.
(e) The solid phase that is stable at the higher pressure is more dense. The more dense phase is diamond.

Section Problems
Properties of Liquids (Section 11.1)

11.28 It's water's surface tension that keeps the needle on top.

11.29 The different liquid behavior is due to a difference in viscosity.

11.30 (a) CH_2Br_2 has the higher surface tension because it is polar while CCl_4 is nonpolar.
(b) Ethylene glycol has the higher surface tension because it can hydrogen bond from both ends of the molecule while ethanol can only hydrogen bond from one end of the molecule.

11.31 (a) 1-Hexanol has the higher viscosity because it can hydrogen bond while hexane cannot.
(b) Pentane has the higher viscosity because the linear pentane has larger dispersion forces than the branched neopentane.

11.32 Oleic acid has a higher viscosity than H_2O because it is a much larger molecule than H_2O with larger dispersion forces. It can also hydrogen bond.

11.33 H_2O has the higher viscosity because it can hydrogen bond while $(CH_3)_2S$ cannot.

Vapor Pressure and Phase Changes (Sections 11.2 and 11.3)

11.34 ΔH_{vap} is usually larger than ΔH_{fusion} because ΔH_{vap} is the heat required to overcome all intermolecular forces.

11.35 Sublimation is the direct conversion of a solid to a gas. A solid can also be converted to a gas in two steps; melting followed by vaporization. The energy to convert a solid to a gas must be the same regardless of the path. Therefore $\Delta H_{subl} = \Delta H_{fusion} + \Delta H_{vap}$.

11.36 (a) Hg(l) $\rightarrow$ Hg(g)
(b) no change of state, Hg remains a liquid
(c) Hg(g) $\rightarrow$ Hg(l) $\rightarrow$ Hg(s)

11.37 (a) solid I_2 melts to form liquid I_2 (b) no change of state, I_2 remains a liquid

11.38 As the pressure over the liquid H_2O is lowered, H_2O vapor is removed by the pump. As H_2O vapor is removed, more of the liquid H_2O is converted to H_2O vapor. This conversion is an endothermic process and the temperature decreases. The combination of both a decrease in pressure and temperature takes the system across the liquid/solid boundary in the phase diagram so the H_2O that remains turns to ice.

11.39 The normal boiling point for ether is relatively low (34.6 °C). As the pressure is reduced by the pump, the relatively high vapor pressure of the ether equals the external pressure produced by the pump, and the liquid boils.

11.40 H_2O, 18.02; $5.00 \text{ g } H_2O \times \dfrac{1 \text{ mol } H_2O}{18.02 \text{ g } H_2O} = 0.2775 \text{ mol } H_2O$

$q_1 = (0.2775 \text{ mol})[36.6 \times 10^{-3} \text{ kJ/(K} \cdot \text{mol)}](273 \text{ K} - 263 \text{ K}) = 0.1016 \text{ kJ}$
$q_2 = (0.2775 \text{ mol})(6.01 \text{ kJ/mol}) = 1.668 \text{ kJ}$
$q_3 = (0.2775 \text{ mol})(75.3 \times 10^{-3} \text{ kJ/(K} \cdot \text{mol)}](303 \text{ K} - 273 \text{ K}) = 0.6269 \text{ kJ}$
$q_{total} = q_1 + q_2 + q_3 = 2.40 \text{ kJ}$; 2.40 kJ of heat is required.

11.41 H_2O, 18.02; $15.3 \text{ g } H_2O \times \dfrac{1 \text{ mol } H_2O}{18.02 \text{ g } H_2O} = 0.8491 \text{ mol } H_2O$

$q_1 = (0.8491 \text{ mol})[33.6 \times 10^{-3} \text{ kJ/(K} \cdot \text{mol)}](373 \text{ K} - 388 \text{ K}) = -0.4279 \text{ kJ}$
$q_2 = -(0.8491 \text{ mol})(40.67 \text{ kJ/mol}) = -34.53 \text{ kJ}$
$q_3 = (0.8491 \text{ mol})[75.3 \times 10^{-3} \text{ kJ/(K} \cdot \text{mol)}](348 \text{ K} - 373 \text{ K}) = -1.598 \text{ kJ}$
$q_{total} = q_1 + q_2 + q_3 = -36.6 \text{ kJ}$
36.6 kJ of heat is released.

11.42 H_2O, 18.02; $7.55 \text{ g } H_2O \times \dfrac{1 \text{ mol } H_2O}{18.02 \text{ g } H_2O} = 0.4190 \text{ mol } H_2O$

$q_1 = (0.4190 \text{ mol})[75.3 \times 10^{-3} \text{ kJ/(K} \cdot \text{mol)}](273.15 \text{ K} - 306.65 \text{ K}) = -1.057 \text{ kJ}$
$q_2 = -(0.4190 \text{ mol})(6.01 \text{ kJ/mol}) = -2.518 \text{ kJ}$
$q_3 = (0.4190 \text{ mol})[36.6 \times 10^{-3} \text{ kJ/(K} \cdot \text{mol)}](263.15 \text{ K} - 273.15 \text{ K}) = -0.1534 \text{ kJ}$
$q_{total} = q_1 + q_2 + q_3 = -3.73 \text{ kJ}$
3.73 kJ of heat is released.

11.43 C_2H_5OH, 46.07; $25.0 \text{ g } C_2H_5OH \times \dfrac{1 \text{ mol } C_2H_5OH}{46.07 \text{ g } C_2H_5OH} = 0.543 \text{ mol } C_2H_5OH$

$q_1 = (0.543 \text{ mol})[65.6 \times 10^{-3} \text{ kJ/(K} \cdot \text{mol)}](351.45 \text{ K} - 366.15 \text{ K}) = -0.524 \text{ kJ}$
$q_2 = -(0.543 \text{ mol})(38.56 \text{ kJ/mol}) = -20.94 \text{ kJ}$
$q_3 = (0.543 \text{ mol})[112.3 \times 10^{-3} \text{ kJ/(K} \cdot \text{mol)}](263.15 \text{ K} - 351.45 \text{ K}) = -5.38 \text{ kJ}$
$q_{total} = q_1 + q_2 + q_3 = -26.8 \text{ kJ}$
26.8 kJ of heat is released.

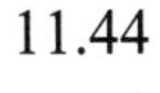

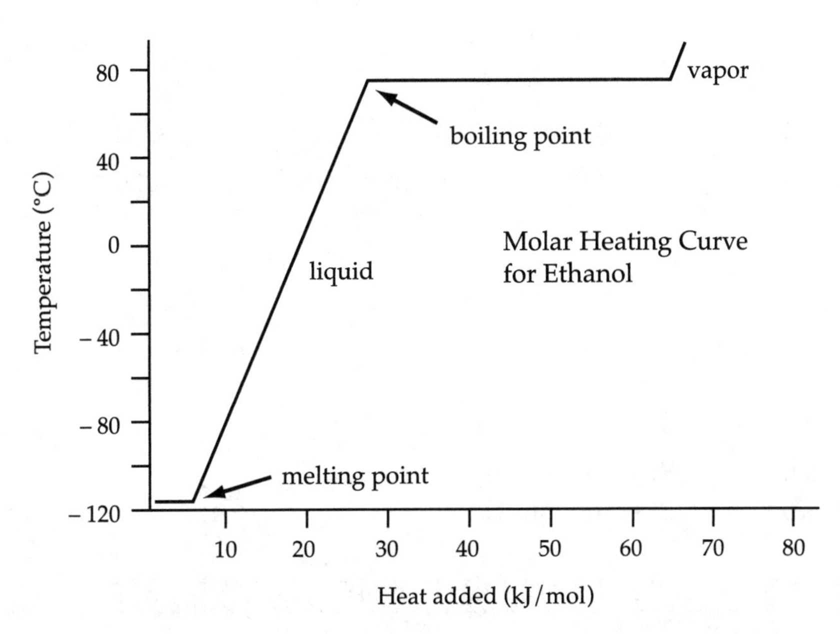

11.45

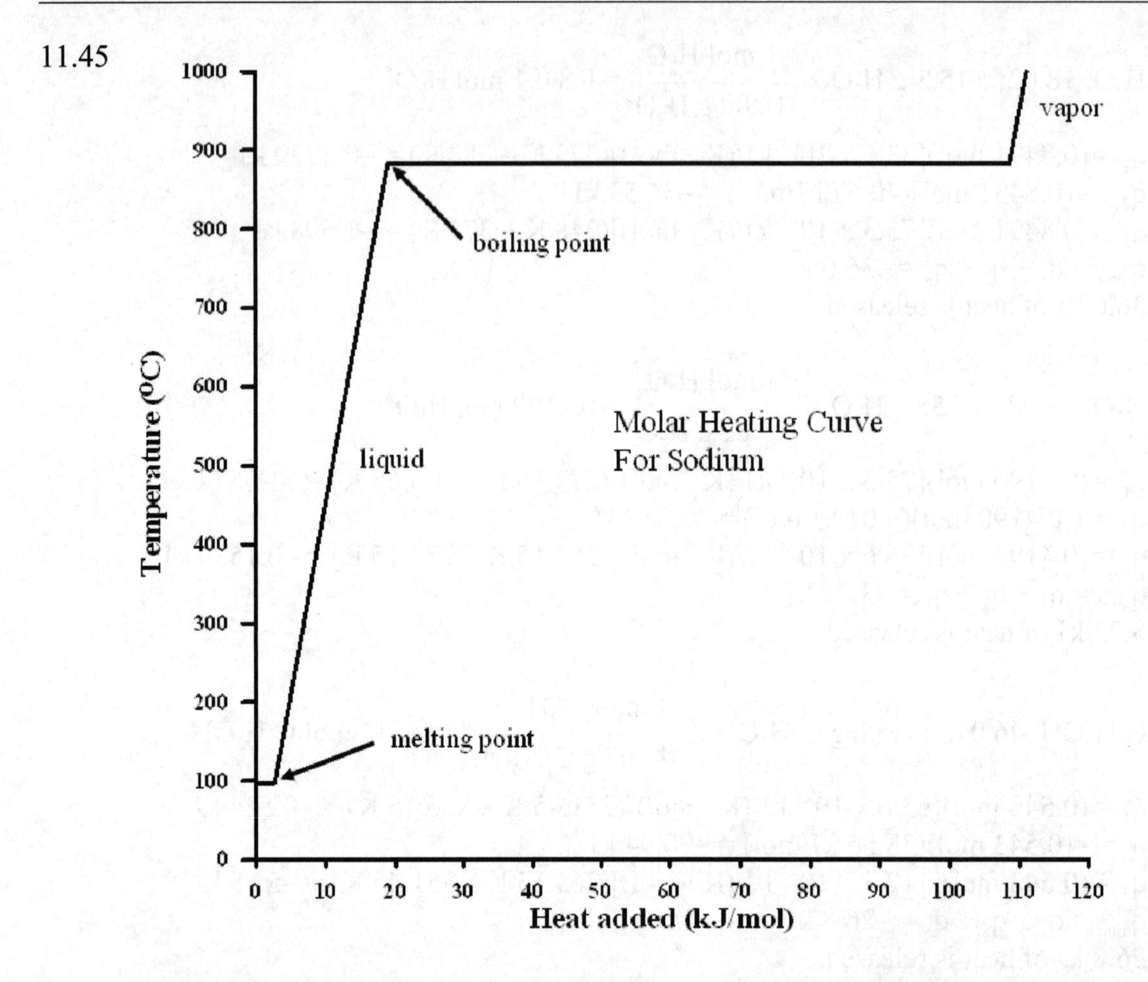

11.46 boiling point = 218 °C = 491 K

$\Delta G = \Delta H_{vap} - T\Delta S_{vap}$; At the boiling point (phase change), $\Delta G = 0$

$$\Delta H_{vap} = T\Delta S_{vap}; \quad \Delta S_{vap} = \frac{\Delta H_{vap}}{T} = \frac{43.3 \text{ kJ/mol}}{491 \text{ K}} = 0.0882 \text{ kJ/(K}\cdot\text{ mol)} = 88.2 \text{ J/(K}\cdot\text{ mol)}$$

11.47 $$\Delta S_{fus} = \frac{\Delta H_{fus}}{T} = \frac{2.64 \text{ kJ/mol}}{371 \text{ K}} = 0.007\ 12 \text{ kJ/(K}\cdot\text{ mol)} = 7.12 \text{ J/(K}\cdot\text{ mol)}$$

11.48 $$\Delta H_{vap} = \frac{(\ln P_2 - \ln P_1)(R)}{\left(\frac{1}{T_1} - \frac{1}{T_2}\right)}$$

$T_1 = -5.1$ °C = 268.0 K; $P_1 = 100$ mm Hg

$T_2 = 46.5$ °C = 319.6 K; $P_2 = 760$ mm Hg

$$\Delta H_{vap} = \frac{[\ln(760) - \ln(100)][8.3145 \times 10^{-3} \text{ kJ/(K}\cdot\text{ mol)}]}{\left(\frac{1}{268.0 \text{ K}} - \frac{1}{319.6 \text{ K}}\right)} = 28.0 \text{ kJ/mol}$$

11.49 $\Delta H_{vap} = \dfrac{(\ln P_2 - \ln P_1)(R)}{\left(\dfrac{1}{T_1} - \dfrac{1}{T_2}\right)}$

$P_1 = 100$ mm Hg; $T_1 = 5.4\ ^\circ C = 278.6$ K
$P_2 = 760$ mm Hg; $T_2 = 57.7\ ^\circ C = 330.8$ K

$$\Delta H_{vap} = \frac{[\ln(760) - \ln(100)][8.3145 \times 10^{-3}\ kJ/(K \cdot mol)]}{\left(\dfrac{1}{278.6\ K} - \dfrac{1}{330.8\ K}\right)} = 29.8\ kJ/mol$$

11.50 $\ln P_2 = \ln P_1 + \dfrac{\Delta H_{vap}}{R}\left(\dfrac{1}{T_1} - \dfrac{1}{T_2}\right)$

$\Delta H_{vap} = 28.0$ kJ/mol
$P_1 = 100$ mm Hg; $T_1 = -5.1\ ^\circ C = 268.0$ K; $T_2 = 20.0\ ^\circ C = 293.2$ K
Solve for P_2.

$$\ln P_2 = \ln(100) + \frac{28.0\ kJ/mol}{[8.3145 \times 10^{-3}\ kJ/(K \cdot mol)]}\left(\frac{1}{268.0\ K} - \frac{1}{293.2\ K}\right)$$

$\ln P_2 = 5.6852$; $P_2 = e^{5.6852} = 294.5$ mm Hg $= 294$ mm Hg

11.51 $\ln P_2 = \ln P_1 + \dfrac{\Delta H_{vap}}{R}\left(\dfrac{1}{T_1} - \dfrac{1}{T_2}\right)$

$\Delta H_{vap} = 29.8$ kJ/mol
$P_1 = 100$ mm Hg; $T_1 = 5.4\ ^\circ C = 278.6$ K; $T_2 = 30.0\ ^\circ C = 303.2$ K
Solve for P_2.

$$\ln P_2 = \ln(100) + \frac{29.8\ kJ/mol}{[8.3145 \times 10^{-3}\ kJ/(K \cdot mol)]}\left(\frac{1}{278.6\ K} - \frac{1}{303.2\ K}\right)$$

$\ln P_2 = 5.6489$; $P_2 = e^{5.6489} = 284$ mm Hg

11.52

T(K)	P_{vap}(mm Hg)	$\ln P_{vap}$	1/T
263	80.1	4.383	0.003 802
273	133.6	4.8949	0.003 663
283	213.3	5.3627	0.003 534
293	329.6	5.7979	0.003 413
303	495.4	6.2054	0.003 300
313	724.4	6.5853	0.003 195

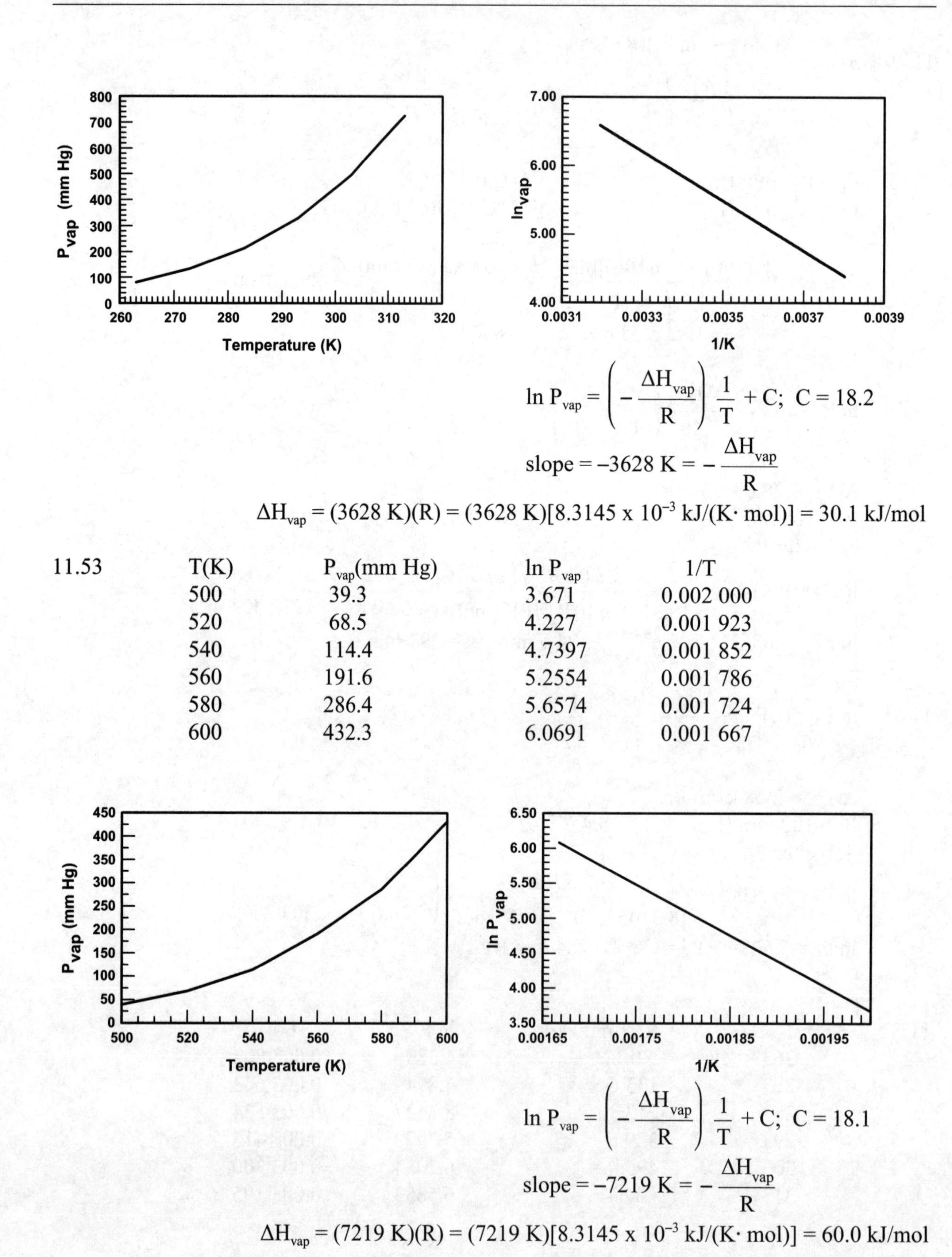

$$\ln P_{vap} = \left(-\frac{\Delta H_{vap}}{R}\right)\frac{1}{T} + C;\ \ C = 18.2$$

$$\text{slope} = -3628\ \text{K} = -\frac{\Delta H_{vap}}{R}$$

$$\Delta H_{vap} = (3628\ \text{K})(R) = (3628\ \text{K})[8.3145 \times 10^{-3}\ \text{kJ/(K}\cdot\text{mol)}] = 30.1\ \text{kJ/mol}$$

11.53

T(K)	P_{vap}(mm Hg)	$\ln P_{vap}$	1/T
500	39.3	3.671	0.002 000
520	68.5	4.227	0.001 923
540	114.4	4.7397	0.001 852
560	191.6	5.2554	0.001 786
580	286.4	5.6574	0.001 724
600	432.3	6.0691	0.001 667

$$\ln P_{vap} = \left(-\frac{\Delta H_{vap}}{R}\right)\frac{1}{T} + C;\ \ C = 18.1$$

$$\text{slope} = -7219\ \text{K} = -\frac{\Delta H_{vap}}{R}$$

$$\Delta H_{vap} = (7219\ \text{K})(R) = (7219\ \text{K})[8.3145 \times 10^{-3}\ \text{kJ/(K}\cdot\text{mol)}] = 60.0\ \text{kJ/mol}$$

11.54 $\Delta H_{vap} = 30.1$ kJ/mol

11.55 $\Delta H_{vap} = 60.0$ kJ/mol

11.56 $$\Delta H_{vap} = \frac{(\ln P_2 - \ln P_1)(R)}{\left(\frac{1}{T_1} - \frac{1}{T_2}\right)}$$

$P_1 = 80.1$ mm Hg; $T_1 = 263$ K
$P_2 = 724.4$ mm Hg; $T_2 = 313$ K

$$\Delta H_{vap} = \frac{[\ln(724.4) - \ln(80.1)][8.3145 \times 10^{-3}\ \text{kJ/(K}\cdot\text{mol)}]}{\left(\frac{1}{263\ \text{K}} - \frac{1}{313\ \text{K}}\right)} = 30.1\ \text{kJ/mol}$$

The calculated ΔH_{vap} and that obtained from the plot in Problem 11.54 are the same.

11.57 $$\Delta H_{vap} = \frac{(\ln P_2 - \ln P_1)(R)}{\left(\frac{1}{T_1} - \frac{1}{T_2}\right)}$$

$P_1 = 39.3$ mm Hg; $T_1 = 500$ K
$P_2 = 432.3$ mm Hg; $T_2 = 600$ K

$$\Delta H_{vap} = \frac{[\ln(432.3) - \ln(39.3)][8.3145 \times 10^{-3}\ \text{kJ/(K}\cdot\text{mol)}]}{\left(\frac{1}{500\ \text{K}} - \frac{1}{600\ \text{K}}\right)} = 59.8\ \text{kJ/mol}$$

The calculated ΔH_{vap} and that obtained from the plot in Problem 11.55 are consistent with each other. The value from the slope is 60.0 kJ/mol.

Structures of Solids (Sections 11.4–11.8)

11.58 molecular solid, CO_2, I_2; metallic solid, any metallic element; covalent network solid, diamond; ionic solid, NaCl

11.59 molecular solid, covalent molecules; metallic solid, metal atoms; covalent network solid, nonmetal atoms; ionic solid, cations and anions

11.60 (a) rubber (b) Na_3PO_4 (c) CBr_4 (d) quartz (e) Au

11.61 (a) glass (b) KCl (c) Cl_2 (d) diamond (e) Hg

11.62 Silicon carbide is a covalent network solid.

11.63 Arsenic tribromide is a molecular solid.

11.64 $d = \dfrac{n\lambda}{2\sin\theta} = \dfrac{(1)(154.2\text{ pm})}{2\sin(22.5^\circ)} = 201\text{ pm}$

11.65 $d = \dfrac{n\lambda}{2\sin\theta} = \dfrac{(1)(154.2\text{ pm})}{2\sin(76.84^\circ)} = 79.2\text{ pm}$

11.66 From Table 11.5.
Hexagonal and cubic closest packing are the most efficient because 74% of the available space is used.
Simple cubic packing is the least efficient because only 52% of the available space is used.

11.67 (a) A unit cell is the smallest repeating unit that makes up a crystal.
(b) A primitive cubic unit cell has 1 atom. A body-centered cubic unit cell has 2 atoms. A face-centered cubic unit cell has 4 atoms.

11.68 Cu is face-centered cubic. d = 362 pm

$r = \sqrt{\dfrac{d^2}{8}} = \sqrt{\dfrac{(362\text{ pm})^2}{8}} = 128\text{ pm}$

$362\text{ pm} = 362 \times 10^{-12}\text{ m} = 3.62 \times 10^{-8}\text{ cm}$

unit cell volume = $(3.62 \times 10^{-8}\text{ cm})^3 = 4.74 \times 10^{-23}\text{ cm}^3$

mass of one Cu atom = $63.55\text{ g/mol} \times \dfrac{1\text{ mol}}{6.022 \times 10^{23}\text{ atom}} = 1.055 \times 10^{-22}\text{ g/atom}$

Cu is face-centered cubic; there are, therefore, four Cu atoms in the unit cell.

unit cell mass = $(4\text{ atoms})(1.055 \times 10^{-22}\text{ g/atom}) = 4.22 \times 10^{-22}\text{ g}$

density = $\dfrac{\text{mass}}{\text{volume}} = \dfrac{4.22 \times 10^{-22}\text{g}}{4.74 \times 10^{-23}\text{ cm}^3} = 8.90\text{ g/cm}^3$

11.69 Pb is face-centered cubic. d = 495 pm = 4.95×10^{-8} cm

$r = \sqrt{\dfrac{d^2}{8}} = \sqrt{\dfrac{(495\text{ pm})^2}{8}} = 175\text{ pm}$

unit cell volume = $(4.95 \times 10^{-8}\text{ cm})^3 = 1.2129 \times 10^{-22}\text{ cm}^3$

mass of one Pb atom = $207.2\text{ g/mol} \times \dfrac{1\text{ mol}}{6.022 \times 10^{23}\text{ atoms}} = 3.4407 \times 10^{-22}\text{ g/atom}$

Pb is face-centered cubic; there are, therefore, four Pb atoms in the unit cell.

density = $\dfrac{\text{mass}}{\text{volume}} = \dfrac{4(3.4407 \times 10^{-22}\text{ g})}{1.2129 \times 10^{-22}\text{ cm}^3} = 11.3\text{ g/cm}^3$

11.70 mass of one Al atom = $26.98 \text{ g/mol} \times \dfrac{1 \text{ mol}}{6.022 \times 10^{23} \text{ atom}} = 4.480 \times 10^{-23}$ g/atom

Al is face-centered cubic; there are, therefore, four Al atoms in the unit cell.
unit cell mass = (4 atoms)(4.480×10^{-23} g/atom) = 1.792×10^{-22} g

$$\text{density} = \frac{\text{mass}}{\text{volume}}$$

$$\text{unit cell volume} = \frac{\text{unit cell mass}}{\text{density}} = \frac{1.792 \times 10^{-22} \text{ g}}{2.699 \text{ g/cm}^3} = 6.640 \times 10^{-23} \text{ cm}^3$$

$$\text{unit cell edge} = d = \sqrt[3]{6.640 \times 10^{-23} \text{ cm}^3} = 4.049 \times 10^{-8} \text{ cm}$$

$$d = 4.049 \times 10^{-8} \text{ cm} \times \frac{1 \text{m}}{100 \text{ cm}} = 4.049 \times 10^{-10} \text{ m} = 404.9 \times 10^{-12} \text{ m} = 404.9 \text{ pm}$$

11.71 W is body-centered cubic. d = 317 pm
a = edge = d; b = face diagonal; c = body diagonal
$b^2 = 2a^2$
$c^2 = a^2 + b^2$
$c^2 = a^2 + 2a^2 = 3a^2$
$c = \sqrt{3}\, a$
unit cell body diagonal = $\sqrt{3}\, d = \sqrt{3}\,(317 \text{ pm}) = 549$ pm

11.72 unit cell body diagonal = 4r = 549 pm
For W, $r = \dfrac{549 \text{ pm}}{4} = 137$ pm

11.73 mass of one Na atom = $23.0 \text{ g/mol} \times \dfrac{1 \text{ mol}}{6.022 \times 10^{23} \text{ atoms}} = 3.82 \times 10^{-23}$ g/atom

Because Na is body-centered cubic; there are two Na atoms in the unit cell.
unit cell mass = 2(3.82×10^{-23} g) = 7.64×10^{-23} g

$$\text{unit cell volume} = \frac{\text{unit cell mass}}{\text{density}} = \frac{7.64 \times 10^{-23} \text{ g}}{0.971 \text{ g/cm}^3} = 7.87 \times 10^{-23} \text{ cm}^3$$

$$\text{unit cell edge} = d = \sqrt[3]{7.87 \times 10^{-23} \text{ cm}^3} = 4.29 \times 10^{-8} \text{ cm} = 429 \text{ pm}$$

$$4r = \sqrt{3}\, d;\ r = \frac{\sqrt{3}\, d}{4} = \frac{\sqrt{3}\,(429 \text{ pm})}{4} = 186 \text{ pm}$$

11.74 mass of one Ti atom = $47.88 \text{ g/mol} \times \dfrac{1 \text{ mol}}{6.022 \times 10^{23} \text{ atoms}} = 7.951 \times 10^{-23}$ g/atom

r = 144.8 pm = 144.8×10^{-12} m

$$r = 144.8 \times 10^{-12} \text{ m} \times \frac{100 \text{ cm}}{1 \text{ m}} = 1.448 \times 10^{-8} \text{ cm}$$

Calculate the volume and then the density for Ti assuming it is primitive cubic, body-centered cubic, and face-centered cubic. Compare the calculated density with the actual density to identify the unit cell.

For primitive cubic:

$d = 2r$; volume $= d^3 = [2(1.448 \times 10^{-8}\ cm)]^3 = 2.429 \times 10^{-23}\ cm^3$

$$density = \frac{unit\ cell\ mass}{volume} = \frac{7.951 \times 10^{-23}\ g}{2.429 \times 10^{-23}\ cm^3} = 3.273\ g/cm^3$$

For face-centered cubic:

$d = 2\sqrt{2}r$; volume $= d^3 = [2\sqrt{2}(1.448 \times 10^{-8}\ cm)]^3 = 6.870 \times 10^{-23}\ cm^3$

$$density = \frac{4(7.951 \times 10^{-23}\ g)}{6.870 \times 10^{-23}\ cm^3} = 4.630\ g/cm^3$$

For body-centered cubic:

$$d = \frac{4r}{\sqrt{3}};\ volume = d^3 = \left[\frac{4(1.448 \times 10^{-8}\ cm)}{\sqrt{3}}\right]^3 = 3.739 \times 10^{-23}\ cm^3$$

$$density = \frac{2(7.951 \times 10^{-23}\ g)}{3.739 \times 10^{-23}\ cm^3} = 4.253\ g/cm^3$$

The calculated density for a face-centered cube (4.630 g/cm^3) is closest to the actual density of 4.54 g/cm^3. Ti crystallizes in the face-centered cubic unit cell.

11.75 mass of one Ca = $40.08\ g/mol \times \dfrac{1\ mol}{6.022 \times 10^{23}\ atom} = 6.656 \times 10^{-23}\ g/atom$

unit cell edge = d = 558.2 pm = 5.582×10^{-8} cm

unit cell volume = $d^3 = (5.582 \times 10^{-8}\ cm)^3 = 1.739 \times 10^{-22}\ cm^3$

unit cell mass = $(1.739 \times 10^{-22}\ cm^3)(1.55\ g/cm^3) = 2.695 \times 10^{-22}$ g

(a) number of Ca atoms in unit cell $= \dfrac{unit\ cell\ mass}{mass\ of\ one\ Ca\ atom}$

$$= \frac{2.695 \times 10^{-22}\ g}{6.656 \times 10^{-23}\ g/atom} = 4.05 = 4\ Ca\ atoms$$

(b) Because the unit cell contains 4 Ca atoms, the unit cell is face-centered cubic.

11.76 Six Na^+ ions touch each H^- ion and six H^- ions touch each Na^+ ion.

11.77 For CsCl: (1/8 x 8 corners), so 1 Cl^- and 1 minus per unit cell
1 Cs^+ inside, so 1 plus per unit cell

11.78 Na^+ H^- Na^+
← 488 pm → unit cell edge = d = 488 pm; Na–H bond = d/2 = 244 pm

11.79 body diagonal = $\sqrt{3}\ d = \sqrt{3}\ (412.3\ pm) = 714.12$ pm

Cs–Cl bond = body diagonal/2 = (714.12 pm)/2 = 357.1 pm

Cs–Cl bond length = $r_{Cs^+} + r_{Cl^-}$

$357.1 \text{ pm} = r_{Cs^+} + r_{Cl^-}$

$357.1 \text{ pm} = r_{Cs^+} + 181 \text{ pm}$

$r_{Cs^+} = 357.1 \text{ pm} - 181 \text{ pm} = 176 \text{ pm}$

11.80 mass of one Pb atom = $207.2 \text{ g/mol} \times \dfrac{1 \text{ mol}}{6.022 \times 10^{23} \text{ atom}} = 3.441 \times 10^{-22}$ g/atom

If the unit cell for Pb is the primitive cube it would contain one Pb atom and weigh 3.441×10^{-22} g.

If the unit cell for Pb is the face-centered cube it would contain four Pb atom and weigh $(4)(3.441 \times 10^{-22} \text{ g}) = 1.376 \times 10^{-21}$ g.

For a primitive cubic unit cell:
unit cell edge = $d = 2r = 2(175 \text{ pm}) = 3.50 \times 10^{-8}$ cm
unit cell volume = $d^3 = (3.50 \times 10^{-8} \text{ cm})^3 = 4.29 \times 10^{-23} \text{ cm}^3$
unit cell mass = $(4.29 \times 10^{-23} \text{ cm}^3)(11.34 \text{ g/cm}^3) = 4.86 \times 10^{-22}$ g

For a face-centered cubic unit cell:
unit cell edge = $d = 2\sqrt{2}\,r = 2\sqrt{2}\,(175 \text{ pm}) = 4.95 \times 10^{-8}$ cm
unit cell volume = $d^3 = (4.95 \times 10^{-8} \text{ cm})^3 = 1.21 \times 10^{-22} \text{ cm}^3$
unit cell mass = $(1.21 \times 10^{-22} \text{ cm}^3)(11.34 \text{ g/cm}^3) = 1.38 \times 10^{-21}$ g

The masses agree for the face-centered cube, therefore it is the unit cell for Pb.

11.81 The unit cell is a face-centered cube. There are four atoms in the unit cell.

unit cell volume = $\left(350.7 \times 10^{-12} \text{ m} \times \dfrac{1 \text{ cm}}{1 \times 10^{-2} \text{ m}}\right)^3 = 4.313 \times 10^{-23} \text{ cm}^3$

unit cell mass = $(4.313 \times 10^{-23} \text{ cm}^3)(6.84 \text{ g/cm}^3) = 2.95 \times 10^{-22}$ g

atom mass = $\dfrac{2.95 \times 10^{-22} \text{ g}}{4 \text{ atoms}} = 7.37 \times 10^{-23}$ g/atom

atomic mass = $(7.37 \times 10^{-23} \text{ g/atom})(6.022 \times 10^{23} \text{ atoms/mol}) = 44.4$ g/mol

atomic radius = $\sqrt{\dfrac{d^2}{8}} = \sqrt{\dfrac{(350.7 \text{ pm})^2}{8}} = 124.0$ pm

The metal is Sc.

Phase Diagrams (Section 11.9)

11.82 (a) gas (b) liquid (c) solid

11.83 (a) $H_2O(l) \rightarrow H_2O(s)$
(b) 380 °C is above the critical temperature; therefore, the water cannot be liquefied. At the higher pressure, it will behave as a supercritical fluid.

11.84

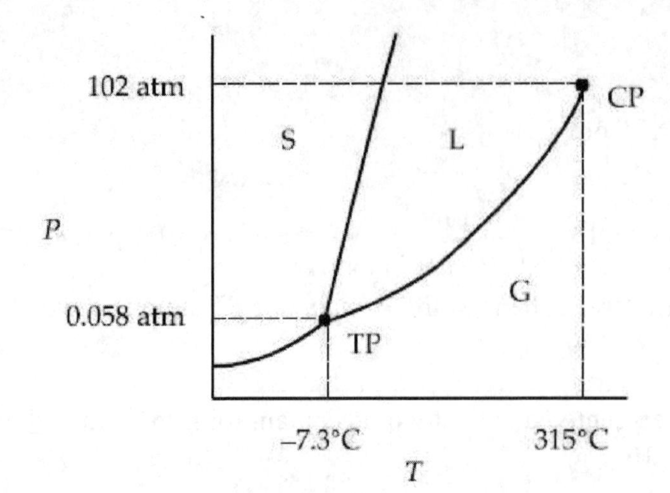

11.85

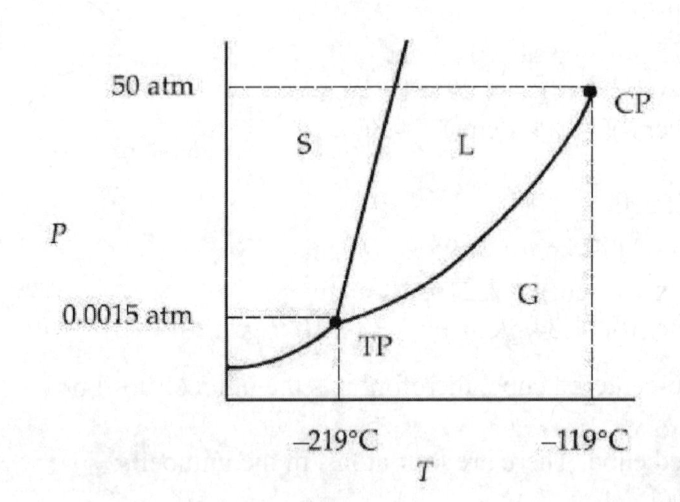

11.86 (a) Br_2(s) (b) Br_2(l)

11.87 (a) O_2(l) (b) O_2 - supercritical fluid

11.88 Solid O_2 does not melt when pressure is applied because the solid is denser than the liquid, and the solid/liquid boundary in the phase diagram slopes to the right.

11.89 Ammonia can be liquefied at 25 °C because this temperature is below T_c (132.5 °C). Methane cannot be liquefied at 25 °C because this temperature is above T_c (–82.1 °C). Sulfur dioxide can be liquefied at 25 °C because this temperature is below T_c (157.8 °C).

11.90

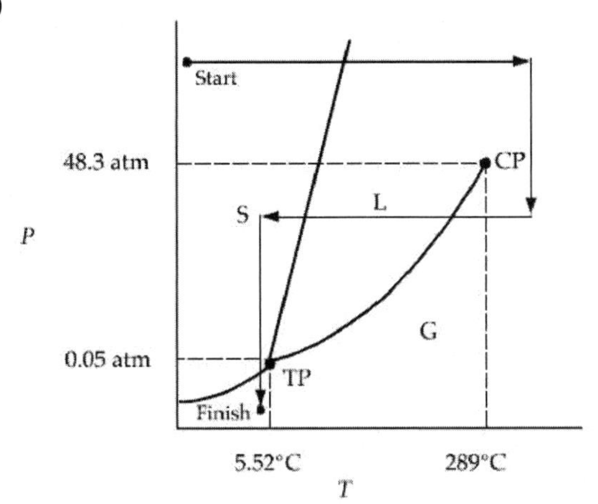

The starting phase is benzene as a solid, and the final phase is benzene as a gas.

11.91

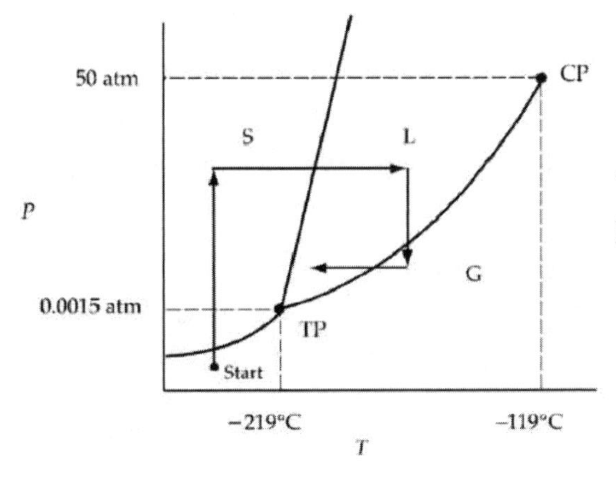

The starting phase is a gas, and the final phase is a liquid.

11.92 solid → liquid → supercritical fluid → liquid → solid → gas

11.93 gas → solid → liquid → gas → liquid

Chapter Problems

11.94 Because Ar crystallizes in a face-centered cubic unit cell, there are four Ar atoms in the unit cell.

$$\text{mass of one Ar atom} = 39.95 \text{ g/mol} \times \frac{1 \text{ mol}}{6.022 \times 10^{23} \text{ atom}} = 6.634 \times 10^{-23} \text{ g/atom}$$

unit cell mass = 4 atoms x mass of one Ar atom

$$= 4 \text{ atoms} \times 6.634 \times 10^{-23} \text{ g/atom} = 2.654 \times 10^{-22} \text{ g}$$

$$\text{density} = \frac{\text{mass}}{\text{volume}}$$

$$\text{unit cell volume} = \frac{\text{unit cell mass}}{\text{density}} = \frac{2.654 \times 10^{-22} \text{ g}}{1.623 \text{ g/cm}^3} = 1.635 \times 10^{-22} \text{ cm}^3$$

$$\text{unit cell edge} = d = \sqrt[3]{1.635 \times 10^{-22} \text{ cm}^3} = 5.468 \times 10^{-8} \text{ cm}$$

$$d = 5.468 \times 10^{-8}\ \text{cm} \times \frac{1 \times 10^{-2}\ \text{m}}{1\ \text{cm}} = 5.468 \times 10^{-10}\ \text{m} = 546.8 \times 10^{-12}\ \text{m} = 546.8\ \text{pm}$$

$$r = \sqrt{\frac{d^2}{8}} = \sqrt{\frac{(546.8\ \text{pm})^2}{8}} = 193.3\ \text{pm}$$

11.95 $7.50\ \text{g} \times \frac{1\ \text{mol}}{200.6\ \text{g}} = 0.037\ 39\ \text{mol Hg}$

$q_1 = (0.037\ 39\ \text{mol})[28.2 \times 10^{-3}\ \text{kJ/(K} \cdot \text{mol)}](234.3\ \text{K} - 223.1\ \text{K}) = 0.011\ 81\ \text{kJ}$

$q_2 = (0.037\ 39\ \text{mol})(2.33\ \text{kJ/mol}) = 0.087\ 12\ \text{kJ}$

$q_3 = (0.037\ 39\ \text{mol})[27.9 \times 10^{-3}\ \text{kJ/(K} \cdot \text{mol)}](323.1\ \text{K} - 234.3\ \text{K}) = 0.092\ 63\ \text{kJ}$

$q_{total} = q_1 + q_2 + q_3 = 0.192\ \text{kJ}$;   0.192 kJ of heat is required.

11.96

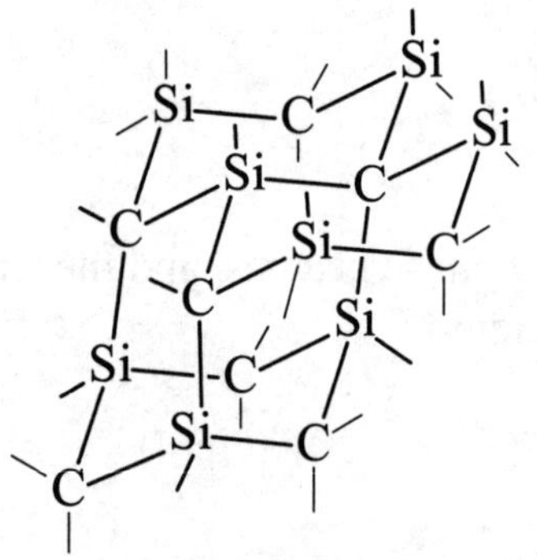

11.97 $\ln P_2 = \ln P_1 + \frac{\Delta H_{vap}}{R}\left(\frac{1}{T_1} - \frac{1}{T_2}\right)$

$\Delta H_{vap} = 40.67\ \text{kJ/mol}$

At 1 atm, H_2O boils at 100 °C; therefore set

$T_1 = 100\ °\text{C} = 373\ \text{K}$, and $P_1 = 1.00\ \text{atm}$.

Let $T_2 = 95\ °\text{C} = 368\ \text{K}$, and solve for P_2. (P_2 is the atmospheric pressure in Denver.)

$$\ln P_2 = \ln(1) + \frac{40.67\ \text{kJ/mol}}{[8.3145 \times 10^{-3}\ \text{kJ/(K} \cdot \text{mol)}]}\left(\frac{1}{373\ \text{K}} - \frac{1}{368\ \text{K}}\right)$$

$\ln P_2 = -0.1782$; $P_2 = e^{-0.1782} = 0.837\ \text{atm}$

11.98 Unit cell volume $= (7900 \times 10^{-12}\ \text{m})^2(3800 \times 10^{-12}\ \text{m})\left(\frac{100\ \text{cm}}{1\ \text{m}}\right)^3 = 2.37 \times 10^{-19}\ \text{cm}^3$

total lysozyme mass $= 8 \times 1.44 \times 10^4\ \text{u} \times \frac{1.660\ 54 \times 10^{-27}\ \text{kg}}{1\ \text{u}} \times \frac{1000\ \text{g}}{1\ \text{kg}} = 1.91 \times 10^{-19}\ \text{g}$

total lysozyme volume $= 1.91 \times 10^{-19}\ \text{g} \times \frac{1\ \text{cm}^3}{1.35\ \text{g}} = 1.42 \times 10^{-19}$ cm3

% occupied unit cell volume $= \frac{1.42 \times 10^{-19}\ \text{cm}^3}{2.37 \times 10^{-19}\ \text{cm}^3} \times 100\% = 60\%$

11.99 Unit cell volume = $(101.9 \times 10^{-27}\ m^3)\left(\dfrac{100\ cm}{1\ m}\right)^3 = 1.019 \times 10^{-19}\ cm^3$

total toxin mass = $16 \times 3336\ u \times \dfrac{1.660\ 54 \times 10^{-27}\ kg}{1\ u} \times \dfrac{1000\ g}{1\ kg} = 8.86 \times 10^{-20}\ g$

total toxin volume = $8.86 \times 10^{-20}\ g \times \dfrac{1\ cm^3}{1.35\ g} = 6.56 \times 10^{-20}\ cm^3$

% occupied unit cell volume = $\dfrac{6.56 \times 10^{-20}\ cm^3}{1.019 \times 10^{-19}\ cm^3} \times 100\% = 64.4\%$

11.100 $\Delta G = \Delta H - T\Delta S$; at the melting point (phase change), $\Delta G = 0$.

$\Delta H = T\Delta S$; $\quad T = \dfrac{\Delta H_{fus}}{\Delta S_{fus}} = \dfrac{9.037\ kJ/mol}{9.79 \times 10^{-3}\ kJ/(K \cdot mol)} = 923\ K = 650\ ^\circ C$

11.101 melting point = –23.2 °C = 250.0 K

$\Delta G = \Delta H_{fusion} - T\Delta S_{fusion}$

At the melting point (phase change), $\Delta G = 0$

$\Delta H_{fusion} = T\Delta S_{fusion}$

$\Delta S_{fusion} = \dfrac{\Delta H_{fusion}}{T} = \dfrac{9.37\ kJ/mol}{250.0\ K} = 0.0375\ kJ/(K \cdot mol) = 37.5\ J/(K \cdot mol)$

11.102 $\Delta H_{vap} = \dfrac{(\ln P_2 - \ln P_1)(R)}{\left(\dfrac{1}{T_1} - \dfrac{1}{T_2}\right)}$

$P_1 = 40.0$ mm Hg; $\quad T_1 = -81.6\ ^\circ C = 191.6\ K$

$P_2 = 400$ mm Hg; $\quad T_2 = -43.9\ ^\circ C = 229.2\ K$

$$\Delta H_{vap} = \frac{[\ln(400) - \ln(40.0)]\left(8.3145 \times 10^{-3} \frac{kJ}{K \cdot mol}\right)}{\left(\dfrac{1}{191.6\ K} - \dfrac{1}{229.2\ K}\right)} = 22.36\ kJ/mol$$

Using $\Delta H_{vap} = 22.36$ kJ/mol

$$\ln P_2 = \ln P_1 + \frac{\Delta H_{vap}}{R}\left(\frac{1}{T_1} - \frac{1}{T_2}\right)$$

$$(\ln P_2 - \ln P_1)\left(\frac{R}{\Delta H_{vap}}\right) = \frac{1}{T_1} - \frac{1}{T_2}$$

$$\frac{1}{T_1} - (\ln P_2 - \ln P_1)\left(\frac{R}{\Delta H_{vap}}\right) = \frac{1}{T_2}$$

$P_1 = 40.0$ mm Hg; $T_1 = 191.6$ K
$P_2 = 760$ mm Hg
Solve for T_2, the normal boiling point.

$$\frac{1}{191.6\text{ K}} - [\ln(760) - \ln(40.0)]\left(\frac{8.3145 \times 10^{-3}\frac{\text{kJ}}{\text{K}\cdot\text{mol}}}{22.36\text{ kJ/mol}}\right) = \frac{1}{T_2}$$

$\frac{1}{T_2} = 0.004\,124\,33$; $T_2 = 242.46\text{ K} = -30.7\ ^\circ\text{C}$

11.103 (a) $$\ln P_2 = \ln P_1 + \frac{\Delta H_{vap}}{R}\left(\frac{1}{T_1} - \frac{1}{T_2}\right)$$

$$(\ln P_2 - \ln P_1)\left(\frac{R}{\Delta H_{vap}}\right) = \frac{1}{T_1} - \frac{1}{T_2}$$

$$\frac{1}{T_1} - (\ln P_2 - \ln P_1)\left(\frac{R}{\Delta H_{vap}}\right) = \frac{1}{T_2}$$

$P_1 = 100.0$ mm Hg; $T_1 = -23\ ^\circ\text{C} = 250$ K
$P_2 = 760.0$ mm Hg
Solve for T_2, the normal boiling point for CCl_3F.

$$\frac{1}{250\text{ K}} - [\ln(760.0) - \ln(100.0)]\left(\frac{8.3145 \times 10^{-3}\frac{\text{kJ}}{\text{K}\cdot\text{mol}}}{24.77\text{ kJ/mol}}\right) = \frac{1}{T_2}$$

$\frac{1}{T_2} = 0.003\,319$; $T_2 = 301.3\text{ K} = 28.1\ ^\circ\text{C}$

(b) $$\Delta S_{vap} = \frac{\Delta H_{vap}}{T} = \frac{24.77\text{ kJ/mol}}{301.3\text{ K}} = 0.082\,21\text{ kJ/(K}\cdot\text{mol)} = 82.2\text{ J/(K}\cdot\text{mol)}$$

11.104 $$\Delta H_{vap} = \frac{(\ln P_2 - \ln P_1)(R)}{\left(\frac{1}{T_1} - \frac{1}{T_2}\right)}$$

$P_1 = 100$ mm Hg; $T_1 = -110.3\ ^\circ\text{C} = 162.85$ K
$P_2 = 760$ mm Hg; $T_2 = -88.5\ ^\circ\text{C} = 184.65$ K

$$\Delta H_{vap} = \frac{[\ln(760) - \ln(100)]\left(8.3145 \times 10^{-3}\frac{\text{kJ}}{\text{K}\cdot\text{mol}}\right)}{\left(\frac{1}{162.85\text{ K}} - \frac{1}{184.65\text{ K}}\right)} = 23.3\text{ kJ/mol}$$

11.105 $\ln P_2 = \ln P_1 + \frac{\Delta H_{vap}}{R}\left(\frac{1}{T_1} - \frac{1}{T_2}\right)$

$$(\ln P_2 - \ln P_1)\left(\frac{R}{\Delta H_{vap}}\right) = \frac{1}{T_1} - \frac{1}{T_2}$$

$$\frac{1}{T_1} - (\ln P_2 - \ln P_1)\left(\frac{R}{\Delta H_{vap}}\right) = \frac{1}{T_2}$$

$P_1 = 760$ mm Hg; $T_1 = 56.1\ °C = 329.2$ K

$P_2 = 105$ mm Hg; Solve for T_2.

$$\frac{1}{329.2\text{ K}} - [\ln(105) - \ln(760)]\left(\frac{8.3145 \times 10^{-3}\frac{\text{kJ}}{\text{K}\cdot\text{mol}}}{29.1\text{ kJ/mol}}\right) = \frac{1}{T_2}$$

$\frac{1}{T_2} = 0.003\ 603$; $T_2 = 277.5\text{ K} = 4.4\ °C$

11.106

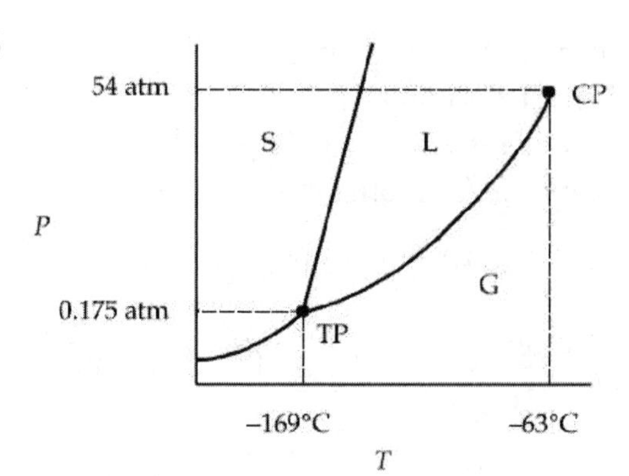

Kr cannot be liquefied at room temperature because room temperature is above T_c (−63 °C).

11.107 (a) Kr(l) (b) supercritical Kr

11.108 For a body-centered cube:

$4r = \sqrt{3}$ edge; edge $= \frac{4r}{\sqrt{3}}$

volume of sphere $= \frac{4}{3}\pi r^3$

volume of unit cell $= \left(\frac{4r}{\sqrt{3}}\right)^3 = \frac{64\,r^3}{3\sqrt{3}}$

volume of 2 spheres $= 2\left(\frac{4}{3}\pi r^3\right) = \frac{8}{3}\pi r^3$

$$\% \text{ volume occupied} = \frac{\left(\frac{8}{3}\pi r^3\right)}{\left(\frac{64r^3}{3\sqrt{3}}\right)} \times 100\% = 68\%$$

11.109 $4r = \sqrt{3}d; \quad r = \frac{\sqrt{3}\,d}{4} = \frac{\sqrt{3}(287 \text{ pm})}{4} = 124 \text{ pm}$

11.110 unit cell edge = d = 287 pm = 287 x 10^{-12} m = 2.87 x 10^{-8} cm
unit cell volume = d^3 = (2.87 x 10^{-8} cm)^3 = 2.364 x 10^{-23} cm^3
unit cell mass = (2.364 x 10^{-23} cm^3)(7.86 g/cm^3) = 1.858 x 10^{-22} g
Fe is body-centered cubic; therefore there are two Fe atoms per unit cell.

$$\text{mass of one Fe atom} = \frac{1.858 \times 10^{-22} \text{ g}}{2 \text{ Fe atoms}} = 9.290 \times 10^{-23} \text{ g/atom}$$

$$\text{Avogadro's number} = 55.85 \text{ g/mol} \times \frac{1 \text{ atom}}{9.290 \times 10^{-23} \text{ g}} = 6.01 \times 10^{23} \text{ atoms/mol}$$

11.111 unit cell edge = d = 408 pm = 408 x 10^{-12} m = 4.08 x 10^{-8} cm
unit cell volume = (4.08 x 10^{-8} cm)^3 = 6.792 x 10^{-23} cm^3
unit cell mass = (10.50 g/cm^3)(6.792 x 10^{-23} cm^3) = 7.132 x 10^{-22} g
Ag is face-centered cubic; therefore there are four Ag atoms in the unit cell.

$$\text{mass of one Ag atom} = \frac{7.132 \times 10^{-22} \text{g}}{4 \text{ Ag atoms}} = 1.783 \times 10^{-22} \text{ g/atom}$$

$$\text{Avogadro's number} = 107.9 \text{ g/mol} \times \frac{1 \text{ atom}}{1.783 \times 10^{-22} \text{ g}} = 6.05 \times 10^{23} \text{ atoms/mol}$$

11.112 (a) unit cell edge = $2r_{Cl^-} + 2r_{Na^+}$ = 2(181 pm) + 2(97 pm) = 556 pm
(b) unit cell edge = d = 556 pm = 556 x 10^{-12} m = 5.56 x 10^{-8} cm
unit cell volume = (5.56 x 10^{-8} cm)^3 = 1.719 x 10^{-22} cm^3
The unit cell contains 4 Na^+ ions and 4 Cl^- ions.

$$\text{mass of one Na}^+ \text{ ion} = 22.99 \text{ g/mol} \times \frac{1 \text{ mol}}{6.022 \times 10^{23} \text{ ions}} = 3.818 \times 10^{-23} \text{ g/Na}^+$$

$$\text{mass of one Cl}^- \text{ ion} = 35.45 \text{ g/mol} \times \frac{1 \text{ mol}}{6.022 \times 10^{23} \text{ ions}} = 5.887 \times 10^{-23} \text{ g/Cl}^-$$

unit cell mass = 4(3.818 x 10^{-23} g) + 4(5.887 x 10^{-23} g) = 3.882 x 10^{-22} g

$$\text{density} = \frac{\text{unit cell mass}}{\text{unit cell volume}} = \frac{3.882 \times 10^{-22} \text{ g}}{1.719 \times 10^{-22} \text{ cm}^3} = 2.26 \text{ g/cm}^3$$

11.113 (a) (1/2 Nb/face)(6 faces) = 3 Nb; (1/4 O/edge)(12 edges) = 3 O
(b) NbO
(c) The oxidation state of Nb is +2.

11.114 Ag_2Te, 343.33; 529 pm = 529×10^{-12} m = 529×10^{-10} cm
unit cell volume = $(529 \times 10^{-10}\text{ cm})^3 = 1.48 \times 10^{-22}\text{ cm}^3$
unit cell mass = $(1.48 \times 10^{-22}\text{ cm}^3)(7.70\text{ g/cm}^3) = 1.14 \times 10^{-21}$ g

$$\text{mass of one } Ag_2Te = \frac{343.33\text{ g } Ag_2Te/\text{mol}}{6.022 \times 10^{23}\ Ag_2Te\text{ formula units/mol}} = 5.70 \times 10^{-22}\text{ g } Ag_2Te/\text{formula unit}$$

$$Ag_2Te\text{ formula units/unit cell} = \frac{1.14 \times 10^{-21}\text{ g/unit cell}}{5.70 \times 10^{-22}\text{ g}/Ag_2Te} = 2\ Ag_2Te/\text{unit cell}$$

$$\text{Ag/unit cell} = \frac{2\ Ag_2Te}{\text{unit cell}} \times \frac{2\text{ Ag}}{Ag_2Te} = 4\text{ Ag/unit cell}$$

11.115 (a)

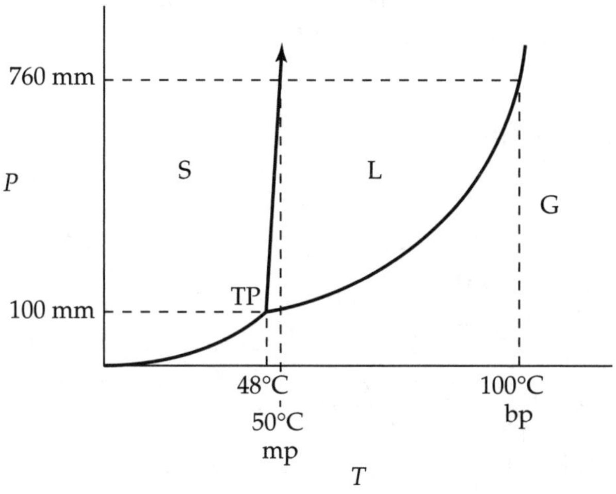

760 mm
100 mm
P
S
L
G
TP
48°C
50°C
mp
100°C
bp
T

(b) (i) solid (ii) gas (iii) liquid (iv) liquid (v) solid

Multiconcept Problems

11.116 Al_2O_3, ionic (greater lattice energy than NaCl because of higher ion charges);
F_2, dispersion; H_2O, H–bonding, dipole-dipole; Br_2, dispersion (larger and more polarizable than F_2), ICl, dipole-dipole, NaCl, ionic
rank according to normal boiling points: $F_2 < Br_2 < ICl < H_2O < NaCl < Al_2O_3$

11.117 $C_2H_5OH(l) \rightarrow C_2H_5OH(g)$
Calculate ΔH and ΔS for this process and assume they do not change as a function of temperature.
$\Delta H° = \Delta H°_f(C_2H_5OH(g)) - \Delta H°_f(C_2H_5OH(l))$
$\Delta H° = [(1\text{ mol})(-235.1\text{ kJ/mol})] - [(1\text{ mol})(-277.7\text{ kJ/mol})] = 42.6\text{ kJ}$
$\Delta S° = S°(C_2H_5OH(g)) - S°(C_2H_5OH(l))$
$\Delta S° = [(1\text{ mol})(282.6\text{ J/(K}\cdot\text{mol))}] - [(1\text{ mol})(161\text{ J/(K}\cdot\text{mol))}] = 122\text{ J/K}$
$\Delta S° = 122 \times 10^{-3}$ kJ/K)
$\Delta G° = \Delta H° - T\Delta S°$ and at the boiling point, $\Delta G = 0$
$0 = \Delta H° - T_{bp}\Delta S°$
$T_{bp}\Delta S° = \Delta H°$

$T_{bp} = \dfrac{\Delta H^o}{\Delta S^o} = \dfrac{42.6\ \text{kJ}}{122 \times 10^{-3}\ \text{kJ/K}} = 349\ \text{K}$

$T_{bp} = 349 - 273 = 76\ ^\circ\text{C}$

$\ln P_2 = \ln P_1 + \dfrac{\Delta H_{vap}}{R}\left(\dfrac{1}{T_1} - \dfrac{1}{T_2}\right)$

$\Delta H_{vap} = 42.6$ kJ/mol

At 1 atm, C_2H_5OH boils at 349 K; therefore set

$T_1 = 349$ K, and $P_1 = 1.00$ atm.

Let $T_2 = 25\ ^\circ\text{C} = 298$ K, and solve for P_2.

P_2 is the vapor pressure of C_2H_5OH at 25 °C.

$\ln P_2 = \ln(1.00) + \dfrac{42.6\ \text{kJ/mol}}{[8.3145 \times 10^{-3}\ \text{kJ/(K} \cdot \text{mol)}]}\left(\dfrac{1}{349\ \text{K}} - \dfrac{1}{298\ \text{K}}\right)$

$\ln P_2 = -2.512;\quad P_2 = e^{-2.512} = 0.0811$ atm

$P_2 = 0.0811\ \text{atm} \times \dfrac{760\ \text{mm Hg}}{1.00\ \text{atm}} = 61.6$ mm Hg

11.118 (a) Let the formula of magnetite be Fe_xO_y, then $Fe_xO_y + y\,CO \rightarrow x\,Fe + y\,CO_2$

$$n_{CO_2} = y = \frac{PV}{RT} = \frac{\left(751\ \text{mm Hg} \times \dfrac{1.00\ \text{atm}}{760\ \text{mm Hg}}\right)(1.136\ \text{L})}{\left(0.082\ 06\ \dfrac{\text{L} \cdot \text{atm}}{\text{K} \cdot \text{mol}}\right)(298\ \text{K})} = 0.04590\ \text{mol}\ CO_2$$

0.04590 mol CO_2 = mol of O in Fe_xO_y

mass of O in Fe_xO_y = $0.04590\ \text{mol O} \times \dfrac{16.0\ \text{g O}}{1\ \text{mol O}} = 0.7345$ g O

mass of Fe in Fe_xO_y = 2.660 g – 0.7345 g = 1.926 g Fe

(b) mol Fe in magnetite = $1.926\ \text{g Fe} \times \dfrac{1\ \text{mol Fe}}{55.85\ \text{g Fe}} = 0.0345$ mol Fe

formula of magnetite: $Fe_{0.0345}O_{0.0459}$ (divide each subscript by the smaller)

$Fe_{0.0345/0.0345}O_{0.0459/0.0345}$

$FeO_{1.33}$ (multiply both subscripts by 3)

$Fe_{(1 \times 3)}O_{(1.33 \times 3)}$; Fe_3O_4

(c) unit cell edge = d = 839 pm = 839×10^{-12} m

$d = 839 \times 10^{-12}\ \text{m} \times \dfrac{100\ \text{cm}}{1\ \text{m}} = 8.39 \times 10^{-8}$ cm

unit cell volume = $d^3 = (8.39 \times 10^{-8}\ \text{cm})^3 = 5.91 \times 10^{-22}\ \text{cm}^3$

unit cell mass = $(5.91 \times 10^{-22}\ \text{cm}^3)(5.20\ \text{g/cm}^3) = 3.07 \times 10^{-21}$ g

mass of Fe in unit cell = $\left(\dfrac{1.926\ \text{g Fe}}{2.660\ \text{g}}\right)(3.07 \times 10^{-21}\ \text{g}) = 2.22 \times 10^{-21}$ g Fe

mass of O in unit cell = $\left(\dfrac{0.7345\ \text{g O}}{2.660\ \text{g}}\right)(3.07 \times 10^{-21}\ \text{g}) = 8.47 \times 10^{-22}$ g O

$$\text{Fe atoms in unit cell} = 2.22 \times 10^{-21}\ \text{g} \times \frac{6.022 \times 10^{23}\ \text{atoms/mol}}{55.847\ \text{g/mol}} = 24\ \text{Fe atoms}$$

$$\text{O atoms in unit cell} = 8.47 \times 10^{-22}\ \text{g} \times \frac{6.022 \times 10^{23}\ \text{atoms/mol}}{16.00\ \text{g/mol}} = 32\ \text{O atoms}$$

11.119 (a) $$n_{H_2} = \frac{PV}{RT} = \frac{\left(740\ \text{mm Hg} \times \dfrac{1.00\ \text{atm}}{760\ \text{mm Hg}}\right)(4.00\ \text{L})}{\left(0.08206\ \dfrac{\text{L}\cdot\text{atm}}{\text{K}\cdot\text{mol}}\right)(296\ \text{K})} = 0.160\ \text{mol}\ H_2$$

M = Group 3A metal; $2\ M(s) + 6\ H^+(aq) \rightarrow 2\ M^{3+}(aq) + 3\ H_2(g)$

$$n_M = 0.160\ \text{mol}\ H_2 \times \frac{2\ \text{mol M}}{3\ \text{mol}\ H_2} = 0.107\ \text{mol M}$$

$\text{mass M} = 1.07\ \text{cm}^3 \times 2.70\ \text{g/cm}^3 = 2.89\ \text{g M}$

$$\text{molar mass M} = \frac{2.89\ \text{g M}}{0.107\ \text{mol M}} = 27.0\ \text{g/mol;}$$ The Group 3A metal is Al

(b) $$\text{mass of one Al atom} = 26.98\ \text{g/mol} \times \frac{1\ \text{mol}}{6.022 \times 10^{23}\ \text{atoms}} = 4.48 \times 10^{-23}\ \text{g/atom}$$

unit cell edge = d = 404 pm = 404×10^{-12} m

$$d = 404 \times 10^{-12}\ \text{m} \times \frac{100\ \text{cm}}{1\ \text{m}} = 4.04 \times 10^{-8}\ \text{cm}$$

unit cell volume = $d^3 = (4.04 \times 10^{-8}\ \text{cm})^3 = 6.59 \times 10^{-23}\ \text{cm}^3$

Calculate the density of Al assuming it is primitive cubic, body-centered cubic, and face-centered cubic. Compare the calculated density with the actual density to identify the unit cell.

For primitive cubic:

$$\text{density} = \frac{\text{unit cell mass}}{\text{unit cell volume}} = \frac{(1\ \text{Al})(4.48 \times 10^{-23}\ \text{g/Al atom})}{6.59 \times 10^{-23}\ \text{cm}^3} = 0.680\ \text{g/cm}^3$$

For body-centered cubic:

$$\text{density} = \frac{\text{unit cell mass}}{\text{unit cell volume}} = \frac{(2\ \text{Al})(4.48 \times 10^{-23}\ \text{g/Al atom})}{6.59 \times 10^{-23}\ \text{cm}^3} = 1.36\ \text{g/cm}^3$$

For face-centered cubic:

$$\text{density} = \frac{\text{unit cell mass}}{\text{unit cell volume}} = \frac{(4\ \text{Al})(4.48 \times 10^{-23}\ \text{g/Al atom})}{6.59 \times 10^{-23}\ \text{cm}^3} = 2.72\ \text{g/cm}^3$$

The calculated density for a face-centered cube (2.72 g/cm^3) is closest to the actual density of 2.70 g/cm^3. Al crystallizes in the face-centered cubic unit cell.

(c) $$r = \sqrt{\frac{d^2}{8}} = \sqrt{\frac{(404\ \text{pm})^2}{8}} = 143\ \text{pm}$$

11.120 (a) M = alkali metal; 500.0 mL = 0.5000 L; 802 °C = 1075 K

$$n_M = \frac{PV}{RT} = \frac{\left(12.5 \text{ mm Hg x } \frac{1.00 \text{ atm}}{760 \text{ mm Hg}}\right)(0.5000 \text{ L})}{\left(0.082\ 06 \frac{\text{L}\cdot\text{atm}}{\text{K}\cdot\text{mol}}\right)(1075 \text{ K})} = 9.32 \times 10^{-5} \text{ mol M}$$

1.62 mm = 1.62×10^{-3} m; crystal volume = $(1.62 \times 10^{-3} \text{ m})^3 = 4.25 \times 10^{-9} \text{ m}^3$

M atoms in crystal = $(9.32 \times 10^{-5} \text{ mol})(6.022 \times 10^{23} \text{ atoms/mol}) = 5.61 \times 10^{19}$ M atoms

Because M is body-centered cubic, only 68% (Table 11.5) of the total volume is occupied by M atoms.

$$\text{volume of M atom} = \frac{(0.68)(4.25 \times 10^{-9} \text{ m})}{5.61 \times 10^{19} \text{ M atoms}} = 5.15 \times 10^{-29} \text{ m}^3\text{/M atom}$$

$$\text{volume of a sphere} = \frac{4}{3}\pi r^3$$

$$r_M = \sqrt[3]{\frac{3(\text{volume})}{4\pi}} = \sqrt[3]{\frac{3(5.15 \times 10^{-29} \text{ m}^3)}{4\pi}} = 2.31 \times 10^{-10} \text{ m} = 231 \times 10^{-12} \text{ m} = 231 \text{ pm}$$

(b) The radius of 231 pm is closest to that of K.

(c) 1.62 mm = 0.162 cm

$$\text{density of solid} = \frac{(9.32 \times 10^{-5} \text{ mol})(39.1 \text{ g/mol})}{(0.162 \text{ cm})^3} = 0.857 \text{ g/cm}^3$$

$$\text{density of vapor} = \frac{(9.32 \times 10^{-5} \text{ mol})(39.1 \text{ g/mol})}{500.0 \text{ cm}^3} = 7.29 \times 10^{-6} \text{ g/cm}^3$$

11.121 (a)

$$n_{X_2} = \frac{PV}{RT} = \frac{\left(755 \text{ mm Hg x } \frac{1.00 \text{ atm}}{760 \text{ mm Hg}}\right)(0.500 \text{ L})}{\left(0.082\ 06 \frac{\text{L}\cdot\text{atm}}{\text{K}\cdot\text{mol}}\right)(298 \text{ K})} = 0.0203 \text{ mol } X_2$$

$M(s) + 1/2\ X_2(g) \rightarrow MX(s)$

$$\text{mol M} = 0.0203 \text{ mol } X_2 \text{ x } \frac{1 \text{ mol M}}{1/2 \text{ mol } X_2} = 0.0406 \text{ mol M}$$

$$\text{molar mass M} = \frac{1.588 \text{ g M}}{0.0406 \text{ mol M}} = 39.1 \text{ g/mol; atomic mass} = 39.1\text{ ; M} = \text{K}$$

(b) From Figure 6.1, the radius for K^+ is ~140 pm.

unit cell edge = 535 pm = $2r_{K^+} + 2r_{X^-}$

$$r_{X^-} = \frac{535 \text{ pm} - 2r_{K^+}}{2} = \frac{535 \text{ pm} - 2(140 \text{ pm})}{2} = 128 \text{ pm}$$

From Figure 6.2, $X^- = F^-$

(c) Because the cation and anion are of comparable size, the anions are not in contact with each other.

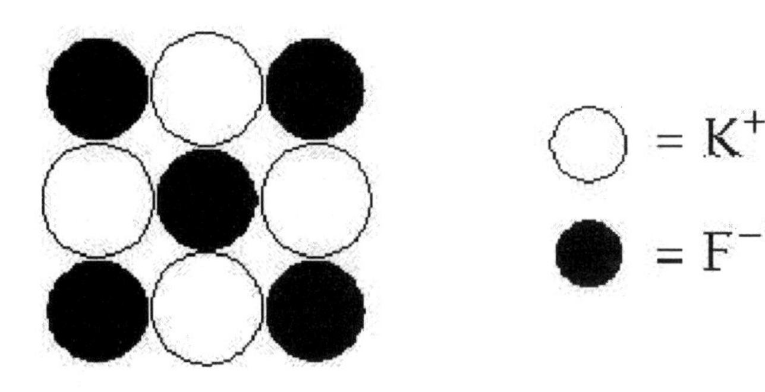

(d) unit cell contents: 1/8 F^- at 8 corners and 1/2 F^- at 6 faces = 4 F^-
1/4 K^+ at 12 edges and 1 K^+ inside = 4 K^+

mass of one $K^+ = \dfrac{39.098 \text{ g/mol}}{6.022 \times 10^{23} \text{ K}^+\text{/mol}} = 6.493 \times 10^{-23} \text{ g/K}^+$

mass of one $F^- = \dfrac{18.998 \text{ g/mol}}{6.022 \times 10^{23} \text{ F}^-\text{/mol}} = 3.155 \times 10^{-23} \text{ g/F}^-$

unit cell mass = $(4\ K^+)(6.493 \times 10^{-23} \text{ g/K}^+) + (4\ F^-)(3.155 \times 10^{-23} \text{ g/F}^-) = 3.859 \times 10^{-22}$ g

unit cell volume = $[(535 \times 10^{-12} \text{ m})(100 \text{ cm/m})]^3 = 1.531 \times 10^{-22} \text{ cm}^3$

density = $\dfrac{\text{mass of unit cell}}{\text{volume of unit cell}} = \dfrac{3.859 \times 10^{-22} \text{ g}}{1.531 \times 10^{-22} \text{ cm}^3} = 2.52 \text{ g/cm}^3$

(e) $K(s) + 1/2\ F_2(g) \rightarrow KF(s)$ is a formation reaction.

$\Delta H^\circ_f(KF) = \dfrac{-22.83 \text{ kJ}}{0.0406 \text{ mol}} = -562 \text{ kJ/mol}$

12 Solutions and Their Properties

12.1 KBr < 1,5 pentanediol < toluene

12.2 Vitamin E is more fat soluble.

12.3 NaCl, 58.44; 1.00 mol NaCl = 58.44 g

1.00 L H_2O = 1000 mL = 1000 g (assuming a density of 1.00 g/mL)

$$\text{mass \% NaCl} = \frac{58.44\text{ g}}{1000\text{ g} + 58.44\text{ g}} \times 100\% = 5.52\text{ mass \%}$$

12.4 $$7.50\% = \left(\frac{15.0\text{ g}}{15.0\text{ g} + x}\right)(100\%)$$

Let x equal the mass of water. Solve for x.

(7.50%)(15.0 g + x) = (15.0 g)(100%)

$$x = \frac{(15.0\text{ g})(100\%)}{7.50\%} - 15.0\text{ g} = 185\text{ g H}_2\text{O}$$

12.5 1.25 μg = 1.25×10^{-6} g; (50.0 mL)(1.00 g/mL) = 50.0 g

$$\frac{1.25 \times 10^{-6}\text{ g}}{50.0\text{ g}} \times 10^9 = 25.0\text{ ppb}$$

This exceeds the 10.0 ppb limit for As.

12.6 $$\text{ppm} = \frac{\text{mass of CO}_2}{\text{total mass of solution}} \times 10^6\text{ ppm}$$

total mass of solution = density x volume = (1.3 g/L)(1.0 L) = 1.3 g

$$35\text{ ppm} = \frac{\text{mass of CO}_2}{1.3\text{ g}} \times 10^6\text{ ppm}$$

$$\text{mass of CO}_2 = \frac{(35\text{ ppm})(1.3\text{ g})}{10^6\text{ ppm}} = 4.6 \times 10^{-5}\text{ g CO}_2$$

12.7 CH_3CO_2Na, 82.03

$$\text{kg H}_2\text{O} = (0.150\text{ mol CH}_3\text{CO}_2\text{Na})\left(\frac{1\text{ kg H}_2\text{O}}{0.500\text{ mol CH}_3\text{CO}_2\text{Na}}\right) = 0.300\text{ kg H}_2\text{O}$$

$$\text{mass CH}_3\text{CO}_2\text{Na} = 0.150\text{ mol CH}_3\text{CO}_2\text{Na} \times \frac{82.03\text{ g CH}_3\text{CO}_2\text{Na}}{1\text{ mol CH}_3\text{CO}_2\text{Na}} = 12.3\text{ g CH}_3\text{CO}_2\text{Na}$$

mass of solution needed = 300 g + 12.3 g = 312 g

12.8 $C_{27}H_{46}O$, 386.7; $CHCl_3$, 119.4; $40.0 \text{ g x } \frac{1 \text{ kg}}{1000 \text{ g}} = 0.0400 \text{ kg}$

$$\text{molality} = \frac{\text{mol } C_{27}H_{46}O}{\text{kg } CHCl_3} = \frac{\left(0.385 \text{ g x } \frac{1 \text{ mol}}{386.7 \text{ g}}\right)}{0.0400 \text{ kg}} = 0.0249 \text{ mol/kg} = 0.0249\ m$$

$$X_{C_{27}H_{46}O} = \frac{\text{mol } C_{27}H_{46}O}{\text{mol } C_{27}H_{46}O + \text{mol } CHCl_3}$$

$$X_{C_{27}H_{46}O} = \frac{\left(0.385 \text{ g x } \frac{1 \text{ mol}}{386.7 \text{ g}}\right)}{\left[\left(0.385 \text{ g x } \frac{1 \text{ mol}}{386.7 \text{ g}}\right) + \left(40.0 \text{ g x } \frac{1 \text{ mol}}{119.4 \text{ g}}\right)\right]} = 2.96 \text{ x } 10^{-3}$$

12.9 H_2O, 18.02

Assume 1.00 L of solution.

mass of 1.00 L = (1.0042 g/mL)(1000 mL) = 1004.2 g of solution

$$0.500 \text{ mol } CH_3CO_2H \text{ x } \frac{60.05 \text{ g } CH_3CO_2H}{1 \text{ mol } CH_3CO_2H} = 30.02 \text{ g } CH_3CO_2H$$

1004.2 g – 30.02 g = 974.2 g = 0.9742 kg of H_2O

$$974.2 \text{ g } H_2O \text{ x } \frac{1 \text{ mol } H_2O}{18.02 \text{ g } H_2O} = 54.06 \text{ mol } H_2O$$

$$X_{CH_3CO_2H} = \frac{\text{mol } CH_3CO_2H}{\text{mol } CH_3CO_2H + \text{mol } H_2O} = \frac{0.500 \text{ mol}}{0.500 \text{ mol} + 54.06 \text{ mol}} = 0.00916$$

$$\text{mass \% } CH_3CO_2H = \frac{30.02 \text{ g}}{30.02 \text{ g} + 974.2 \text{ g}} \text{ x } 100\% = 2.99 \text{ mass \%}$$

$$\text{molality} = \frac{0.500 \text{ mol}}{0.9742 \text{ kg}} = 0.513\ m$$

12.10 Assume you have a solution with 1.000 kg (1000 g) of H_2O. If this solution is 0.258 *m*, then it must also contain 0.258 mol glucose.

$$\text{mass of glucose} = 0.258 \text{ mol x } \frac{180.2 \text{ g}}{1 \text{ mol}} = 46.5 \text{ g glucose}$$

mass of solution = 1000 g + 46.5 g = 1046.5 g

density = 1.0173 g/mL

$$\text{volume of solution} = 1046.5 \text{ g x } \frac{1 \text{ mL}}{1.0173 \text{ g}} = 1028.7 \text{ mL}$$

$$\text{volume} = 1028.7 \text{ mL x } \frac{1 \text{ L}}{1000 \text{ mL}} = 1.029 \text{ L}; \quad \text{molarity} = \frac{0.258 \text{ mol}}{1.029 \text{ L}} = 0.251 \text{ M}$$

12.11 $M = k \cdot P; \ k = \frac{M}{P} = \frac{3.2 \text{ x } 10^{-2} \text{ M}}{1.0 \text{ atm}} = 3.2 \text{ x } 10^{-2} \text{ mol/(L} \cdot \text{atm)}$

12.12 (a) $M = k \cdot P = [3.2 \times 10^{-2}\ \text{mol/(L} \cdot \text{atm)}](2.5\ \text{atm}) = 0.080\ \text{M}$

(b) $P_{CO_2} = (0.0004)(1\ \text{atm}) = 4.0 \times 10^{-4}\ \text{atm}$

$M = k \cdot P = [3.2 \times 10^{-2}\ \text{mol/(L} \cdot \text{atm)}](4.0 \times 10^{-4}\ \text{atm}) = 1.3 \times 10^{-5}\ \text{M}$

12.13 H_2O, 18.02; $CaCl_2$, 110.0

$$100.0\ \text{g H}_2\text{O} \times \frac{1\ \text{mol H}_2\text{O}}{18.02\ \text{g H}_2\text{O}} = 5.549\ \text{mol H}_2\text{O}$$

$$10.00\text{g CaCl}_2 \times \frac{1\ \text{mol CaCl}_2}{110.0\ \text{g CaCl}_2} = 0.090\ 91\ \text{mol CaCl}_2$$

$$X_{H_2O} = \frac{5.549\ \text{mol}}{(2.7)(0.090\ 91\ \text{mol}) + 5.549\ \text{mol}} = 0.9567$$

$$P_{soln} = P_{H_2O} \times X_{H_2O} = (233.7\ \text{mm Hg})(0.9576) = 223.8\ \text{mm Hg}$$

12.14 H_2O, 18.02; $MgCl_2$, 95.21

$$100.0\ \text{g H}_2\text{O} \times \frac{1\ \text{mol H}_2\text{O}}{18.02\ \text{g H}_2\text{O}} = 5.549\ \text{mol H}_2\text{O}$$

$$8.110\text{g MgCl}_2 \times \frac{1\ \text{mol MgCl}_2}{95.21\ \text{g MgCl}_2} = 0.085\ 18\ \text{mol MgCl}_2$$

$$X_{H_2O} = \frac{P_{soln}}{P_{H_2O}} = \frac{224.7\ \text{mm Hg}}{233.7\ \text{mm Hg}} = 0.9615$$

$$X_{H_2O} = 0.9615 = \frac{5.549\ \text{mol}}{(i)(0.085\ 18\ \text{mol}) + 5.549\ \text{mol}};$$ solve for the van't Hoff factor, i.

$$(0.9615)[(i)(0.085\ 18\ \text{mol}) + 5.549\ \text{mol})] = 5.549\ \text{mol}$$

$$i = \frac{\left(\dfrac{5.549\ \text{mol}}{0.9615} - 5.549\ \text{mol}\right)}{0.085\ 18} = 2.6$$

12.15 $P_{soln} = P_{solv} \cdot X_{solv}$; $\quad X_{solv} = \dfrac{P_{soln}}{P_{solv}} = \dfrac{(55.3 - 1.30)\ \text{mm Hg}}{55.3\ \text{mm Hg}} = 0.976$

NaBr dissociates into two ions in aqueous solution.

$$X_{solv} = \frac{\text{mol H}_2\text{O}}{\text{mol H}_2\text{O} + \text{mol Na}^+ + \text{mol Br}^-}$$

$$X_{solv} = 0.976 = \frac{\left(250\ \text{g} \times \dfrac{1\ \text{mol}}{18.02\ \text{g}}\right)}{\left(250\ \text{g} \times \dfrac{1\ \text{mol}}{18.02\ \text{g}}\right) + x\ \text{mol Na}^+ + x\ \text{mol Br}^-}$$

$$0.976 = \frac{13.9\ \text{mol}}{13.9\ \text{mol} + 2x\ \text{mol}};$$ solve for x.

$0.976(13.9 \text{ mol} + 2x \text{ mol}) = 13.9 \text{ mol}$
$13.566 \text{ mol} + 1.952\, x \text{ mol} = 13.9 \text{ mol}$
$1.952\, x \text{ mol} = 13.9 \text{ mol} - 13.566 \text{ mol}$

$$x \text{ mol} = \frac{13.9 \text{ mol} - 13.566 \text{ mol}}{1.952} = 0.171 \text{ mol}$$

$x = 0.171 \text{ mol } Na^+ = 0.171 \text{ mol } Br^- = 0.171 \text{ mol NaBr}$

$$\text{NaBr, 102.9; mass NaBr} = 0.171 \text{ mol} \times \frac{102.9 \text{ g}}{1 \text{ mol}} = 17.6 \text{ g NaBr}$$

12.16 At any given temperature, the vapor pressure of a solution is lower than the vapor pressure of the pure solvent. The upper curve represents the vapor pressure of the pure solvent. The lower curve represents the vapor pressure of the solution.

12.17
$$100 \text{ g } C_2H_5OH \times \frac{1 \text{ mol } C_2H_5OH}{46.07 \text{ g } C_2H_5OH} = 2.171 \text{ mol } C_2H_6O$$

$$25.0 \text{ g } H_2O \times \frac{1 \text{ mol } H_2O}{18.02 \text{ g } H_2O} = 1.387 \text{ mol } H_2O$$

$$X_{C_2H_5OH} = \frac{2.171 \text{ mol}}{2.171 \text{ mol} + 1.387 \text{ mol}} = 0.6102$$

$$X_{H_2O} = \frac{1.387 \text{ mol}}{2.171 \text{ mol} + 1.387 \text{ mol}} = 0.3898$$

$$P_{soln} = X_{C_2H_5OH} P^{o}_{C_2H_5OH} + X_{H_2O} P^{o}_{H_2O}$$

$$P_{soln} = (0.6102)(61.2 \text{ mm Hg}) + (0.3898)(23.8 \text{ mm Hg}) = 46.6 \text{ mm Hg}$$

12.18 The red and blue curves are the pure liquids and the green curve is the mixture.

12.19 $C_2H_6O_2$, 62.07; 500.0 g = 0.5000 kg

$$616.9 \text{ g } C_2H_6O_2 \times \frac{1 \text{ mol } C_2H_6O_2}{62.07 \text{ g } C_2H_6O_2} = 9.939 \text{ mol } C_2H_6O_2$$

$$\Delta T = K_b \cdot m = \left(0.51 \frac{^\circ\text{C}\cdot\text{kg}}{\text{mol}}\right)\left(\frac{9.939 \text{ mol}}{0.5000 \text{ kg}}\right) = 10.1\ ^\circ\text{C}$$

bp = 100.0 °C + 10.1 °C = 110.1 °C

12.20 The red curve represents the vapor pressure of pure chloroform.
(a) The normal boiling point for a liquid is the temperature where the vapor pressure of the liquid equals 1 atm (760 mm Hg). The approximate boiling point of pure chloroform is 62 °C.
(b) The approximate boiling point of the solution is 69 °C.
$\Delta T_b = 69\ ^\circ\text{C} - 62\ ^\circ\text{C} = 7\ ^\circ\text{C}$
$\Delta T_b = K_b \cdot m$

$$m = \frac{\Delta T_b}{K_b} = \frac{7\ ^\circ C}{3.63 \frac{^\circ C \cdot kg}{mol}} = 2\ mol/kg = 2\ m$$

12.21 $C_6H_{12}O_6$, 180.2; 37.0 °C = 310.1 K

$$\Pi = \left(\frac{50.0\ g\ x\ \frac{1 mol}{180.2\ g}}{1.00\ L} \right) \left(0.082\ 06\ \frac{L \cdot atm}{K \cdot mol} \right) (310.1\ K) = 7.06\ atm$$

12.22 (a) $\Pi = iMRT$; $M = \frac{\Pi}{iRT} = \frac{8.0\ atm}{(1.9)\left(0.082\ 06\ \frac{L \cdot atm}{K \cdot mol} \right)(298\ K)} = 0.17\ M$

(b) There would be a net transfer of water from outside the cell to the inside of the cell and the cell would swell.

12.23 $\Pi = MRT$; $M = \frac{\Pi}{RT} = \frac{\left(423.1\ mm\ Hg\ x\ \frac{1\ atm}{760\ mm\ Hg} \right)}{\left(0.082\ 06\ \frac{L \cdot atm}{K \cdot mol} \right)(298\ K)} = 0.0228\ mol/L$

200.0 mL = 0.2000 L

mol = (0.0228 mol/L)(0.2000 L) = 0.004 553 mol

$$\text{molar mass} = \frac{0.8220\ g}{0.004\ 553\ mol} = 180.5\ g/mol$$

12.24 $C_{12}H_{22}O_{11}$, 342.3

$\Pi = MRT$; $M = \frac{\Pi}{RT} = \frac{\left(278\ mm\ Hg\ x\ \frac{1\ atm}{760\ mm\ Hg} \right)}{\left(0.082\ 06\ \frac{L \cdot atm}{K \cdot mol} \right)(298\ K)} = 0.014\ 96\ mol/L$

100.0 mL = 0.1000 L

mol = (0.014 96 mol/L)(0.1000 L) = 0.001 496 mol

$$\text{molar mass} = \frac{0.512\ g}{0.001\ 496\ mol} = 342.3\ g/mol$$

The white powder is sucrose.

12.25 Both solvent molecules and small solute particles can pass through a semipermeable dialysis membrane. Only large colloidal particles such as proteins can't pass through. Only solvent molecules can pass through a semipermeable membrane used for osmosis.

12.26 Solvent–solvent is hydrogen bonding, solvent–solute is hydrogen bonding, and solute–solute is hydrogen bonding.

12.27 (a) Total conc = (137 + 105 + 3.0 + 4.0 + 2.0 + 33 + 0.75 + 11.1) mmol/L = 295.85 mmol/L
295.85 mmol/L = 295.85 x 10^{-3} mol/L = 0.295 85 mol/L
25 °C = 298 K

$$\Pi = MRT = (0.295\ 85\ \text{mol/L})\left(0.082\ 06\ \frac{\text{L}\cdot\text{atm}}{\text{K}\cdot\text{mol}}\right)(298\text{K}) = 7.23\ \text{atm}$$

(b) Solvent moves from the dialysis solution to blood.

12.28 $$12.0\ \text{L} \times \frac{1\ \text{mL}}{1 \times 10^{-3}\ \text{L}} \times \frac{1\ \text{g}}{\text{mL}} = 12{,}000\ \text{g}$$

$$2\ \text{ppm} = \frac{\text{mass of F}^-}{12{,}000\ \text{g}} \times 10^6$$

$$\text{mass of F}^- = \frac{(2)(12{,}000\ \text{g})}{10^6} = 0.024\ \text{g} = 24\ \text{mg}$$

Conceptual Problems

12.29 Vitamin D is more fat soluble.

12.30 (a) < (b) < (c)

12.31 (a) Because the vapor pressure of the solution (red curve) is higher than that of the first liquid (green curve), the vapor pressure of the second liquid must be higher than that of the solution (red curve). Because the second liquid has a higher vapor pressure than the first liquid, the second liquid has a lower boiling point.

(b)

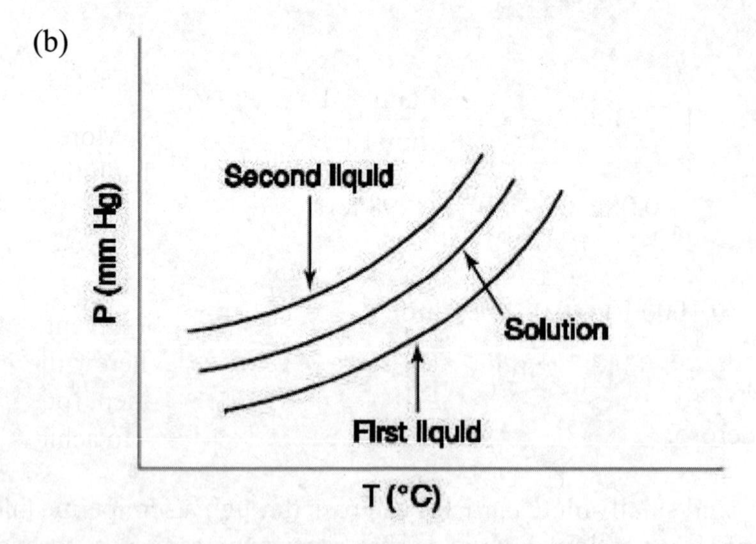

12.32 The upper curve is pure ether.
(a) The normal boiling point for ether is the temperature where the upper curve intersects the 760 mm Hg line, ~ 37 °C.

(b) $\Delta T_b \approx 3\ ^\circ C$

$\Delta T_b = K_b \cdot m;$ $$m = \frac{\Delta T_b}{K_b} = \frac{3\ ^\circ C}{2.02 \frac{^\circ C \cdot kg}{mol}} \approx 1.5\ mol/kg \approx 1.5\ m$$

12.33 At any given temperature, the vapor pressure of a mixture of two pure liquids falls between the individual vapor pressures of the two pure liquids themselves. Because the vapor pressure of the mixture is greater than the vapor pressure of the solvent, the second liquid is more volatile (has a higher vapor pressure) than the solvent.

12.34 (a) The red curve represents the solution of a volatile solute and the green curve represents the solution of a nonvolatile solute.
(b) & (d)

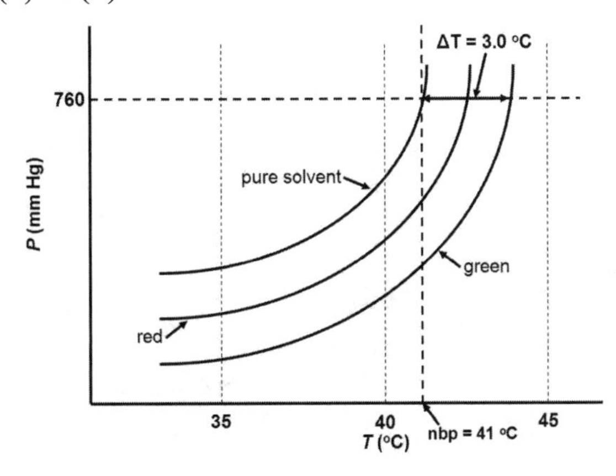

(c) $\Delta T = K_f \cdot m;\ m = \frac{\Delta T}{K_f} = \frac{3.0\ ^\circ C}{2.0\ ^\circ C/m} = 1.5\ m$

12.35 The vapor pressure of the NaCl solution is lower than that of pure H_2O. More H_2O molecules will go into the vapor from the pure H_2O than from the NaCl solution. More H_2O vapor molecules will go into the NaCl solution than into pure H_2O. The result is represented by (b).

12.36

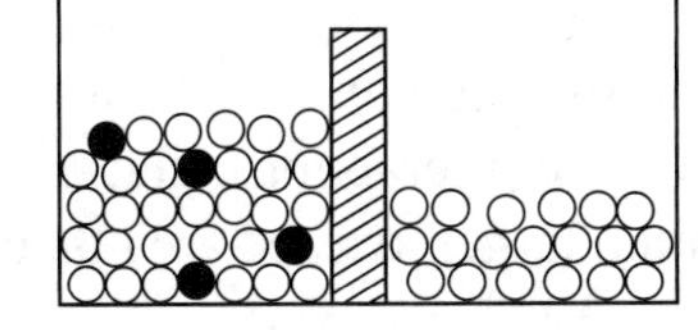

Assume that only the blue (open) spheres (solvent) can pass through the semipermeable membrane. There will be a net transfer of solvent from the right compartment (pure solvent) to the left compartment (solution) to achieve equilibrium.

12.37 (a) Solution B (i = 3) is represented by the red line because its freezing point depressed the most. Solution A (i = 1) is represented by the blue line.
(b) There is a difference of 3.0 °C between the red and blue line. The 3.0 °C equates to 2 i units ($i_{red} - i_{blue} = 3 - 1 = 2$) so 1 i unit is worth 1.5 °C. The blue line (i = 1) is at 14.0 °C, so the freezing point of the pure liquid is 14.0 °C + 1.5 °C = 15.5 °C.

(c) Both the solutions are the same concentration.

$$\Delta T = K_f \cdot m;\ m = \frac{\Delta T}{K_f} = \frac{1.5\ ^\circ C}{3.0\ ^\circ C/m} = 0.50\ m$$

12.38 (a) - (c) (b) ~95 °C

126° C
vapor
b
a
bp
d
liquid
c
69° C

0	0.20	0.40	0.60	0.80	1.00	X_{decane}
1.00	0.80	0.60	0.40	0.20	0	X_{octane}

12.39 (a) and (c)

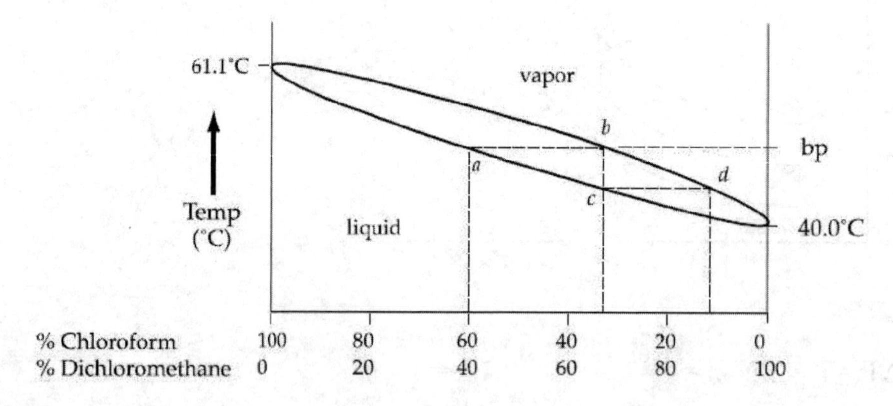

(b) The mixture will begin to boil at ~50 °C.
(d) After two cycles of boiling and condensing, the approximate composition of the liquid would be 90% dichloromethane and 10% chloroform.

Section Problems
Solutions and Energy Changes (Sections 12.1 and 12.2)

12.40 The surface area of a solid plays an important role in determining how rapidly a solid dissolves. The larger the surface area, the more solid–solvent interactions, and the more rapidly the solid will dissolve. Powdered NaCl has a much larger surface area than a large block of NaCl, and it will dissolve more rapidly.

12.41 Energy is required to overcome intermolecular forces holding solute particles together in the crystal. For an ionic solid, this is the lattice energy. Substances with higher lattice energies tend to be less soluble than substances with lower lattice energies.

12.42 (a) Na^+ has the larger hydration energy because of its smaller size and higher charge density.
(b) Ba^{2+} has the larger hydration energy because of its higher charge.

12.43 SO_4^{2-} has the larger hydration energy because of its higher charge. Both SO_4^{2-} and ClO_4^- are comparable in size, so size is not a factor.

12.44 Solvent–solvent is hydrogen bonding, solvent–solute is dispersion, and solute–solute is dispersion. I_2 is not soluble in water.

12.45 Solvent–solvent is hydrogen bonding, solvent–solute is hydrogen bonding, and solute–solute is hydrogen bonding. Glucose would be soluble in water.

12.46 Both Br_2 and CCl_4 are nonpolar, and intermolecular forces for both are dispersion forces. H_2O is a polar molecule with dipole-dipole forces and hydrogen bonding. Therefore, Br_2 is more soluble in CCl_4.

12.47 CH_2O and water are both polar molecules and water can hydrogen bond to CH_2O. CCl_4 is nonpolar. The solubility of CH_2O is greater in water.

12.48 CH_3COOH and water are both polar molecules and they both can hydrogen bond. C_6H_6 is nonpolar. The solubility of CH_3COOH is greater in water.

12.49 C_4H_{10} and C_6H_6 are both nonpolar molecules with only dispersion forces. Water is a polar molecule that can hydrogen bond. The solubility of C_4H_{10} is greater in C_6H_6.

12.50 Toluene is nonpolar and is insoluble in water.
Br_2 is nonpolar but because of its size, is polarizable and is soluble in water.
KBr is an ionic compound and is very soluble in water.
toluene < Br_2 < KBr (solubility in H_2O)

12.51 Hexane, C_6H_{14}, is nonpolar with only dispersion forces.
NaBr is an ionic compound and insoluble in hexane.
CH_3COOH is polar, can hydrogen bond, and has some solubility in hexane.
CCl_4 is nonpolar with only dispersion forces and is very soluble in hexane.
NaBr < CH_3COOH < CCl_4 (solubility in C_6H_{14})

12.52 There are three hydrogen bonds.

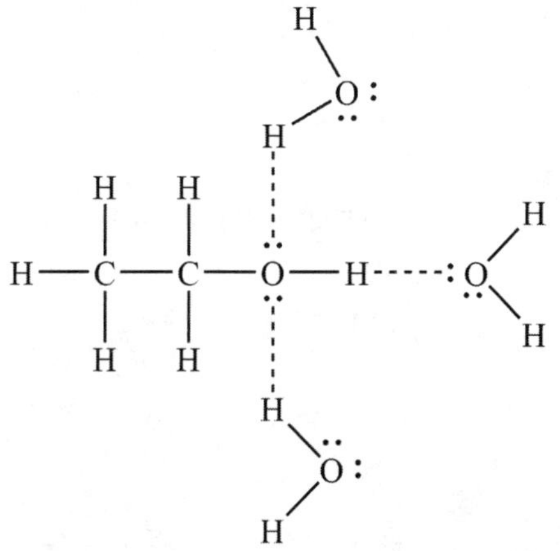

```
                 H
                  \
                   O:
                  /
                 H
     H    H      :
     |    |      :          H
                 ..        /
 H — C —— C —— O —— H·····:O
                 ..        \
     |    |      :          H
     H    H      :
                 H
                  \
                   O:
                  /
                 H
```

12.53 There are six hydrogen bonds.

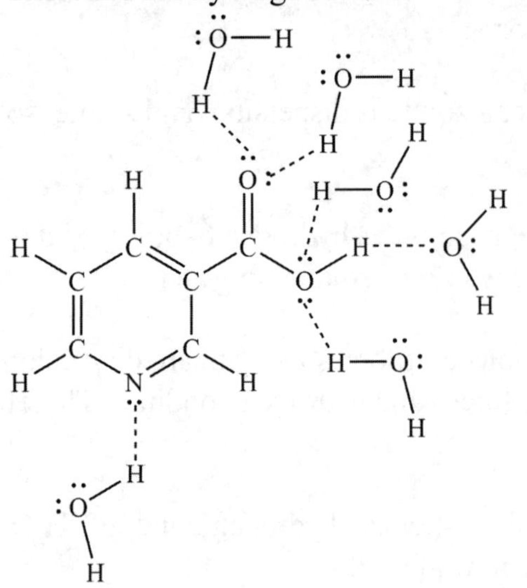

12.54 Ethyl alcohol and water are both polar with small dispersion forces. They both can hydrogen bond, and are miscible.
Pentyl alcohol is slightly polar and can hydrogen bond. It has, however, a relatively large solute-solute dispersion force because of its size, which limits its water solubility.

12.55 The intermolecular forces associated with octane are dispersion forces. Both pentyl alcohol and methyl alcohol can hydrogen bond. Pentyl alcohol has relatively large dispersion forces because of its size. Methyl alcohol does not. Pentyl alcohol is soluble in octane; methyl alcohol is not.

Units of Concentration (Section 12.3)

12.56 $CaCl_2$, 110.98
For a 1.00 *m* solution:
heat released = 81,300 J
mass of solution = 1000 g H_2O + 110.98 g $CaCl_2$ = 1110.98 g

$$\Delta T = \frac{q}{\text{(specific heat)(mass of solution)}} = \frac{81{,}300 \text{ J}}{[4.18 \text{ J/(K}\cdot\text{g)}](1110.98 \text{ g})} = 17.5 \text{ K} = 17.5\ ^\circ\text{C}$$

Final temperature = 25.0 °C + 17.5 °C = 42.5 °C

12.57 NH_4ClO_4, 117.48
For a 1.00 *m* solution: heat absorbed = 33,500 J
mass of solution = 1000 g H_2O + 117.48 g NH_4ClO_4 = 1117.48 g

$$\Delta T = \frac{q}{\text{(specific heat)(mass of solution)}} = \frac{-33{,}500 \text{ J}}{[4.18 \text{ J/(K}\cdot\text{g)}](1117.48 \text{ g})} = -7.2 \text{ K} = -7.2\ ^\circ\text{C}$$

Final temperature = 25.0 °C – 7.2 °C = 17.8 °C

12.58 Assume 1.00 L of seawater.
mass of 1.00 L = (1000 mL)(1.025 g/mL) = 1025 g

$$\frac{\text{mass NaCl}}{1025\text{ g}} \times 100\% = 3.50\text{ mass \%}; \qquad \text{mass NaCl} = \frac{1025\text{ g} \times 3.50}{100} = 35.88\text{ g}$$

There are 35.88 g NaCl per 1.00 L of solution.

$$M = \frac{\left(35.88\text{ g NaCl} \times \frac{1\text{ mol NaCl}}{58.44\text{ g NaCl}}\right)}{1.00\text{ L}} = 0.614\text{ M}$$

12.59 Assume you have 100.0 g of seawater.
mass NaCl = (0.0350)(100.0 g) = 3.50 g NaCl
mass H_2O = 100.0 g – 3.50 g = 96.5 g H_2O

NaCl, 58.44; $\text{mol NaCl} = 3.50\text{ g} \times \frac{1\text{ mol}}{58.44\text{ g}} = 0.0599\text{ mol NaCl}$

$\text{mass } H_2O = 96.5\text{ g} \times \frac{1\text{ kg}}{1000\text{ g}} = 0.0965\text{ kg } H_2O$; $\quad \text{molality} = \frac{0.0599\text{ mol}}{0.0965\text{ kg}} = 0.621\ m$

12.60 $C_{16}H_{21}NO_2$, 259.3; 50 ng = 50×10^{-9} g
(a) mass of solution = (1.025 g/mL)(1000 mL) = 1025 g

$$\text{ppb} = \frac{50 \times 10^{-9}\text{ g}}{1025\text{ g}} \times 10^9 = 0.049\text{ ppb}$$

(b) $50 \times 10^{-9}\text{ g} \times \frac{1\text{ mol } C_{16}H_{21}NO_2}{259.3\text{ g } C_{16}H_{21}NO_2} = 1.9 \times 10^{-10}\text{ mol}$

$$M = \frac{1.9 \times 10^{-10}\text{ mol}}{1.0\text{ L}} = 1.9 \times 10^{-10}\text{ mol/L}$$

12.61 $C_8H_{14}ClN_5$, 215.7; 0.050 μg = 0.050×10^{-6} g
(a) mass of solution = (1.00 g/mL)(1000 mL) = 1000 g

$$\text{ppb} = \frac{0.050 \times 10^{-6}\text{ g}}{1000\text{ g}} \times 10^9 = 0.050\text{ ppb}$$

(b) $0.050 \times 10^{-6}\text{ g} \times \frac{1\text{ mol } C_8H_{14}ClN_5}{215.7\text{ g } C_8H_{14}ClN_5} = 2.3 \times 10^{-10}\text{ mol}$

$$M = \frac{2.3 \times 10^{-10}\text{ mol}}{1.0\text{ L}} = 2.3 \times 10^{-10}\text{ mol/L}$$

12.62 (a) Dissolve 0.150 mol of glucose in water; dilute to 1.00 L.
(b) Dissolve 1.135 mol of KBr in 1.00 kg of H_2O.
(c) Mix together 0.15 mol of CH_3OH with 0.85 mol of H_2O.

12.63 (a) Dissolve 15.5 mg urea in 100 mL water
(b) Choose a K^+ salt, say KCl, and dissolve 0.0075 mol (0.559 g) in water; dilute to100 mL.

12.64 $C_7H_6O_2$, 122.12, 165 mL = 0.165 L
mol $C_7H_6O_2$ = (0.0268 mol/L)(0.165 L) = 0.004 42 mol

$\text{mass } C_7H_6O_2 = 0.004\ 42 \text{ mol x } \frac{122.12 \text{ g}}{1 \text{ mol}} = 0.540 \text{ g}$

Dissolve 4.42×10^{-3} mol (0.540 g) of $C_7H_6O_2$ in enough $CHCl_3$ to make 165 mL of solution.

12.65 $C_7H_6O_2$, 122.12

$$0.325 \text{ mol } C_7H_6O_2 \text{ x } \frac{122.12 \text{ g } C_7H_6O_2}{1 \text{ mol } C_7H_6O_2} = 39.70 \text{ g } C_7H_6O_2$$

Place 39.70 g of $C_7H_6O_2$ in a 1.000 L volumetric flask, fill to the mark with $CHCl_3$, and take 250 mL of the solution.

12.66 (a) KCl, 74.6

A 0.500 M KCl solution contains 37.3 g of KCl per 1.00 L of solution.
A 0.500 mass % KCl solution contains 5.00 g of KCl per 995 g of water.
The 0.500 M KCl solution is more concentrated (that is, it contains more solute per amount of solvent).

(b) Both solutions contain the same amount of solute. The 1.75 M solution contains less solvent than the 1.75 m solution. The 1.75 M solution is more concentrated.

12.67 (a) KI, 166.00; KBr, 119.00; assume 1.000 L = 1000 mL = 1000 g solution

$$10 \text{ ppm} = \frac{\text{mass KI}}{1000 \text{ g}} \text{ x } 10^6; \qquad \text{mass KI} = 0.010 \text{ g}$$

$$10{,}000 \text{ ppb} = \frac{\text{mass KBr}}{1000 \text{ g}} \text{ x } 10^9; \qquad \text{mass KBr} = 0.010 \text{ g}$$

Both solutions contain the same mass of solute in the same amount of solvent. Because the molar mass of KBr is less than that of KI, the number of moles of KBr is larger than the number of moles of KI. The KBr solution has a higher molarity than the KI solution.

(b) Because the mass % of the two solutions is the same, they both contain the same mass of solute and solution. Because the molar mass of KCl is less than that of citric acid, the number of moles of KCl is larger than the number of moles of citric acid. The KCl solution has a higher molarity than the citric acid solution.

12.68 (a) $C_6H_8O_7$, 192.12

$$0.655 \text{ mol } C_6H_8O_7 \text{ x } \frac{192.12 \text{ g } C_6H_8O_7}{1 \text{ mol } C_6H_8O_7} = 126 \text{ g } C_6H_8O_7$$

$$\text{mass \% } C_6H_8O_7 = \frac{126 \text{ g}}{126 \text{ g} + 1000 \text{ g}} \text{ x } 100\% = 11.2 \text{ mass \%}$$

(b) $0.135 \text{ mg} = 0.135 \times 10^{-3} \text{ g}$

$(5.00 \text{ mL } H_2O)(1.00 \text{ g/mL}) = 5.00 \text{ g } H_2O$

$$\text{mass \% KBr} = \frac{0.135 \times 10^{-3} \text{ g}}{(0.135 \times 10^{-3} \text{ g}) + 5.00 \text{ g}} \text{ x } 100\% = 0.002\ 70 \text{ mass \% KBr}$$

(c) $$\text{mass \% aspirin} = \frac{5.50 \text{ g}}{5.50 \text{ g} + 145 \text{ g}} \text{ x } 100\% = 3.65 \text{ mass \% aspirin}$$

12.69 (a) $\text{molality} = \dfrac{0.655\ \text{mol}}{1.00\ \text{kg}} = 0.655\ m$

(b) KBr, 119.00; 5.00 g = 0.005 00 kg

$$\text{molality} = \frac{\left(0.135 \times 10^{-3}\ \text{g} \times \dfrac{1\ \text{mol}}{119.00\ \text{g}}\right)}{0.005\ 00\ \text{kg}} = 2.27 \times 10^{-4}\ \text{mol/kg} = 2.27 \times 10^{-4}\ m$$

(c) $C_9H_8O_4$, 180.16; 145 g = 0.145 kg

$$\text{molality} = \frac{\left(5.50\ \text{g} \times \dfrac{1\ \text{mol}}{180.16\ \text{g}}\right)}{0.145\ \text{kg}} = 0.211\ \text{mol/kg} = 0.211\ m$$

12.70 $P_{O_3} = P_{total} \cdot X_{O_3}$

$$X_{O_3} = \frac{P_{O_3}}{P_{total}} = \frac{1.6 \times 10^{-9}\ \text{atm}}{1.3 \times 10^{-2}\ \text{atm}} = 1.2 \times 10^{-7}$$

Assume one mole of air (29 g/mol)

mol $O_3 = n_{air} \cdot X_{O_3} = (1\ \text{mol})(1.2 \times 10^{-7}) = 1.2 \times 10^{-7}$ mol O_3

O_3, 48.00; mass $O_3 = 1.2 \times 10^{-7}\ \text{mol} \times \dfrac{48.0\ \text{g}}{1\ \text{mol}} = 5.8 \times 10^{-6}\ \text{g}\ O_3$

$$\text{ppm}\ O_3 = \frac{5.8 \times 10^{-6}\ \text{g}}{29\ \text{g}} \times 10^6 = 0.20\ \text{ppm}$$

12.71 Assume 1 mL of blood weighs 1 g. 1 dL = 0.1 L = 100 mL = 100 g

$$\text{ppb} = \frac{10 \times 10^{-6}\ \text{g}}{100\ \text{g}} \times 10^9 = 100\ \text{ppb}$$

12.72 (a) H_2SO_4, 98.08; $\text{molality} = \dfrac{\left(25.0\ \text{g} \times \dfrac{1\ \text{mol}}{98.08\ \text{g}}\right)}{1.30\ \text{kg}} = 0.196\ \text{mol/kg} = 0.196\ m$

(b) $C_{10}H_{14}N_2$, 162.23; CH_2Cl_2, 84.93

$$2.25\ \text{g}\ C_{10}H_{14}N_2 \times \frac{1\ \text{mol}\ C_{10}H_{14}N_2}{162.23\ \text{g}\ C_{10}H_{14}N_2} = 0.0139\ \text{mol}\ C_{10}H_{14}N_2$$

$$80.0\ \text{g}\ CH_2Cl_2 \times \frac{1\ \text{mol}\ CH_2Cl_2}{84.93\ \text{g}\ CH_2Cl_2} = 0.942\ \text{mol}\ CH_2Cl_2$$

$$X_{C_{10}H_{14}N_2} = \frac{0.0139\ \text{mol}}{0.942\ \text{mol} + 0.0139\ \text{mol}} = 0.0145; \quad X_{CH_2Cl_2} = \frac{0.942\ \text{mol}}{0.942\ \text{mol} + 0.0139\ \text{mol}} = 0.985$$

12.73 NaOCl, 74.44

A 5.0 mass % aqueous solution of NaOCl contains 5.0 g NaOCl and 95 g H_2O.

$$\text{molality} = \frac{\left(5.0\ \text{g} \times \dfrac{1\ \text{mol}}{74.44\ \text{g}}\right)}{0.095\ \text{kg}} = 0.71\ \text{mol/kg} = 0.71\ m$$

$$5.0 \text{ g NaOCl x } \frac{1 \text{ mol NaOCl}}{74.44 \text{ g NaOCl}} = 0.0672 \text{ mol NaOCl}$$

$$95 \text{ g } H_2O \text{ x } \frac{1 \text{ mol } H_2O}{18.02 \text{ g } H_2O} = 5.27 \text{ mol } H_2O$$

$$X_{NaOCl} = \frac{0.0672 \text{ mol}}{5.27 \text{ mol} + 0.0672 \text{ mol}} = 0.013$$

12.74 $16.0 \text{ mass } \% = \dfrac{16.0 \text{ g } H_2SO_4}{16.0 \text{ g } H_2SO_4 + 84.0 \text{ g } H_2O}$

H_2SO_4, 98.08; density = 1.1094 g/mL

$$\text{volume of solution} = 100.0 \text{ g x } \frac{1 \text{ mL}}{1.1094 \text{ g}} = 90.14 \text{ mL} = 0.090\ 14 \text{ L}$$

$$\text{molarity} = \frac{\left(16.0 \text{ g x } \frac{1 \text{ mol}}{98.08 \text{ g}}\right)}{0.090\ 14 \text{ L}} = 1.81 \text{ M}$$

12.75 $C_2H_6O_2$, 62.07

A 40.0 mass % aqueous solution of $C_2H_6O_2$ contains 40.0 g $C_2H_6O_2$ and 60.0 g H_2O.

density = 1.0514 g/mL

$$\text{volume of solution} = 100.0 \text{ g x } \frac{1 \text{ mL}}{1.0514 \text{ g}} = 95.1 \text{ mL} = 0.0951 \text{ L}$$

$$\text{molarity} = \frac{\left(40.0 \text{ g x } \frac{1 \text{ mol}}{62.07 \text{ g}}\right)}{0.0951 \text{ L}} = 6.78 \text{ M}$$

12.76 $\text{molality} = \dfrac{\left(40.0 \text{ g x } \frac{1 \text{ mol}}{62.07 \text{ g}}\right)}{0.0600 \text{ kg}} = 10.7 \text{ mol/kg} = 10.7\ m$

12.77 $C_2H_6O_2$, 62.07

$$4.028 \text{ mol } C_2H_6O_2 \text{ x } \frac{62.07 \text{ g } C_2H_6O_2}{1 \text{ mol } C_2H_6O_2} = 250.0 \text{ g } C_2H_6O_2$$

A 4.028 *m* solution contains 4.028 mol (250.0 g) $C_2H_6O_2$ in 1.0 kg (1000.0 g) of water.

solution mass = 1000.0 g + 250.0 g = 1250.0 g

$$\text{solution volume} = 1250.0 \text{ g x } \frac{1 \text{ mL}}{1.0241 \text{ g}} \text{ x } \frac{1 \text{ x } 10^{-3} \text{ L}}{1 \text{ mL}} = 1.221 \text{ L}$$

$$\text{molarity} = \frac{4.028 \text{ mol}}{1.221 \text{ L}} = 3.300 \text{ M}$$

12.78 $C_{19}H_{21}NO_3$, 311.38; 1.5 mg = 1.5×10^{-3} g

$$1.3 \times 10^{-3} \text{ mol/kg} = \frac{\left(1.5 \times 10^{-3} \text{ g} \times \frac{1 \text{ mol}}{311.38 \text{ g}}\right)}{\text{kg of solvent}}; \quad \text{solve for kg of solvent.}$$

$$\text{kg of solvent} = \frac{\left(1.5 \times 10^{-3} \text{ g} \times \frac{1 \text{ mol}}{311.38 \text{ g}}\right)}{1.3 \times 10^{-3} \text{ mol/kg}} = 0.0037 \text{ kg}$$

Because the solution is very dilute, kg of solvent ≈ kg of solution.

$$\text{g of solution} = (0.0037 \text{ kg})\left(\frac{1000 \text{ g}}{1 \text{ kg}}\right) = 3.7 \text{ g}$$

12.79 $C_{12}H_{22}O_{11}$, 342.30

$$32.5 \text{ g } C_{12}H_{22}O_{11} \times \frac{1 \text{ mol } C_{12}H_{22}O_{11}}{342.30 \text{ g } C_{12}H_{22}O_{11}} = 0.0949 \text{ mol } C_{12}H_{22}O_{11}$$

$$0.850\ m = 0.850 \text{ mol/kg} = \frac{0.0949 \text{ mol}}{\text{kg of } H_2O}; \quad \text{kg of } H_2O = \frac{0.0949 \text{ mol}}{0.850 \text{ mol/kg}} = 0.112 \text{ kg}$$

$$\text{mass of } H_2O = 0.112 \text{ kg} \times \frac{1000 \text{ g}}{1 \text{ kg}} = 112 \text{ g } H_2O$$

12.80 $C_6H_{12}O_6$, 180.16; H_2O, 18.02; Assume 1.00 L of solution.

mass of solution = (1000 mL)(1.0624 g/mL) = 1062.4 g

$$\text{mass of solute} = 0.944 \text{ mol} \times \frac{180.16 \text{ g}}{1 \text{ mol}} = 170.1 \text{ g } C_6H_{12}O_6$$

mass of H_2O = 1062.4 g – 170.1 g = 892.3 g H_2O

$$\text{mol } C_6H_{12}O_6 = 0.944 \text{ mol}; \quad \text{mol } H_2O = 892.3 \text{ g} \times \frac{1 \text{ mol}}{18.02 \text{ g}} = 49.5 \text{ mol}$$

$$\text{(a) } X_{C_6H_{12}O_6} = \frac{\text{mol } C_6H_{12}O_6}{\text{mol } C_6H_{12}O_6 + \text{mol } H_2O} = \frac{0.944 \text{ mol}}{0.944 \text{ mol} + 49.5 \text{ mol}} = 0.0187$$

$$\text{(b) mass \%} = \frac{\text{mass } C_6H_{12}O_6}{\text{total mass of solution}} \times 100\% = \frac{170.1 \text{ g}}{1062.4 \text{ g}} \times 100\% = 16.0\%$$

$$\text{(c) molality} = \frac{\text{mol } C_6H_{12}O_6}{\text{kg } H_2O} = \frac{0.944 \text{ mol}}{0.8923 \text{ kg}} = 1.06 \text{ mol/kg} = 1.06\ m$$

12.81 $C_{12}H_{22}O_{11}$, 342.30; Assume 1.00 L of solution.

mass of solution = (1000 mL)(1.0432 g/mL) = 1043.2 g

$$\text{mass of solute} = 0.335 \text{ mol } C_{12}H_{22}O_{11} \times \frac{342.30 \text{ g } C_{12}H_{22}O_{11}}{1 \text{ mol } C_{12}H_{22}O_{11}} = 114.7 \text{ g } C_{12}H_{22}O_{11}$$

mass of H_2O = 1043.2 g – 114.7 g = 928.5 g H_2O

$$\text{mol } C_{12}H_{22}O_{11} = 0.335 \text{ mol}; \quad 928.5 \text{ g } H_2O \times \frac{1 \text{ mol } H_2O}{18.02 \text{ g } H_2O} = 51.53 \text{ mol } H_2O$$

(a) $X_{C_{12}H_{22}O_{11}} = \dfrac{0.335 \text{ mol}}{51.53 \text{ mol} + 0.335 \text{ mol}} = 0.006\ 46$

(b) mass % $C_{12}H_{22}O_{11} = \dfrac{114.7 \text{ g}}{1043.2 \text{ g}} \times 100\% = 11.0$ mass % $C_{12}H_{22}O_{11}$

(c) molality $= \dfrac{0.335 \text{ mol}}{0.9285 \text{ kg}} = 0.361 \text{ mol/kg} = 0.361\ m$

Solubility and Henry's Law (Section 12.4)

12.82 From Figure 12.6, first determine the solubility, in g/100 mL, for each compound.

(a) $CuSO_4$, 159.6, ~42 g/100 mL; NH_4Cl, 53.5, ~56 g/100 mL

$$M = \frac{\left(42 \text{ g} \times \dfrac{1 \text{ mol } CuSO_4}{159.6 \text{ g}}\right)}{0.100 \text{ L}} = 2.6 \text{ mol/L}$$

$$M = \frac{\left(56 \text{ g} \times \dfrac{1 \text{ mol } NH_4Cl}{53.5 \text{ g}}\right)}{0.100 \text{ L}} = 10.5 \text{ mol/L}$$

NH_4Cl has the higher molar solubility.

(b) CH_3CO_2Na, 82.0, ~48 g/100 mL; glucose ($C_6H_{12}O_6$), 180.2, ~90 g/100 mL

$$M = \frac{\left(48 \text{ g} \times \dfrac{1 \text{ mol } CH_3CO_2Na}{82.0 \text{ g}}\right)}{0.100 \text{ L}} = 5.9 \text{ mol/L}$$

$$M = \frac{\left(90 \text{ g} \times \dfrac{1 \text{ mol glucose}}{180.2 \text{ g}}\right)}{0.100 \text{ L}} = 5.0 \text{ mol/L}$$

CH_3CO_2Na has the higher molar solubility.

12.83 From Figure 12.6, first determine the solubility, in g/100 mL, for each compound.

(a) NaCl, 58.4, ~40 g/100 mL; NH_4Cl, 53.5, ~48 g/100 mL

$$M = \frac{\left(40 \text{ g} \times \dfrac{1 \text{ mol NaCl}}{58.4 \text{ g}}\right)}{0.100 \text{ L}} = 6.8 \text{ mol/L}$$

$$M = \frac{\left(48 \text{ g} \times \dfrac{1 \text{ mol } NH_4Cl}{53.5 \text{ g}}\right)}{0.100 \text{ L}} = 9.0 \text{ mol/L}$$

NH_4Cl has the higher molar solubility.

(b) K_2SO_4, 174.3, ~12 g/100 mL; $CuSO_4$, 159.6, ~20 g/100 mL

$$M = \frac{\left(12 \text{ g} \times \dfrac{1 \text{ mol } K_2SO_4}{174.3 \text{ g}}\right)}{0.100 \text{ L}} = 0.7 \text{ mol/L}$$

$$M = \frac{\left(20\text{ g x }\frac{1\text{ mol }CuSO_4}{159.6\text{ g}}\right)}{0.100\text{ L}} = 1.3\text{ mol/L}$$

$CuSO_4$ has the higher molar solubility.

12.84 $M = k \cdot P = (0.091\ \frac{\text{mol}}{\text{L} \cdot \text{atm}})(0.75\text{ atm}) = 0.068\text{ M}$

12.85 $M = k \cdot P;\ k = \frac{M}{P} = \frac{0.195\text{ M}}{1.00\text{ atm}} = 0.195\text{ mol/(L}\cdot\text{atm)}$

$P = 25.5\text{ mm Hg x }\frac{1.00\text{ atm}}{760\text{ mm Hg}} = 0.0336\text{ atm}$

$M = k \cdot P = (0.195\ \frac{\text{mol}}{\text{L} \cdot \text{atm}})(0.0336\text{ atm}) = 6.55\text{ x }10^{-3}\text{ M}$

12.86 $M = k \cdot P$

$k = \frac{M}{P} = \frac{2.21\text{ x }10^{-3}\text{ mol/L}}{1.00\text{ atm}} = 2.21\text{ x }10^{-3}\ \frac{\text{mol}}{\text{L} \cdot \text{atm}}$

Convert 4 mg/L to mol/L:

$4\text{ mg} = 4\text{ x }10^{-3}\text{ g}$

$$O_2\text{ molarity} = \frac{\left(4\text{ x }10^{-3}\text{ g x }\frac{1\text{ mol}}{32.00\text{ g}}\right)}{1.00\text{ L}} = 1.25\text{ x }10^{-4}\text{ M}$$

$$P_{O_2} = \frac{M}{k} = \frac{1.25\text{ x }10^{-4}\ \frac{\text{mol}}{\text{L}}}{2.21\text{ x }10^{-3}\ \frac{\text{mol}}{\text{L} \cdot \text{atm}}} = 0.06\text{ atm}$$

12.87 $k = 1.93\text{ x }10^{-3}\text{ mol/(L} \cdot \text{atm)}$

$M = k \cdot P = [1.93\text{ x }10^{-3}\text{ mol/(L} \cdot \text{atm)}]\left(68\text{ mm Hg x }\frac{1.00\text{ atm}}{760\text{ mm Hg}}\right) = 1.73\text{ x }10^{-4}\text{ mol/L}$

$1.73\text{ x }10^{-4}\text{ mol/L x }\frac{32.00\text{ g }O_2}{1\text{ mol }O_2}\text{ x }\frac{1\text{ mg}}{1\text{ x }10^{-3}\text{ g}} = 5.5\text{ mg/L}$

12.88 $k = 2.4\text{ x }10^{-4}\text{ mol/(L} \cdot \text{atm)}$

$M = k \cdot P = [2.4\text{ x }10^{-4}\text{ mol/(L} \cdot \text{atm)}](2.00\text{ atm}) = 4.8\text{ x }10^{-4}\text{ mol/L}$

12.89 $M = k \cdot P$

$k = \frac{M}{P} = \frac{1.01\text{ x }10^{-3}\text{ mol/L}}{0.63\text{ atm}} = 1.6\text{ x }10^{-3}\ \frac{\text{mol}}{\text{L} \cdot \text{atm}}$

Colligative Properties (Sections 12.5–12.8)

12.90 $FeCl_3$ and $CaCl_2$ are a strong electrolytes with a van't Hoff factors of i = 4 and 3, respectively. Glucose is a nonelectrolyte. The effective molality of $FeCl_3$ is (4)(0.10) = 0.40 *m*. The effective molality of $CaCl_2$ is (3)(0.15) = 0.45 *m*. Glucose = (1)(0.30) = 0.30 *m*.
Freezing point ranking: $CaCl_2$ < $FeCl_3$ < glucose

12.91 HNO_3 is a strong acid with a van't Hoff factor of approximately i = 2. Sucrose is a nonelectrolyte. The effective molality of HNO_3 is (2)(0.35) = 0.70 *m*.
(a) 0.50 *m* $C_{12}H_{22}O_{11}$ (b) and (c) 0.35 *m* HNO_3

12.92 $C_7H_6O_2$, 122.1; C_2H_6O, 46.07

$$X_{solv} = \frac{\text{mol } C_2H_6O}{\text{mol } C_2H_6O + \text{mol } C_7H_6O_2} = \frac{\left(100 \text{ g x } \frac{1 \text{ mol}}{46.07 \text{ g}}\right)}{\left(100 \text{ g x } \frac{1 \text{ mol}}{46.07 \text{ g}}\right) + \left(5.00 \text{ g x } \frac{1 \text{ mol}}{122.1 \text{ g}}\right)} = 0.981$$

$$P_{soln} = P_{solv} \cdot X_{solv} = (100.5 \text{ mm Hg})(0.981) = 98.6 \text{ mm Hg}$$

12.93 $C_9H_8O_4$, 180.2; $CHCl_3$ is the solvent. For $CHCl_3$, $K_b = 3.63 \frac{°C \cdot kg}{mol}$

$$75.00 \text{ g x } \frac{1 \text{ kg}}{1000 \text{ g}} = 0.075\ 00 \text{ kg}$$

$$\Delta T_b = K_b \cdot m = \left(3.63 \frac{°C \cdot kg}{mol}\right)\left(\frac{\left(1.50 \text{ g x } \frac{1 \text{ mol}}{180.2 \text{ g}}\right)}{0.075\ 00 \text{ kg}}\right) = 0.40\ °C$$

Solution boiling point = 61.7 °C + ΔT_b = 61.7 °C + 0.40 °C = 62.1 °C

12.94 $MgCl_2$, 95.21

$$110 \text{ g x } \frac{1 \text{ kg}}{1000 \text{ g}} = 0.110 \text{ kg}$$

$$\Delta T_f = K_f \cdot m \cdot i = \left(1.86 \frac{°C \cdot kg}{mol}\right)\left(\frac{\left(7.40 \text{ g x } \frac{1 \text{ mol}}{95.21 \text{ g}}\right)}{0.110 \text{ kg}}\right)(2.7) = 3.55\ °C$$

Solution freezing point = 0.00 °C – ΔT_f = 0.00 °C – 3.55 °C = –3.55 °C

12.95 $\Delta T_f = K_f \cdot m \cdot i$; For KBr, i = 2.
Solution freezing point = –2.95 °C = 0.00 °C – ΔT_f; ΔT_f = 2.95 °C

$$m = \frac{\Delta T_f}{K_f \cdot i} = \frac{2.95\ °C}{\left(1.86 \frac{°C \cdot kg}{mol}\right)(2)} = 0.793 \text{ mol/kg} = 0.793\ m$$

12.96 HCl, 36.46; $\Delta T_f = K_f \cdot m \cdot i$

$$190 \text{ g} \times \frac{1 \text{ kg}}{1000 \text{ g}} = 0.190 \text{ kg}$$

Solution freezing point = – 4.65 °C = 0.00 °C – ΔT_f; ΔT_f = 4.65 °C

$$i = \frac{\Delta T_f}{K_f \cdot m} = \frac{4.65\ ^\circ\text{C}}{\left(1.86 \frac{^\circ\text{C} \cdot \text{kg}}{\text{mol}}\right)\left(\frac{9.12 \text{ g} \times \frac{1 \text{ mol}}{36.46 \text{ g}}}{0.190 \text{ kg}}\right)} = 1.9$$

12.97 $\Pi = iMRT$; $$i = \frac{\Pi}{MRT} = \frac{8.57 \text{ atm}}{(0.125 \text{ mol/L})\left(0.082\ 06 \frac{\text{L} \cdot \text{atm}}{\text{K} \cdot \text{mol}}\right)(310 \text{ K})} = 2.70$$

12.98 NaCl is a nonvolatile solute. Methyl alcohol is a volatile solute. When NaCl is added to water, the vapor pressure of the solution is decreased, which means that the boiling point of the solution will increase. When methyl alcohol is added to water, the vapor pressure of the solution is increased, which means that the boiling point of the solution will decrease.

12.99 When 100 mL of 9 M H_2SO_4 at 0 °C is added to 100 mL of liquid water at 0 °C, the temperature rises because ΔH_{soln} for H_2SO_4 is exothermic.
When 100 mL of 9 M H_2SO_4 at 0 °C is added to 100 g of solid ice at 0 °C, some of the ice will melt (an endothermic process) and the temperature will fall because the H_2SO_4 (solute) lowers the freezing point of the ice/water mixture.

12.100

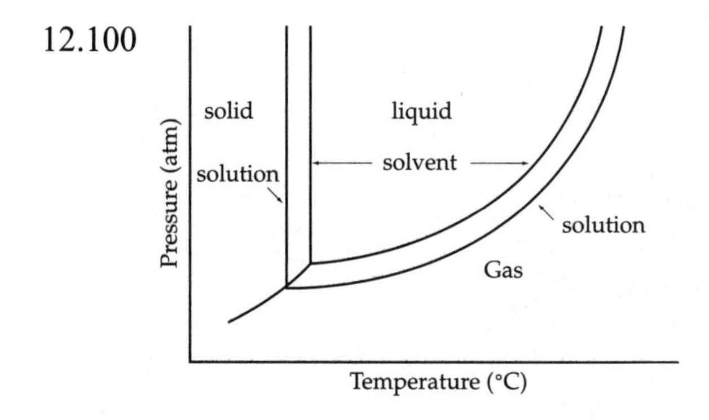

12.101 Molality is a temperature independent concentration unit. For freezing point depression and boiling point elevation, molality is used so that the solute concentration is independent of temperature changes. Molarity is temperature dependent. Molarity can be used for osmotic pressure because osmotic pressure is measured at a fixed temperature.

12.102 (a) CH_4N_2O, 60.06; H_2O, 18.02

$$10.0 \text{ g } CH_4N_2O \times \frac{1 \text{ mol } CH_4N_2O}{60.06 \text{ g } CH_4N_2O} = 0.167 \text{ mol } CH_4N_2O$$

$$150.0 \text{ g } H_2O \times \frac{1 \text{ mol } H_2O}{18.02 \text{ g } H_2O} = 8.32 \text{ mol } H_2O$$

$$X_{H_2O} = \frac{8.32 \text{ mol}}{8.32 \text{ mol} + 0.167 \text{ mol}} = 0.980$$

$$P_{soln} = P^{o}_{H_2O} \cdot X_{H_2O} = (71.93 \text{ mm Hg})(0.980) = 70.5 \text{ mm Hg}$$

(b) LiCl, 42.39; $10.0 \text{ g LiCl} \times \frac{1 \text{ mol LiCl}}{42.39 \text{ g LiCl}} = 0.236 \text{ mol LiCl}$

LiCl dissociates into $Li^+(aq)$ and $Cl^-(aq)$ in H_2O.
mol Li^+ = mol Cl^- = mol LiCl = 0.236 mol

$$150.0 \text{ g } H_2O \times \frac{1 \text{ mol } H_2O}{18.02 \text{ g } H_2O} = 8.32 \text{ mol } H_2O$$

$$X_{H_2O} = \frac{8.32 \text{ mol}}{8.32 \text{ mol} + 0.236 \text{ mol} + 0.236 \text{ mol}} = 0.946$$

$$P_{soln} = P^{o}_{H_2O} \cdot X_{H_2O} = (71.93 \text{ mm Hg})(0.946) = 68.0 \text{ mm Hg}$$

12.103 $C_6H_{12}O_6$, 180.16; CH_3OH, 32.04

$$16.0 \text{ g } C_6H_{12}O_6 \times \frac{1 \text{ mol } C_6H_{12}O_6}{180.16 \text{ g } C_6H_{12}O_6} = 0.0888 \text{ mol } C_6H_{12}O_6$$

$$80.0 \text{ g } CH_3OH \times \frac{1 \text{ mol } CH_3OH}{32.04 \text{ g } CH_3OH} = 2.50 \text{ mol } CH_3OH$$

$$X_{CH_3OH} = \frac{2.50 \text{ mol}}{2.50 \text{ mol} + 0.0888 \text{ mol}} = 0.966$$

$$P_{soln} = P^{o}_{CH_3OH} \cdot X_{CH_3OH} = (140 \text{ mm Hg})(0.966) = 135 \text{ mm Hg}$$

12.104 For H_2O, $K_b = 0.51 \frac{°C \cdot kg}{mol}$; 150.0 g = 0.1500 kg

(a) $\Delta T_b = K_b \cdot m = \left(0.51 \frac{°C \cdot kg}{mol}\right)\left(\frac{0.167 \text{ mol}}{0.1500 \text{ kg}}\right) = 0.57 \text{ °C}$

Solution boiling point = 100.00 °C + ΔT_b = 100.00 °C + 0.57 °C = 100.57 °C

(b) $\Delta T_b = K_b \cdot m = \left(0.51 \frac{°C \cdot kg}{mol}\right)\left(\frac{2(0.236 \text{ mol})}{0.1500 \text{ kg}}\right) = 1.6 \text{ °C}$

Solution boiling point = 100.00 °C + ΔT_b = 100.00 °C + 1.6 °C = 101.6 °C

12.105 For H_2O, $K_f = 1.86 \frac{°C \cdot kg}{mol}$; 150.0 g = 0.1500 kg

(a) $\Delta T_f = K_f \cdot m = \left(1.86 \frac{°C \cdot kg}{mol}\right)\left(\frac{0.167 \text{ mol}}{0.1500 \text{ kg}}\right) = 2.07 \text{ °C}$

Solution freezing point = 0.00 °C – ΔT_f = 0.00 °C – 2.07 °C = –2.07 °C

(b) $\Delta T_f = K_f \cdot m = \left(1.86 \frac{°C \cdot kg}{mol}\right)\left(\frac{2(0.236 \text{ mol})}{0.1500 \text{ kg}}\right) = 5.85\ °C$

Solution freezing point = 0.00 °C – ΔT_f = 0.00 °C – 5.85 °C = –5.85 °C

12.106 Solution freezing point = – 4.3 °C = 0.00 °C – ΔT_f; ΔT_f = 4.3 °C

$\Delta T_f = K_f \cdot m \cdot i; \qquad i = \frac{\Delta T_f}{K_f \cdot m} = \frac{4.3\ °C}{\left(1.86 \frac{°C \cdot kg}{mol}\right)(1.0 \text{ mol/kg})} = 2.3$

12.107 $\Delta T_b = K_b \cdot m \cdot i = \left(0.51 \frac{°C \cdot kg}{mol}\right)(0.75 \text{ mol/kg})(1.85) = 0.71\ °C$

Solution boiling point = 100.00 °C + ΔT_b = 100.00 °C + 0.71 °C = 100.71 °C

12.108 Let $X_{heptane}$ = x and X_{octane} = 1 – x

(428 mm Hg)x + (175 mm Hg)(1 – x) = 305 mm Hg

(428 mm Hg)x + 175 mm Hg – (175 mm Hg)x = 305 mm Hg

(428 mm Hg – 175 mm Hg)x = 305 mm Hg – 175 mm Hg

(253 mm Hg)x = 130 mm Hg

$x = X_{heptane} = \frac{130 \text{ mm Hg}}{253 \text{ mm Hg}} = 0.514$

12.109 Let $X_{cyclopentane}$ = x and $X_{cyclohexane}$ = 1 – x

(385 mm Hg)x + (122 mm Hg)(1 – x) = 212 mm Hg

(385 mm Hg)x + 122 mm Hg – (122 mm Hg)x = 212 mm Hg

(385 mm Hg – 122 mm Hg)x = 212 mm Hg – 122 mm Hg

(263 mm Hg)x = 90 mm Hg

$x = X_{cyclopentane} = \frac{90 \text{ mm Hg}}{263 \text{ mm Hg}} = 0.342$

12.110 Acetone, C_3H_6O, 58.08, $P^o_{C_3H_6O}$ = 285 mm Hg

Ethyl acetate, $C_4H_8O_2$, 88.11, $P^o_{C_4H_8O_2}$ = 118 mm Hg

$25.0 \text{ g } C_3H_6O \times \frac{1 \text{ mol } C_3H_6O}{58.08 \text{ g } C_3H_6O} = 0.430 \text{ mol } C_3H_6O$

$25.0 \text{ g } C_4H_8O_2 \times \frac{1 \text{ mol } C_4H_8O_2}{88.11 \text{ g } C_4H_8O_2} = 0.284 \text{ mol } C_4H_8O_2$

$X_{C_3H_6O} = \frac{0.430 \text{ mol}}{0.430 \text{ mol} + 0.284 \text{ mol}} = 0.602; \quad X_{C_4H_8O_2} = \frac{0.284 \text{ mol}}{0.430 \text{ mol} + 0.284 \text{ mol}} = 0.398$

$P_{soln} = P^o_{C_3H_6O} \cdot X_{C_3H_6O} + P^o_{C_4H_8O_2} \cdot X_{C_4H_8O_2}$

P_{soln} = (285 mm Hg)(0.602) + (118 mm Hg)(0.398) = 219 mm Hg

12.111 $CHCl_3$, 119.38, $P^{o}_{CHCl_3} = 205$ mm Hg; CH_2Cl_2, 84.93, $P^{o}_{CH_2Cl_2} = 415$ mm Hg

$$15.0\text{ g } CHCl_3 \times \frac{1\text{ mol } CHCl_3}{119.38\text{ g } CHCl_3} = 0.126\text{ mol } CHCl_3$$

$$37.5\text{ g } CH_2Cl_2 \times \frac{1\text{ mol } CH_2Cl_2}{84.93\text{ g } CH_2Cl_2} = 0.442\text{ mol } CH_2Cl_2$$

$$X_{CHCl_3} = \frac{0.126\text{ mol}}{0.126\text{ mol} + 0.442\text{ mol}} = 0.222; \quad X_{CH_2Cl_2} = \frac{0.442\text{ mol}}{0.126\text{ mol} + 0.442\text{ mol}} = 0.778$$

$$P_{soln} = P^{o}_{CHCl_3} \cdot X_{CHCl_3} + P^{o}_{CH_2Cl_2} \cdot X_{CH_2Cl_2}$$

$$P_{soln} = (205\text{ mm Hg})(0.222) + (415\text{ mm Hg})(0.778) = 368\text{ mm Hg}$$

12.112 In the liquid, $X_{acetone} = 0.602$ and $X_{ethyl\ acetate} = 0.398$
In the vapor, $P_{Total} = 219$ mm Hg

$$P_{acetone} = P^{o}_{acetone} \cdot X_{acetone} = (285\text{ mm Hg})(0.602) = 172\text{ mm Hg}$$

$$P_{ethyl\ acetate} = P^{o}_{ethyl\ acetate} \cdot X_{ethyl\ acetate} = (118\text{ mm Hg})(0.398) = 47\text{ mm Hg}$$

$$X_{acetone} = \frac{P_{acetone}}{P_{total}} = \frac{172\text{ mm Hg}}{219\text{ mm Hg}} = 0.785; \quad X_{ethyl\ acetate} = \frac{P_{ethyl\ acetate}}{P_{total}} = \frac{47\text{ mm Hg}}{219\text{ mm Hg}} = 0.215$$

12.113 In the liquid, $X_{CHCl_3} = 0.222$ and $X_{CH_2Cl_2} = 0.778$
In the vapor, $P_{total} = 368$ mm Hg

$$P_{CHCl_3} = P^{o}_{CHCl_3} \cdot X_{CHCl_3} = (205\text{ mm Hg})(0.222) = 45.5\text{ mm Hg}$$

$$P_{CH_2Cl_2} = P^{o}_{CH_2Cl_2} \cdot X_{CH_2Cl_2} = (415\text{ mm Hg})(0.778) = 323\text{ mm Hg}$$

$$X_{CHCl_3} = \frac{P_{CHCl_3}}{P_{total}} = \frac{45.5\text{ mm Hg}}{368\text{ mm Hg}} = 0.124; \quad X_{CH_2Cl_2} = \frac{P_{CH_2Cl_2}}{P_{total}} = \frac{323\text{ mm Hg}}{368\text{ mm Hg}} = 0.878$$

12.114 $C_9H_8O_4$, 180.16; 215 g = 0.215 kg

$$\Delta T_b = K_b \cdot m = 0.47\ ^{\circ}C; \quad K_b = \frac{\Delta T_b}{m} = \frac{0.47\ ^{\circ}C}{\left(\dfrac{5.00\text{ g} \times \dfrac{1\text{ mol}}{180.16\text{ g}}}{0.215\text{ kg}}\right)} = 3.6\ \frac{^{\circ}C \cdot kg}{mol}$$

12.115 $C_6H_8O_6$, 176.13; 50.0 g = 0.0500 kg

$$\Delta T_f = K_f \cdot m = 1.33\ ^{\circ}C; \quad K_f = \frac{\Delta T_f}{m} = \frac{1.33\ ^{\circ}C}{\left(\dfrac{3.00\text{ g} \times \dfrac{1\text{ mol}}{176.13\text{ g}}}{0.0500\text{ kg}}\right)} = 3.90\ \frac{^{\circ}C \cdot kg}{mol}$$

12.116 $\Delta T_b = K_b \cdot m = 1.76\ ^\circ C$; $\quad m = \dfrac{\Delta T_b}{K_b} = \dfrac{1.76\ ^\circ C}{3.07\ \dfrac{^\circ C \cdot kg}{mol}} = 0.573\ m$

12.117 $C_6H_{12}O_6$, 180.16

For ethyl alcohol, $K_b = 1.22\ \dfrac{^\circ C \cdot kg}{mol}$; 285 g = 0.285 kg

$$\Delta T_b = K_b \cdot m = \left(1.22\ \frac{^\circ C \cdot kg}{mol}\right)\left(\frac{26.0\ g\ x\ \dfrac{1\ mol}{180.16\ g}}{0.285\ kg}\right) = 0.618\ ^\circ C$$

Solution boiling point = normal boiling point + ΔT_b = 79.1 °C

Normal boiling point = 79.1 °C – ΔT_b = 79.1 °C – 0.618 °C = 78.5 °C

12.118 $\Pi = MRT$

(a) NaCl 58.44; 350.0 mL = 0.3500 L

There are 2 moles of ions/mole of NaCl

$$\Pi = (2)\left(\frac{5.00\ g\ x\ \dfrac{1\ mol}{58.44\ g}}{0.3500\ L}\right)\left(0.082\ 06\ \frac{L \cdot atm}{K \cdot mol}\right)(323\ K) = 13.0\ atm$$

(b) CH_3CO_2Na, 82.03; 55.0 mL = 0.0550 L

There are 2 moles of ions/mole of CH_3CO_2Na

$$\Pi = (2)\left(\frac{6.33\ g\ x\ \dfrac{1\ mol}{82.03\ g}}{0.0550\ L}\right)\left(0.082\ 06\ \frac{L \cdot atm}{K \cdot mol}\right)(283K) = 65.2\ atm$$

12.119 $$\Pi = MRT = \left(\frac{11.5\ x\ 10^{-3}\ g\ x\ \dfrac{1\ mol}{5990\ g}}{0.006\ 60\ L}\right)\left(0.082\ 06\ \frac{L \cdot atm}{K \cdot mol}\right)(298K) = 0.007\ 11\ atm$$

$$\Pi = 0.007\ 11\ atm\ x\ \frac{760\ mm\ Hg}{1\ atm} = 5.41\ mm\ Hg$$

$$\text{height of } H_2O \text{ column} = 5.41\ mm\ Hg\ x\ \frac{13.534\ mm\ H_2O}{1.00\ mm\ Hg} = 73.2\ mm$$

$$\text{height of } H_2O \text{ column} = 73.2\ mm\ x\ \frac{1\ m}{1000\ mm} = 0.0732\ m$$

12.120 $\Pi = MRT$; $\quad M = \dfrac{\Pi}{RT} = \dfrac{4.85\ atm}{\left(0.082\ 06\ \dfrac{L \cdot atm}{K \cdot mol}\right)(300\ K)} = 0.197\ M$

12.121 $\Pi = MRT; \quad M = \frac{\Pi}{RT} = \frac{7.7 \text{ atm}}{\left(0.082\ 06 \frac{L \cdot atm}{K \cdot mol}\right)(310 \text{ K})} = 0.30 \text{ M}$

12.122 K_f for snow (H_2O) is $1.86 \frac{^oC \cdot kg}{mol}$. Reasonable amounts of salt are capable of lowering the freezing point (ΔT_f) of the snow below an air temperature of –2 °C. Reasonable amounts of salt, however, are not capable of causing a ΔT_f of more than 30 °C, which would be required if it is to melt snow when the air temperature is –30 °C.

12.123 $C_6H_{12}O_6$ does not dissociate in aqueous solution. LiCl and NaCl both dissociate into two solute particles per formula unit in aqueous solution. $CaCl_2$ dissociates into three solute particles per formula unit in aqueous solution. Assume that you have 1.00 g of each substance. Calculate the number of moles of solute particles in 1.00 g of each substance.

$C_6H_{12}O_6$, 180.2; moles solute particles $= 1.00 \text{ g x } \frac{1 \text{ mol}}{180.2 \text{ g}} = 0.005\ 55$ moles

LiCl, 42.4; moles solute particles $= 2\left(1.00 \text{ g x } \frac{1 \text{ mol}}{42.4 \text{ g}}\right) = 0.0472$ moles

NaCl, 58.4; moles solute particles $= 2\left(1.00 \text{ g x } \frac{1 \text{ mol}}{58.4 \text{ g}}\right) = 0.0342$ moles

$CaCl_2$, 111.0; moles solute particles $= 3\left(1.00 \text{ g x } \frac{1 \text{ mol}}{111.0 \text{ g}}\right) = 0.0270$ moles

LiCl produces more solute particles/gram than any of the other three substances. LiCl would be the most efficient per unit mass.

12.124 $\Pi = 407.2 \text{ mm Hg x } \frac{1 \text{ atm}}{760 \text{ mm Hg}} = 0.5358 \text{ atm}$

$\Pi = MRT; \quad M = \frac{\Pi}{RT} = \frac{0.5358 \text{ atm}}{\left(0.082\ 06 \frac{L \cdot atm}{K \cdot mol}\right)(298.15 \text{ K})} = 0.021\ 90 \text{ M}$

$200.0 \text{ mL x } \frac{1 \text{ L}}{1000 \text{ mL}} = 0.2000 \text{ L}$

mol cellobiose = (0.2000 L)(0.021 90 mol/L) = 4.380×10^{-3} mol

molar mass of cellobiose $= \frac{1.500 \text{ g cellobiose}}{4.380 \times 10^{-3} \text{ mol cellobiose}} = 342.5$ g/mol

molecular weight = 342.5

12.125 height of Hg column = 32.9 cm H_2O x $\dfrac{1.00 \text{ cm Hg}}{13.534 \text{ cm } H_2O}$ = 2.43 cm Hg

$\Pi = 2.43 \text{ cm Hg} \times \dfrac{1.00 \text{ atm}}{76.0 \text{ cm Hg}} = 0.0320 \text{ atm}$

$\Pi = MRT; \quad M = \dfrac{\Pi}{RT} = \dfrac{0.0320 \text{ atm}}{\left(0.082\ 06 \dfrac{L \cdot atm}{K \cdot mol}\right)(298 \text{ K})} = 0.001\ 31 \text{ M}$

$20.0 \text{ mL} \times \dfrac{1 \text{ L}}{1000 \text{ mL}} = 0.0200 \text{ L}$

$15.0 \text{ mg} \times \dfrac{1.00 \text{ g}}{1000 \text{ mg}} = 0.0150 \text{ g}$

mol met-enkephalin = (0.0200 L)(0.001 31 mol/L) = 2.62×10^{-5} mol

molar mass of met-enkephalin = $\dfrac{0.0150 \text{ g met-enkephalin}}{2.62 \times 10^{-5} \text{ mol met-enkephalin}}$ = 573 g/mol

molecular weight = 573

12.126 HCl is a strong electrolyte in H_2O and completely dissociates into two solute particles per each HCl.
HF is a weak electrolyte in H_2O. Only a few percent of the HF molecules dissociates into ions.

12.127 Na_2SO_4, 142.0; $m = \dfrac{\left(71 \text{ g} \times \dfrac{1 \text{ mol}}{142.0 \text{ g}}\right)}{1.00 \text{ kg}} = 0.50 \text{ mol/kg} = 0.50\ m$

$\Delta T_b = K_b \cdot m = \left(0.51 \dfrac{^\circ C \cdot kg}{mol}\right)(0.50\ m) = 0.26\ ^\circ C$

The experimental ΔT is approximately 3 times that predicted by the equation above because Na_2SO_4 dissociates into three solute particles (2 Na^+ and SO_4^{2-}) in aqueous solution.

12.128 First, determine the empirical formula:
Assume 100.0 g of β-carotene.

10.51% H $\quad$ $10.51 \text{ g H} \times \dfrac{1 \text{ mol H}}{1.008 \text{ g H}} = 10.43 \text{ mol H}$

89.49% C $\quad$ $89.49 \text{ g C} \times \dfrac{1 \text{ mol C}}{12.01 \text{ g C}} = 7.45 \text{ mol C}$

$C_{7.45}H_{10.43}$; divide each subscript by the smaller, 7.45.

$C_{7.45/7.45}H_{10.43/7.45}$

$CH_{1.4}$

Multiply each subscript by 5 to obtain integers.

Empirical formula is C_5H_7, 67.1.

Second, calculate the molecular weight:

$\Delta T_f = K_f \cdot m; \quad m = \dfrac{\Delta T_f}{K_f} = \dfrac{1.17\ ^\circ C}{37.7\ \dfrac{^\circ C \cdot kg}{mol}} = 0.0310\ mol/kg = 0.0310\ m$

$1.50\ g \times \dfrac{1 kg}{1000\ g} = 1.50 \times 10^{-3}\ kg$

mol β-carotene = $(1.50 \times 10^{-3}\ kg)(0.0310\ mol/kg) = 4.65 \times 10^{-5}$ mol

molar mass of β-carotene = $\dfrac{0.0250\ g\ \beta\text{-carotene}}{4.65 \times 10^{-5}\ mol\ \beta\text{-carotene}} = 538\ g/mol$

molecular weight = 538

Finally, determine the molecular formula:
Divide the molecular weight by the empirical formula weight.

$\dfrac{538}{67.1} = 8$

molecular formula is $C_{(8 \times 5)}H_{(8 \times 7)}$, or $C_{40}H_{56}$

12.129 First, determine the empirical formula:
Assume a 100.0 g sample of lysine.

49.29% C $\quad 49.29\ g\ C \times \dfrac{1\ mol\ C}{12.011\ g\ C} = 4.10\ mol\ C$

9.65% H $\quad 9.65\ g\ H \times \dfrac{1\ mol\ H}{1.008\ g\ H} = 9.57\ mol\ H$

19.16% N $\quad 19.16\ g\ N \times \dfrac{1\ mol\ N}{14.007\ g\ N} = 1.37\ mol\ N$

21.89% O $\quad 21.89\ g\ O \times \dfrac{1\ mol\ O}{15.999\ g\ O} = 1.37\ mol\ O$

$C_{4.10}H_{9.57}N_{1.37}O_{1.37}$; divide each subscript by the smallest, 1.37.
$C_{4.10/1.37}H_{9.57/1.37}N_{1.37/1.37}O_{1.37/1.37}$
Empirical formula is C_3H_7NO, 73.09
Second, calculate the molecular weight:

$\Delta T_f = K_f \cdot m = 1.37\ ^\circ C; \quad m = \dfrac{\Delta T_f}{K_f} = \dfrac{1.37\ ^\circ C}{8.00\ \dfrac{^\circ C \cdot kg}{mol}} = 0.171\ mol/kg = 0.171\ m$

$1.200\ g \times \dfrac{1 kg}{1000\ g} = 1.200 \times 10^{-3}\ kg; \quad 30.0\ mg \times \dfrac{1.00\ g}{1000\ mg} = 0.0300\ g$

mol lysine = $(1.200 \times 10^{-3}\ kg)(0.171\ mol/kg) = 2.05 \times 10^{-4}$ mol

molar mass of lysine = $\dfrac{0.0300\ g\ lysine}{2.05 \times 10^{-4}\ mol\ lysine} = 146\ g/mol$

molecular weight = 146

Finally, determine the molecular formula:
Divide the molecular weight by the empirical formula weight.

$$\frac{146}{73.09} = 2$$

molecular formula is $C_{(2 \times 3)}H_{(2 \times 7)}N_{(2 \times 1)}O_{(2 \times 1)}$, or $C_6H_{14}N_2O_2$

Chapter Problems

12.130 $C_2H_6O_2$, 62.07; $\Delta T_f = 22.0\ °C$

$\Delta T_f = K_f \cdot m$

$$m = \frac{\Delta T_f}{K_f} = \frac{22.0\ °C}{1.86\ \frac{°C \cdot kg}{mol}} = 11.8\ mol/kg = 11.8\ m$$

mol $C_2H_6O_2$ = (3.55 kg)(11.8 mol/kg) = 41.9 mol $C_2H_6O_2$

$$\text{mass } C_2H_6O_2 = 41.9 \text{ mol } C_2H_6O_2 \times \frac{62.07 \text{ g } C_2H_6O_2}{1 \text{ mol } C_2H_6O_2} = 2.60 \times 10^3 \text{ g } C_2H_6O_2$$

12.131 The vapor pressure of toluene is lower than the vapor pressure of benzene at the same temperature. When 1 mL of toluene is added to 100 mL of benzene, the vapor pressure of the solution decreases, which means that the boiling point of the solution will increase. When 1 mL of benzene is added to 100 mL of toluene, the vapor pressure of the solution increases, which means that the boiling point of the solution will decrease.

12.132 When solid $CaCl_2$ is added to liquid water, the temperature rises because ΔH_{soln} for $CaCl_2$ is exothermic.
When solid $CaCl_2$ is added to ice at 0 °C, some of the ice will melt (an endothermic process) and the temperature will fall because the $CaCl_2$ lowers the freezing point of an ice/water mixture.

12.133 AgCl, 143.32; there are 2 ions/AgCl; $\Pi = 2MRT$

$$\Pi = 2\left(\frac{0.007 \times 10^{-3} \text{ g} \times \frac{1 \text{ mol}}{143.32 \text{ g}}}{0.001 \text{ L}}\right)\left(0.082\ 06\ \frac{L \cdot atm}{K \cdot mol}\right)(278 \text{ K}) = 0.002 \text{ atm}$$

12.134 Sucrose ($C_{12}H_{22}O_{11}$), 342.3; fructose ($C_6H_{12}O_6$), 180.2

$$\Pi = MRT; \quad M = \frac{\Pi}{RT} = \frac{0.1843 \text{ atm}}{\left(0.082\ 06\ \frac{L \cdot atm}{K \cdot mol}\right)(298.0 \text{ K})} = 0.007\ 537 \text{ M}$$

n_{total} = (0.007 537 mol/L)(1.50 L) = 0.0113 mol
Let x = g of sucrose and 2.850 g – x = g of fructose
0.0113 mol = mol sucrose + mol fructose

$$0.0113 \text{ mol} = (x)\left(\frac{1 \text{ mol sucrose}}{342.3 \text{ g}}\right) + (2.850 \text{ g} - x)\left(\frac{1 \text{ mol fructose}}{180.2 \text{ g}}\right)$$

Solve for x.

$$0.0113 \text{ mol} = (x)\left(\frac{1 \text{ mol}}{342.3 \text{ g}}\right) + (2.850 \text{ g} - x)\left(\frac{1 \text{ mol}}{180.2 \text{ g}}\right)$$

$$0.0113 \text{ mol} = (x)\left(\frac{1 \text{ mol}}{342.3 \text{ g}}\right) + 0.0158 \text{ mol} - (x)\left(\frac{1 \text{ mol}}{180.2 \text{ g}}\right)$$

$$0.0113 \text{ mol} - 0.0158 \text{ mol} = (x)\left(\frac{1 \text{ mol}}{342.3 \text{ g}}\right) - (x)\left(\frac{1 \text{ mol}}{180.2 \text{ g}}\right)$$

$$-0.0045 \text{ mol} = (x)\left[\left(\frac{1 \text{ mol}}{342.3 \text{ g}}\right) - \left(\frac{1 \text{ mol}}{180.2 \text{ g}}\right)\right]$$

$$x = \frac{-0.0045 \text{ mol}}{\left[\left(\frac{1 \text{ mol}}{342.3 \text{ g}}\right) - \left(\frac{1 \text{ mol}}{180.2 \text{ g}}\right)\right]} = 1.71 \text{ g sucrose}$$

$$1.71 \text{ g sucrose x } \frac{1 \text{ mol sucrose}}{342.3 \text{ g sucrose}} = 0.005\ 00 \text{ mol sucrose}$$

$$X_{sucrose} = \frac{n_{sucrose}}{n_{total}} = \frac{0.005\ 00 \text{ mol}}{0.0113 \text{ mol}} = 0.442$$

12.135 Glycerol ($C_3H_8O_3$), 92.09; diethylformamide ($C_5H_{11}NO$), 101.1

$$\Pi = MRT; \quad M = \frac{\Pi}{RT} = \frac{1.466 \text{ atm}}{\left(0.082\ 06 \frac{\text{L} \cdot \text{atm}}{\text{K} \cdot \text{mol}}\right)(298.0 \text{ K})} = 0.059\ 95 \text{ M}$$

n_{total} = (0.059 95 mol/L)(1.75 L) = 0.105 mol

Let x = g of glycerol and 10.208 g – x = g of diethylformamide

0.105 mol = mol glycerol + mol diethylformamide

$$0.105 \text{ mol} = (x)\left(\frac{1 \text{ mol glycerol}}{92.09 \text{ g}}\right) + (10.208 \text{ g} - x)\left(\frac{1 \text{ mol diethylformamide}}{101.1 \text{ g}}\right)$$

Solve for x.

$$0.105 \text{ mol} = (x)\left(\frac{1 \text{ mol}}{92.09 \text{ g}}\right) + (10.208 \text{ g} - x)\left(\frac{1 \text{ mol}}{101.1 \text{ g}}\right)$$

$$0.105 \text{ mol} = (x)\left(\frac{1 \text{ mol}}{92.09 \text{ g}}\right) + 0.101 \text{ mol} - (x)\left(\frac{1 \text{ mol}}{101.1 \text{ g}}\right)$$

$$0.105 \text{ mol} - 0.101 \text{ mol} = (x)\left(\frac{1 \text{ mol}}{92.09 \text{ g}}\right) - (x)\left(\frac{1 \text{ mol}}{101.1 \text{ g}}\right)$$

$$0.004 \text{ mol} = (x)\left[\left(\frac{1 \text{ mol}}{92.09 \text{ g}}\right) - \left(\frac{1 \text{ mol}}{101.1 \text{ g}}\right)\right]$$

$$x = \frac{0.004 \text{ mol}}{\left[\left(\frac{1 \text{ mol}}{92.09 \text{ g}}\right) - \left(\frac{1 \text{ mol}}{101.1 \text{ g}}\right)\right]} = 4.13 \text{ g glycerol}$$

$$4.13 \text{ g glycerol x } \frac{1 \text{ mol glycerol}}{92.09 \text{ g glycerol}} = 0.0448 \text{ mol glycerol}$$

$$X_{glycerol} = \frac{n_{glycerol}}{n_{total}} = \frac{0.0448 \text{ mol}}{0.105 \text{ mol}} = 0.427$$

12.136 $C_{10}H_8$, 128.17; $\Delta T_f = 0.35$ °C

$$\Delta T_f = K_f \cdot m; \qquad m = \frac{\Delta T_f}{K_f} = \frac{0.35 \text{ °C}}{5.12 \frac{\text{°C} \cdot \text{kg}}{\text{mol}}} = 0.0684 \text{ mol/kg} = 0.0684\ m$$

$$150.0 \text{ g x } \frac{1 \text{kg}}{1000 \text{ g}} = 0.1500 \text{ kg}$$

mol $C_{10}H_8$ = (0.1500 kg)(0.0684 mol/kg) = 0.0103 mol $C_{10}H_8$

$$\text{mass } C_{10}H_8 = 0.0103 \text{ mol } C_{10}H_8 \text{ x } \frac{128.17 \text{ g } C_{10}H_8}{1 \text{ mol } C_{10}H_8} = 1.3 \text{ g } C_{10}H_8$$

12.137 Br_2, 159.81; CCl_4, 153.82

$$P^o_{Br_2} = 30.5 \text{ kPa x } \frac{760 \text{ mm Hg}}{101.325 \text{ kPa}} = 228.8 \text{ mm Hg}$$

$$P^o_{CCl_4} = 16.5 \text{ kPa x } \frac{760 \text{ mm Hg}}{101.325 \text{ kPa}} = 123.8 \text{ mm Hg}$$

$$1.50 \text{ g } Br_2 \text{ x } \frac{1 \text{ mol } Br_2}{159.81 \text{ g } Br_2} = 9.39 \text{ x } 10^{-3} \text{ mol } Br_2$$

$$145.0 \text{ g } CCl_4 \text{ x } \frac{1 \text{ mol } CCl_4}{153.82 \text{ g } CCl_4} = 0.943 \text{ mol } CCl_4$$

$$X_{Br_2} = \frac{9.39 \text{ x } 10^{-3} \text{ mol}}{(0.943 \text{ mol}) + (9.39 \text{ x } 10^{-3} \text{ mol})} = 0.009\ 86$$

$$X_{CCl_4} = \frac{0.943 \text{ mol}}{(0.943 \text{ mol}) + (9.39 \text{ x } 10^{-3} \text{ mol})} = 0.990$$

$$P_{soln} = P^o_{Br_2} \cdot X_{Br_2} + P^o_{CCl_4} \cdot X_{CCl_4}$$

P_{soln} = (228.8 mm Hg)(0.009 86) + (123.8 mm Hg)(0.990) = 125 mm Hg

12.138 NaCl, 58.44; there are 2 ions/NaCl

A 3.5 mass % aqueous solution of NaCl contains 3.5 g NaCl and 96.5 g H_2O.

$$\text{molality} = \frac{\left(3.5 \text{ g x } \frac{1 \text{ mol}}{58.44 \text{ g}}\right)}{0.0965 \text{ kg}} = 0.62 \text{ mol/kg} = 0.62\ m$$

$$\Delta T_f = K_f \cdot 2 \cdot m = \left(1.86 \frac{\text{°C} \cdot \text{kg}}{\text{mol}}\right)(2)(0.62 \text{ mol/kg}) = 2.3 \text{ °C}$$

Solution freezing point = 0.0 °C – ΔT_f = 0.0 °C – 2.3 °C = –2.3 °C

$$\Delta T_b = K_b \cdot 2 \cdot m = \left(0.51 \frac{°C \cdot kg}{mol}\right)(2)(0.62 \text{ mol/kg}) = 0.63 \text{ °C}$$

Solution boiling point = 100.00 °C + ΔT_b = 100.00 °C + 0.63 °C = 100.63 °C

12.139 (a) Assume a total mass of solution of 1000.0 g.

$$\text{ppm} = \frac{\text{mass of solute ion}}{\text{total mass of solution}} \times 10^6$$

For each ion: $\text{mass of solute ion} = \frac{(\text{ppm})(1000.0 \text{ g})}{10^6}$

Ion	Mass	Moles
Cl^-	19.0 g	0.536 mol
Na^+	10.5 g	0.457 mol
SO_4^{2-}	2.65 g	0.0276 mol
Mg^{2+}	1.35 g	0.0555 mol
Ca^{2+}	0.400 g	0.009 98 mol
K^+	0.380 g	0.009 72 mol
HCO_3^-	0.140 g	0.002 29 mol
Br^-	0.065 g	0.000 81 mol
Total	34.5 g	1.099 mol

Mass of H_2O = 1000.0 g – 34.5 g = 965.5 g H_2O = 0.9655 kg H_2O

$$\text{molality} = \frac{1.099 \text{ mol}}{0.9655 \text{ kg}} = 1.138 \text{ mol/kg} = 1.138\ m$$

(b) Assume M = m for a dilute solution.

$$\Pi = MRT = (1.138 \text{ mol/L})\left(0.082\ 06 \frac{L \cdot atm}{K \cdot mol}\right)(300 \text{ K}) = 28.0 \text{ atm}$$

12.140 (a) $90 \text{ mass \% isopropyl alcohol} = \frac{10.5 \text{ g}}{10.5 \text{ g} + \text{mass of } H_2O} \times 100\%$

Solve for the mass of H_2O.

$$\text{mass of } H_2O = \left(10.5 \text{ g} \times \frac{100}{90}\right) - 10.5 \text{ g} = 1.2 \text{ g}$$

mass of solution = 10.5 g + 1.2 g = 11.7 g

11.7 g of rubbing alcohol contains 10.5 g of isopropyl alcohol.

(b) C_3H_8O, 60.10

mass C_3H_8O = (0.90)(50.0 g) = 45 g

$$45 \text{ g } C_3H_8O \times \frac{1 \text{ mol } C_3H_8O}{60.10 \text{ g } C_3H_8O} = 0.75 \text{ mol } C_3H_8O$$

12.141 $C_6H_{12}O_6$, 180.16; 50.0 mL = 0.0500 L; 17.5 mg = 17.5 x 10^{-3} g

$\Pi = MRT$

$$T = \frac{\Pi}{MR} = \frac{\left(37.8 \text{ mm Hg x } \frac{1 \text{ atm}}{760 \text{ mm Hg}}\right)}{\left(\frac{\left(17.5 \text{ x } 10^{-3} \text{g x } \frac{1 \text{ mol}}{180.16 \text{ g}}\right)}{0.0500 \text{ L}}\right)\left(0.082\ 06 \frac{\text{L} \cdot \text{atm}}{\text{K} \cdot \text{mol}}\right)} = 312 \text{ K}$$

12.142 $\Delta T_f = K_f \cdot i \cdot m$; $CaCl_2$, 111.0

$$m = \frac{\Delta T_f}{K_f \cdot i} = \frac{1.14\ ^\circ\text{C}}{\left(1.86 \frac{^\circ\text{C} \cdot \text{kg}}{\text{mol}}\right)(2.71)} = 0.226 \text{ mol/kg}$$

$$0.226 \text{ mol CaCl}_2 \text{ x } \frac{111.0 \text{ g CaCl}_2}{1 \text{ mol CaCl}_2} = 25.1 \text{ g CaCl}_2$$

$$\text{mass \% CaCl}_2 = \frac{\text{mass CaCl}_2}{\text{mass CaCl}_2 + \text{mass H}_2\text{O}} \text{ x } 100\%$$

$$= \frac{25.1 \text{ g}}{25.1 \text{ g} + 1000.0 \text{ g}} \text{ x } 100\% = 2.45\%$$

12.143 K_2SO_4, 174.3

Assume 100.0 g of solution. For a 5.00% K_2SO_4 by mass solution there are 5.00 g K_2SO_4 and 95.0 g (= 0.0950 kg) H_2O.

$$5.00 \text{ g K}_2\text{SO}_4 \text{ x } \frac{1 \text{ mol K}_2\text{SO}_4}{174.3 \text{ g K}_2\text{SO}_4} = 0.0287 \text{ mol K}_2\text{SO}_4$$

$$m = \frac{0.0287 \text{ mol}}{0.0950 \text{ kg}} = 0.302\ m$$

$\Delta T_f = K_f \cdot i \cdot m$

$$i = \frac{\Delta T_f}{K_f \cdot m} = \frac{1.21\ ^\circ\text{C}}{\left(1.86 \frac{^\circ\text{C} \cdot \text{kg}}{\text{mol}}\right)(0.302 \text{ mol/kg})} = 2.15$$

12.144 H_2O, 18.02

A 0.62 m LiCl solution contains 0.62 mol of LiCl and 1.00 kg (= 1000 g) of H_2O.

(0.62 mol LiCl)(1.96) = 1.21 mol of solute particles

$$1000 \text{ g H}_2\text{O x } \frac{1 \text{ mol H}_2\text{O}}{18.02 \text{ g H}_2\text{O}} = 55.5 \text{ mol H}_2\text{O}$$

$$P_{soln} = P^\circ_{H_2O} \cdot X_{H_2O} = 23.76 \text{ mm Hg x } \frac{55.5 \text{ mol H}_2\text{O}}{55.5 \text{ mol H}_2\text{O} + 1.21 \text{ mol solute}} = 23.25 \text{ mm Hg}$$

Vapor pressure depression = 23.76 mm Hg – 23.25 mm Hg = 0.51 mm Hg

12.145 H_2O, 18.02

$$1000\ g\ H_2O \times \frac{1\ mol\ H_2O}{18.02\ g\ H_2O} = 55.5\ mol\ H_2O$$

$$P_{soln} = 23.76\ mm\ Hg - 0.734\ mm\ Hg = 23.03\ mm\ Hg$$

$$P_{soln} = P^o_{H_2O} \cdot X_{H_2O} = 23.76\ mm\ Hg \times \frac{55.5\ mol\ H_2O}{55.5\ mol\ H_2O + i(1.00\ mol\ KCl)}$$

$$23.03\ mm\ Hg = 23.76\ mm\ Hg \times \frac{55.5\ mol\ H_2O}{55.5\ mol\ H_2O + i(1.00\ mol\ KCl)}$$

$$0.9693 = \frac{55.5\ mol\ H_2O}{55.5\ mol\ H_2O + i(1.00\ mol\ KCl)}$$

$$55.5\ mol + i(1.00\ mol) = \frac{55.5\ mol}{0.9693} = 57.26\ mol$$

$i\ (1.00\ mol) = 57.26\ mol - 55.5\ mol = 1.76\ mol$
$i = 1.76$

12.146 First, determine the empirical formula.
$3.47\ mg = 3.47 \times 10^{-3}\ g$ sample
$10.10\ mg = 10.10 \times 10^{-3}\ g\ CO_2$
$2.76\ mg = 2.76 \times 10^{-3}\ g\ H_2O$

$$\text{mass C} = 10.10 \times 10^{-3}\ g\ CO_2 \times \frac{12.01\ g\ C}{44.01\ g\ CO_2} = 2.76 \times 10^{-3}\ g\ C$$

$$\text{mass H} = 2.76 \times 10^{-3}\ g\ H_2O \times \frac{2 \times 1.008\ g\ H}{18.02\ g\ H_2O} = 3.09 \times 10^{-4}\ g\ H$$

$$\text{mass O} = 3.47 \times 10^{-3}\ g - 2.76 \times 10^{-3}\ g\ C - 3.09 \times 10^{-4}\ g\ H = 4.01 \times 10^{-4}\ g\ O$$

$$2.76 \times 10^{-3}\ g\ C \times \frac{1\ mol\ C}{12.01\ g\ C} = 2.30 \times 10^{-4}\ mol\ C$$

$$3.09 \times 10^{-4}\ g\ H \times \frac{1\ mol\ H}{1.008\ g\ H} = 3.07 \times 10^{-4}\ mol\ H$$

$$4.01 \times 10^{-4}\ g\ O \times \frac{1\ mol\ O}{16.00\ g\ O} = 2.51 \times 10^{-5}\ mol\ O = 0.251 \times 10^{-4}\ mol\ O$$

To simplify the empirical formula, divide each mol quantity by 10^{-4}.
$C_{2.30}H_{3.07}O_{0.251}$; divide all subscripts by the smallest, 0.251.
$C_{2.30/0.251}H_{3.07/0.251}O_{0.251/0.251}$
$C_{9.16}H_{12.23}O$
Empirical formula is $C_9H_{12}O$, 136.

Second, determine the molecular weight.

$$7.55\ mg = 7.55 \times 10^{-3}\ g\ \text{estradiol};\ 0.500\ g \times \frac{1\ kg}{1000\ g} = 5.00 \times 10^{-4}\ kg\ \text{camphor}$$

$$\Delta T_f = K_f \cdot m;\ m = \frac{\Delta T_f}{K_f} = \frac{2.10\ ^\circ C}{37.7\ \frac{^\circ C \cdot kg}{mol}} = 0.0557\ mol/kg = 0.0557\ m$$

$$m = \frac{\text{mol estradiol}}{\text{kg solvent}}$$

mol estradiol = m x (kg solvent) = (0.0557 mol/kg)(5.00 x 10^{-4} kg) = 2.79 x 10^{-5} mol

$$\text{molar mass} = \frac{7.55 \times 10^{-3}\ \text{g estradiol}}{2.79 \times 10^{-5}\ \text{mol estradiol}} = 271\ \text{g/mol};\ \text{molecular weight} = 271$$

Finally, determine the molecular formula:
Divide the molecular weight by the empirical formula weight.

$$\frac{271}{136} = 2$$

molecular formula is $C_{(2 \times 9)}H_{(2 \times 12)}O_{(2 \times 1)}$, or $C_{18}H_{24}O_2$

12.147 $CCl_3CO_2H(aq) \rightleftarrows H^+(aq) + CCl_3CO_2^-(aq)$
1.00 – x x x

$$\Delta T_f = K_f \cdot m;\ m = \frac{\Delta T_f}{K_f} = \frac{2.53\ ^\circ C}{1.86\ \frac{^\circ C \cdot kg}{mol}} = 1.36\ m$$

1.36 = 1.00 – x + x + x = 1 + x; x = 0.36
36% of the acid molecules are dissociated.

12.148 (a) H_2SO_4, 98.08; 2.238 mol H_2SO_4 x $\frac{98.08\ g\ H_2SO_4}{1\ mol\ H_2SO_4}$ = 219.50 g H_2SO_4

mass of 2.238 m solution = 219.50 g H_2SO_4 + 1000 g H_2O = 1219.50 g

volume of 2.238 m solution = 1219.50 g x $\frac{1.0000\ mL}{1.1243\ g}$ = 1084.68 mL = 1.0847 L

molarity of 2.238 m solution = $\frac{2.238\ mol}{1.0847\ L}$ = 2.063 M

The molarity of the H_2SO_4 solution is less than the molarity of the $BaCl_2$ solution. Because equal volumes of the two solutions are mixed, H_2SO_4 is the limiting reactant and the number of moles of H_2SO_4 determines the number of moles of $BaSO_4$ produced as the white precipitate.

$$(0.05000\ L) \times (2.063\ mol\ H_2SO_4/L) \times \frac{1\ mol\ BaSO_4}{1\ mol\ H_2SO_4} \times \frac{233.39\ g\ BaSO_4}{1\ mol\ BaSO_4} = 24.07\ g\ BaSO_4$$

(b) More precipitate will form because of the excess $BaCl_2$ in the solution.

12.149 KCl, 74.55; KNO_3, 101.10; $Ba(NO_3)_2$, 261.34

$$\Pi = MRT;\; M = \frac{\Pi}{RT} = \frac{\left(744.7\text{ mm Hg} \times \frac{1.00\text{ atm}}{760\text{ mm Hg}}\right)}{\left(0.082\,06\,\frac{L\cdot atm}{K\cdot mol}\right)(298\text{ K})} = 0.040\,07\text{ M}$$

$$0.040\,07\text{ M} = \frac{\text{mol ions}}{0.500\text{ L}};\;\text{mol ions} = (0.040\,07\text{ mol/L})(0.500\text{ L}) = 0.020\,035\text{ mol ions}$$

mass Cl = 1.000 g x 0.2092 = 0.2092 g Cl

$$\text{mass KCl} = 0.2092\text{ g Cl} \times \frac{1\text{ mol Cl}}{35.453\text{ g Cl}} \times \frac{1\text{ mol KCl}}{1\text{ mol Cl}} \times \frac{74.55\text{ g KCl}}{1\text{ mol KCl}} = 0.440\text{ g KCl}$$

$$\text{mol ions from KCl} = 0.440\text{ g KCl} \times \frac{1\text{ mol KCl}}{74.55\text{ g KCl}} \times \frac{2\text{ mol ions}}{1\text{ mol KCl}} = 0.0118\text{ mol ions}$$

mol ions from KNO_3 and $Ba(NO_3)_2$ = 0.020 035 – 0.0118 = 0.008 235 mol ions

Let x = mass KNO_3 and y = mass $Ba(NO_3)_2$

x + y = 1.000 g – 0.440 g = 0.560 g

$$\left(\frac{x}{101.10}\right)2 + \left(\frac{y}{261.34}\right)3 = 0.008\,235\text{ mol ions}$$

x = 0.560 – y

0.0198x + 0.0115y = 0.008 235

0.0198(0.560 – y) + 0.0115y = 0.008 235

0.011 09 – 0.0198y + 0.0115y = 0.008 235

$$0.002\,855 = 0.0083y;\; y = \frac{0.002\,855}{0.0083} = 0.3440;\; x = 0.560 - 0.3440 = 0.216$$

$$\text{mass \% KCl} = \frac{0.440\text{ g}}{1.000\text{ g}} \times 100\% = 44.0\%$$

$$\text{mass \% }KNO_3 = \frac{0.216\text{ g}}{1.000\text{ g}} \times 100\% = 21.6\%$$

$$\text{mass \% }Ba(NO_3)_2 = \frac{0.344\text{ g}}{1.000\text{ g}} \times 100\% = 34.4\%$$

12.150 Let x = X_{H_2O} and y = X_{CH_3OH} and assume n_{total} = 1.00 mol

(14.5 mm Hg)x + (82.5 mm Hg)y = 39.4 mm Hg

(26.8 mm Hg)x + (140.3 mm Hg)y = 68.2 mm Hg

$$x = \frac{68.2 - 140.3y}{26.8}$$

$$\frac{14.5(68.2 - 140.3y)}{26.8} + 82.5y = 39.4$$

$$\frac{(988.9 - 2034.35y)}{26.8} + 82.5y = 39.4$$

$$36.90 - 75.91y + 82.5y = 39.4;\; 6.59y = 2.5;\; y = \frac{2.5}{6.59} = 0.3794$$

$$x = \frac{[68.2 - 140.3(0.3794)]}{26.8} = 0.5586$$

$$X_{LiCl} = 1 - X_{H_2O} - X_{CH_3OH} = 1 - 0.5586 - 0.3794 = 0.0620$$

The mole fraction equals the number of moles of each component because $n_{total} = 1.00$ mol.

$$\text{mass LiCl} = 0.0620 \text{ mol LiCl} \times \frac{42.39 \text{ g LiCl}}{1 \text{ mol LiCl}} = 2.6 \text{ g LiCl}$$

$$\text{mass } H_2O = 0.5588 \text{ mol } H_2O \times \frac{18.02 \text{ g } H_2O}{1 \text{ mol } H_2O} = 10.1 \text{ g } H_2O$$

$$\text{mass } CH_3OH = 0.3794 \text{ mol } CH_3OH \times \frac{32.04 \text{ g } CH_3OH}{1 \text{ mol } CH_3OH} = 12.2 \text{ g } CH_3OH$$

total mass = 2.6 g + 10.1 g + 12.2 g = 24.9 g

$$\text{mass \% LiCl} = \frac{2.6 \text{ g}}{24.9 \text{ g}} \times 100\% = 10\%$$

$$\text{mass \% } H_2O = \frac{10.1 \text{ g}}{24.9 \text{ g}} \times 100\% = 41\%$$

$$\text{mass \% } CH_3OH = \frac{12.2 \text{ g}}{24.9 \text{ g}} \times 100\% = 49\%$$

12.151 KI, 166.00

$$\Delta T_f = K_f \cdot m \cdot i; \; m = \frac{\Delta T_f}{K_f \cdot i} = \frac{1.95 \text{ °C}}{\left(1.86 \frac{\text{°C} \cdot \text{kg}}{\text{mol}}\right)(2)} = 0.524 \text{ mol/kg} = 0.524 \, m$$

$$\Pi = i \cdot MRT; \; M = \frac{\Pi}{i \cdot RT} = \frac{25.0 \text{ atm}}{(2)\left(0.082\,06 \frac{\text{L} \cdot \text{atm}}{\text{K} \cdot \text{mol}}\right)(298 \text{ K})} = 0.511 \text{ M} = 0.511 \text{ mol/L}$$

1.000 L of solution contains 0.511 mol KI and is 0.524 *m*.

$$\text{mass KI} = 0.511 \text{ mol KI} \times \frac{166.00 \text{ g KI}}{1 \text{ mol KI}} = 84.83 \text{ g KI}$$

Calculate the mass of solvent in this solution.

$$0.524 \, m = 0.524 \text{ mol/kg} = \frac{0.511 \text{ mol}}{\text{mass of solvent}}$$

$$\text{mass of solvent} = \frac{0.511 \text{ mol}}{0.524 \text{ mol/kg}} = 0.9752 \text{ kg} = 975.2 \text{ g}$$

mass of solution = mass KI + mass of solvent = 84.83 g + 975.2 g = 1060 g

$$\text{density} = \frac{1060 \text{ g}}{1000 \text{ mL}} = 1.06 \text{ g/mL}$$

12.152 Solution freezing point = –1.03 °C = 0.00 °C – ΔT_f; ΔT_f = 1.03 °C

$$\Delta T_f = K_f \cdot m; \quad m = \frac{\Delta T_f}{K_f} = \frac{1.03\ ^\circ C}{1.86 \frac{^\circ C \cdot kg}{mol}} = 0.554\ mol/kg = 0.554\ m$$

$$\Pi = MRT; \quad M = \frac{\Pi}{RT} = \frac{(12.16\ atm)}{\left(0.082\ 06 \frac{L \cdot atm}{K \cdot mol}\right)(298\ K)} = 0.497\ \frac{mol}{L}$$

Assume 1.000 L = 1000 mL of solution.

mass of solution = (1000 mL)(1.063 g/mL) = 1063 g

$$\text{mass of } H_2O \text{ in 1000 mL of solution} = \frac{1000\ g\ H_2O}{0.554\ mol\ of\ solute} \times 0.497\ mol = 897\ g\ H_2O$$

mass of solute = total mass – mass of H_2O = 1063 g – 897 g = 166 g solute

$$\text{molar mass} = \frac{166\ g}{0.497\ mol} = 334\ g/mol$$

12.153 C_6H_6, 78.11

$$299\ mm\ Hg = P^o_{C_6H_6} \cdot X_{C_6H_6} + P^o_X \cdot X_X$$

$$299\ mm\ Hg = (395\ mm\ Hg) \cdot X_{C_6H_6} + (96\ mm\ Hg) \cdot X_X$$

$$X_{C_6H_6} + X_X = 1; \quad X_X = 1 - X_{C_6H_6}$$

$$299\ mm\ Hg = (395\ mm\ Hg) \cdot X_{C_6H_6} + (96\ mm\ Hg)(1 - X_{C_6H_6})$$

$$299\ mm\ Hg = (395\ mm\ Hg) \cdot X_{C_6H_6} + 96\ mm\ Hg - (96\ mm\ Hg) \cdot X_{C_6H_6}$$

$$299\ mm\ Hg - 96\ mm\ Hg = (395\ mm\ Hg) \cdot X_{C_6H_6} - (96\ mm\ Hg) \cdot X_{C_6H_6}$$

$$203\ mm\ Hg = (299\ mm\ Hg) \cdot X_{C_6H_6}$$

$$X_{C_6H_6} = 203\ mm\ Hg/299\ mm\ Hg = 0.679$$

Assume the mixture contains 1.00 mol (78.11 g) of C_6H_6. Then a 50/50 mixture will also contain 78.11 g of X.

$$X_{C_6H_6} = \frac{1\ mol\ C_6H_6}{1\ mol\ C_6H_6 + mol\ X} = 0.679$$

$$1\ mol\ C_6H_6 = (0.679)(1\ mol\ C_6H_6 + mol\ X)$$

$$\frac{1\ mol\ C_6H_6}{0.679} = 1\ mol\ C_6H_6 + mol\ X$$

$$\frac{1\ mol\ C_6H_6}{0.679} - 1\ mol\ C_6H_6 = mol\ X$$

mol X = 0.473 mol

molar mass X = 78.11 g/0.473 mol = 165 g/mol

12.154 (a) NaCl, 58.44; $CaCl_2$, 110.98; H_2O, 18.02

$$mol\ NaCl = 100.0\ g\ NaCl \times \frac{1\ mol\ NaCl}{58.44\ g\ NaCl} = 1.711\ mol\ NaCl$$

$$\text{mol } CaCl_2 = 100.0 \text{ g } CaCl_2 \times \frac{1 \text{ mol } CaCl_2}{110.98 \text{ g } CaCl_2} = 0.9011 \text{ mol } CaCl_2$$

mass of solution = (1000 mL)(1.15 g/mL) = 1150 g

mass of H_2O in solution = mass of solution – mass NaCl – mass $CaCl_2$

$$= 1150 \text{ g} - 100.0 \text{ g} - 100.0 \text{ g} = 950 \text{ g}$$

$$= 950 \text{ g} \times \frac{1 \text{ kg}}{1000 \text{ g}} = 0.950 \text{ kg}$$

$$\Delta T_b = K_b \cdot (m_{NaCl} \cdot i + m_{CaCl_2} \cdot i)$$

$$\Delta T_b = \left(0.51 \frac{°C \cdot kg}{mol}\right)\left(\frac{(1.711 \text{ mol NaCl} \cdot 2) + (0.9011 \text{ mol } CaCl_2 \cdot 3)}{0.950 \text{ kg}}\right) = 3.3 \text{ °C}$$

solution boiling point = 100.0 °C + ΔT_b = 100.0 °C + 3.3 °C = 103.3 °C

(b) $$\text{mol } H_2O = 950 \text{ g } H_2O \times \frac{1 \text{ mol } H_2O}{18.02 \text{ g } H_2O} = 52.7 \text{ mol } H_2O$$

$$P_{Solution} = P° \cdot X_{H_2O}$$

$$P_{Solution} = P° \cdot \left(\frac{52.7 \text{ mol } H_2O}{(52.7 \text{ mol } H_2O) + (1.711 \text{ mol NaCl} \cdot 2) + (0.9011 \text{ mol } CaCl_2 \cdot 3)}\right)$$

$$P_{Solution} = (23.8 \text{ mm Hg})(0.896) = 21.3 \text{ mm Hg}$$

12.155 HIO_3, 175.91

mass of 1.00 L solution = (1.00 x 10^3 mL)(1.07 g/mL) = 1.07 x 10^3 g

1.00 L of solution contains 1 mol (175.91 g) HIO_3.

mass of H_2O = 1070 g – 175.91 g = 894 g = 0.894 kg

m = 1.00 mol/0.894 kg = 1.12 m

$$\Delta T_f = K_f \cdot m \cdot i$$

$$i = \frac{\Delta T_f}{K_f \cdot m} = \frac{2.78 \text{ °C}}{\left(1.86 \frac{°C \cdot kg}{mol}\right)(1.12 \text{ mol/kg})} = 1.33;$$ The acid is 33% dissociated.

12.156 (a) KI, 166.00

Assume you have 1.000 L of 1.24 M solution.

mass of solution = (1000 mL)(1.15 g/mL) = 1150 g

$$\text{mass of KI in solution} = 1.24 \text{ mol KI} \times \frac{166.00 \text{ g KI}}{1 \text{ mol KI}} = 206 \text{ g KI}$$

mass of H_2O in solution = mass of solution – mass KI = 1150 g – 206 g = 944 g

$$= 944 \text{ g} \times \frac{1 \text{ kg}}{1000 \text{ g}} = 0.944 \text{ kg}$$

$$\text{molality} = \frac{1.24 \text{ mol KI}}{0.944 \text{ kg } H_2O} = 1.31 \; m$$

(b) For KI, i = 2 assuming complete dissociation.

$$\Delta T_f = K_f \cdot m \cdot i = \left(1.86\ \frac{^\circ C \cdot kg}{mol}\right)(1.31\ m)(2) = 4.87\ ^\circ C$$

Solution freezing point = 0.00 °C – ΔT_f = 0.00 °C – 4.87 °C = – 4.87 °C

(c) $$i = \frac{\Delta T_f}{K_f \cdot m} = \frac{4.46\ ^\circ C}{\left(1.86\ \frac{^\circ C \cdot kg}{mol}\right)(1.31\ mol/kg)} = 1.83$$

Because the calculated i is only 1.83 and not 2, the percent dissociation for KI is 83%.

12.157 (a) For NaCl, i = 2 and for $MgCl_2$, i = 3; T = 25 °C = 25 + 273 = 298 K

$$\Pi = i \cdot MRT = [(2)(0.470\ mol/L) + (3)(0.068\ mol/L)]\left(0.082\ 06\ \frac{L \cdot atm}{K \cdot mol}\right)(298\ K) = 28.0\ atm$$

(b) Calculate the molarity for an osmotic pressure = 100.0 atm.

$$\Pi = MRT;\ M = \frac{\Pi}{RT} = \frac{(100.0\ atm)}{\left(0.082\ 06\ \frac{L \cdot atm}{K \cdot mol}\right)(298\ K)} = 4.09\ mol/L$$

$M_{conc} \times V_{conc} = M_{dil} \times V_{dil}$

$$V_{conc} = \frac{M_{dil} \times V_{dil}}{M_{conc}} = \frac{[(2)(0.470\ mol/L) + (3)(0.068\ mol/L)](1.00\ L)}{4.09\ mol/L} = 0.28\ L$$

A volume of 1.00 L of seawater can be reduced to 0.28 L by an osmotic pressure of 100.0 atm. The volume of fresh water that can be obtained is (1.00 L – 0.28 L) = 0.72 L.

12.158 NaCl, 58.44; $C_{12}H_{22}O_{11}$, 342.3

Let X = mass NaCl and Y = mass $C_{12}H_{22}O_{11}$, then X + Y = 100.0 g.

$$500.0\ g \times \frac{1\ kg}{1000\ g} = 0.5000\ kg$$

Solution freezing point = –2.25 °C = 0.00 °C – ΔT_f; ΔT_f = 0.00 °C + 2.25 °C = 2.25 °C

$\Delta T_f = K_f \cdot (m_{NaCl} \cdot i + m_{C_{12}H_{22}O_{11}})$

$$\Delta T_b = \left(1.86\ \frac{^\circ C \cdot kg}{mol}\right)\left(\frac{(mol\ NaCl \cdot 2) + (mol\ C_{12}H_{22}O_{11})}{0.5000\ kg}\right) = 2.25\ ^\circ C$$

$$mol\ NaCl = X\ g\ NaCl \times \frac{1\ mol\ NaCl}{58.44\ g\ NaCl} = X/58.44\ mol$$

$$mol\ C_{12}H_{22}O_{11} = Y\ g\ C_{12}H_{22}O_{11} \times \frac{1\ mol\ C_{12}H_{22}O_{11}}{342.3\ g\ C_{12}H_{22}O_{11}} = Y/342.3\ mol$$

$$\Delta T_b = \left(1.86\ \frac{^\circ C \cdot kg}{mol}\right)\left(\frac{((X/58.44) \cdot 2\ mol) + ((Y/342.3)\ mol)}{0.5000\ kg}\right) = 2.25\ ^\circ C$$

X = 100 – Y

$$\left(1.86\ \frac{^\circ C \cdot kg}{mol}\right)\left(\frac{\{[(100 - Y)/58.44] \cdot 2\ mol]\} + [(Y/342.3)\ mol]}{0.5000\ kg}\right) = 2.25\ ^\circ C$$

$$\left(\frac{[(200/58.44) - (2Y/58.44) + (Y/342.3)]\ \text{mol}}{0.5000\ \text{kg}}\right) = \frac{2.25\ {}^{\circ}\text{C}}{\left(1.86\ \frac{{}^{\circ}\text{C}\cdot\text{kg}}{\text{mol}}\right)} = 1.21\ \text{mol/kg}$$

$$\left(\frac{[(3.42) - (0.0313Y)]\ \text{mol}}{0.5000\ \text{kg}}\right) = 1.21\ \text{mol/kg}$$

$[(3.42) - (0.0313Y)] = (0.5000\ \text{kg})(1.21) = 0.605$
$- 0.0313\ Y = 0.605 - 3.42 = - 2.81$
$Y = (- 2.81)/(- 0.0313) = 89.8$ g of $C_{12}H_{22}O_{11}$
$X = 100.0\ \text{g} - Y = 100.0\ \text{g} - 89.8\ \text{g} = 10.2$ g of NaCl

12.159 $\Delta T_f = K_f \cdot m$; $\quad m = \frac{\Delta T_f}{K_f} = \frac{2.10\ {}^{\circ}\text{C}}{37.7\ \frac{{}^{\circ}\text{C}\cdot\text{kg}}{\text{mol}}} = 0.0557\ \text{mol/kg} = 0.0557\ m$

$$35.00\ \text{g x}\ \frac{1\ \text{kg}}{1000\ \text{g}} = 0.03500\ \text{kg}$$

$$\text{mol} = 0.0557\ \frac{\text{mol}}{\text{kg}}\ \text{x}\ 0.03500\ \text{kg} = 0.001\ 95\ \text{mol naphthalene}$$

$$\text{molar mass of naphthalene} = \frac{0.250\ \text{g}}{0.001\,95\ \text{mol}} = 128\ \text{g/mol}$$

Multiconcept Problems

12.160 (a) 382.6 mL = 0.3826 L; 20.0 °C = 293.2 K
PV = nRT

$$n_{H_2} = \frac{PV}{RT} = \frac{\left(755\ \text{mm Hg x}\ \frac{1.0\ \text{atm}}{760\ \text{mm Hg}}\right)(0.3826\ \text{L})}{\left(0.082\ 06\ \frac{\text{L}\cdot\text{atm}}{\text{K}\cdot\text{mol}}\right)(293.2\ \text{K})} = 0.0158\ \text{mol}\ H_2$$

(b) $M + x\,HCl \rightarrow x/2\ H_2 + MCl_x$

$$\text{moles HCl reacted} = 0.0158\ \text{mol}\ H_2\ \text{x}\ \frac{x\ \text{mol HCl}}{x/2\ \text{mol}\ H_2} = 0.0316\ \text{mol HCl}$$

moles Cl reacted = moles HCl reacted = 0.0316 mol Cl

$$\text{mass Cl} = 0.0316\ \text{mol Cl x}\ \frac{35.453\ \text{g Cl}}{1\ \text{mol Cl}} = 1.120\ \text{g Cl}$$

mass MCl_x = mass M + mass Cl = 1.385 g + 1.120 g = 2.505 g MCl_x

(c) $\Delta T_f = K_f \cdot m$; $\quad m = \frac{\Delta T_f}{K_f} = \frac{3.53\ {}^{\circ}\text{C}}{1.86\ \frac{{}^{\circ}\text{C}\cdot\text{kg}}{\text{mol}}} = 1.90\ \text{mol/kg} = 1.90\ m$

(d) 25.0 g = 0.0250 kg

$$1.90\ m = 1.90\ \frac{\text{mol}}{\text{kg}} = \frac{\text{x mol ions}}{0.0250\ \text{kg}}$$

mol ions = (1.90 mol/kg)(0.0250 kg) = 0.0475 mol ions

(e) mol M = mol ions – mol Cl = 0.0475 mol – 0.0316 mol = 0.0159 mol M

$$\frac{\text{Cl}}{\text{M}} = \frac{0.0316\ \text{mol}}{0.0159\ \text{mol}} = 2\text{, the formula is } MCl_2.$$

$$\text{molar mass} = \frac{2.505\ \text{g}}{0.0159\ \text{mol}} = 157.5\ \text{g/mol};\ \ \text{molecular mass} = 157.5$$

(f) atomic mass of M = 157.5 – 2(35.453) = 86.6; M = Sr

12.161 (a) 20.00 mL = 0.02000 L

mol NaOH = (0.02000 L)(2.00 mol/L) = 0.0400 mol NaOH

$$\text{mol } CO_2 = 0.0400\ \text{mol NaOH} \times \frac{1\ \text{mol } CO_2}{2\ \text{mol NaOH}} = 0.0200\ \text{mol } CO_2$$

$$\text{mol C} = 0.0200\ \text{mol } CO_2 \times \frac{1\ \text{mol C}}{1\ \text{mol } CO_2} = 0.0200\ \text{mol C}$$

$$\text{mass C} = 0.0200\ \text{mol C} \times \frac{12.011\ \text{g C}}{1\ \text{mol C}} = 0.240\ \text{g C}$$

mass H = mass of compound – mass of C = 0.270 g – 0.240 g = 0.030 g H

$$\text{mol H} = 0.030\ \text{g H} \times \frac{1\ \text{mol H}}{1.008\ \text{g H}} = 0.030\ \text{mol H}$$

The mole ratio of C and H in the molecule is $C_{0.0200}\,H_{0.030}$.

$C_{0.0200}\,H_{0.030}$; divide both subscripts by the smaller of the two, 0.0200.

$C_{0.0200\,/\,0.0200}\,H_{0.030\,/\,0.0200}$

$C_1H_{1.5}$, multiply both subscripts by 2.

$C_{(2 \times 1)}H_{(2 \times 1.5)}$

C_2H_3 (27.05) is the empirical formula.

$$\text{(b)}\ \ \Delta T_f = K_f \cdot m;\ m = \frac{\Delta T_f}{K_f} = \frac{(179.8\ ^\circ\text{C} - 177.9\ ^\circ\text{C})}{37.7\ \frac{^\circ\text{C}\cdot\text{kg}}{\text{mol}}} = 0.050\ \text{mol/kg} = 0.050\ m$$

$$50.0\ \text{g} \times \frac{1\ \text{kg}}{1000\ \text{g}} = 0.0500\ \text{kg}$$

mol solute = (0.050 mol/kg)(0.0500 kg) = 0.0025 mol

$$\text{molar mass} = \frac{0.270\ \text{g}}{0.0025\ \text{mol}} = 108\ \text{g/mol};\ \ \text{molecular weight} = 108$$

(c) To find the molecular formula, first divide the molecular weight by the weight of the empirical formula unit.

$$\frac{108}{27} = 4$$

Multiply the subscripts in the empirical formula by the result of this division, 4.

$C_{(4 \times 2)}\,H_{(4 \times 3)}$

C_8H_{12} is the molecular formula of the compound.

12.162 CO_2, 44.01; H_2O, 18.02

$$\text{mol C} = 106.43 \text{ mg } CO_2 \times \frac{1 \text{ g}}{1000 \text{ mg}} \times \frac{1 \text{ mol } CO_2}{44.01 \text{ g } CO_2} \times \frac{1 \text{ mol C}}{1 \text{ mol } CO_2} = 0.002\ 418 \text{ mol C}$$

$$\text{mass C} = 0.002\ 418 \text{ mol C} \times \frac{12.011 \text{ g C}}{1 \text{ mol C}} = 0.029\ 04 \text{ g C}$$

$$\text{mol H} = 32.100 \text{ mg } H_2O \times \frac{1 \text{ g}}{1000 \text{ mg}} \times \frac{1 \text{ mol } H_2O}{18.02 \text{ g } H_2O} \times \frac{2 \text{ mol H}}{1 \text{ mol } H_2O} = 0.003\ 563 \text{ mol H}$$

$$\text{mass H} = 0.003\ 563 \text{ mol H} \times \frac{1.008 \text{ g H}}{1 \text{ mol H}} = 0.003\ 592 \text{ g H}$$

$$\text{mass O} = \left(36.72 \text{ mg} \times \frac{1 \text{ g}}{1000 \text{ mg}}\right) - 0.029\ 04 \text{ g C} - 0.003\ 592 \text{ g H} = 0.004\ 088 \text{ g O}$$

$$\text{mol O} = 0.004\ 088 \text{ g O} \times \frac{1 \text{ mol O}}{16.00 \text{ g O}} = 0.000\ 255\ 5 \text{ mol O}$$

$C_{0.002\ 418}H_{0.003\ 563}O_{0.000\ 255\ 5}$; divide all subscripts by the smallest, 0.000 255 5.

$C_{0.002\ 418\ /\ 0.000\ 255\ 5}H_{0.003\ 563\ /\ 0.000\ 255\ 5}O_{0.000\ 255\ 5\ /\ 0.000\ 255\ 5}$

$C_{9.5}H_{14}O$; multiply all subscripts by 2.

$C_{(9.5 \times 2)}H_{(14 \times 2)}O_{(1 \times 2)}$

Empirical formula is $C_{19}H_{28}O_2$, 288

T = 25 °C = 25 + 273 = 298 K

$$\Pi = MRT;\ M = \frac{\Pi}{RT} = \frac{\left(21.5 \text{ mm Hg} \times \frac{1 \text{ atm}}{760 \text{ mm Hg}}\right)}{\left(0.082\ 06 \frac{\text{L}\cdot\text{atm}}{\text{K}\cdot\text{mol}}\right)(298 \text{ K})} = 0.001\ 16 \text{ mol/L}$$

15.0 mL = 0.0150 L

mol solute = (0.001 16 mol/L)(0.0150 L) = 1.74×10^{-5} mol

$$\text{molar mass} = \frac{\left(5.00 \text{ mg} \times \frac{1 \text{ g}}{1000 \text{ mg}}\right)}{1.74 \times 10^{-5} \text{ mol}} = 287 \text{ g/mol}$$

The molar mass and the empirical formula weight are essentially identical, so the molecular formula and the empirical formula are the same. The molecular formula is $C_{19}H_{28}O_2$.

12.163 AgCl, 143.32

Solution freezing point = – 4.42 °C = 0.00 °C – ΔT_f; ΔT_f = 0.00 °C + 4.42 °C = 4.42 °C

$\Delta T_f = K_f \cdot m$

$$\text{total ion } m = \frac{\Delta T_f}{K_f} = \frac{4.42\ °C}{1.86 \frac{°C\cdot kg}{mol}} = 2.376 \text{ mol/kg} = 2.376\ m$$

$$150.0 \text{ g} \times \frac{1 \text{ kg}}{1000 \text{ g}} = 0.1500 \text{ kg}$$

total mol of ions = (2.376 mol/kg)(0.1500 kg) = 0.3564 mol of ions
An excess of $AgNO_3$ reacts with all Cl^- to produce 27.575 g AgCl.

$$\text{total mol } Cl^- = 27.575 \text{ g AgCl} \times \frac{1 \text{ mol AgCl}}{143.32 \text{ g AgCl}} \times \frac{1 \text{ mol } Cl^-}{1 \text{ mol AgCl}} = 0.1924 \text{ mol } Cl^-$$

Let P = mol XCl and Q = mol YCl_2.
0.3564 mol ions = 2 x mol XCl + 3 x mol YCl_2 = (2 x P) + (3 x Q)
0.1924 mol Cl^- = mol XCl + 2 x mol YCl_2 = P + (2 x Q)
P = 0.1924 – (2 x Q)
0.3564 = 2 x [0.1924 – (2 x Q)] + (3 x Q) = 0.3848 – (4 x Q) + (3 x Q)
Q = 0.3848 – 0.3564 = 0.0284 mol YCl_2
P = 0.1924 – (2 x Q) = 0.1924 – (2 x 0.0284) = 0.1356 mol XCl

$$\text{mass Cl in XCl} = 0.1356 \text{ mol XCl} \times \frac{1 \text{ mol Cl}}{1 \text{ mol XCl}} \times \frac{35.453 \text{ g Cl}}{1 \text{ mol Cl}} = 4.81 \text{ g Cl}$$

$$\text{mass Cl in } YCl_2 = 0.0284 \text{ mol } YCl_2 \times \frac{2 \text{ mol Cl}}{1 \text{ mol } YCl_2} \times \frac{35.453 \text{ g Cl}}{1 \text{ mol Cl}} = 2.01 \text{ g Cl}$$

total mass of XCl and YCl_2 = 8.900 g
mass of X + Y = total mass – mass Cl = 8.900 g – 4.81 g – 2.01 g = 2.08 g
X is an alkali metal and there are 0.1356 mol of X in XCl.
If X = Li, then mass of X = (0.1356 mol)(6.941 g/mol) = 0.941 g
If X = Na, then mass of X = (0.1356 mol)(22.99 g/mol) = 3.12 g but this is not possible because 3.12 g is greater than the total mass of X + Y. Therefore, X is Li.
mass of Y = 2.08 – mass of X = 2.08 g – 0.941 g = 1.14 g
Y is an alkaline earth metal and there are 0.0284 mol of Y in YCl_2.
molar mass of Y = 1.14 g/0.0284 mol = 40.1 g/mol. Therefore, Y is Ca.

$$\text{mass LiCl} = 0.1356 \text{ mol LiCl} \times \frac{42.39 \text{ g LiCl}}{1 \text{ mol LiCl}} = 5.75 \text{ g LiCl}$$

$$\text{mass } CaCl_2 = 0.0284 \text{ mol } CaCl_2 \times \frac{110.98 \text{ g } CaCl_2}{1 \text{ mol } CaCl_2} = 3.15 \text{ g } CaCl_2$$

12.164 CO_2, 44.01; H_2O, 18.02

$$\text{mol C} = 7.0950 \text{ g } CO_2 \times \frac{1 \text{ mol } CO_2}{44.01 \text{ g } CO_2} \times \frac{1 \text{ mol C}}{1 \text{ mol } CO_2} = 0.1612 \text{ mol C}$$

$$\text{mass C} = 0.1612 \text{ mol C} \times \frac{12.011 \text{ g C}}{1 \text{ mol C}} = 1.9362 \text{ g C}$$

$$\text{mol H} = 2.2668 \text{ g } H_2O \times \frac{1 \text{ mol } H_2O}{18.02 \text{ g } H_2O} \times \frac{2 \text{ mol H}}{1 \text{ mol } H_2O} = 0.2516 \text{ mol H}$$

$$\text{mass H} = 0.2516 \text{ mol H} \times \frac{1.008 \text{ g H}}{1 \text{ mol H}} = 0.2536 \text{ g H}$$

mass O = 3.0078 g − 1.9362 g C − 0.2536 g H = 0.8180 g O

$$\text{mol O} = 0.8180 \text{ g O} \times \frac{1 \text{ mol O}}{16.00 \text{ g O}} = 0.05112 \text{ mol O}$$

$C_{0.1612}H_{0.2516}O_{0.05112}$; divide all subscripts by the smallest, 0.05112.
$C_{0.1612\,/\,0.05112}H_{0.2516\,/\,0.05112}O_{0.05112\,/\,0.05112}$
$C_{3.159}H_{4.922}O$, mass of this unit is 58.90

$\Pi = MRT$

$$M = \frac{\Pi}{RT} = \frac{(0.026\,44 \text{ atm})}{\left(0.082\,06 \dfrac{\text{L}\cdot\text{atm}}{\text{K}\cdot\text{mol}}\right)(298 \text{ K})} = 0.001\,081 \text{ mol/L}$$

mol solute = (0.001 081 mol/L)(0.800 L) = 8.65×10^{-4} mol

$$\text{molar mass} = \frac{0.6617 \text{ g}}{8.65 \times 10^{-4} \text{ mol}} = 765 \text{ g/mol}$$

Divide the molar mass by the unit mass to find the appropriate multiplier.

$$\frac{765}{58.9} = 13$$

$C_{3.159}H_{4.922}O$, multiply each subscript by 13 to get the molecular formula.
$C_{(3.159 \times 13)}H_{(4.922 \times 13)}O_{(1 \times 13)}$
$C_{41}H_{64}O_{13}$ is the molecular formula.

13 Chemical Kinetics

13.1 $3\ I^-(aq) + H_3AsO_4(aq) + 2\ H^+(aq) \rightarrow I_3^-(aq) + H_3AsO_3(aq) + H_2O(l)$

(a) $-\dfrac{\Delta[I^-]}{\Delta t} = 4.8 \times 10^{-4}$ M/s

$$\frac{\Delta[I_3^-]}{\Delta t} = \frac{1}{3}\left(-\frac{\Delta[I^-]}{\Delta t}\right) = \left(\frac{1}{3}\right)(4.8 \times 10^{-4}\ \text{M/s}) = 1.6 \times 10^{-4}\ \text{M/s}$$

(b) $-\dfrac{\Delta[H^+]}{\Delta t} = 2\left(\dfrac{\Delta[I_3^-]}{\Delta t}\right) = (2)(1.6 \times 10^{-4}\ \text{M/s}) = 3.2 \times 10^{-4}\ \text{M/s}$

13.2 (a) $\dfrac{-\Delta[A]}{\Delta t} = \dfrac{0.04\ \text{M} - 0.07\ \text{M}}{500\ \text{s}} = 6.0 \times 10^{-5}\ \text{M/s}$

$$\frac{-\Delta[B]}{\Delta t} = \frac{0.01\ \text{M} - 0.07\ \text{M}}{500\ \text{s}} = 1.2 \times 10^{-4}\ \text{M/s}$$

$$\frac{\Delta[C]}{\Delta t} = \frac{0.03\ \text{M} - 0.0\ \text{M}}{500\ \text{s}} = 6.0 \times 10^{-5}\ \text{M/s}$$

(b) $A + 2B \rightarrow C$

13.3 Rate = $k[BrO_3^-][Br^-][H^+]^2$,
1st order in BrO_3^-, 1st order in Br^-, 2nd order in H^+, 4th order overall
Rate = $k[H_2][I_2]$, 1st order in H_2, 1st order in I_2, 2nd order overall

13.4 For NO, when the [NO] doubles, the rate doubles. This indicates that the reaction order for NO is 1. For Cl_2, when the $[Cl_2]$ is halved, the rate decreases by a factor of 4. This indicates that the reaction order for Cl_2 is 2. The reaction is 3rd order overall.

13.5 (a) Rate = $k[NO_2]^m[CO]^n$

$$m = \frac{\ln\left(\dfrac{\text{Rate}_2}{\text{Rate}_1}\right)}{\ln\left(\dfrac{[NO_2]_2}{[NO_2]_1}\right)} = \frac{\ln\left(\dfrac{1.13 \times 10^{-2}}{5.00 \times 10^{-3}}\right)}{\ln\left(\dfrac{0.150}{0.100}\right)} = 2$$

$$n = \frac{\ln\left(\frac{\text{Rate}_3 \cdot [NO_2]_2}{\text{Rate}_2 \cdot [NO_2]_3}\right)}{\ln\left(\frac{[CO]_3}{[CO]_2}\right)} = \frac{\ln\left(\frac{(2.00 \times 10^{-2})(0.150)^2}{(1.13 \times 10^{-2})(0.200)^2}\right)}{\ln\left(\frac{0.200}{0.100}\right)} = 0$$

$\text{Rate} = k[NO_2]^2$

(b) From Experiment 1:

$$k = \frac{\text{Rate}}{[NO_2]^2} = \frac{5.00 \times 10^{-3}\ \text{M/s}}{(0.100\ \text{M})^2} = 0.500\ 1/(\text{M} \cdot \text{s})$$

(c) $\text{Rate} = k[NO_2]^2 = [0.500\ 1/(\text{M} \cdot \text{s})](0.150\ \text{M})^2 = 1.13 \times 10^{-2}\ \text{M/s}$

13.6 (a) $\text{Rate} = k[C_2H_4Br_2]^m[I^-]^n$

$$m = \frac{\ln\left(\frac{\text{Rate}_2}{\text{Rate}_1}\right)}{\ln\left(\frac{[C_2H_4Br_2]_2}{[C_2H_4Br_2]_1}\right)} = \frac{\ln\left(\frac{1.74 \times 10^{-4}}{6.45 \times 10^{-5}}\right)}{\ln\left(\frac{0.343}{0.127}\right)} = 1$$

$$n = \frac{\ln\left(\frac{\text{Rate}_3 \cdot [C_2H_4Br_2]_2}{\text{Rate}_2 \cdot [C_2H_4Br_2]_3}\right)}{\ln\left(\frac{[I^-]_3}{[I^-]_2}\right)} = \frac{\ln\left(\frac{(1.26 \times 10^{-4})(0.343)}{(1.74 \times 10^{-4})(0.203)}\right)}{\ln\left(\frac{0.125}{0.102}\right)} = 1$$

$\text{Rate} = k[C_2H_4Br_2][I^-]$

(b) From Experiment 1:

$$k = \frac{\text{Rate}}{[C_2H_4Br_2][I^-]} = \frac{6.45 \times 10^{-5}\ \text{M/s}}{(0.127\ \text{M})(0.102\ \text{M})} = 4.98 \times 10^{-3}/(\text{M} \cdot \text{s})$$

(c) $\text{Rate} = \frac{\Delta[I_3^-]}{\Delta t} = k[C_2H_4Br_2][I^-] = [4.98 \times 10^{-3}/(\text{M} \cdot \text{s})](0.150\ \text{M})^2 = 1.12 \times 10^{-4}\ \text{M/s}$

(d) $-\frac{1}{3}\frac{\Delta[I^-]}{\Delta t} = \frac{1}{1}\frac{\Delta[I_3^-]}{\Delta t} = 1.12 \times 10^{-4}\ \text{M/s}$

$$-3\frac{\Delta[I_3^-]}{\Delta t} = (-3)(1.12 \times 10^{-4}\ \text{M/s}) = -3.36 \times 10^{-4}\ \text{M/s}$$

13.7 (a) The reactions in vessels (a) and (b) have the same rate, the same number of B molecules, but different numbers of A molecules. Therefore, the rate does not depend on

A and its reaction order is zero. The same conclusion can be drawn from the reactions in vessels (c) and (d).
The rate for the reaction in vessel (c) is four times the rate for the reaction in vessel (a). Vessel (c) has twice as many B molecules than does vessel (a). Because the rate quadruples when the concentration of B doubles, the reaction order for B is two.
(b) rate = $k[B]^2$

13.8 The rate law is Rate = $k[A]^2[B]$. In vessel 1, the Rate = $k(2)^2(4) = 0.01$ M/s

$$k = \frac{0.01\ \text{M/s}}{(2)^2(4)} = 6.25 \times 10^{-4}\ \text{M/s}$$

In vessel 2, Rate = $(6.25 \times 10^{-4}\ \text{M/s})(4)^2(2) = 0.02$ M/s
In vessel 3, Rate = $(6.25 \times 10^{-4}\ \text{M/s})(1)^2(8) = 0.005$ M/s

13.9 $k = 1.50 \times 10^{-6}$ M/s
$[A]_t = -kt + [A]_o$
$kt = [A]_o - [A]_t$

$$t = \frac{[A]_o - [A]_t}{k} = \frac{(5.00 \times 10^{-3}\ \text{M}) - (1.00 \times 10^{-3}\ \text{M})}{1.50 \times 10^{-6}\ \text{M/s}} = 2667\ \text{s} = 2670\ \text{s}$$

$$t = 2667\ \text{s} \times \frac{1\ \text{min}}{60\ \text{s}} = 44.4\ \text{min}$$

13.10

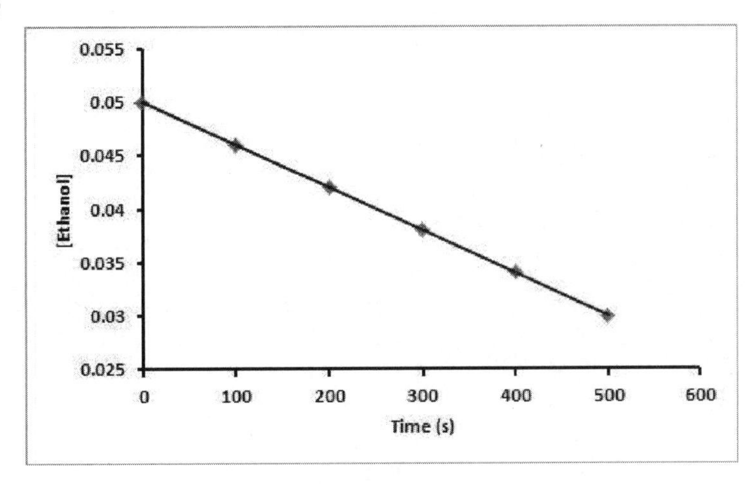

(a) A graph of $[C_2H_5OH]$ vs time is linear (slope = -4.0×10^{-5} M/s), which indicates that the reaction is zeroth order in $[C_2H_5OH]$.
(b) k = –slope = 4.0×10^{-5} M/s
(c) $t = 15.0\ \text{min s} \times \frac{60\ \text{s}}{1\ \text{min}} = 900\ \text{s}$

$[C_2H_5OH]_t = -kt + [C_2H_5OH]_o$
$[C_2H_5OH]_t = -(4.0 \times 10^{-5}\ \text{M/s})(900\ \text{s}) + (5.0 \times 10^{-2}\ \text{M}) = 0.014$ M

13.11 (a) $\ln \dfrac{[Co(NH_3)_5Br^{2+}]_t}{[Co(NH_3)_5Br^{2+}]_o} = -kt$

$k = 6.3 \times 10^{-6}/s$; $\quad t = 10.0\ h \times \dfrac{3600\ s}{1\ h} = 36{,}000\ s$

$\ln[Co(NH_3)_5Br^{2+}]_t = -kt + \ln[Co(NH_3)_5Br^{2+}]_o$

$\ln[Co(NH_3)_5Br^{2+}]_t = -(6.3 \times 10^{-6}/s)(36{,}000\ s) + \ln(0.100)$

$\ln[Co(NH_3)_5Br^{2+}]_t = -2.5294$; $\quad$ After 10.0 h, $[Co(NH_3)_5Br^{2+}] = e^{-2.5294} = 0.080\ M$

(b) $[Co(NH_3)_5Br^{2+}]_o = 0.100\ M$

If 75% of the $Co(NH_3)_5Br^{2+}$ reacts then 25% remains.

$[Co(NH_3)_5Br^{2+}]_t = (0.25)(0.100\ M) = 0.025\ M$

$$\ln \frac{[Co(NH_3)_5Br^{2+}]_t}{[Co(NH_3)_5Br^{2+}]_o} = -kt;\quad t = \frac{\ln \dfrac{[Co(NH_3)_5Br^{2+}]_t}{[Co(NH_3)_5Br^{2+}]_o}}{-k}$$

$$t = \frac{\ln\left(\dfrac{0.025}{0.100}\right)}{-(6.3 \times 10^{-6}/s)} = 2.2 \times 10^5\ s;\quad t = 2.2 \times 10^5\ s \times \frac{1\ h}{3600\ s} = 61\ h$$

13.12 $\ln \dfrac{[A]_t}{[A]_o} = -kt$

(a) Let $[A]_o = 100$ and $[A]_t = 45$ (55% of 100 has reacted, 45 is left)

$$k = -\frac{\ln \dfrac{[A]_t}{[A]_o}}{t} = -\frac{\ln \dfrac{[45]_t}{[100]_o}}{14.2\ h} = 0.0562\ h^{-1}$$

(b) Let $[A]_o = 100$ and $[A]_t = 15$ (85% of 100 has reacted, 15 is left)

$$t = -\frac{\ln \dfrac{[A]_t}{[A]_o}}{k} = -\frac{\ln \dfrac{[15]_t}{[100]_o}}{0.0562\ h^{-1}} = 33.8\ h$$

13.13

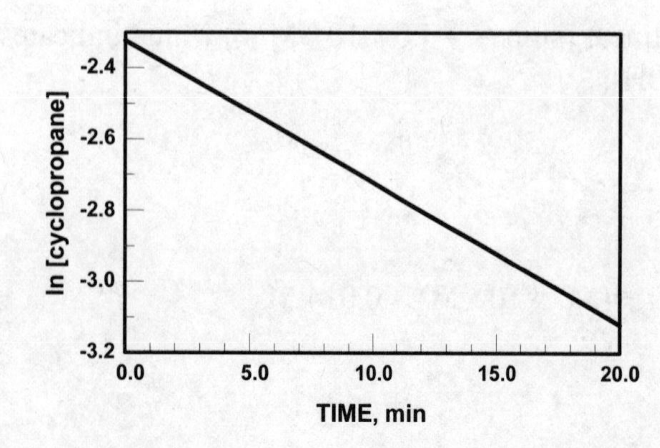

Slope = –0.03989/min = –6.6 x 10^{-4}/s and k = –slope
A plot of ln[cyclopropane] versus time is linear, indicating that the data fit the equation for a first-order reaction. k = 6.6 x 10^{-4}/s (0.040/min)

13.14 $\ln\frac{[N_2O_5]_t}{[N_2O_5]_0} = -kt$, k = – (slope) = – (– 9.8 x 10^{-4}/s) = 9.8 x 10^{-4}/s

(a) $t = 10 \text{ min} \times \frac{60 \text{ s}}{1 \text{ min}} = 600 \text{ s}$

$\ln[N_2O_5]_t = -kt + \ln[N_2O_5]_0 = -(9.8 \times 10^{-4}/s)(600\ s) + \ln(0.100) = -2.891$
$[N_2O_5]_t = e^{-2.891} = 0.055$ M

	$2\ N_2O_5$	$\rightarrow$	$4\ NO_2$	+	O_2
initial (M)	0.100		0		0
change (M)	–2x		+4x		+x
	2x = 0.100 – 0.055 = 0.045				
final (M)	0.055		0.090		0.023

After 10.0 min: $[N_2O_5]$ = 0.055 M, $[NO_2]$ = 0.090 M and $[O_2]$ = 0.023 M

13.15 (a) k = 1.8 x 10^{-5}/s
$t_{1/2} = \frac{0.693}{k} = \frac{0.693}{1.8 \times 10^{-5}/s} = 38{,}500$ s; $\quad t_{1/2} = 38{,}500 \text{ s} \times \frac{1 \text{ h}}{3600 \text{ s}} = 11$ h

(b) 0.30 M $\xrightarrow{t_{1/2}}$ 0.15 M $\xrightarrow{t_{1/2}}$ 0.075 M $\xrightarrow{t_{1/2}}$ 0.0375 M $\xrightarrow{t_{1/2}}$ 0.019 M
(c) Because 25% of the initial concentration corresponds to 1/4 or $(1/2)^2$ of the initial concentration, the time required is two half-lives: $t = 2t_{1/2} = 2(11\text{ h}) = 22$ h

13.16 After one half-life, there would be four A molecules remaining. After two half-lives, there would be two A molecules remaining. This is represented by the drawing at t = 10 min. 10 min is equal to two half-lives, therefore, $t_{1/2}$ = 5 min for this reaction. After 15 min (three half-lives) only one A molecule would remain.

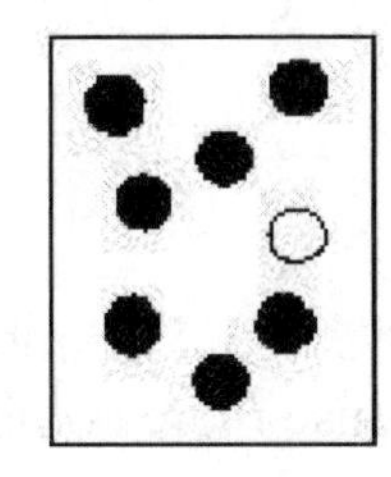

13.17

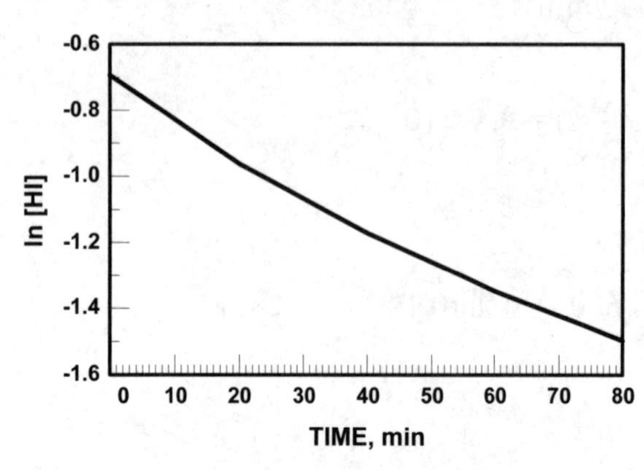

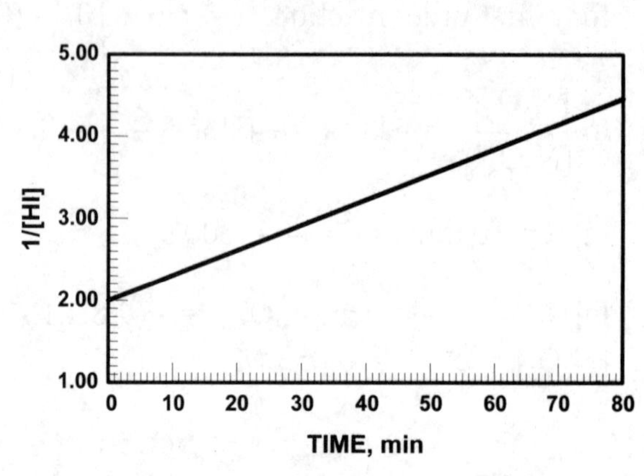

(a) A plot of 1/[HI] versus time is linear. The reaction is second-order.

(b) $k = \text{slope} = 0.0308/(M \cdot \text{min})$

(c) $$t = \frac{1}{k}\left[\frac{1}{[HI]_t} - \frac{1}{[HI]_o}\right] = \frac{1}{0.0308/(M\cdot\text{min})}\left[\frac{1}{0.100\ M} - \frac{1}{0.500\ M}\right] = 260\ \text{min}$$

(d) It requires one half-life ($t_{1/2}$) for the [HI] to drop from 0.400 M to 0.200 M.

$$t_{1/2} = \frac{1}{k[HI]_o} = \frac{1}{[0.0308/(M\cdot\text{min})](0.400\ M)} = 81.2\ \text{min}$$

13.18

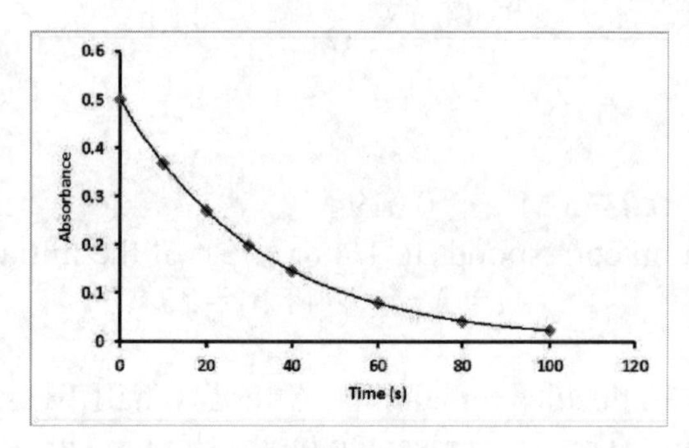

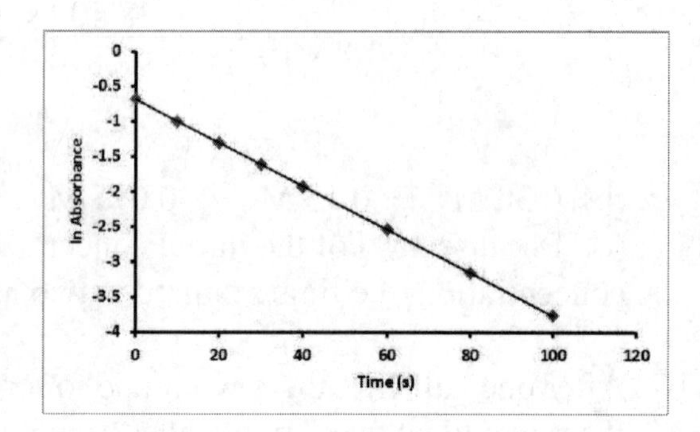

A graph of ln Abs vs time is linear (slope = –0.03071 /s), which indicates that the reaction is first-order in dye.

$k = -\text{slope} = 0.03071\ /s$

$$t_{1/2} = \frac{0.693}{k} = \frac{0.693}{0.03071/s} = 22.6\ s$$

13.19 (a) Because $\Delta E < 0$, the reaction is exothermic.

(b) E_a for the reverse reaction equals 132 kJ/mol + 226 kJ/mol = 358 kJ/mol

(c) The reaction rate increases as temperature increases because more collisions occur with an energy greater than the activation energy.

13.20

unsuccessful

$O{-}N({=}O)\cdots C{\equiv}O$

successful

$O{-}N{=}O\cdots C{\equiv}O$ (O end of NO₂ bent structure toward C)

unsuccessful

$O{-}N({=}O)\cdots O{\equiv}C$

unsuccessful

$O{-}N{=}O\cdots O{\equiv}C$

13.21 (a) $\ln\left(\frac{k_2}{k_1}\right) = \left(\frac{-E_a}{R}\right)\left(\frac{1}{T_2} - \frac{1}{T_1}\right)$

$k_1 = 3.7 \times 10^{-5}/s$, $T_1 = 25\ °C = 298\ K$
$k_2 = 1.7 \times 10^{-3}/s$, $T_2 = 55\ °C = 328\ K$

$$E_a = -\frac{[\ln k_2 - \ln k_1]R}{\left(\frac{1}{T_2} - \frac{1}{T_1}\right)}$$

$$E_a = -\frac{[\ln(1.7 \times 10^{-3}) - \ln(3.7 \times 10^{-5})][8.314 \times 10^{-3}\ kJ/(K\cdot mol)]}{\left(\frac{1}{328\ K} - \frac{1}{298\ K}\right)} = 104\ kJ/mol$$

(b) $k_1 = 3.7 \times 10^{-5}/s$, $T_1 = 25\ °C = 298\ K$
solve for k_2, $T_2 = 35\ °C = 308\ K$

$$\ln k_2 = \left(\frac{-E_a}{R}\right)\left(\frac{1}{T_2} - \frac{1}{T_1}\right) + \ln k_1$$

$$\ln k_2 = \left(\frac{-104\ kJ/mol}{8.314 \times 10^{-3}\ kJ/(K\cdot mol)}\right)\left(\frac{1}{308\ K} - \frac{1}{298\ K}\right) + \ln(3.7 \times 10^{-5})$$

$\ln k_2 = -8.84$; $k_2 = e^{-8.84} = 1.4 \times 10^{-4}/s$

13.22 $\ln\left(\frac{k_2}{k_1}\right) = \left(\frac{-E_a}{R}\right)\left(\frac{1}{T_2} - \frac{1}{T_1}\right)$

$k_1 = 1$, $T_1 = 25\ °C = 298\ K$
$k_2 = 3$, $T_2 = 36\ °C = 309\ K$

$$E_a = -\frac{[\ln k_2 - \ln k_1]R}{\left(\frac{1}{T_2} - \frac{1}{T_1}\right)}$$

$$E_a = -\frac{[\ln(3) - \ln(1)][8.314 \times 10^{-3}\ \text{kJ/(K·mol)}]}{\left(\frac{1}{309\ \text{K}} - \frac{1}{298\ \text{K}}\right)} = 76.5\ \text{kJ/mol}$$

13.23 (a)

$NO_2(g) + F_2(g) \rightarrow NO_2F(g) + F(g)$
$F(g) + NO_2(g) \rightarrow NO_2F(g)$

Overall reaction $2\,NO_2(g) + F_2(g) \rightarrow 2\,NO_2F(g)$

Because F(g) is produced in the first reaction and consumed in the second, it is a reaction intermediate.

(b) In each reaction there are two reactants, so each elementary reaction is bimolecular.

13.24 (a)

$H_2O_2(aq) \rightarrow 2\,OH(aq)$
$H_2O_2(aq) + OH(aq) \rightarrow H_2O(l) + HO_2(aq)$
$HO_2(aq) + OH(aq) \rightarrow H_2O(l) + O_2(g)$

Overall reaction $2\,H_2O_2(aq) \rightarrow 2\,H_2O(l) + O_2(g)$

Because OH(aq) and HO_2(aq) are produced and then consumed, they are reaction intermediates.

(b) Step 1 is unimolecular because it has only one reactant. Steps 2 and 3 are bimolecular because in each reaction there are two reactants.

13.25 (a) Rate = $k[O_3][O]$ (b) Rate = $k[Br]^2[Ar]$ (c) Rate = $k[Co(CN)_5(H_2O)^{2-}]$

13.26 (a) ii) (b) iii) (c) i)

13.27

$NO(g) + Cl_2(g) \rightarrow NOCl(g) + Cl(g)$ (slow)
$NO(g) + Cl(g) \rightarrow NOCl(g)$ (fast)

Overall reaction $2\,NO(g) + Cl_2(g) \rightarrow 2\,NOCl(g)$

The predicted rate law for the overall reaction is the rate law for the first (slow) elementary reaction: Rate = $k[NO][Cl_2]$
The predicted rate law is in accord with the observed rate law.

13.28

$Co(CN)_5(H_2O)^{2-}(aq) \rightarrow Co(CN)_5^{2-}(aq) + H_2O(l)$ (slow)
$Co(CN)_5^{2-}(aq) + I^-(aq) \rightarrow Co(CN)_5I^{3-}(aq)$ (fast)

Overall reaction $Co(CN)_5(H_2O)^{2-}(aq) + I^-(aq) \rightarrow Co(CN)_5I^{3-}(aq) + H_2O(l)$

The predicted rate law for the overall reaction is the rate law for the first (slow) elementary reaction: Rate = $k[Co(CN)_5(H_2O)^{2-}]$
The predicted rate law is in accord with the observed rate law.

13.29
$$NO(g) + O_2(g) \underset{k_{-1}}{\overset{k_1}{\rightleftharpoons}} NO_3(g) \quad \text{fast}$$
$$NO_3(g) + NO(g) \xrightarrow{k_2} 2NO_2(g) \quad \text{slow}$$

(a) $2\ NO(g) + O_2(g) \rightarrow 2\ NO_2(g)$
(b) $Rate_{forward} = k_1[NO][O_2]$ and $Rate_{reverse} = k_{-1}[NO_3]$
Because of the equilibrium, $Rate_{forward} = Rate_{reverse}$, and $k_1[NO][O_2] = k_{-1}[NO_3]$.
$$[NO_3] = \frac{k_1}{k_{-1}}[NO][O_2]$$
The rate law for the rate determining step is $Rate = k_2[NO_3][NO]$. In this rate law substitute for $[NO_3]$.
$Rate = k_2 \dfrac{k_1}{k_{-1}}[NO]^2[O_2]$, which is consistent with the experimental rate law.
(c) $k = \dfrac{k_2 k_1}{k_{-1}}$

13.30
$$I_2(g) \underset{k_{-1}}{\overset{k_1}{\rightleftharpoons}} 2\ I(g) \quad \text{fast}$$
$$H_2(g) + I(g) \underset{k_{-2}}{\overset{k_2}{\rightleftharpoons}} H_2I(g) \quad \text{fast}$$
$$H_2I(g) + I(g) \xrightarrow{k_3} 2\ HI(g) \quad \text{slow}$$

(a) $H_2(g) + I_2(g) \rightarrow 2\ HI(g)$
(b) From step 1, $Rate_{forward} = k_1[I_2]$ and $Rate_{reverse} = k_{-1}[I]^2$
Because of the equilibrium, $Rate_{forward} = Rate_{reverse}$, and $k_1[I_2] = k_{-1}[I]^2$.
$$[I]^2 = \frac{k_1}{k_{-1}}[I_2]$$
From step 2, $Rate_{forward} = k_2[H_2][I]$ and $Rate_{reverse} = k_{-2}[H_2I]$
Because of the equilibrium, $Rate_{forward} = Rate_{reverse}$, and $k_2[H_2][I] = k_{-2}[H_2I]$.
$$[H_2I] = \frac{k_2}{k_{-2}}[H_2][I]$$
The rate law for the rate determining step (step 3) is $Rate = k_3[H_2I][I]$.
Substitute for $[H_2I]$. $Rate = k_3 \dfrac{k_2}{k_{-2}}[H_2][I][I] = k_3 \dfrac{k_2}{k_{-2}}[H_2][I]^2$

Substitute for $[I]^2$. $Rate = k_3 \dfrac{k_2}{k_{-2}}[H_2]\dfrac{k_1}{k_{-1}}[I_2] = k_3 \dfrac{k_1}{k_{-1}}\dfrac{k_2}{k_{-2}}[H_2][I_2]$, which is consistent with the experimental rate law.
(c) $k = \dfrac{k_1 k_2 k_3}{k_{-1} k_{-2}}$

13.31 Assume that concentration is proportional to the number of each molecule in a box.
(a) Comparing boxes (1) and (2), the concentration of A doubles, B and C_2 remain the same and the rate does not change. This means the reaction is zeroth-order in A.
Comparing boxes (1) and (3), the concentration of C_2 doubles, A and B remain the same and the rate doubles. This means the reaction is first-order in C_2.
Comparing boxes (1) and (4), the concentration of B triples, A and C_2 remain the same and the rate triples. This means the reaction is first-order in B.
(b) Rate = k [B][C_2]
(c) The mechanism agrees with the rate law. The rate law for the overall reaction is the rate law for the first (slow) elementary reaction: Rate = k[B][C_2]
(d) B doesn't appear in the overall reaction because it is consumed in the first step and regenerated in the third step. B is therefore a catalyst. C and CB are intermediates because they are formed in one step and then consumed in subsequent steps in the mechanism.

13.32 (a)

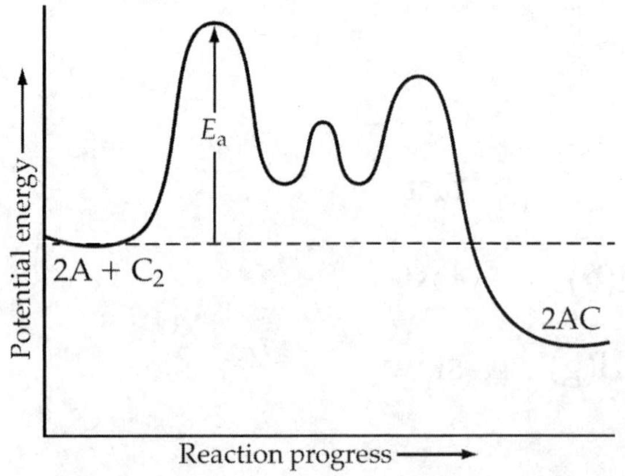

E_a

Potential energy

$2A + C_2$

2AC

Reaction progress

(b)

$$\begin{array}{lcll} B + C_2 & \rightarrow & BC_2 & \text{(slow)} \\ A + BC_2 & \rightarrow & AC + BC & \\ A + BC & \rightarrow & AC + B & \\ \hline 2\,A + C_2 & \rightarrow & 2\,AC & \text{(overall)} \end{array}$$

B doesn't appear in the overall reaction because it is consumed in the first step and regenerated in the third step. B is therefore a catalyst. BC_2 and BC are intermediates because they are formed in one step and then consumed in a subsequent step in the reaction.

13.33 (a) Chlorine radicals are gas phase and homogeneous.
(b) Ice crystals in polar stratospheric clouds are heterogeneous.

13.34 $\ln\left(\frac{k_2}{k_1}\right) = \left(\frac{-E_a}{R}\right)\left(\frac{1}{T_2} - \frac{1}{T_1}\right)$

$k_1 = 8.57 \times 10^{-16}$ cm³/molecules · s, $T_1 = 225$ K
$k_2 = 8.42 \times 10^{-15}$ cm³/molecules · s, $T_2 = 300$ K

$$E_a = -\frac{[\ln k_2 - \ln k_1]R}{\left(\frac{1}{T_2} - \frac{1}{T_1}\right)}$$

$$E_a = -\frac{[\ln(8.42 \times 10^{-15}) - \ln(8.57 \times 10^{-16})][8.314 \times 10^{-3}\ \text{kJ/(K}\cdot\text{mol)}]}{\left(\frac{1}{300\ \text{K}} - \frac{1}{225\ \text{K}}\right)} = 17.1\ \text{kJ/mol}$$

$$k = A e^{-E_a/RT}$$

$$A = \frac{k}{e^{-E_a/RT}} = \frac{8.57 \times 10^{-16}\ \text{cm}^3/\text{molecules}\cdot\text{s}}{e^{-(17.1\ \text{kJ/mol})/[(8.314 \times 10^{-3}\ \text{kJ/mol}\cdot\text{K})(225\ \text{K})]}} = 8.0 \times 10^{-12}\ \text{cm}^3/\text{molecules}\cdot\text{s}$$

13.35 $O_3 + O \rightarrow 2\ O_2$

(a) Rate = $k[O_3][O]$

(b) $k = A e^{-E_a/RT} = (8.0 \times 10^{-12}\ \text{cm}^3/\text{molecules}\cdot\text{s}) e^{-(17.1\ \text{kJ/mol})/[(8.314 \times 10^{-3}\ \text{kJ/mol}\cdot\text{K})(190\ \text{K})]}$

$k = 1.6 \times 10^{-16}\ \text{cm}^3/\text{molecules}\cdot\text{s}$

(c) Rate = $k[O_3][O]$

$= (1.6 \times 10^{-16}\ \text{cm}^3/\text{molecules}\cdot\text{s})(3.5 \times 10^{12}\ \text{molecules/cm}^3)(4.0 \times 10^5\ \text{molecules/cm}^3)$

$= 2.2 \times 10^2\ \text{molecules/cm}^3\cdot\text{s}$

13.36

(1)	$2[Cl + O_3 \rightarrow O_2 + ClO]$	fast
(2)	$2\ ClO \rightarrow Cl_2O_2$	slow, Rate-determining step
(3)	$Cl_2O_2 + h\nu \rightarrow 2\ Cl + O_2$	fast
	$2\ O3 + h\nu \rightarrow 3\ O_2$	overall

(a) Rate = $k[ClO]^2$

(b) Rate = $(7.2 \times 10^{-13}\ \text{cm}^3/\text{molecules}\cdot\text{s})(2.4 \times 10^9\ \text{molecules/cm}^3)^2$

$= 4.1 \times 10^6\ \text{molecules/cm}^3\cdot\text{s}$

(c) 4.1×10^6 molecules/cm^3 · s because the rate of loss of ozone is determined by the rate limiting step and there are two O_3 molecules lost for every 2 ClO molecules.

13.37 $$\frac{\text{Rate}_{(\text{catalytic})}}{\text{Rate}_{(\text{natural})}} = \frac{4.1 \times 10^6\ \text{molecules/cm}^3\cdot\text{s}}{2.2 \times 10^2\ \text{molecules/cm}^3\cdot\text{s}} = 1.9 \times 10^4$$

The catalytic chlorine process has a rate that is approximately 19,000 times greater than the rate of natural loss with oxygen.

13.38

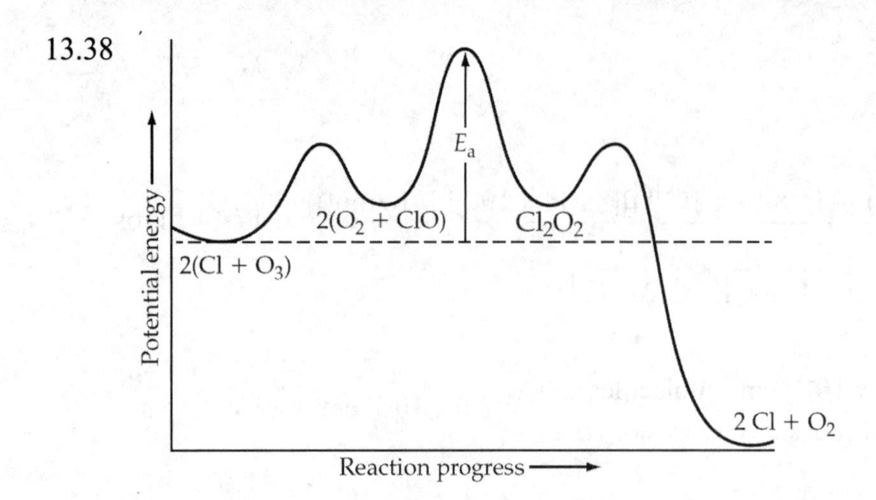

Conceptual Problems

13.39 (a) $\dfrac{-\Delta[A]}{\Delta t} = \dfrac{0.02\ M - 0.07\ M}{250\ s} = 2.0 \times 10^{-4}\ M/s$

$\dfrac{\Delta[B]}{\Delta t} = \dfrac{0.05\ M - 0.0\ M}{250\ s} = 2.0 \times 10^{-4}\ M/s$

$\dfrac{\Delta[C]}{\Delta t} = \dfrac{0.025\ M - 0.0\ M}{250\ s} = 1.0 \times 10^{-4}\ M/s$

(b) $2\ A \rightarrow 2\ B + C$

13.40 (a) Because Rate = k[A][B], the rate is proportional to the product of the number of A molecules and the number of B molecules. The relative rates of the reaction in vessels (a) – (d) are 2 : 1 : 4 : 2.
(b) Because the same reaction takes place in each vessel, the k's are all the same.

13.41 (a) Because Rate = k[A], the rate is proportional to the number of A molecules in each reaction vessel. The relative rates of the reaction are 2 : 4 : 3.
(b) For a first-order reaction, half-lives are independent of concentration. The half-lives are the same.
(c) Concentrations will double, rates will double, and half-lives will be unaffected.

13.42 (a) For the first-order reaction, half of the A molecules are converted to B molecules each minute.

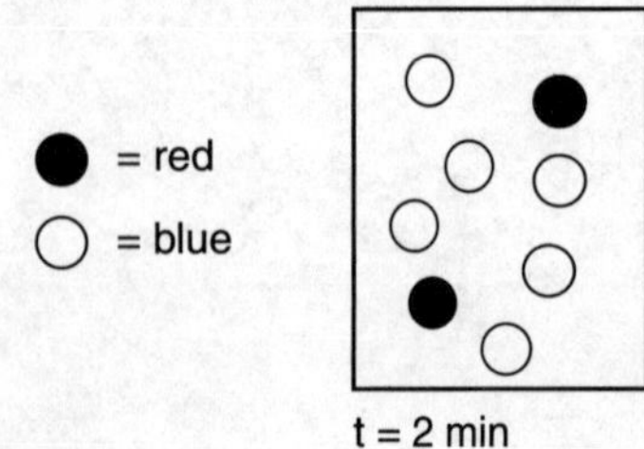

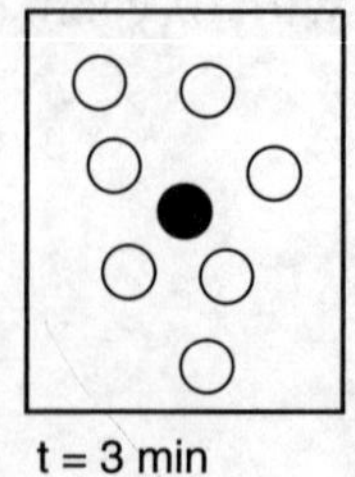

(b) Because half of the A molecules are converted to B molecules in 1 min, the half-life is 1 minute.

13.43 (a) Two molecules of A are converted to two molecules of B every minute. This means the rate is constant throughout the course of the reaction. The reaction is zeroth-order.

(b)

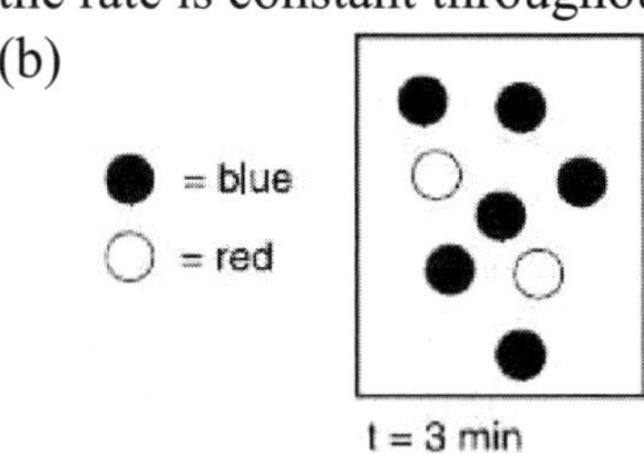

(c) Rate = k

$$k = \frac{(2)\left(\frac{6.0 \times 10^{21}\ \text{molecules}}{1.0\ \text{L}}\right)\left(\frac{1\ \text{mol}}{6.022 \times 10^{23}\ \text{molecules}}\right)}{\text{min}} \times \frac{1\ \text{min}}{60\ \text{s}} = 3.3 \times 10^{-4}\ \text{M/s}$$

13.44 (a) Because the half-life is inversely proportional to the concentration of A molecules, the reaction is second-order in A.

(b) Rate = $k[A]^2$

(c) The second box represents the passing of one half-life, and the third box represents the passing of a second half-life for a second-order reaction. A relative value of k can be calculated.

$$k = \frac{1}{t_{1/2}[A]} = \frac{1}{(1)(16)} = 0.0625$$

$t_{1/2}$ in going from box 3 to box 4 is: $t_{1/2} = \frac{1}{k[A]} = \frac{1}{(0.0625)(4)} = 4\ \text{min}$

(For fourth box, t = 7 min)

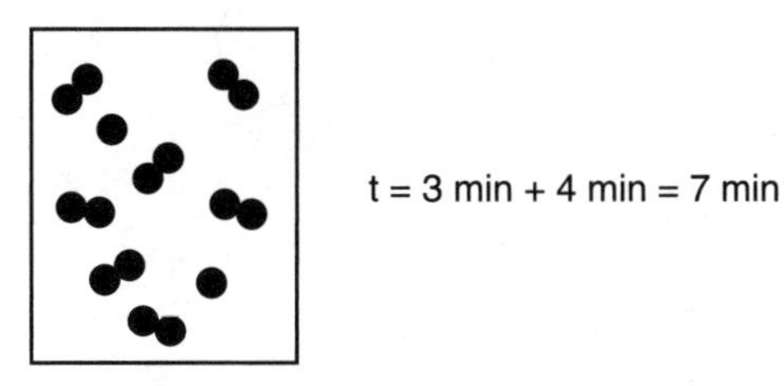

13.45 (a) bimolecular (b) unimolecular (c) termolecular

13.46 Assume that concentration is proportional to the number of each molecule in a box.

(a) Comparing boxes (1) and (2), the concentration of B doubles, A remains the same and the rate does not change. This means the reaction is zeroth-order in B. Comparing boxes (3) and (2), the concentration of A doubles, B remains the same and the rate quadruples. This means the reaction is second-order in A.

(b) Rate = k $[A]^2$.

(c)
$$\begin{array}{lll} 2\,A \rightarrow A_2 & & \text{(slow)} \\ \underline{A_2 + B \rightarrow AB + A} & & \\ A + B \rightarrow AB & & \text{(overall)} \end{array}$$

(d) A_2 is an intermediate because it is formed in one step and then consumed in a subsequent step in the reaction.

13.47 Assume that concentration is proportional to the number of each molecule in a box.
(a) Comparing boxes (2) and (1), the concentration of C doubles, B and AB remain the same and the rate doubles. This means the reaction is first-order in C.
Comparing boxes (2) and (3), the concentration of AB doubles, B and C remain the same and the rate remains the same. This means the reaction is zeroth-order in AB.
Comparing boxes (1) and (4), the concentration of B doubles, C and AB remain the same and the rate doubles. This means the reaction is first-order in B.
Rate = k [B][C]

(b)
$$\begin{array}{lll} B + C \rightarrow BC & & \text{(slow)} \\ \underline{BC + AB \rightarrow A + B_2 + C} & & \\ AB + B \rightarrow A + B_2 & & \text{(overall)} \end{array}$$

(c) C doesn't appear in the overall reaction because it is consumed in the first step and regenerated in the second step. Therefore, C is a catalyst. BC is an intermediate because it is formed in one step and then consumed in a subsequent step in the reaction.

13.48 (a) BC + D → B + CD
(b) 1. B–C + D (reactants), A (catalyst); 2. B---C---A (transition state), D (reactant); 3. A–C (intermediate), B (product), D (reactant); 4. A---C---D (transition state), B (product); 5. A (catalyst), C–D + B (products)
(c) The first step is rate determining because the first maximum in the potential energy curve is greater than the second (relative) maximum; Rate = k[A][BC]
(d) Endothermic

13.49 O–N–O–O–O (zigzag structure: N bonded to O and O; central O bonded to O)

Section Problems
Reaction Rates (Section 13.1)

13.50 $2\,N_2O_5(g) \rightarrow 4\,NO_2(g) + O_2(g)$

time	$[N_2O_5]$	$[O_2]$
200 s	0.0142 M	0.0029 M
300 s	0.0120 M	0.0040 M

$$\text{Rate of decomposition of } N_2O_5 = -\frac{\Delta[N_2O_5]}{\Delta t} = -\frac{0.0120\text{ M} - 0.0142\text{ M}}{300\text{ s} - 200\text{ s}} = 2.2 \times 10^{-5}\text{ M/s}$$

$$\text{Rate of formation of } O_2 = \frac{\Delta[O_2]}{\Delta t} = \frac{0.0040\text{ M} - 0.0029\text{ M}}{300\text{ s} - 200\text{ s}} = 1.1 \times 10^{-5}\text{ M/s}$$

13.51 $2\ N_2O_5(g) \rightarrow 4\ NO_2(g) + O_2(g)$

time	$[N_2O_5]$	$[NO_2]$
500 s	0.0086 M	0.0229 M
600 s	0.0072 M	0.0256 M

$$\text{Rate of decomposition of } N_2O_5 = -\frac{\Delta[N_2O_5]}{\Delta t} = -\frac{0.0072\text{ M} - 0.0086\text{ M}}{600\text{ s} - 500\text{ s}} = 1.4 \times 10^{-5}\text{ M/s}$$

$$\text{Rate of formation of } NO_2 = \frac{\Delta[NO_2]}{\Delta t} = \frac{0.0256\text{ M} - 0.0229\text{ M}}{600\text{ s} - 500\text{ s}} = 2.7 \times 10^{-5}\text{ M/s}$$

13.52 (a) $\text{Rate} = \frac{\Delta[NO_2]}{\Delta t} = \frac{(0.0278\text{ M}) - (0\text{ M})}{700\text{ s} - 0\text{ s}} = 4.0 \times 10^{-5}\text{ M/s}$

(b) $\text{Rate} = \frac{\Delta[NO_2]}{\Delta t} = \frac{(0.0256\text{ M}) - (0.0063\text{ M})}{600\text{ s} - 100\text{ s}} = 3.9 \times 10^{-5}\text{ M/s}$

(c) $\text{Rate} = \frac{\Delta[NO_2]}{\Delta t} = \frac{(0.0229\text{ M}) - (0.0115\text{ M})}{500\text{ s} - 200\text{ s}} = 3.8 \times 10^{-5}\text{ M/s}$

(d) $\text{Rate} = \frac{\Delta[NO_2]}{\Delta t} = \frac{(0.0197\text{ M}) - (0.0160\text{ M})}{400\text{ s} - 300\text{ s}} = 3.7 \times 10^{-5}\text{ M/s}$

Rate (d) is the best estimate of the instantaneous rate because it is determined from measurements taken over the smallest time interval.

13.53 (a) $\text{Rate} = -\frac{\Delta[NO_2]}{\Delta t} = -\frac{(2.53 \times 10^{-3}\text{ M}) - (8.00 \times 10^{-3}\text{ M})}{500\text{ s} - 0\text{ s}} = 1.09 \times 10^{-5}\text{ M/s}$

(b) $\text{Rate} = -\frac{\Delta[NO_2]}{\Delta t} = -\frac{(2.93 \times 10^{-3}\text{ M}) - (5.59 \times 10^{-3}\text{ M})}{400\text{ s} - 100\text{ s}} = 8.87 \times 10^{-6}\text{ M/s}$

(c) $\text{Rate} = -\frac{\Delta[NO_2]}{\Delta t} = -\frac{(3.48 \times 10^{-3}\text{ M}) - (4.29 \times 10^{-3}\text{ M})}{300\text{ s} - 200\text{ s}} = 8.1 \times 10^{-6}\text{ M/s}$

Rate (c) is the best estimate of the instantaneous rate because it is determined from measurements taken over the smallest time interval.

13.54 (a) The instantaneous rate of decomposition of N_2O_5 at t = 200 s is determined from the slope of the curve at t = 200 s.

$$\text{Rate} = -\frac{\Delta[N_2O_5]}{\Delta t} = -\text{slope} = -\frac{(1.20 \times 10^{-2}\text{ M}) - (1.69 \times 10^{-2}\text{ M})}{300\text{ s} - 100\text{ s}} = 2.4 \times 10^{-5}\text{ M/s}$$

(b) The initial rate of decomposition of N_2O_5 is determined from the slope of the curve at t = 0 s. This is equivalent to the slope of the curve from 0 s to 100 s because in this time interval the curve is almost linear.

$$\text{Initial rate} = -\frac{\Delta[N_2O_5]}{\Delta t} = -\text{slope} = -\frac{(1.69 \times 10^{-2}\text{ M}) - (2.00 \times 10^{-2}\text{ M})}{100\text{ s} - 0\text{ s}} = 3.1 \times 10^{-5}\text{ M/s}$$

13.55 (a) The instantaneous rate of decomposition of NO_2 at t = 100 s is determined from the slope of the curve at t = 100 s.

$$\text{Rate} = -\frac{\Delta[NO_2]}{\Delta t} = -\text{slope} = -\frac{(4.00 \times 10^{-3}\ M) - (7.00 \times 10^{-3}\ M)}{190\ s - 20\ s} = 1.8 \times 10^{-5}\ M/s$$

(b) The initial rate of decomposition of NO_2 is determined from the slope of the curve at t = 0 s. This is equivalent to the slope of the curve from 0 s to 50 s because in this time interval the curve is almost linear.

$$\text{Initial rate} = -\frac{\Delta[NO_2]}{\Delta t} = -\text{slope} = -\frac{(6.58 \times 10^{-3}\ M) - (8.00 \times 10^{-3}\ M)}{50\ s - 0\ s} = 2.8 \times 10^{-5}\ M/s$$

13.56 (a) $-\frac{\Delta[H_2]}{\Delta t} = -3\frac{\Delta[N_2]}{\Delta t}$; The rate of consumption of H_2 is 3 times faster.

(b) $\frac{\Delta[NH_3]}{\Delta t} = -2\frac{\Delta[N_2]}{\Delta t}$; The rate of formation of NH_3 is 2 times faster.

13.57 (a) $-\frac{\Delta[O_2]}{\Delta t} = -\frac{5}{4}\frac{\Delta[NH_3]}{\Delta t}$; The rate of consumption of O_2 is 1.25 times faster.

(b) $\frac{\Delta[NO]}{\Delta t} = -\frac{\Delta[NH_3]}{\Delta t}$; The rate of formation of NO is the same.

$\frac{\Delta[H_2O]}{\Delta t} = -\frac{6}{4}\frac{\Delta[NH_3]}{\Delta t}$; The rate of formation of H_2O is 1.5 times faster.

13.58 (a) $-\frac{1}{2}\frac{\Delta[Br_2]}{\Delta t} = \frac{\Delta[ClO_2^-]}{\Delta t} = -2.4 \times 10^{-6}\ M/s$

(b) $\text{Rate} = -\frac{\Delta[Br^-]}{\Delta t} = -4\frac{\Delta[ClO_2^-]}{\Delta t} = -4(-2.4 \times 10^{-6}\ M/s) = 9.6 \times 10^{-6}\ M/s$

13.59 (a) $\frac{\Delta[H^+]}{\Delta t} = \frac{6}{2}\frac{\Delta[CH_3CO_2H]}{\Delta t} = \frac{6}{2}(5.0 \times 10^{-8}\ M/s) = 1.5 \times 10^{-7}\ M/s$

(b) $\text{Rate} = -\frac{\Delta[Ce^{4+}]}{\Delta t} = \frac{6}{2}\frac{\Delta[CH_3CO_2H]}{\Delta t} = \frac{6}{2}(5.0 \times 10^{-8}\ M/s) = 1.5 \times 10^{-7}\ M/s$

Rate Laws (Sections 13.2 and 13.3)

13.60 Rate = $k[H_2][ICl]$; units for k are $\frac{L}{mol \cdot s}$ or $1/(M \cdot s)$

13.61 Rate = $k[NO]^2[H_2]$, units for k are $1/(M^2 \cdot s)$

13.62 (a) Rate = $k[CH_3Br][OH^-]$
(b) Because the reaction is first-order in OH^-, if the $[OH^-]$ is decreased by a factor of 5, the rate will also decrease by a factor of 5.
(c) Because the reaction is first-order in each reactant, if both reactant concentrations are doubled, the rate will increase by a factor of 2 x 2 = 4.

13.63 (a) Rate = $k[Br^-][BrO_3^-][H^+]^2$
(b) The overall reaction order is 1 + 1 + 2 = 4.
(c) Because the reaction is second-order in H^+, if the $[H^+]$ is tripled, the rate will increase by a factor of $3^2 = 9$.
(d) Because the reaction is first-order in both Br^- and BrO_3^-, if both reactant concentrations are halved, the rate will decrease by a factor of 4 (1/2 x 1/2 = 1/4).

13.64 (a) Rate = $k[Cu(C_{10}H_8N_2)_2^+]^2[O_2]$
(b) The overall reaction order is 2 + 1 = 3.
(c) Because the reaction is second-order in $Cu(C_{10}H_8N_2)_2^+$, if the $[Cu(C_{10}H_8N_2)_2^+]$ is decreased by a factor of four, the rate will decrease by a factor of 16 (1/4 x 1/4 = 1/16).

13.65 (a) Rate = $k[NO]^2[O_2]$
(b) The overall reaction order is 2 + 1 = 3.
(c) Because the reaction is first-order in O_2, if the $[O_2]$ is doubled, the rate will double.
(d) Because the reaction is second-order in NO and first-order in O_2, if the [NO] is doubled and the $[O_2]$ is halved, the rate will double (2^2 x 1/2 = 2).

13.66 (a) Rate = $k[NH_4^+]^m[NO_2^-]^n$

$$m = \frac{\ln\left(\frac{\text{Rate}_2}{\text{Rate}_1}\right)}{\ln\left(\frac{[NH_4^+]_2}{[NH_4^+]_1}\right)} = \frac{\ln\left(\frac{3.6 \times 10^{-6}}{7.2 \times 10^{-6}}\right)}{\ln\left(\frac{0.12}{0.24}\right)} = 1; \quad n = \frac{\ln\left(\frac{\text{Rate}_3}{\text{Rate}_2}\right)}{\ln\left(\frac{[NO_2^-]_3}{[NO_2^-]_2}\right)} = \frac{\ln\left(\frac{5.4 \times 10^{-6}}{3.6 \times 10^{-6}}\right)}{\ln\left(\frac{0.15}{0.10}\right)} = 1$$

Rate = $k[NH_4^+][NO_2^-]$

(b) From Experiment 1: $k = \frac{\text{Rate}}{[NH_4^+][NO_2^-]} = \frac{7.2 \times 10^{-6}\ \text{M/s}}{(0.24\ \text{M})(0.10\ \text{M})} = 3.0 \times 10^{-4}/(\text{M} \cdot \text{s})$

(c) Rate = $k[NH_4^+][NO_2^-] = [3.0 \times 10^{-4}/(\text{M} \cdot \text{s})](0.39\ \text{M})(0.052\ \text{M}) = 6.1 \times 10^{-6}$ M/s

13.67 (a) Rate = $k[NO]^m[Cl_2]^n$

$$m = \frac{\ln\left(\frac{\text{Rate}_1}{\text{Rate}_2}\right)}{\ln\left(\frac{[NO]_1}{[NO]_2}\right)} = \frac{\ln\left(\frac{7.2 \times 10^{-6}}{3.6 \times 10^{-6}}\right)}{\ln\left(\frac{0.24}{0.12}\right)} = 1; \quad n = \frac{\ln\left(\frac{\text{Rate}_3}{\text{Rate}_2}\right)}{\ln\left(\frac{[Cl_2]_3}{[Cl_2]_2}\right)} = \frac{\ln\left(\frac{5.4 \times 10^{-6}}{3.6 \times 10^{-6}}\right)}{\ln\left(\frac{0.15}{0.10}\right)} = 1$$

Rate = k[NO][Cl_2]

(b) From Experiment 1: $k = \frac{\text{Rate}}{[NO][Cl_2]} = \frac{7.2 \times 10^{-6}\ M/s}{(0.24\ M)(0.10\ M)} = 3.0 \times 10^{-4}/(M \cdot s)$

(c) $\text{Rate} = k[NO][Cl_2] = [3.0 \times 10^{-4}/(M \cdot s)](0.12\ M)(0.12\ M) = 4.3 \times 10^{-6}\ M/s$

Integrated Rate Law; Half-Life (Sections 13.4–13.6)

13.68 $\ln\frac{[C_3H_6]_t}{[C_3H_6]_0} = -kt,\ k = 6.7 \times 10^{-4}/s$

(a) $t = 30\ \text{min} \times \frac{60\ s}{1\ \text{min}} = 1800\ s$

$\ln[C_3H_6]_t = -kt + \ln[C_3H_6]_0 = -(6.7 \times 10^{-4}/s)(1800\ s) + \ln(0.0500) = -4.202$

$[C_3H_6]_t = e^{-4.202} = 0.015\ M$

(b) $t = \frac{\ln\frac{[C_3H_6]_t}{[C_3H_6]_0}}{-k} = \frac{\ln\left(\frac{0.0100}{0.0500}\right)}{-(6.7 \times 10^{-4}/s)} = 2402\ s;\quad t = 2402\ s \times \frac{1\ \text{min}}{60\ s} = 40\ \text{min}$

(c) $[C_3H_6]_0 = 0.0500\ M$; If 25% of the C_3H_6 reacts then 75% remains.

$[C_3H_6]_t = (0.75)(0.0500\ M) = 0.0375\ M$

$t = \frac{\ln\frac{[C_3H_6]_t}{[C_3H_6]_0}}{-k} = \frac{\ln\left(\frac{0.0375}{0.0500}\right)}{-(6.7 \times 10^{-4}/s)} = 429\ s;\quad t = 429\ s \times \frac{1\ \text{min}}{60\ s} = 7.2\ \text{min}$

13.69 $\ln\frac{[CH_3NC]_t}{[CH_3NC]_0} = -kt,\ k = 5.11 \times 10^{-5}/s$

(a) $t = 2.00\ \text{hr} \times \frac{60\ \text{min}}{1\ \text{hr}} \times \frac{60\ s}{1\ \text{min}} = 7200\ s$

$\ln[CH_3NC]_t = -kt + \ln[CH_3NC]_0 = -(5.11 \times 10^{-5}/s)(7200\ s) + \ln(0.0340) = -3.749$

$[CH_3NC]_t = e^{-3.749} = 0.0235\ M$

(b) $t = \frac{\ln\frac{[CH_3NC]_t}{[CH_3NC]_0}}{-k} = \frac{\ln\left(\frac{0.0300}{0.0340}\right)}{-5.11 \times 10^{-5}/s)} = 2449\ s;\quad t = 2449\ s \times \frac{1\ \text{min}}{60\ s} = 40.8\ \text{min}$

(c) $[CH_3NC]_0 = 0.0340\ M$; If 20% of the CH_3NC reacts then 80% remains.

$[CH_3NC]_t = (0.80)(0.0340\ M) = 0.0272\ M$

$t = \frac{\ln\frac{[CH_3NC]_t}{[CH_3NC]_0}}{-k} = \frac{\ln\left(\frac{0.0272}{0.0340}\right)}{-(5.11 \times 10^{-5}/s)} = 4367\ s;\quad t = 4367\ s \times \frac{1\ \text{min}}{60\ s} = 72.8\ \text{min}$

13.70 $t_{1/2} = \dfrac{0.693}{k} = \dfrac{0.693}{6.7 \times 10^{-4}/s} = 1034 \text{ s} = 17 \text{ min}$

$$t = \frac{\ln\dfrac{[C_3H_6]_t}{[C_3H_6]_o}}{-k} = \frac{\ln\dfrac{(0.0625)(0.0500)}{(0.0500)}}{-6.7 \times 10^{-4}/s} = 4140 \text{ s}$$

$$t = 4140 \text{ s} \times \frac{1 \text{ min}}{60 \text{ s}} = 69 \text{ min}$$

This is also 4 half-lives. $100 \xrightarrow{t_{1/2}} 50 \xrightarrow{t_{1/2}} 25 \xrightarrow{t_{1/2}} 12.5 \xrightarrow{t_{1/2}} 6.25$

13.71 $t_{1/2} = \dfrac{0.693}{k} = \dfrac{0.693}{5.11 \times 10^{-5}/s} = 13{,}562 \text{ s}$

$$t_{1/2} = 13{,}562 \text{ s} \times \frac{1 \text{ min}}{60 \text{ s}} \times \frac{1 \text{ hr}}{60 \text{ min}} = 3.77 \text{ hr}$$

$$t = \frac{\ln\dfrac{[CH_3NC]_t}{[CH_3NC]_o}}{-k} = \frac{\ln\dfrac{(0.125)(0.0340)}{(0.0340)}}{-5.11 \times 10^{-5}/s} = 40{,}694 \text{ s}$$

$$t = 40{,}694 \text{ s} \times \frac{1 \text{ min}}{60 \text{ s}} \times \frac{1 \text{ hr}}{60 \text{ min}} = 11.3 \text{ hr}$$

This is also 3 half-lives. $100 \xrightarrow{t_{1/2}} 50 \xrightarrow{t_{1/2}} 25 \xrightarrow{t_{1/2}} 12.5$

13.72 $kt = \dfrac{1}{[C_4H_6]_t} - \dfrac{1}{[C_4H_6]_0}$, $\quad k = 4.0 \times 10^{-2}/(M \cdot s)$

(a) $t = 1.00 \text{ h} \times \dfrac{60 \text{ min}}{1 \text{ hr}} \times \dfrac{60 \text{ s}}{1 \text{ min}} = 3600 \text{ s}$

$$\frac{1}{[C_4H_6]_t} = kt + \frac{1}{[C_4H_6]_0} = (4.0 \times 10^{-2}/(M \cdot s))(3600 \text{ s}) + \frac{1}{0.0200 \text{ M}}$$

$$\frac{1}{[C_4H_6]_t} = 194/M \text{ and } [C_4H_6] = 5.2 \times 10^{-3} \text{ M}$$

(b) $t = \dfrac{1}{k}\left[\dfrac{1}{[C_4H_6]_t} - \dfrac{1}{[C_4H_6]_0}\right]$

$$t = \frac{1}{4.0 \times 10^{-2}/(M \cdot s)}\left[\frac{1}{(0.0020 \text{ M})} - \frac{1}{(0.0200 \text{ M})}\right] = 11{,}250 \text{ s}$$

$$t = 11{,}250 \text{ s} \times \frac{1 \text{ min}}{60 \text{ s}} \times \frac{1 \text{ hr}}{60 \text{ min}} = 3.1 \text{ h}$$

13.73 $kt = \frac{1}{[HI]_t} - \frac{1}{[HI]_0}$, $k = 9.7 \times 10^{-6}/(M \cdot s)$

(a) $t = 6.00 \text{ day} \times \frac{24 \text{ hr}}{1 \text{ day}} \times \frac{60 \text{ min}}{1 \text{ hr}} \times \frac{60 \text{ s}}{1 \text{ min}} = 518{,}400 \text{ s}$

$$\frac{1}{[HI]_t} = kt + \frac{1}{[HI]_0} = (9.7 \times 10^{-6}/(M \cdot s))(518{,}400 \text{ s}) + \frac{1}{0.100 \text{ M}}$$

$$\frac{1}{[HI]_t} = 15.03/M \text{ and } [HI] = 0.067 \text{ M}$$

(b) $t = \frac{1}{k}\left[\frac{1}{[HI]_t} - \frac{1}{[HI]_0}\right]$

$$t = \frac{1}{9.7 \times 10^{-6}/(M \cdot s)}\left[\frac{1}{(0.085 \text{ M})} - \frac{1}{(0.100 \text{ M})}\right] = 181{,}928 \text{ s}$$

$$t = 181{,}928 \text{ s} \times \frac{1 \text{ min}}{60 \text{ s}} \times \frac{1 \text{ hr}}{60 \text{ min}} \times \frac{1 \text{ day}}{24 \text{ hr}} = 2.1 \text{ days}$$

13.74 $t_{1/2} = \frac{1}{k[C_4H_6]_o} = \frac{1}{[4.0 \times 10^{-2}/(M \cdot s)](0.0200 \text{ M})} = 1250 \text{ s} = 21 \text{ min}$

$$t = t_{1/2} = \frac{1}{k[C_4H_6]_o} = \frac{1}{[4.0 \times 10^{-2}/(M \cdot s)](0.0100 \text{ M})} = 2500 \text{ s} = 42 \text{ min}$$

13.75 $t_{1/2} = \frac{1}{k[HI]_o} = \frac{1}{[9.7 \times 10^{-6}/(M \cdot s)](0.100 \text{ M})} = 1{,}030{,}928 \text{ s}$

$$1{,}030{,}928 \text{ s} \times \frac{1 \text{ min}}{60 \text{ s}} \times \frac{1 \text{ hr}}{60 \text{ min}} \times \frac{1 \text{ day}}{24 \text{ hr}} = 12 \text{ days}$$

$$t = t_{1/2} = \frac{1}{k[HI]_o} = \frac{1}{[9.7 \times 10^{-6}/(M \cdot s)](0.200 \text{ M})} = 515{,}464 \text{ s}$$

$$515{,}464 \text{ s} \times \frac{1 \text{ min}}{60 \text{ s}} \times \frac{1 \text{ hr}}{60 \text{ min}} \times \frac{1 \text{ day}}{24 \text{ hr}} = 6.0 \text{ days}$$

13.76

time (min)	$[N_2O]$	$\ln[N_2O]$	$1/[N_2O]$
0	0.250	−1.386	4.00
60	0.228	−1.478	4.39
90	0.216	−1.532	4.63
300	0.128	−2.056	7.81
600	0.0630	−2.765	15.9

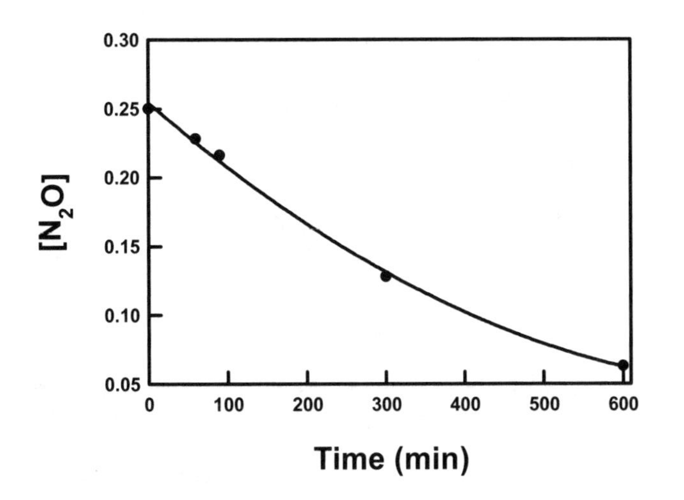

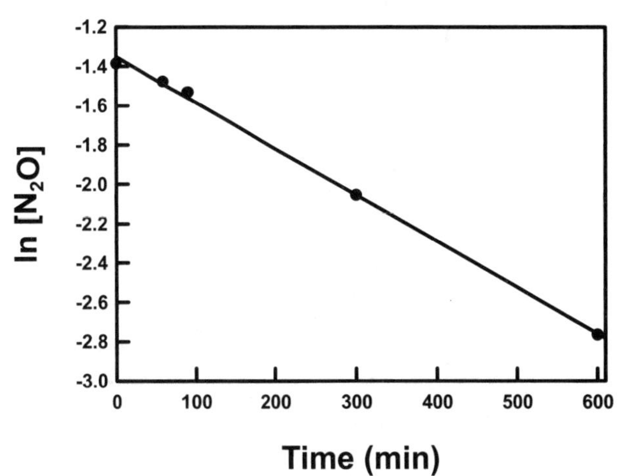

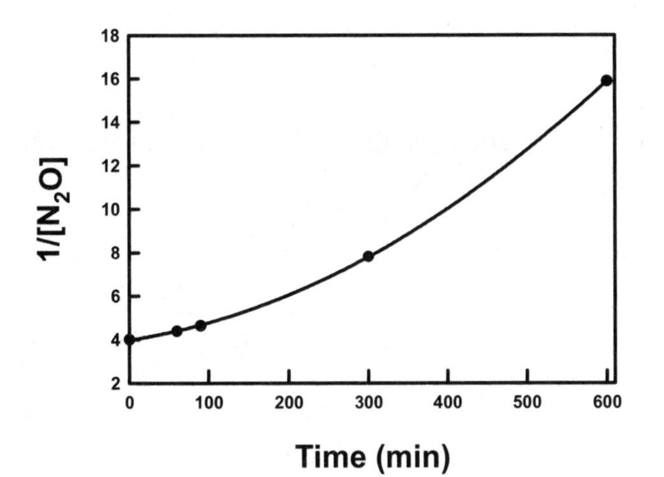

A plot of $\ln[N_2O]$ versus time is linear. The reaction is first-order in N_2O.

$k = -\text{slope} = -(-2.35 \times 10^{-3}/\text{min}) = 2.35 \times 10^{-3}/\text{min}$

$k = 2.35 \times 10^{-3}/\text{min} \times \dfrac{1\ \text{min}}{60\ \text{s}} = 3.92 \times 10^{-5}/\text{s}$

13.77

time (s)	[NOBr]	ln[NOBr]	1/[NOBr]
0	0.0390	–3.244	25.6
10	0.0315	–3.458	31.7
40	0.0175	–4.046	57.1
120	0.00784	–4.849	127.6
320	0.00376	–5.583	266.0

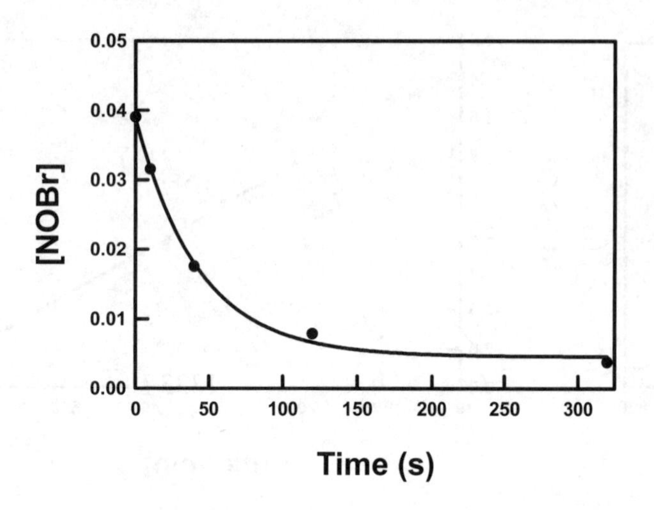

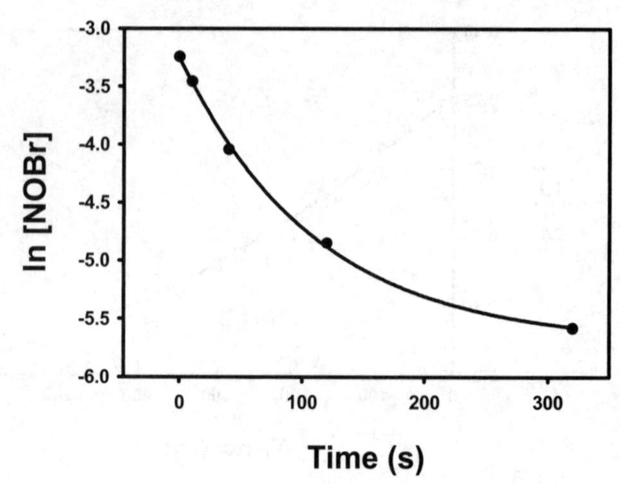

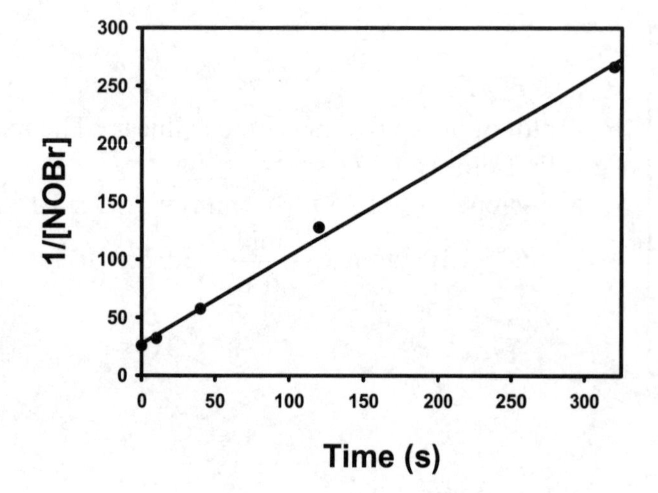

A plot of 1/[NOBr] versus time is linear. The reaction is second-order in NOBr.
k = slope = 0.76/(M · s)

13.78 $k = \dfrac{0.693}{t_{1/2}} = \dfrac{0.693}{248\ s} = 2.79 \times 10^{-3}/s$

13.79 $kt = \dfrac{1}{[C_7H_{12}]_t} - \dfrac{1}{[C_7H_{12}]_0}$, $\quad k = 0.030\ /(M \cdot s)$

(a) $\dfrac{1}{[C_7H_{12}]_t} = kt + \dfrac{1}{[C_7H_{12}]_0} = (0.030/(M \cdot s))(1600\ s) + \dfrac{1}{0.035\ M}$

$\dfrac{1}{[C_7H_{12}]_t} = 79.6/M$ and $[C_7H_{12}]_t = 0.013\ M$

(b) $t = \dfrac{1}{k}\left[\dfrac{1}{[C_7H_{12}]_t} - \dfrac{1}{[C_7H_{12}]_0}\right]$

$$t = \frac{1}{0.030/(M \cdot s)}\left[\frac{1}{(0.035/20\ M)} - \frac{1}{(0.035\ M)}\right] = 1.8 \times 10^4\ s$$

(c) $t_{1/2} = \dfrac{1}{k[C_7H_{12}]_0} = \dfrac{1}{[0.030/(M \cdot s)](0.075\ M)} = 4.4 \times 10^2\ s$

13.80 (a) The units for the rate constant, k, indicate the reaction is zeroth-order.

(b) For a zeroth-order reaction, $[A]_t - [A]_o = -kt$

$t = 30\ min \times \dfrac{60\ s}{1\ min} = 1800\ s$

$[A]_t = -kt + [A]_o = -(3.6 \times 10^{-5}\ M/s)(1800\ s) + 0.096\ M = 0.031\ M$

(c) Let $[A]_t = [A]_o/2$, $t_{1/2} = \dfrac{[A]_o/2 - [A]_o}{-k} = \dfrac{0.096/2\ M - 0.096\ M}{-3.6 \times 10^{-5}\ M/s} = 1333\ s$

$t_{1/2} = 1333\ s \times \dfrac{1\ min}{60\ s} = 22\ min$

13.81 (a)

time (min)	[AB]
0	0.206
20	0.186
40	0.181
120	0.117
220	0.036

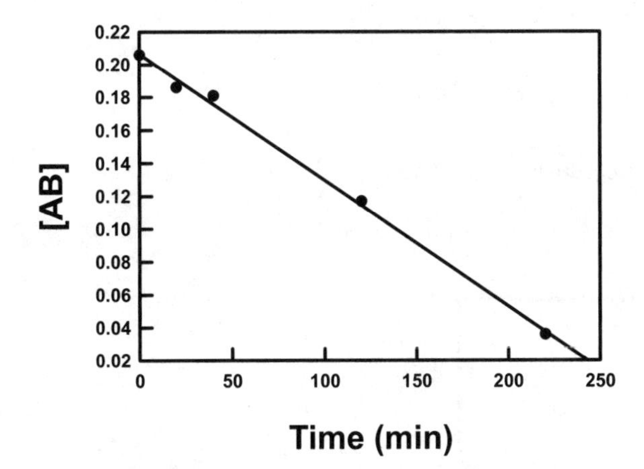

A plot of [AB] versus time is linear.
The reaction is zeroth-order in AB and
$k = -\text{slope} = 7.7 \times 10^{-4}\ M/min$

$k = 7.7 \times 10^{-4}\ M/min \times \dfrac{1\ min}{60\ s} = 1.3 \times 10^{-5}\ M/s$

(b) $[A]_t - [A]_o = -kt$

$[A]_t = -kt + [A]_o = -(7.7 \times 10^{-4}\ M/min)(192\ min) + 0.206\ M = 0.058\ M$

(c) $[A]_t - [A]_o = -kt$

$$t = \frac{[A]_t - [A]_o}{-k} = \frac{0.0250\ M - 0.206\ M}{-7.7 \times 10^{-4}\ M/min} = 2.4 \times 10^2\ min$$

The Arrhenius Equation (Sections 13.7 and 13.8)

13.82 Very few collisions involve a collision energy greater than or equal to the activation energy, and only a fraction of those have the proper orientation for reaction.

13.83 The two reactions have frequency factors that differ by a factor of 10.

13.84 Plot ln k versus 1/T to determine the activation energy, E_a.

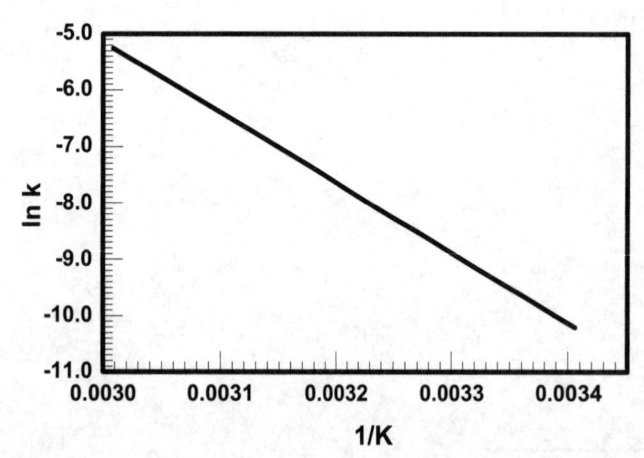

Slope = -1.25×10^4 K

$E_a = -R(\text{slope}) = -[8.314 \times 10^{-3}\ kJ/(K \cdot mol)](-1.25 \times 10^4\ K) = 104\ kJ/mol$

13.85 Plot ln k versus 1/T to determine the activation energy, E_a.

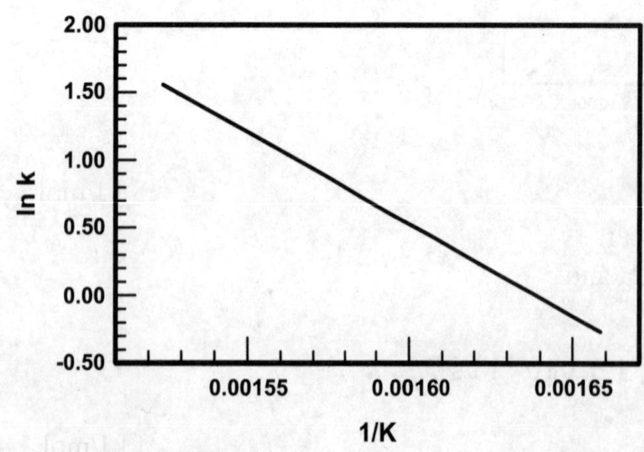

Slope = -1.359×10^4 K

$E_a = -R(\text{slope}) = -[8.314 \times 10^{-3}\ kJ/(K \cdot mol)](-1.359 \times 10^4\ K) = 113\ kJ/mol$

13.86 (a) $\ln\left(\frac{k_2}{k_1}\right) = \left(\frac{-E_a}{R}\right)\left(\frac{1}{T_2} - \frac{1}{T_1}\right)$

$k_1 = 1.3/(M \cdot s)$, $T_1 = 700$ K
$k_2 = 23.0/(M \cdot s)$, $T_2 = 800$ K

$$E_a = -\frac{[\ln k_2 - \ln k_1](R)}{\left(\frac{1}{T_2} - \frac{1}{T_1}\right)}$$

$$E_a = -\frac{[\ln(23.0) - \ln(1.3)][8.314 \times 10^{-3} \text{ kJ/(K}\cdot\text{mol)}]}{\left(\frac{1}{800 \text{ K}} - \frac{1}{700 \text{ K}}\right)} = 134 \text{ kJ/mol}$$

(b) $k_1 = 1.3/(M \cdot s)$, $T_1 = 700$ K
solve for k_2, $T_2 = 750$ K

$$\ln k_2 = \left(\frac{-E_a}{R}\right)\left(\frac{1}{T_2} - \frac{1}{T_1}\right) + \ln k_1$$

$$\ln k_2 = \left(\frac{-133.8 \text{ kJ/mol}}{8.314 \times 10^{-3} \text{ kJ/(K}\cdot\text{mol)}}\right)\left(\frac{1}{750 \text{ K}} - \frac{1}{700 \text{ K}}\right) + \ln(1.3) = 1.795$$

$k_2 = e^{1.795} = 6.0/(M \cdot s)$

13.87 $\ln\left(\frac{k_2}{k_1}\right) = \left(\frac{-E_a}{R}\right)\left[\frac{1}{T_2} - \frac{1}{T_1}\right]$

(a) Because the rate doubles, $k_2 = 2k_1$
$k_1 = 1.0 \times 10^{-3}$/s, $T_1 = 25$ °C $= 298$ K
$k_2 = 2.0 \times 10^{-3}$/s, $T_2 = 35$ °C $= 308$ K

$$E_a = -\frac{[\ln k_2 - \ln k_1](R)}{\left(\frac{1}{T_2} - \frac{1}{T_1}\right)}$$

$$E_a = -\frac{[\ln(2.0 \times 10^{-3}) - \ln(1.0 \times 10^{-3})][8.314 \times 10^{-3} \text{ kJ/(K}\cdot\text{mol)}]}{\left(\frac{1}{308 \text{ K}} - \frac{1}{298 \text{ K}}\right)} = 53 \text{ kJ/mol}$$

(b) Because the rate triples, $k_2 = 3k_1$
$k_1 = 1.0 \times 10^{-3}$/s, $T_1 = 25$ °C $= 298$ K
$k_2 = 3.0 \times 10^{-3}$/s, $T_2 = 35$ °C $= 308$ K

$$E_a = -\frac{[\ln(3.0 \times 10^{-3}) - \ln(1.0 \times 10^{-3})][8.314 \times 10^{-3} \text{ kJ/(K}\cdot\text{mol)}]}{\left(\frac{1}{308 \text{ K}} - \frac{1}{298 \text{ K}}\right)} = 84 \text{ kJ/mol}$$

13.88 $\ln\left(\frac{k_2}{k_1}\right) = \left(\frac{-E_a}{R}\right)\left(\frac{1}{T_2} - \frac{1}{T_1}\right)$

assume $k_1 = 1.0/(M \cdot s)$ at $T_1 = 25\ ^\circ C = 298\ K$
assume $k_2 = 15/(M \cdot s)$ at $T_2 = 50\ ^\circ C = 323\ K$

$$E_a = -\frac{[\ln k_2 - \ln k_1](R)}{\left(\frac{1}{T_2} - \frac{1}{T_1}\right)}$$

$$E_a = -\frac{[\ln(15) - \ln(1.0)][8.314 \times 10^{-3}\ kJ/(K \cdot mol)]}{\left(\frac{1}{323\ K} - \frac{1}{298\ K}\right)} = 87\ kJ/mol$$

13.89 $\ln\left(\frac{k_2}{k_1}\right) = \left(\frac{-E_a}{R}\right)\left(\frac{1}{T_2} - \frac{1}{T_1}\right)$

assume $k_1 = 1.0/(M \cdot s)$ at $T_1 = 15\ ^\circ C = 288\ K$
assume $k_2 = 6.37/(M \cdot s)$ at $T_2 = 45\ ^\circ C = 318\ K$

$$E_a = -\frac{[\ln k_2 - \ln k_1](R)}{\left(\frac{1}{T_2} - \frac{1}{T_1}\right)}$$

$$E_a = -\frac{[\ln(6.37) - \ln(1.0)][8.314 \times 10^{-3}\ kJ/(K \cdot mol)]}{\left(\frac{1}{318\ K} - \frac{1}{288\ K}\right)} = 47.0\ kJ/mol$$

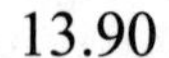

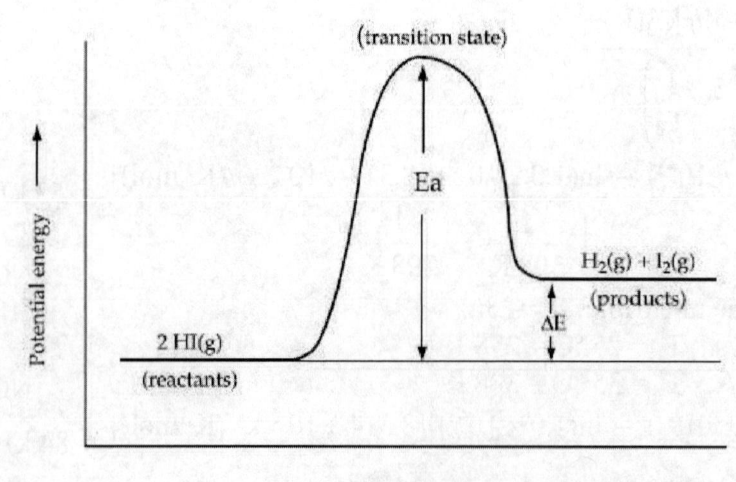

13.91 (a) (b)

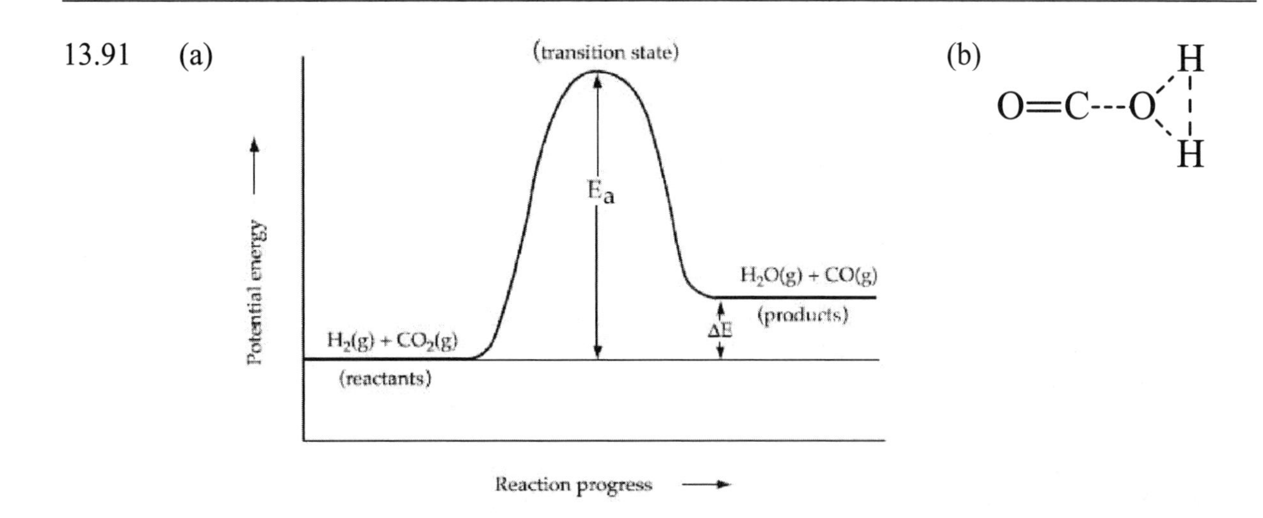

Reaction Mechanisms (Sections 13.9–13.11)

13.92 There is no relationship between the coefficients in a balanced chemical equation for an overall reaction and the exponents in the rate law unless the overall reaction occurs in a single elementary step, in which case the coefficients in the balanced equation are the exponents in the rate law.

13.93 The rate-determining step is the slowest step in a multistep reaction. The coefficients in the balanced equation for the rate-determining step are the exponents in the rate law.

13.94 (a)

$$H_2(g) + ICl(g) \rightarrow HI(g) + HCl(g)$$
$$HI(g) + ICl(g) \rightarrow I_2(g) + HCl(g)$$

Overall reaction $H_2(g) + 2\,ICl(g) \rightarrow I_2(g) + 2\,HCl(g)$

(b) Because HI(g) is produced in the first step and consumed in the second step, it is a reaction intermediate.

(c) In each reaction there are two reactant molecules, so each elementary reaction is bimolecular.

13.95 (a)

$$NO(g) + Cl_2(g) \rightarrow NOCl_2(g)$$
$$NOCl_2(g) + NO(g) \rightarrow 2\,NOCl(g)$$

Overall reaction $2\,NO(g) + Cl_2(g) \rightarrow 2\,NOCl(g)$

(b) Because $NOCl_2$ is produced in the first step and consumed in the second step, $NOCl_2$ is a reaction intermediate.

(c) Each elementary step is bimolecular.

13.96 (a) bimolecular, Rate = $k[O_3][Cl]$ (b) unimolecular, Rate = $k[NO_2]$
(c) bimolecular, Rate = $k[ClO][O]$ (d) termolecular, Rate = $k[Cl]^2[N_2]$

13.97 (a) unimolecular, Rate = $k[I_2]$ (b) termolecular, Rate = $k[NO]^2[Br_2]$
(c) unimolecular, Rate = $k[N_2O_5]$

13.98 (a)

$$\begin{array}{lll} & NO_2Cl(g) \rightarrow & NO_2(g) + Cl(g) \\ & \underline{Cl(g) + NO_2Cl(g) \rightarrow} & \underline{NO_2(g) + Cl_2(g)} \\ \text{Overall reaction} & 2\ NO_2Cl(g) \rightarrow & 2\ NO_2(g) + Cl_2(g) \end{array}$$

(b) 1. unimolecular; 2. bimolecular

(c) Rate = $k[NO_2Cl]$

13.99 (a)

$$\begin{array}{lll} & Mo(CO)_6 \rightarrow & Mo(CO)_5 + CO \\ & \underline{Mo(CO)_5 + L \rightarrow} & \underline{Mo(CO)_5L} \\ \text{Overall reaction} & Mo(CO)_6 + L \rightarrow & Mo(CO)_5L + CO \end{array}$$

(b) 1. unimolecular; 2. bimolecular

(c) Rate = $k[Mo(CO)_6]$

13.100 $NO_2(g) + F_2(g) \rightarrow NO_2F(g) + F(g)$ (slow)

$F(g) + NO_2(g) \rightarrow NO_2F(g)$ (fast)

13.101 $O_3(g) + NO(g) \rightarrow O_2(g) + NO_2(g)$ (slow)

$NO_2(g) + O(g) \rightarrow O_2(g) + NO(g)$ (fast)

13.102 (a) $2\ NO(g) + O_2(g) \rightarrow 2\ NO_2(g)$

(b) $\text{Rate}_{\text{forward}} = k_1[NO]^2$ and $\text{Rate}_{\text{reverse}} = k_{-1}[N_2O_2]$

Because of the equilibrium, $\text{Rate}_{\text{forward}} = \text{Rate}_{\text{reverse}}$, and $k_1[NO]^2 = k_{-1}[N_2O_2]$.

$$[N_2O_2] = \frac{k_1}{k_{-1}}[NO]^2$$

The rate law for the rate determining step is Rate $= 2k_2[N_2O_2][O_2]$ because two NO molecules are consumed in the overall reaction for every N_2O_2 that reacts in the second step. In this rate law substitute for $[N_2O_2]$. Rate $= 2k_2 \dfrac{k_1}{k_{-1}}[NO]^2[O_2]$

(c) $k = \dfrac{2k_2k_1}{k_{-1}}$

13.103 (a) $2\ N_2O_5(g) \rightarrow 4\ NO_2(g) + O_2(g)$

(b) An intermediate is formed in one step and then consumed in a subsequent step in the reaction. NO_2 is an intermediate as well as a product. NO_3 and NO are intermediates.

(c) $\text{Rate}_{\text{forward}} = k_1[N_2O_5]$ and $\text{Rate}_{\text{reverse}} = k_{-1}[NO_2][NO_3]$

Because of the equilibrium, $\text{Rate}_{\text{forward}} = \text{Rate}_{\text{reverse}}$, and $k_1[N_2O_5] = k_{-1}[NO_2][NO_3]$.

$$\frac{k_1}{k_{-1}}[N_2O_5] = [NO_2][NO_3]$$

The rate law for the rate determining step is Rate $= k_2[NO_2][NO_3]$. In this rate law substitute for $[NO_2][NO_3]$.

$$\text{Rate} = k_2\frac{k_1}{k_{-1}}[N_2O_5]$$

(d) $k = \dfrac{k_2k_1}{k_{-1}}$

Catalysis (Sections 13.13–13.15)

13.104 A catalyst does participate in the reaction, but it is not consumed because it reacts in one step of the reaction and is regenerated in a subsequent step.

13.105 A catalyst doesn't appear in the chemical equation for a reaction because a catalyst reacts in one step of the reaction but is regenerated in a subsequent step.

13.106 (a) $O_3(g) + O(g) \rightarrow 2\ O_2(g)$
(b) Cl acts as a catalyst.
(c) ClO is a reaction intermediate.
(d) A catalyst reacts in one step and is regenerated in a subsequent step. A reaction intermediate is produced in one step and consumed in another.

13.107 (a)

$$2\ SO_2(g) + 2\ NO_2(g) \rightarrow 2\ SO_3(g) + 2\ NO(g)$$
$$\underline{2\ NO(g) + O_2(g) \rightarrow 2\ NO_2(g)}$$

Overall reaction $2\ SO_2(g) + O_2(g) \rightarrow 2\ SO_3(g)$

(b) $NO_2(g)$ acts as a catalyst because it is used in the first step and regenerated in the second. NO(g) is a reaction intermediate because it is produced in the first step and consumed in the second.

13.108 (a)

$$NH_2NO_2(aq) + OH^-(aq) \rightarrow NHNO_2^-(aq) + H_2O(l)$$
$$\underline{NHNO_2^-(aq) \rightarrow N_2O(g) + OH^-(aq)}$$

Overall reaction $NH_2NO_2(aq) \rightarrow N_2O(g) + H_2O(l)$

(b) OH^- acts as a catalyst because it is used in the first step and regenerated in the second. $NHNO_2^-$ is a reaction intermediate because it is produced in the first step and consumed in the second.
(c) The rate will decrease because added acid decreases the concentration of OH^-, which appears in the rate law since it is a catalyst.

13.109 The reaction in Problem 13.101 involves a catalyst (NO) because NO is used in the first step and is regenerated in the second step. The reaction also involves an intermediate (NO_2) because NO_2 is produced in the first step and is used up in the second step.

13.110 Soluble enzymes are homogeneous catalysts and membrane-bound enzymes are heterogeneous.

13.111 (a) S $\rightarrow$ P
(b) E is a catalyst. ES is an intermediate.
(c) $\text{Rate}_{\text{forward}} = k_1[E][S]$ and $\text{Rate}_{\text{reverse}} = k_{-1}[ES]$
Because of the equilibrium, $\text{Rate}_{\text{forward}} = \text{Rate}_{\text{reverse}}$, and $k_1[E][S] = k_{-1}[ES]$.

$$\frac{k_1}{k_{-1}}[E][S] = [ES]$$

The rate law for the rate determining step is Rate = k_2[ES]. In this rate law substitute for [ES].

$$\text{Rate} = k_2 \frac{k_1}{k_{-1}}[E][S]$$

(d) $k = \dfrac{k_2 k_1}{k_{-1}}$

13.112

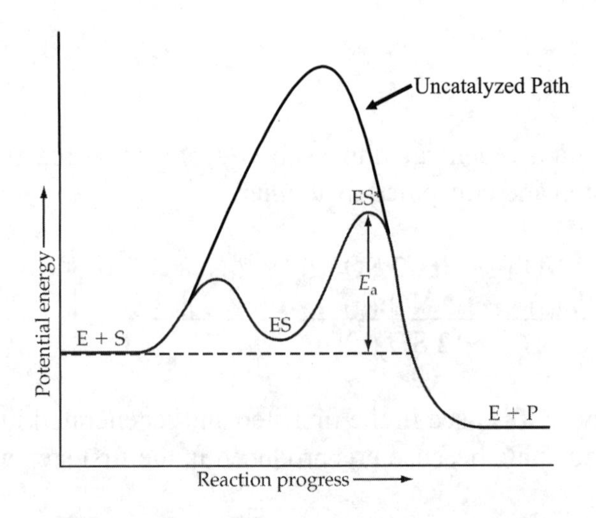

13.113 The substrate concentration is changed by a factor of (3.4 x 10^{-5} M)/(1.4 x 10^{-5} M) = 2.4. Because the reaction is first-order at these (low) substrate concentrations, the enzyme catalyzed rate of product formation will increase by a factor of 2.4.

13.114 (a) Because the reaction is zeroth-order in substrate at high substrate concentration, there is no change in rate when the substrate concentration is changed from 2.8 x 10^{-3} M to 4.8 x 10^{-3} M.
(b) At high substrate concentrations, the enzyme becomes saturated (i.e., completely bound) with the substrate. At this point, the reaction rate reaches a maximum value and becomes independent of substrate concentration (zeroth order in substrate) because only substrate bound to the enzyme can react.

Chapter Problems

13.115 $2\ AB_2 \rightarrow A_2 + 2\ B_2$
(a) Measure the change in the concentration of AB_2 as a function of time.
(b) and (c) If a plot of $[AB_2]$ versus time is linear, the reaction is zeroth-order and k = –slope. If a plot of ln $[AB_2]$ versus time is linear, the reaction is first-order and k = –slope. If a plot of $1/[AB_2]$ versus time is linear, the reaction is second-order and k = slope.

13.116 $A \rightarrow B + C$
(a) Measure the change in the concentration of A as a function of time at several different temperatures.

(b) Plot ln [A] versus time, for each temperature. Straight line graphs will result and k at each temperature equals –slope. Graph ln k versus 1/K, where K is the kelvin temperature. Determine the slope of the line. $E_a = -R(\text{slope})$ where $R = 8.314 \times 10^{-3}$ kJ/(K · mol).

13.117 (a) Rate = $k[B_2][C]$

(b) $B_2 + C \rightarrow CB + B$ (slow)

$CB + A \rightarrow AB + C$ (fast)

(c) C is a catalyst. C does not appear in the chemical equation because it is consumed in the first step and regenerated in the second step.

13.118 (a)

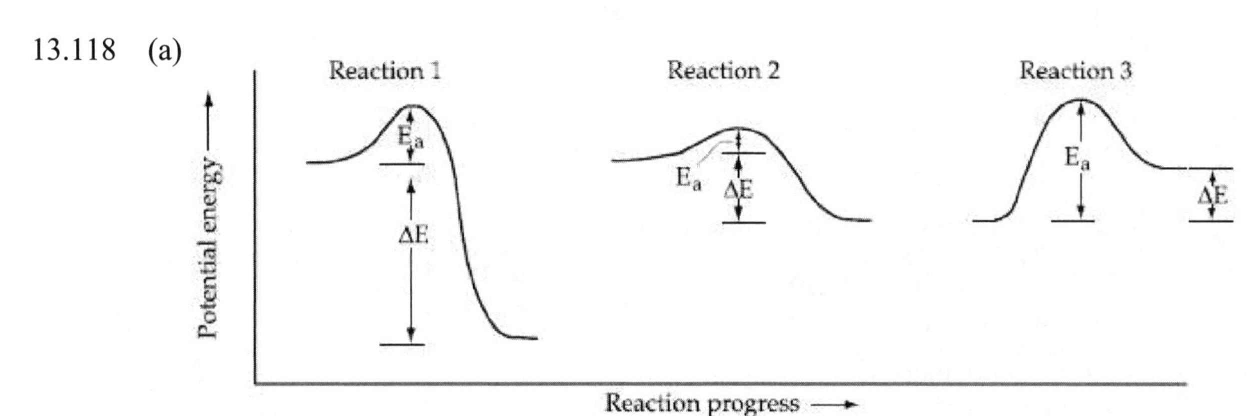

(b) Reaction 2 is the fastest (smallest E_a), and reaction 3 is the slowest (largest E_a).

(c) Reaction 3 is the most endothermic (positive ΔE), and reaction 1 is the most exothermic (largest negative ΔE).

13.119 The first maximum represents the potential energy of the transition state for the first step. The second maximum represents the potential energy of the transition state for the second step. The saddle point between the two maxima represents the potential energy of the intermediate products.

13.120 Because 0.060 M is half of 0.120 M, 5.2 h is the half-life.

For a first-order reaction, the half-life is independent of initial concentration. Because 0.015 M is half of 0.030 M, it will take one half-life, 5.2 h.

$$k = \frac{0.693}{t_{1/2}} = \frac{0.693}{5.2\ \text{h}} = 0.133/\text{h}$$

$$\ln\frac{[N_2O_5]_t}{[N_2O_5]_0} = -kt$$

$$t = \frac{\ln\frac{[N_2O_5]_t}{[N_2O_5]_0}}{-k} = \frac{\ln\left(\frac{0.015}{0.480}\right)}{-(0.133/\text{h})} = 26\ \text{h} \quad \text{(Note that t is five half-lives.)}$$

13.121 (a) The reaction rate will increase with an increase in temperature at constant volume.
(b) The reaction rate will decrease with an increase in volume at constant temperature because reactant concentrations will decrease.
(c) The reaction rate will increase with the addition of a catalyst.
(d) Addition of an inert gas at constant volume will not affect the reaction rate.

13.122 As the temperature of a gas is raised by 10 °C, even though the collision frequency increases by only ~2%, the reaction rate increases by 100% or more because there is an exponential increase in the fraction of the collisions that leads to products.

13.123 $H_2O_2(aq) + 3\,I^-(aq) + 2\,H^+(aq) \rightarrow I_3^-(aq) + 2\,H_2O(l)$

Rate = $k[H_2O_2]^m[I^-]^n$

(a) $\dfrac{\text{Rate}_3}{\text{Rate}_1} = \dfrac{2.30 \times 10^{-4}\ \text{M/s}}{1.15 \times 10^{-4}\ \text{M/s}} = 2 \qquad \dfrac{[H_2O_2]_3}{[H_2O_2]_1} = \dfrac{0.200\ \text{M}}{0.100\ \text{M}} = 2$

Because both ratios are the same, m = 1.

$\dfrac{\text{Rate}_2}{\text{Rate}_1} = \dfrac{2.30 \times 10^{-4}\ \text{M/s}}{1.15 \times 10^{-4}\ \text{M/s}} = 2 \qquad \dfrac{[I^-]_2}{[I^-]_1} = \dfrac{0.200\ \text{M}}{0.100\ \text{M}} = 2$

Because both ratios are the same, n = 1.

The rate law is: Rate = $k[H_2O_2][I^-]$

(b) $k = \dfrac{\text{Rate}}{[H_2O_2][I^-]}$

Using data from Experiment 1: $k = \dfrac{1.15 \times 10^{-4}\ \text{M/s}}{(0.100\ \text{M})(0.100\ \text{M})} = 1.15 \times 10^{-2}\ /(\text{M} \cdot \text{s})$

(c) Rate = $k[H_2O_2][I^-] = [1.15 \times 10^{-2}/(\text{M} \cdot \text{s})](0.300\ \text{M})(0.400\ \text{M}) = 1.38 \times 10^{-3}\ \text{M/s}$

13.124 (a) Rate = $k[(CH_3)_3N]^m[ClO_2]^n$

$$m = \frac{\ln\left(\dfrac{\text{Rate}_3}{\text{Rate}_2}\right)}{\ln\left(\dfrac{[(CH_3)_3N]_3}{[(CH_3)_3N]_2}\right)} = \frac{\ln\left(\dfrac{1.79}{0.90}\right)}{\ln\left(\dfrac{1.3 \times 10^{-2}}{6.5 \times 10^{-3}}\right)} = 1$$

$$n = \frac{\ln\left(\dfrac{\text{Rate}_1 \cdot [(CH_3)_3N]_3}{\text{Rate}_3 \cdot [(CH_3)_3N]_1}\right)}{\ln\left(\dfrac{[ClO_2]_1}{[ClO_2]_3}\right)} = \frac{\ln\left(\dfrac{(0.90)(1.30 \times 10^{-2})}{(1.79)(3.25 \times 10^{-3})}\right)}{\ln\left(\dfrac{4.60 \times 10^{-3}}{2.30 \times 10^{-3}}\right)} = 1$$

$Rate = k[(CH_3)_3N][ClO_2]$

From Experiment 1:

$$k = \frac{Rate}{[(CH_3)_3N][ClO_2]} = \frac{0.90\ M/s}{(3.25\ x\ 10^{-3}\ M)(4.60\ x\ 10^{-3}\ M)} = 6.0\ x\ 10^4/(M \cdot s)$$

(b) $Rate = k[(CH_3)_3N][ClO_2]$

$Rate = [6.0\ x\ 10^4/(M \cdot s)](4.2\ x\ 10^{-2}\ M)(3.4\ x\ 10^{-2}\ M) = 86\ M/s$

13.125 $\ln\frac{[11\text{-cis-retinal}]_t}{[11\text{-cis-retinal}]_0} = -kt,\ k = 1.02\ x\ 10^{-5}/s$

(a) $t = 6\ h\ x\ \frac{60\ min}{1\ h}\ x\ \frac{60\ s}{1\ min} = 21{,}600\ s$

$\ln[11\text{-cis-retinal}]_t = -kt + \ln[11\text{-cis-retinal}]_0$

$\ln[11\text{-cis-retinal}]_t = -(1.02\ x\ 10^{-5}/s)(21{,}600\ s) + \ln(3.50\ x\ 10^{-3}) = -5.875$

$[11\text{-cis-retinal}]_t = e^{-5.875} = 2.81\ x\ 10^{-3}\ M$

(b) $[11\text{-cis-retinal}]_0 = 3.50\ x\ 10^{-3}\ M$

If 25% of the 11-cis-retinal reacts then 75% remains.

$[11\text{-cis-retinal}]_t = (0.75)(3.50\ x\ 10^{-3}\ M) = 2.625\ x\ 10^{-3}\ M$

$$t = \frac{\ln\frac{[11\text{-cis-retinal}]_t}{[11\text{-cis-retinal}]_0}}{-k} = \frac{\ln\left(\frac{2.625\ x\ 10^{-3}\ M}{3.50\ x\ 10^{-3}\ M}\right)}{-(1.02\ x\ 10^{-5}/s)} = 28{,}200\ s$$

$t = 28{,}200\ s\ x\ \frac{1\ min}{60\ s} = 470\ min$

(c) $[11\text{-cis-retinal}]_t = [11\text{-cis-retinal}]_0 - [11\text{-trans-retinal}]_t$

$[11\text{-cis-retinal}]_t = (3.50\ x\ 10^{-3}) - (3.15\ x\ 10^{-3}) = 3.50\ x\ 10^{-4}\ M$

$$t = \frac{\ln\frac{[11\text{-cis-retinal}]_t}{[11\text{-cis-retinal}]_0}}{-k} = \frac{\ln\left(\frac{3.50\ x\ 10^{-4}\ M}{3.50\ x\ 10^{-3}\ M}\right)}{-(1.02\ x\ 10^{-5}/s)} = 225{,}744\ s$$

$t = 225{,}744\ s\ x\ \frac{1\ min}{60\ s}\ x\ \frac{1\ h}{60\ min} = 63\ h$

13.126 (a) From the data in the table for Experiment 1, we see that 0.20 mol of A reacts with 0.10 mol of B to produce 0.10 mol of D. The balanced equation for the reaction is:

$2\ A + B \rightarrow D$

(b) From the data in the table, initial Rates $= -\frac{\Delta A}{\Delta t}$ have been calculated.

For example, from Experiment 1:

$$\text{Initial rate} = -\frac{\Delta A}{\Delta t} = -\frac{(4.80\ M - 5.00\ M)}{60\ s} = 3.33\ x\ 10^{-3}\ M/s$$

Initial concentrations and initial rate data have been collected in the table below.

EXPT	$[A]_o$ (M)	$[B]_o$ (M)	$[C]_o$ (M)	Initial Rate (M/s)
1	5.00	2.00	1.00	3.33×10^{-3}
2	10.00	2.00	1.00	6.66×10^{-3}
3	5.00	4.00	1.00	3.33×10^{-3}
4	5.00	2.00	2.00	6.66×10^{-3}

Rate = $k[A]^m[B]^n[C]^p$
From Expts 1 and 2, [A] doubles and the initial rate doubles; therefore m = 1.
From Expts 1 and 3, [B] doubles but the initial rate does not change; therefore n = 0.
From Expts 1 and 4, [C] doubles and the initial rate doubles; therefore p = 1.
The reaction is: first-order in A; zeroth-order in B; first-order in C; second-order overall.
(c) Rate = k[A][C]
(d) C is a catalyst. C appears in the rate law, but it is not consumed in the reaction.
(e) $A + C \rightarrow AC$ (slow)
$AC + B \rightarrow AB + C$ (fast)
$A + AB \rightarrow D$ (fast)

(f) From data in Expt 1:

$$k = \frac{\text{Rate}}{[A][C]} = \frac{\Delta D/\Delta t}{[A][C]} = \frac{0.10\ \text{M}/60\ \text{s}}{(5.00\ \text{M})(1.00\ \text{M})} = 3.4 \times 10^{-4}/(\text{M} \cdot \text{s})$$

13.127 For E_a = 50 kJ/mol

$$f = e^{-E_a/RT} = \exp\left\{\frac{-50\ \text{kJ/mol}}{[8.314 \times 10^{-3}\ \text{kJ/(K}\cdot\text{mol)}](300\ \text{K})}\right\} = 2.0 \times 10^{-9}$$

For E_a = 100 kJ/mol

$$f = e^{-E_a/RT} = \exp\left\{\frac{-100\ \text{kJ/mol}}{[8.314 \times 10^{-3}\ \text{kJ/(K}\cdot\text{mol)}](300\ \text{K})}\right\} = 3.9 \times 10^{-18}$$

13.128 $$\ln\left(\frac{k_2}{k_1}\right) = \left(\frac{-E_a}{R}\right)\left(\frac{1}{T_2} - \frac{1}{T_1}\right)$$

$k_2 = 2.5k_1$
$k_1 = 1.0$, T_1 = 20 °C = 293 K
$k_2 = 2.5$, T_2 = 30 °C = 303 K

$$E_a = -\frac{[\ln k_2 - \ln k_1](R)}{\left(\frac{1}{T_2} - \frac{1}{T_1}\right)}$$

$$E_a = -\frac{[\ln(2.5) - \ln(1.0)][8.314 \times 10^{-3} \text{ kJ/(K}\cdot\text{mol)}]}{\left(\frac{1}{303 \text{ K}} - \frac{1}{293 \text{ K}}\right)} = 68 \text{ kJ/mol}$$

$k_1 = 1.0$, $T_1 = 120\ ^\circ\text{C} = 393$ K
$k_2 = ?$, $T_2 = 130\ ^\circ\text{C} = 403$ K
Solve for k_2.

$$\ln k_2 = \frac{-E_a}{R}\left(\frac{1}{T_2} - \frac{1}{T_1}\right) + \ln k_1$$

$$\ln k_2 = \frac{-68 \text{ kJ/mol}}{[8.314 \times 10^{-3} \text{ kJ/(K}\cdot\text{mol)}]}\left(\frac{1}{403 \text{ K}} - \frac{1}{393 \text{ K}}\right) + \ln(1.0) = 0.516$$

$k_2 = e^{0.516} = 1.7$; The rate increases by a factor of 1.7.

13.129 $[A] = -kt + [A]_o$
$[A]_o/2 = -kt_{1/2} + [A]_o$
$[A]_o/2 - [A]_o = -kt_{1/2}$
$-[A]_o/2 = -kt_{1/2}$
$[A]_o/2 = kt_{1/2}$

For a zeroth-order reaction, $t_{1/2} = \frac{[A]_o}{2\,k}$

For a zeroth-order reaction, each half-life is half of the previous one.
For a first-order reaction, each half-life is the same as the previous one.
For a second-order reaction, each half-life is twice the previous one.

13.130 (a) $I^-(aq) + OCl^-(aq) \rightarrow Cl^-(aq) + OI^-(aq)$

(b) From the data in the table, initial rates $= -\frac{\Delta[I^-]}{\Delta t}$ have been calculated.

For example, from Experiment 1:

$$\text{Initial rate} = -\frac{\Delta[I^-]}{\Delta t} = -\frac{(2.17 \times 10^{-4}\text{ M} - 2.40 \times 10^{-4}\text{ M})}{10\text{ s}} = 2.30 \times 10^{-6}\text{ M/s}$$

Initial concentrations and initial rate data have been collected in the table below.

EXPT	$[I^-]_o$ (M)	$[OCl^-]_o$ (M)	$[OH^-]_o$ (M)	Initial Rate (M/s)
1	2.40×10^{-4}	1.60×10^{-4}	1.00	2.30×10^{-6}
2	1.20×10^{-4}	1.60×10^{-4}	1.00	1.20×10^{-6}
3	2.40×10^{-4}	4.00×10^{-5}	1.00	6.00×10^{-7}
4	1.20×10^{-4}	1.60×10^{-4}	2.00	6.00×10^{-7}

Rate $= k[I^-]^m[OCl^-]^n[OH^-]^p$

From Expts 1 and 2, $[I^-]$ is cut in half and the initial rate is cut in half; therefore m = 1.
From Expts 1 and 3, $[OCl^-]$ is reduced by a factor of four and the initial rate is reduced

by a factor of four; therefore n = 1.
From Expts 2 and 4, $[OH^-]$ is doubled and the initial rate is cut in half; therefore p = –1.

$$\text{Rate} = k\frac{[I^-][OCl^-]}{[OH^-]}$$

From data in Expt 1:

$$k = \frac{\text{Rate}\,[OH^-]}{[I^-][OCl^-]} = \frac{(2.30 \times 10^{-6}\ M/s)(1.00\ M)}{(2.40 \times 10^{-4}\ M)(1.60 \times 10^{-4}\ M)} = 60/s$$

(c) The reaction does not occur by a single-step mechanism because OH^- appears in the rate law but not in the overall reaction.

(d)

	$OCl^-(aq) + H_2O(l) \rightleftarrows HOCl(aq) + OH^-(aq)$	(fast)
	$HOCl(aq) + I^-(aq) \rightarrow HOI(aq) + Cl^-(aq)$	(slow)
	$HOI(aq) + OH^-(aq) \rightarrow H_2O(l) + OI^-(aq)$	(fast)
Overall reaction	$I^-(aq) + OCl^-(aq) \rightarrow Cl^-(aq) + OI^-(aq)$	

Because the forward and reverse rates in step 1 are equal, $k_1[OCl^-][H_2O] = k_{-1}[HOCl][OH^-]$. Solving for [HOCl] and substituting into the rate law for the second step gives

$$\text{Rate} = k_2[HOCl][I^-] = \frac{k_1 k_2}{k_{-1}}\frac{[OCl^-][H_2O][I^-]}{[OH^-]}$$

$[H_2O]$ is constant and can be combined into k.
Because the rate law for the overall reaction is equal to the rate law for the rate-determining step, the rate law for the overall reaction is

$$\text{Rate} = k\frac{[OCl^-][I^-]}{[OH^-]} \quad \text{where } k = \frac{k_1 k_2 [H_2O]}{k_{-1}}$$

13.131 (a) $\text{Rate}_f = k_f[A]$ and $\text{Rate}_r = k_r[B]$

(b)

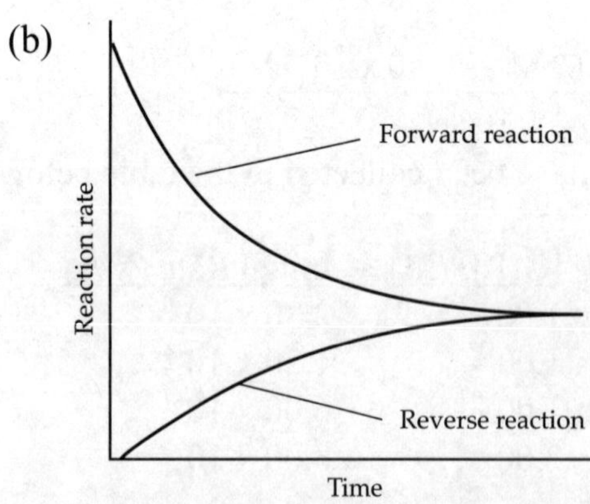

(c) When $\text{Rate}_f = \text{Rate}_r$, $k_f[A] = k_r[B]$, and $\dfrac{[B]}{[A]} = \dfrac{k_f}{k_r} = \dfrac{(3.0 \times 10^{-3})}{(1.0 \times 10^{-3})} = 3$

13.132 X → products is a first-order reaction

$$t = 60\text{ min} \times \frac{60\text{ s}}{1\text{ min}} = 3600\text{ s}$$

$$\ln\frac{[X]_t}{[X]_o} = -kt; \qquad k = \frac{\ln\frac{[X]_t}{[X]_o}}{-t}$$

At 25 °C, calculate k_1: $$k_1 = \frac{\ln\left(\frac{0.600\text{ M}}{1.000\text{ M}}\right)}{-3600\text{ s}} = 1.42 \times 10^{-4}\text{ s}^{-1}$$

At 35 °C, calculate k_2: $$k_2 = \frac{\ln\left(\frac{0.200\text{ M}}{0.600\text{ M}}\right)}{-3600\text{ s}} = 3.05 \times 10^{-4}\text{ s}^{-1}$$

At an unknown temperature calculate k_3: $$k_3 = \frac{\ln\left(\frac{0.010\text{ M}}{0.200\text{ M}}\right)}{-3600\text{ s}} = 8.32 \times 10^{-4}\text{ s}^{-1}$$

$T_1 = 25\text{ °C} = 25 + 273 = 298\text{ K}$
$T_2 = 35\text{ °C} = 35 + 273 = 308\text{ K}$

Calculate E_a using k_1 and k_2.

$$\ln\left(\frac{k_2}{k_1}\right) = \left(\frac{-E_a}{R}\right)\left(\frac{1}{T_2} - \frac{1}{T_1}\right)$$

$$E_a = -\frac{[\ln k_2 - \ln k_1](R)}{\left(\frac{1}{T_2} - \frac{1}{T_1}\right)}$$

$$E_a = -\frac{[\ln(3.05 \times 10^{-4}) - \ln(1.42 \times 10^{-4})][8.314 \times 10^{-3}\text{ kJ/(K}\cdot\text{mol)}]}{\left(\frac{1}{308\text{ K}} - \frac{1}{298\text{ K}}\right)} = 58.3\text{ kJ/mol}$$

Use E_a, k_1, and k_3 to calculate T_3.

$$\frac{1}{T_3} = \frac{\ln\left(\frac{k_3}{k_1}\right)}{\left(\frac{-E_a}{R}\right)} + \frac{1}{T_1} = \frac{\ln\left(\frac{8.32 \times 10^{-4}}{1.42 \times 10^{-4}}\right)}{\left(\frac{-58.3\text{ kJ/mol}}{8.314 \times 10^{-3}\text{ kJ/(K}\cdot\text{mol)}}\right)} + \frac{1}{298\text{ K}} = 0.003104/\text{K}$$

$$T_3 = \frac{1}{0.003104/\text{K}} = 322\text{ K} = 322 - 273 = 49\text{ °C}$$

At 3:00 p.m. raise the temperature to 49 °C to finish the reaction by 4:00 p.m.

13.133

	$N_2O_4(g)$	$\rightarrow$ $2\ NO_2(g)$
before (mm Hg)	17.0	0
change (mm Hg)	–x	+2x
after (mm Hg)	17.0 – x	2x

2x = 1.3 mm Hg

x = 1.3 mm Hg/2 = 0.65 mm Hg

$P_t(N_2O_4) = 17.0 - x = 17.0 - 0.65 = 16.35$ mm Hg

$$k = \frac{0.693}{t_{1/2}} = \frac{0.693}{1.3 \times 10^{-5}\ s} = 5.3 \times 10^4\ s^{-1}$$

$$\ln\frac{P_t}{P_o} = -kt;\quad t = \frac{\ln\frac{P_t}{P_o}}{-k} = \frac{\ln\frac{16.35}{17}}{-5.3 \times 10^4\ s^{-1}} = 7.4 \times 10^{-7}\ s$$

13.134 (a) When equal volumes of two solutions are mixed, both concentrations are cut in half.

$[H_3O^+]_o = [OH^-]_o = 1.0$ M

When 99.999% of the acid is neutralized, $[H_3O^+] = [OH^-] = 1.0\ M - (1.0\ M \times 0.99999)$
$= 1.0 \times 10^{-5}$ M

Using the 2nd order integrated rate law:

$$kt = \frac{1}{[H_3O^+]_t} - \frac{1}{[H_3O^+]_o};\qquad t = \frac{1}{k}\left[\frac{1}{[H_3O^+]_t} - \frac{1}{[H_3O^+]_o}\right]$$

$$t = \frac{1}{(1.3 \times 10^{11}\ M^{-1}s^{-1})}\left[\frac{1}{(1.0 \times 10^{-5}\ M)} - \frac{1}{(1.0\ M)}\right] = 7.7 \times 10^{-7}\ s$$

(b) The rate of an acid-base neutralization reaction would be limited by the speed of mixing, which is much slower than the intrinsic rate of the reaction itself.

13.135

$$\frac{1}{([O_2])^2} = 8\,kt + \frac{1}{([O_2]_o)^2}$$

$$\frac{1}{([O_2])^2} = 8(25\ M^{-2}s^{-1})(100.0\ s) + \frac{1}{(0.0100\ M)^2}$$

$$\frac{1}{([O_2])^2} = 30{,}000\ M^{-2}$$

$$[O_2] = \sqrt{\frac{1}{30{,}000\ M^{-2}}} = 0.005\ 77\ M$$

	$2\ NO(g)$	+ $O_2(g)$	$\rightarrow$ $2\ NO_2(g)$
before (M)	0.0200	0.0100	0
change (M)	–2x	–x	+2x
after (M)	0.0200 – 2x	0.0100 – x	2x

$[O_2] = 0.005\ 77\ M = 0.0100\ M - x$
$x = 0.0100\ M - 0.005\ 77\ M = 0.004\ 23\ M$
$[NO] = 0.0200\ M - 2x = 0.0200\ M - 2(0.004\ 23\ M) = 0.0115\ M$
$[O_2] = 0.005\ 77\ M$
$[NO_2] = 2x = 2(0.004\ 23\ M) = 0.008\ 46\ M$

13.136 Looking at the two experiments at 600 K, when the NO_2 concentration is doubled, the rate increased by a factor of 4. Therefore, the reaction is 2nd order.
Rate = $k\,[NO_2]^2$
Calculate k_1 at 600 K: $k_1 = \text{Rate}/[NO_2]^2 = 5.4 \times 10^{-7}\ M\ s^{-1}/(0.0010\ M)^2 = 0.54\ M^{-1}\ s^{-1}$
Calculate k_2 at 700 K: $k_2 = \text{Rate}/[NO_2]^2 = 5.2 \times 10^{-5}\ M\ s^{-1}/(0.0020\ M)^2 = 13\ M^{-1}\ s^{-1}$
Calculate E_a using k_1 and k_2.

$$\ln\left(\frac{k_2}{k_1}\right) = \left(\frac{-E_a}{R}\right)\left(\frac{1}{T_2} - \frac{1}{T_1}\right)$$

$$E_a = -\frac{[\ln k_2 - \ln k_1](R)}{\left(\frac{1}{T_2} - \frac{1}{T_1}\right)}$$

$$E_a = -\frac{[\ln(13) - \ln(0.54)][8.314 \times 10^{-3}\ kJ/(K\cdot mol)]}{\left(\frac{1}{700\ K} - \frac{1}{600\ K}\right)} = 111\ kJ/mol$$

Calculate k_3 at 650 K using E_a and k_1.
Solve for k_3.

$$\ln k_3 = \frac{-E_a}{R}\left(\frac{1}{T_3} - \frac{1}{T_1}\right) + \ln k_1$$

$$\ln k_3 = \frac{-111\ kJ/mol}{[8.314 \times 10^{-3}\ kJ/(K\cdot mol)]}\left(\frac{1}{650\ K} - \frac{1}{600\ K}\right) + \ln(0.54) = 1.0955$$

$k_3 = e^{1.0955} = 3.0\ M^{-1}\ s^{-1}$

$$k_3 t = \frac{1}{[NO_2]_t} - \frac{1}{[NO_2]_o}; \qquad t = \frac{1}{k_3}\left[\frac{1}{[NO_2]_t} - \frac{1}{[NO_2]_o}\right]$$

$$t = \frac{1}{(3.0\ M^{-1}s^{-1})}\left[\frac{1}{(0.0010\ M)} - \frac{1}{(0.0050\ M)}\right] = 2.7 \times 10^2\ s$$

13.137 Rate = $k\,[A]^x[B]^y$
Comparing Experiments 1 and 2, the concentration of B does not change, the concentration of A doubles, and the rate doubles. This means the reaction is first-order in A ($x = 1$).
Comparing Experiments 1 and 3, the rate would drop to 0.9×10^{-5} M/s as a result of the concentration of A being cut in half. Then with the concentration of B doubling, the rate increases by a factor of 4, to 3.6×10^{-5} M/s. This means the reaction is second-order in B ($y = 2$).

$$\text{At 600 K, } k_1 = \frac{\text{Rate}}{[A][B]^2} = \frac{4.3 \times 10^{-5}\ \text{M/s}}{(0.50\ \text{M})(0.50\ \text{M})^2} = 3.4 \times 10^{-4}\ \text{M}^{-2}\ \text{s}^{-1}$$

$$\text{At 700 K, } k_2 = \frac{\text{Rate}}{[A][B]^2} = \frac{1.8 \times 10^{-5}\ \text{M/s}}{(0.20\ \text{M})(0.10\ \text{M})^2} = 9.0 \times 10^{-3}\ \text{M}^{-2}\ \text{s}^{-1}$$

$$\ln\left(\frac{k_2}{k_1}\right) = \left(\frac{-E_a}{R}\right)\left(\frac{1}{T_2} - \frac{1}{T_1}\right)$$

$$E_a = -\frac{[\ln k_2 - \ln k_1](R)}{\left(\frac{1}{T_2} - \frac{1}{T_1}\right)}$$

$$E_a = -\frac{[\ln(9.0 \times 10^{-3}) - \ln(3.4 \times 10^{-4})][8.314 \times 10^{-3}\ \text{kJ/(K}\cdot\text{mol)}]}{\left(\frac{1}{700\ \text{K}} - \frac{1}{600\ \text{K}}\right)} = 114\ \text{kJ/mol}$$

13.138 A $\rightarrow$ C is a first-order reaction.
The reaction is complete at 200 s when the absorbance of C reaches 1.200.
Because there is a one to one stoichiometry between A and C, the concentration of A must be proportional to 1.200 – absorbance of C. Any two data points can be used to find k. Let $[A]_o \propto 1.200$ and at 100 s, $[A]_t \propto 1.200 - 1.188 = 0.012$

$$\ln\frac{[A]_t}{[A]_o} = -kt; \qquad k = \frac{\ln\frac{[A]_t}{[A]_o}}{-t}; \qquad k = \frac{\ln\left(\frac{0.012\ \text{M}}{1.200\ \text{M}}\right)}{-100\ \text{s}} = 0.0461\ \text{s}^{-1}$$

$$t_{1/2} = \frac{0.693}{k} = \frac{0.693}{0.0461\ \text{s}^{-1}} = 15\ \text{s}$$

13.139 Rate = $k\,[HI]^x$

$$\frac{\text{Rate}_2}{\text{Rate}_1} = \frac{k[0.30]^x}{k[0.10]^x}$$

$$x = \frac{\log\frac{\text{Rate}_2}{\text{Rate}_1}}{\log\frac{(0.30)}{(0.10)}} = \frac{\log\frac{1.6 \times 10^{-4}}{1.8 \times 10^{-5}}}{\log\frac{(0.30)}{(0.10)}} = \frac{0.949}{0.477} = 2$$

Rate = $k\,[HI]^2$

$$\text{At 700 K, } k_1 = \frac{\text{Rate}}{[HI]^2} = \frac{1.8 \times 10^{-5}\ \text{M/s}}{(0.10\ \text{M})^2} = 1.8 \times 10^{-3}\ \text{M}^{-1}\ \text{s}^{-1}$$

At 800 K, $k_2 = \dfrac{\text{Rate}}{[HI]^2} = \dfrac{3.9 \times 10^{-3}\ M/s}{(0.20\ M)^2} = 9.7 \times 10^{-2}\ M^{-1}\ s^{-1}$

$$\ln\left(\frac{k_2}{k_1}\right) = \left(\frac{-E_a}{R}\right)\left(\frac{1}{T_2} - \frac{1}{T_1}\right)$$

$$E_a = -\frac{[\ln k_2 - \ln k_1](R)}{\left(\dfrac{1}{T_2} - \dfrac{1}{T_1}\right)}$$

$$E_a = -\frac{[\ln(9.7 \times 10^{-2}) - \ln(1.8 \times 10^{-3})][8.314 \times 10^{-3}\ kJ/(K \cdot mol)]}{\left(\dfrac{1}{800\ K} - \dfrac{1}{700\ K}\right)} = 186\ kJ/mol$$

Calculate k_4 at 650 K using E_a and k_1.
Solve for k_4.

$$\ln k_4 = \frac{-E_a}{R}\left(\frac{1}{T_4} - \frac{1}{T_1}\right) + \ln k_1$$

$$\ln k_4 = \frac{-186\ kJ/mol}{[8.314 \times 10^{-3}\ kJ/(K \cdot mol)]}\left(\frac{1}{650\ K} - \frac{1}{700\ K}\right) + \ln(1.8 \times 10^{-3}) = -8.778$$

$k_4 = e^{-8.778} = 1.5 \times 10^{-4}\ M^{-1}\ s^{-1}$

$$[HI] = \sqrt{\frac{\text{Rate}}{k_4}} = \sqrt{\frac{1.0 \times 10^{-5}\ M/s}{1.5 \times 10^{-4}\ M^{-1}s^{-1}}} = 0.26\ M$$

13.140 $$\ln\left(\frac{k_2}{k_1}\right) = \left(\frac{-E_a}{R}\right)\left(\frac{1}{T_2} - \frac{1}{T_1}\right)$$

$k_1 = 0.267\ L/(mg \cdot min)$, $T_1 = 5\ ^\circ C = 278\ K$
$k_2 = 3.45\ L/(mg \cdot min)$, $T_2 = 30\ ^\circ C = 303\ K$

$$E_a = -\frac{[\ln k_2 - \ln k_1](R)}{\left(\dfrac{1}{T_2} - \dfrac{1}{T_1}\right)}$$

$$E_a = -\frac{[\ln(3.45) - \ln(0.267)][8.314 \times 10^{-3}\ kJ/(K \cdot mol)]}{\left(\dfrac{1}{303\ K} - \dfrac{1}{278\ K}\right)} = 71.7\ kJ/mol$$

13.141 $k = \dfrac{0.693}{t_{1/2}}$; $500\ y \times \dfrac{365\ d}{1\ y} \times \dfrac{24\ h}{1\ d} \times \dfrac{60\ min}{1\ h} \times \dfrac{60\ s}{1\ min} = 1.58 \times 10^{10}\ s$

$$k_1 = \frac{0.693}{1.58 \times 10^{10}\ s} = 4.39 \times 10^{-11}\ /s \text{ and } k_2 = \frac{0.693}{0.010\ s} = 69.3\ /s$$

$$\frac{k_2}{k_1} = \frac{69.3\ /s}{4.39 \times 10^{-11}\ /s} = 1.6 \times 10^{12}$$

The enzyme increases the rate of the peptide bond breaking reaction by a factor of 1.6×10^{12}.

13.142

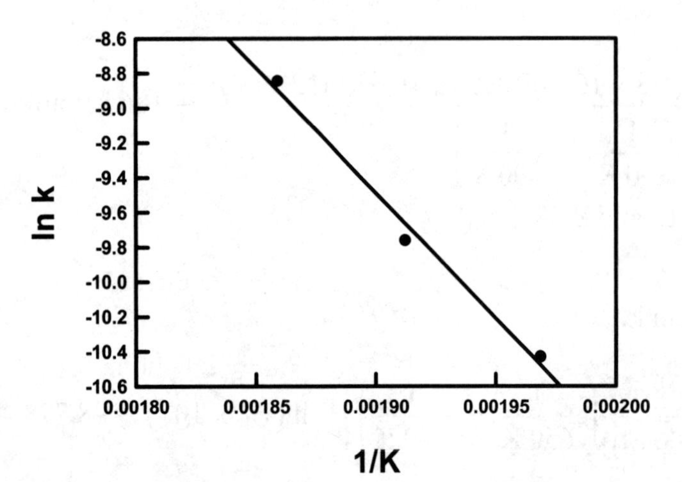

$$\text{Slope} = -\frac{E_a}{R} = -14418.2\ K$$

$$E_a = [8.314 \times 10^{-3}\ kJ/(K\cdot mol)](14418.2\ K) = 120\ kJ/mol$$

13.143 (a)

$$\ln\left(\frac{k_2}{k_1}\right) = \left(\frac{-E_a}{R}\right)\left(\frac{1}{T_2} - \frac{1}{T_1}\right)$$

$k_1 = 2.60 \times 10^{-4}\ /s$, $T_1 = 530\ °C = 803\ K$
$k_2 = 9.45 \times 10^{-3}\ /s$, $T_2 = 620\ °C = 893\ K$

$$E_a = -\frac{[\ln k_2 - \ln k_1](R)}{\left(\frac{1}{T_2} - \frac{1}{T_1}\right)}$$

$$E_a = -\frac{[\ln(9.45 \times 10^{-3}) - \ln(2.60 \times 10^{-4})][8.314 \times 10^{-3}\ kJ/(K\cdot mol)]}{\left(\frac{1}{893\ K} - \frac{1}{803\ K}\right)} = 238\ kJ/mol$$

(b) Calculate k_3 at 580 °C = 853 K using E_a and k_1.
Solve for k_3.

$$\ln k_3 = \frac{-E_a}{R}\left(\frac{1}{T_3} - \frac{1}{T_1}\right) + \ln k_1$$

$$\ln k_3 = \frac{-238 \text{ kJ/mol}}{[8.314 \times 10^{-3} \text{ kJ/(K} \cdot \text{mol)}]}\left(\frac{1}{853 \text{ K}} - \frac{1}{803 \text{ K}}\right) + \ln(2.60 \times 10^{-4}) = -6.165$$

$k_3 = e^{-6.165} = 2.10 \times 10^{-3}$ /s

$$t_{1/2} = \frac{0.693}{k_3} = \frac{0.693}{2.10 \times 10^{-3}/\text{s}} = 330 \text{ s}$$

Multiconcept Problems

13.144 (a) $k = Ae^{-\frac{E_a}{RT}} = (6.0 \times 10^8/(\text{M} \cdot \text{s}))\, e^{-\frac{6.3 \text{ kJ/mol}}{[8.314 \times 10^{-3} \text{ kJ/(K} \cdot \text{mol)}](298 \text{ K})}} = 4.7 \times 10^7/(\text{M} \cdot \text{s})$

(b) :Ö=N̈—F̤: N has 3 electron clouds, is sp^2 hybridized, and the molecule is bent.

(c) O—N---F---F

(d) The reaction has such a low activation energy because the F–F bond is very weak and the N–F bond is relatively strong.

13.145 $2\,HI(g) \rightarrow H_2(g) + I_2(g)$

(a) $\text{mass HI} = 1.50 \text{ L} \times \frac{1000 \text{ mL}}{1 \text{ L}} \times \frac{0.0101 \text{ g}}{1 \text{ mL}} = 15.15 \text{ g HI}$

$$15.15 \text{ g HI} \times \frac{1 \text{ mol HI}}{127.91 \text{ g HI}} = 0.118 \text{ mol HI}$$

$$[HI] = \frac{0.118 \text{ mol}}{1.50 \text{ L}} = 0.0787 \text{ mol/L}$$

$$-\frac{\Delta[HI]}{\Delta t} = k[HI]^2 = (0.031/(\text{M} \cdot \text{min}))(0.0787 \text{ M})^2 = 1.92 \times 10^{-4} \text{ M/min}$$

$2\,HI(g) \rightarrow H_2(g) + I_2(g)$

$$\frac{\Delta[I_2]}{\Delta t} = \frac{1}{2}\left(-\frac{\Delta[HI]}{\Delta t}\right) = \frac{1.92 \times 10^{-4} \text{ M/min}}{2} = 9.60 \times 10^{-5} \text{ M/min}$$

$(9.60 \times 10^{-5} \text{ M/min})(1.50 \text{ L})(6.022 \times 10^{23} \text{ molecules/mol}) = 8.7 \times 10^{19}$ molecules/min

(b) Rate = $k[HI]^2$

$$\frac{1}{[HI]_t} = kt + \frac{1}{[HI]_o} = (0.031/(\text{M} \cdot \text{min}))\left(8.00 \text{ h} \times \frac{60.0 \text{ min}}{1 \text{ h}}\right) + \frac{1}{0.0787 \text{ M}} = 27.59/\text{M}$$

$$[HI]_t = \frac{1}{27.59/\text{M}} = 0.0362 \text{ M}$$

From stoichiometry, $[H_2]_t = 1/2\,([HI]_o - [HI]_t) = 1/2\,(0.0787 \text{ M} - 0.0362 \text{ M}) = 0.0212 \text{ M}$

410 °C = 683 K

PV = nRT

$$P_{H_2} = \left(\frac{n}{V}\right)RT = (0.0212\text{ mol/L})\left(0.082\ 06\ \frac{L\cdot atm}{K\cdot mol}\right)(683\text{ K}) = 1.2\text{ atm}$$

13.146 $2\ NO_2(g) \rightarrow 2\ NO(g) + O_2(g)$

k = 4.7/(M · s)

(a) The units for k indicate a second-order reaction.

(b) 383 °C = 656 K

PV = nRT

$$[NO_2]_o = \frac{n}{V} = \frac{P}{RT} = \frac{\left(746\text{ mm Hg x } \frac{1.000\text{ atm}}{760\text{ mm Hg}}\right)}{\left(0.082\ 06\ \frac{L\cdot atm}{K\cdot mol}\right)(656\text{ K})} = 0.01823\text{ mol/L}$$

initial rate = $k[NO_2]_o^2$ = [4.7/(M · s)](0.01823 mol/L)^2 = 1.56 x 10^{-3} mol/(L · s)

$$\text{initial rate for } O_2 = \frac{\text{initial rate for } NO_2}{2} = \frac{1.56 \times 10^{-3}\text{ mol/(L}\cdot\text{s)}}{2} = 7.80 \times 10^{-4}\text{ mol/(L}\cdot\text{s)}$$

initial rate for O_2 = [7.80 x 10^{-4} mol/(L · s)](32.00 g/mol) = 0.025 g/(L · s)

(c) $$\frac{1}{[NO_2]_t} = kt + \frac{1}{[NO_2]_0} = [4.7/(M\cdot s)](60\text{ s}) + \frac{1}{0.01823\text{ M}}$$

$$\frac{1}{[NO_2]_t} = 336.9/M$$ and $[NO_2]$ = 0.00297 M

	$2\ NO_2(g)$	$\rightarrow$	$2\ NO(g)$	+	$O_2(g)$
before reaction (M)	0.01823		0		0
change (M)	–2x		+2x		+x
after 1.00 min (M)	0.01823 – 2x		2x		x

after 1.00 min $[NO_2]$ = 0.00297 M = 0.01823 – 2x

x = 0.00763 M = $[O_2]$

mass O_2 = (0.00763 mol/L)(5.00 L)(32.00 g/mol) = 1.22 g O_2

13.147 (a) N_2O_5, 108.01

$$[N_2O_5]_o = \frac{\left(2.70\text{ g } N_2O_5 \text{ x } \frac{1\text{ mol } N_2O_5}{108.01\text{ g } N_2O_5}\right)}{2.00\text{ L}} = 0.0125\text{ mol/L}$$

$$\ln [N_2O_5]_t = -kt + \ln [N_2O_5]_o = -(1.7 \times 10^{-3}\text{ s}^{-1})\left(13.0\text{ min x } \frac{60.0\text{ s}}{1\text{ min}}\right) + \ln(0.0125) = -5.71$$

$[N_2O_5]_t = e^{-5.71}$ = 3.31 x 10^{-3} mol/L

After 13.0 min, mol N_2O_5 = (3.31 x 10^{-3} mol/L)(2.00 L) = 6.62 x 10^{-3} mol N_2O_5

	$N_2O_5(g)$	→	$2\ NO_2(g)$	+	$1/2\ O_2(g)$
before reaction (mol)	0.0250		0		0
change (mol)	–x		+2x		+1/2x
after reaction (mol)	0.0250 – x		2x		1/2x

After 13.0 min, mol $N_2O_5 = 6.62 \times 10^{-3} = 0.0250 - x$
x = 0.0184 mol

After 13.0 min, $n_{total} = n_{N_2O_5} + n_{NO_2} + n_{O_2} = (6.62 \times 10^{-3}) + 2(0.0184) + 1/2(0.0184)$

$n_{total} = 0.0526$ mol
55 °C = 328 K

$$PV = nRT; \quad P_{total} = \frac{nRT}{V} = \frac{(0.0526\ \text{mol})\left(0.082\ 06\ \frac{\text{L}\cdot\text{atm}}{\text{K}\cdot\text{mol}}\right)(328\ \text{K})}{2.00\ \text{L}} = 0.71\ \text{atm}$$

(b) $N_2O_5(g) \rightarrow 2\ NO_2(g) + 1/2\ O_2(g)$
$\Delta H^\circ_{rxn} = 2\ \Delta H^\circ_f(NO_2) - \Delta H^\circ_f(N_2O_5)$
$\Delta H^\circ_{rxn} = (2\ \text{mol})(33.2\ \text{kJ/mol}) - (1\ \text{mol})(11\ \text{kJ/mol}) = 55.4\ \text{kJ} = 5.54 \times 10^4\ \text{J}$
initial rate $= k[N_2O_5]_o = (1.7 \times 10^{-3}\ s^{-1})(0.0125\ \text{mol/L}) = 2.125 \times 10^{-5}\ \text{mol/(L} \cdot \text{s)}$
initial rate absorbing heat $= [2.125 \times 10^{-5}\ \text{mol/(L} \cdot \text{s)}](2.00\ \text{L})(5.54 \times 10^4\ \text{J/mol}) = 2.4\ \text{J/s}$

(c)

$$\ln [N_2O_5]_t = -kt + \ln [N_2O_5]_o = -(1.7 \times 10^{-3}\ s^{-1})\left(10.0\ \text{min} \times \frac{60.0\ \text{s}}{1\ \text{min}}\right) + \ln(0.0125) = -5.40$$

$[N_2O_5]_t = e^{-5.40} = 4.52 \times 10^{-3}$ mol/L
After 10.0 min, mol $N_2O_5 = (4.52 \times 10^{-3}\ \text{mol/L})(2.00\ \text{L}) = 9.03 \times 10^{-3}$ mol N_2O_5

	$N_2O_5(g)$	→	$2\ NO_2(g)$	+	$1/2\ O_2(g)$
before reaction (mol)	0.0250		0		0
change (mol)	–x		+2x		+1/2x
after reaction (mol)	0.0250 – x		2x		1/2x

After 10.0 min, mol $N_2O_5 = 9.03 \times 10^{-3} = 0.0250 - x$
x = 0.0160 mol
heat absorbed = (0.0160 mol)(55.4 kJ/mol) = 0.89 kJ

13.148 $2\ N_2O(g) \rightarrow 2\ N_2(g) + O_2(g)$
P_{O_2}(in exit gas) = 1.0 mm Hg; P_{total} = 1.50 atm = 1140 mm Hg
From the reaction stoichiometry:
P_{N_2}(in exit gas) = 2 P_{O_2} = 2.0 mm Hg
P_{N_2O}(in exit gas) = $P_{total} - P_{N_2} - P_{O_2}$ = 1140 – 2.0 – 1.0 = 1137 mm Hg
Assume P_{N_2O}(initial) = P_{total} = 1140 mm Hg (In assuming a constant total pressure in the tube, we are neglecting the slight change in pressure due to the reaction.)

Volume of tube = $\pi r^2 l = \pi(1.25\text{ cm})^2(20\text{ cm}) = 98.2\text{ cm}^3 = 0.0982\text{ L}$

Time, t, gases are in the tube = $\dfrac{\text{volume of tube}}{\text{flow rate}}$ x $\dfrac{0.0982\text{ L}}{0.75\text{ L/min}}$ x $\dfrac{60\text{ s}}{1\text{ min}} = 7.86\text{ s}$

At time t, $\dfrac{[N_2O]_t}{[N_2O]_0} = \dfrac{P_{N_2O}\text{ (in exit gas)}}{P_{N_2O}\text{ (initial)}} = \dfrac{1137\text{ mm Hg}}{1140\text{ mm Hg}} = 0.997\ 37$

Because $k = Ae^{-\frac{E_a}{RT}}$ and $A = 4.2 \times 10^9\text{ s}^{-1}$, k has units of s^{-1}. Therefore, this is a first-order reaction and the appropriate integrated rate law is $\ln\dfrac{[N_2O]_t}{[N_2O]_0} = -kt$.

$$k = \frac{-\ln\dfrac{[N_2O]_t}{[N_2O]_0}}{t} = \frac{-\ln(0.997\,37)}{7.86\text{ s}} = 3.35 \times 10^{-4}\text{ s}^{-1}$$

From the Arrhenius equation, $\ln k = \ln A - \dfrac{E_a}{RT}$

$$T = \frac{E_a}{(R)[\ln A - \ln k]} = \frac{222\text{ kJ/mol}}{(8.314 \times 10^{-3}\text{ kJ/(K}\cdot\text{mol))}[(22.16) - (-8.00)]} = 885\text{ K}$$

13.149 H_2O_2, 34.01

mass H_2O_2 = (0.500 L)(1000 mL/1 L)(1.00 g/ 1 mL)(0.0300) = 15.0 g H_2O_2

mol H_2O_2 = 15.0 g H_2O_2 x $\dfrac{1\text{ mol }H_2O_2}{34.01\text{ g }H_2O_2}$ = 0.441 H_2O_2

$[H_2O_2]_o = \dfrac{0.441\text{ mol}}{0.500\text{L}} = 0.882\text{ mol /L}$

$k = \dfrac{0.693}{t_{1/2}} = \dfrac{0.693}{10.7\text{ h}} = 6.48 \times 10^{-2}\text{/h}$

$\ln [H_2O_2]_t = -kt + \ln [H_2O_2]_o$

$\ln [H_2O_2]_t = -(6.48 \times 10^{-2}\text{/h})(4.02\text{ h}) + \ln (0.882)$

$\ln [H_2O_2]_t = -0.386$; $[H_2O_2]_t = e^{-0.386} = 0.680$ mol/L

mol H_2O_2 = (0.680 mol/L)(0.500 L) = 0.340 mol

	$2\ H_2O_2(aq)$	$\rightarrow$	$2\ H_2O(l)$	+	$O_2(g)$
before reaction (mol)	0.441		0		0
change (mol)	– 2x		+2x		+x
after reaction (mol)	0.441 – 2x		2x		x

After 4.02 h, mol H_2O_2 = 0.340 mol = 0.441 – 2x; solve for x.

2x = 0.101

x = 0.0505 mol = mol O_2

P = 738 mm Hg x $\dfrac{1.00\text{ atm}}{760\text{ mm Hg}}$ = 0.971 atm

$PV = nRT$

$$V = \frac{nRT}{P} = \frac{(0.0505\ \text{mol})\left(0.082\ 06\ \frac{\text{L}\cdot\text{atm}}{\text{K}\cdot\text{mol}}\right)(293\ \text{K})}{0.971\ \text{atm}} = 1.25\ \text{L}$$

$P\Delta V = (0.971\ \text{atm})(1.25\ \text{L}) = 1.21\ \text{L}\cdot\text{atm}$

$$w = -P\Delta V = -1.21\ \text{L}\cdot\text{atm} = (-1.21\ \text{L}\cdot\text{atm})\left(101\ \frac{\text{J}}{\text{L}\cdot\text{atm}}\right) = -122\ \text{J}$$

13.150 (a)

	$CH_3CHO(g)$	$\rightarrow$	$CH_4(g)$	+	$CO(g)$
before (atm)	0.500		0		0
change (atm)	–x		+x		+x
after (atm)	0.500 – x		x		x

At 605 s, $P_{total} = P_{CH_3CHO} + P_{CH_4} + P_{CO} = (0.500\ \text{atm} - x) + x + x = 0.808\ \text{atm}$

$x = 0.808\ \text{atm} - 0.500\ \text{atm} = 0.308\ \text{atm}$

The integrated rate law for a second-order reaction in terms of molar concentrations is $\frac{1}{[A]_t} = kt + \frac{1}{[A]_o}$. The ideal gas law, PV = nRT, can be rearranged to show how P is proportional to the molar concentration of a gas.

$P = \frac{n}{V}RT$ (R and T are constant), so $P \propto \frac{n}{V}$ = molar concentration

Because of this relationship, the second-order integrated rate law can be rewritten in terms of partial pressures.

$$\frac{1}{P_t} = kt + \frac{1}{P_o};\quad \frac{1}{P_t} - \frac{1}{P_o} = kt;\quad \frac{\left(\frac{1}{P_t} - \frac{1}{P_o}\right)}{t} = k$$

P is the partial pressure of CH_3CHO.

At t = 0, $P_o = 0.500$ and at t = 605 s, $P_t = 0.500\ \text{atm} - 0.308\ \text{atm} = 0.192\ \text{atm}$

$$k = \frac{\left(\frac{1}{0.192\ \text{atm}} - \frac{1}{0.500\ \text{atm}}\right)}{605\ \text{s}} = 5.30 \times 10^{-3}\ \text{atm}^{-1}\ \text{s}^{-1}$$

(b) Use the ideal gas law to convert atm^{-1} to M^{-1}.

$$P = \frac{n}{V}RT;\quad \frac{P}{RT} = \frac{n}{V};\quad \frac{1}{P}RT = \frac{V}{n} = M^{-1}$$

So, multiply k by RT to convert $\text{atm}^{-1}\ \text{s}^{-1}$ to $M^{-1}\ \text{s}^{-1}$.

$k = (5.30 \times 10^{-3}\ \text{atm}^{-1}\ \text{s}^{-1})RT$

$$k = (5.30 \times 10^{-3}\ \text{atm}^{-1}\ \text{s}^{-1})\left(0.082\ 06\ \frac{\text{L}\cdot\text{atm}}{\text{K}\cdot\text{mol}}\right)(791\ \text{K}) = 0.344\ \frac{\text{L}}{\text{mol}\cdot\text{s}} = 0.344\ M^{-1}\ \text{s}^{-1}$$

(c) $CH_3CHO(g) \rightarrow CH_4(g) + CO(g)$

$\Delta H^\circ_{rxn} = [\Delta H^\circ_f(CH_4) + \Delta H^\circ_f(CO)] - \Delta H^\circ_f(CH_3CHO)]$

$\Delta H^\circ_{rxn} = [(1\text{ mol})(-74.8\text{ kJ/mol}) + (1\text{ mol})(-110.5\text{ kJ/mol})] - (1\text{ mol})(-166.2\text{ kJ/mol})$
$\Delta H^\circ_{rxn} = -19.1$ kJ per mole of CH_3CHO that decomposes
$PV = nRT$

$$\text{mol } CH_3CHO \text{ reacted} = \frac{PV}{RT} = \frac{(0.308\text{ atm})(1.00\text{ L})}{\left(0.082\ 06\ \frac{\text{L}\cdot\text{atm}}{\text{K}\cdot\text{mol}}\right)(791\text{ K})} = 0.004\ 74\text{ mol}$$

$q = (0.004\ 74\text{ mol})(19.1\text{ kJ/mol})(1000\text{ J/kJ}) = 90.6$ J liberated after a reaction time of 605 s.

14 Chemical Equilibrium

14.1 (a) $K_c = \dfrac{[SO_3]^2}{[SO_2]^2[O_2]}$ (b) $K_c = \dfrac{[H^+][CH_3COO^-]}{[CH_3COOH]}$

(c) $K_c' = \dfrac{[SO_2]^4[O_2]^2}{[SO_3]^4}$; $K_c' = \left(\dfrac{1}{K_c}\right)^2$

14.2 (a) $K_c(\text{overall}) = \dfrac{[NO_2]^2}{[N_2][O_2]^2}$

(b) $K_c(\text{overall}) = K_{c1} \text{ x } K_{c2} = (4.3 \text{ x } 10^{-25})(6.4 \text{ x } 10^9) = 2.8 \text{ x } 10^{-15}$

14.3 (a) $K_c = \dfrac{[SO_3]^2}{[SO_2]^2[O_2]} = \dfrac{(5.0 \text{ x } 10^{-2})^2}{(3.0 \text{ x } 10^{-3})^2(3.5 \text{ x } 10^{-3})} = 7.9 \text{ x } 10^4$

(b) $K_c' = \dfrac{1}{K_c} = \dfrac{[SO_2]^2[O_2]}{[SO_3]^2} = \dfrac{(3.0 \text{ x } 10^{-3})^2(3.5 \text{ x } 10^{-3})}{(5.0 \text{ x } 10^{-2})^2} = 1.3 \text{ x } 10^{-5}$

14.4 (a) $K_c = \dfrac{[H^+][C_3H_5O_3^-]}{[C_3H_6O_3]} = \dfrac{(3.65 \text{ x } 10^{-3})(3.65 \text{ x } 10^{-3})}{(9.64 \text{ x } 10^{-2})} = 1.38 \text{ x } 10^{-4}$

(b) $[C3H6O3] = \dfrac{[H^+][C_3H_5O_3^-]}{K_c} = \dfrac{(1.17 \text{ x } 10^{-2})(1.17 \text{ x } 10^{-2})}{(1.38 \text{ x } 10^{-4})} = 0.992 \text{ M}$

14.5 From (1), $K_c = \dfrac{[AB][B]}{[A][B_2]} = \dfrac{(1)(2)}{(1)(2)} = 1$

For a mixture to be at equilibrium, $\dfrac{[AB][B]}{[A][B_2]}$ must be equal to 1.

For (2), $\dfrac{[AB][B]}{[A][B_2]} = \dfrac{(2)(1)}{(2)(1)} = 1$. This mixture is at equilibrium.

For (3), $\dfrac{[AB][B]}{[A][B_2]} = \dfrac{(1)(1)}{(4)(2)} = 0.125$. This mixture is not at equilibrium.

For (4), $\dfrac{[AB][B]}{[A][B_2]} = \dfrac{(2)(2)}{(4)(1)} = 1$. This mixture is at equilibrium.

14.6 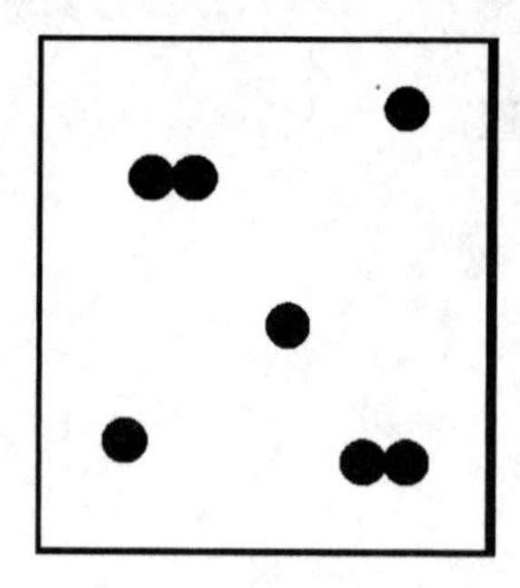

14.7 $K_p = \dfrac{(P_{CO_2})(P_{H_2})}{(P_{CO})(P_{H_2O})} = \dfrac{(6.12)(20.3)}{(1.31)(10.0)} = 9.48$

14.8 $K_p = 25 = \dfrac{(P_{HI})^2}{(P_{H_2})(P_{I_2})}$

$P_{HI} = \sqrt{(25)(P_{H_2})(P_{I_2})} = \sqrt{(25)(0.286)(0.286)} = 1.43$ atm

14.9 $2\ NO(g) + O_2(g) \rightleftarrows 2\ NO_2(g)$; $\Delta n = 2 - 3 = -1$
$K_p = K_c(RT)^{\Delta n}$, $K_c = K_p(1/RT)^{\Delta n}$
at 500 K: $K_p = (6.9 \times 10^5)[(0.082\ 06)(500)]^{-1} = 1.7 \times 10^4$
at 1000 K: $K_c = (1.3 \times 10^{-2})\left(\dfrac{1}{(0.082\ 06)(1000)}\right)^{-1} = 1.1$

14.10 $K_c' = \dfrac{1}{K_c} = \dfrac{1}{245}$

$2\ SO_3(g) \rightleftarrows 2\ SO_2(g) + O_2(g)$; $\Delta n = 3 - 2 = 1$
$K_p = K_c'(RT)^{\Delta n}$
$K_p = (1/245)[(0.082\ 06)(1,000)] = 0.335$

14.11 (a) $K_c = \dfrac{[H_2]^3}{[H_2O]^3}$, $K_p = \dfrac{(P_{H_2})^3}{(P_{H_2O})^3}$, $\Delta n = (3) - (3) = 0$ and $K_p = K_c$

(b) $K_c = [H_2]^2[O_2]$, $K_p = (P_{H_2})^2(P_{O_2})$, $\Delta n = (3) - (0) = 3$ and $K_p = K_c(RT)^3$

(c) $K_c = \dfrac{[HCl]^4}{[SiCl_4][H_2]^2}$, $K_p = \dfrac{(P_{HCl})^4}{(P_{SiCl_4})(P_{H_2})^2}$, $\Delta n = (4) - (3) = 1$ and $K_p = K_c(RT)$

(d) $K_c = \dfrac{1}{[Hg_2^{2+}][Cl^-]^2}$

14.12 $Mg(OH)_2(s) \rightleftarrows Mg^{2+}(aq) + 2\ OH^-(aq)$
$Kc = [Mg^{2+}][OH^-]^2 = (1.65 \times 10^{-4})(3.30 \times 10^{-4})^2 = 1.80 \times 10^{-11}$

14.13 $K_c = 1.2 \times 10^{-42}$. Because K_c is very small, the equilibrium mixture contains mostly H_2 molecules. H is in periodic group 1A. A very small value of K_c is consistent with strong bonding between 2 H atoms, each with one valence electron.

14.14 $K_c = 1.2 \times 10^{82}$ is very large. When equilibrium is reached, very little if any ethanol will remain because the reaction goes to completion.

14.15 The container volume of 5.0 L must be included to calculate molar concentrations.

(a) $$Q_c = \frac{[NO_2]_t^2}{[NO]_t^2[O_2]_t} = \frac{(0.80\text{ mol}/5.0\text{ L})^2}{(0.060\text{ mol}/5.0\text{ L})^2(1.0\text{ mol}/5.0\text{ L})} = 890$$

Because $Q_c < K_c$, the reaction is not at equilibrium. The reaction will proceed to the right to reach equilibrium.

(b) $$Q_c = \frac{[NO_2]_t^2}{[NO]_t^2[O_2]_t} = \frac{(4.0\text{ mol}/5.0\text{ L})^2}{(5.0 \times 10^{-3}\text{ mol}/5.0\text{ L})^2(0.20\text{ mol}/5.0\text{ L})} = 1.6 \times 10^7$$

Because $Q_c > K_c$, the reaction is not at equilibrium. The reaction will proceed to the left to reach equilibrium.

14.16 $K_c = \dfrac{[AB]^2}{[A_2][B_2]} = 4$; For a mixture to be at equilibrium, $\dfrac{[AB]^2}{[A_2][B_2]}$ must be equal to 4.

For (1), $Q_c = \dfrac{[AB]^2}{[A_2][B_2]} = \dfrac{(6)^2}{(1)(1)} = 36$, $Q_c > K_c$

For (2), $Q_c = \dfrac{[AB]^2}{[A_2][B_2]} = \dfrac{(4)^2}{(2)(2)} = 4$, $Q_c = K_c$

For (3), $Q_c = \dfrac{[AB]^2}{[A_2][B_2]} = \dfrac{(2)^2}{(3)(3)} = 0.44$, $Q_c < K_c$

(a) (2) (b) (1), reverse; (3), forward

14.17

	$CO(g)$ +	$H_2O(g)$ ⇄	$CO_2(g)$ +	$H_2(g)$
initial (M)	0.150	0.150	0	0
change (M)	–x	–x	+x	+x
equil (M)	0.150 – x	0.150 – x	x	x

$$K_c = 4.24 = \frac{[CO_2][H_2]}{[CO][H_2O]} = \frac{x^2}{(0.150 - x)^2}$$

Take the square root of both sides and solve for x.

$$\sqrt{4.24} = \sqrt{\frac{x^2}{(0.150 - x)^2}}; \quad 2.06 = \frac{x}{0.150 - x}; \quad x = 0.101$$

At equilibrium, $[CO_2] = [H_2] = x = 0.101$ M

$[CO] = [H_2O] = 0.150 - x = 0.150 - 0.101 = 0.049$ M

14.18

	$CO(g)$ +	$H_2O(g)$ ⇄	$CO_2(g)$ +	$H_2(g)$
initial (M)	0.100	0.100	0.100	0.100
change (M)	–x	–x	+x	+x
equil (M)	0.100 – x	0.100 – x	0.100 + x	0.100 + x

$$K_c = 4.24 = \frac{[CO_2][H_2]}{[CO][H_2O]} = \frac{(0.100 + x)^2}{(0.100 - x)^2}$$

Take the square root of both sides and solve for x.

$$\sqrt{4.24} = \sqrt{\frac{(0.100 + x)^2}{(0.100 - x)^2}}; \quad 2.06 = \frac{0.100 + x}{0.100 - x}; \quad x = 0.035$$

At equilibrium, $[CO_2] = [H_2] = 0.100 + x = 0.100 + 0.035 = 0.135$ M

$[CO] = [H_2O] = 0.100 - x = 0.100 - 0.035 = 0.065$ M

14.19

	$H_2(g)$ +	$I_2(g)$ ⇄	$2\ HI(g)$
initial (M)	0.100	0.300	0
change (M)	–x	–x	+2x
equil (M)	0.100 – x	0.300 – x	2x

$$K_c = \frac{[HI]^2}{[H_2][I_2]} = \frac{(2x)^2}{(0.100 - x)(0.300 - x)} = 57.0$$

$53x^2 - 22.8x + 1.71 = 0$

Use the quadratic formula to solve for x.

$$x = \frac{-(-22.8) \pm \sqrt{(-22.8)^2 - 4(53)(1.71)}}{2(53)} = \frac{22.8 \pm 12.54}{106}$$

x = 0.333 and 0.0968

Discard 0.333 because it leads to negative concentrations and that is impossible.

$[H_2] = 0.100 - x = 0.100 - 0.0968 = 0.0032$ M

$[I_2] = 0.300 - x = 0.300 - 0.0968 = 0.2032$

$[HI] = 2x = (2)(0.0968) = 0.1936$ M

14.20

	$N_2O_4(g)$ ⇄	$2\ NO_2(g)$
initial (M)	0.500	0
change (M)	–x	+2x
equil (M)	0.500 – x	2x

$$K_c = \frac{[NO_2]^2}{[N_2O_4]} = \frac{(2x)^2}{(0.500 - x)} = 4.64 \times 10^{-3}$$

$4x^2 + (4.64 \times 10^{-3})x - (2.32 \times 10^{-3}) = 0$

Use the quadratic formula to solve for x.

$$x = \frac{-(4.64 \times 10^{-3}) \pm \sqrt{(4.64 \times 10^{-3})^2 - 4(4)(-2.32 \times 10^{-3})}}{2(4)} = \frac{-0.004\,64 \pm 0.1927}{8}$$

x = –0.0247 and 0.0235

Discard the negative solution (–0.0247) because it leads to a negative concentration of NO_2 and that is impossible.

$[N_2O_4] = 0.500 - x = 0.500 - 0.0235 = 0.476$ M
$[NO_2] = 2x = 2(0.0235) = 0.0470$ M

14.21

	C(s) +	$H_2O(g)$	⇄	CO(g)	+	$H_2(g)$
initial (atm)		0		1.00		1.40
change (atm)		+x		−x		−x
equil (atm)		x		1.00 − x		1.40 − x

$$K_p = \frac{(P_{CO})(P_{H_2})}{(P_{H_2O})} = 2.44 = \frac{(1.00 - x)(1.40 - x)}{x}$$

$x^2 - 4.84x + 1.40 = 0$

Use the quadratic formula to solve for x.

$$x = \frac{-(-4.84) \pm \sqrt{(-4.84)^2 - 4(1)(1.40)}}{2(1)} = \frac{4.84 \pm 4.22}{2}$$

x = 4.53 and 0.310

Discard 4.53 because it leads to negative partial pressures and that is impossible.

$P_{H_2O} = x = 0.310$ atm

$P_{CO} = 1.00 - x = 1.00 - 0.310 = 0.69$ atm

$P_{H_2} = 1.40 - x = 1.40 - 0.310 = 1.09$ atm

14.22 $K_p = \dfrac{(P_{CO})(P_{H_2})}{(P_{H_2O})} = 2.44,\ Q_p = \dfrac{(1.00)(1.40)}{(1.20)} = 1.17,\ Q_p < K_p$ and the reaction goes to the right to reach equilibrium.

	C(s) +	$H_2O(g)$	⇄	CO(g)	+	$H_2(g)$
initial (atm)		1.20		1.00		1.40
change (atm)		−x		+x		+x
equil (atm)		1.20 − x		1.00 + x		1.40 + x

$$K_p = \frac{(P_{CO})(P_{H_2})}{(P_{H_2O})} = 2.44 = \frac{(1.00 + x)(1.40 + x)}{(1.20 - x)}$$

$x^2 + 4.84x - 1.53 = 0$

Use the quadratic formula to solve for x.

$$x = \frac{-(4.84) \pm \sqrt{(4.84)^2 - 4(1)(-1.53)}}{2(1)} = \frac{-4.84 \pm 5.44}{2}$$

x = −5.14 and 0.300

Discard the negative solution (−5.14) because it leads to negative partial pressures and that is impossible.

$P_{H_2O} = 1.20 - x = 1.20 - 0.300 = 0.90$ atm

$P_{CO} = 1.00 + x = 1.00 + 0.300 = 1.30$ atm

$P_{H_2} = 1.40 + x = 1.40 + 0.300 = 1.70$ atm

14.23 (a) CO (reactant) added, H_2 concentration increases.
(b) CO_2 (product) added, H_2 concentration decreases.
(c) H_2O (reactant) removed, H_2 concentration decreases.
(d) CO_2 (product) removed, H_2 concentration increases.
At equilibrium, $Q_c = K_c = \frac{[CO_2][H_2]}{[CO][H_2O]}$. If some CO_2 is removed from the equilibrium mixture, the numerator in Q_c is decreased, which means that $Q_c < K_c$ and the reaction will shift to the right, increasing the H_2 concentration.

14.24 (a) The concentration of both Ca^{2+} and $C_2O_4^{2-}$ will increase as fluid is lost and the reaction will proceed toward products.
(b) Increasing the concentration of Ca^{2+} will cause the reaction to proceed toward products.
(c) Decreasing the concentration of $C_2O_4^{2-}$ will cause the reaction to proceed toward reactants and the kidney stone will dissolve.
(d) The concentration of both Ca^{2+} and $C_2O_4^{2-}$ will decrease causing the reaction proceed toward reactants and the kidney stone will dissolve.

14.25 (a) Because there are 2 mol of gas on both sides of the balanced equation, the composition of the equilibrium mixture is unaffected by a change in pressure. The number of moles of reaction products remains the same.
(b) Because there are 2 mol of gas on the left side and 1 mol of gas on the right side of the balanced equation, the stress of an increase in pressure is relieved by a shift in the reaction to the side with fewer moles of gas (in this case, to products). The number of moles of reaction products increases.
(c) Because there is 1 mol of gas on the left side and 2 mol of gas on the right side of the balanced equation, the stress of an increase in pressure is relieved by a shift in the reaction to the side with fewer moles of gas (in this case, to reactants). The number of moles of reaction product decreases.

14.26 (a)

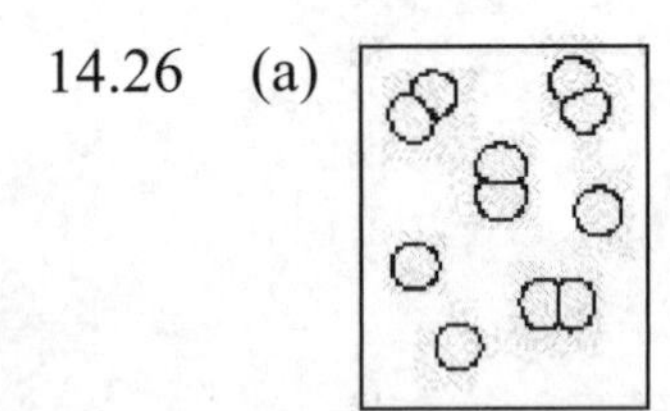

(b) If there is no change in temperature, K_c remains the same.

14.27 Le Châtelier's principle predicts that a stress of added heat will be relieved by net reaction in the direction that absorbs the heat. Since the reaction is endothermic, the equilibrium will shift from left to right (K_c will increase) with an increase in temperature. Therefore, the equilibrium mixture will contain more of the offending NO, the higher the temperature.

14.28 The reaction is exothermic. As the temperature is increased the reaction shifts from right to left. The amount of ethyl acetate decreases.

$$K_c = \frac{[CH_3CO_2C_2H_5][H_2O]}{[CH_3CO_2H][C_2H_5OH]}$$

As the temperature is decreased, the reaction shifts from left to right. The product concentrations increase, and the reactant concentrations decrease. This corresponds to an increase in K_c.

14.29 There are more AB(g) molecules at the higher temperature. The equilibrium shifted to the right at the higher temperature, which means the reaction is endothermic.

14.30 Heat + $BaCO_3(s) \rightleftarrows BaO(s) + CO_2(g)$

(a) 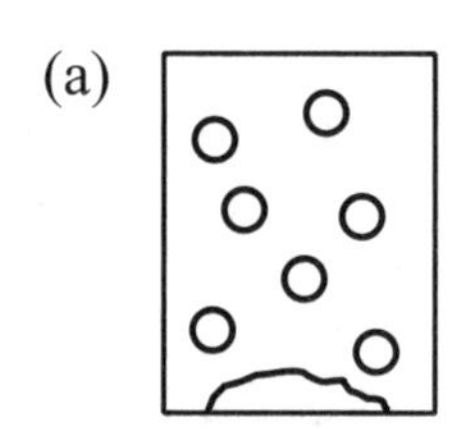 (b) 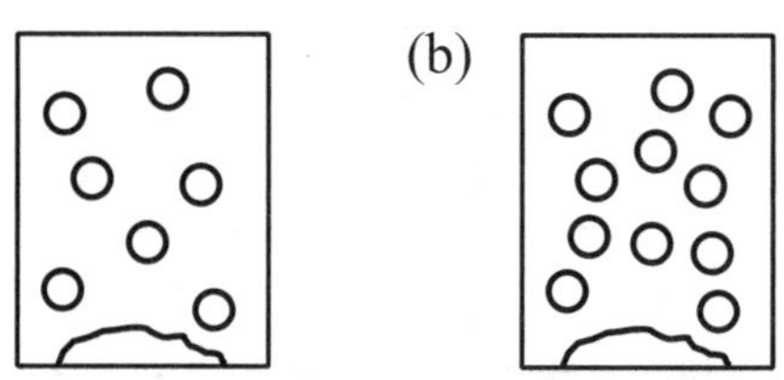

14.31 (a) Because K_c is so large, k_f is larger than k_r.

(b) $K_c = \frac{k_f}{k_r}$; $k_r = \frac{k_f}{K_c} = \frac{8.5 \times 10^6\ M^{-1} s^{-1}}{3.4 \times 10^{34}} = 2.5 \times 10^{-28}\ M^{-1}\ s^{-1}$

(c) Because the reaction is exothermic, E_a (forward) is less than E_a (reverse). Consequently, as the temperature decreases, k_r decreases more than k_f decreases, and therefore $K_c = \frac{k_f}{k_r}$ increases.

14.32 (a) If A is similar for forward and reverse reactions, then k_r is larger because the activation energy for the reverse reaction is lower.

(b) Because $K_c = \frac{k_f}{k_r}$ and $k_r > k_f$, then $K_c < 1$.

(c) The rates of forward and reverse reactions both decrease. However, the rate of the forward reaction has a larger decrease than the rate of reverse reaction. K_c decreases when temperature is lowered in a endothermic process.

14.33 (a) 75% (b) 20%

(c) An efficient unloading of oxygen occurs in working muscle tissue when the partial pressure of oxygen drops.

14.34 An increase in temperature will shift the position of the equilibrium toward reactants, releasing oxygen from hemoglobin and making it available for use by the muscles.

14.35 (a) Equilibrium constants for each sequential step in binding oxygen become larger.

(b) Each step becomes more product favored, meaning that when oxygen binds it increases the affinity of the remaining heme groups for other oxygen molecules.

(c) The oxygen carrying capacity is greatly increased.

14.36 (a) $K_c = K_{c1} \times K_{c2} \times K_{c3} \times K_{c4}$

$K_c = (1.5 \times 10^4)(3.5 \times 10^4)(5.9 \times 10^4)(1.7 \times 10^6) = 5.3 \times 10^{19}$

(b) $Hb + O_2 \rightleftarrows Hb(O_2)_4$

$[O_2] = (1.61\ \mu M/mm\ Hg)(95\ mm\ Hg) = 153\ \mu M = 153 \times 10^{-6}\ M$

$$K_c = \frac{[Hb(O_2)_4]}{[Hb][O_2]^4} = 5.3 \times 10^{19}$$

$$\frac{[Hb(O_2)_4]}{[Hb]} = K_c[O_2]^4 = (5.3 \times 10^{19})(153 \times 10^{-6})^4 = 2.9 \times 10^4$$

14.37 $Hb + O_2 \rightleftarrows Hb(O_2)$

If CO binds to Hb, Hb is removed from the reaction and the reaction will shift to the left, resulting in O_2 being released from $Hb(O_2)$. This will decrease the effectiveness of Hb for carrying O_2.

14.38 $Hb(O_2)(aq) + CO(g) \rightleftarrows Hb(CO)(aq) + O_2(g)$ $\quad K = 207$

(a) $P_{O_2} = 0.20$ atm and $P_{CO} = 0.0015$ atm

$$K = \frac{[Hb(CO)](P_{O_2})}{[Hb(O_2)](P_{CO})} = 207$$

$$\frac{[Hb(CO)]}{[Hb(O_2)]} = \frac{(P_{CO})(207)}{(P_{O_2})} = \frac{(0.0015)(207)}{(0.20)} = 1.6$$

(b) An increase in the concentration of oxygen in the blood will shift the equilibrium toward reactants increasing the concentration of oxygen bound to hemoglobin.

Conceptual Problems

14.39 (a) (1) and (3) because the number of A's and B's are the same in the third and fourth box.

(b) $K_c = \frac{[B]}{[A]} = \frac{6}{4} = 1.5$

(c) Because the same number of molecules appear on both sides of the equation, the volume terms in K_c all cancel. Therefore, we can calculate K_c without including the volume.

14.40 (a) $A_2 + C_2 \rightleftarrows 2\ AC$ (most product molecules)

(b) $A_2 + B_2 \rightleftarrows 2\ AB$ (fewest product molecules)

14.41 (a) Only reaction (3), $K_c = \frac{[A][AB]}{[A_2][B]} = \frac{(2)(4)}{(2)(2)} = 2$, is at equilibrium.

(b) $Q_c = \frac{[A][AB]}{[A_2][B]} = \frac{(3)(5)}{(1)(1)} = 15$ for reaction (1). Because $Q_c > K_c$, the reaction will go in the reverse direction to reach equilibrium.

$Q_c = \frac{[A][AB]}{[A_2][B]} = \frac{(1)(3)}{(3)(3)} = 1/3$ for reaction (2). Because $Q_c < K_c$, the reaction will go in the forward direction to reach equilibrium.

14.42 (a) $A_2 + 2\,B \rightleftarrows 2\,AB$

(b) The number of AB molecules will increase, because as the volume is decreased at constant temperature, the pressure will increase and the reaction will shift to the side of fewer molecules to reduce the pressure.

14.43 When the stopcock is opened, the reaction will go in the reverse direction because there will be initially an excess of AB molecules.

14.44 As the temperature is raised, the reaction proceeds in the reverse direction. This is consistent with an exothermic reaction where "heat" can be considered as a product.

14.45 (a) $AB \rightarrow A + B$

(b) The reaction is endothermic because a stress of added heat (higher temperature) shifts the $AB \rightleftarrows A + B$ equilibrium to the right.

(c) If the volume is increased, the pressure is decreased. The stress of decreased pressure will be relieved by a shift in the equilibrium from left to right, thus increasing the number of A atoms.

14.46 (a) (b) 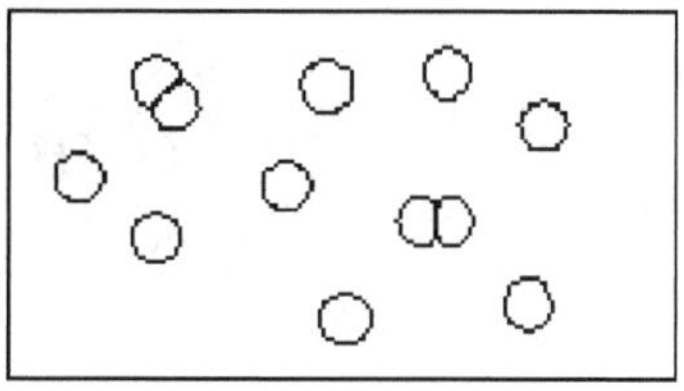 (c)

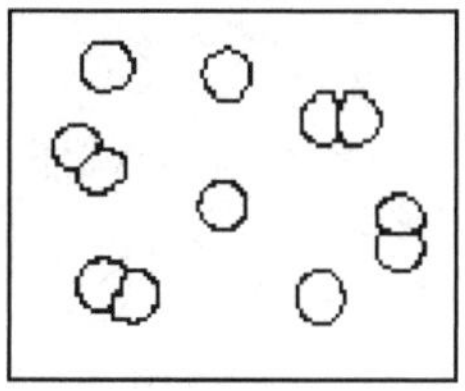

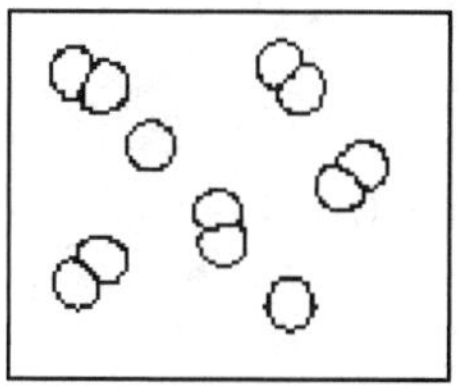

14.47 This equilibrium mixture has a $K_c \propto \frac{(2)(2)}{(3)^2}$ and is less than 1. This means that $k_f < k_r$.

14.48 (a) $A \rightarrow 2\,B$

(b) (1) The reaction is exothermic. As the temperature is increased, the reaction shifts to the left. A increases, B decreases, and K_c decreases.

(2) When the volume decreases, the reaction shifts to the side with fewer gas molecules, which is towards the reactant. The amount of A increases.

(3) If there is no volume change, there is no change in the equilibrium composition, and the amount of A remains the same.

(4) A catalyst does not change the equilibrium composition, and the amount of A remains the same.

14.49 (a) $2\,A + B_2 \rightarrow 2\,AB$

(b) (1) Increasing the partial pressure of B_2 (a reactant) shifts the reaction toward products and the amount of AB increases.

(2) Adding solid A does not change the equilibrium composition, and the amount of AB remains the same.
(3) When the volume increases, the reaction shifts to the side with more gas molecules, which is toward the product. The amount of AB increases.
(4) The reaction is exothermic. As the temperature is increased, the reaction shifts to the left. B_2 increases, AB decreases, and K_c decreases.

Section Problems
Equilibrium Constant Expressions and Equilibrium Constants (Sections 14.1–14.4)

14.50 (d) The rate of the forward reaction is equal to the rate of the reverse reaction.

14.51 (b) The concentrations of A and B are constant.

14.52 $K_c = \frac{[C_2H_5OC_2H_5][H_2O]}{[C_2H_5OH]^2}$

14.53 $K_c = \frac{[HOCH_2CH_2OH]}{[C_2H_4O][H_2O]}$

14.54 (a) $K_c = \frac{[CO][H_2]^3}{[CH_4][H_2O]}$ (b) $K_c = \frac{[ClF_3]^2}{[F_2]^3[Cl_2]}$ (c) $K_c = \frac{[HF]^2}{[H_2][F_2]}$

14.55 (a) $K_c = \frac{[CH_3CHO]^2}{[C_2H_4]^2[O_2]}$ (b) $K_c = \frac{[CO_2][H_2]}{[CO][H_2O]}$ (c) $K_c = \frac{[NO]^4[H_2O]^6}{[NH_3]^4[O_2]^5}$

14.56 (a) $K_p = \frac{(P_{CO})(P_{H_2})^3}{(P_{CH_4})(P_{H_2O})}$, $\Delta n = 2$ and $K_p = K_c(RT)^2$

(b) $K_p = \frac{(P_{ClF_3})^2}{(P_{F_2})^3(P_{Cl_2})}$, $\Delta n = -2$ and $K_p = K_c(RT)^{-2}$

(c) $K_p = \frac{(P_{HF})^2}{(P_{H_2})(P_{F_2})}$, $\Delta n = 0$ and $K_p = K_c$

14.57 (a) $K_p = \frac{(P_{CH_3CHO})^2}{(P_{C_2H_4})^2(P_{O_2})}$, $\Delta n = -1$ and $K_p = K_c(RT)^{-1}$

(b) $K_p = \frac{(P_{CO_2})(P_{H_2})}{(P_{CO})(P_{H_2O})}$, $\Delta n = 0$ and $K_p = K_c$

(c) $K_p = \dfrac{(P_{NO})^4(P_{H_2O})^6}{(P_{NH_3})^4(P_{O_2})^5}$, $\Delta n = 1$ and $K_p = K_c(RT)$

14.58 (a) $K_c = \dfrac{[CO_2]^3}{[CO]^3}$, $K_p = \dfrac{(P_{CO_2})^3}{(P_{CO})^3}$ (b) $K_c = \dfrac{1}{[O_2]^3}$, $K_p = \dfrac{1}{(P_{O_2})^3}$

(c) $K_c = [SO_3]$, $K_p = P_{SO_3}$ (d) $K_c = [Ba^{2+}][SO_4^{2-}]$

14.59 (a) $K_c = \dfrac{[H_2O]^3}{[H_2]^3}$, $K_p = \dfrac{(P_{H_2O})^3}{(P_{H_2})^3}$ (b) $K_c = \dfrac{1}{[Ag^+][Cl^-]}$

(c) $K_c = \dfrac{[HCl]^6}{[H_2O]^3}$, $K_p = \dfrac{(P_{HCl})^6}{(P_{H_2O})^3}$ (d) $K_c = [CO_2]$, $K_p = P_{CO_2}$

14.60 (a) This reaction is the reverse of the original reaction.

$K_c' = \dfrac{1}{K_c} = \dfrac{1}{7.5 \times 10^{-9}} = 1.3 \times 10^8$

(b) This reaction is the reverse of the original reaction multiplied by ½.

$K_c' = \sqrt{\dfrac{1}{K_c}} = \sqrt{\dfrac{1}{7.5 \times 10^{-9}}} = 1.2 \times 10^4$

(c) This reaction is the original reaction multiplied by 2.

$K_c' = (K_c)^2 = (7.5 \times 10^{-9})^2 = 5.6 \times 10^{-17}$

14.61 (a) The two reactions are the reverse of each other.

$K_p(\text{reverse}) = \dfrac{1}{K_p(\text{forward})} = \dfrac{1}{50.2} = 1.99 \times 10^{-2}$

(b) This reaction is the original reaction multiplied by 2.

$K_p' = (K_p)^2 = (50.2)^2 = 2520$

(c) This reaction is the reverse of the original reaction multiplied by ½.

$K_p' = \sqrt{\dfrac{1}{K_p}} = \sqrt{\dfrac{1}{50.2}} = 0.141$

14.62 $K_p(1) = 7.2 \times 10^7$

(a) $K_p(2) = \dfrac{1}{K_p(1)} = \dfrac{1}{7.2 \times 10^7} = 1.4 \times 10^{-8}$

(b) $K_p(3) = (K_p(1))^2 = (7.2 \times 10^7)^2 = 5.2 \times 10^{15}$

(c) $K_p(4) = \dfrac{1}{(K_p(1))^3} = \dfrac{1}{(7.2 \times 10^7)^3} = 2.7 \times 10^{-24}$

14.63 $K_p(1) = 2.9 \times 10^{-5}$

(a) $K_p(2) = \dfrac{1}{K_p(1)} = \dfrac{1}{2.9 \times 10^{-5}} = 3.4 \times 10^4$

(b) $K_p(3) = (K_p(1))^3 = (2.9 \times 10^{-5})^3 = 2.4 \times 10^{-14}$

(c) $K_p(4) = \dfrac{1}{(K_p(1))^2} = \dfrac{1}{(2.9 \times 10^{-5})^2} = 1.2 \times 10^9$

14.64

$2\ Na(l) + O_2(g) \rightleftarrows Na_2O_2(s)$ $\quad K_c' = 1/K_c = 1/(5 \times 10^{-29})$

$\underline{Na_2O(s) \rightleftarrows 2\ Na(l) + \frac{1}{2}\ O_2(g)}$ $\quad K_c = 2 \times 10^{-25}$

Overall $Na_2O(s) + \frac{1}{2}\ O_2(g) \rightleftarrows Na_2O_2(s)$

$K_c(\text{overall}) = K_c' \times K_c = \dfrac{2 \times 10^{-25}}{5 \times 10^{-29}} = 4 \times 10^3$

14.65

$3 \times [2\ H_2(g) + O_2(g) \rightleftarrows 2\ H_2O(g)]$ $\quad K_{c1}' = (K_c)^3 = (3.2 \times 10^{81})^3$

$\underline{2 \times [2\ NH_3(g) \rightleftarrows N_2(g) + 3\ H_2(g)]}$ $\quad K_{c2}' = (1/K_c)^2 = [1/(3.5 \times 10^8)]^2$

Overall $4\ NH_3(g) + 3\ O_2(g) \rightleftarrows 2\ N_2(g) + 6\ H_2O(g)$

$K_c(\text{overall}) = K_{c1}' \times K_{c2}' = \dfrac{(3.2 \times 10^{81})^3}{(3.5 \times 10^8)^2} = 2.7 \times 10^{227}$

14.66 $K_c = \dfrac{[PCl_3][Cl_2]}{[PCl_5]} = \dfrac{(1.5 \times 10^{-2})(3.2 \times 10^{-2})}{(8.3 \times 10^{-3})} = 0.058$

14.67 $K_p = \dfrac{(P_{ClNO})^2}{(P_{NO})^2(P_{Cl_2})} = \dfrac{(1.35)^2}{(0.240)^2(0.608)} = 52.0$

14.68 $\Delta n = (1) - (1 + 1) = -1$

$K_p = K_c(RT)^{\Delta n} = (2.2 \times 10^5)[(0.08206)(298)]^{-1} = 9.0 \times 10^3$

14.69 $\Delta n = (2 + 1) - (2) = 1$

$K_c = K_p\left(\dfrac{1}{RT}\right)^{\Delta n} = (1.6 \times 10^{-6})\left(\dfrac{1}{(0.082\ 06)(298)}\right) = 6.5 \times 10^{-8}$

14.70 $K_p = P_{H_2O} = 0.0313$ atm; $\Delta n = 1$

$K_c = K_p\left(\dfrac{1}{RT}\right)^{\Delta n} = (0.0313)\left(\dfrac{1}{(0.082\ 06)(298)}\right) = 1.28 \times 10^{-3}$

14.71 $P_{C_{10}H_8} = 0.10 \text{ mm Hg} \times \dfrac{1 \text{ atm}}{760 \text{ mm Hg}} = 1.3 \times 10^{-4}$ atm

$K_p = P_{C_{10}H_8} = 1.3 \times 10^{-4}$; $\quad \Delta n = 1 - 0 = 1$, $T = 27\ °C = 300$ K

$$K_c = K_p\left(\frac{1}{RT}\right)^{\Delta n} = (1.3 \times 10^{-4})\left(\frac{1}{(0.082\ 06)(300)}\right) = 5.3 \times 10^{-6}$$

14.72 $2\ ClO(g) \rightleftarrows Cl_2O_2(g) \quad K_c = 4.96 \times 10^{11}$

$$K_c = \frac{[Cl_2O_2]}{[ClO]^2} = \frac{(6.00 \times 10^{-6})}{[ClO]^2} = 4.96 \times 10^{11}$$

$$[ClO]^2 = \frac{(6.00 \times 10^{-6})}{(4.96 \times 10^{11})}; \qquad [ClO] = \sqrt{\frac{(6.00 \times 10^{-6})}{(4.96 \times 10^{11})}} = 3.48 \times 10^{-9}\ M$$

14.73 $2\ SO_2(g) + O_2(g) \rightleftarrows 2\ SO_3(g) \quad K_c = 7.9 \times 10^4$

$$K_c = \frac{[SO_3]^2}{[SO_2]^2[O_2]} = \frac{[SO_3]^2}{(1.5 \times 10^{-3})^2(3.0 \times 10^{-3})} = 7.9 \times 10^4$$

$$[SO_3] = \sqrt{(1.5 \times 10^{-3})^2(3.0 \times 10^{-3})(7.9 \times 10^4)} = 2.3 \times 10^{-2}\ M$$

Using the Equilibrium Constant (Section 14.5)

14.74 (a) Because K_c is very large, the equilibrium mixture contains mostly product.
(b) Because K_c is very small, the equilibrium mixture contains mostly reactants.

14.75 (a) Because K_c is very large, the equilibrium mixture contains mostly product.
(b) Because $K_c = 7.5 \times 10^{-3}$, the equilibrium mixture contains an appreciable concentration of both reactants and products.
(c) Because K_c is very small, the equilibrium mixture contains mostly reactant.

14.76 K_p is large. The equilibrium lies far to the right and much SO_2 will be present at equilibrium.

14.77 Because K_c is very small, pure air will contain very little O_3 (ozone) at equilibrium.

$$3\ O_2(g) \rightleftarrows 2\ O_3(g); \qquad K_c = \frac{[O_3]^2}{[O_2]^3} = 1.7 \times 10^{-56}; \qquad [O_2] = 8 \times 10^{-3}\ M$$

$$[O_3] = \sqrt{[O_2]^3 \times K_c} = \sqrt{(8 \times 10^{-3})^3(1.7 \times 10^{-56})} = 9 \times 10^{-32}\ M$$

14.78 The container volume of 10.0 L must be included to calculate molar concentrations.

$$Q_c = \frac{[CS_2]_t[H_2]_t^4}{[CH_4]_t[H_2S]_t^2} = \frac{(3.0\ mol/10.0\ L)(3.0\ mol/10.0\ L)^4}{(2.0\ mol/10.0\ L)(4.0\ mol/10.0\ L)^2} = 7.6 \times 10^{-2}; \quad K_c = 2.5 \times 10^{-3}$$

The reaction is not at equilibrium because $Q_c > K_c$. The reaction will proceed in the reverse direction to attain equilibrium.

14.79 $Q_c = \frac{[CO]_t[H_2]_t^3}{[H_2O]_t[CH_4]_t} = \frac{(0.15)(0.20)^3}{(0.035)(0.050)} = 0.69; \quad K_c = 4.7$

The reaction is not at equilibrium because $Q_c < K_c$. The reaction will proceed in the forward direction to attain equilibrium.

14.80 (a) $Q_p = \frac{(P_{P_2})(P_{H_2})^3}{(P_{PH_3})^2} = \frac{(0.871)(0.517)^3}{(0.0260)^2} = 178; \quad K_p = 398$

Because $Q_p < K_p$, the reaction will proceed in the forward direction to attain equilibrium.

(b) $K_p = \frac{(P_{P_2})(P_{H_2})^3}{(P_{PH_3})^2} = 398$

$$(P_{PH_3})^2 = \frac{(P_{P_2})(P_{H_2})^3}{398}$$

$$P_{PH_3} = \sqrt{\frac{(P_{P_2})(P_{H_2})^3}{398}} = \sqrt{\frac{(0.412)(0.822)^3}{398}} = 0.0240 \text{ atm}$$

14.81 $H_2(g) \rightleftarrows 2\,H(g) \quad K_c = 1.2 \times 10^{-42}$

(a) $\frac{[H]^2}{[H_2]} = \frac{[H]^2}{(0.10)} = 1.2 \times 10^{-42}$

$[H] = \sqrt{(0.10)(1.2 \times 10^{-42})} = 3.5 \times 10^{-22}$ M

(b) H atoms = $(3.5 \times 10^{-22}$ mol/L)(1.0 L)(6.022×10^{23} atoms/mol) = 210 H atoms

H_2 molecules = (0.10 mol/L)(1.0 L)(6.022×10^{23} molecules/mol) = 6.0×10^{22} H_2 molecules

14.82

	$N_2O_4(g)$	$\rightleftarrows$	$2\,NO_2(g)$
initial (M)	0.0500		0
change (M)	–x		+2x
equil (M)	0.0500 – x		2x

$K_c = 4.64 \times 10^{-3} = \frac{[NO_2]^2}{[N_2O_4]} = \frac{(2x)^2}{(0.0500 - x)}$

$4x^2 + (4.64 \times 10^{-3})x - (2.32 \times 10^{-4}) = 0$

Use the quadratic formula to solve for x.

$$x = \frac{-(4.64 \times 10^{-3}) \pm \sqrt{(4.64 \times 10^{-3})^2 - 4(4)(-2.32 \times 10^{-4})}}{2(4)} = \frac{-0.00464 \pm 0.06110}{8}$$

x = –0.008 22 and 0.007 06

Discard the negative solution (–0.008 22) because it leads to a negative concentration of NO_2 and that is impossible.

$[N_2O_4]$ = 0.0500 – x = 0.0500 – 0.007 06 = 0.0429 M

$[NO_2]$ = 2x = 2(0.007 06) = 0.0141 M

14.83 $N_2O_4(g) \rightleftarrows 2\ NO_2(g)$

$$Q_c = \frac{[NO_2]_t^2}{[N_2O_4]_t} = \frac{(0.0300\ \text{mol/L})^2}{(0.0200\ \text{mol/L})} = 0.0450; \quad Q_c > K_c$$

The reaction will approach equilibrium by going from right to left.

	$N_2O_4(g)$	$\rightleftarrows$	$2\ NO_2(g)$
initial (M)	0.0200		0.0300
change (M)	+x		−2x
equil (M)	0.0200 + x		0.0300 − 2x

$$K_c = 4.64 \times 10^{-3} = \frac{[NO_2]^2}{[N_2O_4]} = \frac{(0.0300 - 2x)^2}{(0.0200 + x)}$$

$4x^2 - 0.1246x + (8.072 \times 10^{-4}) = 0$

Use the quadratic formula to solve for x.

$$x = \frac{-(-0.1246) \pm \sqrt{(-0.1246)^2 - 4(4)(8.072 \times 10^{-4})}}{2(4)} = \frac{0.1246 \pm 0.05109}{8}$$

x = 0.0220 and 0.009 19

Discard the larger solution (0.0220) because it leads to a negative concentration of NO_2, and that is impossible.

$[N_2O_4] = 0.0200 + x = 0.0200 + 0.009\ 19 = 0.0292$ M

$[NO_2] = 0.0300 - 2x = 0.0300 - 2(0.009\ 19) = 0.0116$ M

14.84 The container volume of 2.00 L must be included to calculate molar concentrations.

Initial [HI] = 9.30×10^{-3} mol/2.00 L = 4.65×10^{-3} M = 0.004 65 M

	$H_2(g)$	+	$I_2(g)$	$\rightleftarrows$	$2\ HI(g)$
initial (M)	0		0		0.004 65
change (M)	+x		+x		−2x
equil (M)	x		x		0.004 65 − 2x

$x = [H_2] = [I_2] = 6.29 \times 10^{-4}$ M = 0.000 629 M

[HI] = 0.004 65 − 2x = 0.004 65 − 2(0.000 629) = 0.003 39 M

$$K_c = \frac{[HI]^2}{[H_2][I_2]} = \frac{(0.003\ 39)^2}{(0.000\ 629)^2} = 29.0$$

14.85 (a) $K_c = \dfrac{[CH_3CO_2C_2H_5][H_2O]}{[CH_3CO_2H][C_2H_5OH]}$

(b)

	$CH_3CO_2H(soln)$	+	$C_2H_5OH(soln)$	$\rightleftarrows$	$CH_3CO_2C_2H_5(soln)$	+	$H_2O(soln)$
initial (mol)	1.00		1.00		0		0
change (mol)	−x		−x		+x		+x
equil (mol)	1.00 − x		1.00 − x		x		x

x = 0.65 mol; 1.00 − x = 0.35 mol; $\quad K_c = \dfrac{(0.65)^2}{(0.35)^2} = 3.4$

Because there are the same number of molecules on both sides of the equation, the volume terms in K_c cancel. Therefore, we can calculate K_c without including the volume.

14.86 $CH_3CO_2C_2H_5(soln) + H_2O(soln) \rightleftharpoons CH_3CO_2H(soln) + C_2H_5OH(soln)$

$$K_c(\text{hydrolysis}) = \frac{1}{K_c(\text{forward})} = \frac{1}{3.4} = 0.29$$

14.87 (a) $$Q_p = \frac{(P_{InH_2})}{(P_{In})(P_{H_2})} = \frac{(0.0760)}{(0.0600)(0.0350)} = 36.2; \quad K_p = 1.48$$

Because $Q_p > K_p$, the reaction will proceed in the reverse direction to attain equilibrium.

(b)	$In(g)$ +	$H_2(g)$ ⇄	$InH_2(g)$
initial (atm)	0.0600	0.0350	0.0760
change (atm)	+x	+x	–x
equil (atm)	0.0600 + x	0.0350 + x	0.0760 – x

$$K_p = \frac{(P_{InH_2})}{(P_{In})(P_{H_2})} = 1.48 = \frac{(0.0760 - x)}{(0.0600 + x)(0.0350 + x)}$$

$$1.48 = \frac{(0.0760 - x)}{[(0.0600)(0.0350) + (0.0600x) + (0.0350x) + x^2]}$$

$1.48x^2 + 1.1406x - 0.07289 = 0$

Use the quadratic formula to solve for x.

$$x = \frac{-(1.1406) \pm \sqrt{(1.1406)^2 - 4(1.48)(-0.07289)}}{2(1.48)} = \frac{-1.1406 \pm 1.3162}{2.96}$$

x = –0.830 and 0.0593

Discard the negative solution (–0.835) because it gives negative In and H_2 partial pressures and that is impossible.

$P_{In} = 0.0600 + 0.0593 = 0.1193$ atm

$P_{H_2} = 0.0350 + 0.0593 = 0.0943$ atm

$P_{InH_2} = 0.0760 - 0.0593 = 0.0167$ atm

14.88 (a) $$Q_c = \frac{[Br_2][Cl_2]}{[BrCl]^2} = \frac{(0.035)(0.035)}{(0.050)^2} = 0.49; \quad K_c = 0.145$$

Because $Q_c > K_c$, the reaction will proceed in the reverse direction to attain equilibrium.

(b)	2 $BrCl(\mathit{soln})$ ⇄	$Br_2(\mathit{soln})$ +	$Cl_2(\mathit{soln})$
initial (M)	0.050	0.035	0.035
change (M)	+2x	–x	–x
equil (M)	0.050 + 2x	0.035 – x	0.035 – x

$$K_c = \frac{[Br_2][Cl_2]}{[BrCl]^2} = 0.145 = \frac{(0.035 - x)(0.035 - x)}{(0.050 + 2x)^2}$$

Take the square root of both sides and solve for x.

$$\sqrt{0.145} = \sqrt{\frac{(0.035 - x)^2}{(0.050 + 2x)^2}}$$

$$0.381 = \frac{(0.035 - x)}{(0.050 + 2x)}; \quad x = 0.009$$

$[BrCl] = 0.050 + 2x = 0.050 + 2(0.009) = 0.068$ M

$[Br_2] = [Cl_2] = 0.035 - x = 0.035 - 0.009 = 0.026$ M

14.89 $K_c = \frac{[NH_3]^2}{[N_2][H_2]^3} = 0.29$; At equilibrium, $[N_2] = 0.036$ M and $[H_2] = 0.15$ M

$[NH_3] = \sqrt{[N_2] \times [H_2]^3 \times K_c} = \sqrt{(0.036)(0.15)^3(0.29)} = 5.9 \times 10^{-3}$ M

14.90 $K_c = 2.7 \times 10^2 = \frac{[SO_3]^2}{[SO_2]^2[O_2]}$; Because $[SO_3] = [SO_2]$, then $2.7 \times 10^2 = \frac{1}{[O_2]}$

$[O_2] = 3.7 \times 10^{-3}$ M

14.91

	$N_2(g)$ +	$O_2(g)$ ⇄	$2\ NO(g)$
initial (M)	1.40	1.40	0
change (M)	–x	–x	+2x
equil (M)	1.40 – x	1.40 – x	2x

$$K_c = \frac{[NO]^2}{[N_2][O_2]} = 1.7 \times 10^{-3} = \frac{(2x)^2}{(1.40 - x)^2}$$

Take the square root of both sides and solve for x.

$$\sqrt{1.7 \times 10^{-3}} = \sqrt{\frac{(2x)^2}{(1.40 - x)^2}}; \quad 4.1 \times 10^{-2} = \frac{2x}{1.40 - x}; \quad x = 2.8 \times 10^{-2}$$

At equilibrium, $[NO] = 2x = 2(2.8 \times 10^{-2}) = 0.056$ M

$[N_2] = [O_2] = 1.40 - x = 1.40 - (2.8 \times 10^{-2}) = 1.37$ M

14.92

	$N_2(g)$ +	$O_2(g)$ ⇄	$2\ NO(g)$
initial (M)	2.24	0.56	0
change (M)	–x	–x	+2x
equil (M)	2.24 – x	0.56 – x	2x

$$K_c = \frac{[NO]^2}{[N_2][O_2]} = 1.7 \times 10^{-3} = \frac{(2x)^2}{(2.24 - x)(0.56 - x)}$$

$4x^2 + (4.8 \times 10^{-3})x - (2.1 \times 10^{-3}) = 0$

Use the quadratic formula to solve for x.

$$x = \frac{-(4.8 \times 10^{-3}) \pm \sqrt{(4.8 \times 10^{-3})^2 - 4(4)(-2.1 \times 10^{-3})}}{2(4)} = \frac{-0.0048 \pm 0.1834}{8}$$

$x = -0.0235$ and 0.0223

Discard the negative solution (–0.0235) because it gives a negative NO concentration and that is impossible.

$[N_2] = 2.24 - x = 2.24 - 0.0223 = 2.22$ M; $[O_2] = 0.56 - x = 0.56 - 0.0223 = 0.54$ M

$[NO] = 2x = 2(0.0223) = 0.045$ M

14.93

	L-α-lysine	⇄	L-β-lysine
initial (M)	0.003 00		0
change (M)	−x		+x
equil (M)	0.003 00 − x		x

$$K_c = \frac{[\text{L-}\beta\text{-lysine}]}{[\text{L-}\alpha\text{-lysine}]} = 7.20 = \frac{x}{(0.003\,00 - x)}$$

$0.0216 - 7.20x = x$; $0.0216 = 8.20x$; $x = \frac{0.0216}{8.20} = 0.002\,63$ M

[L-α-lysine] = 0.003 00 − x = 0.003 00 − 0.002 63 = 0.000 37 M

[L-β-lysine] = x = 0.002 63 M

14.94 (a) $K_c = \frac{[CH_3CO_2C_2H_5][H_2O]}{[CH_3CO_2H][C_2H_5OH]} = 3.4 = \frac{(x)(12.0)}{(4.0)(6.0)}$; $x = 6.8$ moles $CH_3CO_2C_2H_5$

Note that the volume cancels because the same number of molecules appear on both sides of the chemical equation.

(b)

	$CH_3CO_2H(soln)$ +	$C_2H_5OH(soln)$ ⇄	$CH_3CO_2C_2H_5(soln)$ +	$H_2O(soln)$
initial (mol)	1.00	10.00	0	0
change (mol)	−x	−x	+x	+x
equil (mol)	1.00 − x	10.00 − x	x	x

$$K_c = 3.4 = \frac{x^2}{(1.00 - x)(10.00 - x)}$$

$2.4x^2 - 37.4x + 34 = 0$

Use the quadratic formula to solve for x.

$$x = \frac{-(-37.4) \pm \sqrt{(-37.4)^2 - 4(2.4)(34)}}{2(2.4)} = \frac{37.4 \pm 32.75}{4.8}$$

x = 0.969 and 14.6

Discard the larger solution (14.6) because it leads to negative concentrations and that is impossible.

mol CH_3CO_2H = 1.00 − x = 1.00 − 0.969 = 0.03 mol

mol C_2H_5OH = 10.00 − x = 10.00 − 0.969 = 9.03 mol

mol $CH_3CO_2C_2H_5$ = mol H_2O = x = 0.97 mol

14.95 When equal volumes of two solutions are mixed together, their concentrations are cut in half.

	$CH_3Cl(aq)$ +	$OH^-(aq)$ ⇄	$CH_3OH(aq)$ +	$Cl^-(aq)$
initial (M)	0.05	0.1	0	0
assume complete reaction (M)	0	0.05	0.05	0.05
assume small back reaction (M)	+x	+x	−x	−x
equil (M)	x	0.05 + x	0.05 − x	0.05 − x

$$K_c = \frac{[CH_3OH][Cl^-]}{[CH_3Cl][OH^-]} = 10^{16} = \frac{(0.05 - x)^2}{x(0.05 + x)}$$

Because K_c is very large, $x << 0.05$.

$10^{16} \approx \dfrac{(0.05)^2}{x(0.05)}$; $x = 5 \times 10^{-18}$

$[CH_3Cl] = x = 5 \times 10^{-18}$ M; $[OH^-] = [CH_3OH] = [Cl^-] \approx 0.05$ M

14.96

	$ClF_3(g)$	⇄ $ClF(g)$	+ $F_2(g)$
initial (atm)	1.47	0	0
change (atm)	–x	+x	+x
equil (atm)	1.47 – x	x	x

$K_p = \dfrac{(P_{ClF})(P_{F_2})}{(P_{ClF_3})} = 0.140 = \dfrac{(x)(x)}{1.47 - x}$; solve for x.

$x^2 + 0.140x - 0.2058 = 0$

Use the quadratic formula to solve for x.

$$x = \frac{-(0.140) \pm \sqrt{(0.140)^2 - (4)(1)(-0.2058)}}{2(1)}$$

$$x = \frac{-0.140 \pm 0.918}{2}$$

x = 0.389 and –0.529

Discard the negative solution (–0.529) because it gives negative partial pressures and that is impossible.

$P_{ClF} = P_{F_2} = x = 0.389$ atm

$P_{ClF_3} = 1.47 - x = 1.47 - 0.389 = 1.08$ atm

14.97

	$Fe_2O_3(s)$ + $3\,CO(g)$	⇄ $2\,Fe(s)$ + $3\,CO_2(g)$
initial (atm)	0.978	0
change (atm)	–3x	+3x
equil (atm)	0.978 – 3x	3x

$K_p = \dfrac{(P_{CO_2})^3}{(P_{CO})^3} = 19.9 = \dfrac{(3x)^3}{(0.978 - 3x)^3}$; take the cube root of both sides and solve for x.

$\sqrt[3]{19.9} = 2.71 = \dfrac{3x}{(0.978 - 3x)}$

$2.65 - 8.13x = 3x$

$2.65 = 11.13x$

$x = 2.65/11.13 = 0.238$ atm

$P_{CO} = 0.978 - 3x = 0.978 - 3(0.238) = 0.264$ atm

$P_{CO_2} = 3x = 3(0.238) = 0.714$ atm

Le Châtelier's Principle (Sections 14.6–14.9)

14.98 (a) Cl^- (reactant) added, AgCl(s) increases

(b) Ag^+ (reactant) added, AgCl(s) increases

(c) Ag^+ (reactant) removed, AgCl(s) decreases
(d) Cl^- (reactant) removed, AgCl(s) decreases

Disturbing the equilibrium by decreasing $[Cl^-]$ increases Q_c $\left(Q_c = \frac{1}{[Ag^+]_t[Cl^-]_t}\right)$ to a value greater than K_c. To reach a new state of equilibrium, Q_c must decrease, which means that the denominator must increase; that is, the reaction must go from right to left, thus decreasing the amount of solid AgCl.

14.99 (a) NOCl (product) added, NO_2 concentration decreases
(b) NO (reactant) added, NO_2 concentration increases
(c) NO (reactant) removed, NO_2 concentration decreases
(d) NO_2Cl (reactant) added, NO_2 concentration increases

Adding NO_2Cl decreases the value of Q_c $\left(Q_c = \frac{[NOCl][NO_2]}{[NO_2Cl][NO]}\right)$. To reach a new state of equilibrium, the reaction must go from left to right, thus increasing the concentration of NO_2.

14.100 (a) Because there are 2 mol of gas on the left side and 3 mol of gas on the right side of the balanced equation, the stress of an increase in pressure is relieved by a shift in the reaction to the side with fewer moles of gas (in this case, to reactants). The number of moles of reaction products decreases.
(b) Because there are 2 mol of gas on both sides of the balanced equation, the composition of the equilibrium mixture is unaffected by a change in pressure. The number of moles of reaction product remains the same.
(c) Because there are 2 mol of gas on the left side and 1 mol of gas on the right side of the balanced equation, the stress of an increase in pressure is relieved by a shift in the reaction to the side with fewer moles of gas (in this case, to products). The number of moles of reaction products increases.

14.101 As the volume increases, the pressure decreases at constant temperature.
(a) Because there is 1 mol of gas on the left side and 2 mol of gas on the right side of the balanced equation, the stress of an increase in volume (decrease in pressure) is relieved by a shift in the reaction to the side with the larger number of moles of gas (in this case, to products).
(b) Because there are 3 mol of gas on the left side and 2 mol of gas on the right side of the balanced equation, the stress of an increase in volume (decrease in pressure) is relieved by a shift in the reaction to the side with the larger number of moles of gas (in this case, to reactants).
(c) Because there are 3 mol of gas on both sides of the balanced equation, the composition of the equilibrium mixture is unaffected by an increase in volume (decrease in pressure). There is no net reaction in either direction.

14.102 $CO(g) + H_2O(g) \rightleftarrows CO_2(g) + H_2(g)$ $\Delta H^o = -41.2$ kJ

The reaction is exothermic. $[H_2]$ decreases when the temperature is increased.
As the temperature is decreased, the reaction shifts to the right. $[CO_2]$ and $[H_2]$ increase, $[CO]$ and $[H_2O]$ decrease, and K_c increases.

14.103 Because ΔH^o is positive, the reaction is endothermic.

heat + $3\,O_2(g) \rightleftarrows 2\,O_3(g)$

$$K_c = \frac{[O_3]^2}{[O_2]^3}$$

As the temperature increases, heat is added to the reaction, causing a shift to the right. The $[O_3]$ increases, and the $[O_2]$ decreases. This results in an increase in K_c.

14.104 (a) HCl is a source of Cl^- (product), the reaction shifts left, the equilibrium $[CoCl_4^{2-}]$ increases.
(b) $Co(NO_3)_2$ is a source of $Co(H_2O)_6^{2+}$ (product), the reaction shifts left, the equilibrium $[CoCl_4^{2-}]$ increases.
(c) All concentrations will initially decrease and the reaction will shift to the right; the equilibrium $[CoCl_4^{2-}]$ decreases.
(d) For an exothermic reaction, the reaction shifts to the left when the temperature is increased; the equilibrium $[CoCl_4^{2-}]$ increases.

14.105 (a) $Fe(NO_3)_3$ is a source of Fe^{3+}. Fe^{3+} (reactant) added; the $FeCl^{2+}$ concentration increases.
(b) Cl^- (reactant) removed; the $FeCl^{2+}$ concentration decreases.
(c) An endothermic reaction shifts to the right as the temperature increases; the $FeCl^{2+}$ concentration increases.
(d) A catalyst does not affect the composition of the equilibrium mixture; no change in $FeCl^{2+}$ concentration.

14.106 (a) The reaction is exothermic. The amount of CH_3OH (product) decreases as the temperature increases.
(b) When the volume decreases, the reaction shifts to the side with fewer gas molecules. The amount of CH_3OH increases.
(c) Addition of an inert gas (He) does not affect the equilibrium composition. There is no change.
(d) Addition of CO (reactant) shifts the reaction toward product. The amount of CH_3OH increases.
(e) Addition or removal of a catalyst does not affect the equilibrium composition. There is no change.

14.107 (a) An endothermic reaction shifts to the right as the temperature increases. The amount of acetone increases.
(b) Because there is 1 mol of gas on the left side and 2 mol of gas on the right side of the balanced equation, the stress of an increase in volume (decrease in pressure) is relieved by a shift in the reaction to the side with the larger number of moles of gas (in this case, to products). The amount of acetone increases.

(c) The addition of Ar (an inert gas) with no volume change does not affect the composition of the equilibrium mixture. The amount of acetone does not change.
(d) H_2 (product) added; the amount of acetone decreases.
(e) A catalyst does not affect the composition of the equilibrium mixture. The amount of acetone does not change.

14.108 (a) add Au (a solid); no shift
(b) OH^- (product) increased; shift toward reactants.
(c) O_2 (reactant) partial pressure increased; shift toward products.
(d) Fe^{3+} decreases CN^- (reactant) by forming $Fe(CN)_6^{3-}$; shift toward reactants.

14.109 (a) The reaction is endothermic. The amount of $CaCO_3$ (product) increases as the temperature increases.
(b) Adding CaO (a solid) does not change the amount of $CaCO_3$.
(c) Removing CH_4 (a reactant) decreases the amount of $CaCO_3$.
(d) Because there are 3 mol of gas on the left side and 4 mol of gas on the right side of the balanced equation, the stress of an increase in volume (decrease in pressure) is relieved by a shift in the reaction to the side with the larger number of moles of gas (in this case, to products). The amount of $CaCO_3$ increases.
(e) Adding Ir (a catalyst) does not affect the equilibrium mixture so there is no change in the amount of $CaCO_3$.

Chemical Equilibrium and Chemical Kinetics (Sections 14.10 and 14.11)

14.110 (a) A catalyst does not affect the equilibrium composition. The amount of CO remains the same.
(b) The reaction is exothermic. An increase in temperature shifts the reaction toward reactants. The amount of CO increases.
(c) Because there are 3 mol of gas on the left side and 2 mol of gas on the right side of the balanced equation, the stress of an increase in pressure is relieved by a shift in the reaction to the side with fewer moles of gas (in this case, to products). The amount of CO decreases.
(d) An increase in pressure as a result of the addition of an inert gas (with no volume change) does not affect the equilibrium composition. The amount of CO remains the same.
(e) Adding O_2 increases the O_2 concentration and shifts the reaction toward products. The amount of CO decreases.

14.111 A decrease in volume (a) and the addition of reactants (c) will affect the composition of the equilibrium mixture, but leave the value of K_c unchanged.
A change in temperature (b) affects the value of K_c.
Addition of a catalyst (d) or an inert gas (e) affects neither the composition of the equilibrium mixture nor the value of K_c.

14.112 Because the rate of the forward reaction is greater than the rate of the reverse reaction, the reaction will proceed in the forward direction to attain equilibrium.

14.113 Because the rate of the forward reaction is less than the rate of the reverse reaction, the reaction will proceed in the reverse direction to attain equilibrium.

14.114 $A + B \rightleftarrows C$

$rate_f = k_f[A][B]$ and $rate_r = k_r[C]$; at equilibrium, $rate_f = rate_r$

$k_f[A][B] = k_r[C]$; $\quad \dfrac{k_f}{k_r} = \dfrac{[C]}{[A][B]} = K_c$

14.115 An equilibrium mixture that contains large amounts of reactants and small amounts of products has a small K_c. A small K_c has $k_f < k_r$ (c).

14.116 $K_c = \dfrac{k_f}{k_r} = \dfrac{0.13}{6.2 \times 10^{-4}} = 210$

14.117 $K_c = \dfrac{k_f}{k_r}$; $\quad k_r = \dfrac{k_f}{K_c} = \dfrac{6 \times 10^{-6}\ M^{-1}s^{-1}}{10^{16}} = 6 \times 10^{-22}\ M^{-1}s^{-1}$

14.118 k_r increases more than k_f, this means that E_a (reverse) is greater than E_a (forward). The reaction is exothermic when E_a (reverse) > E_a (forward).

14.119 k_f increases more than k_r, this means that E_a (forward) is greater than E_a (reverse). The reaction is endothermic when E_a (forward) > E_a (reverse).

Chapter Problems

14.120 $2\ HI(g) \rightleftarrows H_2(g) + I_2(g)$

Calculate K_c. $\quad K_c = \dfrac{[H_2][I_2]}{[HI]^2} = \dfrac{(0.13)(0.70)}{(2.1)^2} = 0.0206$

$[HI] = \dfrac{0.20\ mol}{0.5000\ L} = 0.40\ M$

	$2\ HI(g)$	$\rightleftarrows$	$H_2(g)$	+	$I_2(g)$
initial (M)	0.40		0		0
change (M)	–2x		+x		+x
equil (M)	0.40 – 2x		x		x

$K_c = 0.0206 = \dfrac{[H_2][I_2]}{[HI]^2} = \dfrac{x^2}{(0.40 - 2x)^2}$

Take the square root of both sides, and solve for x.

$\sqrt{0.0206} = \sqrt{\dfrac{x^2}{(0.40 - 2x)^2}}$; $\quad 0.144 = \dfrac{x}{0.40 - 2x}$; $\quad x = 0.045$

At equilibrium, $[H_2] = [I_2] = x = 0.045\ M$; $\quad [HI] = 0.40 - 2x = 0.40 - 2(0.045) = 0.31\ M$

14.121 Note the container volume is 5.00 L

$[H_2] = [I_2] = 1.00\ \text{mol}/5.00\ \text{L} = 0.200\ \text{M}$; $[HI] = 2.50\ \text{mol}/5.00\ \text{L} = 0.500\ \text{M}$

	$H_2(g)$	+	$I_2(g)$	⇄	2 HI(g)
initial (M)	0.200		0.200		0.500
change (M)	–x		–x		+2x
equil (M)	0.200 – x		0.200 – x		0.500 + 2x

$$K_c = \frac{[HI]^2}{[H_2][I_2]} = 129 = \frac{(0.500 + 2x)^2}{(0.200 - x)^2}$$

Take the square root of both sides, and solve for x.

$$\sqrt{129} = \sqrt{\frac{(0.500 + 2x)^2}{(0.200 - x)^2}};\quad 11.4 = \frac{0.500 + 2x}{0.200 - x};\quad x = 0.133$$

$[H_2] = [I_2] = 0.200 - x = 0.200 - 0.133 = 0.067\ \text{M}$

$[HI] = 0.500 + 2x = 0.500 + 2(0.133) = 0.766\ \text{M}$

14.122 $[H_2O] = \dfrac{6.00\ \text{mol}}{5.00\ \text{L}} = 1.20\ \text{M}$

	C(s)	+	$H_2O(g)$	⇄	CO(g)	+	$H_2(g)$
initial (M)			1.20		0		0
change (M)			–x		+x		+x
equil (M)			1.20 – x		x		x

$$K_c = \frac{[CO][H_2]}{[H_2O]} = 3.0 \times 10^{-2} = \frac{x^2}{1.20 - x}$$

$x^2 + (3.0 \times 10^{-2})x - 0.036 = 0$

Use the quadratic formula to solve for x.

$$x = \frac{-(0.030) \pm \sqrt{(0.030)^2 - 4(-0.036)}}{2(1)} = \frac{-0.030 \pm 0.381}{2}$$

x = 0.176 and –0.206

Discard the negative solution (–0.206) because it leads to negative concentrations and that is impossible.

$[CO] = [H_2] = x = 0.18\ \text{M}$

$[H_2O] = 1.20 - x = 1.20 - 0.18 = 1.02\ \text{M}$

14.123 (a) Because K_p is larger at the higher temperature, the reaction has shifted toward products at the higher temperature, which means the reaction is endothermic.

(b) (i) Increasing the volume causes the reaction to shift toward the side with more mol of gas (product side). The equilibrium amounts of PCl_3 and Cl_2 increase while that of PCl_5 decreases.

(ii) If there is no volume change, there is no change in equilibrium concentrations.

(iii) Addition of a catalyst does not affect the equilibrium concentrations.

14.124 (a) $[PCl_5] = 1.000 \text{ mol}/5.000 \text{ L} = 0.2000 \text{ M}$

	$PCl_5(g)$	⇄	$PCl_3(g)$	+	$Cl_2(g)$
initial (M)	0.2000		0		0
change (M)	–(0.2000)(0.7850)		+(0.2000)(0.7850)		+(0.2000)(0.7850)
equil (M)	0.0430		0.1570		0.1570

$$K_c = \frac{[PCl_3][Cl_2]}{[PCl_5]} = \frac{(0.1570)(0.1570)}{(0.0430)} = 0.573$$

$\Delta n = 1$ and $K_p = K_c(RT) = (0.573)(0.082\ 06)(500) = 23.5$

(b) $$Q_c = \frac{[PCl_3][Cl_2]}{[PCl_5]} = \frac{(0.150)(0.600)}{(0.500)} = 0.18$$

Because $Q_c < K_c$, the reaction proceeds to the right to reach equilibrium.

	$PCl_5(g)$	⇄	$PCl_3(g)$	+	$Cl_2(g)$
initial (M)	0.500		0.150		0.600
change (M)	– x		+x		+x
equil (M)	0.500 – x		0.150 + x		0.600 + x

$$K_c = \frac{[PCl_3][Cl_2]}{[PCl_5]} = 0.573 = \frac{(0.150 + x)(0.600 + x)}{(0.500 - x)}; \text{ solve for x.}$$

$x^2 + 1.323x - 0.1965 = 0$

$$x = \frac{-(1.323) \pm \sqrt{(1.323)^2 - (4)(1)(-0.1965)}}{2(1)} = \frac{-1.323 \pm 1.593}{2}$$

$x = -1.458$ and 0.135

Discard the negative solution (–1.458) because it will lead to negative concentrations and that is impossible.

$[PCl_5] = 0.500 - x = 0.500 - 0.135 = 0.365 \text{ M}$

$[PCl_3] = 0.150 + x = 0.150 + 0.135 = 0.285 \text{ M}$

$[Cl_2] = 0.600 + x = 0.600 + 0.135 = 0.735 \text{ M}$

14.125

	H_2O	+	D_2O	⇄	2 HDO	$K_c = 3.86$
initial (mol)	1.00		1.00		0	
change (mol)	–x		–x		+2x	
equil (mol)	1.00 – x		1.00 – x		2x	

$$K_c = \frac{[HDO]^2}{[H_2O][D_2O]} = 3.86 = \frac{(2x)^2}{(1.00 - x)^2}$$

$$\sqrt{3.86} = \sqrt{\frac{(2x)^2}{(1.00 - x)^2}}$$

$$1.96 = \frac{2x}{(1.00 - x)}$$

$1.96 - 1.96x = 2x$

$1.96 = 3.96x$

$x = \dfrac{1.96}{3.96} = 0.495$ mol

mol H_2O = mol D_2O = 1.00 – x = 1.00 – 0.495 = 0.50 mol

mol HDO = 2x = 2(0.495) = 0.99 mol

14.126 (a) $K_c = \dfrac{[C_2H_6][C_2H_4]}{[C_4H_{10}]}$ $\qquad K_p = \dfrac{(P_{C_2H_6})(P_{C_2H_4})}{P_{C_4H_{10}}}$

(b) $K_p = 12$; $\Delta n = 1$; $K_c = K_p\left(\dfrac{1}{RT}\right) = (12)\left(\dfrac{1}{(0.082\,06)(773)}\right) = 0.19$

(c)	$C_4H_{10}(g)$	⇄	$C_2H_6(g)$	+	$C_2H_4(g)$
initial (atm)	50		0		0
change (atm)	–x		+x		+x
equil (atm)	50 – x		x		x

$K_p = 12 = \dfrac{x^2}{50 - x}$; $\qquad x^2 + 12x - 600 = 0$

Use the quadratic formula to solve for x.

$$x = \frac{(-12) \pm \sqrt{(12)^2 - 4(1)(-600)}}{2(1)} = \frac{-12 \pm 50.44}{2}$$

x = –31.22 and 19.22

Discard the negative solution (–31.22) because it leads to negative concentrations and that is impossible.

% C_4H_{10} converted = $\dfrac{19.22}{50}$ x 100% = 38%

$P_{total} = P_{C_4H_{10}} + P_{C_2H_6} + P_{C_2H_4} = (50 - x) + x + x = (50 - 19) + 19 + 19 = 69$ atm

(d) A decrease in volume would decrease the % conversion of C_4H_{10}.

14.127	C(s)	+	$CO_2(g)$	⇄	2 CO(g)
initial (M)	excess		1.50 mol/20.0 L		0
change (M)			–x		+2x
equil (M)			0.0750 – x		2x

[CO] = 2x = 7.00 x 10^{-2} M; $\qquad$ x = 0.0350 M

(a) $[CO_2]$ = 0.0750 – x = 0.0750 – 0.0350 = 0.0400 M

(b) $K_c = \dfrac{[CO]^2}{[CO_2]} = \dfrac{(7.00 \text{ x } 10^{-2})^2}{(0.0400)} = 0.122$

14.128 (a) $K_p = 3.45$; $\Delta n = 1$; $K_c = K_p\left(\dfrac{1}{RT}\right) = (3.45)\left(\dfrac{1}{(0.082\,06)(500)}\right) = 0.0840$

(b) $[(CH_3)_3CCl]$ = 1.00 mol/5.00 L = 0.200 M

	$(CH_3)_3CCl(g)$	⇄	$(CH_3)_2C{=}CH_2(g)$	+	HCl(g)
initial (M)	0.200		0		0
change (M)	–x		+x		+x
equil (M)	0.200 – x		x		x

$K_c = 0.0840 = \dfrac{x^2}{0.200 - x}$; $x^2 + 0.0840x - 0.0168 = 0$

Use the quadratic formula to solve for x.

$$x = \frac{(-0.0840) \pm \sqrt{(0.0840)^2 - 4(1)(-0.0168)}}{2(1)} = \frac{-0.0840 \pm 0.272}{2}$$

x = –0.178 and 0.094

Discard the negative solution (–0.178) because it leads to negative concentrations and that is impossible.

$[(CH_3)_2C{=}CCH_2] = [HCl] = x = 0.094$ M

$[(CH_3)_3CCl] = 0.200 - x = 0.200 - 0.094 = 0.106$ M

(c) $K_p = 3.45$

	$(CH_3)_3CCl(g)$	⇄	$(CH_3)_2C{=}CH_2(g)$	+	$HCl(g)$
initial (atm)	0		0.400		0.600
change (atm)	+x		–x		–x
equil (atm)	x		0.400 – x		0.600 – x

$$K_p = 3.45 = \frac{(0.400 - x)(0.600 - x)}{x}$$

$x^2 - 4.45x + 0.240 = 0$

Use the quadratic formula to solve for x.

$$x = \frac{-(-4.45) \pm \sqrt{(-4.45)^2 - 4(1)(0.240)}}{2(1)} = \frac{4.45 \pm 4.34}{2}$$

x = 0.055 and 4.40

Discard the larger solution (4.40) because it leads to negative partial pressures and that is impossible.

$P_{\text{t-butyl chloride}} = x = 0.055$ atm; $P_{\text{isobutylene}} = 0.400 - x = 0.400 - 0.055 = 0.345$ atm

$P_{HCl} = 0.600 - x = 0.600 - 0.055 = 0.545$ atm

14.129 (a) The Arrhenius equation gives for the forward and reverse reactions:

$$k_f = A_f e^{-E_{a,f}/RT} \quad \text{and} \quad k_r = A_r e^{-E_{a,r}/RT}$$

Addition of a catalyst decreases the activation energies by ΔE_a, so

$$k_f = A_f e^{-(E_{a,f} - \Delta E_a)/RT} = A_f e^{-E_{a,f}/RT} \times e^{\Delta E_a/RT}$$

and

$$k_r = A_r e^{-(E_{a,r} - \Delta E_a)/RT} = A_r e^{-E_{a,r}/RT} \times e^{\Delta E_a/RT}$$

Therefore, the rate constants for both the forward and reverse reactions increase by the same factor, $e^{\Delta E_a/RT}$.

(b) The equilibrium constant is given by $K_c = \dfrac{k_f}{k_r} = \dfrac{A_f e^{-E_{a,f}/RT}}{A_r e^{-E_{a,r}/RT}} = \dfrac{A_f}{A_r} e^{-\Delta E/RT}$

where $\Delta E = E_{a,f} - E_{a,r}$. Addition of a catalyst decreases the activation energies by ΔE_a.

So, $$K_c = \frac{k_f}{k_r} = \frac{A_f e^{-E_{a,f}/RT} \times e^{\Delta E_a/RT}}{A_r e^{-E_{a,r}/RT} \times e^{\Delta E_a/RT}} = \frac{A_f}{A_r} e^{-\Delta E/RT}$$

K_c is unchanged because of cancellation of $e^{\Delta E_a/RT}$, the factor by which the two rate constants increase.

14.130 The activation energy (E_a) is positive, and for an exothermic reaction, $E_{a,r} > E_{a,f}$.

$k_f = A_f\, e^{-E_{a,f}/RT}$, $k_r = A_r\, e^{-E_{a,r}/RT}$

$$K_c = \frac{k_f}{k_r} = \frac{A_f e^{-E_{a,f}/RT}}{A_r e^{-E_{a,r}/RT}} = \frac{A_f}{A_r} e^{(E_{a,r} - E_{a,f})/RT}$$

$(E_{a,r} - E_{a,f})$ is positive, so the exponent is always positive. As the temperature increases, the exponent, $(E_{a,r} - E_{a,f})/RT$, decreases and the value for K_c decreases as well.

14.131 (a) $PV = nRT$

$$P^o_{Br_2} = \frac{nRT}{V} = \frac{(0.974 \text{ mol})\left(0.082\ 06 \frac{L \cdot atm}{K \cdot mol}\right)(1000 \text{ K})}{1.00 \text{ L}} = 80 \text{ atm}$$

$$P^o_{H_2} = \frac{nRT}{V} = \frac{(1.22 \text{ mol})\left(0.082\ 06 \frac{L \cdot atm}{K \cdot mol}\right)(1000 \text{ K})}{1.00 \text{ L}} = 100 \text{ atm}$$

Because K_p is very large, assume first that the reaction goes to completion and then is followed by a small back reaction.

	$H_2(g)$	+	$Br_2(g)$	⇄	$2\ HBr(g)$
before (atm)	100		80		0
100% reaction (atm)	–80		–80		+2(80)
after (atm)	20		0		160
back reaction (atm)	+x		+x		–2x
after (atm)	20 + x		x		160 – 2x

$P_{H_2} = 20 + x \approx 20$ atm

$P_{HBr} = 160 - 2x \approx 160$ atm

$$K_p = \frac{(P_{HBr})^2}{(P_{H_2})(P_{Br_2})} = 2.1 \times 10^6 = \frac{(160)^2}{(20)(x)};\quad P_{Br_2} = x = \frac{(160)^2}{(20)(2.1 \times 10^6)} = 6.1 \times 10^{-4} \text{ atm}$$

(b) (ii) Adding Br_2 will cause the greatest increase in the pressure of HBr. The very large value of K_p means that the reaction goes essentially to completion. Therefore, the reaction stops when the limiting reactant, Br_2, is essentially consumed. No matter how much H_2 is added or how far the equilibrium is shifted (by lowering the temperature) to favor the formation of HBr, the amount of HBr will ultimately be limited by the amount of Br_2 present. Therefore, more Br_2 must be added in order to produce more HBr.

14.132 (a) $PV = nRT$, $n_{total} = \frac{PV}{RT} = \frac{(0.588 \text{ atm})(1.00 \text{ L})}{\left(0.082\ 06 \frac{\text{L} \cdot \text{atm}}{\text{K} \cdot \text{mol}}\right)(300 \text{ K})} = 0.0239 \text{ mol}$

	$2 NOBr(g)$	⇄	$2 NO(g)$	+	$Br_2(g)$
initial (mol)	0.0200		0		0
change (mol)	–2x		+2x		+x
equil (mol)	0.0200 – 2x		2x		x

$n_{total} = 0.0239 \text{ mol} = (0.0200 - 2x) + 2x + x = 0.0200 + x$

$x = 0.0239 - 0.0200 = 0.0039 \text{ mol}$

Because the volume is 1.00 L, the molarity equals the number of moles.

$[NOBr] = 0.0200 - 2x = 0.0200 - 2(0.0039) = 0.0122 \text{ M}$

$[NO] = 2x = 2(0.0039) = 0.0078 \text{ M}$

$[Br_2] = x = 0.0039 \text{ M}$

$$K_c = \frac{[NO]^2[Br_2]}{[NOBr]^2} = \frac{(0.0078)^2(0.0039)}{(0.0122)^2} = 1.6 \times 10^{-3}$$

(b) $\Delta n = (3) - (2) = 1$, $K_p = K_c(RT) = (1.6 \times 10^{-3})(0.082\ 06)(300) = 0.039$

14.133 NO_2, 46.01

$$\text{mol } NO_2 = 4.60 \text{ g } NO_2 \times \frac{1 \text{ mol } NO_2}{46.01 \text{ g } NO_2} = 0.100 \text{ mol } NO_2$$

$[NO_2] = 0.100 \text{ mol}/10.0 \text{ L} = 0.0100 \text{ M}$

	$2 NO_2(g)$	⇄	$N_2O_4(g)$
initial (M)	0.0100		0
change (M)	–2x		+x
equil (M)	0.0100 – 2x		x

$$K_c = \frac{[N_2O_4]}{[NO_2]^2} = 4.72 = \frac{x}{(0.0100 - 2x)^2}$$

$18.88x^2 - 1.189x + 0.000\ 472 = 0$

Use the quadratic formula to solve for x.

$$x = \frac{-(-1.189) \pm \sqrt{(-1.189)^2 - 4(18.88)(0.000\ 472)}}{2(18.88)} = \frac{1.189 \pm 1.1739}{37.76}$$

$x = 0.0626$ and 4.00×10^{-4}

Discard the larger solution (0.0626) because it leads to a negative concentration of NO_2 and that is impossible.

$n_{total} = (0.0100 - 2x + x)(10.0 \text{ L}) = (0.0100 - x)(10.0 \text{ L})$

$= [0.0100 \text{ mol/L} - (4.00 \times 10^{-4} \text{ mol/L})](10.0 \text{ L}) = 0.0960 \text{ mol}$

$100\ °C = 100 + 273 = 373 \text{ K}$

$$P_{total} = \frac{n_{total}RT}{V} = \frac{(0.0960 \text{ mol})\left(0.082\ 06 \frac{\text{L} \cdot \text{atm}}{\text{K} \cdot \text{mol}}\right)(373 \text{ K})}{10.0 \text{ L}} = 0.294 \text{ atm}$$

14.134 (a) $W(s) + 4\,Br(g) \rightleftarrows WBr_4(g)$

$$K_p = \frac{P_{WBr_4}}{(P_{Br})^4} = 100,\ P_{WBr_4} = (P_{Br})^4(100) = (0.010\ atm)^4(100) = 1.0 \times 10^{-6}\ atm$$

(b) Because K_p is smaller at the higher temperature, the reaction has shifted toward reactants at the higher temperature, which means the reaction is exothermic.

(c) At 2800 K, $Q_p = \frac{(1.0 \times 10^{-6})}{(0.010)^4} = 100$, $Q_p > K_p$ so the reaction will go from products to reactants, depositing tungsten back onto the filament.

14.135 (a)

	$(NH_4)(NH_2CO_2)(s)$	$\rightleftarrows$	$2\,NH_3(g)$	+	$CO_2(g)$
initial (atm)			0		0
change (atm)			+2x		+x
equil (atm)			2x		x

$P_{total} = 2x + x = 3x = 0.116$ atm
$x = 0.116\ atm/3 = 0.0387$ atm
$P_{NH_3} = 2x = 2(0.0387\ atm) = 0.0774$ atm
$P_{CO_2} = x = 0.0387$ atm
$K_p = (P_{NH_3})^2(P_{CO_2}) = (0.0774)^2(0.0387) = 2.32 \times 10^{-4}$

(b) (i) The total quantity of NH_3 would decrease. When product, CO_2, is added, the equilibrium will shift to the left.
(ii) The total quantity of NH_3 would remain unchanged. Adding a pure solid, $(NH_4)(NH_2CO_2)$, to a heterogeneous equilibrium will not affect the position of the equilibrium.
(iii) The total quantity of NH_3 would increase. When product, CO_2, is removed, the equilibrium will shift to the right.
(iv) The total quantity of NH_3 would increase. When the total volume is increased, the reaction will shift to the side with more total moles of gas, which in this case is the product side.
(v) The total quantity of NH_3 would remain unchanged. Neon is an inert gas that will have no effect on the reaction or on the position of equilibrium.
(vi) The total quantity of NH_3 would increase. Because the reaction is endothermic, an increase in temperature will shift the equilibrium to the right.

14.136 $2\,NO_2(g) \rightleftarrows N_2O_4(g)$
$\Delta n = (1) - (2) = -1$ and $K_p = K_c(RT)^{-1} = (216)[(0.082\ 06)(298)]^{-1} = 8.83$

$$K_p = \frac{P_{N_2O_4}}{(P_{NO_2})^2} = 8.83$$

Let $X = P_{N_2O_4}$ and $Y = P_{NO_2}$

$P_{total} = 1.50 \text{ atm} = X + Y$ and $\dfrac{X}{Y^2} = 8.83$. Use these two equations to solve for X and Y.

$X = 1.50 - Y$

$$\frac{1.50 - Y}{Y^2} = 8.83$$

$8.83Y^2 + Y - 1.50 = 0$

Use the quadratic formula to solve for Y.

$$Y = \frac{-(1) \pm \sqrt{(1)^2 - 4(8.83)(-1.50)}}{2(8.83)} = \frac{-1 \pm 7.35}{17.7}$$

$Y = -0.472$ and 0.359

Discard the negative solution (–0.472) because it leads to a negative partial pressure of NO_2 and that is impossible.

$Y = P_{NO_2} = 0.359$ atm; $X = P_{N_2O_4} = 1.50 \text{ atm} - Y = 1.50 \text{ atm} - 0.359 \text{ atm} = 1.14$ atm

14.137 500 °C = 500 + 273 = 773 K and 840 °C = 840 + 273 = 1113 K

Calculate the undissociated pressure of F_2 at 1113 K.

$$\frac{P_2}{T_2} = \frac{P_1}{T_1}; \; P_2 = \frac{P_1 T_2}{T_1} = \frac{(0.600 \text{ atm})(1113 \text{ K})}{773 \text{ K}} = 0.864 \text{ atm}$$

	$F_2(g)$	⇄	$2\,F(g)$
initial (atm)	0.864		0
change (atm)	–x		+2x
equil (atm)	0.864 – x		2x

$P_{total} = (0.864 \text{ atm} - x) + 2x = 0.864 \text{ atm} + x = 0.984$ atm

$x = 0.984 \text{ atm} - 0.864 \text{ atm} = 0.120$ atm

$P_{F_2} = (0.864 \text{ atm} - x) = (0.864 \text{ atm} - 0.120 \text{ atm}) = 0.744$ atm

$P_F = 2x = 2(0.120 \text{ atm}) = 0.240$ atm

$$K_p = \frac{(P_F)^2}{P_{F_2}} = \frac{(0.240)^2}{0.744} = 0.0774$$

14.138 (a) $P_{NO} = X_{NO} \cdot P_{total} = \dfrac{2.0 \text{ mol NO}}{(2.0 \text{ mol NO} + 3.0 \text{ mol } NO_2)} \times (1.65 \text{ atm}) = 0.66$ atm

$P_{NO_2} = X_{NO_2} \cdot P_{total} = \dfrac{3.0 \text{ mol } NO_2}{(2.0 \text{ mol NO} + 3.0 \text{ mol } NO_2)} \times (1.65 \text{ atm}) = 0.99$ atm

	NO(g)	+	$NO_2(g)$	⇄	$N_2O_3(g)$
initial (atm)	0.66		0.99		0
change (atm)	–x		–x		+x
equil (atm)	0.66 – x		0.99 – x		x

$K_c = 13$; $\Delta n = (1) - (1 + 1) = -1$; $K_p = K_c(RT)^{-1} = (13)[(0.082\,06)(298)]^{-1} = 0.53$

$$K_p = \frac{P_{N_2O_3}}{(P_{NO})(P_{NO_2})} = 0.53 = \frac{x}{(0.66 - x)(0.99 - x)}$$

$0.53x^2 - 1.87x + 0.35 = 0$

Use the quadratic formula to solve for x.

$$x = \frac{-(-1.87) \pm \sqrt{(-1.87)^2 - 4(0.53)(0.35)}}{2(0.53)} = \frac{1.87 \pm 1.66}{1.06}$$

x = 3.33 and 0.20

Discard the larger solution (3.33) because it leads to a negative concentrations of NO and NO_2 and that is impossible.

$P_{NO} = 0.66 - x = 0.66 - 0.20 = 0.46$ atm

$P_{NO_2} = 0.99 - x = 0.99 - 0.20 = 0.79$ atm

$P_{N_2O_3} = x = 0.20$ atm

(b) PV = nRT

$$V = \frac{nRT}{P} = \frac{(2.0\text{ mol} + 3.0\text{ mol})\left(0.082\ 06\ \frac{L \cdot atm}{K \cdot mol}\right)(298\text{ K})}{1.65\text{ atm}} = 74\text{ L}$$

14.139 [COCl] = 1.00 mol/50.0 L = 0.0200 M

	$COCl_2(g)$	⇄	CO(g)	+	$Cl_2(g)$
initial (M)	0.0200		0		0
change (M)	–x		+x		+x
equil (M)	0.0200 – x		x		x

$$K_c = \frac{[CO][Cl_2]}{[COCl_2]} = 8.4 \times 10^{-4} = \frac{x^2}{0.0200 - x}$$

$x^2 + 8.4 \times 10^{-4}x - 1.68 \times 10^{-5} = 0$

Use the quadratic formula to solve for x.

$$x = \frac{-(8.4 \times 10^{-4}) \pm \sqrt{(8.4 \times 10^{-4})^2 - 4(1)(-1.68 \times 10^{-5})}}{2(1)} = \frac{(-8.4 \times 10^{-4}) \pm (8.2 \times 10^{-3})}{2}$$

x = 0.0037 and –0.0045

Discard the negative solution (–0.0045) because it leads to a negative concentrations of CO and Cl_2 and that is impossible.

$[COCl_2] = 0.0200 - x = 0.0200 - 0.0037 = 0.0163$ M

$[CO] = [Cl_2] = x = 0.0037$ M

mol $COCl_2$ = (0.0163 mol/L)(50.0 L) = 0.815 mol

mol CO = mol Cl_2 = (0.0037 mol/L)(50.0 L) = 0.185 mol

360 °C = 360 + 273 = 633 K

$$P = \frac{nRT}{V} = \frac{(0.815\text{ mol} + 0.185\text{ mol} + 0.185\text{ mol})\left(0.082\ 06\ \frac{L \cdot atm}{K \cdot mol}\right)(633\text{ K})}{50.0\text{ L}} = 1.23\text{ atm}$$

14.140

	$N_2(g)$ +	$3\,H_2(g)$ ⇄	$2\,NH_3(g)$
initial (mol)	0	0	X
change (mol)	+y	+3y	−2y
equil (mol)	y	3y	X − 2y

y = 0.200 mol

Because the volume is 1.00 L, the molarity equals the number of moles.

$[N_2] = y = 0.200$ M; $[H_2] = 3y = 3(0.200) = 0.600$ M

$$K_c = \frac{[NH_3]^2}{[N_2][H_2]^3} = \frac{[NH_3]^2}{(0.200)(0.600)^3} = 4.20\text{, solve for } [NH_3]_{eq}$$

$$[NH_3]_{eq}^2 = [N_2][H_2]^3(4.20) = (0.200)(0.600)^3(4.20)$$

$$[NH_3]_{eq} = \sqrt{[N_2][H_2]^3(4.20)} = \sqrt{(0.200)(0.600)^3(4.20)} = 0.426\text{ M}$$

$[NH_3]_{eq} = 0.426\text{ M} = X - 2(0.200) = [NH_3]_o - 2(0.200)$

$[NH_3]_o = 0.426 + 2(0.200) = 0.826$ M

0.826 mol of NH_3 were placed in the 1.00 L reaction vessel.

14.141 N_2O_4, 92.01

$$\text{mol } N_2O_4 = 46.0\text{ g x } \frac{1\text{ mol } N_2O_4}{92.01\text{ g}} = 0.500\text{ mol } N_2O_4$$

45 °C = 318 K

$K_p = K_c(RT)^{\Delta n}$; $\Delta n = 2 - 1 = 1$; $K_p = K_c(RT) = (0.619)(0.08\,206)(318) = 16.2$

$$P_{N_2O_4} = \frac{nRT}{V} = \frac{(0.500\text{ mol})\left(0.082\ 06\ \frac{L\cdot atm}{K\cdot mol}\right)(318\text{ K})}{2.00\text{ L}} = 6.52\text{ atm}$$

	$N_2O_4(g)$ ⇄	$2\,NO_2(g)$
initial (atm)	6.52	0
change (atm)	−x	+2x
equil (atm)	6.52 − x	2x

$$K_p = \frac{(P_{NO_2})^2}{P_{N_2O_4}} = 16.2 = \frac{(2x)^2}{6.52 - x}$$

$(16.2)(6.52 - x) = 4x^2$

$105.6 - 16.2x = 4x^2$

$4x^2 + 16.2x - 105.6 = 0$

Use the quadratic formula to solve for x.

$$x = \frac{-(16.2) \pm \sqrt{(16.2)^2 - 4(4)(-105.6)}}{2(4)} = \frac{-16.2 \pm 44.2}{8}$$

x = −7.55 and 3.50

Discard the negative solution (−7.55) because it leads to a negative partial pressure and that is impossible.

$P_{N_2O_4} = 6.52 - x = 6.52 - 3.50 = 3.02$ atm; $P_{NO_2} = 2x = 2(3.50) = 7.00$ atm

14.142 ClF_3, 92.45

(a) mol $ClF_3 = 9.25 \text{ g} \times \dfrac{1 \text{ mol } ClF_3}{92.45 \text{ g}} = 0.100 \text{ mol } ClF_3$

$[ClF_3] = \dfrac{0.100 \text{ mol } ClF_3}{2.00 \text{ L}} = 0.0500 \text{ M}$

	$ClF_3(g)$	⇄	$ClF(g)$	+	$F_2(g)$	
initial (M)	0.0500		0		0	
change (M)	−x		+x		+x	where x = 0.0500 x 0.198 = 0.009 90
equil (M)	0.0500 − x		x		x	
	0.0401		0.009 90		0.009 90	

$K_c = \dfrac{[ClF][F_2]}{[ClF_3]} = \dfrac{(0.009\,90)^2}{0.0401} = 0.002\,44$

(b) $K_p = K_c(RT)^{\Delta n}$; $\Delta n = 2 - 1 = 1$; $K_p = K_c(RT) = (0.002\,44)(0.082\,06)(700) = 0.140$

(c) mol $ClF_3 = 39.4 \text{ g} \times \dfrac{1 \text{ mol } ClF_3}{92.45 \text{ g}} = 0.426 \text{ mol } ClF_3$

$[ClF_3] = \dfrac{0.426 \text{ mol } ClF_3}{2.00 \text{ L}} = 0.213 \text{ M}$

	$ClF_3(g)$	⇄	$ClF(g)$	+	$F_2(g)$
initial (M)	0.213		0		0
change (M)	−x		+x		+x
equil (M)	0.213 − x		x		x

$K_c = \dfrac{[ClF][F_2]}{[ClF_3]} = 0.00\,244 = \dfrac{x^2}{0.213 - x}$

$(0.002\,44)(0.213 - x) = x^2$

$5.20 \times 10^{-4} - 0.002\,44x = x^2$

$x^2 + 0.002\,44x - (5.20 \times 10^{-4}) = 0$

Use the quadratic formula to solve for x.

$x = \dfrac{-(0.002\,44) \pm \sqrt{(0.002\,44)^2 - 4(1)(-5.20 \times 10^{-4})}}{2(1)} = \dfrac{-0.002\,44 \pm 0.0457}{2}$

x = −0.0241 and 0.0216

Discard the negative solution (−0.0241) because it leads to a negative partial pressure and that is impossible.

$[ClF_3] = 0.213 - x = 0.213 - 0.0216 = 0.191$ M; $[ClF] = [F_2] = x = 0.0216$ M

14.143

	L-Aspartate	⇄	Fumarate	+	NH_4^+
initial (M)	0.008 32		0		0
change (M)	−x		+x		+x
equil (M)	0.008 32 − x		x		x

$$K_c = \frac{[\text{Fumarate}][NH_4^+]}{[\text{L-Aspartate}]} = 6.95 \times 10^{-3} = \frac{x^2}{0.008\ 32 - x}$$

$x^2 + 0.006\ 95x - 5.78 \times 10^{-5} = 0$

Use the quadratic formula to solve for x.

$$x = \frac{-(0.006\ 95) \pm \sqrt{(0.006\ 95)^2 - 4(1)(-5.78 \times 10^{-5})}}{2(1)} = \frac{-0.006\ 95 \pm 0.0167}{2}$$

x = 0.004 87 and –0.0118

Discard the negative solution (–0.0118) because it leads to a negative concentrations and that is impossible.

[L-Aspartate] = 0.008 32 – x = 0.008 32 – 0.004 87 = 0.003 45 M

$[\text{Fumarate}] = [NH_4^+] = x = 0.004\ 87$ M

14.144

	Fumarate	⇄	L-Malate
initial (M)	0.001 56		0.002 27
change (M)	–x		+x
equil (M)	0.001 56 – x		0.002 27 + x

$$K_c = \frac{[\text{L-Malate}]}{[\text{Fumarate}]} = 3.3 = \frac{0.002\ 27 + x}{0.001\ 56 - x}$$

(3.3)(0.001 56 – x) = 0.002 27 + x

0.005 15 – 3.3x = 0.002 27 + x

0.002 88 = 4.3x

$$x = \frac{0.002\ 88}{4.3} = 6.7 \times 10^{-4} = 0.000\ 67$$

[Fumarate] = 0.001 56 – x = 0.001 56 – 0.000 67 = 8.9×10^{-4} M

[L-Malate] = 0.002 27 + x = 0.002 27 + 0.000 67 = 2.94×10^{-3} M

4.145 (a) $K_c = \frac{[H^+][CH_3CO_2^-]}{[CH_3CO_2H]}$

(b)

	$CH_3CO_2H(aq)$	⇄	$H^+(aq)$	+	$CH_3CO_2^-(aq)$
initial (M)	1.0		0		0
change (M)	–0.0042		+0.0042		+0.0042
equil (M)	1.0 – 0.0042		0.0042		0.0042

$$K_c = \frac{(0.0042)(0.0042)}{(1.0 - 0.0042)} = 1.8 \times 10^{-5}$$

14.146 (a) Addition of a solid does not affect the equilibrium composition. There is no change in the number of moles of CO_2.

(b) Adding a product causes the reaction to shift toward reactants. The number of moles of CO_2 decreases.

(c) Decreasing the volume causes the reaction to shift toward the side with fewer mol of gas (reactant side). The number of moles of CO_2 decreases.

(d) The reaction is endothermic. An increase in temperature shifts the reaction toward products. The number of moles of CO_2 increases.

Multiconcept Problems

14.147 (a) $K_p = \frac{(P_F)^2}{P_{F_2}} = 7.83$ at 1500 K

$P_F = \sqrt{K_p P_{F_2}} = \sqrt{(7.83)(0.200)} = 1.25$ atm

(b)	$F_2(g)$	⇄	2 F(g)
initial (atm)	x		0
change (atm)	−y		+2y
equil (atm)	x − y		2y

2y = 1.25 so y = 0.625

x − y = 0.200; x = 0.200 + y = 0.200 + 0.625 = 0.825

$f = \frac{0.625}{0.825} = 0.758$

(c) The shorter bond in F_2 is expected to be stronger. However, because of the small size of F, repulsion between the lone pairs of the two halogen atoms are much greater in F_2 than in Cl_2.

14.148 (a) $[N_2O_4] = \frac{0.500 \text{ mol}}{4.00 \text{ L}} = 0.125$ M

	$N_2O_4(g)$	⇄	$2\ NO_2(g)$
initial (M)	0.125		0
change (M)	−(0.793)(0.125)		+(2)(0.793)(0.125)
equil (M)	0.125 − (0.793)(0.125)		(2)(0.793)(0.125)

At equilibrium, $[N_2O_4]$ = 0.125 − (0.793)(0.125) = 0.0259 M

$[NO_2]$ = (2)(0.793)(0.125) = 0.198 M

$K_c = \frac{[NO_2]^2}{[N_2O_4]} = \frac{(0.198)^2}{(0.0259)} = 1.51$

$\Delta n = 2 - 1 = 1$ and $K_p = K_c(RT)^{\Delta n}$; $K_p = K_c(RT) = (1.51)(0.082\ 06)(400) = 49.6$

(b) 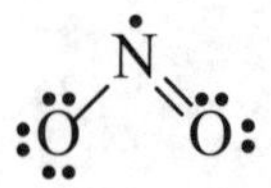 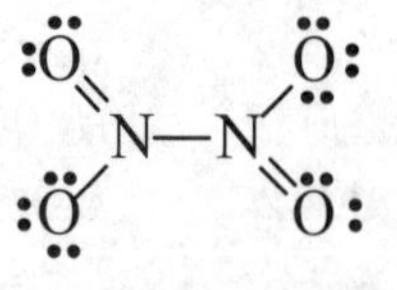 [Lewis structures of NO_2 and N_2O_4]

14.149 (a) Because $\Delta n = 0$, $K_p = K_c = 1.0 \times 10^5$

(b) $K_p = 1.0 \times 10^5 = \frac{P_{propene}}{P_{cyclopropane}}$; $P_{cyclopropane} = \frac{P_{propene}}{1.0 \times 10^5} = \frac{5.0 \text{ atm}}{1.0 \times 10^5} = 5.0 \times 10^{-5}$ atm

(c) The ratio of the two concentrations is equal to K_c. The ratio (K_c) cannot be changed by adding cyclopropane. The individual concentrations can change, but the ratio of the concentrations cannot.

Because there is one mole of gas on each side of the balanced equation, the composition of the equilibrium mixture is unaffected by a decrease in volume. The ratio of the two concentrations will not change.

(d) Because K_c is large, $k_f > k_r$.
(e) Because the C–C–C bond angles are 60° and the angles between sp^3 hybrid orbitals are 109.5°, the hybrid orbitals are not oriented along the bond directions. Their overlap is therefore poor, and the C–C bonds are correspondingly weak.

14.150 2 monomer ⇄ dimer
(a) In benzene, $K_c = 1.51 \times 10^2$

	2 monomer	⇄	dimer
initial (M)	0.100		0
change (M)	−2x		+x
equil (M)	0.100 − 2x		x

$$K_c = \frac{[\text{dimer}]}{[\text{monomer}]^2} = 1.51 \times 10^2 = \frac{x}{(0.100 - 2x)^2}$$

$604x^2 - 61.4x + 1.51 = 0$

Use the quadratic formula to solve for x.

$$x = \frac{-(-61.4) \pm \sqrt{(-61.4)^2 - (4)(604)(1.51)}}{2(604)} = \frac{61.4 \pm 11.04}{1208}$$

x = 0.0600 and 0.0417

Discard the larger solution (0.0600) because it gives a negative concentration of the monomer and that is impossible.

[monomer] = 0.100 − 2x = 0.100 − 2(0.0417) = 0.017 M; [dimer] = x = 0.0417 M

$$\frac{[\text{dimer}]}{[\text{monomer}]} = \frac{0.0417\ \text{M}}{0.017\ \text{M}} = 2.5$$

(b) In H_2O, $K_c = 3.7 \times 10^{-2}$

	2 monomer	⇄	dimer
initial (M)	0.100		0
change (M)	−2x		+x
equil (M)	0.100 − 2x		x

$$K_c = \frac{[\text{dimer}]}{[\text{monomer}]^2} = 3.7 \times 10^{-2} = \frac{x}{(0.100 - 2x)^2}$$

$0.148x^2 - 1.0148x + 0.000\,37 = 0$

Use the quadratic formula to solve for x.

$$x = \frac{-(-1.0148) \pm \sqrt{(-1.0148)^2 - (4)(0.148)(0.00037)}}{2(0.148)} = \frac{1.0148 \pm 1.0147}{0.296}$$

x = 6.86 and 3.4×10^{-4}

Discard the larger solution (6.86) because it gives a negative concentration of the monomer and that is impossible.

[monomer] = 0.100 − 2x = 0.100 − 2(3.4×10^{-4}) = 0.099 M; [dimer] = x = 3.4×10^{-4} M

$$\frac{[\text{dimer}]}{[\text{monomer}]} = \frac{3.4 \times 10^{-4}\ \text{M}}{0.099\ \text{M}} = 0.0034$$

(c) K_c for the water solution is so much smaller than K_c for the benzene solution because H_2O can hydrogen bond with acetic acid, thus preventing acetic acid dimer formation. Benzene cannot hydrogen bond with acetic acid.

14.151 (a) H_2O, 18.015; $125.4 \text{ g } H_2O \times \dfrac{1 \text{ mol } H_2O}{18.015 \text{ g } H_2O} = 6.96 \text{ mol } H_2O$

Given that mol CO = mol H_2O = 6.96 mol

$$P_{CO} = P_{H_2O} = \frac{nRT}{V} = \frac{(6.96 \text{ mol})\left(0.082\,06 \dfrac{L\cdot atm}{K\cdot mol}\right)(700 \text{ K})}{10.0 \text{ L}} = 40.0 \text{ atm}$$

	$CO(g)$	+	$H_2O(g)$	⇄	$CO_2(g)$	+	$H_2(g)$
initial (atm)	40.0		40.0		0		0
equil (atm)	9.80		9.80		40.0 – 9.80 = 30.2		40.0 – 9.80 = 30.2

$$K_p = \frac{(P_{CO_2})(P_{H_2})}{(P_{CO})(P_{H_2O})} = \frac{(30.2)(30.2)}{(9.80)(9.80)} = 9.50$$

(b) $31.4 \text{ g } H_2O \times \dfrac{1 \text{ mol } H_2O}{18.015 \text{ g } H_2O} = 1.743 \text{ mol } H_2O$

$$P_{H_2O} = \frac{nRT}{V} = \frac{(1.743 \text{ mol})\left(0.082\,06 \dfrac{L\cdot atm}{K\cdot mol}\right)(700 \text{ K})}{10.0 \text{ L}} = 10.0 \text{ atm}$$

P_{H_2O} has been increased by 10.0 atm; a new equilibrium will be established.

	$CO(g)$	+	$H_2O(g)$	⇄	$CO_2(g)$	+	$H_2(g)$
initial (atm)	9.80		9.80 +10.0		30.2		30.2
change (atm)	–x		–x		+x		+x
equil (atm)	9.80 – x		19.8 – x		30.2 + x		30.2 + x

$$K_p = \frac{(P_{CO_2})(P_{H_2})}{(P_{CO})(P_{H_2O})} = 9.50 = \frac{(30.2 + x)(30.2 + x)}{(9.80 - x)(19.80 - x)}$$

$8.50x^2 - 341.6x + 931.34 = 0$

Use the quadratic formula to solve for x.

$$x = \frac{-(-341.6) \pm \sqrt{(-341.6)^2 - 4(8.50)(931.34)}}{2(8.50)} = \frac{341.6 \pm 291.6}{17.0}$$

x = 37.25 and 2.94

Discard the larger solution (37.25) because it leads to negative partial pressures and that is impossible.

$P_{CO} = 9.80 - x = 9.80 - 2.94 = 6.86$ atm

$P_{H_2O} = 19.8 - x = 19.8 - 2.94 = 16.9$ atm

$P_{CO_2} = P_{H_2} = 30.2 + x = 30.2 + 2.94 = 33.1$ atm

$$n_{H_2} = \frac{PV}{RT} = \frac{(33.1\ \text{atm})(10.0\ \text{L})}{\left(0.082\ 06\ \frac{\text{L}\cdot\text{atm}}{\text{K}\cdot\text{mol}}\right)(700\ \text{K})} = 5.76\ \text{mol}\ H_2$$

$$\frac{5.76\ \text{mol}\ H_2}{10.0\ \text{L}} \times \frac{1\ \text{L}}{1000\ \text{mL}} \times \frac{1\ \text{mL}}{1\ \text{cm}^3} \times \frac{6.022 \times 10^{23}\ H_2\ \text{molecules}}{1\ \text{mol}\ H_2} = 3.47 \times 10^{20}\ H_2\ \text{molecules/cm}^3$$

14.152 (a) CO_2, 44.01; CO, 28.01

$$79.2\ \text{g}\ CO_2 \times \frac{1\ \text{mol}\ CO_2}{44.01\ \text{g}\ CO_2} = 1.80\ \text{mol}\ CO_2$$

	$CO_2(g)$	+	C(s)	⇄	2 CO(g)
initial (mol)	1.80				0
change (mol)	–x				+2x
equil (mol)	1.80 – x				2x

total mass of gas in flask = (16.3 g/L)(5.00 L) = 81.5 g

81.5 = (1.80 – x)(44.01) + (2x)(28.01)

81.5 = 79.22 – 44.01x + 56.02x; 2.28 = 12.01x; x = 2.28/12.01 = 0.19

n_{CO_2} = 1.80 – x = 1.80 – 0.19 = 1.61 mol CO_2; n_{CO} = 2x = 2(0.19) = 0.38 mol CO

$$P_{CO_2} = \frac{nRT}{V} = \frac{(1.61\ \text{mol})\left(0.082\ 06\ \frac{\text{L}\cdot\text{atm}}{\text{K}\cdot\text{mol}}\right)(1000\ \text{K})}{5.0\ \text{L}} = 26.4\ \text{atm}$$

$$P_{CO} = \frac{nRT}{V} = \frac{(0.38\ \text{mol})\left(0.082\ 06\ \frac{\text{L}\cdot\text{atm}}{\text{K}\cdot\text{mol}}\right)(1000\ \text{K})}{5.0\ \text{L}} = 6.24\ \text{atm}$$

$$K_p = \frac{(P_{CO})^2}{(P_{CO_2})} = \frac{(6.24)^2}{(26.4)} = 1.47$$

(b) At 1100K, the total mass of gas in flask = (16.9 g/L)(5.00 L) = 84.5 g

84.5 = (1.80 – x)(44.01) + (2x)(28.01)

84.5 = 79.22 – 44.01x + 56.02x; 5.28 = 12.01x; x = 5.28/12.01 = 0.44

n_{CO_2} = 1.80 – x = 1.80 – 0.44 = 1.36 mol CO_2; n_{CO} = 2x = 2(0.44) = 0.88 mol CO

$$P_{CO_2} = \frac{nRT}{V} = \frac{(1.36\ \text{mol})\left(0.082\ 06\ \frac{\text{L}\cdot\text{atm}}{\text{K}\cdot\text{mol}}\right)(1100\ \text{K})}{5.0\ \text{L}} = 24.6\ \text{atm}$$

$$P_{CO} = \frac{nRT}{V} = \frac{(0.88\ \text{mol})\left(0.082\ 06\ \frac{\text{L}\cdot\text{atm}}{\text{K}\cdot\text{mol}}\right)(1100\ \text{K})}{5.0\ \text{L}} = 15.9\ \text{atm}$$

$$K_p = \frac{(P_{CO})^2}{(P_{CO_2})} = \frac{(15.9)^2}{(24.6)} = 10.3$$

(c) In agreement with Le Châtelier's principle, the reaction is endothermic because K_p increases with increasing temperature.

14.153 CO_2, 44.01; CO, 28.01; $BaCO_3$, 197.34

$$1.77 \text{ g } CO_2 \times \frac{1 \text{ mol } CO_2}{44.01 \text{ g } CO_2} = 0.0402 \text{ mol } CO_2$$

	$CO_2(g)$ + $C(s)$	⇄ $2\,CO(g)$
initial (mol)	0.0402	0
change (mol)	–x	+2x
equil (mol)	0.0402 – x	2x

$$3.41 \text{ g } BaCO_3 \times \frac{1 \text{ mol } BaCO_3}{197.34 \text{ g } BaCO_3} \times \frac{1 \text{ mol } CO_2}{1 \text{ mol } BaCO_3} = 0.0173 \text{ mol } CO_2$$

mol CO_2 = 0.0173 = 0.0402 – x; x = 0.0229

mol CO = 2x = 2(0.0229) = 0.0458 mol CO

$$P_{CO_2} = \frac{nRT}{V} = \frac{(0.0173 \text{ mol})\left(0.082\,06 \frac{L\cdot atm}{K\cdot mol}\right)(1100 \text{ K})}{1.000 \text{ L}} = 1.562 \text{ atm}$$

$$P_{CO} = \frac{nRT}{V} = \frac{(0.0458 \text{ mol})\left(0.082\,06 \frac{L\cdot atm}{K\cdot mol}\right)(1100 \text{ K})}{1.000 \text{ L}} = 4.134 \text{ atm}$$

$$K_p = \frac{(P_{CO})^2}{(P_{CO_2})} = \frac{(4.134)^2}{(1.562)} = 11.0$$

14.154 (a) N_2O_4, 92.01

$$14.58 \text{ g } N_2O_4 \times \frac{1 \text{ mol } N_2O_4}{92.01 \text{ g } N_2O_4} = 0.1585 \text{ mol } N_2O_4$$

$$PV = nRT \qquad P_{N_2O_4} = \frac{nRT}{V} = \frac{(0.1585 \text{ mol})\left(0.082\,06 \frac{L\cdot atm}{K\cdot mol}\right)(400 \text{ K})}{1.000 \text{ L}} = 5.20 \text{ atm}$$

	$N_2O_4(g)$	⇄ $2\,NO_2(g)$
initial (atm)	5.20	0
change (atm)	–x	+2x
equil (atm)	5.20 – x	2x

$P_{total} = P_{N_2O_4} + P_{NO_2} = (5.20 - x) + (2x) = 9.15$ atm

5.20 + x = 9.15 atm

x = 3.95 atm

$P_{N_2O_4} = 5.20 - x = 5.20 - 3.95 = 1.25$ atm

$P_{NO_2} = 2x = 2(3.95) = 7.90$ atm

$$K_p = \frac{(P_{NO_2})^2}{(P_{N_2O_4})} = \frac{(7.90)^2}{(1.25)} = 49.9$$

$\Delta n = 1$ and $K_c = K_p\left(\frac{1}{RT}\right) = \frac{(49.9)}{(0.082\ 06)(400)} = 1.52$

(b) $\Delta H^\circ_{rxn} = [2\ \Delta H^\circ_f(NO_2)] - \Delta H^\circ_f(N_2O_4)$

$\Delta H^\circ_{rxn} = [(2\text{ mol})(33.2\text{ kJ/mol})] - [(1\text{mol})(11.1\text{ kJ/mol})] = 55.3\text{ kJ}$

moles N_2O_4 reacted $= n = \frac{PV}{RT} = \frac{(3.95\text{ atm})(1.000\text{ L})}{\left(0.082\ 06\ \frac{L\cdot atm}{K\cdot mol}\right)(400\text{ K})} = 0.1203\text{ mol } N_2O_4$

$q = (55.3\text{ kJ/mol } N_2O_4)(0.1203\text{ mol } N_2O_4) = 6.65\text{ kJ}$

14.155 $C_{10}H_8(s) \rightleftarrows C_{10}H_8(g)$

(a) $K_c = [C_{10}H_8] = 5.40 \times 10^{-6}$

room volume $= 8.0\text{ ft} \times 10.0\text{ ft} \times 8.0\text{ ft} = 640\text{ ft}^3$

room volume $= 640\text{ ft}^3 \times \left(\frac{12\text{ in.}}{1\text{ ft}}\right)^3 \times \left(\frac{2.54\text{ cm}}{1\text{ in.}}\right)^3 \times \left(\frac{1.0\text{ L}}{1000\text{ cm}^3}\right) = 18{,}122.8\text{ L}$

$C_{10}H_8$ molecules $= (5.40 \times 10^{-6}\text{ mol/L})(18{,}122.8\text{ L})(6.022 \times 10^{23}\text{ molecules/mol})$
$= 5.89 \times 10^{22}\ C_{10}H_8$ molecules

(b) $C_{10}H_8$, 128.17

mol $C_{10}H_8 = (5.40 \times 10^{-6}\text{ mol/L})(18{,}122.8\text{ L}) = 0.0979\text{ mol } C_{10}H_8$

mass $C_{10}H_8 = 0.0979\text{ mol } C_{10}H_8 \times \frac{128.17\text{ g } C_{10}H_8}{1\text{ mol } C_{10}H_8} = 12.55\text{ g } C_{10}H_8$

moth ball: $r = 12.0\text{ mm}/2 = 6.0\text{ mm} = 0.60\text{ cm}$

volume of moth ball $= (4/3)\pi r^3 = (4/3)\pi(0.60\text{cm})^3 = 0.905\text{ cm}^3$

mass of moth ball $= (0.905\text{ cm}^3/\text{moth ball})(1.16\text{ g/cm}^3) = 1.05\text{ g/moth ball}$

number of moth balls $= \frac{12.55\text{ g } C_{10}H_8}{1.05\text{ g } C_{10}H_8/\text{moth ball}} = 12$ moth balls

14.156 The atmosphere is 21% (0.21) O_2; $P_{O_2} = (0.21)\left(720\text{ mm Hg} \times \frac{1\text{ atm}}{760\text{ mm Hg}}\right) = 0.199\text{ atm}$

$2\ O_3(g) \rightleftarrows 3\ O_2(g)$

$K_p = \frac{(P_{O_2})^3}{(P_{O_3})^2}$; $\quad P_{O_3} = \sqrt{\frac{(P_{O_2})^3}{K_p}} = \sqrt{\frac{(0.199)^3}{1.3 \times 10^{57}}} = 2.46 \times 10^{-30}\text{ atm}$

vol $= 10 \times 10^6\text{ m}^3 \times \left(\frac{100\text{ cm}}{1\text{ m}}\right)^3 \times \frac{1\text{ L}}{1000\text{ cm}^3} = 1.0 \times 10^{10}\text{ L}$

$n_{O_3} = \frac{PV}{RT} = \frac{(2.46 \times 10^{-30}\text{ atm})(1.0 \times 10^{10}\text{ L})}{\left(0.082\ 06\ \frac{L\cdot atm}{K\cdot mol}\right)(298\text{ K})} = 1.0 \times 10^{-21}\text{ mol } O_3$

O_3 molecules $= 1.0 \times 10^{-21}\text{ mol } O_3 \times \frac{6.022 \times 10^{23}\ O_3\text{ molecules}}{1\text{ mol } O_3} = 6.0 \times 10^2\ O_3$ molecules

14.157 250.0 mL = 0.2500 L

$[CH_3CO_2H]$ = 0.0300 mol/0.2500 L = 0.120 M

	$2\ CH_3CO_2H$ ⇄	$(CH_3CO_2H)_2$
initial (M)	0.120	0
change (M)	–2x	+x
equil (M)	0.120 – 2x	x

$$K_c = \frac{[\text{dimer}]}{[\text{monomer}]^2} = 1.51 \times 10^2 = \frac{x}{(0.120 - 2x)^2}$$

$604x^2 - 73.48x + 2.1744 = 0$

Use the quadratic formula to solve for x.

$$x = \frac{-(-73.48) \pm \sqrt{(-73.48)^2 - 4(604)(2.1744)}}{2(604)} = \frac{73.48 \pm 12.08}{1208}$$

x = 0.0708 and 0.0508

Discard the larger solution (0.0708) because it leads to a negative concentration and that is impossible.

[monomer] = 0.120 – 2x = 0.120 – 2(0.0508) = 0.0184 M

[dimer] = x = 0.0508 M

(b) 25 °C = 25 + 273 = 298 K

$$\Pi = MRT = (0.0184\text{ M} + 0.0508\text{ M})\left(0.082\ 06\ \frac{\text{L}\cdot\text{atm}}{\text{K}\cdot\text{mol}}\right)(298\text{K}) = 1.69\text{ atm}$$

14.158 $PCl_5(g) \rightleftarrows PCl_3(g) + Cl_2(g)$

$\Delta n = (2) - (1) = 1$ and at 700 K, $K_p = K_c(RT) = (46.9)(0.082\ 06)(700) = 2694$

(a) Because K_p is larger at the higher temperature, the reaction has shifted toward products at the higher temperature, which means the reaction is endothermic. Because the reaction involves breaking two P–Cl bonds and forming just one Cl–Cl bond, it should be endothermic.

(b) PCl_5, 208.24

$$\text{mol } PCl_5 = 1.25\text{ g } PCl_5 \times \frac{1\text{ mol } PCl_5}{208.24\text{ g } PCl_5} = 6.00 \times 10^{-3}\text{ mol}$$

$$PV = nRT,\quad P_{PCl_5} = \frac{nRT}{V} = \frac{(6.00 \times 10^{-3}\text{ mol})\left(0.082\ 06\ \frac{\text{L}\cdot\text{atm}}{\text{K}\cdot\text{mol}}\right)(700\text{ K})}{0.500\text{ L}} = 0.689\text{ atm}$$

Because K_p is so large, first assume the reaction goes to completion and then allow for a small back reaction.

	$PCl_5(g)$ ⇄	$PCl_3(g)$ +	$Cl_2(g)$
before rxn (atm)	0.689	0	0
100% rxn (atm)	–0.689	+0.689	+0.689
after rxn (atm)	0	0.689	0.689
back rxn (atm)	+x	–x	–x
equil (atm)	x	0.689 – x	0.689 – x

$$K_p = \frac{(P_{PCl_3})(P_{Cl_2})}{P_{PCl_5}} = 2694 = \frac{(0.689 - x)^2}{x} \approx \frac{(0.689)^2}{x}$$

$$x = P_{PCl_5} = \frac{(0.689)^2}{2694} = 1.76 \text{ x } 10^{-4} \text{ atm}$$

$$P_{total} = P_{PCl_5} + P_{PCl_3} + P_{Cl_2}$$

$$P_{total} = x + (0.689 - x) + (0.689 - x) = 0.689 + 0.689 - 1.76 \text{ x } 10^{-4} = 1.38 \text{ atm}$$

$$\% \text{ dissociation} = \frac{(P_{PCl_5})_o - (P_{PCl_5})}{(P_{PCl_5})_o} \text{ x } 100\% = \frac{0.689 - (1.76 \text{ x } 10^{-4})}{0.689} \text{ x } 100\% = 99.97\%$$

(c) [Lewis structure of PCl_5]

The molecular geometry is trigonal bipyramidal. There is no dipole moment because of a symmetrical distribution of Cl's around the central P.

[Lewis structure of PCl_3]

The molecular geometry is trigonal pyramidal. There is a dipole moment because of the lone pair of electrons on the P and an unsymmetrical distribution of Cl's around the central P.

14.159 (a) Hydrogen bonding

(b) [Structure: $H_3C-CH_2-CH_2-O-H\cdots O=C(OCH_3)-C(CH_3)=CH_2$]

(c)

	PrOH	+	MMA	⇄	PrOH · MMA
initial (M)	0.100		0.0500		0
change (M)	−x		−x		+x
equil (M)	0.100 − x		0.0500 − x		x

$$K_c = \frac{[PrOH \cdot MMA]}{[PrOH][MMA]} = 0.701 = \frac{x}{(0.100 - x)(0.0500 - x)}$$

$0.701x^2 - 1.105x + 0.0035 = 0$

Use the quadratic formula to solve for x.

$$x = \frac{-(-1.105) \pm \sqrt{(-1.105)^2 - 4(0.701)(0.0035)}}{2(0.701)} = \frac{1.105 \pm 1.101}{1.402}$$

x = 1.573 and 0.003
Discard the larger solution (1.573) because it leads to a negative concentrations and that is impossible.
[PrOH] = 0.100 – x = 0.100 – 0.003 = 0.097 M
[MMA] = 0.0500 – x = 0.0500 – 0.003 = 0.047 M
[PrOH · MMA] = x = 0.003 M

15 Aqueous Equilibria: Acids and Bases

15.1 (a) $H_2SO_4(aq) + H_2O(l) \rightleftharpoons H_3O^+(aq) + HSO_4^-(aq)$
conjugate base

(b) $HSO_4^-(aq) + H_2O(l) \rightleftharpoons H_3O^+(aq) + SO_4^{2-}(aq)$
conjugate base

(c) $H_3O^+(aq) + H_2O(l) \rightleftharpoons H_3O^+(aq) + H_2O(l)$
conjugate base

(d) $NH_4^+(aq) + H_2O(l) \rightleftharpoons H_3O^+(aq) + NH_3(aq)$
conjugate base

15.2 (a) $HCO_3^-(aq) + H_2O(l) \rightleftharpoons H_2CO_3(aq) + OH^-(aq)$
conjugate acid

(b) $CO_3^{2-}(aq) + H_2O(l) \rightleftharpoons HCO_3^-(aq) + OH^-(aq)$
conjugate acid

(c) $OH^-(aq) + H_2O(l) \rightleftharpoons H_2O(l) + OH^-(aq)$
conjugate acid

(d) $H_2PO_4^-(aq) + H_2O(l) \rightleftharpoons H_3PO_4(aq) + OH^-(aq)$
conjugate acid

15.3 $HCl(aq) + NH_3(aq) \rightleftharpoons NH_4^+(aq) + Cl^-(aq)$
acid base acid base
conjugate acid-base pairs

15.4 (a) $CH_3NH_2(aq) + H_2O(l) \rightleftharpoons OH^-(aq) + CH_3NH_3^+(aq)$
base acid base acid
conjugate acid-base pairs

(b) $HNO_3(aq) + H_2O(l) \rightleftharpoons H_3O^+(aq) + NO_3^-(aq)$
acid base acid base
conjugate acid-base pairs

15.5 (a) $HF(aq) + NO_3^-(aq) \rightleftharpoons HNO_3(aq) + F^-(aq)$
HNO_3 is a stronger acid than HF, and F^- is a stronger base than NO_3^- (see Table 15.1). Because proton transfer occurs from the stronger acid to the stronger base, the reaction proceeds from right to left.

(b) $NH_4^+(aq) + CO_3^{2-}(aq) \rightleftharpoons HCO_3^-(aq) + NH_3(aq)$
NH_4^+ is a stronger acid than HCO_3^-, and CO_3^{2-} is a stronger base than NH_3 (see Table 15.1). Because proton transfer occurs from the stronger acid to the stronger base, the reaction proceeds from left to right.

15.6 (a) Both HX and HY have the same initial concentration. HY is more dissociated than HX. Therefore, HY is the stronger acid.
(b) The conjugate base (X^-) of the weaker acid (HX) is the stronger base.
(c) $HX + Y^- \rightleftharpoons HY + X^-$; Proton transfer occurs from the stronger acid to the stronger base. The reaction proceeds to the left.

15.7 (a) H_2Se is a stronger acid than H_2S because Se is below S in the 6A group and the H–Se bond is weaker than the H–S bond.
(b) HI is a stronger acid than H_2Te because I is to the right of Te in the same row of the periodic table, I is more electronegative than Te, and the H–I bond is more polar.
(c) HNO_3 is a stronger acid than HNO_2 because acid strength increases with increasing oxidation number of N. The oxidation number for N is +5 in HNO_3 and +3 in HNO_2.
(d) H_2SO_3 is a stronger acid than H_2SeO_3 because acid strength increases with increasing electronegativity of the central atom. S is more electronegative than Se.

15.8 The weaker acid has the stronger conjugate base.
(a) H_2Se is a stronger acid than H_2S; $HS^- > HSe^-$
(b) HI is a stronger acid than H_2Te; $HTe^- > I^-$
(c) HNO_3 is a stronger acid than HNO_2; $NO_2^- > NO_3^-$
(d) H_2SO_3 is a stronger acid than H_2SeO_3; $HSeO_3^- > HSO_3^-$

15.9 $$[OH^-] = \frac{K_w}{[H_3O^+]} = \frac{1.0 \times 10^{-14}}{1.4 \times 10^{-4}} = 7.1 \times 10^{-11}\ M$$
Because $[H_3O^+] > [OH^-]$, the solution is acidic.

15.10 $K_w = [H_3O^+][OH^-]$; In a neutral solution, $[H_3O^+] = [OH^-]$
At 50 °C, $[H_3O^+] = [OH^-] = \sqrt{K_w} = \sqrt{5.5 \times 10^{-14}} = 2.3 \times 10^{-7}\ M$

15.11 (a) $$[H_3O^+] = \frac{K_w}{[OH^-]} = \frac{1.0 \times 10^{-14}}{1.58 \times 10^{-6}} = 6.3 \times 10^{-9}\ M$$
$pH = -\log[H_3O^+] = -\log(6.3 \times 10^{-9}) = 8.20$
(b) $pH = -\log[H_3O^+] = -\log(6.0 \times 10^{-5}) = 4.22$

15.12 $pH = -\log[H_3O^+] = -\log(6.3) = -0.80$

15.13 (a) $[H_3O^+] = 10^{-pH} = 10^{-7.40} = 4.0 \times 10^{-8}\ M$
$$[OH^-] = \frac{K_w}{[H_3O^+]} = \frac{1.0 \times 10^{-14}}{4.0 \times 10^{-8}} = 2.5 \times 10^{-7}\ M$$

(b) $[H_3O^+] = 10^{-pH} = 10^{-2.8} = 2 \times 10^{-3}$ M

$$[OH^-] = \frac{K_w}{[H_3O^+]} = \frac{1.0 \times 10^{-14}}{2 \times 10^{-3}} = 5 \times 10^{-12} \text{ M}$$

15.14 milk: $[H_3O^+] = 10^{-pH} = 10^{-6.6} = 2.5 \times 10^{-7}$ M

black coffee: $[H_3O^+] = 10^{-pH} = 10^{-5.0} = 1.0 \times 10^{-5}$ M

$$\frac{[H_3O^+]_{\text{black coffee}}}{[H_3O^+]_{\text{milk}}} = \frac{1.0 \times 10^{-5}}{2.5 \times 10^{-7}} = 40$$

The $[H_3O^+]$ in coffee is 40 times greater.

15.15 (a) Because $HClO_4$ is a strong acid, $[H_3O^+] = 0.050$ M.

$pH = -\log[H_3O^+] = -\log(0.050) = 1.30$

(b) Because HCl is a strong acid, $[H_3O^+] = 6.0$ M.

$pH = -\log[H_3O^+] = -\log(6.0) = -0.78$

(c) Because KOH is a strong base, $[OH^-] = 4.0$ M.

$$[H_3O^+] = \frac{K_w}{[OH^-]} = \frac{1.0 \times 10^{-14}}{4.0} = 2.5 \times 10^{-15} \text{ M}$$

$pH = -\log[H_3O^+] = -\log(2.5 \times 10^{-15}) = 14.60$

(d) Because $Ba(OH)_2$ is a strong base, $[OH^-] = 2(0.010 \text{ M}) = 0.020$ M.

$$[H_3O^+] = \frac{K_w}{[OH^-]} = \frac{1.0 \times 10^{-14}}{0.020} = 5.0 \times 10^{-13} \text{ M}$$

$pH = -\log[H_3O^+] = -\log(5.0 \times 10^{-13}) = 12.30$

15.16 $CaO(s) + H_2O(l) \rightarrow Ca(OH)_2(aq)$; CaO, 56.08

$$0.25 \text{ g CaO} \times \frac{1 \text{ mol CaO}}{56.08 \text{ g CaO}} \times \frac{1 \text{ mol } Ca(OH)_2}{1 \text{ mol CaO}} \times \frac{2 \text{ mol } OH^-}{1 \text{ mol } Ca(OH)_2} = 8.92 \times 10^{-3} \text{ mol } OH^-$$

$$[OH^-] = \frac{8.92 \times 10^{-3} \text{ mol } OH^-}{1.50 \text{ L}} = 5.94 \times 10^{-3} \text{ M}$$

$$[H_3O^+] = \frac{K_w}{[OH^-]} = \frac{1.0 \times 10^{-14}}{5.94 \times 10^{-3}} = 1.68 \times 10^{-12} \text{ M}$$

$pH = -\log[H_3O^+] = -\log(1.68 \times 10^{-12}) = 11.77$

15.17

	$HOCl(aq) + H_2O(l)$	$\rightleftarrows$	$H_3O^+(aq)$	$+ OCl^-(aq)$
initial (M)	0.10		~0	0
change (M)	–x		+x	+x
equil (M)	0.10 – x		x	x

$x = [H_3O^+] = 10^{-pH} = 10^{-4.23} = 5.9 \times 10^{-5}$ M

$[OCl^-] = x = 5.9 \times 10^{-5}$ M; $[HOCl] = 0.10 - x = (0.10 - 5.9 \times 10^{-5})$ M

$$K_a = \frac{[H_3O^+][OCl^-]}{[HOCl]} = \frac{(5.9 \times 10^{-5})(5.9 \times 10^{-5})}{(0.10 - 5.9 \times 10^{-5})} = 3.5 \times 10^{-8}$$

This value of K_a agrees with the value in Table 15.2.

15.18 (a) HZ is completely dissociated. HX and HY are at the same concentration and HX is more dissociated than HY. The strongest acid is HZ, the weakest is HY.
K_a (HY) < K_a (HX) < K_a (HZ)
(b) HZ
(c) HY has the highest pH; HX has the lowest pH (highest $[H_3O^+]$).

15.19 (a)

	$CH_3CO_2H(aq) + H_2O(l)$	$\rightleftarrows$	$H_3O^+(aq)$	$+ CH_3CO_2^-(aq)$
initial (M)	1.00		~0	0
change (M)	−x		+x	+x
equil (M)	1.00 − x		x	x

$$K_a = \frac{[H_3O^+][CH_3CO_2^-]}{[CH_3CO_2H]} = 1.8 \times 10^{-5} = \frac{x^2}{1.00 - x} \approx \frac{x^2}{1.00}$$

Solve for x. $x = [H_3O^+] = 4.2 \times 10^{-3}$ M
$pH = -\log[H_3O^+] = -\log(4.2 \times 10^{-3}) = 2.38$
$[CH_3CO_2^-] = x = 4.2 \times 10^{-3}$ M; $[CH_3CO_2H] = 1.00 - x = 1.00$ M

$$[OH^-] = \frac{K_w}{[H_3O^+]} = \frac{1.0 \times 10^{-14}}{4.2 \times 10^{-3}} = 2.4 \times 10^{-12} \text{ M}$$

(b)

	$CH_3CO_2H(aq) + H_2O(l)$	$\rightleftarrows$	$H_3O^+(aq)$	$+ CH_3CO_2^-(aq)$
initial (M)	0.0100		~0	0
change (M)	−x		+x	+x
equil (M)	0.0100 − x		x	x

$$K_a = \frac{[H_3O^+][CH_3CO_2^-]}{[CH_3CO_2H]} = 1.8 \times 10^{-5} = \frac{x^2}{0.0100 - x}$$

$x^2 + (1.8 \times 10^{-5})x - (1.8 \times 10^{-7}) = 0$
Use the quadratic formula to solve for x.

$$x = \frac{-(1.8 \times 10^{-5}) \pm \sqrt{(1.8 \times 10^{-5})^2 - 4(-1.8 \times 10^{-7})}}{2(1)} = \frac{(-1.8 \times 10^{-5}) \pm (8.5 \times 10^{-4})}{2}$$

$x = 4.2 \times 10^{-4}$ and -4.3×10^{-4}
Of the two solutions for x, only the positive value of x has physical meaning because x is the $[H_3O^+]$.
$x = [H_3O^+] = 4.2 \times 10^{-4}$ M
$pH = -\log[H_3O^+] = -\log(4.2 \times 10^{-4}) = 3.38$
$[CH_3CO_2^-] = x = 4.2 \times 10^{-4}$ M
$[CH_3CO_2H] = 0.0100 - x = 0.0100 - (4.2 \times 10^{-4}) = 0.0096$ M

$$[OH^-] = \frac{K_w}{[H_3O^+]} = \frac{1.0 \times 10^{-14}}{4.2 \times 10^{-4}} = 2.4 \times 10^{-11} \text{ M}$$

15.20 $[H_3O^+] = 10^{-pH} = 10^{-2.00} = 0.010$ M

	$HCO_2H(aq) + H_2O(l)$	$\rightleftarrows$	$H_3O^+(aq)$	$+ HCO_2^-(aq)$	$Ka = 1.8 \times 10^{-4}$
(M)	x − 0.010		0.010	0.010	

$$K_a = \frac{[H_3O^+][HCO_2^-]}{[HCO_2H]} = 1.8 \times 10^{-4} = \frac{(0.010)^2}{(x - 0.010)}$$

$1.8 \times 10^{-4}x - 1.8 \times 10^{-6} = 1.0 \times 10^{-4}$

$x = (1.0 \times 10^{-4} + 1.8 \times 10^{-6})/(1.8 \times 10^{-4}) = 0.57$ M

15.21

	$H_2SO_3(aq) + H_2O(l)$	$\rightleftarrows$	$H_3O^+(aq)$	+ $HSO_3^-(aq)$
initial (M)	0.10		~0	0
change (M)	–x		+x	+x
equil (M)	0.10 – x		x	x

$$K_{a1} = \frac{[H_3O^+][HSO_3^-]}{[H_2SO_3]} = 1.5 \times 10^{-2} = \frac{x^2}{0.10 - x}$$

$x^2 + 0.015x - 0.0015 = 0$

Use the quadratic formula to solve for x.

$$x = \frac{-(0.015) \pm \sqrt{(0.015)^2 - (4)(-0.0015)}}{2(1)} = \frac{-0.015 \pm 0.079}{2}$$

x = 0.032 and –0.047

Of the two solutions for x, only the positive value of x has physical meaning since x is the $[H_3O^+]$.

$x = [H_3O^+] = [HSO_3^-] = 0.032$ M; $[H_2SO_3] = 0.10 - x = 0.10 - 0.032 = 0.07$ M

The second dissociation of H_2SO_3 produces a negligible amount of H_3O^+ compared with that from the first dissociation.

$HSO_3^-(aq) + H_2O(l) \rightleftarrows H_3O^+(aq) + SO_3^{2-}(aq)$

$$K_{a2} = \frac{[H_3O^+][SO_3^{2-}]}{[HSO_3^-]} = 6.3 \times 10^{-8} = \frac{(0.032)[SO_3^{2-}]}{(0.032)}$$

$[SO_3^{2-}] = K_{a2} = 6.3 \times 10^{-8}$ M

$$[OH^-] = \frac{K_w}{[H_3O^+]} = \frac{1.0 \times 10^{-14}}{0.032} = 3.1 \times 10^{-13} \text{ M}$$

$pH = -\log[H_3O^+] = -\log(0.032) = 1.49$

15.22 Solubility $= k \cdot P = [3.2 \times 10^{-2}$ mol/(L · atm)](4.5 atm) = 0.14 M

	$H_2CO_3(aq) + H_2O(l)$	$\rightleftarrows$	$H_3O^+(aq)$	+ $HCO_3^-(aq)$
initial (M)	0.14		~0	0
change (M)	–x		+x	+x
equil (M)	0.14 – x		x	x

$$K_{a1} = \frac{[H_3O^+][HCO_3^-]}{[H_2CO_3]} = 4.3 \times 10^{-7} = \frac{x^2}{0.14 - x} \approx \frac{x^2}{0.14}$$

Solve for x. $x = 2.5 \times 10^{-4}$

$[H_3O^+] = [HCO_3^-] = x = 2.5 \times 10^{-4}$ M; $[H_2CO_3] = 0.14 - x = 0.14$ M

The second dissociation of H_2CO_3 produces a negligible amount of H_3O^+ compared with that from the first dissociation.

$HCO_3^-(aq) + H_2O(l) \rightleftarrows H_3O^+(aq) + CO_3^{2-}(aq)$

$$K_{a2} = \frac{[H_3O^+][CO_3^{2-}]}{[HCO_3^-]} = 5.6 \times 10^{-11} = \frac{(2.5 \times 10^{-4})[CO_3^{2-}]}{(2.5 \times 10^{-4})}$$

$[CO_3^{2-}] = K_{a2} = 5.6 \times 10^{-11}$ M

$$[OH^-] = \frac{K_w}{[H_3O^+]} = \frac{1.0 \times 10^{-14}}{2.5 \times 10^{-4}} = 4.0 \times 10^{-11} \text{ M}$$

$pH = -\log[H_3O^+] = -\log(2.5 \times 10^{-4}) = 3.60$

15.23

	$NH_3(aq)$ + $H_2O(l)$	$\rightleftarrows$	$NH_4^+(aq)$	+ $OH^-(aq)$
initial (M)	0.40		0	~0
change (M)	–x		+x	+x
equil (M)	0.40 – x		x	x

$$K_b = \frac{[NH_4^+][OH^-]}{[NH_3]} = 1.8 \times 10^{-5} = \frac{x^2}{0.40 - x} \approx \frac{x^2}{0.40}$$

Solve for x. $x = [OH^-] = 2.7 \times 10^{-3}$ M

$[NH_4^+] = x = 2.7 \times 10^{-3}$ M; $[NH_3] = 0.40 - x = 0.40$ M

$$[H_3O^+] = \frac{K_w}{[OH^-]} = \frac{1.0 \times 10^{-14}}{2.7 \times 10^{-3}} = 3.7 \times 10^{-12} \text{ M}$$

$pH = -\log[H_3O^+] = -\log(3.7 \times 10^{-12}) = 11.43$

15.24 $[H_3O^+] = 10^{-pH} = 10^{-8.15} = 7.1 \times 10^{-9}$ M

$$[OH^-] = \frac{K_w}{[H_3O^+]} = \frac{1.0 \times 10^{-14}}{7.1 \times 10^{-9}} = 1.4 \times 10^{-6} \text{ M}$$

	$C_3H_5O_3^-(aq)$ + $H_2O(l)$	$\rightleftarrows$	$C_3H_5O_3H(aq)$	+ $OH^-(aq)$
(M)	$0.028 - 1.4 \times 10^{-6}$		1.4×10^{-6}	1.4×10^{-6}

$$K_b = \frac{[C_3H_5O_3H][OH^-]}{[C_3H_5O_3^-]} = \frac{(1.4 \times 10^{-6})^2}{(0.028 - 1.4 \times 10^{-6})} \approx \frac{(1.4 \times 10^{-6})^2}{0.028} = 7.0 \times 10^{-11}$$

15.25 (a) $$K_a = \frac{K_w}{K_b \text{ for } C_5H_{11}N} = \frac{1.0 \times 10^{-14}}{1.3 \times 10^{-3}} = 7.7 \times 10^{-12}$$

(b) $$K_b = \frac{K_w}{K_a \text{ for HOCl}} = \frac{1.0 \times 10^{-14}}{3.5 \times 10^{-8}} = 2.9 \times 10^{-7}$$

(c) pK_b for $HCO_2^- = 14.00 - pK_a = 14.00 - 3.74 = 10.26$

15.26 (a) The strongest acid has the largest K_a. HX is the strongest acid because it is the most dissociated. Therefore, HX has the largest K_a.

(b) The weakest acid has the strongest conjugate base. HY is the weakest acid because it is the least dissociated. Therefore, Y^- has the largest K_b.
(c) The strongest acid has the weakest conjugate base. HX is the strongest acid because it is the most dissociated. Therefore, X^- is the weakest base.

15.27 (a) 0.25 M NH_4Br

NH_4^+ is an acidic cation. Br^- is a neutral anion. The salt solution is acidic.

For NH_4^+, $K_a = \dfrac{K_w}{K_b \text{ for } NH_3} = \dfrac{1.0 \times 10^{-14}}{1.8 \times 10^{-5}} = 5.6 \times 10^{-10}$

	$NH_4^+(aq)$ + $H_2O(l)$	⇄	$H_3O^+(aq)$	+ $NH_3(aq)$
initial (M)	0.25		~0	0
change (M)	−x		+x	+x
equil (M)	0.25 − x		x	x

$$K_a = \frac{[H_3O^+][NH_3]}{[NH_4^+]} = 5.6 \times 10^{-10} = \frac{x^2}{0.25 - x} \approx \frac{x^2}{0.25}$$

Solve for x. $x = [H_3O^+] = 1.2 \times 10^{-5}$ M

$pH = -\log[H_3O^+] = -\log(1.2 \times 10^{-5}) = 4.92$

(b) 0.20 M $NaNO_2$

For NO_2^-, $K_b = \dfrac{K_w}{K_a \text{ for } HNO_2} = \dfrac{1.0 \times 10^{-14}}{4.6 \times 10^{-4}} = 2.2 \times 10^{-11}$

	$NO_2^-(aq)$ + $H_2O(l)$	⇄	$HNO_2(aq)$	+ $OH^-(aq)$
initial (M)	0.20		0	~0
change (M)	−x		+x	+x
equil (M)	0.20 − x		x	x

$$K_b = \frac{[HNO_2][OH^-]}{[NO_2^-]} = 2.2 \times 10^{-11} = \frac{x^2}{0.20 - x} \approx \frac{x^2}{0.20}$$

Solve for x. $x = [OH^-] = 2.1 \times 10^{-6}$ M

$[H_3O^+] = \dfrac{K_w}{[OH^-]} = \dfrac{1.0 \times 10^{-14}}{2.1 \times 10^{-6}} = 4.8 \times 10^{-9}$ M

$pH = -\log[H_3O^+] = -\log(4.8 \times 10^{-9}) = 8.32$

15.28 (a) Zn^{2+} is an acidic cation. Cl^- is a neutral anion. The salt solution is acidic.

	$Zn(H_2O)_6^{2+}(aq)$ + $H_2O(l)$	⇄	$H_3O^+(aq)$	+ $Zn(H_2O)_5(OH)^+(aq)$
initial (M)	0.40		~0	0
change (M)	−x		+x	+x
equil(M)	0.40 − x		x	x

$$K_a = \frac{[H_3O^+][Zn(H_2O)_5(OH)^+]}{[Zn(H_2O)_6^{2+}]} = 2.5 \times 10^{-10} = \frac{x^2}{0.40 - x} \approx \frac{x^2}{0.40}$$

Solve for x. $x = [H_3O^+] = 1.0 \times 10^{-5}$ M

$pH = -\log[H_3O^+] = -\log(1.0 \times 10^{-5}) = 5.00$

$$\% \text{ dissociation} = \frac{[Zn(H_2O)_6{}^{2+}]_{diss}}{[Zn(H_2O)_6{}^{2+}]_{initial}} \times 100\% = \frac{1.0 \times 10^{-5}\text{ M}}{0.40\text{ M}} \times 100\% = 2.5 \times 10^{-3}\,\%$$

(b) $Fe(H_2O)_6{}^{3+}$ is a stronger acid than $Zn(H_2O)_6{}^{2+}$ because of the higher 3+ charge. The stronger acid, $Fe(H_2O)_6{}^{3+}$, has the higher percent dissociation.

15.29 (a) Lewis acid, $AlCl_3$; Lewis base, Cl^-
(b) Lewis acid, Ag^+; Lewis base, NH_3
(c) Lewis acid, SO_2; Lewis base, OH^-
(d) Lewis acid, Cr^{3+}; Lewis base, H_2O

15.30

$$:\ddot{\underset{..}{Cl}}—Be—\ddot{\underset{..}{Cl}}: \; + \; 2\,:\ddot{\underset{..}{Cl}}:^- \longrightarrow \left[\begin{array}{c} :\ddot{Cl}: \\ | \\ :\ddot{\underset{..}{Cl}}—Be—\ddot{\underset{..}{Cl}}: \\ | \\ :\underset{..}{Cl}: \end{array}\right]^{2-}$$

15.31 (a) $\Delta pH = 5.6 - 3.6 = 2.0$
A decrease of 2 pH units represents an increase in $[H_3O^+]$ by a factor of 100.
(b) $[H_3O^+] = 10^{-pH} = 10^{-5.10} = 7.9 \times 10^{-6}$ M

$$[OH^-] = \frac{K_w}{[H_3O^+]} = \frac{1.0 \times 10^{-14}}{7.9 \times 10^{-6}} = 1.3 \times 10^{-9}\text{ M}$$

15.32 (a) $P_{CO_2} = \left(\frac{750}{10^6} \cdot 1.0\text{ atm}\right) = 7.5 \times 10^{-4}$ atm

Solubility $= k \cdot P = [3.2 \times 10^{-2}\text{ mol/(L} \cdot \text{atm)}](7.5 \times 10^{-4}\text{ atm}) = 2.4 \times 10^{-5}$ M

(b)	$H_2CO_3(aq)$ + $H_2O(l)$	⇄	$H_3O^+(aq)$	+ $HCO_3^-(aq)$
initial (M)	2.4×10^{-5}		~0	0
change (M)	$-x$		$+x$	$+x$
equil (M)	$2.4 \times 10^{-5} - x$		x	x

$$K_{a1} = \frac{[H_3O^+][HCO_3^-]}{[H_2CO_3]} = 4.3 \times 10^{-7} = \frac{x^2}{2.4 \times 10^{-5} - x}$$

$x^2 + 4.3 \times 10^{-7}x - 1.0 \times 10^{-11} = 0$

Use the quadratic formula to solve for x.

$$x = \frac{-(4.3 \times 10^{-7}) \pm \sqrt{(4.3 \times 10^{-7})^2 - (4)(1)(-1.0 \times 10^{-11})}}{2(1)} = \frac{(-4.3 \times 10^{-7}) \pm (6.3 \times 10^{-6})}{2}$$

$x = -3.4 \times 10^{-6}$ and 3.0×10^{-6}

Of the two solutions for x, only the positive value of x has physical meaning because x is the $[H_3O^+]$.

$x = [H_3O^+] = 3.0 \times 10^{-6}$ M

The second dissociation of H_2CO_3 produces a negligible amount of H_3O^+ compared with that from the first dissociation.

$pH = -\log[H_3O^+] = -\log(3.0 \times 10^{-6}) = 5.52$

(c) The acidity of rain will increase but only slightly.

15.33 NO_2, 46.0

$$5.50 \text{ mg} \times \frac{1 \times 10^{-3} \text{ g}}{1 \text{ mg}} \times \frac{1 \text{ mol } NO_2}{46.0 \text{ g } NO_2} = 1.20 \times 10^{-4} \text{ mol } NO_2$$

$$4\,NO_2(g) + 2\,H_2O(l) + O_2(g) \rightarrow 4\,HNO_3(aq)$$

mol HNO_3 = mol NO_2 = 1.20×10^{-4} mol
$[HNO_3] = 1.20 \times 10^{-4}$ mol/1.00 L = 1.20×10^{-4} M
Because HNO_3 is a strong acid, $[H_3O^+] = [HNO_3] = 1.20 \times 10^{-4}$ M
$pH = -\log [H_3O^+] = -\log (1.20 \times 10^{-4}) = 3.921$

15.34 Lewis acids include not only H^+ but also other cations and neutral molecules having vacant valence orbitals that can accept a share in a pair of electrons donated by a Lewis base. The O^{2-} from CaO is the Lewis base and SO_2 is the Lewis acid.

$$:\ddot{O}-\ddot{S}-\ddot{O}: \; + \; :\ddot{O}:^{2-} \longrightarrow :SO_3^{2-}$$

15.35 For NH_4^+, $K_a = \dfrac{K_w}{K_b \text{ for } NH_3} = \dfrac{1.0 \times 10^{-14}}{1.8 \times 10^{-5}} = 5.6 \times 10^{-10}$

For SO_4^{2-}, $K_b = \dfrac{K_w}{K_a \text{ for } HSO_4^-} = \dfrac{1.0 \times 10^{-14}}{1.2 \times 10^{-2}} = 8.3 \times 10^{-13}$

NH_4^+ is acidic, NO_3^- is neutral. NH_4NO_3 is an acidic salt.
NH_4^+ is acidic, SO_4^{2-} is basic. $K_a > K_b$, $(NH_4)_2SO_4$ is an acidic salt.

Conceptual Problems

15.36 (a) acids, HCO_3^- and H_3O^+; bases, H_2O and CO_3^{2-}
(b) acids, HF and H_2CO_3; bases HCO_3^- and F^-

15.37 (a) X^-, Y^-, Z^- (b) HX < HZ < HY (c) HY (d) HX (e) (2/10) x 100% = 20%

15.38 (c) represents a solution of a weak diprotic acid, H_2A. Because K_{a2} is always less than K_{a1}, (a) and (d) represent impossible situations. (b) contains no H_2A.

15.39 For H_2SO_4, there is complete dissociation of only the first H^+. At equilibrium, there should be H_3O^+, HSO_4^-, and a small amount of SO_4^{2-}. This is best represented by (b).

15.40 (a) $Y^- < Z^- < X^-$
(b) The weakest base, Y^-, has the strongest conjugate acid.
(c) X^- is the strongest conjugate base and has the smallest pK_b.
(d) The numbers of HA molecules and OH^- ions are equal because the reaction of A^- with water has a 1:1 stoichiometry: $A^- + H_2O \rightleftharpoons HA + OH^-$

15.41 (a) $M(H_2O)_6^{n+}(aq) + H_2O(l) \rightleftharpoons H_3O^+(aq) + M(H_2O)_5(OH)^{(n-1)+}(aq)$

$$K_a = \frac{[H_3O^+][M(H_2O)_5(OH)^{(n-1)+}]}{[M(H_2O)_6^{n+}]}$$

(b) As the charge of the metal cation increases, the equilibrium constant increases because the OH bonds in coordinated water molecules are more polar, facilitating the dissociation of a H^+.

(c) $M(H_2O)_6^{3+}$ is the strongest acid. $M(H_2O)_6^+$, the weakest acid, has the strongest conjugate base.

15.42

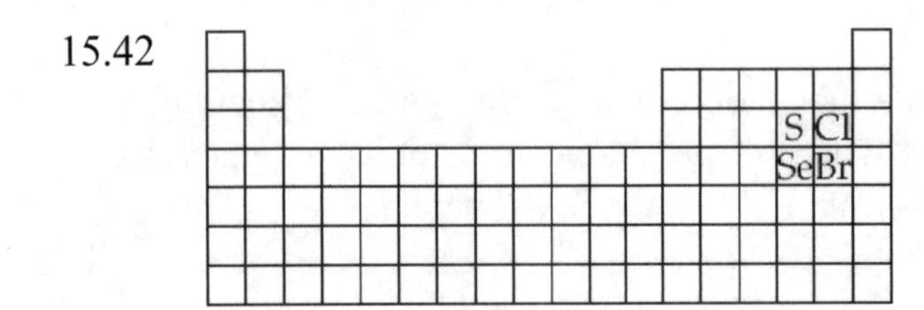

(a) H_2S, weakest; HBr, strongest. Acid strength for H_nX increases with increasing polarity of the H–X bond and with increasing size of X.

(b) H_2SeO_3, weakest; $HClO_3$, strongest. Acid strength for H_nYO_3 increases with increasing electronegativity of Y.

15.43 (a) Brønsted-Lowry acids: NH_4^+, $H_2PO_4^-$; Brønsted-Lowry bases: SO_3^{2-}, OCl^-, $H_2PO_4^-$

(b) Lewis acids: Fe^{3+}, BCl_3; Lewis bases: SO_3^{2-}, OCl^-, $H_2PO_4^-$

15.44 (a) $H_3BO_3(aq) + H_2O(l) \rightleftharpoons H_3O^+(aq) + H_2BO_3^-(aq)$

(b) $H_3BO_3(aq) + 2\ H_2O(l) \rightleftharpoons H_3O^+(aq) + B(OH)_4^-(aq)$

15.45

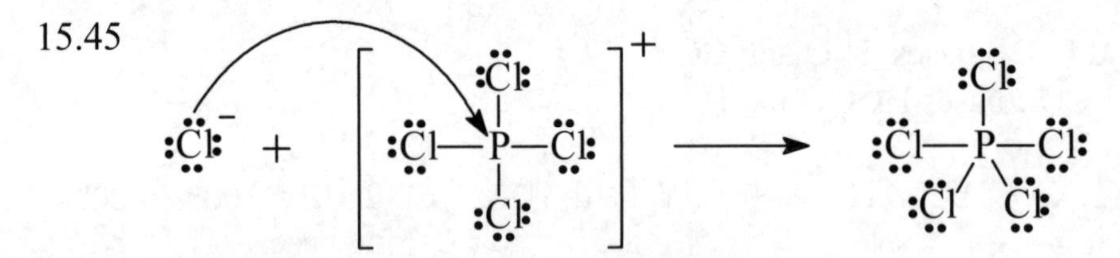

Section Problems

Acid–Base Concepts (Sections 15.1 and 15.2)

15.46 A Brønsted-Lowry base is a proton acceptor. An Arrhenius base dissociates producing OH^-.
(a) NH_3 and (c) HCO_3^-

15.47 A Brønsted-Lowry acid is a proton donor. A Brønsted-Lowry base is a proton acceptor.
(a) HCO_3^- and (b) H_2O

15.48 (a) SO_4^{2-} (b) HSO_3^- (c) HPO_4^{2-} (d) NH_3 (e) OH^- (f) NH_2^-

15.49 (a) HSO_3^- (b) H_3O^+ (c) $CH_3NH_3^+$ (d) H_2O (e) H_2CO_3 (f) H_2

15.50 (a) $CH_3CO_2H(aq)$ + $NH_3(aq)$ ⇄ $NH_4^+(aq)$ + $CH_3CO_2^-(aq)$
acid base —— acid base

(b) $CO_3^{2-}(aq)$ + $H_3O^+(aq)$ ⇄ $H_2O(l)$ + $HCO_3^-(aq)$
base acid —— base acid

(c) $HSO_3^-(aq)$ + $H_2O(l)$ ⇄ $H_3O^+(aq)$ + $SO_3^{2-}(aq)$
acid base —— acid base

(d) $HSO_3^-(aq)$ + $H_2O(l)$ ⇄ $H_2SO_3(aq)$ + $OH^-(aq)$
base acid acid base

15.51 (a) $CN^-(aq)$ + $H_2O(l)$ ⇄ $HCN(aq)$ + $OH^-(aq)$
base acid acid base

(b) $H_2PO_4^-(aq)$ + $H_2O(l)$ ⇄ $H_3O^+(aq)$ + $HPO_4^{2-}(aq)$
acid base —— acid base

(c) $HPO_4^{2-}(aq)$ + $H_2O(l)$ ⇄ $H_2PO_4^-(aq)$ + $OH^-(aq)$
base acid acid base

(d) $NH_4^+(aq)$ + $NO_2^-(aq)$ ⇄ $HNO_2(aq)$ + $NH_3(aq)$
acid base —— acid base

15.52 From data in Table 15.1: Strong acids: HNO_3 and H_2SO_4; Strong bases: H^- and O^{2-}

15.53 The weaker acid of the pair has the stronger conjugate base.
(a) H_2CO_3 is a weak acid. H_2SO_4 is a strong acid. H_2CO_3 has the stronger conjugate base.
(b) HCl is a strong acid. HF is a weak acid. HF has the stronger conjugate base.
(c) NH_4^+ is a weaker acid than HF. NH_4^+ has the stronger conjugate base.
(d) HCN is a weaker acid than HSO_4^-. HCN has the stronger conjugate base.

15.54 The direction of the reaction for $K_c > 1$, is proton transfer from the stronger acid to the stronger base to give the weaker base and the weaker acid:
HSO_4^- + NO_2^- → SO_4^{2-} + HNO_2

15.55 The direction of the reaction for $K_c > 1$, is proton transfer to the stronger base from the stronger acid to give the weaker acid and the weaker base:
$NH_3 + H_2CO_3 \rightarrow NH_4^+ + HCO_3^-$

Factors That Affect Acid Strength (Section 15.3)

15.56 (a) $PH_3 < H_2S < HCl$; electronegativity increases from P to Cl
(b) $NH_3 < PH_3 < AsH_3$; X–H bond strength decreases from N to As (down a group)
(c) $HBrO < HBrO_2 < HBrO_3$; acid strength increases with the number of O atoms

15.57 (a) $H_2Se > H_2S > H_2O$; X–H bond strength decreases from O to Se (down a group)
(b) $HClO_3 > HBrO_3 > HIO_3$; electronegativity increases from I to Cl
(c) $HCl > H_2S > PH_3$; electronegativity increases from P to Cl

15.58 (a) HCl; The strength of a binary acid H_nA increases as A moves from left to right and from top to bottom in the periodic table.
(b) $HClO_3$; The strength of an oxoacid increases with increasing electronegativity and increasing oxidation state of the central atom.
(c) HBr; The strength of a binary acid H_nA increases as A moves from left to right and from top to bottom in the periodic table.

15.59 (a) H_2SO_3; $HClO_3$ is weaker than $HClO_4$ because of the lower oxidation number of the Cl, and H_2SO_3 is weaker than $HClO_3$ because S is less electronegative than Cl.
(b) NH_3; H_2O is weaker than H_2S because of the smaller size of O and the stronger O–H bond. NH_3 is weaker than H_2O because N is less electronegative than O, and the N–H bond is therefore less polar.
(c) $Ga(OH)_3$; The acid strength of an oxoacid decreases with decreasing electronegativity of the central atom.

15.60 (a) H_2Te, weaker X–H bond
(b) H_3PO_4, P has higher electronegativity
(c) $H_2PO_4^-$, lower negative charge
(d) NH_4^+, higher positive charge and N is more electronegative than C

15.61 (a) ClO_2^- (conjugate base of $HClO_2$) is a stronger base than ClO_3^- (conjugate base of $HClO_3$) because $HClO_2$ is the weaker acid.
(b) $HSeO_4^-$ (conjugate base of H_2SeO_4) is a stronger base than HSO_4^- (conjugate base of H_2SO_4) because H_2SeO_4 is the weaker acid.
(c) OH^- (conjugate base of H_2O) is a stronger base than HS^- (conjugate base of H_2S) because H_2O is the weaker acid.
(d) HS^- is the stronger base because Br^- has no basic properties. HBr is a strong acid.

15.62 (b)

```
      O
      ‖
H—C—O—H
```

The HCO group is more electronegative than the CH_3 group, (b) is more acidic.

15.63 For RCO_2H, the more electronegative the R, the more acidic is the acid.
acetic acid < chloroacetic acid < trichloroacetic acid

Dissociation of Water; pH (Sections 15.4–15.6)

15.64 $$[H_3O^+] = \frac{K_w}{[OH^-]} = \frac{1.0 \times 10^{-14}}{2.0 \times 10^{-6}} = 5.0 \times 10^{-9} \text{ M}$$
Because $[OH^-] > [H_3O^+]$, the solution is basic.

15.65 $$[H_3O^+] = \frac{K_w}{[OH^-]} = \frac{1.0 \times 10^{-14}}{2.24 \times 10^{-7}} = 4.5 \times 10^{-8} \text{ M}$$
Because $[H_3O^+] < 1.0 \times 10^{-7}$ M, the solution is basic.

15.66 If $[H_3O^+] > 1.0 \times 10^{-7}$ M, solution is acidic.
If $[H_3O^+] < 1.0 \times 10^{-7}$ M, solution is basic.
If $[H_3O^+] = [OH^-] = 1.0 \times 10^{-7}$ M, solution is neutral.
If $[OH^-] > 1.0 \times 10^{-7}$ M, solution is basic
If $[OH^-] < 1.0 \times 10^{-7}$ M, solution is acidic.

(a) $$[OH^-] = \frac{K_w}{[H_3O^+]} = \frac{1.0 \times 10^{-14}}{3.4 \times 10^{-9}} = 2.9 \times 10^{-6} \text{ M, basic}$$

(b) $$[H_3O^+] = \frac{K_w}{[OH^-]} = \frac{1.0 \times 10^{-14}}{0.010} = 1.0 \times 10^{-12} \text{ M, basic}$$

(c) $$[H_3O^+] = \frac{K_w}{[OH^-]} = \frac{1.0 \times 10^{-14}}{1.0 \times 10^{-10}} = 1.0 \times 10^{-4} \text{ M, acidic}$$

(d) $$[OH^-] = \frac{K_w}{[H_3O^+]} = \frac{1.0 \times 10^{-14}}{1.0 \times 10^{-7}} = 1.0 \times 10^{-7} \text{ M, neutral}$$

(e) $$[OH^-] = \frac{K_w}{[H_3O^+]} = \frac{1.0 \times 10^{-14}}{8.6 \times 10^{-5}} = 1.2 \times 10^{-10} \text{ M, acidic}$$

15.67 If $[H_3O^+] > 1.0 \times 10^{-7}$ M, solution is acidic.
If $[H_3O^+] < 1.0 \times 10^{-7}$ M, solution is basic.
If $[H_3O^+] = [OH^-] = 1.0 \times 10^{-7}$ M, solution is neutral.
If $[OH^-] > 1.0 \times 10^{-7}$ M, solution is basic
If $[OH^-] < 1.0 \times 10^{-7}$ M, solution is acidic.

(a) $$[OH^-] = \frac{K_w}{[H_3O^+]} = \frac{1.0 \times 10^{-14}}{2.5 \times 10^{-4}} = 4.0 \times 10^{-11} \text{ M, acidic}$$

(b) $$[OH^-] = \frac{K_w}{[H_3O^+]} = \frac{1.0 \times 10^{-14}}{2.0} = 5.0 \times 10^{-15} \text{ M, acidic}$$

(c) $[H_3O^+] = \frac{K_w}{[OH^-]} = \frac{1.0 \times 10^{-14}}{5.6 \times 10^{-9}} = 1.8 \times 10^{-6}$ M, acidic

(d) $[H_3O^+] = \frac{K_w}{[OH^-]} = \frac{1.0 \times 10^{-14}}{1.5 \times 10^{-3}} = 6.7 \times 10^{-12}$ M, basic

(e) $[H_3O^+] = \frac{K_w}{[OH^-]} = \frac{1.0 \times 10^{-14}}{1.0 \times 10^{-7}} = 1.0 \times 10^{-7}$ M, neutral

15.68 At 200 °C and 750 atm, $K_w = [H_3O^+][OH^-] = 1.5 \times 10^{-11}$ and in pure water $[H_3O^+] = [OH^-]$.
$[H_3O^+]^2 = K_w$
$[H_3O^+] = [OH^-] = \sqrt{K_w} = \sqrt{1.5 \times 10^{-11}} = 3.9 \times 10^{-6}$ M
Because $[H_3O^+] = [OH^-]$, the solution is neutral.

15.69 At 500 °C and 250 atm, $K_w = [H_3O^+][OH^-] = 1.7 \times 10^{-19}$ and in pure water $[H_3O^+] = [OH^-]$.
$[H_3O^+]^2 = K_w$
$[H_3O^+] = [OH^-] = \sqrt{K_w} = \sqrt{1.7 \times 10^{-19}} = 4.1 \times 10^{-10}$ M
Because $[H_3O^+] = [OH^-]$, the solution is neutral.

15.70 (a) $pH = -\log[H_3O^+] = -\log(2.0 \times 10^{-5}) = 4.70$
(b) $[H_3O^+] = \frac{K_w}{[OH^-]} = \frac{1.0 \times 10^{-14}}{4 \times 10^{-3}} = 2.5 \times 10^{-12}$ M
$pH = -\log[H_3O^+] = -\log(2.5 \times 10^{-12}) = 11.6$
(c) $pH = -\log[H_3O^+] = -\log(3.56 \times 10^{-9}) = 8.449$
(d) $pH = -\log[H_3O^+] = -\log(10^{-3}) = 3$
(e) $[H_3O^+] = \frac{K_w}{[OH^-]} = \frac{1.0 \times 10^{-14}}{12} = 8.3 \times 10^{-16}$ M
$pH = -\log[H_3O^+] = -\log(8.3 \times 10^{-16}) = 15.08$

15.71 (a) $[H_3O^+] = \frac{K_w}{[OH^-]} = \frac{1.0 \times 10^{-14}}{7.6 \times 10^{-3}} = 1.3 \times 10^{-12}$ M
$pH = -\log[H_3O^+] = -\log(1.3 \times 10^{-12}) = 11.88$
(b) $pH = -\log[H_3O^+] = -\log(10^{-8}) = 8$
(c) $pH = -\log[H_3O^+] = -\log(5.0) = -0.70$
(d) $[H_3O^+] = \frac{K_w}{[OH^-]} = \frac{1.0 \times 10^{-14}}{1.0 \times 10^{-7}} = 1.0 \times 10^{-7}$ M
$pH = -\log[H_3O^+] = -\log(1.0 \times 10^{-7}) = 7.00$
(e) $pH = -\log[H_3O^+] = -\log(2.18 \times 10^{-10}) = 9.662$

15.72 $[H_3O^+] = 10^{-pH}$; (a) 8×10^{-5} M (b) 1.5×10^{-11} M (c) 1.0 M
(d) 5.6×10^{-15} M (e) 10 M (f) 5.78×10^{-6} M

15.73 $[H_3O^+] = 10^{-pH}$; (a) 1×10^{-9} M (b) 1.0×10^{-7} M (c) 2 M (d) 6.6×10^{-16} M (e) 2.3×10^{-3} M (f) 1.75×10^{-11} M

15.74 (a) chlorphenol red (b) thymol blue (c) methyl orange

15.75 (a) bromcresol green (b) thymolphthalein (c) methyl violet

Strong Acids and Strong Bases (Section 15.7)

15.76 (a) Because $Sr(OH)_2$ is a strong base, $[OH^-] = 2(1.0 \times 10^{-3}\ M) = 2.0 \times 10^{-3}$ M.

$$[H_3O^+] = \frac{K_w}{[OH^-]} = \frac{1.0 \times 10^{-14}}{2.0 \times 10^{-3}} = 5.0 \times 10^{-12}\ M$$

$pH = -\log[H_3O^+] = -\log(5.0 \times 10^{-12}) = 11.30$

(b) Because HNO_3 is a strong acid, $[H_3O^+] = 0.015$ M.

$pH = -\log[H_3O^+] = -\log(0.015) = 1.82$

(c) Because NaOH is a strong base, $[OH^-] = 0.035$ M.

$$[H_3O^+] = \frac{K_w}{[OH^-]} = \frac{1.0 \times 10^{-14}}{0.035} = 2.9 \times 10^{-13}\ M$$

$pH = -\log[H_3O^+] = -\log(2.9 \times 10^{-13}) = 12.54$

15.77 (b) Because HCl is a strong acid, $[H_3O^+] = 0.48$ M.

$pH = -\log[H_3O^+] = -\log(0.48) = 0.32$

(a) Because $Ba(OH)_2$ is a strong base, $[OH^-] = 2(2.5 \times 10^{-3}\ M) = 5.0 \times 10^{-3}$ M.

$$[H_3O^+] = \frac{K_w}{[OH^-]} = \frac{1.0 \times 10^{-14}}{5.0 \times 10^{-3}} = 2.0 \times 10^{-12}\ M$$

$pH = -\log[H_3O^+] = -\log(2.0 \times 10^{-12}) = 11.70$

(c) Because NaOH is a strong base, $[OH^-] = 0.075$ M.

$$[H_3O^+] = \frac{K_w}{[OH^-]} = \frac{1.0 \times 10^{-14}}{0.075} = 1.3 \times 10^{-13}\ M$$

$pH = -\log[H_3O^+] = -\log(1.3 \times 10^{-13}) = 12.89$

15.78 (a) LiOH, 23.95; 250 mL = 0.250 L

$$\text{molarity of LiOH(aq)} = \frac{\left(4.8\ g \times \frac{1\ mol}{23.95\ g}\right)}{0.250\ L} = 0.80\ M$$

LiOH is a strong base; therefore $[OH^-] = 0.80$ M

$$[H_3O^+] = \frac{K_w}{[OH^-]} = \frac{1.0 \times 10^{-14}}{0.80} = 1.25 \times 10^{-14}\ M$$

$pH = -\log[H_3O^+] = -\log(1.25 \times 10^{-14}) = 13.90$

(b) HCl, 36.46

$$\text{molarity of HCl(aq)} = \frac{\left(0.93 \text{ g} \times \frac{1 \text{ mol}}{36.46 \text{ g}}\right)}{0.40 \text{ L}} = 0.064 \text{ M}$$

HCl is a strong acid; therefore $[H_3O^+] = 0.064$ M
$pH = -\log[H_3O^+] = -\log(0.064) = 1.19$

(c) $M_f \cdot V_f = M_i \cdot V_i$

$$M_f = \frac{M_i \cdot V_i}{V_f} = \frac{(0.10 \text{ M})(50 \text{ mL})}{(1000 \text{ mL})} = 5.0 \times 10^{-3} \text{ M}$$

$pH = -\log[H_3O^+] = -\log(5.0 \times 10^{-3}) = 2.30$

$$\text{(d) For HCl, } M_f = \frac{M_i \cdot V_i}{V_f} = \frac{(2.0 \times 10^{-3} \text{ M})(100 \text{ mL})}{(500 \text{ mL})} = 4.0 \times 10^{-4} \text{ M}$$

$$\text{For } HClO_4,\ M_f = \frac{M_i \cdot V_i}{V_f} = \frac{(1.0 \times 10^{-3} \text{ M})(400 \text{ mL})}{(500 \text{ mL})} = 8.0 \times 10^{-4} \text{ M}$$

$[H_3O^+] = (4.0 \times 10^{-4} \text{ M}) + (8.0 \times 10^{-4} \text{ M}) = 1.2 \times 10^{-3}$ M
$pH = -\log[H_3O^+] = -\log(1.2 \times 10^{-3}) = 2.92$

15.79 (a) Na_2O, 61.98; 100 mL = 0.100 L

$$\text{moles of } Na_2O = 0.20 \text{ g} \times \frac{1 \text{ mol}}{61.98 \text{ g}} = 0.0032 \text{ mol}$$

$$O^{2-}(aq) + H_2O(l) \xrightarrow{100\%} 2\ OH^-(aq)$$

moles of OH^- = (2)(0.0032 mol) = 0.0064 mol

$$[OH^-] = \frac{0.0064 \text{ mol}}{0.100 \text{ L}} = 0.064 \text{ M}$$

$$[H_3O^+] = \frac{K_w}{[OH^-]} = \frac{1.0 \times 10^{-14}}{0.064} = 1.6 \times 10^{-13} \text{ M}$$

$pH = -\log[H_3O^+] = -\log(1.6 \times 10^{-13}) = 12.80$

(b) HNO_3, 63.01

$$\text{molarity of } HNO_3(aq) = \frac{\left(1.26 \text{ g} \times \frac{1 \text{ mol}}{63.01 \text{ g}}\right)}{0.500 \text{ L}} = 0.0400 \text{ M}$$

$pH = -\log[H_3O^+] = -\log(0.0400) = 1.398$

(c) $[OH^-] = 2(0.075 \text{ M}) = 0.15$ M
$M_f \cdot V_f = M_i \cdot V_i$

$$M_f = \frac{M_i \cdot V_i}{V_f} = \frac{(0.15 \text{ M})(40.0 \text{ mL})}{(300.0 \text{ mL})} = 0.020 \text{ M}$$

$$[H_3O^+] = \frac{K_w}{[OH^-]} = \frac{1.0 \times 10^{-14}}{0.020} = 5.0 \times 10^{-13} \text{ M}$$

$pH = -\log[H_3O^+] = -\log(5.0 \times 10^{-13}) = 12.30$

(d) On mixing equal volumes of the two strong acids, both acid concentrations are cut in half.
$[H_3O^+] = 0.10\ M + 0.25\ M = 0.35\ M$
$pH = -\log[H_3O^+] = -\log(0.35) = 0.46$

15.80 CaO, 56.08
$[H_3O^+] = 10^{-pH} = 10^{-(10.50)} = 3.16 \times 10^{-11}\ M$

$$[OH^-] = \frac{K_w}{[H_3O^+]} = \frac{1.0 \times 10^{-14}}{3.16 \times 10^{-11}} = 3.16 \times 10^{-4}\ M = 3.16 \times 10^{-4}\ mol/L$$

$CaO(s) + H_2O(l) \rightarrow Ca^{2+}(aq) + 2\ OH^-(aq)$

$$3.16 \times 10^{-4}\ mol\ OH^- \times \frac{1\ mol\ CaO}{2\ mol\ OH^-} \times \frac{56.08\ g\ CaO}{1\ mol\ CaO} = 0.0089\ g\ CaO$$

15.81 SrO, 103.6
$[H_3O^+] = 10^{-pH} = 10^{-(10.00)} = 1.0 \times 10^{-10}\ M$

$$[OH^-] = \frac{K_w}{[H_3O^+]} = \frac{1.0 \times 10^{-14}}{1.0 \times 10^{-10}} = 1.0 \times 10^{-4}\ M = 1.0 \times 10^{-4}\ mol/L$$

$(1.0 \times 10^{-4}\ mol/L)(2.0\ L) = 2.0 \times 10^{-4}\ mol\ OH^-$
$SrO(s) + H_2O(l) \rightarrow Sr^{2+}(aq) + 2\ OH^-(aq)$

$$2.0 \times 10^{-4}\ mol\ OH^- \times \frac{1\ mol\ SrO}{2\ mol\ OH^-} \times \frac{103.6\ g\ SrO}{1\ mol\ SrO} = 0.0104\ g\ SrO$$

Weak Acids (Sections 15.8–15.10)

15.82 (a) The larger the K_a, the stronger the acid.
$C_6H_5OH < HOCl < CH_3CO_2H < HNO_3$
(b) The larger the K_a, the larger the percent dissociation for the same concentration.
$HNO_3 > CH_3CO_2H > HOCl > C_6H_5OH$
1 M HNO_3, $[H_3O^+] = 1\ M$

1 M CH_3CO_2H, $[H_3O^+] = \sqrt{[HA] \times K_a} = \sqrt{(1\ M)(1.8 \times 10^{-5})} = 4 \times 10^{-3}\ M$

1 M HOCl, $[H_3O^+] = \sqrt{[HA] \times K_a} = \sqrt{(1\ M)(3.5 \times 10^{-8})} = 2 \times 10^{-4}\ M$

1 M C_6H_5OH, $[H_3O^+] = \sqrt{[HA] \times K_a} = \sqrt{(1\ M)(1.3 \times 10^{-10})} = 1 \times 10^{-5}\ M$

15.83 (a) The larger the K_a, the stronger the acid.
$HCN < HOBr < HCO_2H < HClO_4$
(b) The larger the K_a, the larger the percent dissociation for the same concentration.
$HClO_4 > HCO_2H > HOBr > HCN$
1 M $HClO_4$, $[H_3O^+] = 1\ M$

1 M HCO_2H, $[H_3O^+] = \sqrt{[HA] \times K_a} = \sqrt{(1\ M)(1.8 \times 10^{-4})} = 1 \times 10^{-2}\ M$

1 M HOBr, $[H_3O^+] = \sqrt{[HA] \times K_a} = \sqrt{(1\ M)(2.0 \times 10^{-9})} = 4 \times 10^{-5}\ M$

1 M HCN, $[H_3O^+] = \sqrt{[HA] \times K_a} = \sqrt{(1\ M)(4.9 \times 10^{-10})} = 2 \times 10^{-5}\ M$

15.84

	$HOBr(aq)$ + $H_2O(l)$ ⇄	$H_3O^+(aq)$ +	$OBr^-(aq)$
initial (M)	0.040	~0	0
change (M)	–x	+x	+x
equil (M)	0.040 – x	x	x

$x = [H_3O^+] = 10^{-pH} = 10^{-5.05} = 8.9 \times 10^{-6}$ M

$$K_a = \frac{[H_3O^+][OBr^-]}{[HOBr]} = \frac{x^2}{0.040 - x} = \frac{(8.9 \times 10^{-6})^2}{0.040 - (8.9 \times 10^{-6})} = 2.0 \times 10^{-9}$$

15.85

	$C_3H_6O_3(aq)$ + $H_2O(l)$ ⇄	$H_3O^+(aq)$ +	$C_3H_5O_3^-(aq)$
initial (M)	0.10	~0	0
change (M)	–x	+x	+x
equil (M)	0.10 – x	x	x

$x = [H_3O^+] = 10^{-pH} = 10^{-2.43} = 3.7 \times 10^{-3}$ M

$[C_3H_5O_3^-] = x = 3.7 \times 10^{-3}$ M

$[C_3H_6O_3] = 0.10 - x = 0.10 - (3.7 \times 10^{-3})$ M $= 0.10$ M

$$K_a = \frac{[H_3O^+][C_3H_5O_3^-]}{[C_3H_6O_3]} = \frac{(3.7 \times 10^{-3})(3.7 \times 10^{-3})}{0.10} = 1.4 \times 10^{-4}$$

$pK_a = -\log K_a = -\log(1.4 \times 10^{-4}) = 3.85$

15.86 $C_6H_8O_6$, 176.13; 250 mg = 0.250 g; 250 mL = 0.250 L

$$[C_6H_8O_6] = \frac{\left(0.250\text{ g} \times \frac{1\text{ mol}}{176.13\text{ g}}\right)}{0.250\text{ L}} = 5.68 \times 10^{-3}\text{ M}$$

	$C_6H_8O_6(aq)$ + $H_2O(l)$ ⇄	$H_3O^+(aq)$ +	$C_6H_7O_6^-(aq)$
initial (M)	5.68×10^{-3}	~0	0
change (M)	–x	+x	+x
equil (M)	$(5.68 \times 10^{-3}) - x$	x	x

$$K_a = \frac{[H_3O^+][C_6H_7O_6^-]}{[C_6H_8O_6]} = 8.0 \times 10^{-5} = \frac{x^2}{(5.68 \times 10^{-3}) - x}$$

$x^2 + (8.0 \times 10^{-5})x - (4.54 \times 10^{-7}) = 0$

Use the quadratic formula to solve for x.

$$x = \frac{-(8.0 \times 10^{-5}) \pm \sqrt{(8.0 \times 10^{-5})^2 - (4)(-4.54 \times 10^{-7})}}{2(1)} = \frac{(-8.0 \times 10^{-5}) \pm 0.001\ 35}{2}$$

$x = 6.35 \times 10^{-4}$ and -7.15×10^{-4}

Of the two solutions for x, only the positive value of x has physical meaning because x is the $[H_3O^+]$.

$x = [H_3O^+] = 6.35 \times 10^{-4}$ M

$pH = -\log[H_3O^+] = -\log(6.35 \times 10^{-4}) = 3.20$

15.87 CH_3COOH, 60.05

Assume 1.00 L of vinegar solution.

Vinegar solution mass = $(1.00 \times 10^3\text{ mL})(1.02\text{ g/mL}) = 1020$ g

CH_3COOH mass in vinegar solution = (1020 g)(0.0350) = 35.7 g CH_3COOH

$$[CH_3COOH] = \frac{\left(35.7 \text{ g x } \frac{1 \text{ mol}}{60.05 \text{ g}}\right)}{1.00 \text{ L}} = 0.595 \text{ M}$$

	$CH_3COOH(aq)$ + $H_2O(l)$	⇄ $H_3O^+(aq)$	+ $CH_3COO^-(aq)$
initial (M)	0.595	~0	0
change (M)	–x	+x	+x
equil (M)	0.595 – x	x	x

$$K_a = \frac{[H_3O^+][CH_3COO^-]}{[CH_3COOH]} = 1.8 \times 10^{-5} = \frac{x^2}{0.595 - x} \approx \frac{x^2}{0.595}$$

Solve for x. x = 0.0033 M = $[H_3O^+]$

pH = –log$[H_3O^+]$ = –log(0.0033) = 2.49

15.88 $K_a = 10^{-pK_a} = 10^{-4.25} = 5.6 \times 10^{-5}$

	$HC_3H_3O_2(aq)$ + $H_2O(l)$	⇄ $H_3O^+(aq)$	+ $C_3H_3O_2^-(aq)$
initial (M)	0.150	~0	0
change (M)	–x	+x	+x
equil (M)	0.150 – x	x	x

$$K_a = \frac{[H_3O^+][C_3H_3O_2^-]}{[HC_3H_3O_2]} = 5.6 \times 10^{-5} = \frac{x^2}{0.150 - x} \approx \frac{x^2}{0.150}$$

Solve for x. x = 0.0029 M = $[H_3O^+]$ = $[C_3H_3O_2^-]$

$[HC_3H_3O_2]$ = 0.150 – x = 0.150 – 0.0029 = 0.147 M

pH = –log$[H_3O^+]$ = –log(0.0029) = 2.54

$$[OH^-] = \frac{K_w}{[H_3O^+]} = \frac{1.0 \times 10^{-14}}{0.0029} = 3.4 \times 10^{-12} \text{ M}$$

(b)	$HC_3H_3O_2(aq)$ + $H_2O(l)$	⇄ $H_3O^+(aq)$	+ $C_3H_3O_2^-(aq)$
initial (M)	0.0500	~0	0
change (M)	–x	+x	+x
equil (M)	0.0500 – x	x	x

$$K_a = \frac{[H_3O^+][C_3H_3O_2^-]}{[HC_3H_3O_2]} = 5.6 \times 10^{-5} = \frac{x^2}{0.0500 - x} \approx \frac{x^2}{0.0500}$$

Solve for x. x = 0.001 67 M = $[H_3O^+]$ = $[HC_3H_3O_2]_{diss}$

$$\% \text{ dissociation} = \frac{[HC_3H_3O_2]_{diss}}{[HC_3H_3O_2]_{initial}} \times 100\% = \frac{0.001\ 67 \text{ M}}{0.0500 \text{ M}} \times 100\% = 3.3\%$$

15.89 $K_a = 10^{-pK_a} = 10^{-3.62} = 2.4 \times 10^{-4}$

	$HC_9H_8NO_3(aq)$ + $H_2O(l)$	⇄ $H_3O^+(aq)$	+ $HC_9H_8NO_3^-(aq)$
initial (M)	0.100	~0	0
change (M)	–x	+x	+x
equil (M)	0.100 – x	x	x

$$K_a = \frac{[H_3O^+][C_9H_8NO_3^-]}{[HC_9H_8NO_3]} = 2.4 \times 10^{-4} = \frac{x^2}{0.100 - x}$$

$x^2 + (2.4 \times 10^{-4})x - (2.4 \times 10^{-5}) = 0$

Use the quadratic formula to solve for x.

$$x = \frac{-(2.4 \times 10^{-4}) \pm \sqrt{(2.4 \times 10^{-4})^2 - (4)(-2.4 \times 10^{-5})}}{2(1)} = \frac{(-2.4 \times 10^{-4}) \pm (9.8 \times 10^{-3})}{2}$$

$x = 4.8 \times 10^{-3}$ and -5.0×10^{-3}

Of the two solutions for x, only the positive value of x has physical meaning because x is the $[H_3O^+]$.

$x = 4.8 \times 10^{-3}\ M = [H_3O^+] = [HC_9H_8NO_3^-]$

$[HC_9H_8NO_3] = 0.100 - x = 0.100 - 0.0048 = 0.095\ M$

$pH = -\log[H_3O^+] = -\log(4.8 \times 10^{-3}) = 2.32$

$$[OH^-] = \frac{K_w}{[H_3O^+]} = \frac{1.0 \times 10^{-14}}{0.0048} = 2.1 \times 10^{-12}\ M$$

(b)	$HC_9H_8NO_3(aq)$ + $H_2O(l)$	⇄	$H_3O^+(aq)$	+ $HC_9H_8NO_3^-(aq)$
initial (M)	0.0750		~0	0
change (M)	−x		+x	+x
equil (M)	0.0750 − x		x	x

$$K_a = \frac{[H_3O^+][C_9H_8NO_3^-]}{[HC_9H_8NO_3]} = 2.4 \times 10^{-4} = \frac{x^2}{0.0750 - x}$$

$x^2 + (2.4 \times 10^{-4})x - (1.8 \times 10^{-5}) = 0$

Use the quadratic formula to solve for x.

$$x = \frac{-(2.4 \times 10^{-4}) \pm \sqrt{(2.4 \times 10^{-4})^2 - (4)(-1.8 \times 10^{-5})}}{2(1)} = \frac{(-2.4 \times 10^{-4}) \pm (8.5 \times 10^{-3})}{2}$$

$x = 4.1 \times 10^{-3}$ and -4.4×10^{-3}

Of the two solutions for x, only the positive value of x has physical meaning because x is the $[H_3O^+]$.

$x = 4.1 \times 10^{-3}\ M = [H_3O^+] = [HC_9H_8NO_3]_{diss}$

$$\%\ \text{dissociation} = \frac{[HC_9H_8NO_3]_{diss}}{[HC_9H_8NO_3]_{initial}} \times 100\% = \frac{0.0041\ M}{0.0750\ M} \times 100\% = 5.5\%$$

15.90	$HNO_2(aq)$ + $H_2O(l)$	⇄	$H_3O^+(aq)$	+ $NO_2^-(aq)$
initial (M)	1.5		~0	0
change (M)	−x		+x	+x
equil (M)	1.5 − x		x	x

$$K_a = \frac{[H_3O^+][NO_2^-]}{[HNO_2]} = 4.5 \times 10^{-4} = \frac{x^2}{1.5 - x} \approx \frac{x^2}{1.5}$$

Solve for x. $x = 0.026\ M = [H_3O^+]$; $pH = -\log[H_3O^+] = -\log(0.026) = 1.59$

$$\%\ \text{dissociation} = \frac{[HNO_2]_{diss}}{[HNO_2]_{initial}} \times 100\% = \frac{0.026\ M}{1.5\ M} \times 100\% = 1.7\%$$

15.91 $C_9H_8O_4$, 180.16; 300 mL = 0.300 L; 324 mg = 0.324 g

$$[C_9H_8O_4] = \frac{\left((2)(0.324\text{ g}) \times \frac{1\text{ mol}}{180.16\text{ g}}\right)}{0.300\text{ L}} = 0.0120\text{ M}$$

	$C_9H_8O_4(aq)$ + $H_2O(l)$	⇄ $H_3O^+(aq)$	+ $C_9H_7O_4^-(aq)$
initial (M)	0.0120	~0	0
change (M)	–x	+x	+x
equil (M)	0.0120 – x	x	x

$$K_a = \frac{[H_3O^+][C_9H_7O_4^-]}{[C_9H_8O_4]} = 3.0 \times 10^{-4} = \frac{x^2}{0.0120 - x}$$

$x^2 + (3.0 \times 10^{-4})x - (3.6 \times 10^{-6}) = 0$

Use the quadratic formula to solve for x.

$$x = \frac{-(3.0 \times 10^{-4}) \pm \sqrt{(3.0 \times 10^{-4})^2 - (4)(-3.6 \times 10^{-6})}}{2(1)} = \frac{(-3.0 \times 10^{-4}) \pm (3.8 \times 10^{-3})}{2}$$

$x = 1.8 \times 10^{-3}$ and -2.1×10^{-3}

Of the two solutions for x, only the positive value of x has physical meaning because x is the $[H_3O^+]$.

$[H_3O^+] = x = 1.8 \times 10^{-3}$ M

$pH = -\log[H_3O^+] = -\log(1.8 \times 10^{-3}) = 2.74$

$$\%\text{ dissociation} = \frac{[C_9H_8O_4]_{diss}}{[C_9H_8O_4]_{initial}} \times 100\% = \frac{1.8 \times 10^{-3}\text{ M}}{0.0120\text{ M}} \times 100\% = 15\%$$

15.92 (a) From Example 15.10 in the text:

$[H_3O^+] = [HF]_{diss} = 4.0 \times 10^{-3}$ M

$$\%\text{ dissociation} = \frac{[HF]_{diss}}{[HF]_{initial}} \times 100\% = \frac{4.0 \times 10^{-3}\text{ M}}{0.050\text{ M}} \times 100\% = 8.0\%\text{ dissociation}$$

(b)

	HF(aq) + $H_2O(l)$	⇄ $H_3O^+(aq)$	+ $F^-(aq)$
initial (M)	0.50	~0	0
change (M)	–x	+x	+x
equil (M)	0.50 – x	x	x

$$K_a = \frac{[H_3O^+][F^-]}{[HF]} = 3.5 \times 10^{-4} = \frac{x^2}{0.50 - x}$$

$x^2 + (3.5 \times 10^{-4})x - (1.75 \times 10^{-4}) = 0$

Use the quadratic formula to solve for x.

$$x = \frac{-(3.5 \times 10^{-4}) \pm \sqrt{(3.5 \times 10^{-4})^2 - 4(1)(-1.75 \times 10^{-4})}}{2(1)} = \frac{(-3.5 \times 10^{-4}) \pm 0.0265}{2}$$

x = 0.0131 and –0.0134

Of the two solutions for x, only the positive value of x has physical meaning, because x is the $[H_3O^+]$. $[H_3O^+] = [HF]_{diss} = 0.013$ M

$$\%\text{ dissociation} = \frac{[HF]_{diss}}{[HF]_{initial}} \times 100\% = \frac{0.013\text{ M}}{0.50\text{ M}} \times 100\% = 2.6\%\text{ dissociation}$$

15.93 (a)

	$HNO_2(aq)$	+	$H_2O(l)$	⇄	$H_3O^+(aq)$	+	$NO_2^-(aq)$
initial (M)	0.010				~0		0
change (M)	–x				+x		+x
equil (M)	0.010 – x				x		x

$$K_a = \frac{[H_3O^+][NO_2^-]}{[HNO_2]} = 4.5 \text{ x } 10^{-4} = \frac{x^2}{0.010 - x}$$

$x^2 + (4.5 \text{ x } 10^{-4})x - (4.5 \text{ x } 10^{-6}) = 0$

Use the quadratic formula to solve for x.

$$x = \frac{-(4.5 \text{ x } 10^{-4}) \pm \sqrt{(4.5 \text{ x } 10^{-4})^2 - 4(1)(-4.5 \text{ x } 10^{-6})}}{2(1)} = \frac{(-4.5 \text{ x } 10^{-4}) \pm 0.00427}{2}$$

x = 0.0019 and –0.0024

Of the two solutions for x, only the positive value of x has physical meaning, because x is the $[H_3O^+]$.

$[H_3O^+] = [HNO_2]_{diss} = 0.0019$ M

$$\% \text{ dissociation} = \frac{[HNO_2]_{diss}}{[HNO_2]_{initial}} \text{ x } 100\% = \frac{0.0019 \text{ M}}{0.010 \text{ M}} \text{ x } 100\% = 19\% \text{ dissociation}$$

(b)

	$HNO_2(aq)$	+	$H_2O(l)$	⇄	$H_3O^+(aq)$	+	$NO_2^-(aq)$
initial (M)	1.00				~0		0
change (M)	–x				+x		+x
equil (M)	1.00 – x				x		x

$$K_a = \frac{[H_3O^+][NO_2^-]}{[HNO_2]} = 4.5 \text{ x } 10^{-4} = \frac{x^2}{1.00 - x}$$

$x^2 + (4.5 \text{ x } 10^{-4})x - (4.5 \text{ x } 10^{-4}) = 0$

Use the quadratic formula to solve for x.

$$x = \frac{-(4.5 \text{ x } 10^{-4}) \pm \sqrt{(4.5 \text{ x } 10^{-4})^2 - 4(1)(-4.5 \text{ x } 10^{-4})}}{2(1)} = \frac{(-4.5 \text{ x } 10^{-4}) \pm 0.0424}{2}$$

x = 0.021 and –0.021

Of the two solutions for x, only the positive value of x has physical meaning, because x is the $[H_3O^+]$.

$[H_3O^+] = [HNO_2]_{diss} = 0.021$ M

$$\% \text{ dissociation} = \frac{[HNO_2]_{diss}}{[HNO_2]_{initial}} \text{ x } 100\% = \frac{0.021 \text{ M}}{1.00 \text{ M}} \text{ x } 100\% = 2.1\% \text{ dissociation}$$

Polyprotic Acids (Section 15.11)

15.94 $H_2SeO_4(aq) + H_2O(l) \rightleftharpoons H_3O^+(aq) + HSeO_4^-(aq)$; $K_{a1} = \dfrac{[H_3O^+][HSeO_4^-]}{[H_2SeO_4]}$

$HSeO_4^-(aq) + H_2O(l) \rightleftharpoons H_3O^+(aq) + SeO_4^{2-}(aq)$; $K_{a2} = \dfrac{[H_3O^+][SeO_4^{2-}]}{[HSeO_4^-]}$

15.95 (a) $H_3PO_4(aq) + H_2O(l) \rightleftarrows H_3O^+(aq) + H_2PO_4^-(aq)$; $K_{a1} = \dfrac{[H_3O^+][H_2PO_4^-]}{[H_3PO_4]}$

(b) $H_2PO_4^-(aq) + H_2O(l) \rightleftarrows H_3O^+(aq) + HPO_4^{2-}(aq)$; $K_{a2} = \dfrac{[H_3O^+][HPO_4^{2-}]}{[H_2PO_4^-]}$

(c) $HPO_4^{2-}(aq) + H_2O(l) \rightleftarrows H_3O^+(aq) + PO_4^{3-}(aq)$; $K_{a3} = \dfrac{[H_3O^+][PO_4^{3-}]}{[HPO_4^{2-}]}$

15.96

	$H_2CO_3(aq)$ + $H_2O(l)$	$\rightleftarrows$	$H_3O^+(aq)$	+ $HCO_3^-(aq)$
initial (M)	0.010		~0	0
change (M)	−x		+x	+x
equil (M)	0.010 − x		x	x

$$K_{a1} = \frac{[H_3O^+][HCO_3^-]}{[H_2CO_3]} = 4.3 \times 10^{-7} = \frac{x^2}{0.010 - x} \approx \frac{x^2}{0.010}$$

Solve for x. $x = 6.6 \times 10^{-5}$

$[H_3O^+] = [HCO_3^-] = x = 6.6 \times 10^{-5}$ M; $[H_2CO_3] = 0.010 - x = 0.010$ M

The second dissociation of H_2CO_3 produces a negligible amount of H_3O^+ compared with that from the first dissociation.

$HCO_3^-(aq) + H_2O(l) \rightleftarrows H_3O^+(aq) + CO_3^{2-}(aq)$

$$K_{a2} = \frac{[H_3O^+][CO_3^{2-}]}{[HCO_3^-]} = 5.6 \times 10^{-11} = \frac{(6.6 \times 10^{-5})[CO_3^{2-}]}{(6.6 \times 10^{-5})}$$

$[CO_3^{2-}] = K_{a2} = 5.6 \times 10^{-11}$ M

$$[OH^-] = \frac{K_w}{[H_3O^+]} = \frac{1.0 \times 10^{-14}}{6.6 \times 10^{-5}} = 1.5 \times 10^{-10}\ M$$

$pH = -\log[H_3O^+] = -\log(6.6 \times 10^{-5}) = 4.18$

15.97

	$H_2SO_3(aq)$ + $H_2O(l)$	$\rightleftarrows$	$H_3O^+(aq)$	+ $HSO_3^-(aq)$
initial (M)	0.025		~0	0
change (M)	−x		+x	+x
equil (M)	0.025 − x		x	x

$$K_{a1} = \frac{[H_3O^+][HSO_3^-]}{[H_2SO_3]} = 1.5 \times 10^{-2} = \frac{x^2}{0.025 - x}$$

$x^2 + 0.015x - (3.75 \times 10^{-4}) = 0$

Use the quadratic formula to solve for x.

$$x = \frac{-(0.015) \pm \sqrt{(0.015)^2 - (4)(-3.75 \times 10^{-4})}}{2(1)} = \frac{-0.015 \pm 0.0415}{2}$$

$x = 0.013$ and -0.028

Of the two solutions for x, only the positive value of x has physical meaning because x is the $[H_3O^+]$.
$x = [H_3O^+] = [HSO_3^-] = 0.013$ M
$[H_2SO_3] = 0.025 - x = 0.025 - 0.013 = 0.012$ M

The second dissociation of H_2SO_3 produces a negligible amount of H_3O^+ compared with that from the first dissociation.

$HSO_3^-(aq) + H_2O(l) \rightleftarrows H_3O^+(aq) + SO_3^{2-}(aq)$

$$K_{a2} = \frac{[H_3O^+][SO_3^{2-}]}{[HSO_3^-]} = 6.3 \times 10^{-8} = \frac{(0.013)[SO_3^{2-}]}{(0.013)}$$

$[SO_3^{2-}] = K_{a2} = 6.3 \times 10^{-8}$ M
$pH = -\log[H_3O^+] = -\log(0.013) = 1.89$

15.98 For the dissociation of the first proton, the following equilibrium must be considered:

	$H_2C_2O_4(aq)$ + $H_2O(l)$	$\rightleftarrows$ $H_3O^+(aq)$	+ $HC_2O_4^-(aq)$
initial (M)	0.20	~0	0
change (M)	$-x$	$+x$	$+x$
equil (M)	$0.20 - x$	x	x

$$K_{a1} = \frac{[H_3O^+][HC_2O_4^-]}{[H_2C_2O_4]} = 5.9 \times 10^{-2} = \frac{x^2}{0.20 - x}$$

$x^2 + 0.059x - 0.0118 = 0$
Use the quadratic formula to solve for x.

$$x = \frac{-(0.059) \pm \sqrt{(0.059)^2 - 4(1)(-0.0118)}}{2(1)} = \frac{-0.059 \pm 0.225}{2}$$

$x = 0.083$ and -0.142
Of the two solutions for x, only the positive value of x has physical meaning, because x is the $[H_3O^+]$.
$[H_3O^+] = [HC_2O_4^-] = 0.083$ M

For the dissociation of the second proton, the following equilibrium must be considered:

	$HC_2O_4^-(aq)$ + $H_2O(l)$	$\rightleftarrows$ $H_3O^+(aq)$	+ $C_2O_4^{2-}(aq)$
initial (M)	0.083	0.083	0
change (M)	$-x$	$+x$	$+x$
equil (M)	$0.083 - x$	$0.083 + x$	x

$$K_{a2} = \frac{[H_3O^+][C_2O_4^{2-}]}{[HC_2O_4^-]} = 6.4 \times 10^{-5} = \frac{(0.083 + x)(x)}{0.083 - x} \approx \frac{(0.083)(x)}{0.083} = x$$

$[H_3O^+] = 0.083 + x = 0.083$ M
$pH = -\log[H_3O^+] = -\log(0.083) = 1.08$
$[C_2O_4^{2-}] = x = 6.4 \times 10^{-5}$ M

15.99 $C_4H_6O_6$; $K_{a1} = 10^{-pK_{a1}} = 10^{-2.89} = 1.3 \times 10^{-3}$; $K_{a2} = 10^{-pK_{a2}} = 10^{-4.40} = 4.0 \times 10^{-5}$

For the dissociation of the first proton, the following equilibrium must be considered:

	$C_4H_6O_6(aq)$ + $H_2O(l)$	⇄ $H_3O^+(aq)$	+ $C_4H_5O_6^-(aq)$
initial (M)	0.50	~0	0
change (M)	–x	+x	+x
equil (M)	0.50 – x	x	x

$$K_{a1} = \frac{[H_3O^+][C_4H_5O_6^-]}{[C_4H_6O_6]} = 1.3 \times 10^{-3} = \frac{x^2}{0.50 - x}$$

$x^2 + (1.3 \times 10^{-3})x - (6.5 \times 10^{-4}) = 0$

Use the quadratic formula to solve for x.

$$x = \frac{-(1.3 \times 10^{-3}) \pm \sqrt{(1.3 \times 10^{-3})^2 - 4(1)(-6.5 \times 10^{-4})}}{2(1)} = \frac{-(1.3 \times 10^{-3}) \pm 0.051}{2}$$

x = 0.025 and –0.026

Of the two solutions for x, only the positive value of x has physical meaning, because x is the $[H_3O^+]$.

$[H_3O^+] = [C_4H_5O_6^-] = 0.025$ M

For the dissociation of the second proton, the following equilibrium must be considered:

	$C_4H_5O_6^-(aq)$ + $H_2O(l)$	⇄ $H_3O^+(aq)$	+ $C_4H_4O_6^{2-}(aq)$
initial (M)	0.025	0.025	0
change (M)	–x	+x	+x
equil (M)	0.025 – x	0.025 + x	x

$$K_{a2} = \frac{[H_3O^+][C_4H_4O_4^{2-}]}{[C_4H_5O_6^-]} = 4.0 \times 10^{-5} = \frac{(0.025 + x)(x)}{0.025 - x} \approx \frac{(0.025)(x)}{0.025} = x$$

$[H_3O^+] = 0.025 + x = 0.025$ M

$pH = -\log[H_3O^+] = -\log(0.025) = 1.60$

15.100 From the complete dissociation of the first proton, $[H_3O^+] = [HSeO_4^-] = 0.50$ M.

For the dissociation of the second proton, the following equilibrium must be considered:

	$HSeO_4^-(aq)$ + $H_2O(l)$	⇄ $H_3O^+(aq)$	+ $SeO_4^{2-}(aq)$
initial (M)	0.50	0.50	0
change (M)	–x	+x	+x
equil (M)	0.50 – x	0.50 + x	x

$$K_{a2} = \frac{[H_3O^+][SeO_4^{2-}]}{[HSeO_4^-]} = 1.2 \times 10^{-2} = \frac{(0.50 + x)(x)}{0.50 - x}$$

$x^2 + 0.512x - 0.0060 = 0$

Use the quadratic formula to solve for x.

$$x = \frac{-(0.512) \pm \sqrt{(0.512)^2 - 4(-0.0060)}}{2(1)} = \frac{-0.512 \pm 0.535}{2}$$

x = 0.011 and –0.524

Of the two solutions for x, only the positive value of x has physical meaning, since x is the $[SeO_4^{2-}]$.
$[H_2SeO_4] = 0$ M; $[HSeO_4^-] = 0.50 - x = 0.49$ M; $[SeO_4^{2-}] = x = 0.011$ M
$[H_3O^+] = 0.50 + x = 0.51$ M
$pH = -\log[H_3O^+] = -\log(0.51) = 0.29$

$$[OH^-] = \frac{K_w}{[H_3O^+]} = \frac{1.0 \times 10^{-14}}{0.51} = 2.0 \times 10^{-14}\ M$$

15.101 Mixing equal volumes of the two strong acids cuts the initial acid concentrations in half.
$[H_3O^+] = 0.1$ M (from HCl) + 0.3 M (from H_2SO_4) = 0.4 M
For the dissociation of the second proton from H_2SO_4, the following equilibrium must be considered:

	$HSO_4^-(aq)$	+	$H_2O(l)$	⇄	$H_3O^+(aq)$	+	$SO_4^{2-}(aq)$
initial (M)	0.3				0.4		0
change (M)	−x				+x		+x
equil (M)	0.3 − x				0.4 + x		x

$$K_{a2} = \frac{[H_3O^+][SO_4^{2-}]}{[HSO_4^-]} = 1.2 \times 10^{-2} = \frac{(0.4 + x)(x)}{0.3 - x}$$

$x^2 + 0.412x - 0.0036 = 0$
Use the quadratic formula to solve for x.

$$x = \frac{-(0.412) \pm \sqrt{(0.412)^2 - 4(1)(-0.0036)}}{2(1)} = \frac{-0.412 \pm 0.429}{2}$$

x = 0.0085 and −0.420
Of the two solutions for x, only the positive value of x has physical meaning, because x is the $[SO_4^{2-}]$.
$[H_3O^+] = 0.4 + x = 0.4$ M
$[SO_4^{2-}] = x = 0.008$ M

Weak Bases; Relation Between K_a and K_b (Sections 15.12 and 15.13)

15.102 (a) $(CH_3)_2NH(aq) + H_2O(l) \rightleftarrows (CH_3)_2NH_2^+(aq) + OH^-(aq)$; $K_b = \dfrac{[(CH_3)_2NH_2^+][OH^-]}{[(CH_3)_2NH]}$

(b) $C_6H_5NH_2(aq) + H_2O(l) \rightleftarrows C_6H_5NH_3^+(aq) + OH^-(aq)$; $K_b = \dfrac{[C_6H_5NH_3^+][OH^-]}{[C_6H_5NH_2]}$

(c) $CN^-(aq) + H_2O(l) \rightleftarrows HCN(aq) + OH^-(aq)$; $K_b = \dfrac{[HCN][OH^-]}{[CN^-]}$

15.103 (a) $C_5H_5N(aq) + H_2O(l) \rightleftarrows C_5H_5NH^+(aq) + OH^-(aq)$; $K_b = \dfrac{[C_5H_5NH^+][OH^-]}{[C_5H_5N]}$

(b) $C_2H_5NH_2(aq) + H_2O(l) \rightleftarrows C_2H_5NH_3^+(aq) + OH^-(aq)$; $K_b = \dfrac{[C_2H_5NH_3^+][OH^-]}{[C_2H_5NH_2]}$

(c) $CH_3CO_2^-(aq) + H_2O(l) \rightleftarrows CH_3CO_2H(aq) + OH^-(aq)$; $K_b = \dfrac{[CH_3CO_2H][OH^-]}{[CH_3CO_2^-]}$

15.104 $C_{21}H_{22}N_2O_2$, 334.42; 16 mg = 0.016 g

$$\text{molarity} = \frac{\left(0.016 \text{ g x } \dfrac{1 \text{ mol}}{334.42 \text{ g}}\right)}{0.100 \text{ L}} = 4.8 \times 10^{-4} \text{ M}$$

	$C_{21}H_{22}N_2O_2(aq)$ + $H_2O(l)$	$\rightleftarrows$	$C_{21}H_{23}N_2O_2^+(aq)$	+ $OH^-(aq)$
initial (M)	4.8×10^{-4}		0	~0
change (M)	–x		+x	+x
equil (M)	$(4.8 \times 10^{-4}) - x$		x	x

$$K_b = \frac{[C_{21}H_{23}N_2O_2^+][OH^-]}{[C_{21}H_{22}N_2O_2]} = 1.8 \times 10^{-6} = \frac{x^2}{(4.8 \times 10^{-4}) - x}$$

$x^2 + (1.8 \times 10^{-6})x - (8.6 \times 10^{-10}) = 0$

Use the quadratic formula to solve for x.

$$x = \frac{-(1.8 \times 10^{-6}) \pm \sqrt{(1.8 \times 10^{-6})^2 - (4)(-8.6 \times 10^{-10})}}{2(1)} = \frac{(-1.8 \times 10^{-6}) \pm (5.87 \times 10^{-5})}{2}$$

$x = 2.84 \times 10^{-5}$ and -3.02×10^{-5}

Of the two solutions for x, only the positive value of x has physical meaning, because x is the $[OH^-]$.

$[OH^-] = 2.84 \times 10^{-5}$ M

$$[H_3O^+] = \frac{K_w}{[OH^-]} = \frac{1.0 \times 10^{-14}}{2.84 \times 10^{-5}} = 3.52 \times 10^{-10} \text{ M}$$

$pH = -\log[H_3O^+] = -\log(3.52 \times 10^{-10}) = 9.45$

15.105

	$NH_3(aq)$ + $H_2O(l)$	$\rightleftarrows$	$NH_4^+(aq)$	+ $OH^-(aq)$
initial (M)	0.5		0	~0
change (M)	–x		+x	+x
equil (M)	0.5 – x		x	x

$$K_b = \frac{[NH_4^+][OH^-]}{[NH_3]} = 1.8 \times 10^{-5} = \frac{x^2}{0.5 - x} \approx \frac{x^2}{0.5}$$

Solve for x. $x = [OH^-] = 3.0 \times 10^{-3}$ M

$$[H_3O^+] = \frac{K_w}{[OH^-]} = \frac{1.0 \times 10^{-14}}{3.0 \times 10^{-3}} = 3.3 \times 10^{-12} \text{ M}$$

$pH = -\log[H_3O^+] = -\log(3.3 \times 10^{-12}) = 11.5$

15.106 $[H_3O^+] = 10^{-pH} = 10^{-9.5} = 3.16 \times 10^{-10}$ M

$$[OH^-] = \frac{K_w}{[H_3O^+]} = \frac{1.0 \times 10^{-14}}{3.16 \times 10^{-10}} = 3.16 \times 10^{-5} \text{ M}$$

	$C_{17}H_{19}NO_3(aq)$ + $H_2O(l)$	⇌ $C_{17}H_{20}NO_3^+(aq)$	+ $OH^-(aq)$
initial (M)	7.0×10^{-4}	0	~0
change (M)	$-x$	$+x$	$+x$
equil (M)	$(7.0 \times 10^{-4}) - x$	x	x

$x = [OH^-] = 3.16 \times 10^{-5}$ M

$$K_b = \frac{[C_{17}H_{20}NO_3^+][OH^-]}{[C_{17}H_{19}NO_3]} = \frac{x^2}{(7.0 \times 10^{-4}) - x} = \frac{(3.16 \times 10^{-5})^2}{(7.0 \times 10^{-4}) - (3.16 \times 10^{-5})} = 1.49 \times 10^{-6}$$

$K_b = 1 \times 10^{-6}$

$pK_b = -\log K_b = -\log(1.49 \times 10^{-6}) = 5.827 = 5.8$

15.107 $[H_3O^+] = 10^{-pH} = 10^{-9.75} = 1.78 \times 10^{-10}$ M

$$[OH^-] = \frac{K_w}{[H_3O^+]} = \frac{1.0 \times 10^{-14}}{1.78 \times 10^{-10}} = 5.62 \times 10^{-5} \text{ M}$$

	quinine(aq) + $H_2O(l)$	⇌ quinine$H^+(aq)$	+ $OH^-(aq)$
initial (M)	1.00×10^{-3}	0	~0
change (M)	$-x$	$+x$	$+x$
equil (M)	$(1.00 \times 10^{-3}) - x$	x	x

$x = [OH^-] = 5.62 \times 10^{-5}$ M

$$K_b = \frac{[\text{quinineH}^+][OH^-]}{[\text{quinine}]} = \frac{x^2}{(1.00 \times 10^{-3}) - x} = \frac{(5.62 \times 10^{-5})^2}{[(1.00 \times 10^{-3}) - (5.62 \times 10^{-5})]} = 3.3 \times 10^{-6}$$

$pK_b = -\log K_b = -\log(3.3 \times 10^{-6}) = 5.48$

15.108 $K_b = 10^{-pK_b} = 10^{-5.47} = 3.4 \times 10^{-6}$

	$C_{18}H_{21}NO_4(aq)$ + $H_2O(l)$	⇌ $HC_{18}H_{21}NO_4^+(aq)$	+ $OH^-(aq)$
initial (M)	2.50×10^{-3}	0	~0
change (M)	$-x$	$+x$	$+x$
equil (M)	$(2.50 \times 10^{-3}) - x$	x	x

$$K_b = \frac{[HC_{18}H_{21}NO_4^+][OH^-]}{[C_{18}H_{21}NO_4]} = 3.4 \times 10^{-6} = \frac{x^2}{(2.50 \times 10^{-3}) - x}$$

$x^2 + (3.4 \times 10^{-6})x - (8.5 \times 10^{-9}) = 0$

Use the quadratic formula to solve for x.

$$x = \frac{-(3.4 \times 10^{-6}) \pm \sqrt{(3.4 \times 10^{-6})^2 - 4(1)(-8.5 \times 10^{-9})}}{2(1)}$$

$$x = = \frac{-(3.4 \times 10^{-6}) \pm (1.84 \times 10^{-4})}{2}$$

$x = 9.0 \times 10^{-5}$ and -1.9×10^{-4}

Of the two solutions for x, only the positive value of x has physical meaning, because x is the $[OH^-]$.

$x = 9.0 \times 10^{-5}$ M $= [OH^-] = [HC_{18}H_{21}NO_4^+]$

$[C_{18}H_{21}NO_4] = (2.50 \times 10^{-3}) - x = (2.50 \times 10^{-3}) - (9.0 \times 10^{-5}) = 0.0024$ M

$[H_3O^+] = \dfrac{K_w}{[OH^-]} = \dfrac{1.0 \times 10^{-14}}{9.0 \times 10^{-5}} = 1.1 \times 10^{-10}$ M

pH = $-\log[H_3O^+] = -\log(1.1 \times 10^{-10}) = 9.96$

15.109 $K_b = 10^{-pK_b} = 10^{-5.68} = 2.1 \times 10^{-6}$

	$C_4H_9NO(aq)$ + $H_2O(l)$	⇄	$HC_4H_9NO^+(aq)$	+ $OH^-(aq)$
initial (M)	0.0100		0	~0
change (M)	–x		+x	+x
equil (M)	0.0100 – x		x	x

$K_b = \dfrac{[HC_4H_9NO^+][OH^-]}{[C_4H_9NO]} = 2.1 \times 10^{-6} = \dfrac{x^2}{0.0100 - x}$

$x^2 + (2.1 \times 10^{-6})x - (2.1 \times 10^{-8}) = 0$

Use the quadratic formula to solve for x.

$x = \dfrac{-(2.1 \times 10^{-6}) \pm \sqrt{(2.1 \times 10^{-6})^2 - 4(1)(-2.1 \times 10^{-8})}}{2(1)}$

$x = = \dfrac{-(2.1 \times 10^{-6}) \pm (2.9 \times 10^{-4})}{2}$

$x = 1.4 \times 10^{-4}$ and -1.5×10^{-4}

Of the two solutions for x, only the positive value of x has physical meaning, because x is the $[OH^-]$.

$x = 1.4 \times 10^{-4}$ M $= [OH^-] = [HC_4H_9NO^+]$

$[C_4H_9NO] = 0.0100 - x = 0.0100 - 1.4 \times 10^{-4} = 0.0099$ M

$[H_3O^+] = \dfrac{K_w}{[OH^-]} = \dfrac{1.0 \times 10^{-14}}{1.4 \times 10^{-4}} = 7.1 \times 10^{-11}$ M

pH = $-\log[H_3O^+] = -\log(7.1 \times 10^{-11}) = 10.15$

15.110 (a) $K_a = \dfrac{K_w}{K_b \text{ for } C_3H_7NH_2} = \dfrac{1.0 \times 10^{-14}}{5.1 \times 10^{-4}} = 2.0 \times 10^{-11}$

(b) $K_a = \dfrac{K_w}{K_b \text{ for } NH_2OH} = \dfrac{1.0 \times 10^{-14}}{9.1 \times 10^{-9}} = 1.1 \times 10^{-6}$

(c) $K_a = \dfrac{K_w}{K_b \text{ for } C_6H_5NH_2} = \dfrac{1.0 \times 10^{-14}}{4.3 \times 10^{-10}} = 2.3 \times 10^{-5}$

(d) $K_a = \dfrac{K_w}{K_b \text{ for } C_5H_5N} = \dfrac{1.0 \times 10^{-14}}{1.8 \times 10^{-9}} = 5.6 \times 10^{-6}$

15.111 (a) $K_b = \dfrac{K_w}{K_a \text{ for HF}} = \dfrac{1.0 \times 10^{-14}}{3.5 \times 10^{-4}} = 2.9 \times 10^{-11}$

(b) $K_b = \dfrac{K_w}{K_a \text{ for HOBr}} = \dfrac{1.0 \times 10^{-14}}{2.0 \times 10^{-9}} = 5.0 \times 10^{-6}$

(c) $K_b = \dfrac{K_w}{K_a \text{ for } H_2S} = \dfrac{1.0 \times 10^{-14}}{1.0 \times 10^{-7}} = 1.0 \times 10^{-7}$

(d) $K_b = \dfrac{K_w}{K_a \text{ for } HS^-} = \dfrac{1.0 \times 10^{-14}}{1 \times 10^{-19}} = 1 \times 10^{5}$

Acid–Base Properties of Salts (Section 15.14)

15.112 (a) $CH_3NH_3^+(aq) + H_2O(l) \rightleftarrows H_3O^+(aq) + CH_3NH_2(aq)$

acid base — acid base

(b) $Cr(H_2O)_6^{3+}(aq) + H_2O(l) \rightleftarrows H_3O^+(aq) + Cr(H_2O)_5(OH)^{2+}(aq)$

acid base — acid base

(c) $CH_3CO_2^-(aq) + H_2O(l) \rightleftarrows CH_3CO_2H(aq) + OH^-(aq)$

base acid acid base

(d) $PO_4^{3-}(aq) + H_2O(l) \rightleftarrows HPO_4^{2-}(aq) + OH^-(aq)$

base acid acid base

15.113 (a) Na_2CO_3: Na^+ is a neutral cation. CO_3^{2-} is a basic anion.

$CO_3^{2-}(aq) + H_2O(l) \rightleftarrows HCO_3^-(aq) + OH^-(aq)$

base acid acid base

(b) NH_4NO_3: NH_4^+ is an acidic cation. NO_3^- is a neutral anion.

$NH_4^+(aq) + H_2O(l) \rightleftarrows H_3O^+(aq) + NH_3(aq)$

acid base — acid base

(c) NaCl: Na^+ is a neutral cation. Cl^- is a neutral anion.

$H_2O(l) + H_2O(l) \rightleftarrows H_3O^+(aq) + OH^-(aq)$

acid base — acid base

(d) $ZnCl_2$: Zn^{2+} is an acidic cation. Cl^- is a neutral anion.

$Zn(H_2O)_6^{2+}(aq) + H_2O(l) \rightleftarrows H_3O^+(aq) + Zn(H_2O)_5(OH)^+(aq)$

acid base — acid base

15.114 (a) F^- (conjugate base of a weak acid), basic solution
(b) Br^- (anion of a strong acid), neutral solution
(c) NH_4^+ (conjugate acid of a weak base), acidic solution
(d) $K(H_2O)_6^+$ (neutral cation), neutral solution
(e) SO_3^{2-} (conjugate base of a weak acid), basic solution
(f) $Cr(H_2O)_6^{3+}$ (acidic cation), acidic solution

15.115 (a) $Fe(NO_3)_3$: Fe^{3+}, acidic cation; NO_3^-, neutral anion; solution is acidic
(b) $Ba(NO_3)_2$: Ba^{2+}, neutral cation; NO_3^-, neutral anion; solution is neutral
(c) NaOCl: Na^+, neutral cation; OCl^-, basic anion; solution is basic
(d) NH_4I: NH_4^+, acidic cation; I^-, neutral anion; solution is acidic
(e) NH_4NO_2; For NH_4^+, $K_a = 5.6 \times 10^{-10}$; for NO_2^-, $K_b = 2.2 \times 10^{-11}$
Because $K_a > K_b$, the solution is acidic.
(f) $(CH_3NH_3)Cl$: $CH_3NH_3^+$, acidic cation; Cl^-, neutral anion; solution is acidic

15.116 (a) $(C_2H_5NH_3)NO_3$: $C_2H_5NH_3^+$, acidic cation; NO_3^-, neutral anion
$C_2H_5NH_2$, $K_b = 6.4 \times 10^{-4}$

$$C_2H_5NH_3^+,\ K_a = \frac{K_w}{K_b \text{ for } C_2H_5NH_2} = \frac{1.0 \times 10^{-14}}{6.4 \times 10^{-4}} = 1.56 \times 10^{-11}$$

	$C_2H_5NH_3^+(aq)$ + $H_2O(l)$	⇄ $H_3O^+(aq)$	+ $C_2H_5NH_2(aq)$
initial (M)	0.10	~0	0
change (M)	–x	+x	+x
equil (M)	0.10 – x	x	x

$$K_a = \frac{[H_3O^+][C_2H_5NH_2]}{[C_2H_5NH_3^+]} = 1.56 \times 10^{-11} = \frac{x^2}{0.10 - x} \approx \frac{x^2}{0.10}$$

Solve for x. $x = 1.25 \times 10^{-6}\ M = 1.2 \times 10^{-6}\ M = [H_3O^+] = [C_2H_5NH_2]$
$pH = -\log[H_3O^+] = -\log(1.25 \times 10^{-6}) = 5.90$
$[C_2H_5NH_3^+] = 0.10 - x = 0.10\ M$; $[NO_3^-] = 0.10\ M$

$$[OH^-] = \frac{K_w}{[H_3O^+]} = \frac{1.0 \times 10^{-14}}{1.25 \times 10^{-6}\ M} = 8.0 \times 10^{-9}$$

(b) $Na(CH_3CO_2)$: Na^+, neutral cation; $CH_3CO_2^-$, basic anion
CH_3CO_2H, $K_a = 1.8 \times 10^{-5}$

$$CH_3CO_2^-,\ K_b = \frac{K_w}{K_a \text{ for } CH_3CO_2H} = \frac{1.0 \times 10^{-14}}{1.8 \times 10^{-5}} = 5.6 \times 10^{-10}$$

	$CH_3CO_2^-(aq)$ + $H_2O(aq)$	⇄ $CH_3CO_2H(aq)$	+ $OH^-(aq)$
initial (M)	0.10	0	~0
change (M)	–x	+x	+x
equil (M)	0.10 – x	x	x

$$K_b = \frac{[CH_3CO_2H][OH^-]}{[CH_3CO_2^-]} = 5.6 \times 10^{-10} = \frac{x^2}{0.10 - x} \approx \frac{x^2}{0.10}$$

Solve for x. $x = 7.5 \times 10^{-6}\ M = [CH_3CO_2H] = [OH^-]$

$[CH_3CO_2^-] = 0.10 - x = 0.10$ M; $[Na^+] = 0.10$ M

$$[H_3O^+] = \frac{K_w}{[OH^-]} = \frac{1.0 \times 10^{-14}}{7.5 \times 10^{-6}} = 1.3 \times 10^{-9} \text{ M}$$

$pH = -\log[H_3O^+] = -\log(1.3 \times 10^{-9}) = 8.89$

(c) $NaNO_3$: Na^+, neutral cation; NO_3^-, neutral anion

$[Na^+] = [NO_3^-] = 0.10$ M

$[H_3O^+] = [OH^-] = 1.0 \times 10^{-7}$ M; pH = 7.00

15.117 (a)

	$Fe(H_2O)_6^{2+}(aq)$ + $H_2O(l)$	⇄	$H_3O^+(aq)$	+	$Fe(H_2O)_5(OH)^+(aq)$
initial (M)	0.020		~0		0
change (M)	–x		+x		+x
equil (M)	0.020 – x		x		x

$$K_a = \frac{[H_3O^+][Fe(H_2O)_5(OH)^+]}{[Fe(H_2O)_6^{2+}]} = 3.2 \times 10^{-10} = \frac{x^2}{0.020 - x} \approx \frac{x^2}{0.020}$$

Solve for x. $x = [H_3O^+] = 2.5 \times 10^{-6}$ M

$pH = -\log[H_3O^+] = -\log(2.5 \times 10^{-6}) = 5.60$

$$\% \text{ dissociation} = \frac{[Fe(H_2O)_6^{2+}]_{diss}}{[Fe(H_2O)_6^{2+}]_{initial}} \times 100\% = \frac{2.5 \times 10^{-6} \text{ M}}{0.020 \text{ M}} \times 100\% = 0.012\%$$

(b)

	$Fe(H_2O)_6^{3+}(aq)$ + $H_2O(l)$	⇄	$H_3O^+(aq)$	+	$Fe(H_2O)_5(OH)^{2+}(aq)$
initial (M)	0.020		~0		0
change (M)	–x		+x		+x
equil (M)	0.020 – x		x		x

$$K_a = \frac{[H_3O^+][Fe(H_2O)_5(OH)^{2+}]}{[Fe(H_2O)_6^{3+}]} = 6.3 \times 10^{-3} = \frac{x^2}{0.020 - x}$$

$x^2 + (6.3 \times 10^{-3})x - (1.26 \times 10^{-4}) = 0$

Use the quadratic formula to solve for x.

$$x = \frac{-(6.3 \times 10^{-3}) \pm \sqrt{(6.3 \times 10^{-3})^2 - (4)(-1.26 \times 10^{-4})}}{2(1)} = \frac{(-6.3 \times 10^{-3}) \pm 0.0233}{2}$$

x = 0.0085 and –0.0148

Of the two solutions for x, only the positive value of x has physical meaning because x is the $[H_3O^+]$.

$x = [H_3O^+] = 0.0085$ M

$pH = -\log[H_3O^+] = -\log(0.0085) = 2.07$

$$\% \text{ dissociation} = \frac{[Fe(H_2O)_6^{3+}]_{diss}}{[Fe(H_2O)_6^{3+}]_{initial}} \times 100\% = \frac{0.0085 \text{ M}}{0.020 \text{ M}} \times 100\% = 42\%$$

15.118 For NH_4^+, $K_a = \dfrac{K_w}{K_b \text{ for } NH_3} = \dfrac{1.0 \times 10^{-14}}{1.8 \times 10^{-5}} = 5.6 \times 10^{-10}$

For CN^-, $K_b = \dfrac{K_w}{K_a \text{ for HCN}} = \dfrac{1.0 \times 10^{-14}}{4.9 \times 10^{-10}} = 2.0 \times 10^{-5}$

Because $K_b > K_a$, the solution is basic.

15.119 (a) KBr: K^+, neutral cation; Br^-, neutral anion; solution is neutral
(b) $NaNO_2$: Na^+, neutral cation; NO_2^-, basic anion; solution is basic
(c) NH_4Br: NH_4^+, acidic cation; Br^-, neutral anion; solution is acidic
(d) $ZnCl_2$: Zn^{2+}, acidic cation; Cl^-, neutral anion; solution is acidic
(e) NH_4F

For NH_4^+, $K_a = \dfrac{K_w}{K_b \text{ for } NH_3} = \dfrac{1.0 \times 10^{-14}}{1.8 \times 10^{-5}} = 5.6 \times 10^{-10}$

For F^-, $K_b = \dfrac{K_w}{K_a \text{ for HF}} = \dfrac{1.0 \times 10^{-14}}{3.5 \times 10^{-4}} = 2.9 \times 10^{-11}$

Because $K_a > K_b$, the solution is acidic.

Lewis Acids and Bases (Section 15.16)

15.120 (a) Lewis acid, SiF_4; Lewis base, F^-
(b) Lewis acid, Zn^{2+}; Lewis base, NH_3
(c) Lewis acid, $HgCl_2$; Lewis base, Cl^-
(d) Lewis acid, CO_2; Lewis base, H_2O

15.121 (a) Lewis acid, $BeCl_2$; Lewis base, Cl^-
(b) Lewis acid, Mg^{2+}; Lewis base, H_2O
(c) Lewis acid, SO_3; Lewis base, OH^-
(d) Lewis acid, BF_3; Lewis base, F^-

15.122 (a) $2\ :\ddot{\underset{\cdot\cdot}{F}}:^- + SiF_4 \longrightarrow SiF_6^{2-}$
(b) $4\ \ddot{N}H_3 + Zn^{2+} \longrightarrow Zn(NH_3)_4^{2+}$
(c) $2\ :\ddot{\underset{\cdot\cdot}{Cl}}:^- + HgCl_2 \longrightarrow HgCl_4^{2-}$
(d) $H_2\ddot{O}: + CO_2 \longrightarrow H_2CO_3$

15.123 (a) $2\ :\ddot{\underset{\cdot\cdot}{Cl}}:^- + BeCl_2 \longrightarrow BeCl_4^{2-}$
(b) $6\ H_2\ddot{O}: + Mg^{2+} \longrightarrow Mg(H_2O)_6^{2+}$
(c) $:\ddot{\underset{\cdot\cdot}{O}}H^- + SO_3 \longrightarrow HSO_4^-$
(d) $:\ddot{\underset{\cdot\cdot}{F}}:^- + BF_3 \longrightarrow BF_4^-$

15.124 (a) CN^-, Lewis base
(b) H^+, Lewis acid
(c) H_2O, Lewis base
(d) Fe^{3+}, Lewis acid
(e) OH^-, Lewis base
(f) CO_2, Lewis acid
(g) $P(CH_3)_3$, Lewis base
(h) $B(CH_3)_3$, Lewis acid

15.125 (a) BF_3, F is more electronegative than H
(b) SO_3, more O atoms draw electron density away from S
(c) Sn^{4+}, higher charge
(d) CH_3^+, electron deficient

Chapter Problems

15.126 In aqueous solution:
H_2S acts as an acid only.
HS^- can act as both an acid and a base.
S^{2-} can act as a base only.
H_2O can act as both an acid and a base.
H_3O^+ acts as an acid only.
OH^- acts as a base only.

15.127 OH^- because stronger bases accept a proton from water, yielding OH^-.

15.128 $HCO_3^-(aq) + Al(H_2O)_6^{3+}(aq) \rightarrow H_2O(l) + CO_2(g) + Al(H_2O)_5(OH)^{2+}(aq)$

15.129 (a) $Zn(NO_3)_2$, weakly acidic salt (b) Na_2O, strong base
(c) NaOCl, weakly basic salt (d) $NaClO_4$, neutral salt
(e) $HClO_4$, strong acid
Na_2O < NaOCl < $NaClO_4$ < $Zn(NO_3)_2$ < $HClO_4$

15.130 H_2O, 18.02

$$\text{at } 0\ ^\circ\text{C},\ [H_2O] = \frac{\left(0.9998\text{ g x }\dfrac{1\text{ mol}}{18.02\text{ g}}\right)}{0.001\text{ L}} = 55.48\text{ M}$$

$K_w = [H_3O^+][OH^-]$, for a neutral solution $[H_3O^+] = [OH^-]$

$[H_3O^+] = \sqrt{K_w} = \sqrt{1.14 \text{ x } 10^{-15}} = 3.376 \text{ x } 10^{-8}\text{ M}$

$pH = -\log[H_3O^+] = -\log(3.376 \text{ x } 10^{-8}) = 7.472$

$$\text{fraction dissociated} = \frac{[H_2O]_{diss}}{[H_2O]_{initial}} = \frac{3.376 \text{ x } 10^{-8}\text{ M}}{55.48\text{ M}} = 6.09 \text{ x } 10^{-10}$$

$$\%\text{ dissociation} = \frac{[H_2O]_{diss}}{[H_2O]_{initial}} \text{ x } 100\% = \frac{3.376 \text{ x } 10^{-8}\text{ M}}{55.48\text{ M}} \text{ x } 100\% = 6.09 \text{ x } 10^{-8}\%$$

15.131 $HCN(aq) + H_2O(l) \rightleftarrows H_3O^+(aq) + CN^-(aq);\quad K_a = \dfrac{[H_3O^+][CN^-]}{[HCN]}$

$CN^-(aq) + H_2O(l) \rightleftarrows HCN(aq) + OH^-(aq);\quad K_b = \dfrac{[HCN][OH^-]}{[CN^-]}$

$$K_a \text{ x } K_b = \left(\frac{[H_3O^+][CN^-]}{[HCN]}\right)\left(\frac{[HCN][OH^-]}{[CN^-]}\right) = [H_3O^+][OH^-] = K_w$$

15.132

	$HA(aq)$	+	$H_2O(l)$	$\rightleftarrows$	$H_3O^+(aq)$	+	$A^-(aq)$
initial (M)	0.050				~0		0
change (M)	–x				+x		+x
equil (M)	0.050 – x				x		x

$x = [H_3O^+] = 10^{-pH} = 10^{-2.86} = 1.38 \times 10^{-3}$ M

$$K_a = \frac{[H_3O^+][A^-]}{[HA]} = \frac{x^2}{0.050 - x} = \frac{(1.38 \times 10^{-3})^2}{0.050 - (1.38 \times 10^{-3})} = 3.92 \times 10^{-5} = 3.9 \times 10^{-5}$$

$pK_a = -\log K_a = -\log(3.92 \times 10^{-5}) = 4.41$

15.133

	$HA(aq)$	+	$H_2O(l)$	$\rightleftarrows$	$H_3O^+(aq)$	+	$A^-(aq)$
initial (M)	0.040				~0		0
change (M)	–x				+x		+x
equil (M)	0.040 – x				x		x

$x = [H_3O^+] = 10^{-pH} = 10^{-1.96} = 0.0110$ M

$$K_a = \frac{[H_3O^+][A^-]}{[HA]} = \frac{x^2}{0.040 - x} = \frac{(0.0110)^2}{0.040 - 0.0110} = 4.17 \times 10^{-3} = 4.2 \times 10^{-3}$$

$pK_a = -\log K_a = -\log(4.17 \times 10^{-3}) = 2.38$

15.134 For $C_{10}H_{14}N_2H^+$, $K_{a1} = \dfrac{K_w}{K_{b1} \text{ for } C_{10}H_{14}N_2} = \dfrac{1.0 \times 10^{-14}}{1.0 \times 10^{-6}} = 1.0 \times 10^{-8}$

For $C_{10}H_{14}N_2H_2^{2+}$, $K_{a2} = \dfrac{K_w}{K_{b2} \text{ for } C_{10}H_{14}N_2H^+} = \dfrac{1.0 \times 10^{-14}}{1.3 \times 10^{-11}} = 7.7 \times 10^{-4}$

15.135 $C_6H_5CO_2Na$: Na^+, neutral cation, $C_6H_5CO_2^-$, basic anion

$C_6H_5CO_2H$, $K_a = 6.5 \times 10^{-5}$

$$C_6H_5CO_2^-,\ K_b = \frac{K_w}{K_a \text{ for } C_6H_6CO_2H} = \frac{1.0 \times 10^{-14}}{6.5 \times 10^{-5}} = 1.54 \times 10^{-10}$$

	$C_6H_5CO_2^-(aq)$	+	$H_2O(l)$	$\rightleftarrows$	$C_6H_5CO_2H(aq)$	+	$OH^-(aq)$
initial (M)	0.050				0		~0
change (M)	–x				+x		+x
equil (M)	0.050 – x				x		x

$$K_b = \frac{[C_6H_5CO_2H][OH^-]}{[C_6H_5CO_2^-]} = 1.54 \times 10^{-10} = \frac{x^2}{0.050 - x} \approx \frac{x^2}{0.050}$$

Solve for x. $x = 2.77 \times 10^{-6} = 2.8 \times 10^{-6}$ M $= [C_6H_5CO_2H] = [OH^-]$

$[Na^+] = 0.050$ M; $[C_6H_5CO_2^-] = 0.050 - x = 0.050$ M

$$[H_3O^+] = \frac{K_w}{[OH^-]} = \frac{1.0 \times 10^{-14}}{2.77 \times 10^{-6}} = 3.61 \times 10^{-9} \text{ M} = 3.6 \times 10^{-9} \text{ M}$$

$pH = -\log[H_3O^+] = -\log(3.61 \times 10^{-9}) = 8.44$

15.136 (a) $A^-(aq) + H_2O(l) \rightleftharpoons HA(aq) + OH^-(aq)$; basic

(b) $M(H_2O)_6^{3+}(aq) + H_2O(l) \rightleftharpoons H_3O^+(aq) + M(H_2O)_5(OH)^{2+}(aq)$; acidic

(c) $2\ H_2O(l) \rightleftharpoons H_3O^+(aq) + OH^-(aq)$; neutral

(d) $M(H_2O)_6^{3+}(aq) + A^-(aq) \rightleftharpoons HA(aq) + M(H_2O)_5(OH)^{2+}(aq)$;
acidic because K_a for $M(H_2O)_6^{3+}$ (10^{-4}) is greater than K_b for A^- (10^{-9})

15.137 HCl is a strong acid. $[H_3O^+]_{initial} = 0.10$ M
The dissociation of HF produces a negligible amount of H_3O^+ compared with that produced from HCl.

$HF(aq) + H_2O(l) \rightleftharpoons H_3O^+(aq) + F^-(aq)$

$$K_a = \frac{[H_3O^+][F^-]}{[HF]} = 3.5 \times 10^{-4} = \frac{(0.10)[F^-]}{(0.10)}$$

$[F^-] = K_a = 3.5 \times 10^{-4}$ M; $[HF] = 0.10$ M; $[H_3O^+] = 0.10$ M

$$[OH^-] = \frac{K_w}{[H_3O^+]} = \frac{1.0 \times 10^{-14}}{0.10} = 1.0 \times 10^{-13} \text{ M}$$

$pH = -\log[H_3O^+] = -\log(0.10) = 1.00$

15.138

	$HIO_3(aq)$	+	$H_2O(l)$	$\rightleftharpoons$	$H_3O^+(aq)$	+	$IO_3^-(aq)$
initial (M)	0.0500				~0		0
change (M)	−x				+x		+x
equil (M)	0.0500 − x				x		x

$$K_a = \frac{[H_3O^+][IO_3^-]}{[HIO_3]} = 1.7 \times 10^{-1} = \frac{x^2}{0.0500 - x}$$

$x^2 + 0.17x - 0.0085 = 0$

Use the quadratic formula to solve for x.

$$x = \frac{-(0.17) \pm \sqrt{(0.17)^2 - (4)(1)(-0.0085)}}{2(1)} = \frac{(-0.17) \pm 0.251}{2}$$

x = −0.210 and 0.0405

Of the two solutions for x, only the positive value of x has physical meaning because x is the $[H_3O^+]$.

$x = [H_3O^+] = 0.0405$ M $= 0.040$ M

$pH = -\log[H_3O^+] = -\log(0.040) = 1.39$

$[HIO_3] = 0.0500 - x = 0.0500 - 0.040 = 0.010$ M

$[IO_3^-] = x = 0.040$ M

$$[OH^-] = \frac{K_w}{[H_3O^+]} = \frac{1.0 \times 10^{-14}}{0.040} = 2.5 \times 10^{-13} \text{ M}$$

15.139 $K_{a1} = 10^{-pK_{a1}} = 10^{-2.43} = 3.7 \times 10^{-3}$; $K_{a2} = 10^{-pK_{a2}} = 10^{-4.78} = 1.7 \times 10^{-5}$

	$H_2C_7H_3NO_4(aq)$ + $H_2O(l)$	⇄ $H_3O^+(aq)$	+ $HC_7H_3NO_4^-(aq)$
initial (M)	0.050	~0	0
change (M)	−x	+x	+x
equil (M)	0.050 − x	x	x

$$K_{a1} = \frac{[H_3O^+][HC_7H_3NO_4^-]}{[H_2C_7H_3NO_4]} = 3.7 \times 10^{-3} = \frac{x^2}{0.050 - x}$$

$x^2 + (3.7 \times 10^{-3})x - (1.85 \times 10^{-4}) = 0$

Use the quadratic formula to solve for x.

$$x = \frac{-(3.7 \times 10^{-3}) \pm \sqrt{(3.7 \times 10^{-3})^2 - (4)(1)(-1.85 \times 10^{-4})}}{2(1)} = \frac{-(3.7 \times 10^{-3}) \pm 0.0275}{2}$$

x = 0.012 and −0.016

Of the two solutions for x, only the positive value of x has physical meaning because x is the $[H_3O^+]$.

$x = 0.012\ M = [H_3O^+] = [HC_7H_3NO_4^-]$

$[H_2C_7H_3NO_4] = 0.050 - x = 0.050 - 0.012 = 0.038\ M$

The second dissociation of $H_2C_7H_3NO_4$ produces a negligible amount of H_3O^+ compared with that from the first dissociation.

$HC_7H_3NO_4^-(aq) + H_2O(l) \rightleftarrows H_3O^+(aq) + C_7H_3NO_4^{2-}(aq)$

$$K_{a2} = \frac{[H_3O^+][C_7H_3NO_4^{2-}]}{[HC_7H_3NO_4^-]} = 1.7 \times 10^{-5} = \frac{(0.012)[C_7H_3NO_4^{2-}]}{(0.012)}$$

$[C_7H_3NO_4^{2-}] = K_{a2} = 1.7 \times 10^{-5}\ M$

$$[OH^-] = \frac{K_w}{[H_3O^+]} = \frac{1.0 \times 10^{-14}}{0.012} = 8.3 \times 10^{-13}\ M$$

$pH = -\log[H_3O^+] = -\log(0.012) = 1.92$

15.140 $K_{a1} = 10^{-pK_{a1}} = 10^{-2.89} = 1.3 \times 10^{-3}$; $K_{a2} = 10^{-pK_{a2}} = 10^{-5.51} = 3.1 \times 10^{-6}$

	$H_2C_8H_4O_4(aq)$ + $H_2O(l)$	⇄ $H_3O^+(aq)$	+ $HC_8H_4O_4^-(aq)$
initial (M)	0.0250	~0	0
change (M)	−x	+x	+x
equil (M)	0.0250 − x	x	x

$$K_{a1} = \frac{[H_3O^+][HC_8H_4O_4^-]}{[H_2C_8H_4O_4]} = 1.3 \times 10^{-3} = \frac{x^2}{0.0250 - x}$$

$x^2 + (1.3 \times 10^{-3})x - (3.25 \times 10^{-5}) = 0$

Use the quadratic formula to solve for x.

$$x = \frac{-(1.3 \times 10^{-3}) \pm \sqrt{(1.3 \times 10^{-3})^2 - (4)(1)(-3.25 \times 10^{-5})}}{2(1)} = \frac{-(1.3 \times 10^{-3}) \pm 0.0115}{2}$$

x = 0.0051 and −0.0064

Of the two solutions for x, only the positive value of x has physical meaning because x is the $[H_3O^+]$.

$x = 0.0051\ M = [H_3O^+] = [HC_8H_4O_4^-]$
$[H_2C_8H_4O_4] = 0.0250 - x = 0.0250 - 0.0051 = 0.020\ M$
The second dissociation of $H_2C_7H_3NO_4$ produces a negligible amount of H_3O^+ compared with that from the first dissociation.

$HC_8H_4O_4^-(aq) + H_2O(l) \rightleftarrows H_3O^+(aq) + C_8H_4O_4^{2-}(aq)$

$$K_{a2} = \frac{[H_3O^+][C_8H_4O_4^{2-}]}{[HC_8H_4O_4^-]} = 3.1 \times 10^{-6} = \frac{(0.0051)[C_8H_4O_4^{2-}]}{(0.0051)}$$

$[C_8H_4O_4^{2-}] = K_{a2} = 3.1 \times 10^{-6}\ M$

$$[OH^-] = \frac{K_w}{[H_3O^+]} = \frac{1.0 \times 10^{-14}}{0.0051} = 2.0 \times 10^{-12}\ M$$

$pH = -\log[H_3O^+] = -\log(0.0051) = 2.29$

15.141 $K = 1.33 = \dfrac{[H_2SO_3]}{P_{SO_2}}$

$[H_2SO_3] = 1.33 \times P_{SO_2}$, and because P_{SO_2} is maintained at 1.00 atm, the $[H_2SO_3]$ remains constant at 1.33 M.

	$H_2SO_3(aq) + H_2O(l)$	$\rightleftarrows$	$H_3O^+(aq)$	$+\ HSO_3^-(aq)$
equil (M)	1.33		x	x

$$K_{a1} = \frac{[H_3O^+][HSO_3^-]}{[H_2SO_3]} = 1.5 \times 10^{-2} = \frac{x^2}{1.33}$$

$x = [H_3O^+] = [HSO_3^-] = \sqrt{(1.5 \times 10^{-2})(1.33)} = 0.14\ M$

	$HSO_3^-(aq) + H_2O(l)$	$\rightleftarrows$	$H_3O^+(aq)$	$+\ SO_3^{2-}(aq)$
initial (M)	0.14		0.14	0
change (M)	−y		+y	+y
equil (M)	0.14 − y		0.14 + y	y

$$K_{a2} = \frac{[H_3O^+][SO_3^{2-}]}{[HSO_3^-]} = 6.3 \times 10^{-8} = \frac{(0.14 + y)y}{(0.14 - y)} \approx \frac{(0.14)y}{0.14} = y$$

$y = [SO_3^{2-}] = 6.3 \times 10^{-8}\ M$
$pH = -\log[H_3O^+] = -\log(0.14) = 0.85$

15.142 (a) NH_4F; For NH_4^+, $K_a = 5.6 \times 10^{-10}$ and for F^-, $K_b = 2.9 \times 10^{-11}$
Because $K_a > K_b$, the salt solution is acidic.
(b) $(NH_4)_2SO_3$; For NH_4^+, $K_a = 5.6 \times 10^{-10}$ and for SO_3^{2-}, $K_b = 1.6 \times 10^{-7}$
Because $K_b > K_a$, the salt solution is basic.

15.143 (a)

	$Cr(H_2O)_6^{3+}(aq) + H_2O(l)$	$\rightleftarrows$	$H_3O^+(aq)$	$+\ Cr(H_2O)_5(OH)^{2+}(aq)$
initial (M)	0.010		~0	0
change (M)	−x		+x	+x
equil (M)	0.010 − x		x	x

$$K_a = \frac{[H_3O^+][Cr(H_2O)_5(OH)^{2+}]}{[Cr(H_2O)_6{}^{3+}]} = 1.6 \times 10^{-4} = \frac{x^2}{0.010 - x}$$

$x^2 + (1.6 \times 10^{-4})x - (1.6 \times 10^{-6}) = 0$

Use the quadratic formula to solve for x.

$$x = \frac{-(1.6 \times 10^{-4}) \pm \sqrt{(1.6 \times 10^{-4})^2 - (4)(1)(-1.6 \times 10^{-6})}}{2(1)} = \frac{-(1.6 \times 10^{-4}) \pm 0.00253}{2}$$

x = 0.001 18 and –0.001 34

Of the two solutions for x, only the positive value of x has physical meaning because x is the $[H_3O^+]$.

$x = 0.001\ 18\ M = [H_3O^+] = [Cr(H_2O)_6{}^{3+}]_{diss}$

$pH = -\log[H_3O^+] = -\log(0.001\ 18) = 2.93$

$$\%\ \text{dissociation} = \frac{[Cr(H_2O)_6{}^{3+}]_{diss}}{[Cr(H_2O)_6{}^{3+}]_{initial}} \times 100\% = \frac{0.001\ 18\ M}{0.010\ M} \times 100\% = 12\%$$

(b)	$Cr(H_2O)_6{}^{3+}(aq)$ + $H_2O(l)$	⇄	$H_3O^+(aq)$	+	$Cr(H_2O)_5(OH)^{2+}(aq)$
initial (M)	0.0050		~0		0
change (M)	–x		+x		+x
equil (M)	0.0050 – x		x		x

$$K_a = \frac{[H_3O^+][Cr(H_2O)_5(OH)^{2+}]}{[Cr(H_2O)_6{}^{3+}]} = 1.6 \times 10^{-4} = \frac{x^2}{0.0050 - x}$$

$x^2 + (1.6 \times 10^{-4})x - (8.0 \times 10^{-7}) = 0$

Use the quadratic formula to solve for x.

$$x = \frac{-(1.6 \times 10^{-4}) \pm \sqrt{(1.6 \times 10^{-4})^2 - (4)(1)(-8.0 \times 10^{-7})}}{2(1)} = \frac{-(1.6 \times 10^{-4}) \pm 0.00180}{2}$$

$x = 8.2 \times 10^{-4}$ and -9.8×10^{-4}

Of the two solutions for x, only the positive value of x has physical meaning because x is the $[H_3O^+]$.

$x = 8.2 \times 10^{-4}\ M = [H_3O^+] = [Cr(H_2O)_6{}^{3+}]_{diss}$

$pH = -\log[H_3O^+] = -\log(8.2 \times 10^{-4}) = 3.09$

$$\%\ \text{dissociation} = \frac{[Cr(H_2O)_6{}^{3+}]_{diss}}{[Cr(H_2O)_6{}^{3+}]_{initial}} \times 100\% = \frac{8.2 \times 10^{-4}\ M}{0.0050\ M} \times 100\% = 16\%$$

(c)	$Cr(H_2O)_6{}^{3+}(aq)$ + $H_2O(l)$	⇄	$H_3O^+(aq)$	+	$Cr(H_2O)_5(OH)^{2+}(aq)$
initial (M)	0.0010		~0		0
change (M)	–x		+x		+x
equil (M)	0.0010 – x		x		x

$$K_a = \frac{[H_3O^+][Cr(H_2O)_5(OH)^{2+}]}{[Cr(H_2O)_6{}^{3+}]} = 1.6 \times 10^{-4} = \frac{x^2}{0.0010 - x}$$

$x^2 + (1.6 \times 10^{-4})x - (1.6 \times 10^{-7}) = 0$

Use the quadratic formula to solve for x.

$$x = \frac{-(1.6 \text{ x } 10^{-4}) \pm \sqrt{(1.6 \text{ x } 10^{-4})^2 - (4)(1)(-1.6 \text{ x } 10^{-7})}}{2(1)}$$

$$x = \frac{-(1.6 \text{ x } 10^{-4}) \pm (8.2 \text{ x } 10^{-4})}{2}$$

$x = 3.3 \text{ x } 10^{-4}$ and $-4.9 \text{ x } 10^{-4}$

Of the two solutions for x, only the positive value of x has physical meaning because x is the $[H_3O^+]$.

$x = 3.3 \text{ x } 10^{-4}\text{ M} = [H_3O^+] = [Cr(H_2O)_6^{\ 3+}]_{diss}$

$pH = -\log[H_3O^+] = -\log(3.3 \text{ x } 10^{-4}) = 3.48$

$$\% \text{ dissociation} = \frac{[Cr(H_2O)_6^{\ 3+}]_{diss}}{[Cr(H_2O)_6^{\ 3+}]_{initial}} \text{ x } 100\% = \frac{3.3 \text{ x } 10^{-4}\text{ M}}{0.0010\text{ M}} \text{ x } 100\% = 33\%$$

15.144 Fraction dissociated $= \dfrac{[HA]_{diss}}{[HA]_{initial}}$

For a weak acid, $[HA]_{diss} = [H_3O^+] = [A^-]$

$$K_a = \frac{[H_3O^+][A^-]}{[HA]} = \frac{[H_3O^+]^2}{[HA]}; \qquad [H_3O^+] = \sqrt{K_a[HA]}$$

$$\text{Fraction dissociated} = \frac{[HA]_{diss}}{[HA]} = \frac{[H_3O^+]}{[HA]} = \frac{\sqrt{K_a[HA]}}{[HA]} = \sqrt{\frac{K_a}{[HA]}}$$

When the concentration of HA that dissociates is negligible compared with its initial concentration, the equilibrium concentration, [HA], equals the initial concentration, $[HA]_{initial}$.

$$\% \text{ dissociation} = \sqrt{\frac{K_a}{[HA]_{initial}}} \text{ x } 100\%$$

15.145

	Reaction	K
	$NH_4^+(aq) + H_2O(l) \rightleftarrows H_3O^+(aq) + NH_3(aq)$	$K_1 = K_a$
	$F^-(aq) + H_2O(l) \rightleftarrows HF(aq) + OH^-(aq)$	$K_2 = K_b$
	$H_3O^+(aq) + OH^-(aq) \rightleftarrows 2\ H_2O(l)$	$K_3 = 1/K_w$
Overall reaction	$NH_4^+(aq) + F^-(aq) \rightleftarrows HF(aq) + NH_3(aq)$	$K = K_1K_2K_3$

$$K = K_1K_2K_3 = \frac{K_aK_b}{K_w}$$

The equilibrium constant for the transfer of a proton from the cation of a salt to the anion of the salt is equal to $(K_aK_b)/K_w$.

(a)

	$NH_4^+(aq)$	+ $F^-(aq)$	$\rightleftarrows$ $HF(aq)$	+ $NH_3(aq)$
initial (M)	0.25	0.25	0	0
change (M)	–x	–x	+x	+x
equil (M)	0.25 – x	0.25 – x	x	x

$$K = \frac{K_aK_b}{K_w} = \frac{(5.56 \times 10^{-10})(2.86 \times 10^{-11})}{(1.0 \times 10^{-14})} = \frac{[HF][NH_3]}{[NH_4^+][F^-]} = \frac{x^2}{(0.25 - x)^2}$$

where K_a is K_a for NH_4^+ and K_b is K_b for F^-.
Take the square root of both sides and solve for x.
$x = 3.15 \times 10^{-4}$
$[NH_4^+] = [F^-] = 0.25 - x = 0.25$ M; $[NH_3] = [HF] = x = 3.15 \times 10^{-4}$ M

$$K_a(NH_4^+) = \frac{[H_3O^+][NH_3]}{[NH_4^+]} = \frac{[H_3O^+](3.15 \times 10^{-4})}{0.25} = 5.56 \times 10^{-10}$$

$[H_3O^+] = 4.4 \times 10^{-7}$ M

$$[OH^-] = \frac{K_w}{[H_3O^+]} = \frac{1.0 \times 10^{-14}}{4.4 \times 10^{-7}} = 2.3 \times 10^{-8}\ M$$

$pH = -\log[H_3O^+] = -\log(4.4 \times 10^{-7}) = 6.36$

(b)

	$NH_4^+(aq)$ +	$SO_3^{2-}(aq)$	⇄	$HSO_3^-(aq)$ +	$NH_3(aq)$
initial (M)	0.50	0.25		0	0
change (M)	–x	–x		+x	+x
equil (M)	0.50 – x	0.25 – x		x	x

$$K = \frac{K_aK_b}{K_w} = \frac{(5.56 \times 10^{-10})(1.59 \times 10^{-7})}{(1.0 \times 10^{-14})} = \frac{[HSO_3^-][NH_3]}{[NH_4^+][SO_3^{2-}]} = \frac{x^2}{(0.50 - x)(0.25 - x)}$$

where K_a is K_a for NH_4^+ and K_b is K_b for SO_3^{2-}.
$0.9912x^2 + 0.006\,63x - 0.001\,11 = 0$
Use the quadratic formula to solve for x.

$$x = \frac{-(0.006\,63) \pm \sqrt{(0.006\,63)^2 - 4(0.9912)(-0.001\,11)}}{2(0.9912)} = \frac{-0.006\,63 \pm 0.066\,67}{2(0.9912)}$$

x = 0.0303 and –0.0370
Of the two solutions for x, only the positive value of x has physical meaning, because x is the $[NH_3]$ and $[HSO_3^-]$.
x = 0.030
$[NH_4^+] = 0.50 - x = 0.47$ M; $[SO_3^{2-}] = 0.25 - x = 0.22$ M
$[NH_3] = [HSO_3^-] = x = 0.030$ M

$$K_b(SO_3^{2-}) = \frac{[HSO_3^-][OH^-]}{[SO_3^{2-}]} = \frac{(0.030)[OH^-]}{0.22} = 1.59 \times 10^{-7}$$

$[OH^-] = 1.166 \times 10^{-6}\ M = 1.2 \times 10^{-6}$ M

$$[H_3O^+] = \frac{K_w}{[OH^-]} = \frac{1.0 \times 10^{-14}}{1.166 \times 10^{-6}} = 8.6 \times 10^{-9}\ M$$

$pH = -\log[H_3O^+] = -\log(8.6 \times 10^{-9}) = 8.07$

15.146 Both reactions occur together.
Let $x = [H_3O^+]$ from CH_3CO_2H and $y = [H_3O^+]$ from $C_6H_5CO_2H$
The following two equilibria must be considered:

	$CH_3CO_2H(aq)$ + $H_2O(l)$	⇄ $H_3O^+(aq)$	+ $CH_3CO_2^-(aq)$
initial (M)	0.10	y	0
change (M)	−x	+x	+x
equil (M)	0.10 − x	x + y	x

	$C_6H_5CO_2H(aq)$ + $H_2O(l)$	⇄ $H_3O^+(aq)$	+ $C_6H_5CO_2^-(aq)$
initial (M)	0.10	x	0
change (M)	−y	+y	+y
equil (M)	0.10 − y	x + y	y

$$K_a(\text{for } CH_3CO_2H) = \frac{[H_3O^+][CH_3CO_2^-]}{[CH_3CO_2H]} = 1.8 \times 10^{-5} = \frac{(x+y)(x)}{0.10-x} \approx \frac{(x+y)(x)}{0.10}$$

$1.8 \times 10^{-6} = (x + y)(x)$

$$K_a(\text{for } C_6H_5CO_2H) = \frac{[H_3O^+][C_6H_5CO_2^-]}{[C_6H_5CO_2H]} = 6.5 \times 10^{-5} = \frac{(x+y)(y)}{0.10-y} \approx \frac{(x+y)(y)}{0.10}$$

$6.5 \times 10^{-6} = (x + y)(y)$

$1.8 \times 10^{-6} = (x + y)(x)$

$6.5 \times 10^{-6} = (x + y)(y)$

These two equations must be solved simultaneously for x and y. Divide the first equation by the second.

$$\frac{x}{y} = \frac{1.8 \times 10^{-6}}{6.5 \times 10^{-6}};\ x = 0.277y$$

$6.5 \times 10^{-6} = (x + y)(y)$; substitute $x = 0.277y$ into this equation and solve for y.

$6.5 \times 10^{-6} = (0.277y + y)(y) = 1.277y^2$

$y = 0.002\,256$

$x = 0.277y = (0.277)(0.002\,256) = 0.000\,624\,9$

$[H_3O^+] = (x + y) = (0.000\,624\,9 + 0.002\,256) = 0.002\,881$ M

$pH = -\log[H_3O^+] = -\log(0.002\,881) = 2.54$

15.147 For 1.0×10^{-10} M HCl, pH = 7.00, and the principal source of H_3O^+ is the dissociation of H_2O.

For 1.0×10^{-7} M HCl,

	$2\,H_2O(l)$ ⇄	$H_3O^+(aq)$	+ $OH^-(aq)$
initial (M)		1.0×10^{-7}	~0
change (M)		+x	+x
equil (M)		$(1.0 \times 10^{-7}) + x$	x

$K_w = 1.0 \times 10^{-14} = [H_3O^+][OH^-] = [(1.0 \times 10^{-7}) + x](x)$

$x^2 + (1.0 \times 10^{-7})x - (1.0 \times 10^{-14}) = 0$

Solve for x using the quadratic formula.

$$x = \frac{-(1.0 \times 10^{-7}) \pm \sqrt{(1.0 \times 10^{-7})^2 - (4)(-1.0 \times 10^{-14})}}{2(1)} = \frac{(-1.0 \times 10^{-7}) \pm (2.236 \times 10^{-7})}{2}$$

$x = 6.18 \times 10^{-8}$ and -1.62×10^{-7}

Of the two solutions for x, only the positive value of x has physical meaning, because x is the $[OH^-]$.

$[H_3O^+] = (1.0 \times 10^{-7}) + x = (1.0 \times 10^{-7}) + (6.18 \times 10^{-8}) = 1.618 \times 10^{-7}$ M

$pH = -\log[H_3O^+] = -\log(1.618 \times 10^{-7}) = 6.79$

15.148 $2\ NO_2(g) + H_2O(l) \rightarrow HNO_3(aq) + HNO_2(aq)$

$$\text{mol } HNO_3 = 0.0500 \text{ mol } NO_2 \times \frac{1 \text{ mol } HNO_3}{2 \text{ mol } NO_2} = 0.0250 \text{ mol } HNO_3$$

$$\text{mol } HNO_2 = 0.0500 \text{ mol } NO_2 \times \frac{1 \text{ mol } HNO_2}{2 \text{ mol } NO_2} = 0.0250 \text{ mol } HNO_2$$

Because the volume is 1.00 L, mol and molarity are the same.

HNO_3 is a strong acid and completely dissociated. From HNO_3, $[NO_3^-] = [H_3O^+] = 0.0250$ M.

	$HNO_2(aq)$ + $H_2O(l)$ ⇄	$H_3O^+(aq)$ +	$NO_2^-(aq)$
initial (M)	0.0250	0.0250	0
change (M)	−x	+x	+x
equil (M)	0.0250 − x	0.0250 + x	x

$$K_a = \frac{[H_3O^+][NO_2^-]}{[HNO_2]} = 4.5 \times 10^{-4} = \frac{(0.0250 + x)x}{0.0250 - x}$$

$x^2 + 0.02545x - 1.125 \times 10^{-5} = 0$

Solve for x using the quadratic formula.

$$x = \frac{-(0.025\,45) \pm \sqrt{(0.025\,45)^2 - (4)(-1.125 \times 10^{-5})}}{2(1)} = \frac{(-0.025\,45) \pm (0.026\,32)}{2}$$

$x = -0.0259$ and 4.35×10^{-4}

Of the two solutions for x, only the positive value of x has physical meaning, because x is the $[NO_2^-]$.

$[H_3O^+] = (0.0250) + x = (0.0250) + (4.35 \times 10^{-4}) = 0.0254$ M

$pH = -\log[H_3O^+] = -\log(0.02543) = 1.59$

$$[OH^-] = \frac{K_w}{[H_3O^+]} = \frac{1.0 \times 10^{-14}}{0.0254} = 3.9 \times 10^{-13} \text{ M}$$

$[NO_3^-] = 0.0250$ M

$[HNO_2] = 0.0250 - x = 0.0250 - 4.35 \times 10^{-4} = 0.0246$ M; $[NO_2^-] = x = 4.3 \times 10^{-4}$ M

15.149 (a) $2\ HSol \rightleftarrows H_2Sol^+ + Sol^-$

$K_{HSol} = [H_2Sol^+][Sol^-] = 1 \times 10^{-35}$

(b) $$K_b(CN^-) = \frac{K_{HSol}}{K_a(HCN)} = \frac{1 \times 10^{-35}}{1.3 \times 10^{-13}} = 7.7 \times 10^{-23}$$

	CN^- + HSol ⇄	HCN +	Sol^-
initial (M)	0.010	0	0
change (M)	−x	+x	+x
equil (M)	0.010 − x	x	x

$$K_b = \frac{[HCN][Sol^-]}{[CN^-]} = 7.7 \times 10^{-23} = \frac{x^2}{0.010 - x} \approx \frac{x^2}{0.010}$$

$x = [Sol^-] = \sqrt{(7.7 \times 10^{-23})(0.010)} = 8.8 \times 10^{-13}$ M

$$[H_2Sol^+] = \frac{K_{HSol}}{[Sol^-]} = \frac{1 \times 10^{-35}}{8.8 \times 10^{-13}} = 1 \times 10^{-23} \text{ M}$$

Multiconcept Problems

15.150 H_3PO_4, 98.00

Assume 1.000 L of solution.

$$\text{Mass of solution} = 1.000 \text{ L} \times \frac{1000 \text{ mL}}{1 \text{ L}} \times \frac{1.0353 \text{ g}}{1 \text{ mL}} = 1035.3 \text{ g}$$

Mass H_3PO_4 = (0.070)(1035.3 g) = 72.47 g H_3PO_4

$$\text{mol } H_3PO_4 = 72.47 \text{ g } H_3PO_4 \times \frac{1 \text{ mol } H_3PO_4}{98.00 \text{ g } H_3PO_4} = 0.740 \text{ mol } H_3PO_4$$

$$[H_3PO_4] = \frac{0.740 \text{ mol } H_3PO_4}{1.000 \text{ L}} = 0.740 \text{ M}$$

For the dissociation of the first proton, the following equilibrium must be considered:

	$H_3PO_4(aq)$ + $H_2O(l)$	⇄ $H_3O^+(aq)$	+ $H_2PO_4^-(aq)$
initial (M)	0.740	~0	0
change (M)	–x	+x	+x
equil (M)	0.740 – x	x	x

$$K_{a1} = \frac{[H_3O^+][H_2PO_4^-]}{[H_3PO_4]} = 7.5 \times 10^{-3} = \frac{x^2}{0.740 - x}$$

$x^2 + (7.5 \times 10^{-3})x - (5.55 \times 10^{-3}) = 0$

Solve for x using the quadratic formula.

$$x = \frac{-(7.5 \times 10^{-3}) \pm \sqrt{(7.5 \times 10^{-3})^2 - (4)(-5.55 \times 10^{-3})}}{2(1)} = \frac{(-7.5 \times 10^{-3}) \pm 0.149}{2}$$

x = 0.0708 and –0.0783

Of the two solutions for x, only the positive value of x has physical meaning, because x is the $[H_3O^+]$.

$x = 0.0708 \text{ M} = [H_2PO_4^-] = [H_3O^+]$

For the dissociation of the second proton, the following equilibrium must be considered:

	$H_2PO_4^-(aq)$ + $H_2O(l)$	⇄ $H_3O^+(aq)$	+ $HPO_4^{2-}(aq)$
initial (M)	0.0708	0.0708	0
change (M)	–y	+y	+y
equil (M)	0.0708 – y	0.0708 + y	y

$$K_{a2} = \frac{[H_3O^+][HPO_4^{2-}]}{[H_2PO_4^-]} = 6.2 \times 10^{-8} = \frac{(0.0708 + y)(y)}{0.0708 - y} \approx \frac{(0.0708)(y)}{0.0708} = y$$

$y = 6.2 \times 10^{-8} \text{ M} = [HPO_4^{2-}]$

For the dissociation of the third proton, the following equilibrium must be considered:

	$HPO_4^{2-}(aq)$ + $H_2O(l)$ ⇄	$H_3O^+(aq)$ +	$PO_4^{3-}(aq)$
initial (M)	6.2×10^{-8}	0.0708	0
change (M)	$-z$	$+z$	$+z$
equil (M)	$(6.2 \times 10^{-8}) - z$	$0.0708 + z$	z

$$K_{a3} = \frac{[H_3O^+][PO_4^{3-}]}{[HPO_4^{2-}]} = 4.8 \times 10^{-13} = \frac{(0.0708 + z)(z)}{(6.2 \times 10^{-8}) - z} \approx \frac{(0.0708)(z)}{6.2 \times 10^{-8}}$$

$z = 4.2 \times 10^{-19}\ M = [PO_4^{3-}]$

$[H_3PO_4] = 0.740 - x = 0.740 - 0.0708 = 0.67\ M$

$[H_2PO_4^-] = [H_3O^+] = 0.0708\ M = 0.071\ M$

$[HPO_4^{2-}] = 6.2 \times 10^{-8}\ M$

$[PO_4^{3-}] = 4.2 \times 10^{-19}\ M$

$$[OH^-] = \frac{K_w}{[H_3O^+]} = \frac{1.0 \times 10^{-14}}{0.0708} = 1.4 \times 10^{-13}\ M$$

$pH = -\log[H_3O^+] = -\log(0.0708) = 1.15$

15.151 $C_7H_5NO_3S$, 183.19; 348 mg = 0.348 g

$$[C_7H_5NO_3S] = \frac{\left(0.348\ g \times \frac{1\ mol}{183.19\ g}\right)}{0.100\ L} = 0.0190\ M$$

Let $C_7H_5NO_3S$ = HSac

Let $x = [H_3O^+]$ from HSac and $y = [H_3O^+]$ from H_2O

	HSac(aq) + $H_2O(l)$ ⇄	$H_3O^+(aq)$ +	$Sac^-(aq)$
initial (M)	0.0190	y	0
change (M)	$-x$	$+x$	$+x$
equil (M)	$0.0190 - x$	$x + y$	x

$$K_a = \frac{[H_3O^+][Sac^-]}{[HSac]} = 2.1 \times 10^{-12} = \frac{(x + y)(x)}{0.0190 - x} \approx \frac{(x + y)(x)}{0.0190}$$

$4.0 \times 10^{-14} = (x + y)(x)$

	2 $H_2O(l)$ ⇄ $H_3O^+(aq)$ +	$OH^-(aq)$
initial (M)	x	~0
change (M)	$+y$	$+y$
equil (M)	$x + y$	y

$K_w = [H_3O^+][OH^-] = 1.0 \times 10^{-14} = (x + y)(y)$

$4.0 \times 10^{-14} = (x + y)(x)$

$1.0 \times 10^{-14} = (x + y)(y)$

Solve these two equations simultaneously for x and y. Divide the first equation by the second.

$$\frac{x}{y} = \frac{4.0 \times 10^{-14}}{1.0 \times 10^{-14}} = 4.0; \qquad x = 4.0y$$

$1.0 \times 10^{-14} = (x + y)(y) = (4.0y + y)y = 5.0y^2$

$y = \sqrt{\dfrac{1.0 \times 10^{-14}}{5.0}} = 4.5 \times 10^{-8}$

$x = 4.0y = (4.0)(4.5 \times 10^{-8}) = 1.8 \times 10^{-7}$

$[H_3O^+] = (x + y) = (1.8 \times 10^{-7}) + (4.5 \times 10^{-8}) = 2.25 \times 10^{-7}\ M = 2.2 \times 10^{-7}\ M$

$pH = -\log[H_3O^+] = -\log(2.25 \times 10^{-7}) = 6.65$

15.152 $[H_3O^+] = 10^{-pH} = 10^{-9.07} = 8.51 \times 10^{-10}\ M$

$[H_3O^+][OH^-] = K_w$

$[OH^-] = \dfrac{K_w}{[H_3O^+]} = \dfrac{1.0 \times 10^{-14}}{8.51 \times 10^{-10}} = 1.18 \times 10^{-5}\ M = x$ below.

$K_a = 1.8 \times 10^{-5}$ for CH_3CO_2H and $K_b = \dfrac{K_w}{K_a} = \dfrac{1.0 \times 10^{-14}}{1.8 \times 10^{-5}} = 5.56 \times 10^{-10}$

Use the equilibrium associated with a weak base to solve for $[CH_3CO_2^-] = y$ below.

	$CH_3CO_2^-(aq)$ + $H_2O(l)$	⇄	$CH_3CO_2H(aq)$	+ $OH^-(aq)$
initial (M)	y		~0	0
change (M)	–x		+x	+x
equil (M)	y – x		x	x

$K_b = \dfrac{[CH_3CO_2H][OH^-]}{[CH_3CO_2^-]} = 5.56 \times 10^{-10} = \dfrac{(1.18 \times 10^{-5})^2}{[y - (1.18 \times 10^{-5})]}$

Solve for y.

$(5.56 \times 10^{-10})[y - (1.18 \times 10^{-5})] = (1.18 \times 10^{-5})^2$

$(5.56 \times 10^{-10})y - 6.56 \times 10^{-15} = 1.39 \times 10^{-10}$

$(5.56 \times 10^{-10})y = 1.39 \times 10^{-10}$

$y = (1.39 \times 10^{-10})/(5.56 \times 10^{-10}) = [CH_3CO_2^-] = 0.25\ M$

In 1.00 L of solution, the mass of CH_3CO_2Na solute $= (0.25\ mol/L)\left(\dfrac{82.035\ g\ CH_3CO_2Na}{1\ mol\ CH_3CO_2Na}\right) = 20.5\ g$

mass of solution $= (1000\ mL)\left(\dfrac{1.0085\ g}{1\ mL}\right) = 1008.5\ g$

mass of solvent $= 1008.5\ g - 20.5\ g = 988\ g = 0.988\ kg$

$m = \dfrac{0.25\ mol\ CH_3CO_2Na}{0.988\ kg} = 0.25\ m$

Because CH_3CO_2Na is a strong electrolyte, the ionic compound is completely dissociated and $[CH_3CO_2^-] = [Na^+]$. The contribution of CH_3CO_2H and OH^- to the total molality of the solution is negligible.

$\Delta T_f = K_f \cdot (2 \cdot m) = (1.86\ °C/m)(2)(0.25\ m) = 0.93\ °C$

Solution freezing point $= 0.00\ °C - \Delta T_f = 0.00\ °C - 0.93\ °C = -0.93\ °C$

15.153 (a) SO_2, 64.06; $CaCO_3$, 100.09

$$\text{mass } SO_2 \text{ in g} = 4.0 \times 10^6 \text{ tons} \times \frac{2000 \text{ lb}}{1 \text{ ton}} \times \frac{453.59 \text{ g}}{1 \text{ lb}} = 3.63 \times 10^{12} \text{ g } SO_2$$

$$\text{mol } SO_2 = 3.63 \times 10^{12} \text{ g } SO_2 \times \frac{1 \text{ mol } SO_2}{64.06 \text{ g } SO_2} = 5.67 \times 10^{10} \text{ mol } SO_2$$

$$\text{mol } CaCO_3 \text{ reacted} = 5.67 \times 10^{10} \text{ mol } SO_2 \times \frac{1 \text{ mol } SO_3}{1 \text{ mol } SO_2} \times \frac{1 \text{ mol } H_2SO_4}{1 \text{ mol } SO_3} \times \frac{1 \text{ mol } CaCO_3}{1 \text{ mol } H_2SO_4}$$

$$= 5.67 \times 10^{10} \text{ mol } CaCO_3$$

$$\text{mass } CaCO_3 = 5.67 \times 10^{10} \text{ mol } CaCO_3 \times \frac{100.09 \text{ g } CaCO_3}{1 \text{ mol } CaCO_3} \times \frac{1 \text{ lb}}{453.59 \text{ g}} = 1.25 \times 10^{10} \text{ lbs}$$

(500 lbs)(0.03) = 15 lbs

$$\text{statues damaged} = \frac{1.25 \times 10^{10} \text{ lbs}}{15 \text{ lbs/statue}} = 8.3 \times 10^8 \text{ statues}$$

(b) $$\text{mol } CO_2 = 5.67 \times 10^{10} \text{ mol } CaCO_3 \times \frac{1 \text{ mol } CO_2}{1 \text{ mol } CaCO_3} = 5.67 \times 10^{10} \text{ mol } CO_2$$

$$PV = nRT; \quad V = \frac{nRT}{P} = \frac{(5.67 \times 10^{10} \text{ mol})\left(0.082\ 06 \frac{\text{L} \cdot \text{atm}}{\text{K} \cdot \text{mol}}\right)(293 \text{ K})}{\left(735 \text{ mm Hg} \times \frac{1.00 \text{ atm}}{760 \text{ mm Hg}}\right)} = 1.4 \times 10^{12} \text{ L } CO_2$$

(c) The cation is H_3O^+. O has four charge clouds (one being a lone pair) and is sp^3 hybridized. The geometry of H_3O^+ is trigonal pyramidal.

15.154 Na_3PO_4, 163.94

$$3.28 \text{ g } Na_3PO_4 \times \frac{1 \text{ mol } Na_3PO_4}{163.94 \text{ g } Na_3PO_4} = 0.0200 \text{ mol} = 20.0 \text{ mmol } Na_3PO_4$$

300.0 mL x 0.180 mmol/mL = 54.0 mmol HCl

	$H_3O^+(aq)$ +	$PO_4^{3-}(aq)$ ⇄	$HPO_4^{2-}(aq)$ + $H_2O(l)$
before (mmol)	54.0	20.0	0
change (mmol)	–20.0	–20.0	+20.0
after (mmol)	34.0	0	20.0

	$H_3O^+(aq)$ +	$HPO_4^{2-}(aq)$ ⇄	$H_2PO_4^-(aq)$ + $H_2O(l)$
before (mmol)	34.0	20.0	0
change (mmol)	–20.0	–20.0	+20.0
after (mmol)	14.0	0	20.0

	$H_3O^+(aq)$ +	$H_2PO_4^-(aq)$ ⇄	$H_3PO_4(aq)$ + $H_2O(l)$
before (mmol)	14.0	20.0	0
change (mmol)	–14.0	–14.0	+14.0
after (mmol)	0	6.0	14.0

$[H_3PO_4] = \dfrac{14.0\ \text{mmol}}{300.0\ \text{mL}} = 0.047\ \text{M}$; $[H_2PO_4^-] = \dfrac{6.0\ \text{mmol}}{300.0\ \text{mL}} = 0.020\ \text{M}$

	$H_3PO_4(aq)$ + $H_2O(l)$	⇄ $H_3O^+(aq)$	+ $H_2PO_4^-(aq)$
initial (M)	0.047	~0	0.020
change (M)	–x	+x	+x
equil (M)	0.047 – x	x	0.020 + x

$$K_a = \frac{[H_3O^+][H_2PO_4^{2-}]}{[H_3PO_4]} = 7.5 \times 10^{-3} = \frac{x(0.020 + x)}{(0.047 - x)}$$

$x^2 + 0.0275x - (3.525 \times 10^{-4}) = 0$

Solve for x using the quadratic formula.

$$x = \frac{-(0.0275) \pm \sqrt{(0.0275)^2 - (4)(-3.525 \times 10^{-4})}}{2(1)} = \frac{-0.0275 \pm 0.0465}{2}$$

$x = 0.009\ 52$ and -0.0370

Of the two solutions for x, only the positive value of x has physical meaning, because x is the $[H_3O^+]$.

$pH = -\log[H_3O^+] = -\log(0.009\ 52) = 2.02$

15.155 Let $x = [H_3O^+]$ from $Li(H_2O)_4^+$ and $y = [H_3O^+]$ from H_2O

	$Li(H_2O)_4^+(aq)$ + $H_2O(l)$	⇄ $H_3O^+(aq)$	+ $Li(H_2O)_3(OH)(aq)$
initial (M)	0.10	y	0
change (M)	–x	+x	+x
equil (M)	0.10 – x	x	x

$$K_a = \frac{[H_3O^+][Li(H_2O)_3(OH)]}{[Li(H_2O)_4^+]} = 2.5 \times 10^{-14} = \frac{(x+y)(x)}{0.10 - x} \approx \frac{(x+y)(x)}{0.10}$$

$2.5 \times 10^{-15} = (x + y)(x)$

	$2\ H_2O(l)$ ⇄	$H_3O^+(aq)$	+ $OH^-(aq)$
initial (M)		x	~0
change (M)		+y	+y
equil (M)		x + y	y

$K_w = [H_3O^+][OH^-] = 1.0 \times 10^{-14} = (x + y)(y)$

$2.5 \times 10^{-15} = (x + y)(x)$
$1.0 \times 10^{-14} = (x + y)(y)$

Solve these two equations simultaneously for x and y. Divide the first equation by the second.

$$\frac{x}{y} = \frac{2.5 \times 10^{-15}}{1.0 \times 10^{-14}} = 0.25 \qquad x = 0.25y$$

$1.0 \times 10^{-14} = (x + y)(y) = (0.25y + y)y = (1.25y)y = 1.25y^2$

$$y = \sqrt{\frac{1.0 \times 10^{-14}}{1.25}} = 8.9 \times 10^{-8}$$

$x = 0.25y = (0.25)(8.9 \times 10^{-8}) = 2.24 \times 10^{-8}$

$[H_3O^+] = (x + y) = (2.2 \times 10^{-8}) + (8.9 \times 10^{-8}) = 1.11 \times 10^{-7}$ M
$pH = -\log[H_3O^+] = -\log(1.11 \times 10^{-7}) = 6.95$

15.156 (a) $PV = nRT; \quad n = \dfrac{PV}{RT} = \dfrac{(0.601 \text{ atm})(1.000 \text{ L})}{\left(0.082\ 06 \dfrac{\text{L}\cdot\text{atm}}{\text{K}\cdot\text{mol}}\right)(293.1 \text{ K})} = 0.0250 \text{ mol HF}$

$$50.0 \text{ mL} \times \frac{1.00 \text{ L}}{1000 \text{ mL}} = 0.0500 \text{ L}$$

$$[HF] = \frac{0.0250 \text{ mol HF}}{0.0500 \text{ L}} = 0.500 \text{ M}$$

	$HF(aq)$ +	$H_2O(l)$ ⇄	$H_3O^+(aq)$ +	$F^-(aq)$
initial (M)	0.500		~0	0
change (M)	–x		+x	+x
equil (M)	0.500 – x		x	x

$$K_a = \frac{[H_3O^+][F^-]}{[HF]} = 3.5 \times 10^{-4} = \frac{x^2}{0.500 - x}$$

$x^2 + (3.5 \times 10^{-4})x - (1.75 \times 10^{-4}) = 0$
Solve for x using the quadratic formula.

$$x = \frac{-(3.5 \times 10^{-4}) \pm \sqrt{(3.5 \times 10^{-4})^2 - (4)(-1.75 \times 10^{-4})}}{2(1)} = \frac{(-3.5 \times 10^{-4}) \pm 0.0265}{2}$$

x = –0.0134 and 0.0131
Of the two solutions for x, only the positive value of x has physical meaning, because x is the $[H_3O^+]$.
$pH = -\log[H_3O^+] = -\log(0.0131) = 1.883 = 1.88$

(b) $\% \text{ dissociation} = \dfrac{0.0131 \text{ M}}{0.500} \times 100\% = 2.62\% = 2.6\%$

New % dissociation = (3)(2.62%) = 7.86%
Let X equal the concentration of HF dissociated and Y the new volume (in liters) that would triple the % dissociation.

$$K_a = \frac{X^2}{(0.0250/Y) - X} = 3.5 \times 10^{-4}$$

$\% \text{ dissociation} = \dfrac{X}{(0.0250/Y)} \times 100\% = 7.86\%$ and $\dfrac{X}{(0.0250/Y)} = 0.0786$

$X = 1.965 \times 10^{-3}/Y$
Substitute X into the K_a equation.

$$\frac{(1.965 \times 10^{-3}/Y)^2}{(0.0250/Y) - (1.965 \times 10^{-3}/Y)} = 3.5 \times 10^{-4}$$

$$\frac{3.861 \times 10^{-6}/Y^2}{0.0230/Y} = 3.5 \times 10^{-4}$$

$$\frac{3.861 \times 10^{-6}/Y}{0.0230} = 3.5 \times 10^{-4}$$

$$\frac{3.861 \times 10^{-6}}{8.05 \times 10^{-6}} = Y = 0.48 \text{ L}$$

The result in Problem 15.144 can't be used here because the concentration of HF that dissociates can't be neglected compared with the initial HF concentration.

15.157 PV = nRT; 25 °C = 25 + 273 = 298 K

$$n_{NH_3} = \frac{PV}{RT} = \frac{\left(650.8 \text{ mm Hg} \times \frac{1.00 \text{ atm}}{760 \text{ mm Hg}}\right)(2.000 \text{ L})}{\left(0.082\ 06 \frac{\text{L}\cdot\text{atm}}{\text{K}\cdot\text{mol}}\right)(298 \text{ K})} = 0.0700 \text{ mol}$$

200.0 mL = 0.2000 L; mol CH_3CO_2H = (0.350 mol/L)(0.2000 L) = 0.0700 mol

There are equal moles of NH_3 and CH_3CO_2H.

$[NH_3]$ = 0.0700 mol/0.2000 L = 0.350 M

$CH_3CO_2H(aq) + H_2O(l) \rightleftarrows H_3O^+(aq) + CH_3CO_2^-(aq)$ $\quad K_a = 1.8 \times 10^{-5}$

$NH_3(aq) + H_2O(l) \rightleftarrows NH_4^+(aq) + OH^-(aq)$ $\quad K_b = 1.8 \times 10^{-5}$

$\underline{H_3O^+(aq) + OH^-(aq) \rightleftarrows 2\ H_2O(l)}$ $\quad 1/K_w$

$CH_3CO_2H(aq) + NH_3(aq) \rightleftarrows NH_4^+(aq) + CH_3CO_2^-(aq)$ $\quad K = K_aK_b/K_w = 3.2 \times 10^4$

Because of the magnitude of K, assume 100% neutralization followed by a small amount of the back reaction.

	$CH_3CO_2H(aq)$ +	$NH_3(aq)$ ⇄	$NH_4^+(aq)$ +	$CH_3CO_2^-(aq)$
before reaction (M)	0.350	0.350	0	0
assume 100% reaction	–0.350	–0.350	+0.350	+0.350
after reaction (M)	0	0	0.350	0.350
assume small back rxn	+x	+x	–x	–x
equil (M)	x	x	0.350 – x	0.350 – x

$$K = K_aK_b/K_w = \frac{[NH_4^+][CH_3CO_2^-]}{[CH_3CO_2H][NH_3]} = 3.2 \times 10^4 = \frac{(0.350 - x)^2}{x^2}$$

$$\sqrt{3.2 \times 10^4} = 179 = \frac{0.350 - x}{x}$$

179x = 0.350 – x

180x = 0.350

x = 0.350/180 = 1.9×10^{-3}

$[CH_3CO_2H] = [NH_3] = x = 1.9 \times 10^{-3}$ M

$[NH_4^+] = [CH_3CO_2^-] = 0.350 - x = 0.350 - (1.9 \times 10^{-3}) = 0.348$ M

Because K_a and K_b are equal, the solution is neutral and $[H_3O^+] = [OH^-] = 1.0 \times 10^{-7}$ M

$pH = -\log[H_3O^+] = -\log(1.0 \times 10^{-7}) = 7.00$

15.158 (a) Rate = $k[OCl^-]^x[NH_3]^y[OH^-]^z$

From experiments 1 & 2, the $[OCl^-]$ doubles and the rate doubles, therefore x = 1.
From experiments 2 & 3, the $[NH_3]$ triples and the rate triples, therefore y = 1.
From experiments 3 & 4, the $[OH^-]$ goes up by a factor of 10 and the rate goes down by a factor of 10, therefore z = –1.

$$\text{Rate} = k\frac{[OCl^-][NH_3]}{[OH^-]}$$

$$[H_3O^+] = 10^{-pH} = 10^{-12} = 1 \times 10^{-12}\ M$$

$$[OH^-] = \frac{K_w}{[H_3O^+]} = \frac{1.0 \times 10^{-14}}{1 \times 10^{-12}} = 0.01\ M$$

Using experiment 1: $k = \dfrac{(\text{Rate})[OH^-]}{[OCl^-][NH_3]} = \dfrac{(0.017\ M/s)(0.01\ M)}{(0.001\ M)(0.01\ M)} = 17\ s^{-1}$

(b) $K_1 = \dfrac{[HOCl][OH^-]}{[OCl^-]} = K_b(OCl^-) = \dfrac{K_w}{K_a(HOCl)} = \dfrac{1.0 \times 10^{-14}}{3.5 \times 10^{-8}} = 2.9 \times 10^{-7}$

For second step, Rate = k_2 [HOCl][NH_3]

Multiply the Rate by $\dfrac{K_1}{K_1} = \dfrac{K_1}{\left(\dfrac{[HOCl][OH^-]}{[OCl^-]}\right)}$

$$\text{Rate} = K_1 k_2 \frac{[HOCl][NH_3]}{\left(\dfrac{[HOCl][OH^-]}{[OCl^-]}\right)} = K_1 k_2 \frac{[HOCl][NH_3][OCl^-]}{[HOCl][OH^-]} = K_1 k_2 \frac{[OCl^-][NH_3]}{[OH^-]}$$

$K_1 k_2 = k = 17\ s^{-1}$

$$k_2 = \frac{17\ s^{-1}}{2.9 \times 10^{-7}\ M} = 5.9 \times 10^7\ M^{-1}s^{-1}$$

16 Applications of Aqueous Equilibria

16.1 (a) $HNO_2(aq) + OH^-(aq) \rightleftarrows NO_2^-(aq) + H_2O(l)$; NO_2^- (basic anion), pH > 7.00
(b) $H_3O^+(aq) + NH_3(aq) \rightleftarrows NH_4^+(aq) + H_2O(l)$; NH_4^+ (acidic cation), pH < 7.00
(c) $OH^-(aq) + H_3O^+(aq) \rightleftarrows 2\,H_2O(l)$; pH = 7.00

16.2 (a) $HF(aq) + OH^-(aq) \rightleftarrows H_2O(l) + F^-(aq)$

$$K_n = \frac{K_a}{K_w} = \frac{3.5 \times 10^{-4}}{1.0 \times 10^{-14}} = 3.5 \times 10^{10}$$

(b) $H_3O^+(aq) + OH^-(aq) \rightleftarrows 2\,H_2O(l)$

$$K_n = \frac{1}{K_w} = \frac{1}{1.0 \times 10^{-14}} = 1.0 \times 10^{14}$$

(c) $HF(aq) + NH_3(aq) \rightleftarrows NH_4^+(aq) + F^-(aq)$

$$K_n = \frac{K_a K_b}{K_w} = \frac{(3.5 \times 10^{-4})(1.8 \times 10^{-5})}{1.0 \times 10^{-14}} = 6.3 \times 10^5$$

The tendency to proceed to completion is determined by the magnitude of K_n. The larger the value of K_n, the further does the reaction proceed to completion.
The tendency to proceed to completion is: reaction (c) < reaction (a) < reaction (b)

16.3

	$HCN(aq)$	+ $H_2O(l)$	$\rightleftarrows$ $H_3O^+(aq)$	+ $CN^-(aq)$
initial (M)	0.025		~0	0.010
change (M)	–x		+x	+x
equil (M)	0.025 – x		x	0.010 + x

$$K_a = \frac{[H_3O^+][CN^-]}{[HCN]} = 4.9 \times 10^{-10} = \frac{x(0.010 + x)}{0.025 - x} \approx \frac{x(0.010)}{0.025}$$

Solve for x. $x = 1.23 \times 10^{-9}$ M = 1.2×10^{-9} M = $[H_3O^+]$

$pH = -\log[H_3O^+] = -\log(1.23 \times 10^{-9}) = 8.91$

$$[OH^-] = \frac{K_w}{[H_3O^+]} = \frac{1.0 \times 10^{-14}}{1.23 \times 10^{-9}} = 8.1 \times 10^{-6}\ M$$

$[Na^+] = [CN^-] = 0.010$ M; $[HCN] = 0.025$ M

$$\%\ \text{dissociation} = \frac{[HCN]_{diss}}{[HCN]_{initial}} \times 100\% = \frac{1.23 \times 10^{-9}\ M}{0.025\ M} \times 100\% = 4.9 \times 10^{-6}\ \%$$

16.4 On mixing equal volumes of two solutions, both concentrations are cut in half.
$[CH_3NH_2] = 0.10$ M; $[CH_3NH_3Cl] = 0.30$ M

	$CH_3NH_2(aq)$ + $H_2O(l)$	⇄ $CH_3NH_3^+(aq)$	+ $OH^-(aq)$
initial (M)	0.10	0.30	~0
change (M)	–x	+x	+x
equil (M)	0.10 – x	0.30 + x	x

$$K_b = \frac{[CH_3NH_3^+][OH^-]}{[CH_3NH_2]} = 3.7 \times 10^{-4} = \frac{(0.30 + x)x}{0.10 - x} \approx \frac{(0.30)x}{0.10}$$

Solve for x. $x = [OH^-] = 1.2 \times 10^{-4}$ M

$$[H_3O^+] = \frac{K_w}{[OH^-]} = \frac{1.0 \times 10^{-14}}{1.2 \times 10^{-4}} = 8.3 \times 10^{-11}\ M$$

$pH = -\log[H_3O^+] = -\log(8.3 \times 10^{-11}) = 10.08$

16.5 Both solutions contain the same number of HF molecules but solution 2 also contains five F^- ions. The dissociation equilibrium lies farther to the left for solution 2, and therefore solution 2 has the lower $[H_3O^+]$ and the higher pH. For solution 1, no common ion is present to suppress the dissociation of HF, and therefore solution 1 has the larger percent dissociation.

16.6 Each solution contains the same number of B molecules. The presence of BH^+ from BHCl lowers the percent dissociation of B. Solution (2) contains no BH^+, therefore it has the largest percent dissociation. BH^+ is the conjugate acid of B. Solution (1) has the largest amount of BH^+, and it would be the most acidic solution and have the lowest pH.

16.7

	$HF(aq)$ + $H_2O(l)$	⇄ $H_3O^+(aq)$	+ $F^-(aq)$
initial (M)	0.25	~0	0.50
change (M)	–x	+x	+x
equil (M)	0.25 – x	x	0.50 + x

$$K_a = \frac{[H_3O^+][F^-]}{[HF]} = 3.5 \times 10^{-4} = \frac{x(0.50 + x)}{0.25 - x} \approx \frac{x(0.50)}{0.25}$$

Solve for x. $x = 1.75 \times 10^{-4}\ M = [H_3O^+]$

For the buffer, $pH = -\log[H_3O^+] = -\log(1.75 \times 10^{-4}) = 3.76$

(a) mol HF = 0.025 mol; mol F^- = 0.050 mol; vol = 0.100 L

	$F^-(aq)$	+ $H_3O^+(aq)$	$\xrightarrow{100\%}$ $HF(aq)$ + $H_2O(l)$
before (mol)	0.050	0.002	0.025
change (mol)	–0.002	–0.002	+0.002
after (mol)	0.048	0	0.027

$$[H_3O^+] = K_a \frac{[HF]}{[F^-]} = (3.5 \times 10^{-4})\left(\frac{0.27}{0.48}\right) = 1.97 \times 10^{-4}\ M$$

$pH = -\log[H_3O^+] = -\log(1.97 \times 10^{-4}) = 3.71$

(b) mol HF = 0.025 mol; mol F^- = 0.050 mol; vol = 0.100 L

	$HF(aq)$	+ $OH^-(aq)$	$\xrightarrow{100\%}$ $F^-(aq)$	+ $H_2O(l)$
before (mol)	0.025	0.004	0.050	
change (mol)	–0.004	–0.004	+0.004	
after (mol)	0.021	0	0.054	

$$[H_3O^+] = K_a \frac{[HF]}{[F^-]} = (3.5 \times 10^{-4})\left(\frac{0.21}{0.54}\right) = 1.36 \times 10^{-4}\ M$$

$pH = -\log[H_3O^+] = -\log(1.36 \times 10^{-4}) = 3.87$

16.8 (a)

	$HF(aq)$	+ $H_2O(l)$ ⇄	$H_3O^+(aq)$	+ $F^-(aq)$
initial (M)	0.050		~0	0.100
change (M)	–x		+x	+x
equil (M)	0.050 – x		x	0.100 + x

$$K_a = \frac{[H_3O^+][F^-]}{[HF]} = 3.5 \times 10^{-4} = \frac{x(0.100 + x)}{0.050 - x} \approx \frac{x(0.100)}{0.050}$$

Solve for x. x = $[H_3O^+] = 1.75 \times 10^{-4}$ M

$pH = -\log[H_3O^+] = -\log(1.75 \times 10^{-4}) = 3.76$

mol HF = 0.050 mol/L x 0.100 L = 0.0050 mol HF

mol F^- = 0.100 mol/L x 0.100 L = 0.0100 mol F^-

mol HNO_3 = mol H_3O^+ = 0.002 mol

Neutralization reaction:	$F^-(aq)$	+ $H_3O^+(aq)$	$\xrightarrow{100\%}$ $HF(aq)$	+ $H_2O(l)$
before reaction (mol)	0.0100	0.002	0.0050	
change (mol)	–0.002	–0.002	+0.002	
after reaction (mol)	0.008	0	0.007	

$$[HF] = \frac{0.007\ mol}{0.100\ L} = 0.07\ M; \qquad [F^-] = \frac{0.008\ mol}{0.100\ L} = 0.08\ M$$

$$[H_3O^+] = K_a \frac{[HF]}{[F^-]} = (3.5 \times 10^{-4})\frac{(0.07)}{(0.08)} = 3 \times 10^{-4}\ M$$

$pH = -\log[H_3O^+] = -\log(3 \times 10^{-4}) = 3.5$

(b) When the solution is diluted, the acid to conjugate base ratio remains the same and therefore the pH does not change.

16.9 (a) (1) and (3). Both pictures show equal concentrations of HA and A^-.
(b) (3). It contains a higher concentration of HA and A^-.

16.10 When equal volumes of two solutions are mixed together, the concentration of each solution is cut in half.

$$pH = pK_a + \log\frac{[base]}{[acid]} = pK_a + \log\frac{[CO_3^{2-}]}{[HCO_3^-]}$$

For HCO_3^-, $K_a = 5.6 \times 10^{-11}$, $pK_a = -\log K_a = -\log(5.6 \times 10^{-11}) = 10.25$

$$pH = 10.25 + \log\left(\frac{0.050}{0.10}\right) = 10.25 - 0.30 = 9.95$$

16.11 (a) $pH = pK_a + \log\frac{[base]}{[acid]} = 9.15 + \log\left(\frac{1}{50}\right) = 7.45$

(b) $\frac{[base]}{[acid]} = \frac{1}{50}$; % dissociation $= \frac{1}{(50+1)} \times 100\% = 1.96\%$

16.12 $pH = pK_a + \log\frac{[base]}{[acid]} = pK_a + \log\frac{[CO_3^{2-}]}{[HCO_3^-]}$

For HCO_3^-, $K_a = 5.6 \times 10^{-11}$, $pK_a = -\log K_a = -\log(5.6 \times 10^{-11}) = 10.25$

$10.40 = 10.25 + \log\frac{[CO_3^{2-}]}{[HCO_3^-]}$; $\log\frac{[CO_3^{2-}]}{[HCO_3^-]} = 10.40 - 10.25 = 0.15$

$$\frac{[CO_3^{2-}]}{[HCO_3^-]} = 10^{0.15} = 1.4$$

To obtain a buffer solution with pH 10.40, make the Na_2CO_3 concentration 1.4 times the concentration of $NaHCO_3$.

16.13 (a) HOCl, $K_a = 3.5 \times 10^{-8}$, $pK_a = 7.46$; HOBr, $K_a = 2.0 \times 10^{-9}$, $pK_a = 8.70$
Choose an acid with a pK_a within 1 pH unit of the desired pH. The desired pH is 7.00, so HOCl–NaOCl is the buffer of choice.

(b) $pH = 7.00 = pK_a + \log\frac{[base]}{[acid]} = 7.46 + \log\frac{[OCl^-]}{[HOCl]}$

$7.00 = 7.46 + \log\frac{[OCl^-]}{[HOCl]}$; $\log\frac{[OCl^-]}{[HOCl]} = 7.00 - 7.46 = -0.46$

$$\frac{[OCl^-]}{[HOCl]} = 10^{-0.46} = 0.35$$

16.14 (a) mol HCl = mol H_3O^+ = 0.100 mol/L x 0.0400 L = 0.004 00 mol
mol NaOH = mol OH^- = 0.100 mol/L x 0.0350 L = 0.003 50 mol

Neutralization reaction:	$H_3O^+(aq)$ +	$OH^-(aq)$ →	$2\,H_2O(l)$
before reaction (mol)	0.004 00	0.003 50	
change (mol)	−0.003 50	−0.003 50	
after reaction (mol)	0.000 50	0	

$$[H_3O^+] = \frac{0.000\ 50\ mol}{(0.0400\ L + 0.0350\ L)} = 6.7 \times 10^{-3}\ M$$

$pH = -\log[H_3O^+] = -\log(6.7 \times 10^{-3}) = 2.17$

(b) mol HCl = mol H_3O^+ = 0.100 mol/L x 0.0400 L = 0.004 00 mol
mol NaOH = mol OH^- = 0.100 mol/L x 0.0450 L = 0.004 50 mol

Neutralization reaction:	$H_3O^+(aq)$	+ $OH^-(aq)$	→ $2\ H_2O(l)$
before reaction (mol)	0.004 00	0.004 50	
change (mol)	–0.004 00	–0.004 00	
after reaction (mol)	0	0.000 50	

$$[OH^-] = \frac{0.000\ 50\ \text{mol}}{(0.0400\ \text{L} + 0.0450\ \text{L})} = 5.9 \times 10^{-3}\ \text{M}$$

$$[H_3O^+] = \frac{K_w}{[OH^-]} = \frac{1.0 \times 10^{-14}}{5.9 \times 10^{-3}} = 1.7 \times 10^{-12}\ \text{M}$$

$pH = -\log[H_3O^+] = -\log(1.7 \times 10^{-12}) = 11.77$

The results obtained here are consistent with the pH data in Table 16.1.

16.15 (a) mol NaOH = mol OH^- = 0.100 mol/L x 0.0400 L = 0.004 00 mol

mol HCl = mol H_3O^+ = 0.0500 mol/L x 0.0600 L = 0.003 00 mol

Neutralization reaction:	$H_3O^+(aq)$	+ $OH^-(aq)$	→ $2\ H_2O(l)$
before reaction (mol)	0.003 00	0.004 00	
change (mol)	–0.003 00	–0.003 00	
after reaction (mol)	0	0.001 00	

$$[OH^-] = \frac{0.001\ 00\ \text{mol}}{(0.0400\ \text{L} + 0.0600\ \text{L})} = 1.0 \times 10^{-2}\ \text{M}$$

$$[H_3O^+] = \frac{K_w}{[OH^-]} = \frac{1.0 \times 10^{-14}}{1.0 \times 10^{-2}} = 1.0 \times 10^{-12}\ \text{M}$$

$pH = -\log[H_3O^+] = -\log(1.0 \times 10^{-12}) = 12.00$

(b) mol NaOH = mol OH^- = 0.100 mol/L x 0.0400 L = 0.004 00 mol

mol HCl = mol H_3O^+ = 0.0500 mol/L x 0.0802 L = 0.004 01 mol

Neutralization reaction:	$H_3O^+(aq)$	+ $OH^-(aq)$	→ $2\ H_2O(l)$
before reaction (mol)	0.004 01	0.004 00	
change (mol)	–0.004 00	–0.004 00	
after reaction (mol)	0.000 01	0	

$$[H_3O^+] = \frac{0.000\ 01\ \text{mol}}{(0.0400\ \text{L} + 0.0802\ \text{L})} = 8.3 \times 10^{-5}\ \text{M}$$

$pH = -\log[H_3O^+] = -\log(8.3 \times 10^{-5}) = 4.08$

(c) mol NaOH = mol OH^- = 0.100 mol/L x 0.0400 L = 0.004 00 mol

mol HCl = mol H_3O^+ = 0.0500 mol/L x 0.1000 L = 0.005 00 mol

Neutralization reaction:	$H_3O^+(aq)$	+ $OH^-(aq)$	→ $2\ H_2O(l)$
before reaction (mol)	0.005 00	0.004 00	
change (mol)	–0.004 00	–0.004 00	
after reaction (mol)	0.001 00	0	

$$[H_3O^+] = \frac{0.001\ 00\ \text{mol}}{(0.0400\ \text{L} + 0.1000\ \text{L})} = 7.1 \times 10^{-3}\ \text{M}$$

$pH = -\log[H_3O^+] = -\log(7.1 \times 10^{-3}) = 2.15$

16.16 mol NaOH required = $\left(\frac{0.016 \text{ mol HOCl}}{\text{L}}\right)(0.100 \text{ L})\left(\frac{1 \text{ mol NaOH}}{1 \text{ mol HOCl}}\right) = 0.0016 \text{ mol}$

vol NaOH required = $(0.0016 \text{ mol})\left(\frac{1 \text{ L}}{0.0400 \text{ mol}}\right) = 0.040 \text{ L} = 40 \text{ mL}$

40 mL of 0.0400 M NaOH are required to reach the equivalence point.

(a) mmol HOCl = 0.016 mmol/mL x 100.0 mL = 1.6 mmol

mmol NaOH = mmol OH^- = 0.0400 mmol/mL x 10.0 mL = 0.400 mmol

Neutralization reaction:	$HOCl(aq)$	+ $OH^-(aq)$	→ $OCl^-(aq)$	+ $H_2O(l)$
before reaction (mmol)	1.6	0.400	0	
change (mmol)	–0.400	–0.400	+0.400	
after reaction (mmol)	1.2	0	0.400	

$[HOCl] = \frac{1.2 \text{ mmol}}{(100.0 \text{ mL} + 10.0 \text{ mL})} = 1.09 \times 10^{-2} \text{ M}$

$[OCl^-] = \frac{0.400 \text{ mmol}}{(100.0 \text{ mL} + 10.0 \text{ mL})} = 3.64 \times 10^{-3} \text{ M}$

	$HOCl(aq) + H_2O(l)$	⇄ $H_3O^+(aq)$	+ $OCl^-(aq)$
initial (M)	0.0109	~0	0.003 64
change (M)	–x	+x	+x
equil (M)	0.0109 – x	x	0.003 64 + x

$K_a = \frac{[H_3O^+][OCl^-]}{[HOCl]} = 3.5 \times 10^{-8} = \frac{x(0.003\ 64 + x)}{0.0109 - x} \approx \frac{x(0.003\ 64)}{0.0109}$

Solve for x. x = $[H_3O^+]$ = 1.05×10^{-7} M

pH = $-\log[H_3O^+] = -\log(1.05 \times 10^{-7}) = 6.98$

(b) Halfway to the equivalence point, $[OCl^-] = [HOCl]$

pH = $pK_a = -\log K_a = -\log(3.5 \times 10^{-8}) = 7.46$

(c) At the equivalence point the solution contains the salt, NaOCl.

mol NaOCl = initial mol HOCl = 0.0016 mol = 1.6 mmol

$[OCl^-] = \frac{1.6 \text{ mmol}}{(100.0 \text{ mL} + 40.0 \text{ mL})} = 1.1 \times 10^{-2} \text{ M}$

For OCl^-, $K_b = \frac{K_w}{K_a \text{ for HOCl}} = \frac{1.0 \times 10^{-14}}{3.5 \times 10^{-8}} = 2.9 \times 10^{-7}$

	$OCl^-(aq) + H_2O(l)$	⇄ $HOCl(aq)$	+ $OH^-(aq)$
initial (M)	0.011	0	~0
change (M)	–x	+x	+x
equil (M)	0.011 – x	x	x

$K_b = \frac{[HOCl][OH^-]}{[OCl^-]} = 2.9 \times 10^{-7} = \frac{x^2}{0.011 - x} \approx \frac{x^2}{0.011}$

Solve for x. x = $[OH^-]$ = 5.65×10^{-5} M

$[H_3O^+] = \frac{K_w}{[OH^-]} = \frac{1.0 \times 10^{-14}}{5.65 \times 10^{-5}} = 1.77 \times 10^{-10} = 1.8 \times 10^{-10} \text{ M}$

pH = $-\log[H_3O^+] = -\log(1.77 \times 10^{-10}) = 9.75$

(d) pH = 9.75 at the equivalence point.
Use thymolphthalein (pH 9.4–10.6). Bromthymol blue (6.0–7.6) is unacceptable because it changes color halfway to the equivalence point. Alizarin yellow (10.1–12.0) could be used, but thymolphthalein is better.

16.17 (a) (3), only HA present (b) (1), HA and A^- present
(c) (4), only A^- present (d) (2), A^- and OH^- present

16.18 (a) mol NaOH required to reach first equivalence point

$$= \left(\frac{0.0800 \text{ mol } H_2SO_3}{L}\right)(0.0400 \text{ L})\left(\frac{1 \text{ mol NaOH}}{1 \text{ mol } H_2SO_3}\right) = 0.003\ 20 \text{ mol}$$

vol NaOH required to reach first equivalence point

$$= (0.003\ 20 \text{ mol})\left(\frac{1 \text{ L}}{0.160 \text{ mol}}\right) = 0.020 \text{ L} = 20.0 \text{ mL}$$

20.0 mL is enough NaOH solution to reach the first equivalence point for the titration of the diprotic acid, H_2SO_3.
For H_2SO_3,

$K_{a1} = 1.5 \times 10^{-2}$, $pK_{a1} = -\log K_{a1} = -\log(1.5 \times 10^{-2}) = 1.82$

$K_{a2} = 6.3 \times 10^{-8}$, $pK_{a2} = -\log K_{a2} = -\log(6.3 \times 10^{-8}) = 7.20$

At the first equivalence point, $pH = \frac{pK_{a1} + pK_{a2}}{2} = \frac{1.82 + 7.20}{2} = 4.51$

(b) mol NaOH required to reach second equivalence point

$$= \left(\frac{0.0800 \text{ mol } H_2SO_3}{L}\right)(0.0400 \text{ L})\left(\frac{2 \text{ mol NaOH}}{1 \text{ mol } H_2SO_3}\right) = 0.006\ 40 \text{ mol}$$

vol NaOH required to reach second equivalence point

$$= (0.006\ 40 \text{ mol})\left(\frac{1 \text{ L}}{0.160 \text{ mol}}\right) = 0.040 \text{ L} = 40.0 \text{ mL}$$

30.0 mL is enough NaOH solution to reach halfway to the second equivalent point.
Halfway to the second equivalence point
$pH = pK_{a2} = -\log K_{a2} = -\log(6.3 \times 10^{-8}) = 7.20$

(c) mmol HSO_3^- = 0.0800 mmol/mL x 40.0 mL = 3.20 mmol
volume NaOH added after first equivalence point = 35.0 mL – 20.0 mL = 15.0 mL
mmol NaOH = mmol OH^- = 0.160 mmol/L x 15.0 mL = 2.40 mmol

Neutralization reaction:	$HSO_3^-(aq)$ +	$OH^-(aq)$ ⇄	$SO_3^{2-}(aq) + H_2O(l)$
before reaction (mmol)	3.20	2.40	0
change (mmol)	–2.40	–2.40	+2.40
after reaction (mmol)	0.80	0	2.40

$$[HSO_3^-] = \frac{0.80 \text{ mmol}}{(40.0 \text{ mL} + 35.0 \text{ mL})} = 0.0107 \text{ M}$$

$$[SO_3^{2-}] = \frac{2.40 \text{ mmol}}{(40.0 \text{ mL} + 35.0 \text{ mL})} = 0.0320 \text{ M}$$

	$HSO_3^-(aq)$ + $H_2O(l)$	⇄ $H_3O^+(aq)$	+ $SO_3^{2-}(aq)$
initial (M)	0.0107	~0	0.0320
change (M)	–x	+x	+x
equil (M)	0.0107 – x	x	0.0320 + x

$$K_a = \frac{[H_3O^+][SO_3^{2-}]}{[HSO_3^-]} = 6.3 \times 10^{-8} = \frac{x(0.0320 + x)}{0.0107 - x} \approx \frac{x(0.0320)}{0.0107}$$

Solve for x. $x = [H_3O^+] = 2.1 \times 10^{-8}$ M

$pH = -\log[H_3O^+] = -\log(2.1 \times 10^{-8}) = 7.68$

16.19 Let H_2A^+ = valine cation

(a) mol NaOH required to reach first equivalence point

$$= \left(\frac{0.0250 \text{ mol } H_2A^+}{L}\right)(0.0400 \text{ L})\left(\frac{1 \text{ mol NaOH}}{1 \text{ mol } H_2A^+}\right) = 0.001\ 00 \text{ mol}$$

vol NaOH required to reach first equivalence point

$$= (0.001\ 00 \text{ mol})\left(\frac{1 \text{ L}}{0.100 \text{ mol}}\right) = 0.0100 \text{ L} = 10.0 \text{ mL}$$

10.0 mL is enough NaOH solution to reach the first equivalence point for the titration of the diprotic acid, H_2A^+.

For H_2A^+,

$K_{a1} = 4.8 \times 10^{-3}$, $pK_{a1} = -\log K_{a1} = -\log(4.8 \times 10^{-3}) = 2.32$

$K_{a2} = 2.4 \times 10^{-10}$, $pK_{a2} = -\log K_{a2} = -\log(2.4 \times 10^{-10}) = 9.62$

At the first equivalence point, $pH = \frac{pK_{a1} + pK_{a2}}{2} = \frac{2.32 + 9.62}{2} = 5.97$

(b) mol NaOH required to reach second equivalence point

$$= \left(\frac{0.0250 \text{ mol } H_2A^+}{L}\right)(0.0400 \text{ L})\left(\frac{2 \text{ mol NaOH}}{1 \text{ mol } H_2A^+}\right) = 0.002\ 00 \text{ mol}$$

vol NaOH required to reach second equivalence point

$$= (0.002\ 00 \text{ mol})\left(\frac{1 \text{ L}}{0.100 \text{ mol}}\right) = 0.0200 \text{ L} = 20.0 \text{ mL}$$

15.0 mL is enough NaOH solution to reach halfway to the second equivalent point.

Halfway to the second equivalence point

$pH = pK_{a2} = -\log K_{a2} = -\log(2.4 \times 10^{-10}) = 9.62$

(c) 20.0 mL is enough NaOH to reach the second equivalence point.

At the second equivalence point

mmol A^- = (0.0250 mmol/mL)(40.0 mL) = 1.00 mmol A^-

solution volume = 40.0 mL + 20.0 mL = 60.0 mL

$$[A^-] = \frac{1.00 \text{ mmol}}{60.0 \text{ mL}} = 0.0167 \text{ M}$$

	$A^-(aq)$ + $H_2O(l)$	⇄	$HA(aq)$	+ $OH^-(aq)$
initial (M)	0.0167		0	~0
change (M)	–x		+x	+x
equil (M)	0.0167 – x		x	x

$$K_b = \frac{K_w}{K_a \text{ for HA}} = \frac{K_w}{K_{a2}} = \frac{1.0 \times 10^{-14}}{2.4 \times 10^{-10}} = 4.17 \times 10^{-5}$$

$$K_b = \frac{[HA][OH^-]}{[A^-]} = 4.17 \times 10^{-5} = \frac{x^2}{0.0167 - x}$$

$x^2 + (4.17 \times 10^{-5})x - (6.964 \times 10^{-7}) = 0$

Use the quadratic formula to solve for x.

$$x = \frac{-(4.17 \times 10^{-5}) \pm \sqrt{(4.17 \times 10^{-5})^2 - (4)(1)(-6.964 \times 10^{-7})}}{2(1)} = \frac{(-4.17 \times 10^{-5}) \pm (1.67 \times 10^{-3})}{2}$$

$x = 8.14 \times 10^{-4}$ and -8.56×10^{-4}

Of the two solutions for x, only the positive value has physical meaning because x is the $[OH^-]$.

$x = [OH^-] = 8.14 \times 10^{-4}$ M

$$[H_3O^+] = \frac{K_w}{[OH^-]} = \frac{1.0 \times 10^{-14}}{8.14 \times 10^{-4}} = 1.23 \times 10^{-11} \text{ M}$$

$pH = -\log[H_3O^+] = -\log(1.23 \times 10^{-11}) = 10.91$

16.20 (a) $K_{sp} = [Ag^+][Cl^-]$ (b) $K_{sp} = [Pb^{2+}][I^-]^2$
(c) $K_{sp} = [Ca^{2+}]^3[PO_4^{3-}]^2$ (d) $K_{sp} = [Cr^{3+}][OH^-]^3$

16.21 Let the number of ions be proportional to its concentration.
For AgX, $K_{sp} = [Ag^+][X^-] \propto (4)(4) = 16$
For AgY, $K_{sp} = [Ag^+][Y^-] \propto (1)(9) = 9$
For AgZ, $K_{sp} = [Ag^+][Z^-] \propto (3)(6) = 18$
(a) AgZ (b) AgY

16.22 $K_{sp} = [Ca^{2+}]^3[PO_4^{3-}]^2 = (2.01 \times 10^{-8})^3(1.6 \times 10^{-5})^2 = 2.1 \times 10^{-33}$

16.23 CaC_2O_4, $K_{sp} = 2.3 \times 10^{-9}$; $[Ca^{2+}] = 3.0 \times 10^{-8}$ M
$K_{sp} = [Ca^{2+}][C_2O_4^{2-}] = (3.0 \times 10^{-8})[C_2O_4^{2-}] = 2.3 \times 10^{-9}$

$$[C_2O_4^{2-}] = \frac{2.3 \times 10^{-9}}{3.0 \times 10^{-8}} = 0.077 \text{ M}$$

minimum $[Na_2C_2O_4] = 0.077$ M

16.24 (a)

	$AgCl(s)$ ⇄	$Ag^+(aq)$	+ $Cl^-(aq)$
equil (M)		x	x

$K_{sp} = [Ag^+][Cl^-] = 1.8 \times 10^{-10} = (x)(x)$

molar solubility $= x = \sqrt{K_{sp}} = 1.3 \times 10^{-5}$ mol/L

AgCl, 143.32, $\text{solubility} = \dfrac{\left(1.3 \times 10^{-5}\text{ mol} \times \dfrac{143.32\text{ g}}{1\text{ mol}}\right)}{1\text{ L}} = 0.0019\text{ g/L}$

(b) $Ag_2CrO_4(s) \rightleftarrows 2\,Ag^+(aq) + CrO_4^{2-}(aq)$

equil (M) 2x x

$K_{sp} = [Ag^+]^2[CrO_4^{2-}] = 1.1 \times 10^{-12} = (2x)^2(x) = 4x^3$

$\text{molar solubility} = x = \sqrt[3]{\dfrac{1.1 \times 10^{-12}}{4}} = 6.5 \times 10^{-5}\text{ mol/L}$

Ag_2CrO_4, 331.73, $\text{solubility} = \dfrac{\left(6.5 \times 10^{-5}\text{ mol} \times \dfrac{331.73\text{ g}}{1\text{ mol}}\right)}{1\text{ L}} = 0.022\text{ g/L}$

Ag_2CrO_4 has both the higher molar and gram solubility, despite its smaller value of K_{sp}.

16.25 $[Ba^{2+}] = [SO_4^{2-}] = 1.05 \times 10^{-5}$ M; $K_{sp} = [Ba^{2+}][SO_4^{2-}] = (1.05 \times 10^{-5})^2 = 1.10 \times 10^{-10}$

16.26 $[Mg^{2+}]_0$ is from 0.10 M $MgCl_2$.

	$MgF_2(s)$ ⇄	$Mg^{2+}(aq)$ +	$2\,F^-(aq)$
initial (M)		0.10	0
change (M)		+x	+2x
equil (M)		0.10 + x	2x

$K_{sp} = 7.4 \times 10^{-11} = [Mg^{2+}][F^-]^2 = (0.10 + x)(2x)^2 \approx (0.10)(4x^2)$

$x = 1.4 \times 10^{-5}$, molar solubility = x = 1.4×10^{-5} M

16.27 pH = 11; $[H_3O^+] = 10^{-pH} = 10^{-11} = 1.0 \times 10^{-11}$ M

$[OH^-] = \dfrac{K_w}{[H_3O^+]} = \dfrac{1.0 \times 10^{-14}}{1.0 \times 10^{-11}} = 0.0010\text{ M}$

	Zn(OH)2(s) ⇄	$Zn^{2+}(aq)$ +	$2\,OH^-(aq)$
equil (M)		x	0.0010

$K_{sp} = 4.1 \times 10^{-17} = [Zn^{2+}][OH^-]^2 = (x)(0.0010)^2$

$x = 4.1 \times 10^{-11}$, molar solubility = x = 4.1×10^{-11} M

16.28 Compounds that contain basic anions are more soluble in acidic solution than in pure water. AgCN, $Al(OH)_3$, and ZnS all contain basic anions.

16.29 $[Cu^{2+}] = (5.0 \times 10^{-3}\text{ mol})/(0.500\text{ L}) = 0.010$ M

	$Cu^{2+}(aq)$ +	$4\,NH_3(aq)$ ⇄	$Cu(NH_3)_4^{2+}(aq)$
before reaction (M)	0.010	0.40	0
assume 100% reaction (M)	–0.010	– 4(0.010)	+0.010
after reaction (M)	0	0.36	0.010
assume small back reaction (M)	+x	+4x	–x
equil (M)	x	0.36 + 4x	0.010 – x

$$K_f = \frac{[Cu(NH_3)_4^{2+}]}{[Cu^{2+}][NH_3]^4} = 5.6 \times 10^{11} = \frac{(0.010 - x)}{(x)(0.36 + 4x)^4} \approx \frac{0.010}{x(0.36)^4}$$

Solve for x. $x = [Cu^{2+}] = 1.1 \times 10^{-12}$ M

16.30 Total solution volume = 25.0 mL + 35.0 mL = 60.0 mL

$$[Au^{3+}] = \frac{(25.0 \text{ mL})(3.0 \times 10^{-2} \text{ M})}{(60.0 \text{ mL})} = 0.0125 \text{ M}$$

$$[CN^-] = \frac{(35.0 \text{ mL})(1.0 \text{ M})}{(60.0 \text{ mL})} = 0.583 \text{ M}$$

	$Au^{3+}(aq)$ +	$2\ CN^-(aq)$ ⇄	$Au(CN)_2^-(aq)$
before reaction (M)	0.0125	0.583	0
assume 100% reaction (M)	−0.0125	−2(0.0125)	+0.0125
after reaction (M)	0	0.558	0.0125
assume small back reaction (M)	+x	+2x	−x
equil (M)	x	0.558 + 2x	0.0125 – x

$$K_f = \frac{[Au(CN)_2^-]}{[Au^{3+}][CN^-]^2} = 2 \times 10^{38} = \frac{(0.0125 - x)}{(x)(0.558 + 2x)^2} \approx \frac{0.0125}{x(0.558)^2}$$

Solve for x. $x = [Au^{3+}] = 2 \times 10^{-40}$ M

16.31

$AgBr(s) \rightleftarrows Ag^+(aq) + Br^-(aq)$ $\quad K_{sp} = 5.4 \times 10^{-13}$

$Ag^+(aq) + 2\ S_2O_3^{2-} \rightarrow Ag(S_2O_3)_2^{3-}(aq)$ $\quad K_f = 4.7 \times 10^{13}$

dissolution reaction $\quad AgBr(s) + 2\ S_2O_3^{2-}(aq) \rightleftarrows Ag(S_2O_3)_2^{3-}(aq) + Br^-(aq)$

$K = (K_{sp})(K_f) = (5.4 \times 10^{-13})(4.7 \times 10^{13}) = 25.4$

	$AgBr(s) + 2\ S_2O_3^{2-}(aq)$ ⇄	$Ag(S_2O_3)_2^{3-}(aq)$ +	$Br^-(aq)$
initial (M)	0.10	0	0
change (M)	−2x	+x	+x
equil (M)	0.10 – 2x	x	x

$$K = \frac{[Ag(S_2O_3)_2^{3-}][Br^-]}{[S_2O_3^{2-}]^2} = 25.4 = \frac{x^2}{(0.10 - 2x)^2}$$

Take the square root of both sides and solve for x.

$$\sqrt{25.4} = \sqrt{\frac{x^2}{(0.10 - 2x)^2}}; \quad 5.04 = \frac{x}{0.10 - 2x}; \quad x = \text{molar solubility} = 0.045 \text{ mol/L}$$

16.32 Step 1: The precipitate is $Cu(OH)_2(s)$. NH_3 is a base and OH^- ions are present in aqueous solution to react with the Cu^{2+} ions.

$$Cu^{2+}(aq) + 2\ OH^-(aq) \rightleftarrows Cu(OH)_2(s)$$

Step 2: In the presence of additional NH_3, the complex ion $Cu(NH_3)_4^{2+}$ forms.

$$Cu(OH)_2(s) + 4\ NH_3(aq) \rightleftarrows Cu(NH_3)_4^{2+}(aq) + 2\ OH^-(aq)$$

16.33 On mixing equal volumes of two solutions, the concentrations of both solutions are cut in half.

For $BaCO_3$, $K_{sp} = 2.6 \times 10^{-9}$

(a) $IP = [Ba^{2+}][CO_3^{2-}] = (1.5 \times 10^{-3})(1.0 \times 10^{-3}) = 1.5 \times 10^{-6}$

$IP > K_{sp}$; a precipitate of $BaCO_3$ will form.

(b) $IP = [Ba^{2+}][CO_3^{2-}] = (5.0 \times 10^{-6})(2.0 \times 10^{-5}) = 1.0 \times 10^{-10}$

$IP < K_{sp}$; no precipitate will form.

16.34 $$pH = pK_a + \log\frac{[base]}{[acid]} = pK_a + \log\frac{[NH_3]}{[NH_4^+]}$$

For NH_4^+, $K_a = 5.6 \times 10^{-10}$, $pK_a = -\log K_a = -\log(5.6 \times 10^{-10}) = 9.25$

$$pH = 9.25 + \log\frac{(0.20)}{(0.20)} = 9.25; \quad [H_3O^+] = 10^{-pH} = 10^{-9.25} = 5.6 \times 10^{-10}\ M$$

$$[OH^-] = \frac{K_w}{[H_3O^+]} = \frac{1.0 \times 10^{-14}}{5.6 \times 10^{-10}} = 1.8 \times 10^{-5}\ M$$

$$[Fe^{2+}] = [Mn^{2+}] = \frac{(25\ mL)(1.0 \times 10^{-3}\ M)}{250\ mL} = 1.0 \times 10^{-4}\ M$$

For $Mn(OH)_2$, $K_{sp} = 2.1 \times 10^{-13}$

$IP = [Mn^{2+}][OH^-]^2 = (1.0 \times 10^{-4})(1.8 \times 10^{-5})^2 = 3.2 \times 10^{-14}$

$IP < K_{sp}$; no precipitate will form.

For $Fe(OH)_2$, $K_{sp} = 4.9 \times 10^{-17}$

$IP = [Fe^{2+}][OH^-]^2 = (1.0 \times 10^{-4})(1.8 \times 10^{-5})^2 = 3.2 \times 10^{-14}$

$IP > K_{sp}$; a precipitate of $Fe(OH)_2$ will form.

16.35 $MS(s) + 2\ H_3O^+(aq) \rightleftarrows M^{2+}(aq) + H_2S(aq) + 2\ H_2O(l)$

$$K_{spa} = \frac{[M^{2+}][H_2S]}{[H_3O^+]^2}$$

For ZnS, $K_{spa} = 3 \times 10^{-2}$; for CdS, $K_{spa} = 8 \times 10^{-7}$

$[Cd^{2+}] = [Zn^{2+}] = 0.005\ M$

Because the two cation concentrations are equal, Q_c is the same for both.

$$Q_c = \frac{[M^{2+}]_t[H_2S]_t}{[H_3O^+]_t^2} = \frac{(0.005)(0.10)}{(0.3)^2} = 6 \times 10^{-3}$$

$Q_c > K_{spa}$ for CdS; CdS will precipitate. $Q_c < K_{spa}$ for ZnS; Zn^{2+} will remain in solution.

16.36 According to Le Châtelier's Principle removing a product will shift the equilibrium toward the products. In the three equilibrium reactions shown, CO_3^{2-} is a product in the last reaction. As organisms use CO_3^{2-} to make their shells, all three equilibrium reactions shift toward the products.

16.37 (a) pH = 8.2; $[H_3O^+] = 10^{-pH} = 10^{-8.2} = 6.3 \times 10^{-9}$ M

$$150 = \frac{[H_3O^+]_{final} - [H_3O^+]_{initial}}{[H_3O^+]_{initial}} \times 100 = \frac{[H_3O^+]_{final} - 6.3 \times 10^{-9}\ M}{6.3 \times 10^{-9}\ M} \times 100$$

$$[H_3O^+]_{final} = 1.6 \times 10^{-8}\ M$$

$pH = -\log[H_3O^+] = -\log(1.6 \times 10^{-8}) = 7.8$

(b) HCO_3^-, $K_a = 5.6 \times 10^{-11}$, $pK_a = 10.25$

$$pH = pK_a + \log\frac{[base]}{[acid]} = 10.25 - \log\frac{[HCO_3^-]}{[CO_3^{2-}]}$$

$$pH = 8.2 = pK_a + \log\frac{[base]}{[acid]} = 8.2 = 10.25 - \log\frac{[HCO_3^-]}{[CO_3^{2-}]}$$

$$\log\frac{[HCO_3^-]}{[CO_3^{2-}]} = 10.25 - 8.2 = 2.05; \quad \frac{[HCO_3^-]}{[CO_3^{2-}]} = 10^{2.05} = 112$$

$$pH = 7.8 = pK_a + \log\frac{[base]}{[acid]} = 7.8 = 10.25 - \log\frac{[HCO_3^-]}{[CO_3^{2-}]}$$

$$\log\frac{[HCO_3^-]}{[CO_3^{2-}]} = 10.25 - 7.8 = 2.45; \quad \frac{[HCO_3^-]}{[CO_3^{2-}]} = 10^{2.45} = 282$$

16.38

	$CaCO_3(s)$ ⇄	$Ca^{2+}(aq)$ +	$CO_3^{2-}(aq)$
equil (M)		x	x

$K_{sp} = [Ca^{2+}][CO_3^{2-}] = 5.0 \times 10^{-9} = (x)(x)$

molar solubility = $x = \sqrt{K_{sp}} = 7.1 \times 10^{-5}$ mol/L

16.39 (a)

$CaCO_3(s) \rightleftarrows Ca^{2+}(aq) + CO_3^{2-}(aq)$ $\quad K_{sp} = 5.0 \times 10^{-9}$

$H_2CO_3(aq) + H_2O(l) \rightleftarrows H_3O^+(aq) + HCO_3^-(aq)$ $\quad K_{a1} = 4.3 \times 10^{-7}$

$CO_3^{2-}(aq) + H_3O^+(aq) \rightleftarrows HCO_3^-(aq) + H_2O(l)$ $\quad 1/K_{a2} = 1/5.6 \times 10^{-11}$

$CaCO_3(s) + H_2CO_3(aq) \rightleftarrows Ca^{2+}(aq) + 2\ HCO_3^-(aq)$ $\quad K = K_{sp} \cdot K_{a1} \cdot (1/K_{a2})$

$K = K_{sp} \cdot K_{a1} \cdot (1/K_{a2}) = (5.0 \times 10^{-9})(4.3 \times 10^{-7})(1/5.6 \times 10^{-11}) = 3.8 \times 10^{-5}$

(b) As atmospheric CO_2 levels rise, the concentration of H_2CO_3 will increase, shifting the equilibrium of the overall reaction to the products. The molar solubility of $CaCO_3$ will increase.

Conceptual Problems

16.40 (4); only A^- and water should be present

16.41 (a) (2) has the highest pH, $[A^-] > [HA]$
(3) has the lowest pH, $[HA] > [A^-]$

(b)
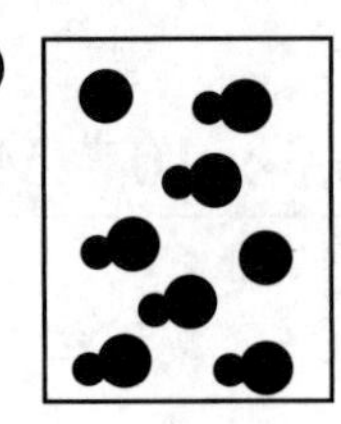
(c)
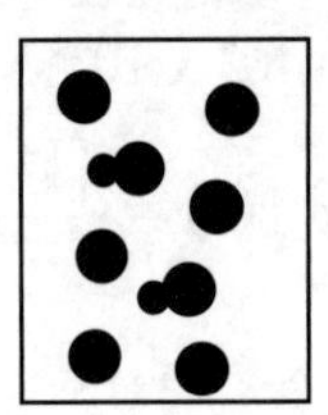

16.42 A buffer solution contains a conjugate acid-base pair in about equal concentrations.
(a) (1), (3), and (4)
(b) (4) because it has the highest buffer concentration.

16.43 (a) (1) corresponds to (iii); (2) to (i); (3) to (iv); and (4) to (ii)
(b) Solution (3) has the highest pH; solution (2) has the lowest pH.

16.44 (a) (i) (1), only B present (ii) (4), equal amounts of B and BH^+ present
(iii) (3), only BH^+ present (iv) (2), BH^+ and H_3O^+ present
(b) The pH is less than 7 because BH^+ is an acidic cation.

16.45 (a) (1) corresponds to (iii); (2) to (i); (3) to (ii); and (4) to (iv)
(b)
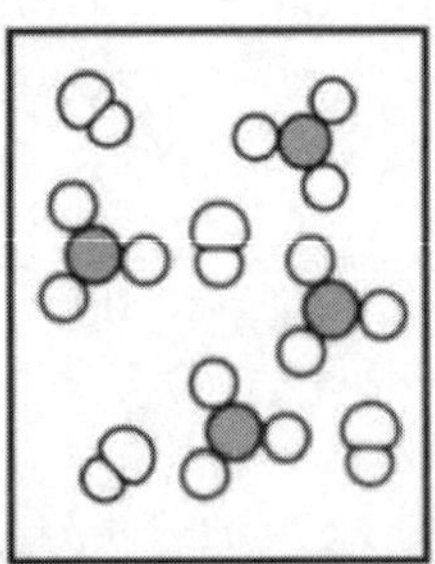

16.46 (a) The lower curve represents the titration of a strong acid; the upper curve represents the titration of a weak acid.
(b) pH = 7 for titration of the strong acid; pH = 10 for titration of the weak acid.
(c) Halfway to the equivalence point, the pH = $pK_a \sim 6.3$.

16.47 (a) There are two equivalence points. It takes 40.0 mL of base to reach the first equivalence point and 80.0 mL to reach the second.
(b) At the first equivalence point the pH is approximately 7.5. At the second equivalence point the pH is approximately 11.
(c) pK_{a1} is equal to the pH halfway to the first equivalence point, $pK_{a1} = 5$. pK_{a2} is equal to the pH halfway between the first and second equivalence point, $pK_{a2} = 10$.

16.48 (2) is supersaturated; (3) is unsaturated; (4) is unsaturated

16.49 Let the number of ions be proportional to its concentration.
For Ag_2CrO_4, $K_{sp} = [Ag^+]^2[CrO_4^{2-}] \propto (4)^2(2) = 32$
For (2), $IP = [Ag^+]^2[CrO_4^{2-}] \propto (2)^2(4) = 16$
For (3), $IP = [Ag^+]^2[CrO_4^{2-}] \propto (6)^2(2) = 72$
For (4), $IP = [Ag^+]^2[CrO_4^{2-}] \propto (2)^2(6) = 24$
A precipitate will form when $IP > K_{sp}$. A precipitate will form only in (3).

Section Problems
Neutralization Reactions (Section 16.1)

16.50 (a) $HNO_3(aq) + KOH(aq) \rightarrow H_2O(l) + KNO_3(aq)$
net ionic equation: $H_3O^+(aq) + OH^-(aq) \rightarrow 2\,H_2O(l)$
The solution at neutralization contains a neutral salt (KNO_3); pH = 7.00.
(b) $2\,HOI(aq) + Ba(OH)_2(aq) \rightarrow 2\,H_2O(l) + Ba(OI)_2(aq)$
net ionic equation: $HOI(aq) + OH^-(aq) \rightarrow H_2O(l) + OI^-(aq)$
The solution at neutralization contains a basic anion (OI^-); pH > 7.00
(c) $HBr(aq) + C_6H_5NH_2(aq) \rightarrow C_6H_5NH_3Br(aq)$
net ionic equation: $H_3O^+(aq) + C_6H_5NH_2(aq) \rightarrow H_2O(l) + C_6H_5NH_3^+(aq)$
The solution at neutralization contains an acidic cation ($C_6H_5NH_3^+$); pH < 7.00.
(d) $HNO_2(aq) + KOH(aq) \rightarrow H_2O(l) + KNO_2(aq)$
net ionic equation: $HNO_2(aq) + OH^-(aq) \rightarrow H_2O(l) + NO_2^-(aq)$
The solution at neutralization contains a basic anion (NO_2^-); pH > 7.00.

16.51 (a) $C_6H_5CO_2H(aq) + NaOH(aq) \rightarrow H_2O(l) + C_6H_5CO_2Na(aq)$
net ionic equation: $C_6H_5CO_2H(aq) + OH^-(aq) \rightarrow H_2O(l) + C_6H_5CO_2^-(aq)$
The solution at neutralization contains a basic anion ($C_6H_5CO_2^-$); pH > 7.00
(b) $HClO_4(aq) + NH_3(aq) \rightarrow NH_4ClO_4(aq)$
net ionic equation: $H_3O^+(aq) + NH_3(aq) \rightarrow H_2O(l) + NH_4^+(aq)$
The solution at neutralization contains an acidic cation (NH_4^+); pH < 7.00
(c) $HI(aq) + KOH(aq) \rightarrow H_2O(l) + KI(aq)$
net ionic equation: $H_3O^+(aq) + OH^-(aq) \rightarrow 2\,H_2O(l)$
The solution at neutralization contains a neutral salt (KI); pH = 7.00
(d) $HOBr(aq) + (CH_3)_3N(aq) \rightarrow (CH_3)_3NHOBr(aq)$
net ionic equation: $HOBr(aq) + (CH_3)_3N(aq) \rightarrow (CH_3)_3NH^+(aq) + OBr^-(aq)$
The solution at neutralization contains the salt $(CH_3)_3NHOBr(aq)$.
$K_a((CH_3)_3NH^+) = 1.5 \times 10^{-10}$ and $K_b(OBr^-) = 5.0 \times 10^{-5}$
$K_b(OBr^-) > K_a((CH_3)_3NH^+)$; pH > 7.00

16.52 (a) After mixing, the solution contains the basic salt, NaCN; pH > 7.00
(b) After mixing, the solution contains the neutral salt, $NaClO_4$; pH = 7.00
Solution (a) has the higher pH.

16.53 (a) After mixing, the solution contains the neutral salt, KI; pH = 7.00
(b) After mixing, the solution contains the acidic salt, NH_4I; pH < 7.00
Solution (b) has the lower pH.

16.54 Weak acid - weak base reaction $$K_n = \frac{K_aK_b}{K_w} = \frac{(1.3 \times 10^{-10})(1.8 \times 10^{-9})}{1.0 \times 10^{-14}} = 2.3 \times 10^{-5}$$
K_n is small so the neutralization reaction does not proceed very far to completion.

16.55 Weak acid - weak base reaction $K_n = \frac{K_aK_b}{K_w} = \frac{(8.0 \times 10^{-5})(4.3 \times 10^{-10})}{1.0 \times 10^{-14}} = 3.4$

Because K_n is close to 1, there will be an appreciable amount of aniline present at equilibrium.

16.56 $K_n = \frac{K_aK_b}{K_w} = 2.1 \times 10^{-4}$; $K_b = \frac{K_nK_w}{K_a} = \frac{(2.1 \times 10^{-4})(1.0 \times 10^{-14})}{(1.4 \times 10^{-4})} = 1.5 \times 10^{-14}$

16.57 $K_n = \frac{K_aK_b}{K_w} = 24$; $K_b = \frac{K_nK_w}{K_a} = \frac{(24)(1.0 \times 10^{-14})}{(5.8 \times 10^{-10})} = 4.1 \times 10^{-4}$

The Common-Ion Effect (Section 16.2)

16.58 (a) $HF(aq) + H_2O(l) \rightleftarrows H_3O^+(aq) + F^-(aq)$

LiF is a source of F^- (reaction product). The equilibrium shifts toward reactants, and the $[H_3O^+]$ decreases. The pH increases.

(b) Because HI is a strong acid, addition of KI, a neutral salt, does not change the pH.

(c) $NH_3(aq) + H_2O(l) \rightleftarrows NH_4^+(aq) + OH^-(aq)$

NH_4Cl is a source of NH_4^+ (reaction product). The equilibrium shifts toward reactants, and the $[OH^-]$ decreases. The pH decreases.

16.59 (a) $NH_3(aq) + H_2O(l) \rightleftarrows NH_4^+(aq) + OH^-(aq)$

NH_4NO_3 is a source of NH_4^+ (reaction product). The equilibrium shifts toward reactants, and the $[OH^-]$ decreases. The pH decreases.

(b) $HCO_3^-(aq) + H_2O(l) \rightleftarrows H_3O^+(aq) + CO_3^{2-}(aq)$

Na_2CO_3 is a source of CO_3^{2-} (reaction product). The equilibrium shifts toward reactants, and the $[H_3O^+]$ decreases. The pH increases.

(c) Because NaOH is a strong base, addition of $NaClO_4$, a neutral salt, does not change the pH.

16.60 For 0.25 M HF and 0.10 M NaF

	$HF(aq) + H_2O(l)$	$\rightleftarrows$	$H_3O^+(aq)$	+ $F^-(aq)$
initial (M)	0.25		~0	0.10
change (M)	–x		+x	+x
equil (M)	0.25 – x		x	0.10 + x

$$K_a = \frac{[H_3O^+][F^-]}{[HF]} = 3.5 \times 10^{-4} = \frac{x(0.10 + x)}{0.25 - x} \approx \frac{x(0.10)}{0.25}$$

Solve for x. $x = [H_3O^+] = 8.8 \times 10^{-4}$ M

$pH = -\log[H_3O^+] = -\log(8.8 \times 10^{-4}) = 3.06$

16.61 From $NH_4Cl(s)$, $[NH_4^+]_{initial} = \frac{0.10 \text{ mol}}{0.500 \text{ L}} = 0.20 \text{ M}$

	$NH_3(aq) + H_2O(l)$	$\rightleftarrows$	$NH_4^+(aq)$	$+ OH^-(aq)$
initial (M)	0.40		0.20	~0
change (M)	–x		+x	+x
equil (M)	0.40 – x		0.20 + x	x

$$K_b = \frac{[NH_4^+][OH^-]}{[NH_3]} = 1.8 \times 10^{-5} = \frac{(0.20 + x)(x)}{(0.40 - x)} \approx \frac{(0.20)(x)}{(0.40)}$$

Solve for x. $x = [OH^-] = 3.6 \times 10^{-5}$ M

$$[H_3O^+] = \frac{K_w}{[OH^-]} = \frac{1.0 \times 10^{-14}}{3.6 \times 10^{-5}} = 2.8 \times 10^{-10} \text{ M}$$

$pH = -\log[H_3O^+] = -\log(2.8 \times 10^{-10}) = 9.55$

16.62 $pH = 4.86$; $[H_3O^+] = 10^{-pH} = 10^{-4.86} = 1.38 \times 10^{-5}$ M

$HN_3(aq) + H_2O(l) \rightleftarrows H_3O^+(aq) + N_3^-(aq)$

$$K_a = \frac{[H_3O^+][N_3^-]}{[HN_3]} = 1.9 \times 10^{-5} = \frac{(1.38 \times 10^{-5})[N_3^-]}{(0.016)}$$

$[N_3^-] = 0.022$ M

16.63 $pH = 8.90$; $[H_3O^+] = 10^{-pH} = 10^{-8.90} = 1.26 \times 10^{-9}$ M

$$[OH^-] = \frac{K_w}{[H_3O^+]} = \frac{1.0 \times 10^{-14}}{1.26 \times 10^{-9}} = 7.9 \times 10^{-6} \text{ M}$$

$NH_3(aq) + H_2O(l) \rightleftarrows NH_4^+(aq) + OH^-(aq)$

$$K_b = \frac{[NH_4^+][OH^-]}{[NH_3]} = 1.8 \times 10^{-5}$$

$$[NH_4^+] = \frac{(K_b)[NH_3]}{[OH^-]} = \frac{(1.8 \times 10^{-5})(0.016)}{(7.9 \times 10^{-6})} = 0.036 \text{ M}$$

16.64 For 0.10 M HN_3:

	$HN_3(aq)$ + $H_2O(l)$	$\rightleftarrows$	$H_3O^+(aq)$	+ $N_3^-(aq)$
initial (M)	0.10		~0	0
change (M)	–x		+x	+x
equil (M)	0.10 – x		x	x

$$K_a = \frac{[H_3O^+][N_3^-]}{[HN_3]} = 1.9 \times 10^{-5} = \frac{x^2}{0.10 - x} \approx \frac{x^2}{0.10}$$

Solve for x. $x = 1.4 \times 10^{-3}$ M

$$\% \text{ dissociation} = \frac{[HN_3]_{diss}}{[HN_3]_{initial}} \times 100\% = \frac{1.4 \times 10^{-3} \text{ M}}{0.10 \text{ M}} \times 100\% = 1.4\%$$

For 0.10 M HN_3 in 0.10 M HCl:

	$HN_3(aq)$ + $H_2O(l)$	⇌ $H_3O^+(aq)$	+ $N_3^-(aq)$
initial (M)	0.10	0.10	0
change (M)	–x	+x	+x
equil (M)	0.10 – x	0.10 + x	x

$$K_a = \frac{[H_3O^+][N_3^-]}{[HN_3]} = 1.9 \times 10^{-5} = \frac{(0.10 + x)(x)}{0.10 - x} \approx \frac{(0.10)(x)}{0.10} = x$$

Solve for x. x = 1.9×10^{-5} M

$$\% \text{ dissociation} = \frac{[HN_3]_{diss}}{[HN_3]_{initial}} \times 100\% = \frac{1.9 \times 10^{-5}\text{ M}}{0.10\text{ M}} \times 100\% = 0.019\%$$

The % dissociation is less because of the common ion (H_3O^+) effect.

16.65

	$NH_3(aq)$ + $H_2O(l)$	⇌ $NH_4^+(aq)$	+ $OH^-(aq)$
initial (M)	0.30	0	~0
change (M)	–x	+x	+x
equil (M)	0.30 – x	x	x

$$K_b = \frac{[NH_4^+][OH^-]}{[NH_3]} = 1.8 \times 10^{-5} = \frac{x^2}{0.30 - x} \approx \frac{x^2}{0.30}$$

Solve for x. x = $[OH^-] = 2.3 \times 10^{-3}$ M

$$[H_3O^+] = \frac{K_w}{[OH^-]} = \frac{1.0 \times 10^{-14}}{2.3 \times 10^{-3}} = 4.3 \times 10^{-12}\text{ M}$$

pH = $-\log[H_3O^+] = -\log(4.3 \times 10^{-12}) = 11.37$

Add 4.0 g of NH_4NO_3.

NH_4NO_3, 80.04; $[NH_4^+]$ = molarity of NH_4NO_3 = $\dfrac{\left(4.0\text{ g} \times \dfrac{1\text{ mol}}{80.04\text{ g}}\right)}{0.100\text{ L}} = 0.50$ M

	$NH_3(aq)$ + $H_2O(l)$	⇌ $NH_4^+(aq)$	+ $OH^-(aq)$
initial (M)	0.30	0.50	~0
change (M)	–x	+x	+x
equil (M)	0.30 – x	0.50 + x	x

$$K_b = \frac{[NH_4^+][OH^-]}{[NH_3]} = 1.8 \times 10^{-5} = \frac{(0.50 + x)x}{0.30 - x} \approx \frac{(0.50)x}{0.30}$$

Solve for x. x = $[OH^-] = 1.1 \times 10^{-5}$ M

$$[H_3O^+] = \frac{K_w}{[OH^-]} = \frac{1.0 \times 10^{-14}}{1.1 \times 10^{-5}} = 9.1 \times 10^{-10}\text{ M}; \quad pH = -\log[H_3O^+] = -\log(9.1 \times 10^{-10}) = 9.04$$

The % dissociation decreases because of the common ion (NH_4^+) effect.

Buffer Solutions (Sections 16.3 and 16.4)

16.66 Solutions (a), (c), and (d) are buffer solutions. Neutralization reactions for (c) and (d) result in solutions with equal concentrations of HF and F^-.

16.67 Solutions (b), (c), and (d) are buffer solutions. Neutralization reactions for (b) and (d) result in solutions with equal concentrations of NH_3 and NH_4^+.

16.68 Both solutions buffer at the same pH because in both cases the $[NO_2^-]/[HNO_2] = 1$. Solution (a), however, has a higher concentration of both HNO_2 and NO_2^-, and therefore it has the greater buffer capacity.

16.69 Both solutions buffer at the same pH because in both cases the $[NH_3]/[NH_4^+] = 1.5$. Solution (b), however, has a higher concentration of both NH_3 and NH_4^+, therefore it has the greater buffer capacity.

16.70 $$pH = pK_a + \log\frac{[base]}{[acid]} = pK_a + \log\frac{[CN^-]}{[HCN]}$$

For HCN, $K_a = 4.9 \times 10^{-10}$, $pK_a = -\log K_a = -\log(4.9 \times 10^{-10}) = 9.31$

$$pH = 9.31 + \log\left(\frac{0.12}{0.20}\right) = 9.09$$

The pH of a buffer solution will not change on dilution because the acid and base concentrations will change by the same amount and their ratio will remain the same.

16.71 $NaHCO_3$, 84.01; Na_2CO_3, 105.99

$$[HCO_3^-] = \text{molarity of } NaHCO_3 = \frac{\left(4.2\text{ g} \times \frac{1\text{ mol}}{84.01\text{ g}}\right)}{0.20\text{ L}} = 0.25\text{ M}$$

$$[CO_3^{2-}] = \text{molarity of } Na_2CO_3 = \frac{\left(5.3\text{ g} \times \frac{1\text{ mol}}{105.99\text{ g}}\right)}{0.20\text{ L}} = 0.25\text{ M}$$

$$pH = pK_a + \log\frac{[base]}{[acid]} = pK_a + \log\frac{[CO_3^{2-}]}{[HCO_3^-]}$$

For HCO_3^-, $K_{a2} = 5.6 \times 10^{-11}$, $pK_{a2} = -\log K_{a2} = -\log(5.6 \times 10^{-11}) = 10.25$

$$pH = 10.25 + \log\frac{[0.25]}{[0.25]} = 10.25$$

The pH of a buffer solution will not change on dilution because the acid and base concentrations will change by the same amount and their ratio will remain the same.

16.72

	$HCO_2H(aq)$ + $H_2O(l)$	⇄	$H_3O^+(aq)$	+ $HCO_2^-(aq)$
initial (M)	0.36		~0	0.30
change (M)	–x		+x	+x
equil (M)	0.36 – x		x	0.30 + x

$$K_a = \frac{[H_3O^+][HCO_2^-]}{[HCO_2H]} = 1.8 \times 10^{-4} = \frac{x(0.30 + x)}{0.36 - x} \approx \frac{x(0.30)}{0.36}$$

Solve for x. $x = 2.16 \times 10^{-4}$ M = $[H_3O^+]$

For the buffer, pH = $-\log[H_3O^+] = -\log(2.16 \times 10^{-4}) = 3.67$

mol HCO_2H = (0.36 mol/L)(0.250 L) = 0.090 mol HCO_2H

mol HCO_2^- = (0.30 mol/L)(0.250 L) = 0.075 mol HCO_2^-

(a)

	$HCO_2H(aq)$	+ $OH^-(aq)$	$\xrightarrow{100\%}$ $HCO_2^-(aq)$	+ $H_2O(l)$
before (mol)	0.090	0.0050	0.075	
change (mol)	–0.0050	–0.0050	+0.0050	
after (mol)	0.085	0	0.080	

$$[H_3O^+] = K_a \frac{[HCO_2H]}{[HCO_2^-]} = (1.8 \times 10^{-4})\left(\frac{0.085}{0.080}\right) = 1.91 \times 10^{-4}\text{ M}$$

pH = $-\log[H_3O^+] = -\log(1.91 \times 10^{-4}) = 3.72$

(b)

	$HCO_2^-(aq)$	+ $H_3O^+(aq)$	$\xrightarrow{100\%}$ $HCO_2H(aq)$	+ $H_2O(l)$
before (mol)	0.075	0.0050	0.090	
change (mol)	–0.0050	–0.0050	+0.0050	
after (mol)	0.070	0	0.095	

$$[H_3O^+] = K_a \frac{[HCO_2H]}{[HCO_2^-]} = (1.8 \times 10^{-4})\left(\frac{0.095}{0.070}\right) = 2.44 \times 10^{-4}\text{ M}$$

pH = $-\log[H_3O^+] = -\log(2.44 \times 10^{-4}) = 3.61$

16.73

	$CH_3CO_2H(aq)$ + $H_2O(l)$	⇄	$H_3O^+(aq)$	+ $CH_3CO_2^-(aq)$
initial (M)	0.18		~0	0.29
change (M)	–x		+x	+x
equil (M)	0.18 – x		x	0.29 + x

$$K_a = \frac{[H_3O^+][CH_3CO_2^-]}{[CH_3CO_2H]} = 1.8 \times 10^{-5} = \frac{x(0.29 + x)}{0.18 - x} \approx \frac{x(0.29)}{0.18}$$

Solve for x. $x = 1.12 \times 10^{-5}$ M = $[H_3O^+]$

For the buffer, pH = $-\log[H_3O^+] = -\log(1.12 \times 10^{-5}) = 4.95$

mol CH_3CO_2H = (0.18 mol/L)(0.375 L) = 0.0675 mol CH_3CO_2H

mol $CH_3CO_2^-$ = (0.29 mol/L)(0.375 L) = 0.109 mol $CH_3CO_2^-$

(a)

	$CH_3CO_2H(aq)$	+ $OH^-(aq)$	$\xrightarrow{100\%}$ $CH_3CO_2^-(aq)$	+ $H_2O(l)$
before (mol)	0.0675	0.0060	0.109	
change (mol)	–0.0060	–0.0060	+0.0060	
after (mol)	0.0615	0	0.115	

$[H_3O^+] = K_a \frac{[CH_3CO_2H]}{[CH_3CO_2^-]} = (1.8 \times 10^{-5})\left(\frac{0.0615}{0.115}\right) = 9.63 \times 10^{-6}\ M$

$pH = -\log[H_3O^+] = -\log(9.63 \times 10^{-6}) = 5.02$

(b)

	$CH_3CO_2^-(aq)$	+	$H_3O^+(aq)$	$\xrightarrow{100\%}$	$CH_3CO_2H(aq)$	+	$H_2O(l)$
before (mol)	0.109		0.0060		0.0675		
change (mol)	–0.0060		–0.0060		+0.0060		
after (mol)	0.103		0		0.0735		

$[H_3O^+] = K_a \frac{[CH_3CO_2H]}{[CH_3CO_2^-]} = (1.8 \times 10^{-5})\left(\frac{0.0735}{0.103}\right) = 1.28 \times 10^{-5}\ M$

$pH = -\log[H_3O^+] = -\log(1.28 \times 10^{-5}) = 4.89$

16.74 $pH = pK_a + \log\frac{[base]}{[acid]} = pK_a + \log\frac{[HCO_2^-]}{[HCO_2H]}$

For HCO_2H, $K_a = 1.8 \times 10^{-4}$; $pK_a = -\log K_a = -\log(1.8 \times 10^{-4}) = 3.74$

$pH = 3.74 + \log\frac{(0.50)}{(0.25)} = 4.04$

16.75 $pH = pK_a + \log\frac{[base]}{[acid]} = pK_a + \log\frac{[C_3H_5O_3^-]}{[HC_3H_5O_3]}$

For $HC_3H_5O_3$, $K_a = 1.4 \times 10^{-4}$; $pK_a = -\log K_a = -\log(1.4 \times 10^{-4}) = 3.85$

$pH = 3.85 + \log\frac{(0.36)}{(0.58)} = 3.64$

16.76 $pH = pK_a + \log\frac{[base]}{[acid]} = pK_a + \log\frac{[HCO_3^-]}{[H_2CO_3]}$

For H_2CO_3, at 37 °C, $K_a = 7.9 \times 10^{-7}$; $pK_a = -\log K_a = -\log(7.9 \times 10^{-7}) = 6.10$

$7.40 = 6.10 + \log\frac{[HCO_3^-]}{[H_2CO_3]}$; $1.30 = \log\frac{[HCO_3^-]}{[H_2CO_3]}$

$\frac{[HCO_3^-]}{[H_2CO_3]} = 10^{1.30} = 20.0$

16.77 $pH = pK_a + \log\frac{[base]}{[acid]} = pK_a + \log\frac{[HCO_3^-]}{[H_2CO_3]}$

For H_2CO_3, at 37 °C, $K_a = 7.9 \times 10^{-7}$; $pK_a = -\log K_a = -\log(7.9 \times 10^{-7}) = 6.10$

$pH = 6.10 + \log(10) = 7.10$

16.78 $\text{pH} = \text{p}K_a + \log\frac{[\text{base}]}{[\text{acid}]} = \text{p}K_a + \log\frac{[NH_3]}{[NH_4^+]}$

For NH_4^+, $K_a = 5.6 \times 10^{-10}$; $\text{p}K_a = -\log K_a = -\log(5.6 \times 10^{-10}) = 9.25$

$9.80 = 9.25 + \log\frac{[NH_3]}{[NH_4^+]}$; $0.550 = \log\frac{[NH_3]}{[NH_4^+]}$; $\frac{[NH_3]}{[NH_4^+]} = 10^{0.55} = 3.5$

The volume of the 1.0 M NH_3 solution should be 3.5 times the volume of the 1.0 M NH_4Cl solution so that the mixture will buffer at pH 9.80.

16.79 $\text{pH} = \text{p}K_a + \log\frac{[\text{base}]}{[\text{acid}]} = \text{p}K_a + \log\frac{[CH_3CO_2^-]}{[CH_3CO_2H]}$

For CH_3CO_2H, $K_a = 1.8 \times 10^{-5}$; $\text{p}K_a = -\log K_a = -\log(1.8 \times 10^{-5}) = 4.74$

$4.44 = 4.74 + \log\frac{[CH_3CO_2^-]}{[CH_3CO_2H]}$; $-0.30 = \log\frac{[CH_3CO_2^-]}{[CH_3CO_2H]}$

$\frac{[CH_3CO_2^-]}{[CH_3CO_2H]} = 10^{-0.30} = 0.50$

The solution should have 0.50 mol of $CH_3CO_2^-$ per mole of CH_3CO_2H. For example, you could dissolve 41g of CH_3CO_2Na in 1.00 L of 1.00 M CH_3CO_2H.

16.80 H_3PO_4, $K_{a1} = 7.5 \times 10^{-3}$; $\text{p}K_{a1} = -\log K_{a1} = 2.12$

$H_2PO_4^-$, $K_{a2} = 6.2 \times 10^{-8}$; $\text{p}K_{a2} = -\log K_{a2} = 7.21$

HPO_4^{2-}, $K_{a3} = 4.8 \times 10^{-13}$; $\text{p}K_{a3} = -\log K_{a3} = 12.32$

The buffer system of choice for pH 7.00 is (b) $H_2PO_4^- - HPO_4^{2-}$ because the $\text{p}K_a$ for $H_2PO_4^-$ (7.21) is closest to 7.00.

16.81 HSO_4^-, $K_{a2} = 1.2 \times 10^{-2}$; $\text{p}K_{a2} = -\log K_{a2} = 1.92$

HOCl, $K_a = 3.5 \times 10^{-8}$; $\text{p}K_a = -\log K_a = 7.46$

$C_6H_5CO_2H$, $K_a = 6.5 \times 10^{-5}$; $\text{p}K_a = -\log K_a = 4.19$

The buffer system of choice for pH = 4.50 is (c) $C_6H_5CO_2H$ - $C_6H_5CO_2^-$ because the $\text{p}K_a$ for $C_6H_5CO_2H$ (4.19) is closest to 4.50.

16.82 $\text{pH} = \text{p}K_a + \log\frac{[\text{base}]}{[\text{acid}]}$

$9.46 = \text{p}K_a + \log\left(\frac{34.5}{100 - 34.5}\right)$

$\text{p}K_a = 9.46 - \log\left(\frac{34.5}{100 - 34.5}\right) = 9.74$

$K_a = 10^{-\text{p}K_a} = 10^{-9.74} = 1.8 \times 10^{-10}$

16.83 $pH = pK_a + \log \frac{[base]}{[acid]}$

$$9.95 = pK_a + \log\left(\frac{86.3}{100 - 86.3}\right)$$

$$pK_a = 9.95 - \log\left(\frac{86.3}{100 - 86.3}\right) = 9.15$$

$$K_a = 10^{-pK_a} = 10^{-9.15} = 7.1 \times 10^{-10}$$

pH Titration Curves (Sections 16.5–16.9)

16.84 (a) (0.060 L)(0.150 mol/L)(1000 mmol/mol) = 9.00 mmol HNO_3

(b) $\text{vol NaOH} = (9.00 \text{ mmol } HNO_3)\left(\frac{1 \text{ mmol NaOH}}{1 \text{ mmol } HNO_3}\right)\left(\frac{1 \text{ mL NaOH}}{0.450 \text{ mmol NaOH}}\right) = 20.0 \text{ mL NaOH}$

(c) At the equivalence point the solution contains the neutral salt $NaNO_3$. The pH is 7.00.

(d)

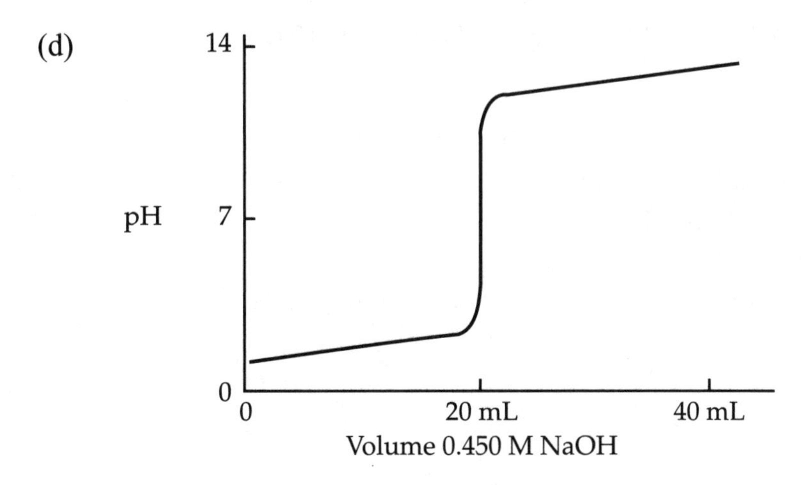

16.85

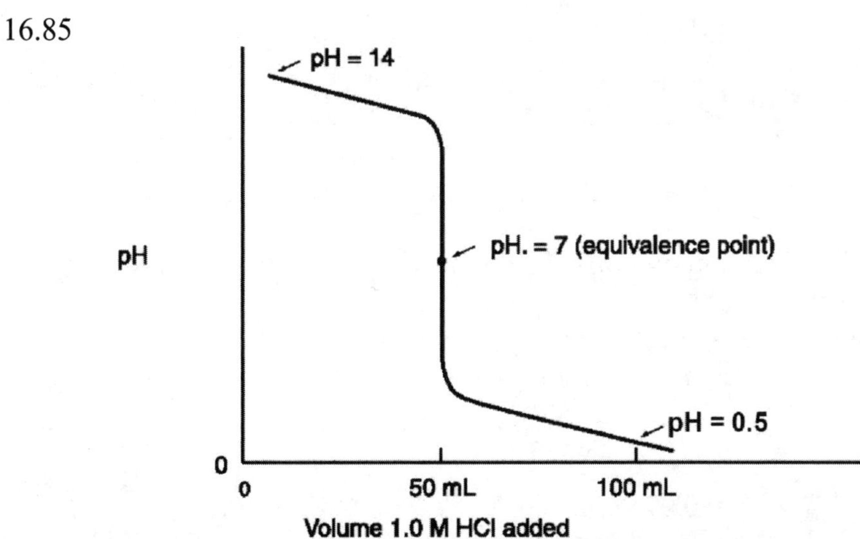

mmol NaOH = (50.0 mL)(1.0 mmol/mL) = 50 mmol
mmol HCl = mmol NaOH = 50 mmol

$$\text{vol HCl} = (50 \text{ mmol})\left(\frac{1.0 \text{ mL}}{1.0 \text{ mmol}}\right) = 50 \text{ mL}$$

50 mL of 1.0 M HCl is needed to reach the equivalence point.

16.86 (0.0250 L)(0.125 mol/L)(1000 mmol/mol) = 3.125 mmol HCl

(a) (0.0030 L)(0.100 mol/L)(1000 mmol/mol) = 0.300 mmol NaOH

Neutralization reaction:	$H_3O^+(aq)$	+	$OH^-(aq)$	→	$H_2O(l)$
before reaction (mmol)	3.125		0.300		
change (mmol)	–0.300		–0.300		
after reaction (mmol)	2.825		0		

$$[H_3O^+] = \frac{2.825 \text{ mmol}}{(25.0 \text{ mL} + 3.0 \text{ mL})} = 0.1009 \text{ M}$$

$pH = -\log[H_3O^+] = -\log(0.1009) = 0.996$

(b) (0.020 L)(0.100 mol/L)(1000 mmol/mol) = 2.0 mmol NaOH

Neutralization reaction:	$H_3O^+(aq)$	+	$OH^-(aq)$	→	$H_2O(l)$
before reaction (mmol)	3.125		2.0		
change (mmol)	–2.0		–2.0		
after reaction (mmol)	1.125		0		

$$[H_3O^+] = \frac{1.125 \text{ mmol}}{(25.0 \text{ mL} + 20 \text{ mL})} = 0.025 \text{ M}$$

$pH = -\log[H_3O^+] = -\log(0.025) = 1.60$

(c) (0.065 L)(0.100 mol/L)(1000 mmol/mol) = 6.5 mmol NaOH

Neutralization reaction:	$H_3O^+(aq)$	+	$OH^-(aq)$	→	$H_2O(l)$
before reaction (mmol)	3.125		6.5		
change (mmol)	–3.125		–3.125		
after reaction (mmol)	0		3.375		

$$[OH^-] = \frac{3.375 \text{ mmol}}{(25.0 \text{ mL} + 65 \text{ mL})} = 0.0375 \text{ M}$$

$$[H_3O^+] = \frac{K_w}{[OH^-]} = \frac{1.0 \times 10^{-14}}{0.0375} = 2.7 \times 10^{-13} \text{ M}$$

$pH = -\log[H_3O^+] = -\log(2.7 \times 10^{-13}) = 12.57$

16.87 (0.0500 L)(0.116 mol/L)(1000 mmol/mol) = 5.80 mmol NaOH

(a) (0.0050 L)(0.0750 mol/L)(1000 mmol/mol) = 0.375 mmol HCl

Neutralization reaction:	$H_3O^+(aq)$	+	$OH^-(aq)$	→	$H_2O(l)$
before reaction (mmol)	0.375		5.80		
change (mmol)	–0.375		–0.375		
after reaction (mmol)	0		5.425		

$$[OH^-] = \frac{5.425 \text{ mmol}}{(50.0 \text{ mL} + 5.0 \text{ mL})} = 0.0986 \text{ M}$$

$$[H_3O^+] = \frac{K_w}{[OH^-]} = \frac{1.0 \times 10^{-14}}{0.0986} = 1.01 \times 10^{-13}\ M$$

$pH = -\log[H_3O^+] = -\log(1.01 \times 10^{-13}) = 13.00$

(b) (0.050 L)(0.0750 mol/L)(1000 mmol/mol) = 3.75 mmol HCl

Neutralization reaction:	$H_3O^+(aq)$ +	$OH^-(aq)$ →	$H_2O(l)$
before reaction (mmol)	3.75	5.80	
change (mmol)	−3.75	−3.75	
after reaction (mmol)	0	2.05	

$$[OH^-] = \frac{2.05\ \text{mmol}}{(50.0\ \text{mL} + 50\ \text{mL})} = 0.0205\ M$$

$$[H_3O^+] = \frac{K_w}{[OH^-]} = \frac{1.0 \times 10^{-14}}{0.0205} = 4.88 \times 10^{-13}\ M$$

$pH = -\log[H_3O^+] = -\log(4.88 \times 10^{-13}) = 12.31$

(c) (0.10 L)(0.0750 mol/L)(1000 mmol/mol) = 7.50 mmol HCl

Neutralization reaction:	$H_3O^+(aq)$ +	$OH^-(aq)$ →	$H_2O(l)$
before reaction (mmol)	7.50	5.80	
change (mmol)	−5.80	−5.80	
after reaction (mmol)	1.70	0	

$$[H_3O^+] = \frac{1.70\ \text{mmol}}{(50.0\ \text{mL} + 100\ \text{mL})} = 0.0113\ M$$

$pH = -\log[H_3O^+] = -\log(0.0113) = 1.95$

16.88 mmol HF = (40.0 mL)(0.250 mmol/mL) = 10.0 mmol

mmol NaOH required = mmol HF = 10.0 mmol

$$\text{mL NaOH required} = (10.0\ \text{mmol})\left(\frac{1.00\ \text{mL}}{0.200\ \text{mmol}}\right) = 50.0\ \text{mL}$$

50.0 mL of 0.200 M NaOH is required to reach the equivalence point.

For HF, $K_a = 3.5 \times 10^{-4}$; $pK_a = -\log K_a = -\log(3.5 \times 10^{-4}) = 3.46$

(a) mmol HF = 10.0 mmol

mmol NaOH = (0.200 mmol/mL)(10.0 mL) = 2.00 mmol

Neutralization reaction:	$HF(aq)$ +	$OH^-(aq)$ →	$F^-(aq)$ +	$H_2O(l)$
before reaction (mmol)	10.0	2.00	0	
change (mmol)	−2.00	−2.00	+2.00	
after reaction (mmol)	8.0	0	2.00	

$$[HF] = \frac{8.0\ \text{mmol}}{(40.0\ \text{mL} + 10.0\ \text{mL})} = 0.16\ M; \quad [F^-] = \frac{2.00\ \text{mmol}}{(40.0\ \text{mL} + 10.0\ \text{mL})} = 0.0400\ M$$

	$HF(aq)$ +	$H_2O(l)$ ⇄	$H_3O^+(aq)$ +	$F^-(aq)$
initial (M)	0.16		~0	0.0400
change (M)	−x		+x	+x
equil (M)	0.16 − x		x	0.0400 + x

$$K_a = \frac{[H_3O^+][F^-]}{[HF]} = 3.5 \times 10^{-4} = \frac{x(0.0400 + x)}{0.16 - x} \approx \frac{x(0.0400)}{0.16}$$

Solve for x. x = $[H_3O^+]$ = 1.4 x 10^{-3} M

pH = $-\log[H_3O^+] = -\log(1.4 \times 10^{-3}) = 2.85$

(b) Halfway to the equivalence point,

$pH = pK_a = -\log K_a = -\log(3.5 \times 10^{-4}) = 3.46$

(c) At the equivalence point only the salt NaF is in solution.

$$[F^-] = \frac{10.0 \text{ mmol}}{(40.0 \text{ mL} + 50.0 \text{ mL})} = 0.111 \text{ M}$$

	$F^-(aq)$ + $H_2O(l)$	⇄	$HF(aq)$	+ $OH^-(aq)$
initial (M)	0.111		0	~0
change (M)	–x		+x	+x
equil (M)	0.111 – x		x	x

For F^-, $K_b = \frac{K_w}{K_a \text{ for HF}} = \frac{1.0 \times 10^{-14}}{3.5 \times 10^{-4}} = 2.9 \times 10^{-11}$

$$K_b = \frac{[HF][OH^-]}{[F^-]} = 2.9 \times 10^{-11} = \frac{x^2}{0.111 - x} \approx \frac{x^2}{0.111}$$

Solve for x. x = $[OH^-]$ = 1.8 x 10^{-6} M

$$[H_3O^+] = \frac{K_w}{[OH^-]} = \frac{1.0 \times 10^{-14}}{1.8 \times 10^{-6}} = 5.6 \times 10^{-9} \text{ M}$$

pH = $-\log[H_3O^+] = -\log(5.6 \times 10^{-9}) = 8.25$

(d) mmol HF = 10.0 mmol

mmol NaOH = (0.200 mmol/mL)(80.0 mL) = 16.0 mmol

Neutralization reaction:	$HF(aq)$ +	$OH^-(aq)$ →	$F^-(aq)$ + $H_2O(l)$
before reaction (mmol)	10.0	16.0	0
change (mmol)	–10.0	–10.0	+10.0
after reaction (mmol)	0	6.0	10.0

After the equivalence point, the pH of the solution is determined by the $[OH^-]$.

$$[OH^-] = \frac{6.0 \text{ mmol}}{(40.0 \text{ mL} + 80.0 \text{ mL})} = 5.0 \times 10^{-2} \text{ M}$$

$$[H_3O^+] = \frac{K_w}{[OH^-]} = \frac{1.0 \times 10^{-14}}{5.0 \times 10^{-2}} = 2.0 \times 10^{-13} \text{ M}$$

pH = $-\log[H_3O^+] = -\log(2.0 \times 10^{-13}) = 12.70$

16.89 mmol HCO_2H = (25.0 mL)(0.200 mmol/mL) = 5.0 mmol

mmol NaOH required = mmol HCO_2H = 5.0 mmol

$$\text{mL NaOH required} = (5.0 \text{ mmol})\left(\frac{1.00 \text{ mL}}{0.250 \text{ mmol}}\right) = 20.0 \text{ mL}$$

20.0 mL of 0.250 M NaOH is required to reach the equivalence point.

For HCO_2H, $K_a = 1.8 \times 10^{-4}$; $pK_a = -\log K_a = -\log(1.8 \times 10^{-4}) = 3.74$

(a) mmol HCO_2H = 5.0 mmol

mmol NaOH = (0.250 mmol/mL)(7.0 mL) = 1.75 mmol

Neutralization reaction:	$HCO_2H(aq)$ +	$OH^-(aq)$ →	$HCO_2^-(aq)$ + $H_2O(l)$
before reaction (mmol)	5.0	1.75	0
change (mmol)	–1.75	–1.75	+1.75
after reaction (mmol)	3.25	0	1.75

$$[HCO_2H] = \frac{3.25 \text{ mmol}}{(25.0 \text{ mL} + 7.0 \text{ mL})} = 0.102 \text{ M};$$

$$[HCO_2^-] = \frac{1.75 \text{ mmol}}{(25.0 \text{ mL} + 7.0 \text{ mL})} = 0.0547 \text{ M}$$

	$HCO_2H(aq)$ + $H_2O(l)$ ⇄	$H_3O^+(aq)$ +	$HCO_2^-(aq)$
initial (M)	0.102	~0	0.0547
change (M)	–x	+x	+x
equil (M)	0.102 – x	x	0.0547 + x

$$K_a = \frac{[H_3O^+][HCO_2^-]}{[HCO_2H]} = 1.8 \times 10^{-4} = \frac{x(0.0547 + x)}{0.102 - x} \approx \frac{x(0.0547)}{0.102}$$

Solve for x. $x = [H_3O^+] = 3.4 \times 10^{-4}$ M

$pH = -\log[H_3O^+] = -\log(3.4 \times 10^{-4}) = 3.47$

(b) Halfway to the equivalence point,

$pH = pK_a = -\log K_a = -\log(1.8 \times 10^{-4}) = 3.74$

(c) At the equivalence point only the salt $NaHCO_2$ is in solution.

$$[HCO_2^-] = \frac{5.0 \text{ mmol}}{(25.0 \text{ mL} + 20.0 \text{ mL})} = 0.111 \text{ M}$$

	$HCO_2^-(aq)$ + $H_2O(l)$ ⇄	$HCO_2H(aq)$ +	$OH^-(aq)$
initial (M)	0.111	0	~0
change (M)	–x	+x	+x
equil (M)	0.111 – x	x	x

$$\text{For } HCO_2^-,\ K_b = \frac{K_w}{K_a \text{ for } HCO_2H} = \frac{1.0 \times 10^{-14}}{1.8 \times 10^{-4}} = 5.6 \times 10^{-11}$$

$$K_b = \frac{[HCO_2H][OH^-]}{[HCO_2^-]} = 5.6 \times 10^{-11} = \frac{x^2}{0.111 - x} \approx \frac{x^2}{0.111}$$

Solve for x. $x = [OH^-] = 2.5 \times 10^{-6}$ M

$$[H_3O^+] = \frac{K_w}{[OH^-]} = \frac{1.0 \times 10^{-14}}{2.5 \times 10^{-6}} = 4.0 \times 10^{-9} \text{ M}$$

$pH = -\log[H_3O^+] = -\log(4.0 \times 10^{-9}) = 8.40$

(d) mmol HCO_2H = 5.0 mmol

mmol NaOH = (0.250 mmol/mL)(25.0 mL) = 6.25 mmol

Neutralization reaction:	$HCO_2H(aq)$ +	$OH^-(aq)$ →	$HCO_2^-(aq)$ + $H_2O(l)$
before reaction (mmol)	5.0	6.25	0
change (mmol)	–5.0	–5.0	+5.0
after reaction (mmol)	0	1.25	5.0

After the equivalence point, the pH of the solution is determined by the $[OH^-]$.

$$[OH^-] = \frac{1.25 \text{ mmol}}{(25.0 \text{ mL} + 25.0 \text{ mL})} = 2.5 \times 10^{-2} \text{ M}$$

$$[H_3O^+] = \frac{K_w}{[OH^-]} = \frac{1.0 \times 10^{-14}}{2.5 \times 10^{-2}} = 4.0 \times 10^{-13} \text{ M}$$

$pH = -\log[H_3O^+] = -\log(4.0 \times 10^{-13}) = 12.40$

16.90 mmol CH_3NH_2 = (100.0 mL)(0.100 mmol/mL) = 10.0 mmol

mmol HNO_3 required = mmol CH_3NH_2 = 10.0 mmol

$$\text{vol } HNO_3 \text{ required} = (10.0 \text{ mmol})\left(\frac{1.00 \text{ mL}}{0.250 \text{ mmol}}\right) = 40.0 \text{ mL}$$

40.0 mL of 0.250 M HNO_3 are required to reach the equivalence point.

(a)	$CH_3NH_2(aq)$ + $H_2O(l)$ ⇄	$CH_3NH_3^+(aq)$ +	$OH^-(aq)$
initial (M)	0.100	0	~0
change (M)	–x	+x	+x
equil (M)	0.100 – x	x	x

$$K_b = \frac{[CH_3NH_3^+][OH^-]}{[CH_3NH_2]} = 3.7 \times 10^{-4} = \frac{x^2}{0.100 - x}$$

$x^2 + (3.7 \times 10^{-4})x - (3.7 \times 10^{-5}) = 0$

Use the quadratic formula to solve for x.

$$x = \frac{-(3.7 \times 10^{-4}) \pm \sqrt{(3.7 \times 10^{-4})^2 - (4)(-3.7 \times 10^{-5})}}{2(1)} = \frac{-3.7 \times 10^{-4} \pm 0.0122}{2}$$

x = 0.0059 and –0.0063

Of the two solutions for x, only the positive value of x has physical meaning because x is the $[OH^-]$.

$[OH^-]$ = x = 0.0059 M

$$[H_3O^+] = \frac{K_w}{[OH^-]} = \frac{1.0 \times 10^{-14}}{5.9 \times 10^{-3}} = 1.7 \times 10^{-12} \text{ M}$$

$pH = -\log[H_3O^+] = -\log(1.7 \times 10^{-12}) = 11.77$

(b) 20.0 mL of HNO_3 is halfway to the equivalence point.

$$\text{For } CH_3NH_3^+, K_a = \frac{K_w}{K_b \text{ for } CH_3NH_2} = \frac{1.0 \times 10^{-14}}{3.7 \times 10^{-4}} = 2.7 \times 10^{-11}$$

$pH = pK_a = -\log(2.7 \times 10^{-11}) = 10.57$

(c) At the equivalence point only the salt $CH_3NH_3NO_3$ is in solution.

mmol $CH_3NH_3NO_3$ = (0.100 mmol/mL)(100.0 mL) = 10.0 mmol

$$[CH_3NH_3^+] = \frac{10.0 \text{ mmol}}{(100.0 \text{ mL} + 40.0 \text{ mL})} = 0.0714 \text{ M}$$

	$CH_3NH_3^+(aq)$ + $H_2O(l)$ ⇄	$H_3O^+(aq)$ +	$CH_3NH_2(aq)$
initial (M)	0.0714	~0	0
change (M)	–x	+x	+x
equil (M)	0.0714 – x	x	x

$$K_a = \frac{[H_3O^+][CH_3NH_2]}{[CH_3NH_3^+]} = 2.7 \times 10^{-11} = \frac{x^2}{0.0714 - x} \approx \frac{x^2}{0.0714}$$

Solve for x. x = $[H_3O^+] = 1.4 \times 10^{-6}$ M; pH = $-\log[H_3O^+] = -\log(1.4 \times 10^{-6}) = 5.85$

(d) mmol CH_3NH_2 = (0.100 mmol/mL)(100.0 mL) = 10.0 mmol

mmol HNO_3 = (0.250 mmol/mL)(60.0 mL) = 15.0 mmol

Neutralization reaction:	$CH_3NH_2(aq)$ +	$H_3O^+(aq)$ →	$CH_3NH_3^+(aq)$ + $H_2O(l)$
before reaction (mmol)	10.0	15.0	0
change (mmol)	−10.0	−10.0	+10.0
after reaction (mmol)	0	5.0	10.0

After the equivalence point the pH of the solution is determined by the $[H_3O^+]$.

$$[H_3O^+] = \frac{5.0 \text{ mmol}}{(100.0 \text{ mL} + 60.0 \text{ mL})} = 3.1 \times 10^{-2} \text{ M}$$

pH = $-\log[H_3O^+] = -\log(3.1 \times 10^{-2}) = 1.51$

16.91 mmol NH_3 = (50.0 mL)(0.250 mmol/mL) = 12.5 mmol

mmol HNO_3 required = mmol NH_3 = 12.5 mmol

$$\text{vol } HNO_3 \text{ required} = (12.5 \text{ mmol})\left(\frac{1.00 \text{ mL}}{0.250 \text{ mmol}}\right) = 50.0 \text{ mL}$$

50.0 mL of 0.250 M HNO_3 are required to reach the equivalence point.

(a)	$NH_3(aq)$ + $H_2O(l)$ ⇄	$NH_4^+(aq)$ +	$OH^-(aq)$
initial (M)	0.250	0	~0
change (M)	−x	+x	+x
equil (M)	0.250 − x	x	x

$$K_b = \frac{[NH_4^+][OH^-]}{[NH_3]} = 1.8 \times 10^{-5} = \frac{x^2}{0.250 - x} \approx \frac{x^2}{0.250}$$

Solve for x. x = $[OH^-] = 2.1 \times 10^{-3}$ M

$$[H_3O^+] = \frac{K_w}{[OH^-]} = \frac{1.0 \times 10^{-14}}{2.1 \times 10^{-3}} = 4.7 \times 10^{-12} \text{ M}$$

pH = $-\log[H_3O^+] = -\log(4.7 \times 10^{-12}) = 11.33$

(b) 25.0 mL of HNO_3 is halfway to the equivalence point.

$$\text{For } NH_4^+, K_a = \frac{K_w}{K_b \text{ for } NH_3} = \frac{1.0 \times 10^{-14}}{1.8 \times 10^{-5}} = 5.6 \times 10^{-10}$$

pH = $pK_a = -\log(5.6 \times 10^{-10}) = 9.26$

(c) At the equivalence point only the salt NH_4NO_3 is in solution.

mmol NH_4NO_3 = (0.250 mmol/mL)(50.0 mL) = 12.5 mmol

$$[NH_4^+] = \frac{12.5 \text{ mmol}}{(50.0 \text{ mL} + 50.0 \text{ mL})} = 0.125 \text{ M}$$

	$NH_4^+(aq)$ + $H_2O(l)$ ⇄	$H_3O^+(aq)$ +	$NH_3(aq)$
initial (M)	0.125	~0	0
change (M)	−x	+x	+x
equil (M)	0.125 − x	x	x

$$K_a = \frac{[H_3O^+][NH_3]}{[NH_4^+]} = 5.6 \times 10^{-10} = \frac{x^2}{0.125 - x} \approx \frac{x^2}{0.125}$$

Solve for x. $x = [H_3O^+] = 8.4 \times 10^{-6}$ M

$pH = -\log[H_3O^+] = -\log(8.4 \times 10^{-6}) = 5.08$

(d) mmol NH_3 = (50.0 mL)(0.250 mmol/mL) = 12.5 mmol

mmol HNO_3 = (0.250 mmol/mL)(60.0 mL) = 15.0 mmol

Neutralization reaction:	$NH_3(aq)$ +	$H_3O^+(aq)$ →	$NH_4^+(aq)$ + $H_2O(l)$
before reaction (mmol)	12.5	15.0	0
change (mmol)	−12.5	−12.5	+12.5
after reaction (mmol)	0	2.5	12.5

After the equivalence point the pH of the solution is determined by the $[H_3O^+]$.

$$[H_3O^+] = \frac{2.5 \text{ mmol}}{(50.0 \text{ mL} + 60.0 \text{ mL})} = 2.3 \times 10^{-2} \text{ M}$$

$pH = -\log[H_3O^+] = -\log(2.3 \times 10^{-2}) = 1.64$

16.92 For H_2A^+, $K_{a1} = 4.6 \times 10^{-3}$ and $K_{a2} = 2.0 \times 10^{-10}$

(a) (10.0 mL)(0.100 mmol/mL) = 1.00 mmol NaOH added = 1.00 mmol HA produced.

(50.0 mL)(0.100 mmol/mL) = 5.00 mmol H_2A^+

5.00 mmol H_2A^+ – 1.00 mmol NaOH = 4.00 mmol H_2A^+ after neutralization

$$[H_2A^+] = \frac{4.00 \text{ mmol}}{(50.0 \text{ mL} + 10.0 \text{ mL})} = 6.67 \times 10^{-2} \text{ M}$$

$$[HA] = \frac{1.00 \text{ mmol}}{(50.0 \text{ mL} + 10.0 \text{ mL})} = 1.67 \times 10^{-2} \text{ M}$$

$$pH = pK_{a1} + \log\frac{[HA]}{[H_2A^+]} = -\log(4.6 \times 10^{-3}) + \log\left(\frac{1.67 \times 10^{-2}}{6.67 \times 10^{-2}}\right) = 1.74$$

(b) Halfway to the first equivalence point, $pH = pK_{a1} = 2.34$

(c) At the first equivalence point, $pH = \frac{pK_{a1} + pK_{a2}}{2} = 6.02$

(d) Halfway between the first and second equivalence points, $pH = pK_{a2} = 9.70$

(e) At the second equivalence point only the basic salt, NaA, is in solution.

$$K_b = \frac{K_w}{K_a \text{ for HA}} = \frac{K_w}{K_{a2}} = \frac{1.0 \times 10^{-14}}{2.0 \times 10^{-10}} = 5.0 \times 10^{-5}$$

mmol A^- = (50.0 mL)(0.100 mmol/mL) = 5.00 mmol

$$[A^-] = \frac{5.0 \text{ mmol}}{(50.0 \text{ mL} + 100.0 \text{ mL})} = 3.3 \times 10^{-2} \text{ M}$$

	$A^-(aq)$ + $H_2O(l)$ ⇄	$HA(aq)$ +	$OH^-(aq)$
initial (M)	0.033	0	~0
change (M)	−x	+x	+x
equil (M)	0.033 – x	x	x

$$K_b = \frac{[HA][OH^-]}{[A^-]} = 5.0 \times 10^{-5} = \frac{(x)(x)}{0.033 - x} \approx \frac{x^2}{0.033}$$

Solve for x.

$x = [OH^-] = \sqrt{(5.0 \times 10^{-5})(0.033)} = 1.3 \times 10^{-3}$ M

$$[H_3O^+] = \frac{K_w}{[OH^-]} = \frac{1.0 \times 10^{-14}}{1.3 \times 10^{-3}} = 7.7 \times 10^{-12} \text{ M}$$

$pH = -\log[H_3O^+] = -\log(7.7 \times 10^{-12}) = 11.11$

16.93 For H_2CO_3, $K_{a1} = 4.3 \times 10^{-7}$ and $K_{a2} = 5.6 \times 10^{-11}$

(a) (25.0 mL)(0.0200 mmol/mL) = 0.500 mmol H_2CO_3

(10.0 mL)(0.0250 mmol/mL) = 0.250 mmol KOH added

0.500 mmol H_2CO_3 – 0.250 mmol KOH = 0.250 mmol HCO_3^- produced

This is halfway to the first equivalence point where $pH = pK_{a1} = -\log(4.3 \times 10^{-7}) = 6.37$

(b) At the first equivalence point, $pH = \frac{pK_{a1} + pK_{a2}}{2} = 8.31$

(c) Halfway between the first and second equivalence points, $pH = pK_{a2} = 10.25$

(d) At the second equivalence point only the basic salt, K_2CO_3, is in solution.

$$K_b = \frac{K_w}{K_a \text{ for } HCO_3^-} = \frac{K_w}{K_{a2}} = \frac{1.0 \times 10^{-14}}{5.6 \times 10^{-11}} = 1.8 \times 10^{-4}$$

mmol CO_3^{2-} = mmol H_2CO_3 = 0.500 mmol

$$[CO_3^{2-}] = \frac{0.500 \text{ mmol}}{(25.0 \text{ mL} + 40.0 \text{ mL})} = 0.00769 \text{ M}$$

	$CO_3^{2-}(aq)$ + $H_2O(l)$	⇄	$HCO_3^-(aq)$	+	$OH^-(aq)$
initial (M)	0.00769		0		~0
change (M)	–x		+x		+x
equil (M)	0.00769 – x		x		x

$$K_b = \frac{[HCO_3^-][OH^-]}{[CO_3^{2-}]} = 1.8 \times 10^{-4} = \frac{(x)(x)}{0.00769 - x}; \quad x^2 + (1.8 \times 10^{-4})x - (1.4 \times 10^{-6}) = 0$$

Use the quadratic formula to solve for x.

$$x = \frac{-(1.8 \times 10^{-4}) \pm \sqrt{(1.8 \times 10^{-4})^2 - (4)(1)(-1.4 \times 10^{-6})}}{2(1)} = \frac{-(1.8 \times 10^{-4}) \pm (2.37 \times 10^{-3})}{2}$$

$x = -1.28 \times 10^{-3}$ and 1.10×10^{-3}

Of the two solutions for x, only the positive value of x has physical meaning because x is the $[OH^-]$.

$[OH^-] = x = 1.10 \times 10^{-3}$ M

$$[H_3O^+] = \frac{K_w}{[OH^-]} = \frac{1.0 \times 10^{-14}}{1.10 \times 10^{-3}} = 9.1 \times 10^{-12} \text{ M}$$

$pH = -\log[H_3O^+] = -\log(9.1 \times 10^{-12}) = 11.04$

(e) excess KOH

(50.0 mL – 40.0 mL)(0.025 mmol/mL) = 0.250 mmol KOH = 0.250 mmol OH^-

$$[OH^-] = \frac{0.250 \text{ mmol}}{(25.0 \text{ mL} + 50.0 \text{ mL})} = 3.33 \times 10^{-3} \text{ M}$$

$$[H_3O^+] = \frac{K_w}{[OH^-]} = \frac{1.0 \times 10^{-14}}{3.33 \times 10^{-3}} = 3.0 \times 10^{-12}\ M$$

$pH = -\log[H_3O^+] = -\log(3.0 \times 10^{-12}) = 11.52$

16.94 (a) The strongest acid has the lowest pH at the equivalence point, C is the strongest acid.
(b) The weakest acid has the highest pH at the equivalence point, A is the weakest acid.

16.95 (a) The strongest base has the highest pH at the equivalence point, C is the strongest base.
(b) The weakest base has the lowest pH at the equivalence point, B is the weakest base.

16.96 When equal volumes of acid and base react, all concentrations are cut in half.
(a) At the equivalence point, only the salt $NaNO_2$ is in solution.
$[NO_2^-] = 0.050\ M$

$$\text{For } NO_2^-,\ K_b = \frac{K_w}{K_a \text{ for } HNO_2} = \frac{1.0 \times 10^{-14}}{4.5 \times 10^{-4}} = 2.2 \times 10^{-11}$$

	$NO_2^-(aq)$ + $H_2O(l)$	⇄	$HNO_2(aq)$	+ $OH^-(aq)$
initial (M)	0.050		0	~0
change (M)	–x		+x	+x
equil (M)	0.050 – x		x	x

$$K_b = \frac{[HNO_2][OH^-]}{[NO_2^-]} = 2.2 \times 10^{-11} = \frac{(x)(x)}{0.050 - x} \approx \frac{x^2}{0.050}$$

Solve for x. $x = [OH^-] = 1.1 \times 10^{-6}\ M$

$$[H_3O^+] = \frac{K_w}{[OH^-]} = \frac{1.0 \times 10^{-14}}{1.1 \times 10^{-6}} = 9.1 \times 10^{-9}\ M$$

$pH = -\log[H_3O^+] = -\log(9.1 \times 10^{-9}) = 8.04$

Phenol red would be a suitable indicator. (see Figure 15.5)

(b) The pH is 7.00 at the equivalence point for the titration of a strong acid (HI) with a strong base (NaOH).
Bromthymol blue or phenol red would be suitable indicators. (Any indicator that changes color in the pH range 4 – 10 is satisfactory for a strong acid – strong base titration.)

(c) At the equivalence point only the salt CH_3NH_3Cl is in solution.
$[CH_3NH_3^+] = 0.050\ M$

$$\text{For } CH_3NH_3^+,\ K_a = \frac{K_w}{K_b \text{ for } CH_3NH_2} = \frac{1.0 \times 10^{-14}}{3.7 \times 10^{-4}} = 2.7 \times 10^{-11}$$

	$CH_3NH_3^+(aq)$ + $H_2O(l)$	⇄	$H_3O^+(aq)$	+ $CH_3NH_2(aq)$
initial (M)	0.050		~0	0
change (M)	–x		+x	+x
equil (M)	0.050 – x		x	x

$$K_a = \frac{[H_3O^+][CH_3NH_2]}{[CH_3NH_3^+]} = 2.7 \times 10^{-11} = \frac{(x)(x)}{0.050 - x} \approx \frac{x^2}{0.050}$$

Solve for x. x = $[H_3O^+]$ = 1.2 x 10^{-6} M
pH = –log$[H_3O^+]$ = –log(1.2 x 10^{-6}) = 5.92
Chlorphenol red would be a suitable indicator.

16.97 When equal volumes of acid and base react, all concentrations are cut in half.
(a) At the equivalence point only the salt $C_5H_{11}NHNO_3$ is in solution.
$[C_5H_{11}NH^+]$ = 0.10 M

$$\text{For } C_5H_{11}NH^+,\ K_a = \frac{K_w}{K_b \text{ for } C_5H_{11}N} = \frac{1.0 \times 10^{-14}}{1.3 \times 10^{-3}} = 7.7 \times 10^{-12}$$

	$C_5H_{11}NH^+(aq)$ + $H_2O(l)$	⇄	$H_3O^+(aq)$ +	$C_5H_{11}N(aq)$
initial (M)	0.10		~0	0
change (M)	–x		+x	+x
equil (M)	0.10 – x		x	x

$$K_a = \frac{[H_3O^+][C_5H_{11}N]}{[C_5H_{11}NH^+]} = 7.7 \times 10^{-12} = \frac{(x)(x)}{0.10 - x} \approx \frac{x^2}{0.10}$$

Solve for x. x = $[H_3O^+]$ = 8.8 x 10^{-7} M
pH = –log$[H_3O^+]$ = –log(8.8 x 10^{-7}) = 6.06
Chlorphenol red would be a suitable indicator.
(b) At the equivalence point only the salt Na_2SO_3 is in solution.
$[SO_3^{2-}]$ = 0.10 M

$$\text{For } SO_3^{2-},\ K_b = \frac{K_w}{K_a \text{ for } HSO_3^-} = \frac{1.0 \times 10^{-14}}{6.3 \times 10^{-8}} = 1.6 \times 10^{-7}$$

	$SO_3^{2-}(aq)$ + $H_2O(l)$	⇄	$HSO_3^-(aq)$ +	$OH^-(aq)$
Initial (M)	0.10		0	~0
change (M)	–x		+x	+x
equil (M)	0.10 – x		x	x

$$K_b = \frac{[HSO_3^-][OH^-]}{[SO_3^{2-}]} = 1.6 \times 10^{-7} = \frac{(x)(x)}{0.10 - x} \approx \frac{x^2}{0.10}$$

Solve for x. x = $[OH^-]$ = 1.26 x 10^{-4} M

$$[H_3O^+] = \frac{K_w}{[OH^-]} = \frac{1.0 \times 10^{-14}}{1.26 \times 10^{-4}} = 7.9 \times 10^{-11}\ M$$

pH = –log$[H_3O^+]$ = –log(7.9 x 10^{-11}) = 10.10
Thymolphthalein would be a suitable indicator.
(c) The pH is 7.00 at the equivalence point for the titration of a strong acid (HBr) with a strong base ($Ba(OH)_2$).
Bromthymol blue or phenol red would be suitable indicators. (Any indicator that changes color in the pH range 4–10 is satisfactory for a strong acid–strong base titration.)

Solubility Equilibria (Sections 16.10 and 16.11)

16.98 (a) $Ag_2CO_3(s) \rightleftarrows 2\ Ag^+(aq) + CO_3^{2-}(aq)$ $\quad K_{sp} = [Ag^+]^2[CO_3^{2-}]$

(b) $PbCrO_4(s) \rightleftarrows Pb^{2+}(aq) + CrO_4^{2-}(aq)$ $\quad K_{sp} = [Pb^{2+}][CrO_4^{2-}]$

(c) $Al(OH)_3(s) \rightleftarrows Al^{3+}(aq) + 3\ OH^-(aq)$ $\quad K_{sp} = [Al^{3+}][OH^-]^3$

(d) $Hg_2Cl_2(s) \rightleftarrows Hg_2^{2+}(aq) + 2\ Cl^-(aq)$ $\quad K_{sp} = [Hg_2^{2+}][Cl^-]^2$

16.99 (a) $K_{sp} = [Ca^{2+}][OH^-]^2$ (b) $K_{sp} = [Ag^+]^3[PO_4^{3-}]$

(c) $K_{sp} = [Ba^{2+}][CO_3^{2-}]$ (d) $K_{sp} = [Ca^{2+}]^5[PO_4^{3-}]^3[OH^-]$

16.100 (a) $K_{sp} = [Pb^{2+}][I^-]^2 = (5.0 \times 10^{-3})(1.3 \times 10^{-3})^2 = 8.5 \times 10^{-9}$

(b) $$[I^-] = \sqrt{\frac{K_{sp}}{[Pb^{2+}]}} = \sqrt{\frac{(8.5 \times 10^{-9}}{(2.5 \times 10^{-4})}} = 5.8 \times 10^{-3}\ M$$

(c) $$[Pb^{2+}] = \frac{K_{sp}}{[I^-]^2} = \frac{(8.5 \times 10^{-9})}{(2.5 \times 10^{-4})^2} = 0.14\ M$$

16.101 (a) $K_{sp} = [Ca^{2+}]^3[PO_4^{2-}]^2 = (2.9 \times 10^{-7})^3(2.9 \times 10^{-7})^2 = 2.1 \times 10^{-33}$

(b) $$[Ca^{2+}] = \sqrt[3]{\frac{K_{sp}}{[PO_4^{2-}]^2}} = \sqrt[3]{\frac{2.1 \times 10^{-33}}{(0.010)^2}} = 2.8 \times 10^{-10}\ M$$

(c) $$[PO_4^{2-}] = \sqrt{\frac{K_{sp}}{[Ca^{2+}]^3}} = \sqrt{\frac{2.1 \times 10^{-33}}{(0.010)^3}} = 4.6 \times 10^{-14}\ M$$

16.102 $Ag_2CO_3(s) \rightleftarrows 2\ Ag^+(aq) + CO_3^{2-}(aq)$

equil (M) $\quad 2x \quad x$

$[Ag^+] = 2x = 2.56 \times 10^{-4}\ M$; $[CO_3^{2-}] = x = (2.56 \times 10^{-4}\ M)/2 = 1.28 \times 10^{-4}\ M$

$K_{sp} = [Ag^+]^2[CO_3^{2-}] = (2.56 \times 10^{-4})^2(1.28 \times 10^{-4}) = 8.39 \times 10^{-12}$

16.103 $Pb(N_3)_2(s) \rightleftarrows Pb^{2+}(aq) + 2\ N_3^-(aq)$

equil (M) $\quad x \quad 2x$

$[Pb^{2+}] = x = 8.5 \times 10^{-4}\ M$; $[N_3^-] = 2x = 2(8.5 \times 10^{-4}\ M) = 1.7 \times 10^{-3}\ M$

$K_{sp} = [Pb^{2+}][N_3^-]^2 = (8.5 \times 10^{-4})(1.7 \times 10^{-3})^2 = 2.5 \times 10^{-9}$

16.104 (a) $SrF_2(s) \rightleftarrows Sr^{2+}(aq) + 2\ F^-(aq)$

equil (M) $\quad x \quad 2x$

$[Sr^{2+}] = x = 1.03 \times 10^{-3}\ M$; $[F^-] = 2x = 2(1.03 \times 10^{-3}\ M) = 2.06 \times 10^{-3}\ M$

$K_{sp} = [Sr^{2+}][F^-]^2 = (1.03 \times 10^{-3})(2.06 \times 10^{-3})^2 = 4.37 \times 10^{-9}$

(b) $CuI(s) \rightleftarrows Cu^{+}(aq) + I^{-}(aq)$

equil (M) x x

$[Cu^{+}] = [I^{-}] = x = 1.05 \times 10^{-6}$ M

$K_{sp} = [Cu^{+}][I^{-}] = (1.05 \times 10^{-6})^2 = 1.10 \times 10^{-12}$

(c) MgC_2O_4, 112.32

$$[Mg^{2+}] = \text{molarity of } MgC_2O_4 = \frac{\left(0.094 \text{ g} \times \frac{1 \text{ mol}}{112.32 \text{ g}}\right)}{1 \text{ L}} = 8.37 \times 10^{-4} \text{ M}$$

$MgC_2O_4(s) \rightleftarrows Mg^{2+}(aq) + C_2O_4^{-}(aq)$

equil (M) x x

$[Mg^{2+}] = [C_2O_4^{-}] = x = 8.37 \times 10^{-4}$ M

$K_{sp} = [Mg^{2+}][C_2O_4^{-}] = (8.37 \times 10^{-4})^2 = 7.0 \times 10^{-7}$

(d) $Zn(CN)_2$, 117.41

$$[Zn^{2+}] = \text{molarity of } Zn(CN)_2 = \frac{\left(4.95 \times 10^{-4} \text{ g} \times \frac{1 \text{ mol}}{117.41 \text{ g}}\right)}{1 \text{ L}} = 4.22 \times 10^{-6} \text{ M}$$

$Zn(CN)_2(s) \rightleftarrows Zn^{2+}(aq) + 2\ CN^{-}(aq)$

equil (M) x 2x

$[Zn^{2+}] = x = 4.22 \times 10^{-6}$ M; $[CN^{-}] = 2x = 2(4.22 \times 10^{-6} \text{ M}) = 8.44 \times 10^{-6}$ M

$K_{sp} = [Zn^{2+}][CN^{-}]^2 = (4.22 \times 10^{-6})(8.44 \times 10^{-6})^2 = 3.01 \times 10^{-16}$

16.105 (a) $CdCO_3(s) \rightleftarrows Cd^{2+}(aq) + CO_3^{-}(aq)$

equil (M) x x

$[Cd^{2+}] = [CO_3^{2-}] = x = 1.0 \times 10^{-6}$ M

$K_{sp} = [Cd^{2+}][CO_3^{2-}] = (1.0 \times 10^{-6})^2 = 1.0 \times 10^{-12}$

(b) $Ca(OH)_2(s) \rightleftarrows Ca^{2+}(aq) + 2\ OH^{-}(aq)$

equil (M) x 2x

$[Ca^{2+}] = x = 1.06 \times 10^{-2}$ M; $[OH^{-}] = 2x = 2(1.06 \times 10^{-2} \text{ M}) = 2.12 \times 10^{-2}$ M

$K_{sp} = [Ca^{2+}][OH^{-}]^2 = (1.06 \times 10^{-2})(2.12 \times 10^{-2})^2 = 4.76 \times 10^{-6}$

(c) $PbBr_2$, 367.01

$$[Pb^{2+}] = \text{molarity of } PbBr_2 = \frac{\left(4.34 \text{ g} \times \frac{1 \text{ mol}}{367.01 \text{ g}}\right)}{1 \text{ L}} = 1.18 \times 10^{-2} \text{ M}$$

$PbBr_2(s) \rightleftarrows Pb^{2+}(aq) + 2\ Br^{-}(aq)$

equil (M) x 2x

$[Pb^{2+}] = x = 1.18 \times 10^{-2}$ M; $[Br^{-}] = 2x = 2(1.18 \times 10^{-2} \text{ M}) = 2.36 \times 10^{-2}$ M

$K_{sp} = [Pb^{2+}][Br^{-}]^2 = (1.18 \times 10^{-2})(2.36 \times 10^{-2})^2 = 6.57 \times 10^{-6}$

(d) $BaCrO_4$, 253.32

$$[Ba^{2+}] = [CrO_4^{2-}] = \text{molarity of } BaCrO_4 = \frac{\left(2.8 \times 10^{-3}\text{ g} \times \dfrac{1\text{ mol}}{253.32\text{ g}}\right)}{1\text{ L}} = 1.1 \times 10^{-5}\text{ M}$$

$$BaCrO_4(s) \rightleftarrows Ba^{2+}(aq) + CrO_4^{2-}(aq)$$

equil (M) x x

$[Ba^{2+}] = [CrO_4^{2-}] = x = 1.1 \times 10^{-5}$ M

$K_{sp} = [Ba^{2+}][CrO_4^{2-}] = (1.1 \times 10^{-5})^2 = 1.2 \times 10^{-10}$

16.106 (a) $BaCrO_4(s) \rightleftarrows Ba^{2+}(aq) + CrO_4^{2-}(aq)$

equil (M) x x

$K_{sp} = [Ba^{2+}][CrO_4^{2-}] = 1.2 \times 10^{-10} = (x)(x)$

molar solubility $= x = \sqrt{1.2 \times 10^{-10}} = 1.1 \times 10^{-5}$ M

(b) $Mg(OH)_2(s) \rightleftarrows Mg^{2+}(aq) + 2\,OH^-(aq)$

equil (M) x 2x

$K_{sp} = [Mg^{2+}][OH^-]^2 = 5.6 \times 10^{-12} = x(2x)^2 = 4x^3$

$$\text{molar solubility} = x = \sqrt[3]{\frac{5.6 \times 10^{-12}}{4}} = 1.1 \times 10^{-4}\text{ M}$$

(c) $Ag_2SO_3(s) \rightleftarrows 2\,Ag^+(aq) + SO_3^{2-}(aq)$

equil (M) 2x x

$K_{sp} = [Ag^+]^2[SO_3^{2-}] = 1.5 \times 10^{-14} = (2x)^2x = 4x^3$

$$\text{molar solubility} = x = \sqrt[3]{\frac{1.5 \times 10^{-14}}{4}} = 1.6 \times 10^{-5}\text{ M}$$

16.107 (a) $Ag_2CO_3(s) \rightleftarrows 2\,Ag^+(aq) + CO_3^{2-}(aq)$

equil (M) 2x x

$K_{sp} = [Ag^+]^2[CO_3^{2-}] = 8.4 \times 10^{-12} = (2x)^2(x) = 4x^3$

$$\text{molar solubility} = x = \sqrt[3]{\frac{8.4 \times 10^{-12}}{4}} = 1.3 \times 10^{-4}\text{ M}$$

Ag_2CO_3, 275.75

solubility = $(1.3 \times 10^{-4}$ mol/L)(275.75 g/mol) = 0.036 g/L

(b) $CuBr(s) \rightleftarrows Cu^+(aq) + Br^-(aq)$

equil (M) x x

$K_{sp} = [Cu^+][Br^-] = 6.3 \times 10^{-9} = (x)(x)$

molar solubility $= x = \sqrt{6.3 \times 10^{-9}} = 7.9 \times 10^{-5}$ M

CuBr, 143.45

solubility = $(7.9 \times 10^{-5}$ mol/L)(143.45 g/mol) = 0.011 g/L

(c) $Cu_3(PO_4)_2(s) \rightleftharpoons 3\ Cu^{2+}(aq) + 2\ PO_4^{3-}(aq)$

equil (M) 3x 2x

$K_{sp} = [Cu^{2+}]^3[PO_4^{3-}]^2 = 1.4 \times 10^{-37} = (3x)^3(2x)^2 = 108x^5$

$$\text{molar solubility} = x = \sqrt[5]{\frac{1.4 \times 10^{-37}}{108}} = 1.7 \times 10^{-8}$$

$Cu_3(PO_4)_2$, 380.58

solubility = $(1.7 \times 10^{-8}\ \text{mol/L})(380.58\ \text{g/mol}) = 6.5 \times 10^{-6}\ \text{g/L}$

Factors That Affect Solubility (Section 16.12)

16.108 $Ag_2CO_3(s) \rightleftharpoons 2\ Ag^+(aq) + CO_3^{2-}(aq)$

(a) $AgNO_3$, source of Ag^+; equilibrium shifts left
(b) HNO_3, source of H_3O^+, removes CO_3^{2-}; equilibrium shifts right
(c) Na_2CO_3, source of CO_3^{2-}; equilibrium shifts left
(d) NH_3, forms $Ag(NH_3)_2^+$, removes Ag^+; equilibrium shifts right

16.109 $BaF_2(s) \rightleftharpoons Ba^{2+}(aq) + 2\ F^-(aq)$

(a) H^+ from HCl reacts with F^- forming the weak acid HF. The equilibrium shifts to the right increasing the solubility of BaF_2.
(b) KF, source of F^-; equilibrium shifts left, solubility of BaF_2 decreases.
(c) No change in solubility.
(d) $Ba(NO_3)_2$, source of Ba^{2+}; equilibrium shifts left, solubility of BaF_2 decreases.

16.110 (a) $PbCrO_4(s) \rightleftharpoons Pb^{2+}(aq) + CrO_4^{2-}(aq)$

equil (M) x x

$K_{sp} = [Pb^{2+}][CrO_4^{2-}] = 2.8 \times 10^{-13} = (x)(x)$

molar solubility = $x = \sqrt{2.8 \times 10^{-13}} = 5.3 \times 10^{-7}\ M$

(b) $PbCrO_4(s) \rightleftharpoons Pb^{2+}(aq) + CrO_4^{2-}(aq)$

initial(M) 0 1.0×10^{-3}

equil (M) x $1.0 \times 10^{-3} + x$

$K_{sp} = [Pb^{2+}][CrO_4^{2-}] = 2.8 \times 10^{-13} = (x)(1.0 \times 10^{-3} + x) \approx (x)(1.0 \times 10^{-3})$

$$\text{molar solubility} = x = \frac{2.8 \times 10^{-13}}{1 \times 10^{-3}} = 2.8 \times 10^{-10}\ M$$

16.111 (a) $SrF_2(s) \rightleftharpoons Sr^{2+}(aq) + 2\ F^-(aq)$

initial (M) 0.010 0

equil (M) 0.010 + x 2x

$K_{sp} = [Sr^{2+}][F^-]^2 = 4.3 \times 10^{-9} = (0.010 + x)(2x)^2 \approx (0.010)(2x)^2 = 0.040\ x^2$

$$\text{molar solubility} = x = \sqrt{\frac{4.3 \times 10^{-9}}{0.040}} = 3.3 \times 10^{-4}\ M$$

(b) $SrF_2(s) \rightleftarrows Sr^{2+}(aq) + 2\,F^-(aq)$

	Sr^{2+}	F^-
initial (M)	0	0.010
equil (M)	x	0.010 + 2x

$K_{sp} = [Sr^{2+}][F^-]^2 = 4.3 \times 10^{-9} = (x)(0.010 + 2x)^2 \approx (x)(0.010)^2 = x(0.00010)$

molar solubility $= x = \dfrac{4.3 \times 10^{-9}}{0.00010} = 4.3 \times 10^{-5}$ M

16.112 (b), (c), and (d) are more soluble in acidic solution.

(a) $AgBr(s) \rightleftarrows Ag^+(aq) + Br^-(aq)$

(b) $CaCO_3(s) + H_3O^+(aq) \rightleftarrows Ca^{2+}(aq) + HCO_3^-(aq) + H_2O(l)$

(c) $Ni(OH)_2(s) + 2\,H_3O^+(aq) \rightleftarrows Ni^{2+}(aq) + 4\,H_2O(l)$

(d) $Ca_3(PO_4)_2(s) + 2\,H_3O^+(aq) \rightleftarrows 3\,Ca^{2+}(aq) + 2\,HPO_4^{2-}(aq) + 2\,H_2O(l)$

16.113 (a), (b), and (d) are more soluble in acidic solution.

(a) $MnS(s) + 2\,H_3O^+(aq) \rightleftarrows Mn^{2+}(aq) + H_2S(aq) + 2\,H_2O(l)$

(b) $Fe(OH)_3(s) + 3\,H_3O^+(aq) \rightleftarrows Fe^{3+}(aq) + 6\,H_2O(l)$

(c) $AgCl(s) \rightleftarrows Ag^+(aq) + Cl^-(aq)$

(d) $BaCO_3(s) + H_3O^+(aq) \rightleftarrows Ba^{2+}(aq) + HCO_3^-(aq) + H_2O(l)$

16.114 (a) Because $Zn(OH)_2$ contains a basic anion, it becomes more soluble as the acidity of the solution increases. $Zn(OH)_2(s) + 2\,H_3O^+(aq) \rightleftarrows Zn^{2+}(aq) + 4\,H_2O(l)$

(b) Because $Zn(OH)_2$ forms the complex anion, $Zn(OH)_4^{2-}$, $Zn(OH)_2$ becomes more soluble in basic solution. $Zn(OH)_2(s) + 2\,OH^-(aq) \rightleftarrows Zn(OH)_4^{2-}(aq)$

(c) Because $Zn(OH)_2$ forms the complex anion, $Zn(CN)_4^{2-}$, $Zn(OH)_2$ becomes more soluble in the presence of CN^-.

$Zn(OH)_2(s) + 4\,CN^-(aq) \rightleftarrows Zn(CN)_4^{2-}(aq) + 2\,OH^-(aq)$

16.115 (a) Because $Fe(OH)_3$ contains a basic anion, it becomes more soluble as the acidity of the solution increases. $Fe(OH)_3(s) + 3\,H_3O^+(aq) \rightleftarrows Fe^{3+}(aq) + 6\,H_2O(l)$

(b) $Fe(OH)_3$ does not form a complex anion with OH^-; the solubility of $Fe(OH)_3$ is decreased by the presence of a common ion (OH^-) in the solution.

$Fe(OH)_3(s) \rightleftarrows Fe^{3+}(aq) + 3\,OH^-(aq)$

(c) Because $Fe(OH)_3$ forms the complex anion, $Fe(CN)_6^{3-}$, $Fe(OH)_3$ becomes more soluble in the presence of CN^-.

$Fe(OH)_3(s) + 6\,CN^-(aq) \rightleftarrows Fe(CN)_6^{3-}(aq) + 3\,OH^-(aq)$

16.116 On mixing equal volumes of two solutions, the concentrations of both solutions are cut in half.

	$Ag^+(aq)$ +	$2\,CN^-(aq)$ $\rightleftarrows$	$Ag(CN)_2^-(aq)$
before reaction (M)	0.0010	0.10	0
assume 100% reaction	–0.0010	–2(0.0010)	0.0010
after reaction (M)	0	0.098	0.0010
assume small back rxn	+x	+2x	–x
equil (M)	x	0.098 + 2x	0.0010 – x

$$K_f = 3.0 \times 10^{20} = \frac{[Ag(CN)_2^-]}{[Ag^+][CN^-]^2} = \frac{(0.0010 - x)}{x(0.098 + 2x)^2} \approx \frac{0.0010}{x(0.098)^2}$$

Solve for x. $x = [Ag^+] = 3.5 \times 10^{-22}$ M

16.117

	$Cr^{3+}(aq)$	+ $4\ OH^-(aq)$	⇄ $Cr(OH)_4^-(aq)$
before reaction (M)	0.0050	1.0	0
assume 100% reaction	–0.0050	–(4)(0.0050)	+0.0050
after reaction(M)	0	0.98	0.0050
assume small back rxn	+x	+4x	–x
equil (M)	x	0.98 + 4x	0.0050 – x

$$K_f = \frac{[Cr(OH)_4^-]}{[Cr^{3+}][OH^-]^4} = 8 \times 10^{29} = \frac{(0.0050 - x)}{(x)(0.98 + 4x)^4} \approx \frac{(0.0050)}{(x)(0.98)^4}$$

Solve for x. $x = [Cr^{3+}] = 6.8 \times 10^{-33}\ M = 7 \times 10^{-33}$ M

$$\text{fraction uncomplexed } Cr^{3+} = \frac{[Cr^{3+}]}{[Cr(OH)_4^-]} = \frac{7 \times 10^{-33}\ M}{0.0050\ M} = 1.4 \times 10^{-30} = 1 \times 10^{-30}$$

16.118 (a)

$AgI(s) \rightleftarrows Ag^+(aq) + I^-(aq)$ $\quad K_{sp} = 8.5 \times 10^{-17}$

$\underline{Ag^+(aq) + 2\ CN^-(aq) \rightarrow Ag(CN)_2^-(aq)}$ $\quad K_f = 3.0 \times 10^{20}$

dissolution rxn $AgI(s) + 2\ CN^-(aq) \rightleftarrows Ag(CN)_2^-(aq) + I^-(aq)$

$K = (K_{sp})(K_f) = (8.5 \times 10^{-17})(3.0 \times 10^{20}) = 2.6 \times 10^4$

(b)

$Al(OH)_3(s) \rightleftarrows Al^{3+}(aq) + 3\ OH^-(aq)$ $\quad K_{sp} = 1.9 \times 10^{-33}$

$\underline{Al^{3+}(aq) + 4\ OH^-(aq) \rightarrow Al(OH)_4^-(aq)}$ $\quad K_f = 3 \times 10^{33}$

dissolution rxn $Al(OH)_3(s) + OH^-(aq) \rightleftarrows Al(OH)_4^-(aq)$

$K = (K_{sp})(K_f) = (1.9 \times 10^{-33})(3 \times 10^{33}) = 6$

(c)

$Zn(OH)_2(s) \rightleftarrows Zn^{2+}(aq) + 2\ OH^-(aq)$ $\quad K_{sp} = 4.1 \times 10^{-17}$

$\underline{Zn^{2+}(aq) + 4\ NH_3(aq) \rightarrow Zn(NH_3)_4^{2+}(aq)}$ $\quad K_f = 7.8 \times 10^8$

dissolution rxn $Zn(OH)_2(s) + 4\ NH_3(aq) \rightleftarrows Zn(NH_3)_4^{2+} + 2\ OH^-(aq)$

$K = (K_{sp})(K_f) = (4.1 \times 10^{-17})(7.8 \times 10^8) = 3.2 \times 10^{-8}$

16.119 (a)

$Zn(OH)_2(s) \rightleftarrows Zn^{2+}(aq) + 2\ OH^-(aq)$ $\quad K_{sp} = 4.1 \times 10^{-17}$

$\underline{Zn^{2+}(aq) + 4\ OH^-(aq) \rightarrow Zn(OH)_4^{2-}(aq)}$ $\quad K_f = 3 \times 10^{15}$

dissolution rxn $Zn(OH)_2(s) + 2\ OH^-(aq) \rightleftarrows Zn(OH)_4^{2-}(aq)$

$K = (K_{sp})(K_f) = (4.1 \times 10^{-17})(3 \times 10^{15}) = 0.1$

(b)

$Cu(OH)_2(s) \rightleftarrows Cu^{2+}(aq) + 2\ OH^-(aq)$ $\quad K_{sp} = 1.6 \times 10^{-19}$

$\underline{Cu^{2+}(aq) + 4\ NH_3(aq) \rightarrow Cu(NH_3)_4^{2+}(aq)}$ $\quad K_f = 5.6 \times 10^{11}$

dissolution rxn $Cu(OH)_2(s) + 4\ NH_3(aq) \rightleftarrows Cu(NH_3)_4^{2+}(aq) + 2\ OH^-(aq)$

$K = (K_{sp})(K_f) = (1.6 \times 10^{-19})(5.6 \times 10^{11}) = 9.0 \times 10^{-8}$

(c)

$AgBr(s) \rightleftarrows Ag^+(aq) + Br^-(aq)$ $\quad K_{sp} = 5.4 \times 10^{-13}$

$\underline{Ag^+(aq) + 2\ NH_3(aq) \rightarrow Ag(NH_3)_2^+(aq)}$ $\quad K_f = 1.7 \times 10^7$

dissolution rxn $AgBr(s) + 2\ NH_3(aq) \rightleftarrows Ag(NH_3)_2^+(aq) + Br^-(aq)$

$K = (K_{sp})(K_f) = (5.4 \times 10^{-13})(1.7 \times 10^7) = 9.2 \times 10^{-6}$

16.120 (a) $AgI(s) \rightleftarrows Ag^+(aq) + I^-(aq)$

		Ag^+	I^-
equil (M)		x	x

$K_{sp} = [Ag^+][I^-] = 8.5 \times 10^{-17} = (x)(x)$

molar solubility = $x = \sqrt{8.5 \times 10^{-17}} = 9.2 \times 10^{-9}$ M

(b) $AgI(s) + 2\,CN^-(aq) \rightleftarrows Ag(CN)_2^-(aq) + I^-(aq)$

	CN^-	$Ag(CN)_2^-$	I^-
initial (M)	0.10	0	0
change (M)	–2x	+x	+x
equil (M)	0.10 – 2x	x	x

$K = (K_{sp})(K_f) = (8.5 \times 10^{-17})(3.0 \times 10^{20}) = 2.6 \times 10^4$

$$K = 2.6 \times 10^4 = \frac{[Ag(CN)_2^-][I^-]}{[CN^-]^2} = \frac{x^2}{(0.10 - 2x)^2}$$

Take the square root of both sides and solve for x.

molar solubility = x = 0.050 M

16.121 $Cr(OH)_3(s) + OH^-(aq) \rightleftarrows Cr(OH)_4^-(aq)$

	OH^-	$Cr(OH)_4^-$
initial (M)	0.50	0
change (M)	–x	+x
equil (M)	0.50 – x	x

$K = (K_{sp})(K_f) = (6.7 \times 10^{-31})(8 \times 10^{29}) = 0.54$

$$K = 0.54 = \frac{[Cr(OH)_4^-]}{[OH^-]} = \frac{x}{0.50 - x}$$

$0.27 - 0.54x = x$

$0.27 = 1.54x$

molar solubility = $x = \dfrac{0.27}{1.54} = 0.2$ M

Precipitation; Qualitative Analysis (Sections 16.13–16.15)

16.122 $BaSO_4$, $K_{sp} = 1.1 \times 10^{-10}$; $Fe(OH)_3$, $K_{sp} = 2.6 \times 10^{-39}$

Total volume = 80 mL + 20 mL = 100 mL

$$[Ba^{2+}] = \frac{(1.0 \times 10^{-5}\text{ M})(80\text{ mL})}{(100\text{ mL})} = 8.0 \times 10^{-6}\text{ M}$$

$[OH^-] = 2[Ba^{2+}] = 2(8.0 \times 10^{-6}) = 1.6 \times 10^{-5}$ M

$$[Fe^{3+}] = \frac{2(1.0 \times 10^{-5}\text{ M})(20\text{ mL})}{(100\text{ mL})} = 4.0 \times 10^{-6}\text{ M}$$

$$[SO_4^{2-}] = \frac{3(1.0 \times 10^{-5}\text{ M})(20\text{ mL})}{(100\text{ mL})} = 6.0 \times 10^{-6}\text{ M}$$

For $BaSO_4$, IP = $[Ba^{2+}]_t[SO_4^{2-}]_t = (8.0 \times 10^{-6})(6.0 \times 10^{-6}) = 4.8 \times 10^{-11}$

IP < K_{sp}; $BaSO_4$ will not precipitate.

For $Fe(OH)_3$, IP = $[Fe^{3+}]_t[OH^-]_t^3 = (4.0 \times 10^{-6})(1.6 \times 10^{-5})^3 = 1.6 \times 10^{-20}$

IP > K_{sp}; $Fe(OH)_3(s)$ will precipitate.

16.123 (a) $[CO_3^{2-}] = \dfrac{(2.0 \times 10^{-3}\ M)(0.10\ mL)}{(250\ mL)} = 8.0 \times 10^{-7}\ M$

$K_{sp} = 5.0 \times 10^{-9} = [Ca^{2+}][CO_3^{2-}]$

$IP = [Ca^{2+}][CO_3^{2-}] = (8.0 \times 10^{-4})(8.0 \times 10^{-7}) = 6.4 \times 10^{-10}$

$IP < K_{sp}$; no precipitate will form.

(b) Na_2CO_3, 106; 10 mg = 0.010 g

$$[CO_3^{2-}] = \frac{\left(0.010\ g \times \dfrac{1\ mol}{106\ g}\right)}{0.250\ L} = 3.8 \times 10^{-4}\ M$$

$IP = [Ca^{2+}][CO_3^{2-}] = (8.0 \times 10^{-4})(3.8 \times 10^{-4}) = 3.0 \times 10^{-7}$

$IP > K_{sp}$; $CaCO_3(s)$ will precipitate.

16.124 pH = 10.80; $[H_3O^+] = 10^{-pH} = 10^{-10.80} = 1.6 \times 10^{-11}\ M$

$$[OH^-] = \frac{K_w}{[H_3O^+]} = \frac{1.0 \times 10^{-14}}{1.6 \times 10^{-11}} = 6.3 \times 10^{-4}\ M$$

For $Mg(OH)_2$, $K_{sp} = 5.6 \times 10^{-12}$

$IP = [Mg^{2+}]_t[OH^-]_t^2 = (2.5 \times 10^{-4})(6.3 \times 10^{-4})^2 = 9.9 \times 10^{-11}$

$IP > K_{sp}$; $Mg(OH)_2(s)$ will precipitate

16.125 $Mg(OH)_2$, $K_{sp} = 5.6 \times 10^{-12}$; $Al(OH)_3$, $K_{sp} = 1.9 \times 10^{-33}$

pH = 8; $[H_3O^+] = 10^{-pH} = 10^{-8} = 1 \times 10^{-8}\ M$

$$[OH^-] = \frac{K_w}{[H_3O^+]} = \frac{1.0 \times 10^{-14}}{1 \times 10^{-8}} = 1 \times 10^{-6}\ M$$

For $Mg(OH)_2$, $IP = [Mg^{2+}][OH^-]^2 = (0.01)(1 \times 10^{-6})^2 = 1 \times 10^{-14}$

$IP < K_{sp}$; $Mg(OH)_2$ will not precipitate.

For $Al(OH)_3$, $IP = [Al^{3+}][OH^-]^3 = (0.01)(1 \times 10^{-6})^3 = 1 \times 10^{-20}$

$IP > K_{sp}$; $Al(OH)_3$ will precipitate.

16.126 $K_{spa} = \dfrac{[M^{2+}][H_2S]}{[H_3O^+]^2}$; FeS, $K_{spa} = 6 \times 10^2$; SnS, $K_{spa} = 1 \times 10^{-5}$

Fe^{2+} and Sn^{2+} can be separated by bubbling H_2S through an acidic solution containing the two cations because their K_{spa} values are so different.

For FeS and SnS, $Q_c = \dfrac{(0.01)(0.10)}{(0.3)^2} = 1.1 \times 10^{-2}$

For FeS, $Q_c < K_{spa}$, and FeS will not precipitate.

For SnS, $Q_c > K_{spa}$, and SnS will precipitate.

16.127 CoS, $K_{spa} = 3$; ZnS, $K_{spa} = 3 \times 10^{-2}$; In 0.3 M HCl, $[H_3O^+] = 0.3\ M$

$$Q_c = \frac{[M^{2+}]_t[H_2S]_t}{[H_3O^+]_t^2} = \frac{(0.01)(0.10)}{(0.3)^2} = 0.011$$

For CoS, $Q_c < K_{spa}$; CoS will not precipitate
For ZnS, $Q_c < K_{spa}$; ZnS will not precipitate
Co^{2+} and Zn^{2+} cannot be separated from each other under these conditions.

16.128 FeS, $K_{spa} = \dfrac{[Fe^{2+}][H_2S]}{[H_3O^+]^2} = 6 \times 10^2$

(i) In 0.4 M HCl, $[H_3O^+] = 0.4$ M

$$Q_c = \frac{[Fe^{2+}]_t[H_2S]_t}{[H_3O^+]_t^2} = \frac{(0.10)(0.10)}{(0.4)^2} = 0.0625; \quad Q_c < K_{spa}; \text{ FeS will not precipitate}$$

(ii) pH = 8; $[H_3O^+] = 10^{-pH} = 10^{-8} = 1 \times 10^{-8}$ M

$$Q_c = \frac{[Fe^{2+}]_t[H_2S]_t}{[H_3O^+]_t^2} = \frac{(0.10)(0.10)}{(1 \times 10^{-8})^2} = 1 \times 10^{14}; \quad Q_c > K_{spa}; \text{ FeS(s) will precipitate}$$

16.129 CoS, $K_{spa} = \dfrac{[Co^{2+}][H_2S]}{[H_3O^+]^2} = 3$

(i) In 0.5 M HCl, $[H_3O^+] = 0.5$ M

$$Q_c = \frac{[Co^{2+}]_t[H_2S]_t}{[H_3O^+]_t^2} = \frac{(0.10)(0.10)}{(0.5)^2} = 0.04; \quad Q_c < K_{spa}; \text{ CoS will not precipitate}$$

(ii) pH = 8; $[H_3O^+] = 10^{-pH} = 10^{-8} = 1 \times 10^{-8}$ M

$$Q_c = \frac{[Co^{2+}]_t[H_2S]_t}{[H_3O^+]_t^2} = \frac{(0.10)(0.10)}{(1 \times 10^{-8})^2} = 1 \times 10^{14}; \quad Q_c > K_{spa}; \text{ CoS(s) will precipitate}$$

16.130 (a) add Cl^- to precipitate AgCl
(b) add CO_3^{2-} to precipitate $CaCO_3$
(c) add H_2S to precipitate MnS
(d) add NH_3 and NH_4Cl to precipitate $Cr(OH)_3$
(Need buffer to control $[OH^-]$; excess OH^- produces the soluble $Cr(OH)_4^-$.)

16.131 (a) add Cl^- to precipitate Hg_2Cl_2
(b) add $(NH_4)_2HPO_4$ to precipitate $MgNH_4PO_4$
(c) add HCl and H_2S to precipitate HgS
(d) add Cl^- to precipitate $PbCl_2$

Chapter Problems

16.132 $$Ca_5(PO_4)_3(OH)(s) \rightleftarrows 5\,Ca^{2+}(aq) + 3\,PO_4^{3-}(aq) + OH^-(aq)$$

equil (M): 5x, 3x, x

$K_{sp} = [Ca^{2+}]^5[PO_4^{3-}]^3[OH^-] = 2.3 \times 10^{-59} = (5x)^5(3x)^3(x) = 84{,}375\, x^9$

Solve for x, x = molar solubility = 8.6×10^{-8} M

$$Ca_5(PO_4)_3(F)(s) \rightleftarrows 5\ Ca^{2+}(aq) + 3\ PO_4^{3-}(aq) + F^-(aq)$$

equil (M) 5x 3x x

$K_{sp} = [Ca^{2+}]^5[PO_4^{3-}]^3[F^-] = 3.2 \times 10^{-60} = (5x)^5(3x)^3(x) = 84,375\ x^9$

Solve for x, x = molar solubility = 7.0×10^{-8} M

16.133 Assume 1.00 L of solution with a density of 1.00 g/mL. The solution weighs 1000 g.

$$1\text{ ppm } F^- = \frac{\text{mass } F^-}{1000\text{ g}} \times 10^6$$

$$\text{mass } F^- = \frac{1000\text{ g}}{10^6} = 0.0010\text{ g } F^-$$

$$[F^-] = \frac{\left(0.0010\text{ g} \times \frac{1\text{ mol}}{19.0\text{ g}}\right)}{1.00\text{ L}} = 5.3 \times 10^{-5}\text{ M}$$

$[Ca^{2+}] = 5.0 \times 10^{-4}$ M

For CaF_2, $K_{sp} = [Ca^{2+}][F^-]^2 = 3.5 \times 10^{-11}$

For CaF_2, IP = $[Ca^{2+}][F^-]^2 = (5.0 \times 10^{-4})(5.3 \times 10^{-5})^2 = 1.4 \times 10^{-12}$

IP < K_{sp}; CaF_2 will not precipitate.

16.134 Prepare aqueous solutions of the three salts. Add a solution of $(NH_4)_2HPO_4$. If a white precipitate forms, the solution contains Mg^{2+}. Perform flame test on the other two solutions. A yellow flame test indicates Na^+. A violet flame test indicates K^+.

16.135

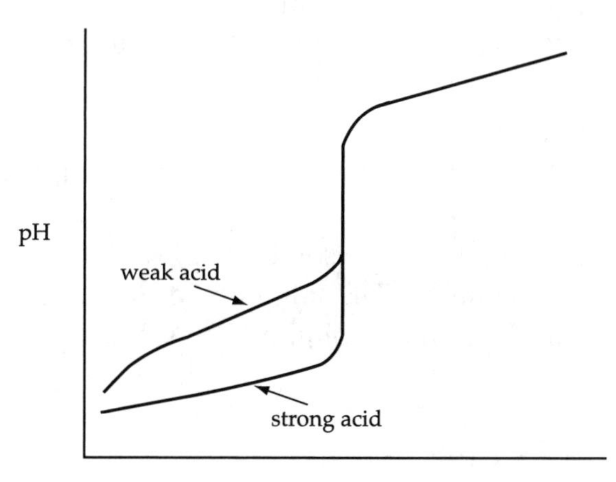

(a) The pH for the weak acid is higher.
(b) Initially, the pH rises more quickly for the weak acid, but then the curve becomes more level in the region halfway to the equivalence point.
(c) The pH is higher at the equivalence point for the weak acid.
(d) Both curves are identical beyond the equivalence point because the pH is determined by the excess $[OH^-]$.
(e) If the acid concentrations are the same, the volume of base needed to reach the equivalence point is the same.

16.136 (a)

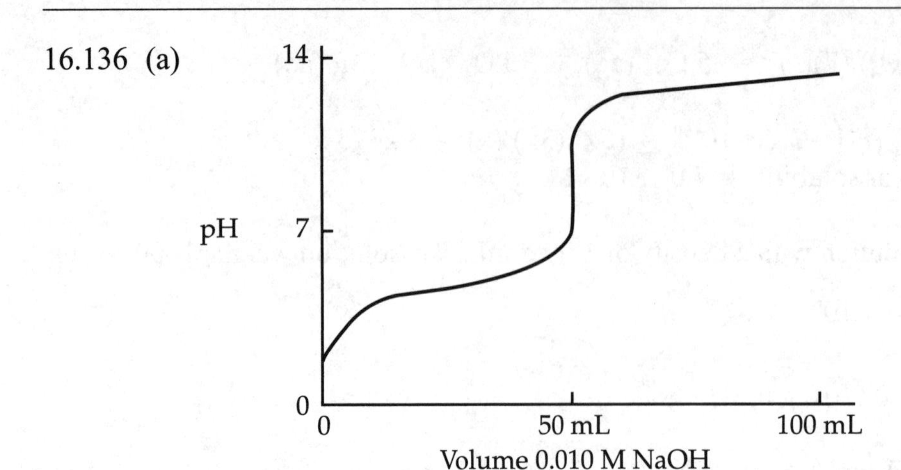

(b) $\text{mol NaOH required} = \left(\dfrac{0.010 \text{ mol HA}}{\text{L}}\right)(0.0500 \text{ L})\left(\dfrac{1 \text{ mol NaOH}}{1 \text{ mol HA}}\right) = 0.000\ 50 \text{ mol}$

$\text{vol NaOH required} = (0.000\ 50 \text{ mol})\left(\dfrac{1 \text{ L}}{0.010 \text{ mol}}\right) = 0.050 \text{ L} = 50 \text{ mL}$

(c) A basic salt is present at the equivalence point; pH > 7.00
(d) Halfway to the equivalence point, the $pH = pK_a = 4.00$

16.137 (a) $AgBr(s) \rightleftarrows Ag^+(aq) + Br^-(aq)$
(i) HBr is a source of Br^- (reaction product). The solubility of AgBr is decreased.
(ii) unaffected
(iii) $AgNO_3$ is a source of Ag^+ (reaction product). The solubility of AgBr is decreased.
(iv) NH_3 forms a complex with Ag^+, removing it from solution. The solubility of AgBr is increased.
(b) $BaCO_3(s) \rightleftarrows Ba^{2+}(aq) + CO_3^{2-}(aq)$
(i) HNO_3 reacts with CO_3^{2-}, removing it from the solution. The solubility of $BaCO_3$ is increased.
(ii) $Ba(NO_3)_2$ is a source of Ba^{2+} (reaction product). The solubility of $BaCO_3$ is decreased.
(iii) Na_2CO_3 is a source of CO_3^{2-} (reaction product). The solubility of $BaCO_3$ is decreased.
(iv) CH_3CO_2H reacts with CO_3^{2-}, removing it from the solution. The solubility of $BaCO_3$ is increased.

16.138 $pH = 10.35$; $[H_3O^+] = 10^{-pH} = 10^{-10.35} = 4.5 \times 10^{-11}$ M

$$[OH^-] = \frac{K_w}{[H_3O^+]} = \frac{1.0 \times 10^{-14}}{4.5 \times 10^{-11}} = 2.2 \times 10^{-4} \text{ M}$$

$$[Mg^{2+}] = \frac{[OH^-]}{2} = \frac{2.2 \times 10^{-4}}{2} = 1.1 \times 10^{-4} \text{ M}$$

$K_{sp} = [Mg^{2+}][OH^-]^2 = (1.1 \times 10^{-4})(2.2 \times 10^{-4})^2 = 5.3 \times 10^{-12}$

16.139 $H_2PO_4^-(aq) + H_2O(l) \rightleftarrows H_3O^+(aq) + HPO_4^{2-}(aq)$
(a) Na_2HPO_4, source of HPO_4^{2-}, equilibrium shifts left, pH increases.
(b) Addition of the strong acid, HBr, decreases the pH.

(c) Addition of the strong base, KOH, increases the pH.
(d) There is no change in the pH with the addition of the neutral salt KI.
(e) H_3PO_4, source of $H_2PO_4^-$, equilibrium shifts right, pH decreases.
(f) Na_3PO_4, source of PO_4^{3-}, decreases $[H_3O^+]$ by forming HPO_4^{2-}, pH increases.

16.140 NaOH, 40.0; $20 \text{ g x } \dfrac{1 \text{ mol}}{40.0 \text{ g}} = 0.50 \text{ mol NaOH}$

$(0.500 \text{ L})(1.5 \text{ mol/L}) = 0.75 \text{ mol } NH_4Cl$

	$NH_4^+(aq)$ +	$OH^-(aq)$ ⇄	$NH_3(aq)$ + $H_2O(l)$
before reaction (mol)	0.75	0.50	0
change (mol)	–0.50	–0.50	+0.50
after reaction (mol)	0.25	0	0.50

This reaction produces a buffer solution.

$[NH_4^+] = 0.25 \text{ mol}/0.500 \text{ L} = 0.50 \text{ M}$; $[NH_3] = 0.50 \text{ mol}/0.500 \text{ L} = 1.0 \text{ M}$

$$pH = pK_a + \log\frac{[\text{base}]}{[\text{acid}]} = pK_a + \log\frac{[NH_3]}{[NH_4^+]}$$

$$\text{For } NH_4^+,\ K_a = \frac{K_w}{K_b \text{ for } NH_3} = \frac{1.0 \times 10^{-14}}{1.8 \times 10^{-5}} = 5.6 \times 10^{-10};\ pK_a = -\log K_a = 9.25$$

$$pH = 9.25 + \log\left(\frac{1.0}{0.5}\right) = 9.55$$

16.141 (a) AgCl, $K_{sp} = [Ag^+][Cl^-] = 1.8 \times 10^{-10}$

$$[Cl^-] = \frac{K_{sp}}{[Ag^+]} = \frac{1.8 \times 10^{-10}}{0.030} = 6.0 \times 10^{-9} \text{ M}$$

(b) Hg_2Cl_2, $K_{sp} = [Hg_2^{2+}][Cl^-]^2 = 1.4 \times 10^{-18}$

$$[Cl^-] = \sqrt{\frac{K_{sp}}{[Hg_2^{2+}]}} = \sqrt{\frac{1.4 \times 10^{-18}}{0.030}} = 6.8 \times 10^{-9} \text{ M}$$

(c) $PbCl_2$, $K_{sp} = [Pb^{2+}][Cl^-]^2 = 1.2 \times 10^{-5}$

$$[Cl^-] = \sqrt{\frac{K_{sp}}{[Pb^{2+}]}} = \sqrt{\frac{1.2 \times 10^{-5}}{0.030}} = 0.020 \text{ M}$$

AgCl(s) will begin to precipitate when the $[Cl^-]$ just exceeds 6.0×10^{-9} M. At this Cl^- concentration, IP < K_{sp} for $PbCl_2$ so all of the Pb^{2+} will remain in solution.

16.142 $$\text{For } NH_4^+,\ K_a = \frac{K_w}{K_b \text{ for } NH_3} = \frac{1.0 \times 10^{-14}}{1.8 \times 10^{-5}} = 5.6 \times 10^{-10};\ pK_a = -\log K_a = 9.25$$

$$pH = pK_a + \log\frac{[NH_3]}{[NH_4^+]} = 9.25 + \log\frac{(0.50)}{(0.30)} = 9.47$$

$[H_3O^+] = 10^{-pH} = 10^{-9.47} = 3.4 \times 10^{-10}$ M

For MnS, $K_{spa} = \dfrac{[Mn^{2+}][H_2S]}{[H_3O^+]^2} = 3 \times 10^7$

molar solubility = $[Mn^{2+}] = \dfrac{K_{spa}[H_3O^+]^2}{[H_2S]} = \dfrac{(3 \times 10^7)(3.4 \times 10^{-10})^2}{(0.10)} = 3.5 \times 10^{-11}$ M

MnS, 87.00; solubility = $(3.5 \times 10^{-11}$ mol/L)(87.00 g/mol) = 3×10^{-9} g/L

16.143 mol H_2A = mol NaOH/2 = (0.1000 mol/L)(0.03472/2 L) = 0.001 736 mol H_2A

H_2A molar mass = $\dfrac{0.2015 \text{ g } H_2A}{0.001\ 736 \text{ mol } H_2A} = 116.1$ g/mol

At the second equivalence point, the dominant reaction is:

$A^{2-}(aq) + H_2O(l) \rightleftarrows OH^-(aq) + HA^-(aq)$

$K_{b1} = \dfrac{[OH^-][HA^-]}{[A^{2-}]}$ and $[OH^-] = [HA^-]$

$K_{b1} = \dfrac{[OH^-]^2}{[A^{2-}]}$

$[A^{2-}] = \dfrac{0.001\ 736 \text{ mol}}{(0.02500 \text{ L} + 0.03472 \text{ L})} = 0.02907$ M

$[H_3O^+] = 10^{-pH} = 10^{-9.27} = 5.37 \times 10^{-10}$ M

$[OH^-] = \dfrac{K_w}{[H_3O^+]} = \dfrac{1.00 \times 10^{-14}}{5.37 \times 10^{-10}} = 1.86 \times 10^{-5}$ M

$K_{b1} = \dfrac{[OH^-]^2}{[A^{2-}]} = \dfrac{(1.86 \times 10^{-5})^2}{(0.02907)} = 1.19 \times 10^{-8}$

$K_{a2} \cdot K_{b1} = K_w$

$K_{a2} = \dfrac{K_w}{K_{b1}} = \dfrac{1.00 \times 10^{-14}}{1.19 \times 10^{-8}} = 8.40 \times 10^{-7}$; $pK_{a2} = -\log(8.40 \times 10^{-7}) = 6.08$

At the first equivalence point, pH = $\dfrac{pK_{a1} + pK_{a2}}{2} = 3.95$

$3.95 = \dfrac{pK_{a1} + 6.08}{2}$

$pK_{a1} = (2)(3.95) - 6.08 = 1.82$

16.144 60.0 mL = 0.0600 L

mol H_3PO_4 = 0.0600 L x $\dfrac{1.00 \text{ mol } H_3PO_4}{1.00 \text{ L}}$ = 0.0600 mol H_3PO_4

mol LiOH = 1.00 L x $\dfrac{0.100 \text{ mol LiOH}}{1.00 \text{ L}}$ = 0.100 mol LiOH

	$H_3PO_4(aq)$ +	$OH^-(aq)$ →	$H_2PO_4^-(aq)$ + $H_2O(l)$
before reaction (mol)	0.0600	0.100	0
change (mol)	–0.0600	–0.0600	+0.0600
after reaction (mol)	0	0.040	0.0600

	$H_2PO_4^-(aq)$ +	$OH^-(aq)$ →	$HPO_4^{2-}(aq)$ + $H_2O(l)$
before reaction (mol)	0.0600	0.040	0
change (mol)	–0.040	–0.040	+0.040
after reaction (mol)	0.020	0	0.040

The resulting solution is a buffer because it contains the conjugate acid-base pair, $H_2PO_4^-$ and HPO_4^{2-}, at acceptable buffer concentrations.
For $H_2PO_4^-$, $K_{a2} = 6.2 \times 10^{-8}$ and $pK_{a2} = -\log K_{a2} = -\log(6.2 \times 10^{-8}) = 7.21$

$$pH = pK_{a2} + \log\frac{[HPO_4^{2-}]}{[H_2PO_4^-]} = 7.21 + \log\frac{(0.040\ mol/1.06\ L)}{(0.020\ mol/1.06\ L)}$$

$$pH = 7.21 + \log\frac{(0.040)}{(0.020)} = 7.21 + 0.30 = 7.51$$

16.145 (a) The mixture of 0.100 mol H_3PO_4 and 0.150 mol NaOH is a buffer and contains mainly $H_2PO_4^-$ and HPO_4^{2-} from the reactions:

	$H_3PO_4(aq)$ +	$OH^-(aq)$ →	$H_2PO_4^-(aq)$ + $H_2O(l)$
before (mol)	0.100	0.150	0
change (mol)	–0.100	–0.100	+0.100
after (mol)	0	0.050	0.100

	$H_2PO_4^-(aq)$ +	$OH^-(aq)$ →	$HPO_4^{2-}(aq)$ + $H_2O(l)$
before (mol)	0.100	0.050	0
change (mol)	–0.050	–0.050	+0.050
after (mol)	0.050	0	0.050

If water were used to dilute the solution instead of HCl, the pH would be equal to pK_{a2} because $[H_2PO_4^-] = [HPO_4^{2-}] = 0.050\ mol/1.00\ L = 0.050\ M$

$H_2PO_4^-(aq) + H_2O(l) \rightleftarrows H_3O^+(aq) + HPO_4^{2-}(aq)$ $\quad K_{a2} = 6.2 \times 10^{-8}$

$pK_{a2} = -\log K_{2a} = -\log(6.2 \times 10^{-8}) = 7.21$

$$pH = pK_{a2} + \log\frac{[HPO_4^{2-}]}{[H_2PO_4^-]} = pK_{a2} + \log(1) = pK_{a2} = 7.21$$

The pH is lower (6.73) because the added HCl converts some HPO_4^{2-} to $H_2PO_4^-$.

	$HPO_4^{2-}(aq)$ +	$H_3O^+(aq)$ →	$H_2PO_4^-(aq)$ + $H_2O(l)$
before (M)	0.050	x	0.050
change (M)	–x	–x	+x
after (M)	0.050 – x	0	0.050 + x

$[HPO_4^{2-}] + [H_2PO_4^-] = (0.050 - x) + (0.050 + x) = 0.100\ M$

$$pH = pK_{a2} + \log \frac{[HPO_4^{2-}]}{[H_2PO_4^-]}$$

$[HPO_4^{2-}] = 0.100 - [H_2PO_4^-]$

$$6.73 = 7.21 + \log \frac{(0.100 - [H_2PO_4^-])}{[H_2PO_4^-]}$$

$$6.73 - 7.21 = -0.48 = \log \frac{(0.100 - [H_2PO_4^-])}{[H_2PO_4^-]}$$

$$10^{-0.48} = 0.331 = \frac{(0.100 - [H_2PO_4^-])}{[H_2PO_4^-]}$$

$(0.331)[H_2PO_4^-] = 0.100 - [H_2PO_4^-]$
$(1.331)[H_2PO_4^-] = 0.100$
$[H_2PO_4^-] = 0.100/1.331 = 0.075\ M$
$[HPO_4^{2-}] = 0.100 - [H_2PO_4^-] = 0.100 - 0.075 = 0.025\ M$
$H_3PO_4(aq) + H_2O(l) \rightleftharpoons H_3O^+(aq) + H_2PO_4^-(aq) \qquad K_{a1} = 7.5 \times 10^{-3}$

$$K_{a1} = \frac{[H_3O^+][H_2PO_4^-]}{[H_3PO_4]}$$

$$[H_3PO_4] = \frac{[H_3O^+][H_2PO_4^-]}{K_{a1}}$$

$[H_3O^+] = 10^{-pH} = 10^{-6.73} = 1.86 \times 10^{-7}\ M$

$$[H_3PO_4] = \frac{(1.86 \times 10^{-7})(0.075)}{7.5 \times 10^{-3}} = 1.9 \times 10^{-6}\ M$$

(b) If distilled water were used and not HCl, the mole amounts of both $H_2PO_4^-$ and HPO_4^{2-} would be 0.050 mol. The HCl converted some HPO_4^{2-} to $H_2PO_4^-$.

	$HPO_4^{2-}(aq)$	+ $H_3O^+(aq)$	→ $H_2PO_4^-(aq)$	+ $H_2O(l)$
before (mol)	0.050	x	0.050	
change (mol)	−x	−x	+x	
after (mol)	0.050 − x	0	0.050 + x	

From part (a), $[HPO_4^{2-}] = 0.025\ M$
mol HPO_4^{2-} = (0.025 mol/L)(1.00 L) = 0.025 mol = 0.050 − x
x = mol H_3O^+ = mol HCl inadvertently added = 0.050 − 0.025 = 0.025 mol HCl

16.146 For CH_3CO_2H, $K_a = 1.8 \times 10^{-5}$ and $pK_a = -\log K_a = -\log(1.8 \times 10^{-5}) = 4.74$
The mixture will be a buffer solution containing the conjugate acid-base pair, CH_3CO_2H and $CH_3CO_2^-$, having a pH near the pK_a of CH_3CO_2H.

$$pH = pK_a + \log \frac{[CH_3CO_2^-]}{[CH_3CO_2H]}$$

$$4.85 = 4.74 + \log \frac{[CH_3CO_2^-]}{[CH_3CO_2H]}; \quad 4.85 - 4.74 = \log \frac{[CH_3CO_2^-]}{[CH_3CO_2H]}$$

$$0.11 = \log \frac{[CH_3CO_2^-]}{[CH_3CO_2H]}; \quad \frac{[CH_3CO_2^-]}{[CH_3CO_2H]} = 10^{0.11} = 1.3$$

In the Henderson-Hasselbalch equation, moles can be used in place of concentrations because both components are in the same volume so the volume terms cancel.
20.0 mL = 0.0200 L

Let X equal the volume of 0.10 M CH_3CO_2H and Y equal the volume of 0.15 M $CH_3CO_2^-$. Therefore, X + Y = 0.0200 L and

$$\frac{Y \text{ x } [CH_3CO_2^-]}{X \text{ x } [CH_3CO_2H]} = \frac{Y(0.15 \text{ mol/L})}{X(0.10 \text{ mol/L})} = 1.3$$

$X = 0.0200 - Y$

$$\frac{Y(0.15 \text{ mol/L})}{(0.020 - Y)(0.10 \text{ mol/L})} = 1.3$$

$$\frac{0.15Y}{0.0020 - 0.10Y} = 1.3$$

$0.15Y = 1.3(0.0020 - 0.10Y)$
$0.15Y = 0.0026 - 0.13Y$
$0.15Y + 0.13Y = 0.0026$
$0.28Y = 0.0026$
$Y = 0.0026/0.28 = 0.0093$ L
$X = 0.0200 - Y = 0.0200 - 0.0093 = 0.0107$ L
$X = 0.0107$ L = 10.7 mL and $Y = 0.0093$ L = 9.3 mL
You need to mix together 10.7 mL of 0.10 M CH_3CO_2H and 9.3 mL of 0.15 M $NaCH_3CO_2$ to prepare 20.0 mL of a solution with a pH of 4.85.

16.147 $[H_3O^+] = 10^{-pH} = 10^{-2.37} = 0.004\ 27$ M

$H_3Cit(aq) + H_2O(l) \rightleftarrows H_3O^+(aq) + H_2Cit^-(aq)$

$$K_{a1} = 7.1 \text{ x } 10^{-4} = \frac{[H_3O^+][H_2Cit^-]}{[H_3Cit]}$$

$(7.1 \text{ x } 10^{-4})[H_3Cit] = (0.004\ 27)[H_2Cit^-]$

$[H_3Cit] = (0.004\ 27)[H_2Cit^-]/(7.1 \text{ x } 10^{-4}) = (6.01)[H_2Cit^-]$

$H_2Cit^-(aq) + H_2O(l) \rightleftarrows H_3O^+(aq) + HCit^{2-}(aq)$

$$K_{a2} = 1.7 \text{ x } 10^{-5} = \frac{[H_3O^+][HCit^{2-}]}{[H_2Cit^-]}$$

$(1.7 \text{ x } 10^{-5})[H_2Cit^-] = (0.004\ 27)[HCit^{2-}]$
$[HCit^{2-}] = (1.7 \text{ x } 10^{-5})[H_2Cit^-]/(0.004\ 27) = (0.003\ 98)[H_2Cit^-]$
$[H_3Cit] + [H_2Cit^-] + [HCit^{2-}] + [Cit^{3-}] = 0.350$ M

Now assume $[Cit^{3-}] \approx 0$, so $[H_3Cit] + [H_2Cit^-] + [HCit^{2-}] = 0.350$ M and then by substitution:

$(6.01)[H_2Cit^-] + [H_2Cit^-] + (0.003\ 98)[H_2Cit^-] = 0.350$ M

$(7.01)[H_2Cit^-] = 0.350$ M

$[H_2Cit^-] = 0.350\ M/7.01 = 0.050$ M

$[H_3Cit] = (6.01)[H_2Cit^-] = (6.01)(0.050\ M) = 0.30$ M

$[HCit^{2-}] = (0.003\ 98)[H_2Cit^-] = (0.003\ 98)(0.050\ M) = 2.0 \times 10^{-4}$ M

$$HCit^{2-}(aq) + H_2O(l) \rightleftarrows H_3O^+(aq) + Cit^{3-}(aq)$$

$$K_{a3} = 4.1 \times 10^{-7} = \frac{[H_3O^+][Cit^{3-}]}{[HCit^{2-}]}$$

$$[Cit^{3-}] = \frac{(K_{a3})[HCit^{2-}]}{[H_3O^+]} = \frac{(4.1 \times 10^{-7})(2.0 \times 10^{-4})}{(0.004\ 27)} = 1.9 \times 10^{-8}\ M$$

16.148 (a) HCl is a strong acid. HCN is a weak acid with $K_a = 4.9 \times 10^{-10}$. Before the titration, the $[H_3O^+] = 0.100$ M. The HCN contributes an insignificant amount of additional H_3O^+, so the pH $= -\log[H_3O^+] = -\log(0.100) = 1.00$

(b) 100.0 mL = 0.1000 L

$$\text{mol } H_3O^+ = 0.1000\ L \times \frac{0.100\ \text{mol HCl}}{1.00\ L} = 0.0100\ \text{mol } H_3O^+$$

add 75.0 mL of 0.100 M NaOH; 75.0 mL = 0.0750 L

$$\text{mol } OH^- = 0.0750\ L \times \frac{0.100\ \text{mol NaOH}}{1.00\ L} = 0.00750\ \text{mol } OH^-$$

	$H_3O^+(aq)$	+ $OH^-(aq)$	→ $2\ H_2O(l)$
before reaction (mol)	0.0100	0.0075	
change (mol)	–0.0075	–0.0075	
after reaction (mol)	0.0025	0	

$$[H_3O^+] = \frac{0.0025\ \text{mol } H_3O^+}{0.1000\ L + 0.0750\ L} = 0.0143\ M$$

pH $= -\log[H_3O^+] = -\log(0.0143) = 1.84$

(c) 100.0 mL of 0.100 M NaOH will completely neutralize all of the H_3O^+ from 100.0 mL of 0.100 M HCl. Only NaCl and HCN remain in the solution. NaCl is a neutral salt and does not affect the pH of the solution. [HCN] changes because of dilution. Because the solution volume is doubled, [HCN] is cut in half.

[HCN] = 0.100 M/2 = 0.0500 M

	HCN(aq)	+ $H_2O(l)$	⇄ $H_3O^+(aq)$	+ $CN^-(aq)$
initial (M)	0.0500		~0	0
change (M)	–x		+x	+x
equil (M)	0.0500 – x		x	x

$$K_a = \frac{[H_3O^+][CN^-]}{HCN} = 4.9 \times 10^{-10} = \frac{x^2}{0.0500 - x} \approx \frac{x^2}{0.0500}$$

$[H_3O^+] = x = \sqrt{(0.0500)(4.9 \times 10^{-10})} = 4.95 \times 10^{-6}$ M

$pH = -\log[H_3O^+] = -\log(4.95 \times 10^{-6}) = 5.31$

(d) Add an additional 25.0 mL of 0.100 M NaOH.

25.0 mL = 0.0250 L

$$\text{additional mol } OH^- = 0.0250 \text{ L} \times \frac{0.100 \text{ mol NaOH}}{1.00 \text{ L}} = 0.00250 \text{ mol } OH^-$$

$$\text{mol HCN} = 0.200 \text{ L} \times \frac{0.0500 \text{ mol HCN}}{1.00 \text{ L}} = 0.0100 \text{ mol HCN}$$

	$HCN(aq)$	+ $OH^-(aq)$	→ $CN^-(aq)$	+ $H_2O(l)$
before reaction (mol)	0.0100	0.00250	0	
change (mol)	−0.00250	−0.00250	+0.00250	
after reaction (mol)	0.0075	0	0.00250	

The resulting solution is a buffer because it contains the conjugate acid-base pair, HCN and CN^-, at acceptable buffer concentrations.

For HCN, $K_a = 4.9 \times 10^{-10}$ and $pK_a = -\log K_a = -\log(4.9 \times 10^{-10}) = 9.31$

$$pH = pK_a + \log\frac{[CN^-]}{[HCN]} = 9.31 + \log\frac{(0.00250 \text{ mol}/0.2250 \text{ L})}{(0.0075 \text{ mol}/0.2250 \text{ L})}$$

$$pH = 9.31 + \log\frac{(0.00250)}{(0.0075)} = 9.31 - 0.48 = 8.83$$

16.149 (a)

	$Cd(OH)_2(s)$	⇄ $Cd^{2+}(aq)$	+ $2\ OH^-(aq)$
initial (M)		0	~0
equil (M)		x	2x

$K_{sp} = [Cd^{2+}][OH^-]^2 = 5.3 \times 10^{-15} = (x)(2x)^2 = 4x^3$

$$\text{molar solubility} = x = \sqrt[3]{\frac{5.3 \times 10^{-15}}{4}} = 1.1 \times 10^{-5} \text{ M}$$

$[OH^-] = 2x = 2(1.1 \times 10^{-5}\text{ M}) = 2.2 \times 10^{-5}$ M

$$[H_3O^+] = \frac{1.0 \times 10^{-14}}{2.2 \times 10^{-5}} = 4.5 \times 10^{-10} \text{ M}$$

$pH = -\log[H_3O^+] = -\log(4.5 \times 10^{-10}) = 9.35$

(b) 90.0 mL = 0.0900 L

mol HNO_3 = (0.100 mol/L)(0.0900 L) = 0.009 00 mol HNO_3

The addition of HNO_3 dissolves some $Cd(OH)_2(s)$.

	$Cd(OH)_2(s)$	+ $2\ HNO_3(aq)$	→ $Cd^{2+}(aq)$	+ $2\ H_2O(l)$
before (mol)	0.100	0.009 00	1.1×10^{-5}	
change (mol)	−0.0045	−2(0.0045)	$1.1 \times 10^{-5} + 0.0045$	
after (mol)	0.0955	0	~0.0045	

total volume = 100.0 mL + 90.0 mL = 190.0 mL = 0.1900 L

$[Cd^{2+}]$ = 0.0045 mol/0.1900 L = 0.024 M

$K_{sp} = 5.3 \times 10^{-15} = [Cd^{2+}][OH^-]^2 = (0.024)[OH^-]^2$

$[OH^-] = \sqrt{\frac{5.3 \times 10^{-15}}{0.024}} = 4.7 \times 10^{-7}$ M; $[H_3O^+] = \frac{1.0 \times 10^{-14}}{4.7 \times 10^{-7}} = 2.1 \times 10^{-8}$ M

pH = $-\log[H_3O^+] = -\log(2.1 \times 10^{-8}) = 7.68$

(c) volume $HNO_3 = 0.0100 \text{ mol } Cd(OH)_2 \times \frac{2 \text{ mol } HNO_3}{1 \text{ mol } Cd(OH)_2} \times \frac{1.00 \text{ L}}{0.100 \text{ mol}} \times \frac{1000 \text{ mL}}{1.00 \text{ L}} = 200 \text{ mL}$

16.150 (a)

	$Zn(OH)_2(s) \rightleftarrows$	$Zn^{2+}(aq)$	+ $2\,OH^-(aq)$
initial (M)		0	~0
equil (M)		x	2x

$K_{sp} = [Zn^{2+}][OH^-]^2 = 4.1 \times 10^{-17} = (x)(2x)^2 = 4x^3$

molar solubility = $x = \sqrt[3]{\frac{4.1 \times 10^{-17}}{4}} = 2.2 \times 10^{-6}$ M

(b) $[OH^-] = 2x = 2(2.2 \times 10^{-6} \text{ M}) = 4.4 \times 10^{-6}$ M

$[H_3O^+] = \frac{1.0 \times 10^{-14}}{4.4 \times 10^{-6}} = 2.3 \times 10^{-9}$ M; pH = $-\log[H_3O^+] = -\log(2.3 \times 10^{-9}) = 8.64$

(c)

$Zn(OH)_2(s) \rightleftarrows Zn^{2+}(aq) + 2\,OH^-(aq)$ $K_{sp} = 4.1 \times 10^{-17}$

$Zn^{2+}(aq) + 4\,OH^-(aq) \rightleftarrows Zn(OH)_4^{2-}(aq)$ $K_f = 3 \times 10^{15}$

$Zn(OH)_2(s) + 2\,OH^-(aq) \rightleftarrows Zn(OH)_4^{2-}(aq)$ $K = K_{sp} \cdot K_f = 0.123$

	$2\,OH^-(aq)$	$Zn(OH)_4^{2-}(aq)$
initial (M)	0.10	0
change (M)	–2x	+x
equil (M)	0.10 – 2x	x

$K = \frac{[Zn(OH)_4^{2-}]}{[OH^-]^2} = 0.123 = \frac{x}{(0.10 - 2x)^2}$

$0.492x^2 - 1.0492x + 0.00123 = 0$

Use the quadratic formula to solve for x.

$x = \frac{-(-1.0492) \pm \sqrt{(-1.0492)^2 - (4)(0.492)(0.00123)}}{2(0.492)} = \frac{1.0492 \pm 1.0480}{0.984}$

$x = 2.1$ and 1.2×10^{-3}

Of the two solutions for x, only 1.2×10^{-3} has physical meaning because the other solution leads to a negative $[OH^-]$.

molar solubility of $Zn(OH)_4^{2-}$ in 0.10 M NaOH = x = 1.2×10^{-3} M

16.151 (a)

$Fe(OH)_3(s) \rightleftarrows Fe^{3+}(aq) + 3\,OH^-(aq)$ $K_{sp} = 2.6 \times 10^{-39}$

$H_3Cit(aq) + H_2O(l) \rightleftarrows H_3O^+(aq) + H_2Cit^-(aq)$ $K_{a1} = 7.1 \times 10^{-4}$

$H_2Cit^-(aq) + H_2O(l) \rightleftarrows H_3O^+(aq) + HCit^{2-}(aq)$ $K_{a2} = 1.7 \times 10^{-5}$

$HCit^{2-}(aq) + H_2O(l) \rightleftarrows H_3O^+(aq) + Cit^{3-}(aq)$ $K_{a3} = 4.1 \times 10^{-7}$

$Fe^{3+}(aq) + Cit^{3-}(aq) \rightleftarrows Fe(Cit)(aq)$ $K_f = 6.3 \times 10^{11}$

$3\,[H_3O^+(aq) + OH^-(aq) \rightleftarrows 2\,H_2O(l)]$ $(1/K_w)^3 = 1.0 \times 10^{42}$

$Fe(OH)_3(s) + H_3Cit(aq) \rightleftarrows Fe(Cit)(aq) + 3\,H_2O(l)$

$K = K_{sp}K_{a1}K_{a2}K_{a3}K_f(1/K_w)^3 = 8.1$

(b)	$Fe(OH)_3(s)$ +	$H_3Cit(aq)$	⇄	$Fe(Cit)(aq)$	+ $3\,H_2O(l)$
initial (M)		0.500		0	
change (M)		−x		+x	
equil (M)		0.500 − x		x	

$$K = \frac{[Fe(Cit)]}{[H_3Cit]} = 8.1 = \frac{x}{0.500 - x}$$

$8.1(0.500 - x) = x$

$4.05 - 8.1x = x$

$4.05 = 9.1x$

x = molar solubility = 4.05/9.1 = 0.45 M

Multiconcept Problems

16.152 (a) pH = 5.5; $[H_3O^+] = 10^{-pH} = 10^{-5.5} = 3.2 \times 10^{-6}$ M

	$H_2C_2O_4(aq)$	+ $H_2O(l)$ ⇄	$H_3O^+(aq)$	+ $HC_2O_4^-(aq)$
equil (M)	$1.1\times10^{-4} - 3.2\times10^{-6}$		3.2×10^{-6}	3.2×10^{-6}

The second dissociation of $H_2C_2O_3$ produces a negligible amount of H_3O^+ compared with that from the first dissociation.

$HC_2O_4^-(aq)$ + $H_2O(l)$ ⇄	$H_3O^+(aq)$ +	$C_2O_4^{2-}(aq)$
3.2×10^{-6}	3.2×10^{-6}	x

$$K_{a2} = \frac{[H_3O^+][C_2O_4^{2-}]}{[HC_2O_4^-]} = 6.4 \times 10^{-5} = \frac{(3.2 \times 10^{-6})[C_2O_4^{2-}]}{(3.2 \times 10^{-6})}$$

$x = [C_2O_4^{2-}] = K_{a2} = 6.4 \times 10^{-5}$

$Ksp = [Ca^{2+}][C_2O_4^{2-}] = 2.3 \times 10^{-9}$

For CaC_2O_4, $IP = [Ca^{2+}][C_2O_4^{2-}] = (2.5 \times 10^{-3})(6.4 \times 10^{-5}) = 1.6 \times 10^{-7}$

$IP > K_{sp}$; CaC_2O_4 will precipitate.

(b) An ionic compound that contains a basic anion becomes less soluble as the acidity of the solution decreases (i.e., pH increases). Kidney stones are made of CaC_2O_4. CaC_2O_4 contains a basic anion. Kidney stones would be more likely to form in urine with a higher pH.

16.153 (a) $HA^-(aq) + H_2O(l) \rightleftarrows H_3O^+(aq) + A^{2-}(aq)$ $\quad K_{a2} = 10^{-10}$

$HA^-(aq) + H_2O(l) \rightleftarrows H_2A(aq) + OH^-(aq)$ $\quad K_b = \dfrac{K_w}{K_{a1}} = 10^{-10}$

$2\,HA^-(aq) \rightleftarrows H_2A(aq) + A^{2-}(aq)$ $\quad K = \dfrac{K_{a2}}{K_{a1}} = 10^{-6}$

$2\,H_2O(l) \rightleftarrows H_3O^+(aq) + OH^-(aq)$ $\quad K_w = 1.0 \times 10^{-14}$

The principal reaction of the four is the one with the largest K, and that is the third reaction.

(b) $K_{a1} = \dfrac{[H_3O^+][HA^-]}{[H_2A]}$ and $K_{a2} = \dfrac{[H_3O^+][A^{2-}]}{[HA^-]}$

$[H_3O^+] = \dfrac{K_{a1}\,[H_2A]}{[HA^-]}$ and $[H_3O^+] = \dfrac{K_{a2}\,[HA^-]}{[A^{2-}]}$

$\dfrac{K_{a1}\,[H_2A]}{[HA^-]} \times \dfrac{K_{a2}\,[HA^-]}{[A^{2-}]} = [H_3O^+]^2$; $\quad \dfrac{K_{a1}K_{a2}\,[H_2A]}{[A^{2-}]} = [H_3O^+]^2$

Because the principal reaction is $2\ HA^-(aq) \rightleftarrows H_2A(aq) + A^{2-}(aq)$, $[H_2A] = [A^{2-}]$.

$K_{a1}\,K_{a2} = [H_3O^+]^2$

$\log K_{a1} + \log K_{a2} = 2 \log [H_3O^+]$

$\dfrac{\log K_{a1} + \log K_{a2}}{2} = \log[H_3O^+]$; $\quad \dfrac{-\log K_{a1} + (-\log K_{a2})}{2} = -\log[H_3O^+]$

$\dfrac{pK_{a1} + pK_{a2}}{2} = pH$

(c)

	$2\ HA^-(aq)$	$\rightleftarrows$	$H_2A(aq)$	+	$A^{2-}(aq)$
initial (M)	1.0		0		0
change (M)	–2x		+x		+x
equil (M)	1.0 – 2x		x		x

$K = \dfrac{[H_2A][A^{2-}]}{[HA^-]^2} = 1 \times 10^{-6} = \dfrac{x^2}{(1.0 - 2x)^2}$

Take the square root of both sides and solve for x.

$x = [A^{2-}] = 1 \times 10^{-3}$ M

mol A^{2-} = $(1 \times 10^{-3}$ mol/L)(0.0500 L) = 5×10^{-5} mol A^{2-}

number of A^{2-} ions = $(5 \times 10^{-5}$ mol $A^{2-})(6.022 \times 10^{23}$ ions/mol) = 3×10^{19} A^{2-} ions

16.154 (a) (i)

	en(aq)	+	$H_2O(l)$	$\rightleftarrows$	$enH^+(aq)$	+	$OH^-(aq)$
initial (M)	0.100				0		~0
change (M)	–x				+x		+x
equil (M)	0.100 – x				x		x

$K_b = \dfrac{[enH^+][OH^-]}{[en]} = 5.2 \times 10^{-4} = \dfrac{(x)(x)}{0.100 - x}$

$x^2 + (5.2 \times 10^{-4})x - (5.2 \times 10^{-5}) = 0$

Use the quadratic formula to solve for x.

$x = \dfrac{-(5.2 \times 10^{-4}) \pm \sqrt{(5.2 \times 10^{-4})^2 - 4(1)(-5.2 \times 10^{-5})}}{2(1)} = \dfrac{-5.2 \times 10^{-4} \pm 0.01443}{2}$

x = –0.0075 and 0.0070

Of the two solutions for x, only the positive value of x has physical meaning because x is the $[OH^-]$.

$[OH^-] = x = 0.0070$ M; $\quad [H_3O^+] = \dfrac{K_w}{[OH^-]} = \dfrac{1.0 \times 10^{-14}}{0.0070} = 1.43 \times 10^{-12}$ M

$pH = -\log[H_3O^+] = -\log(1.43 \times 10^{-12}) = 11.84$

(ii) (30.0 mL)(0.100 mmol/mL) = 3.00 mmol en
(15.0 mL)(0.100 mmol/mL) = 1.50 mmol HCl
Halfway to the first equivalence point, $[OH^-] = K_{b1}$

$$[H_3O^+] = \frac{K_w}{[OH^-]} = \frac{1.0 \times 10^{-14}}{5.2 \times 10^{-4}} = 1.92 \times 10^{-11} \text{ M}$$

$pH = -\log[H_3O^+] = -\log(1.92 \times 10^{-11}) = 10.72$

(iii) At the first equivalence point $pH = \dfrac{pK_{a1} + pK_{a2}}{2} = 9.14$

(iv) Halfway between the first and second equivalence points, $[OH^-] = K_{b2} = 3.7 \times 10^{-7}$ M

$$[H_3O^+] = \frac{K_w}{[OH^-]} = \frac{1.0 \times 10^{-14}}{3.7 \times 10^{-7}} = 2.70 \times 10^{-8} \text{ M}$$

$pH = -\log[H_3O^+] = -\log(2.70 \times 10^{-8}) = 7.57$

(v) At the second equivalence point only the acidic enH_2Cl_2 is in solution.

For enH_2^{2+}, $K_a = \dfrac{K_w}{K_b \text{ for } enH^+} = \dfrac{K_w}{K_{b2}} = \dfrac{1.0 \times 10^{-14}}{3.7 \times 10^{-7}} = 2.70 \times 10^{-8}$

$$[enH_2^{2+}] = \frac{3.00 \text{ mmol}}{(30.0 \text{ mL} + 60.0 \text{ mL})} = 0.0333 \text{ M}$$

	$enH_2^{2+}(aq)$ + $H_2O(l)$	⇄ $H_3O^+(aq)$	+ $enH^+(aq)$
initial (M)	0.0333	~0	0
change (M)	–x	+x	+x
equil (M)	0.0333 – x	x	x

$$K_a = \frac{[H_3O^+][enH^+]}{[enH_2^{2+}]} = 2.70 \times 10^{-8} = \frac{(x)(x)}{0.0333 - x} \approx \frac{x^2}{0.0333}$$

Solve for x. $x = [H_3O^+] = \sqrt{(2.70 \times 10^{-8})(0.0333)} = 3.00 \times 10^{-5}$ M
$pH = -\log[H_3O^+] = -\log(3.00 \times 10^{-5}) = 4.52$

(vi) excess HCl
(75.0 mL – 60.0 mL)(0.100 mmol/mL) = 1.50 mmol HCl = 1.50 mmol H_3O^+

$$[H_3O^+] = \frac{1.50 \text{ mmol}}{(30.0 \text{ mL} + 75.0 \text{ mL})} = 0.0143 \text{ M}$$

$pH = -\log[H_3O^+] = -\log(0.0143) = 1.84$

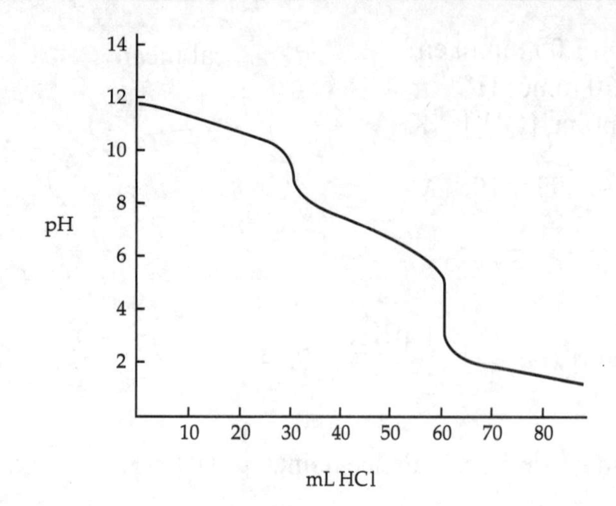

(b) $H-\ddot{N}(H)-C(H)(H)-C(H)(H)-\ddot{N}(H)-H$ Each of the two nitrogens in ethylenediamine can accept a proton.

(c) Each nitrogen is sp^3 hybridized.

16.155 (a) The first equivalence point is reached when all the H_3O^+ from the HCl, and the H_3O^+ from the first ionization of H_3PO_4, is consumed.

At the first equivalence point $pH = \dfrac{pK_{a1} + pK_{a2}}{2} = 4.66$

$[H_3O^+] = 10^{-pH} = 10^{(-4.66)} = 2.2 \times 10^{-5}$ M

(88.0 mL)(0.100 mmol/mL) = 8.80 mmol NaOH are used to get to the first equivalence point

(b) mmol ($HCl + H_3PO_4$) = mmol NaOH = 8.8 mmol

mmol H_3PO_4 = (126.4 mL – 88.0 mL)(0.100 mmol/mL) = 3.84 mmol

mmol HCl = (8.8 – 3.84) = 4.96 mmol

$[HCl] = \dfrac{4.96 \text{ mmol}}{40.0 \text{ mL}} = 0.124$ M; $[H_3PO_4] = \dfrac{3.84 \text{ mmol}}{40.0 \text{ mL}} = 0.0960$ M

(c) 100% of the HCl is neutralized at the first equivalence point.

(d)

	$H_3PO_4(aq)$ + $H_2O(l)$	⇄	$H_3O^+(aq)$	+ $H_2PO_4^-(aq)$
initial (M)	0.0960		0.124	0
change (M)	–x		+x	+x
equil (M)	0.0960 – x		0.124 + x	x

$$K_{a1} = \frac{[H_3O^+][H_2PO_4^-]}{[H_3PO_4]} = 7.5 \times 10^{-3} = \frac{(0.124 + x)(x)}{0.0960 - x}$$

$x^2 + 0.132x - (7.2 \times 10^{-4}) = 0$

Use the quadratic formula to solve for x.

$$x = \frac{-(0.132) \pm \sqrt{(0.132)^2 - 4(1)(-7.2 \times 10^{-4})}}{2(1)} = \frac{-0.132 \pm 0.142}{2}$$

x = –0.137 and 0.005

Of the two solutions for x, only the positive value of x has physical meaning because the other solution would give a negative $[H_3O^+]$.
$[H_3O^+] = 0.124 + x = 0.124 + 0.005 = 0.129$ M
$pH = -\log[H_3O^+] = -\log(0.129) = 0.89$
(e)

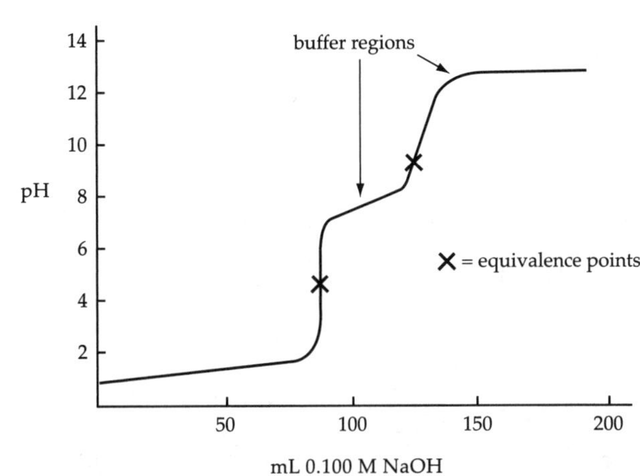

(f) Bromcresol green or methyl orange are suitable indicators for the first equivalence point. Thymolphthalein is a suitable indicator for the second equivalence point.

16.156 (a) PV = nRT; 25 °C = 298 K

$$n_{HCl} = \frac{PV}{RT} = \frac{\left(732 \text{ mm Hg x } \dfrac{1.00 \text{ atm}}{760 \text{ mm Hg}}\right)(1.000 \text{ L})}{\left(0.082\ 06 \dfrac{\text{L} \cdot \text{atm}}{\text{K} \cdot \text{mol}}\right)(298 \text{ K})} = 0.0394 \text{ mol HCl}$$

Na_2CO_3, 105.99

$$\text{mol } Na_2CO_3 = 6.954 \text{ g } Na_2CO_3 \text{ x } \frac{1 \text{ mol } Na_2CO_3}{105.99 \text{ g } Na_2CO_3} = 0.0656 \text{ mol } Na_2CO_3$$

	$CO_3^{2-}(aq)$ +	$H_3O^+(aq)$ →	$HCO_3^-(aq)$ +	$H_2O(l)$
before reaction (mol)	0.0656	0.0394	0	
change (mol)	−0.0394	−0.0394	+0.0394	
after reaction (mol)	0.0656 – 0.0394	0	0.0394	

mol CO_3^{2-} = 0.0656 – 0.0394 = 0.0262 mol and mol HCO_3^- = 0.0394 mol
Therefore, we have an HCO_3^-/CO_3^{2-} buffer solution.

$$pH = pK_{a2} + \log\frac{[CO_3^{2-}]}{[HCO_3^-]} = -\log(5.6 \text{ x } 10^{-11}) + \log\frac{0.0262 \text{ mol/V}}{0.0394 \text{ mol/V}}$$

pH = 10.25 – 0.177 = 10.08

(b) mol Na^+ = 2(0.0656 mol) = 0.1312 mol
mol CO_3^{2-} = 0.0262 mol
mol HCO_3^- = 0.0394 mol
mol Cl^- = 0.0394 mol
total ion moles = 0.2362 mol

$$\Delta T_f = K_f \cdot m; \qquad \Delta T_f = \left(1.86\ \frac{°C \cdot kg}{mol}\right)\left(\frac{0.2362\ mol}{0.2500\ kg}\right) = 1.76\ °C$$

Solution freezing point = 0 °C – ΔT_f = –1.76 °C

(c) H_2O, 18.02

$$mol\ H_2O = 250.0\ g \times \frac{1\ mol\ H_2O}{18.02\ g\ H_2O} = 13.87\ mol\ H_2O$$

$$X_{solv} = \frac{mol\ H_2O}{mol\ H_2O + mol\ ions} = \frac{13.87\ mol}{13.87\ mol + 0.2362\ mol} = 0.9833$$

$$P_{soln} = P_{solv} \cdot X_{solv} = (23.76\ mm\ Hg)(0.9833) = 23.36\ mm\ Hg$$

16.157 25 °C = 298 K

$$\Pi = 2MRT; \qquad M = \frac{\Pi}{2RT} = \frac{\left(74.4\ mm\ Hg \times \frac{1.00\ atm}{760\ mm\ Hg}\right)}{(2)\left(0.082\ 06\ \frac{L \cdot atm}{K \cdot mol}\right)(298\ K)} = 0.00200\ M$$

$[M^+] = [X^-] = 0.00200$ M

$K_{sp} = [M^+][X^-] = (0.00200)^2 = 4.00 \times 10^{-6}$

16.158 (a) $HCO_3^-(aq) + OH^-(aq) \rightarrow CO_3^{2-}(aq) + H_2O(l)$

(b) mol HCO_3^- = (0.560 mol/L)(0.0500 L) = 0.0280 mol HCO_3^-

mol OH^- = (0.400 mol/L)(0.0500 L) = 0.0200 mol OH^-

	$HCO_3^-(aq)$ +	$OH^-(aq)$ →	$CO_3^{2-}(aq)$ +	$H_2O(l)$
before reaction (mol)	0.0280	0.0200	0	
change (mol)	–0.0200	–0.0200	+0.0200	
after reaction (mol)	0.0280 – 0.0200	0	0.0200	

mol HCO_3^- = 0.0280 – 0.0200 = 0.0080 mol

$$[HCO_3^-] = \frac{0.0080\ mol}{0.1000\ L} = 0.080\ M; \qquad [CO_3^{2-}] = \frac{0.0200\ mol}{0.1000\ L} = 0.200\ M$$

	$HCO_3^-(aq)$ +	$H_2O(l)$ ⇄	$H_3O^+(aq)$ +	$CO_3^{2-}(aq)$
initial (M)	0.080		~0	0.200
change (M)	–x		+x	+x
equil (M)	0.080 – x		x	0.200 + x

$$K_a = \frac{[H_3O^+][CO_3^{2-}]}{[HCO_3^-]} = 5.6 \times 10^{-11} = \frac{x(0.200 + x)}{0.080 - x} \approx \frac{x(0.200)}{0.080}$$

Solve for x. x = $[H_3O^+] = 2.24 \times 10^{-11}$ M

pH = $-\log[H_3O^+] = -\log(2.24 \times 10^{-11}) = 10.65$

Because this solution contains both a weak acid (HCO_3^-) and its conjugate base, the solution is a buffer.

(c) $HCO_3^-(aq) + OH^-(aq) \rightarrow CO_3^{2-}(aq) + H_2O(l)$

$\Delta H^\circ_{rxn} = [\Delta H^\circ_f(CO_3^{2-}) + \Delta H^\circ_f(H_2O)] - [\Delta H^\circ_f(HCO_3^-) + \Delta H^\circ_f(OH^-)]$

$\Delta H^\circ_{rxn} = [(1\ mol)(-677.1\ kJ/mol) + (1\ mol)(-285.8\ kJ/mol)]$
$\qquad - [(1\ mol)(-692.0\ kJ/mol) + (1\ mol)(-230\ kJ/mol)]$

$\Delta H^\circ_{rxn} = -40.9\ kJ$

0.0200 moles each of HCO_3^- and OH^- reacted.

heat produced = q = (0.0200 mol)(40.9 kJ/mol) = 0.818 kJ = 818 J

(d) q = m x specific heat x ΔT

$$\Delta T = \frac{q}{m \text{ x specific heat}} = \frac{818\ J}{(100.0\ g)[4.18\ J/(g\cdot{}^\circ C)]} = 2.0\ ^\circ C$$

Final temperature = 25 °C + 2.0 °C = 27 °C

16.159 (a) species present initially:

NH_4^+	CO_3^{2-}	H_2O
acid	base	acid or base

$2H_2O(l) \rightleftharpoons H_3O^+(aq) + OH^-(aq)$

$NH_4^+(aq) + H_2O(l) \rightleftharpoons NH_3(aq) + H_3O^+(aq)$

$CO_3^{2-}(aq) + H_2O(l) \rightleftharpoons HCO_3^-(aq) + OH^-(aq)$

NH_3, $K_b = 1.8 \times 10^{-5}$
NH_4^+, $K_a = 5.6 \times 10^{-10}$
CO_3^{2-}, $K_b = 1.8 \times 10^{-4}$
HCO_3^-, $K_a = 5.6 \times 10^{-11}$

In the mixture, proton transfer takes place from the stronger acid to the stronger base, so the principal reaction is $NH_4^+(aq) + CO_3^{2-}(aq) \rightleftharpoons HCO_3^-(aq) + NH_3(aq)$

(b)

$NH_4^+(aq) + OH^-(aq) \rightleftharpoons NH_3(aq) + H_2O(l)$ $\quad K_1 = 1/K_b(NH_3)$

$CO_3^{2-}(aq) + H_2O(l) \rightleftharpoons HCO_3^-(aq) + OH^-(aq)$ $\quad K_2 = K_b(CO_3^{2-})$

$NH_4^+(aq) + CO_3^{2-}(aq) \rightleftharpoons HCO_3^-(aq) + NH_3(aq)$ $\quad K = K_1 \cdot K_2$

	$NH_4^+(aq)$	$CO_3^{2-}(aq)$	$HCO_3^-(aq)$	$NH_3(aq)$
initial (M)	0.16	0.080	0	0.16
change (M)	–x	–x	+x	+x
equil (M)	0.16 – x	0.080 – x	x	0.16 + x

$$K = \frac{[HCO_3^-][NH_3]}{[NH_4^+][CO_3^{2-}]} = \frac{1.8 \times 10^{-4}}{1.8 \times 10^{-5}} = 10 = \frac{x(0.16 + x)}{(0.16 - x)(0.080 - x)}$$

$9x^2 - 2.56x + 0.128 = 0$

Use the quadratic formula to solve for x.

$$x = \frac{-(-2.56) \pm \sqrt{(-2.56)^2 - (4)(9)(0.128)}}{2(9)} = \frac{2.56 \pm 1.395}{18}$$

x = 0.220 and 0.0647

Of the two solutions for x, only 0.00647 has physical meaning because 0.220 leads to negative concentrations.

$[NH_4^+] = 0.16 - x = 0.16 - 0.0647 = 0.0953\ M = 0.095\ M$
$[NH_3] = 0.16 + x = 0.16 + 0.0647 = 0.225\ M = 0.23\ M$
$[CO_3^{2-}] = 0.080 - x = 0.080 - 0.0647 = 0.0153\ M = 0.015\ M$
$[HCO_3^-] = x = 0.0647\ M = 0.065\ M$
The solution is a buffer containing two different sets of conjugate acid-base pairs. Either pair can be used to calculate the pH.
For NH_4^+, $K_a = 5.6 \times 10^{-10}$ and $pK_a = 9.25$

$$pH = pK_a + \log \frac{[NH_3]}{[NH_4^+]} = 9.25 + \log \frac{(0.225)}{(0.0953)} = 9.62$$

$[H_3O^+] = 10^{-pH} = 10^{-9.62} = 2.4 \times 10^{-10}\ M$

$$[OH^-] = \frac{1.0 \times 10^{-14}}{2.4 \times 10^{-10}} = 4.2 \times 10^{-5}\ M$$

$$[H_2CO_3] = \frac{[HCO_3^-][H_3O^+]}{K_a} = \frac{(0.647)(2.4 \times 10^{-10})}{(4.3 \times 10^{-7})} = 3.6 \times 10^{-4}\ M$$

(c) For MCO_3, $IP = [M^{2+}][CO_3^{2-}] = (0.010)(0.0153) = 1.5 \times 10^{-4}$
$K_{sp}(CaCO_3) = 5.0 \times 10^{-9}$, $10^3\ K_{sp} = 5.0 \times 10^{-6}$
$K_{sp}(BaCO_3) = 2.6 \times 10^{-9}$, $10^3\ K_{sp} = 2.6 \times 10^{-6}$
$K_{sp}(MgCO_3) = 6.8 \times 10^{-6}$, $10^3\ K_{sp} = 6.8 \times 10^{-3}$
$IP > 10^3\ K_{sp}$ for $CaCO_3$ and $BaCO_3$, but $IP < 10^3\ K_{sp}$ for $MgCO_3$ so the $[CO_3^{2-}]$ is large enough to give observable precipitation of $CaCO_3$ and $BaCO_3$, but not $MgCO_3$.
(d) For $M(OH)_2$, $IP = [M^{2+}][OH^-]^2 = (0.010)(4.17 \times 10^{-5})^2 = 1.7 \times 10^{-11}$
$K_{sp}(Ca(OH)_2) = 4.7 \times 10^{-6}$, $10^3\ K_{sp} = 4.7 \times 10^{-3}$
$K_{sp}(Ba(OH)_2) = 5.0 \times 10^{-3}$, $10^3\ K_{sp} = 5.0$
$K_{sp}(Mg(OH)_2) = 5.6 \times 10^{-12}$, $10^3\ K_{sp} = 5.6 \times 10^{-9}$
$IP < 10^3\ K_{sp}$ for all three $M(OH)_2$. None precipitate.

(e)	$CO_3^{2-}(aq)$	+ $H_2O(l)$	$\rightleftarrows$ $HCO_3^-(aq)$	+ $OH^-(aq)$
initial (M)	0.08		0	~0
change (M)	$-x$		$+x$	$+x$
equil (M)	$0.08 - x$		x	x

$$K_b = \frac{[HCO_3^-][OH^-]}{[CO_3^{2-}]} = 1.8 \times 10^{-4} = \frac{x^2}{(0.08 - x)}$$

$x^2 + (1.8 \times 10^{-4})x - (1.44 \times 10^{-5}) = 0$
Use the quadratic formula to solve for x.

$$x = \frac{-(1.8 \times 10^{-4}) \pm \sqrt{(1.8 \times 10^{-4})^2 - (4)(1)(-1.44 \times 10^{-5})}}{2(1)} = \frac{-(1.8 \times 10^{-4}) \pm 7.59 \times 10^{-3}}{2}$$

$x = 0.0037$ and -0.0039
Of the two solutions for x, only 0.0037 has physical meaning because -0.0039 leads to negative concentrations.
$[OH^-] = x = 3.7 \times 10^{-3}\ M$
For MCO_3, $IP = [M^{2+}][CO_3^{2-}] = (0.010)(0.08) = 8.0 \times 10^{-4}$
For $M(OH)_2$, $IP = [M^{2+}][OH^-]^2 = (0.010)(3.7 \times 10^{-3})^2 = 1.4 \times 10^{-7}$

Comparing IP's here and 10^3 K_{sp}'s in (c) and (d) above, Ca^{2+} and Ba^{2+} cannot be separated from Mg^{2+} using 0.08 M Na_2CO_3. Na_2CO_3 is more basic than $(NH_4)_2CO_3$ and $Mg(OH)_2$ would precipitate along with $CaCO_3$ and $BaCO_3$.

16.160 (a) H_2SO_4, 98.08

Assume 1.00 L = 1000 mL of solution.

mass of solution = (1000 mL)(1.836 g/mL) = 1836 g

mass H_2SO_4 = (0.980)(1836 g) = 1799 g H_2SO_4

$$\text{mol } H_2SO_4 = 1799 \text{ g } H_2SO_4 \times \frac{1 \text{ mol } H_2SO_4}{98.08 \text{ g } H_2SO_4} = 18.3 \text{ mol } H_2SO_4$$

$[H_2SO_4]$ = 18.3 mol/ 1.00 L = 18.3 M

(b) Na_2CO_3, 105.99; 1 kg = 1000 g = 2.2046 lb

$$H_2SO_4(aq) + Na_2CO_3(s) \rightarrow Na_2SO_4(aq) + H_2O(l) + CO_2(g)$$

$$\text{mass } H_2SO_4 = (0.980)(36 \text{ tons}) \times \frac{2000 \text{ lb}}{1 \text{ ton}} \times \frac{1000 \text{ g}}{2.2046 \text{ lb}} = 3.20 \times 10^7 \text{ g } H_2SO_4$$

$$\text{mol } H_2SO_4 = 3.20 \times 10^7 \text{ g } H_2SO_4 \times \frac{1 \text{ mol } H_2SO_4}{98.08 \text{ g } H_2SO_4} = 3.26 \times 10^5 \text{ mol } H_2SO_4$$

$$\text{mass } Na_2CO_3 = 3.26 \times 10^5 \text{ mol } H_2SO_4 \times \frac{1 \text{ mol } Na_2CO_3}{1 \text{ mol } H_2SO_4} \times \frac{105.99 \text{ g } Na_2CO_3}{1 \text{ mol } Na_2CO_3} \times \frac{1 \text{ kg}}{1000 \text{ g}} = 3.5 \times 10^4 \text{ kg } Na_2CO_3$$

(c)
$$\text{mol } CO_2 = 3.26 \times 10^5 \text{ mol } H_2SO_4 \times \frac{1 \text{ mol } CO_2}{1 \text{ mol } H_2SO_4} = 3.26 \times 10^5 \text{ mol } CO_2$$

18 °C = 18 + 273 = 291 K

PV = nRT

$$V = \frac{nRT}{P} = \frac{(3.26 \times 10^5 \text{ mol})\left(0.082\ 06 \frac{L \cdot atm}{K \cdot mol}\right)(291 \text{ K})}{\left(745 \text{ mm Hg} \times \frac{1.00 \text{ atm}}{760 \text{ mm Hg}}\right)} = 7.9 \times 10^6 \text{ L}$$

16.161 $Pb(CH_3CO_2)_2$, 325.29; PbS, 239.27

$$\text{(a) mass PbS} = (2 \text{ mL})(1 \text{ g/mL})(0.003) \times \frac{1 \text{ mol } Pb(CH_3CO_2)_2}{325.29 \text{ g } Pb(CH_3CO_2)_2} \times \frac{1 \text{ mol PbS}}{1 \text{ mol } Pb(CH_3CO_2)_2} \times \frac{239.27 \text{ g PbS}}{1 \text{ mol PbS}} \times (30/100) = 0.0013 \text{ g}$$

= 1.3 mg PbS per dye application

(b) $[H_3O^+] = 10^{-pH} = 10^{-5.50} = 3.16 \times 10^{-6}$ M

	$PbS(s)$ +	$2\ H_3O^+(aq)$	⇄ $Pb^{2+}(aq)$	+ $H_2S(aq)$	+ $2\ H_2O(l)$
initial (M)		3.16×10^{-6}	0	0	
change (M)		$-2x$	$+x$	$+x$	
equil (M)		$3.16 \times 10^{-6} - 2x$	x	x	

$$K_{spa} = \frac{[Pb^{2+}][H_2S]}{[H_3O^+]^2} = \frac{x^2}{(3.16 \times 10^{-6} - 2x)^2} \approx \frac{x^2}{(3.16 \times 10^{-6})^2} = 3 \times 10^{-7}$$

$x^2 = (3.16 \times 10^{-6})^2(3 \times 10^{-7}) = 3.0 \times 10^{-18}$

$x = 1.7 \times 10^{-9}$ M = $[Pb^{2+}]$ for a saturated solution.

mass of PbS dissolved per washing =

$$(3\ \text{gal})(3.7854\ \text{L/1 gal})(1.7 \times 10^{-9}\ \text{mol/L}) \times \frac{239.27\ \text{g PbS}}{1\ \text{mol PbS}} = 4.6 \times 10^{-6}\ \text{g PbS/washing}$$

Number of washings required to remove 50% of the PbS from one application =

$$\frac{(0.0013\ \text{g PbS})(50/100)}{(4.6 \times 10^{-6}\ \text{g PbS/washing})} = 1.4 \times 10^2\ \text{washings}$$

(c) The number of washings does not look reasonable. It seems too high considering that frequent dye application is recommended. If the PbS is located mainly on the surface of the hair, as is believed to be the case, solid particles of PbS can be lost by abrasion during shampooing.

17 Thermodynamics: Entropy, Free Energy, and Equilibrium

17.1 (a) spontaneous; (b), (c), and (d) $Q_p > K_p$, are nonspontaneous

17.2 (a) $H_2O(g) \rightarrow H_2O(l)$
A liquid has less randomness than a gas. Therefore, ΔS is negative.
(b) $I_2(g) \rightarrow 2\ I(g)$
ΔS is positive because the reaction increases the number of gaseous particles from 1 mol to 2 mol.
(c) $CaCO_3(s) \rightarrow CaO(s) + CO_2(g)$
ΔS is positive because the reaction increases the number of gaseous molecules.
(d) $Ag^+(aq) + Br^-(aq) \rightarrow AgBr(s)$
A solid has less randomness than +1 and −1 charged ions in an aqueous solution. Therefore, ΔS is negative.
(e) Deposition of frost on a cold morning, $H_2O(g) \rightarrow H_2O(s)$
Deposition is the formation of a solid from a gas. A solid has less randomness than a gas. Therefore, ΔS is negative.

17.3 (a) $A_2(g) + 2\ B(g) \rightarrow 2\ AB(g)$.
(b) Because the reaction decreases the number of gaseous particles from 3 mol to 2 mol, the entropy change is negative.

17.4 $S = k \ln W$, $k = 1.38 \times 10^{-23}$ J/K
(a) $S = (1.38 \times 10^{-23}\ \text{J/K}) \ln (3^{100}) = 1.52 \times 10^{-21}$ J/K
(b) $S = (1.38 \times 10^{-23}\ \text{J/K}) \ln (3^{6.02 \times 10^{23}}) = (1.38 \times 10^{-23}\ \text{J/K})(6.022 \times 10^{23}) \ln 3 = 9.13$ J/K

17.5 $S = k \ln W$, $k = 1.38 \times 10^{-23}$ J/K
(a) $W = 1$; $S = (1.38 \times 10^{-23}\ \text{J/K}) \ln (1) = 0$
(b) $W = 3^{10}$; $S = (1.38 \times 10^{-23}\ \text{J/K}) \ln (3^{10}) = 1.52 \times 10^{-22}$ J/K

17.6 (a) 1 mole N_2 at STP (larger volume, more randomness)
(b) $\Delta S = R \ln \left(\dfrac{V_{\text{State B}}}{V_{\text{State A}}}\right) = R \ln \left(\dfrac{11.2\ \text{L}}{22.4\ \text{L}}\right) = (8.314\ \text{J/K}) \ln(1/2) = -5.76$ J/K

17.7 $CaCO_3(s) \rightarrow CaO(s) + CO_2(g)$
$\Delta S^\circ = [S^\circ(CaO) + S^\circ(CO_2)] - S^\circ(CaCO_3)$
$\Delta S^\circ = [(1\ \text{mol})(38.1\ \text{J/(K·mol)}) + (1\ \text{mol})(213.6\ \text{J/(K·mol)})]$
$- (1\ \text{mol})(91.7\ \text{J/(K·mol)}) = +160.0$ J/K

17.8 (a) $C_3H_8(g) + 5\ O_2(g) \rightarrow 3\ CO_2(g) + 4\ H_2O(l)$

$\Delta S°$ is negative because 6 mol of gas in the reactants are converted to 3 mol of gas in the products.

(b) $\Delta S° = -376.6\ J/K = [3\ S°(CO2) + 4\ S°(H_2O(l))] - [S°(C_3H_8) + 5\ S°(O_2)]$

$S°(C_3H_8) = [3\ S°(CO2) + 4\ S°(H_2O(l))] - [5\ S°(O_2)] + 376.6\ J/K$

$S°(C_3H_8) = [(3\ mol)(213.6\ J/(K \cdot mol)) + (4\ mol)(69.9\ J/(K \cdot mol))]$
$- [(5\ mol)(205.0\ J/(K \cdot mol))] + 376.6\ J/K$

$S°(C_3H_8) = 272.0\ J/(K \cdot mol)$

17.9 From Problem 17.7, $\Delta S_{sys} = \Delta S° = 160.0\ J/K$

$CaCO_3(s) \rightarrow CaO(s) + CO_2(g)$

$\Delta H° = [\Delta H°_f(CaO) + \Delta H°_f(CO_2)] - \Delta H°_f(CaCO_3)$

$\Delta H° = [(1\ mol)(-634.9\ kJ/mol) + (1\ mol)(-393.5\ kJ/mol)]$
$- (1\ mol)(-1207.6\ kJ/mol) = +179.2\ kJ$

$$\Delta S_{surr} = \frac{-\Delta H°}{T} = \frac{-179{,}200\ J}{298\ K} = -601\ J/K$$

$\Delta S_{total} = \Delta S_{sys} + \Delta S_{surr} = 160.0\ J/K + (-601\ J/K) = -441\ J/K$

Because ΔS_{total} is negative, the reaction is not spontaneous under standard-state conditions at 25 °C.

17.10 $Br_2(l) \rightarrow Br_2(g)$

$\Delta H° = \Delta H°_f(Br_2(g)) = (1\ mol)(30.9\ kJ/mol) = 30.9\ kJ$

$\Delta S° = S°(Br_2(g)) - S°(Br_2(l))$

$\Delta S° = (1\ mol)(245.4\ J/(K \cdot mol)) - (1\ mol)(152.2\ J/(K \cdot mol)) = 93.2\ J/K = 93.2 \times 10^{-3}\ kJ/K$

$\Delta G = \Delta H - T\Delta S$

The boiling point (phase change) is associated with an equilibrium. Set $\Delta G = 0$ and solve for T, the boiling point.

$$0 = \Delta H - T\Delta S; \quad T_{bp} = \frac{\Delta H}{\Delta S} = \frac{30.9\ kJ}{93.2 \times 10^{-3}\ kJ/K} = 331.5\ K = 58.4\ °C$$

17.11 (a) $\Delta G = \Delta H - T\Delta S = 55.3\ kJ - (298\ K)(0.1757\ kJ/K) = +2.9\ kJ$

Because $\Delta G > 0$, the reaction is nonspontaneous at 25 °C (298 K)

(b) Set $\Delta G = 0$ and solve for T.

$$0 = \Delta H - T\Delta S; \quad T = \frac{\Delta H}{\Delta S} = \frac{55.3\ kJ}{0.1757\ kJ/K} = 315\ K = 42\ °C$$

17.12 $\Delta H < 0$ (reaction involves bond making - exothermic)

$\Delta S < 0$ (the reaction has less randomness in going from reactants (2 atoms) to products (1 molecule)

$\Delta G < 0$ (the reaction is spontaneous)

17.13 From Problems 17.7 and 17.9: $\Delta H° = 179.2\ kJ$ and $\Delta S° = 160.0\ J/K = 0.1600\ kJ/K$

(a) $\Delta G° = \Delta H° - T\Delta S° = 179.2\ kJ - (298\ K)(0.1600\ kJ/K) = +131.5\ kJ$

(b) Because $\Delta G > 0$, the reaction is nonspontaneous at 25 °C (298 K).

(c) Set $\Delta G = 0$ and solve for T, the temperature above which the reaction becomes spontaneous.

$0 = \Delta H - T\Delta S; \qquad T = \dfrac{\Delta H}{\Delta S} = \dfrac{179.2\ kJ}{0.1600\ kJ/K} = 1120\ K = 847\ ^\circ C$

17.14 $2\ AB_2 \rightarrow A_2 + 2\ B_2$

(a) ΔS° is positive because the reaction increases the number of molecules.

(b) ΔH° is positive because the reaction is endothermic.

$\Delta G^\circ = \Delta H^\circ - T\Delta S^\circ$

For the reaction to be spontaneous, ΔG° must be negative. This will only occur at high temperature where $T\Delta S^\circ$ is greater than ΔH°.

17.15 (a) $CaC_2(s) + 2\ H_2O(l) \rightarrow C_2H_2(g) + Ca(OH)_2(s)$

$\Delta G^\circ = [\Delta G^\circ_f(C_2H_2) + \Delta G^\circ_f(Ca(OH)_2)] - [\Delta G^\circ_f(CaC_2) + 2\ \Delta G^\circ_f(H_2O)]$

$\Delta G^\circ = [(1\ mol)(209.9\ kJ/mol) + (1\ mol)(-897.5\ kJ/mol)]$
$\quad - [(1\ mol)(-64.8\ kJ/mol) + (2\ mol)(-237.2\ kJ/mol)] = -148.4\ kJ$

This reaction can be used for the synthesis of C_2H_2 because $\Delta G < 0$.

(b) It is not possible to synthesize acetylene from solid graphite and gaseous H_2 at 25 °C and 1 atm because $\Delta G^\circ_f(C_2H_2) > 0$.

17.16 $C_{diamond}(s) \rightarrow C_{graphite}(s)$

(a) $\Delta G^\circ = \Delta G^\circ_f(C_{graphite}) - \Delta G^\circ_f(C_{diamond}) = 0 - 2.9\ kJ = -2.9\ kJ$

(b) The reaction is spontaneous because $\Delta G^\circ < 0$.

(c) The reaction is spontaneous, but the reaction rate is extremely low.

17.17 $C(s) + 2\ H_2(g) \rightarrow C_2H_4(g)$

$Q_p = \dfrac{P_{C_2H_4}}{(P_{H_2})^2} = \dfrac{(0.10)}{(100)^2} = 1.0 \times 10^{-5}$

$\Delta G = \Delta G^\circ + RT \ln Q_p$

$\Delta G = 68.1\ kJ/mol + [8.314 \times 10^{-3}\ kJ/(K \cdot mol)](298\ K)\ln(1.0 \times 10^{-5}) = +39.6\ kJ/mol$

Because $\Delta G > 0$, the reaction is spontaneous in the reverse direction.

17.18 $\Delta G = \Delta G^\circ + RT \ln Q$ and $\Delta G^\circ = 15\ kJ$

For $A_2(g) + B_2(g) \rightleftarrows 2\ AB(g)$, $Q_p = \dfrac{(P_{AB})^2}{(P_{A_2})(P_{B_2})}$

Let the number of molecules be proportional to the partial pressure.

(1) $Q_p = 1.0$ (2) $Q_p = 0.0667$ (3) $Q_p = 18$

(a) Reaction (3) has the largest ΔG because Q_p is the largest. Reaction (2) has the smallest ΔG because Q_p is the smallest.

(b) $\Delta G = \Delta G^\circ = 15\ kJ$ because $Q_p = 1$ and $\ln(1) = 0$.

17.19 From Problem 17.13, $\Delta G^\circ = +131.5\ kJ$

$\Delta G^\circ = -RT \ln K_p$

$$\ln K_p = \frac{-\Delta G^o}{RT} = \frac{-131.5 \text{ kJ/mol}}{[8.314 \times 10^{-3} \text{ kJ/(K} \cdot \text{mol)}](298 \text{ K})} = -53.1$$

$K_p = e^{-53.1} = 9 \times 10^{-24}$

17.20 $\Delta G^o = -RT \ln K = -[8.314 \times 10^{-3} \text{ kJ/(K} \cdot \text{mol)}](298 \text{ K}) \ln (1.0 \times 10^{-14}) = 80 \text{ kJ/mol}$

17.21 $H_2O(l) \rightleftarrows H_2O(g)$

$K_p = P_{H_2O}$; K_p is equal to the vapor pressure for H_2O.

$\Delta G^o = \Delta G^o_f(H_2O(g)) - \Delta G^o_f(H_2O(l))$

$\Delta G^o = (1 \text{ mol})(-228.6 \text{ kJ/mol}) - (1 \text{ mol})(-237.2 \text{ kJ/mol}) = +8.6 \text{ kJ}$

$\Delta G^o = -RT \ln K_p$

$$\ln K_p = \frac{-\Delta G^o}{RT} = \frac{-8.6 \text{ kJ/mol}}{[8.314 \times 10^{-3} \text{ kJ/(K} \cdot \text{mol)}](298 \text{ K})} = -3.5$$

$K_p = P_{H_2O} = e^{-3.5} = 0.03 \text{ atm}$

7.22 $P_{EtOH} = 60.6 \text{ mm Hg} \times \frac{1.00 \text{ atm}}{760 \text{ mm Hg}} = 0.0797 \text{ atm}$

$K_p = P_{EtOH} = 0.0797 \text{ atm}$

$\Delta G^o = -RT \ln K_p = -[8.314 \text{ J/K}](298 \text{ K}) \ln (0.0797) = 6267 \text{ J} = 6.27 \text{ kJ}$

17.23 The growth of a human adult from a single cell does not violate the second law of thermodynamics. The energy an human obtains from glucose is used to build and organize complex molecules, resulting in a decrease in entropy for the human. At the same time, however, the entropy of the surroundings increases as the human releases small, simple waste products such as CO_2 and H_2O. Furthermore, heat is released by the human, further increasing the entropy of the surroundings. Thus, an organism pays for its decrease in entropy by increasing the entropy of the rest of the universe.

17.24 (a) $C_6H_{12}O_6(s) + 6\,O_2(g) \rightleftarrows 6\,CO_2(g) + 6\,H_2O(l)$ $\Delta G^{o'} = -2870 \text{ kJ}$

$ADP^{3-}(aq) + H_2PO_4^-(aq) \rightleftarrows ATP^{4-}(aq) + H_2O(l)$ $\Delta G^{o'} = +30.5 \text{ kJ}$

$C_6H_{12}O_6(s) + 6\,O_2(g) \rightleftarrows 6\,CO_2(g) + 6\,H_2O(l)$

$\underline{32\,[ADP^{3-}(aq) + H_2PO_4^-(aq) \rightleftarrows ATP^{4-}(aq) + H_2O(l)]}$

$C_6H_{12}O_6(s) + 6\,O_2(g) + 32\,ADP^{3-}(aq) + 32\,H_2PO_4^-(aq) \rightleftarrows 6\,CO_2(g) + 32\,ATP^{4-}(aq) + 38\,H_2O(l)$

$\Delta G^{o'} = (-2870 \text{ kJ}) + 32(30.5 \text{ kJ}) = -1894 \text{ kJ}$

(b) Because $\Delta G^{o'} < 0$, the reaction is spontaneous.

17.25 (a) Because $\Delta G^{o'} > 0$, the reaction is not spontaneous.

(b) $\Delta G^{o'} = (13.8 \text{ kJ}) + (-30.5 \text{ kJ}) = -16.7 \text{ kJ}$

Because the overall $\Delta G^{o'} < 0$, the reaction is spontaneous.

(c) $\Delta G^o = -RT \ln K$

$$\ln K = \frac{-\Delta G^o}{RT} = \frac{-(-16.7\text{ kJ})}{[8.314 \times 10^{-3}\text{ kJ/K}](310\text{ K})} = 6.48; \quad K = e^{6.48} = 652$$

17.26 (a) $ATP^{4-}(aq) + H_2O(l) \rightleftharpoons ADP^{3-}(aq) + H_2PO_4^-(aq)$ $\quad \Delta G^{o'} = -30.5$ kJ

$$Q = \frac{[ADP^{3-}][H_2PO_4^-]}{[ATP^{4-}]} = \frac{(8.0 \times 10^{-3})(0.9 \times 10^{-3})}{(8.0 \times 10^{-3})}$$

$$\Delta G = \Delta G^{o'} + RT \ln Q$$

$$\Delta G = -30.5\text{ kJ} + (8.314 \times 10^{-3}\text{ kJ/K})(310\text{ K}) \ln\left(\frac{(8.0 \times 10^{-3})(0.9 \times 10^{-3})}{(8.0 \times 10^{-3})}\right) = -48.6\text{ kJ}$$

(b) The amount of free-energy released increases at the concentrations present in muscle cells.

17.27 creatine phosphate(aq) + $H_2O(l) \rightleftharpoons$ creatine(aq) + $H_2PO_4^-(aq)$ $\quad \Delta G^{o'} = -43.1$ kJ

$ADP^{3-}(aq) + H_2PO_4^-(aq) \rightleftharpoons ATP^{4-}(aq) + H_2O(l)$ $\quad \Delta G^{o'} = +30.5$ kJ

creatine phosphate(aq) + $ADP^{3-}(aq) \rightleftharpoons$ creatine(aq) + $ATP^{4-}(aq)$

$\Delta G^{o'} = (-43.1\text{ kJ}) + (30.5\text{ kJ}) = -12.6$ kJ

Conceptual Problems

17.28 (a) $A_2 + AB_3 \rightarrow 3\ AB$

(b) ΔS is positive because the reaction increases the number of gaseous molecules.

17.29 (a)

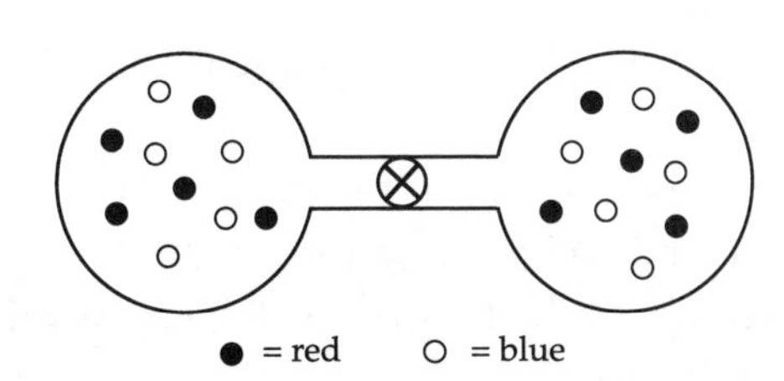

(b) $\Delta H = 0$ (no heat is gained or lost in the mixing of ideal gases)

$\Delta S > 0$ (the mixture of the two gases has more randomness)

$\Delta G < 0$ (the mixing of the two gases is spontaneous)

(c) For an isolated system, $\Delta S_{surr} = 0$ and $\Delta S_{sys} = \Delta S_{Total} > 0$ for the spontaneous process.

(d) $\Delta G > 0$ and the process is nonspontaneous.

17.30 $\Delta H > 0$ (heat is absorbed during sublimation)

$\Delta S > 0$ (gas has more randomness than solid)

$\Delta G < 0$ (the reaction is spontaneous)

17.31 $\Delta H < 0$ (heat is lost during condensation)

$\Delta S < 0$ (liquid has less randomness than vapor)

$\Delta G < 0$ (the reaction is spontaneous)

17.32 $\Delta H = 0$ (system is an ideal gas at constant temperature)
$\Delta S < 0$ (there is less randomness in the smaller volume)
$\Delta G > 0$ (compression of a gas is not spontaneous)

17.33 (a) $2\ A_2 + B_2 \rightarrow 2\ A_2B$
(b) $\Delta H < 0$ (because ΔS is negative, ΔH must also be negative in order for ΔG to be negative)
$\Delta S < 0$ (the mixture becomes less random going from reactants (3 molecules) to products (2 molecules))
$\Delta G < 0$ (the reaction is spontaneous)

17.34 (a) For initial state 1, $Q_p < K_p$
(more reactant (A_2) than product (A) compared to the equilibrium state)
For initial state 2, $Q_p > K_p$
(more product (A) than reactant (A_2) compared to the equilibrium state)
(b) $\Delta H > 0$ (reaction involves bond breaking - endothermic)
$\Delta S > 0$ (equilibrium state has more randomness than initial state 1)
$\Delta G < 0$ (reaction spontaneously proceeds toward equilibrium)
(c) $\Delta H < 0$ (reaction involves bond making - exothermic)
$\Delta S < 0$ (equilibrium state has less randomness than initial state 2)
$\Delta G < 0$ (reaction spontaneously proceeds toward equilibrium)
(d) State 1 lies to the left of the minimum in Figure 17.11. State 2 lies to the right of the minimum.

17.35 (a) $\Delta H^\circ > 0$ (reaction involves bond breaking - endothermic)
$\Delta S^\circ > 0$ (2 A's have more randomness than A_2)
(b) ΔS° is for the complete conversion of 1 mole of A_2 in its standard state to 2 moles of A in its standard state.
(c) There is not enough information to say anything about the sign of ΔG°. ΔG° decreases (becomes less positive or more negative) as the temperature increases.
(d) K_p increases as the temperature increases. As the temperature increases there will be more A and less A_2.
(e) $\Delta G = 0$ at equilibrium.

17.36 (a) Because the free energy decreases as pure reactants form products and also decreases as pure products form reactants, the free energy curve must go through a minimum somewhere between pure reactants and pure products. At the minimum point, $\Delta G = 0$ and the system is at equilibrium.
(b) The minimum in the plot is on the left side of the graph because $\Delta G^\circ > 0$ and the equilibrium composition is rich in reactants.

17.37 $\Delta G^\circ = -RT \ln K$ where $K = \dfrac{[X]}{[A]}$ or $\dfrac{[Y]}{[A]}$ or $\dfrac{[Z]}{[A]}$
Let the number of molecules be proportional to the concentration.
(1) K = 1, ln K = 0, and $\Delta G^\circ = 0$.

(2) K > 1, ln K is positive, and ΔG° is negative.
(3) K < 1, ln K is negative, and ΔG° is positive.

17.38 The equilibrium mixture is richer in reactant A at the higher temperature. This means the reaction is exothermic ($\Delta H < 0$). At 25 °C, $\Delta G^\circ < 0$ because K > 1 and at 45 °C, $\Delta G^\circ > 0$ because K < 1. Using the relationship. $\Delta G^\circ = \Delta H^\circ - T\Delta S^\circ$, with $\Delta H^\circ < 0$, ΔG° will become positive at the higher temperature only if ΔS° is negative.

17.39 $A_2 + B_2 \rightleftarrows 2\,AB$

(a) At equilibrium, $P_{A_2} = P_{B_2} = 1.0$ atm and $P_{AB} = 2.0$ atm

$$K_p = \frac{(P_{AB})^2}{(P_{A_2})(P_{B_2})} = \frac{(2.0)^2}{(1.0)(1.0)} = 4.0$$

$\Delta G^\circ = -RT \ln K_p = -(8.314 \times 10^{-3}\ \text{kJ/(K} \cdot \text{mol))}(298\ \text{K}) \ln(4.0) = -3.4\ \text{kJ/mol}$

(b) $\Delta G = \Delta G^\circ + RT \ln Q_p$

$$Q_p = \frac{(P_{AB})^2}{(P_{A_2})(P_{B_2})} = \frac{(3.0)^2}{(0.5)(0.5)} = 36.0$$

$\Delta G = (-3.4\ \text{kJ/mol}) + (8.314 \times 10^{-3}\ \text{kJ/(K} \cdot \text{mol))}(298\ \text{K}) \ln(36.0) = 5.5\ \text{kJ/mol}$

Because $\Delta G > 0$, the reaction proceeds toward reactants.

Section Problems
Spontaneous Processes (Section 17.1)

17.40 (a) and (d) nonspontaneous; (b) and (c) spontaneous

17.41 (a) and (c) spontaneous; (b) and (d) nonspontaneous

17.42 (b) and (d) spontaneous (because of the large positive K_p's)

17.43 (a) and (d) nonspontaneous (because of the small K's)

Entropy (Sections 17.2–17.4)

17.44 Molecular randomness is called entropy. For the following reaction, the entropy increases: $H_2O(s) \rightarrow H_2O(l)$ at 25 °C.

17.45 Exothermic reactions can become nonspontaneous at high temperatures if ΔS is negative. Endothermic reactions can become spontaneous at high temperatures if ΔS is positive.

17.46 (a) + (solid $\rightarrow$ gas)
(b) – (liquid $\rightarrow$ solid)
(c) – (aqueous ions $\rightarrow$ solid)
(d) + ($CO_2(aq) \rightarrow CO_2(g)$)

17.47 (a) + (increase in moles of gas)
(b) – (decrease in moles of gas and formation of liquid)
(c) + (aqueous ions to gas)
(d) – (decrease in moles of gas)

17.48 (a) – (liquid → solid)
(b) – (decrease in number of O_2 molecules)
(c) + (gas has more randomness in larger volume)
(d) – (aqueous ions → solid)

17.49 (a) + (solid dissolved in water)
(b) + (increase in moles of gas)
(c) + (mixed gases have more randomness)
(d) + (liquid to gas)

17.50 $S = k \ln W$, $k = 1.38 \times 10^{-23}$ J/K
(a) $S = (1.38 \times 10^{-23}\text{ J/K}) \ln (4^{12}) = 2.30 \times 10^{-22}$ J/K
(b) $S = (1.38 \times 10^{-23}\text{ J/K}) \ln (4^{120}) = 2.30 \times 10^{-21}$ J/K
(c) $S = (1.38 \times 10^{-23}\text{ J/K}) \ln (4^{6.02 \times 10^{23}}) = (1.38 \times 10^{-23}\text{ J/K})(6.022 \times 10^{23})\ln 4 = 11.5$ J/K
If all C–D bonds point in the same direction, S = 0.

17.51 $S = k \ln W$, $k = 1.38 \times 10^{-23}$ J/K
(a) $W = 1$; $S = (1.38 \times 10^{-23}\text{ J/K}) \ln (1) = 0$
(b) $W = 3^2, = 9$; $S = (1.38 \times 10^{-23}\text{ J/K}) \ln (3^2) = 3.03 \times 10^{-23}$ J/K
(c) $W = 1$; $S = (1.38 \times 10^{-23}\text{ J/K}) \ln (1) = 0$
(d) $W = 3^3 = 27$; $S = (1.38 \times 10^{-23}\text{ J/K}) \ln (3^3) = 4.55 \times 10^{-23}$ J/K
(e) $W = 1$; $S = (1.38 \times 10^{-23}\text{ J/K}) \ln (1) = 0$
(f) $W = 3^{6.02 \times 10^{23}}$; $S = (1.38 \times 10^{-23}\text{ J/K}) \ln (3^{6.02 \times 10^{23}}) = 9.13$ J/K

$\Delta S = R \ln\left(\frac{V_f}{V_i}\right) = (8.314\text{ J/K})\ln 3 = 9.13$ J/K   The results are the same.

17.52 $S = k \ln W$, $k = 1.38 \times 10^{-23}$ J/K
$W = 1000^{100}$; $S = (1.38 \times 10^{-23}\text{ J/K}) \ln (1000^{100}) = 9.53 \times 10^{-21}$ J/K

17.53 $S = k \ln W$, $k = 1.38 \times 10^{-23}$ J/K
$W = 10{,}000^{100}$; $S = (1.38 \times 10^{-23}\text{ J/K}) \ln (10{,}000^{100}) = 1.27 \times 10^{-20}$ J/K

17.54 $S = k \ln W$
$S_i = k \ln W_i = k \ln (1.00 \times 10^6)^{1000}$ and $S_f = k \ln W_f = k \ln (1.00 \times 10^7)^{1000}$

$$\frac{S_f}{S_i} = \frac{k \ln (1.00 \times 10^7)^{1000}}{k \ln (1.00 \times 10^6)^{1000}} = \frac{k (1000) \ln (1.00 \times 10^7)}{k (1000) \ln (1.00 \times 10^6)} = \frac{\ln(1.00 \times 10^7)}{\ln(1.00 \times 10^6)} = 1.17$$

and

$S_i = k \ln W_i = k \ln (1.00 \times 10^{16})^{1000}$ and $S_f = k \ln W_f = k \ln (1.00 \times 10^{17})^{1000}$

$$\frac{S_f}{S_i} = \frac{k \ln (1.00 \times 10^{17})^{1000}}{k \ln (1.00 \times 10^{16})^{1000}} = \frac{k (1000) \ln (1.00 \times 10^{17})}{k (1000) \ln (1.00 \times 10^{16})} = \frac{\ln (1.00 \times 10^{17})}{\ln (1.00 \times 10^{16})} = 1.06$$

17.55 $S = k \ln W$, $k = 1.38 \times 10^{-23}$ J/K
Y = # of molecules
$S = 3.73 \times 10^{-20}$ J/K $= (1.38 \times 10^{-23}$ J/K$) \ln (50{,}000^Y)$
3.73×10^{-20} J/K $= (1.38 \times 10^{-23}$ J/K$)(Y) \ln (50{,}000)$

$$Y = \frac{(3.73 \times 10^{-20} \text{ J/K})}{(1.38 \times 10^{-23} \text{ J/K})[\ln(50{,}000)]} = 250 \text{ molecules}$$

17.56 (a) disordered N_2O (more randomness)
(b) quartz glass (amorphous solid, more randomness)

17.57 (a) 1 mol of N_2 gas at 273 K and 0.10 atm (larger volume, more randomness)
(b) NaCl crystal at 50 °C (higher average kinetic energy, more randomness)

17.58 (a) H_2 at 25 °C in 50 L (larger volume)
(b) O_2 at 25 °C, 1 atm (larger volume)
(c) H_2 at 100 °C, 1 atm (larger volume and higher T)
(d) CO_2 at 100 °C, 0.1 atm (larger volume and higher T)

17.59 (a) Ice at 0 °C, because of the higher temperature.
(b) N_2 at STP, because it has the larger volume.
(c) N_2 at 0 °C and 50 L, because it has the larger volume.
(d) Water vapor at 150 °C and 1 atm, because it has a larger volume and higher temperature.

17.60 $$\Delta S = nR \ln\left(\frac{V_f}{V_i}\right) = (0.050 \text{ mol})(8.314 \text{ J/K} \cdot \text{mol})\ln\left(\frac{3.5 \text{ L}}{2.5 \text{ L}}\right) = 0.14 \text{ J/K}$$

17.61 $$\Delta S = nR \ln\left(\frac{V_f}{V_i}\right) = (0.15 \text{ mol})(8.314 \text{ J/K} \cdot \text{mol})\ln\left(\frac{20.0 \text{ L}}{30.0 \text{ L}}\right) = -0.51 \text{ J/K}$$

Standard Molar Entropies and Standard Entropies of Reaction (Section 17.5)

17.62 (a) $C_2H_6(g)$; more atoms/molecule
(b) $CO_2(g)$; more atoms/molecule
(c) $I_2(g)$; gas has more randomness than the solid
(d) $CH_3OH(g)$; gas has more randomness than the liquid

17.63 (a) $NO_2(g)$; more atoms/molecule
(b) $CH_3CO_2H(l)$; more atoms/molecule
(c) $Br_2(l)$; liquid has more randomness than the solid
(d) $SO_3(g)$; gas has more randomness than the solid

17.64 $2\ CO(g) + O_2(g) \rightarrow 2\ CO_2(g)$
$\Delta S^\circ = [2\ S^\circ(CO_2)] - [2\ S^\circ(CO) + S^\circ(O_2)]$
$\Delta S^\circ = [(2\ mol)(213.6\ J/(K \cdot mol))]$
$- [(2\ mol)(197.6\ J/(K \cdot mol)) + (1\ mol)(205.0\ J/(K \cdot mol))] = -173.0\ J/K$

17.65 $C(s) + O_2(g) \rightarrow 2\ CO_2(g)$
$\Delta S^\circ = [2\ S^\circ(CO_2)] - [S^\circ(C) + S^\circ(O_2)]$
$\Delta S^\circ = [(2\ mol)(213.6\ J/(K \cdot mol))]$
$- [(1\ mol)(5.7\ J/(K \cdot mol)) + (1\ mol)(205.0\ J/(K \cdot mol))] = +216.5\ J/K$

17.66 (a) $2\ H_2O_2(l) \rightarrow 2\ H_2O(l) + O_2(g)$
$\Delta S^\circ = [2\ S^\circ(H_2O(l)) + S^\circ(O_2)] - 2\ S^\circ(H_2O_2)$
$\Delta S^\circ = [(2\ mol)(69.9\ J/(K \cdot mol)) + (1\ mol)(205.0\ J/(K \cdot mol))]$
$- (2\ mol)(110\ J/(K \cdot mol)) = +125\ J/K$ (+, because moles of gas increase)
(b) $2\ Na(s) + Cl_2(g) \rightarrow 2\ NaCl(s)$
$\Delta S^\circ = 2\ S^\circ(NaCl) - [2\ S^\circ(Na) + S^\circ(Cl_2)]$
$\Delta S^\circ = (2\ mol)(72.1\ J/(K \cdot mol)) - [(2\ mol)(51.2\ J/(K \cdot mol)) + (1\ mol)(223.0\ J/(K \cdot mol))]$
$\Delta S^\circ = -181.2\ J/K$ (–, because moles of gas decrease)
(c) $2\ O_3(g) \rightarrow 3\ O_2(g)$
$\Delta S^\circ = 3\ S^\circ(O_2) - 2\ S^\circ(O_3)$
$\Delta S^\circ = (3\ mol)(205.0\ J/(K \cdot mol)) - (2\ mol)(238.8\ J/(K \cdot mol))$
$\Delta S^\circ = +137.4\ J/K$ (+, because moles of gas increase)
(d) $4\ Al(s) + 3\ O_2(g) \rightarrow 2\ Al_2O_3(s)$
$\Delta S^\circ = 2\ S^\circ(Al_2O_3) - [4\ S^\circ(Al) + 3\ S^\circ(O_2)]$
$\Delta S^\circ = (2\ mol)(50.9\ J/(K \cdot mol)) - [(4\ mol)(28.3\ J/(K \cdot mol)) + (3\ mol)(205.0\ J/(K \cdot mol))]$
$\Delta S^\circ = -626.4\ J/K$ (–, because moles of gas decrease)

17.67 (a) $2\ S(s) + 3\ O_2(g) \rightarrow 2\ SO_3(g)$
$\Delta S^\circ = 2\ S^\circ(SO_3) - [2\ S^\circ(S) + 3\ S^\circ(O_2)]$
$\Delta S^\circ = (2\ mol)(256.6\ J/(K \cdot mol)) - [(2\ mol)(31.8\ J/(K \cdot mol)) + (3\ mol)(205.0\ J/(K \cdot mol))]$
$\Delta S^\circ = -165.4\ J/K$ (–, because moles of gas decrease)
(b) $SO_3(g) + H_2O(l) \rightarrow H_2SO_4(aq)$
$\Delta S^\circ = S^\circ(H_2SO_4) - [S^\circ(SO_3) + S^\circ(H_2O)]$
$\Delta S^\circ = (1\ mol)(20\ J/(K \cdot mol)) - [(1\ mol)(256.6\ J/(K \cdot mol)) + (1\ mol)(69.9\ J/(K \cdot mol))]$
$\Delta S^\circ = -307\ J/K$ (–, because of the conversion of a gas and water to an aqueous solution)
(c) $AgCl(s) \rightarrow Ag^+(aq) + Cl^-(aq)$
$\Delta S^\circ = [S^\circ(Ag^+) + S^\circ(Cl^-)] - S^\circ(AgCl)$
$\Delta S^\circ = [(1\ mol)(72.7\ J/(K \cdot mol)) + (1\ mol)(56.5\ J/(K \cdot mol))] - (1\ mol)(96.2\ J/(K \cdot mol))$
$\Delta S^\circ = +33.0\ J/K$ (+, because a solid is converted to ions in aqueous solution)
(d) $NH_4NO_3(s) \rightarrow N_2O(g) + 2\ H_2O(g)$
$\Delta S^\circ = [S^\circ(N_2O) + 2\ S^\circ(H_2O)] - S^\circ(NH_4NO_3)$
$\Delta S^\circ = [(1\ mol)(219.7\ J/(K \cdot mol)) + (2\ mol)(188.7\ J/(K \cdot mol))] - (1\ mol)(151.1\ J/(K \cdot mol))$
$\Delta S^\circ = +446.0\ J/K$ (+, because moles of gas increase)

Entropy and the Second Law of Thermodynamics (Section 17.6)

17.68 In any spontaneous process, the total entropy of a system and its surroundings always increases.

17.69 For a spontaneous process, $\Delta S_{total} = \Delta S_{sys} + \Delta S_{surr} > 0$. For an isolated system, $\Delta S_{surr} = 0$, and so $\Delta S_{sys} > 0$ is the criterion for spontaneous change. An example of a spontaneous process in an isolated system is the mixing of two gases.

17.70 $\Delta S_{surr} = \dfrac{-\Delta H}{T}$; the temperature (T) is always positive.

(a) For an exothermic reaction, ΔH is negative and ΔS_{surr} is positive.
(b) For an endothermic reaction, ΔH is positive and ΔS_{surr} is negative.

17.71 $\Delta S_{surr} \propto \dfrac{1}{T}$

Consider the surroundings as an infinitely large constant-temperature bath to which heat can be added without changing its temperature. If the surroundings has a low temperature, it has only a small amount of randomness, in which case addition of a given quantity of heat results in a substantial increase in the amount of randomness (a relatively large value of ΔS_{surr}). If the surroundings has a high temperature, it already has a large amount of randomness, and addition of the same quantity of heat produces only a marginal increase in the randomness (a relatively small value of ΔS_{surr}). Thus, we expect ΔS_{surr} to vary inversely with temperature.

17.72 $HgO(s) + Zn(s) \rightarrow ZnO(s) + Hg(l)$

(a) $\Delta S_{surr} = \dfrac{-\Delta H^\circ}{T} = \dfrac{-(-259.7 \text{ x } 10^3 \text{ J})}{298 \text{ K}} = +871.5 \text{ J/K}$

$\Delta S_{total} = \Delta S^\circ + \Delta S_{surr} = +7.8 \text{ J/K} + 871.5 \text{ J/K} = +879.3 \text{ J/K}$
The reaction is spontaneous because ΔS_{total} is > 0.
(b) Because $\Delta S^\circ > 0$ and $\Delta H^\circ < 0$, there is no temperature at which the reaction is not spontaneous.

17.73 $2\ ZnS(s) + O_2(g) \rightarrow 2\ ZnO(s) + 2\ S(s)$

(a) $\Delta S_{surr} = \dfrac{-\Delta H^\circ}{T} = \dfrac{-(-289.0 \text{ x } 10^3 \text{ J})}{298 \text{ K}} = +969.8 \text{ J/K}$

$\Delta S_{total} = \Delta S^\circ + \Delta S_{surr} = -169.4 \text{ J/K} + 969.8 \text{ J/K} = +800.4 \text{ J/K}$
Because $\Delta S_{total} > 0$, the reaction is spontaneous under standard-state conditions at 25°C.

(b) $\Delta S_{total} = \Delta S^\circ + \Delta S_{surr} = \Delta S^\circ + \dfrac{-\Delta H^\circ}{T}$

To find the temperature where the reaction becomes nonspontaneous, set $\Delta S_{total} = 0$ and solve for T.

$0 = \Delta S^\circ + \dfrac{-\Delta H^\circ}{T} = -169.4 \text{ J/K} + \dfrac{-(-289.0 \text{ x } 10^3 \text{ J})}{T}$; $\quad T = \dfrac{289.0 \text{ x } 10^3 \text{ J}}{169.4 \text{ J/K}} = 1706 \text{ K}$

17.74 $3\ O_2(g) \rightarrow 2\ O_3(g)$

$\Delta H^o = 2\ \Delta H^o_f(O_3) = (2\ mol)(143\ kJ/mol) = 286\ kJ = 286 \times 10^3\ J$

$\Delta S^o = 2\ S^o(O_3) - 3\ S^o(O_2)]$

$\Delta S^o = (2\ mol)(238.8\ J/(K \cdot mol)) - (3\ mol)(205.0\ J/(K \cdot mol))]$

$\Delta S^o = -137.4\ J/K$

$$\Delta S_{total} = \Delta S^o + \Delta S_{surr} = \Delta S^o + \frac{-\Delta H^o}{T} = -137.4\ J/K + \frac{-(286 \times 10^3\ J)}{298\ K} = -1097\ J/K$$

Because $\Delta S_{total} < 0$, the reaction is not spontaneous under standard-state conditions at 25 °C.

17.75 $2\ SO_2(g) + O_2(g) \rightarrow 2\ SO_3(g)$

$\Delta H^o = 2\ \Delta H^o_f(SO_3) - 2\ \Delta H^o_f(SO_2)$

$\Delta H^o = (2\ mol)(-395.7\ kJ/mol) - (2\ mol)(-296.8\ kJ/mol) = -197.8\ kJ = -197.8 \times 10^3\ J$

$\Delta S^o = 2\ S^o(SO_3) - [2\ S^o(SO_2) + S^o(O_2)]$

$\Delta S^o = (2\ mol)(256.6\ J/(K \cdot mol)) - [(2\ mol)(248.1\ J/(K \cdot mol)) + (1\ mol)(205.0\ J/(K \cdot mol))]$

$\Delta S^o = -188.0\ J/K$

$$\Delta S_{total} = \Delta S^o + \Delta S_{surr} = \Delta S^o + \frac{-\Delta H^o}{T} = -188.0\ J/K + \frac{-(-197.8 \times 10^3\ J)}{298\ K} = +475.8\ J/K$$

Because $\Delta S_{total} > 0$, the reaction is spontaneous under standard-state conditions at 25 °C.

17.76 $2\ HgO(s) \rightarrow 2\ Hg(l) + O_2(g)$

$\Delta H^o = 0 - 2\ \Delta H^o_f(HgO)$

$\Delta H^o = -(2\ mol)(-90.8\ kJ/mol) = +181.6\ kJ = +181.6 \times 10^3\ J$

$\Delta S^o = [2\ S^o(Hg) + S^o(O_2)] - 2\ S^o(HgO)$

$\Delta S^o = [(2\ mol)(76.0\ J/(K \cdot mol)) + (1\ mol)(205.0\ J/(K \cdot mol))] - (2\ mol)(70.3\ J/(K \cdot mol))$

$\Delta S^o = \Delta S_{sys} = +216.4\ J/K$

$$\Delta S_{surr} = \frac{-\Delta H^o}{T} = \frac{-(181.6 \times 10^3\ J)}{298\ K} = -609.4\ J/K$$

$\Delta S_{total} = \Delta S^o + \Delta S_{surr} = 216.4\ J/K + (-609.4\ J/K) = -393.0\ J/K$

Because $\Delta S_{total} < 0$, the reaction is not spontaneous under standard-state conditions at 25 °C.

(b) $\Delta S_{total} = \Delta S^o + \Delta S_{surr} = \Delta S^o + \dfrac{-\Delta H^o}{T}$

To find the temperature where the reaction becomes spontaneous, set $\Delta S_{total} = 0$ and solve for T.

$$0 = \Delta S^o + \frac{-\Delta H^o}{T} = +216.4\ J/K + \frac{-(181.6 \times 10^3\ J)}{T}$$

$$T = \frac{-(181.6 \times 10^3\ J)}{-216.4\ J/K} = 839.2\ K$$

17.77 $PCl_3(g) + Cl_2(g) \rightarrow PCl_5(g)$

$\Delta H^o = \Delta H^o_f(PCl_5) - \Delta H^o_f(PCl_3)$

$\Delta H^o = (1\ mol)(-374.9\ kJ/mol) - (1\ mol)(-287.0\ kJ/mol) = -87.9\ kJ = -87.9 \times 10^3\ J$

$\Delta S^o = S^o(PCl_5) - [S^o(PCl_3) + S^o(Cl_2)]$

$\Delta S^o = (1\ mol)(364.5\ J/(K \cdot mol)) - [(1\ mol)(311.7\ J/(K \cdot mol)) + (1\ mol)(223.0\ J/(K \cdot mol))]$

$\Delta S^o = \Delta S_{sys} = -170.2\ J/K$

$$\Delta S_{surr} = \frac{-\Delta H^o}{T} = \frac{-(-87.9 \times 10^3\ J)}{298\ K} = +295.0\ J/K$$

$\Delta S_{total} = \Delta S^o + \Delta S_{surr} = -170.2\ J/K + 295.0\ J/K = +124.8\ J/K$

Because $\Delta S_{total} > 0$, the reaction is spontaneous under standard-state conditions at 25 °C.

(b) $\Delta S_{total} = \Delta S^o + \Delta S_{surr} = \Delta S^o + \dfrac{-\Delta H^o}{T}$

To find the temperature where the reaction becomes nonspontaneous, set $\Delta S_{total} = 0$ and solve for T.

$$0 = \Delta S^o + \frac{-\Delta H^o}{T} = -170.2\ J/K + \frac{-(-87.9 \times 10^3\ J)}{T}$$

$$T = \frac{87.9 \times 10^3\ J}{170.2\ J/K} = 516.5\ K$$

17.78 (a) $\Delta S_{surr} = \dfrac{-\Delta H_{vap}}{T} = \dfrac{-30{,}700\ J/mol}{343\ K} = -89.5\ J/(K \cdot mol)$

$\Delta S_{total} = \Delta S_{vap} + \Delta S_{surr} = 87.0\ J/(K \cdot mol) + (-89.5\ J/(K \cdot mol)) = -2.5\ J/(K \cdot mol)$

(b) $\Delta S_{surr} = \dfrac{-\Delta H_{vap}}{T} = \dfrac{-30{,}700\ J/mol}{353\ K} = -87.0\ J/(K \cdot mol)$

$\Delta S_{total} = \Delta S_{vap} + \Delta S_{surr} = 87.0\ J/(K \cdot mol) + (-87.0\ J/(K \cdot mol)) = 0$

(c) $\Delta S_{surr} = \dfrac{-\Delta H_{vap}}{T} = \dfrac{-30{,}700\ J/mol}{363\ K} = -84.6\ J/(K \cdot mol)$

$\Delta S_{total} = \Delta S_{vap} + \Delta S_{surr} = 87.0\ J/(K \cdot mol) + (-84.6\ J/(K \cdot mol)) = +2.4\ J/(K \cdot mol)$

Benzene does not boil at 70 °C (343 K) because ΔS_{total} is negative.

The normal boiling point for benzene is 80 °C (353 K), where $\Delta S_{total} = 0$.

17.79 (a) $\Delta S_{surr} = \dfrac{-\Delta H_{fusion}}{T} = \dfrac{-28{,}160\ J/mol}{1050\ K} = -26.8\ J/(K \cdot mol)$

$\Delta S_{total} = \Delta S_{sys} + \Delta S_{surr} = 26.22\ J/(K \cdot mol) + (-26.8\ J/(K \cdot mol)) = -0.6\ J/(K \cdot mol)$

(b) $\Delta S_{surr} = \dfrac{-\Delta H_{fusion}}{T} = \dfrac{-28{,}160\ J/mol}{1074\ K} = -26.22\ J/(K \cdot mol)$

$\Delta S_{total} = \Delta S_{sys} + \Delta S_{surr} = 26.22\ J/(K \cdot mol) + (-26.22\ J/(K \cdot mol)) = 0$

(c) $\Delta S_{surr} = \dfrac{-\Delta H_{fusion}}{T} = \dfrac{-28{,}160\ J/mol}{1100\ K} = -25.6\ J/(K \cdot mol)$

$\Delta S_{total} = \Delta S_{sys} + \Delta S_{surr} = 26.22\ J/(K \cdot mol) + (-25.6\ J/(K \cdot mol)) = +0.6\ J/(K \cdot mol)$

NaCl melts at 1100 K because $\Delta S_{total} > 0$.

The melting point of NaCl is 1074 K, where $\Delta S_{total} = 0$.

Free Energy (Section 17.7)

17.80

ΔH	ΔS	$\Delta G = \Delta H - T\Delta S$	Reaction Spontaneity
–	+	–	Spontaneous at all temperatures
–	–	– or +	Spontaneous at low temperatures where $\lvert\Delta H\rvert > \lvert T\Delta S\rvert$ Nonspontaneous at high temperatures where $\lvert\Delta H\rvert < \lvert T\Delta S\rvert$
+	–	+	Nonspontaneous at all temperatures
+	+	– or +	Spontaneous at high temperatures where $T\Delta S > \Delta H$ Nonspontaneous at low temperature where $T\Delta S < \Delta H$

17.81 When ΔH and ΔS are both positive or both negative, the temperature determines the direction of spontaneous reaction. See Problem 17.80 for an explanation.

17.82 (e) a negative free-energy change.

17.83 (d) $\Delta H = -,\ \Delta S = +$

17.84 (a) 0 °C (temperature is below mp); $\Delta H > 0,\ \Delta S > 0,\ \Delta G > 0$
(b) 15 °C (temperature is above mp); $\Delta H > 0,\ \Delta S > 0,\ \Delta G < 0$

17.85 (a) $\Delta H = 0$

$$\Delta S = R \ln \frac{V_{final}}{V_{initial}} = (8.314\ J/K) \ln 2 = 5.76\ J/K$$

$\Delta G = \Delta H - T\Delta S$

Because $\Delta H = 0$, $\Delta G = -T\Delta S = -(298\ K)(5.76\ J/K) = -1717\ J = -1.72\ kJ$

(b) For a process in an isolated system, $\Delta S_{surr} = 0$. Therefore, $\Delta S_{total} = \Delta S_{sys} > 0$, and the process is spontaneous.

17.86 $\Delta H_{vap} = 30.7\ kJ/mol$

$\Delta S_{vap} = 87.0\ J/(K \cdot mol) = 87.0 \times 10^{-3}\ kJ/(K \cdot mol)$

$\Delta G_{vap} = \Delta H_{vap} - T\Delta S_{vap}$

(a) $\Delta G_{vap} = 30.7\ kJ/mol - (343\ K)(87.0 \times 10^{-3}\ kJ/(K \cdot mol)) = +0.9\ kJ/mol$

At 70 °C (343 K), benzene does not boil because ΔG_{vap} is positive.

(b) $\Delta G_{vap} = 30.7\ kJ/mol - (353\ K)(87.0 \times 10^{-3}\ kJ/(K \cdot mol)) = 0$

80 °C (353 K) is the boiling point for benzene because $\Delta G_{vap} = 0$

(c) $\Delta G_{vap} = 30.7\ kJ/mol - (363\ K)(87.0 \times 10^{-3}\ kJ/(K \cdot mol)) = -0.9\ kJ/mol$

At 90 °C (363 K), benzene boils because ΔG_{vap} is negative.

17.87 $\Delta H_{fusion} = 28.16\ kJ/mol$; $\Delta S_{fusion} = 26.22 \times 10^{-3}\ kJ/(K \cdot mol)$

$\Delta G_{fusion} = \Delta H_{fusion} - T\Delta S_{fusion}$

(a) $\Delta G_{fusion} = 28.16\ kJ/mol - (1050\ K)(26.22 \times 10^{-3}\ kJ/(K \cdot mol)) = +0.63\ kJ/mol$

At 1050 K, NaCl does not melt because ΔG_{fusion} is positive.
(b) ΔG_{fusion} = 28.16 kJ/mol – (1074 K)(26.22 x 10^{-3} kJ/(K · mol)) = 0.0
1074 K is the melting point for NaCl because ΔG_{fusion} = 0.
(c) ΔG_{fusion} = 28.16 kJ/mol – (1100 K)(26.22 x 10^{-3} kJ/(K · mol)) = –0.68 kJ/mol
At 1100 K, NaCl does melt because ΔG_{fusion} is negative.

17.88 At the melting point (phase change), ΔG_{fusion} = 0.
$\Delta G_{fusion} = \Delta H_{fusion} - T\Delta S_{fusion}$

$$0 = \Delta H_{fusion} - T\Delta S_{fusion}; \quad T = \frac{\Delta H_{fusion}}{\Delta S_{fusion}} = \frac{18.02 \text{ kJ/mol}}{45.56 \times 10^{-3} \text{ kJ/(K} \cdot \text{mol)}} = 395.5 \text{ K} = 122.4 \text{ °C}$$

17.89 128 °C = 401 K
At the melting point (phase change), ΔG_{fusion} = 0.
$\Delta G_{fusion} = \Delta H_{fusion} - T\Delta S_{fusion}$
$0 = \Delta H_{fusion} - T\Delta S_{fusion}$
$\Delta H_{fusion} = T\Delta S_{fusion}$ = (401 K)[47.7 x 10^{-3} kJ/(K · mol)] = 19.1 kJ/mol

Standard Free-Energy Changes and Standard Free Energies of Formation (Sections 17.8 and 17.9)

17.90 (a) $\Delta G°$ is the change in free energy that occurs when reactants in their standard states are converted to products in their standard states.
(b) $\Delta G°_f$ is the free-energy change for formation of one mole of a substance in its standard state from the most stable form of the constituent elements in their standard states.

17.91 The standard state of a substance (solid, liquid, or gas) is the most stable form of a pure substance at 25 °C and 1 atm pressure. For solutes, the condition is 1 M at 25 °C.

17.92 (a) $N_2(g) + 2\ O_2(g) \rightarrow 2\ NO_2(g)$
$\Delta H° = 2\ \Delta H°_f(NO_2)$ = (2 mol)(33.2 kJ/mol) = 66.4 kJ
$\Delta S° = 2\ S°(NO_2) - [S°(N_2) + 2\ S°(O_2)]$
$\Delta S°$ = (2 mol)(240.0 J/(K · mol)) – [(1 mol)(191.5 J/(K · mol)) + (2 mol)(205.0 J/(K · mol))]
$\Delta S°$ = –121.5 J/K = –121.5 x 10^{-3} kJ/K
$\Delta G° = \Delta H° - T\Delta S°$ = 66.4 kJ – (298 K)(–121.5 x 10^{-3} kJ/K) = +102.6 kJ
Because $\Delta G°$ is positive, the reaction is nonspontaneous under standard-state conditions at 25 °C.
(b) $2\ KClO_3(s) \rightarrow 2\ KCl(s) + 3\ O_2(g)$
$\Delta H° = 2\ \Delta H°_f(KCl) - 2\ \Delta H°_f(KClO_3)$
$\Delta H°$ = (2 mol)(–436.5 kJ/mol) – (2 mol)(–397.7 kJ/mol) = –77.6 kJ
$\Delta S° = [2\ S°(KCl) + 3\ S°(O_2)] - 2\ S°(KClO_3)$
$\Delta S°$ = [(2 mol)(82.6 J/(K · mol)) + (3 mol)(205.0 J/(K · mol))] – (2 mol)(143.1 J/(K · mol))
$\Delta S°$ = 494.0 J/K = 494.0 x 10^{-3} kJ/K
$\Delta G° = \Delta H° - T\Delta S°$ = –77.6 kJ – (298 K)(494.0 x 10^{-3} kJ/K) = –224.8 kJ
Because $\Delta G°$ is negative, the reaction is spontaneous under standard-state conditions at 25 °C.
(c) $CH_3CH_2OH(l) + O_2(g) \rightarrow CH_3CO_2H(l) + H_2O(l)$
$\Delta H° = [\Delta H°_f(CH_3CO_2H) + \Delta H°_f(H_2O)] - \Delta H°_f(CH_3CH_2OH)$
$\Delta H°$ = [(1 mol)(–484.5 kJ/mol) + (1 mol)(–285.8 kJ/mol)] – (1 mol)(–277.7 kJ/mol) = –492.6 kJ
$\Delta S° = [S°(CH_3CO_2H) + S°(H_2O)] - [S°(CH_3CH_2OH) + S°(O_2)]$

$\Delta S° = [(1\ mol)(160\ J/(K \cdot mol)) + (1\ mol)(69.9\ J/(K \cdot mol))]$
$- [(1\ mol)(161\ J/(K \cdot mol)) + (1\ mol)(205.0\ J/(K \cdot mol))]$
$\Delta S° = -136.1\ J/K = -136.1 \times 10^{-3}\ kJ/K$
$\Delta G° = \Delta H° - T\Delta S° = -492.6\ kJ - (298\ K)(-136.1 \times 10^{-3}\ kJ/K) = -452.0\ kJ$
Because $\Delta G°$ is negative, the reaction is spontaneous under standard-state conditions at 25 °C.

17.93 (a) $2\ SO_2(g) + O_2(g) \rightarrow 2\ SO_3(g)$
$\Delta H° = 2\ \Delta H°_f(SO_3) - 2\ \Delta H°_f(SO_2)$
$\Delta H° = (2\ mol)(-395.7\ kJ/mol) - (2\ mol)(-296.8\ kJ/mol) = -197.8\ kJ$
$\Delta S° = 2\ S°(SO_3) - [2\ S°(SO_2) + S°(O_2)]$
$\Delta S° = (2\ mol)(256.6\ J/(K \cdot mol)) - [(2\ mol)(248.1\ J/(K \cdot mol)) + (1\ mol)(205.0\ J/(K \cdot mol))]$
$\Delta S° = -188.0\ J/K = -188.0 \times 10^{-3}\ kJ/K$
$\Delta G° = \Delta H° - T\Delta S° = -197.8\ kJ - (298\ K)(-188.0 \times 10^{-3}\ kJ/K) = -141.8\ kJ$
Because $\Delta G°$ is negative, the reaction is spontaneous under standard-state conditions at 25 °C.
(b) $N_2(g) + 2\ H_2(g) \rightarrow N_2H_4(l)$
$\Delta H° = \Delta H°_f(N_2H_4)$
$\Delta H° = (1\ mol)(50.6\ kJ/mol) = 50.6\ kJ$
$\Delta S° = S°(N_2H_4) - [S°(N_2) + 2\ S°(H_2)]$
$\Delta S° = (1\ mol)(121.2\ J/(K \cdot mol)) - [(1\ mol)(191.5\ J/(K \cdot mol)) + (2\ mol)(130.6\ J/(K \cdot mol))]$
$\Delta S° = -331.5\ J/K = -331.5 \times 10^{-3}\ kJ/K$
$\Delta G° = \Delta H° - T\Delta S° = 50.6\ kJ - (298\ K)(-331.5 \times 10^{-3}\ kJ/K) = +149.4\ kJ$
Because $\Delta G°$ is positive, the reaction is nonspontaneous under standard-state conditions at 25 °C.
(c) $CH_3OH(l) + O_2(g) \rightarrow HCO_2H(l) + H_2O(l)$
$\Delta H° = [\Delta H°_f(HCO_2H) + \Delta H°_f(H_2O)] - \Delta H°_f(CH_3OH)$
$\Delta H° = [(1\ mol)(-424.7\ kJ/mol) + (1\ mol)(-285.8\ kJ/mol)] - (1\ mol)(-239.2\ kJ/mol) = -471.3\ kJ$
$\Delta S° = [S°(HCO_2H) + S°(H_2O)] - [S°(CH_3OH) + S°(O_2)]$
$\Delta S° = [(1\ mol)(129.0\ J/(K \cdot mol)) + (1\ mol)(69.9\ J/(K \cdot mol))]$
$- [(1\ mol)(127\ J/(K \cdot mol)) + (1\ mol)(205.0\ J/(K \cdot mol))]$
$\Delta S° = -133.1\ J/K = -133.1 \times 10^{-3}\ kJ/K$
$\Delta G° = \Delta H° - T\Delta S° = -471.3\ kJ - (298\ K)(-133.1 \times 10^{-3}\ kJ/K) = -431.6\ kJ$
Because $\Delta G°$ is negative, the reaction is spontaneous under standard-state conditions at 25 °C.

17.94 (a) $N_2(g) + 2\ O_2(g) \rightarrow 2\ NO_2(g)$
$\Delta G° = 2\ \Delta G°_f(NO_2) = (2\ mol)(51.3\ kJ/mol) = +102.6\ kJ$
(b) $2\ KClO_3(s) \rightarrow 2\ KCl(s) + 3\ O_2(g)$
$\Delta G° = 2\ \Delta G°_f(KCl) - 2\ \Delta G°_f(KClO_3)$
$\Delta G° = (2\ mol)(-408.5\ kJ/mol) - (2\ mol)(-296.3\ kJ/mol) = -224.4\ kJ$
(c) $CH_3CH_2OH(l) + O_2(g) \rightarrow CH_3CO_2H(l) + H_2O(l)$
$\Delta G° = [\Delta G°_f(CH_3CO_2H) + \Delta G°_f(H_2O)] - \Delta G°_f(CH_3CH_2OH)$
$\Delta G° = [(1\ mol)(-390\ kJ/mol) + (1\ mol)(-237.2\ kJ/mol)] - (1\ mol)(-174.9\ kJ/mol) = -452\ kJ$

17.95 (a) $2\ SO_2(g) + O_2(g) \rightarrow 2\ SO_3(g)$
$\Delta G° = 2\ \Delta G°_f(SO_3) - 2\ \Delta G°_f(SO_2)$
$\Delta G° = (2\ mol)(-371.1\ kJ/mol) - (2\ mol)(-300.2\ kJ/mol) = -141.8\ kJ$
(b) $N_2(g) + 2\ H_2(g) \rightarrow N_2H_4(l)$
$\Delta G° = \Delta G°_f(N_2H_4) = (1\ mol)(149.2\ kJ/mol) = 149.2\ kJ$

(c) $CH_3OH(l) + O_2(g) \rightarrow HCO_2H(l) + H_2O(l)$
$\Delta G^o = [\Delta G^o_f(HCO_2H) + \Delta G^o_f(H_2O)] - \Delta G^o_f(CH_3OH)$
$\Delta G^o = [(1\ mol)(-361.4\ kJ/mol) + (1\ mol)(-237.2\ kJ/mol)] - (1\ mol)(-166.6\ kJ/mol)$
$\Delta G^o = -432.0\ kJ$

17.96 A compound is thermodynamically stable with respect to its constituent elements at 25 °C if ΔG^o_f is negative.

	ΔG^o_f (kJ/mol)	Stable
(a) $BaCO_3(s)$	–1134.4	yes
(b) $HBr(g)$	–53.4	yes
(c) $N_2O(g)$	+104.2	no
(d) $C_2H_4(g)$	+68.1	no

17.97 A compound is thermodynamically stable with respect to its constituent elements at 25 °C if ΔG^o_f is negative.

	ΔG^o_f (kJ/mol)	Stable
(a) $C_6H_6(l)$	+124.5	no
(b) $NO(g)$	+87.6	no
(c) $PH_3(g)$	+13.5	no
(d) $FeO(s)$	–255	yes

17.98 $C_2H_4(g) + Cl_2(g) \rightarrow CH_2ClCH_2Cl(l)$
$\Delta G^o = \Delta G^o_f(CH_2ClCH_2Cl) - \Delta G^o_f(C_2H_4)$
$\Delta G^o = (1\ mol)(-79.6\ kJ/mol) - (1\ mol)(68.1\ kJ/mol) = -147.7\ kJ$
Because $\Delta G^o < 0$, dichloroethane can be synthesized from gaseous C_2H_4 and Cl_2, each at 25 °C and 1 atm pressure.

17.99 $2\ NH_3(g) \rightarrow H_2(g) + N_2H_4(l)$
$\Delta G^o = \Delta G^o_f(N_2H_4) - 2\ \Delta G^o_f(NH_3)$
$\Delta G^o = (1\ mol)(149.2\ kJ/mol) - (2\ mol)(-16.5\ kJ/mol) = +182.2\ kJ$
Because $\Delta G^o > 0$, it is not worth trying to find a catalyst for the reaction because the reaction is not spontaneous in the forward direction.

17.100 $CH_2{=}CH_2(g) + H_2O(l) \rightarrow CH_3CH_2OH(l)$
$\Delta H^o = \Delta H^o_f(CH_3CH_2OH) - [\Delta H^o_f(CH_2{=}CH_2) + \Delta H^o_f(H_2O)]$
$\Delta H^o = (1\ mol)(-277.7\ kJ/mol) - [(1\ mol)(52.3\ kJ/mol) + (1\ mol)(-285.8\ kJ/mol)]$
$\Delta H^o = -44.2\ kJ$
$\Delta S^o = S^o(CH_3CH_2OH) - [S^o(CH_2{=}CH_2) + S^o(H_2O)]$
$\Delta S^o = (1\ mol)(161\ J/(K \cdot mol)) - [(1\ mol)(219.5\ J/(K \cdot mol)) + (1\ mol)(69.9\ J/(K \cdot mol))]$
$\Delta S^o = -128\ J/(K \cdot mol) = -128 \times 10^{-3}\ kJ/(K \cdot mol)$
$\Delta G^o = \Delta H^o - T\Delta S^o = -44.2\ kJ - (298\ K)(-128 \times 10^{-3}\ kJ/K) = -6.1\ kJ$
Because ΔG^o is negative, the reaction is spontaneous under standard-state conditions at 25 °C. The reaction becomes nonspontaneous at high temperatures because ΔS^o is negative. To find the crossover temperature, set $\Delta G = 0$ and solve for T.

$$T = \frac{\Delta H^o}{\Delta S^o} = \frac{-44{,}200 \text{ J}}{-128 \text{ J/K}} = 345 \text{ K} = 72 \text{ °C}$$

The reaction becomes nonspontaneous at 72 °C.

17.101 $2\ H_2S(g) + SO_2(g) \rightarrow 3\ S(s) + 2\ H_2O(g)$

$\Delta H^o = 2\ \Delta H^o_f(H_2O) - [2\ \Delta H^o_f(H_2S) + \Delta H^o_f(SO_2)]$

$\Delta H^o = (2 \text{ mol})(-241.8 \text{ kJ/mol}) - [(2 \text{ mol})(-20.6 \text{ kJ/mol}) + (1 \text{ mol})(-296.8 \text{ kJ/mol})] = -145.6 \text{ kJ}$

$\Delta S^o = [3\ S^o(S) + 2\ S^o(H_2O)] - [2\ S^o(H_2S) + S^o(SO_2)]$

$\Delta S^o = [(3 \text{ mol})(31.8 \text{ J/(K} \cdot \text{mol})) + (2 \text{ mol})(188.7 \text{ J/(K} \cdot \text{mol}))]$
$\quad - [(2 \text{ mol})(205.7 \text{ J/(K} \cdot \text{mol})) + (1 \text{ mol})(248.1 \text{ J/(K} \cdot \text{mol}))]$

$\Delta S^o = -186.7 \text{ J/K} = -186.7 \times 10^{-3} \text{ kJ/K}$

$\Delta G^o = \Delta H^o - T\Delta S^o = -145.6 \text{ kJ} - (298 \text{ K})(-186.7 \times 10^{-3} \text{ kJ/K}) = -90.0 \text{ kJ}$

Because ΔG^o is negative, the reaction is spontaneous under standard-state conditions at 25 °C.
The reaction becomes nonspontaneous at high temperatures because ΔS^o is negative.
To find the crossover temperature set $\Delta G = 0$ and solve for T.

$$T = \frac{\Delta H^o}{\Delta S^o} = \frac{-145{,}600 \text{ J}}{-186.7 \text{ J/K}} = 780 \text{ K} = 507 \text{ °C.}$$

The reaction becomes nonspontaneous at 507 °C.

17.102 $3\ C_2H_2(g) \rightarrow C_6H_6(l)$

$\Delta G^o = \Delta G^o_f(C_6H_6) - 3\ \Delta G^o_f(C_2H_2)$

$\Delta G^o = (1 \text{ mol})(124.5 \text{ kJ/mol}) - (3 \text{ mol})(209.9 \text{ kJ/mol}) = -505.2 \text{ kJ}$

Because ΔG^o is negative, the reaction is possible. Look for a catalyst.
Because ΔG^o_f for benzene is positive (+124.5 kJ/mol), the synthesis of benzene from graphite and gaseous H_2 at 25 °C and 1 atm pressure is not possible.

17.103 $CH_2ClCH_2Cl(l) \rightarrow CH_2{=}CHCl(g) + HCl(g)$

$\Delta G^o = [\Delta G^o_f(CH_2{=}CHCl) + \Delta G^o_f(HCl)] - \Delta G^o_f(CH_2ClCH_2Cl)$

$\Delta G^o = [(1 \text{ mol})(51.9 \text{ kJ/mol}) + (1 \text{ mol})(-95.3 \text{ kJ/mol})] - (1 \text{ mol})(-79.6 \text{ kJ/mol}) = +36.2 \text{ kJ}$

Because ΔG^o is positive, the reaction is nonspontaneous under standard-state conditions at 25 °C.

$CH_2ClCH_2Cl(l) \rightarrow CH_2{=}CHCl(g) + HCl(g)$

$\underline{NaOH(aq) + HCl(g) \rightarrow Na^+(aq) + Cl^-(aq) + H_2O(l)}$

Sum: $CH_2ClCH_2Cl(l) + NaOH(aq) \rightarrow CH_2{=}CHCl(g) + Na^+(aq) + Cl^-(aq) + H_2O(l)$

$\Delta G^o = [\Delta G^o_f(CH_2{=}CHCl) + \Delta G^o_f(Na^+) + \Delta G^o_f(Cl^-) + \Delta G^o_f(H_2O)]$
$\quad - [\Delta G^o_f(CH_2ClCH_2Cl) + \Delta G^o_f(NaOH)]$

$\Delta G^o = [(1 \text{ mol})(51.9 \text{ kJ/mol}) + (1 \text{ mol})(-261.9 \text{ kJ/mol})$
$\quad + (1 \text{ mol})(-131.3 \text{ kJ/mol}) + (1 \text{ mol})(-237.2 \text{ kJ/mol})]$
$\quad - [(1 \text{ mol})(-79.6 \text{ kJ/mol}) + (1 \text{ mol})(-419.2 \text{ kJ/mol})] = -79.7 \text{ kJ}$

Using NaOH(aq), $\Delta G^o = -79.7$ kJ and the reaction is spontaneous. (More generally, base removes HCl, driving the reaction to the right.)
The synthesis of a compound from its constituent elements is thermodynamically feasible at 25 °C and 1 atm pressure if ΔG^o_f is negative.
Because $\Delta G^o_f(CH_2{=}CHCl) = +51.9$ kJ, the synthesis of vinyl chloride from its elements is not possible at 25 °C and 1 atm pressure.

Free Energy, Composition, and Chemical Equilibrium (Sections 17.10 and 17.11)

17.104 $\Delta G = \Delta G^\circ + RT \ln Q$

17.105 $\Delta G = \Delta G^\circ + RT \ln Q$
(a) If $Q < 1$, then $RT \ln Q$ is negative and $\Delta G < \Delta G^\circ$.
(b) If $Q = 1$, then $RT \ln Q = 0$ and $\Delta G = \Delta G^\circ$.
(c) If $Q > 1$, then $RT \ln Q$ is positive and $\Delta G > \Delta G^\circ$.
As Q increases the thermodynamic tendency for the reaction to occur decreases.

17.106 $2\ NO(g) + Cl_2(g) \rightarrow 2\ NOCl(g)$
$\Delta G^\circ = 2\ \Delta G^\circ_f(NOCl) - 2\ \Delta G^\circ_f(NO)$
$\Delta G^\circ = (2\text{ mol})(66.1\text{ kJ/mol}) - (2\text{ mol})(87.6\text{ kJ/mol}) = -43.0\text{ kJ}$

$$\Delta G = \Delta G^\circ + RT \ln\left[\frac{(P_{NOCl})^2}{(P_{NO})^2(P_{Cl_2})}\right]$$

$$\Delta G = (-43.0\text{ kJ/mol}) + [8.314 \times 10^{-3}\text{ kJ/(K}\cdot\text{mol)}](298\text{ K})\ln\left[\frac{(2.00)^2}{(1.00 \times 10^{-3})^2(1.00 \times 10^{-3})}\right]$$

$\Delta G = +11.8\text{ kJ/mol}$
The reaction is spontaneous in the reverse direction.

17.107 $U(s) + 3\ F_2(g) \rightarrow UF_6(s)$

$$\Delta G = \Delta G^\circ + RT \ln\left[\frac{1}{(P_{F_2})^3}\right]$$

$$\Delta G = (-2068\text{ kJ/mol}) + [8.314 \times 10^{-3}\text{ kJ/(K}\cdot\text{mol)}](298\text{ K})\ln\left[\frac{1}{(0.045)^3}\right] = -2045\text{ kJ/mol}$$

The reaction is spontaneous in the forward direction.

17.108
$$\Delta G = \Delta G^\circ + RT \ln\left[\frac{(P_{SO_3})^2}{(P_{SO_2})^2(P_{O_2})}\right]$$

(a) $$\Delta G = (-141.8\text{ kJ/mol}) + [8.314 \times 10^{-3}\text{ kJ/(K}\cdot\text{mol)}](298\text{ K})\ln\left[\frac{(1.0)^2}{(100)^2(100)}\right] = -176.0\text{ kJ/mol}$$

(b) $$\Delta G = (-141.8\text{ kJ/mol}) + [8.314 \times 10^{-3}\text{ kJ/(K}\cdot\text{mol)}](298\text{ K})\ln\left[\frac{(10)^2}{(2.0)^2(1.0)}\right] = -133.8\text{ kJ/mol}$$

(c) $Q = 1$, $\ln Q = 0$, $\Delta G = \Delta G^\circ = -141.8\text{ kJ/mol}$

17.109
$$\Delta G = \Delta G^\circ + RT \ln\left[\frac{[NH_2CONH_2]}{(P_{NH_3})^2(P_{CO_2})}\right]$$

(a) $\Delta G = -13.6 \text{ kJ/mol} + [8.314 \times 10^{-3} \text{ kJ/(K} \cdot \text{mol)}](298 \text{ K})\ln\left[\dfrac{1.0}{(10)^2(10)}\right] = -30.7 \text{ kJ/mol}$

(b) $\Delta G = -13.6 \text{ kJ/mol} + [8.314 \times 10^{-3} \text{ kJ/(K} \cdot \text{mol)}](298 \text{ K})\ln\left[\dfrac{1.0}{(0.10)^2(0.10)}\right] = +3.5 \text{ kJ/mol}$

(c) The reaction is spontaneous for conditions in part (a) because ΔG is negative.

17.110 $\Delta G^\circ = -RT \ln K$
(a) If $K > 1$, ΔG° is negative. (b) If $K = 1$, $\Delta G^\circ = 0$. (c) If $K < 1$, ΔG° is positive.

17.111 $K = e^{\frac{-\Delta G^\circ}{RT}}$ (a) If ΔG° is positive, K is small. (b) If ΔG° is negative, K is large.

17.112 $\Delta G^\circ = -RT \ln K_p = -141.8 \text{ kJ}$

$\ln K_p = \dfrac{-\Delta G^\circ}{RT} = \dfrac{-(-141.8 \text{ kJ/mol})}{[8.314 \times 10^{-3} \text{ kJ/(K} \cdot \text{mol)}](298 \text{ K})} = 57.23$

$K_p = e^{57.23} = 7.2 \times 10^{24}$

17.113 $\Delta G^\circ = -RT \ln K = -13.6 \text{ kJ}$

$\ln K = \dfrac{-\Delta G^\circ}{RT} = \dfrac{-(-13.6 \text{ kJ/mol})}{[8.314 \times 10^{-3} \text{ kJ/(K} \cdot \text{mol)}](298 \text{ K})} = 5.49$

$K = e^{5.49} = 2.4 \times 10^2$

17.114 $C_2H_5OH(l) \rightleftarrows C_2H_5OH(g)$

$\Delta G^\circ = \Delta G^\circ_f(C_2H_5OH(g)) - \Delta G^\circ_f(C_2H_5OH(l))$

$\Delta G^\circ = (1 \text{ mol})(-167.9 \text{ kJ/mol}) - (1 \text{ mol})(-174.9 \text{ kJ/mol}) = +7.0 \text{ kJ}$

$\Delta G^\circ = -RT \ln K$

$\ln K = \dfrac{-\Delta G^\circ}{RT} = \dfrac{-(7.0 \text{ kJ/mol})}{[8.314 \times 10^{-3} \text{ kJ/(K} \cdot \text{mol)}](298 \text{ K})} = -2.83$

$K = e^{-2.83} = 0.059$; $K = K_p = P_{C_2H_5OH} = 0.059 \text{ atm}$

17.115 $\Delta G^\circ = -RT \ln K_a$

$\Delta G^\circ = -[8.314 \times 10^{-3} \text{ kJ/(K} \cdot \text{mol)}](298 \text{ K})\ln(3.0 \times 10^{-4}) = +20.1 \text{ kJ/mol}$

17.116 $Br_2(l) \rightleftarrows Br_2(g)$

$\Delta G^\circ = \Delta G^\circ_f(Br_2(g)) = 3.14 \text{ kJ/mol}$

$\Delta G^\circ = -RT \ln K$

$\ln K = \dfrac{-\Delta G^\circ}{RT} = \dfrac{-(3.14 \text{ kJ/mol})}{[8.314 \times 10^{-3} \text{ kJ/(K} \cdot \text{mol)}](298 \text{ K})} = -1.267$

$K = e^{-1.267} = 0.282$; $K = K_p = P_{Br_2} = 0.282 \text{ atm} = 0.28 \text{ atm}$

17.117 $I_2(s) \rightleftarrows I_2(g)$

$\Delta G^\circ = \Delta G^\circ_f(I_2(g)) = 19.4$ kJ/mol

$\Delta G^\circ = -RT \ln K$

$$\ln K = \frac{-\Delta G^\circ}{RT} = \frac{-(19.4 \text{ kJ/mol})}{[8.314 \times 10^{-3} \text{ kJ/(K}\cdot\text{mol)}](298 \text{ K})} = -7.830$$

$K = e^{-7.830} = 3.98 \times 10^{-4}$; $\quad K = K_p = P_{I_2} = 3.98 \times 10^{-4}$ atm $= 4.0 \times 10^{-4}$ atm

17.118 $2\ CH_2{=}CH_2(g) + O_2(g) \rightarrow 2\ C_2H_4O(g)$

$\Delta G^\circ = 2\ \Delta G^\circ_f(C_2H_4O) - 2\ \Delta G^\circ_f(CH_2{=}CH_2)$

$\Delta G^\circ = (2 \text{ mol})(-13.1 \text{ kJ/mol}) - (2 \text{ mol})(68.1 \text{ kJ/mol}) = -162.4$ kJ

$\Delta G^\circ = -RT \ln K$

$$\ln K = \frac{-\Delta G^\circ}{RT} = \frac{-(-162.4 \text{ kJ/mol})}{[8.314 \times 10^{-3} \text{ kJ/(K}\cdot\text{mol)}](298 \text{ K})} = 65.55$$

$K = K_p = e^{65.55} = 2.9 \times 10^{28}$

17.119 $TiO_2(s) + 2\ Cl_2(g) + 2\ C(s) \rightarrow TiCl_4(l) + 2\ CO(g)$

$\Delta G^\circ = [\Delta G^\circ_f(TiCl_4) + 2\ \Delta G^\circ_f(CO)] - \Delta G^\circ_f(TiO_2)$

$\Delta G^\circ = [(1 \text{ mol})(-737.2 \text{ kJ/mol}) + (2 \text{ mol})(-137.2 \text{ kJ/mol})] - (1 \text{ mol})(-888.8 \text{ kJ/mol})$

$\Delta G^\circ = -122.8$ kJ

$\Delta G^\circ = -RT \ln K$

$$\ln K = \frac{-\Delta G^\circ}{RT} = \frac{-(-122.8 \text{ kJ/mol})}{[8.314 \times 10^{-3} \text{ kJ/(K}\cdot\text{mol)}](298 \text{ K})} = 49.56$$

$K = K_p = e^{49.56} = 3.3 \times 10^{21}$

Chapter Problems

17.120 C_3H_8, 44.10; 20 °C = 293 K

$$\text{mol } C_3H_8 = 1.32 \text{ g} \times \frac{1 \text{ mol } C_3H_8}{44.10 \text{ g}} = 0.0300 \text{ mol } C_3H_8$$

$$V = \frac{nRT}{P} = \frac{(0.0300 \text{ mol})\left(0.082\ 06 \frac{\text{L}\cdot\text{atm}}{\text{K}\cdot\text{mol}}\right)(293 \text{ K})}{0.100 \text{ atm}} = 7.21 \text{ L}$$

Compress 7.21 L by a factor of 5 (7.21/5) to 1.44 L

$$\Delta S = nR \ln\left(\frac{V_f}{V_i}\right) = (0.0300 \text{ mol})(8.314 \text{ J/K}\cdot\text{mol})\ln\left(\frac{1.44 \text{ L}}{7.21 \text{ L}}\right) = -0.402 \text{ J/K}$$

17.121 (a), (c), and (d) are nonspontaneous; (b) is spontaneous.

17.122 (a) Spontaneous does not mean fast, just possible.

(b) For a spontaneous reaction $\Delta S_{total} > 0$. ΔS_{sys} can be positive or negative.

(c) An endothermic reaction can be spontaneous if $\Delta S_{sys} > 0$.

(d) True, because the sign of ΔG changes when the direction of a reaction is reversed.

17.123

Point Total	Possible Ways	Number of Ways
2	(1+1)	1
3	(2+1)(1+2)	2
4	(1+3)(2+2)(3+1)	3
5	(1+4)(2+3)(3+2)(4+1)	4
6	(1+5)(2+4)(3+3)(4+2)(5+1)	5
7	(1+6)(2+5)(3+4)(4+3)(5+2)(6+1)	6
8	(2+6)(3+5)(4+4)(5+3)(6+2)	5
9	(3+6)(4+5)(5+4)(6+3)	4
10	(4+6)(5+5)(6+4)	3
11	(6+5)(5+6)	2
12	(6+6)	1

Because a point total of 7 can be rolled in the most ways, it is the most probable point total.

17.124

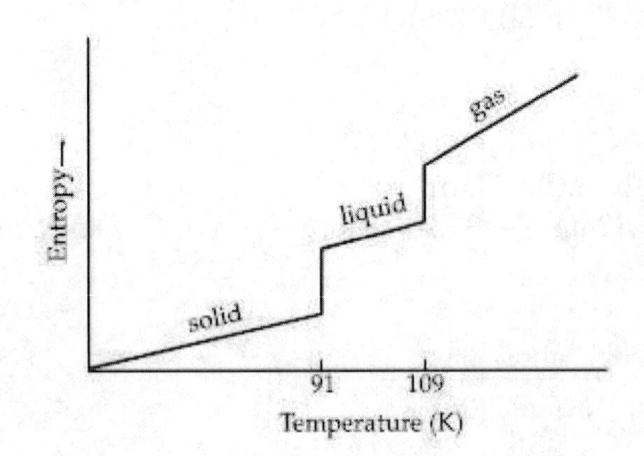

17.125 (a) $Q = 1$, $\ln Q = 0$, $\Delta G = \Delta G^\circ = +79.9$ kJ

Because ΔG is positive, the reaction is spontaneous in the reverse direction.

(b) $\Delta G = \Delta G^\circ + RT \ln Q$; $Q = [H_3O^+][OH^-] = (1.0 \times 10^{-7})^2 = 1.0 \times 10^{-14}$

$\Delta G = 79.9\text{ kJ/mol} + [8.314 \times 10^{-3}\text{ kJ/(K}\cdot\text{mol)}](298\text{ K}) \ln(1.0 \times 10^{-14}) = 0$

Because $\Delta G = 0$, the reaction is at equilibrium.

(c) $\Delta G = \Delta G^\circ + RT \ln Q$

$Q = [H_3O^+][OH^-] = (1.0 \times 10^{-7})(1.0 \times 10^{-10}) = 1.0 \times 10^{-17}$

$\Delta G = 79.9\text{ kJ/mol} + [8.314 \times 10^{-3}\text{ kJ/(K}\cdot\text{mol)}](298\text{ K}) \ln(1.0 \times 10^{-17}) = -17.1\text{ kJ/mol}$

Because ΔG is negative, the reaction is spontaneous in the forward direction.

The results are consistent with Le Châtelier's principle. When the $[H_3O^+]$ and $[OH^-]$ are larger than the equilibrium concentrations (a), the reverse reaction takes place. When the product of $[H_3O^+]$ and the $[OH^-]$ is less than the equilibrium value, the forward reaction is spontaneous.

$\Delta G^\circ = -RT \ln K$

$$\ln K = \frac{-\Delta G^\circ}{RT} = \frac{-79.9\text{ kJ/mol}}{[8.314 \times 10^{-3}\text{ kJ/(K}\cdot\text{mol)}](298\text{ K})} = -32.25$$

$K = K_a = e^{-32.25} = 9.9 \times 10^{-15}$

17.126 At the normal boiling point, $\Delta G = 0$.

$\Delta G_{vap} = \Delta H_{vap} - T\Delta S_{vap}$; $\quad T = \dfrac{\Delta H_{vap}}{\Delta S_{vap}} = \dfrac{38{,}600\ J}{110\ J/K} = 351\ K = 78\ ^\circ C$

17.127 At the normal boiling point, $\Delta G_{vap} = 0$. $\quad 61\ ^\circ C = 334\ K$

$\Delta G_{vap} = \Delta H_{vap} - T\Delta S_{vap}$; $\quad \Delta S_{vap} = \dfrac{\Delta H_{vap}}{T} = \dfrac{29{,}240\ J}{334\ K} = 87.5\ J/K$

17.128 $\Delta G = \Delta H - T\Delta S$

(a) ΔH must be positive (endothermic) and greater than $T\Delta S$ in order for ΔG to be positive (nonspontaneous reaction).

(b) Set $\Delta G = 0$ and solve for ΔH.

$\Delta G = 0 = \Delta H - T\Delta S = \Delta H - (323\ K)(104\ J/K) = \Delta H - (33592\ J) = \Delta H - (33.6\ kJ)$

$\Delta H = 33.6\ kJ$

ΔH must be greater than 33.6 kJ.

17.129 $NH_4NO_3(s) \rightarrow N_2O(g) + 2\ H_2O(g)$

(a) $\Delta G^\circ = [\Delta G^\circ_f(N_2O) + 2\ \Delta G^\circ_f(H_2O)] - \Delta G^\circ_f(NH_4NO_3)$

$\Delta G^\circ = [(1\ mol)(104.2\ kJ/mol) + (2\ mol)(-228.6\ kJ/mol)] - (1\ mol)(-184.0\ kJ/mol)$

$\Delta G^\circ = -169.0\ kJ$

Because ΔG° is negative, the reaction is spontaneous.

(b) Because the reaction increases the number of moles of gas, ΔS° is positive.

$\Delta G^\circ = \Delta H^\circ - T\Delta S^\circ$

As the temperature is raised, ΔG° becomes more negative.

(c) $\Delta G^\circ = -RT \ln K$

$$\ln K = \frac{-\Delta G^\circ}{RT} = \frac{-(-169.0\ kJ/mol)}{[8.314 \times 10^{-3}\ kJ/(K \cdot mol)](298\ K)} = 68.21$$

$K = K_p = e^{68.21} = 4.2 \times 10^{29}$

(d) $Q = (P_{N_2O})(P_{H_2O})^2 = (30)(30)^2 = (30)^3$

$\Delta G = \Delta G^\circ + RT \ln Q$

$\Delta G = -169.0\ kJ/mol + [8.314 \times 10^{-3}\ kJ/(K \cdot mol)](298\ K) \ln[(30)^3] = -143.7\ kJ/mol$

17.130 For $PbCrO_4$, $K_{sp} = 2.8 \times 10^{-13}$

$\Delta G^\circ = -RT \ln K_{sp}$

$\Delta G^\circ = -[8.314 \times 10^{-3}\ kJ/K](298\ K) \ln(2.8 \times 10^{-13}) = +71.6\ kJ$

17.131 (a) $2\ Mg(s) + O_2(g) \rightarrow 2\ MgO(s)$

$\Delta H^\circ = 2\ \Delta H^\circ_f(MgO) = (2\ mol)(-601.7\ kJ/mol) = -1203.4\ kJ$

$\Delta S^\circ = 2\ S^\circ(MgO) - [2\ S^\circ(Mg) + S^\circ(O_2)]$

$\Delta S^\circ = (2\ mol)(26.9\ J/(K \cdot mol)) - [(2\ mol)(32.7\ J/(K \cdot mol)) + (1\ mol)(205.0\ J/(K \cdot mol))]$

$\Delta S^\circ = -216.6\ J/K = -216.6 \times 10^{-3}\ kJ/K$

$\Delta G^\circ = \Delta H^\circ - T\Delta S^\circ = -1203.4\ kJ - (298\ K)(-216.6 \times 10^{-3}\ kJ/K) = -1138.8\ kJ$

Because ΔG° is negative, the reaction is spontaneous at 25 °C. ΔG° becomes less negative as the temperature is raised.

(b) $MgCO_3(s) \rightarrow MgO(s) + CO_2(g)$

$\Delta H^o = [\Delta H^o_f(MgO) + \Delta H^o_f(CO_2)] - \Delta H^o_f(MgCO_3)$

$\Delta H^o = [(1\ mol)(-601.1\ kJ/mol) + (1\ mol)(-393.5\ kJ/mol)] - (1\ mol)(-1096\ kJ/mol) = +101\ kJ$

$\Delta S^o = [S^o(MgO) + S^o(CO_2)] - S^o(MgCO_3)$

$\Delta S^o = [(1\ mol)(26.9\ J/(K \cdot mol)) + (1\ mol)(213.6\ J/(K \cdot mol))] - (1\ mol)(65.7\ J/(K \cdot mol))$

$\Delta S^o = 174.8\ J/K = 174.8 \times 10^{-3}\ kJ/K$

$\Delta G^o = \Delta H^o - T\Delta S^o = 101\ kJ - (298\ K)(174.8 \times 10^{-3}\ kJ/K) = +49\ kJ$

Because ΔG^o is positive, the reaction is not spontaneous at 25 °C. ΔG^o becomes less positive as the temperature is raised.

(c) $Fe_2O_3(s) + 2\ Al(s) \rightarrow Al_2O_3(s) + 2\ Fe(s)$

$\Delta H^o = \Delta H^o_f(Al_2O_3) - \Delta H^o_f(Fe_2O_3)$

$\Delta H^o = (1\ mol)(-1676\ kJ/mol) - (1\ mol)(-824.2\ kJ/mol) = -852\ kJ$

$\Delta S^o = [S^o(Al_2O_3) + 2\ S^o(Fe)] - [S^o(Fe_2O_3) + 2\ S^o(Al)]$

$\Delta S^o = [(1\ mol)(50.9\ J/(K \cdot mol)) + (2\ mol)(27.3\ J/(K \cdot mol))] - [(1\ mol)(87.4\ J/(K \cdot mol)) + (2\ mol)(28.3\ J/(K \cdot mol))]$

$\Delta S^o = -38.5\ J/K = -38.5 \times 10^{-3}\ kJ/K$

$\Delta G^o = \Delta H^o - T\Delta S^o = -852\ kJ - (298\ K)(-38.5 \times 10^{-3}\ kJ/K) = -841\ kJ$

Because ΔG^o is negative, the reaction is spontaneous at 25 °C. ΔG^o becomes less negative as the temperature is raised.

(d) $2\ NaHCO_3(s) \rightarrow Na_2CO_3(s) + CO_2(g) + H_2O(g)$

$\Delta H^o = [\Delta H^o_f(Na_2CO_3) + \Delta H^o_f(CO_2) + \Delta H^o_f(H_2O)] - 2\ \Delta H^o_f(NaHCO_3)$

$\Delta H^o = [(1\ mol)(-1130.7\ kJ/mol) + (1\ mol)(-393.5\ kJ/mol) + (1\ mol)(-241.8\ kJ/mol)] - (2\ mol)(-950.8\ kJ/mol) = +135.6\ kJ$

$\Delta S^o = [S^o(Na_2CO_3) + S^o(CO_2) + S^o(H_2O)] - 2\ S^o(NaHCO_3)$

$\Delta S^o = [(1\ mol)(135.0\ J/(K \cdot mol)) + (1\ mol)(213.6\ J/(K \cdot mol)) + (1\ mol)(188.7\ J/(K \cdot mol))] - (2\ mol)(102\ J/(K \cdot mol))$

$\Delta S^o = +333\ J/K = +333 \times 10^{-3}\ kJ/K$

$\Delta G^o = \Delta H^o - T\Delta S^o = +135.6\ kJ - (298\ K)(+333 \times 10^{-3}\ kJ/K) = +36.4\ kJ$

Because ΔG^o is positive, the reaction is not spontaneous at 25 °C. ΔG^o becomes less positive as the temperature is raised.

17.132 (a)

	$\Delta H_{vap}/T_{bp}$
ammonia	98 J/K
benzene	87 J/K
carbon tetrachloride	85 J/K
chloroform	87 J/K
mercury	94 J/K

(b) All processes are the conversion of a liquid to a gas at the boiling point. They should all have similar ΔS values. $\Delta H_{vap}/T_{bp}$ is equal to ΔS_{vap}.

(c) NH_3 deviates from Trouton's rule because of hydrogen bonding. $NH_3(l)$ has less randomness and ΔS_{vap} is larger. Hg has metallic bonding which also leads to less randomness of the liquid.

17.133 $NH_4HS(s) \rightarrow NH_3(g) + H_2S(g)$

$P_{total} = P_{NH_3} + P_{H_2S} = 0.658$ atm, and $P_{NH_3} = P_{H_2S} = 0.658\ atm/2 = 0.329$ atm

$\Delta G = \Delta G^\circ + RT \ln[(P_{NH_3})(P_{H_2S})]$

At equilibrium, $\Delta G = 0$

$\Delta G^\circ = -RT \ln[(P_{NH_3})(P_{H_2S})]$

$\Delta G^\circ = -[8.314 \times 10^{-3}\ kJ/(K \cdot mol)](298\ K) \ln[(0.329)(0.329)] = +5.51\ kJ/mol$

$\Delta G^\circ = [\Delta G^\circ_f(NH_3) + \Delta G^\circ_f(H_2S)] - \Delta G^\circ_f(NH_4HS)$

$5.51\ kJ = [(1\ mol)(-16.5\ kJ/mol) + (1\ mol)(-33.6\ kJ/mol)] - \Delta G^\circ_f(NH_4HS)$

$\Delta G^\circ_f(NH_4HS) = [(1\ mol)(-16.5\ kJ/mol) + (1\ mol)(-33.6\ kJ/mol)] - 5.51\ kJ$

$\Delta G^\circ_f(NH_4HS) = -55.6\ kJ/mol$

17.134 $Ni(s) + 4\ CO(g) \rightarrow Ni(CO)_4(l)$

(a) $\Delta H^\circ = \Delta H^\circ_f(Ni(CO)_4) - 4\ \Delta H^\circ_f(CO)$

$\Delta H^\circ = (1\ mol)(-633.0\ kJ/mol) - (4\ mol)(-110.5\ kJ/mol) = -191.0\ kJ$

$\Delta S^\circ = S^\circ(Ni(CO)_4) - [S^\circ(Ni) + 4\ S^\circ(CO)]$

$\Delta S^\circ = (1\ mol)(313.4\ J/(K \cdot mol)) - [(1\ mol)(29.9\ J/(K \cdot mol)) + (4\ mol)(197.6\ J/(K \cdot mol))]$

$\Delta S^\circ = -506.9\ J/K = -506.9 \times 10^{-3}\ kJ/K$

$\Delta G^\circ = \Delta H^\circ - T\Delta S^\circ = -191.0\ kJ - (298\ K)(-506.9 \times 10^{-3}\ kJ/K) = -39.9\ kJ$

(b) To find the crossover temperature set $\Delta G = 0$ and solve for T.

$$T = \frac{\Delta H^\circ}{\Delta S^\circ} = \frac{-191.0\ kJ}{-506.9 \times 10^{-3}\ kJ/K} = 376.8\ K$$

The reaction becomes nonspontaneous at 376.8 K

(c) $\Delta G^\circ = \Delta G^\circ_f(Ni(CO)_4) - 4\ \Delta G^\circ_f(CO)$

$-39.9\ kJ = (1\ mol)\Delta G^\circ_f(Ni(CO)_4) - (4\ mol)(-137.2\ kJ/mol)$

$\Delta G^\circ_f(Ni(CO)_4) = (-39.9\ kJ - 548.8\ kJ)/mol = -588.7\ kJ/mol$

17.135 (a) $6\ C(s) + 3\ H_2(g) \rightarrow C_6H_6(l)$

$\Delta S^\circ_f = S^\circ(C_6H_6) - [6\ S^\circ(C) + 3\ S^\circ(H_2)]$

$\Delta S^\circ_f = (1\ mol)(173.4\ J/(K \cdot mol)) - [(6\ mol)(5.7\ J/(K \cdot mol)) + (3\ mol)(130.6\ J/(K \cdot mol))]$

$\Delta S^\circ_f = -253\ J/K = -253\ J/(K \cdot mol)$

$\Delta G^\circ_f = \Delta H^\circ_f - T\Delta S^\circ_f$

$$\Delta S^\circ_f = \frac{\Delta H^\circ_f - \Delta G^\circ_f}{T} = \frac{49.0\ kJ/mol - 124.5\ kJ/mol}{298\ K} = -0.2533\ kJ/(K \cdot mol)$$

$\Delta S^\circ_f = -253.3\ J/(K \cdot mol)$

Both calculations lead to the same value of ΔS°_f.

(b) $Ca(s) + S(s) + 2\ O_2(g) \rightarrow CaSO_4(s)$

$\Delta S^\circ_f = S^\circ(CaSO_4) - [S^\circ(Ca) + S^\circ(S) + 2\ S^\circ(O_2)]$

$\Delta S^\circ_f = (1\ mol)(107\ J/(K \cdot mol))$

$- [(1\ mol)(41.4\ J/(K \cdot mol)) + (1\ mol)(31.8\ J/(K \cdot mol)) + (2\ mol)(205.0\ J/(K \cdot mol))]$

$\Delta S^\circ_f = -376\ J/K = -376\ J/(K \cdot mol)$

$\Delta G^\circ_f = \Delta H^\circ_f - T\Delta S^\circ_f$

$$\Delta S^\circ_f = \frac{\Delta H^\circ_f - \Delta G^\circ_f}{T} = \frac{-1434.1\ kJ/mol - (-1321.9\ kJ/mol)}{298\ K} = -0.377\ kJ/(K \cdot mol)$$

$\Delta S^\circ_f = -377\ J/(K \cdot mol)$

Both calculations lead to the same value of ΔS°_f.

(c) $2\ C(s)\ +\ 3\ H_2(g)\ +\ 1/2\ O_2(g)\ \rightarrow\ C_2H_5OH(l)$
$\Delta S^\circ_f = S^\circ(C_2H_5OH) - [S^\circ(C) + S^\circ(H_2) + 1/2\ S^\circ(O_2)]$
$\Delta S^\circ_f = (1\text{ mol})(161\text{ J/(K}\cdot\text{mol))}$
$-[(2\text{ mol})(5.7\text{ J/(K}\cdot\text{mol))} + (3\text{ mol})(130.6\text{ J/(K}\cdot\text{mol))} + (0.5\text{ mol})(205.0\text{ J/(K}\cdot\text{mol))}]$
$\Delta S^\circ_f = -345\text{ J/K} = -345\text{ J/(K}\cdot\text{mol)}$
$\Delta G^\circ_f = \Delta H^\circ_f - T\Delta S^\circ_f$

$$\Delta S^\circ_f = \frac{\Delta H^\circ_f - \Delta G^\circ_f}{T} = \frac{-277.7\text{ kJ/mol} - (-174.9\text{ kJ/mol})}{298\text{ K}} = -0.345\text{ kJ/(K}\cdot\text{mol)}$$

$\Delta S^\circ_f = -345\text{ J/(K}\cdot\text{mol)}$
Both calculations lead to the same value of ΔS°_f.

17.136 $MgCO_3(s) \rightarrow MgO(s) + CO_2(g)$
From Problem 17.131(b)
$\Delta H^\circ = +101\text{ kJ}$; $\Delta S^\circ = 174.8\text{ J/K} = 174.8 \times 10^{-3}\text{ kJ/K}$
The equilibrium pressure of CO_2 is equal to $K_p = P_{CO_2}$. K_p is not affected by the quantities of $MgCO_3$ and MgO present. K_p can be calculated from ΔG°.
$\Delta G^\circ = \Delta H^\circ - T\Delta S^\circ$
$\Delta G^\circ = -RT \ln K_p$
(a) $\Delta G^\circ = 101\text{ kJ} - (298\text{ K})(174.8 \times 10^{-3}\text{ kJ/K}) = +49\text{ kJ}$

$$\ln K_p = \frac{-\Delta G^\circ}{RT} = \frac{-49\text{ kJ/mol}}{[8.314 \times 10^{-3}\text{ kJ/(K}\cdot\text{mol)}](298\text{ K})} = -19.8$$

$K_p = P_{CO_2} = e^{-19.8} = 3 \times 10^{-9}\text{ atm}$
(b) $\Delta G^\circ = 101\text{ kJ} - (553\text{ K})(174.8 \times 10^{-3}\text{ kJ/K}) = 4.3\text{ kJ}$

$$\ln K_p = \frac{-\Delta G^\circ}{RT} = \frac{-4.3\text{ kJ/mol}}{[8.314 \times 10^{-3}\text{ kJ/(K}\cdot\text{mol)}](553\text{ K})} = -0.94$$

$K_p = P_{CO_2} = e^{-0.94} = 0.39\text{ atm}$
(c) $P_{CO_2} = 0.39$ atm because the temperature is the same as in (b).

17.137 $\Delta G^\circ = -RT \ln K_b$
At 20 °C: $\Delta G^\circ = -[8.314 \times 10^{-3}\text{ kJ/(K}\cdot\text{mol)}](293\text{ K}) \ln(1.710 \times 10^{-5}) = +26.74\text{ kJ/mol}$
At 50 °C: $\Delta G^\circ = -[8.314 \times 10^{-3}\text{ kJ/(K}\cdot\text{mol)}](323\text{ K}) \ln(1.892 \times 10^{-5}) = +29.20\text{ kJ/mol}$
$\Delta G^\circ = \Delta H^\circ - T\Delta S^\circ$

$26.74 = \Delta H^\circ - 293\Delta S^\circ$
$29.20 = \Delta H^\circ - 323\Delta S^\circ$ Solve these two equations simultaneously for ΔH° and ΔS°.

$26.74 + 293\Delta S^\circ = \Delta H^\circ$
$29.20 + 323\Delta S^\circ = \Delta H^\circ$ Set these two equations equal to each other.

$26.74 + 293\Delta S^\circ = 29.20 + 323\Delta S^\circ$
$26.74 - 29.20 = 323\Delta S^\circ - 293\ \Delta S^\circ$
$-2.46 = 30\Delta S^\circ$

$\Delta S^\circ = -2.46/30 = -0.0820 = -0.0820\ kJ/K = -82.0\ J/K$
$\Delta H^\circ = 26.74 + 293\Delta S^\circ = 26.74 + 293(-0.0820) = +2.71\ kJ$

17.138 (a) $\Delta H^\circ = 2\ \Delta H^\circ_f(NH_3) = (2\ mol)(-46.1\ kJ/mol) = -92.2\ kJ$
$\Delta G^\circ = 2\ \Delta G^\circ_f(NH_3) = (2\ mol)(-16.5\ kJ/mol) = -33.0\ kJ$
$\Delta G^\circ = \Delta H^\circ - T\Delta S^\circ$
$\Delta H^\circ - \Delta G^\circ = T\Delta S^\circ$

$$\Delta S^\circ = \frac{\Delta H^\circ - \Delta G^\circ}{T} = \frac{-92.2\ kJ - (-33.0\ kJ)}{298\ K} = -0.199\ kJ/K = -199\ J/K$$

(b) ΔS° is negative because the number of mol of gas molecules decreases from 4 mol to 2 mol on going from reactants to products.
(c) The reaction is spontaneous because ΔG° is negative.
(d) $\Delta G^\circ = \Delta H^\circ - T\Delta S^\circ = -92.2\ kJ - (350\ K)(-0.199\ kJ/K) = -22.55\ kJ$
$\Delta G^\circ = -RT \ln K_p$

$$\ln K_p = \frac{-\Delta G^\circ}{RT} = \frac{-(-22.55\ kJ/mol)}{[8.314 \times 10^{-3}\ kJ/(K\cdot mol)](350\ K)} = 7.749$$

$K_p = e^{7.749} = 2.3 \times 10^3$
$\Delta n = 2 - (1 + 3) = -2$

$$K_c = K_p\left(\frac{1}{RT}\right)^{\Delta n} = (2.3 \times 10^3)\left(\frac{1}{RT}\right)^{-2} = (2.3 \times 10^3)(RT)^2$$

$K_c = (2.3 \times 10^3)[(0.082\ 06)(350)]^2 = 1.9 \times 10^6$

17.139 (a) $\Delta H^\circ = [\Delta H^\circ_f(Ag^+(aq)) + \Delta H^\circ_f(Br^-(aq))] - \Delta H^\circ_f(AgBr(s))$
$\Delta H^\circ = [(1\ mol)(105.6\ kJ/mol) + (1\ mol)(-121.5\ kJ/mol)] - (1\ mol)(-100.4\ kJ/mol) = +84.5\ kJ$
$\Delta S^\circ = [S^\circ(Ag^+(aq)) + S^\circ(Br^-(aq))] - S^\circ(AgBr(s))$
$\Delta S^\circ = [(1\ mol)(72.7\ J/(K\cdot mol)) + (1\ mol)(82.4\ J/(K\cdot mol))]$
$- (1\ mol)(107.1\ J/(K\cdot mol)) = +48.0\ J$
$\Delta G^\circ = \Delta H^\circ - T\Delta S^\circ = 84.5\ kJ - (298\ K)(48.0 \times 10^{-3}\ kJ/K) = +70.2\ kJ$
(b) $\Delta G^\circ = -RT \ln K_{sp}$

$$\ln K_{sp} = \frac{-\Delta G^\circ}{RT} = \frac{-70.2\ kJ/mol}{[8.314 \times 10^{-3}\ kJ/(K\cdot mol)](298\ K)} = -28.3$$

$K_{sp} = e^{-28.3} = 5 \times 10^{-13}$
(c) $Q = [Ag^+][Br^-] = (1.00 \times 10^{-5})(1.00 \times 10^{-5}) = 1.00 \times 10^{-10}$
$\Delta G = \Delta G^\circ + RT\ln Q$
$\Delta G = 70.2\ kJ/mol + [8.314 \times 10^{-3}\ kJ/(K\cdot mol)](298\ K)\ln(1.00 \times 10^{-10}) = 13.2\ kJ/mol$
A positive value of ΔG means that the forward reaction is nonspontaneous under these conditions. The reverse reaction is therefore spontaneous, which is consistent with the fact that $Q > K_{sp}$.

17.140 (a) $\Delta G^\circ = \Delta H^\circ - T\Delta S^\circ$ and $\Delta G^\circ = -RT \ln K$
Set the two equations equal to each other.
$-RT \ln K = \Delta H^\circ - T\Delta S^\circ$

$$\ln K = \frac{\Delta H^\circ - T\Delta S^\circ}{-RT}$$

$$\ln K = \frac{-\Delta H^\circ}{RT} + \frac{T\Delta S^\circ}{RT}$$

$$\ln K = \frac{-\Delta H^\circ}{RT} + \frac{\Delta S^\circ}{R}$$

$$\ln K = \frac{-\Delta H^\circ}{R}\left(\frac{1}{T}\right) + \frac{\Delta S^\circ}{R}$$ This is the equation for a straight line ($y = mx + b$).

$y = \ln K$; $m = -\frac{\Delta H^\circ}{R}$ = slope; $x = \frac{1}{T}$; $b = \frac{\Delta S^\circ}{R}$ = intercept

(b) Plot ln K versus 1/T
$\Delta H^\circ = -R(\text{slope})$ $\Delta S^\circ = R(\text{intercept})$
(c) For a reaction where K increases with increasing temperature, the following plot would be obtained:

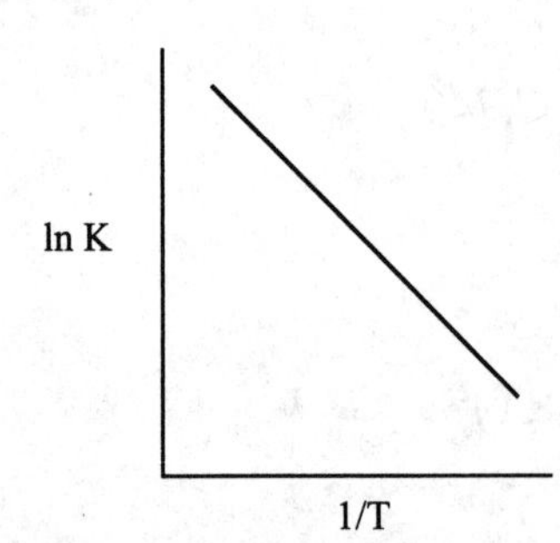

The slope is negative.
Because $\Delta H^\circ = -R(\text{slope})$, ΔH° is positive, and the reaction is endothermic.

This prediction is in accord with Le Châtelier's principle because when you add heat (raise the temperature) for an endothermic reaction, the reaction in the forward direction takes place, the product concentrations increase and the reactant concentrations decrease. This results in an increase in K.

17.141 $Br_2(l) \rightleftarrows Br_2(g)$
$\Delta S^\circ = S^\circ(Br_2(g)) - S^\circ(Br_2(l))$
$\Delta S^\circ = (1\text{ mol})(245.4\text{ J/(K}\cdot\text{mol)}) - (1\text{ mol})(152.2\text{ J/(K}\cdot\text{mol)}) = 93.2\text{ J/K} = 93.2 \times 10^{-3}\text{ kJ/K}$
$\Delta G = \Delta H^\circ - T\Delta S^\circ$
At the boiling point, $\Delta G = 0$.
$0 = \Delta H^\circ - T_{bp}\Delta S^\circ$

$$T_{bp} = \frac{\Delta H^\circ}{\Delta S^\circ}$$

$\Delta H^\circ = T_{bp}\,\Delta S^\circ = (332\text{ K})(93.2 \times 10^{-3}\text{ kJ/K}) = 30.9\text{ kJ}$

$$K_p = P_{Br_2} = \left(227\text{ mm Hg} \times \frac{1\text{ atm}}{760\text{ mm Hg}}\right) = 0.299\text{ atm}$$

$\Delta G^\circ = -RT \ln K_p$ and $\Delta G^\circ = \Delta H^\circ - T\Delta S^\circ$ (set equations equal to each other)
$\Delta H^\circ - T\Delta S^\circ = -RT \ln K_p$ (rearrange)

$$\ln K_p = \frac{-\Delta H^\circ}{R}\frac{1}{T} + \frac{\Delta S^\circ}{R}$$ (solve for T)

$$T = \frac{\left(\frac{-\Delta H^\circ}{R}\right)}{\left(\ln K_p - \frac{\Delta S^\circ}{R}\right)} = \frac{\left(\frac{-30.9 \text{ kJ/mol}}{8.314 \times 10^{-3} \text{ kJ/(K}\cdot\text{mol)}}\right)}{\left(\ln(0.299) - \frac{93.2 \times 10^{-3} \text{ kJ/(K}\cdot\text{mol)}}{8.314 \times 10^{-3} \text{ kJ/(K}\cdot\text{mol)}}\right)} = 299 \text{ K} = 26\ ^\circ\text{C}$$

$Br_2(l)$ has a vapor pressure of 227 mm Hg at 26 °C.

17.142 For PbI_2, $K_{sp} = [Pb^{2+}][I^-]^2$

	$PbI_2(s)$ ⇄	$Pb^{2+}(aq)$ +	$2\ I^-(aq)$
initial (M)		0	0
equil (M)		x	2x

$K_{sp} = x(2x)^2 = 4x^3$, where x = molar solubility

At 20 °C = 20 + 273 = 293 K, $K_{sp} = 4(1.45 \times 10^{-3})^3 = 1.22 \times 10^{-8}$

At 80 °C = 80 + 273 = 353 K, $K_{sp} = 4(6.85 \times 10^{-3})^3 = 1.29 \times 10^{-6}$

From problem 17.140, $\ln K = \frac{-\Delta H^\circ}{RT} + \frac{\Delta S^\circ}{R}$

$$\ln K_1 - \ln K_2 = \frac{-\Delta H^\circ}{RT_1} + \frac{\Delta S^\circ}{R} - \left(\frac{-\Delta H^\circ}{RT_2} + \frac{\Delta S^\circ}{R}\right)$$

$$\ln\frac{K_1}{K_2} = \frac{-\Delta H^\circ}{RT_1} - \frac{-\Delta H^\circ}{RT_2} = \frac{-\Delta H^\circ}{R}\left(\frac{1}{T_1} - \frac{1}{T_2}\right) = \frac{\Delta H^\circ}{R}\left(\frac{1}{T_2} - \frac{1}{T_1}\right)$$

$$\Delta H^\circ = \frac{[\ln K_1 - \ln K_2]R}{\left(\frac{1}{T_2} - \frac{1}{T_1}\right)}$$

$$\Delta H^\circ = \frac{[\ln(1.22 \times 10^{-8}) - \ln(1.29 \times 10^{-6})][8.314 \times 10^{-3} \text{ kJ/(K}\cdot\text{mol)}]}{\left(\frac{1}{353 \text{ K}} - \frac{1}{293 \text{ K}}\right)} = 66.8 \text{ kJ/mol}$$

$\Delta G^\circ = -RT \ln K_{sp} = -[8.314 \times 10^{-3} \text{ kJ/(K}\cdot\text{mol)}](293 \text{ K}) \ln(1.22 \times 10^{-8}) = 44.4 \text{ kJ/mol}$

$\Delta G^\circ = \Delta H^\circ - T\Delta S^\circ$; $\Delta H^\circ - \Delta G^\circ = T\Delta S^\circ$; $\Delta S^\circ = \frac{\Delta H^\circ - \Delta G^\circ}{T}$

$$\Delta S^\circ = \frac{66.8 \text{ kJ/mol} - 44.4 \text{ kJ/mol}}{293 \text{ K}} = 0.0765 \text{ kJ/(K}\cdot\text{mol)} = 76.5 \text{ J/(K}\cdot\text{mol)}$$

17.143 $\Delta H^\circ = [2\ \Delta H^\circ_f(Cl^-(aq))] - [2\ \Delta H^\circ_f(Br^-(aq))]$

$\Delta H^\circ = [(2 \text{ mol})(-167.2 \text{ kJ/mol})] - [(2 \text{ mol})(-121.5 \text{ kJ/mol})] = -91.4 \text{ kJ}$

$\Delta S^\circ = [S^\circ(Br_2(l)) + 2\ S^\circ(Cl^-(aq))] - [2\ S^\circ(Br^-(aq)) + S^\circ(Cl_2(g))]$

$\Delta S^\circ = [(1 \text{ mol})(152.2 \text{ J/(K}\cdot\text{mol)}) + (2 \text{ mol})(56.5 \text{ J/(K}\cdot\text{mol)})]$
$\quad - [(2 \text{ mol})(82.4 \text{ J/(K}\cdot\text{mol)}) + (1 \text{ mol})(223.0 \text{ J/(K}\cdot\text{mol)})] = -122.6 \text{ J/K}$

80 °C = 80 + 273 = 353 K

$\Delta G^\circ = \Delta H^\circ - T\Delta S^\circ = -91.4 \text{ kJ} - (353 \text{ K})(-122.6 \times 10^{-3} \text{ kJ/K}) = -48.1 \text{ kJ}$

$\Delta G^\circ = -RT \ln K$

$$\ln K = \frac{-\Delta G^o}{RT} = \frac{-(-48.1 \text{ kJmol})}{[8.314 \times 10^{-3} \text{ kJ/(K·mol)}](353 \text{ K})} = 16.4$$

$K = e^{16.4} = 1.3 \times 10^7$

17.144 $CS_2(l) \rightleftarrows CS_2(g)$

$\Delta H^o = \Delta H^o_f(CS_2(g)) - \Delta H^o_f(CS_2(l))$

$\Delta H^o = [(1 \text{ mol})(116.7 \text{ kJ/mol})] - [(1 \text{ mol})(89.0 \text{ kJ/mol})] = 27.7 \text{ kJ}$

$\Delta S^o = S^o(CS_2(g)) - S^o(CS_2(l))$

$\Delta S^o = [(1 \text{ mol})(237.7 \text{ J/(K·mol)})] - [(1 \text{ mol})(151.3 \text{ J/(K·mol)})] = 86.4 \text{ J/K}$

$\Delta G = \Delta H^o - T\Delta S^o$

At the boiling point, $\Delta G = 0$.

$0 = \Delta H^o - T_{bp}\Delta S^o$

$$T_{bp} = \frac{\Delta H^o}{\Delta S^o} = \frac{27.7 \text{ kJ}}{86.4 \times 10^{-3} \text{ kJ/K}} = 321 \text{ K}$$

$T_{bp} = 321 \text{ K} = 321 - 273 = 48\ ^oC$

17.145 $35\ ^oC = 35 + 273 = 308 \text{ K}$

$\Delta G^o = \Delta H^o - T\Delta S^o = -352 \text{ kJ} - (308 \text{ K})(-899 \times 10^{-3} \text{ kJ/K}) = -75.1 \text{ kJ}$

$\Delta G^o = -RT \ln K_p$

$$\ln K_p = \frac{-\Delta G^o}{RT} = \frac{-(-75.1 \text{ kJ/mol})}{[8.314 \times 10^{-3} \text{ kJ/(K·mol)}](308 \text{ K})} = 29.33$$

$K_p = e^{29.33} = 5.5 \times 10^{12}$

$$K_p = \frac{1}{(P_{H_2O})^6} = 5.5 \times 10^{12}$$

$$P_{H_2O} = \sqrt[6]{\frac{1}{5.5 \times 10^{12}}} = 0.0075 \text{ atm}$$

$$P_{H_2O} = 0.0075 \text{ atm} \times \frac{760 \text{ mm Hg}}{1 \text{ atm}} = 5.7 \text{ mm Hg}$$

17.146 $2\ KClO_3(s) \rightarrow 2\ KCl(s) + 3\ O_2(g)$

$\Delta H^o = 2\ \Delta H^o_f(KCl) - 2\ \Delta H^o_f(KClO_3)$

$\Delta H^o = (2 \text{ mol})(-436.5 \text{ kJ}) - (2 \text{ mol})(-397.7 \text{ kJ}) = -77.6 \text{ kJ}$

$25\ ^oC = 25 + 273 = 298 \text{ K}$

$\Delta G^o = \Delta H^o - T\Delta S^o$

$\Delta H^o - \Delta G^o = T\Delta S^o$

$$\Delta S^o = \frac{\Delta H^o - \Delta G^o}{T} = \frac{-77.6 \text{ kJ} - (-224.4 \text{ kJ})}{298 \text{ K}} = 0.493 \text{ kJ/K} = 493 \text{ J/K}$$

$\Delta S^o = [2\ S^o(KCl) + 3\ S^o(O_2)] - 2\ S^o(KClO_3)$

$493 \text{ J/K} = [(2 \text{ mol})(82.6 \text{ J/(K·mol)}) + (3 \text{ mol})S^o(O_2)] - (2 \text{ mol})(143.1 \text{ J/(K·mol)})$

$(3 \text{ mol})S^o(O_2) = 493 \text{ J/K} - (2 \text{ mol})(82.6 \text{ J/(K·mol)}) + (2 \text{ mol})(143.1 \text{ J/(K·mol)})$

$(3 \text{ mol})S^o(O_2) = 614 \text{ J/K}$

$S^o(O_2) = (614 \text{ J/K})/(3 \text{ mol}) = 204.7 \text{ J/(K·mol)} = 205 \text{ J/(K·mol)}$

17.147 $N_2O_4(g) \rightleftarrows 2\ NO_2(g)$

$\Delta H° = 2\ \Delta H°_f(NO_2) - \Delta H°_f(N_2O_4) = (2\ mol)(33.2\ kJ) - (1\ mol)(11.1\ kJ) = 55.3\ kJ$

$\Delta S° = 2\ S°(NO_2) - S°(N_2O_4) = (2\ mol)(240.0\ J/(K \cdot mol)) - (1\ mol)(304.3\ J/(K \cdot mol))$

$\Delta S° = 175.7\ J/K = 175.7 \times 10^{-3}\ kJ/K$

$\Delta G° = \Delta H° - T\Delta S°$ and $\Delta G° = -RT \ln K_p$; Set these two equations equal to each other and solve for T.

$\Delta H° - T\Delta S° = -RT \ln K_p$

$\Delta H° = T\Delta S° - RT \ln K_p = T(\Delta S° - R \ln K_p)$

$$T = \frac{\Delta H°}{\Delta S° - R \ln K_p}$$

(a) $P_{N_2O_4} + P_{NO_2} = 1.00$ atm and $P_{NO_2} = 2P_{N_2O_4}$

$P_{N_2O_4} + 2P_{N_2O_4} = 3P_{N_2O_4} = 1.00$ atm

$P_{N_2O_4} = 1.00\ atm/3 = 0.333$ atm

$P_{NO_2} = 1.00\ atm - P_{N_2O_4} = 1.00 - 0.333 = 0.667$ atm

$$K_p = \frac{(P_{NO_2})^2}{P_{N_2O_4}} = \frac{(0.667)^2}{(0.333)} = 1.34$$

$$T = \frac{\Delta H°}{\Delta S° - R \ln K_p}$$

$$T = \frac{55.3\ kJ/mol}{[175.7 \times 10^{-3}\ kJ/(K \cdot mol)] - [8.314 \times 10^{-3}\ kJ/(K \cdot mol)] \ln(1.34)} = 319\ K$$

$T = 319\ K = 319 - 273 = 46\ °C$

(b) $P_{N_2O_4} + P_{NO_2} = 1.00$ atm and $P_{NO_2} = P_{N_2O_4}$ so $P_{NO_2} = P_{N_2O_4} = 0.50$ atm

$$K_p = \frac{(P_{NO_2})^2}{P_{N_2O_4}} = \frac{(0.500)^2}{(0.500)} = 0.500$$

$$T = \frac{\Delta H°}{\Delta S° - R \ln K_p}$$

$$T = \frac{55.3\ kJ/mol}{[175.7 \times 10^{-3}\ kJ/(K \cdot mol)] - [8.314 \times 10^{-3}\ kJ/(K \cdot mol)] \ln(0.500)} = 305\ K$$

$T = 305\ K = 305 - 273 = 32\ °C$

Multiconcept Problems

17.148 The kinetic parameters [(a), (b), and (h)] are affected by a catalyst. The thermodynamic and equilibrium parameters [(c), (d), (e), (f), and (g)] are not affected by a catalyst.

17.149 $N_2(g) + 3\ H_2(g) \rightleftarrows 2\ NH_3(g)$

$\Delta H° = 2\ \Delta H°_f(NH_3) - [\Delta H°_f(N_2) + 3\ \Delta H°_f(H_2)] = (2\ mol)(-46.1\ kJ) - [0] = -92.2\ kJ$

$\Delta S° = 2\ S°(NH_3) - [S°(N_2) + 3\ S°(H_2)]$

$\Delta S° = (2\ mol)(192.3\ J/(K \cdot mol))$
$\quad - [(1\ mol)(191.5\ J/(K \cdot mol)) + (3\ mol)(130.6\ J/(K \cdot mol))] = -198.7\ J/K$
$\Delta G° = \Delta H° - T\Delta S° = -92.2\ kJ - (673\ K)(-198.7 \times 10^{-3}\ kJ/K) = 41.5\ kJ$
$\Delta G° = -RT \ln K_p$

$$\ln K_p = \frac{-\Delta G°}{RT} = \frac{-41.5\ kJ/mol}{[8.314 \times 10^{-3}\ kJ/(K \cdot mol)](673\ K)} = -7.42$$

$K_p = e^{-7.42} = 6.0 \times 10^{-4}$
Because $K_p = K_c(RT)^{\Delta n}$, $K_c = K_p(RT)^{-\Delta n}$
$K_c = K_p(RT)^2 = (6.0 \times 10^{-4})[(0.082\ 06)(673)]^2 = 1.83$
N_2, 28.01; H_2, 2.016
Initial concentrations:

$$[N_2] = \frac{(14.0\ g)\left(\frac{1\ mol}{28.01\ g}\right)}{5.00\ L} = 0.100\ M \text{ and } [H_2] = \frac{(3.024\ g)\left(\frac{1\ mol}{2.016\ g}\right)}{5.00\ L} = 0.300\ M$$

	$N_2(g)$	+ $3\ H_2(g)$	⇄ $2\ NH_3(g)$
initial (M)	0.100	0.300	0
change (M)	−x	−3x	+2x
equil (M)	0.100 − x	0.300 − 3x	2x

$$K_c = \frac{[NH_3]^2}{[N_2][H_2]^3} = \frac{(2x)^2}{(0.100 - x)(0.300 - 3x)^3} = \frac{4x^2}{27(0.100 - x)^4} = 1.83$$

$$\left(\frac{x}{(0.100 - x)^2}\right)^2 = \frac{(27)(1.83)}{4} = 12.35; \quad \frac{x}{(0.100 - x)^2} = \sqrt{12.35} = 3.514$$

$3.514x^2 - 1.703x + 0.03514 = 0$
Use the quadratic formula to solve for x.

$$x = \frac{-(-1.703) \pm \sqrt{(-1.703)^2 - (4)(3.514)(0.03514)}}{2(3.514)} = \frac{1.703 \pm 1.551}{7.028}$$

x = 0.463 and 0.0216
Of the two solutions for x, only 0.0216 has physical meaning because 0.463 would lead to negative concentrations of N_2 and H_2.
$[N_2] = 0.100 - x = 0.100 - 0.0216 = 0.078\ M$
$[H_2] = 0.300 - 3x = 0.300 - 3(0.0216) = 0.235\ M$
$[NH_3] = 2x = 2(0.0216) = 0.043\ M$

17.150 (a) $2\ SO_2(g) + O_2(g) \rightleftarrows 2\ SO_3(g)$
$\Delta H° = 2\ \Delta H°_f(SO_3) - 2\ \Delta H°_f(SO_2)$
$\Delta H° = (2\ mol)(-395.7\ kJ/mol) - (2\ mol)(-296.8\ kJ/mol) = -197.8\ kJ$
$\Delta S° = 2\ S°(SO_3) - [2\ S°(SO_2) + S°(O_2)]$
$\Delta S° = (2\ mol)(256.6\ J/(K \cdot mol)) - [(2\ mol)(248.1\ J/(K \cdot mol)) + (1\ mol)(205.0\ J/(K \cdot mol))]$
$\Delta S° = -188.0\ J/K = -188.0 \times 10^{-3}\ kJ/K$
$\Delta G° = \Delta H° - T\Delta S° = -197.8\ kJ - (800\ K)(-188.0 \times 10^{-3}\ kJ/K) = -47.4\ kJ$

$\Delta G^\circ = -RT \ln K_p$

$$\ln K_p = \frac{-\Delta G^\circ}{RT} = \frac{-(-47.4 \text{ kJ/mol})}{[8.314 \times 10^{-3} \text{ kJ/(K}\cdot\text{mol)}](800 \text{ K})} = 7.13$$

$K_p = e^{7.13} = 1249$

SO_2, 64.06; O_2, 32.00

At 800 K:

$$P_{SO_2} = \frac{nRT}{V} = \frac{\left(192 \text{ g} \times \frac{1 \text{ mol}}{64.06 \text{ g}}\right)\left(0.082\,06 \frac{\text{L}\cdot\text{atm}}{\text{K}\cdot\text{mol}}\right)(800 \text{ K})}{15.0 \text{ L}} = 13.1 \text{ atm}$$

$$P_{O_2} = \frac{nRT}{V} = \frac{\left(48.0 \text{ g} \times \frac{1 \text{ mol}}{32.00 \text{ g}}\right)\left(0.082\,06 \frac{\text{L}\cdot\text{atm}}{\text{K}\cdot\text{mol}}\right)(800 \text{ K})}{15.0 \text{ L}} = 6.56 \text{ atm}$$

	$2\,SO_2(g)$	+	$O_2(g)$	⇄	$2\,SO_3(g)$
initial (atm)	13.1		6.56		0
assume complete rxn (atm)	0		0		13.1
assume a small back rxn	+2x		+x		−2x
equil (atm)	2x		x		13.1 − 2x

$$K_p = 1249 = \frac{[SO_3]^2}{[SO_2]^2[O_2]} = \frac{(13.1-2x)^2}{(2x)^2(x)} \approx \frac{(13.1)^2}{(2x)^2(x)}$$

Solve for x. $x^3 = 0.0343$; $x = 0.325$

Use successive approximations to solve for x because 2x is not negligible compared with 13.1.

Second approximation:

$$1249 = \frac{[13.1-(2)(0.325)]^2}{(2x)^2(x)};$$ Solve for x. $x^3 = 0.0310$; $x = 0.314$

Third approximation:

$$1249 = \frac{[13.1-(2)(0.314)]^2}{(2x)^2(x)};$$ Solve for x. $x^3 = 0.0311$; $x = 0.315$ (x has converged)

$P_{SO_2} = 2x = 2(0.315) = 0.63$ atm

$P_{O_2} = x = 0.32$ atm

$P_{SO_3} = 13.1 - 2x = 13.1 - 2(0.315) = 12.5$ atm

(b) The % yield of SO_3 decreases with increasing temperature because ΔS° is negative. ΔG° becomes less negative and K_p gets smaller as the temperature increases.

(c) At 1000 K:

$\Delta G^\circ = \Delta H^\circ - T\Delta S^\circ = -197.8 \text{ kJ} - (1000 \text{ K})(-188.0 \times 10^{-3} \text{ kJ/K}) = -9.8 \text{ kJ}$

$\Delta G^\circ = -RT \ln K_p$

$$\ln K_p = \frac{-\Delta G^\circ}{RT} = \frac{-(-9.8 \text{ kJ/mol})}{[8.314 \times 10^{-3} \text{ kJ/(K}\cdot\text{mol)}](1000 \text{ K})} = 1.179$$

$K_p = e^{1.179} = 3.25$

$$P_{SO_2} = \frac{nRT}{V} = \frac{\left(192\ g \times \frac{1\ mol}{64.06\ g}\right)\left(0.082\ 06\ \frac{L\cdot atm}{K\cdot mol}\right)(1000\ K)}{15.0\ L} = 16.4\ atm$$

$$P_{O_2} = \frac{nRT}{V} = \frac{\left(48.0\ g \times \frac{1\ mol}{32.00\ g}\right)\left(0.082\ 06\ \frac{L\cdot atm}{K\cdot mol}\right)(1000\ K)}{15.0\ L} = 8.2\ atm$$

	$2\ SO_2(g)$	+ $O_2(g)$	⇄ $2\ SO_3(g)$
initial (atm)	16.4	8.2	0
assume complete rxn (atm)	0	0	16.4
assume a small back rxn	+2x	+x	–2x
equil (atm)	2x	x	16.4 – 2x

$$K_p = 3.25 = \frac{[SO_3]^2}{[SO_2]^2[O_2]} = \frac{(16.4 - 2x)^2}{(2x)^2(x)} \approx \frac{(16.4)^2}{(2x)^2(x)}$$

Solve for x. $x^3 = 20.7$; $x = 2.7$

Use successive approximations to solve for x because 2x is not negligible compared with 16.4.

Second approximation:

$$3.25 = \frac{[16.4 - (2)(2.7)]^2}{(2x)^2(x)};\ \text{Solve for x.}\ x^3 = 9.31;\ x = 2.1$$

Third approximation:

$$3.25 = \frac{[16.4 - (2)(2.1)]^2}{(2x)^2(x)};\ \text{Solve for x.}\ x^3 = 11.4;\ x = 2.3$$

Fourth approximation:

$$3.25 = \frac{[16.4 - (2)(2.3)]^2}{(2x)^2(x)};\ \text{Solve for x.}\ x^3 = 10.7;\ x = 2.2\ \text{(x has converged)}$$

$P_{SO_2} = 2x = 2(2.2) = 4.4$ atm

$P_{O_2} = x = 2.2$ atm

$P_{SO_3} = 16.4 - 2x = 16.4 - 2(2.2) = 12.0$ atm

$P_{total} = P_{SO_2} + P_{O_2} + P_{SO_3} = 4.4 + 2.2 + 12.0 = 18.6$ atm

On going from 800 K to 1000 K, P_{total} increases to 18.6 atm (because K_p decreases, but P increases with temperature at constant volume).

17.151 $Pb(s) + PbO_2(s) + 2\ H^+(aq) + 2\ HSO_4^-(aq) \rightarrow 2\ PbSO_4(s) + 2\ H_2O(l)$

(a) $\Delta G^\circ = [2\ \Delta G^\circ_f(PbSO_4) + 2\ \Delta G^\circ_f(H_2O)] - [\Delta G^\circ_f(PbO_2) + 2\ \Delta G^\circ_f(HSO_4^-)]$

ΔG° = (2 mol)(–813.2 kJ/mol) + (2 mol)(–237.2 kJ/mol)]

– [(1 mol)(–217.4 kJ/mol) + (2 mol)(–756.0 kJ/mol)] = –371.4 kJ

(b) °C = 5/9(°F – 32) = 5/9(10 – 32) = –12.2 °C; –12.2 °C = 261 K

$\Delta H^\circ = [2\ \Delta H^\circ_f(PbSO_4) + 2\ \Delta H^\circ_f(H_2O)] - [\Delta H^\circ_f(PbO_2) + 2\ \Delta H^\circ_f(HSO_4^-)]$
$\Delta H^\circ = [(2\text{ mol})(-919.9\text{ kJ/mol}) + (2\text{ mol})(-285.8\text{ kJ/mol})]$
$- [(1\text{ mol})(-277\text{ kJ/mol}) + (2\text{ mol})(-887.3\text{ kJ/mol})] = -359.8\text{ kJ}$
$\Delta S^\circ = [2\ S^\circ(PbSO_4) + 2\ S^\circ(H_2O)] - [S^\circ(Pb) + S^\circ(PbO_2) + 2\ S^\circ(H^+) + 2\ S^\circ(HSO_4^-)]$
$\Delta S^\circ = [(2\text{ mol})(148.6\text{ J/(K}\cdot\text{mol)}) + (2\text{ mol})(69.9\text{ J/(K}\cdot\text{mol)})]$
$- [(1\text{ mol})(64.8\text{ J/(K}\cdot\text{mol)}) + (1\text{ mol})(68.6\text{ J/(K}\cdot\text{mol)})$
$+ (2\text{ mol})(132\text{ J/(K}\cdot\text{mol)})] = 39.6\text{ J/K} = 39.6 \times 10^{-3}\text{ kJ/K}$
$\Delta G^\circ = \Delta H^\circ - T\Delta S^\circ = -359.8\text{ kJ} - (261\text{ K})(39.6 \times 10^{-3}\text{ kJ/K}) = -370.1\text{ kJ at 261 K}$

	$HSO_4^-(aq)$ + $H_2O(l)$ ⇄	$H_3O^+(aq)$ +	$SO_4^{2-}(aq)$
initial (M)	0.100	0.100	0
change (M)	−x	+x	+x
equil (M)	0.100 − x	0.100 + x	x

$$K_{a2} = \frac{[H_3O^+][SO_4^{2-}]}{[HSO_4^-]} = 1.2 \times 10^{-2} = \frac{(0.100 + x)x}{0.100 - x}$$

$x^2 + 0.112x - (1.2 \times 10^{-3}) = 0$
Use the quadratic formula to solve for x.

$$x = \frac{-(0.112) \pm \sqrt{(0.112)^2 - (4)(1)(-1.2 \times 10^{-3})}}{2(1)} = \frac{-0.112 \pm 0.132}{2}$$

x = −0.122 and 0.010
Of the two solutions for x, only 0.010 has physical meaning because −0.122 would lead to negative concentrations of H_3O^+ and SO_4^{2-}.
$[H^+] = 0.100 + x = 0.100 + 0.010 = 0.110\text{ M}$
$[HSO_4^-] = 0.100 - x = 0.100 - 0.010 = 0.090\text{ M}$

$$\Delta G = \Delta G^\circ + RT \ln \frac{1}{[H^+]^2[HSO_4^-]^2}$$

$$\Delta G = (-370.1\text{ kJ/mol}) + [8.314 \times 10^{-3}\text{ kJ/(K}\cdot\text{mol)}](261\text{ K}) \ln \frac{1}{(0.110)^2(0.090)^2}$$

$\Delta G = -350.1\text{ kJ/mol}$

17.152 $CaCO_3(s) \rightleftarrows Ca^{2+}(aq) + CO_3^{2-}(aq)$
$\Delta H^\circ = [\Delta H^\circ_f(Ca^{2+}) + \Delta H^\circ_f(CO_3^{2-})] - \Delta H^\circ_f(CaCO_3)$
$\Delta H^\circ = [(1\text{ mol})(-542.8\text{ kJ/mol}) + (1\text{ mol})(-677.1\text{ kJ/mol})] - (1\text{ mol})(-1207.6\text{ kJ/mol})$
$\Delta H^\circ = -12.3\text{ kJ}$
$\Delta S^\circ = [S^\circ(Ca^{2+}) + S^\circ(CO_3^{2-})] - S^\circ(CaCO_3)$
$\Delta S^\circ = [(1\text{ mol})(-53.1\text{ J/(K}\cdot\text{mol)}) + (1\text{ mol})(-56.9\text{ J/(K}\cdot\text{mol)})] - (1\text{ mol})(91.7\text{ J/(K}\cdot\text{mol)})$
$\Delta S^\circ = -201.7\text{ J/K} = -201.7 \times 10^{-3}\text{ kJ/K}$
$50\ ^\circ C = 50 + 273 = 323\text{ K}$
$\Delta G = \Delta H^\circ - T\Delta S^\circ = -12.3\text{ kJ} - (323\text{ K})(-201.7 \times 10^{-3}\text{ kJ/K}) = +52.85\text{ kJ}$
$\Delta G = -RT \ln K_{sp}$

$$\ln K_{sp} = \frac{-\Delta G}{RT} = \frac{-52.85\text{ J/mol}}{[8.314 \times 10^{-3}\text{ kJ/(K}\cdot\text{mol)}](323\text{ K})} = -19.68$$

$K_{sp} = e^{-19.68} = 2.8 \times 10^{-9}$

20 °C = 20 +273 = 293 K

$$n_{CO_2} = \frac{PV}{RT} = \frac{\left(731 \text{ mm Hg} \times \frac{1.00 \text{ atm}}{760 \text{ mm Hg}}\right)(1.000 \text{ L})}{\left(0.082\ 06 \frac{\text{L} \cdot \text{atm}}{\text{K} \cdot \text{mol}}\right)(293 \text{ K})} = 0.0400 \text{ mol } CO_2$$

$Ca(OH)_2$, 74.09

$$\text{mol } Ca(OH)_2 = 3.335 \text{ g } Ca(OH)_2 \times \frac{1 \text{ mol } Ca(OH)_2}{74.09 \text{ g } Ca(OH)_2} = 0.0450 \text{ mol } Ca(OH)_2$$

$CO_2(g) + H_2O(l) \rightarrow H_2CO_3(aq)$

	$Ca(OH)_2(aq)$	+ $H_2CO_3(aq)$	→ $CaCO_3(s)$	+ $2\ H_2O(l)$
before (mol)	0.0450	0.0400	0	
change (mol)	−0.0400	−0.0400	+0.0400	
after (mol)	0.0050	0	0.0400	

500.0 mL = 0.5000 L
$[Ca(OH)_2] = [Ca^{2+}] = 0.0050 \text{ mol}/0.5000 \text{ L} = 0.010 \text{ M}$

	$CaCO_3(s)$ ⇄	$Ca^{2+}(aq)$	+ $CO_3^{2-}(aq)$
initial (M)		0.010	0
change (M)		+x	+x
equil (M)		0.010 + x	x

$K_{sp} = [Ca^{2+}][CO_3^{2-}] = 2.8 \times 10^{-9} = (0.010 + x)x \approx 0.010x$
$x = \text{molar solubility} = 2.8 \times 10^{-9}/0.010 = 2.8 \times 10^{-7}$ M
Because ΔH° is negative (exothermic), the solubility of $CaCO_3$ is lower at 50 °C.

17.153 PV = nRT

$$n_{NH_3} = \frac{PV}{RT} = \frac{\left(744 \text{ mm Hg} \times \frac{1.00 \text{ atm}}{760 \text{ mm Hg}}\right)(1.00 \text{ L})}{\left(0.082\ 06 \frac{\text{L} \cdot \text{atm}}{\text{K} \cdot \text{mol}}\right)(298.1 \text{ K})} = 0.0400 \text{ mol } NH_3$$

500.0 mL = 0.5000 L
$[NH_3] = 0.0400 \text{ mol}/0.5000 \text{ L} = 0.0800 \text{ M}$
$NH_3(aq) + H_2O(l) \rightleftharpoons NH_4^+(aq) + OH^-(aq)$
$\Delta H^\circ = [\Delta H^\circ_f(NH_4^+) + \Delta H^\circ_f(OH^-)] - [\Delta H^\circ_f(NH_3) + \Delta H^\circ_f(H_2O)]$
$\Delta H^\circ = [(1 \text{ mol})(-132.5 \text{ kJ/mol}) + (1 \text{ mol})(-230.0 \text{ kJ/mol})]$
$- [(1 \text{ mol})(-80.3 \text{ kJ/mol}) + (1 \text{ mol})(-285.8 \text{ kJ/mol})] = +3.6 \text{ kJ}$
$\Delta S^\circ = [S^\circ(NH_4^+) + S^\circ(OH^-)] - [S^\circ(NH_3) + S^\circ(H_2O)]$
$\Delta S^\circ = [(1 \text{ mol})(113 \text{ J/(K} \cdot \text{mol)}) + (1 \text{ mol})(-10.8 \text{ J/(K} \cdot \text{mol)})]$
$- [(1 \text{ mol})(111 \text{ J/(K} \cdot \text{mol)}) + (1 \text{ mol})(69.9 \text{ J/(K} \cdot \text{mol)})] = -78.7 \text{ J/K}$

$T = 2.0\ °C = 2.0 + 273.1 = 275.1\ K$

$\Delta G° = \Delta H° - T\Delta S° = 3.6\ kJ - (275.1\ K)(-78.7 \times 10^{-3}\ kJ/K) = 25.3\ kJ$

$\Delta G° = -RT \ln K_b$

$$\ln K_b = \frac{-\Delta G°}{RT} = \frac{-25.3\ kJ/mol}{[8.314 \times 10^{-3}\ kJ/(K \cdot mol)](275.1\ K)} = -11.06$$

$K_b = e^{-11.06} = 1.6 \times 10^{-5}$

	$NH_3(aq)$ + $H_2O(l)$	$\rightleftarrows$	$NH_4^+(aq)$	+ $OH^-(aq)$
initial (M)	0.0800		0	~0
change (M)	−x		+x	+x
equil (M)	0.0800 − x		x	x

$$\text{at } 2\ °C,\ K_b = \frac{[NH_4^+][OH^-]}{[NH_3]} = 1.6 \times 10^{-5} = \frac{x^2}{0.0800 - x} \approx \frac{x^2}{0.0800}$$

$x^2 = (1.6 \times 10^{-5})(0.0800)$

$x = [OH^-] = \sqrt{(1.6 \times 10^{-5})(0.0800)} = 1.13 \times 10^{-3}\ M$

$$[H_3O^+] = \frac{1.0 \times 10^{-14}}{1.13 \times 10^{-3}} = 8.85 \times 10^{-12}\ M$$

$pH = -\log[H_3O^+] = -\log(8.85 \times 10^{-12}) = 11.05$

17.154 (a) $I_2(s) \rightarrow 2\ I^-(aq)$

$[I_2(s) + 2\ e^- \rightarrow 2\ I^-(aq)] \times 5$ reduction half reaction

$I_2(s) \rightarrow 2\ IO_3^-(aq)$

$I_2(s) + 6\ H_2O(l) \rightarrow 2\ IO_3^-(aq)$

$I_2(s) + 6\ H_2O(l) \rightarrow 2\ IO_3^-(aq) + 12\ H^+(aq)$

$I_2(s) + 6\ H_2O(l) \rightarrow 2\ IO_3^-(aq) + 12\ H^+(aq) + 10\ e^-$ oxidation half reaction

Combine the two half reactions.

$6\ I_2(s) + 6\ H_2O(l) \rightarrow 10\ I^-(aq) + 2\ IO_3^-(aq) + 12\ H^+(aq)$

Divide all coefficients by 2.

$3\ I_2(s) + 3\ H_2O(l) \rightarrow 5\ I^-(aq) + IO_3^-(aq) + 6\ H^+(aq)$

$3\ I_2(s) + 3\ H_2O(l) + 6\ OH^-(aq) \rightarrow 5\ I^-(aq) + IO_3^-(aq) + 6\ H^+(aq) + 6\ OH^-(aq)$

$3\ I_2(s) + 3\ H_2O(l) + 6\ OH^-(aq) \rightarrow 5\ I^-(aq) + IO_3^-(aq) + 6\ H_2O(l)$

$3\ I_2(s) + 6\ OH^-(aq) \rightarrow 5\ I^-(aq) + IO_3^-(aq) + 3\ H_2O(l)$

(b) $\Delta G° = [5\ \Delta G°_f(I^-) + \Delta G°_f(IO_3^-) + 3\ \Delta G°_f(H_2O(l))] - 6\ \Delta G°_f(OH^-)$

$\Delta G° = [(5\ mol)(-51.6\ kJ/mol) + (1\ mol)(-128.0\ kJ/mol) + (3\ mol)(-237.2\ kJ/mol)] - (6\ mol)(-157.3\ kJ/mol) = -153.8\ kJ$

(c) The reaction is spontaneous because $\Delta G°$ is negative.

(d) $25\ °C = 25 + 273 = 298\ K$

$\Delta G° = -RT \ln K_c$

$$\ln K_c = \frac{-\Delta G^\circ}{RT} = \frac{-(-153.8 \text{ kJ/mol})}{[8.314 \times 10^{-3} \text{ kJ/(K}\cdot\text{mol)}](298 \text{ K})} = 62.077$$

$K_c = e^{62.077} = 9.1 \times 10^{26}$

$$K_c = \frac{[I^-]^5[IO_3^-]}{[OH^-]^6} = 9.1 \times 10^{26} = \frac{(0.10)^5(0.50)}{[OH^-]^6}$$

$$[OH^-] = \sqrt[6]{\frac{(0.10)^5(0.50)}{9.1 \times 10^{26}}} = 4.2 \times 10^{-6} \text{ M}; \quad [H_3O^+] = \frac{1.0 \times 10^{-14}}{4.2 \times 10^{-6}} = 2.38 \times 10^{-9} \text{ M}$$

$pH = -\log[H_3O^+] = -\log(2.38 \times 10^{-9}) = 8.62$

17.155 (a) $N_2O_4(g) \rightleftarrows 2\ NO_2(g)$

$\Delta H^\circ = 2\ \Delta H^\circ_f(NO_2) - \Delta H^\circ_f(N_2O_4) = (2 \text{ mol})(33.2 \text{ kJ/mol}) - (1 \text{ mol})(11.1 \text{ kJ/mol}) = 55.3 \text{ kJ}$

$\Delta S^\circ = 2\ S^\circ(NO_2) - S^\circ(N_2O_4) = (2 \text{ mol})(240.0 \text{ J/(K}\cdot\text{mol)}) - (1 \text{ mol})(304.3 \text{ J/(K}\cdot\text{mol)})$

$\Delta S^\circ = 175.7 \text{ J/K} = 175.7 \times 10^{-3} \text{ kJ/K}$

$\Delta G^\circ = \Delta H^\circ - T\Delta S^\circ = 55.3 \text{ kJ} - (373 \text{ K})(175.7 \times 10^{-3} \text{ kJ/K}) = -10.2 \text{ kJ}$

$$K_p = \frac{(P_{NO_2})^2}{P_{N_2O_4}}$$

$$\Delta G^\circ = -RT \ln K_p; \quad \ln K_p = \frac{-\Delta G^\circ}{RT} = \frac{-(-10.2 \text{ kJ/mol})}{[8.314 \times 10^{-3} \text{ kJ/(K}\cdot\text{mol)}](373 \text{ K})} = 3.29$$

$K_p = e^{3.29} = 27$

	$N_2O_4(g)$	$\rightleftarrows$	$2\ NO_2(g)$
initial (atm)	1.00		1.00
change (atm)	–x		+2x
equil (atm)	1.00 – x		1.00 + 2x

$$K_p = \frac{(P_{NO_2})^2}{P_{N_2O_4}} = 27 = \frac{(1.00 + 2x)^2}{(1.00 - x)}$$

$4x^2 + 31x - 26 = 0$

Use the quadratic formula to solve for x.

$$x = \frac{-(31) \pm \sqrt{(31)^2 - (4)(4)(-26)}}{2(4)} = \frac{-31 \pm 37.1}{8}$$

x = 0.76 and –8.5

Of the two solutions for x, only 0.76 has physical meaning because –8.5 would lead to a negative partial pressure for NO_2.

$P_{N_2O_4} = 1.00 - x = 1.00 - 0.76 = 0.24$ atm; $P_{NO_2} = 1.00 + 2x = 1.00 + 2(0.76) = 2.52$ atm

(b) One resonance structure is shown here.

:Ö\ /Ö:
N—N
:O// \Ö:

Each N is sp^2 hybridized.
There is a trigonal planar geometry about each N.

18 Electrochemistry

18.1 (a) $MnO_4^-(aq) \rightarrow MnO_2(s)$ (reduction)

$IO_3^-(aq) \rightarrow IO_4^-(aq)$ (oxidation)

(b) $NO_3^-(aq) \rightarrow NO_2(g)$ (reduction)

$SO_2(aq) \rightarrow SO_4^{2-}(aq)$ (oxidation)

18.2 $ClO^-(aq) \rightarrow Cl^-(aq)$ Cl goes from +1 to −1 (reduction)

$I^-(aq) \rightarrow I_2(aq)$ I goes from −1 to 0 (oxidation)

$ClO^-(aq) + I^-(aq) \rightarrow Cl^-(aq) + I_2(aq)$ overall reaction (unbalanced)

18.3 $Fe(OH)_2(s) + O_2(g) \rightarrow Fe(OH)_3(s)$

$[Fe(OH)_2(s) + OH^-(aq) \rightarrow Fe(OH)_3(s) + e^-] \times 4$ (oxidation half reaction)

$O_2(g) \rightarrow 2\ H_2O(l)$

$4\ H^+(aq) + O_2(g) \rightarrow 2\ H_2O(l)$

$4\ e^- + 4\ H^+(aq) + O_2(g) \rightarrow 2\ H_2O(l)$

$4\ e^- + 4\ H^+(aq) + 4\ OH^-(aq) + O_2(g) \rightarrow 2\ H_2O(l) + 4\ OH^-(aq)$

$4\ e^- + 4\ H_2O(l) + O_2(g) \rightarrow 2\ H_2O(l) + 4\ OH^-(aq)$

$4\ e^- + 2\ H_2O(l) + O_2(g) \rightarrow 4\ OH^-(aq)$ (reduction half reaction)

Combine the two half reactions.

$4\ Fe(OH)_2(s) + 4\ OH^-(aq) + 2\ H_2O(l) + O_2(g) \rightarrow 4\ Fe(OH)_3(s) + 4\ OH^-(aq)$

$4\ Fe(OH)_2(s) + 2\ H_2O(l) + O_2(g) \rightarrow 4\ Fe(OH)_3(s)$

18.4 $NO_3^-(aq) + Cu(s) \rightarrow NO(g) + Cu^{2+}(aq)$

$[Cu(s) \rightarrow Cu^{2+}(aq) + 2\ e^-] \times 3$ (oxidation half reaction)

$NO_3^-(aq) \rightarrow NO(g)$

$NO_3^-(aq) \rightarrow NO(g) + 2\ H_2O(l)$

$4\ H^+(aq) + NO_3^-(aq) \rightarrow NO(g) + 2\ H_2O(l)$

$[3\ e^- + 4\ H^+(aq) + NO_3^-(aq) \rightarrow NO(g) + 2\ H_2O(l)] \times 2$ (reduction half reaction)

Combine the two half reactions.

$2\ NO_3^-(aq) + 8\ H^+(aq) + 3\ Cu(s) \rightarrow 3\ Cu^{2+}(aq) + 2\ NO(g) + 4\ H_2O(l)$

18.5 $2\ Ag^+(aq) + Ni(s) \rightarrow 2\ Ag(s) + Ni^{2+}(aq)$

There is a Ni anode in an aqueous solution of Ni^{2+}, and a Ag cathode in an aqueous solution of Ag^+. A salt bridge connects the anode and cathode compartment. The electrodes are connected through an external circuit.

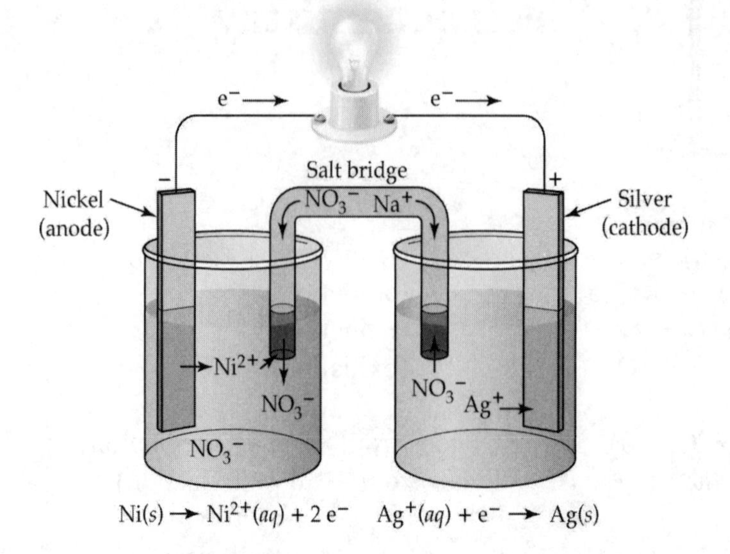

18.6

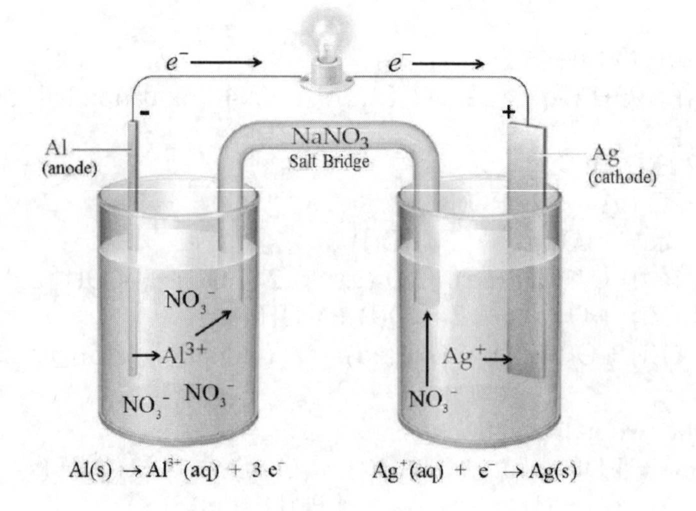

anode reaction	$Al(s) \rightarrow Al^{3+}(aq) + 3\ e^-$
cathode reaction	$\underline{3\ Ag^+(aq) + 3\ e^- \rightarrow 3\ Ag(s)}$
overall reaction	$Al(s) + 3\ Ag^+(aq) \rightarrow Al^{3+}(aq) + 3\ Ag(s)$

18.7 $Pb(s) + Br_2(l) \rightarrow Pb^{2+}(aq) + 2\ Br^-(aq)$

There is a Pb anode in an aqueous solution of Pb^{2+}. The cathode is a Pt wire that dips into a pool of liquid Br_2 and an aqueous solution that is saturated with Br_2. A salt bridge connects the anode and cathode compartment. The electrodes are connected through an external circuit.

18.8 (a) and (b)

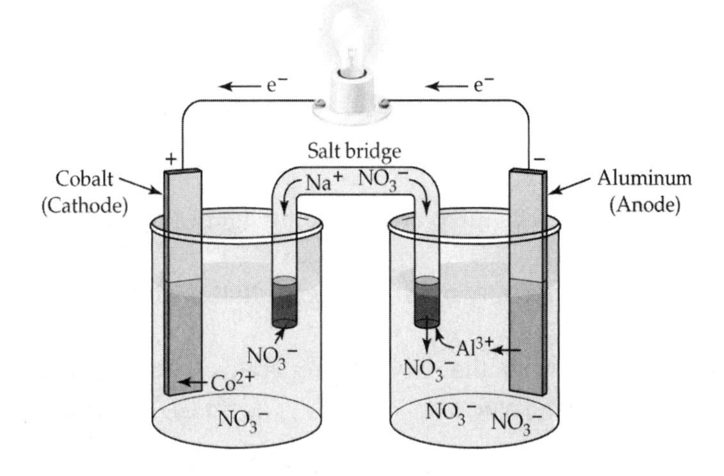

(c) $2\ Al(s) + 3\ Co^{2+}(aq) \rightarrow 2\ Al^{3+}(aq) + 3\ Co(s)$

(d) $Al(s) | Al^{3+}(aq) \| Co^{2+}(aq) | Co(s)$

18.9 $Cr_2O_7^{2-}(aq) + 3\ Sn^{2+}(aq) + 14\ H^+(aq) \rightarrow 2\ Cr^{3+}(aq) + 3\ Sn^{4+}(aq) + 7\ H_2O(l)$

$n = 6$ mol e^-

$$\Delta G^o = -nFE^o = -(6 \text{ mol } e^-)\left(\frac{96{,}500\ C}{1 \text{ mol } e^-}\right)(1.21\ V)\left(\frac{1\ J}{1\ C \cdot V}\right) = -700{,}590\ J = -701\ kJ$$

18.10 $Hg(l) + I_2(s) \rightarrow Hg^{2+}(aq) + 2\ I^-(aq)$

$n = 2$ mol e^- and $1\ J/C = 1V$

$\Delta G^o = 59.8\ kJ = 59{,}800\ J = -nFE^o$

$$E^o = \frac{-\Delta G^o}{nF} = \frac{-(59{,}800\ J)}{(2 \text{ mol } e^-)\left(\frac{96{,}500\ C}{1 \text{ mol } e^-}\right)} = -0.310\ J/C = -0.310\ V$$

Because $E^o < 0$, the reaction is nonspontaneous.

18.11

oxidation:	$Al(s) \rightarrow Al^{3+}(aq) + 3\ e^-$	$E^o = 1.66\ V$
reduction:	$Cr^{3+}(aq) + 3\ e^- \rightarrow Cr(s)$	$E^o = ?$
overall	$Al(s) + Cr^{3+}(aq) \rightarrow Al^{3+}(aq) + Cr(s)$	$E^o = 0.92\ V$

The standard reduction potential for the Cr^{3+}/Cr half cell is:

$E^o = 0.92 - 1.66 = -0.74\ V$

18.12 (a) $Cl_2(g) + 2\ e^- \rightarrow 2\ Cl^-(aq)$ $E^o = 1.36\ V$

$Ag^+(aq) + e^- \rightarrow Ag(s)$ $E^o = 0.80\ V$

Cl_2 has the greater tendency to be reduced (larger E^o). The species that has the greater tendency to be reduced is the stronger oxidizing agent. Cl_2 is the stronger oxidizing agent.

(b) $Fe^{2+}(aq) + 2\ e^- \rightarrow Fe(s)$ $E^o = -0.45\ V$

$Mg^{2+}(aq) + 2\ e^- \rightarrow Mg(s)$ $E^o = -2.37\ V$

The second half-reaction has the lesser tendency to occur in the forward direction (more negative E°) and the greater tendency to occur in the reverse direction. Therefore, Mg is the stronger reducing agent.

18.13 (a) D is the strongest reducing agent. D^+ has the most negative standard reduction potential. A^{3+} is the strongest oxidizing agent. It has the most positive standard reduction potential.
(b) An oxidizing agent can oxidize any reducing agent that is below it in the table. B^{2+} can oxidize C and D.
A reducing agent can reduce any oxidizing agent that is above it in the table. C can reduce A^{3+} and B^{2+}.
(c) Use the two half-reactions that have the most positive and the most negative standard reduction potentials, respectively.

$$A^{3+} + 2e^- \rightarrow A^+ \qquad 1.47\text{ V}$$
$$2 \times (D \rightarrow D^+ + e^-) \qquad 1.38\text{ V}$$
$$A^{3+} + 2D \rightarrow A^+ + 2D^+ \qquad 2.85\text{ V}$$

18.14 (a) $2\,Fe^{3+}(aq) + 2\,I^-(aq) \rightarrow 2\,Fe^{2+}(aq) + I_2(s)$
reduction: $Fe^{3+}(aq) + e^- \rightarrow Fe^{2+}(aq)$ E° = 0.77 V
oxidation: $2\,I^-(aq) \rightarrow I_2(s) + 2\,e^-$ E° = −0.54 V
overall E° = 0.23 V
Because E° for the overall reaction is positive, this reaction can occur under standard-state conditions.
(b) $3\,Ni(s) + 2\,Al^{3+}(aq) \rightarrow 3\,Ni^{2+}(aq) + 2\,Al(s)$
oxidation: $Ni(s) \rightarrow Ni^{2+}(aq) + 2\,e^-$ E° = 0.26 V
reduction: $Al^{3+}(aq) + 3\,e^- \rightarrow Al(s)$ E° = −1.66 V
overall E° = −1.40 V
Because E° for the overall reaction is negative, this reaction cannot occur under standard-state conditions. This reaction can occur in the reverse direction.

18.15 (a) $Ni(s) + 2\,Ag^+(aq) \rightarrow Ni^{2+}(aq) + 2\,Ag(s)$
oxidation: $Ni(s) \rightarrow Ni^{2+}(aq) + 2\,e^-$ E° = 0.26 V
reduction: $Ag^+(aq) + e^- \rightarrow Ag(s)$ E° = 0.80 V
overall E° = 1.06 V
(b) $Ni(s)|Ni^{2+}(aq)(1.0\text{ M})\|Ag^+(aq)(1.0\text{ M})|Ag(s)$
(c) Ni(s) is the anode and Ag(s) is the cathode.

18.16 $Cu(s) + 2\,Fe^{3+}(aq) \rightarrow Cu^{2+}(aq) + 2\,Fe^{2+}(aq)$

$E^\circ = E^\circ_{Cu \rightarrow Cu^{2+}} + E^\circ_{Fe^{3+} \rightarrow Fe^{2+}} = -0.34\text{ V} + 0.77\text{ V} = 0.43\text{ V}$; n = 2 mol e^-

$$E = E^\circ - \frac{0.0592\text{ V}}{n}\log\frac{[Cu^{2+}][Fe^{2+}]^2}{[Fe^{3+}]^2} = 0.43\text{ V} - \frac{(0.0592\text{ V})}{2}\log\frac{(0.25)(0.20)^2}{(1.0 \times 10^{-4})^2} = 0.25\text{ V}$$

18.17 (a) anode: $4[Al(s) \rightarrow Al^{3+}(aq) + 3\,e^-]$ E° = 1.66 V
cathode: $3[O_2(g) + 4\,H^+(aq) + 4\,e^- \rightarrow 2\,H_2O(l)]$ E° = 1.23 V
overall: $4\,Al(s) + 3\,O_2(g) + 12\,H^+(aq) \rightarrow 4\,Al^{3+}(aq) + 6\,H_2O(l)$ E° = 2.89 V

(b) & (c) $E = E^\circ - \dfrac{2.303\,RT}{nF}\log\dfrac{[Al^{3+}]^4}{(P_{O_2})^3[H^+]^{12}}$

$$E = 2.89\ V - \frac{(2.303)\left(8.314\ \frac{J}{K\cdot mol}\right)(310\ K)}{(12\ mol\ e^-)(96{,}500\ C/mol\ e^-)}\log\left(\frac{(1.0\ x\ 10^{-9})^4}{(0.20)^3(1.0\ x\ 10^{-7})^{12}}\right)$$

$E = 2.89\ V - 0.257\ V = 2.63\ V$

18.18 $5\,[Cu(s) \rightarrow Cu^{2+}(aq) + 2\,e^-]$ (oxidation half reaction)
$2\,[5\,e^- + 8\,H^+(aq) + MnO_4^-(aq) \rightarrow Mn^{2+}(aq) + 4\,H_2O(l)]$ (reduction half reaction)

$5\,Cu(s) + 16\,H^+(aq) + 2\,MnO_4^-(aq) \rightarrow 5\,Cu^{2+}(aq) + 2\,Mn^{2+}(aq) + 8\,H_2O(l)$

$$\Delta E = -\frac{0.0592\ V}{n}\log\frac{[Cu^{2+}]^5[Mn^{2+}]^2}{[MnO_4^-]^2[H^+]^{16}}$$

(a) The anode compartment contains Cu^{2+}.

$$\Delta E = -\frac{0.0592\ V}{10}\log\frac{(0.01)^5(1)^2}{(1)^2(1)^{16}} = +0.059\ V$$

(b) The cathode compartment contains Mn^{2+}, MnO_4^-, and H^+.

$$\Delta E = -\frac{0.0592\ V}{10}\log\frac{(1)^5(0.01)^2}{(0.01)^2(0.01)^{16}} = -0.19\ V$$

18.19 $Zn(s) + Cu^{2+}(aq) \rightarrow Zn^{2+}(aq) + Cu(s)$

oxidation: $Zn(s) \rightarrow Zn^{2+}(aq) + 2\,e^-$ $E^\circ = 0.76\ V$
reduction: $Cu^{2+}(aq) + 2\,e^- \rightarrow Cu(s)$ $\underline{E^\circ = 0.34\ V}$
overall $E^\circ = 1.10\ V$

$$E = 1.16\ V = E^\circ - \frac{0.0592\ V}{n}\log\frac{[Zn^{2+}]}{[Cu^{2+}]} = 1.10\ V - \frac{(0.0592\ V)}{2}\log\frac{[Zn^{2+}]}{[Cu^{2+}]}$$

$$\frac{(1.16\ V - 1.10\ V)}{(-0.0592\ V/2)} = \log\frac{[Zn^{2+}]}{[Cu^{2+}]}$$

$$\log\frac{[Zn^{2+}]}{[Cu^{2+}]} = -2.03 \quad\text{and}\quad \frac{[Zn^{2+}]}{[Cu^{2+}]} = 10^{-2.03} = 9.3\ x\ 10^{-3}$$

18.20 $H_2(g) + Pb^{2+}(aq) \rightarrow 2\,H^+(aq) + Pb(s)$

$E^\circ = E^\circ_{H_2 \rightarrow H^+} + E^\circ_{Pb^{2+} \rightarrow Pb} = 0\ V + (-0.13\ V) = -0.13\ V$; $n = 2$ mol e^-

$$E = E^\circ - \frac{0.0592\ V}{n}\log\frac{[H_3O^+]^2}{[Pb^{2+}](P_{H_2})}$$

$$0.28\ V = -0.13\ V - \frac{(0.0592\ V)}{2}\log\frac{[H_3O^+]^2}{(1)(1)} = -0.13\ V - (0.0592\ V)\log[H_3O^+]$$

$pH = -\log[H_3O^+]$ therefore $0.28\ V = -0.13\ V + (0.0592\ V)\ pH$

$$pH = \frac{(0.28\ V + 0.13\ V)}{0.0592\ V} = 6.9$$

18.21 $H_2(g) + Hg_2Cl_2(s) \rightarrow 2\ H^+(aq) + 2\ Hg(l) + 2Cl^-(aq)$

$E^o = E^o_{H_2 \rightarrow H^+} + E^o_{Hg_2Cl_2 \rightarrow Hg} = 0\ V + 0.28\ V = 0.28\ V$; $n = 2$ mol e^-

$$E = E^o - \frac{0.0592\ V}{n} \log \frac{[H_3O^+]^2[Cl^-]^2}{(P_{H_2})}$$

$$E = 0.28\ V - \frac{0.0592\ V}{2} \log \frac{(1.0 \times 10^{-7})^2(1.0)^2}{(1.0)} = 0.69\ V$$

18.22 $4\ Fe^{2+}(aq) + O_2(g) + 4\ H^+(aq) \rightarrow 4\ Fe^{3+}(aq) + 2\ H_2O(l)$

$E^o = E^o_{Fe^{2+} \rightarrow Fe^{3+}} + E^o_{O_2 \rightarrow H_2O} = -0.77\ V + 1.23\ V = 0.46\ V$; $n = 4$ mol e^-

$$E^o = \frac{0.0592\ V}{n} \log K; \quad \log K = \frac{nE^o}{0.0592\ V} = \frac{(4)(0.46\ V)}{0.0592\ V} = 31; \quad K = 10^{31} \text{ at } 25\ ^\circ C$$

18.23 $$E^o = \frac{0.0592\ V}{n} \log K = \frac{0.0592\ V}{2} \log(1.8 \times 10^{-5}) = -0.140\ V$$

18.24 (a) $Zn(s) + 2\ MnO_2(s) + 2\ NH_4^+(aq) \rightarrow Zn^{2+}(aq) + Mn_2O_3(s) + 2\ NH_3(aq) + H_2O(l)$

(b) $Zn(s) + 2\ MnO_2(s) \rightarrow ZnO(s) + Mn_2O_3(s)$

(c) $Cd(s) + 2\ NiO(OH)(s) + 2\ H_2O(l) \rightarrow Cd(OH)_2(s) + 2\ Ni(OH)_2(s)$

(d) $x\ Li(s) + MnO_2(s) \rightarrow Li_xMnO_2(s)$

(e) $Li_xC_6(s) + Li_{1-x}CoO_2(s) \rightarrow 6\ C(s) + LiCoO_2(s)$

18.25 (a) $[Mg(s) \rightarrow Mg^{2+}(aq) + 2\ e^-] \times 2$

$\underline{O_2(g) + 4\ H^+(aq) + 4\ e^- \rightarrow 2\ H_2O(l)}$

$2\ Mg(s) + O_2(g) + 4\ H^+(aq) \rightarrow 2\ Mg^{2+}(aq) + 2\ H_2O(l)$

(b) $[Fe(s) \rightarrow Fe^{2+}(aq) + 2\ e^-] \times 4$

$[O_2(g) + 4\ H^+(aq) + 4\ e^- \rightarrow 2\ H_2O(l)] \times 2$

$4\ Fe^{2+}(aq) + O_2(g) + 4\ H^+(aq) \rightarrow 4\ Fe^{3+}(aq) + 2\ H_2O(l)$

$\underline{[2\ Fe^{3+}(aq) + 4\ H_2O(l) \rightarrow Fe_2O_3 \cdot H_2O(s) + 6\ H^+(aq)] \times 2}$

$4\ Fe(s) + 3\ O_2(g) + 2\ H_2O(l) \rightarrow 2\ Fe_2O_3 \cdot H_2O(s)$

18.26 (a)

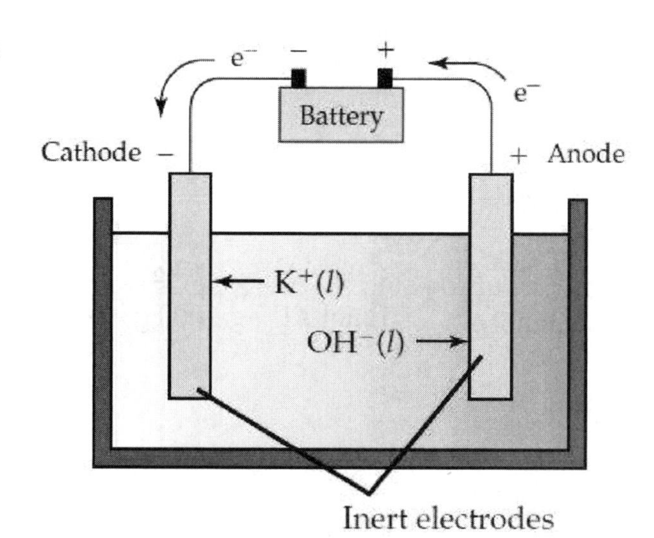

(b)

anode reaction	$4\ OH^-(l) \rightarrow O_2(g) + 2\ H_2O(l) + 4\ e^-$
cathode reaction	$4\ K^+(l) + 4\ e^- \rightarrow 4\ K(l)$
overall reaction	$4\ K^+(l) + 4\ OH^-(l) \rightarrow 4\ K(l) + O_2(g) + 2\ H_2O(l)$

18.27 (a)

anode reaction	$2\ Cl^-(aq) \rightarrow Cl_2(g) + 2\ e^-$
cathode reaction	$2\ H_2O(l) + 2\ e^- \rightarrow H_2(g) + 2\ OH^-(aq)$
overall reaction	$2\ Cl^-(aq) + 2\ H_2O(l) \rightarrow Cl_2(g) + H_2(g) + 2\ OH^-(aq)$

(b)

anode reaction	$2\ H_2O(l) \rightarrow O_2(g) + 4\ H^+(aq) + 4\ e^-$
cathode reaction	$2\ Cu^{2+}(aq) + 4\ e^- \rightarrow 2\ Cu(s)$
overall reaction	$2\ Cu^{2+}(aq) + 2\ H_2O(l) \rightarrow 2\ Cu(s) + O_2(g) + 4\ H^+(aq)$

(c)

anode reaction	$2\ H_2O(l) \rightarrow O_2(g) + 4\ H^+(aq) + 4\ e^-$
cathode reaction	$4\ H_2O(l) + 4\ e^- \rightarrow 2\ H_2(g) + 4\ OH^-(aq)$
overall reaction	$2\ H_2O(l) \rightarrow 2\ H_2(g) + O_2(g)$

18.28

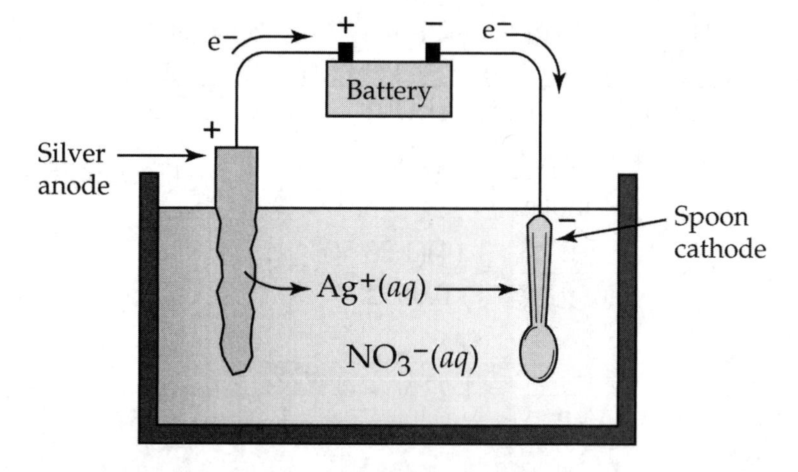

anode reaction	$Ag(s) \rightarrow Ag^+(aq) + e^-$
cathode reaction	$Ag^+(aq) + e^- \rightarrow Ag(s)$

The overall reaction is transfer of silver metal from the silver anode to the spoon.

18.29 $\text{Charge} = \left(1.00 \times 10^5\ \frac{C}{s}\right)(8.00\ h)\left(\frac{60\ min}{h}\right)\left(\frac{60\ s}{min}\right) = 2.88 \times 10^9\ C$

$\text{Moles of } e^- = (2.88 \times 10^9\ C)\left(\frac{1\ mol\ e^-}{96{,}500\ C}\right) = 2.98 \times 10^4\ mol\ e^-$

cathode reaction: $Al^{3+} + 3\ e^- \rightarrow Al$

$\text{mass Al} = (2.98 \times 10^4\ mol\ e^-) \times \frac{1\ mol\ Al}{3\ mol\ e^-} \times \frac{26.98\ g\ Al}{1\ mol\ Al} \times \frac{1\ kg}{1000\ g} = 268\ kg\ Al$

18.30 $3.00\ g\ Ag \times \frac{1\ mol\ Ag}{107.9\ g\ Ag} = 0.0278\ mol\ Ag$

cathode reaction: $Ag^+(aq) + e^- \rightarrow Ag(s)$

$\text{Charge} = (0.0278\ mol\ Ag)\left(\frac{1\ mol\ e^-}{1\ mol\ Ag}\right)\left(\frac{96{,}500\ C}{1\ mol\ e^-}\right) = 2682.7\ C$

$\text{Time} = \frac{C}{A} = \left(\frac{2682.7\ C}{0.100\ C/s} \times \frac{1\ h}{3600\ s}\right) = 7.45\ h$

18.31 A fuel cell and a battery are both galvanic cells that convert chemical energy into electrical energy utilizing a spontaneous redox reaction. A fuel cell differs from an ordinary battery in that the reactants are not contained within the cell but instead are continuously supplied from an external reservoir.

18.32 (a)

anode reaction	$2\ H_2(g) \rightarrow 4\ H^+(aq) + 4\ e^-$	$E^\circ = 0.00\ V$
cathode reaction	$O_2(g) + 4\ H^+(aq) + 4\ e^- \rightarrow 2\ H_2O(l)$	$E^\circ = 1.23\ V$
overall reaction	$2\ H_2(g) + O_2(g) \rightarrow 2\ H_2O(l)$	$E^\circ = 1.23\ V$

(b) $E = E^\circ - \frac{0.0592\ V}{n} \log \frac{1}{(P_{H_2})^2(P_{O_2})} = 1.23\ V - \frac{0.0592\ V}{4} \log \frac{1}{(6)^2(0.2)} = 1.24\ V$

18.33 (a)

anode reaction	$2\ H_2(g) \rightarrow 4\ H^+(aq) + 4\ e^-$	$E^\circ = 0.00\ V$
cathode reaction	$O_2(g) + 4\ H^+(aq) + 4\ e^- \rightarrow 2\ H_2O(l)$	$E^\circ = 1.23\ V$
overall reaction	$2\ H_2(g) + O_2(g) \rightarrow 2\ H_2O(l)$	$E^\circ = 1.23\ V$

(b) $\Delta G^\circ = -nFE^\circ = -(4\ mol\ e^-)\left(\frac{96{,}500\ C}{1\ mol\ e^-}\right)(1.23\ V)\left(\frac{1\ J}{1\ C \cdot V}\right) = -474{,}780\ J = -475\ kJ$

$E^\circ = \frac{0.0592\ V}{n} \log K;\ \log K = \frac{nE^\circ}{0.0592\ V} = \frac{(4)(1.23\ V)}{0.0592\ V} = 83.1$

$K = 10^{83.1} = 1.28 \times 10^{83}$

(c) $E = E^\circ - \frac{0.0592\ V}{n} \log \frac{1}{(P_{H_2})^2(P_{O_2})} = 1.23\ V - \frac{0.0592\ V}{4} \log \frac{1}{(25)^2(25)} = 1.29\ V$

18.34 $2\ CH_3OH(l) + 3\ O_2(g) \rightarrow 2\ CO_2(g) + 4\ H_2O(l)$

$\Delta G^\circ = [2\ \Delta G^\circ_f(CO_2) + 4\ \Delta G^\circ_f(H_2O)] - [2\ \Delta G^\circ_f(CH_3OH)]$

$\Delta G^\circ = [(2\ mol)(-394.4\ kJ/mol) + (4\ mol)(-237.2\ kJ/mol)] - (2\ mol)(-166.6\ kJ/mol)$

$\Delta G^\circ = -1404$ kJ

anode: $2\ CH_3OH(l) + 2\ H_2O(l) \rightarrow 2\ CO_2(g) + 12\ H^+(aq) + 12\ e^-$

cathode: $3\ O_2(g) + 12\ H^+(aq) + 12\ e^- \rightarrow 6\ H_2O(l)$

n = 12 mol e^- and 1 J = 1 C x 1 V

$$\Delta G^\circ = -nFE^\circ \qquad E^\circ = \frac{-\Delta G^\circ}{nF} = \frac{-(-1{,}404{,}000\ J)}{(12\ mol\ e^-)\left(\frac{96{,}500\ C}{1\ mol\ e^-}\right)} = +1.21\ J/C = +1.21\ V$$

$$E^\circ = \frac{0.0592\ V}{n}\log K;\ \log K = \frac{nE^\circ}{0.0592\ V} = \frac{(12)(1.21\ V)}{0.0592\ V} = 245;\ K = 10^{245} = 1 \times 10^{245}$$

18.35 (a) H is reduced and C is oxidized.

(b) H is reduced and C is oxidized. H_2O is the oxidizing agent and CO is the reducing agent.

(c) The drawback of the process is the high reaction temperatures, which require a lot of energy and the greenhouse gas CO_2 is also produced.

18.36 $6\ H_2O(l) \rightarrow 2\ H_2(g) + O_2(g) + 4\ H^+(aq) + 4\ OH^-(aq)$

1 A = 1 C/s

(a) $$\text{mol } e^- = 250.0\ \frac{C}{s} \times 30\ min \times \frac{60\ s}{min} \times \frac{1\ mol\ e^-}{96{,}500\ C} = 4.66\ mol\ e^-$$

$$\text{mass } H_2 = 4.66\ mol\ e^- \times \frac{2\ mol\ H_2}{4\ mol\ e^-} \times \frac{2.02\ g\ H_2}{1\ mol\ H_2} = 4.71\ g\ H_2$$

(b) $$\text{charge} = 25\ mol\ O_2 \times \frac{4\ mol\ e^-}{1\ mol\ O_2} \times \frac{96{,}500\ C}{1\ mol\ e^-} = 9.65 \times 10^6\ C$$

$$\text{time} = \frac{9.65 \times 10^6\ C}{500.0\ C/s} \times \frac{1\ min}{60\ s} \times \frac{1\ h}{60\ min} = = 5.36\ h$$

Conceptual Problems

18.37 (a) - (d)

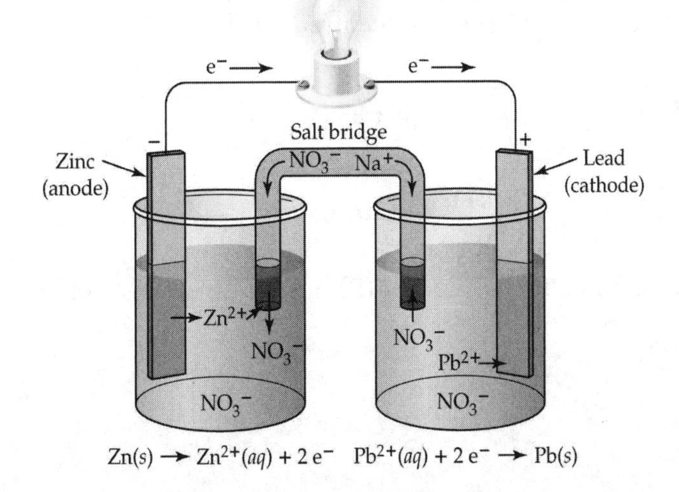

(e) anode reaction $Zn(s) \rightarrow Zn^{2+}(aq) + 2\ e^-$
cathode reaction $Pb^{2+}(aq) + 2\ e^- \rightarrow Pb(s)$
overall reaction $Zn(s) + Pb^{2+}(aq) \rightarrow Zn^{2+}(aq) + Pb(s)$

18.38 (a) anode is Ni; cathode is Pt
(b) anode reaction $3\ Ni(s) \rightarrow 3\ Ni^{2+}(aq) + 6\ e^-$
cathode reaction $Cr_2O_7^{2-}(aq) + 14\ H^+(aq) + 6\ e^- \rightarrow 2\ Cr^{3+}(aq) + 7\ H_2O(l)$
overall reaction $Cr_2O_7^{2-}(aq) + 3\ Ni(s) + 14\ H^+(aq) \rightarrow$
$2\ Cr^{3+}(aq) + 3\ Ni^{2+}(aq) + 7\ H_2O(l)$
(c) $Ni(s)|Ni^{2+}(aq)\ \|Cr_2O_7^{2-}(aq), Cr^{3+}|Pt(s)$

18.39 (a) The three cell reactions are the same except for cation concentrations.

anode reaction	$Cu(s) \rightarrow Cu^{2+}(aq) + 2\ e^-$	$E^\circ = -0.34$ V
cathode reaction	$2\ Fe^{3+}(aq) + 2\ e^- \rightarrow 2\ Fe^{2+}(aq)$	$E^\circ = 0.77$ V
overall reaction	$Cu(s) + 2\ Fe^{3+}(aq) \rightarrow Cu^{2+}(aq) + 2\ Fe^{2+}(aq)$	$E^\circ = 0.43$ V

(b)

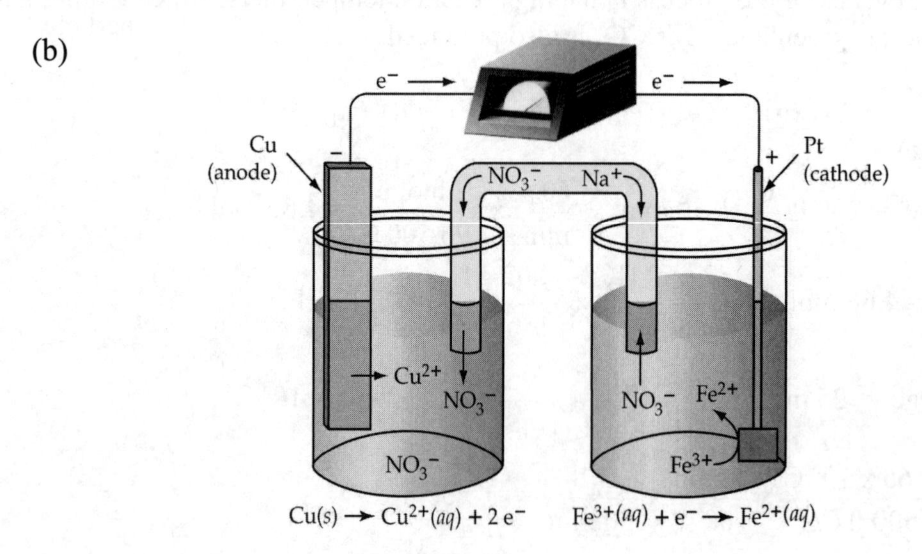

(c) $E = E^\circ - \dfrac{0.0592\ V}{n} \log \dfrac{[Cu^{2+}][Fe^{2+}]^2}{[Fe^{2+}]^2}$; n = 2 mol e^-

(1) $E = E^\circ = 0.43$ V because all cation concentrations are 1 M.

(2) $E = E^\circ - \dfrac{0.0592\ V}{2} \log \dfrac{(1)(5)^2}{(1)^2} = 0.39$ V

(3) $E = E^\circ - \dfrac{0.0592\ V}{2} \log \dfrac{(0.1)(0.1)^2}{(0.1)^2} = 0.46$ V

Cell (3) has the largest potential, while cell (2) has the smallest as calculated from the Nernst equation.

18.40 (a) - (b)

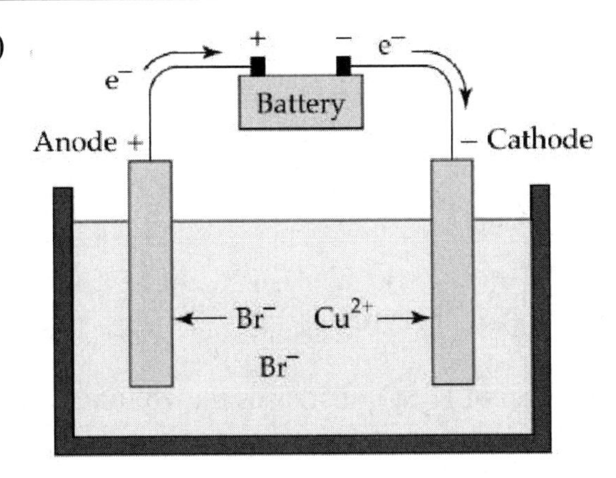

(c) anode reaction $2\ Br^-(aq) \rightarrow Br_2(aq) + 2\ e^-$

cathode reaction $Cu^{2+}(aq) + 2\ e^- \rightarrow Cu(s)$

overall reaction $Cu^{2+}(aq) + 2\ Br^-(aq) \rightarrow Cu(s) + Br_2(aq)$

18.41 (a) This is an electrolytic cell that has a battery connected between two inert electrodes.

(b)

+ − e⁻

e⁻

Battery

Anode + − Cathode

$Ni^{2+}(aq)$

$SO_4^{2-}(aq)$

Inert electrodes

(c) anode reaction $2\ H_2O(l) \rightarrow O_2(g) + 4\ H^+(aq) + 4\ e^-$

cathode reaction $Ni^{2+}(aq) + 2\ e^- \rightarrow Ni(s)$

overall reaction $2\ Ni^{2+}(aq) + 2\ H_2O(l) \rightarrow 2\ Ni(s) + O_2(g) + 4\ H^+(aq)$

18.42 (a) & (b)

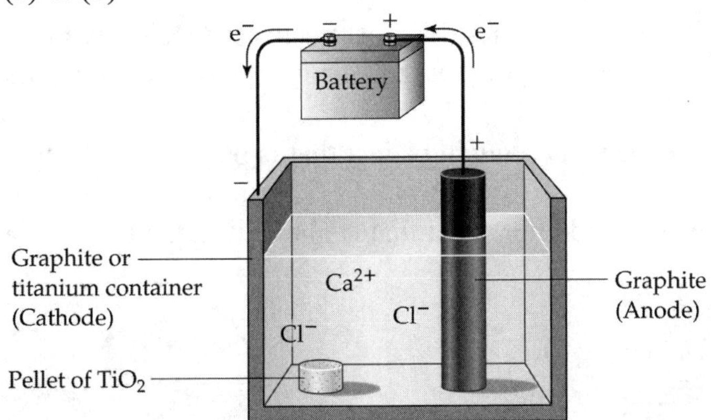

(c) anode reaction $2\,O^{2-} \rightarrow O_2(g) + 4\,e^-$
cathode reaction $TiO_2(s) + 4\,e^- \rightarrow Ti(s) + 2\,O^{2-}$
overall reaction $TiO_2(s) \rightarrow Ti(s) + O_2(g)$

18.43 $Zn(s) + Cu^{2+}(aq) \rightarrow Zn^{2+}(aq) + Cu(s)$; $E = E^\circ - \frac{0.0592\text{ V}}{2}\log\frac{[Zn^{2+}]}{[Cu^{2+}]}$

(a) E increases because increasing $[Cu^{2+}]$ decreases $\log\frac{[Zn^{2+}]}{[Cu^{2+}]}$.

(b) E will decrease because addition of H_2SO_4 increases the volume which, decreases $[Cu^{2+}]$ and increases $\log\frac{[Zn^{2+}]}{[Cu^{2+}]}$.

(c) E decreases because increasing $[Zn^{2+}]$ increases $\log\frac{[Zn^{2+}]}{[Cu^{2+}]}$.

(d) Because there is no change in $[Zn^{2+}]$, there is no change in E.

18.44 $Cu(s) + 2\,Ag^+(aq) \rightarrow Cu^{2+}(aq) + 2\,Ag(s)$; $E = E^\circ - \frac{0.0592\text{ V}}{2}\log\frac{[Cu^{2+}]}{[Ag^+]^2}$

(a) E decreases because addition of NaCl precipitates AgCl, which decreases $[Ag^+]$ and increases $\log\frac{[Cu^{2+}]}{[Ag^+]^2}$.

(b) E increases because addition of NaCl increases the volume, which decreases $[Cu^{2+}]$ and decreases $\log\frac{[Cu^{2+}]}{[Ag^+]^2}$.

(c) E decreases because addition of NH_3 complexes Ag^+, yielding $Ag(NH_3)_2^+$, which decreases $[Ag^+]$ and increases $\log\frac{[Cu^{2+}]}{[Ag^+]^2}$.

(d) E increases because addition of NH_3 complexes Cu^{2+}, yielding $Cu(NH_3)_4^{2+}$, which decreases $[Cu^{2+}]$ and decreases $\log\frac{[Cu^{2+}]}{[Ag^+]^2}$.

18.45 (a) A^+ is the strongest oxidizing agent. It has the most positive standard reduction potential. D is the strongest reducing agent. D^{3+} has the most negative standard reduction potential.

(b) An oxidizing agent can oxidize any reducing agent that is below it in the table. B^{2+} can oxidize C^- and D.

A reducing agent can reduce any oxidizing agent that is above it in the table. D can reduce A^+, B^{2+}, and C_2.

(c) $3 \times (C_2 + 2\,e^- \rightarrow 2\,C^-)$ 0.17 V
$2 \times (D \rightarrow D^{3+} + 3\,e^-)$ 1.36 V
$3\,C_2 + 2\,D \rightarrow 6\,C^- + 2\,D^{3+}$ 1.53 V

Section Problems
Balancing Redox Reactions (Section 18.1)

18.46 (a) N oxidation number decreases from +5 to +2; reduction.
(b) Zn oxidation number increases from 0 to +2; oxidation.
(c) Ti oxidation number increases from +3 to +4; oxidation.
(d) Sn oxidation number decreases from +4 to +2; reduction.

18.47 (a) O oxidation number decreases from 0 to –2; reduction.
(b) O oxidation number increases from –1 to 0; oxidation.
(c) Mn oxidation number decreases from +7 to +6; reduction.
(d) C oxidation number increases from –2 to 0; oxidation.

18.48 (a) $NO_3^-(aq) \rightarrow NO(g)$
$NO_3^-(aq) \rightarrow NO(g) + 2\ H_2O(l)$
$4\ H^+(aq) + NO_3^-(aq) \rightarrow NO(g) + 2\ H_2O(l)$
$3\ e^- + 4\ H^+(aq) + NO_3^-(aq) \rightarrow NO(g) + 2\ H_2O(l)$
(b) $Zn(s) \rightarrow Zn^{2+}(aq) + 2\ e^-$
(c) $Ti^{3+}(aq) \rightarrow TiO_2(s)$
$Ti^{3+}(aq) + 2\ H_2O(l) \rightarrow TiO_2(s)$
$Ti^{3+}(aq) + 2\ H_2O(l) \rightarrow TiO_2(s) + 4\ H^+(aq)$
$Ti^{3+}(aq) + 2\ H_2O(l) \rightarrow TiO_2(s) + 4\ H^+(aq) + e^-$
(d) $Sn^{4+}(aq) + 2\ e^- \rightarrow Sn^{2+}(aq)$

18.49 (a) $O_2(g) \rightarrow OH^-(aq)$
$O_2(g) \rightarrow OH^-(aq) + H_2O(l)$
$3\ H^+(aq) + O_2(g) \rightarrow OH^-(aq) + H_2O(l)$
$3\ H^+(aq) + 3\ OH^-(aq) + O_2(g) \rightarrow 4\ OH^-(aq) + H_2O(l)$
$3\ H_2O(l) + O_2(g) \rightarrow 4\ OH^-(aq) + H_2O(l)$
$4\ e^- + 2\ H_2O(l) + O_2(g) \rightarrow 4\ OH^-(aq)$
(b) $H_2O_2(aq) \rightarrow O_2(g)$
$H_2O_2(aq) \rightarrow O_2(g) + 2\ H^+(aq)$
$2\ OH^-(aq) + H_2O_2(aq) \rightarrow O_2(g) + 2\ H^+(aq) + 2\ OH^-(aq)$
$2\ OH^-(aq) + H_2O_2(aq) \rightarrow O_2(g) + 2\ H_2O(l) + 2\ e^-$
(c) $MnO_4^-(aq) \rightarrow MnO_4^{2-}(aq)$
$MnO_4^-(aq) + e^- \rightarrow MnO_4^{2-}(aq)$
(d) $CH_3OH(aq) \rightarrow CH_2O(aq)$
$CH_3OH(aq) \rightarrow CH_2O(aq) + 2\ H^+(aq)$
$CH_3OH(aq) + 2\ OH^-(aq) \rightarrow CH_2O(aq) + 2\ H^+(aq) + 2\ OH^-(aq)$
$CH_3OH(aq) + 2\ OH^-(aq) \rightarrow CH_2O(aq) + 2\ H_2O(l)$
$CH_3OH(aq) + 2\ OH^-(aq) \rightarrow CH_2O(aq) + 2\ H_2O(l) + 2\ e^-$

18.50 (a) $Te(s) + NO_3^-(aq) \rightarrow TeO_2(s) + NO(g)$
oxidation: $Te(s) \rightarrow TeO_2(s)$
reduction: $NO_3^-(aq) \rightarrow NO(g)$

(b) $H_2O_2(aq) + Fe^{2+}(aq) \rightarrow Fe^{3+}(aq) + H_2O(l)$
oxidation: $Fe^{2+}(aq) \rightarrow Fe^{3+}(aq)$
reduction: $H_2O_2(aq) \rightarrow H_2O(l)$

18.51 (a) $Mn(s) + NO_3^-(aq) \rightarrow Mn^{2+}(aq) + NO_2(g)$
oxidation: $Mn(s) \rightarrow Mn^{2+}(aq)$
reduction: $NO_3^-(aq) \rightarrow NO_2(g)$
(b) $Mn^{3+}(aq) \rightarrow MnO_2(s) + Mn^{2+}(aq)$
oxidation: $Mn^{3+}(aq) \rightarrow MnO_2(s)$
reduction: $Mn^{3+}(aq) \rightarrow Mn^{2+}(aq)$

18.52 (a) $Cr_2O_7^{2-}(aq) \rightarrow Cr^{3+}(aq)$
$Cr_2O_7^{2-}(aq) \rightarrow 2\ Cr^{3+}(aq)$
$Cr_2O_7^{2-}(aq) \rightarrow 2\ Cr^{3+}(aq) + 7\ H_2O(l)$
$14\ H^+(aq) + Cr_2O_7^{2-}(aq) \rightarrow 2\ Cr^{3+}(aq) + 7\ H_2O(l)$
$14\ H^+(aq) + Cr_2O_7^{2-}(aq) + 6\ e^- \rightarrow 2\ Cr^{3+}(aq) + 7\ H_2O(l)$
(b) $CrO_4^{2-}(aq) \rightarrow Cr(OH)_4^-(aq)$
$4\ H^+(aq) + CrO_4^{2-}(aq) \rightarrow Cr(OH)_4^-(aq)$
$4\ H^+(aq) + 4\ OH^-(aq) + CrO_4^{2-}(aq) \rightarrow Cr(OH)_4^-(aq) + 4\ OH^-(aq)$
$4\ H_2O(l) + CrO_4^{2-}(aq) \rightarrow Cr(OH)_4^-(aq) + 4\ OH^-(aq)$
$4\ H_2O(l) + CrO_4^{2-}(aq) + 3\ e^- \rightarrow Cr(OH)_4^-(aq) + 4\ OH^-(aq)$
(c) $Bi^{3+}(aq) \rightarrow BiO_3^-(aq)$
$Bi^{3+}(aq) + 3\ H_2O(l) \rightarrow BiO_3^-(aq)$
$Bi^{3+}(aq) + 3\ H_2O(l) \rightarrow BiO_3^-(aq) + 6\ H^+(aq)$
$Bi^{3+}(aq) + 3\ H_2O(l) + 6\ OH^-(aq) \rightarrow BiO_3^-(aq) + 6\ H^+(aq) + 6\ OH^-(aq)$
$Bi^{3+}(aq) + 3\ H_2O(l) + 6\ OH^-(aq) \rightarrow BiO_3^-(aq) + 6\ H_2O(l)$
$Bi^{3+}(aq) + 6\ OH^-(aq) \rightarrow BiO_3^-(aq) + 3\ H_2O(l)$
$Bi^{3+}(aq) + 6\ OH^-(aq) \rightarrow BiO_3^-(aq) + 3\ H_2O(l) + 2\ e^-$
(d) $ClO^-(aq) \rightarrow Cl^-(aq)$
$ClO^-(aq) \rightarrow Cl^-(aq) + H_2O(l)$
$2\ H^+(aq) + ClO^-(aq) \rightarrow Cl^-(aq) + H_2O(l)$
$2\ H^+(aq) + 2\ OH^-(aq) + ClO^-(aq) \rightarrow Cl^-(aq) + H_2O(l) + 2\ OH^-(aq)$
$2\ H_2O(l) + ClO^-(aq) \rightarrow Cl^-(aq) + H_2O(l) + 2\ OH^-(aq)$
$H_2O(l) + ClO^-(aq) \rightarrow Cl^-(aq) + 2\ OH^-(aq)$
$H_2O(l) + ClO^-(aq) + 2\ e^- \rightarrow Cl^-(aq) + 2\ OH^-(aq)$

18.53 (a) $VO^{2+}(aq) \rightarrow V^{3+}(aq)$
$VO^{2+}(aq) \rightarrow V^{3+}(aq) + H_2O(l)$
$2\ H^+(aq) + VO^{2+}(aq) \rightarrow V^{3+}(aq) + H_2O(l)$
$2\ H^+(aq) + VO^{2+}(aq) + e^- \rightarrow V^{3+}(aq) + H_2O(l)$
(b) $Ni(OH)_2(s) \rightarrow Ni_2O_3(s)$
$2\ Ni(OH)_2(s) \rightarrow Ni_2O_3(s) + H_2O(l)$
$2\ Ni(OH)_2(s) \rightarrow Ni_2O_3(s) + H_2O(l) + 2\ H^+(aq)$
$2\ Ni(OH)_2(s) + 2\ OH^-(aq) \rightarrow Ni_2O_3(s) + H_2O(l) + 2\ H^+(aq) + 2\ OH^-(aq)$
$2\ Ni(OH)_2(s) + 2\ OH^-(aq) \rightarrow Ni_2O_3(s) + 3\ H_2O(l) + 2\ e^-$

(c) $NO_3^-(aq) \rightarrow NO_2(g)$
$NO_3^-(aq) \rightarrow NO_2(g) + H_2O(l)$
$2\,H^+(aq) + NO_3^-(aq) \rightarrow NO_2(g) + H_2O(l)$
$2\,H^+(aq) + NO_3^-(aq) + e^- \rightarrow NO_2(g) + H_2O(l)$

(d) $Br_2(aq) \rightarrow BrO_3^-(aq)$
$Br_2(aq) \rightarrow 2\,BrO_3^-(aq)$
$Br_2(aq) + 6\,H_2O(l) \rightarrow 2\,BrO_3^-(aq)$
$Br_2(aq) + 6\,H_2O(l) \rightarrow 2\,BrO_3^-(aq) + 12\,H^+(aq)$
$Br_2(aq) + 6\,H_2O(l) + 12\,OH^-(aq) \rightarrow 2\,BrO_3^-(aq) + 12\,H^+(aq) + 12\,OH^-(aq)$
$Br_2(aq) + 6\,H_2O(l) + 12\,OH^-(aq) \rightarrow 2\,BrO_3^-(aq) + 12\,H_2O(l)$
$Br_2(aq) + 12\,OH^-(aq) \rightarrow 2\,BrO_3^-(aq) + 6\,H_2O(l) + 10\,e^-$

18.54 (a) $MnO_4^-(aq) \rightarrow MnO_2(s)$
$MnO_4^-(aq) \rightarrow MnO_2(s) + 2\,H_2O(l)$
$4\,H^+(aq) + MnO_4^-(aq) \rightarrow MnO_2(s) + 2\,H_2O(l)$
$[4\,H^+(aq) + MnO_4^-(aq) + 3\,e^- \rightarrow MnO_2(s) + 2\,H_2O(l)] \times 2$ (reduction half reaction)

$IO_3^-(aq) \rightarrow IO_4^-(aq)$
$H_2O(l) + IO_3^-(aq) \rightarrow IO_4^-(aq)$
$H_2O(l) + IO_3^-(aq) \rightarrow IO_4^-(aq) + 2\,H^+(aq)$
$[H_2O(l) + IO_3^-(aq) \rightarrow IO_4^-(aq) + 2\,H^+(aq) + 2\,e^-] \times 3$ (oxidation half reaction)

Combine the two half reactions.
$8\,H^+(aq) + 3\,H_2O(l) + 2\,MnO_4^-(aq) + 3\,IO_3^-(aq) \rightarrow$
$6\,H^+(aq) + 4\,H_2O(l) + 2\,MnO_2(s) + 3\,IO_4^-(aq)$
$2\,H^+(aq) + 2\,MnO_4^-(aq) + 3\,IO_3^-(aq) \rightarrow 2\,MnO_2(s) + 3\,IO_4^-(aq) + H_2O(l)$
$2\,H^+(aq) + 2\,OH^-(aq) + 2\,MnO_4^-(aq) + 3\,IO_3^-(aq) \rightarrow$
$2\,MnO_2(s) + 3\,IO_4^-(aq) + H_2O(l) + 2\,OH^-(aq)$
$2\,H_2O(l) + 2\,MnO_4^-(aq) + 3\,IO_3^-(aq) \rightarrow$
$2\,MnO_2(s) + 3\,IO_4^-(aq) + H_2O(l) + 2\,OH^-(aq)$
$H_2O(l) + 2\,MnO_4^-(aq) + 3\,IO_3^-(aq) \rightarrow 2\,MnO_2(s) + 3\,IO_4^-(aq) + 2\,OH^-(aq)$

(b) $Cu(OH)_2(s) \rightarrow Cu(s)$
$Cu(OH)_2(s) \rightarrow Cu(s) + 2\,H_2O(l)$
$2\,H^+(aq) + Cu(OH)_2(s) \rightarrow Cu(s) + 2\,H_2O(l)$
$[2\,H^+(aq) + Cu(OH)_2(s) + 2\,e^- \rightarrow Cu(s) + 2\,H_2O(l)] \times 2$ (reduction half reaction)

$N_2H_4(aq) \rightarrow N_2(g)$
$N_2H_4(aq) \rightarrow N_2(g) + 4\,H^+(aq)$
$N_2H_4(aq) \rightarrow N_2(g) + 4\,H^+(aq) + 4\,e^-$ (oxidation half reaction)

Combine the two half reactions.
$4\,H^+(aq) + 2\,Cu(OH)_2(s) + N_2H_4(aq) \rightarrow 2\,Cu(s) + 4\,H_2O(l) + N_2(g) + 4\,H^+(aq)$
$2\,Cu(OH)_2(s) + N_2H_4(aq) \rightarrow 2\,Cu(s) + 4\,H_2O(l) + N_2(g)$

(c) $Fe(OH)_2(s) \rightarrow Fe(OH)_3(s)$
$Fe(OH)_2(s) + H_2O(l) \rightarrow Fe(OH)_3(s)$
$Fe(OH)_2(s) + H_2O(l) \rightarrow Fe(OH)_3(s) + H^+(aq)$
$[Fe(OH)_2(s) + H_2O(l) \rightarrow Fe(OH)_3(s) + H^+(aq) + e^-] \times 3$ (oxidation half reaction)

$CrO_4^{2-}(aq) \rightarrow Cr(OH)_4^-(aq)$
$4\,H^+(aq) + CrO_4^{2-}(aq) \rightarrow Cr(OH)_4^-(aq)$
$4\,H^+(aq) + CrO_4^{2-}(aq) + 3\,e^- \rightarrow Cr(OH)_4^-(aq)$ (reduction half reaction)

Combine the two half reactions.
$3\,Fe(OH)_2(s) + 3\,H_2O(l) + 4\,H^+(aq) + CrO_4^{2-}(aq) \rightarrow$
$3\,Fe(OH)_3(s) + 3\,H^+(aq) + Cr(OH)_4^-(aq)$
$3\,Fe(OH)_2(s) + 3\,H_2O(l) + H^+(aq) + CrO_4^{2-}(aq) \rightarrow 3\,Fe(OH)_3(s) + Cr(OH)_4^-(aq)$
$3\,Fe(OH)_2(s) + 3\,H_2O(l) + H^+(aq) + OH^-(aq) + CrO_4^{2-}(aq) \rightarrow$
$3\,Fe(OH)_3(s) + Cr(OH)_4^-(aq) + OH^-(aq)$
$3\,Fe(OH)_2(s) + 4\,H_2O(l) + CrO_4^{2-}(aq) \rightarrow 3\,Fe(OH)_3(s) + Cr(OH)_4^-(aq) + OH^-(aq)$

(d) $ClO_4^-(aq) \rightarrow ClO_2^-(aq)$
$ClO_4^-(aq) \rightarrow ClO_2^-(aq) + 2\,H_2O(l)$
$4\,H^+(aq) + ClO_4^-(aq) \rightarrow ClO_2^-(aq) + 2\,H_2O(l)$
$4\,H^+(aq) + ClO_4^-(aq) + 4\,e^- \rightarrow ClO_2^-(aq) + 2\,H_2O(l)$ (reduction half reaction)

$H_2O_2(aq) \rightarrow O_2(g)$
$H_2O_2(aq) \rightarrow O_2(g) + 2\,H^+(aq)$
$[H_2O_2(aq) \rightarrow O_2(g) + 2\,H^+(aq) + 2\,e^-] \times 2$ (oxidation half reaction)

Combine the two half reactions.
$4\,H^+(aq) + ClO_4^-(aq) + 2\,H_2O_2(aq) \rightarrow ClO_2^-(aq) + 2\,H_2O(l) + 2\,O_2(g) + 4\,H^+(aq)$
$ClO_4^-(aq) + 2\,H_2O_2(aq) \rightarrow ClO_2^-(aq) + 2\,H_2O(l) + 2\,O_2(g)$

18.55 (a) $S_2O_3^{2-}(aq) \rightarrow S_4O_6^{2-}(aq)$
$2\,S_2O_3^{2-}(aq) \rightarrow S_4O_6^{2-}(aq)$
$2\,S_2O_3^{2-}(aq) \rightarrow S_4O_6^{2-}(aq) + 2\,e^-$ (oxidation half reaction)

$I_2(aq) \rightarrow I^-(aq)$
$I_2(aq) \rightarrow 2\,I^-(aq)$
$I_2(aq) + 2\,e^- \rightarrow 2\,I^-(aq)$ (reduction half reaction)

Combine the two half reactions.
$2\,S_2O_3^{2-}(aq) + I_2(aq) \rightarrow S_4O_6^{2-}(aq) + 2\,I^-(aq)$

(b) $Mn^{2+}(aq) \rightarrow MnO_2(s)$
$Mn^{2+}(aq) + 2\,H_2O(l) \rightarrow MnO_2(s)$
$Mn^{2+}(aq) + 2\,H_2O(l) \rightarrow MnO_2(s) + 4\,H^+(aq)$
$Mn^{2+}(aq) + 2\,H_2O(l) \rightarrow MnO_2(s) + 4\,H^+(aq) + 2\,e^-$ (oxidation half reaction)

$H_2O_2(aq) \rightarrow 2\ H_2O(l)$
$2\ H^+(aq) + H_2O_2(aq) \rightarrow 2\ H_2O(l)$
$2\ H^+(aq) + H_2O_2(aq) + 2\ e^- \rightarrow 2\ H_2O(l)$ (reduction half reaction)

Combine the two half reactions.
$Mn^{2+}(aq) + 2\ H_2O(l) + 2\ H^+(aq) + H_2O_2(aq) \rightarrow MnO_2(s) + 4\ H^+(aq) + 2\ H_2O(l)$
$Mn^{2+}(aq) + H_2O_2(aq) \rightarrow MnO_2(s) + 2\ H^+(aq)$
$Mn^{2+}(aq) + H_2O_2(aq) + 2\ OH^-(aq) \rightarrow MnO_2(s) + 2\ H^+(aq) + 2\ OH^-(aq)$
$Mn^{2+}(aq) + H_2O_2(aq) + 2\ OH^-(aq) \rightarrow MnO_2(s) + 2\ H_2O(l)$

(c) $Zn(s) \rightarrow Zn(OH)_4^{2-}(aq)$
$4\ H_2O(l) + Zn(s) \rightarrow Zn(OH)_4^{2-}(aq)$
$4\ H_2O(l) + Zn(s) \rightarrow Zn(OH)_4^{2-}(aq) + 4\ H^+(aq)$
$[4\ H_2O(l) + Zn(s) \rightarrow Zn(OH)_4^{2-}(aq) + 4\ H^+(aq) + 2\ e^-] \times 4$ (oxidation half reaction)

$NO_3^-(aq) \rightarrow NH_3(aq)$
$NO_3^-(aq) \rightarrow NH_3(aq) + 3\ H_2O(l)$
$9\ H^+(aq) + NO_3^-(aq) \rightarrow NH_3(aq) + 3\ H_2O(l)$
$9\ H^+(aq) + NO_3^-(aq) + 8\ e^- \rightarrow NH_3(aq) + 3\ H_2O(l)$ (reduction half reaction)

Combine the two half reactions.
$16\ H_2O(l) + 4\ Zn(s) + 9\ H^+(aq) + NO_3^-(aq) \rightarrow$
$4\ Zn(OH)_4^{2-}(aq) + 16\ H^+(aq) + NH_3(aq) + 3\ H_2O(l)$
$13\ H_2O(l) + 4\ Zn(s) + NO_3^-(aq) \rightarrow 4\ Zn(OH)_4^{2-}(aq) + 7\ H^+(aq) + NH_3(aq)$
$13\ H_2O(l) + 4\ Zn(s) + NO_3^-(aq) + 7\ OH^-(aq) \rightarrow$
$4\ Zn(OH)_4^{2-}(aq) + 7\ H^+(aq) + 7\ OH^-(aq) + NH_3(aq)$
$13\ H_2O(l) + 4\ Zn(s) + NO_3^-(aq) + 7\ OH^-(aq) \rightarrow$
$4\ Zn(OH)_4^{2-}(aq) + 7\ H_2O(l) + NH_3(aq)$
$6\ H_2O(l) + 4\ Zn(s) + NO_3^-(aq) + 7\ OH^-(aq) \rightarrow 4\ Zn(OH)_4^{2-}(aq) + NH_3(aq)$

(d) $Bi(OH)_3(s) \rightarrow Bi(s)$
$Bi(OH)_3(s) \rightarrow Bi(s) + 3\ H_2O(l)$
$3\ H^+(aq) + Bi(OH)_3(s) \rightarrow Bi(s) + 3\ H_2O(l)$
$[3\ H^+(aq) + Bi(OH)_3(s) + 3\ e^- \rightarrow Bi(s) + 3\ H_2O(l)] \times 2$ (reduction half reaction)

$Sn(OH)_3^-(aq) \rightarrow Sn(OH)_6^{2-}(aq)$
$Sn(OH)_3^-(aq) + 3\ H_2O(l) \rightarrow Sn(OH)_6^{2-}(aq)$
$Sn(OH)_3^-(aq) + 3\ H_2O(l) \rightarrow Sn(OH)_6^{2-}(aq) + 3\ H^+(aq)$
$[Sn(OH)_3^-(aq) + 3\ H_2O(l) \rightarrow Sn(OH)_6^{2-}(aq) + 3\ H^+(aq) + 2\ e^-] \times 3$
(oxidation half reaction)

Combine the two half reactions.
$6\ H^+(aq) + 2\ Bi(OH)_3(s) + 3\ Sn(OH)_3^-(aq) + 9\ H_2O(l) \rightarrow$
$2\ Bi(s) + 6\ H_2O(l) + 3\ Sn(OH)_6^{2-}(aq) + 9\ H^+(aq)$
$2\ Bi(OH)_3(s) + 3\ Sn(OH)_3^-(aq) + 3\ H_2O(l) \rightarrow 2\ Bi(s) + 3\ Sn(OH)_6^{2-}(aq) + 3\ H^+(aq)$
$2\ Bi(OH)_3(s) + 3\ Sn(OH)_3^-(aq) + 3\ H_2O(l) + 3\ OH^-(aq) \rightarrow$
$2\ Bi(s) + 3\ Sn(OH)_6^{2-}(aq) + 3\ H^+(aq) + 3\ OH^-(aq)$

$2\ Bi(OH)_3(s) + 3\ Sn(OH)_3^-(aq) + 3\ H_2O(l) + 3\ OH^-(aq) \rightarrow$
$2\ Bi(s) + 3\ Sn(OH)_6^{2-}(aq) + 3\ H_2O(l)$
$2\ Bi(OH)_3(s) + 3\ Sn(OH)_3^-(aq) + 3\ OH^-(aq) \rightarrow 2\ Bi(s) + 3\ Sn(OH)_6^{2-}(aq)$

18.56 (a) $Zn(s) \rightarrow Zn^{2+}(aq)$
$Zn(s) \rightarrow Zn^{2+}(aq) + 2\ e^-$ (oxidation half reaction)

$VO^{2+}(aq) \rightarrow V^{3+}(aq)$
$VO^{2+}(aq) \rightarrow V^{3+}(aq) + H_2O(l)$
$2\ H^+(aq) + VO^{2+}(aq) \rightarrow V^{3+}(aq) + H_2O(l)$
$[2\ H^+(aq) + VO^{2+}(aq) + e^- \rightarrow V^{3+}(aq) + H_2O(l)] \times 2$ (reduction half reaction)

Combine the two half reactions.
$Zn(s) + 2\ VO^{2+}(aq) + 4\ H^+(aq) \rightarrow Zn^{2+}(aq) + 2\ V^{3+}(aq) + 2\ H_2O(l)$

(b) $Ag(s) \rightarrow Ag^+(aq)$
$Ag(s) \rightarrow Ag^+(aq) + e^-$ (oxidation half reaction)

$NO_3^-(aq) \rightarrow NO_2(g)$
$NO_3^-(aq) \rightarrow NO_2(g) + H_2O(l)$
$2\ H^+(aq) + NO_3^-(aq) \rightarrow NO_2(g) + H_2O(l)$
$2\ H^+(aq) + NO_3^-(aq) + e^- \rightarrow NO_2(g) + H_2O(l)$ (reduction half reaction)

Combine the two half reactions.
$2\ H^+(aq) + Ag(s) + NO_3^-(aq) \rightarrow Ag^+(aq) + NO_2(g) + H_2O(l)$

(c) $Mg(s) \rightarrow Mg^{2+}(aq)$
$[Mg(s) \rightarrow Mg^{2+}(aq) + 2\ e^-] \times 3$ (oxidation half reaction)

$VO_4^{3-}(aq) \rightarrow V^{2+}(aq)$
$VO_4^{3-}(aq) \rightarrow V^{2+}(aq) + 4\ H_2O(l)$
$8\ H^+(aq) + VO_4^{3-}(aq) \rightarrow V^{2+}(aq) + 4\ H_2O(l)$
$[8\ H^+(aq) + VO_4^{3-}(aq) + 3\ e^- \rightarrow V^{2+}(aq) + 4\ H_2O(l)] \times 2$ (reduction half reaction)

Combine the two half reactions.
$3\ Mg(s) + 16\ H^+(aq) + 2\ VO_4^{3-}(aq) \rightarrow 3\ Mg^{2+}(aq) + 2\ V^{2+}(aq) + 8\ H_2O(l)$

(d) $I^-(aq) \rightarrow I_3^-(aq)$
$3\ I^-(aq) \rightarrow I_3^-(aq)$
$[3\ I^-(aq) \rightarrow I_3^-(aq) + 2\ e^-] \times 8$ (oxidation half reaction)

$IO_3^-(aq) \rightarrow I_3^-(aq)$
$3\ IO_3^-(aq) \rightarrow I_3^-(aq)$
$3\ IO_3^-(aq) \rightarrow I_3^-(aq) + 9\ H_2O(l)$
$18\ H^+(aq) + 3\ IO_3^-(aq) \rightarrow I_3^-(aq) + 9\ H_2O(l)$
$18\ H^+(aq) + 3\ IO_3^-(aq) + 16\ e^- \rightarrow I_3^-(aq) + 9\ H_2O(l)$ (reduction half reaction)

Combine the two half reactions.
$18\ H^+(aq) + 3\ IO_3^-(aq) + 24\ I^-(aq) \rightarrow 9\ I_3^-(aq) + 9\ H_2O(l)$
Divide each coefficient by 3.
$6\ H^+(aq) + IO_3^-(aq) + 8\ I^-(aq) \rightarrow 3\ I_3^-(aq) + 3\ H_2O(l)$

18.57 (a) $MnO_4^-(aq) \rightarrow Mn^{2+}(aq)$
$MnO_4^-(aq) \rightarrow Mn^{2+}(aq) + 4\ H_2O(l)$
$8\ H^+(aq) + MnO_4^-(aq) \rightarrow Mn^{2+}(aq) + 4\ H_2O(l)$
$[8\ H^+(aq) + MnO_4^-(aq) + 5\ e^- \rightarrow Mn^{2+}(aq) + 4\ H_2O(l)] \times 4$
(reduction half reaction)
$C_2H_5OH(aq) \rightarrow CH_3CO_2H(aq)$
$C_2H_5OH(aq) + H_2O(l) \rightarrow CH_3CO_2H(aq)$
$C_2H_5OH(aq) + H_2O(l) \rightarrow CH_3CO_2H(aq) + 4\ H^+(aq)$
$[C_2H_5OH(aq) + H_2O(l) \rightarrow CH_3CO_2H(aq) + 4\ H^+(aq) + 4\ e^-] \times 5$
(oxidation half reaction)
Combine the two half reactions.
$32\ H^+(aq) + 4\ MnO_4^-(aq) + 5\ C_2H_5OH(aq) + 5\ H_2O(l) \rightarrow$
$4\ Mn^{2+}(aq) + 16\ H_2O(l) + 5\ CH_3CO_2H(aq) + 20\ H^+(aq)$
$12\ H^+(aq) + 4\ MnO_4^-(aq) + 5\ C_2H_5OH(aq) \rightarrow$
$4\ Mn^{2+}(aq) + 11\ H_2O(l) + 5\ CH_3CO_2H(aq)$

(b) $Cr_2O_7^{2-}(aq) \rightarrow Cr^{3+}(aq)$
$Cr_2O_7^{2-}(aq) \rightarrow 2\ Cr^{3+}(aq)$
$Cr_2O_7^{2-}(aq) \rightarrow 2\ Cr^{3+}(aq) + 7\ H_2O(l)$
$14\ H^+(aq) + Cr_2O_7^{2-}(aq) \rightarrow 2\ Cr^{3+}(aq) + 7\ H_2O(l)$
$14\ H^+(aq) + Cr_2O_7^{2-}(aq) + 6\ e^- \rightarrow 2\ Cr^{3+}(aq) + 7\ H_2O(l)$ (reduction half reaction)

$H_2O_2(aq) \rightarrow O_2(g)$
$H_2O_2(aq) \rightarrow O_2(g) + 2\ H^+(aq)$
$[H_2O_2(aq) \rightarrow O_2(g) + 2\ H^+(aq) + 2\ e^-] \times 3$ (oxidation half reaction)

Combine the two half reactions.
$14\ H^+(aq) + Cr_2O_7^{2-}(aq) + 3\ H_2O_2(aq) \rightarrow$
$2\ Cr^{3+}(aq) + 7\ H_2O(l) + 3\ O_2(g) + 6\ H^+(aq)$
$8\ H^+(aq) + Cr_2O_7^{2-}(aq) + 3\ H_2O_2(aq) \rightarrow 2\ Cr^{3+}(aq) + 7\ H_2O(l) + 3\ O_2(g)$

(c) $Sn^{2+}(aq) \rightarrow Sn^{4+}(aq)$
$[Sn^{2+}(aq) \rightarrow Sn^{4+}(aq) + 2\ e^-] \times 4$ (oxidation half reaction)

$IO_4^-(aq) \rightarrow I^-(aq)$
$IO_4^-(aq) \rightarrow I^-(aq) + 4\ H_2O(l)$
$8\ H^+(aq) + IO_4^-(aq) \rightarrow I^-(aq) + 4\ H_2O(l)$
$8\ H^+(aq) + IO_4^-(aq) + 8\ e^- \rightarrow I^-(aq) + 4\ H_2O(l)$ (reduction half reaction)

Combine the two half reactions.
$4\ Sn^{2+}(aq) + 8\ H^+(aq) + IO_4^-(aq) \rightarrow 4\ Sn^{4+}(aq) + I^-(aq) + 4\ H_2O(l)$

(d) $PbO_2(s) + Cl^-(aq) \rightarrow PbCl_2(s)$
$PbO_2(s) + 2\,Cl^-(aq) \rightarrow PbCl_2(s)$
$PbO_2(s) + 2\,Cl^-(aq) \rightarrow PbCl_2(s) + 2\,H_2O(l)$
$PbO_2(s) + 4\,H^+(aq) + 2\,Cl^-(aq) \rightarrow PbCl_2(s) + 2\,H_2O(l)$
$[PbO_2(s) + 4\,H^+(aq) + 2\,Cl^-(aq) + 2\,e^- \rightarrow PbCl_2(s) + 2\,H_2O(l)] \times 2$
(reduction half reaction)

$H_2O(l) \rightarrow O_2(g)$
$2\,H_2O(l) \rightarrow O_2(g)$
$2\,H_2O(l) \rightarrow O_2(g) + 4\,H^+(aq)$
$2\,H_2O(l) \rightarrow O_2(g) + 4\,H^+(aq) + 4\,e^-$ (oxidation half reaction)

Combine the two half reactions.
$2\,PbO_2(s) + 8\,H^+(aq) + 4\,Cl^-(aq) + 2\,H_2O(l) \rightarrow$
$2\,PbCl_2(s) + 4\,H_2O(l) + O_2(g) + 4\,H^+(aq)$
$2\,PbO_2(s) + 4\,H^+(aq) + 4\,Cl^-(aq) \rightarrow 2\,PbCl_2(s) + 2\,H_2O(l) + O_2(g)$

Galvanic Cells (Sections 18.2 and 18.3)

18.58 The cathode of a galvanic cell is considered to be the positive electrode because electrons flow through the external circuit toward the positive electrode (the cathode).

18.59 The salt bridge maintains charge neutrality in both the anode and cathode compartments of a galvanic cell.

18.60 (a) $Cd(s) + Sn^{2+}(aq) \rightarrow Cd^{2+}(aq) + Sn(s)$

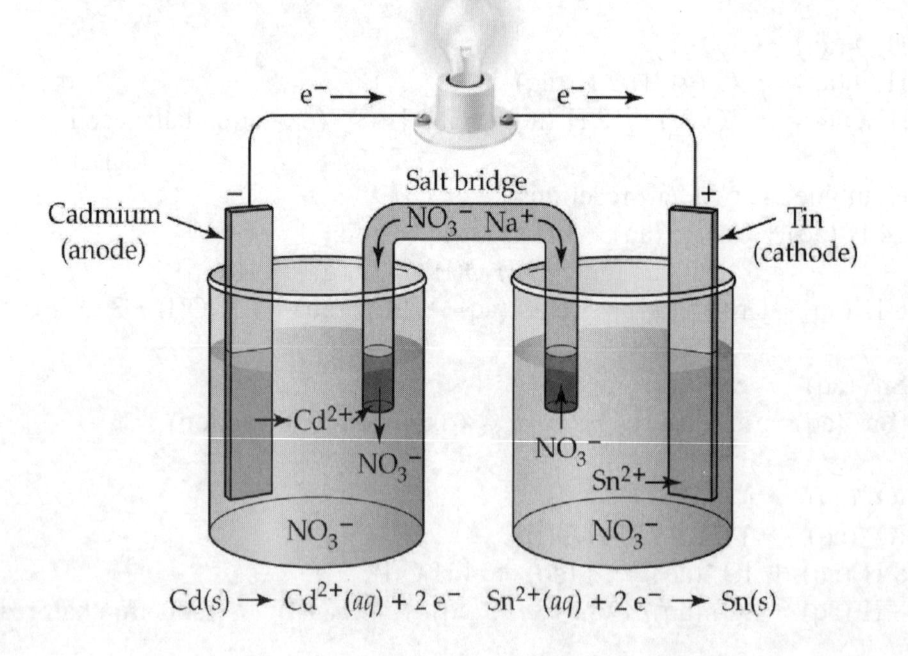

(b) $2\ Al(s) + 3\ Cd^{2+}(aq) \rightarrow 2\ Al^{3+}(aq) + 3\ Cd(s)$

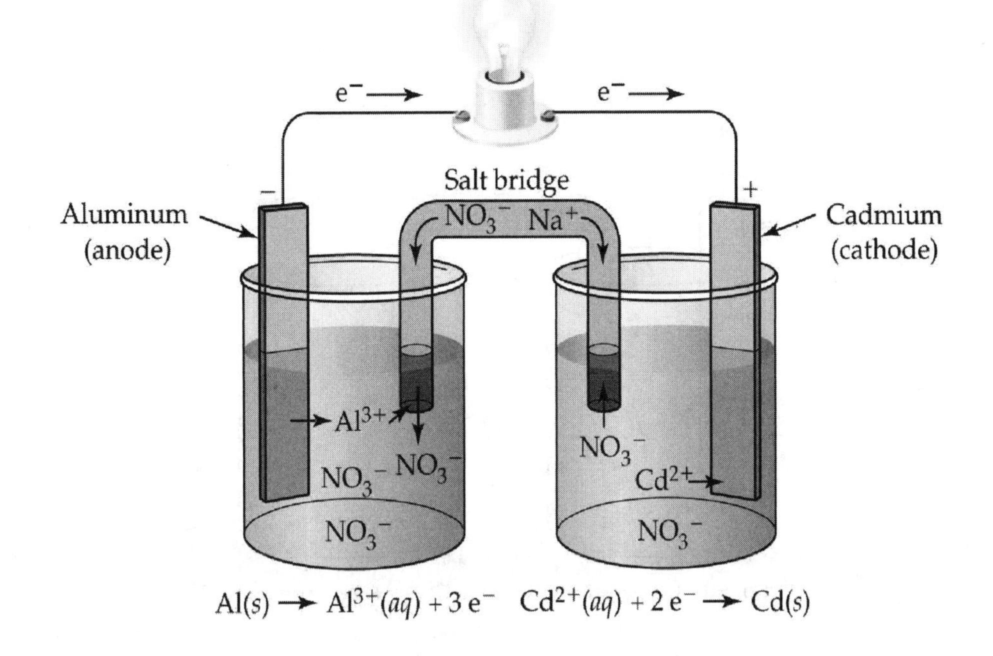

(c) $6\ Fe^{2+}(aq) + Cr_2O_7^{2-}(aq) + 14\ H^+(aq) \rightarrow 6\ Fe^{3+}(aq) + 2\ Cr^{3+}(aq) + 7\ H_2O(l)$

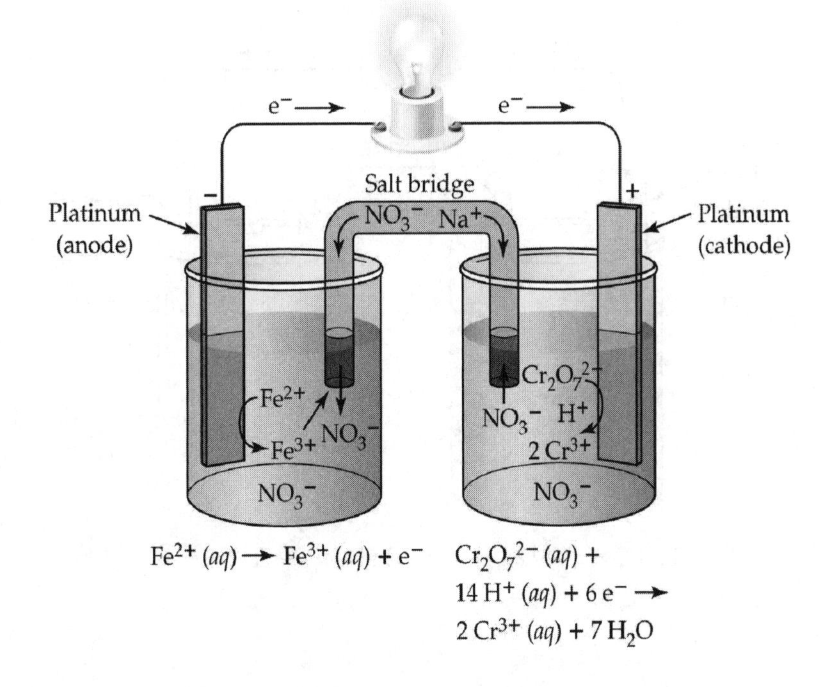

18.61 (a) $3\ Cu^{2+}(aq) + 2\ Cr(s) \rightarrow 3\ Cu(s) + 2\ Cr^{3+}(aq)$

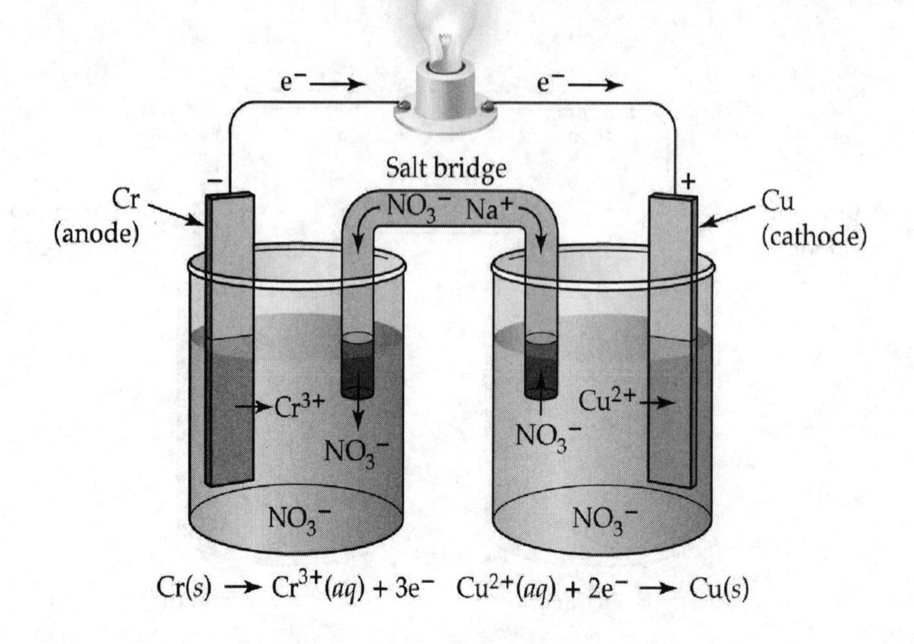

(b) $Pb(s) + 2\ H^{+}(aq) \rightarrow Pb^{2+}(aq) + H_2(g)$

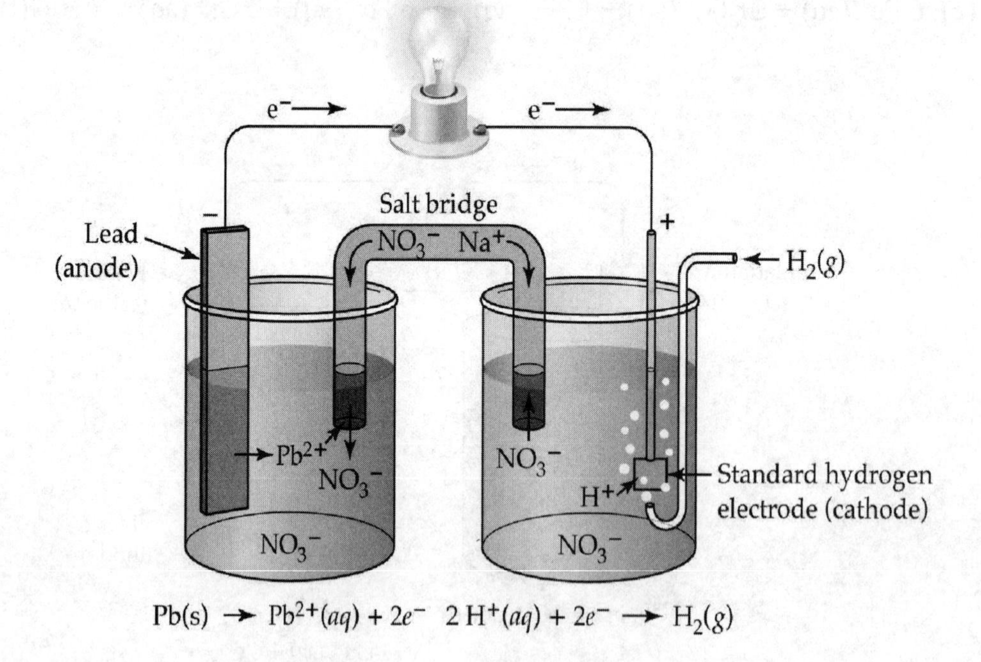

(c) $Cl_2(g) + Sn^{2+}(aq) \rightarrow Sn^{4+}(aq) + 2\ Cl^-(aq)$

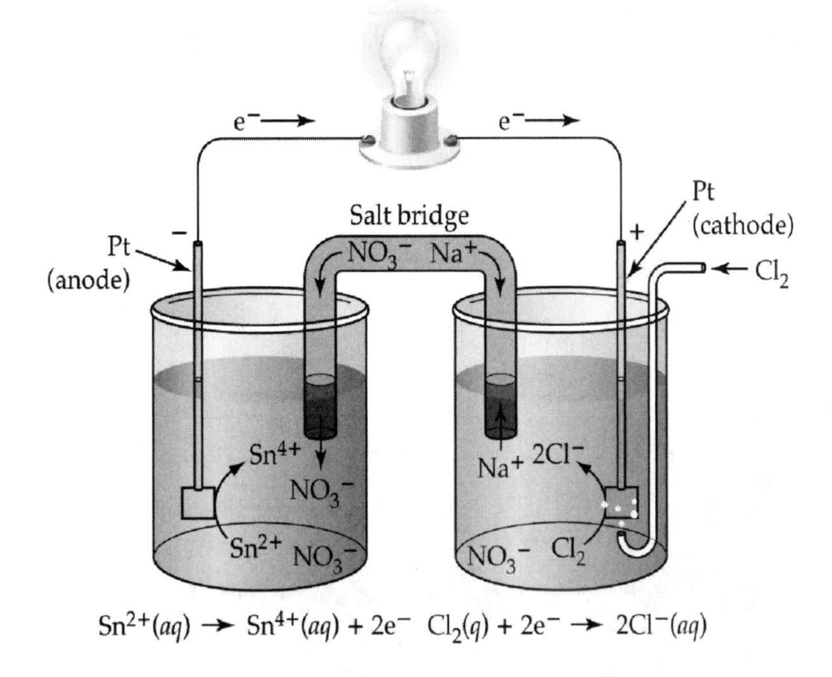

18.62 $2\ Br^-(aq) + Cl_2(g) \rightarrow Br_2(l) + 2\ Cl^-(aq)$
Inert electrodes are required because none of the reactants or products is an electrical conductor.

18.63 $Fe(s)|Fe^{3+}(aq)\|Cr_2O_7^{2-}(aq), Cr^{3+}(aq)|Pt(s)$

18.64 $Al(s)|Al^{3+}(aq)\|Cd^{2+}|Cd(s)$

18.65 $Fe(s)|Fe^{2+}(aq)\|I_2(s)|I^-(aq)|Pt(s)$

18.66 (a)

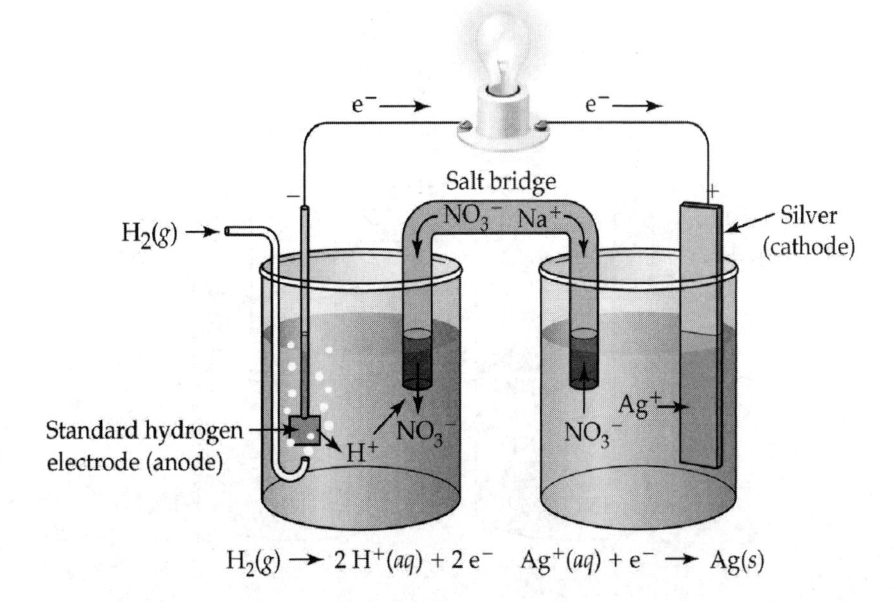

(b) anode reaction $H_2(g) \rightarrow 2\ H^+(aq) + 2\ e^-$
cathode reaction $2\ Ag^+(aq) + 2\ e^- \rightarrow 2\ Ag(s)$
overall reaction $H_2(g) + 2\ Ag^+(aq) \rightarrow 2\ H^+(aq) + 2\ Ag(s)$

(c) $Pt(s)|H_2(g)|H^+(aq)\|Ag^+(aq)|Ag(s)$

18.67 (a)

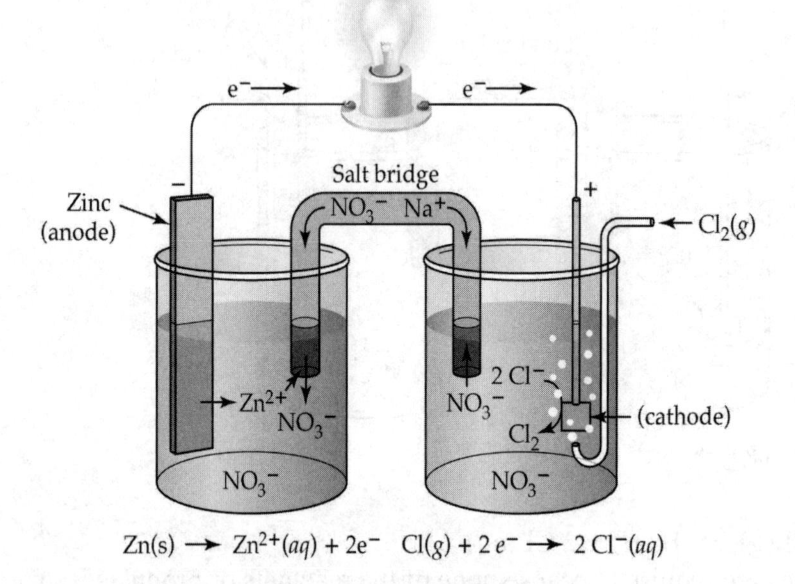

(b) anode reaction $Zn(s) \rightarrow Zn^{2+}(aq) + 2\ e^-$
cathode reaction $Cl_2(g) + 2\ e^- \rightarrow 2\ Cl^-(aq)$
overall reaction $Zn(s) + Cl_2(g) \rightarrow Zn^{2+}(aq) + 2\ Cl^-(aq)$

(c) $Zn(s)|Zn^{2+}(aq)\|Cl_2(g)|Cl^-(aq)|C(s)$

18.68 (a) anode reaction $Co(s) \rightarrow Co^{2+}(aq) + 2\ e^-$
cathode reaction $Cu^{2+}(aq) + 2\ e^- \rightarrow Cu(s)$
overall reaction $Co(s) + Cu^{2+}(aq) \rightarrow Co^{2+}(aq) + Cu(s)$

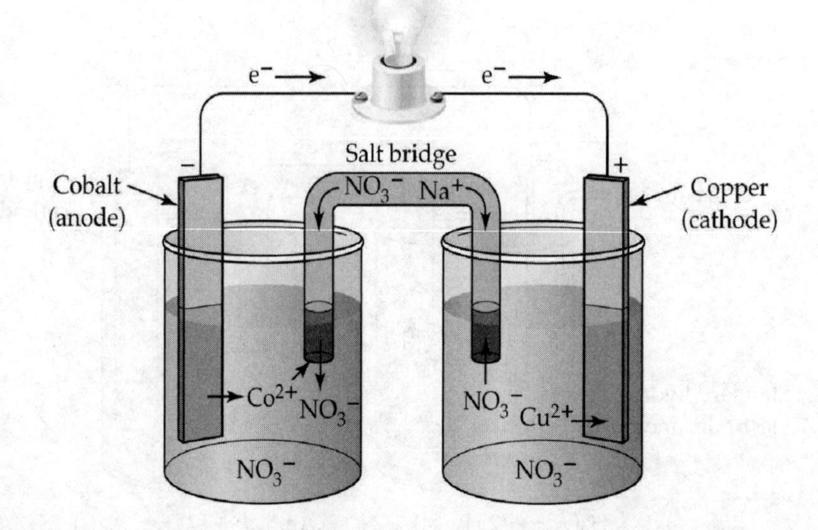

(b) anode reaction $2\ Fe(s) \rightarrow 2\ Fe^{2+}(aq) + 4\ e^-$
cathode reaction $O_2(g) + 4\ H^+(aq) + 4\ e^- \rightarrow 2\ H_2O(l)$
overall reaction $2\ Fe(s) + O_2(g) + 4\ H^+(aq) \rightarrow 2\ Fe^{2+}(aq) + 2\ H_2O(l)$

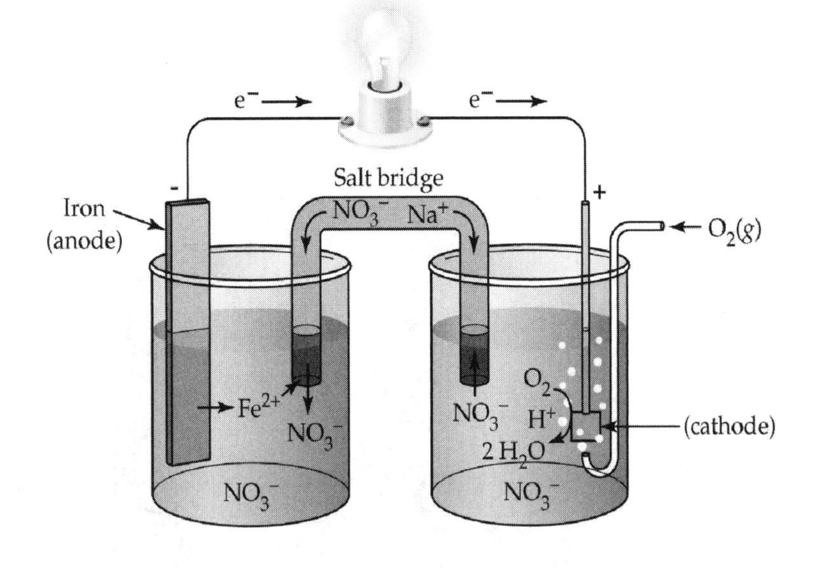

18.69 (a) anode reaction $Mn(s) \rightarrow Mn^{2+}(aq) + 2\ e^-$
cathode reaction $Pb^{2+}(aq) + 2\ e^- \rightarrow Pb(s)$
overall reaction $Mn(s) + Pb^{2+}(aq) \rightarrow Mn^{2+}(aq) + Pb(s)$

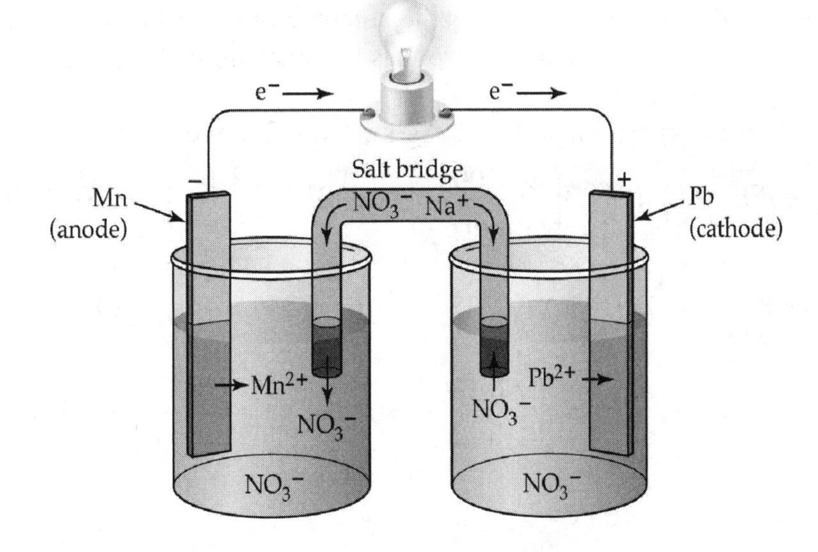

(b) anode reaction $H_2(g) \rightarrow 2\,H^+(aq) + 2\,e^-$
cathode reaction $2\,AgCl(s) + 2\,e^- \rightarrow 2\,Ag(s) + 2\,Cl^-(aq)$
overall reaction $H_2(g) + 2\,AgCl(s) \rightarrow 2\,Ag(s) + 2\,H^+(aq) + 2\,Cl^-(aq)$

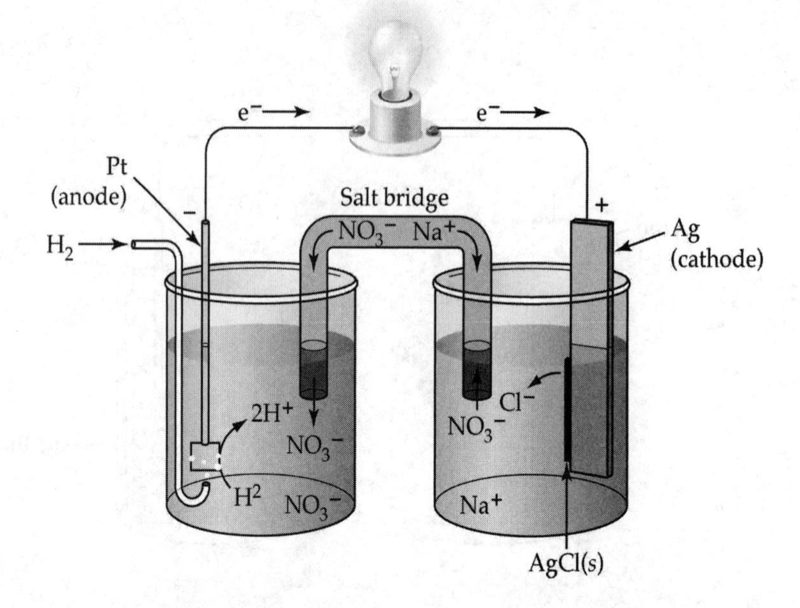

Cell Potentials and Free-Energy Changes; Standard Reduction Potentials (Sections 18.4–18.6)

18.70 E is the standard cell potential (E°) when all reactants and products are in their standard states–solutes at 1 M concentrations, gases at a partial pressure of 1 atm, solids and liquids in pure form, all at 25 °C.

18.71 The standard reduction potential is the potential of the reduction half reaction in a galvanic cell where the other electrode is the standard hydrogen electrode.

18.72 $Zn(s) + Ag_2O(s) \rightarrow ZnO(s) + 2\,Ag(s);\quad n = 2\text{ mol e}^-$

$$\Delta G = -nFE = -(2\text{ mol e}^-)\left(\frac{96{,}500\text{ C}}{1\text{ mol e}^-}\right)(1.60\text{ V})\left(\frac{1\text{ J}}{1\text{ C}\cdot\text{V}}\right) = -308{,}800\text{ J} = -309\text{ kJ}$$

18.73 $Pb(s) + PbO_2(s) + 2\,H^+(aq) + 2\,HSO_4^-(aq) \rightarrow 2\,PbSO_4(s) + 2\,H_2O(l)$
$n = 2\text{ mol e}^-$

$$\Delta G^\circ = -nFE^\circ = -(2\text{ mol e}^-)\left(\frac{96{,}500\text{ C}}{1\text{ mol e}^-}\right)(1.924\text{ V})\left(\frac{1\text{ J}}{1\text{ C}\cdot\text{V}}\right) = -371{,}300\text{ J} = -371\text{ kJ}$$

18.74 $\Delta G^\circ = -nFE^\circ;\ 1\text{ J} = \text{C}\cdot\text{V}$

$$n = \frac{-\Delta G^\circ}{FE^\circ} = \frac{-(-414{,}000\text{ J})}{\left(\frac{96{,}500\text{ C}}{1\text{ mol e}^-}\right)(1.43\text{ V})} = \frac{-(-414{,}000\text{ C}\cdot\text{V})}{\left(\frac{96{,}500\text{ C}}{1\text{ mol e}^-}\right)(1.43\text{ V})} = 3\text{ mol e}^-$$

18.75 $\Delta G^\circ = -nFE^\circ$; $1\ J = C\cdot V$

$$n = \frac{-\Delta G^\circ}{FE^\circ} = \frac{-(-527{,}000\ J)}{\left(\frac{96{,}500\ C}{1\ mol\ e^-}\right)(0.91\ V)} = \frac{-(-527{,}000\ C\cdot V)}{\left(\frac{96{,}500\ C}{1\ mol\ e^-}\right)(0.91\ V)} = 6\ mol\ e^-$$

$2x = 6$, $x = 3$ and $3y = 6$, $y = 2$

18.76 $2\ H_2(g) + O_2(g) \rightarrow 2\ H_2O(l)$; $n = 4\ mol\ e^-$ and $1\ V = 1\ J/C$

$\Delta G^\circ = 2\ \Delta G^\circ_f(H_2O(l)) = (2\ mol)(-237.2\ kJ/mol) = -474.4\ kJ$

$$\Delta G^\circ = -nFE^\circ \qquad E^\circ = \frac{-\Delta G^\circ}{nF} = \frac{-(-474{,}400\ J)}{(4\ mol\ e^-)\left(\frac{96{,}500\ C}{1\ mol\ e^-}\right)} = +1.23\ J/C = +1.23\ V$$

18.77 $CH_4(g) + 2\ O_2(g) \rightarrow CO_2(g) + 2\ H_2O(l)$; $n = 8\ mol\ e^-$ and $1\ V = 1\ J/C$

$\Delta G^\circ = [\Delta G^\circ_f(CO_2) + 2\ \Delta G^\circ_f(H_2O(l))] - \Delta G^\circ_f(CH_4)$

$\Delta G^\circ = [(1\ mol)(-394.4\ kJ/mol) + (2\ mol)(-237.2\ kJ/mol)]$
$\qquad - (1\ mol)(-50.8\ kJ/mol) = -818.0\ kJ$

$$\Delta G^\circ = -nFE^\circ \qquad E^\circ = \frac{-\Delta G^\circ}{nF} = \frac{-(-818{,}000\ J)}{(8\ mol\ e^-)\left(\frac{96{,}500\ C}{1\ mol\ e^-}\right)} = +1.06\ J/C = +1.06\ V$$

18.78 oxidation: $Zn(s) \rightarrow Zn^{2+}(aq) + 2\ e^-$ $\quad E^\circ = 0.76\ V$
reduction: $Eu^{3+}(aq) + e^- \rightarrow Eu^{2+}(aq)$ $\quad E^\circ = ?$
overall $Zn(s) + 2\ Eu^{3+}(aq) \rightarrow Zn^{2+}(aq) + 2\ Eu^{2+}(aq)$ $\quad E^\circ = 0.40\ V$
The standard reduction potential for the Eu^{3+}/Eu^{2+} half cell is:
$E^\circ = 0.40 - 0.76 = -0.36\ V$

18.79 oxidation: $2\ Ag(s) + 2\ Br^-(aq) \rightarrow 2\ AgBr(s) + 2\ e^-$ $\quad E^\circ = ?$
reduction: $Cu^{2+}(aq) + 2\ e^- \rightarrow Cu(s)$ $\quad E^\circ = 0.34\ V$
overall $Cu^{2+}(aq) + 2\ Ag(s) + 2\ Br^-(aq) \rightarrow Cu(s) + 2\ AgBr(s)$ $\quad E^\circ = 0.27\ V$
E° for the oxidation half reaction $= 0.27 - 0.34 = -0.07\ V$
For $AgBr(s) + e^- \rightarrow Ag(s) + Br^-(aq)$, $E^\circ = -(-0.07\ V) = +0.07\ V$

18.80 $Sn^{4+}(aq) < Br_2(aq) < MnO_4^-$

18.81 $Pb(s) < Fe(s) < Al(s)$

18.82 $Cr_2O_7^{2-}(aq)$ is highest in the table of standard reduction potentials, therefore it is the strongest oxidizing agent.
$Fe^{2+}(aq)$ is lowest in the table of standard reduction potentials, therefore it is the weakest oxidizing agent.

18.83 From Table 18.1:
Sn^{2+} is the strongest reducing agent and Fe^{2+} is the weakest reducing agent.

18.84 oxidation: $Co(s) \rightarrow Co^{2+}(aq) + 2\ e^-$ $E° = 0.28$ V
reduction: $I_2(s) + 2\ e^- \rightarrow 2\ I^-(aq)$ $E° = 0.54$ V
overall $I_2(s) + Co(s) \rightarrow Co^{2+}(aq) + 2\ I^-(aq)$ $E° = 0.82$ V
n = 2 mol e⁻

$$\Delta G° = -nFE° = -(2 \text{ mol e}^-)\left(\frac{96{,}500 \text{ C}}{1 \text{ mol e}^-}\right)(0.82 \text{ V})\left(\frac{1 \text{ J}}{1 \text{ C} \cdot \text{V}}\right) = -158{,}260 \text{ J} = -1.6 \times 10^2 \text{ kJ}$$

18.85 oxidation: $Cu(s) \rightarrow Cu^{2+}(aq) + 2\ e^-$ $E° = -0.34$ V
reduction: $I_2(s) + 2\ e^- \rightarrow 2\ I^-(aq)$ $E° = 0.54$ V
overall $I_2(s) + Cu(s) \rightarrow Cu^{2+}(aq) + 2\ I^-(aq)$ $E° = 0.20$ V
n = 2 mol e⁻

$$\Delta G° = -nFE° = -(2 \text{ mol e}^-)\left(\frac{96{,}500 \text{ C}}{1 \text{ mol e}^-}\right)(0.20 \text{ V})\left(\frac{1 \text{ J}}{1 \text{ C} \cdot \text{V}}\right) = -38{,}600 \text{ J} = -39 \text{ kJ}$$

18.86 oxidation: $2\ Al(s) \rightarrow 2\ Al^{3+}(aq) + 6\ e^-$ $E° = 1.66$ V
reduction: $3\ Cd^{2+}(aq) + 6\ e^- \rightarrow 3\ Cd(s)$ $E° = -0.40$ V
overall $2\ Al(s) + 3\ Cd^{2+}(aq) \rightarrow Al^{3+}(aq) + 3\ Cd(s)$ $E° = 1.26$ V
n = 6 mol e⁻

$$\Delta G° = -nFE° = -(6 \text{ mol e}^-)\left(\frac{96{,}500 \text{ C}}{1 \text{ mol e}^-}\right)(1.26 \text{ V})\left(\frac{1 \text{ J}}{1 \text{ C} \cdot \text{V}}\right) = -729{,}540 \text{ J} = -730 \text{ kJ}$$

18.87 oxidation: $6\ Fe^{2+}(aq) \rightarrow 6\ Fe^{3+}(aq) + 6\ e^-$ $E° = -0.77$ V
reduction: $Cr_2O_7^{2-}(aq) + 14\ H^+(aq) + 6\ e^- \rightarrow 2\ Cr^{3+}(aq) + 7\ H_2O(l)$ $E° = 1.36$ V
overall $Cr_2O_7^{2-}(aq) + 6\ Fe^{2+}(aq) + 14\ H^+(aq)$
$\rightarrow 2\ Cr^{3+}(aq) + 6\ Fe^{3+}(aq) + 7\ H_2O(l)$ $E° = 0.59$ V
n = 6 mol e⁻

$$\Delta G° = -nFE° = -(6 \text{ mol e}^-)\left(\frac{96{,}500 \text{ C}}{1 \text{ mol e}^-}\right)(0.59 \text{ V})\left(\frac{1 \text{ J}}{1 \text{ C} \cdot \text{V}}\right) = -341{,}610 \text{ J} = -342 \text{ kJ}$$

18.88 (a) $2\ Fe^{2+}(aq) + Pb^{2+}(aq) \rightarrow 2\ Fe^{3+}(aq) + Pb(s)$
oxidation: $2\ Fe^{2+}(aq) \rightarrow 2\ Fe^{3+}(aq) + 2\ e^-$ $E° = -0.77$ V
reduction: $Pb^{2+}(aq) + 2\ e^- \rightarrow Pb(s)$ $E° = -0.13$ V
overall $E° = -0.90$ V
Because the overall E° is negative, this reaction is nonspontaneous.

(b) $Mg(s) + Ni^{2+}(aq) \rightarrow Mg^{2+}(aq) + Ni(s)$
oxidation: $Mg(s) \rightarrow Mg^{2+}(aq) + 2\ e^-$ $E° = 2.37$ V
reduction: $Ni^{2+}(aq) + 2\ e^- \rightarrow Ni(s)$ $E° = -0.26$ V
overall $E° = 2.11$ V
Because the overall E° is positive, this reaction is spontaneous.

18.89 (a) $5\ Ag^+(aq) + Mn^{2+}(aq) + 4\ H_2O(l) \rightarrow 5\ Ag(s) + MnO_4^-(aq) + 8\ H^+(aq)$

oxidation: $Mn^{2+}(aq) + 4\ H_2O(l) \rightarrow MnO_4^-(aq) + 8\ H^+(aq) + 5\ e^-$ $E° = -1.51\ V$

reduction: $5\ Ag^+(aq) + 5\ e^- \rightarrow 5\ Ag(s)$ $E° = 0.80\ V$

overall $E° = -0.71\ V$

Because the overall E° is negative, this reaction is nonspontaneous.

(b) $2\ H_2O_2(aq) \rightarrow O_2(g) + 2\ H_2O(l)$

oxidation: $H_2O_2(aq) \rightarrow O_2(g) + 2\ H^+(aq) + 2\ e^-$ $E° = -0.70\ V$

reduction: $H_2O_2(aq) + 2\ H^+(aq) + 2\ e^- \rightarrow 2\ H_2O(l)$ $E° = 1.78\ V$

overall $E° = 1.08\ V$

Because the overall E° is positive, this reaction is spontaneous.

18.90 (a) oxidation: $Sn^{2+}(aq) \rightarrow Sn^{4+}(aq) + 2\ e^-$ $E° = -0.15\ V$

reduction: $Br_2(aq) + 2\ e^- \rightarrow 2\ Br^-(aq)$ $E° = 1.09\ V$

overall $E° = +0.94\ V$

Because the overall E° is positive, $Sn^{2+}(aq)$ can be oxidized by $Br_2(aq)$.

(b) oxidation: $Sn^{2+}(aq) \rightarrow Sn^{4+}(aq) + 2\ e^-$ $E° = -0.15\ V$

reduction: $Ni^{2+}(aq) + 2\ e^- \rightarrow Ni(s)$ $E° = -0.26\ V$

overall $E° = -0.41\ V$

Because the overall E° is negative, $Ni^{2+}(aq)$ cannot be reduced by $Sn^{2+}(aq)$.

(c) oxidation: $2\ Ag(s) \rightarrow 2\ Ag^+(aq) + 2\ e^-$ $E° = -0.80\ V$

reduction: $Pb^{2+}(aq) + 2\ e^- \rightarrow Pb(s)$ $E° = -0.13\ V$

overall $E° = -0.93\ V$

Because the overall E° is negative, Ag(s) cannot be oxidized by $Pb^{2+}(aq)$.

(d) oxidation: $H_2SO_3(aq) + H_2O(l) \rightarrow SO_4^{2-}(aq) + 4\ H^+(aq) + 2\ e^-$ $E° = -0.17\ V$

reduction: $I_2(s) + 2\ e^- \rightarrow 2\ I^-(aq)$ $E° = 0.54\ V$

overall $E° = +0.37\ V$

Because the overall E° is positive, $I_2(s)$ can be reduced by H_2SO_3.

18.91 (a) oxidation: $Ni(s) \rightarrow Ni^{2+}(aq) + 2\ e^-$ $E° = 0.26\ V$

reduction: $Pb^{2+}(aq) + 2\ e^- \rightarrow Pb(s)$ $E° = -0.13\ V$

overall $E° = +0.13\ V$

Because the overall E° is positive, $Pb^{2+}(aq)$ can be reduced by Ni(s).

(b) oxidation: $Au^+(aq) \rightarrow Au^{3+}(aq) + 2\ e^-$ $E° = -1.40\ V$

reduction: $Mn^{2+}(aq) + 2\ e^- \rightarrow Mn(s)$ $E° = -1.18\ V$

overall $E° = -2.58\ V$

Because the overall E° is negative, $Au^+(aq)$ cannot be oxidized by $Mn^{2+}(aq)$.

(c) oxidation: $Mn(s) \rightarrow Mn^{2+}(aq) + 2\ e^-$ $E° = 1.18\ V$

reduction: $I_2(s) + 2\ e^- \rightarrow 2\ I^-(aq)$ $E° = 0.54\ V$

overall $E° = +1.72\ V$

Because the overall E° is positive, $I_2(s)$ can be reduced by Mn(s).

(d) oxidation: $2\ Fe^{2+}(aq) \rightarrow 2\ Fe^{3+}(aq) + 2\ e^-$ $E° = -0.77\ V$
reduction: $Br_2(aq) + 2\ e^- \rightarrow 2\ Br^-(aq)$ $E° = \underline{1.09\ V}$
overall $E° = +0.32\ V$

Because the overall E° is positive, $Fe^{2+}(aq)$ can be oxidized by $Br_2(l)$.

18.92 (a) oxidation: $2\ Cr^{3+}(aq) + 7\ H_2O(l) \rightarrow Cr_2O_7^{2-}(aq) + 14\ H^+(aq) + 6\ e^-$ $E° = -1.36\ V$
reduction: $O_2(g) + 4\ H^+(aq) + 4\ e^- \rightarrow 2\ H_2O(l)$ $E° = \underline{1.23\ V}$
overall $E° = -0.13\ V$

There is no reaction because the overall E° is negative.

(b) oxidation: $Pb(s) \rightarrow Pb^{2+}(aq) + 2\ e^-$ $E° = 0.13\ V$
reduction: $2\ Ag^+(aq) + 2\ e^- \rightarrow 2\ Ag(s)$ $E° = \underline{0.80\ V}$
overall $E° = +0.93\ V$

$Pb(s) + 2\ Ag^+(aq) \rightarrow Pb^{2+}(aq) + 2\ Ag(s)$
The reaction is spontaneous because the overall E° is positive.

(c) oxidation: $H_2C_2O_4(aq) \rightarrow 2\ CO_2(g) + 2\ H^+(aq) + 2\ e^-$ $E° = 0.49\ V$
reduction: $Cl_2(g) + 2\ e^- \rightarrow 2\ Cl^-(aq)$ $E° = \underline{1.36\ V}$
overall $E° = +1.85\ V$

$Cl_2(g) + H_2C_2O_4(aq) \rightarrow 2\ Cl^-(aq) + 2\ CO_2(g) + 2\ H^+(aq)$
The reaction is spontaneous because the overall E° is positive.

(d) oxidation: $Ni(s) \rightarrow Ni^{2+}(aq) + 2\ e^-$ $E° = 0.26\ V$
reduction: $2\ HClO(aq) + 2\ H^+(aq) + 2\ e^- \rightarrow Cl_2(g) + H_2O(l)$ $E° = \underline{1.61\ V}$
overall $E° = +1.87\ V$

$Ni(s) + 2\ HClO(aq) + 2\ H^+(aq) \rightarrow Ni^{2+}(aq) + Cl_2(g) + H_2O(l)$
The reaction is spontaneous because the overall E° is positive.

18.93 (a) oxidation: $Zn(s) \rightarrow Zn^{2+}(aq) + 2\ e^-$ $E° = 0.76\ V$
reduction: $Pb^{2+}(aq) + 2\ e^- \rightarrow Pb(s)$ $E° = \underline{-0.13\ V}$
overall $E° = 0.63\ V$

$Zn(s) + Pb^{2+}(aq) \rightarrow Zn^{2+}(aq) + Pb(s)$
The reaction is spontaneous because the overall E° is positive.

(b) oxidation: $4\ Fe^{2+}(aq) \rightarrow 4\ Fe^{3+}(aq) + 4\ e^-$ $E° = -0.77\ V$
reduction: $O_2(g) + 4\ H^+(aq) + 4\ e^- \rightarrow 2\ H_2O(l)$ $E° = \underline{1.23\ V}$
overall $E° = 0.46\ V$

$4\ Fe^{2+}(aq) + O_2(g) + 4\ H^+(aq) \rightarrow 4\ Fe^{3+}(aq) + 2\ H_2O(l)$
The reaction is spontaneous because the overall E° is positive.

(c) oxidation: $2\ Ag(s) \rightarrow 2\ Ag^+(aq) + 2\ e^-$ $E° = -0.80\ V$
reduction: $Ni^{2+}(aq) + 2\ e^- \rightarrow Ni(s)$ $E° = \underline{-0.26\ V}$
overall $E° = -1.06\ V$

There is no reaction because the overall E° is negative.

(d) oxidation: $H_2(g) \rightarrow 2\,H^+(aq) + 2\,e^-$ $E° = 0.00$ V

reduction: $Cd^{2+}(aq) + 2\,e^- \rightarrow Cd(s)$ $E° = -0.40$ V

overall $E° = -0.40$ V

There is no reaction because the overall E° is negative.

The Nernst Equation (Sections 18.7 and 18.8)

18.94 $2\,Ag^+(aq) + Sn(s) \rightarrow 2\,Ag(s) + Sn^{2+}(aq)$

oxidation: $Sn(s) \rightarrow Sn^{2+}(aq) + 2\,e^-$ $E° = 0.14$ V

reduction: $2\,Ag^+(aq) + 2\,e^- \rightarrow 2\,Ag(s)$ $E° = 0.80$ V

overall $E° = 0.94$ V

$$E = E° - \frac{0.0592\,V}{n}\log\frac{[Sn^{2+}]}{[Ag^+]^2} = 0.94\,V - \frac{(0.0592\,V)}{2}\log\frac{(0.020)}{(0.010)^2} = 0.87\,V$$

18.95 $2\,Fe^{2+}(aq) + Cl_2(g) \rightarrow 2\,Fe^{3+}(aq) + 2\,Cl^-(aq)$

oxidation: $2\,Fe^{2+}(aq) \rightarrow 2\,Fe^{3+}(aq) + 2\,e^-$ $E° = -0.77$ V

reduction: $Cl_2(g) + 2\,e^- \rightarrow 2\,Cl^-(aq)$ $E° = 1.36$ V

overall $E° = 0.59$ V

$$E = E° - \frac{0.0592\,V}{n}\log\frac{[Fe^{3+}]^2[Cl^-]^2}{[Fe^{2+}]^2 P_{Cl_2}} = 0.59\,V - \frac{(0.0592\,V)}{2}\log\frac{(0.0010)^2(0.0030)^2}{(1.0)^2(0.50)} = 0.91\,V$$

18.96 $Pb(s) + Cu^{2+}(aq) \rightarrow Pb^{2+}(aq) + Cu(s)$

oxidation: $Pb(s) \rightarrow Pb^{2+}(aq) + 2\,e^-$ $E° = 0.13$ V

reduction: $Cu^{2+}(aq) + 2\,e^- \rightarrow Cu(s)$ $E° = 0.34$ V

overall $E° = 0.47$ V

$$E = E° - \frac{0.0592\,V}{n}\log\frac{[Pb^{2+}]}{[Cu^{2+}]} = 0.47\,V - \frac{(0.0592\,V)}{2}\log\frac{1.0}{(1.0 \times 10^{-4})} = 0.35\,V$$

When E = 0, $0 = E° - \frac{0.0592\,V}{n}\log\frac{[Pb^{2+}]}{[Cu^{2+}]} = 0.47\,V - \frac{(0.0592\,V)}{2}\log\frac{1.0}{[Cu^{2+}]}$

$$0 = 0.47\,V + \frac{(0.0592\,V)}{2}\log[Cu^{2+}]$$

$$\log[Cu^{2+}] = (-0.47\,V)\left(\frac{2}{0.0592\,V}\right) = -15.88$$

$$[Cu^{2+}] = 10^{-15.88} = 1 \times 10^{-16}\,M$$

18.97 $Fe(s) + Cu^{2+}(aq) \rightarrow Fe^{2+}(aq) + Cu(s)$

oxidation: $Fe(s) \rightarrow Fe^{2+}(aq) + 2\,e^-$ $E° = 0.45$ V

reduction: $Cu^{2+}(aq) + 2\,e^- \rightarrow Cu(s)$ $E° = 0.34$ V

overall $E° = 0.79$ V

$$E = 0.67\,V = E° - \frac{0.0592\,V}{n}\log\frac{[Fe^{2+}]}{[Cu^{2+}]} = 0.79\,V - \frac{(0.0592\,V)}{2}\log\left(\frac{0.10}{[Cu^{2+}]}\right)$$

$$0.67\ V = 0.79\ V - \frac{(0.0592\ V)}{2}(\log(0.10) - \log[Cu^{2+}])$$

$\log[Cu^{2+}] = -5.05$; $[Cu^{2+}] = 10^{-5.05} = 8.9 \times 10^{-6}$ M

18.98 $Zn(s) + Cu^{2+}(aq) \rightarrow Zn^{2+}(aq) + Cu(s)$

oxidation: $Zn(s) \rightarrow Zn^{2+}(aq) + 2\ e^-$ $E° = 0.76$ V

reduction: $Cu^{2+}(aq) + 2\ e^- \rightarrow Cu(s)$ $E° = 0.34$ V

overall $E° = 1.10$ V

$$E = 1.07\ V = E° - \frac{0.0592\ V}{n}\log\frac{[Zn^{2+}]}{[Cu^{2+}]} = 1.10\ V - \frac{(0.0592\ V)}{2}\log\left(\frac{[Zn^{2+}]}{[Cu^{2+}]}\right)$$

$$1.07\ V - 1.10\ V = -\frac{(0.0592\ V)}{2}\log\left(\frac{[Zn^{2+}]}{[Cu^{2+}]}\right)$$

$$\frac{0.03\ V}{\frac{(0.0592\ V)}{2}} = \log\left(\frac{[Zn^{2+}]}{[Cu^{2+}]}\right) = 1; \quad \frac{[Zn^{2+}]}{[Cu^{2+}]} = 10^1 = 10$$

18.99 $Fe(s) + Sn^{2+}(aq) \rightarrow Fe^{2+}(aq) + Sn(s)$

oxidation: $Fe(s) \rightarrow Fe^{2+}(aq) + 2\ e^-$ $E° = 0.45$ V

reduction: $Sn^{2+}(aq) + 2\ e^- \rightarrow Sn(s)$ $E° = -0.14$ V

overall $E° = 0.31$ V

$$E = 0.35\ V = E° - \frac{0.0592\ V}{n}\log\frac{[Fe^{2+}]}{[Sn^{2+}]} = 0.31\ V - \frac{(0.0592\ V)}{2}\log\left(\frac{[Fe^{2+}]}{[Sn^{2+}]}\right)$$

$$0.35\ V - 0.31\ V = -\frac{(0.0592\ V)}{2}\log\left(\frac{[Fe^{2+}]}{[Sn^{2+}]}\right)$$

$$\frac{0.04\ V}{-\frac{(0.0592\ V)}{2}} = \log\left(\frac{[Fe^{2+}]}{[Sn^{2+}]}\right) = -1.35; \quad \frac{[Fe^{2+}]}{[Sn^{2+}]} = 10^{-1.35} = 0.04$$

18.100 (a) $E = E° - \frac{0.0592\ V}{n}\log[I^-]^2 = 0.54\ V - \frac{(0.0592\ V)}{2}\log(0.020)^2 = 0.64\ V$

(b) $E = E° - \frac{0.0592\ V}{n}\log\frac{[Fe^{2+}]}{[Fe^{3+}]} = 0.77\ V - \frac{(0.0592\ V)}{1}\log\left(\frac{0.10}{0.10}\right) = 0.77\ V$

(c) $E = E° - \frac{0.0592\ V}{n}\log\frac{[Sn^{4+}]}{[Sn^{2+}]} = -0.15\ V - \frac{(0.0592\ V)}{2}\log\left(\frac{0.40}{0.0010}\right) = -0.23\ V$

(d) $E = E° - \frac{0.0592\ V}{n}\log\frac{[Cr_2O_7^{2-}][H^+]^{14}}{[Cr^{3+}]^2} = -1.36\ V - \frac{(0.0592\ V)}{6}\log\left(\frac{(1.0)(0.010)^{14}}{1.0}\right)$

$$E = -1.36\ V - \frac{(0.0592\ V)}{6}(14)\log(0.010) = -1.08\ V$$

18.101 $E = E^\circ - \frac{0.0592\ V}{n} \log \frac{P_{H_2}}{[H_3O^+]^2}$; $E^\circ = 0$, $n = 2$ mol e^-, and $P_{H_2} = 1$ atm

(a) $[H_3O^+] = 1.0$ M; $E = -\frac{0.0592\ V}{2} \log \frac{1}{(1.0)^2} = 0$

(b) pH = 4.00, $[H_3O^+] = 10^{-4.00} = 1.0 \times 10^{-4}$ M

$$E = -\frac{0.0592\ V}{2} \log \frac{1}{(1.0 \times 10^{-4})^2} = -0.24\ V$$

(c) $[H_3O^+] = 1.0 \times 10^{-7}$ M; $E = -\frac{0.0592\ V}{2} \log \frac{1}{(1.0 \times 10^{-7})^2} = -0.41\ V$

(d) $[OH^-] = 1.0$ M; $[H_3O^+] = \frac{K_w}{[OH^-]} = \frac{1.0 \times 10^{-14}}{1.0} = 1.0 \times 10^{-14}$ M

$$E = -\frac{0.0592\ V}{2} \log \frac{1}{(1.0 \times 10^{-14})^2} = -0.83\ V$$

18.102 $H_2(g) + Ni^{2+}(aq) \rightarrow 2\ H^+(aq) + Ni(s)$

$$E^\circ = E^\circ_{H_2 \rightarrow H^+} + E^\circ_{Ni^{2+} \rightarrow Ni} = 0\ V + (-0.26\ V) = -0.26\ V$$

$$E = E^\circ - \frac{0.0592\ V}{n} \log \frac{[H_3O^+]^2}{[Ni^{2+}](P_{H_2})}$$

$$0.27\ V = -0.26\ V - \frac{(0.0592\ V)}{2} \log \frac{[H_3O^+]^2}{(1)(1)}$$

$$0.27\ V = -0.26\ V - (0.0592\ V) \log [H_3O^+]$$

pH = – log $[H_3O^+]$ therefore $0.27\ V = -0.26\ V + (0.0592\ V)$ pH

$$pH = \frac{(0.27\ V + 0.26\ V)}{0.0592\ V} = 9.0$$

18.103 $Zn(s) + 2\ H^+(aq) \rightarrow Zn^{2+}(aq) + H_2(g)$

$$E^\circ = E^\circ_{H^+ \rightarrow H_2} + E^\circ_{Zn \rightarrow Zn^{2+}} = 0\ V + 0.76\ V = 0.76\ V$$

$$E = E^\circ - \frac{0.0592\ V}{n} \log \frac{[Zn^{2+}](P_{H_2})}{[H_3O^+]^2}$$

$$0.58\ V = 0.76\ V - \frac{(0.0592\ V)}{2} \log \frac{(1)(1)}{[H_3O^+]^2}$$

$$0.58\ V = 0.76\ V + (0.0592\ V) \log [H_3O^+]$$

pH = – log $[H_3O^+]$ therefore $0.58\ V = 0.76\ V - (0.0592\ V)$ pH

$$pH = \frac{-(0.58\ V - 0.76\ V)}{0.0592\ V} = 3.0$$

Standard Cell Potentials and Equilibrium Constants (Section 18.9)

18.104 $\Delta G^\circ = -nFE^\circ$

Because n and F are always positive, ΔG° is negative when E° is positive because of the negative sign in the equation.

$$E^\circ = \frac{0.0592\ V}{n}\log K; \quad \log K = \frac{nE^\circ}{0.0592\ V}; \qquad K = 10^{\frac{nE^\circ}{0.0592}}$$

If E° is positive, the exponent is positive (because n is positive), and K is greater than 1.

18.105 If $K < 1$, $E^\circ < 0$. When $E^\circ = 0$, $K = 1$.

18.106 $Ni(s) + 2\ Ag^+(aq) \rightarrow Ni^{2+}(aq) + 2\ Ag(s)$

oxidation: $Ni(s) \rightarrow Ni^{2+}(aq) + 2\ e^-$ $\qquad E^\circ = 0.26\ V$

reduction: $2\ Ag^+(aq) + 2\ e^- \rightarrow 2\ Ag(s)$ $\qquad \underline{E^\circ = 0.80\ V}$

overall $E^\circ = 1.06\ V$

$$E^\circ = \frac{0.0592\ V}{n}\log K; \ \log K = \frac{nE^\circ}{0.0592\ V} = \frac{(2)(1.06\ V)}{0.0592\ V} = 35.8; \ \ K = 10^{35.8} = 6 \times 10^{35}$$

18.107 $2\ MnO_4^-(aq) + 10\ Cl^-(aq) + 16\ H^+(aq) \rightarrow 2\ Mn^{2+}(aq) + 5\ Cl_2(g) + 8\ H_2O(l)$

oxidation: $10\ Cl^-(aq) \rightarrow 5\ Cl_2(g) + 10\ e^-$ $\qquad E^\circ = -1.36\ V$

reduction: $2\ MnO_4^-(aq) + 16\ H^+(aq) + 10\ e^- \rightarrow 2\ Mn^{2+}(aq) + 8\ H_2O(l)$ $\qquad \underline{E^\circ = \ \ 1.51\ V}$

overall $E^\circ = \ 0.15\ V$

$$E^\circ = \frac{0.0592\ V}{n}\log K; \ \log K = \frac{nE^\circ}{0.0592\ V} = \frac{(10)(0.15\ V)}{0.0592\ V} = 25.3; \ \ K = 10^{25.3} = 2 \times 10^{25}$$

18.108 $Cd(s) + Sn^{2+}(aq) \rightarrow Cd^{2+}(aq) + Sn(s)$

oxidation $\quad Cd(s) \rightarrow Cd^{2+}(aq) + 2\ e^-$ $\qquad E^\circ = \ 0.40\ V$

reduction: $\quad Sn^{2+}(aq) + 2\ e^- \rightarrow Sn(s)$ $\qquad \underline{E^\circ = -0.14\ V}$

overall $\quad E^\circ = \ 0.26\ V$

$$E^\circ = \frac{0.0592\ V}{n}\log K; \ \log K = \frac{nE^\circ}{0.0592\ V} = \frac{(2)(0.26\ V)}{0.0592\ V} = 8.8; \ K = 10^{8.8} = 6.3 \times 10^{8}$$

18.109 $Cl_2(g) + Sn^{2+}(aq) \rightarrow Sn^{4+}(aq) + 2\ Cl^-(aq)$

oxidation $\quad Sn^{2+}(aq) \rightarrow Sn^{4+}(aq) + 2\ e^-$ $\qquad E^\circ = -0.15\ V$

reduction: $\quad Cl_2(g) + 2\ e^- \rightarrow 2\ Cl^-(aq)$ $\qquad \underline{E^\circ = \ \ 1.36\ V}$

overall $\quad E^\circ = \ 1.21\ V$

$$E^\circ = \frac{0.0592\ V}{n}\log K; \ \log K = \frac{nE^\circ}{0.0592\ V} = \frac{(2)(1.21\ V)}{0.0592\ V} = 40.9; \ K = 10^{40.9} = 8.0 \times 10^{40}$$

18.110 $Hg_2^{2+}(aq) \rightarrow Hg(l) + Hg^{2+}(aq)$

oxidation: $\quad ½[Hg_2^{2+}(aq) \rightarrow 2\ Hg^{2+}(aq) + 2\ e^-]$ $\qquad E^\circ = -0.92\ V$

reduction: $\quad ½[Hg_2^{2+}(aq) + 2\ e^- \rightarrow 2\ Hg(l)]$ $\qquad \underline{E^\circ = \ \ 0.80\ V}$

overall $E^\circ = -0.12\ V$

$$E^\circ = \frac{0.0592\ V}{n}\log K$$

$$\log K = \frac{nE^\circ}{0.0592\ V} = \frac{(1)(-0.12\ V)}{0.0592\ V} = -2.027; \qquad K = 10^{-2.027} = 9 \times 10^{-3}$$

18.111 $2\ H_2O_2(aq) \rightarrow 2\ H_2O(l) + O_2(g)$

oxidation: $H_2O_2(aq) \rightarrow O_2(g) + 2\ H^+(aq) + 2e^-$ $E^\circ = -0.70\ V$

reduction: $H_2O_2(aq) + 2\ H^+(aq) + 2e^- \rightarrow 2\ H_2O(l)$ $\underline{E^\circ = \ 1.78\ V}$

overall $E^\circ = \ 1.08\ V$

$$E^\circ = \frac{0.0592\ V}{n}\log K; \ \log K = \frac{nE^\circ}{0.0592\ V} = \frac{(2)(1.08\ V)}{0.0592\ V} = 36.5; \quad K = 10^{36.5} = 3 \times 10^{36}$$

Batteries; Corrosion (Sections 18.10 and 18.11)

18.112 (a)

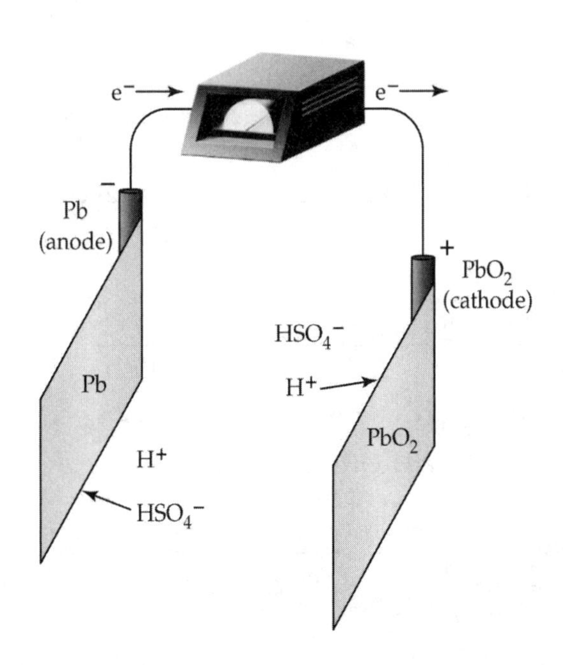

(b) anode: $Pb(s) + HSO_4^-(aq) \rightarrow PbSO_4(s) + H^+(aq) + 2\ e^-$ $E^\circ = 0.296\ V$

cathode: $\underline{PbO_2(s) + 3\ H^+(aq) + HSO_4^-(aq) + 2\ e^- \rightarrow PbSO_4(s) + 2\ H_2O(l)}$ $\underline{E^\circ = 1.628\ V}$

overall $Pb(s) + PbO_2(s) + 2\ H^+(aq) + 2\ HSO_4^-(aq) \rightarrow 2\ PbSO_4(s) + 2\ H_2O(l)$ $E^\circ = 1.924\ V$

(c) $E^\circ = \frac{0.0592\ V}{n}\log K; \quad \log K = \frac{nE^\circ}{0.0592\ V} = \frac{(2)(1.924\ V)}{0.0592\ V} = 65.0; \quad K = 1 \times 10^{65}$

(d) When the cell reaction reaches equilibrium the cell voltage = 0.

18.113 anode: $2\ H_2(g) + 4\ OH^-(aq) \rightarrow 4\ H_2O(l) + 4\ e^-$ $E^\circ = 0.83\ V$

cathode: $\underline{O_2(g) + 2\ H_2O(l) + 4\ e^- \rightarrow 4\ OH^-(aq)}$ $\underline{E^\circ = 0.40\ V}$

overall: $H_2(g) + O_2(g) \rightarrow 2\ H_2O(l)$ $E^\circ = 1.23\ V$

$n = 4$ mol e^- and 1 J = 1 C x 1 V

$$\Delta G^\circ = -nFE^\circ = -(4 \text{ mol e}^-)\left(\frac{96{,}500 \text{ C}}{1 \text{ mol e}^-}\right)(1.23 \text{ V})\left(\frac{1 \text{ J}}{1 \text{ C} \cdot \text{V}}\right) = -474{,}780 \text{ J} = -475 \text{ kJ}$$

$$E^\circ = \frac{0.0592 \text{ V}}{n}\log K; \quad \log K = \frac{nE^\circ}{0.0592 \text{ V}} = \frac{(4)(1.23 \text{ V})}{0.0592 \text{ V}} = 83.1; \; K = 10^{83.1} = 1 \times 10^{83}$$

$$E = E^\circ - \frac{0.0592 \text{ V}}{n}\log\frac{1}{(P_{H_2})^2(P_{O_2})} = 1.23 \text{ V} - \frac{0.0592 \text{ V}}{4}\log\frac{1}{(25)^2(25)} = 1.29 \text{ V}$$

18.114 Rust is a hydrated form of iron(III) oxide ($Fe_2O_3 \cdot H_2O$). Rust forms from the oxidation of Fe in the presence of O_2 and H_2O. Rust can be prevented by coating Fe with Zn (galvanizing).

18.115 The potential for the reduction of O_2 becomes more positive the lower the pH of the solution. Rust should form more readily at lower pH.

18.116 Cr forms a protective oxide coating similar to Al.

18.117 (c) Steel is coated with zinc because zinc is more easily oxidized than iron.

18.118 (d) A strip of magnesium is attached to steel because the magnesium is more easily oxidized than iron.

18.119 (a) $[Zn(s) \rightarrow Zn^{2+}(aq) + 2\,e^-] \times 2$

$\underline{O_2(g) + 4\,H^+(aq) + 4\,e^- \rightarrow 2\,H_2O(l)}$

$2\,Zn(s) + O_2(g) + 4\,H^+(aq) \rightarrow 2\,Zn^{2+}(aq) + 2\,H_2O(l)$

(b) $[Fe(s) \rightarrow Fe^{2+}(aq) + 2\,e^-] \times 4$

$[O_2(g) + 4\,H^+(aq) + 4\,e^- \rightarrow 2\,H_2O(l)] \times 2$

$4\,Fe^{2+}(aq) + O_2(g) + 4\,H^+(aq) \rightarrow 4\,Fe^{3+}(aq) + 2\,H_2O(l)$

$\underline{[2\,Fe^{3+}(aq) + 4\,H_2O(l) \rightarrow Fe_2O_3 \cdot H_2O(s) + 6\,H^+(aq)] \times 2}$

$4\,Fe(s) + 3\,O_2(g) + 2\,H_2O(l) \rightarrow 2\,Fe_2O_3 \cdot H_2O(s)$

18.120 Mn and Al

18.121 Mn, Mg and Cr

Electrolysis (Sections 18.12–18.14)

18.122 (a)

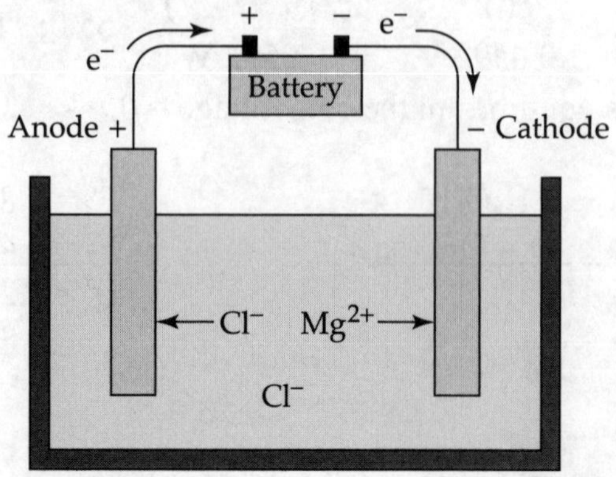

+ −
e^- e^-
Battery
Anode +
− Cathode
$\leftarrow Cl^-$ $Mg^{2+} \rightarrow$
Cl^-

(b) anode: $2\,Cl^-(l) \rightarrow Cl_2(g) + 2\,e^-$
cathode: $Mg^{2+}(l) + 2\,e^- \rightarrow Mg(l)$
overall: $Mg^{2+}(l) + 2\,Cl^-(l) \rightarrow Mg(l) + Cl_2(g)$

18.123 (a)

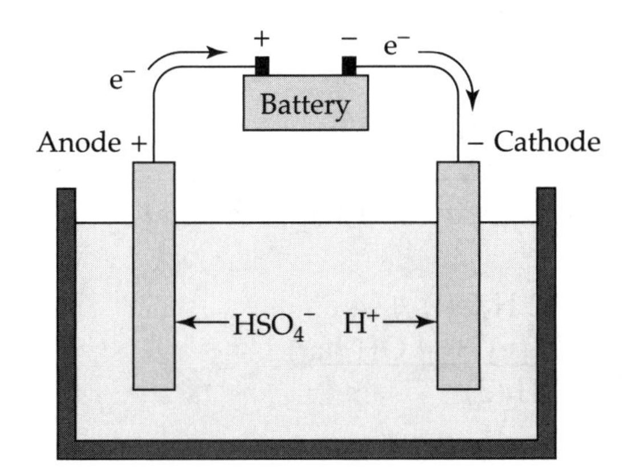

(b) anode: $2\,H_2O(l) \rightarrow O_2(g) + 4\,H^+(aq) + 4\,e^-$
cathode: $4\,H^+(aq) + 4\,e^- \rightarrow 2\,H_2(g)$
overall: $2\,H_2O(l) \rightarrow O_2(g) + 2\,H_2(g)$

18.124 Possible anode reactions:

$2\,Cl^-(aq) \rightarrow Cl_2(g) + 2\,e^-$
$2\,H_2O(l) \rightarrow O_2(g) + 4\,H^+(aq) + 4\,e^-$

Possible cathode reactions:

$2\,H_2O(l) + 2\,e^- \rightarrow H_2(g) + 2\,OH^-(aq)$
$Mg^{2+}(aq) + 2\,e^- \rightarrow Mg(s)$

Actual reactions:

anode: $2\,Cl^-(aq) \rightarrow Cl_2(g) + 2\,e^-$
cathode: $2\,H_2O(l) + 2\,e^- \rightarrow H_2(g) + 2\,OH^-(aq)$

This anode reaction takes place instead of $2\,H_2O(l) \rightarrow O_2(g) + 4\,H^+(aq) + 4\,e^-$ because of a high overvoltage for formation of gaseous O_2.
This cathode reaction takes place instead of $Mg^{2+}(aq) + 2\,e^- \rightarrow Mg(s)$ because H_2O is easier to reduce than Mg^{2+}.

18.125 (a) K(l) and $Cl_2(g)$ (b) $H_2(g)$ and $Cl_2(g)$. Solvent H_2O is reduced in preference to K^+.

18.126 (a) NaBr

anode: $2\,Br^-(aq) \rightarrow Br_2(l) + 2\,e^-$
cathode: $2\,H_2O(l) + 2\,e^- \rightarrow H_2(g) + 2\,OH^-(aq)$
overall: $2\,H_2O(l) + 2\,Br^-(aq) \rightarrow Br_2(l) + H_2(g) + 2\,OH^-(aq)$

(b) $CuCl_2$

anode: $2\,Cl^-(aq) \rightarrow Cl_2(g) + 2\,e^-$
cathode: $Cu^{2+}(aq) + 2\,e^- \rightarrow Cu(s)$
overall: $Cu^{2+}(aq) + 2\,Cl^-(aq) \rightarrow Cu(s) + Cl_2(g)$

(c) LiOH

anode: $4\ OH^-(aq) \rightarrow O_2(g) + 2\ H_2O(l) + 4\ e^-$

cathode: $4\ H_2O(l) + 4\ e^- \rightarrow 2\ H_2(g) + 4\ OH^-(aq)$

overall: $2\ H_2O(l) \rightarrow O_2(g) + 2\ H_2(g)$

18.127 (a) Ag_2SO_4

anode: $2\ H_2O(l) \rightarrow O_2(g) + 4\ H^+(aq) + 4\ e^-$

cathode: $4\ Ag^+(aq) + 4\ e^- \rightarrow 4\ Ag(s)$

overall: $4\ Ag^+(aq) + 2\ H_2O(l) \rightarrow O_2(g) + 4\ H^+(aq) + 4\ Ag(s)$

(b) $Ca(OH)_2$

anode: $4\ OH^-(aq) \rightarrow O_2(g) + 2\ H_2O(l) + 4\ e^-$

cathode: $4\ H_2O(l) + 4\ e^- \rightarrow 2\ H_2(g) + 4\ OH^-(aq)$

overall: $2\ H_2O(l) \rightarrow O_2(g) + 2\ H_2(g)$

(c) KI

anode: $2\ I^-(aq) \rightarrow I_2(s) + 2\ e^-$

cathode: $2\ H_2O(l) + 2\ e^- \rightarrow H_2(g) + 2\ OH^-(aq)$

overall: $2\ I^-(aq) + 2\ H_2O(l) \rightarrow I_2(s) + H_2(g) + 2\ OH^-(aq)$

18.128 $Ag^+(aq) + e^- \rightarrow Ag(s)$; $1\ A = 1\ C/s$

$$\text{mass Ag} = 2.40\ \frac{C}{s} \times 20.0\ \text{min} \times \frac{60\ s}{1\ \text{min}} \times \frac{1\ \text{mol}\ e^-}{96{,}500\ C} \times \frac{1\ \text{mol Ag}}{1\ \text{mol}\ e^-} \times \frac{107.87\ \text{g Ag}}{1\ \text{mol Ag}} = 3.22\ g$$

18.129 $Cu^{2+}(aq) + 2\ e^- \rightarrow Cu(s)$

$$\text{mol}\ e^- = 100.0\ \frac{C}{s} \times 24.0\ h \times \frac{60\ \text{min}}{h} \times \frac{60\ s}{\text{min}} \times \frac{1\ \text{mol}\ e^-}{96{,}500\ C} = 89.5\ \text{mol}\ e^-$$

$$\text{mass Cu} = 89.5\ \text{mol}\ e^- \times \frac{1\ \text{mol Cu}}{2\ \text{mol}\ e^-} \times \frac{63.54\ \text{g Cu}}{1\ \text{mol Cu}} \times \frac{1\ \text{kg}}{1000\ g} = 2.84\ \text{kg Cu}$$

18.130 $2\ Na^+(l) + 2\ Cl^-(l) \rightarrow 2\ Na(l) + Cl_2(g)$

$Na^+(l) + e^- \rightarrow Na(l)$; $1\ A = 1\ C/s$; $1.00 \times 10^3\ kg = 1.00 \times 10^6\ g$

$$\text{Charge} = 1.00 \times 10^6\ \text{g Na} \times \frac{1\ \text{mol Na}}{22.99\ \text{g Na}} \times \frac{1\ \text{mol}\ e^-}{1\ \text{mol Na}} \times \frac{96{,}500\ C}{1\ \text{mol}\ e^-} = 4.20 \times 10^9\ C$$

$$\text{Time} = \frac{4.20 \times 10^9\ C}{30{,}000\ C/s} \times \frac{1\ h}{3600\ s} = 38.9\ h$$

$$1.00 \times 10^6\ \text{g Na} \times \frac{1\ \text{mol Na}}{22.99\ \text{g Na}} \times \frac{1\ \text{mol}\ Cl_2}{2\ \text{mol Na}} = 21{,}748.6\ \text{mol}\ Cl_2$$

$PV = nRT$

$$V = \frac{nRT}{P} = \frac{(21{,}748.6\ \text{mol})\left(0.082\ 06\ \frac{L \cdot atm}{K \cdot mol}\right)(273.15\ K)}{1.00\ atm} = 4.87 \times 10^5\ L\ Cl_2$$

18.131 $Al^{3+} + 3\ e^- \rightarrow Al$; 40.0 kg = 40,000 g; 1 h = 3600 s

$$\text{Charge} = 40{,}000\text{ g Al} \times \frac{1\text{ mol Al}}{26.98\text{ g Al}} \times \frac{3\text{ mol e}^-}{1\text{ mol Al}} \times \frac{96{,}500\text{ C}}{1\text{ mol e}^-} = 4.29 \times 10^8\text{ C}$$

$$\text{Current} = \frac{4.29 \times 10^8\text{ C}}{3600\text{ s}} = 1.19 \times 10^5\text{ A}$$

18.132 $M^{2+} + 2\ e^- \rightarrow M$; 20.0 A = 20.0 C/s

$$\text{mol e}^- = 20.0\ \frac{\text{C}}{\text{s}} \times 325\text{ min} \times \frac{60\text{ s}}{\text{min}} \times \frac{1\text{ mol e}^-}{96{,}500\text{ C}} = 4.04\text{ mol e}^-$$

$$4.04\text{ mol e}^- \times \frac{1\text{ mol M}}{2\text{ mol e}^-} = 2.02\text{ mol M}$$

$$\text{molar mass} = \frac{111\text{ g M}}{2.02\text{ mol M}} = 54.9\text{ g/mol},\ M^{2+} = Mn^{2+}$$

18.133 $M^{3+} + 3\ e^- \rightarrow M$; 35.0 A = 35.0 C/s

$$\text{mol e}^- = 35.0\ \frac{\text{C}}{\text{s}} \times 4.0\text{ h} \times \frac{60\text{ min}}{\text{h}} \times \frac{60\text{ s}}{\text{min}} \times \frac{1\text{ mol e}^-}{96{,}500\text{ C}} = 5.22\text{ mol e}^-$$

$$5.22\text{ mol e}^- \times \frac{1\text{ mol M}}{3\text{ mol e}^-} = 1.74\text{ mol M}$$

$$\text{molar mass} = \frac{90.52\text{ g M}}{1.74\text{ mol M}} = 52.0\text{ g/mol},\ M^{3+} = Cr^{3+}$$

Chapter Problems

18.134 $$\text{volume} = \left(0.0100\text{ mm} \times \frac{1\text{ cm}}{10\text{ mm}}\right)(10.0\text{ cm})^2 = 0.100\text{ cm}^3$$

$$\text{mol } Al_2O_3 = (0.100\text{ cm}^3)(3.97\text{ g/cm}^3)\frac{1\text{ mol } Al_2O_3}{102.0\text{ g } Al_2O_3} = 3.892 \times 10^{-3}\text{ mol } Al_2O_3$$

$$\text{mole e}^- = 3.892 \times 10^{-3}\text{ mol } Al_2O_3 \times \frac{6\text{ mol e}^-}{1\text{ mol } Al_2O_3} = 0.02335\text{ mol e}^-$$

$$\text{coulombs} = 0.02335\text{ mol e}^- \times \frac{96{,}500\text{ C}}{1\text{ mol e}^-} = 2253\text{ C}$$

$$\text{time} = \frac{\text{C}}{\text{A}} = \frac{2253\text{ C}}{0.600\text{ C/s}} \times \frac{1\text{ min}}{60\text{ s}} = 62.6\text{ min}$$

18.135

anode reaction	$Ti(s) + 2\ H_2O(l) \rightarrow TiO_2(s) + 4\ H^+(aq) + 4\ e^-$	$E^\circ = 1.066$ V
cathode reaction	$4\ H^+(aq) + 4\ e^- \rightarrow 2\ H_2(g)$	$E^\circ = 0.000$ V
overall reaction	$Ti(s) + 2\ H_2O(l) \rightarrow TiO_2(s) + 2\ H_2(g)$	$E^\circ = 1.066$ V

18.136 $PbSO_4(s) + 2\ H_2O(l) \rightarrow PbO_2(s) + 4\ H^+(aq) + SO_4^-(aq) + 2\ e^-$

$PbSO_4(s) + 2\ e^- \rightarrow Pb(s) + SO_4^-(aq)$

$2\ PbSO_4(s) + 2\ H_2O(l) \rightarrow Pb(s) + PbO_2(s) + 2\ H^+(aq) + 2\ HSO_4^-(aq)$

(a) The reaction represents an electrolytic cell.
(b) During the recharging (electrolysis) 250.0 g of $PbSO_4$ are oxidized to PbO_2 and 250.0 g of $PbSO_4$ are reduced to Pb.

$$\text{mol PbSO}_4 = 250.0 \text{ g PbSO}_4 \times \frac{1 \text{ mol PbSO}_4}{303.3 \text{ g PbSO}_4} = 0.824 \text{ mol PbSO}_4$$

$$\text{mole e}^- = 0.824 \text{ mol PbSO}_4 \times \frac{2 \text{ mol e}^-}{1 \text{ mol PbSO}_4} = 1.648 \text{ mol e}^-$$

$$\text{coulombs} = 1.648 \text{ mol e}^- \times \frac{96{,}500 \text{ C}}{1 \text{ mol e}^-} = 1.591 \times 10^5 \text{ C}$$

(c) $\text{time} = \dfrac{\text{C}}{\text{A}} = \dfrac{1.591 \times 10^5 \text{ C}}{500 \text{ C/s}} \times \dfrac{1 \text{ min}}{60 \text{ s}} = 5.3 \text{ min}$

18.137 (a) $2\ MnO_4^-(aq) + 16\ H^+(aq) + 5\ Sn^{2+}(aq) \rightarrow 2\ Mn^{2+}(aq) + 5\ Sn^{4+}(aq) + 8\ H_2O(l)$
(b) MnO_4^- is the oxidizing agent; Sn^{2+} is the reducing agent.
(c) $E° = 1.51\ V + (-0.15\ V) = 1.36\ V$

18.138 $2\ Mn^{3+}(aq) + 2\ H_2O(l) \rightarrow Mn^{2+}(aq) + MnO_2(s) + 4\ H^+(aq)$
$E° = 1.54\ V + (-0.95\ V) = +0.59\ V$
Because E° is positive, the disproportionation is spontaneous under standard-state conditions.

18.139 (a) Ag^+ is the strongest oxidizing agent because Ag^+ has the most positive standard reduction potential.
Pb is the strongest reducing agent because Pb^{2+} has the most negative standard reduction potential.
(b)

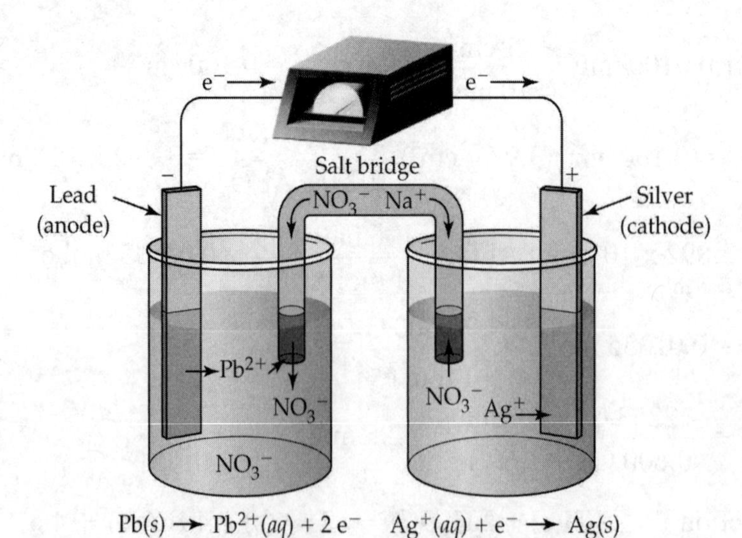

(c) $Pb(s) + 2\ Ag^+(aq) \rightarrow Pb^{2+}(aq) + 2\ Ag(s)$; n = 2 mol e⁻
$E° = E°_{ox} + E°_{red} = 0.13\ V + 0.80\ V = 0.93\ V$

$$\Delta G° = -nFE° = -(2 \text{ mol e}^-)\left(\frac{96{,}500 \text{ C}}{1 \text{ mol e}^-}\right)(0.93 \text{ V})\left(\frac{1 \text{ J}}{1 \text{ C} \cdot \text{V}}\right) = -179{,}490 \text{ J} = -180 \text{ kJ}$$

$$E^\circ = \frac{0.0592\ V}{n}\log K; \quad \log K = \frac{nE^\circ}{0.0592\ V} = \frac{(2)(0.93\ V)}{0.0592\ V} = 31; \quad K = 10^{31}$$

(d) $E = E^\circ - \dfrac{0.0592\,V}{n}\log\dfrac{[Pb^{2+}]}{[Ag^+]^2} = 0.93\ V - \dfrac{0.0592\,V}{2}\log\left(\dfrac{0.01}{(0.01)^2}\right) = 0.87\ V$

18.140 For Pb^{2+}, $E = -0.13 - \dfrac{0.0592\,V}{2}\log\dfrac{1}{[Pb^{2+}]}$

For Cd^{2+}, $E = -0.40 - \dfrac{0.0592\,V}{2}\log\dfrac{1}{[Cd^{2+}]}$

Set these two equations for E equal to each other and solve for $[Cd^{2+}]/[Pb^{2+}]$.

$$-0.13 - \frac{0.0592\,V}{2}\log\frac{1}{[Pb^{2+}]} = -0.40 - \frac{0.0592\,V}{2}\log\frac{1}{[Cd^{2+}]}$$

$$0.27 = \frac{0.0592\,V}{2}(\log[Cd^{2+}] - \log[Pb^{2+}]) = \frac{0.0592\,V}{2}\log\frac{[Cd^{2+}]}{[Pb^{2+}]}$$

$$\log\frac{[Cd^{2+}]}{[Pb^{2+}]} = \frac{(0.27)(2)}{0.0592} = 9.1; \qquad \frac{[Cd^{2+}]}{[Pb^{2+}]} = 10^{9.1} = 1 \times 10^9$$

18.141 (a)

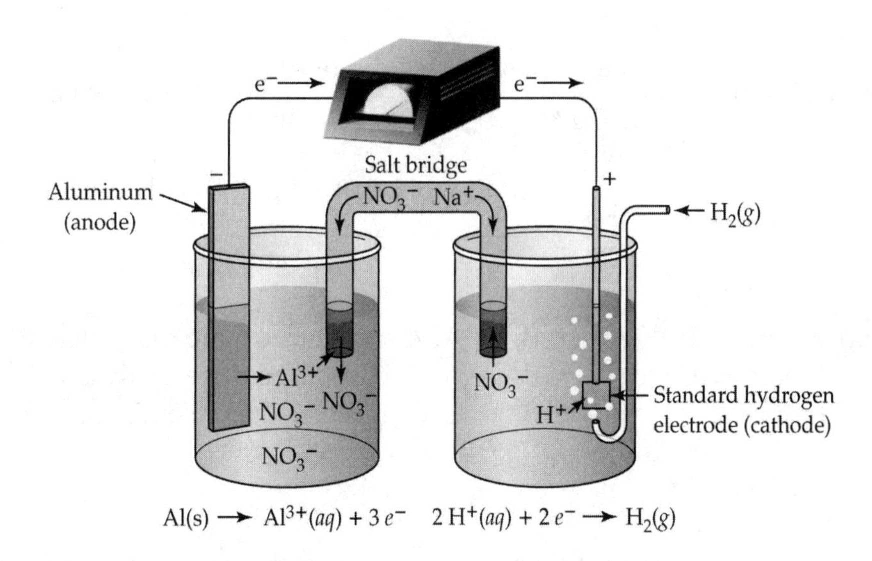

(b) $2\ Al(s) + 6\ H^+(aq) \rightarrow 2\ Al^{3+}(aq) + 3\ H_2(g)$

$E^\circ = E^\circ_{ox} + E^\circ_{red} = 1.66\ V + 0.00\ V = 1.66\ V$

(c)

$$E = E^\circ - \frac{0.0592\,V}{n}\log\frac{[Al^{3+}]^2(P_{H_2})^3}{[H^+]^6} = 1.66\ V - \frac{(0.0592\ V)}{6}\log\left(\frac{(0.10)^2(10.0)^3}{(0.10)^6}\right) = 1.59\ V$$

(d) $\Delta G^\circ = -nFE^\circ = -(6\ mol\ e^-)\left(\dfrac{96{,}500\ C}{1\ mol\ e^-}\right)(1.66\ V)\left(\dfrac{1\ J}{1\ C\cdot V}\right) = -961{,}140\ J = -961\ kJ$

$$E^\circ = \frac{0.0592\ V}{n}\log K; \quad \log K = \frac{nE^\circ}{0.0592\ V} = \frac{(6)(1.66\ V)}{0.0592\ V} = 168; \quad K = 10^{168}$$

(e) $$\text{mass Al} = 10.0\ \frac{C}{s} \times 25.0\ \text{min} \times \frac{60\ s}{1\ \text{min}} \times \frac{1\ \text{mol e}^-}{96{,}500\ C} \times \frac{1\ \text{mol Al}}{3\ \text{mol e}^-} \times \frac{26.98\ \text{g Al}}{1\ \text{mol Al}} = 1.40\ g$$

18.142 $Ni(OH)_2$, 92.71

(a) $$3.35 \times 10^{-2}\ \text{g}\ Ni(OH)_2 \times \frac{1\ \text{mol}\ Ni(OH)_2}{92.71\ \text{g}\ Ni(OH)_2} \times \frac{1\ \text{mol Zn}}{2\ \text{mol}\ Ni(OH)_2} \times \frac{65.38\ \text{g Zn}}{1\ \text{mol Zn}} = 0.0118\ \text{g Zn}$$

(b) 0.100 A = 0.100 C/s

$$6.17 \times 10^{-2}\ \text{g Zn} \times \frac{1\ \text{mol Zn}}{65.38\ \text{g Zn}} \times \frac{2\ \text{mol e}^-}{1\ \text{mol Zn}} = 1.89 \times 10^{-3}\ \text{mol e}^-$$

$$1.89 \times 10^{-3}\ \text{mol e}^- \times \frac{96{,}500\ C}{1\ \text{mol e}^-} \times \frac{1\ s}{0.100\ C} \times \frac{1\ \text{min}}{60\ s} = 30.4\ \text{min}$$

18.143 (a)

$$5 \times [2\ Cl^-(aq) \rightarrow Cl_2(g) + 2\ e^-] \qquad E^\circ = -1.36\ V$$
$$2 \times [MnO_4^-(aq) + 8\ H^+(aq) + 5\ e^- \rightarrow Mn^{2+}(aq) + 4\ H_2O(l)] \qquad E^\circ = 1.51\ V$$
$$10\ Cl^-(aq) + 2\ MnO_4^-(aq) + 16\ H^+(aq) \rightarrow 5\ Cl_2(g) + 2\ Mn^{2+}(aq) + 8\ H_2O(l) \qquad E^\circ = 0.15\ V$$

(b) $E^\circ = 0.15\ V$

$$\Delta G^\circ = -nFE^\circ = -(10\ \text{mol e}^-)\left(\frac{96{,}500\ C}{1\ \text{mol e}^-}\right)(0.15\ V)\left(\frac{1\ J}{1\ C \cdot V}\right) = -144{,}750\ J = -1.4 \times 10^2\ kJ$$

(c) $KMnO_4$, 158.0

$$179\ \text{g}\ KMnO_4 \times \frac{1\ \text{mol}\ KMnO_4}{158.0\ \text{g}\ KMnO_4} \times \frac{5\ \text{mol}\ Cl_2}{2\ \text{mol}\ KMnO_4} = 2.83\ \text{mol}\ Cl_2$$

$$PV = nRT; \quad V = \frac{nRT}{P} = \frac{(2.83\ \text{mol})\left(0.082\ 06\ \frac{L \cdot atm}{K \cdot mol}\right)(298\ K)}{1.0\ atm} = 69\ L$$

18.144 (a) From: $B + A^+ \rightarrow B^+ + A$, A^+ is reduced more easily than B^+

From: $C + A^+ \rightarrow C^+ + A$, A^+ is reduced more easily than C^+

From: $B + C^+ \rightarrow B^+ + C$, C^+ is reduced more easily than B^+

$A^+ + e^- \rightarrow A$

$C^+ + e^- \rightarrow C$

$B^+ + e^- \rightarrow B$

(b) A^+ is the strongest oxidizing agent; B is the strongest reducing agent

(c) $A^+ + B \rightarrow B^+ + A$

18.145 (a) oxidizing agents: PbO_2, H^+, $Cr_2O_7^{2-}$; reducing agents: Al, Fe, Ag

(b) PbO_2 is the strongest oxidizing agent. H^+ is the weakest oxidizing agent.

(c) Al is the strongest reducing agent. Ag is the weakest reducing agent.

(d) oxidized by Cu^{2+}: Fe and Al; reduced by H_2O_2: PbO_2 and $Cr_2O_7^{2-}$

18.146 (a) $2\ Na(soln) + S(soln) \rightarrow 2\ Na^+(soln) + S^{2-}(soln)$

(b) $1\ W = \dfrac{1\ C \cdot 1\ V}{s}$; $\dfrac{1\ W}{1\ V} = 1\ C/s$; $\dfrac{25\ kW}{2.0\ V} = \dfrac{25{,}000\ W}{2.0\ V} = 12{,}500\ C/s$

$$12{,}500\ C/s \times 32\ min \times \frac{60\ s}{1\ min} \times \frac{1\ mol\ e^-}{96{,}500\ C} \times \frac{1\ mol\ Na^+}{1\ mol\ e^-} \times \frac{22.99\ g\ Na}{1\ mol\ Na} = 5.7 \times 10^3\ g = 5.7\ kg$$

18.147 (a) $3\ CH_3CH_2OH(aq) + 2\ Cr_2O_7^{2-}(aq) + 16\ H^+(aq) \rightarrow$
$3\ CH_3CO_2H(aq) + 4\ Cr^{3+}(aq) + 11\ H_2O(l)$

oxidation:
$3\ CH_3CH_2OH(aq) + 3\ H_2O(l) \rightarrow 3\ CH_3CO_2H(aq) + 12\ H^+(aq) + 12\ e^-$ $E° = -0.058V$
reduction:
$2\ Cr_2O_7^{2-}(aq) + 28\ H^+(aq) + 12\ e^- \rightarrow 4\ Cr^{3+}(aq) + 14\ H_2O(l)$ $E° = 1.36\ V$
overall $E° = 1.30\ V$

(b) $$E = E° - \frac{0.0592\ V}{n} \log \frac{[CH_3CO_2H]^3[Cr^{3+}]^4}{[CH_3CH_2OH]^3[Cr_2O_7^{2-}]^2[H^+]^{16}}$$

$pH = 4.00$, $[H^+] = 0.000\ 10\ M$

$$E = 1.30\ V - \frac{(0.0592\ V)}{12} \log\left(\frac{(1.0)^3(1.0)^4}{(1.0)^3(1.0)^2(0.000\ 10)^{16}}\right)$$

$$E = 1.30\ V - \frac{(0.0592\ V)}{12} \log \frac{1}{(0.000\ 10)^{16}} = 0.98\ V$$

18.148 (a) $\Delta G° = -nFE°$

$\Delta G°_3 = \Delta G°_1 + \Delta G°_2$ therefore $-n_3FE°_3 = -n_1FE°_1 + (-n_2FE°_2)$

$n_3E°_3 = n_1E°_1 + n_2E°_2$

$$E°_3 = \frac{n_1E°_1 + n_2E°_2}{n_3}$$

(b) $$E°_3 = \frac{(3)(-0.04\ V) + (2)(0.45\ V)}{1} = 0.78\ V$$

(c) $E°$ values would be additive ($E°_3 = E°_1 + E°_2$) if reaction (3) is an overall cell reaction because the electrons in the two half reactions, (1) and (2), cancel. That is, $n_1 = n_2 = n_3$ in the equation for $E°_3$.

18.149 anode: $Ag(s) + Cl^-(aq) \rightarrow AgCl(s) + e^-$
cathode: $Ag^+(aq) + e^- \rightarrow Ag(s)$
overall: $Ag^+(aq) + Cl^-(aq) \rightarrow AgCl(s)$ $E° = 0.578\ V$

For $AgCl(s) \rightleftharpoons Ag^+(aq) + Cl^-(aq)$ $E° = -0.578\ V$

$$E° = \frac{0.0592\ V}{n} \log K; \quad \log K = \frac{nE°}{0.0592\ V} = \frac{(1)(-0.578\ V)}{0.0592\ V} = -9.76$$

$K = K_{sp} = 10^{-9.76} = 1.7 \times 10^{-10}$

18.150 (a)

anode:	$Cu(s) \rightarrow Cu^{2+}(aq) + 2\ e^-$	$E° = -0.34\ V$
cathode:	$2\ Ag^+(aq) + 2\ e^- \rightarrow 2\ Ag(s)$	$E° = 0.80\ V$
overall:	$2\ Ag^+(aq) + Cu(s) \rightarrow Cu^{2+}(aq) + 2\ Ag(s)$	$E° = 0.46\ V$

$$E = E° - \frac{0.0592\ V}{n}\log\frac{[Cu^{2+}]}{[Ag^+]^2} = 0.46\ V - \frac{(0.0592\ V)}{2}\log\left(\frac{1.0}{(0.050)^2}\right) = 0.38\ V$$

(b) $$[Ag^+] = \frac{K_{sp}}{[Br^-]} = \frac{5.4 \times 10^{-13}}{1.0\ M} = 5.4 \times 10^{-13}\ M$$

$$E = E° - \frac{0.0592\ V}{n}\log\frac{[Cu^{2+}]}{[Ag^+]^2} = 0.46\ V - \frac{(0.0592\ V)}{2}\log\left(\frac{1.0}{(5.4 \times 10^{-13})^2}\right) = -0.27\ V$$

The cell potential for the spontaneous reaction is E = 0.27 V.

The spontaneous reaction is: $Cu^{2+}(aq) + 2\ Ag(s) + 2\ Br^-(aq) \rightarrow 2\ AgBr(s) + Cu(s)$

(c)

$Cu^{2+}(aq) + 2\ e^- \rightarrow Cu(s)$	$E° = 0.34\ V$
$2\ Ag(s) + 2\ Br^-(aq) \rightarrow 2\ AgBr(s) + 2\ e^-$	$E° = ?$
$Cu^{2+}(aq) + 2\ Ag(s) + 2\ Br^-(aq) \rightarrow 2\ AgBr(s) + Cu(s)$	$E° = 0.27\ V$

$E° = ? = 0.27\ V - 0.34\ V = -0.07\ V$

For: $AgBr(s) + e^- \rightarrow Ag(s) + Br^-(aq)$

the standard reduction potential is $E° = 0.07\ V$

18.151 $4\ Fe^{2+}(aq) + O_2(g) + 4\ H^+(aq) \rightarrow 4\ Fe^{3+}(aq) + 2\ H_2O(l)$

oxidation	$4\ Fe^{2+}(aq) \rightarrow 4\ Fe^{3+}(aq) + 4\ e^-$	$E° = -0.77\ V$
reduction	$O_2(g) + 4\ H^+(aq) + 4\ e^- \rightarrow 2\ H_2O(l)$	$E° = 1.23\ V$
	overall	$E° = 0.46\ V$

$$P_{O_2} = 160\ mm\ Hg \times \frac{1.00\ atm}{760\ mm\ Hg} = 0.211\ atm$$

$$E = E° - \frac{0.0592\ V}{n}\log\frac{[Fe^{3+}]^4}{[Fe^{2+}]^4[H^+]^4(P_{O_2})}$$

$$E = 0.46\ V - \frac{0.0592\ V}{4}\log\frac{(1 \times 10^{-7})^4}{(1 \times 10^{-7})^4(1 \times 10^{-7})^4(0.211)}$$

$E = 0.46\ V - 0.42\ V = 0.04\ V$

Because E is positive, the reaction is spontaneous.

18.152 $H_2MoO_4(aq) + As(s) \rightarrow Mo^{3+}(aq) + H_3AsO_4(aq)$

$H_2MoO_4(aq) \rightarrow Mo^{3+}(aq)$

$H_2MoO_4(aq) \rightarrow Mo^{3+}(aq) + 4\ H_2O(l)$

$6\ H^+(aq) + H_2MoO_4(aq) \rightarrow Mo^{3+}(aq) + 4\ H_2O(l)$

$[3\ e^- + 6\ H^+(aq) + H_2MoO_4(aq) \rightarrow Mo^{3+}(aq) + 4\ H_2O(l)] \times 5$

(reduction half reaction)

$As(s) \rightarrow H_3AsO_4(aq)$

$As(s) + 4\ H_2O(l) \rightarrow H_3AsO_4(aq)$

$As(s) + 4\ H_2O(l) \rightarrow H_3AsO_4(aq) + 5\ H^+(aq)$

$[As(s) + 4\,H_2O(l) \rightarrow H_3AsO_4(aq) + 5\,H^+(aq) + 5\,e^-] \times 3$ (oxidation half reaction)

Combine the two half reactions.

$30\,H^+(aq) + 5\,H_2MoO_4(aq) + 3\,As(s) + 12\,H_2O(l) \rightarrow$
$5\,Mo^{3+}(aq) + 3\,H_3AsO_4(aq) + 15\,H^+(aq) + 20\,H_2O(l)$

$15\,H^+(aq) + 5\,H_2MoO_4(aq) + 3\,As(s) \rightarrow 5\,Mo^{3+}(aq) + 3\,H_3AsO_4(aq) + 8\,H_2O(l)$

$5 \times [H_2MoO_4(aq) + 2\,H^+(aq) + 2\,e^- \rightarrow MoO_2(s) + 2\,H_2O(l)]$ $E° = +0.646$ V
$5 \times [MoO_2(s) + 4\,H^+(aq) + e^- \rightarrow Mo^{3+}(aq) + 2\,H_2O(l)]$ $E° = -0.008$ V
$3 \times [As(s) + 3\,H_2O(l) \rightarrow H_3AsO_3(aq) + 3\,H^+(aq) + 3\,e^-]$ $E° = -0.240$ V
$3 \times [H_3AsO_3(aq) + H_2O(l) \rightarrow H_3AsO_4(aq) + 2\,H^+(aq) + 2\,e^-]$ $E° = -0.560$ V

$15\,H^+(aq) + 5\,H_2MoO_4(aq) + 3\,As(s) \rightarrow 5\,Mo^{3+}(aq) + 3\,H_3AsO_4(aq) + 8\,H_2O(l)$

$$\Delta G° = -nFE° = -(10 \text{ mol e}^-)\left(\frac{96{,}500 \text{ C}}{1 \text{ mol e}^-}\right)(0.646 \text{ V})\left(\frac{1 \text{ J}}{1 \text{ C}\cdot\text{V}}\right) = -623{,}390 \text{ J} = -623.4 \text{ kJ}$$

$$\Delta G° = -nFE° = -(5 \text{ mol e}^-)\left(\frac{96{,}500 \text{ C}}{1 \text{ mol e}^-}\right)(-0.008 \text{ V})\left(\frac{1 \text{ J}}{1 \text{ C}\cdot\text{V}}\right) = 3{,}860 \text{ J} = +3.9 \text{ kJ}$$

$$\Delta G° = -nFE° = -(9 \text{ mol e}^-)\left(\frac{96{,}500 \text{ C}}{1 \text{ mol e}^-}\right)(-0.240 \text{ V})\left(\frac{1 \text{ J}}{1 \text{ C}\cdot\text{V}}\right) = 208{,}440 \text{ J} = +208.4 \text{ kJ}$$

$$\Delta G° = -nFE° = -(6 \text{ mol e}^-)\left(\frac{96{,}500 \text{ C}}{1 \text{ mol e}^-}\right)(-0.560 \text{ V})\left(\frac{1 \text{ J}}{1 \text{ C}\cdot\text{V}}\right) = 324{,}240 \text{ J} = +324.2 \text{ kJ}$$

$\Delta G°(\text{total}) = -623.4 \text{ kJ} + 3.9 \text{ kJ} + 208.4 \text{ kJ} + 324.2 \text{ kJ} = -86.9 \text{ kJ} = -86{,}900 \text{ J}$

$1 \text{ V} = 1 \text{ J/C}$

$$\Delta G° = -nFE°;\ E° = \frac{-\Delta G°}{nF} = \frac{-(-86{,}900 \text{ J})}{(15 \text{ mol e}^-)\left(\frac{96{,}500 \text{ C}}{1 \text{ mol e}^-}\right)} = +0.060 \text{ J/C} = +0.060 \text{ V}$$

18.153 First calculate E° for the galvanic cell in order to determine $E°_1$.

anode: $5\,[2\,Hg(l) + 2\,Br^-(aq) \rightarrow Hg_2Br_2(s) + 2\,e^-]$ $E°_1 = ?$
cathode: $2\,[MnO_4^-(aq) + 8\,H^+(aq) + 5\,e^- \rightarrow Mn^{2+}(aq) + 4\,H_2O(l)]$ $E°_2 = 1.51$ V
overall: $2\,MnO_4^-(aq) + 10\,Hg(l) + 10\,Br^-(aq) + 16\,H^+(aq) \rightarrow$
$2\,Mn^{2+}(aq) + 5\,Hg_2Br_2(s) + 8\,H_2O(l)$

$n = 10$ mol e⁻

$$E = E° - \frac{0.0592 \text{ V}}{n} \log \frac{[Mn^{2+}]^2}{[Br^-]^{10}[MnO_4^-]^2[H^+]^{16}}$$

$$1.214 \text{ V} = E° - \frac{(0.0592 \text{ V})}{10} \log\left(\frac{(0.10)^2}{(0.10)^{10}(0.10)^2(0.10)^{16}}\right)$$

$$1.214 \text{ V} = E° - \frac{(0.0592 \text{ V})}{10} \log \frac{1}{(0.10)^{26}} = E° - 0.154 \text{ V}$$

$E° = 1.214 + 0.154 = 1.368$ V

$E°_1 + E°_2 = 1.368$ V; $E°_1 + 1.51$ V $= 1.368$ V; $E°_1 = 1.368$ V $- 1.51$ V $= -0.142$ V

oxidation:	$2\ Hg(l) \rightarrow Hg_2^{2+}(aq) + 2\ e^-$	$E° = -0.80$ V (Appendix D)
reduction:	$Hg_2Br_2(s) + 2\ e^- \rightarrow 2\ Hg(l) + 2\ Br^-(aq)$	$E° = +0.142$ V (from $E°_1$)
overall:	$Hg_2Br_2(s) \rightarrow Hg_2^{2+}(aq) + 2\ Br^-(aq)$	$E° = -0.658$ V

$$E° = \frac{0.0592\text{ V}}{n}\log K;\quad \log K = \frac{nE°}{0.0592\text{ V}} = \frac{(2)(-0.658\text{ V})}{0.0592\text{ V}} = -22.2$$

$K = K_{sp} = 10^{-22.2} = 6 \times 10^{-23}$

18.154

oxidation:	$Cu^+(aq) \rightarrow Cu^{2+}(aq) + e^-$	$E° = -0.15$ V
reduction:	$Cu^{2+}(aq) + 2\ CN^-(aq) + e^- \rightarrow Cu(CN)_2^-(aq)$	$E° = 1.103$ V
overall:	$Cu^+(aq) + 2\ CN^-(aq) \rightarrow Cu(CN)_2^-(aq)$	$E° = 0.953$ V

$$E° = \frac{0.0592\text{ V}}{n}\log K;\quad \log K = \frac{nE°}{0.0592\text{ V}} = \frac{(1)(0.953\text{ V})}{0.0592\text{ V}} = 16.1$$

$K = K_f = 10^{16.1} = 1 \times 10^{16}$

18.155 $$E_1 = E°_1 - \frac{0.0592\text{ V}}{6}\log\frac{[Cu^{2+}]^3(P_{NO})^2}{[NO_3^-]^2[H^+]^8} \text{ and } E_2 = E°_2 - \frac{0.0592\text{ V}}{2}\log\frac{[Cu^{2+}](P_{NO_2})^2}{[NO_3^-]^2[H^+]^4}$$

(a) $$E_1 = 0.62\text{ V} - \frac{0.0592\text{ V}}{6}\log\frac{(0.10)^3(1.0 \times 10^{-3})^2}{(1.0)^2(1.0)^8} = 0.71\text{ V}$$

$$E_2 = 0.45\text{ V} - \frac{0.0592\text{ V}}{2}\log\frac{(0.10)(1.0 \times 10^{-3})^2}{(1.0)^2(1.0)^4} = 0.66\text{ V}$$

Reaction (1) has the greater thermodynamic tendency to occur because of the larger positive potential.

(b) $$E_1 = 0.62\text{ V} - \frac{0.0592\text{ V}}{6}\log\frac{(0.10)^3(1.0 \times 10^{-3})^2}{(10.0)^2(10.0)^8} = 0.81\text{ V}$$

$$E_2 = 0.45\text{ V} - \frac{0.0592\text{ V}}{2}\log\frac{(0.10)(1.0 \times 10^{-3})^2}{(10.0)^2(10.0)^4} = 0.83\text{ V}$$

Reaction (2) has the greater thermodynamic tendency to occur because of the larger positive potential.

(c) Set the two equations equal to each other and solve for x.

$$0.62\text{ V} - \frac{0.0592\text{ V}}{6}\log\frac{(0.10)^3(1.0 \times 10^{-3})^2}{(x)^2(x)^8} = 0.45\text{ V} - \frac{0.0592\text{ V}}{2}\log\frac{(0.10)(1.0 \times 10^{-3})^2}{(x)^2(x)^4}$$

$$0.17 - \frac{0.0592\text{ V}}{6}[(-9) - 10\log x] = -\frac{0.0592\text{ V}}{2}[(-7) - 6\log x]$$

$$0.0516 = 0.0789\log x;\quad \frac{0.0516}{0.0789} = \log x;\quad 0.654 = \log x$$

$[HNO_3] = x = 10^{0.654} = 4.5$ M

Multiconcept Problems

18.156 (a) $4\ CH_2{=}CHCN + 2\ H_2O \rightarrow 2\ NC(CH_2)_4CN + O_2$

(b) $\text{mol } e^- = 3000\ C/s \times 10.0\ h \times \frac{3600\ s}{1\ h} \times \frac{1\ mol\ e^-}{96{,}500\ C} = 1119.2\ mol\ e^-$

mass adiponitrile =

$$1119.2\ mol\ e^- \times \frac{1\ mol\ adiponitrile}{2\ mol\ e^-} \times \frac{108.14\ g\ adiponitrile}{1\ mol\ adiponitrile} \times \frac{1.0\ kg}{1000\ g} = 60.5\ kg$$

(c) $1119.2\ mol\ e^- \times \frac{1\ mol\ O_2}{4\ mol\ e^-} = 279.8\ mol\ O_2$

$$PV = nRT; \quad V = \frac{nRT}{P} = \frac{(279.8\ mol)\left(0.082\ 06\ \frac{L \cdot atm}{K \cdot mol}\right)(298\ K)}{\left(740\ mm\ Hg \times \frac{1\ atm}{760\ mm\ Hg}\right)} = 7030\ L\ O_2$$

18.157 (a) $2\ MnO_4^-(aq) + 5\ H_2C_2O_4(aq) + 6\ H^+(aq) \rightarrow 2\ Mn^{2+}(aq) + 10\ CO_2(g) + 8\ H_2O(l)$

(b) oxidation: $5[H_2C_2O_4(aq) \rightarrow 2\ CO_2(g) + 2\ H^+(aq) + 2\ e^-]$ $\quad E° = 0.49\ V$

reduction: $2[MnO_4^-(aq) + 8\ H^+(aq) + 5\ e^- \rightarrow Mn^{2+}(aq) + 4\ H_2O(l)]$ $\quad \underline{E° = 1.51\ V}$

overall $E° = 2.00\ V$

(c)

$$\Delta G° = -nFE° = -(10\ mol\ e^-)\left(\frac{96{,}500\ C}{1\ mol\ e^-}\right)(2.00\ V)\left(\frac{1\ J}{1\ C \cdot V}\right) = -1{,}930{,}000\ J = -1{,}930\ kJ$$

$$E° = \frac{0.0592\ V}{n}\log K; \quad \log K = \frac{n E°}{0.0592\ V} = \frac{(10)(2.00\ V)}{0.0592\ V} = 338; \quad K = 10^{338}$$

(d) $Na_2C_2O_4$, 134.0

$$1.200\ g\ Na_2C_2O_4 \times \frac{1\ mol\ Na_2C_2O_4}{134.0\ g\ Na_2C_2O_4} \times \frac{2\ mol\ KMnO_4}{5\ mol\ Na_2C_2O_4} = 3.582 \times 10^{-3}\ mol\ KMnO_4$$

$$\text{molarity} = \frac{3.582 \times 10^{-3}\ mol}{0.032\ 50\ L} = 0.1102\ M$$

18.158 $Ba(s) + Cl_2(g) \rightarrow BaCl_2(s)$ $\quad \Delta G°_f = -806.7\ kJ/mol$

$\underline{BaCl_2(s) \rightarrow Ba^{2+}(aq) + 2\ Cl^-(aq)}$ $\quad \underline{\Delta G°_1 = -16.7\ kJ/mol}$

$Ba(s) + Cl_2(g) \rightarrow Ba^{2+}(aq) + 2\ Cl^-(aq)$ $\quad \Delta G°_2 = \Delta G°_f + \Delta G°_1 = -823.4\ kJ/mol$

$$E° = \frac{-\Delta G°}{nF} = \frac{-(-823{,}400\ J)}{(2\ mol\ e^-)\left(\frac{96{,}500\ C}{1\ mol\ e^-}\right)} = +4.266\ J/C = +4.266\ V$$

$Ba(s) + Cl_2(g) \rightarrow Ba^{2+}(aq) + 2\ Cl^-(aq)$ $E^\circ = +4.266\ V$
$2\ Cl^-(aq) \rightarrow Cl_2(g) + 2\ e^-$ $E^\circ = -1.36\ V$
$Ba(s) \rightarrow Ba^{2+}(aq) + 2\ e^-$ $E^\circ = 2.91\ V$
$Ba^{2+}(aq) + 2\ e^- \rightarrow Ba(s)$ $E^\circ = -2.91\ V$

18.159 For a concentration cell $E^\circ = 0$.

(a) $$E = 0.0965\ V = -\frac{0.0592\ V}{n}\log\frac{[Cu^{2+}]}{[Cu^{2+}]} = -\frac{(0.0592\ V)}{2}\log\left(\frac{x}{0.10}\right)$$

$$\frac{(0.0965\ V)(2)}{-0.0592\ V} = \log(x) - \log(0.10)$$

$\log(x) = -4.26;\ x = [Cu^{2+}] = 10^{-4.26}\ M = 5.5 \times 10^{-5}\ M$

(b) $$E = 0.179\ V = -\frac{0.0592\ V}{n}\log\frac{[Cu^{2+}]}{[Cu^{2+}]} = -\frac{(0.0592\ V)}{2}\log\left(\frac{x}{0.10}\right)$$

$$\frac{(0.179\ V)(2)}{-0.0592\ V} = \log(x) - \log(0.10)$$

$\log(x) = -7.05;\ x = [Cu^{2+}] = 10^{-7.05}\ M = 8.9 \times 10^{-8}\ M$

	$Cu^{2+}(aq)$	+	$2\ en(aq)$	$\rightleftarrows$	$Cu(en)_2^{2+}(aq)$
before reaction (M)	0.10		0.20		0
assume 100% reaction	−0.10		−0.20		+0.10
after reaction (M)	0		0		0.10
assume small back rxn	+x		+2x		−x
equil (M)	x		2x		0.10 − x

$x = 8.9 \times 10^{-8}\ M$

$$K_f = \frac{[Cu(en)_2^{2+}]}{[Cu^{2+}][en]^2} = \frac{(0.10 - x)}{(x)(2x)^2} \approx \frac{0.10}{(x)(2x)^2} = \frac{0.10}{(8.9 \times 10^{-8})(2(8.9 \times 10^{-8}))^2} = 3.5 \times 10^{19}$$

18.160 (a) $Cr_2O_7^{2-}(aq) + 6\ Fe^{2+}(aq) + 14\ H^+(aq) \rightarrow 2\ Cr^{3+}(aq) + 6\ Fe^{3+}(aq) + 7\ H_2O(l)$

(b) The two half reactions are:

oxidation: $Fe^{2+}(aq) \rightarrow Fe^{3+}(aq) + e^-$ $E^\circ = -0.77\ V$
reduction: $Cr_2O_7^{2-}(aq) + 14\ H^+(aq) + 6\ e^- \rightarrow 2\ Cr^{3+}(aq) + 7\ H_2O(l)$ $E^\circ = 1.36\ V$

At the equivalence point the potential is given by either of the following expressions:

(1) $$E = 1.36\ V - \frac{0.0592\ V}{6}\log\frac{[Cr^{3+}]^2}{[Cr_2O_7^{2-}][H^+]^{14}}$$

(2) $$E = 0.77\ V - \frac{0.0592\ V}{1}\log\frac{[Fe^{2+}]}{[Fe^{3+}]}$$

where E is the same in both because equilibrium is reached and the solution can have only one potential. Multiplying (1) by 6, adding it to (2), and using some stoichiometric relationships at the equivalence point will simplify the log term.

$$7E = [(6 \times 1.36\ V) + 0.77\ V] - (0.0592\ V)\log\frac{[Fe^{2+}][Cr^{3+}]^2}{[Fe^{3+}][Cr_2O_7^{2-}][H^+]^{14}}$$

At the equivalence point, $[Fe^{2+}] = 6[Cr_2O_7^{2-}]$ and $[Fe^{3+}] = 3[Cr^{3+}]$. Substitute these equalities into the previous equation.

$$7E = [(6 \times 1.36\ V) + 0.77\ V] - (0.0592\ V)\log\frac{6[Cr_2O_7^{2-}][Cr^{3+}]^2}{3[Cr^{3+}][Cr_2O_7^{2-}][H^+]^{14}}$$

Cancel identical terms.

$$7E = [(6 \times 1.36\ V) + 0.77\ V] - (0.0592\ V)\log\frac{6[Cr^{3+}]}{3[H^+]^{14}}$$

mol Fe^{2+} = (0.120 L)(0.100 mol/L) = 0.0120 mol Fe^{2+}

$$\text{mol } Cr_2O_7^{2-} = 0.0120 \text{ mol } Fe^{2+} \times \frac{1 \text{ mol } Cr_2O_7^{2-}}{6 \text{ mol } Fe^{2+}} = 0.002\ 00 \text{ mol } Cr_2O_7^{2-}$$

$$\text{volume } Cr_2O_7^{2-} = 0.002\ 00 \text{ mol} \times \frac{1\ L}{0.120 \text{ mol}} = 0.0167\ L$$

At the equivalence point assume mol Fe^{3+} = initial mol Fe^{2+} = 0.0120 mol

Total volume at the equivalence point is 0.120 L + 0.0167 L = 0.1367 L

$$[Fe^{3+}] = \frac{0.0120 \text{ mol}}{0.1367\ L} = 0.0878\ M; \quad [Cr^{3+}] = [Fe^{3+}]/3 = (0.0878\ M)/3 = 0.0293\ M$$

$[H^+] = 10^{-pH} = 10^{-2.00} = 0.010\ M$

$$7E = [(6 \times 1.36\ V) + 0.77\ V] - (0.0592\ V)\log\frac{6(0.0293)}{3(0.010)^{14}} = 8.93 - 1.585 = 7.345\ V$$

$$E = \frac{7.345\ V}{7} = 1.05\ V \text{ at the equivalence point.}$$

18.161 $2\ H_2(g) + O_2(g) \rightarrow 2\ H_2O(l)$

(a) $\Delta H^\circ = 2\ \Delta H^\circ_f(H_2O) = (2\ mol)(-285.8\ kJ/mol) = -571.6\ kJ$

$\Delta S^\circ = 2\ S^\circ(H_2O) - [2\ S^\circ(H_2) + S^\circ(O_2)]$

$\Delta S^\circ = (2\ mol)(69.9\ J/(K \cdot mol)) - [(2\ mol)(130.6\ J/(K \cdot mol)) + (1\ mol)(205.0\ J/(K \cdot mol))]$

$\Delta S^\circ = -326.4\ J/K = -0.3264\ kJ/K$

95 °C = 368 K

$\Delta G^\circ = \Delta H^\circ - T\Delta S^\circ = -571.6\ kJ - (368\ K)(-0.3264\ kJ/K) = -451.5\ kJ$

1 V = 1 J/C

$$\Delta G^\circ = -nFE^\circ; \quad E^\circ = -\frac{\Delta G^\circ}{nF} = -\frac{-451.5 \times 10^3\ J}{(4)(96{,}500\ C)} = 1.17\ J/C = 1.17\ V$$

(b) $$E = E^\circ - \frac{2.303\,RT}{nF}\log\frac{1}{(P_{H_2})^2(P_{O_2})}$$

$$E = 1.17\ V - \frac{(2.303)\left(8.314\ \frac{J}{mol \cdot K}\right)(368\ K)}{(4\ mol\ e^-)(96{,}500\ C/mol\ e^-)}\log\left(\frac{1}{(25)^2(25)}\right)$$

E = 1.17 V + 0.077 V = 1.25 V

18.162 (a) $Zn(s) + 2\ Ag^+(aq) + H_2O(l) \rightarrow ZnO(s) + 2\ Ag(s) + 2\ H^+(aq)$

$\Delta H^\circ_{rxn} = \Delta H^\circ_f(ZnO) - [2\ \Delta H^\circ_f(Ag^+) + \Delta H^\circ_f(H_2O)]$

$\Delta H^\circ_{rxn} = [(1\ mol)(-350.5\ kJ/mol)] - [(2\ mol)(105.6\ kJ/mol) + (1\ mol)(-285.8\ kJ/mol)]$
$\Delta H^\circ_{rxn} = -275.9\ kJ$
$\Delta S^\circ = [S^\circ(ZnO) + 2\ S^\circ(Ag)] - [S^\circ(Zn) + 2\ S^\circ(Ag^+) + S^\circ(H_2O)]$
$\Delta S^\circ = [(1\ mol)(43.7\ J/(K\cdot mol)) + (2\ mol)(42.6\ J/(K\cdot mol))]$
$\quad - [(1\ mol)(41.6\ J/(K\cdot mol)) + (2\ mol)(72.7\ J/(K\cdot mol)) + (1\ mol)(69.9\ J/(K\cdot mol))$
$\Delta S^\circ = -128.0\ J/K$
$\Delta G^\circ = \Delta H^\circ - T\Delta S^\circ = -275.9\ kJ - (298\ K)(-128.0 \times 10^{-3}\ kJ/K) = -237.8\ kJ$
(b) $1\ V = 1\ J/C$

$$\Delta G^\circ = -nFE^\circ \quad E^\circ = \frac{-\Delta G^\circ}{nF} = \frac{-(-237.8 \times 10^3\ J)}{(2\ mol\ e^-)\left(\frac{96{,}500\ C}{1\ mol\ e^-}\right)} = 1.232\ J/C = 1.232\ V$$

$$E^\circ = \frac{0.0592\ V}{n}\log K; \quad \log K = \frac{nE^\circ}{0.0592\ V} = \frac{(2)(1.232\ V)}{0.0592\ V} = 41.62$$

$K = 10^{41.62} = 4 \times 10^{41}$

(c) $E = E^\circ - \frac{0.0592\ V}{n}\log\frac{[H^+]^2}{[Ag^+]^2}$

The addition of NH_3 to the cathode compartment would result in the formation of the $Ag(NH_3)_2^+$ complex ion, which results in a decrease in Ag^+ concentration. The log term in the Nernst equation becomes larger and the cell voltage decreases.

On mixing equal volumes of two solutions, the concentrations of both solutions are cut in half.

	$Ag^+(aq)$	+	$2\ NH_3(aq)$	⇄	$Ag(NH_3)_2^+(aq)$
before reaction (M)	0.0500		2.00		0
assume 100% reaction	–0.0500		–2(0.0500)		+0.0500
after reaction (M)	0		1.90		0.0500
assume small back rxn	+x		+2x		–x
equil (M)	x		1.90 + 2x		0.0500 – x

$$K_f = 1.7 \times 10^7 = \frac{[Ag(NH_3)_2^+]}{[Ag^+][NH_3]^2} = \frac{(0.0500 - x)}{(x)(1.90 + 2x)^2} \approx \frac{0.0500}{(x)(1.90)^2}$$

Solve for x. $x = [Ag^+] = 8.15 \times 10^{-10}\ M$

$$E = E^\circ - \frac{0.0592\ V}{n}\log\frac{[H^+]^2}{[Ag^+]^2} = 1.232\ V - \frac{0.0592\ V}{2}\log\frac{(1.00\ M)^2}{(8.15 \times 10^{-10}\ M)^2} = 0.694\ V$$

(d) Calculate new initial concentrations because of dilution to 110.0 mL.

$$M_i \times V_i = M_f \times V_f; \quad M_f = [Cl^-] = \frac{M_i \times V_i}{V_f} = \frac{0.200\ M \times 10.0\ mL}{110.0\ mL} = 0.0182\ M$$

$$M_i \times V_i = M_f \times V_f; \quad M_f = [Ag^+] = \frac{M_i \times V_i}{V_f} = \frac{0.0500\ M \times 100.0\ mL}{110.0\ mL} = 0.0455\ M$$

$$M_i \times V_i = M_f \times V_f; \quad M_f = [NH_3] = \frac{M_i \times V_i}{V_f} = \frac{2.00\ M \times 100.0\ mL}{110.0\ mL} = 1.82\ M$$

Now calculate the $[Ag^+]$ as a result of the following equilibrium:

	$Ag^+(aq)$	+ $2\ NH_3(aq)$	⇄ $Ag(NH_3)_2^+(aq)$
before reaction (M)	0.0455	1.82	0
assume 100% reaction	–0.0455	–2(0.0455)	+0.0455
after reaction (M)	0	1.73	0.0455
assume small back rxn	+x	+2x	–x
equil (M)	x	1.73 + 2x	0.0455 – x

$$K_f = 1.7 \times 10^7 = \frac{[Ag(NH_3)_2^+]}{[Ag^+][NH_3]^2} = \frac{(0.0455 - x)}{(x)(1.73 + 2x)^2} \approx \frac{0.0455}{(x)(1.73)^2}$$

Solve for x. $x = [Ag^+] = 8.94 \times 10^{-10}$ M

For AgCl, $K_{sp} = 1.8 \times 10^{-10}$

$IP = [Ag^+][Cl^-] = (8.94 \times 10^{-10}\ M)(0.0182\ M) = 1.6 \times 10^{-11}$

$IP < K_{sp}$, AgCl will not precipitate.

Now calculate new initial concentrations because of dilution to 120.0 mL.

$$M_i \times V_i = M_f \times V_f;\quad M_f = [Br^-] = \frac{M_i \times V_i}{V_f} = \frac{0.200\ M \times 10.0\ mL}{120.0\ mL} = 0.0167\ M$$

$$M_i \times V_i = M_f \times V_f;\quad M_f = [Ag^+] = \frac{M_i \times V_i}{V_f} = \frac{0.0500\ M \times 100.0\ mL}{120.0\ mL} = 0.0417\ M$$

$$M_i \times V_i = M_f \times V_f;\quad M_f = [NH_3] = \frac{M_i \times V_i}{V_f} = \frac{2.00\ M \times 100.0\ mL}{120.0\ mL} = 1.67\ M$$

Now calculate the $[Ag^+]$ as a result of the following equilibrium:

	$Ag^+(aq)$	+ $2\ NH_3(aq)$	⇄ $Ag(NH_3)_2^+(aq)$
before reaction (M)	0.0417	1.67	0
assume 100% reaction	–0.0417	–2(0.0417)	+0.0417
after reaction (M)	0	1.59	0.0417
assume small back rxn	+x	+2x	–x
equil (M)	x	1.59 + 2x	0.0417 – x

$$K_f = 1.7 \times 10^7 = \frac{[Ag(NH_3)_2^+]}{[Ag^+][NH_3]^2} = \frac{(0.0417 - x)}{(x)(1.59 + 2x)^2} \approx \frac{0.0417}{(x)(1.59)^2}$$

Solve for x. $x = [Ag^+] = 9.70 \times 10^{-10}$ M

For AgBr, $K_{sp} = 5.4 \times 10^{-13}$

$IP = [Ag^+][Br^-] = (9.70 \times 10^{-10}\ M)(0.0167\ M) = 1.6 \times 10^{-11}$

$IP > K_{sp}$, AgBr will precipitate.

18.163 (a) anode: $Fe(s) + 2\ OH^-(aq) \rightarrow Fe(OH)_2(s) + 2\ e^-$

cathode: $2 \times [NiO(OH)(s) + H_2O(l) + e^- \rightarrow Ni(OH)_2(s) + OH^-(aq)]$

overall: $Fe(s) + 2\ NiO(OH)(s) + 2\ H_2O(l) \rightarrow Fe(OH)_2(s) + 2\ Ni(OH)_2(s)$

(b)

$$\Delta G^\circ = -nFE^\circ = -(2\ mol\ e^-)\left(\frac{96{,}500\ C}{1\ mol\ e^-}\right)(1.37\ V)\left(\frac{1\ J}{1\ C \cdot V}\right) = -264{,}410\ J = -264\ kJ$$

$E^\circ = \frac{0.0592\ V}{n} \log K; \quad \log K = \frac{nE^\circ}{0.0592\ V} = \frac{(2)(1.37\ V)}{0.0592\ V} = 46.3$

$K = 10^{46.3} = 2 \times 10^{46}$

(c) It would still be 1.37 V because OH^- does not appear in the overall cell reaction. The overall cell reaction contains only solids and one liquid, therefore the cell voltage does not change because there are no concentration changes.

(d) $Fe(OH)_2$, 89.86; 1 A = 1 C/s

$$\text{mol } e^- = (0.250\ C/s)(40.0\ min)\left(\frac{60\ s}{1\ min}\right)\left(\frac{1\ mol\ e^-}{96{,}500\ C}\right) = 6.22 \times 10^{-3}\ mol\ e^-$$

$$\text{mass } Fe(OH)_2 = (6.22 \times 10^{-3}\ mol\ e^-) \times \frac{1\ mol\ Fe(OH)_2}{2\ mol\ e^-} \times \frac{89.86\ g\ Fe(OH)_2}{1\ mol\ Fe(OH)_2} = 0.279\ g$$

H_2O molecules consumed =

$$(6.22 \times 10^{-3}\ mol\ e^-) \times \frac{2\ mol\ H_2O}{2\ mol\ e^-} \times \frac{6.022 \times 10^{23}\ H_2O\ molecules}{1\ mol\ H_2O} = 3.75 \times 10^{21}\ H_2O\ molecules$$

18.164 (a) Oxidation half reaction: $2\ [C_4H_{10}(g) + 13\ O^{2-}(s) \rightarrow 4\ CO_2(g) + 5\ H_2O(l) + 26\ e^-]$

Reduction half reaction: $13\ [O_2(g) + 4\ e^- \rightarrow 2\ O^{2-}(s)]$

Cell reaction: $2\ C_4H_{10}(g) + 13\ O_2(g) \rightarrow 8\ CO_2(g) + 10\ H_2O(l)$

(b) $\Delta H^\circ = [8\ \Delta H^\circ_f(CO_2) + 10\ \Delta H^\circ_f(H_2O)] - [2\ \Delta H^\circ_f(C_4H_{10})]$

$\Delta H^\circ = [(8\ mol)(-393.5\ kJ/mol) + (10\ mol)(-285.8\ kJ/mol)] - [(2\ mol)(-126\ kJ/mol)] = -5754\ kJ$

$\Delta S^\circ = [8\ S^\circ(CO_2) + 10\ S^\circ(H_2O)] - [2\ S^\circ(C_4H_{10}) + 13\ S^\circ(O_2)]$

$\Delta S^\circ = [(8\ mol)(213.6\ J/(K \cdot mol)) + (10\ mol)(69.9\ J/(K \cdot mol))] - [(2\ mol)(310\ J/(K \cdot mol)) + (13\ mol)(205\ J/(K \cdot mol))] = -877.2\ J/K$

$\Delta G^\circ = \Delta H^\circ - T\Delta S^\circ = -5754\ kJ - (298\ K)(-877.2 \times 10^{-3}\ kJ/K) = -5493\ kJ$

1 V = 1 J/C

$$\Delta G^\circ = -nFE^\circ; \quad E^\circ = -\frac{\Delta G^\circ}{nF} = -\frac{-5493 \times 10^3\ J}{(52)(96{,}500\ C)} = 1.09\ J/C = 1.09\ V$$

$\Delta G^\circ = -RT \ln K$

$$\ln K = \frac{-\Delta G^\circ}{RT} = \frac{-(-5493\ kJ)}{(8.314 \times 10^{-3}\ kJ/K)(298\ K)} = 2217$$

$K = e^{2217} = 7 \times 10^{962}$

On raising the temperature, both K and E° will decrease because the reaction is exothermic ($\Delta H^\circ < 0$).

(c) C_4H_{10}, 58.12; 10.5 A = 10.5 C/s

$$\text{mass } C_4H_{10} = 10.5\ C/s \times 8\ hr \times \frac{60\ min}{1\ hr} \times \frac{60\ s}{1\ min} \times \frac{1\ mol\ e^-}{96{,}500\ C} \times \frac{2\ mol\ C_4H_{10}}{52\ mol\ e^-} \times \frac{58.12\ g\ C_4H_{10}}{1\ mol\ C_4H_{10}} = 7.00\ g\ C_4H_{10}$$

$$n = 7.00 \text{ g } C_4H_{10} \times \frac{1 \text{ mol } C_4H_{10}}{58.12 \text{ g } C_4H_{10}} = 0.120 \text{ mol } C_4H_{10}$$

20 °C = 20 + 273 = 293 K

$$PV = nRT \qquad V = \frac{nRT}{P} = \frac{(0.120 \text{ mol})\left(0.082\ 06\ \frac{L\cdot atm}{K\cdot mol}\right)(293 \text{ K})}{\left(815 \text{ mm Hg} \times \frac{1.00 \text{ atm}}{760 \text{ mm Hg}}\right)} = 2.69 \text{ L}$$

18.165 (a) cathode:

(1) $MnO_2(s) + 4\,H^+(aq) + 2\,e^- \rightarrow Mn^{2+}(aq) + 2\,H_2O(l)$ $\qquad E° = +1.22$ V

(2) $Mn(OH)_2(s) + OH^-(aq) \rightarrow MnO(OH)(s) + H_2O(l) + e^-$ $\qquad E° = +0.380$ V

(3) $Mn^{2+}(aq) + 2\,OH^-(aq) \rightarrow Mn(OH)_2(s)$ $\qquad K = 1/K_{sp} = 1/(2.1 \times 10^{-13}) = 4.8 \times 10^{12}$

(4) $4 \times [H_2O(l) \rightarrow H^+(aq) + OH^-(aq)]$ $\qquad K = (K_w)^4 = (1.0 \times 10^{-14})^4 = 1.0 \times 10^{-56}$

$MnO_2(s) + H_2O(l) + e^- \rightarrow MnO(OH)(s) + OH^-(aq)$

$$\Delta G°_1 = -nFE° = -(2 \text{ mol } e^-)\left(\frac{96{,}500 \text{ C}}{1 \text{ mol } e^-}\right)(1.22 \text{ V})\left(\frac{1 \text{ J}}{1 \text{ C}\cdot\text{V}}\right) = -235{,}460 \text{ J} = -235.5 \text{ kJ}$$

$$\Delta G°_2 = -nFE° = -(1 \text{ mol } e^-)\left(\frac{96{,}500 \text{ C}}{1 \text{ mol } e^-}\right)(0.380 \text{ V})\left(\frac{1 \text{ J}}{1 \text{ C}\cdot\text{V}}\right) = -36{,}670 \text{ J} = -36.7 \text{ kJ}$$

$\Delta G°_3 = -RT \ln K = -(8.314 \times 10^{-3} \text{ kJ/K})(298 \text{ K}) \ln (4.8 \times 10^{12}) = -72.3$ kJ

$\Delta G°_4 = -RT \ln K = -(8.314 \times 10^{-3} \text{ kJ/K})(298 \text{ K}) \ln (1.0 \times 10^{-56}) = +319.5$ kJ

$\Delta G°(\text{total}) = -235.5 \text{ kJ} - 36.7 \text{ kJ} - 72.3 \text{ kJ} + 319.5 \text{ kJ} = -25.0 \text{ kJ} = -25{,}000$ J

1 V = 1 J/C

$$\Delta G° = -nFE°;\ E° = \frac{-\Delta G°}{nF} = \frac{-(-25{,}000 \text{ J})}{(1 \text{ mol } e^-)\left(\frac{96{,}500 \text{ C}}{1 \text{ mol } e^-}\right)} = +0.259 \text{ J/C} = +0.259 \text{ V}$$

$E°_{cathode} = +0.259$ V

anode:

(1) $Zn(s) \rightarrow Zn^{2+}(aq) + 2\,e^-$ $\qquad E° = +0.76$ V

(2) $Zn^{2+}(aq) + 2\,OH^-(aq) \rightarrow Zn(OH)_2(s)$ $\qquad K = 1/K_{sp} = 1/(4.1 \times 10^{-17}) = 2.4 \times 10^{16}$

$Zn(s) + 2\,OH^-(aq) \rightarrow Zn(OH)_2(s) + 2\,e^-$

$$\Delta G°_1 = -nFE° = -(2 \text{ mol } e^-)\left(\frac{96{,}500 \text{ C}}{1 \text{ mol } e^-}\right)(0.76 \text{ V})\left(\frac{1 \text{ J}}{1 \text{ C}\cdot\text{V}}\right) = -146{,}680 \text{ J} = -146.7 \text{ kJ}$$

$\Delta G°_2 = -RT \ln K = -(8.314 \times 10^{-3} \text{ kJ/K})(298 \text{ K}) \ln (2.4 \times 10^{16}) = -93.4$ kJ

$\Delta G°(\text{total}) = -146.7 \text{ kJ} - 93.4 \text{ kJ} = -240.1 \text{ kJ} = -240{,}100$ J

1 V = 1 J/C

$$\Delta G° = -nFE°;\ E° = \frac{-\Delta G°}{nF} = \frac{-(-240{,}100 \text{ J})}{(2 \text{ mol } e^-)\left(\frac{96{,}500 \text{ C}}{1 \text{ mol } e^-}\right)} = +1.24 \text{ J/C} = +1.24 \text{ V}$$

$E^\circ_{anode} = +1.24$ V
$E^\circ_{cell} = E^\circ_{cathode} + E^\circ_{anode} = 0.259$ V + 1.24 V = 1.50 V

(b) $FeO_4^{2-}(aq) \rightarrow Fe(OH)_3(s)$
$FeO_4^{2-}(aq) \rightarrow Fe(OH)_3(s) + H_2O(l)$
$FeO_4^{2-}(aq) + 5\,H^+(aq) \rightarrow Fe(OH)_3(s) + H_2O(l)$
$FeO_4^{2-}(aq) + 5\,H^+(aq) + 3\,e^- \rightarrow Fe(OH)_3(s) + H_2O(l)$
$FeO_4^{2-}(aq) + 5\,H^+(aq) + 5\,OH^-(aq) + 3\,e^- \rightarrow Fe(OH)_3(s) + H_2O(l) + 5\,OH^-(aq)$
$FeO_4^{2-}(aq) + 5\,H_2O(l) + 3\,e^- \rightarrow Fe(OH)_3(s) + H_2O(l) + 5\,OH^-(aq)$
$FeO_4^{2-}(aq) + 4\,H_2O(l) + 3\,e^- \rightarrow Fe(OH)_3(s) + 5\,OH^-(aq)$

(c) K_2FeO_4, 198.04; MnO_2, 86.94

$$\text{coulombs} = 10.00\text{ g } K_2FeO_4 \times \frac{1\text{ mol } K_2FeO_4}{198.04\text{ g } K_2FeO_4} \times \frac{3\text{ mol e}^-}{1\text{ mol } K_2FeO_4} \times \frac{96{,}500\text{ C}}{1\text{ mol e}^-} =$$

1.46×10^4 C from 10.00 g K_2FeO_4

$$\text{coulombs} = 10.00\text{ g } MnO_2 \times \frac{1\text{ mol } MnO_2}{86.94\text{ g } MnO_2} \times \frac{1\text{ mol e}^-}{1\text{ mol } MnO_2} \times \frac{96{,}500\text{ C}}{1\text{ mol e}^-} =$$

1.11×10^4 C from 10.00 g MnO_2

18.166 (a) $4\,[Au(s) + 2\,CN^-(aq) \rightarrow Au(CN)_2^-(aq) + e^-]$ (oxidation half reaction)

$O_2(g) \rightarrow 2\,H_2O(l)$
$O_2(g) + 4\,H^+(aq) \rightarrow 2\,H_2O(l)$
$4\,e^- + O_2(g) + 4\,H^+(aq) \rightarrow 2\,H_2O(l)$ (reduction half reaction)

Combine the two half reactions.
$4\,Au(s) + 8\,CN^-(aq) + O_2(g) + 4\,H^+(aq) \rightarrow 4\,Au(CN)_2^-(aq) + 2\,H_2O(l)$
$4\,Au(s) + 8\,CN^-(aq) + O_2(g) + 4\,H^+(aq) + 4\,OH^-(aq)$
$\rightarrow 4\,Au(CN)_2^-(aq) + 2\,H_2O(l) + 4\,OH^-(aq)$
$4\,Au(s) + 8\,CN^-(aq) + O_2(g) + 4\,H_2O(l)$
$\rightarrow 4\,Au(CN)_2^-(aq) + 2\,H_2O(l) + 4\,OH^-(aq)$
$4\,Au(s) + 8\,CN^-(aq) + O_2(g) + 2\,H_2O(l) \rightarrow 4\,Au(CN)_2^-(aq) + 4\,OH^-(aq)$

(b) Add the following five reactions together. ΔG° is calculated below each reaction.
$4\,[Au^+(aq) + 2\,CN^-(aq) \rightarrow Au(CN)_2^-(aq)]$ $K = (K_f)^4$
$\Delta G^\circ = -RT \ln K = -(8.314 \times 10^{-3}\text{ kJ/K})(298\text{ K}) \ln (6.2 \times 10^{38})^4 = -885.2$ kJ

$O_2(g) + 4\,H^+(aq) + 4\,e^- \rightarrow 2\,H_2O(l)$ $E^\circ = 1.229$ V

$$\Delta G^\circ = -nFE^\circ = -(4\text{ mol e}^-)\left(\frac{96{,}500\text{ C}}{1\text{ mol e}^-}\right)(1.229\text{ V})\left(\frac{1\text{ J}}{1\text{ C}\cdot\text{V}}\right) = -474{,}394\text{ J} = -474.4\text{ kJ}$$

$4\,[H_2O(l) \rightleftarrows H^+(aq) + OH^-(aq)]$ $K = (K_w)^4$
$\Delta G^\circ = -RT \ln K = -(8.314 \times 10^{-3}\text{ kJ/K})(298\text{ K}) \ln (1.0 \times 10^{-14})^4 = +319.5$ kJ

$4\,[Au(s) \rightarrow Au^{3+}(aq) + 3\,e^-]$ $E^\circ = -1.498$ V

$$\Delta G^\circ = -nFE^\circ = -(12 \text{ mol e}^-)\left(\frac{96{,}500 \text{ C}}{1 \text{ mol e}^-}\right)(-1.498 \text{ V})\left(\frac{1 \text{ J}}{1 \text{ C}\cdot\text{V}}\right) = +1{,}734{,}684 \text{ J} = +1{,}734.7 \text{ kJ}$$

$4\,[Au^{3+}(aq) + 2\,e^- \rightarrow Au^+(aq)]$ $\quad E^\circ = 1.401$ V

$$\Delta G^\circ = -\,nFE^\circ = -(8 \text{ mol e}^-)\left(\frac{96{,}500 \text{ C}}{1 \text{ mol e}^-}\right)(1.401 \text{ V})\left(\frac{1 \text{ J}}{1 \text{ C}\cdot\text{V}}\right) = -1{,}081{,}572 \text{ J} = -1{,}081.6 \text{ kJ}$$

Overall reaction:
$4\,Au(s) + 8\,CN^-(aq) + O_2(g) + 2\,H_2O(l) \rightarrow 4\,Au(CN)_2^-(aq) + 4\,OH^-(aq)$
$\Delta G^\circ = -\,885.2$ kJ $-$ 474.4 kJ + 319.5 kJ + 1,734.7 kJ $-$ 1,081.6 kJ = -387.0 kJ

18.167 The overall cell reaction is:
$2\,Fe^{3+}(aq) + 2\,Hg(l) + 2\,Cl^-(aq) \rightarrow 2\,Fe^{2+}(aq) + Hg_2Cl_2(s)$
The Nernst equation can be applied to separate half reactions.
One half reaction is for the calomel reference electrode.
$2\,Hg(l) + 2\,Cl^-(aq) \rightarrow Hg_2Cl_2(s) + 2\,e^-$ $\quad E^\circ = -0.28$ V
When $[Cl^-] = 2.9$ M,

$$E_{calomel} = E^\circ - \frac{0.0592\,\text{V}}{n}\log\frac{1}{[Cl^-]^2} = -0.28\text{ V} - \frac{0.0592\,\text{V}}{2}\log\frac{1}{(2.9)^2} = -0.25\text{ V}$$

Balance the titration redox reaction: $MnO_4^-(aq) + Fe^{2+}(aq) \rightarrow Mn^{2+}(aq) + Fe^{3+}(aq)$
$[Fe^{2+}(aq) \rightarrow Fe^{3+}(aq) + e^-] \times 5$

$MnO_4^-(aq) \rightarrow Mn^{2+}(aq)$
$MnO_4^-(aq) \rightarrow Mn^{2+}(aq) + 4\,H_2O(l)$
$MnO_4^-(aq) + 8\,H^+(aq) \rightarrow Mn^{2+}(aq) + 4\,H_2O(l)$
$MnO_4^-(aq) + 8\,H^+(aq) + 5\,e^- \rightarrow Mn^{2+}(aq) + 4\,H_2O(l)$
Combine the two half reactions.
$MnO_4^-(aq) + 5\,Fe^{2+}(aq) + 8\,H^+(aq) \rightarrow Mn^{2+}(aq) + 5\,Fe^{3+}(aq) + 4\,H_2O(l)$

initial mol Fe^{2+} = (0.010 mol/L)(0.1000 L) = 0.0010 mol Fe^{2+}
mL MnO_4^- needed to reach endpoint =

$$0.0010 \text{ mol Fe}^{2+} \times \frac{1 \text{ mol MnO}_4^-}{5 \text{ mol Fe}^{2+}} \times \frac{1.00 \text{ L}}{0.010 \text{ mol MnO}_4^-} \times \frac{1000 \text{ mL}}{1.00 \text{ L}} = 20.0 \text{ mL}$$

(a) initial mol Fe^{2+} = (0.010 mol/L)(0.1000 L) = 0.0010 mol Fe^{2+}
mol MnO_4^- in 5.0 mL = (0.010 mol/L)(0.0050 L) = 0.000 050 mol MnO_4^-

	$MnO_4^-(aq)$	+ $5\,Fe^{2+}(aq)$	+ $8\,H^+(aq)$ $\rightarrow$ $Mn^{2+}(aq)$	+ $5\,Fe^{3+}(aq)$	+ $4\,H_2O(l)$
before (mol)	0.000 050	0.0010		0	
change (mol)	−0.000 050	−5(0.000 050)		(0.000 050)	
after (mol)	0	0.000 75		0.000 25	

Again, the Nernst equation can be applied to separate half reactions.
The other half reaction is: $Fe^{3+}(aq) + e^- \rightarrow 2\,Fe^{2+}(aq)$ $\quad E^\circ = +0.77$ V
E for the half reaction after adding 5.0 mL of MnO_4^- is

$$E_{Fe^{3+}/Fe^{2+}} = E^\circ - \frac{0.0592\,V}{n}\log\frac{[Fe^{2+}]}{[Fe^{3+}]} = 0.77\,V - \frac{0.0592\,V}{1}\log\frac{(0.000\,75)}{(0.000\,25)} = 0.74\,V$$

(Note in the Nernst equation above, we are taking a ratio of Fe^{2+} to Fe^{3+} so we can ignore volumes and just use moles instead of molarity.)

$E_{cell} = E_{Fe^{3+}/Fe^{2+}} + E_{calomel} = 0.74\,V + (-0.25\,V) = 0.49\,V$

(b) initial mol Fe^{2+} = (0.010 mol/L)(0.1000 L) = 0.0010 mol Fe^{2+}
mol MnO_4^- in 10.0 mL = (0.010 mol/L)(0.0100 L) = 0.000 10 mol MnO_4^-

	$MnO_4^-(aq)$ +	$5\,Fe^{2+}(aq)$ +	$8\,H^+(aq)$ →	$Mn^{2+}(aq)$ +	$5\,Fe^{3+}(aq)$ +	$4\,H_2O(l)$
before (mol)	0.000 10	0.0010			0	
change (mol)	−0.000 10	−5(0.000 10)			+5(0.000 10)	
after (mol)	0	0.000 50			0.000 50	

E for the half reaction after adding 10.0 mL of MnO_4^- is

$$E_{Fe^{3+}/Fe^{2+}} = E^\circ - \frac{0.0592\,V}{n}\log\frac{[Fe^{2+}]}{[Fe^{3+}]} = 0.77\,V - \frac{0.0592\,V}{1}\log\frac{(0.000\,50)}{(0.000\,50)} = 0.77\,V$$

$E_{cell} = E_{Fe^{3+}/Fe^{2+}} + E_{calomel} = 0.77\,V + (-0.25\,V) = 0.52\,V$

(c) initial mol Fe^{2+} = (0.010 mol/L)(0.1000 L) = 0.0010 mol Fe^{2+}
mol MnO_4^- in 19.0 mL = (0.010 mol/L)(0.0190 L) = 0.000 19 mol MnO_4^-

	$MnO_4^-(aq)$ +	$5\,Fe^{2+}(aq)$ +	$8\,H^+(aq)$ →	$Mn^{2+}(aq)$ +	$5\,Fe^{3+}(aq)$ +	$4\,H_2O(l)$
before (mol)	0.000 19	0.0010			0	
change (mol)	−0.000 19	−5(0.000 19)			+5(0.000 19)	
after (mol)	0	0.000 05			0.000 95	

E for the half reaction after adding 19.0 mL of MnO_4^- is

$$E_{Fe^{3+}/Fe^{2+}} = E^\circ - \frac{0.0592\,V}{n}\log\frac{[Fe^{2+}]}{[Fe^{3+}]} = 0.77\,V - \frac{0.0592\,V}{1}\log\frac{(0.000\,05)}{(0.000\,95)} = 0.85\,V$$

$E_{cell} = E_{Fe^{3+}/Fe^{2+}} + E_{calomel} = 0.85\,V + (-0.25\,V) = 0.60\,V$

(d) 21.0 mL is past the endpoint so the MnO_4^- is in excess and all of the Fe^{2+} is consumed.
initial mol Fe^{2+} = (0.010 mol/L)(0.1000 L) = 0.0010 mol Fe^{2+}
mol MnO_4^- in 21.0 mL = (0.010 mol/L)(0.0210 L) = 0.000 21 mol MnO_4^-

	$MnO_4^-(aq)$ +	$5\,Fe^{2+}(aq)$ +	$8\,H^+(aq)$ →	$Mn^{2+}(aq)$ +	$5\,Fe^{3+}(aq)$ +	$4\,H_2O(l)$
before (mol)	0.000 21	0.0010		0	0	
change (mol)	−0.000 20	−5(0.000 20)		+0.000 20	+5(0.000 20)	
after (mol)	0.000 01	0		0.000 20	0.0010	

Because the Fe^{2+} is totally consumed, there is a new half reaction:

$MnO_4^-(aq) + 8\,H^+(aq) + 5\,e^- \rightarrow Mn^{2+}(aq) + 4\,H_2O(l)$ $\quad E^\circ = 1.51\,V$

The total volume = 100.0 mL + 21.0 mL = 121.0 mL = 0.1210 L
$[MnO_4^-]$ = 0.000 01 mol/0.1210 L = 0.000 083 M
$[Mn^{2+}]$ = 0.000 20 mol/0.1210 L = 0.001 65 M
We need to determine $[H^+]$ in order to determine the half reaction potential.
$[H_2SO_4]_{dil} \cdot 121.0\ mL = [H_2SO_4]_{conc} \cdot 100.0\ mL$
$[H_2SO_4]_{dil}$ = [(1.50 M)(100.0 mL)]/121.0 mL = 1.24 M
We can ignore the small amount of H^+ consumed by the titration itself, because the H_2SO_4 concentration is so large.

Consider the dissociation of H_2SO_4. From the complete dissociation of the first proton, $[H^+] = [HSO_4^-]$ = 1.24 M.
For the dissociation of the second proton, the following equilibrium must be considered:

	$HSO_4^-(aq)$	⇄	$H^+(aq)$	+	$SO_4^{2-}(aq)$
initial (M)	1.24		1.24		0
change (M)	–x		+x		+x
equil (M)	1.24 – x		1.24 + x		x

$$K_{a2} = \frac{[H^+][SO_4^{2-}]}{[HSO_4^-]} = 1.2 \times 10^{-2} = \frac{(1.24 + x)(x)}{1.24 - x}$$

$x^2 + 1.252x - 0.0149 = 0$
Use the quadratic formula to solve for x.

$$x = \frac{-(1.252) \pm \sqrt{(1.252)^2 - 4(1)(-0.0149)}}{2(1)} = \frac{-1.252 \pm 1.276}{2}$$

x = –1.264 and 0.012
Of the two solutions for x, only the positive value of x has physical meaning, since x is the $[SO_4^{2-}]$.
$[H^+]$ = 1.24 + x = 1.24 + 0.012 = 1.25 M

$$E_{MnO_4^-/Mn^{2+}} = E^\circ - \frac{0.0592\ V}{n} \log \frac{[Mn^{2+}]}{[MnO_4^-][H^+]^8}$$

$$= 1.51\ V - \frac{0.0592\ V}{5} \log \frac{(0.001\ 65)}{(0.000\ 083)(1.25)^8} = 1.50\ V$$

$E_{cell} = E_{MnO_4^-/Mn^{2+}} + E_{calomel}$ = 1.50 V + (–0.25 V) = 1.25 V
Notice that there is a dramatic change in the potential at the equivalence point.

19 Nuclear Chemistry

19.1 (a) In beta emission, the mass number is unchanged, and the atomic number increases by one. $^{106}_{44}Ru \rightarrow {}^{0}_{-1}e + {}^{106}_{45}Rh$

(b) In alpha emission, the mass number decreases by four, and the atomic number decreases by two. $^{189}_{83}Bi \rightarrow {}^{4}_{2}He + {}^{185}_{81}Tl$

(c) In electron capture, the mass number is unchanged, and the atomic number decreases by one. $^{204}_{84}Po + {}^{0}_{-1}e \rightarrow {}^{204}_{83}Bi$

19.2 $^{148}_{69}Tm$ decays to $^{148}_{68}Er$ by either positron emission or electron capture.

19.3 (a) ^{199}Au has a higher neutron/proton ratio and decays by beta emission. ^{173}Au has a lower neutron/proton ratio and decays by alpha emission.

(b) ^{196}Pb has a lower neutron/proton ratio and decays by positron emission. ^{206}Pb is nonradioactive.

19.4 $$^{238}_{92}U \rightarrow {}^{238-(8\times4)-(6\times0)}_{92-(8\times2)-(6\times-1)}X = {}^{206}_{82}Pb$$

19.5 $$t_{1/2} = \frac{0.693}{k} = \frac{0.693}{1.08 \times 10^{-2}\ h^{-1}} = 64.2\ h$$

19.6 $$k = \frac{0.693}{t_{1/2}} = \frac{0.693}{3.82\ d} = 0.181\ d^{-1}$$

19.7 $$\ln\left(\frac{N}{N_0}\right) = -0.693\left(\frac{t}{t_{1/2}}\right) = -0.693\left(\frac{16{,}230\ y}{5715\ y}\right) = -1.968$$

$$\frac{N}{N_0} = e^{-1.968} = 0.140; \qquad \frac{N}{100\%} = 0.140; \qquad N = 14.0\%$$

19.8 Assume $N_0 = 100\%$ and $N = 89.2\%$ at $t = 5.00$ y

$$\ln\left(\frac{N}{N_0}\right) = (-0.693)\left(\frac{t}{t_{1/2}}\right); \qquad \frac{N}{N_0} = \frac{\text{Decay rate at time t}}{\text{Decay rate at time t} = 0}$$

$$\ln\left(\frac{89.2}{100}\right) = (-0.693)\left(\frac{5.00 \text{ y}}{t_{1/2}}\right); \qquad t_{1/2} = 30.3 \text{ y}$$

19.9

$$\ln\left(\frac{N}{N_0}\right) = (-0.693)\left(\frac{t}{t_{1/2}}\right); \qquad \frac{N}{N_0} = \frac{\text{Decay rate at time t}}{\text{Decay rate at t} = 0}$$

$$\ln\left(\frac{10{,}860}{16{,}800}\right) = (-0.693)\left(\frac{28.0 \text{ d}}{t_{1/2}}\right); \qquad t_{1/2} = 44.5 \text{ d}$$

19.10

$$\ln\left(\frac{N}{N_0}\right) = (-0.693)\left(\frac{t}{t_{1/2}}\right); \qquad \frac{N}{N_0} = \frac{\text{Decay rate at time t}}{\text{Decay rate at t} = 0}$$

$t_{1/2} = 44.5$ d

$$\ln\left(\frac{N}{16{,}800}\right) = (-0.693)\left(\frac{40.0 \text{ d}}{44.5 \text{ d}}\right)$$

$$\ln N - \ln(16{,}800) = (-0.693)\left(\frac{40.0 \text{ d}}{44.5 \text{ d}}\right)$$

$$\ln N = (-0.693)\left(\frac{40.0 \text{ d}}{44.5 \text{ d}}\right) + \ln(16{,}800) = 9.106$$

$N = e^{9.106} = 9011$ disintegrations/min

19.11 For ${}^{16}_{8}O$:

First, calculate the total mass of the nucleons (8 n + 8 p)

Mass of 8 neutrons = (8)(1.008 66) = 8.069 28	
Mass of 8 protons = (8)(1.007 28) = 8.058 24	
Mass of 8 n + 8 p	= 16.127 52

Next, calculate the mass of a ^{16}O nucleus by subtracting the mass of 8 electrons from the mass of a ^{16}O atom.

Mass of ^{16}O atom	= 15.994 91
–Mass of 8 electrons = $-(8)(5.486 \times 10^{-4})$	= –0.004 39
Mass of ^{16}O nucleus	= 15.990 52

Then subtract the mass of the ^{16}O nucleus from the mass of the nucleons to find the mass defect:

Mass defect = mass of nucleons – mass of nucleus
= (16.127 52) – (15.990 52) = 0.137 00 u
Mass defect in grams = (0.137 00 u)($1.660\ 54 \times 10^{-24}$ g/u) = 2.2749×10^{-25} g
Mass defect in g/mol = (2.2749×10^{-25} g)(6.022×10^{23} mol^{-1}) = 0.136 99 g/mol
Now, use the Einstein equation to convert the mass defect into the binding energy.
$\Delta E = \Delta mc^2 = (0.136\ 99\ g/mol)(10^{-3}\ kg/g)(3.00 \times 10^8\ m/s)^2$
$\Delta E = 1.233 \times 10^{13}$ J/mol = 1.233×10^{10} kJ/mol

$$\Delta E = \frac{1.233 \times 10^{13}\ J/mol}{6.022 \times 10^{23}\ nuclei/mol} \times \frac{1\ MeV}{1.60 \times 10^{-13}\ J} \times \frac{1\ nucleus}{16\ nucleons} = 8.00\ \frac{MeV}{nucleon}$$

19.12 ^{6}Li atomic mass = (3 proton mass) + (3 neutron mass) + (3 electron mass) – (mass defect)
= (3 x 1.007 28) + (3 x 1.008 66) + ($3 \times 5.486 \times 10^{-4}$) – (0.034 37) = 6.0151 u
Mass defect in g/mol = 0.034 37 g/mol
Now, use the Einstein equation to convert the mass defect into the binding energy.
$\Delta E = \Delta mc^2 = (0.034\ 37\ g/mol)(10^{-3}\ kg/g)(3.00 \times 10^8\ m/s)^2$
$\Delta E = 3.093 \times 10^{12}$ J/mol = 3.093×10^9 kJ/mol

$$\Delta E = \frac{3.093 \times 10^{12}\ J/mol}{6.022 \times 10^{23}\ nuclei/mol} \times \frac{1\ MeV}{1.60 \times 10^{-13}\ J} \times \frac{1\ nucleus}{6\ nucleons} = 5.35\ \frac{MeV}{nucleon}$$

19.13 $$\Delta m = \frac{\Delta E}{c^2} = \frac{-820 \times 10^3\ J}{(3.00 \times 10^8\ m/s)^2} = \frac{-820 \times 10^3\ kg \cdot m^2/s^2}{(3.00 \times 10^8\ m/s)^2} = -9.11 \times 10^{-12}\ kg$$

$$\Delta m = \frac{-9.11 \times 10^{-9}\ g}{2\ mol\ NaCl} = 4.56 \times 10^{-9}\ g/mol$$

19.14 $$\Delta m = \frac{\Delta E}{c^2} = \frac{-3.9 \times 10^{10}\ J}{(3.00 \times 10^8\ m/s)^2} = \frac{-3.9 \times 10^{10}\ kg \cdot m^2/s^2}{(3.00 \times 10^8\ m/s)^2} = -4.3 \times 10^{-7}\ kg$$
$\Delta m = -4.3 \times 10^{-4}$ g

19.15 $${}^{1}_{0}n + {}^{235}_{92}U \rightarrow {}^{137}_{52}Te + {}^{97}_{40}Zr + 2\ {}^{1}_{0}n$$

mass ${}^{235}_{92}U$	235.0439
mass ${}^{1}_{0}n$	1.008 66
–mass ${}^{137}_{52}Te$	–136.9254
–mass ${}^{97}_{40}Zr$	-96.9110
–mass 2 ${}^{1}_{0}n$	–(2)(1.008 66)
mass change	0.1988 u

$(0.1988\ u)(1.660\ 54 \times 10^{-24}\ g/u)(6.022 \times 10^{23}\ mol^{-1}) = 0.1988\ g/mol$
$\Delta E = \Delta mc^2 = (0.1988\ g/mol)(10^{-3}\ kg/g)(3.00 \times 10^8\ m/s)^2$
$\Delta E = 1.79 \times 10^{13}\ J/mol = 1.79 \times 10^{10}\ kJ/mol$

19.16 (a) $^{238}_{92}U \rightarrow ^{232}_{90}Th + ^{4}_{2}He$

(b)

mass ^{238}U	238.0508
mass ^{232}Th	–234.0436
mass ^{4}He	–4.0026
mass change	0.0046 u

$(0.0046\ u/atom)(1.660\ 54 \times 10^{-24}\ g/u) = 7.6 \times 10^{-27}\ g/atom$

$(0.0046\ u)(1.660\ 54 \times 10^{-24}\ g/u)(6.022 \times 10^{23}\ mol^{-1}) = 0.0046\ g/mol$
$\Delta E = \Delta mc^2 = (0.0046\ g/mol)(10^{-3}\ kg/g)(3.00 \times 10^8\ m/s)^2$
$\Delta E = 4.1 \times 10^{11}\ J/mol = 4.1 \times 10^8\ kJ/mol$

(c) Mass is lost and energy is released.

19.17 $^{1}_{1}H + ^{2}_{1}H \rightarrow ^{3}_{2}He$

mass ^{1}H	1.007 83
mass ^{2}H	2.014 10
–mass ^{3}He	–3.016 03
mass change	0.005 90 u

$(0.005\ 90\ u)(1.660\ 54 \times 10^{-24}\ g/u)(6.022 \times 10^{23}\ mol^{-1}) = 0.005\ 90\ g/mol$
$\Delta E = \Delta mc^2 = (0.005\ 90\ g/mol)(10^{-3}\ kg/g)(3.00 \times 10^8\ m/s)^2$
$\Delta E = 5.31 \times 10^{11}\ J/mol = 5.31 \times 10^8\ kJ/mol$

19.18 $^{40}_{18}Ar + ^{1}_{1}p \rightarrow ^{40}_{19}K + ^{1}_{0}n$

19.19 $^{238}_{92}U + ^{12}_{6}C \rightarrow ^{246}_{98}Cf + 4\ ^{1}_{0}n$

19.20 $\ln\left(\frac{N}{N_0}\right) = (-0.693)\left(\frac{t}{t_{1/2}}\right)$; $\frac{N}{N_0} = \frac{\text{Decay rate at time t}}{\text{Decay rate at time t} = 0}$

$\ln\left(\frac{2.4}{15.3}\right) = (-0.693)\left(\frac{t}{5730\ y}\right)$; $t = 1.53 \times 10^4\ y$

19.21

	^{40}K	$\rightarrow$	^{40}Ar
then (mmol)	N_0		0
change (mmol)	–0.95		+0.95
now (mmol)	N = 1.20		0.95

$N_0 - 0.95 = 1.20$ mmol therefore, $N_0 = 1.20 + 0.95 = 2.15$ mmol and N = 1.20 mmol

$$\ln\left(\frac{N}{N_0}\right) = (-0.693)\left(\frac{t}{t_{1/2}}\right)$$

$$\ln\left(\frac{1.20}{2.15}\right) = (-0.693)\left(\frac{t}{1.25 \text{ x } 10^9 \text{ y}}\right); \qquad t = 1.05 \text{ x } 10^9 \text{ y}$$

19.22 $^{235}_{92}U$

mass defect = (92 proton mass) + (143 neutron mass) + (92 electron mass) – $^{235}_{92}U$ mass

mass defect = (92 x 1.007 28) + (143 x 1.008 67) + (92 x 5.486 x 10^{-4}) – (235.043 929 9) = 1.916 11 u

(1.916 11 u/atom)(1.660 54 x 10^{-24} g/u) = 3.1818 x 10^{-24} g/atom

(1.916 11 u)(1.660 54 x 10^{-24} g/u)(6.022 x 10^{23} mol^{-1}) = 1.9161 g/mol

$\Delta E = \Delta mc^2$ = (1.9161 g/mol)(10^{-3} kg/g)(3.00 x 10^8 m/s)^2

ΔE = 1.724 x 10^{14} J/mol = 1.233 x 10^{11} kJ/mol

$$\Delta E = \frac{1.724 \text{ x } 10^{14} \text{ J/mol}}{6.022 \text{ x } 10^{23} \text{ nuclei/mol}} \text{ x } \frac{1 \text{ MeV}}{1.60 \text{ x } 10^{-13} \text{ J}} \text{ x } \frac{1 \text{ nucleus}}{235 \text{ nucleons}} = 7.62 \frac{\text{MeV}}{\text{nucleon}}$$

19.23

$$^{235}_{92}U \rightarrow {}^{231}_{90}Th + {}^{4}_{2}He$$
$$^{231}_{90}Th \rightarrow {}^{231}_{91}Pa + {}^{0}_{-1}e$$
$$^{231}_{91}Pa \rightarrow {}^{227}_{89}Ac + {}^{4}_{2}He$$
$$^{227}_{89}Ac \rightarrow {}^{223}_{87}Fr + {}^{4}_{2}He$$
$$^{223}_{87}Fr \rightarrow {}^{219}_{85}At + {}^{4}_{2}He$$
$$^{219}_{85}At \rightarrow {}^{215}_{83}Bi + {}^{4}_{2}He$$
$$^{215}_{83}Bi \rightarrow {}^{215}_{84}Po + {}^{0}_{-1}e$$
$$^{215}_{84}Po \rightarrow {}^{211}_{82}Pb + {}^{4}_{2}He$$

$$^{211}_{82}Pb \rightarrow \ ^{211}_{83}Bi + \ ^{0}_{-1}e$$
$$^{211}_{83}Bi \rightarrow \ ^{207}_{81}Tl + \ ^{4}_{2}He$$
$$^{207}_{81}Tl \rightarrow \ ^{207}_{82}Pb + \ ^{0}_{-1}e$$

19.24 $N_0 = 3.00$ and $N = 0.72$

$$\ln\left(\frac{N}{N_0}\right) = (-0.693)\left(\frac{t}{t_{1/2}}\right)$$

$$\ln\left(\frac{0.72}{3.00}\right) = (-0.693)\left(\frac{t}{7.03 \times 10^8 \text{ y}}\right); \qquad t = 1.45 \times 10^9 \text{ y}$$

19.25 $^{1}_{0}n + \ ^{235}_{92}U \rightarrow \ ^{140}_{56}Ba + \ ^{93}_{36}Kr + 3\ ^{1}_{0}n$

mass ^{1}n	1.00867
mass ^{235}U	235.0439
mass ^{140}Ba	–139.9106
mass ^{93}Kr	–92.9313
mass 3 ^{1}n	–3(1.00867)
mass change	0.1847

$(0.1847 \text{ u})(1.660\ 54 \times 10^{-24} \text{ g/u})(6.022 \times 10^{23} \text{ mol}^{-1}) = 0.1847 \text{ g/mol}$
$\Delta E = \Delta mc^2 = (0.1847 \text{ g/mol})(10^{-3} \text{ kg/g})(3.00 \times 10^8 \text{ m/s})^2$
$\Delta E = 1.66 \times 10^{13} \text{ J/mol} = 1.66 \times 10^{10} \text{ kJ/mol}$

19.26 2 billion years ago unusually rich uranium deposits, with the fissionable ^{235}U isotope at about 3% abundance, were flooded by groundwater, which acted as a moderator to slow the neutrons released by fission of ^{235}U, thereby allowing a nuclear chain reaction to take place. Today, however, the natural abundance of ^{235}U is only about 0.7% because of nuclear decay over the past few billion years. Because a chain-reaction is no longer self-sustaining at the present 0.7% level of ^{235}U, the conditions needed for natural reactors are no longer present on Earth.

Conceptual Problems

19.27 The shorter arrow pointing right is for beta emission. The longer arrow pointing left is for alpha emission.

$$A = {}^{147+94}_{94}X = {}^{241}_{94}Pu$$
$$B = {}^{146+95}_{95}X = {}^{241}_{95}Am$$
$$C = {}^{144+93}_{93}X = {}^{237}_{93}Np$$
$$D = {}^{142+91}_{91}X = {}^{233}_{91}Pa$$
$$E = {}^{141+92}_{92}X = {}^{233}_{92}U$$

Section Problems
Nuclear Reactions and Radioactivity (Sections 19.1 and 19.2)

19.28 Positron emission is the conversion of a proton in the nucleus into a neutron plus an ejected positron.
Electron capture is the process in which a proton in the nucleus captures an inner-shell electron and is thereby converted into a neutron.

19.29 An alpha particle $\left({}^{4}_{2}He^{2+}\right)$ is a helium nucleus. The He atom has two electrons and is neutral.

19.30 In beta emission a neutron is converted to a proton and the atomic number increases. In positron emission a proton is converted to a neutron and the atomic number decreases.

19.31 Gamma emission (radiation) has no mass. It is high-energy electromagnetic radiation. It typically accompanies α and β emission as a mechanism for the release of energy.

19.32 (a) ${}^{126}_{50}Sn \rightarrow {}^{0}_{-1}e + {}^{126}_{51}Sb$ (b) ${}^{210}_{88}Ra \rightarrow {}^{4}_{2}He + {}^{206}_{86}Rn$

(c) ${}^{77}_{37}Rb \rightarrow {}^{0}_{1}e + {}^{77}_{36}Kr$ (d) ${}^{76}_{36}Kr + {}^{0}_{-1}e \rightarrow {}^{76}_{35}Br$

19.33 (a) ${}^{90}_{38}Sr \rightarrow {}^{0}_{-1}e + {}^{90}_{39}Y$ (b) ${}^{247}_{100}Fm \rightarrow {}^{4}_{2}He + {}^{243}_{98}Cf$

(c) ${}^{49}_{25}Mn \rightarrow {}^{0}_{1}e + {}^{49}_{24}Cr$ (d) ${}^{37}_{18}Ar + {}^{0}_{-1}e \rightarrow {}^{37}_{17}Cl$

19.34 The mass number decreases by four, and the atomic number decreases by two. This is characteristic of alpha emission. $^{214}_{90}Th \rightarrow ^{210}_{88}Ra + ^{4}_{2}He$

19.35 The mass number does not change, and the atomic number increases by one. This is characteristic of beta emission. $^{239}_{92}U \rightarrow ^{239}_{93}Np + ^{0}_{-1}e$

19.36 (a) $^{188}_{80}Hg \rightarrow ^{188}_{79}Au + ^{0}_{1}e$ (b) $^{218}_{85}At \rightarrow ^{214}_{83}Bi + ^{4}_{2}He$

(c) $^{234}_{90}Th \rightarrow ^{234}_{91}Pa + ^{0}_{-1}e$

19.37 (a) $^{24}_{11}Na \rightarrow ^{24}_{12}Mg + ^{0}_{-1}e$ (b) $^{135}_{60}Nd \rightarrow ^{135}_{59}Pr + ^{0}_{1}e$

(c) $^{170}_{78}Pt \rightarrow ^{166}_{76}Os + ^{4}_{2}He$

19.38 (a) $^{162}_{75}Re \rightarrow ^{158}_{73}Ta + ^{4}_{2}He$ (b) $^{138}_{62}Sm + ^{0}_{-1}e \rightarrow ^{138}_{61}Pm$

(c) $^{188}_{74}W \rightarrow ^{188}_{75}Re + ^{0}_{-1}e$ (d) $^{165}_{73}Ta \rightarrow ^{165}_{72}Hf + ^{0}_{1}e$

19.39 (a) $^{157}_{63}Eu \rightarrow ^{157}_{64}Gd + ^{0}_{-1}e$ (b) $^{126}_{56}Ba + ^{0}_{-1}e \rightarrow ^{126}_{55}Cs$

(c) $^{146}_{62}Sm \rightarrow ^{142}_{60}Nd + ^{4}_{2}He$ (d) $^{125}_{56}Ba \rightarrow ^{125}_{55}Cs + ^{0}_{1}e$

19.40 $^{100}_{43}Tc \rightarrow ^{0}_{1}e + ^{100}_{42}Mo$ (positron emission)

$^{100}_{43}Tc + ^{0}_{-1}e \rightarrow ^{100}_{42}Mo$ (electron capture)

19.41 α emission: $^{226}_{89}Ac \rightarrow ^{222}_{87}Fr + ^{4}_{2}He$

β emission: $^{226}_{89}Ac \rightarrow ^{226}_{90}Th + ^{0}_{-1}e$

electron capture: $^{226}_{89}Ac + ^{0}_{-1}e \rightarrow ^{226}_{88}Ra$

Nuclear Stability (Section 19.3)

19.42 ^{160}W is neutron poor and decays by alpha emission. ^{185}W is neutron rich and decays by beta emission.

19.43 $^{136}_{53}I$ is neutron rich and decays by beta emission.

$^{122}_{53}I$ is neutron poor and decays by positron emission.

19.44 "Neutron rich" nuclides emit beta particles to decrease the number of neutrons and increase the number of protons in the nucleus.

19.45 "Neutron poor" nuclides decrease the number of protons and increase the n/p ratio by either alpha emission, positron emission, or electron capture.

19.46
$${}^{241}_{95}\text{Am} \rightarrow {}^{237}_{93}\text{Np} + {}^{4}_{2}\text{He}$$
$${}^{237}_{93}\text{Np} \rightarrow {}^{233}_{91}\text{Pa} + {}^{4}_{2}\text{He}$$
$${}^{233}_{91}\text{Pa} \rightarrow {}^{233}_{92}\text{U} + {}^{0}_{-1}\text{e}$$
$${}^{233}_{92}\text{U} \rightarrow {}^{229}_{90}\text{Th} + {}^{4}_{2}\text{He}$$
$${}^{229}_{90}\text{Th} \rightarrow {}^{225}_{88}\text{Ra} + {}^{4}_{2}\text{He}$$
$${}^{225}_{88}\text{Ra} \rightarrow {}^{225}_{89}\text{Ac} + {}^{0}_{-1}\text{e}$$
$${}^{225}_{89}\text{Ac} \rightarrow {}^{221}_{87}\text{Fr} + {}^{4}_{2}\text{He}$$
$${}^{221}_{87}\text{Fr} \rightarrow {}^{217}_{85}\text{At} + {}^{4}_{2}\text{He}$$
$${}^{217}_{85}\text{At} \rightarrow {}^{213}_{83}\text{Bi} + {}^{4}_{2}\text{He}$$
$${}^{213}_{83}\text{Bi} \rightarrow {}^{213}_{84}\text{Po} + {}^{0}_{-1}\text{e}$$
$${}^{213}_{84}\text{Po} \rightarrow {}^{209}_{82}\text{Pb} + {}^{4}_{2}\text{He}$$
$${}^{209}_{82}\text{Pb} \rightarrow {}^{209}_{83}\text{Bi} + {}^{0}_{-1}\text{e}$$

19.47
$${}^{222}_{86}\text{Rn} \rightarrow {}^{218}_{84}\text{Po} + {}^{4}_{2}\text{He}$$
$${}^{218}_{84}\text{Po} \rightarrow {}^{214}_{82}\text{Pb} + {}^{4}_{2}\text{He}$$
$${}^{214}_{82}\text{Pb} \rightarrow {}^{210}_{80}\text{Hg} + {}^{4}_{2}\text{He}$$
$${}^{210}_{80}\text{Hg} \rightarrow {}^{210}_{81}\text{Tl} + {}^{0}_{-1}\text{e}$$
$${}^{210}_{81}\text{Tl} \rightarrow {}^{210}_{82}\text{Pb} + {}^{0}_{-1}\text{e}$$

19.48 Each alpha emission decreases the mass number by four and the atomic number by two. Each beta emission increases the atomic number by one.

$${}^{232}_{90}\text{Th} \rightarrow {}^{208}_{82}\text{Pb}$$

$$\text{Number of } \alpha \text{ emissions} = \frac{\text{Th mass number} - \text{Pb mass number}}{4}$$

$$= \frac{232 - 208}{4} = 6\ \alpha \text{ emissions}$$

The atomic number decreases by 12 as a result of 6 alpha emissions. The resulting atomic number is (90 – 12) = 78.
Number of β emissions = Pb atomic number – 78 = 82 – 78 = 4 β emissions

19.49 Each alpha emission decreases the mass number by four and the atomic number by two. Each beta emission increases the atomic number by one.

$$^{235}_{92}U \rightarrow \; ^{207}_{82}Pb$$

$$\text{Number of } \alpha \text{ emissions} = \frac{\text{U mass number} - \text{Pb mass number}}{4}$$

$$= \frac{235 - 207}{4} = 7\ \alpha \text{ emissions}$$

The atomic number decreases by 14 as a result of 7 alpha emissions. The resulting atomic number is (92 – 14) = 78.
Number of β emissions = Pb atomic number – 78 = 82 – 78 = 4 β emissions

Radioactive Decay Rates (Section 19.4)

19.50 $$k = \frac{0.693}{t_{1/2}} = \frac{0.693}{2.805\ d} = 0.247\ d^{-1}$$

19.51 $$t_{1/2} = \frac{0.693}{k} = \frac{0.693}{2.88 \times 10^{-5}\ y^{-1}} = 2.41 \times 10^4\ y$$

19.52 $$t_{1/2} = \frac{0.693}{k} = \frac{0.693}{7.95 \times 10^{-3}\ d^{-1}} = 87.17 d$$

$$\ln\left(\frac{N}{N_0}\right) = (-0.693)\left(\frac{t}{t_{1/2}}\right) = (-0.693)\left(\frac{185\ d}{87.17\ d}\right) = -1.4707$$

$$\frac{N}{N_0} = e^{-1.4707} = 0.2298; \qquad \frac{N}{100\%} = 0.2298; \qquad N = 23.0\%$$

19.53 $$t_{1/2} = \frac{0.693}{k} = \frac{0.693}{2.88 \times 10^{-5}\ y^{-1}} = 2.41 \times 10^4\ y$$

After 1000 y: $$\ln\left(\frac{N}{N_0}\right) = (-0.693)\left(\frac{t}{t_{1/2}}\right) = (-0.693)\left(\frac{1000\ y}{2.41 \times 10^4\ y}\right) = -0.028\ 76$$

$$\frac{N}{N_0} = e^{-0.02876} = 0.9717; \qquad \frac{N}{100\%} = 0.9717; \qquad N = 97.17\%$$

After 25,000 y: $\ln\left(\frac{N}{N_0}\right) = (-0.693)\left(\frac{t}{t_{1/2}}\right) = (-0.693)\left(\frac{25{,}000\text{ y}}{2.41 \times 10^4\text{ y}}\right) = -0.7189$

$\frac{N}{N_0} = e^{-0.7189} = 0.4873$; $\frac{N}{100\%} = 0.4873$; $N = 48.73\%$

After 100,000 y: $\ln\left(\frac{N}{N_0}\right) = (-0.693)\left(\frac{t}{t_{1/2}}\right) = (-0.693)\left(\frac{100{,}000\text{ y}}{2.41 \times 10^4\text{ y}}\right) = -2.876$

$\frac{N}{N_0} = e^{-2.876} = 0.0564$; $\frac{N}{100\%} = 0.0564$; $N = 5.64\%$

19.54 $t_{1/2} = (102\text{ y})(365\text{ d/y})(24\text{ h/d})(3600\text{ s/h}) = 3.2167 \times 10^9\text{ s}$

$k = \frac{0.693}{t_{1/2}} = \frac{0.693}{3.2167 \times 10^9\text{ s}} = 2.1544 \times 10^{-10}\text{ s}^{-1}$

$N = (1.0 \times 10^{-9}\text{ g})\left(\frac{1\text{ mol Po}}{209\text{ g Po}}\right)(6.022 \times 10^{23}\text{ atoms/mol}) = 2.881 \times 10^{12}\text{ atoms}$

Decay rate = kN = $(2.1544 \times 10^{-10}\text{ s}^{-1})(2.881 \times 10^{12}\text{ atoms}) = 6.21 \times 10^2\text{ s}^{-1}$

621 α particles are emitted in 1.0 s.

19.55 $t_{1/2} = (3.0 \times 10^5\text{ y})(365\text{ d/y})(24\text{ h/d})(60\text{ min/h}) = 1.6 \times 10^{11}\text{ min}$

$k = \frac{0.693}{t_{1/2}} = \frac{0.693}{1.6 \times 10^{11}\text{ min}} = 4.3 \times 10^{-12}\text{ min}^{-1}$

$N = (5.0 \times 10^{-3}\text{ g})\left(\frac{1\text{ mol }^{36}\text{Cl}}{36\text{ g}}\right)(6.022 \times 10^{23}\text{ atoms/mol}) = 8.4 \times 10^{19}\text{ atoms}$

Decay rate = kN = $(4.3 \times 10^{-12}\text{ min}^{-1})(8.4 \times 10^{19}\text{ atoms}) = 3.6 \times 10^8\text{ min}^{-1}$

19.56 Decay rate = kN

$N = (1.0 \times 10^{-3}\text{ g})\left(\frac{1\text{ mol }^{79}\text{Se}}{79\text{ g}}\right)(6.022 \times 10^{23}\text{ atoms/mol}) = 7.6 \times 10^{18}\text{ atoms}$

$k = \frac{\text{Decay rate}}{N} = \frac{1.5 \times 10^5\text{/s}}{7.6 \times 10^{18}} = 2.0 \times 10^{-14}\text{ s}^{-1}$

$t_{1/2} = \frac{0.693}{k} = \frac{0.693}{2.0 \times 10^{-14}\text{ s}^{-1}} = 3.5 \times 10^{13}\text{ s}$

$t_{1/2} = (3.5 \times 10^{13}\text{ s})\left(\frac{1\text{ h}}{3600\text{ s}}\right)\left(\frac{1\text{ d}}{24\text{ h}}\right)\left(\frac{1\text{ y}}{365\text{ d}}\right) = 1.1 \times 10^6\text{ y}$

19.57 Decay rate = kN

$$N = (1.0 \times 10^{-9}\ g)\left(\frac{1\ mol\ Ti}{44\ g\ Ti}\right)(6.022 \times 10^{23}\ atoms/mol) = 1.37 \times 10^{13}\ atoms$$

$$k = \frac{Decay\ rate}{N} = \frac{4.8 \times 10^{3}\ s^{-1}}{1.37 \times 10^{13}} = 3.50 \times 10^{-10}\ s^{-1}$$

$$k = (3.50 \times 10^{-10}\ s^{-1})(3600\ s/h)(24\ h/d)(365\ d/y) = 1.10 \times 10^{-2}\ y^{-1}$$

$$t_{1/2} = \frac{0.693}{k} = \frac{0.693}{1.10 \times 10^{-2}\ y^{-1}} = 63\ y$$

19.58

$$\ln\left(\frac{N}{N_0}\right) = (-0.693)\left(\frac{t}{t_{1/2}}\right); \quad \frac{N}{N_0} = \frac{Decay\ rate\ at\ time\ t}{Decay\ rate\ at\ time\ t = 0}$$

$$\ln\left(\frac{6990}{8540}\right) = (-0.693)\left(\frac{10.0\ d}{t_{1/2}}\right); \quad t_{1/2} = 34.6\ d$$

19.59

$$\ln\left(\frac{N}{N_0}\right) = (-0.693)\left(\frac{t}{t_{1/2}}\right); \quad \frac{N}{N_0} = \frac{Decay\ rate\ at\ time\ t}{Decay\ rate\ at\ time\ t = 0}$$

$$\ln\left(\frac{10,980}{53,500}\right) = (-0.693)\left(\frac{48.0\ h}{t_{1/2}}\right); \quad t_{1/2} = 21.0\ h$$

19.60 (a) $1 \rightarrow 1/2 \rightarrow 1/4 \rightarrow 1/8$

After three half-lives, 1/8 of the strontium-90 will remain.

(b) $$k = \frac{0.693}{t_{1/2}} = \frac{0.693}{29\ y} = 0.0239\ y^{-1} = 0.024\ y^{-1}$$

(c) $$t = \frac{\ln\frac{N}{N_o}}{-k} = \frac{\ln\frac{(Sr\text{–}90)_t}{(Sr\text{–}90)_o}}{-k} = \frac{\ln\frac{(0.01)}{(1)}}{-0.0239y^{-1}} = 193\ y$$

19.61 Decay rate = kN

$$k = \frac{0.693}{t_{1/2}} = \frac{0.693}{1.25 \times 10^{9}\ y} = 5.54 \times 10^{-10}\ y^{-1}$$

KCl, 74.55

N = number of $^{40}K^{+}$ ions in a 1.00 g sample of KCl

$$N = (0.000\ 117)(1.00\ g)\left(\frac{1\ mol\ KCl}{74.55\ g}\right)\left(\frac{1\ mol\ K^+}{1\ mol\ KCl}\right)(6.022 \times 10^{23}\ mol^{-1})$$

$N = 9.45 \times 10^{17}\ {}^{40}K^+$ ions

Decay rate = $kN = (5.54 \times 10^{-10}\ y^{-1})(9.45 \times 10^{17}) = 5.24 \times 10^8/y$

$$\text{Disintegration/s} = (5.24 \times 10^8/y)\left(\frac{1\ y}{365\ d}\right)\left(\frac{1\ d}{24\ h}\right)\left(\frac{1\ h}{3600\ s}\right) = 16.6/s$$

Energy Changes during Nuclear Reactions (Section 19.5)

19.62 The loss in mass that occurs when protons and neutrons combine to form a nucleus is called the mass defect. The lost mass is converted into the binding energy that is used to hold the nucleons together.

19.63 Energy (heat) is absorbed in an endothermic reaction. The energy is converted to mass. The mass of the products is slightly larger than the mass of the reactants.

19.64 (a) For ${}^{52}_{26}Fe$:

First, calculate the total mass of the nucleons (26 n + 26 p)
Mass of 26 neutrons = (26)(1.008 66) = 26.225 16
Mass of 26 protons = (26)(1.007 28) = 26.189 28
Mass of 26 n + 26 p = 52.414 44

Next, calculate the mass of a ${}^{52}Fe$ nucleus by subtracting the mass of 26 electrons from the mass of a ${}^{52}Fe$ atom.
Mass of ${}^{52}Fe$ atom = 51.948 11
–Mass of 26 electrons = $-(26)(5.486 \times 10^{-4})$ = –0.014 26
Mass of ${}^{52}Fe$ nucleus = 51.933 85
Then subtract the mass of the ${}^{52}Fe$ nucleus from the mass of the nucleons to find the mass defect:
Mass defect = mass of nucleons – mass of nucleus
= (52.414 44) – (51.933 85) = 0.480 59 u
Mass defect in g/mol:
$(0.480\ 59\ u)(1.660\ 54 \times 10^{-24}\ g/u)(6.022 \times 10^{23}\ mol^{-1}) = 0.480\ 58\ g/mol$

(b) For ${}^{92}_{42}Mo$:

First, calculate the total mass of the nucleons (50 n + 42 p)
Mass of 50 neutrons = (50)(1.008 66) = 50.433 00
Mass of 42 protons = (42)(1.007 28) = 42.305 76
Mass of 50 n + 42 p = 92.738 76

Next, calculate the mass of a ^{92}Mo nucleus by subtracting the mass of 42 electrons from the mass of a ^{92}Mo atom.

Mass of ^{92}Mo atom = 91.906 81
–Mass of 42 electrons = $-(42)(5.486 \times 10^{-4})$ = –0.023 04
Mass of ^{92}Mo nucleus = 91.883 77

Then subtract the mass of the ^{92}Mo nucleus from the mass of the nucleons to find the mass defect:

Mass defect = mass of nucleons – mass of nucleus
= (92.738 76) – (91.883 77) = 0.854 97 u

Mass defect in g/mol:
$(0.854\ 99\ \text{u})(1.660\ 54 \times 10^{-24}\ \text{g/u})(6.022 \times 10^{23}\ \text{mol}^{-1}) = 0.854\ 99\ \text{g/mol}$

19.65 (a) For $^{32}_{16}$S:

First, calculate the total mass of the nucleons (16 n + 16 p)
Mass of 16 neutrons = (16)(1.008 66) = 16.138 56
Mass of 16 protons = (16)(1.007 28) = 16.116 48
Mass of 16 n + 16 p = 32.255 04

Next, calculate the mass of a ^{32}S nucleus by subtracting the mass of 16 electrons from the mass of a ^{32}S atom.

Mass of ^{32}S = 31.972 07
Mass of 16 electrons = $-(16)(5.486 \times 10^{-4})$ = –0.008 78
Mass of ^{32}S nucleus = 31.963 29

Then subtract the mass of the ^{32}S nucleus from the mass of the nucleons to find the mass defect:

Mass defect = mass of nucleons – mass of nucleus
= (32.255 04) – (31.963 29) = 0.291 75 u

Mass defect in g/mol:
$(0.291\ 75\ \text{u})(1.660\ 54 \times 10^{-24}\ \text{g/u})(6.022 \times 10^{23}\ \text{mol}^{-1}) = 0.291\ 74\ \text{g/mol}$

(b) For $^{40}_{20}$Ca:

First, calculate the total mass of the nucleons (20 n + 20 p)
Mass of 20 neutrons = (20)(1.008 66) = 20.173 20
Mass of 20 protons = (20)(1.007 28) = 20.145 60
Mass of 20 n + 20 p = 40.318 80

Next, calculate the mass of a ^{40}Ca nucleus by subtracting the mass of 20 electrons from the mass of a ^{40}Ca atom.

Mass of ^{40}Ca = 39.962 59
–Mass of 20 electrons = $-(20)(5.486 \times 10^{-4})$ = –0.010 97
Mass of ^{40}Ca nucleus = 39.951 62

Then subtract the mass of the ^{40}Ca nucleus from the mass of the nucleons to find the mass defect:

Mass defect = mass of nucleons – mass of nucleus

= (40.318 80) – (39.951 62) = 0.367 18 u

Mass defect in g/mol:

(0.367 18 u)(1.660 54 x 10^{-24} g/u)(6.022 x 10^{23} mol^{-1}) = 0.367 17 g/mol

19.66 (a) For $^{58}_{28}$Ni:

First, calculate the total mass of the nucleons (30 n + 28 p)

Mass of 30 neutrons = (30)(1.008 66) = 30.259 80

Mass of 28 protons = (28)(1.007 28) = 28.203 84

Mass of 30 n + 28 p = 58.463 64

Next, calculate the mass of a ^{58}Ni nucleus by subtracting the mass of 28 electrons from the mass of a ^{58}Ni atom.

Mass of ^{58}Ni atom = 57.935 35

–Mass of 28 electrons = –(28)(5.486 x 10^{-4}) = –0.015 36

Mass of ^{58}Ni nucleus = 57.919 99

Then subtract the mass of the ^{58}Ni nucleus from the mass of the nucleons to find the mass defect:

Mass defect = mass of nucleons – mass of nucleus

= (58.463 64) – (57.919 99) = 0.543 65 u

Mass defect in g/mol:

(0.543 65 u)(1.660 54 x 10^{-24} g/u)(6.022 x 10^{23} mol^{-1}) = 0.543 64 g/mol

Now, use the Einstein equation to convert the mass defect into the binding energy.

$\Delta E = \Delta mc^2 = (0.543\ 64\ \text{g/mol})(10^{-3}\ \text{kg/g})(3.00 \times 10^8\ \text{m/s})^2$

$\Delta E = 4.893 \times 10^{13}$ J/mol = 4.893×10^{10} kJ/mol

$$\Delta E = \frac{4.893 \times 10^{13}\ \text{J/mol}}{6.022 \times 10^{23}\ \text{nuclei/mol}} \times \frac{1\ \text{MeV}}{1.60 \times 10^{-13}\ \text{J}} \times \frac{1\ \text{nucleus}}{58\ \text{nucleons}} = 8.76\ \text{MeV/nucleon}$$

(b) For $^{84}_{36}$Kr:

First, calculate the total mass of the nucleons (48 n + 36 p)

Mass of 48 neutrons = (48)(1.008 66) = 48.415 68

Mass of 36 protons = (36)(1.007 28) = 36.262 08

Mass of 48 n + 36 p = 84.677 76

Next, calculate the mass of a ^{84}Kr nucleus by subtracting the mass of 36 electrons from the mass of a ^{84}Kr atom.

Mass of ^{84}Kr atom = 83.911 51

–Mass of 36 electrons = –(36)(5.486 x 10^{-4}) = –0.019 75

Mass of ^{84}Kr nucleus = 83.891 76

Then subtract the mass of the ^{84}Kr nucleus from the mass of the nucleons to find the mass defect:
Mass defect = mass of nucleons – mass of nucleus
= (84.677 76) – (83.891 76) = 0.786 00 u
Mass defect in g/mol:
$(0.786\ 00\ u)(1.660\ 54 \times 10^{-24}\ g/u)(6.022 \times 10^{23}\ mol^{-1}) = 0.785\ 98\ g/mol$
Now, use the Einstein equation to convert the mass defect into the binding energy.
$\Delta E = \Delta mc^2 = (0.785\ 98\ g/mol)(10^{-3}\ kg/g)(3.00 \times 10^8\ m/s)^2$
$\Delta E = 7.074 \times 10^{13}\ J/mol = 7.074 \times 10^{10}\ kJ/mol$

$$\Delta E = \frac{7.074 \times 10^{13}\ J/mol}{6.022 \times 10^{23}\ nuclei/mol} \times \frac{1\ MeV}{1.60 \times 10^{-13}\ J} \times \frac{1\ nucleus}{84\ nucleons} = 8.74\ MeV/nucleon$$

19.67 (a) For $^{63}_{29}Cu$:
First, calculate the total mass of the nucleons (34 n + 29 p)
Mass of 34 neutrons = (34)(1.008 66) = 34.294 44
Mass of 29 protons = (29)(1.007 28) = 29.211 12
Mass of 34 n + 29 p = 63.505 56
Next calculate the mass of a ^{63}Cu nucleus by subtracting the mass of 29 electrons from the mass of a ^{63}Cu atom.
Mass of ^{63}Cu atom = 62.939 60
–Mass of 29 electrons = $-(29)(5.486 \times 10^{-4})$ = –0.015 91
Mass of ^{63}Cu nucleus = 62.923 69
Then subtract the mass of the ^{63}Cu nucleus from the mass of the nucleons to find the mass defect:
Mass defect = mass of nucleons – mass of nucleus
= (63.505 56) – (62.923 69) = 0.581 87 u
Mass defect in g/mol:
$(0.581\ 87\ u)(1.660\ 54 \times 10^{-24}\ g/u)(6.022 \times 10^{23}\ mol^{-1}) = 0.581\ 86\ g/mol$
Now, use the Einstein equation to convert the mass defect into the binding energy.
$\Delta E = \Delta mc^2 = (0.581\ 86\ g/mol)(10^{-3}\ kg/g)(3.00 \times 10^8\ m/s)^2$
$\Delta E = 5.237 \times 10^{13}\ J/mol = 5.237 \times 10^{10}\ kJ/mol$

$$\Delta E = \frac{5.237 \times 10^{13}\ J/mol}{6.022 \times 10^{23}\ nuclei/mol} \times \frac{1\ MeV}{1.60 \times 10^{-13}\ J} \times \frac{1\ nucleus}{63\ nucleons} = 8.63\ MeV/nucleon$$

(b) For $^{84}_{38}Sr$:
First, calculate the total mass of the nucleons (46 n + 38 p)
Mass of 46 neutrons = (46)(1.008 66) = 46.398 36
Mass of 38 protons = (38)(1.007 28) = 38.276 64
Mass of 46 n + 38 p = 84.675 00

Next, calculate the mass of a ^{84}Sr nucleus by subtracting the mass of 38 electrons from the mass of a ^{84}Sr atom.

Mass of ^{84}Sr atom = 83.913 43
–Mass of 38 electrons = $-(38)(5.486 \times 10^{-4})$ = –0.020 85
Mass of ^{84}Sr nucleus = 83.892 58

Then subtract the mass of the ^{84}Sr nucleus from the mass of the nucleons to find the mass defect:

Mass defect = mass of nucleons – mass of nucleus
= (84.675 00) – (83.892 58) = 0.782 42 u

Mass defect in g/mol:
$(0.782\ 42\ u)(1.660\ 54 \times 10^{-24}\ g/u)(6.022 \times 10^{23}\ mol^{-1}) = 0.782\ 40\ g/mol$

Now, use the Einstein equation to convert the mass defect into the binding energy.
$\Delta E = \Delta mc^2 = (0.782\ 40\ g/mol)(10^{-3}\ kg/g)(3.00 \times 10^8\ m/s)^2$
$\Delta E = 7.042 \times 10^{13}\ J/mol = 7.042 \times 10^{10}\ kJ/mol$

$$\Delta E = \frac{7.042 \times 10^{13}\ J/mol}{6.022 \times 10^{23}\ nuclei/mol} \times \frac{1\ MeV}{1.60 \times 10^{-13}\ J} \times \frac{1\ nucleus}{84\ nucleons} = 8.70\ MeV/nucleon$$

19.68 $^{174}_{77}Ir \rightarrow \, ^{170}_{75}Re + \, ^{4}_{2}He$

mass $^{174}_{77}Ir$ 173.966 66
–mass $^{170}_{75}Re$ –169.958 04
–mass $^{4}_{2}He$ – 4.002 60
mass change 0.006 02 u

$(0.006\ 02\ u)(1.660\ 54 \times 10^{-24}\ g/u)(6.022 \times 10^{23}\ mol^{-1}) = 0.006\ 02\ g/mol$
$\Delta E = \Delta mc^2 = (0.006\ 02\ g/mol)(10^{-3}\ kg/g)(3.00 \times 10^8\ m/s)^2$
$\Delta E = 5.42 \times 10^{11}\ J/mol = 5.42 \times 10^8\ kJ/mol$

19.69 $^{28}_{12}Mg \rightarrow \, ^{28}_{13}Al + \, ^{0}_{-1}e$

Reactant: $^{28}_{12}Mg$ nucleus = $^{28}_{12}Mg$ atom – 12 e^-
Product: $^{28}_{13}Al$ nucleus + e^- = ($^{28}_{13}Al$ nucleus – 13 e^-) + e^- = $^{28}_{13}Al$ nucleus – 12 e^-

Change : ($^{28}_{12}Mg$ atom – 12 e^-) – ($^{28}_{13}Al$ nucleus – 12 e^-) = $^{28}_{12}Mg$ atom – $^{28}_{13}Al$ atom
(clectrons cancel)

Mass change = (27.983 88) – (27.981 91) = 0.001 97 u

$(0.001\ 97\ u)(1.660\ 54 \times 10^{-24}\ g/u)(6.022 \times 10^{23}\ mol^{-1}) = 0.001\ 97\ g/mol$
$\Delta E = \Delta mc^2 = (0.001\ 97\ g/mol)(10^{-3}\ kg/g)(3.00 \times 10^8\ m/s)^2$
$\Delta E = 1.77 \times 10^{11}\ J/mol = 1.77 \times 10^8\ kJ/mol$

19.70 $$\Delta m = \frac{\Delta E}{c^2} = \frac{-92.2 \times 10^3\ J}{(3.00 \times 10^8\ m/s)^2} = \frac{-92.2 \times 10^3\ kg \cdot m^2/s^2}{(3.00 \times 10^8\ m/s)^2} = -1.02 \times 10^{-12}\ kg$$

$$\Delta m = \frac{-1.02 \times 10^{-9}\ g}{2\ mol\ NH_3} = 5.10 \times 10^{-10}\ g/mol$$

19.71 $$\Delta m = \frac{\Delta E}{c^2} = \frac{131 \times 10^3\ J}{(3.00 \times 10^8\ m/s)^2} = \frac{131 \times 10^3\ kg \cdot m^2/s^2}{(3.00 \times 10^8\ m/s)^2} = 1.46 \times 10^{-12}\ kg$$

$\Delta m = 1.46 \times 10^{-9}\ g$

Fission and Fusion (Section 19.6)

19.72 (a) Nuclear fission is induced by bombarding a U-235 sample with beta particles.

19.73 Neutrons are used to initiate the fission of U-235 and also sustain the nuclear chain reaction.

19.74 $${}^{10}_{5}B + {}^{1}_{0}n \rightarrow {}^{7}_{3}Li + {}^{4}_{2}He$$

19.75 Uranium fuel rods for a nuclear reactor contain U-235 enriched to ~3%. Uranium fuel for atomic weapons contain U-235 enriched to greater than 85%.

19.76 Fuel rods in a power plant cannot be used to make an atomic weapon unless the fuel rod is processed and significantly enriched in the fissionable U-235.

19.77 The appeal of nuclear fusion as a power source is that the hydrogen isotopes used as fuel are cheap and plentiful and that the fusion products are nonradioactive and nonpolluting. Many technical problems must be solved before achieving a practical and controllable fusion method. Not the least of which is that a temperature of approximately 40 million K is needed to initiate the fusion process.

19.78 $2\ {}^{2}_{1}H \rightarrow {}^{3}_{2}He + {}^{1}_{0}n$

mass 2 ${}^{2}_{1}H$	2(2.0141)
–mass ${}^{3}_{2}He$	–3.0160
–mass ${}^{1}_{0}n$	–1.008 66
mass change	0.003 54 u

(0.003 54 u)(1.660 54 x 10^{-24} g/u)(6.022 x 10^{23} mol^{-1}) = 0.003 54 g/mol
$\Delta E = \Delta mc^2$ = (0.003 54 g/mol)(10^{-3} kg/g)(3.00 x 10^8 m/s)^2
ΔE = 3.2 x 10^{11} J/mol = 3.2 x 10^8 kJ/mol

19.79 ${}^{1}_{0}n + {}^{239}_{94}Pu \rightarrow {}^{146}_{56}Ba + {}^{91}_{38}Sr + 3\ {}^{1}_{0}n$

mass ${}^{1}n$	1.00867
mass ${}^{239}Pu$	239.05216
mass ${}^{146}Ba$	–145.93022
mass ${}^{91}Sr$	–90.91020
mass 3 ${}^{1}n$	–3(1.00867)
mass change	0.1944

(0.1944 u)(1.660 54 x 10^{-24} g/u)(6.022 x 10^{23} mol^{-1}) = 0.1944 g/mol
$\Delta E = \Delta mc^2$ = (0.1944 g/mol)(10^{-3} kg/g)(3.00 x 10^8 m/s)^2
ΔE = 1.75 x 10^{13} J/mol = 1.75 x 10^{10} kJ/mol

Nuclear Transmutation (Section 19.7)

19.80 (a) ${}^{109}_{47}Ag + {}^{4}_{2}He \rightarrow {}^{113}_{49}In$

(b) ${}^{10}_{5}B + {}^{4}_{2}He \rightarrow {}^{13}_{7}N + {}^{1}_{0}n$

19.81 (a) ${}^{235}_{92}U \rightarrow {}^{160}_{62}Sm + {}^{72}_{30}Zn + 3\ {}^{1}_{0}n$

(b) ${}^{235}_{92}U \rightarrow {}^{87}_{35}Br + {}^{146}_{57}La + 2\ {}^{1}_{0}n$

19.82 ${}^{209}_{83}Bi + {}^{58}_{26}Fe \rightarrow {}^{266}_{109}Mt + {}^{1}_{0}n$

19.83 $^{98}_{42}Mo + ^{1}_{0}n \rightarrow ^{99}_{42}Mo$

19.84 $^{238}_{92}U + ^{2}_{1}H \rightarrow ^{238}_{93}Np + 2\ ^{1}_{0}n$

19.85 (a) $^{246}_{96}Cm + ^{12}_{6}C \rightarrow ^{254}_{102}No + 4\ ^{1}_{0}n$

(b) $^{253}_{99}Es + ^{4}_{2}He \rightarrow ^{256}_{101}Md + ^{1}_{0}n$

(c) $^{250}_{98}Cf + ^{11}_{5}B \rightarrow ^{257}_{103}Lr + 4\ ^{1}_{0}n$

Detecting and Measuring Radioactivity (Section 19.8)

19.86 1 Sv = 1 Gy and 1 Gy = 1 J/kg

$5000\ \mu Sv = 5000 \times 10^{-6}\ Sv = 5000 \times 10^{-6}\ Gy = 5000 \times 10^{-6}\ J/kg$

joules absorbed = $(5000 \times 10^{-6}\ J/kg)(60\ kg) = 0.3\ J$

19.87 $23.2\ \text{rads} \times \frac{0.01\ Gy}{1\ rad} = 0.232\ Gy$

$(0.232\ Gy)(255\ g) \times \frac{1\ kg}{1000\ g} \times \frac{1\ J}{kg \cdot Gy} = 0.0592\ J$

19.88 $4.0\ pCi = 4.0 \times 10^{-12}\ Ci$

$1\ Ci = 3.7 \times 10^{10}\ Bq = 3.7 \times 10^{10}$ disintegrations/s

(a) $4.0 \times 10^{-12}\ Ci \times \frac{3.7 \times 10^{10}\ \text{disintegrations/s}}{1\ Ci} \times \frac{60\ s}{1\ min} = 8.9$ disintegrations/min

(b) $\text{Rate} = \frac{8.9\ dis}{1\ min} \times \frac{60\ min}{1\ h} \times \frac{24\ h}{1\ d} = 1.3 \times 10^4\ dis/d$

$\text{Rate} = 1.3 \times 10^4\ dis/d = kN = \left(\frac{0.693}{3.8\ d}\right)N$; solve for N

$N = 7.1 \times 10^4\ ^{222}Rn$

19.89 By definition, 1 Ci is the decay rate of 1 g of ^{226}Ra.

$1\ Ci = 3.7 \times 10^{10}\ Bq = 3.7 \times 10^{10}$ disintegrations/s

$1600\ y \times \frac{365\ d}{1\ y} \times \frac{24\ h}{1\ d} \times \frac{60\ min}{1\ h} \times \frac{60\ s}{1\ min} = 5.0 \times 10^{10}\ s$

For ^{226}Ra, $k = \frac{0.693}{t_{1/2}} = \frac{0.693}{5.0 \times 10^{10}\ s} = 1.4 \times 10^{-11}\ s^{-1}$

Decay rate = kN

$$N = (10.0 \text{ x } 10^{-3} \text{ g})\left(\frac{1 \text{ mol Ra-226}}{226 \text{ g Ra}}\right)(6.022 \text{ x } 10^{23} \text{ atoms/mol}) = 2.66 \text{ x } 10^{19} \text{ atoms}$$

activity = decay rate = kN = $(1.4 \text{ x } 10^{-11} \text{ s}^{-1})(2.66 \text{ x } 10^{19} \text{ atoms}) = 3.7 \text{ x } 10^{8}$ dis/s

decay rate = $3.7 \text{ x } 10^{8}$ Bq

$$\text{decay rate} = 3.7 \text{ x } 10^{8} \text{ Bq x } \frac{1 \text{ Ci}}{3.7 \text{ x } 10^{10} \text{ Bq}} = 0.010 \text{ Ci}$$

19.90 1 Ci = $3.7 \text{ x } 10^{10}$ disintegrations/s and 1.0 rad = $2.2 \text{ x } 10^{11}$ disintegrations of ^{60}Co

$$1800 \text{ rad x } \frac{2.2 \text{ x } 10^{11} \text{ disintegrations}}{1 \text{ rad}} = 3.96 \text{ x } 10^{14} \text{ disintegrations}$$

30 Ci = $(30)(3.7 \text{ x } 10^{10}$ disintegrations/s$) = 1.11 \text{ x } 10^{12}$ disintegrations/s

$$\frac{1.11 \text{ x } 10^{12} \text{ disintegrations/s}}{1 \text{ source}} \text{ x } 201 \text{ sources} = 2.23 \text{ x } 10^{14} \text{ disintegrations/s}$$

$$\text{time} = \frac{3.96 \text{ x } 10^{14} \text{ disintegrations}}{2.23 \text{ x } 10^{14} \text{ disintegrations/s}} = 1.8 \text{ s}$$

19.91 First find the activity of the ^{51}Cr after 17.0 days.

$$\ln\left(\frac{N}{N_0}\right) = (-0.693)\left(\frac{t}{t_{1/2}}\right); \qquad \frac{N}{N_0} = \frac{\text{Decay rate at time t}}{\text{Decay rate at time t = 0}}$$

$$\ln\left(\frac{N}{4.10}\right) = (-0.693)\left(\frac{17.0 \text{ d}}{27.7 \text{ d}}\right)$$

$\ln N - \ln(4.10) = -0.4253$

$\ln N = -0.4253 + \ln(4.10) = 0.9857$

$N = e^{0.9857} = 2.68$ μCi/mL

(20.0 mL)(2.68 μCi/mL) = (total blood volume)(0.009 35 μCi/mL)

total blood volume = 5732 mL = 5.73 L

Some Applications of Nuclear Chemistry (Section 19.9)

19.92
$$\ln\left(\frac{N}{N_0}\right) = (-0.693)\left(\frac{t}{t_{1/2}}\right); \qquad \frac{N}{N_0} = \frac{\text{Decay rate at time t}}{\text{Decay rate at t = 0}}$$

$t_{1/2} = 5715$ y

$$\ln\left(\frac{2.3}{15.3}\right) = (-0.693)\left(\frac{t}{5715 \text{ y}}\right)$$

age of bone = t = $1.6 \text{ x } 10^{4}$ y

19.93 $\ln\left(\frac{N}{N_0}\right) = (-0.693)\left(\frac{t}{t_{1/2}}\right)$; $\frac{N}{N_0} = \frac{\text{Decay rate at time t}}{\text{Decay rate at t = 0}}$

$t_{1/2} = 5715 \text{ y}$

$\ln\left(\frac{N}{15.3}\right) = (-0.693)\left(\frac{3000 \text{ y}}{5715 \text{ y}}\right)$

$\ln N - \ln(15.3) = (-0.693)\left(\frac{3000 \text{ y}}{5715 \text{ y}}\right)$

$\ln N = (-0.693)\left(\frac{3000 \text{ y}}{5715 \text{ y}}\right) + \ln(15.3) = 2.364$

$N = e^{2.364} = 10.6$ decays/min

19.94 U-238, $t_{1/2} = 4.47 \times 10^9$ yr

At time t, U-238 = 105 μmol and Pb-206 = 33 μmol

U-238 at time t_o = (105 + 33) μmol = 138 μmol

$\ln\left(\frac{N}{N_0}\right) = (-0.693)\left(\frac{t}{t_{1/2}}\right)$

$\ln\left(\frac{^{238}U_t}{^{238}U_0}\right) = \ln\left(\frac{105}{138}\right) = (-0.693)\left(\frac{t}{4.47 \times 10^9 \text{ yr}}\right)$

age of rock = t = 1.8×10^9 yr

19.95

	^{40}K	→	^{40}Ar
then (mmol)	N_0		0
change (mmol)	–0.25		+0.25
now (mmol)	N = 3.35		0.25

$N_0 - 0.25 = 3.35$ mmol therefore, $N_0 = 3.35 + 0.25 = 3.60$ mmol and N = 3.35 mmol

$\ln\left(\frac{N}{N_0}\right) = (-0.693)\left(\frac{t}{t_{1/2}}\right)$

$\ln\left(\frac{3.35}{3.60}\right) = (-0.693)\left(\frac{t}{1.25 \times 10^9 \text{ y}}\right)$; $t = 1.30 \times 10^8$ y

Chapter Problems

19.96 $E = (1.50 \text{ MeV})\left(\dfrac{1.60 \times 10^{-13} \text{ J}}{1 \text{ MeV}}\right) = 2.40 \times 10^{-13} \text{ J}$

$$\lambda = \frac{hc}{E} = \frac{(6.626 \times 10^{-34} \text{ J·s})(3.00 \times 10^{8} \text{ m/s})}{2.40 \times 10^{-13} \text{ J}} = 8.28 \times 10^{-13} \text{ m} = 0.000\ 828 \text{ nm}$$

19.97 $E = (6.82 \text{ keV})\left(\dfrac{1 \text{ MeV}}{10^3 \text{ keV}}\right)\left(\dfrac{1.60 \times 10^{-13} \text{J}}{1 \text{ MeV}}\right) = 1.09 \times 10^{-15} \text{ J}$

$$\nu = \frac{E}{h} = \frac{1.09 \times 10^{-15} \text{J}}{6.626 \times 10^{-34} \text{J} \cdot \text{s}} = 1.65 \times 10^{18}/\text{s} = 1.65 \times 10^{18} \text{ Hz}$$

19.98 Mass of positron and electron

$= 2(9.109 \times 10^{-31} \text{ kg})(6.022 \times 10^{23} \text{ mol}^{-1}) = 1.097 \times 10^{-6} \text{ kg/mol}$

$\Delta E = \Delta mc^2 = (1.097 \times 10^{-6} \text{ kg/mol})(3.00 \times 10^{8} \text{ m/s})^2$

$\Delta E = 9.87 \times 10^{10} \text{ J/mol} = 9.87 \times 10^{7} \text{ kJ/mol}$

19.99 $^{10}B + {}^{1}n \rightarrow {}^{4}He + {}^{7}Li + \gamma$

mass ^{10}B	10.012 937
mass ^{1}n	1.008 665
–mass ^{4}He	– 4.002 603
–mass ^{7}Li	–7.016 004
mass change	0.002 995 u

$(0.002\ 995 \text{ u})(1.660\ 54 \times 10^{-24} \text{ g/u}) = 4.973 \times 10^{-27} \text{ g}$

$\Delta E = \Delta mc^2 = (4.973 \times 10^{-27} \text{ g})(10^{-3} \text{ kg/g})(3.00 \times 10^{8} \text{ m/s})^2 = 4.476 \times 10^{-13} \text{ J}$

$\text{Kinetic energy} = 2.31 \text{ MeV} \times \dfrac{1.60 \times 10^{-13} \text{ J}}{1 \text{ MeV}} = 3.696 \times 10^{-13} \text{ J}$

$\gamma \text{ photon energy} = \Delta E - KE = 4.476 \times 10^{-13} \text{ J} - 3.696 \times 10^{-13} \text{ J} = 7.80 \times 10^{-14} \text{ J}$

$= 7.80 \times 10^{-14} \text{ J} \times \dfrac{1 \text{ MeV}}{1.60 \times 10^{-13} \text{ J}} = 0.488 \text{ MeV}$

19.100 For radioactive decay, $\ln \dfrac{N}{N_o} = -kt$

For ^{235}U, $k_1 = \dfrac{0.693}{t_{1/2}} = \dfrac{0.693}{7.04 \times 10^{8} \text{ y}} = 9.84 \times 10^{-10} \text{ y}^{-1}$

For ^{238}U, $k_2 = \frac{0.693}{t_{1/2}} = \frac{0.693}{4.47 \times 10^9\ y} = 1.55 \times 10^{-10}\ y^{-1}$

For ^{235}U, $\ln \frac{N_1}{N_{o1}} = -k_1t$ and $\ln \frac{N_1}{N_{o1}} + k_1t = 0$

For ^{238}U, $\ln \frac{N_2}{N_{o2}} = -k_2t$ and $\ln \frac{N_2}{N_{o2}} + k_2t = 0$

Set the two equations that are equal to zero equal to each other and solve for t.

$$\ln \frac{N_1}{N_{o1}} + k_1t = \ln \frac{N_2}{N_{o2}} + k_2t;$$

$$\ln \frac{N_1}{N_{o1}} - \ln \frac{N_2}{N_{o2}} = k_2t - k_1t = (k_2 - k_1)t$$

$$\ln \frac{\left(\frac{N_1}{N_{o1}}\right)}{\left(\frac{N_2}{N_{o2}}\right)} = (k_2 - k_1)t, \text{ now } N_{o1} = N_{o2}, \text{ so } \ln \frac{N_1}{N_2} = (k_2 - k_1)t$$

$$\frac{N_1}{N_2} = 7.25 \times 10^{-3}, \text{ so } \ln(7.25 \times 10^{-3}) = (1.55 \times 10^{-10}\ y^{-1} - 9.84 \times 10^{-10}\ y^{-1})t$$

$$t = \frac{-4.93}{-8.29 \times 10^{-10}\ y^{-1}} = 5.9 \times 10^9\ y$$

The age of the elements is 5.9×10^9 y (6 billion years).

19.101 $\ln\left(\frac{N}{N_0}\right) = -0.693\left(\frac{t}{t_{1/2}}\right) = -0.693\left(\frac{9.5\ y}{87.7\ y}\right) = -0.0751$

$$\frac{N}{N_0} = e^{-0.0751} = 0.928$$

power output = 240 W x 0.928 = 223 W = 220 W

19.102 $^{232}_{90}Th \rightarrow\ ^{208}_{82}Pb + 6\ ^{4}_{2}He + 4\ ^{0}_{-1}e$

Reactant: $^{232}_{90}Th$ nucleus = $^{232}_{90}Th$ atom – 90 e^-

Product: ${}^{208}_{82}Pb$ nucleus + (6)(${}^{4}_{2}He$ nucleus) + 4 e^-

= (${}^{208}_{82}Pb$ atom – 82 e^-) + (6)(${}^{4}_{2}He$ atom – 2 e^-) + 4 e^-

= ${}^{208}_{82}Pb$ atom + (6)(${}^{4}_{2}He$ atom) – 90 e^-

Change: (${}^{232}_{90}Th$ atom – 90 e^-) – [${}^{208}_{82}Pb$ atom + (6)(${}^{4}_{2}He$ atom) – 90 e^-]

= ${}^{232}_{90}Th$ atom – [${}^{208}_{82}Pb$ atom + (6)(${}^{4}_{2}He$ atom)] (electrons cancel)

Mass change = (232.038 054) – [(207.976 627) + (6)(4.002 603)]
= 0.045 809 u

(0.045 809 u)(1.660 54 x 10^{-24} g/u)(6.022 x 10^{23} mol^{-1}) = 0.045 808 g/mol
$\Delta E = \Delta mc^2$ = (0.045 808 g/mol)(10^{-3} kg/g)(3.00 x 10^8 m/s)^2
ΔE = 4.12 x 10^{12} J/mol = 4.12 x 10^9 kJ/mol

19.103 ${}^{293}_{118}X$, ${}^{289}_{116}Y$, and ${}^{285}_{114}Z$

19.104 (a) For ${}^{50}_{24}Cr$:

First, calculate the total mass of the nucleons (26 n + 24 p)
Mass of 26 neutrons = (26)(1.008 66) = 26.225 16
Mass of 24 protons = (24)(1.007 28) = 24.174 72
Mass of 26 n + 24 p = 50.399 88
Next, calculate the mass of a ^{50}Cr nucleus by subtracting the mass of 24 electrons from the mass of a ^{50}Cr atom.
Mass of ^{50}Cr atom = 49.946 05
–Mass of 24 electrons = –(24)(5.486 x 10^{-4}) = –0.013 17
Mass of ^{50}Cr nucleus = 49.932 88
Then subtract the mass of the ^{50}Cr nucleus from the mass of the nucleons to find the mass defect:
Mass defect = mass of nucleons – mass of nucleus
= (50.399 88) – (49.932 88) = 0.467 00 u
Mass defect in g/mol:
(0.467 00 u)(1.660 54 x 10^{-24} g/u)(6.022 x 10^{23} mol^{-1}) = 0.466 99 g/mol
Now, use the Einstein equation to convert the mass defect into the binding energy.
$\Delta E = \Delta mc^2$ = (0.466 99 g/mol)(10^{-3} kg/g)(3.00 x 10^8 m/s)^2
ΔE = 4.203 x 10^{13} J/mol = 4.203 x 10^{10} kJ/mol

$$\Delta E = \frac{4.203 \times 10^{13}\ \text{J/mol}}{6.022 \times 10^{23}\ \text{nuclei/mol}} \times \frac{1\ \text{MeV}}{1.60 \times 10^{-13}\ \text{J}} \times \frac{1\ \text{nucleus}}{50\ \text{nucleons}} = 8.72\ \text{MeV/nucleon}$$

(b) For $^{64}_{30}Zn$:

First, calculate the total mass of the nucleons (34 n + 30 p)

Mass of 34 neutrons = (34)(1.008 66) = 34.294 44

Mass of 30 protons = (30)(1.007 28) = 30.218 40

Mass of 34 n + 30 p = 64.512 84

Next, calculate the mass of a ^{64}Zn nucleus by subtracting the mass of 30 electrons from the mass of a ^{64}Zn atom.

Mass of ^{64}Zn atom = 63.929 15

–Mass of 30 electrons = $-(30)(5.486 \times 10^{-4})$ = –0.016 46

Mass of ^{64}Zn nucleus = 63.912 69

Then subtract the mass of the ^{64}Zn nucleus from the mass of the nucleons to find the mass defect:

Mass defect = mass of nucleons – mass of nucleus

= (64.512 84) – (63.912 69) = 0.600 15 u

Mass defect in g/mol:

$(0.600\ 15\ \text{u})(1.660\ 54 \times 10^{-24}\ \text{g/u})(6.022 \times 10^{23}\ \text{mol}^{-1}) = 0.600\ 14\ \text{g/mol}$

Now, use the Einstein equation to convert the mass defect into the binding energy.

$\Delta E = \Delta mc^2 = (0.600\ 14\ \text{g/mol})(10^{-3}\ \text{kg/g})(3.00 \times 10^8\ \text{m/s})^2$

$\Delta E = 5.401 \times 10^{13}\ \text{J/mol} = 5.401 \times 10^{10}\ \text{kJ/mol}$

$$\Delta E = \frac{5.401 \times 10^{13}\ \text{J/mol}}{6.022 \times 10^{23}\ \text{nuclei/mol}} \times \frac{1\ \text{MeV}}{1.60 \times 10^{-13}\ \text{J}} \times \frac{1\ \text{nucleus}}{64\ \text{nucleons}} = 8.76\ \text{MeV/nucleon}$$

The ^{64}Zn is more stable because ΔE is larger.

19.105 $\ln\left(\frac{N}{N_0}\right) = (-0.693)\left(\frac{t}{t_{1/2}}\right)$; $\frac{N}{N_0} = \frac{\text{Decay rate at time t}}{\text{Decay rate at time t} = 0}$

$\ln\left(\frac{2.9}{15.3}\right) = (-0.693)\left(\frac{t}{5730\ \text{y}}\right)$; $t = 1.38 \times 10^4\ \text{y}$

19.106 $^{2}_{1}H + ^{3}_{2}He \rightarrow ^{4}_{2}He + ^{1}_{1}H$

mass $^{2}_{1}H$	2.0141
mass $^{3}_{2}He$	3.0160
–mass $^{4}_{2}He$	– 4.0026
–mass $^{1}_{1}H$	–1.0078
mass change	0.0197 u

$(0.0197 \text{ u})(1.660\ 54 \times 10^{-24} \text{ g/u})(6.022 \times 10^{23} \text{ mol}^{-1}) = 0.0197 \text{ g/mol}$
$\Delta E = \Delta mc^2 = (0.0197 \text{ g/mol})(10^{-3} \text{ kg/g})(3.00 \times 10^8 \text{ m/s})^2$
$\Delta E = 1.77 \times 10^{12} \text{ J/mol} = 1.77 \times 10^9 \text{ kJ/mol}$

19.107 Assume a sample of ${}^{40}_{19}K$ containing 100 atoms.

	${}^{40}_{19}K$	+ ${}^{0}_{-1}e$ →	${}^{40}_{18}Ar$
before decay (atoms)	100		0
after decay (atoms)	100 – x		x

$\dfrac{^{40}Ar}{^{40}K} = \dfrac{x}{100 - x} = 1.15;$ Solve for x. x = 53.5

$$\ln\left(\frac{N}{N_0}\right) = (-0.693)\left(\frac{t}{t_{1/2}}\right)$$

N = 100 – x = 100 – 53.5 = 46.5, the amount of ^{40}K at time t.
N_0 = 100, the original amount of ^{40}K.

$$\ln\left(\frac{46.5}{100}\right) = (-0.693)\left(\frac{t}{1.28 \times 10^9 \text{ y}}\right); \qquad t = 1.41 \times 10^9 \text{ y}$$

19.108 ${}^{238}_{92}U + {}^{1}_{0}n \rightarrow {}^{239}_{94}Pu + 2\ {}^{0}_{-1}e$

19.109 $3.9 \times 10^{23} \text{ kJ} = 3.9 \times 10^{26} \text{ J} = 3.9 \times 10^{26} \text{ kg} \cdot \text{m}^2/\text{s}^2$

$$\Delta E = \Delta mc^2; \quad \Delta m = \frac{\Delta E}{c^2} = \frac{3.9 \times 10^{26} \text{ kg} \cdot \text{m}^2/\text{s}^2}{(3.00 \times 10^8 \text{ m/s})^2} = 4.3 \times 10^9 \text{ kg}$$

The sun loses mass at a rate of 4.3×10^9 kg/s.

19.110 Each alpha emission decreases the mass number by four and the atomic number by two. Each beta emission increases the atomic number by one.

${}^{237}_{93}Np \rightarrow {}^{209}_{83}Bi$

$$\text{Number of } \alpha \text{ emissions} = \frac{\text{Np mass number} - \text{Bi mass number}}{4}$$

$$= \frac{237 - 209}{4} = 7\ \alpha \text{ emissions}$$

The atomic number decreases by 14 as a result of 7 alpha emissions. The resulting atomic number is (93 – 14) = 79.
Number of β emissions = Bi atomic number – 79 = 83 – 79 = 4 β emissions

19.111 (a) ${}^{100}_{43}Tc \rightarrow {}^{0}_{1}e + {}^{100}_{42}Mo$ (positron emission)

${}^{100}_{43}Tc + {}^{0}_{-1}e \rightarrow {}^{100}_{42}Mo$ (electron capture)

(b) Positron emission

Reactant: ${}^{100}_{43}Tc$ nucleus = ${}^{100}_{43}Tc$ atom – 43 e^-

Product: ${}^{100}_{42}Mo$ nucleus + e^+ = ${}^{100}_{42}Mo$ atom – 42 e^- + 1 e^+

Change: (${}^{100}_{43}Tc$ atom – 43 e^-) – (${}^{100}_{42}Mo$ atom – 42 e^- + 1 e^+)

= ${}^{100}_{43}Tc$ atom – ${}^{100}_{42}Mo$ atom – 2 e^-

Mass change = (99.907 657) – (99.907 48) – (2)(0.000 5486)

= –0.000 92 u (energy is absorbed)

$(-0.000\ 92\ u)(1.660\ 54 \times 10^{-24}\ g/u)(6.022 \times 10^{23}\ mol^{-1}) = -0.000\ 92\ g/mol$

$\Delta E = \Delta mc^2 = (-0.000\ 92\ g/mol)(10^{-3}\ kg/g)(3.00 \times 10^8\ m/s)^2$

$\Delta E = -8.3 \times 10^{10}\ J/mol = -8.3 \times 10^7\ kJ/mol$

Electron Capture

Reactant: ${}^{100}_{43}Tc$ nucleus + e^- = ${}^{100}_{43}Tc$ atom – 42 e^-

Product: ${}^{100}_{42}Mo$ nucleus = ${}^{100}_{42}Mo$ atom – 42 e^-

---

Change: (${}^{100}_{43}Tc$ atom – 42 e^-) – (${}^{100}_{42}Mo$ atom – 42 e^-)

= ${}^{100}_{43}Tc$ atom – ${}^{100}_{42}Mo$ atom (electrons cancel)

Mass change = (99.907 657) – (99.907 48) = 0.000 177 u

$(0.000\ 177\ u)(1.660\ 54 \times 10^{-24}\ g/u)(6.022 \times 10^{23}\ mol^{-1}) = 0.000\ 177\ g/mol$

$\Delta E = \Delta mc^2 = (0.000\ 177\ g/mol)(10^{-3}\ kg/g)(3.00 \times 10^8\ m/s)^2$

$\Delta E = 1.6 \times 10^{10}\ J/mol = 1.6 \times 10^7\ kJ/mol$

Only electron capture is observed because only this process involves a mass decrease and a release of energy.

19.112 (a) α emission: ${}^{226}_{89}Ac \rightarrow {}^{222}_{87}Fr + {}^{4}_{2}He$

β emission: ${}^{226}_{89}Ac \rightarrow {}^{226}_{90}Th + {}^{0}_{-1}e$

electron capture: ${}^{226}_{89}Ac + {}^{0}_{-1}e \rightarrow {}^{226}_{88}Ra$

(b) $t_{1/2} = \frac{0.693}{k} = \frac{0.693}{0.556\ d^{-1}} = 1.25\ d$

If 80% reacts, then 20% is left.

$$\ln\left(\frac{N}{N_0}\right) = (-0.693)\left(\frac{t}{t_{1/2}}\right); \quad \ln\left(\frac{20}{100}\right) = (-0.693)\left(\frac{t}{1.25\ d}\right)$$

$$t = \frac{\ln\left(\frac{20}{100}\right)(1.25\ d)}{(-0.693)} = 2.90\ d$$

Multiconcept Problems

19.113 $BaCO_3$, 197.34

$$1.000\ g\ BaCO_3 \times \frac{1\ mol\ BaCO_3}{197.34\ g\ BaCO_3} \times \frac{1\ mol\ C}{1\ mol\ BaCO_3} \times \frac{12.011\ g\ C}{1\ mol\ C} = 0.060\ 86\ g\ C$$

4.0×10^{-3} Bq = 4.0×10^{-3} disintegrations/s

(4.0×10^{-3} Bq = 4.0×10^{-3} disintegrations/s)(60 s/min) = 0.24 disintegrations/min

$$\text{sample radioactivity} = \frac{0.24\ \text{disintegrations/min}}{0.060\ 86\ g\ C} = 3.94\ \text{disintegrations/min per gram of C}$$

$$\ln\left(\frac{N}{N_0}\right) = (-0.693)\left(\frac{t}{t_{1/2}}\right); \qquad \frac{N}{N_0} = \frac{\text{Decay rate at time t}}{\text{Decay rate at time t} = 0}$$

$$\ln\left(\frac{3.94}{15.3}\right) = (-0.693)\left(\frac{t}{5730\ y}\right); \qquad t = 11{,}000\ y$$

19.114 $t_{1/2} = 138\ d = 138\ d \times \frac{1\ y}{365\ d} = 0.378\ y$

$$k = \frac{0.693}{t_{1/2}} = \frac{0.693}{0.378\ y} = 1.83\ y^{-1}$$

$$0.700\ mg \times \frac{1 \times 10^{-3}\ g}{1\ mg} = 7.00 \times 10^{-4}\ g$$

$$N_o = (7.00 \times 10^{-4}\ g)\left(\frac{1\ mol\ Po}{210\ g\ Po}\right)(6.022 \times 10^{23}\ atoms/mol) = 2.01 \times 10^{18}\ atoms$$

$$\ln\left(\frac{N}{N_o}\right) = -kt = -(1.83\ y^{-1})(1\ y) = -1.83; \qquad \frac{N}{N_o} = e^{-1.83} = 0.160$$

$N = 0.160\ N_o = (0.160)(2.01 \times 10^{18}\ \text{atoms}) = 0.322 \times 10^{18}\ \text{atoms}$
atoms He = atoms Po decayed
atoms He = 2.01×10^{18} atoms – 0.322×10^{18} atoms = 1.688×10^{18} atoms

$$\text{mol He} = \frac{1.688 \times 10^{18}\ \text{He atoms}}{6.022 \times 10^{23}\ \text{atoms/mol}} = 2.80 \times 10^{-6}\ \text{mol He}$$

20 °C = 293 K

$$P = \frac{nRT}{V} = \frac{(2.80 \times 10^{-6}\ \text{mol})\left(0.082\ 06\ \frac{\text{L}\cdot\text{atm}}{\text{K}\cdot\text{mol}}\right)(293\ \text{K})}{0.2500\ \text{L}} = 2.69 \times 10^{-4}\ \text{atm}$$

$$P = 2.69 \times 10^{-4}\ \text{atm} \times \frac{760\ \text{mm Hg}}{1.00\ \text{atm}} = 2.04\ \text{mm Hg}$$

19.115 First find the activity (N) that the ^{28}Mg would have after 2.4 hours assuming that none of it was removed by precipitation as $MgCO_3$.

$$\ln\left(\frac{N}{N_0}\right) = (-0.693)\left(\frac{t}{t_{1/2}}\right)$$

$$\frac{N}{N_0} = \frac{\text{Decay rate at time t}}{\text{Decay rate at time t = 0}}$$

$$\ln\left(\frac{N}{0.112}\right) = (-0.693)\left(\frac{2.40\ \text{h}}{20.91\ \text{h}}\right)$$

$\ln N - \ln(0.112) = -0.0795$
$\ln N = -0.0795 + \ln(0.112) = -2.27$
$N = e^{-2.27} = 0.103\ \mu\text{Ci/mL}$

20.00 mL = 0.020 00 L and 15.00 mL = 0.015 00 L
mol $MgCl_2$ = (0.007 50 mol/L)(0.020 00 L) = 1.50×10^{-4} mol $MgCl_2$
mol Na_2CO_3 = (0.012 50 mol/L)(0.015 00 L) = 1.88×10^{-4} mol Na_2CO_3
The mol of CO_3^{2-} are in excess, so assume that all of the Mg^{2+} precipitates as $MgCO_3$ according to the reaction:

	$Mg^{2+}(aq)$	+	$CO_3^{2-}(aq)$	→	$MgCO_3(s)$
initial (mol)	0.000 150		0.000 188		0
change (mol)	– 0.000 150		– 0.000 150		+ 0.000 150
final (mol)	0		0.000 038		0.000 150

$[CO_3^{2-}]$ = 0.000 038 mol/(0.020 00 L + 0.015 00 L) = 0.001 09 M

Now consider the dissolution of $MgCO_3$ in the presence of CO_3^{2-}.

	$MgCO_3(s)$ ⇄	$Mg^{2+}(aq)$	+	$CO_3^{2-}(aq)$
initial (M)		0		0.001 09
change (M)		+x		+x
equil (M)		x		0.001 09 + x

The $[Mg^{2+}]$ in the filtrate is proportional to its activity after 2.40 h.

$$x = [Mg^{2+}] = 0.029\ \mu Ci/mL \times \frac{0.007\ 50\ M}{0.103\ \mu Ci/mL} = 0.002\ 11\ M$$

$[CO_3^{2-}] = 0.001\ 09 + x = 0.001\ 09 + 0.002\ 11 = 0.003\ 20\ M$

$K_{sp} = [Mg^{2+}][CO_3^{2-}] = (0.002\ 11)(0.003\ 20) = 6.8 \times 10^{-6}$

20 Transition Elements and Coordination Chemistry

20.1 (a) V, [Ar] $3d^3 4s^2$ (b) Co^{2+}, [Ar] $3d^7$
(c) Mn^{4+} in MnO_2, [Ar] $3d^3$ (d) Cu^{2+} in $CuCl_4^{2-}$, [Ar] $3d^9$

20.2 (a) Mn (b) Ni^{2+}

(c) Ag (d) Mo^{3+}

20.3 Z_{eff} increases from left to right across the first transition series.
(a) The transition metal with the lowest Z_{eff} (Ti) should be the strongest reducing agent because it is easier for Ti to lose its valence electrons. The transition metal with the highest Z_{eff} (Zn) should be the weakest reducing agent because it is more difficult for Zn to lose its valence electrons.
(b) The oxoanion with the highest Z_{eff} (FeO_4^{2-}) should be the strongest oxidizing agent because of the greater attraction for electrons. The oxoanion with the lowest Z_{eff} (VO_4^{3-}) should be the weakest oxidizing agent because of the lower attraction for electrons.

20.4 (a) $Cr_2O_7^{2-}$ (b) Cr^{3+} (c) Cr^{2+} (d) Fe^{2+} (e) Cu^{2+}

20.5 (a) $Cr(OH)_2$ (b) $Cr(OH)_4^-$ (c) CrO_4^{2-} (d) $Fe(OH)_2$ (e) $Fe(OH)_3$

20.6 $[Cr(NH_3)_2(SCN)_4]^-$

20.7 In $Na_4[Fe(CN)_6]$ each sodium is in the +1 oxidation state (+4 total); each cyanide (CN^-) has a −1 charge (−6 total). The compound is neutral; therefore, the oxidation state of the iron is +2.

20.8 (a)

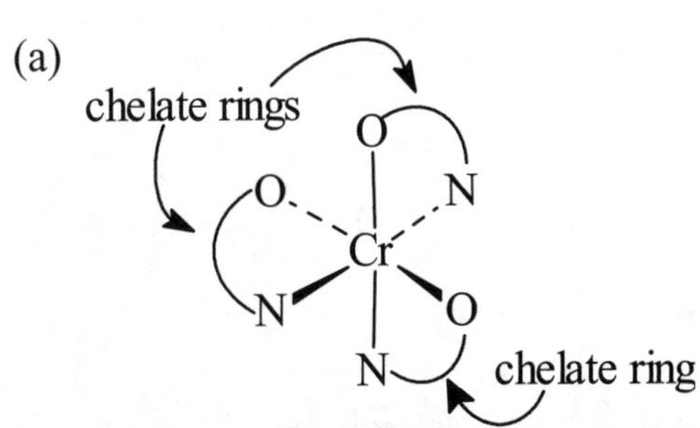

(b) Cr^{3+} is the Lewis acid. The glycinate ligand is the Lewis base. Nitrogen and oxygen are the ligand donor atoms. The chelate rings are identified in the drawing.
(c) The coordination number is 6. The coordination geometry is octahedral. The chromium is in the +3 oxidation state.

20.9 (a) tetraamminecopper(II) sulfate (b) sodium tetrahydroxochromate(III)
(c) triglycinatocobalt(III) (d) pentaaquaisothiocyanatoiron(III) ion

20.10 triamminetrichlorocobalt(III)

20.11 (a) $[Zn(NH_3)_4](NO_3)_2$ (b) $Ni(CO)_4$ (c) $K[Pt(NH_3)Cl_3]$ (d) $[Au(CN)_2]^-$

20.12 $[Pt(NH_3)_3Cl]Cl$, triamminechloroplatinum(II)chloride

20.13 Structures (1) and (4) are identical and are the cis isomer. Structures (2) and (3) are identical and are the trans isomer.

20.14 (1) and (2) are the same. (3) and (4) are the same. (1) and (2) are different from (3) and (4).

20.15 (a) Two diastereoisomers are possible.

NCS SCN Pt H_3N NH_3
cis

NCS NH_3 Pt H_3N SCN
trans

(b) No isomers are possible for a tetrahedral complex of the type MA_2B_2.
(c) Two diastereoisomers are possible.

NH_3 H_3N NH_3 Co O_2N NO_2 NO_2

NO_2 H_3N NH_3 Co O_2N NH_3 NO_2

20.16 (a) No isomers are possible for a complex of this type.

N N Pt Cl Cl

(b) Two diastereoisomers are possible.

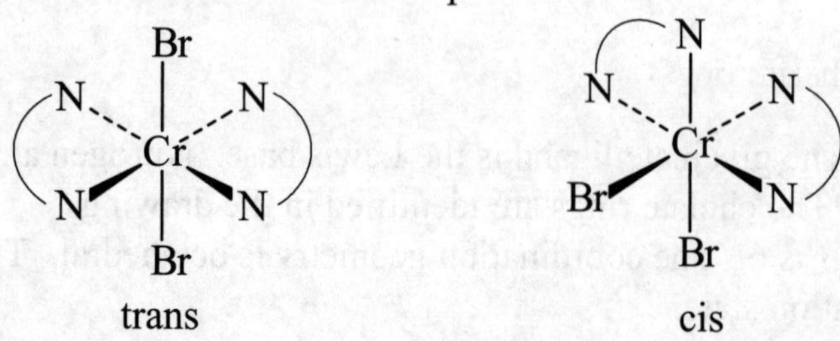

20.17 (a) $[Fe(C_2O_4)_3]^{3-}$ can exist as enantiomers.

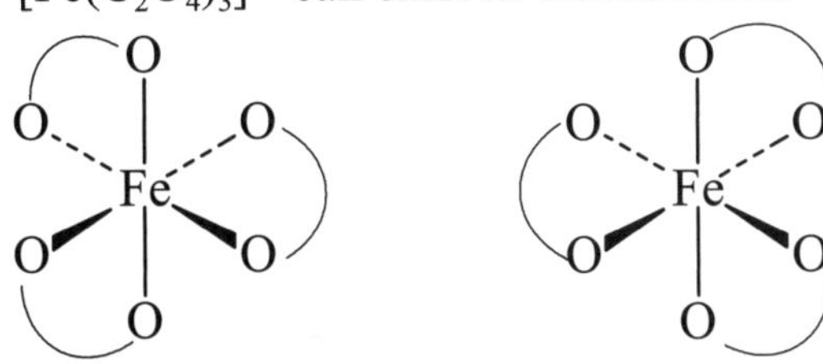

(b) $[Co(NH_3)_4en]^{3+}$ cannot exist as enantiomers.

(c) $[Co(NH_3)_2(en)_2]^{3+}$ can exist as enantiomers.

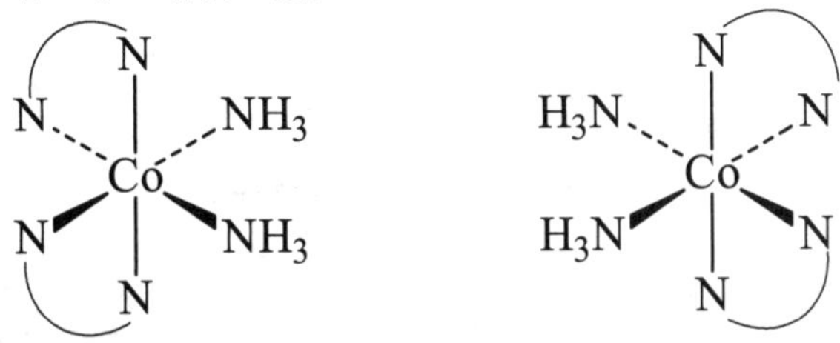

(d) $[Cr(H_2O)_4Cl_2]^+$ cannot exist as enantiomers.

20.18 (a) (2) and (3) are chiral and (1) and (4) are achiral.

(b) enantiomer of (2) enantiomer of (3)

Br, Br, N, Rh, Cl, N, Cl

N, N, N, Rh, Cl, N, Br

20.19 (a) The ion is absorbing in the red (625 nm), so the most likely color for the ion is blue.

(b) 625 nm = 625 x 10^{-9} m

$$E = h\frac{c}{\lambda} = (6.626 \text{ x } 10^{-34} \text{ J}\cdot\text{s})\left(\frac{3.00 \text{ x } 10^{8} \text{ m/s}}{625 \text{ x } 10^{-9} \text{ m}}\right) = 3.18 \text{ x } 10^{-19} \text{ J}$$

20.20 (a)

V^{3+} [Ar] ↑ ↑ _ _ _ (3d) _ (4s) _ _ _ (4p)

$[VCl_4]^-$ [Ar] ↑ ↑ _ _ _ (3d) ↑↓ (4s) ↑↓ ↑↓ ↑↓ (4p)

sp^3 2 unpaired e^-

(b)

Pt^{2+} [Xe] ↑↓ ↑↓ ↑↓ ↑ ↑ (5d) _ (6s) _ _ _ (6p)

$[PtCl_4]^{2-}$ [Xe] ↑↓ ↑↓ ↑↓ ↑↓ ↑↓ (5d) ↑↓ (6s) ↑↓ ↑↓ _ (6p)

dsp^2 no unpaired e^-

20.21 (a) Fe^{3+} [Ar] ↑ ↑ ↑ ↑ ↑ (3d) — (4s) — — — (4p)

$[Fe(CN)_6]^{3-}$ [Ar] ↑↓ ↑↓ ↑ | ↑↓ ↑↓ (3d) ↑↓ (4s) ↑↓ ↑↓ ↑↓ (4p)

d^2sp^3 1 unpaired e^-

(b) Co^{2+} [Ar] ↑↓ ↑↓ ↑ ↑ ↑ (3d) — (4s) — — — (4p)

$[Co(H_2O)_6]^{2+}$ [Ar] ↑↓ ↑↓ ↑ ↑ ↑ (3d) | ↑↓ (4s) ↑↓ ↑↓ ↑↓ (4p) ↑↓ ↑↓ | — — — (4d)

sp^3d^2 3 unpaired e^-

20.22

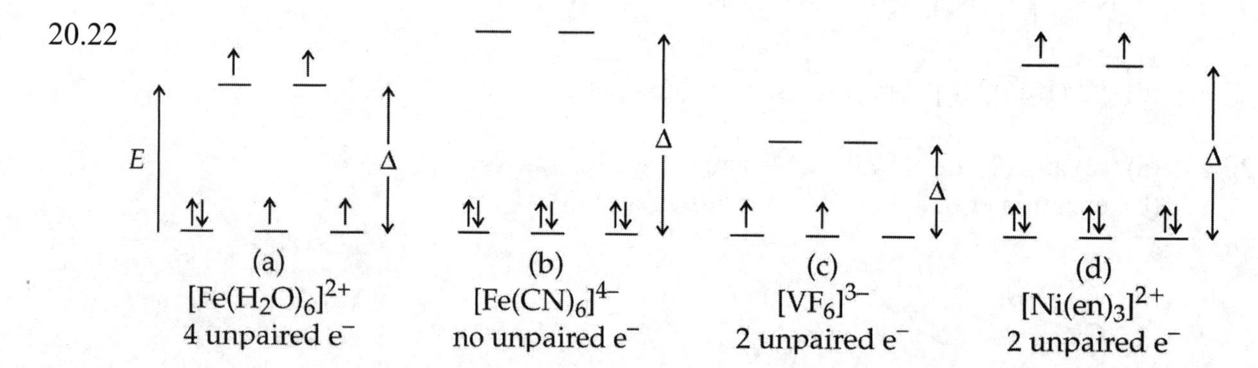

20.23 Cl^- is a weak field ligand and therefore $[FeCl_6]^{3-}$ is a high-spin complex with 5 unpaired electrons. CN^- is a strong field ligand and therefore $[Fe(CN)_6]^{3-}$ is a low-spin complex with 1 unpaired electron. $[FeCl_6]^{3-}$ is more paramagnetic because it has more unpaired electrons.

20.24 Both $[NiCl_4]^{2-}$ and $[Ni(CN)_4]^{2-}$ contain Ni^{2+} with a [Ar] $3d^8$ electron configuration.

(a) $[NiCl_4]^{2-}$ (tetrahedral)

↑↓ ↑ ↑
xy xz yz

↑↓ ↑↓
z^2 x^2-y^2

2 unpaired electrons

(b) $[Ni(CN)_4]^{2-}$ (square planar)

—
x^2-y^2

↑↓
xy

↑↓
z^2

↑↓ ↑↓
xz yz

no unpaired electrons

20.25 A diamagnetic four coordinate d^8 complex is most likely square planar.

20.26 (a) diamminedichloroplatinum(II)
(b) oxidation state = +2 and coordination number is 4
(c) Lewis acid is Pt^{2+} and Lewis bases are Cl^- and NH_3
(d) [Xe] $4f^{14}$ $5d^8$

20.27 $Pt(NH_3)_2Cl_2$

$\overline{}$
x^2-y^2

↑↓
xy
↑↓
z^2
↑↓ ↑↓
xz yz
no unpaired electrons

20.28 (a) The chloride concentration is relatively high in blood plasma and, according to Le Chatelier's Principle, a high concentration of product shifts the equilibrium position toward the reactants. Inside the cell, the chloride concentration is lower, thus shifting the equilibrium positions toward the products.
(b) diammineaquachloroplatinum(II)
(c) According to the spectrochemical series, H_2O is a stronger field ligand than Cl^-. The crystal field splitting energy is larger, which corresponds to shorter wavelength of maximum absorption.

20.29 (a) +4
(b) diamminetetrachloroplatinum(IV)
(c) cis and trans isomers.

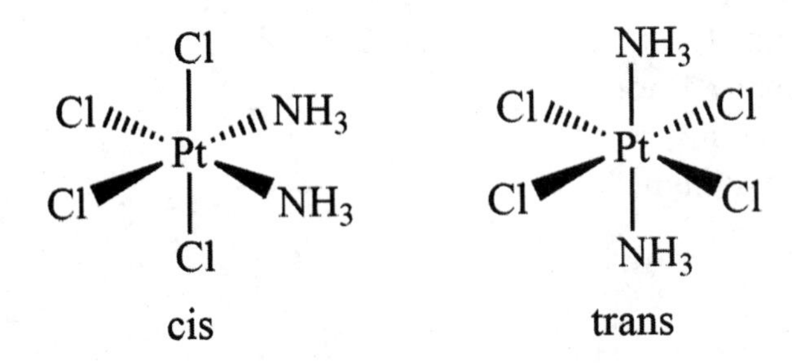

(d) Both the cis and trans isomers have symmetry planes and are achiral.

Conceptual Problems

20.30

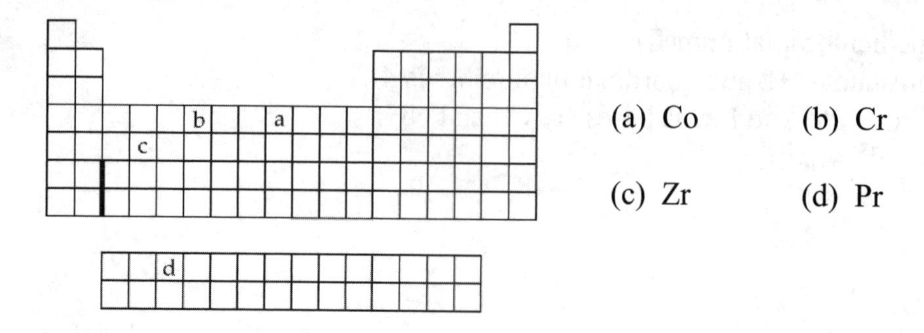

(a) Co (b) Cr

(c) Zr (d) Pr

20.31 (a) A^{2+}, [Ar] $3d^8$ (b) B^+, [Kr] $4d^{10}$ (c) C^{3+}, [Kr] $4d^3$ (d) DO_4^{2-}, [Ar] $3d^2$

20.32 (a) The atomic radii decrease, at first markedly and then more gradually. Toward the end of the series, the radii increase again. The decrease in atomic radii is a result of an increase in Z_{eff}. The increase is due to electron-electron repulsions in doubly occupied d orbitals.
(b) The densities of the transition metals are inversely related to their atomic radii. The densities initially increase from left to right and then decrease toward the end of the series.
(c) Ionization energies generally increase from left to right across the series. The general trend correlates with an increase in Z_{eff} and a decrease in atomic radii.
(d) The standard oxidation potentials generally decrease from left to right across the first transition series. This correlates with the general trend in ionization energies.

20.33 The carbonate ion would be a monodentate ligand. A bidentate carbonato metal complex would not be stable since a 4-membered ring is highly strained.

20.34 (a) **NH_2**–CH_2–CH_2–**NH_2** is a bidentate ligand. It can form a chelate ring using the atoms indicated in bold.
(b) CH_3–CH_2–CH_2–**NH_2** is a monodentate ligand.
(c) **NH_2**–CH_2–CH_2–**NH**–CH_2–**CO_2^-** is a tridentate ligand. It can form chelate rings using the atoms indicated in bold.
(d) **NH_2**–CH_2–CH_2–NH_3^+ is a monodentate ligand. The first N can coordinate to a metal.

20.35 (1) dichloroethylenediamineplatinum(II)
(2) trans-diammineaquachloroplatinate(II) ion
(3) amminepentachloroplatinate(IV) ion
(4) cis-diaquabis(ethylenediamine)platinum(IV) ion

20.36 (a) $Na[Au(CN)_2]$
1 Na^+ 2 CN^-
The oxidation state of the Au is +1.
Coordination number = 2; Linear

$$[CN{-}Au{-}NC]^-$$

(b) $[Co(NH_3)_5Br]SO_4$
1 Br^- 1 SO_4^{2-} 5 NH_3 (no charge)
The oxidation state of the Co is +3.
Coordination number = 6; Octahedral

2+
NH_3
H_3N NH_3
Co
H_3N NH_3
Br

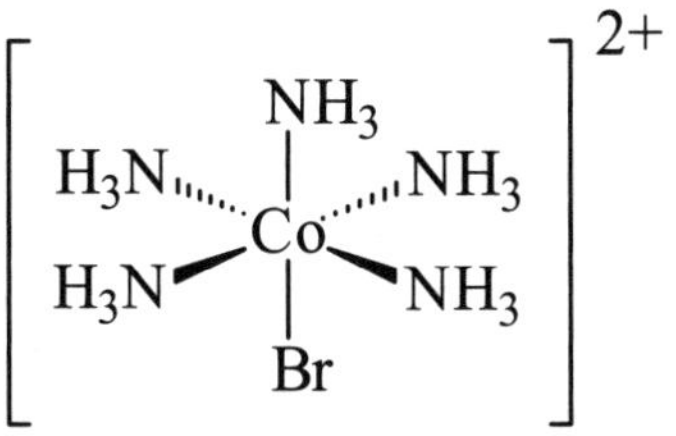

(c) $Pt(en)Cl_2$
2 Cl^- en = $NH_2CH_2CH_2NH_2$ (no charge)
The oxidation state of the Pt is +2.
Coordination number = 4; Square planar

Cl N
Pt
Cl N

(d) $(NH_4)_2[PtCl_2(C_2O_4)_2]$
2 NH_4^+ 2 Cl^- 2 $C_2O_4^{2-}$
The oxidation state of the Pt is +4.
Coordination number = 6; Octahedral

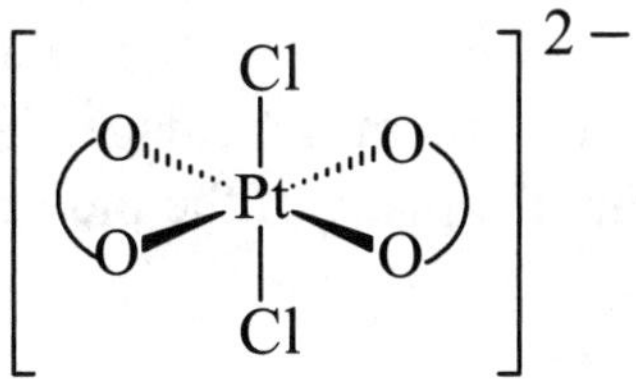

2 –
Cl
O O
Pt
O O
Cl

or

2 –
Cl
Cl O
Pt
O O
O

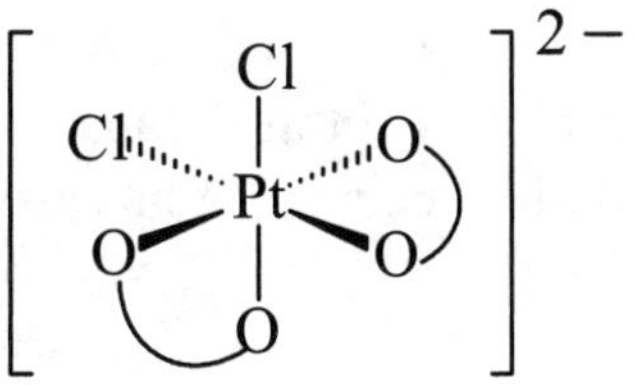

20.37. (a) (1) cis; (2) trans; (3) trans; (4) cis
(b) (1) and (4) are the same. (2) and (3) are the same.
(c) None of the isomers exist as enantiomers because their mirror images are identical.

20.38 (a) (1) chiral; (2) achiral; (3) chiral; (4) chiral

(b)

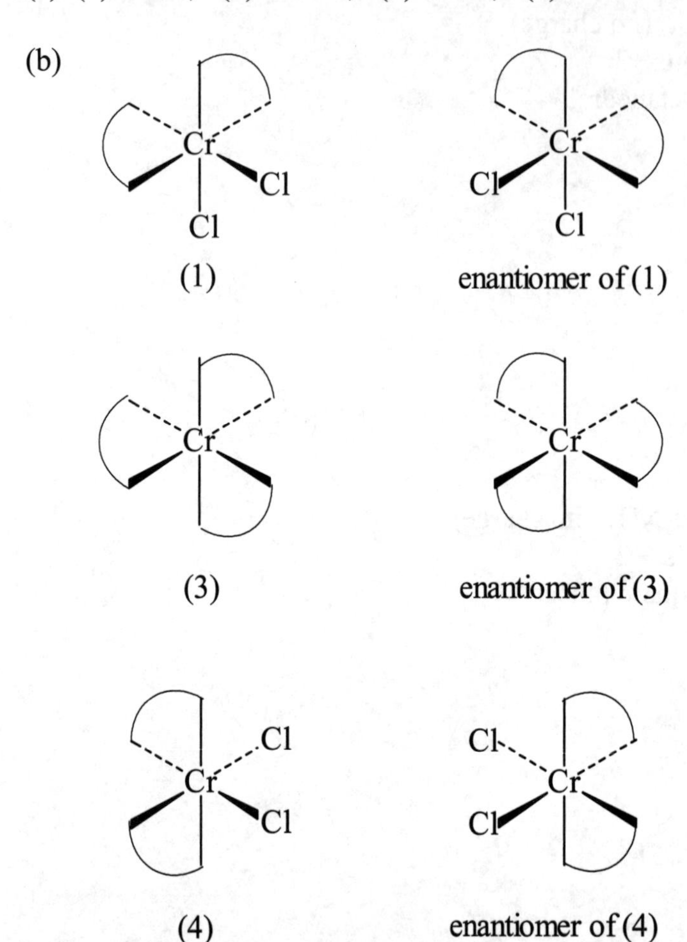

(c) (1) and (4) are enantiomers.

20.39 The tetrahedral complex is chiral because it is not identical to its mirror image. The square planar complex is achiral because it has a symmetry plane (the plane of the molecule).

Section Problems

Electron Configurations and Properties of Transition Elements (Sections 20.1 and 20.2)

20.40 (a) Cr, [Ar] $3d^5 4s^1$ (b) Zr, [Kr] $4d^2 5s^2$ (c) Co^{2+}, [Ar] $3d^7$
(d) Fe^{3+}, [Ar] $3d^5$ (e) Mo^{3+}, [Kr] $4d^3$ (f) Cr(VI), [Ar] $3d^0$

20.41 (a) Mn, [Ar] $3d^5 4s^2$ (b) Cd, [Kr] $4d^{10} 5s^2$ (c) Cr^{3+}, [Ar] $3d^3$
(d) Ag^+, [Kr] $4d^{10}$ (e) Rh^{3+}, [Kr] $4d^6$ (f) Mn(VI), [Ar] $3d^1$

20.42 (a) Cu^{2+}, [Ar] $3d^9$ ⇅ ⇅ ⇅ ⇅ ↑ (3d) 1 unpaired e^-

(b) Ti^{2+}, [Ar] $3d^2$ ↑ ↑ _ _ _ (3d) 2 unpaired e^-

(c) Zn^{2+}, [Ar] $3d^{10}$ ↑↓ ↑↓ ↑↓ ↑↓ ↑↓ (3d) 0 unpaired e^-

(d) Cr^{3+}, [Ar] $3d^3$ ↑ ↑ ↑ _ _ (3d) 3 unpaired e^-

20.43 (a) Sc^{3+}, [Ar] $3d^0$ _ _ _ _ _ (3d) no unpaired e^-

(b) Co^{2+}, [Ar] $3d^7$ ↑↓ ↑↓ ↑ ↑ ↑ (3d) 3 unpaired e^-

(c) Mn^{3+}, [Ar] $3d^4$ ↑ ↑ ↑ ↑ _ (3d) 4 unpaired e^-

(d) Cr^{2+}, [Ar] $3d^4$ ↑ ↑ ↑ ↑ _ (3d) 4 unpaired e^-

20.44 Ti is harder than K and Ca largely because the sharing of d, as well as s, electrons results in stronger metallic bonding.

20.45 Mo has a higher melting point than Y or Cd. Melting points increase as the number of unpaired d electrons available for metallic bonding increases, and then decrease as the d electrons pair up and become less available for bonding.

20.46 (a) The decrease in radii with increasing atomic number is expected because the added d electrons only partially shield the added nuclear charge. As a result, Z_{eff} increases. With increasing Z_{eff}, the electrons are more strongly attracted to the nucleus, and atomic size decreases.

(b) The densities of the transition metals are inversely related to their atomic radii.

20.47 Ti > V > Cr > Mn Atomic radius decreases with increasing Z_{eff}.

20.48 The smaller than expected sizes of the third-transition series atoms are associated with what is called the lanthanide contraction, the general decrease in atomic radii of the f-block lanthanide elements. The lanthanide contraction is due to the increase in Z_{eff} as the 4f subshell is filled.

20.49 Zr and Hf are about the same size. The lanthanide contraction for Hf is due to the increase in Z_{eff} as the 4f subshell is filled.

20.50
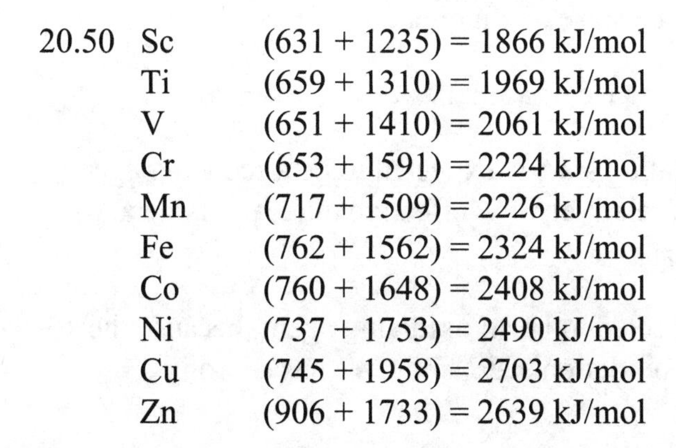

Sc	(631 + 1235) = 1866 kJ/mol
Ti	(659 + 1310) = 1969 kJ/mol
V	(651 + 1410) = 2061 kJ/mol
Cr	(653 + 1591) = 2224 kJ/mol
Mn	(717 + 1509) = 2226 kJ/mol
Fe	(762 + 1562) = 2324 kJ/mol
Co	(760 + 1648) = 2408 kJ/mol
Ni	(737 + 1753) = 2490 kJ/mol
Cu	(745 + 1958) = 2703 kJ/mol
Zn	(906 + 1733) = 2639 kJ/mol

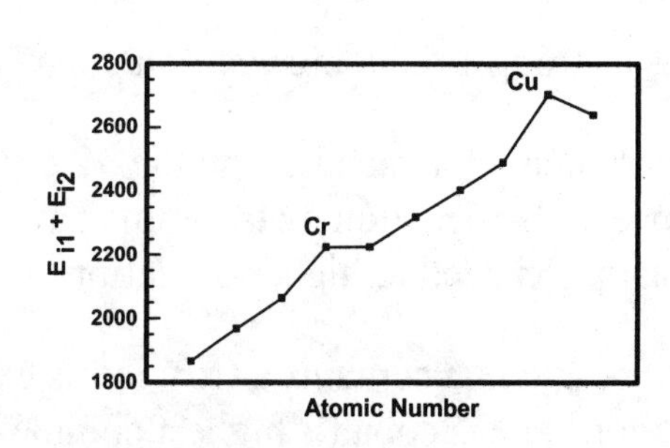

Across the first transition element series, Z_{eff} increases and there is an almost linear increase in the sum of the first two ionization energies. This is what is expected if the two electrons are removed from the 4s orbital. Higher than expected values for the sum of the first two ionization energies are observed for Cr and Cu because of their anomalous electron configurations (Cr $3d^5\ 4s^1$; Cu $3d^{10}\ 4s^1$). An increasing Z_{eff} affects 3d orbitals more than the 4s orbital and the second ionization energy for an electron from the 3d orbital is higher than expected.

20.51 Oxidation potentials generally decrease from left to right across the transition series. The general trend correlates with an increase in ionization energy, which is due in turn to an increase in Z_{eff} and a decrease in atomic radius.

20.52 Ti is more easily oxidized than is Zn because of a smaller Z_{eff}.

20.53 Sc is more easily oxidized than is Co because of a smaller Z_{eff}.

20.54 (a) $Cr(s) + 2\ H^+(aq) \rightarrow Cr^{2+}(aq) + H_2(g)$
(b) $Zn(s) + 2\ H^+(aq) \rightarrow Zn^{2+}(aq) + H_2(g)$
(c) N.R.
(d) $Fe(s) + 2\ H^+(aq) \rightarrow Fe^{2+}(aq) + H_2(g)$

20.55 (a) $Mn(s) + 2\ H^+(aq) \rightarrow Mn^{2+}(aq) + H_2(g)$
(b) N.R.
(c) $2\ Sc(s) + 6\ H^+(aq) \rightarrow 2\ Sc^{3+}(aq) + 3\ H_2(g)$
(d) $Ni(s) + 2\ H^+(aq) \rightarrow Ni^{2+}(aq) + H_2(g)$

Oxidation States (Section 20.3)

20.56 (b) Mn (d) Cu

20.57 (b) Al (d) Sc

20.58 Sc(III), Ti(IV), V(V), Cr(VI), Mn(VII), Fe(VI), Co(III), Ni(II), Cu(II), Zn(II)

20.59 The highest oxidation state for the group 3B-7B metals is the group number, corresponding to the loss of all valence s and d electrons. For the later transition metals, loss of all valence electrons is energetically prohibitive because of the increasing Z_{eff}. Therefore, only lower oxidation states are accessible for the later transition metals.

20.60 Cu^{2+} is a stronger oxidizing agent than Cr^{2+} because of a higher Z_{eff}.

20.61 Ti^{2+} is a stronger reducing agent than Ni^{2+} because of a smaller Z_{eff}.

20.62 A compound with vanadium in the +2 oxidation state is expected to be a reducing agent, because early transition metal atoms have a relatively low effective nuclear charge and are easily oxidized to higher oxidation states.

20.63 A compound that contains a Co^{3+} ion is expected to be an oxidizing agent, because the +3 oxidation state is cobalt's highest common oxidation state, and Co^{3+} has a strong tendency to be reduced.

20.64 Mn^{2+} < MnO_2 < MnO_4^- because of increasing oxidation state of Mn.

20.65 $Cr_2O_7^{2-}$ < Cr^{3+} < Cr^{2+} because of decreasing oxidation state of the Cr.

Chemistry of Selected Transition Elements (Section 20.4)

20.66 (a) $Cr_2O_3(s) + 2\ Al(s) \rightarrow 2\ Cr(s) + Al_2O_3(s)$
(b) $Cu_2S(l) + O_2(g) \rightarrow 2\ Cu(l) + SO_2(g)$

20.67 (a) $Fe(s) + NO_3^-(aq) + 4\ H^+(aq) \rightarrow Fe^{3+}(aq) + NO(g) + 2\ H_2O(l)$
(b) $3\ Cu(s) + 2\ NO_3^-(aq) + 8\ H^+(aq) \rightarrow 3\ Cu^{2+}(aq) + 2\ NO(g) + 4\ H_2O(l)$
(c) $Cr(s) + NO_3^-(aq) + 4\ H^+(aq) \rightarrow Cr^{3+}(aq) + NO(g) + 2\ H_2O(l)$

20.68 $Cr(OH)_3(s) + OH^-(aq) \rightarrow Cr(OH)_4^-(aq)$
The Cr in $Cr(OH)_4^-$ is in the +3 oxidation state. $Cr(OH)_4^-$ is deep green.

20.69 acid: $Cr(OH)_3(s) + OH^-(aq) \rightarrow Cr(OH)_4^-(aq)$
base: $Cr(OH)_3(s) + 3\ H_3O^+(aq) \rightarrow Cr^{3+}(aq) + 6\ H_2O(l)$

20.70 (c) $Cr(OH)_3$

20.71 (c) Cu^+ $2\ Cu^+(aq) \rightarrow Cu(s) + Cu^{2+}(aq)$

20.72 (a) Add excess KOH(aq) and Fe^{3+} will precipitate as $Fe(OH)_3(s)$. $Na^+(aq)$ will remain in solution.
(b) Add excess NaOH(aq) and Fe^{3+} will precipitate as $Fe(OH)_3(s)$. $Cr(OH)_4^-(aq)$ will remain in solution.
(c) Add excess $NH_3(aq)$ and Fe^{3+} will precipitate as $Fe(OH)_3(s)$. $Cu(NH_3)_4^{2+}(aq)$ will remain in solution.

20.73 (a) Add H_2S to precipitate CuS, or add NaOH(aq) to precipitate $Cu(OH)_2$. $K^+(aq)$ will remain in solution.
(b) Add excess NaOH(aq) to precipitate $Cu(OH)_2$. $Cr(OH)_4^-(aq)$ will remain in solution.
(c) Add excess NaOH(aq) to precipitate $Fe(OH)_3$. $Al(OH)_4^-(aq)$ will remain in solution.

20.74 (a) $Cr_2O_7^{2-}(aq) + 6\ Fe^{2+}(aq) + 14\ H^+(aq) \rightarrow 2\ Cr^{3+}(aq) + 6\ Fe^{3+}(aq) + 7\ H_2O(l)$
(b) $4\ Fe^{2+}(aq) + O_2(g) + 4\ H^+(aq) \rightarrow 4\ Fe^{3+}(aq) + 2\ H_2O(l)$
(c) $Cu_2O(s) + 2\ H^+(aq) \rightarrow Cu(s) + Cu^{2+}(aq) + H_2O(l)$
(d) $Fe(s) + 2\ H^+(aq) \rightarrow Fe^{2+}(aq) + H_2(g)$

20.75 (a) $6\ Cr^{2+}(aq) + Cr_2O_7^{2-}(aq) + 14\ H^+(aq) \rightarrow 8\ Cr^{3+}(aq) + 7\ H_2O(l)$
(b) $3\ Cu(s) + 2\ NO_3^-(aq) + 8\ H^+(aq) \rightarrow 3\ Cu^{2+}(aq) + 2\ NO(g) + 4\ H_2O(l)$
(c) $Cu^{2+}(aq) + 4\ NH_3(aq) \rightarrow Cu(NH_3)_4^{2+}(aq)$
(d) $Cr(OH)_4^-(aq) + 4\ H^+(aq) \rightarrow Cr^{3+}(aq) + 4\ H_2O(l)$

20.76 (a) $2\ CrO_4^{2-}(aq) + 2\ H_3O^+(aq) \rightarrow Cr_2O_7^{2-}(aq) + 3\ H_2O(l)$
(yellow) (orange)

(b) $[Fe(H_2O)_6]^{3+}(aq) + SCN^-(aq) \rightarrow [Fe(H_2O)_5(SCN)]^{2+}(aq) + H_2O(l)$
(red)

(c) $3\ Cu(s) + 2\ NO_3^-(aq) + 8\ H^+(aq) \rightarrow 3\ Cu^{2+}(aq) + 2\ NO(g) + 4\ H_2O(l)$
(blue)

(d) $Cr(OH)_3(s) + OH^-(aq) \rightarrow Cr(OH)_4^-(aq)$
$2\ Cr(OH)_4^-(aq) + 3\ HO_2^-(aq) \rightarrow 2\ CrO_4^{2-}(aq) + 5\ H_2O(l) + OH^-(aq)$
(yellow)

20.77 (a) $Cu^{2+}(aq) + 4\ NH_3(aq) \rightarrow [Cu(NH_3)_4]^{2+}(aq)$
(dark blue)

(b) $Cr_2O_7^{2-}(aq) + 2\ OH^-(aq) \rightarrow 2\ CrO_4^{2-}(aq) + H_2O(l)$
(orange) (yellow)

(c) $Fe^{3+}(aq) + 3\ OH^-(aq) \rightarrow Fe(OH)_3(s)$
(red-brown)

(d) $3\ CuS(s) + 8\ H^+(aq) + 2\ NO_3^-(aq) \rightarrow 3\ Cu^{2+}(aq) + 3\ S(s) + 2\ NO(g) + 4\ H_2O(l)$
(blue) (yellow)

Coordination Compounds; Ligands (Sections 20.5 and 20.6)

20.78 (a) Ni^{2+} is the Lewis acid. Ethylenediamine is the Lewis base.
(b) Ethylenediamine is the ligand and the two N's are the donor atoms.
(c) $[Ni(en)_3]^{2+}$ is octahedral with a coordination number of 6.

20.79 (a) Fe^{3+} is the Lewis acid. Oxalate ion is the Lewis base.
(b) Oxalate ion is the ligand and two O's on adjacent carbons are the donor atoms.
(c) $[Fe(C_2O_4)_3]^{3-}$ is octahedral with a coordination number of 6.

20.80

	Coordination Number
(a) $[AgCl_2]^-$	2
(b) $[Cr(H_2O)_5Cl]^{2+}$	6
(c) $[Co(NCS)_4]^{2-}$	4
(d) $[ZrF_8]^{4-}$	8
(e) $[Fe(EDTA)(H_2O)]^-$	7

20.81

	Coordination Number
(a) $[Ni(CN)_5]^{3-}$	5
(b) $Ni(CO)_4$	4
(c) $[Co(en)_2(H_2O)Br]^{2+}$	6
(d) $[Cu(H_2O)_2(C_2O_4)_2]^{2-}$	6
(e) $Co(NH_3)_3(NO_2)_3$	6

20.82 (a) $AgCl_2^-$
$2\ Cl^-$
The oxidation state of the Ag is +1.
(b) $[Cr(H_2O)_5Cl]^{2+}$
$4\ H_2O$ (no charge) $1\ Cl^-$
The oxidation state of the Cr is +3.
(c) $[Co(NCS)_4]^{2-}$
$4\ NCS^-$
The oxidation state of the Co is +2.
(d) $[ZrF_8]^{4-}$
$8\ F^-$
The oxidation state of the Zr is +4.
(e) $[Fe(EDTA)(H_2O)]^-$
H_2O (no charge) $EDTA^{4-}$
The oxidation state of the Fe is +3.

20.83 (a) $[Ni(CN)_5]^{3-}$
$5\ CN^-$
The oxidation state of the Ni is +2.
(b) $Ni(CO)_4$
4 CO (no charge)
The oxidation state of the Ni is 0.
(c) $[Co(en)_2(H_2O)Br]^{2+}$
2 en (no charge) H_2O (no charge) Br^-
The oxidation state of the Co is +3.
(d) $[Cu(H_2O)_2(C_2O_4)_2]^{2-}$
$2\ H_2O$ (no charge) $2\ C_2O_4^{2-}$
The oxidation state of the Cu is +2.
(e) $Co(NH_3)_3(NO_2)_3$
$3\ NH_3$ (no charge) $3\ NO_2^-$
The oxidation state of the Co is +3.

20.84 (a) $Co(NH_3)_3(NO_2)_3$
$3\ NH_3$ (no charge) $3\ NO_2^-$
The oxidation state of the Co is +3.
(b) $[Ag(NH_3)_2]NO_3$
$2\ NH_3$ (no charge) $1\ NO_3^-$
The oxidation state of the Ag is +1.
(c) $K_3[Cr(C_2O_4)_2Cl_2]$
$3\ K^+$ $2\ C_2O_4^{2-}$ $2\ Cl^-$
The oxidation state of the Cr is +3.
(d) $Cs[CuCl_2]$
$1\ Cs^+$ $2\ Cl^-$
The oxidation state of the Cu is +1.

20.85 (a) $(NH_4)_3[RhCl_6]$
$3\ NH_4^+$ $6\ Cl^-$
The oxidation state of the Rh is +3.
(b) $[Cr(NH_3)_4(SCN)_2]Br$
$4\ NH_3$ (no charge) $2\ SCN^-$ $1\ Br^-$
The oxidation state of the Cr is +3.
(c) $[Cu(en)_2]SO_4$
2 en ($NH_2CH_2CH_2NH_2$, no charge) $1\ SO_4^{2-}$
The oxidation state of the Cu is +2.
(d) $Na_2[Mn(EDTA)]$
$2\ Na^+$ $1\ EDTA^{4-}$
The oxidation state of the Mn is +2.

20.86 (a) $Ir(NH_3)_3Cl_3$
(b) $Ni(en)_2Br_2$
(c) $[Pt(en)_2(SCN)_2]^{2+}$

20.87 (a) $[Fe(H_2O)_6]^{3+}$
(b) $[Cr(H_2O)_2(C_2O_4)_2]^-$
(c) $Pt(NH_3)_2Cl_2$

20.88

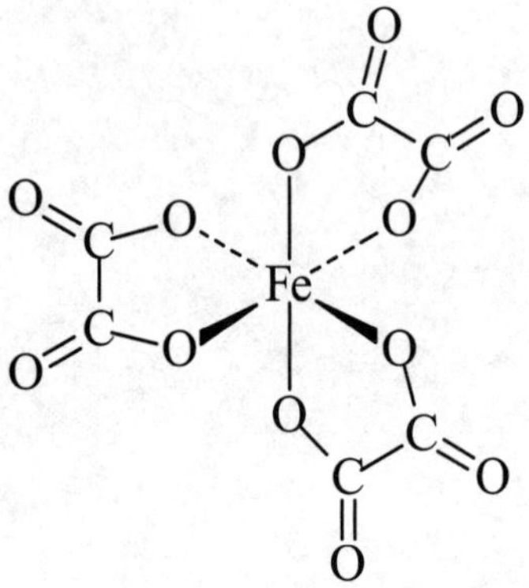

The iron is in the +3 oxidation state, and the coordination number is six. The geometry about the Fe is octahedral. The oxalate ligand is behaving as a bidentate chelating ligand. There are three chelate rings, one formed by each oxalate ligand.

20.89

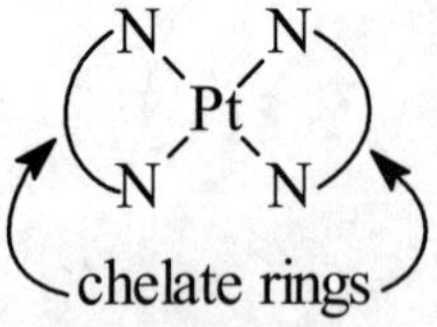

chelate rings

$[Pt(en)_2]^{2+}$ is square planar. Ethylenediamine is a neutral bidentate chelating ligand. The coordination number of the Pt is four, and the oxidation number of the Pt is +2.

Naming Coordination Compounds (Section 20.7)

20.90 (a) tetrachloromanganate(II)
(b) hexaamminenickel(II)
(c) tricarbonatocobaltate(III)
(d) bis(ethylenediamine)dithiocyanatoplatinum(IV)

20.91 (a) tetrachloroaurate(III)
(b) hexacyanoferrate(II)
(c) pentaaquaisothiocyanatoiron(III)
(d) diamminedioxalatochromate(III)

20.92 (a) cesium tetrachloroferrate(III)
(b) hexaaquavanadium(III) nitrate
(c) tetraamminedibromocobalt(III) bromide
(d) diglycinatocopper(II)

20.93 (a) tetraamminecopper(II) sulfate
(b) hexacarbonylchromium(0)
(c) potassium trioxalatoferrate(III)
(d) amminecyanatobis(ethylenediamine)cobalt(III) chloride

20.94 (a) $[Pt(NH_3)_4]Cl_2$ (b) $Na_3[Fe(CN)_6]$
(c) $[Pt(en)_3](SO_4)_2$ (d) $Rh(NH_3)_3(SCN)_3$

20.95 (a) $[Ag(NH_3)_2]NO_3$ (b) $K[Co(H_2O)_2(C_2O_4)_2]$
(c) $Mo(CO)_6$ (d) $[Cr(NH_3)_2(en)_2]Cl_3$

Isomers (Sections 20.8 and 20.9)

20.96 (a) (1) $[Ru(NH_3)_5(NO_2)]Cl$, tetraamminenitroruthenium(II) chloride
(2) $[Ru(NH_3)_5(ONO)]Cl$, tetraamminenitritoruthenium(II) chloride
(3) $[Ru(NH_3)_5Cl]NO_2$, tetraamminechlororuthenium(II) nitrite
(b) (1) and (2) are linkage isomers.
(c) (1) and (2) are ionization isomers with (3).

20.97 (a) (1 and 3), (2 and 5), and (4 and 6) are cis-trans isomers.
(b) (1, 2 and 4) and (3, 5 and 6) are linkage isomers.

20.98 (a) $[Cr(NH_3)_2Cl_4]^-$ can exist as cis and trans diastereoisomers.

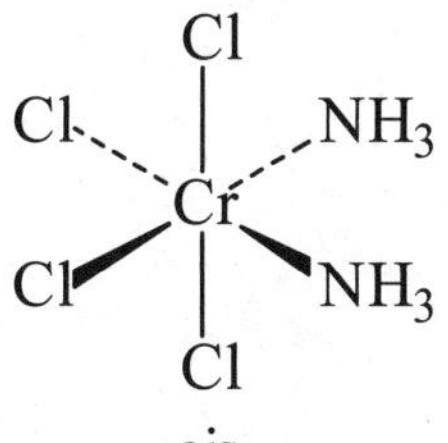

cis 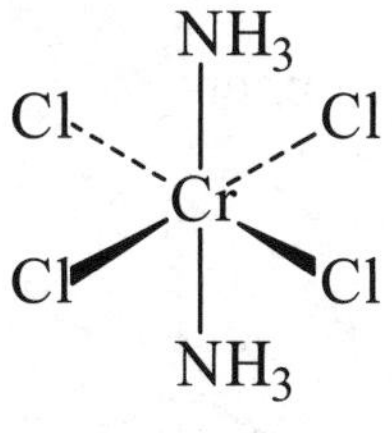 trans

(b) $[Co(NH_3)_5Br]^{2+}$ cannot exist as diastereoisomers.
(c) $[FeCl_2(NCS)_2]^{2-}$ (tetrahedral) cannot exist as diastereoisomers.

(d) $[PtCl_2Br_2]^{2-}$ (square planar) can exist as cis and trans diastereoisomers.

cis

trans

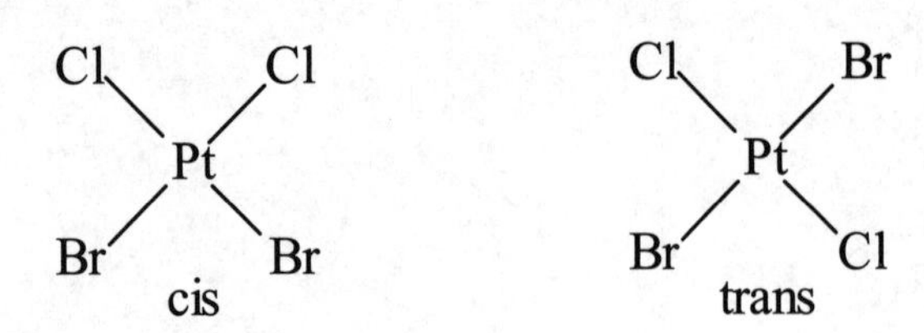

20.99 (a) $Pt(NH_3)_2(CN)_2$ can exist as two (cis and trans) diastereoisomers.

cis

trans

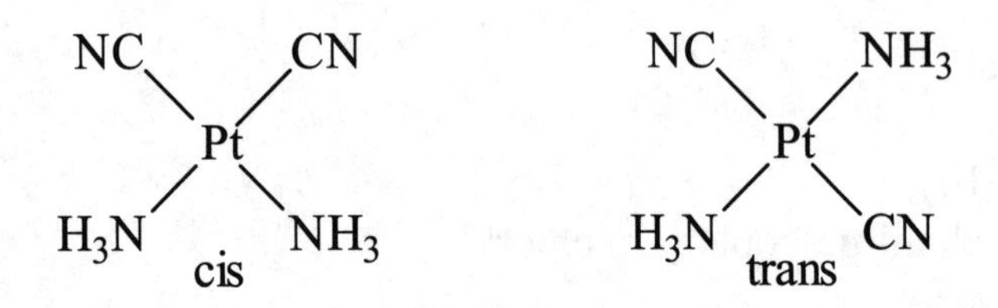

(b) $[Co(en)(SCN)_4]^-$ cannot exist as diastereoisomers.
(c) $[Cr(H_2O)_4Cl_2]^+$ can exist as two (cis and trans) diastereoisomers.

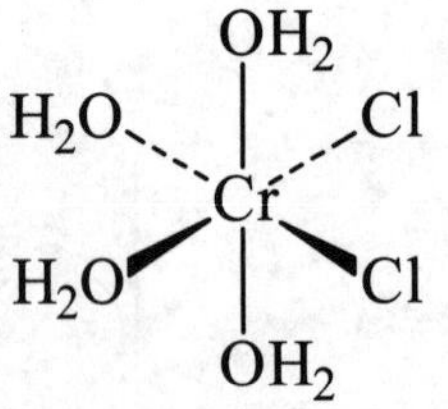

cis

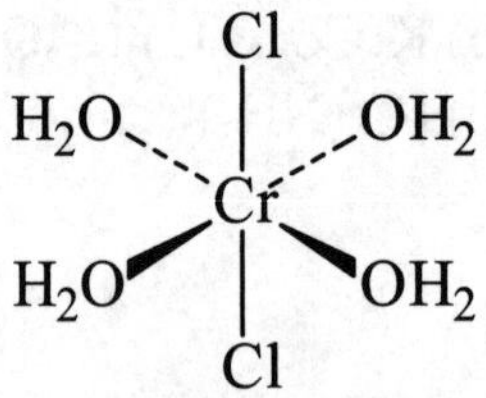

trans

(d) $Ru(NH_3)_3I_3$ can exist as two diastereoisomers.

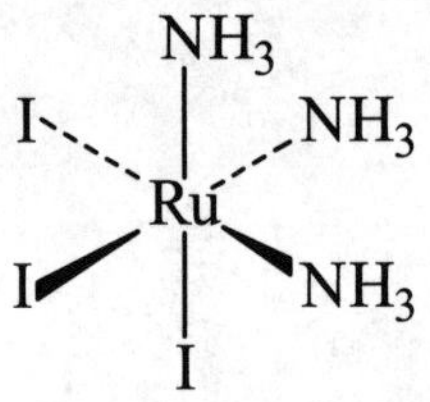

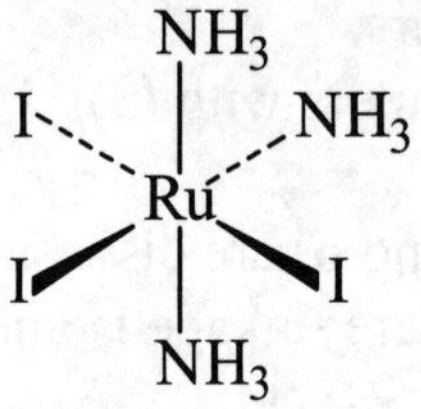

20.100 (c) cis-$[Cr(en)_2(H_2O)_2]^{3+}$ (d) $[Cr(C_2O_4)_3]^{3-}$

20.101 (a) $[Cr(en)_3]^{3+}$ can exist as enantiomers.

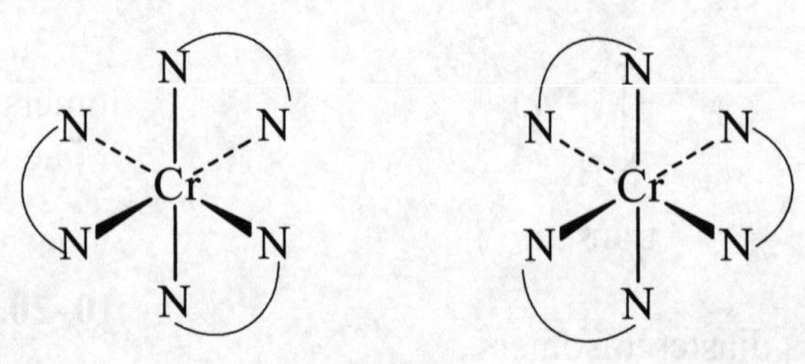

(b) cis-$[Co(en)_2(NH_3)Cl]^{2+}$ can exist as enantiomers.

(c) trans-$[Co(en)_2(NH_3)Cl]^{2+}$ cannot exist as enantiomers.
(d) $[Pt(NH_3)_3Cl_3]^{+}$ cannot exist as enantiomers.

20.102

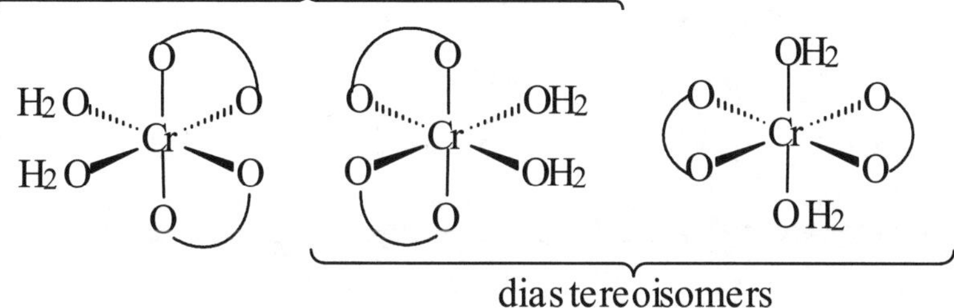

20.103

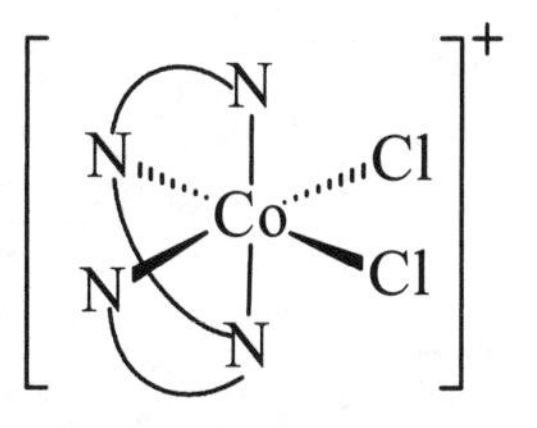

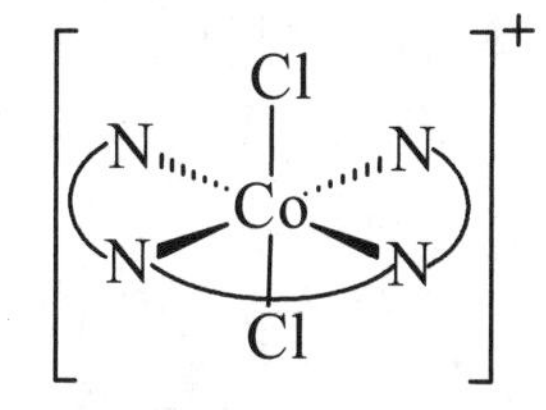

exists as a pair of enantiomers — exists as a pair of enantiomers — has no enantiomers

20.104 Plane-polarized light is light in which the electric vibrations of the light wave are restricted to a single plane. The following chromium complex can rotate the plane of plane-polarized light.

$[Cr(en)_3]^{3+}$

20.105 A racemic mixture is a mixture that contains equal amounts of two enantiomers. A racemic mixture does not affect plane-polarized light because the effect of one enantiomer is canceled by that of the other enantiomer.

Color of Complexes; Valence Bond and Crystal Field Theories (Sections 20.10–20.12)

20.106 The measure of the amount of light absorbed by a substance is called the absorbance, and a graph of absorbance versus wavelength is called an absorption spectrum. If a complex absorbs at 455 nm, its color is orange (use the color wheel in Figure 20.26).

20.107 ~525 nm (use the color wheel in Figure 20.26)

20.108 (a) $[Ti(H_2O)_6]^{3+}$

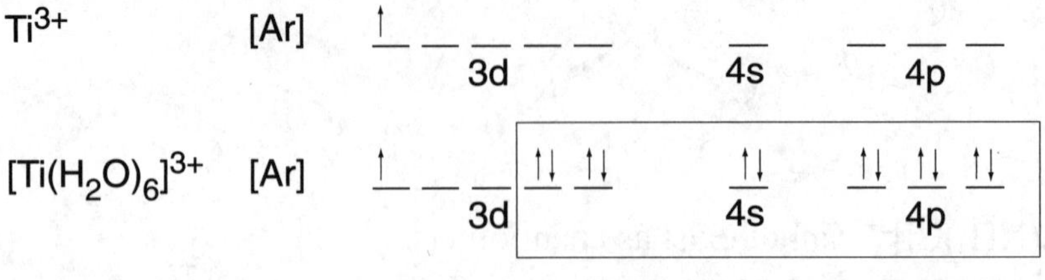

Ti^{3+} [Ar] ↑ _ _ _ _ (3d) _ (4s) _ _ _ (4p)

$[Ti(H_2O)_6]^{3+}$ [Ar] ↑ _ _ ↑↓ ↑↓ (3d) ↑↓ (4s) ↑↓ ↑↓ ↑↓ (4p)

d^2sp^3 1 unpaired e^-

(b) $[NiBr_4]^{2-}$

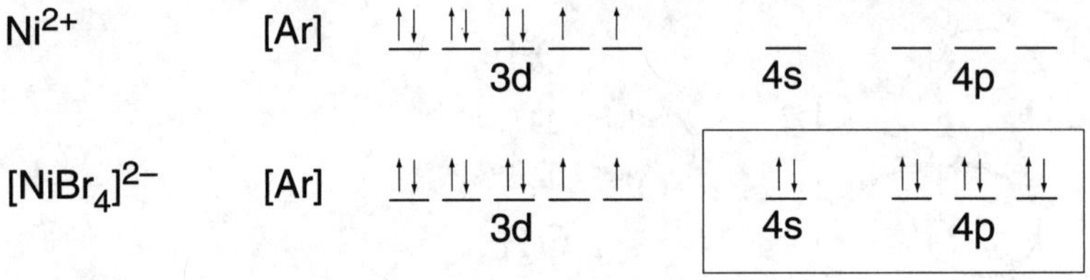

Ni^{2+} [Ar] ↑↓ ↑↓ ↑↓ ↑ ↑ (3d) _ (4s) _ _ _ (4p)

$[NiBr_4]^{2-}$ [Ar] ↑↓ ↑↓ ↑↓ ↑ ↑ (3d) ↑↓ (4s) ↑↓ ↑↓ ↑↓ (4p)

sp^3 2 unpaired e^-

(c) $[Fe(CN)_6]^{3-}$ (low-spin)

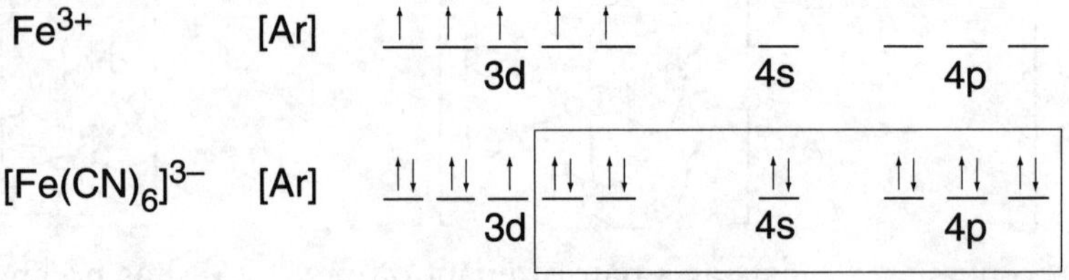

Fe^{3+} [Ar] ↑ ↑ ↑ ↑ ↑ (3d) _ (4s) _ _ _ (4p)

$[Fe(CN)_6]^{3-}$ [Ar] ↑↓ ↑↓ ↑ ↑↓ ↑↓ (3d) ↑↓ (4s) ↑↓ ↑↓ ↑↓ (4p)

d^2sp^3 1 unpaired e^-

(d) $[MnCl_6]^{3-}$ (high-spin)

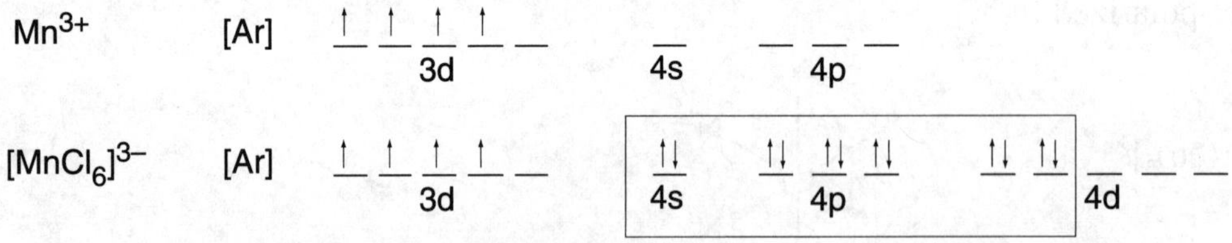

Mn^{3+} [Ar] ↑ ↑ ↑ ↑ _ (3d) _ (4s) _ _ _ (4p)

$[MnCl_6]^{3-}$ [Ar] ↑ ↑ ↑ ↑ _ (3d) ↑↓ (4s) ↑↓ ↑↓ ↑↓ (4p) ↑↓ ↑↓ _ _ _ (4d)

sp^3d^2 4 unpaired e^-

20.109 (a) $[AuCl_4]^-$

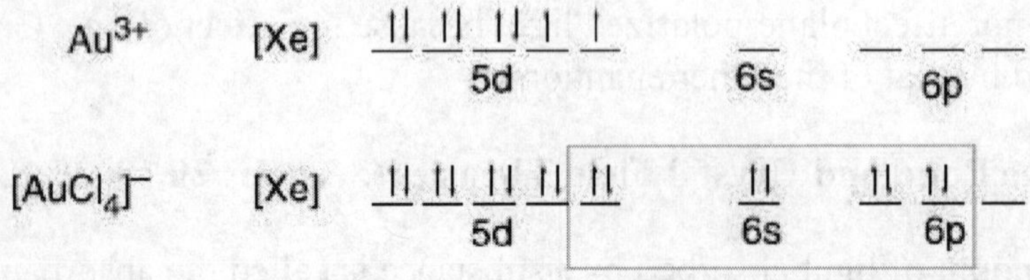

Au^{3+} [Xe] ↑↓ ↑↓ ↑↓ ↑ ↑ (5d) _ (6s) _ _ _ (6p)

$[AuCl_4]^-$ [Xe] ↑↓ ↑↓ ↑↓ ↑↓ ↑↓ (5d) ↑↓ (6s) ↑↓ ↑↓ _ (6p)

dsp^2 no unpaired e^-

(b) $[Ag(NH_3)_2]^+$

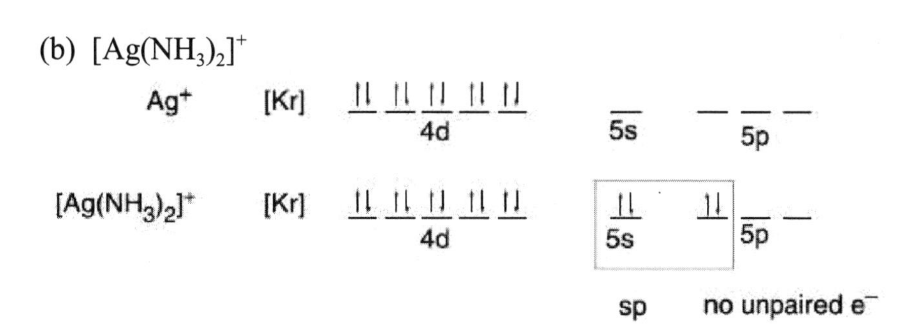

(c) $[Fe(H_2O)_6]^{2+}$ (high-spin)

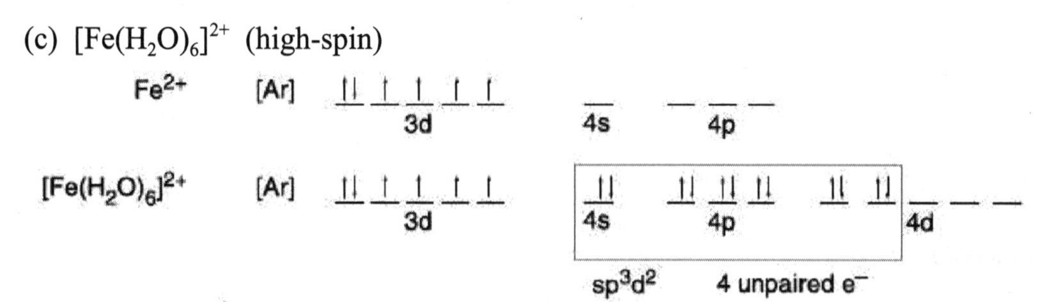

(d) $[Fe(CN)_6]^{4-}$ (low-spin)

Fe^{2+} [Ar] ⥮ ↿ ↿ ↿ ↿ 3d __ 4s __ __ __ 4p

$[Fe(CN)_6]^{4-}$ [Ar] ⥮ ⥮ ⥮ ⥮ ⥮ 3d ⥮ 4s ⥮ ⥮ ⥮ 4p

d^2sp^3 no unpaired e^-

20.110 (a) +3, M = Cr or Ni

(b)

$[Cr(OH)_4]^-$: [Ar] ↑ ↑ ↑ __ __ 3d ⇅ 4s ⇅ ⇅ ⇅ 4p

Four sp^3 bonds to the ligands

$[Ni(OH)_4]^-$: [Ar] ⇅ ⇅ ↑ ↑ ↑ 3d ⇅ 4s ⇅ ⇅ ⇅ 4p

Four sp^3 bonds to the ligands

(c) $[Cr(OH)_4]^-$

20.111 (a) +2, M = V or Co

(b)

$[VCl_4]^{2-}$: [Ar] ↑ ↑ ↑ __ __ (3d) | ↑↓ (4s) ↑↓ ↑↓ ↑↓ (4p)

Four sp^3 bonds to the ligands

$[CoCl_4]^{2-}$: [Ar] ↑↓ ↑↓ ↑ ↑ ↑ (3d) | ↑↓ (4s) ↑↓ ↑↓ ↑↓ (4p)

Four sp^3 bonds to the ligands

(c) $[CoCl_4]^{2-}$

20.112 $[Ti(H_2O)_6]^{3+}$ Ti^{3+} $3d^1$

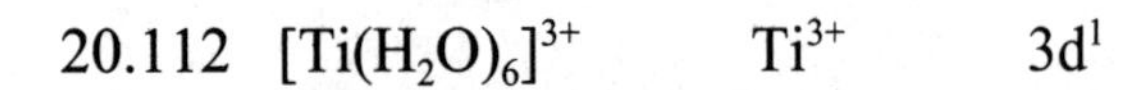

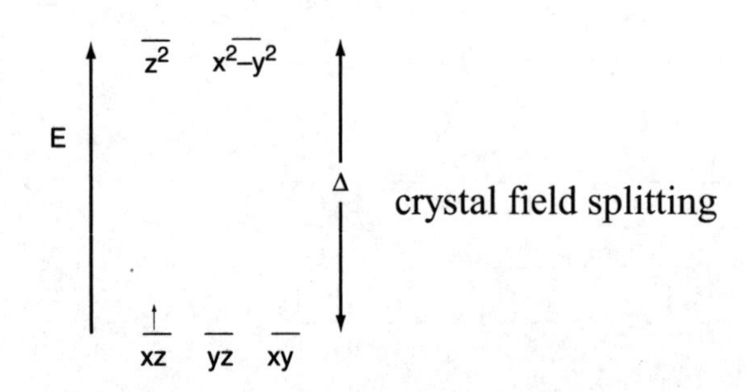

crystal field splitting

$[Ti(H_2O)_6]^{3+}$ is colored because it can absorb light in the visible region, exciting the electron to the higher-energy set of orbitals.

20.113

$d_{x^2-y^2}$ d_{xy}

The $d_{x^2-y^2}$ orbital is higher in energy because its lobes are pointing directly at the ligands.

20.114 $\lambda = 544 \text{ nm} = 544 \times 10^{-9} \text{ m}$

$$\Delta = \frac{hc}{\lambda} = \frac{(6.626 \times 10^{-34} \text{ J} \cdot \text{s})(3.00 \times 10^8 \text{ m/s})}{(544 \times 10^{-9} \text{ m})} = 3.65 \times 10^{-19} \text{ J}$$

$\Delta = (3.65 \times 10^{-19} \text{ J/ion})(6.022 \times 10^{23} \text{ ion/mol}) = 219{,}803 \text{ J/mol} = 220 \text{ kJ/mol}$

For $[Ti(H_2O)_6]^{3+}$, $\Delta = 240$ kJ/mol

Because $\Delta_{NCS^-} < \Delta_{H_2O}$ for the Ti complex, NCS^- is a weaker-field ligand than H_2O. If $[Ti(NCS)_6]^{3-}$ absorbs at 544 nm, its color should be red (use the color wheel in Figure 20.26).

20.115 $[Cr(H_2O)_6]^{3+}$ absorbs at ~580 nm while $[Cr(CN)_6]^{3-}$ absorbs at ~415 nm. CN^- is a stronger-field ligand, and H_2O is a weaker-field ligand.

20.116 (a) $[CrF_6]^{3-}$

— —

↿ ↿ ↿
3 unpaired e⁻

(b) $[V(H_2O)_6]^{3+}$

— —

↿ ↿ —
2 unpaired e⁻

(c) $[Fe(CN)_6]^{3-}$

— —

↿⇂ ↿⇂ ↿
1 unpaired e⁻

20.117 (a) $[Cu(en)_3]^{2+}$

↿⇂ ↿

↿⇂ ↿⇂ ↿⇂
1 unpaired e⁻

(b) $[FeF_6]^{3-}$

↿ ↿

↿ ↿ ↿
5 unpaired e⁻

(c) $[Co(en)_3]^{3+}$

— —

↿⇂ ↿⇂ ↿⇂
no unpaired e⁻

20.118 $Ni^{2+}(aq)$

↿ ↿

↿⇂ ↿⇂ ↿⇂

$Zn^{2+}(aq)$

↿⇂ ↿⇂

↿⇂ ↿⇂ ↿⇂

$Ni^{2+}(aq)$ is green because the Ni^{2+} ion can absorb light, which promotes electrons from the filled d orbitals to the higher energy half-filled d orbitals. $Zn^{2+}(aq)$ is colorless because the d orbitals are completely filled and no electrons can be promoted, so no light is absorbed.

20.119 Cr^{3+} is $3d^3$ and should be colored because absorption of light can promote electrons from the lower-energy to the higher-energy d orbitals. Y^{3+} is $4d^0$ and should be colorless because it can't exhibit d-d transitions.

20.120 Weak-field ligands produce a small Δ. Strong-field ligands produce a large Δ. For a metal complex with weak-field ligands, Δ < P, where P is the pairing energy, and it is easier to place an electron in either d_{z^2} or $d_{x^2-y^2}$ than to pair up electrons; high-spin complexes result. For a metal complex with strong-field ligands, Δ > P and it is easier to pair up electrons than to place them in either d_{z^2} or $d_{x^2-y^2}$; low-spin complexes result.

20.121 Ligands do not point directly at any d-orbitals, and the crystal field splitting is small. Because $\Delta < P$, high-spin complexes result.

20.122

__ $x^2\text{–}y^2$

↑↓ xy

↑↓ z^2

↑↓ ↑↓ xz, yz

Square planar geometry is most common for metal ions with d^8 configurations because this configuration favors low-spin complexes in which all four lower energy d orbitals are filled, and the higher energy $d_{x^2-y^2}$ orbital is vacant.

20.123 (a) $[Pt(NH_3)_4]^{2+}$ (square planar)

__ $x^2\text{–}y^2$

↑↓ xy

↑↓ z^2

↑↓ ↑↓ xz, yz

no unpaired e^-

(b) $[MnCl_4]^{2-}$ (tetrahedral)

↑ ↑ ↑

xy xz yz

↑ ↑

$x^2\text{–}y^2$ z^2

5 unpaired e^-

(c) $[Co(NCS)_4]^{2-}$ (tetrahedral)

↑ ↑ ↑

xy xz yz

↑↓ ↑↓

$x^2\text{–}y^2$ z^2

3 unpaired e^-

(d) $[Cu(en)_2]^{2+}$ (square planar)

↑ $x^2–y^2$

↑↓ xy

↑↓ z^2

↑↓ ↑↓ xz, yz

1 unpaired e^-

Chapter Problems

20.124 (a) $[Mn(CN)_6]^{3-}$ Mn^{3+} [Ar] $3d^4$

CN^- is a strong-field ligand. The Mn^{3+} complex is low-spin.

— —

↑↓ ↑ ↑ 2 unpaired e^-, paramagnetic

(b) $[Zn(NH_3)_4]^{2+}$ Zn^{2+} [Ar] $3d^{10}$

$[Zn(NH_3)_4]^{2+}$ is tetrahedral.

↑↓ ↑↓ ↑↓

↑↓ ↑↓ no unpaired e^-, diamagnetic

(c) $[Fe(CN)_6]^{4-}$ Fe^{2+} [Ar] $3d^6$

CN^- is a strong-field ligand. The Fe^{2+} complex is low-spin.

— —

↑↓ ↑↓ ↑↓ no unpaired e^-, diamagnetic

(d) $[FeF_6]^{4-}$ Fe^{2+} [Ar] $3d^6$

F^- is a weak-field ligand. The Fe^{2+} complex is high-spin.

↑ ↑

↑↓ ↑ ↑ 4 unpaired e^-, paramagnetic

20.125 (a) $[Ni(H_2O)_6]^{2+}$ Ni^{2+} [Ar] $3d^8$

H_2O is a weak-field ligand.

↑ ↑

↑↓ ↑↓ ↑↓ 2 unpaired e^-, paramagnetic

(b) $[Co(CN)_6]^{3-}$ Co^{2+} [Ar] $3d^6$
CN^- is a strong-field ligand. The Co^{3+} complex is low-spin.

— —

↑↓ ↑↓ ↑↓ no unpaired e^-, diamagnetic

(c) $[HgI_4]^{2-}$ Hg^{2+} [Xe] $5d^{10}$
$[HgI_4]^{2-}$ is tetrahedral.

↑↓ ↑↓ ↑↓

↑↓ ↑↓ no unpaired e^-, diamagnetic

(d) $[Cu(NH_3)_4]^{2+}$ Cu^{2+} [Ar] $3d^9$
$[Cu(NH_3)_4]^{2+}$ is square-planar.

↑

↑↓

↑↓
↑↓ ↑↓ 1 unpaired e^-, paramagnetic

20.126 (a) $4\,[Co^{3+}(aq) + e^- \rightarrow Co^{2+}(aq)]$
$\underline{2\,H_2O(l) \rightarrow O_2(g) + 4\,H^+(aq) + 4\,e^-}$
$4\,Co^{3+}(aq) + 2\,H_2O(l) \rightarrow 4\,Co^{2+}(aq) + O_2(g) + 4\,H^+(aq)$

(b) $4\,Cr^{2+}(aq) + O_2(g) + 4\,H^+(aq) \rightarrow 4\,Cr^{3+}(aq) + 2\,H_2O(l)$

(c) $3\,[Cu(s) \rightarrow Cu^{2+}(aq) + 2\,e^-]$
$\underline{Cr_2O_7^{2-}(aq) + 14\,H^+(aq) + 6\,e^- \rightarrow 2\,Cr^{3+}(aq) + 7\,H_2O(l)}$
$3\,Cu(s) + Cr_2O_7^{2-}(aq) + 14\,H^+(aq) \rightarrow 3\,Cu^{2+}(aq) + 2\,Cr^{3+}(aq) + 7\,H_2O(l)$

(d) $2\,CrO_4^{2-}(aq) + 2\,H^+(aq) \rightarrow Cr_2O_7^{2-}(aq) + H_2O(l)$

20.127 (a) $2\,[5\,e^- + 8\,H^+(aq) + MnO_4^-(aq) \rightarrow Mn^{2+}(aq) + 4\,H_2O(l)]$
$\underline{5\,[C_2O_4^{2-}(aq) \rightarrow 2\,CO_2(g) + 2\,e^-]}$
$16\,H^+(aq) + 2\,MnO_4^-(aq) + 5\,C_2O_4^{2-}(aq) \rightarrow 2\,Mn^{2+}(aq) + 8\,H_2O(l) + 10\,CO_2(g)$

(b) $6\,[Ti^{3+}(aq) + H_2O(l) \rightarrow TiO^{2+}(aq) + 2\,H^+(aq) + e^-]$
$\underline{6\,e^- + Cr_2O_7^{2-}(aq) + 14\,H^+(aq) \rightarrow 2\,Cr^{3+}(aq) + 7\,H_2O(l)}$
$6\,Ti^{3+}(aq) + Cr_2O_7^{2-}(aq) + 2\,H^+(aq) \rightarrow 6\,TiO^{2+}(aq) + 2\,Cr^{3+}(aq) + H_2O(l)$

(c) $2\,[MnO_4^-(aq) + e^- \rightarrow MnO_4^{2-}(aq)]$
$\underline{H_2O(l) + SO_3^{2-}(aq) + 2\,OH^-(aq) \rightarrow SO_4^{2-}(aq) + 2\,H_2O(l) + 2\,e^-}$
$2\,MnO_4^-(aq) + SO_3^{2-}(aq) + 2\,OH^-(aq) \rightarrow 2\,MnO_4^{2-}(aq) + SO_4^{2-}(aq) + H_2O(l)$

(d) $4\ [H_2O(l) + Fe(OH)_2(s) \rightarrow Fe(OH)_3(s) + H^+(aq) + e^-]$
$4\ H^+(aq) + 4\ e^- + O_2(g) \rightarrow 2\ H_2O(l)$
$2\ H_2O(l) + 4\ Fe(OH)_2(s) + O_2(g) \rightarrow 4\ Fe(OH)_3(s)$

20.128 $EDTA^{4-}$ in mayonnaise will complex any metal cations that are present in trace amounts. Free metal ions can catalyze the oxidation of oils, causing the mayonnaise to become rancid. The bidentate ligand $H_2NCH_2CO_2^-$ will not bind to metal ions as strongly as does the hexadentate $EDTA^{4-}$ and so would not be an effective substitute for $EDTA^{4-}$.

20.129

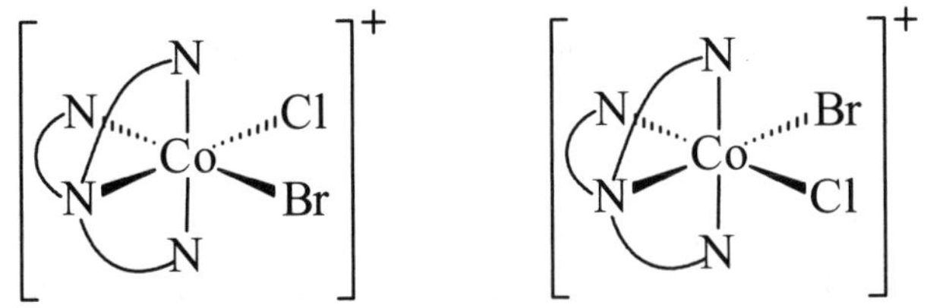

20.130 (a)

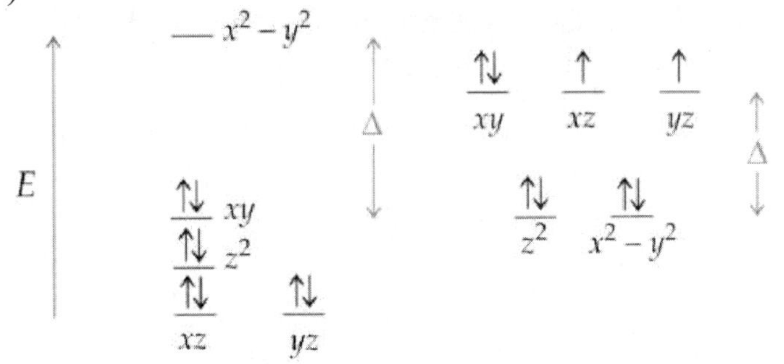

square planar nickel(II) tetrahedral nickel(II)

(b) $NiCl_2L_2$ is tetrahedral, $Ni(NCS)_2L_2$ is square planar.

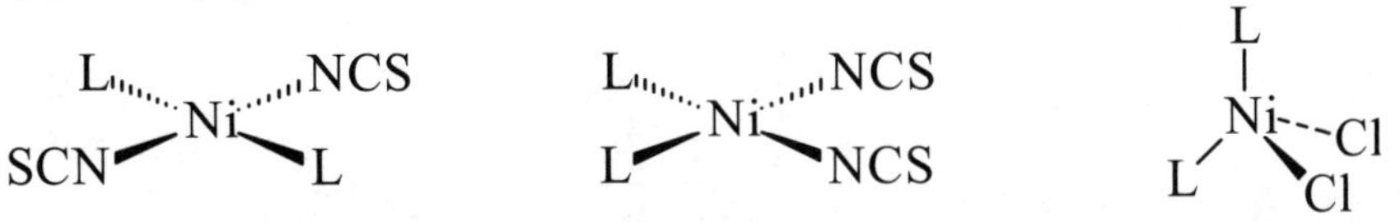

(c) Square planar cis-$Ni(NCS)_2L_2$ and tetrahedral $NiCl_2L_2$ have a dipole moment.

20.131 $[Mn(CN)_6]^{3-}$ Mn^{3+} $3d^4$

valence bond theory

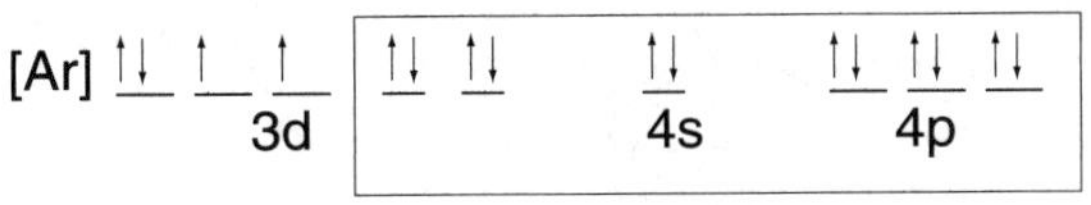

[Ar] ↿⇂ ↿ ↿ | ↿⇂ ↿⇂ ↿⇂ ↿⇂ ↿⇂ ↿⇂
3d 4s 4p

d^2sp^3

crystal field theory

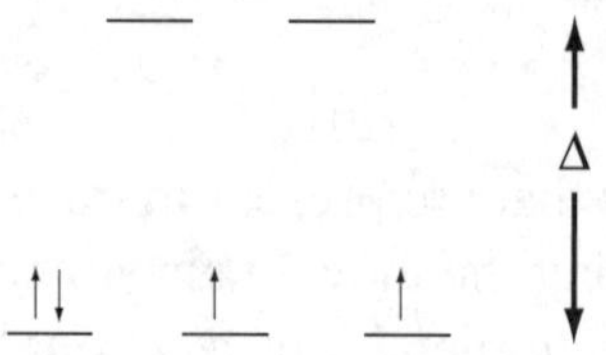

— —

Δ

↑↓ ↑ ↑

Crystal field model predicts 2 unpaired electrons.

20.132 Cr^{3+} is a $3d^3$ ion. Regardless of the crystal field splitting energy, the three electrons singly occupy the three lower energy d orbitals.

20.133 A choice between high-spin and low-spin electron configurations arises only for complexes of metal ions with four to seven d electrons, the so-called d^4-d^7 complexes. For d^1-d^3 and d^8-d^{10} complexes, only one ground-state electron configuration is possible. In d^1-d^3 complexes, all the electrons occupy the lower-energy d orbitals, independent of the value of Δ. In d^8-d^{10} complexes, the lower-energy set of d orbitals is filled with three pairs of electrons, while the higher-energy set contains two, three, or four electrons, again independent of the value of Δ.

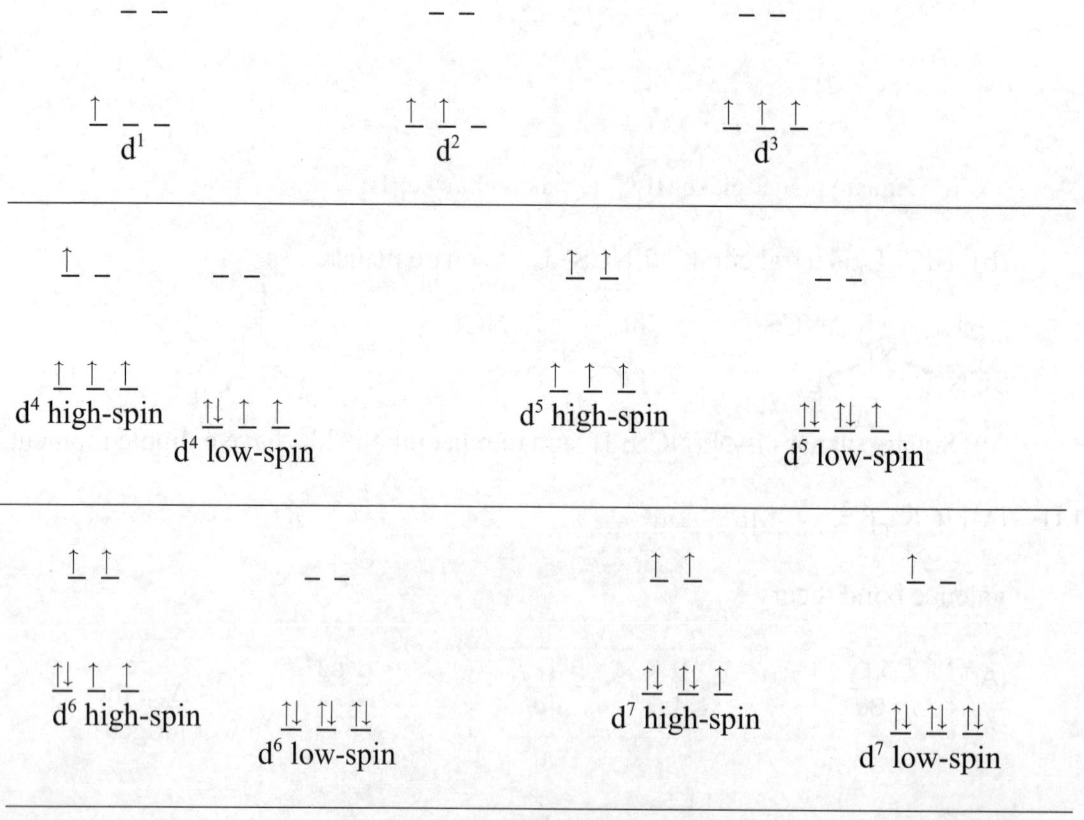

d^1 d^2 d^3

d^4 high-spin d^4 low-spin d^5 high-spin d^5 low-spin

d^6 high-spin d^6 low-spin d^7 high-spin d^7 low-spin

↿⇂ ↿⇂ ↿⇂ / ↿ ↿ — d^8

↿⇂ ↿⇂ ↿⇂ / ↿⇂ ↿ — d^9

↿⇂ ↿⇂ ↿⇂ / ↿⇂ ↿⇂ — d^{10}

20.134 (a) $[Mn(H_2O)_6]^{2+}$ high-spin Mn^{2+}, $3d^5$

↿ ↿

↿ ↿ ↿

5 unpaired e^-

(b) $Pt(NH_3)_2Cl_2$ square-planar Pt^{2+}, $5d^8$

_

↿⇂

↿⇂

↿⇂ ↿⇂

no unpaired e^-

(c) $[FeO_4]^{2-}$ tetrahedral Fe(VI), $3d^2$

_ _ _

↿ ↿

2 unpaired e^-

(d) $[Ru(NH_3)_6]^{2+}$ low-spin Ru^{2+}, $4d^6$

_ _

↿⇂ ↿⇂ ↿⇂

no unpaired e^-

20.135 $[CoCl_4]^{2-}$ is tetrahedral. $[Co(H_2O)_6]^{2+}$ is octahedral. Because $\Delta_{tet} < \Delta_{oct}$, these complexes have different colors. $[CoCl_4]^{2-}$ has absorption bands at longer wavelengths.

20.136 Linkage isomers:

Linkage isomers:

20.137 $Co(gly)_3$

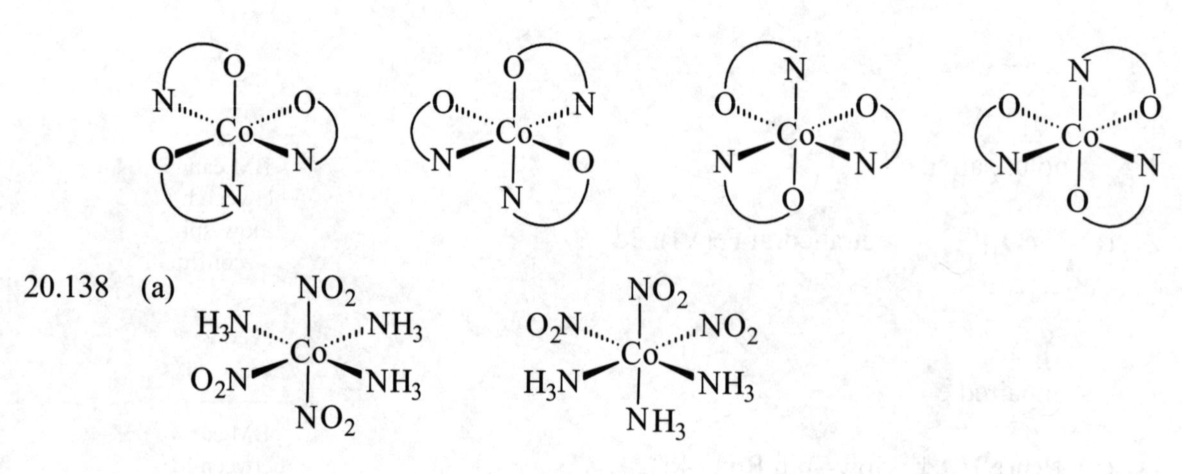

(b) $[Co(NH_3)_4(NO_2)_2][Co(NH_3)_2(NO_2)_4]$

20.139

1 can exist as enantiomers.

20.140 The nitro (–NO_2) complex is orange, which means it absorbs in the blue region (see color wheel) of the visible spectrum. The nitrito (–ONO) is red, which means it absorbs in the green region. The energy of the absorbed light is related to ligand field strength. Blue is higher energy than green, therefore nitro (–NO_2) is the stronger field ligand.

20.141

	weak-field ligands	strong-field ligands	
Ti^{2+} [Ar] $3d^2$	— — ↑ ↑ —	— — ↑ ↑ —	BM cannot distinguish between high-spin and low-spin electron configurations
	BM = $\sqrt{2(2+2)} = 2.83$	BM = $\sqrt{2(2+2)} = 2.83$	
V^{2+} [Ar] $3d^3$	— — ↑ ↑ ↑	— — ↑ ↑ ↑	BM cannot distinguish between high-spin and low-spin electron configurations
	BM = $\sqrt{3(3+2)} = 3.87$	BM = $\sqrt{3(3+2)} = 3.87$	
Cr^{2+} [Ar] $3d^4$	↑ — ↑ ↑ ↑	— — ↑↓ ↑ ↑	BM can distinguish between high-spin and low-spin electron configurations
	BM = $\sqrt{4(4+2)} = 4.90$	BM = $\sqrt{2(2+2)} = 2.83$	
Mn^{2+} [Ar] $3d^5$	↑ ↑ ↑ ↑ ↑	— — ↑↓ ↑↓ ↑	BM can distinguish between high-spin and low-spin electron configurations
	BM = $\sqrt{5(5+2)} = 5.92$	BM = $\sqrt{1(1+2)} = 1.73$	
Fe^{2+} [Ar] $3d^6$	↑ ↑ ↑↓ ↑ ↑	— — ↑↓ ↑↓ ↑↓	BM can distinguish between high-spin and low-spin electron configurations
	BM = $\sqrt{4(4+2)} = 4.90$	BM = 0	

Co^{2+} [Ar] $3d^7$	↑ ↑ ↑↓ ↑↓ ↑	↑ _ ↑↓ ↑↓ ↑↓	BM can distinguish between high-spin and low-spin electron configurations
	$BM = \sqrt{3(3+2)} = 3.87$	$BM = \sqrt{1(1+2)} = 1.73$	
Ni^{2+} [Ar] $3d^8$	↑ ↑ ↑↓ ↑↓ ↑↓	↑ ↑ ↑↓ ↑↓ ↑↓	BM cannot distinguish between high-spin and low-spin electron configurations
	$BM = \sqrt{2(2+2)} = 2.83$	$BM = \sqrt{2(2+2)} = 2.83$	
Cu^{2+} [Ar] $3d^9$	↑↓ ↑ ↑↓ ↑↓ ↑↓	↑↓ ↑ ↑↓ ↑↓ ↑↓	BM cannot distinguish between high-spin and low-spin electron configurations
	$BM = \sqrt{1(1+2)} = 1.73$	$BM = \sqrt{1(1+2)} = 1.73$	
Zn^{2+} [Ar] $3d^{10}$	↑↓ ↑↓ ↑↓ ↑↓ ↑↓	↑↓ ↑↓ ↑↓ ↑↓ ↑↓	BM cannot distinguish between high-spin and low-spin electron configurations
	BM = 0	BM = 0	

20.142 ML_2 (linear)

_ z^2

_ _ xz, yz

_ _ $x^2–y^2$, xy

20.143 ML_5 (square pyramid)

_ $x^2–y^2$

_ z^2

_ xy

_ _ xz, yz

20.144 (a)

NH_3, O_2N, NH_3, Co, O_2N, NH_3, NO_2

1

NH_3, O_2N, NH_3, Co, O_2N, NO_2, NH_3

2

(b) Isomer **2** would give rise to the desired product because it has two trans NO_2 groups.

20.145 (a) A tridentate ligand bonds to a metal using electron pairs on three donor atoms. Diethylenetriamine (dien) has three nitrogens that it can use to bond to a metal.

(b)

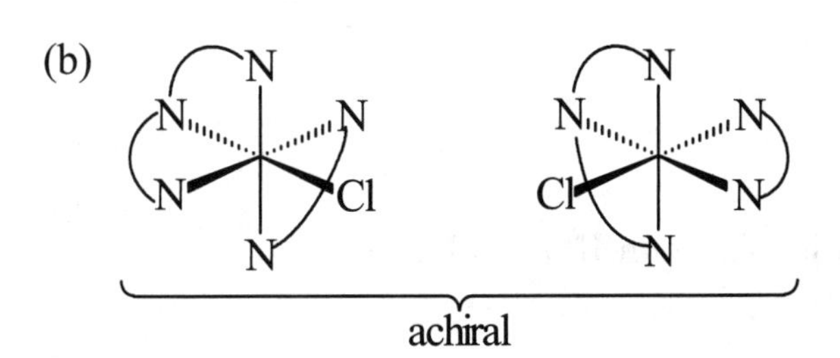

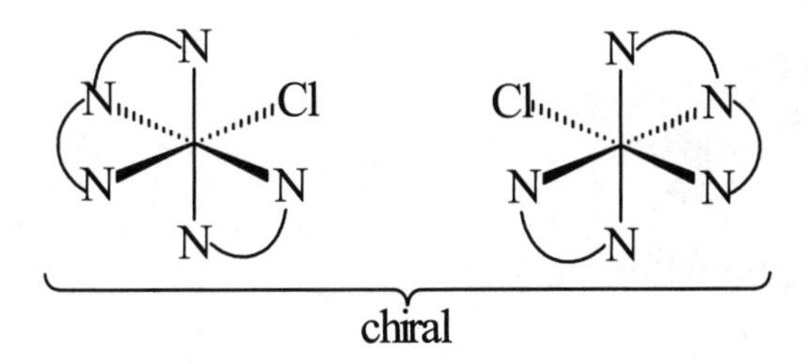

20.146 (a) $(NH_4)[Cr(H_2O)_6](SO_4)_2$, ammonium hexaaquachromium(III) sulfate

Cr^{3+}

_ _

↑ ↑ ↑

3 unpaired e^-

(b) $Mo(CO)_6$, hexacarbonylmolybdenum(0)

Mo^0 — —

↑↓ ↑↓ ↑↓
low-spin, no unpaired e^-

(c) $[Ni(NH_3)_4(H_2O)_2](NO_3)_2$, tetraamminediaquanickel(II) nitrate

Ni^{2+} ↑ ↑

↑↓ ↑↓ ↑↓
2 unpaired e^-

(d) $K_4[Os(CN)_6]$, potassium hexacyanoosmate(II)

Os^{2+} — —

↑↓ ↑↓ ↑↓
low-spin, no unpaired e^-

(e) $[Pt(NH_3)_4](ClO_4)_2$, tetraammineplatinum(II) perchlorate

Pt^{2+} —
↑↓
↑↓
↑↓ ↑↓
low spin, no unpaired e^-

(f) $Na_2[Fe(CO)_4]$, sodium tetracarbonylferrate(–II)

Fe^{2-} ↑↓ ↑↓ ↑↓

↑↓ ↑↓
no unpaired e^-

20.147 (a) Fe^{2+}, sodium pentacyanonitrosylferrate(II)
(b) — —

↑↓ ↑↓ ↑↓
low-spin, no unpaired e^-

20.148 For transition metal complexes, observed colors and absorbed colors are generally complementary. Using the color wheel (Figure 20.26), the absorbed colors in the table are complementary colors to those observed.

	Observed Color	Absorbed Color	Approximate λ (nm)
$Cr(acac)_3$	red	green	530
$[Cr(H_2O)_6]^{3+}$	violet	yellow	580
$[CrCl_2(H_2O)_4]^+$	green	red	700
$[Cr(urea)_6]^{3+}$	green	red	700
$[Cr(NH_3)_6]^{3+}$	yellow	violet	420
$Cr(acetate)_3(H_2O)_3$	blue-violet	orange-yellow	600

The magnitude of Δ is comparable to the energy of the absorbed light from the low energy red end to the high energy violet end (ROYGBIV). The red of $[CrCl_2(H_2O)_4]^+$ is lower energy than the yellow of $[Cr(H_2O)_6]^{3+}$, so $Cl^- < H_2O$. Because $[CrCl_2(H_2O)_4]^+$ and $[Cr(urea)_6]^{3+}$ are both red, Δ for 6 urea's is approximately equal to Δ for 2 Cl^-'s and 4 H_2O's. Therefore, urea is between Cl^- and H_2O.
The spectrochemical series is: $Cl^- <$ urea $<$ acetate $< H_2O <$ acac $< NH_3$

Multiconcept Problems

20.149 (a) $Fe^{3+}(aq) + 3\,C_2O_4^{2-}(aq) \rightleftarrows [Fe(C_2O_4)_3]^{3-}(aq) \qquad K_f = 3.3 \times 10^{20}$

	$[Fe(C_2O_4)_3]^{3-}(aq)$	$\rightleftarrows$	$Fe^{3+}(aq)$	+	$3\,C_2O_4^{2-}(aq)$	$K = 1/K_f = 3.0 \times 10^{-21}$
initial (M)	0.100		0		0	
change (M)	−x		+x		+3x	
equil (M)	0.100 − x		x		3x	

$$K = \frac{[Fe^{3+}][C_2O_4^{2-}]^3}{[[Fe(C_2O_4)_3]^{3-}]} = 3.0 \times 10^{-21} = \frac{x(3x)^3}{0.100 - x} \approx \frac{x(3x)^3}{0.100}$$

$3.0 \times 10^{-22} = 27x^4$

$x = [Fe^{3+}] = \sqrt[4]{3.0 \times 10^{-22}/27} = 1.8 \times 10^{-6}$ M

(b) $[Fe(C_2O_4)_3]^{3-}(aq) \rightleftarrows Fe^{3+}(aq) + 3\,C_2O_4^{2-}(aq) \qquad K_1 = 3.0 \times 10^{-21}$

$3\,H_3O^+(aq) + 3\,C_2O_4^{2-}(aq) \rightleftarrows 3\,HC_2O_4^-(aq) + 3\,H_2O(l) \qquad K_2 = (1/K_{a2})^3$

$\underline{3\,H_3O^+(aq) + 3\,HC_2O_4^-(aq) \rightleftarrows 3\,H_2C_2O_4(aq) + 3\,H_2O(l) \qquad K_3 = (1/K_{a1})^3}$

$[Fe(C_2O_4)_3]^{3-}(aq) + 6\,H_3O^+(aq) \rightleftarrows Fe^{3+}(aq) + 3\,H_2C_2O_4(aq) + 6\,H_2O(l)$

$$K_{overall} = K_1K_2K_3 = (3.0 \times 10^{-21})\left(\frac{1}{6.4 \times 10^{-5}}\right)^3\left(\frac{1}{5.9 \times 10^{-2}}\right)^3 = 5.6 \times 10^{-5}$$

$\Delta G = -RT\ln K = -[8.314 \times 10^{-3}\ kJ/(K \cdot mol)](298\ K) \ln(5.6 \times 10^{-5}) = 24.3$ kJ/mol
The reaction is nonspontaneous because ΔG is positive.

(c) ↑ ↑

↑ ↑ ↑
5 unpaired e^-

(d)

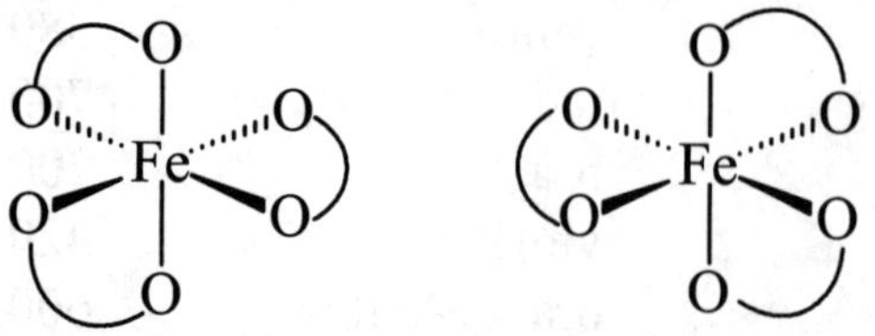

The complex is chiral. Enantiomers are shown.

20.150 (1) $Ni(H_2O)_6^{2+}(aq) + 6\ NH_3(aq) \rightleftarrows Ni(NH_3)_6^{2+}(aq) + 6\ H_2O(l)$ $K_f = 2.0 \times 10^8$

(2) $Ni(H_2O)_6^{2+}(aq) + 3\ en(aq) \rightleftarrows Ni(en)_3^{2+}(aq) + 6\ H_2O(l)$ $K_f = 4 \times 10^{17}$

(a) Reaction (2) should have the larger entropy change because three bidentate en ligands displace six water molecules.

(b) $\Delta G° = \Delta H° - T\Delta S°$

Because $\Delta H°_1$ and $\Delta H°_2$ are almost the same, the difference in $\Delta G°$ is determined by the difference in $\Delta S°$. Because $\Delta S°_2$ is larger than $\Delta S°_1$, $\Delta G°_2$ is more negative than $\Delta G°_1$ which is consistent with the greater stability of $Ni(en)_3^{2+}$.

(c) $\Delta H° - T\Delta S° = \Delta G° = -RT \ln K_f$

$$\Delta H°_1 - T\Delta S°_1 - (\Delta H°_2 - T\Delta S°_2) = -RT \ln K_f(1) - [-RT \ln K_f(2)]$$

$$T\Delta S°_2 - T\Delta S°_1 = RT \ln K_f(2) - RT \ln K_f(1) = RT \ln \frac{K_f(2)}{K_f(1)}$$

$$\Delta S°_2 - \Delta S°_1 = R \ln \frac{K_f(2)}{K_f(1)} = [8.314\ J/(K \cdot mol)] \ln \frac{4 \times 10^{17}}{2.0 \times 10^8}$$

$$\Delta S°_2 - \Delta S°_1 = 178\ J/(K \cdot mol) \text{ or } 180\ J/(K \cdot mol)$$

20.151 (a) $[Fe^{2+}(aq) \rightarrow Fe^{3+}(aq) + e^-] \times 5$ $E° = -0.77\ V$
(oxidation half reaction)

$MnO_4^-(aq) \rightarrow Mn^{2+}(aq)$

$MnO_4^-(aq) \rightarrow Mn^{2+}(aq) + 4\ H_2O(l)$

$8\ H^+(aq) + MnO_4^-(aq) \rightarrow Mn^{2+}(aq) + 4\ H_2O(l)$

$8\ H^+(aq) + MnO_4^-(aq) + 5\ e^- \rightarrow Mn^{2+}(aq) + 4\ H_2O(l)$ $E° = 1.51\ V$
(reduction half reaction)

Combine the two half reactions.

$5\ Fe^{2+}(aq) + MnO_4^-(aq) + 8\ H^+(aq) \rightarrow 5\ Fe^{3+}(aq) + Mn^{2+}(aq) + 4\ H_2O(l)$

(b) $E° = -0.77\ V + 1.51\ V = 0.74\ V$

$$\Delta G° = -nFE° = -(5\ mol\ e^-)\left(\frac{96{,}500\ C}{mol\ e^-}\right)(0.74\ V)\left(\frac{1\ J}{1\ C \cdot V}\right)\left(\frac{1\ kJ}{1000\ J}\right) = -3.6 \times 10^2\ kJ$$

$$E^\circ = \frac{0.0592\ V}{n} \log K$$

$$\log K = \frac{nE^\circ}{0.0592\ V} = \frac{(5)(0.74\ V)}{0.0592\ V} = 62.5; \quad K = 10^{62.5} = 3 \times 10^{62}$$

(c)

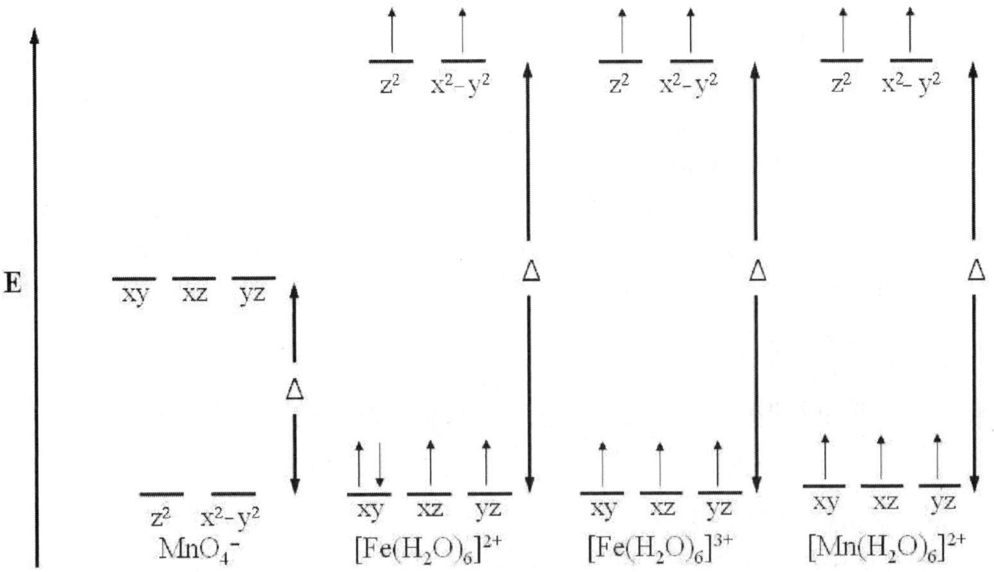

(d) The paramagnetism of the solution increases as the reaction proceeds. When $Fe^{2+}(aq)$ is oxidized to $Fe^{3+}(aq)$, the number of unpaired electrons goes from 4 to 5. In addition, the $Mn^{2+}(aq)$ produced has 5 unpaired electrons.

(e) 34.83 mL = 0.03483 L

mol $MnO_4^-(aq)$ = (0.051 32 mol/L)(0.034 83 L) = 1.787×10^{-3} mol MnO_4^-

$$\text{mass Fe} = 1.787 \times 10^{-3}\ \text{mol MnO}_4^- \times \frac{5\ \text{mol Fe}^{2+}}{1\ \text{mol MnO}_4^-} \times \frac{55.847\ \text{g Fe}}{1\ \text{mol Fe}^{2+}} = 0.4990\ \text{g Fe}$$

$$\text{mass \% Fe} = \frac{0.4990\ \text{g Fe}}{1.265\ \text{g sample}} \times 100\% = 39.45\%\ \text{Fe}$$

20.152 (a) $Cr(s) + 2\,H^+(aq) \rightarrow Cr^{2+}(aq) + H_2(g)$

(b) $$\text{mol Cr} = 2.60\ \text{g Cr} \times \frac{1\ \text{mol Cr}}{52.00\ \text{g Cr}} = 0.0500\ \text{mol Cr}$$

mol H_2SO_4 = (0.050 00 L)(1.200 mol/L) = 0.060 00 mol H_2SO_4

The stoichiometry between Cr and H_2SO_4 is one to one, therefore Cr is the limiting reagent because of the smaller number of moles.

$$\text{mol H}_2 = 0.0500\ \text{mol Cr} \times \frac{1\ \text{mol H}_2}{1\ \text{mol Cr}} = 0.0500\ \text{mol H}_2$$

25 °C = 298 K

PV = nRT

$$V = \frac{nRT}{P} = \frac{(0.0500\ \text{mol})\left(0.082\ 06\ \frac{\text{L}\cdot\text{atm}}{\text{K}\cdot\text{mol}}\right)(298\ \text{K})}{\left(735\ \text{mm Hg x}\ \frac{1.00\ \text{atm}}{760\ \text{mm Hg}}\right)} = 1.26\ \text{L of } H_2$$

(c) 0.060 00 mol H_2SO_4 can provide 0.1200 mol H^+. 0.0500 mol Cr reacts with 2 x (0.0500 mol H^+) = 0.100 mol H^+. This leaves 0.0200 mol H^+ and 0.0600 mol SO_4^{2-}, which will give, after neutralization, 0.0200 mol HSO_4^- and 0.0400 mol SO_4^{2-}.
$[HSO_4^-]$ = 0.0200 mol/0.050 00 L = 0.400 M
$[SO_4^{2-}]$ = 0.0400 mol/0.050 00 L = 0.800 M
The pH of this solution can be determined from the following equilibrium:

	$HSO_4^-(aq)$ + $H_2O(l)$	⇄	$H_3O^+(aq)$	+ $SO_4^{2-}(aq)$
initial (M)	0.400		~0	0.800
change (M)	–x		+x	+x
equil (M)	0.400 – x		x	0.800 + x

$$K_{a2} = \frac{[H_3O^+][SO_4^{2-}]}{[HSO_4^-]} = 1.2 \times 10^{-2} = \frac{(x)(0.800 + x)}{0.400 - x}$$

$x^2 + 0.812x - 0.0048 = 0$
Use the quadratic formula to solve for x.

$$x = \frac{-(0.812) \pm \sqrt{(0.812)^2 - 4(1)(-0.0048)}}{2(1)} = \frac{-0.812 \pm 0.8237}{2}$$

x = 0.005 85 and –0.818
Of the two solutions for x, only the positive value of x has physical meaning, because x is the $[H_3O^+]$.
$[H_3O^+]$ = x = 0.005 85 M
pH = $-\log[H_3O^+] = -\log(0.005\ 85) = 2.23$
(d) Crystal field d-orbital energy-level diagram

E
↑ z^2 $x^2 - y^2$
Δ
↑ xz ↑ yz ↑ xy

Valence bond orbital diagram

$Cr(H_2O)_6^{2+}$ [Ar] ↑ ↑ ↑ ↑ __ (3d) | ⇅ (4s) | ⇅ ⇅ ⇅ (4p) | ⇅ ⇅ __ __ __ (4d)

sp^3d^2 4 unpaired e^-

(e) The addition of excess KCN converts $Cr(H_2O)_6^{2+}(aq)$ to $Cr(CN)_6^{4-}(aq)$. CN^- is a strong field ligand and increases Δ changing the chromium complex from high spin, with 4 unpaired electrons, to low spin, with only 2 unpaired electrons.

20.153 (a) $t_{1/2} = \dfrac{0.693}{k} = \dfrac{0.693}{3.2 \times 10^{-5}\ s^{-1}} = 2.16 \times 10^4\ s$

$t_{1/2} = (2.16 \times 10^4\ s)(1\ h/3600\ s) = 6.0\ h$

(b) Let A = trans-$[Co(en)_2Cl_2]^+$

$\ln \dfrac{[A]_t}{[A]_o} = -kt$; $\quad \ln [A]_t - \ln [A]_o = -kt$; $\quad \ln [A]_t = \ln [A]_o - kt$

$\ln [A]_t = \ln(0.138) - (3.2 \times 10^{-5}\ s^{-1})(16.5\ h)(3600\ s/h) = -3.88$

$[A]_t = e^{-3.88} = 0.021\ M$

(c) trans-$[Co(en)_2Cl_2]^+(aq) \rightarrow$ trans-$[Co(en)_2Cl]^{2+}(aq) + Cl^-(aq)$ (slow)

trans-$[Co(en)_2Cl]^{2+}(aq) + H_2O(l) \rightarrow$ trans-$[Co(en)_2(H_2O)Cl]^{2+}(aq)$ (fast)

(d)

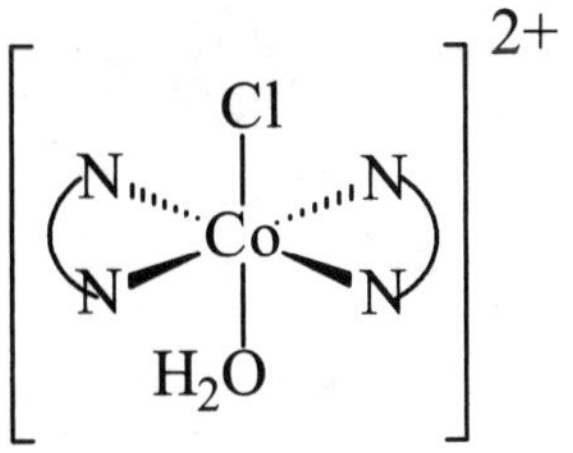

2+
Cl
N N
Co
N N
H_2O

The reaction product is achiral because it has several mirror planes.

(e)

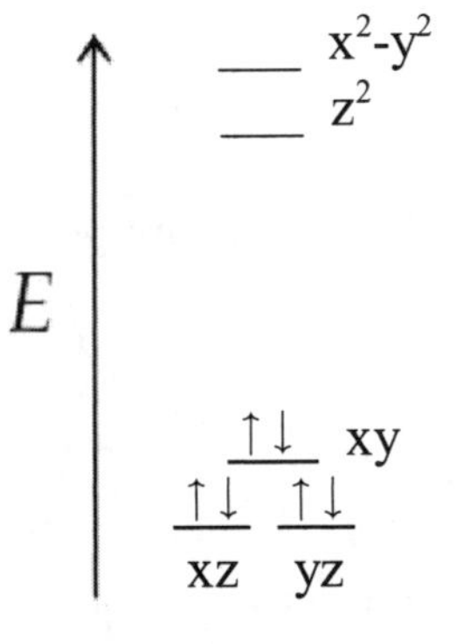

x^2-y^2
z^2

E

↑↓ xy
↑↓ ↑↓
xz yz

20.154 (a) Assume a 100.0 g sample of the chromium compound.

$19.52\ g\ Cr \times \dfrac{1\ mol\ Cr}{51.996\ g\ Cr} = 0.3754\ mol\ Cr$

$39.91\ g\ Cl \times \dfrac{1\ mol\ Cl}{35.453\ g\ Cl} = 1.126\ mol\ Cl$

$40.57\ g\ H_2O \times \dfrac{1\ mol\ H_2O}{18.015\ g\ H_2O} = 2.252\ mol\ H_2O$

$Cr_{0.3754}Cl_{1.126}(H_2O)_{2.252}$, divide each subscript by the smallest, 0.3754.

$Cr_{0.3754 / 0.3754}Cl_{1.126 / 0.3754}(H_2O)_{2.252 / 0.3754}$

$CrCl_3(H_2O)_6$

(b) $Cr(H_2O)_6Cl_3$, 266.45; AgCl, 143.32

For **A**: mol Cr complex = mol Cr = $0.225 \text{ g Cr complex} \times \dfrac{1 \text{ mol Cr complex}}{266.45 \text{ g Cr complex}} = 8.44 \times 10^{-4}$ mol Cr

mol Cl = mol AgCl = $0.363 \text{ g AgCl} \times \dfrac{1 \text{ mol AgCl}}{143.32 \text{ g AgCl}} = 2.53 \times 10^{-3}$ mol Cl

$$\frac{\text{mol Cl}}{\text{mol Cr}} = \frac{2.53 \times 10^{-3} \text{ mol Cl}}{8.44 \times 10^{-4} \text{ mol Cr}} = 3 \text{ Cl/Cr}$$

For **B**: mol Cr complex = mol Cr = $0.263 \text{ g Cr complex} \times \dfrac{1 \text{ mol Cr complex}}{266.45 \text{ g Cr complex}} = 9.87 \times 10^{-4}$ mol Cr

mol Cl = mol AgCl = $0.283 \text{ g AgCl} \times \dfrac{1 \text{ mol AgCl}}{143.32 \text{ g AgCl}} = 1.97 \times 10^{-3}$ mol Cl

$$\frac{\text{mol Cl}}{\text{mol Cr}} = \frac{1.97 \times 10^{-3} \text{ mol Cl}}{9.87 \times 10^{-4} \text{ mol Cr}} = 2 \text{ Cl/Cr}$$

For **C**: mol Cr complex = mol Cr = $0.358 \text{ g Cr complex} \times \dfrac{1 \text{ mol Cr complex}}{266.45 \text{ g Cr complex}} = 1.34 \times 10^{-3}$ mol Cr

mol Cl = mol AgCl = $0.193 \text{ g AgCl} \times \dfrac{1 \text{ mol AgCl}}{143.32 \text{ g AgCl}} = 1.34 \times 10^{-3}$ mol Cl

$$\frac{\text{mol Cl}}{\text{mol Cr}} = \frac{1.34 \times 10^{-3} \text{ mol Cl}}{1.34 \times 10^{-3} \text{ mol Cr}} = 1 \text{ Cl/Cr}$$

Because only the free Cl^- ions (those not bonded to the Cr^{3+}) give an immediate precipitate of AgCl, the probable structural formulas are:

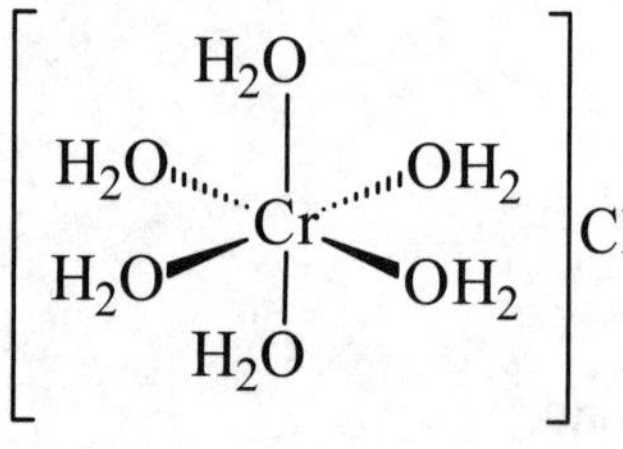

A

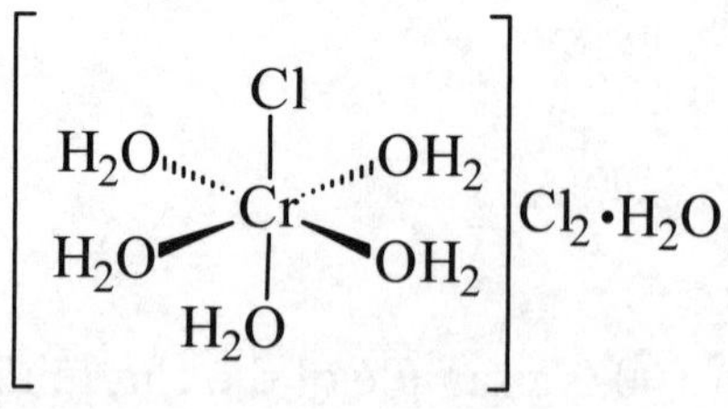

H_2O, H_2O, OH_2, H_2O, OH_2, H_2O around Cr; Cl_3

B

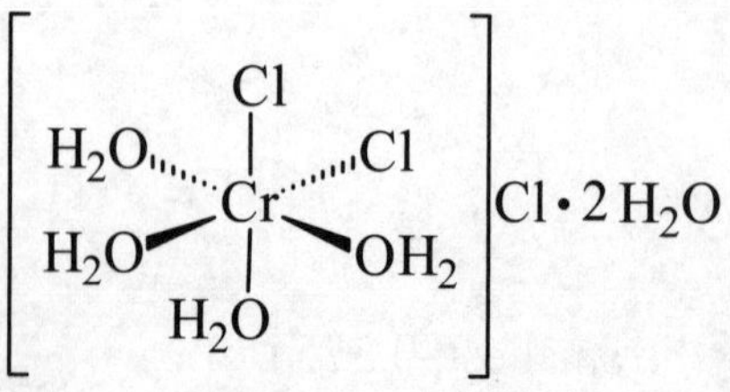

Cl, H_2O, OH_2, H_2O, OH_2, H_2O around Cr; $Cl_2 \cdot H_2O$

C

Cl, H_2O, Cl, H_2O, OH_2, H_2O around Cr; $Cl \cdot 2\,H_2O$

Structure **C** can exist as either cis or trans diastereoisomers.

(c) H_2O is a stronger field ligand than Cl^-. Compound **A** is likely to be violet absorbing in the yellow. Compounds **B** and **C** have weaker field ligands and would appear blue or green absorbing in the orange or red, respectively.

(d) $\Delta T = K_f \cdot m \cdot i$
For **A**, i = 4; for **B**, i = 3; and for **C**, i = 2.
For **A**, $\Delta T = K_f \cdot m \cdot i = (1.86\ °C/m)(0.25\ m)(4) = 1.9\ °C$
freezing point = 0 °C − ΔT = 0 °C − 1.9 °C = −1.9 °C
For **B**, $\Delta T = K_f \cdot m \cdot i = (1.86\ °C/m)(0.25\ m)(3) = 1.4\ °C$
freezing point = 0 °C − ΔT = 0 °C − 1.4 °C = −1.4 °C
For **C**, $\Delta T = K_f \cdot m \cdot i = (1.86\ °C/m)(0.25\ m)(2) = 0.93\ °C$
freezing point = 0 °C − ΔT = 0 °C − 0.93 °C = −0.93 °C

20.155 (a) & (b) Let X⌒Y represent the $tfac^-$ ligand.

A =
X X X
Co
Y Y Y

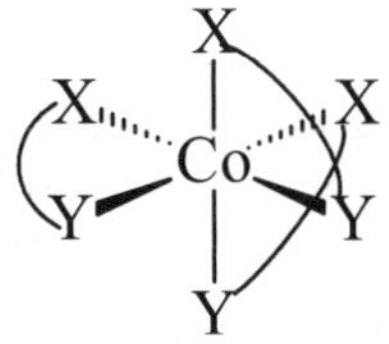

B =
Y X X Co X Y Y
Y X X Co X Y Y

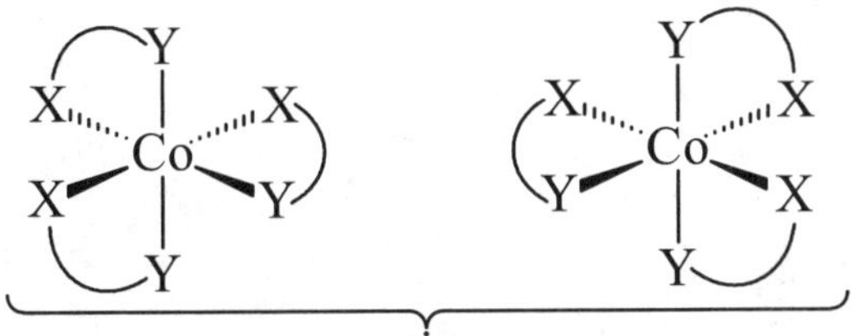

enantiomers

(c) $k_1 = 0.0889\ h^{-1}$, $T_1 = 66.1\ °C = 66.1 + 273.15 = 339.2\ K$
$k_2 = (0.0870\ min^{-1})(60\ min/1\ h) = 5.22\ h^{-1}$, $T_2 = 99.2\ °C = 99.2 + 273.15 = 372.3\ K$

$$\ln\left(\frac{k_2}{k_1}\right) = \left(\frac{-E_a}{R}\right)\left(\frac{1}{T_2} - \frac{1}{T_1}\right)$$

$$E_a = -\frac{[\ln k_2 - \ln k_1](R)}{\left(\frac{1}{T_2} - \frac{1}{T_1}\right)}$$

$$E_a = -\frac{[\ln(5.22) - \ln(0.0889)][8.314 \times 10^{-3}\ kJ/(K \cdot mol)]}{\left(\frac{1}{372.3\ K} - \frac{1}{339.2\ K}\right)} = 129\ kJ/mol$$

(d) — —

↑↓ ↑↓ ↑↓

no unpaired e^-, diamagnetic

20.156 (a) $K = \dfrac{[Cr_2O_7^{2-}]}{[CrO_4^{2-}]^2[H^+]^2} = 1.00 \times 10^{14}$

$[Cr_2O_7^{2-}]/[CrO_4^{2-}]^2 = 1.00 \times 10^{14}\ [H^+]^2$

In neutral solution, $[H^+] = 1.0 \times 10^{-7}$ and $[Cr_2O_7^{2-}]/[CrO_4^{2-}]^2 = 1$, so $[Cr_2O_7^{2-}]$ and $[CrO_4^{2-}]$ are comparable.

In basic solution, $[H^+] < 1.0 \times 10^{-7}$ and $[Cr_2O_7^{2-}]/[CrO_4^{2-}]^2 < 1$, so $[CrO_4^{2-}]$ predominates.

In acidic solution, $[H^+] > 1.0 \times 10^{-7}$ and $[Cr_2O_7^{2-}]/[CrO_4^{2-}]^2 > 1$, so $[Cr_2O_7^{2-}]$ predominates.

(b) At pH = 4.000, the $[H^+] = 1.00 \times 10^{-4}$ M

Let $x = [Cr_2O_7^{2-}]$ and $y = [CrO_4^{2-}]$

$$\frac{[Cr_2O_7^{2-}]}{[CrO_4^{2-}]^2} = [H^+]^2(1.00 \times 10^{14})$$

$$\frac{[Cr_2O_7^{2-}]}{[CrO_4^{2-}]^2} = (1.00 \times 10^{-4})^2(1.00 \times 10^{14})$$

$$\frac{[Cr_2O_7^{2-}]}{[CrO_4^{2-}]^2} = 1.00 \times 10^{6} = \frac{x}{y^2}$$

Because there are 2 Cr atoms per $Cr_2O_7^{2-}$, the total Cr concentration is $2[Cr_2O_7^{2-}] + [CrO_4^{2-}]$, and therefore $2x + y = 0.100$.

$\dfrac{x}{y^2} = 1.00 \times 10^6$ and $2x + y = 0.100$ M; solve these simultaneous equations.

$x = (1.00 \times 10^6)y^2$ and $x = (0.100 - y)/2$; substitute $(0.100 - y)/2$ for x

$(0.100 - y)/2 = (1.00 \times 10^6)y^2$

$(2.00 \times 10^6)y^2 + y - 0.100 = 0$

Use the quadratic formula to solve for y.

$$y = \frac{-(1) \pm \sqrt{(1)^2 - 4(2.00 \times 10^6)(-0.100)}}{2(2.00 \times 10^6)} = \frac{(-1) \pm (894.2)}{4.00 \times 10^6}$$

$y = -2.24 \times 10^{-4}$ and $2.233 \times 10^{-4} = 2.23 \times 10^{-4}$

Of the two solutions for y, only the positive value of y has physical meaning because y is the $[CrO_4^{2-}]$.

$[CrO_4^{2-}] = 2.23 \times 10^{-4}$ M

$[Cr_2O_7^{2-}] = x = (1.00 \times 10^6)y^2 = (1.00 \times 10^6)(2.233 \times 10^{-4}\ M)^2 = 4.99 \times 10^{-2}$ M

(c) At pH = 2.000, the $[H^+] = 1.00 \times 10^{-2}$ M

Let x = $[Cr_2O_7^{2-}]$ and y = $[CrO_4^{2-}]$

$$\frac{[Cr_2O_7^{2-}]}{[CrO_4^{2-}]^2} = [H^+]^2(1.00 \times 10^{14})$$

$$\frac{[Cr_2O_7^{2-}]}{[CrO_4^{2-}]^2} = (1.00 \times 10^{-2})^2(1.00 \times 10^{14})$$

$$\frac{[Cr_2O_7^{2-}]}{[CrO_4^{2-}]^2} = 1.00 \times 10^{10} = \frac{x}{y^2}$$

Because there are 2 Cr atoms per $Cr_2O_7^{2-}$, the total Cr concentration is $2[Cr_2O_7^{2-}] + [CrO_4^{2-}]$, and therefore 2x + y = 0.100.

$\frac{x}{y^2} = 1.00 \times 10^{10}$ and 2x + y = 0.100 M; solve these simultaneous equations.

$x = (1.00 \times 10^{10})y^2$ and x = (0.100 – y)/2; substitute (0.100 – y)/2 for x

$(0.100 - y)/2 = (1.00 \times 10^{10})y^2$

$(2.00 \times 10^{10})y^2 + y - 0.100 = 0$

Use the quadratic formula to solve for y.

$$y = \frac{-(1) \pm \sqrt{(1)^2 - 4(2.00 \times 10^{10})(-0.100)}}{2(2.00 \times 10^{10})} = \frac{(-1) \pm (8.944 \times 10^4)}{4.00 \times 10^{10}}$$

$y = -2.24 \times 10^{-6}$ and $2.236 \times 10^{-6} = 2.24 \times 10^{-6}$

Of the two solutions for y, only the positive value of y has physical meaning because y is the $[CrO_4^{2-}]$.

$[CrO_4^{2-}] = 2.24 \times 10^{-6}$ M

$[Cr_2O_7^{2-}] = x = (1.00 \times 10^{10})y^2 = (1.00 \times 10^{10})(2.236 \times 10^{-6}\ M)^2 = 5.00 \times 10^{-2}$ M

20.157 (a)

	$Au^{3+}(aq)$	+ 4 $SCN^-(aq)$	⇄ $Au(SCN)_4^-(aq)$
assume 100% reaction (M)	0	0	0.050
assume small back reaction (M)	+x	+4x	–x
equil (M)	x	4x	0.050 – x

$$K_f = \frac{[Au(SCN)_4^-]}{[Au^{3+}][SCN^-]^4} = 10^{37} = \frac{(0.050 - x)}{(x)(4x)^4} \approx \frac{0.050}{x(4x)^4}$$

Solve for x. $x = [Au^{3+}] = \sqrt[5]{\frac{0.05}{(256)(10^{37})}} = 7 \times 10^{-9}$ M

(b)

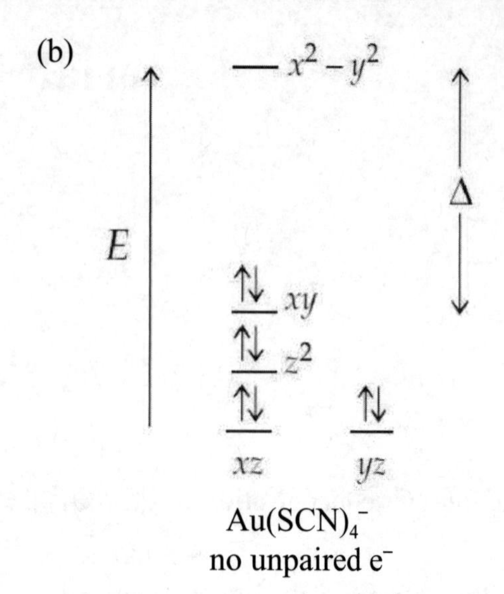

21 Metals and Solid-State Materials

21.1 (a) $Cr_2O_3(s) + 2\ Al(s) \rightarrow 2\ Cr(s) + Al_2O_3(s)$
(b) $Cu_2S(s) + O_2(g) \rightarrow 2\ Cu(s) + SO_2(g)$
(c) $PbO(s) + C(s) \rightarrow Pb(s) + CO(g)$
(d) $2\ K^+(l) + 2\ Cl^-(l) \xrightarrow{\text{electrolysis}} 2\ K(l) + Cl_2(g)$

21.2 $CaO(s) + SiO_2(s) \rightarrow CaSiO_3(l)$ (slag)
The O^{2-} in CaO behaves as a Lewis base and SiO_2 is the Lewis acid. They react with each other in a Lewis acid-base reaction to yield $CaSiO_3$ (Ca^{2+} and SiO_3^{2-}).

21.3 The electron configuration for Hg is [Xe] $4f^{14}\ 5d^{10}\ 6s^2$. Assuming the 5d and 6s bands overlap, the composite band can accommodate 12 valence electrons per metal atom. Weak bonding and a low melting point are expected for Hg because both the bonding and antibonding MOs are occupied.

21.4 (a) The composite s-d band can accommodate 12 valence electrons per metal atom.
Hf [Xe] $6s^2\ 4f^{14}\ 5d^2$, 4 valence electrons (4 bonding, 0 antibonding)
The s-d band is 1/4 full, so Hf is picture (1).
Pt [Xe] $6s^2\ 4f^{14}\ 5d^8$, 10 valence electrons (6 bonding, 4 antibonding)
The s-d band is 5/6 full, so Pt is picture (2).
Re [Xe] $6s^2\ 4f^{14}\ 5d^5$, 7 valence electrons (6 bonding, 1 antibonding)
The s-d band is 7/12 full, so Re is picture (3).
(b) Re has an excess of 5 bonding electrons and it has the highest melting point and is the hardest of the three.
(c) Pt has an excess of only 2 bonding electrons and it has the lowest melting point and is the softest of the three.

21.5 Ge doped with As is an n-type semiconductor because As has an additional valence electron. The extra electrons are in the conduction band. The number of electrons in the conduction band of the doped Ge is much higher than for pure Ge, and the conductivity of the doped semiconductor is higher.

21.6 (a) (1), silicon; (2), white tin; (3), diamond; (4), silicon doped with aluminum
(b) (3) < (1) < (4) < (2)
Diamond (3) is an insulator with a large band gap. Silicon (1) is a semiconductor with a band gap smaller than diamond. The conduction band is partially occupied with a few electrons and the valence band is partially empty. Silicon doped with aluminum (4) is a p-type semiconductor that has fewer electrons than needed for bonding and has vacancies (positive holes) in the valence band. White tin (2) has a partially filled s-p composite band and is a metallic conductor.

21.7 $E = 222\ \text{kJ/mol} \times \dfrac{1000\ \text{J}}{1\ \text{kJ}} \times \dfrac{1\ \text{mol}}{6.02 \times 10^{23}} = 3.69 \times 10^{-19}\ \text{J}$

$$\nu = \frac{E}{h} = \frac{3.69 \times 10^{-19}\ \text{J}}{6.626 \times 10^{-34}\ \text{J·s}} = 5.57 \times 10^{14}\ \text{s}^{-1}$$

$$\lambda = \frac{c}{\nu} = \frac{3.00 \times 10^{8}\ \text{m/s}}{5.57 \times 10^{14}\ \text{s}^{-1}} = 5.39 \times 10^{-7}\ \text{m} = 539 \times 10^{-9}\ \text{m} = 539\ \text{nm}$$

21.8 $Si(OCH_3)_4 + 4\ H_2O \rightarrow Si(OH)_4 + 4\ HOCH_3$

21.9 $Ba[OCH(CH_3)_2]_2 + Ti[OCH(CH_3)_2]_4 + 6\ H_2O \rightarrow BaTi(OH)_6(s) + 6\ HOCH(CH_3)_2$

$$BaTi(OH)_6(s) \xrightarrow{\text{heat}} BaTiO_3(s) + 3\ H_2O(g)$$

21.10 (a) cobalt/tungsten carbide is a ceramic-metal composite.
(b) silicon carbide/zirconia is a ceramic-ceramic composite.
(c) boron nitride/epoxy is a ceramic-polymer composite.
(d) boron carbide/titanium is a ceramic-metal composite.

21.11 The color of the quantum dots depends on the wavelength of light they absorb, which is determined by band-gap energy. Different sizes of CdSe nanoparticles have different band-gap energies.

21.12 (a) 5.0 nm (b) 2.2 nm (c) 3.5 nm

21.13 CdSe diameter = 2 nm = 2×10^{-9} m and human hair diameter = 50 μm = 50×10^{-6} m

$$\left(\frac{50 \times 10^{-6}\ \text{m}}{2 \times 10^{-9}\ \text{m}}\right) = 25{,}000 \text{ CdSe nanoparticles can fit across a human hair.}$$

21.14 (a) Size (a) absorbs red light so it appears green, size (b) absorbs orange light so it appears blue, size (c) absorbs yellow light so it appears violet, and size (d) absorbs green light so it appears red.
(b) Particle sizes from smallest to largest are: $d < c < b < a$

21.15 The smaller the particle, the larger the band gap and the greater the shift in the color of the emitted light from the red to the violet. The yellow quantum dot is larger because yellow is closer to the red than is the blue.

Conceptual Problems

21.16 A – metal oxide; B – metal sulfide; C – metal carbonate; D – free metal

21.17 (a) electrolysis (b) roasting a metal sulfide
(c) A = Li, electrolysis.
B = Hg, roasting of the metal sulfide.
C = Mn, reduction of the metal oxide.
D = Ca, reduction of the metal oxide.

21.18 (a) (2), bonding MO's are filled.
(b) (3), bonding and antibonding MO's are filled.
(c) (3) < (1) < (2). Hardness increases with increasing MO bond order.

21.19 (a) (1) is Mo, $5s^1\ 4d^5$; (2) is Y, $5s^2\ 4d^1$; and (3) is Ag, $5s^1\ 4d^{10}$. The size of the red area is proportional to the number of electrons in the composite s-d band.
(b) (1) has the highest melting point because the bonding MO's are filled and (3) has the lowest because bonding and most antibonding MO's are filled.
(c) Hardness increases with increasing MO bond order. Mo has the maximum MO bond whereas Ag has a minimum MO bond order.

21.20 (a) (1) and (4) are semiconductors; (2) is a metal; (3) is an insulator
(b) (3) < (1) < (4) < (2). The conductivity increases with decreasing band gap.
(c) (1) and (4) increases; (2) decreases; (3) not much change.

21.21

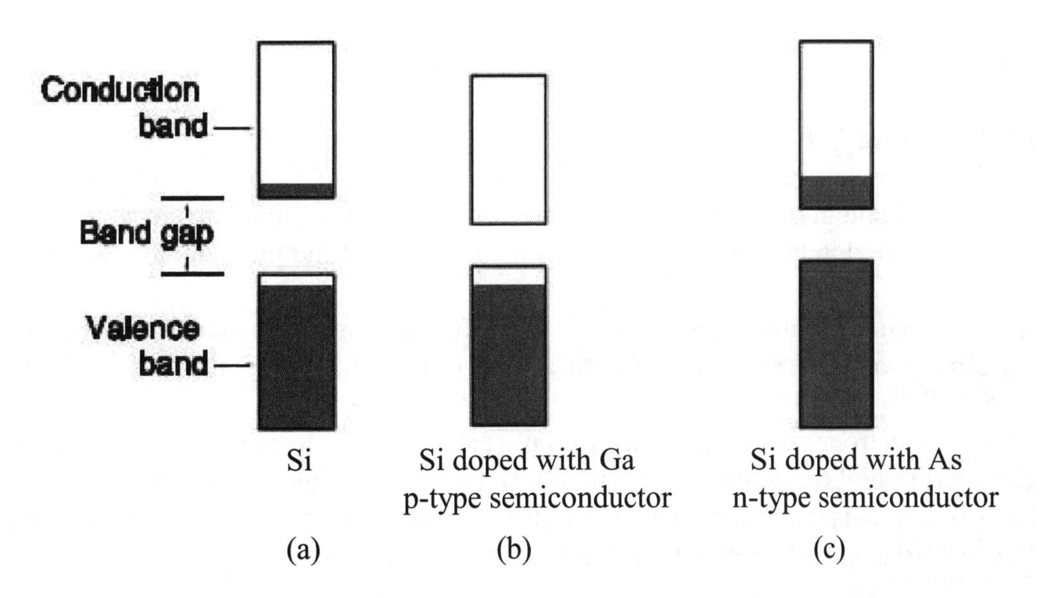

(d) The electrical conductivity of the doped silicon in both cases is higher than for pure silicon. Si doped with Ga is a p-type semiconductor with many more positive holes in the conduction band than in pure Si. This results in a higher conductivity. Si doped with As is an n-type semiconductor with many more electrons in the conduction band than in pure Si. This also results in a higher conductivity.

21.22

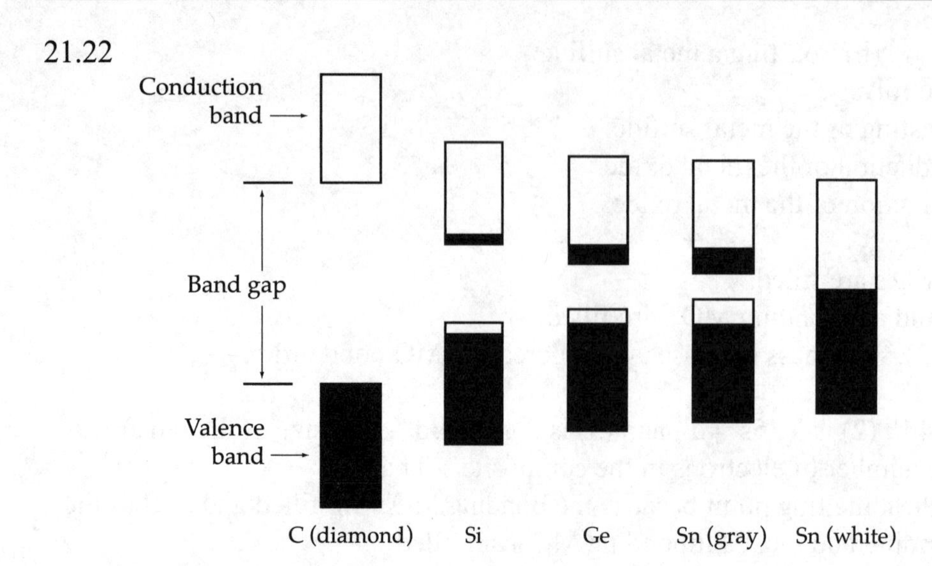

21.23 For LEDs the energy of the light emitted is roughly equal to the band gap energy, the higher the energy the shorter the wavelength, and the smaller the energy the longer the wavelength. LED (1) has the larger band gap energy so it emits blue light; LED (2) has the smaller band gap energy so it emits red light.

Section Problems
Sources of the Metallic Elements (Section 21.1)

21.24 TiO_2, MnO_2, and Fe_2O_3

21.25 Pd, Pt, and Au

21.26 (a) Cu is found in nature as a sulfide. (b) Zr is found in nature as an oxide.
(c) Pd is found in nature uncombined. (d) Bi is found in nature as a sulfide.

21.27 (a) V is found in nature as an oxide. (b) Ag is found in nature as a sulfide.
(c) Rh is found in nature uncombined. (d) Hf is found in nature as an oxide.

21.28 The less electronegative early transition metals tend to form ionic compounds by losing electrons to highly electronegative nonmetals such as oxygen. The more electronegative late transition metals tend to form compounds with more covalent character by bonding to the less electronegative nonmetals such as sulfur.

21.29 The early transition metals, which include Fe, generally occur as oxides because these less electronegative metals tend to form ionic compounds by losing electrons to the very electronegative oxygen. On the other hand, oxides of the s-block metals are strongly basic and far too reactive to exist in an environment that contains acidic oxides such as CO_2 and SiO_2. Consequently, s-block metals, including Ca, are found in nature as carbonates, as silicates, and in the case of Na and K, as chlorides.

21.30 (a) Fe_2O_3, hematite (b) PbS, galena
(c) TiO_2, rutile (d) $CuFeS_2$, chalcopyrite

21.31 (a) cinnabar, HgS (b) bauxite, $Al_2O_3 \cdot x\ H_2O$
(c) sphalerite, ZnS (d) chromite, $FeCr_2O_4$

Metallurgy (Section 21.2)

21.32 The flotation process exploits the differences in the ability of water and oil to wet the surfaces of the mineral and the gangue. The gangue, which contains ionic silicates, is moistened by the polar water molecules and sinks to the bottom of the tank. The mineral particles, which contain the less polar metal sulfide, are coated by the oil and become attached to the soapy air bubbles created by the detergent. The metal sulfide particles are carried to the surface in the soapy froth, which is skimmed off at the top of the tank. This process would not work well for a metal oxide because it is too polar and will be wet by the water and sink with the gangue.

21.33 Sometimes, ores are concentrated by chemical treatment. In the Bayer process, Ga_2O_3 is separated from Fe_2O_3 by treating the ore with hot aqueous NaOH. The amphoteric Ga_2O_3 dissolves as $Ga(OH)_4^-$, but the basic Fe_2O_3 does not.
$Ga_3O_3(s) + 2\ OH^-(aq) + 3\ H_2O(l) \rightarrow 2\ Ga(OH)_4^-(aq)$

21.34 Hg^{2+} in HgS is reduced. S^{2-} in HgS is oxidized. O_2 is reduced.
Hg^{2+} in HgS and O_2 are oxidizing agents. S^{2-} in HgS is the reducing agent.

21.35 W^{6+} in WO_3 is reduced. H_2 is oxidized.
W^{6+} in WO_3 is the oxidizing agent. H_2 is the reducing agent.

21.36 Because $E° < 0$ for Zn^{2+}, the reduction of Zn^{2+} is not favored.
Because $E° > 0$ for Hg^{2+}, the reduction of Hg^{2+} is favored.
The roasting of CdS should yield CdO because, like Zn^{2+}, $E° < 0$ for the reduction of Cd^{2+}.

21.37 $Mg^{2+}(aq) + 2\ e^- \rightarrow Mg(s)$ $E° = -2.37\ V$
$Zn^{2+}(aq) + 2\ e^- \rightarrow Zn(s)$ $E° = -0.76\ V$

The very negative reduction potential for Mg^{2+} indicates that Mg^{2+} is much more difficult to reduce than the Zn^{2+}.

21.38 (a) $V_2O_5(s) + 5\ Ca(s) \rightarrow 2\ V(s) + 5\ CaO(s)$
(b) $2\ PbS(s) + 3\ O_2(g) \rightarrow 2\ PbO(s) + 2\ SO_2(g)$
(c) $MoO_3(s) + 3\ H_2(g) \rightarrow Mo(s) + 3\ H_2O(g)$
(d) $3\ MnO_2(s) + 4\ Al(s) \rightarrow 3\ Mn(s) + 2\ Al_2O_3(s)$

(e) $MgCl_2(l) \xrightarrow{\text{electrolysis}} Mg(l) + Cl_2(g)$

21.39 (a) $Fe_2O_3(s) + 3\ H_2(g) \rightarrow 2\ Fe(s) + 3\ H_2O(g)$
(b) $Cr_2O_3(s) + 2\ Al(s) \rightarrow 2\ Cr(s) + Al_2O_3(s)$
(c) $Ag_2S(s) + O_2(g) \rightarrow 2\ Ag(s) + SO_2(g)$
(d) $TiCl_4(g) + 2\ Mg(l) \rightarrow Ti(l) + 2\ MgCl_2(l)$
(e) $2\ LiCl(l) \xrightarrow{\text{electrolysis}} 2\ Li(l) + Cl_2(g)$

21.40 $2\ ZnS(s) + 3\ O_2(g) \rightarrow 2\ ZnO(s) + 2\ SO_2(g)$
$\Delta H^\circ = [2\ \Delta H^\circ_f(ZnO) + 2\ \Delta H^\circ_f(SO_2)] - [2\ \Delta H^\circ_f(ZnS)]$
$\Delta H^\circ = [(2\ mol)(-350.5\ kJ/mol) + (2\ mol)(-296.8\ kJ/mol)]$
$- (2\ mol)(-206.0\ kJ/mol) = -882.6\ kJ$
$\Delta G^\circ = [2\ \Delta G^\circ_f(ZnO) + 2\ \Delta G^\circ_f(SO_2)] - [2\ \Delta G^\circ_f(ZnS)]$
$\Delta G^\circ = [(2\ mol)(-320.5\ kJ/mol) + (2\ mol)(-300.2\ kJ/mol)]$
$- (2\ mol)(-201.3\ kJ/mol) = -838.8\ kJ$
ΔH° and ΔG° are different because of the entropy change associated with the reaction. The minus sign for $(\Delta H^\circ - \Delta G^\circ)$ indicates that the entropy is negative, which is consistent with a decrease in the number of moles of gas from 3 mol to 2 mol.

21.41 $HgS(s) + O_2(g) \rightarrow Hg(l) + SO_2(g)$
$\Delta H^\circ = \Delta H^\circ_f(SO_2) - \Delta H^\circ_f(HgS)$
$\Delta H^\circ = (1\ mol)(-296.8\ kJ/mol) - (1\ mol)(-58.2\ kJ/mol) = -238.6\ kJ$
$\Delta G^\circ = \Delta G^\circ_f(SO_2) - \Delta G^\circ_f(HgS)$
$\Delta G^\circ = (1\ mol)(-300.2\ kJ/mol) - (1\ mol)(-50.6\ kJ/mol) = -249.6\ kJ$
ΔH° and ΔG° are different because of the entropy change associated with the reaction. The plus sign for $(\Delta H^\circ - \Delta G^\circ)$ indicates that the entropy is positive, which is consistent with a solid going to liquid and no change in the number of moles of gas.

21.42 $FeCr_2O_4(s) + 4\ C(s) \rightarrow \underbrace{Fe(s) + 2\ Cr(s)}_{\text{ferrochrome}} + 4\ CO(g)$

(a) $FeCr_2O_4$, 223.84; Cr, 52.00; 236 kg = 236 x 10^3 g

$$\text{mass Cr} = 236 \times 10^3\ g \times \frac{1\ mol\ FeCr_2O_4}{223.84\ g} \times \frac{2\ mol\ Cr}{1\ mol\ FeCr_2O_4} \times \frac{52.00\ g\ Cr}{1\ mol\ Cr} \times \frac{1.00\ kg}{1000\ g} = 110\ kg\ Cr$$

(b) $$\text{mol CO} = 236 \times 10^3\ g \times \frac{1\ mol\ FeCr_2O_4}{223.84\ g} \times \frac{4\ mol\ CO}{1\ mol\ FeCr_2O_4} = 4217.3\ mol\ CO$$

$$PV = nRT;\quad V = \frac{nRT}{P} = \frac{(4217.3\ mol)\left(0.082\ 06\ \frac{L \cdot atm}{K \cdot mol}\right)(298\ K)}{\left(740\ mm\ Hg \times \frac{1.00\ atm}{760\ mm\ Hg}\right)} = 1.06 \times 10^5\ L\ CO$$

21.43 $Cu_2S(l) + O_2(g) \rightarrow 2\ Cu(l) + SO_2(g)$

Cu_2S, 159.16; Cu, 63.546; SO_2, 64.06

(a) $\text{mass of Cu} = 1.4 \times 10^{10}\ \text{kg Cu} \times \dfrac{1000\ \text{g}}{1\ \text{kg}} = 1.4 \times 10^{13}\ \text{g Cu}$

$$\text{mass Cu}_2\text{S} = 1.4 \times 10^{13}\ \text{g Cu} \times \frac{1\ \text{mol Cu}}{63.546\ \text{g Cu}} \times \frac{1\ \text{mol Cu}_2\text{S}}{2\ \text{mol Cu}} \times \frac{159.16\ \text{g Cu}_2\text{S}}{1\ \text{mol Cu}_2\text{S}} \times \frac{1\ \text{kg}}{1000\ \text{g}}$$

$\text{mass Cu}_2\text{S} = 1.8 \times 10^{10}\ \text{kg Cu}_2\text{S}$

(b) $\text{mol SO}_2 = 1.8 \times 10^{10}\ \text{kg Cu}_2\text{S} \times \dfrac{1000\ \text{g}}{1\ \text{kg}} \times \dfrac{1\ \text{mol Cu}_2\text{S}}{159.16\ \text{g Cu}_2\text{S}} \times \dfrac{1\ \text{mol SO}_2}{1\ \text{mol Cu}_2\text{S}}$

$\text{mol SO}_2 = 1.1 \times 10^{11}\ \text{mol SO}_2$

PV = nRT

$$V = \frac{nRT}{P} = \frac{(1.1 \times 10^{11}\ \text{mol})\left(0.082\ 06\ \dfrac{\text{L} \cdot \text{atm}}{\text{K} \cdot \text{mol}}\right)(273\ \text{K})}{1.0\ \text{atm}} = 2.5 \times 10^{12}\ \text{L SO}_2$$

(c) H_2SO_4, 98.08

$$\text{mass H}_2\text{SO}_4 = 1.1 \times 10^{11}\ \text{mol SO}_2 \times \frac{1\ \text{mol H}_2\text{SO}_4}{1\ \text{mol SO}_2} \times \frac{98.08\ \text{g H}_2\text{SO}_4}{1\ \text{mol H}_2\text{SO}_4} \times \frac{1\ \text{kg}}{1000\ \text{g}}$$

$\text{mass H}_2\text{SO}_4 = 1.1 \times 10^{10}\ \text{kg H}_2\text{SO}_4$

21.44 $Ni^{2+}(aq) + 2\ e^- \rightarrow Ni(s)$; 1 A = 1 C/s

$$\text{mass Ni} = 52.5\ \frac{\text{C}}{\text{s}} \times 8.00\ \text{h} \times \frac{3600\ \text{s}}{1\ \text{h}} \times \frac{1\ \text{mol e}^-}{96{,}500\ \text{C}} \times \frac{1\ \text{mol Ni}}{2\ \text{mol e}^-} \times \frac{58.69\ \text{g Ni}}{1\ \text{mol Ni}} \times \frac{1.00\ \text{kg}}{1000\ \text{g}}$$

mass Ni = 0.460 kg Ni

21.45 $Cu^{2+}(aq) + 2\ e^- \rightarrow Cu(s)$

$$\text{Charge} = 7.50\ \text{kg} \times \frac{1000\ \text{g}}{1\ \text{kg}} \times \frac{1\ \text{mol Cu}}{63.546\ \text{g}} \times \frac{2\ \text{mol e}^-}{1\ \text{mol Cu}} \times \frac{96{,}500\ \text{C}}{1\ \text{mol e}^-} = 2.28 \times 10^7\ \text{C}$$

$$\text{Time} = \frac{2.28 \times 10^7\ \text{C}}{40.0\ \text{C/s}} \times \frac{1\ \text{h}}{3600\ \text{s}} = 158\ \text{h}$$

Iron and Steel (Section 21.3)

21.46 $Fe_2O_3(s) + 3\ CO(g) \rightarrow 2\ Fe(l) + 3\ CO_2(g)$

Fe_2O_3 is the oxidizing agent. CO is the reducing agent.

21.47 Coke (C) is converted to CO, the reducing agent used in the production of Fe.

$C(s) + CO_2(g) \rightarrow 2\ CO(g)$

$2\ C(s) + O_2(g) \rightarrow 2\ CO(g)$

21.48 Slag is a by-product of iron production, consisting mainly of $CaSiO_3$. It is produced from the gangue in iron ore.

21.49 Limestone is added to the blast furnace in the commercial process for producing iron to remove the gangue from the iron ore. At the high temperatures of the blast furnace, the limestone decomposes to lime (CaO), a basic oxide that reacts with SiO_2 and other acidic oxides present in the gangue. The product, called slag, is a molten material consisting mainly of calcium silicate.

$CaCO_3(s) \rightarrow CaO(s) + CO_2(g)$

$CaO(s) + SiO_2(s) \rightarrow CaSiO_3(l)$ (slag)

21.50 Molten iron from a blast furnace is exposed to a jet of pure oxygen gas for about 20 minutes. The impurities are oxidized to yield a molten slag that can be poured off.

$P_4(l) + 5\ O_2(g) \rightarrow P_4O_{10}(l)$

$6\ CaO(s) + P_4O_{10}(l) \rightarrow 2\ Ca_3(PO_4)_2(l)$ (slag)

$2\ Mn(l) + O_2(g) \rightarrow 2\ MnO(s)$

$MnO(s) + SiO_2(s) \rightarrow MnSiO_3(l)$ (slag)

21.51 Slag forms in the basic oxygen process because the impurities are oxidized, and the acidic oxides that form react with the basic CaO to yield a molten slag that can be poured off. Phosphorus, for example, is oxidized to P_4O_{10}, which then reacts with CaO to give molten calcium phosphate:

$P_4(l) + 5\ O_2(g) \rightarrow P_4O_{10}(l)$

$6\ CaO(s) + P_4O_{10}(l) \rightarrow 2\ Ca_3(PO_4)_2(l)$ (slag)

Manganese also passes into the slag because its oxide is basic and reacts with added SiO_2, yielding molten manganese silicate.

$2\ Mn(l) + O_2(g) \rightarrow 2\ MnO(s)$

$MnO(s) + SiO_2(s) \rightarrow MnSiO_3(l)$ (slag)

21.52 $SiO_2(s) + 2\ C(s) \rightarrow Si(s) + 2\ CO(g)$

$Si(s) + O_2(g) \rightarrow SiO_2(s)$

$CaO(s) + SiO_2(s) \rightarrow CaSiO_3(l)$ (slag)

21.53 $CaSO_4(s) + 3\ C(s) \rightarrow CaO(s) + S(l) + 3\ CO(g)$

$2\ S(l) + 3\ O_2(g) \rightarrow 2\ SO_3(g)$

$CaO(s) + SO_3(g) \rightarrow CaSO_4(l)$ (slag)

21.54 $3\ Fe_2O_3(s) + CO(g) \rightarrow 2\ Fe_3O_4(s) + CO_2(g)$ $\quad \Delta H° = -46.4$ kJ

$2 \times [Fe_3O_4(s) + CO(g) \rightarrow 3\ FeO(s) + CO_2(g)]$ $\quad \Delta H° = 2(19.0\ kJ) = 38.0$ kJ

$\underline{6 \times [FeO(s) + CO(g) \rightarrow Fe(s) + CO_2(g)]}$ $\quad \Delta H° = 6(-11.0\ kJ) = -66.0$ kJ

$3\ Fe_2O_3(s) + 9\ CO(g) \rightarrow 6\ Fe(s) + 9\ CO_2(g)$ $\quad \Delta H° = (-46.4 + 38.0 - 66.0) = -74.4$ kJ

divide each coefficient by 3

$Fe_2O_3(s) + 3\ CO(g) \rightarrow 2\ Fe(s) + 3\ CO_2(g)$ $\quad \Delta H° = -74.4\ kJ/3 = -24.8$ kJ

21.55 (a) $CO_2(g) + C(s) \rightarrow 2\ CO(g)$

$\Delta H°_{rxn} = [2\ \Delta H°_f(CO)] - \Delta H°_f(CO_2)$

$\Delta H°_{rxn} = [(2\ mol)(-110.5\ kJ/mol)] - [(1\ mol)(-393.5\ kJ/mol)] = +172.5$ kJ

$\Delta S^\circ = [2\ S^\circ(CO)] - [S^\circ(CO_2) + S^\circ(C)]$
$\Delta S^\circ = [(2\text{ mol})(197.6\text{ J/(K}\cdot\text{mol))}]$
$- [(1\text{ mol})(213.6\text{ J/(K}\cdot\text{mol)}) + (1\text{ mol})(5.7\text{ J/(K}\cdot\text{mol)})] = +175.9\text{ J/K}$

(b) $\Delta G = \Delta H - T\Delta S$; Set $\Delta G = 0$ and solve for T (the crossover temperature).

$$T = \frac{\Delta H}{\Delta S} = \frac{172.5\text{ kJ}}{0.1759\text{ kJ/K}} = 980.7\text{ K}$$

The reaction would become spontaneous at 980.7 K and above.

21.56 No. In a blast furnace tungsten carbide (WC) would be formed.

21.57 No. C and CO are not strong enough reducing agents to reduce aluminum oxide. ΔG° for the reaction of Al_2O_3 with C and with CO is large and positive.

Bonding in Metals (Section 21.4)

21.58

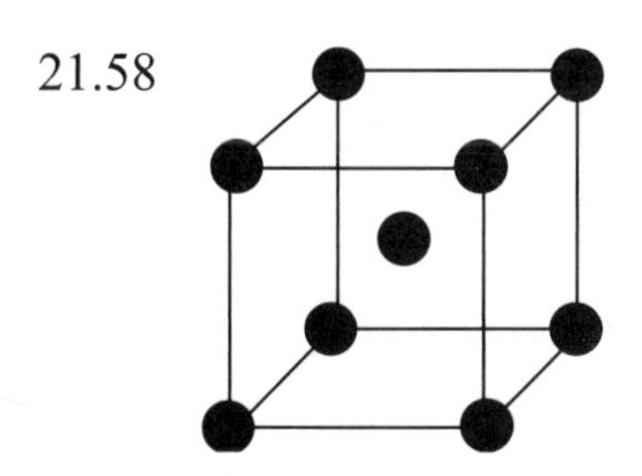

Each K has a single valence electron and has eight nearest neighbor K atoms. The valence electrons cannot be localized in an electron-pair bond between any particular pair of K atoms.

21.59 Cesium has just one valence electron per Cs atom and crystallizes in a body-centered cubic structure in which each Cs atom is surrounded by eight other Cs atoms. Consequently, the valence electrons cannot be localized in a bond between any particular pair of Cs atoms. Instead, the electrons are delocalized and belong to the crystal as a whole. In the electron-sea model, we visualize the crystal as a three-dimensional array of metal cations immersed in a sea of delocalized electrons that are free to move throughout the crystal. The continuum of delocalized, mobile valence electrons acts as an electrostatic glue that holds the metal cations together.

21.60 Malleability and ductility of metals follow from the fact that the delocalized bonding extends in all directions. When a metallic crystal is deformed, no localized bonds are broken. Instead, the electron sea simply adjusts to the new distribution of cations, and the energy of the deformed structure is similar to that of the original. Thus, the energy required to deform a metal is relatively small.

21.61 The electron-sea model affords a simple qualitative explanation for the electrical and thermal conductivity of metals. Because the electrons are mobile, they are free to move away from a negative electrode and toward a positive electrode when a metal is subjected to an electrical potential. The mobile electrons can also conduct heat by carrying kinetic energy from one part of the crystal to another.

21.62 Ionic bonding is much stronger than metallic bonding.

21.63 Ionic bonding is much stronger than metallic bonding.

21.64 The energy required to deform a transition metal like W is greater than that for Cs because W has more valence electrons and hence more electrostatic "glue".

21.65 Na has one valence electron. Mg has two valence electrons. There are more electrons per cation in the electron sea for Mg than for Na. There is more electrostatic glue for Mg, and hence the higher melting point.

21.66 The difference in energy between successive MOs in a metal decreases as the number of metal atoms increases so that the MOs merge into an almost continuous band of energy levels. Consequently, MO theory for metals is often called band theory.

21.67

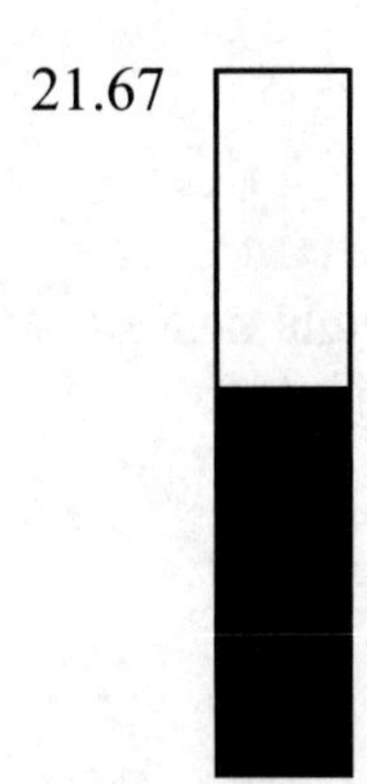

21.68 The energy levels within a band occur in degenerate pairs; one set of energy levels applies to electrons moving to the right, and the other set applies to electrons moving to the left. In the absence of an electrical potential, the two sets of levels are equally populated. As a result there is no net electric current. In the presence of an electrical potential those electrons moving to the right are accelerated, those moving to the left are slowed down, and some change direction. Thus, the two sets of energy levels are now unequally populated. The number of electrons moving to the right is now greater than the number moving to the left, and so there is a net electric current.

21.69 An electrical potential can shift electrons from one set of energy levels to the other only if the band is partially filled. If the band is completely filled, there are no available vacant energy levels to which electrons can be excited, and therefore the two sets of levels must remain equally populated, even in the presence of an electrical potential. This means that an electrical potential cannot accelerate the electrons in a completely filled band. Materials that have only completely filled bands are therefore electrical insulators. By contrast, materials that have partially filled bands are metals.

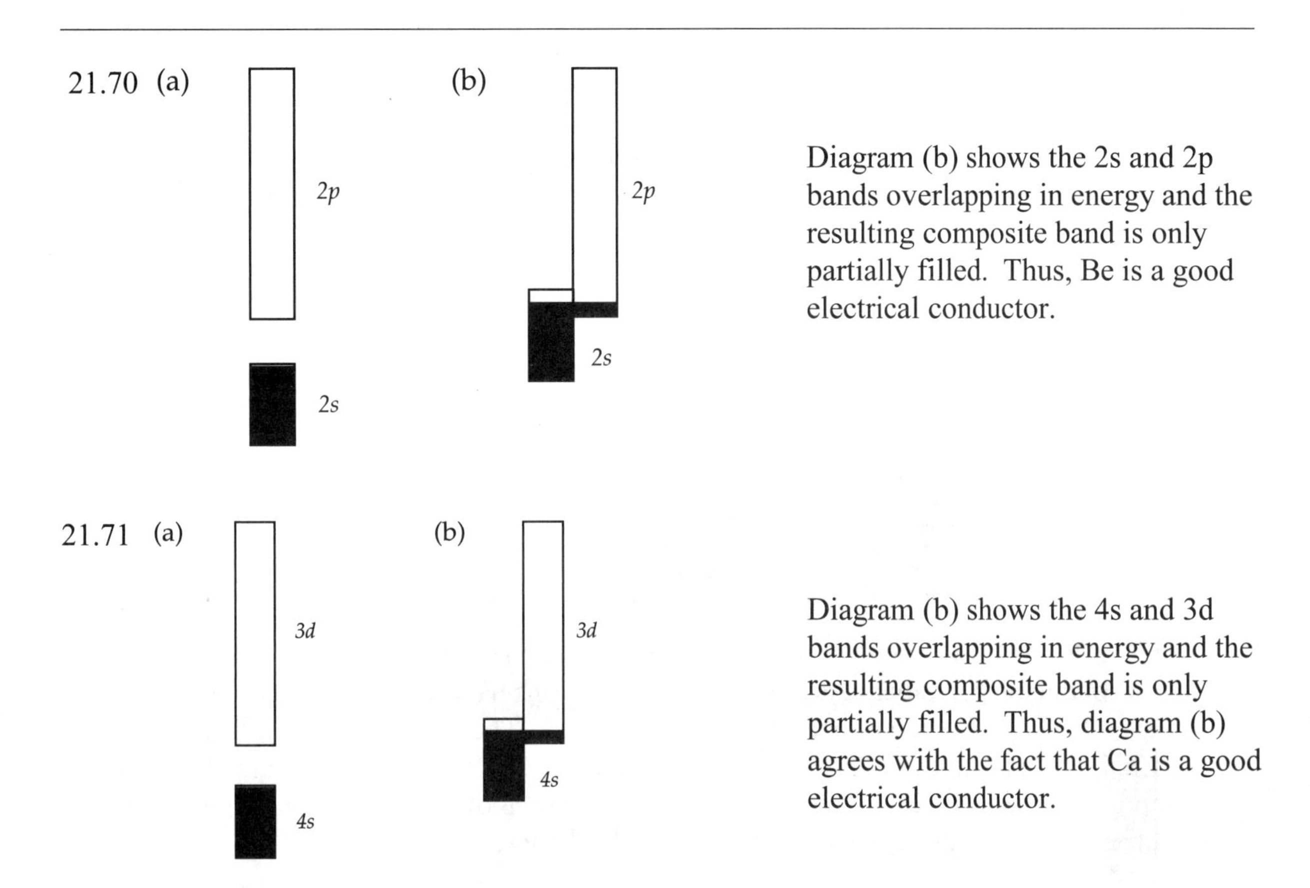

21.70 (a) (b) Diagram (b) shows the 2s and 2p bands overlapping in energy and the resulting composite band is only partially filled. Thus, Be is a good electrical conductor.

21.71 (a) (b) Diagram (b) shows the 4s and 3d bands overlapping in energy and the resulting composite band is only partially filled. Thus, diagram (b) agrees with the fact that Ca is a good electrical conductor.

21.72 Transition metals have a d band that can overlap the s band to give a composite band consisting of six MOs per metal atom. Half of the MOs are bonding and half are antibonding, and thus one expects maximum bonding for metals that have six valence electrons per metal atom. Accordingly, the melting points of the transition metals go through a maximum at or near group 6B.

21.73 Transition metals have a d band that can overlap the s band to give a composite band consisting of six MOs per metal atom. Half of the MOs are bonding and half are antibonding, and thus we might expect maximum bonding and hardness for metals that have six valence electrons per metal atom. The hardness will decrease as the number of valence electrons (beyond six) increases because they go into antibonding MOs. Cu has three more valence electrons than Fe, and Cu is softer.

Semiconductors and Semiconductor Applications (Sections 21.5 and 21.6)

21.74 A semiconductor is a material that has an electrical conductivity intermediate between that of a metal and that of an insulator. Si, Ge, and Sn (gray) are semiconductors.

21.75 (a) The valence band is the band of bonding molecular orbitals in a semiconductor.
(b) The conduction band is the band of antibonding molecular orbitals in a semiconductor.

(c) The band gap is the energy difference between the valence band and the conduction band in a semiconductor.
(d) Doping is the addition of a small amount of impurities to increase the conductivity of a semiconductor.

21.76 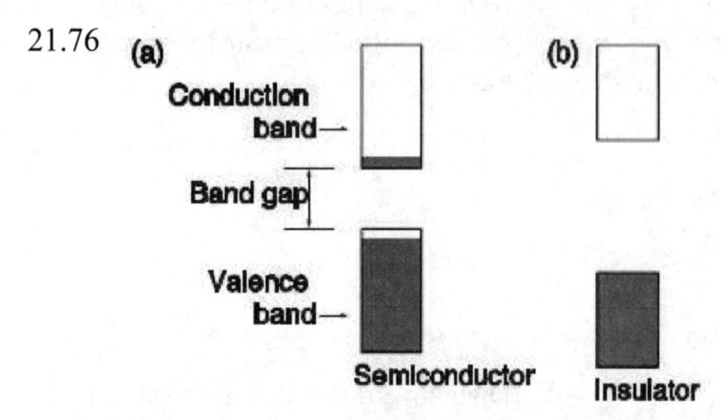

The MOs of a semiconductor are similar to those of an insulator, but the band gap in a semiconductor is smaller. As a result, a few electrons have enough energy to jump the gap and occupy the higher-energy, conduction band. The conduction band is thus partially filled, and the valence band is partially empty. When an electrical potential is applied to a semiconductor, it conducts a small amount of current because the potential can accelerate the electrons in the partially filled bands.

21.77 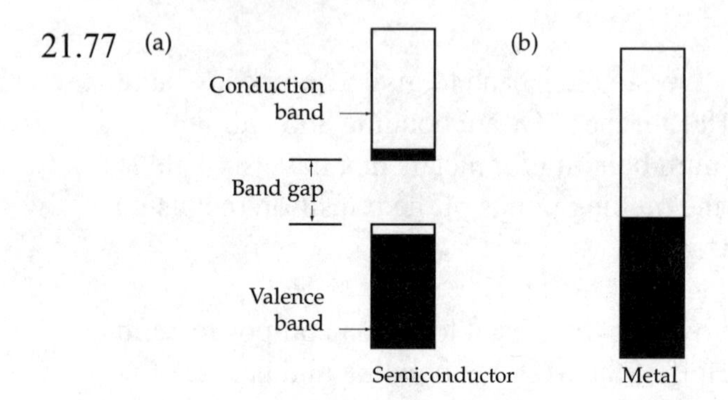

Metals have partially filled bands with no band gap between the highest occupied and lowest unoccupied MOs, which makes them conductors. Semiconductors have a moderate band gap between the valence and conduction bands. A few electrons have enough energy to jump the gap and occupy the conduction band. The result is a partially filled valence and conduction band. When an electrical potential is applied to a semiconductor, it conducts a small amount of current.

21.78 As the band gap increases, the number of electrons able to jump the gap and occupy the higher-energy conduction band decreases, and thus the conductivity decreases.

21.79 The electrical conductivity of a semiconductor increases with increasing temperature because the number of electrons with sufficient energy to occupy the conduction band increases as the temperature rises. At higher temperatures, there are more charge carriers (electrons) in the conduction band and more vacancies in the valence band.

21.80 An n-type semiconductor is a semiconductor doped with a substance with more valence electrons than the semiconductor itself. Si doped with P is an example.

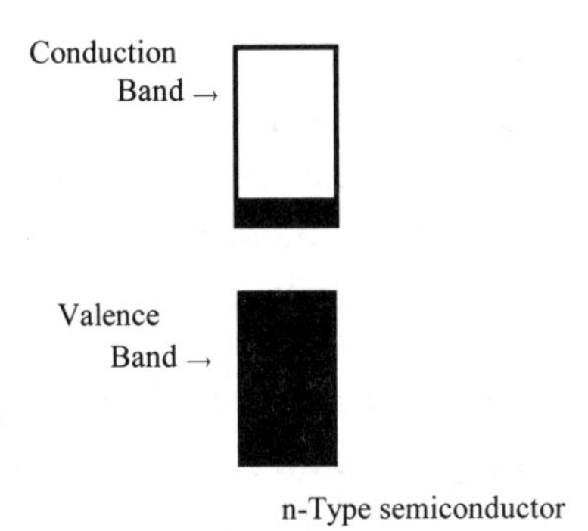

n-Type semiconductor

21.81 A p-type semiconductor is a semiconductor doped with a substance with fewer valence electrons than the semiconductor itself. Si doped with B is an example.

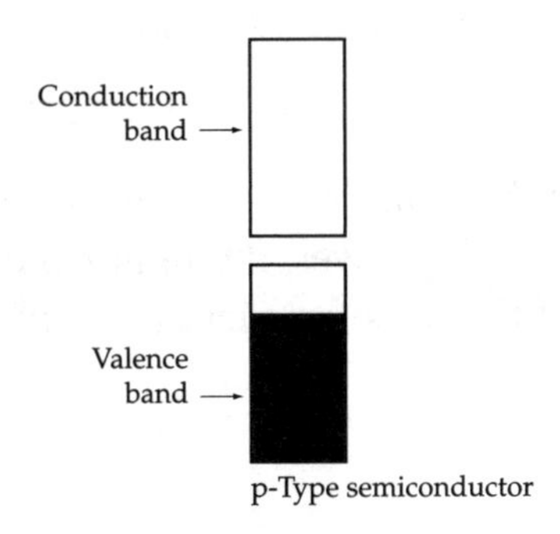

p-Type semiconductor

21.82 In the MO picture, the extra electrons occupy the conduction band. The number of electrons in the conduction band of the doped Ge is much greater than for pure Ge, and the conductivity of the doped semiconductor is correspondingly higher.

21.83 Si doped with Ga is a p-type semiconductor because each Ga atom has one less valence electron than Si. The valence band is thus partially filled, which accounts for the electrical conductivity. The conductivity is greater than that of pure Si because the doped Si has many more positive holes in the valence band. That is, doped Si has more vacant MOs available to which electrons can be excited by an electrical potential.

21.84 (a) p-type (In is electron deficient with respect to Si)
(b) n-type (Sb is electron rich with respect to Ge)
(c) n-type (As is electron rich with respect to gray Sn)

21.85 (a) n-type (As is electron rich with respect to Ge)
(b) p-type (B is electron deficient with respect to Ge)
(c) n-type (Sb is electron rich with respect to Si)

21.86 $Cd(CH_3)_2(g) + H_2Se(g) \rightarrow CdSe(s) + 2\ CH_4(g)$

21.87 $Zn(CH_3)_2(g) + Te(CH_2CH_3)_2(g) \rightarrow ZnTe(s) + 2\ CH_3CH_2CH_3(g)$

21.88 Al_2O_3 < Ge < Ge doped with In < Fe < Cu

21.89 NaCl < Si < gray Sn < gray Sn doped with Sb < Ag

21.90 In a diode, current flows only when the junction is under a forward bias (negative battery terminal on the n-type side). A p-n junction that is part of a circuit and subjected to an alternating potential acts as a rectifier, allowing current to flow in only one direction, thereby converting alternating current to direct current.

21.91 LEDs produce light as a result of the combination of electrons and holes in the region of a p-n junction. This only occurs under a forward bias (negative battery terminal on the n-type side). Under a reverse bias (negative battery terminal on the p-type side), almost no current flows because the electrons and holes move away from one another.

21.92 Both an LED and a photovoltaic cell contain p-n junctions, but the two devices involve opposite processes. An LED converts electrical energy to light; a photovoltaic, or solar, cell converts light to electricity.

21.93 Both LEDs and diode lasers produce light as a result of the combination of electrons and holes in the region of a p-n junction. Compared to an LED, however, the light from a diode laser is more intense, more highly directional, and all of the same frequency and phase.

21.94 $$E = 193 \text{ kJ/mol} \times \frac{1000 \text{ J}}{1 \text{ kJ}} \times \frac{1 \text{ mol}}{6.02 \times 10^{23}} = 3.21 \times 10^{-19} \text{ J}$$

$$\nu = \frac{E}{h} = \frac{3.21 \times 10^{-19} \text{ J}}{6.626 \times 10^{-34} \text{ J·s}} = 4.84 \times 10^{14} \text{ s}^{-1}$$

$$\lambda = \frac{c}{\nu} = \frac{3.00 \times 10^{8} \text{ m/s}}{4.84 \times 10^{14} \text{ s}^{-1}} = 6.20 \times 10^{-7} \text{ m} = 620 \times 10^{-9} \text{ m} = 620 \text{ nm, orange light}$$

21.95 $\lambda = 470 \text{ nm} = 470 \times 10^{-9} \text{ m}$

$$E = h\frac{c}{\lambda} = (6.626 \times 10^{-34} \text{ J·s})\left(\frac{3.00 \times 10^{8} \text{ m/s}}{470 \times 10^{-9} \text{ m}}\right)(6.02 \times 10^{23}/\text{mol}) = 2.55 \times 10^{5} \text{ J/mol}$$

$$E = 2.55 \times 10^{5} \text{ J/mol} \times \frac{1 \text{ kJ}}{1000 \text{ J}} = 255 \text{ kJ/mol}$$

21.96 (a) InN has the smaller band-gap energy because In is larger than Ga.
(b) GaN would emit ultraviolet light and InN would emit red light.

21.97 (a) GaN has the larger band-gap energy because N is smaller than P.
(b) GaN would emit ultraviolet light and GaP would emit green light.

21.98 $GaP_{0.50}As_{0.50}$ < $GaP_{0.80}As_{0.20}$ < $GaP_{1.00}As_{0.00}$

21.99 $Al_{0.05}Ga_{0.95}As$ < $Al_{0.25}Ga_{0.75}As$ < $Al_{0.40}Ga_{0.60}As$

21.100 (a) $E = 107 \text{ kJ/mol} \times \dfrac{1000 \text{ J}}{1 \text{ kJ}} \times \dfrac{1 \text{ mol}}{6.02 \times 10^{23}} = 1.78 \times 10^{-19} \text{ J}$

$$\nu = \frac{E}{h} = \frac{1.78 \times 10^{-19} \text{ J}}{6.626 \times 10^{-34} \text{ J·s}} = 2.68 \times 10^{14} \text{ s}^{-1}$$

$$\lambda = \frac{c}{\nu} = \frac{3.00 \times 10^{8} \text{ m/s}}{2.68 \times 10^{14} \text{ s}^{-1}} = 1.12 \times 10^{-6} \text{ m} = 1120 \times 10^{-9} \text{ m} = 1120 \text{ nm}$$

(b) The wavelength is in the near IR and does not correspond to the highest intensity wavelength in the solar emission spectrum.

21.101 (a) blue-green at ~500 nm = 500×10^{-9} m

(b) $E = \dfrac{hc}{\lambda} = (6.626 \times 10^{-34} \text{ J·s}) \left(\dfrac{3.00 \times 10^{8} \text{ m/s}}{500 \times 10^{-9} \text{ m}} \right) (6.022 \times 10^{23} \text{/mol})$

$E = 2.39 \times 10^{5}$ J/mol = 239 kJ/mol

ZnSe with a band-gap energy of 248 kJ/mol

Superconductors (Section 21.7)

21.102 (1) A superconductor is able to levitate a magnet.
(2) In a superconductor, once an electric current is started, it flows indefinitely without loss of energy. A superconductor has no electrical resistance.

21.103

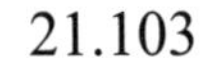

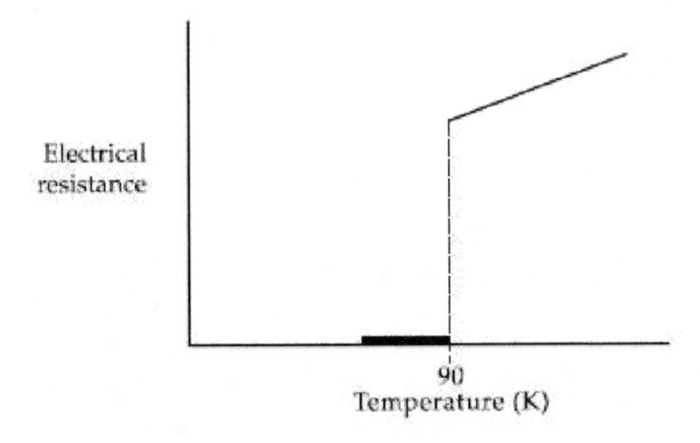

21.104 Some K^{+} ions are surrounded octahedrally by six C_{60}^{3-} ions; others are surrounded tetrahedrally by four C_{60}^{3-} ions.

21.105 In $YBa_2Cu_3O_7$;

atom	coordination numbers
Cu	4 and 5
Y	8
Ba	10

Ceramics and Composites (Sections 21.8 and 21.9)

21.106 Ceramics are inorganic, nonmetallic, nonmolecular solids, including both crystalline and amorphous materials. Ceramics have higher melting points, and they are stiffer, harder, and more resistant to wear and corrosion than are metals.

21.107 The bonding in ceramics consists of either directed covalent bonds or ionic bonds. This is very different from the electron-sea model of bonding for metals.

21.108 Ceramics have higher melting points, and they are stiffer, harder, and more wear resistant than metals because they have stronger bonding. They maintain much of their strength at high temperatures, where metals either melt or corrode because of oxidation.

21.109 Oxide ceramics do not react with oxygen because they are already fully oxidized.

21.110 The brittleness of ceramics is due to strong chemical bonding. In silicon nitride each Si atom is bonded to four N atoms and each N atom is bonded to three Si atoms. The strong, highly directional covalent bonds prevent the planes of atoms from sliding over one another when the solid is subjected to a stress. As a result, the solid cannot deform to relieve the stress. It maintains its shape up to a point, but then the bonds give way suddenly and the material fails catastrophically when the stress exceeds a certain threshold value. By contrast, metals are able to deform under stress because their planes of metal cations can slide easily in the electron sea.

21.111

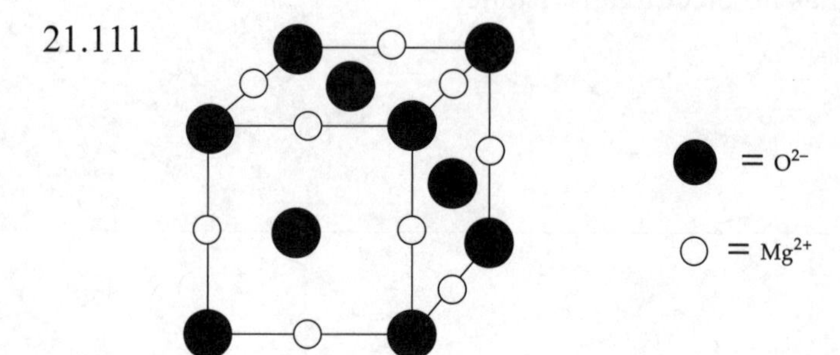

The brittleness of oxide ceramics, as well as their hardness, stiffness, and high melting points, is due to strong ionic bonding. The strong ionic bonding prevents planes of ions from sliding over one another when the solid is subjected to a stress. As a result, the solid cannot deform to relieve the stress. It maintains its shape up to a point, but then the bonds give way suddenly and the material fails catastrophically. For example, if the stress displaces a plane of ions by one-half a unit-cell edge length, the resulting Mg^{2+}–Mg^{2+} and O^{2-}–O^{2-} repulsions cause the crystal to shatter. By contrast, metals are able to deform under stress because their planes of cations can slide easily in the electron sea.

21.112 Ceramic processing is the series of steps that leads from raw material to the finished ceramic object.

21.113 Sintering, which occurs below the melting point, is a process in which the particles of the powder are "welded" together without completely melting. During sintering, the crystal grains grow larger and the density of the material increases as the void spaces between particles disappear.

21.114 $Zr[OCH(CH_3)_2]_4 + 4\ H_2O \rightarrow Zr(OH)_4 + 4\ HOCH(CH_3)_2$

21.115 $Zn(OCH_2CH_3)_2 + 2\ H_2O \rightarrow Zn(OH)_2 + 2\ HOCH_2CH_3$

21.116 $(HO)_3Si–O–H + H–O–Si(OH)_3 \rightarrow (HO)_3Si–O–Si(OH)_3 + H_2O$
Further reactions of this sort give a three-dimensional network of Si–O–Si bridges. On heating, SiO_2 is obtained.

21.117 $(HO)_2Y–O–H + H–O–Y(OH)_2 \rightarrow (HO)_2Y–O–Y(OH)_2 + H_2O$
On drying and sintering, Y_2O_3 is obtained.

21.118 $2\ Ti(BH_4)_3(soln) \rightarrow 2\ TiB_2(s) + B_2H_6(g) + 9\ H_2(g)$

21.119 $Ge(OCH_2CH_3)_4(soln) + 2\ H_2S(soln) \rightarrow GeS_2(s) + 4\ CH_3CH_2OH(soln)$

21.120 $3\ SiCl_4(g) + 4\ NH_3(g) \rightarrow Si_3N_4(s) + 12\ HCl(g)$

21.121 $2\ BCl_3(g) + 3\ H_2(g) \xrightarrow{W} 2\ B(s) + 6\ HCl(g)$

21.122 Graphite/epoxy composites are good materials for making tennis rackets and golf clubs because of their high strength-to-weight ratios.

21.123 Silicon carbide-reinforced alumina is stronger and tougher than pure alumina. The silicon carbide whiskers have great strength along their axis because most of the chemical bonds are aligned in that direction. In addition the silicon carbide whiskers can prevent microscopic cracks from propagating.

Chapter Problems

21.124 $\Delta G = \Delta H - T\Delta S = -160.8\text{ kJ} - (298\text{ K})(-0.410\text{ kJ/K}) = -38.6\text{ kJ}$
Because $\Delta G < 0$, the reaction is spontaneous at 25 °C (298 K).
Set $\Delta G = 0$ and solve for T to find the temperature at which the reaction becomes nonspontaneous.

$$0 = \Delta H - T\Delta S; \quad T = \frac{\Delta H}{\Delta S} = \frac{-160.8\text{ kJ}}{-0.410\text{ kJ/K}} = 392\text{ K} = 119\text{ °C}$$

21.125 Fe_2O_3, 412.1 kJ; Al_2O_3, 838.0 kJ; Fe and Al cannot be obtained by decomposition of their oxides under standard state conditions because $\Delta G°$ is positive for both reactions.

21.126 $2\ Eu^{3+}(aq) + Zn(s) \rightarrow 2\ Eu^{2+}(aq) + Zn^{2+}(aq)$
$Eu^{2+}(aq) + SO_4^{2-}(aq) \rightarrow EuSO_4(s)$

21.127 (a) $E = 3.44\ \text{eV} \times \frac{96.485\ \text{kJ/mol}}{1\ \text{eV}} = 331.9\ \text{kJ/mol}$

$$E = 331.9\ \text{kJ/mol} \times \frac{1000\ \text{J}}{1\ \text{kJ}} \times \frac{1\ \text{mol}}{6.02 \times 10^{23}} = 5.51 \times 10^{-19}\ \text{J}$$

$$\nu = \frac{E}{h} = \frac{5.51 \times 10^{-19}\ \text{J}}{6.626 \times 10^{-34}\ \text{J}\cdot\text{s}} = 8.32 \times 10^{14}\ \text{s}^{-1}$$

$$\lambda = \frac{c}{\nu} = \frac{3.00 \times 10^{8}\ \text{m/s}}{8.32 \times 10^{14}\ \text{s}^{-1}} = 3.61 \times 10^{-7}\ \text{m} = 361 \times 10^{-9}\ \text{m} = 361\ \text{nm}$$

(b) Blue light has a longer wavelength than 361 nm, and so the band of $Ga_xIn_{1-x}N$ is less than the band gap of GaN.

21.128 The chemical composition of the alkaline earth minerals is that of metal sulfates and sulfites, MSO_4 and MSO_3.

21.129 77 K is the boiling point of the readily available liquid N_2.

21.130 Band theory better explains how the number of valence electrons affects properties such as melting point and hardness.

21.131 W [Xe] $4f^{14}\ 5d^4\ 6s^2$ Au [Xe] $4f^{14}\ 5d^{10}\ 6s^1$
These facts are better explained with the MO band model. Transition metals have a d band that can overlap the s band to give a composite band consisting of six MOs per metal atom. Half of the MOs are bonding and half are antibonding, and thus one expects maximum bonding for metals that have six valence electrons per metal atom. Accordingly, the hardness and melting points of the transition metals go through a maximum at or near group 6B.

21.132 V [Ar] $3d^3\ 4s^2$ Zn [Ar] $3d^{10}\ 4s^2$
Transition metals have a d band that can overlap the s band to give a composite band consisting of six MOs per metal atom. Half of the MOs are bonding and half are antibonding. Strong bonding and a high enthalpy of vaporization are expected for V because almost all of the bonding MOs are occupied and all of the antibonding MOs are empty. Weak bonding and a low enthalpy of vaporization are expected for Zn because both the bonding and the antibonding MOs are occupied.

21.133 (a) MgO, insulator
(b) Si doped with Sb (Sb is electron rich with respect to Si), n-type semiconductor
(c) white tin, metallic conductor
(d) Ge doped with Ga (Ga is electron deficient with respect to Ge), p-type semiconductor
(e) stainless steel, metallic conductor

21.134 With a band gap of 130 kJ/mol, GaAs is a semiconductor. Because Ge lies between Ga and As in the periodic table, GaAs is isoelectronic with Ge.

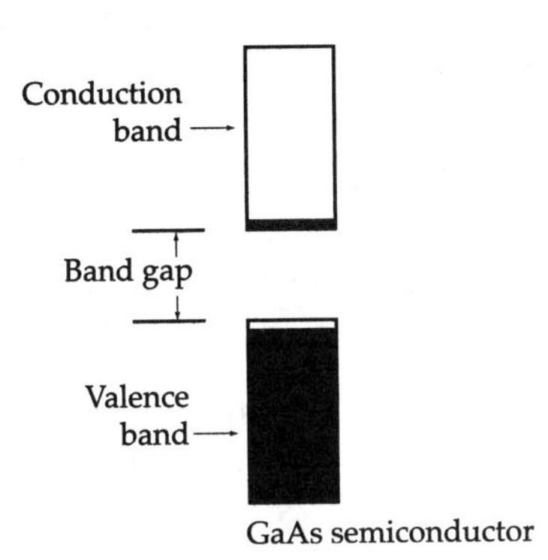

21.135 (a) $P_4(l) + 5\ O_2(g) \rightarrow P_4O_{10}(l)$
$6\ CaO(s) + P_4O_{10}(l) \rightarrow 2\ Ca_3(PO_4)_2(l)$ (slag)
(b) mass $P_4 = (0.0045)(3.4 \times 10^3\ kg) = 15.3\ kg\ P_4$

$$\text{mol } P_4O_{10} = 15.3 \times 10^3\ g \times \frac{1\ mol\ P_4}{123.9\ g} \times \frac{1\ mol\ P_4O_{10}}{1\ mol\ P_4} = 123.5\ mol\ P_4O_{10}$$

$$\text{mass CaO} = 123.5\ mol\ P_4O_{10} \times \frac{6\ mol\ CaO}{1\ mol\ P_4O_{10}} \times \frac{56.08\ g\ CaO}{1\ mol\ CaO} \times \frac{1\ kg}{1000\ g} = 42\ kg\ CaO$$

21.136 $YBa_2Cu_3O_7$, 666.20; $Cu(OCH_2CH_3)_2$, 153.67
$Y(OCH_2CH_3)_3$, 224.09; $Ba(OCH_2CH_3)_2$, 227.45

$$\text{mol } Cu(OCH_2CH_3)_2 = 75.4\ g \times \frac{1\ mol}{153.67\ g} = 0.4907\ mol\ Cu(OCH_2CH_3)_2$$

mass $Y(OCH_2CH_3)_3 = 0.4907\ mol\ Cu(OCH_2CH_3)_2 \times$

$$\frac{1\ mol\ Y(OCH_2CH_3)_3}{3\ mol\ Cu(OCH_2CH_3)_2} \times \frac{224.09\ g\ Y(OCH_2CH_3)_3}{1\ mol\ Y(OCH_2CH_3)_3} = 36.7\ g\ Y(OCH_2CH_3)_3$$

mass $Ba(OCH_2CH_3)_2 = 0.4907\ mol\ Cu(OCH_2CH_3)_2 \times$

$$\frac{2\ mol\ Ba(OCH_2CH_3)_2}{3\ mol\ Cu(OCH_2CH_3)_2} \times \frac{227.45\ g\ Ba(OCH_2CH_3)_2}{1\ mol\ Ba(OCH_2CH_3)_2} = 74.4\ g\ Ba(OCH_2CH_3)_2$$

mass $YBa_2Cu_3O_7 = 0.4907\ mol\ Cu(OCH_2CH_3)_2 \times$

$$\frac{1\ mol\ YBa_2Cu_3O_7}{3\ mol\ Cu(OCH_2CH_3)_2} \times \frac{666.20\ g\ YBa_2Cu_3O_7}{1\ mol\ YBa_2Cu_3O_7} = 109\ g\ YBa_2Cu_3O_7$$

21.137 (a) In an n-type InP semiconductor the valence band is completely filled and the conduction band is partially full. Cd has only two 5s and no 5p electrons. Adding Cd to the n-type InP semiconductor would add positive holes that would combine with free electrons. This results in a decrease in the number of electrons in the conduction band and a decrease in the conductivity.

(b) In a p-type InP semiconductor the valence band is partially filled and the conduction band is empty. The charge carriers are positive holes. Se has 6 valence electrons. Adding Se to the p-type InP semiconductor would add electrons that would combine with positive holes. This results in a decrease in the number of positive holes and a decrease in the conductivity.

21.138 (a) $6\ Al(OCH_2CH_3)_3 + 2\ Si(OCH_2CH_3)_4 + 26\ H_2O \rightarrow$

$$6\ Al(OH)_3(s) + 2\ Si(OH)_4(s) + 26\ HOCH_2CH_3$$

sol

(b) H_2O is eliminated from the sol through a series of reactions linking the sol particles together through a three-dimensional network of O bridges to form the gel.

$(HO)_2Al–O–H + H–O–Si(OH)_3 \rightarrow (HO)_2Al–O–Si(OH)_3 + H_2O$

(c) The remaining H_2O and solvent are removed from the gel by heating to produce the ceramic, $3\ Al_2O_3 \cdot 2\ SiO_2$.

21.139 (a)

$H_2C{=}CH(CN)$ + $H_2C{=}CH(CN)$ ⟶ —CH_2—$CH(CN)$—CH_2—$CH(CN)$—

(b) $\Delta H° = D_{C=C} - 2\ D_{C–C} = 611\ kJ - 2(350)\ kJ = -89\ kJ/unit$; exothermic

21.141

Zn has 2 valence electrons. Ga has 3 valence electrons. When ZnSe is doped with Ga, there are extra electrons in the semiconductor. Extra electrons lead to n-type semiconductors.

21.142 (a)

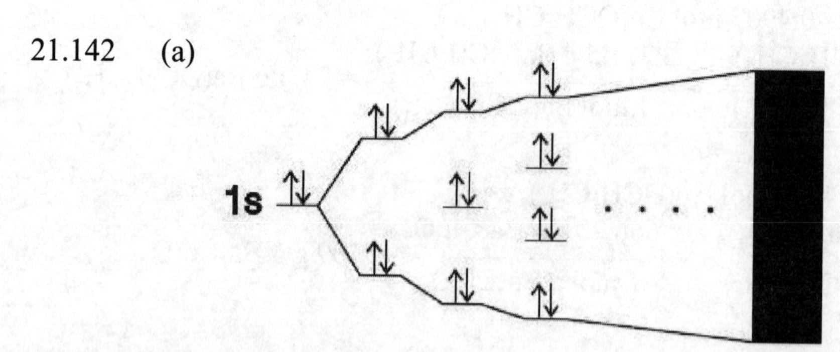

This material is an insulator because all MOs are filled, preventing the movement of electrons.

(b)

Neutral hydrogen atoms have only 1 valence electron, compared with 2 in H^-. Partially empty antibonding MOs will allow the movement of electrons, so the doped material will be a conductor.

(c) The missing electrons in the doped material create "holes" that are positive charge carriers. This type of doped material is a p-type semiconductor.

21.143 (a)

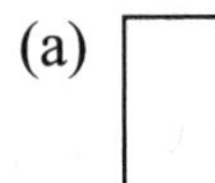

(b) The electrical conductivity of a semiconductor increases with increasing temperature because the number of electrons with sufficient energy to occupy the conduction band increases as the temperature rises.

(c) (i) The conductivity of GaAs would increase when doped with Zn because it would produce positive holes in the valence band. Zn doped GaAs would be a p-type semiconductor.

(ii) The conductivity of GaAs would increase when doped with S because it would put extra electrons in the conduction band. S doped GaAs would be an n-type semiconductor.

21.144 (a) Because nitrogen has one more valence electron than carbon, nitrogen-doped diamond would be an n-type semiconductor.

(b)

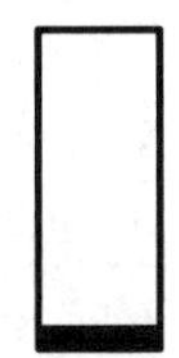

diamond (insulator)

nitrogen-doped diamond (n-type semiconductor)

(c) 425 nm = 425 x 10^{-9} m

$$E = \frac{hc}{\lambda} = (6.626 \times 10^{-34}\ J \cdot s)\left(\frac{3.00 \times 10^{8}\ m/s}{425 \times 10^{-9}\ m}\right)\left(\frac{1\ kJ}{1000\ J}\right)(6.022 \times 10^{23}\ /mol)$$

E = 282 kJ/mol

Multiconcept Problems

21.145 (a) $8.894\ g\ Al \times \frac{1\ cm^3}{2.699\ g\ Al} = 3.295\ cm^3$

$3.295\ cm^3 = (36.5\ cm)^2 \times h$

$$h = \frac{3.295\ cm^3}{(36.5\ cm)^2} = 2.47 \times 10^{-3}\ cm$$

$$h = 2.47 \times 10^{-3}\ cm \times \frac{1\ m}{100\ cm} = 2.47 \times 10^{-5}\ m$$

(b) d = 143 pm = 143 x 10^{-12} m

For a face-centered cube, the unit cell face diagonal = 4r = $\sqrt{2} \cdot d$

$$\text{unit cell edge} = d = \frac{4r}{\sqrt{2}} = \frac{(4)(143 \times 10^{-12}\ m)}{\sqrt{2}} = 4.04 \times 10^{-10}\ m$$

$$\frac{2.47 \times 10^{-5}\ m}{4.04 \times 10^{-10}\ m/\text{unit cell}} = 6.11 \times 10^{4}\ \text{unit cells thick}$$

21.146 (a) Eu^{2+}, [Xe] $4f^7$ (b) $BM = \sqrt{n(n+2)} = \sqrt{7(7+2)} = 7.94$ BM

21.147 $Cr_2O_7^{2-}(aq) + 6\ Fe^{2+}(aq) + 14\ H^+(aq) \rightarrow 6\ Fe^{3+}(aq) + 2\ Cr^{3+}(aq) + 7\ H_2O(l)$

mol $Cr_2O_7^{2-}$ = (0.038 89 L)(0.018 54 mol/L) = 7.210 x 10^{-4} mol $Cr_2O_7^{2-}$

$$\text{mass Fe} = 7.210 \times 10^{-4}\ mol\ Cr_2O_7^{2-} \times \frac{6\ mol\ Fe^{2+}}{1\ mol\ Cr_2O_7^{2-}} \times \frac{55.847\ g\ Fe^{2+}}{1\ mol\ Fe^{2+}} = 0.2416\ g\ Fe^{2+}$$

$$\text{mass \% Fe} = \frac{0.2416\ g}{0.3249\ g} \times 100\% = 74.36\%\ Fe$$

21.148 660 nm = 660 x 10^{-9} m and 3.0 mW = 3.0 x 10^{-3} W = 3.0 x 10^{-3} J/s

$$E = h\frac{c}{\lambda} = (6.626 \times 10^{-34}\ J \cdot s)\left(\frac{3.00 \times 10^{8}\ m/s}{660 \times 10^{-9}\ m}\right) = 3.0 \times 10^{-19}\ J/photon$$

$$\#\ \text{of photons/s} = \frac{3.0 \times 10^{-3}\ J/s}{3.0 \times 10^{-19}\ J/photon} = 1.0 \times 10^{16}\ photons/s$$

# of electrons/s = # of photons/s = 1.0 x 10^{16} electrons/s

$$\#\ \text{of moles of electrons/s} = 1.0 \times 10^{16}\ electrons/s \times \frac{1\ mol\ e^-}{6.02 \times 10^{23}\ e^-} = 1.7 \times 10^{-8}\ mol\ e^-/s$$

$$A = 1.7 \times 10^{-8}\ mol\ e^-/s \times \frac{96{,}500\ C}{1\ mol\ e^-} = 0.0016\ C/s = 0.0016\ A = 1.6 \times 10^{-3}\ A = 1.6\ mA$$

21.149 (a) 431 pm = 431×10^{-12} m

There are 4 oxygen atoms in the face-centered cubic unit cell.

mass of unit cell = $(5.75\ g/cm^3)(431 \times 10^{-12}\ m)^3(100\ cm/1\ m)^3 = 4.604 \times 10^{-22}$ g

$$\text{mass of Fe in unit cell} = (4.604 \times 10^{-22}\ g) - 4\ \text{O atoms} \times \frac{15.9994\ g\ O}{6.022 \times 10^{23}\ \text{O atoms}}$$

$$= 3.541 \times 10^{-22}\ g\ Fe$$

$$\text{number of Fe atoms in unit cell} = 3.541 \times 10^{-22}\ g\ Fe \times \frac{6.022 \times 10^{23}\ \text{Fe atoms}}{55.847\ g\ Fe}$$

$$= 3.818\ \text{Fe atoms}$$

$$\text{For } Fe_xO,\ x = \frac{3.818\ \text{Fe atoms}}{4\ \text{O atoms}} = 0.955$$

(b) The average oxidation state of Fe = $\frac{+2}{0.955} = 2.094$

(c) Let X equal the fraction of Fe^{3+} and Y equal the fraction of Fe^{2+} in wustite.

So, X + Y = 1 and 3X + 2Y = 2.094

Y = 1 – X

3X + 2(1 –X) = 2.094

3X + 2 – 2X = 2.094

X + 2 = 2.094

X = 2.094 – 2 = 0.094

9.4% of the Fe in wustite is Fe^{3+}.

(d) d = 431 pm

$$d = \frac{n\lambda}{2\sin\theta} = \frac{3 \cdot 70.93\,\text{pm}}{2\sin\theta} = 431\ \text{pm}$$

$$\sin\theta = \frac{3 \cdot 70.93\,\text{pm}}{2 \cdot 431\ \text{pm}} = 0.247 \text{ and } \theta = 14.3^\circ$$

(e) The presence of Fe^{3+} in the semiconductor leads to missing electrons that create "holes", which are positive charge carriers. This type of doped material is a p-type semiconductor.

21.150 $SiO_2(s) + 2\ C(s) \rightarrow Si(s) + 2\ CO(g)$

(a) $\Delta H^\circ = 2\ \Delta H^\circ_f(CO) - \Delta H^\circ_f(SiO_2)$

ΔH° = (2 mol)(–110.5 kJ/mol) – (1 mol)(–910.7 kJ/mol) = 689.7 kJ

$\Delta S^\circ = [S^\circ(Si) + 2\ S^\circ(CO)] - [S^\circ(SiO_2) + 2\ S^\circ(C)]$

ΔS° = [(1 mol)(18.8 J/(K· mol)) + (2 mol)(197.6 J/(K· mol))]
– [(1 mol)(41.5 J/(K · mol)) + (2 mol)(5.7 J/(K · mol))]

$\Delta S^\circ = 361.1\ J/K = 361.1 \times 10^{-3}$ kJ/K

$\Delta G^\circ = \Delta H^\circ - T\Delta S^\circ = 689.7\ kJ - (298.15\ K)(361.1 \times 10^{-3}\ kJ/K) = 582.0$ kJ

(b) The reaction is endothermic because $\Delta H^\circ > 0$.

(c) The number of moles of gas increases from 0 to 2 mol, therefore, $\Delta S^\circ > 0$.

(d) Because $\Delta G^\circ > 0$, the reaction is nonspontaneous at 25 °C and 1 atm pressure of CO.

(e) To determine the crossover temperature, set $\Delta G^o = 0$ and solve for T.
$\Delta G^o = 0 = \Delta H^o - T\Delta S^o$
$$\Delta H^o = T\Delta S^o; \qquad T = \frac{\Delta H^o}{\Delta S^o} = \frac{689.7 \text{ kJ}}{361.1 \times 10^{-3} \text{ kJ/K}} = 1910 \text{ K} = 1637 \text{ °C}$$

21.151 $Ni(s) + 4\,CO(g) \rightleftarrows Ni(CO)_4(g)$
$\Delta H^o = -160.8$ kJ; $\Delta S^o = -410$ J/K $= -410 \times 10^{-3}$ kJ/K
(a) 150 °C = 423 K
$\Delta G^o = \Delta H^o - T\Delta S^o = -160.8 \text{ kJ} - (423 \text{ K})(-410 \times 10^{-3} \text{ kJ/K}) = +12.6 \text{ kJ}$
$\Delta G^o = -RT \ln K$
$$\ln K = \frac{-\Delta G^o}{RT} = \frac{-12.6 \text{ kJ/mol}}{[8.314 \times 10^{-3} \text{ kJ/(K}\cdot\text{mol)}](423 \text{ K})} = -3.58$$
$K = K_p = e^{-3.58} = 0.028$
(b) 230 °C = 503 K
$\Delta G^o = \Delta H^o - T\Delta S^o = -160.8 \text{ kJ} - (503 \text{ K})(-410 \times 10^{-3} \text{ kJ/K}) = +45.4 \text{ kJ}$
$\Delta G^o = -RT \ln K$
$$\ln K = \frac{-\Delta G^o}{RT} = \frac{-45.4 \text{ kJ/mol}}{[8.314 \times 10^{-3} \text{ kJ/(K}\cdot\text{mol)}](503 \text{ K})} = -10.86$$
$K = K_p = e^{-10.86} = 1.9 \times 10^{-5}$
(c) ΔS^o is large and negative because as the reaction proceeds in the forward direction, the number of moles of gas decrease from four to one.
Because ΔS^o is negative, $-T\Delta S^o$ is positive, and as T increases, ΔG^o becomes more positive because $\Delta G^o = \Delta H^o - T\Delta S^o$.
(d) The reaction is exothermic because ΔH^o is negative.
$Ni(s) + 4\,CO(g) \rightleftarrows Ni(CO)_4(g) + \text{heat}$
Heat is added as the temperature is raised and the reaction proceeds in the reverse direction to relieve this stress, as predicted by Le Châtelier's principle. As the reverse reaction proceeds, the partial pressure of CO increases and the partial pressure of $Ni(CO)_4$ decreases. K_p decreases as calculated because $K_p = \dfrac{P_{Ni(CO)_4}}{(P_{CO})^4}$.

21.152 $C(s) + CO_2(g) \rightarrow 2\,CO(g)$
(a) CO_2, 44.01
$$\text{mol } CO_2 = 100.0 \text{ g } CO_2 \times \frac{1 \text{ mol } CO_2}{44.01 \text{ g } CO_2} = 2.272 \text{ mol } CO_2$$
$\Delta H^o = [2\,\Delta H^o_f(CO)] - \Delta H^o_f(CO_2)$
$\Delta H^o = (2 \text{ mol})(-110.5 \text{ kJ/mol}) - (1 \text{ mol})(-393.5 \text{ kJ/mol}) = 172.5 \text{ kJ}$
$\Delta S^o = [2\,S^o(CO)] - [S^o(C) + S^o(CO_2)]$
$\Delta S^o = (2 \text{ mol})(197.6 \text{ J/(K}\cdot\text{mol)}) - [(1 \text{ mol})(5.7 \text{ J/(K}\cdot\text{mol)}) + (1 \text{ mol})(213.6 \text{ J/(K}\cdot\text{mol)})]$
$\Delta S^o = 175.9 \text{ J/K} = 175.9 \times 10^{-3} \text{ kJ/K}$
at 500 °C (773 K):
$\Delta G^o = \Delta H^o - T\Delta S^o = 172.5 \text{ kJ} - (773 \text{ K})(175.9 \times 10^{-3} \text{ kJ/K}) = 36.5 \text{ kJ}$

$$\ln K_p = \frac{-\Delta G^o}{RT} = \frac{-36.5 \text{ kJ/mol}}{[8.314 \times 10^{-3} \text{ kJ/(K} \cdot \text{mol)}](773 \text{ K})} = -5.68$$

$K_p = e^{-5.68} = 3.4 \times 10^{-3}$

$$P_{CO_2} = \frac{nRT}{V} = \frac{(2.272 \text{ mol})\left(0.082\ 06 \frac{\text{L} \cdot \text{atm}}{\text{K} \cdot \text{mol}}\right)(773 \text{ K})}{50.00 \text{ L}} = 2.88 \text{ atm}$$

	$C(s) + CO_2(g)$	$\rightleftarrows$	$2\ CO(g)$
initial (atm)	2.88		0
change (atm)	$-x$		$+2x$
equil (atm)	$2.88 - x$		$2x$

$$K_p = \frac{(P_{CO})^2}{P_{CO_2}} = 3.4 \times 10^{-3} = \frac{(2x)^2}{2.88 - x}$$

$4x^2 + (3.4 \times 10^{-3})x - 9.79 \times 10^{-3} = 0$

Use the quadratic formula to solve for x.

$$x = \frac{-(3.4 \times 10^{-3}) \pm \sqrt{(3.4 \times 10^{-3})^2 - (4)(4)(-9.79 \times 10^{-3})}}{2(4)} = \frac{(-3.4 \times 10^{-3}) \pm (0.396)}{8}$$

$x = 0.049\ 08$ and $-0.049\ 93$

Of the two solutions for x, only the positive value of x has physical meaning because 2x is the partial pressure of CO.

$P_{CO_2} = 2.88 - x = 2.88 - 0.049\ 08 = 2.831$ atm

$P_{CO} = 2x = 2(0.049\ 08) = 0.0982$ atm

$P_{total} = P_{CO_2} + P_{CO} = 2.831 + 0.0982 = 2.93$ atm

$$[CO] = \frac{n}{V} = \frac{P}{RT} = \frac{(0.0982 \text{ atm})}{\left(0.082\ 06 \frac{\text{L} \cdot \text{atm}}{\text{K} \cdot \text{mol}}\right)(773 \text{ K})} = 1.5 \times 10^{-3} \text{ M}$$

$$[CO_2] = \frac{n}{V} = \frac{P}{RT} = \frac{(2.831 \text{ atm})}{\left(0.082\ 06 \frac{\text{L} \cdot \text{atm}}{\text{K} \cdot \text{mol}}\right)(773 \text{ K})} = 4.46 \times 10^{-2} \text{ M}$$

(b) at 1000 °C (1273 K):

$\Delta G^o = \Delta H^o - T\Delta S^o = 172.5 \text{ kJ} - (1273 \text{ K})(175.9 \times 10^{-3} \text{ kJ/K}) = -51.4 \text{ kJ}$

$$\ln K_p = \frac{-\Delta G^o}{RT} = \frac{-(-51.4 \text{ kJ/mol})}{[8.314 \times 10^{-3} \text{ kJ/(K} \cdot \text{mol)}](1273 \text{ K})} = 4.86$$

$K_p = e^{4.86} = 1.3 \times 10^2$

$$P_{CO_2} = \frac{nRT}{V} = \frac{(2.272 \text{ mol})\left(0.082\ 06 \frac{\text{L} \cdot \text{atm}}{\text{K} \cdot \text{mol}}\right)(1273 \text{ K})}{50.00 \text{ L}} = 4.75 \text{ atm}$$

	$C(s) + CO_2(g) \rightleftarrows 2\,CO(g)$	
initial (atm)	4.75	0
change (atm)	–x	+2x
equil (atm)	4.75 – x	2x

$$K_p = \frac{(P_{CO})^2}{P_{CO_2}} = 1.3 \times 10^2 = \frac{(2x)^2}{4.75 - x}$$

$4x^2 + (1.3 \times 10^2)x - 617.5 = 0$

Use the quadratic formula to solve for x.

$$x = \frac{-(1.3 \times 10^2) \pm \sqrt{(1.3 \times 10^2)^2 - (4)(4)(-617.5)}}{2(4)} = \frac{(-1.3 \times 10^2) \pm (163.6)}{8}$$

x = 4.200 and –36.70

Of the two solutions for x, only the positive value of x has physical meaning because 2x is the partial pressure of CO.

$P_{CO_2} = 4.75 - x = 4.75 - 4.200 = 0.55$ atm

$P_{CO} = 2x = 2(4.200) = 8.40$ atm

$P_{total} = P_{CO_2} + P_{CO} = 0.55 + 8.40 = 8.95$ atm

$$[CO] = \frac{n}{V} = \frac{P}{RT} = \frac{(8.40\text{ atm})}{\left(0.082\ 06\ \frac{L \cdot atm}{K \cdot mol}\right)(1273\text{ K})} = 8.04 \times 10^{-2}\text{ M}$$

$$[CO_2] = \frac{n}{V} = \frac{P}{RT} = \frac{(0.55\text{ atm})}{\left(0.082\ 06\ \frac{L \cdot atm}{K \cdot mol}\right)(1273\text{ K})} = 5.3 \times 10^{-3}\text{ M}$$

(c) $\Delta G^\circ = \Delta H^\circ - T\Delta S^\circ$; The equilibrium shifts to the right with increasing temperature because ΔH° is positive (endothermic reaction) and ΔS° is positive. Therefore, ΔG° is more negative at higher temperatures.

21.153 (a) $(NH_4)_2Zn(CrO_4)_2(s) \rightarrow ZnCr_2O_4(s) + N_2(g) + 4\,H_2O(g)$

(b) $\text{mol } (NH_4)_2Zn(CrO_4)_2 = 10.36\text{ g } (NH_4)_2Zn(CrO_4)_2 \times \frac{1\text{ mol } (NH_4)_2Zn(CrO_4)_2}{333.45\text{ g } (NH_4)_2Zn(CrO_4)_2}$

$= 0.03107\text{ mol } (NH_4)_2Zn(CrO_4)_2$

$\text{mass } ZnCr_2O_4 = 0.03107\text{ mol } (NH_4)_2Zn(CrO_4)_2 \times \frac{1\text{ mol } ZnCr_2O_4}{1\text{ mol } (NH_4)_2Zn(CrO_4)_2}$

$\times \frac{233.38\text{ g } ZnCr_2O_4}{1\text{ mol } ZnCr_2O_4} = 7.251\text{ g } ZnCr_2O_4$

(c) $\text{mol } N_2 + \text{mol } H_2O = 0.03107\text{ mol } (NH_4)_2Zn(CrO_4)_2 \times \frac{5\text{ mol gas}}{1\text{ mol } (NH_4)_2Zn(CrO_4)_2}$

= 0.1554 mol gaseous by-products

292 °C = 565 K;

$$PV = nRT; \qquad V = \frac{nRT}{P} = \frac{(0.1554\ \text{mol})\left(0.082\ 06\ \frac{\text{L}\cdot\text{atm}}{\text{K}\cdot\text{mol}}\right)(565\ \text{K})}{\left(745\ \text{mm Hg} \times \frac{1.00\ \text{atm}}{760\ \text{mm Hg}}\right)} = 7.35\ \text{L}$$

(d) A face-centered cubic unit cell has four octahedral holes and eight tetrahedral holes. This unit cell contains one Zn^{2+} ion in a tetrahedral hole and two Cr^{3+} ions in octahedral holes, therefore 1/8 of the tetrahedral holes and 1/2 of the octahedral holes are filled.

(e)

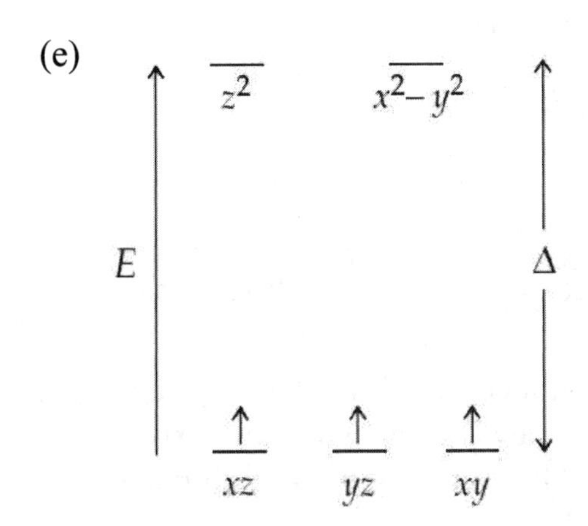

Octahedral Cr^{3+} has three unpaired electrons in the lower-energy d orbitals (xy, xz, yz). Cr^{3+} can absorb visible light to promote one of these d electrons to one of the higher-energy d orbitals making this compound colored. All of the d orbitals in Zn^{2+} are filled and no d electrons can be promoted, consequently the Zn^{2+} ion does not contribute to the color.

21.154 (a)

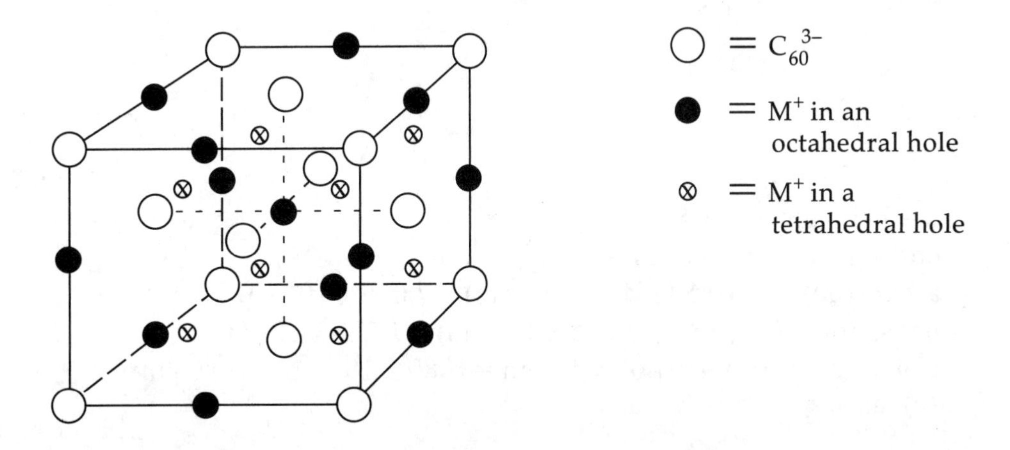

(b) There are 4 C_{60}^{3-} ions, 4 octahedral holes, and 8 tetrahedral holes per unit cell.

(c) Octahedral holes: (1/2,1/2,1/2), (1/2,0,0), (0,1/2,0), (0,0,1/2)
Tetrahedral holes: (1/4,1/4,1/4), (3/4,1/4,1/4), (1/4,3/4,1/4), (3/4,3/4,1/4), (1/4,1/4,3/4), (3/4,1/4,3/4), (1/4,3/4,3/4), (3/4,3/4,3/4)

(d) Let the unit cell edge = a.
The face diagonal is equal to 4R = 4(500 pm) = 2000 pm
$a^2 + a^2 = (2000)^2$; $2a^2 = 4 \times 10^6$; $a^2 = 2 \times 10^6$; $a = \sqrt{2 \times 10^6} = 1414$ pm
$a = 2R(C_{60}^{3-}) + 2R(\text{octahedral hole}) = 1414$ pm

$$R(\text{octahedral hole}) = \frac{1414 \text{ pm} - 2R(C_{60}^{3-})}{2} = \frac{1414 \text{ pm} - 2(500 \text{ pm})}{2} = 207 \text{ pm}$$

The tetrahedron that defines the tetrahedral hole can be thought of as being found inside a cube with edge = a/2 = 707 pm. This cube is located in one corner of the unit cell. The face diagonal of this cube = $2R(C_{60}^{3-})$ = 1000 pm.
The body diagonal of this cube = $\sqrt{707^2 + 1000^2} = 1225$ pm
Body diagonal = $2R(C_{60}^{3-}) + 2R(\text{tetrahedral hole}) = 1225$ pm

$$R(\text{tetrahedral hole}) = \frac{1225 \text{ pm} - 2R(C_{60}^{3-})}{2} = \frac{1225 \text{ pm} - 2(500 \text{ pm})}{2} = 112 \text{ pm}$$

(e) Na^+ will fit into the octahedral and tetrahedral holes without expanding the C_{60}^{3-} framework. K^+ and Rb^+ will fit into the octahedral holes without expanding the C_{60}^{3-} framework but will fit into the tetrahedral holes only if the C_{60}^{3-} framework is expanded.

21.155

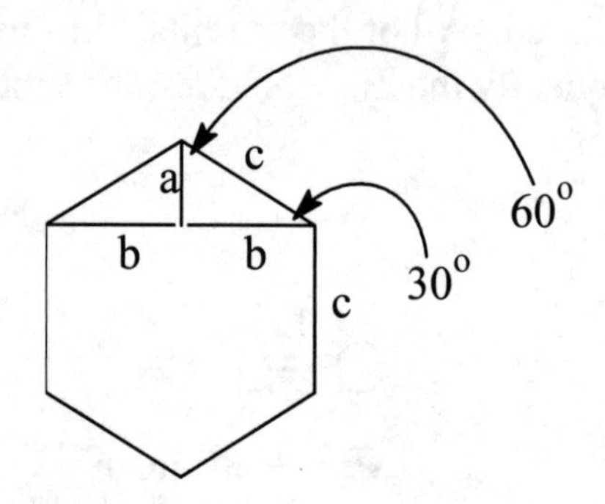

c = 141.5 pm = 141.5×10^{-12} m
cos(60) = a/c and sin(60) = b/c
$a = \cos(60) \cdot c = (0.5)(141.5 \times 10^{-12} \text{ m}) = 7.075 \times 10^{-11}$ m
$b = \sin(60) \cdot c = (0.866)(141.5 \times 10^{-12} \text{ m}) = 1.225 \times 10^{-10}$ m
diameter = 1.08 nm = 1.08×10^{-9} m = 1080×10^{-12} m = 1080 pm

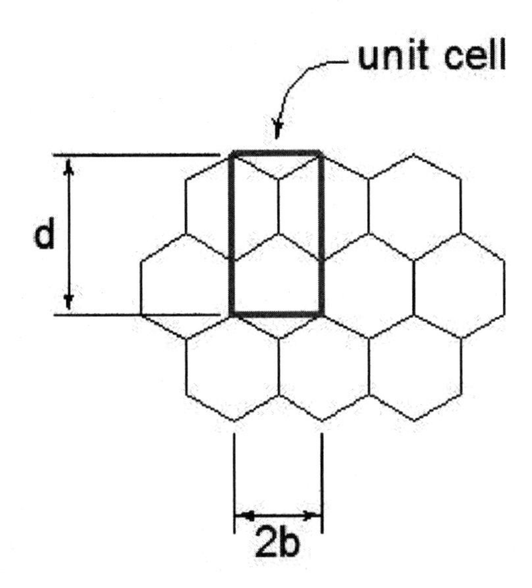

$d = 2a + 2c = 2(7.075 \times 10^{-11}\ m) + 2(141.5 \times 10^{-12}\ m) = 4.245 \times 10^{-10}\ m$
$d = 424.5 \times 10^{-12}\ m = 424.5\ pm$
$2b = 2(1.225 \times 10^{-10}\ m) = 2.450 \times 10^{-10}\ m = 245.0 \times 10^{-12}\ m = 245.0\ pm$
Area of unit cell = (424.5 pm)(245.0 pm) = $1.04 \times 10^5\ pm^2$/cell
No. of C atoms/cell = (4)(1/4) + (2)(1/2) + 2 = 4 C atoms/cell
Surface area of nanotube = $\pi d l = \pi(1080\ pm)(1.0 \times 10^9\ pm) = 3.39 \times 10^{12}\ pm^2$

$$\text{No. of C atoms} = (3.39 \times 10^{12}\ pm^2)\left(\frac{1\ \text{cell}}{1.04 \times 10^5\ pm^2}\right)\left(\frac{4\ \text{C atoms}}{\text{cell}}\right) = 1.3 \times 10^8\ \text{C atoms}$$

22 The Main-Group Elements

22.1 (a) B is above Al in group 3A, and therefore B is more nonmetallic than Al.
(b) Ge and Br are in the same row of the periodic table, but Br (group 7A) is to the right of Ge (group 4A). Therefore, Br is more nonmetallic.
(c) Se (group 6A) is more nonmetallic than In because it is above and to the right of In (group 3A).
(d) Cl (group 7A) is more nonmetallic than Te because it is above and to the right of Te (group 6A).

22.2 Element A

22.3 (a) HNO_3 $\qquad\qquad$ H_3PO_4

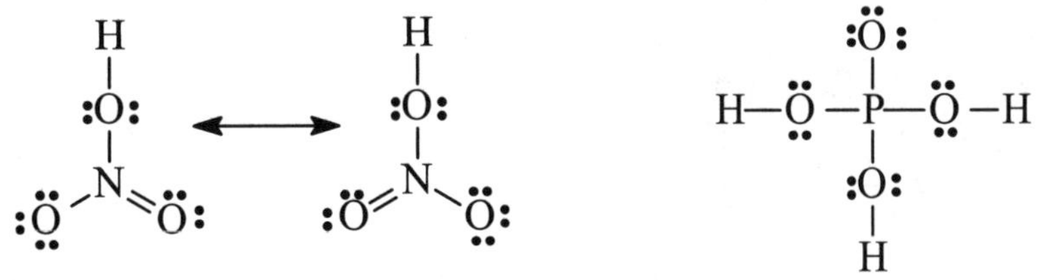

Nitrogen can form very strong pπ - pπ bonds. Phosphorus forms weaker pπ - pπ bonds, so it tends to form more single bonds.
(b) The larger S atom can accommodate six bond pairs in its valence shell, but the smaller O atom is limited to two bond pairs and two lone pairs.

22.4 Carbon forms strong π bonds with oxygen. Silicon does not form strong π bonds with oxygen, and what results are chains of alternating silicon and oxygen singly bonded to each other.

22.5 (a) SiH_4, covalent (b) KH, ionic (c) H_2Se, covalent

22.6 (a) $SrH_2(s) + 2\ H_2O(l) \rightarrow 2\ H_2(g) + Sr^{2+}(aq) + 2\ OH^-(aq)$
(b) $KH(s) + H_2O(l) \rightarrow H_2(g) + K^+(aq) + OH^-(aq)$

22.7 (a) A, KH; B, MgH_2; C, H_2O; D, HCl
(b) HCl
(c) $KH(s) + H_2O(l) \rightarrow H_2(g) + K^+(aq) + OH^-(aq)$
$MgH_2(s) + 2\ H_2O(l) \rightarrow 2\ H_2(g) + Mg^{2+}(aq) + 2\ OH^-(aq)$
(d) HCl reacts with water to give an acidic solution. KH and MgH_2 react with water to give a basic solution.

22.8 (a) (1) ZrH_x, interstitial (2) PH_3, covalent (3) HBr, covalent (4) LiH, ionic
(b) (1) and (4) are likely to be solids at 25 °C. (2) and (3) are likely to be gases at 25 °C. Covalent hydrides, like (2) and (3), form discrete molecules and have only relatively weak intermolecular forces, resulting in gases. (4) is an ionic metal hydride with strong ion-ion forces holding the 3-dimensional lattice together in the solid state. (1) is an interstitial hydride with the metal atoms in a solid crystal lattice and H's occupying holes.
(c) $LiH(s) + H_2O(l) \rightarrow H_2(g) + Li^+(aq) + OH^-(aq)$

22.9 (a) A, NaH; B, PdH_x; C, H_2S; D, HI
(b) NaH (ionic); PdH_x (interstitial); H_2S and HI (covalent)
(c) H_2S and HI (molecular); NaH and PdH_x (3-dimensional crystal)
(d) NaH: Na +1, H −1
H_2S: S −2, H +1
HI: I −1, H +1

22.10 (a) O^{2-} (b) O_2^{2-} (c) O_2^-

22.11 (a) $2\ Cs(s) + 2\ H_2O(l) \rightarrow 2\ Cs^+(aq) + 2\ OH^-(aq) + H_2(g)$
(b) $Rb(s) + O_2(g) \rightarrow RbO_2(s)$

22.12 (a) $Be(s) + Br_2(l) \rightarrow BeBr_2(s)$
(b) $Sr(s) + 2\ H_2O(l) \rightarrow Sr(OH)_2(aq) + H_2(g)$
(c) $2\ Mg(s) + O_2(g) \rightarrow 2\ MgO(s)$

22.13 An ethane-like structure is unlikely for diborane because it would require 14 valence electrons and diborane only has 12. The result is two three-center, two-electron bonds between the borons and the bridging hydrogen atoms.

22.14 (a) Each C atom is sp^2 hybridized with trigonal planar geometry. (b) The unhybridized p orbitals on each carbon atom are perpendicular to the hexagonal arrangement of carbon atoms on the sheet. The p orbitals have sideways overlap and electrons are mobile in the extended bonding system.

22.15 :C≡O: :Ö=C=Ö: [:Ö: | C (=O:, –Ö:)]$^{2-}$

Carbon monoxide will have the strongest carbon-oxygen bond because it is a triple bond.

22.16 $Hb\text{–}O_2 + CO \rightleftharpoons Hb\text{–}CO + O_2$
Mild cases of carbon monoxide poisoning can be treated with O_2. Le Châtelier's principle says that adding a product (O_2) will cause the reaction to proceed in the reverse direction, back to $Hb\text{–}O_2$.

22.17 (a) $Si_8O_{24}{}^{16-}$ (b) $Si_2O_5{}^{2-}$

22.18

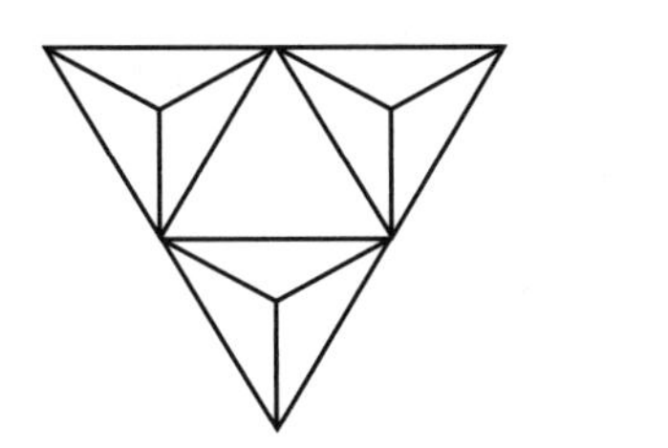

$Si_3O_9{}^{6-}$

22.19 (a) –3 (b) –2 (c) +1 (d) +4

22.20

$$\overset{-1}{:\ddot{N}}=\overset{+1}{N}=\overset{0}{\ddot{O}:} \longleftrightarrow \overset{0}{:N}\equiv\overset{+1}{N}-\overset{-1}{:\ddot{O}:} \longleftrightarrow \overset{-2}{:\ddot{N}}-\overset{+1}{N}\equiv\overset{+1}{O:}$$

The middle resonance structure makes the greatest contribution to the resonance hybrid because formal charges are minimized and the negative formal charge resides on the most electronegative element, oxygen.

22.21 NO_2 because reactants are favored for an exothermic reaction at high temperatures.

22.22 A is Li; B is Ga; C is C
(a) Li_2O, Ga_2O_3, CO_2
(b) Li_2O is the most ionic. CO_2 is the most covalent.
(c) CO_2 is the most acidic. Li_2O is the most basic.
(d) Ga_2O_3 is amphoteric and can react with both $H^+(aq)$ and $OH^-(aq)$.

22.23 (a) $Li_2O(s) + H_2O(l) \rightarrow 2\ Li^+(aq) + 2\ OH^-(aq)$
(b) $SO_3(l) + H_2O(l) \rightarrow H^+(aq) + HSO_4{}^-(aq)$
(c) $Cr_2O_3(s) + 6\ H^+(aq) \rightarrow 2\ Cr^{3+}(aq) + 3\ H_2O(l)$
(d) $Cr_2O_3(s) + 2\ OH^-(aq) + 3\ H_2O(l) \rightarrow 2\ Cr(OH)_4{}^-(aq)$

22.24 (a) $SO_3{}^{2-}$, $HSO_3{}^-$, $SO_4{}^{2-}$, $HSO_4{}^-$ (b) $HSO_4{}^-$ (c) $SO_3{}^{2-}$ (d) $HSO_4{}^-$

22.25 (a) $H-\ddot{\underset{\cdot\cdot}{S}}-H$, bent.
(b) $:\ddot{\underset{\cdot\cdot}{O}}-\ddot{S}=\ddot{O}: \longleftrightarrow :\ddot{O}=\ddot{S}-\ddot{\underset{\cdot\cdot}{O}}:$, bent, S is sp^2 hybridized.
(c)

$$:\ddot{\underset{\cdot\cdot}{O}}-\overset{\overset{:\ddot{O}:}{|}}{S}=\ddot{O}: \longleftrightarrow :\ddot{O}=\overset{\overset{:\ddot{O}:}{|}}{S}-\ddot{\underset{\cdot\cdot}{O}}: \longleftrightarrow :\ddot{\underset{\cdot\cdot}{O}}-\overset{\overset{:\ddot{O}}{\|}}{S}-\ddot{\underset{\cdot\cdot}{O}}:$$

trigonal planar, S is sp^2 hybridized.

22.26 $H_2(g) + 1/2\ O_2(g) \rightarrow H_2O(g)$ $\Delta H^\circ = -242$ kJ

$$\text{mol } H_2 = 1.45 \times 10^6 \text{ L} \times \frac{0.088 \text{ kg}}{1 \text{ L}} \times \frac{1000 \text{ g}}{1 \text{ kg}} \times \frac{1 \text{ mol } H_2}{2.016 \text{ g } H_2} = 6.33 \times 10^7 \text{ mol } H_2$$

$$q = 6.33 \times 10^7 \text{ mol } H_2 \times \frac{242 \text{ kJ}}{1 \text{ mol } H_2} = 1.5 \times 10^{10} \text{ kJ}$$

$$\text{mass } O_2 = 6.33 \times 10^7 \text{ mol } H_2 \times \frac{0.5 \text{ mol } O_2}{1 \text{ mol } H_2} \times \frac{32.00 \text{ g } O_2}{1 \text{ mol } O_2} \times \frac{1 \text{ kg}}{1000 \text{ g}} = 1.0 \times 10^6 \text{ kg } O_2$$

22.27 The steam-hydrocarbon reforming process is the most important industrial preparation of hydrogen.

$$CH_4(g) + H_2O(g) \xrightarrow[\text{Ni catalyst}]{1100\ °C} CO(g) + 3\ H_2(g)$$

$$CO(g) + H_2O(g) \xrightarrow{400\ °C} CO_2(g) + H_2(g)$$

$$CO_2(g) + 2\ OH^-(aq) \rightarrow CO_3^{2-}(aq) + H_2O(l)$$

22.28 (a)

$$H_2O(g) + C(s) \xrightarrow{1000\ °C} CO(g) + H_2(g)$$

(b) $C_3H_8(g) + 3\ H_2O(g) \rightarrow 7\ H_2(g) + 3\ CO(g)$

22.29 Hydrogen can be stored as a solid in the form of solid interstitial hydrides or in the recently discovered tube-shaped molecules called carbon nanotubes.

22.30 Assume 12.0 g of Pd with a volume of 1.0 cm^3.

$V_{H_2} = 935 \text{ cm}^3 = 935 \text{ mL} = 0.935 \text{ L}$

$$PV = nRT; \quad n_{H_2} = \frac{PV}{RT} = \frac{(1.00 \text{ atm})(0.935 \text{ L})}{\left(0.082\ 06 \frac{\text{L} \cdot \text{atm}}{\text{K} \cdot \text{mol}}\right)(273 \text{ K})} = 0.0417 \text{ mol } H_2$$

$n_H = 2 n_{H_2} = 0.0834 \text{ mol H}$

$$12.0 \text{ g Pd} \times \frac{1 \text{ mol Pd}}{106.42 \text{ g Pd}} = 0.113 \text{ mol Pd}$$

$Pd_{0.113}H_{0.0834}$

$Pd_{0.113/0.113}H_{0.0834/0.113}$

$PdH_{0.74}$

g H = (0.0834 mol H)(1.008 g/mol) = 0.0841 g H

$$d_H = 0.0841 \text{ g/cm}^3; \quad M_H = \frac{0.0834 \text{ mol}}{0.001 \text{ L}} = 83.4 \text{ M}$$

22.31 (a) TiH_2, 49.88; Assume 1.0 cm^3 of TiH_2, which has a mass of 3.75 g.

$$3.75 \text{ g } TiH_2 \times \frac{1 \text{ mol } TiH_2}{49.88 \text{ g } TiH_2} = 0.0752 \text{ mol } TiH_2$$

$$0.0752 \text{ mol } TiH_2 \times \frac{2 \text{ mol H}}{1 \text{ mol } TiH_2} = 0.150 \text{ mol H}$$

$$0.150 \text{ mol H x } \frac{1.008 \text{ g H}}{1 \text{ mol H}} = 0.151 \text{ g H}$$

$d_H = 0.15 \text{ g/cm}^3$; the density of H in TiH_2 is about 2.1 times the density of liquid H_2.

(b)

$$PV = nRT; \quad V = \frac{nRT}{P} = \frac{\left(0.15 \text{ g x } \frac{1 \text{ mol}}{2.016 \text{ g}}\right)\left(0.082\ 06 \frac{\text{L} \cdot \text{atm}}{\text{K} \cdot \text{mol}}\right)(273 \text{ K})}{1.00 \text{ atm}} = 1.7 \text{ L } H_2$$

$1.7 \text{ L} = 1.7 \times 10^3 \text{ mL} = 1.7 \times 10^3 \text{ cm}^3$

22.32 (a) NH_3 is the Lewis base and BH_3 is the Lewis acid.

(b) Nitrogen and boron both have sp^3 hybrid orbitals with bond angels of close to 109.5°.

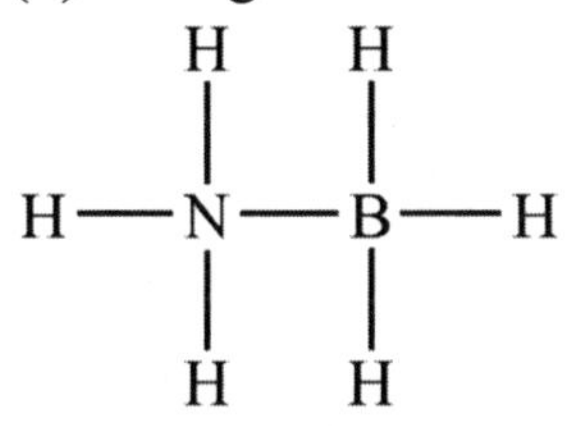

Conceptual Problems

22.33 (a)

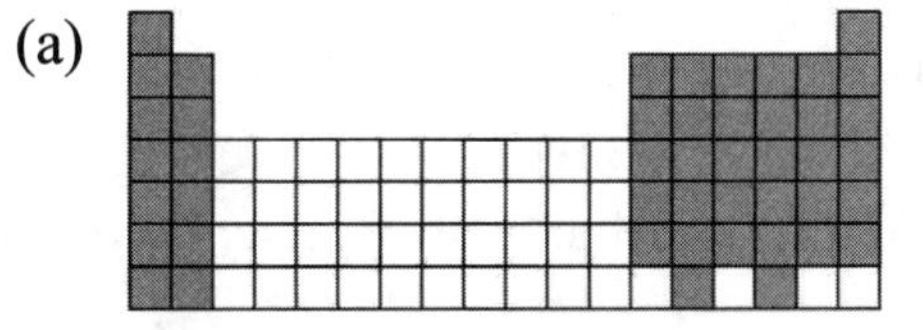

main-group elements

(b)

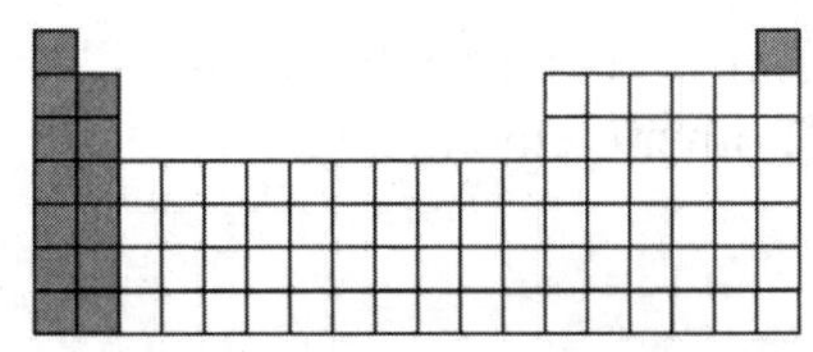

s-block elements

(c)

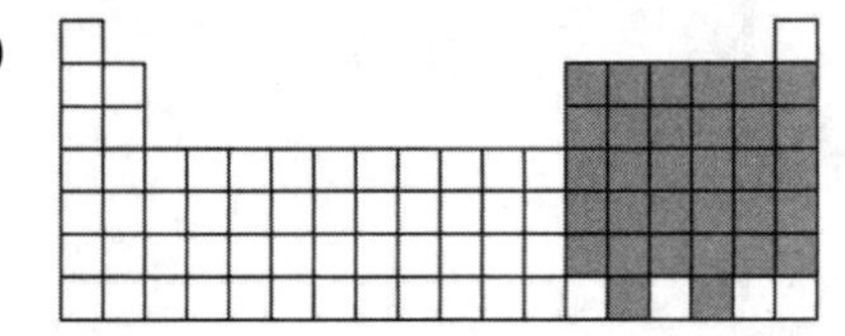

p-block elements

(d)

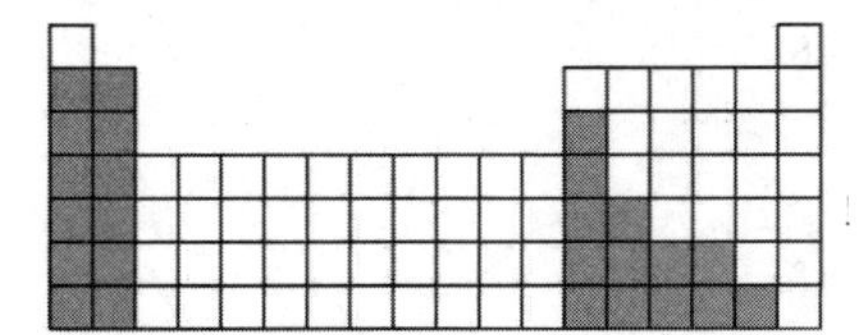

main-group metals

(e) nonmetals

(f) 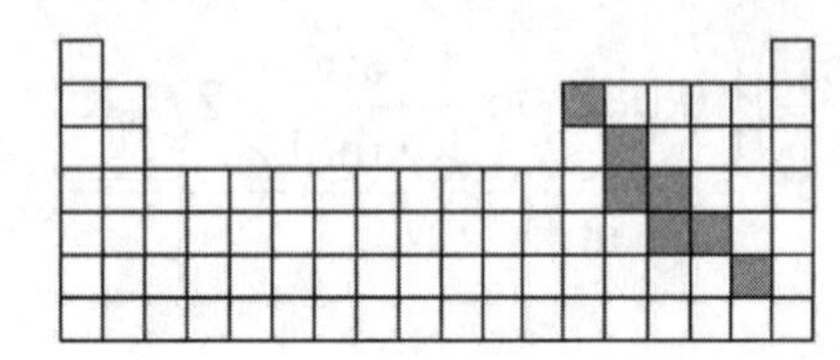 semimetals

22.34 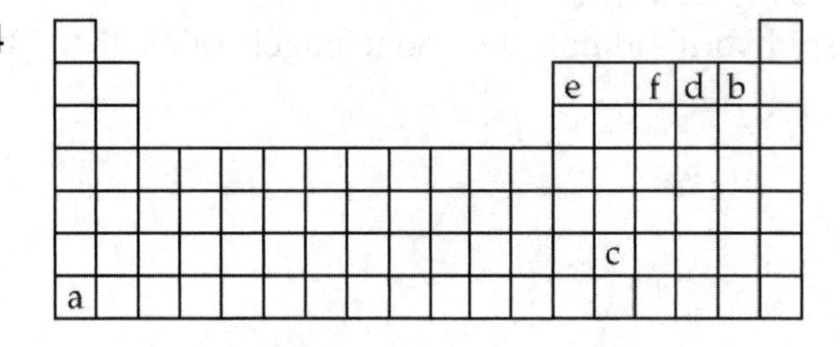

22.35 (a) A, KH; B, MgH_2; C, H_2O; D, HCl
(b) HCl
(c) $KH(s) + H_2O(l) \rightarrow H_2(g) + K^+(aq) + OH^-(aq)$
$MgH_2(s) + 2\ H_2O(l) \rightarrow 2\ H_2(g) + Mg^{2+}(aq) + 2\ OH^-(aq)$
(d) HCl reacts with water to give an acidic solution. KH and MgH_2 react with water to give a basic solution.

22.36 (a) (1) covalent (2) ionic (3) covalent (4) interstitial
(b) (1) H, +1; other element, –3 (2) H, –1; other element, +1
(3) H, +1; other element, –2

22.37 (a)

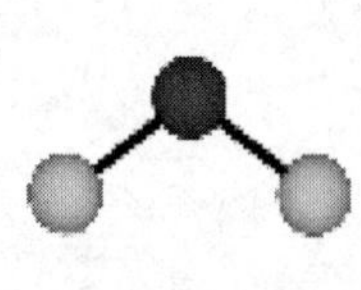 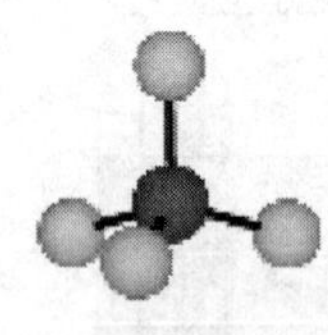

H_2O NH_3 CH_4

2nd row elements cannot form expanded octets.

(b) 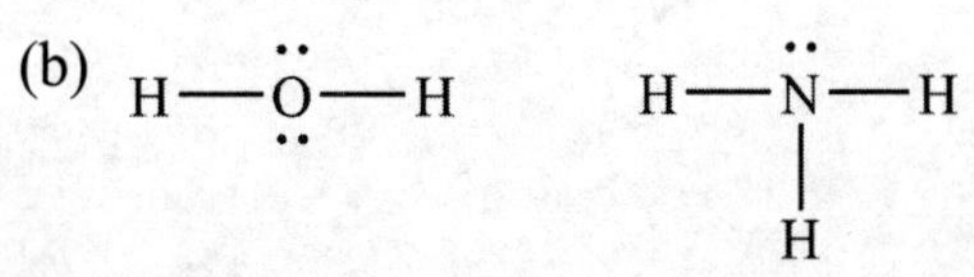 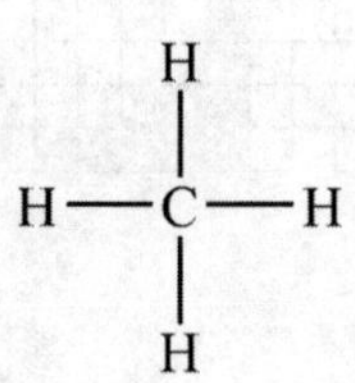

22.38 (a) The ionic hydride (4) has the highest melting point.
(b) (1), (2), and (3) are covalent hydrides. (1) and (2) can hydrogen bond, (3) cannot. Consequently, (3) has the lowest boiling point.
(c) (1), water, and (4), the ionic hydride react together to form $H_2(g)$.

22.39 (a) (1) −2, +4; (2) −2, +6; (3) −2, +2
(b) (1) covalent; (2) covalent; (3) ionic
(c) (1) acidic; (2) acidic; (3) basic
(d) (1) carbon; (2) sulfur

22.40 (a) (1) −2, +2; (2) −2, +1; (3) −2, +5
(b) (1) three-dimensional; (2) molecular; (3) molecular
(c) (1) solid; (2) gas or liquid; (3) gas or liquid
(d) (2) hydrogen; (3) nitrogen

22.41 (a) A, CaO; B, Al_2O_3; C, SO_3; D, SeO_3
(b) CaO (basic); Al_2O_3 (amphoteric); SO_3 and SeO_3 (acidic)
(c) CaO (most ionic); SO_3 (most covalent)
(d) CaO and Al_2O_3 (3-dimensional crystal); SO_3 and SeO_3 (molecular)
(e) CaO (highest melting point); SO_3 (lowest melting point)

22.42 (1) is CO_2. The molecule is linear because C has two charge clouds. There are two C=O double bonds.
(2) is SO_2. The molecule is bent because S has three charge clouds. There is one S–O single bond and one S=O double bond. SO_2 has two resonance structures so each S–O bond appears to be a bond and a half.
CO_2 has the stronger bonds.

22.43 (a) PF_5 and SF_6 (b) CH_4 and NH_4^+ (c) CO and NO_2 (d) P_4O_{10}

22.44 (a) N_2, O_2, F_2, P_4 (tetrahedral), S_8 (crown-shaped ring), Cl_2
(b) :N≡N: :Ö=Ö: :F̤̈–F̤̈: :C̤̈l—C̤̈l:

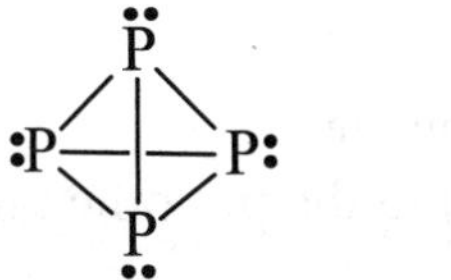

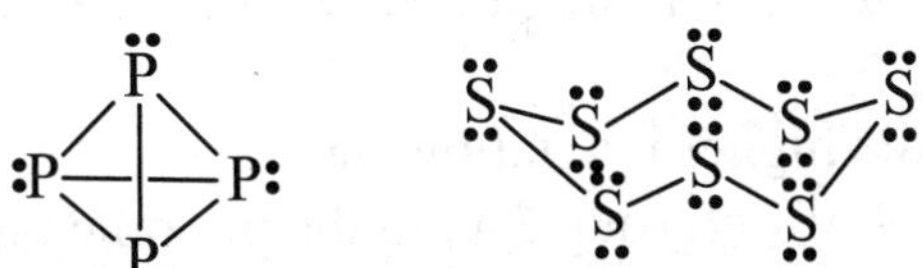

(c) The smaller N and O can form strong π bonds, whereas P and S cannot. In both F_2 and Cl_2, the atoms are joined by a single bond.

22.45 (a)

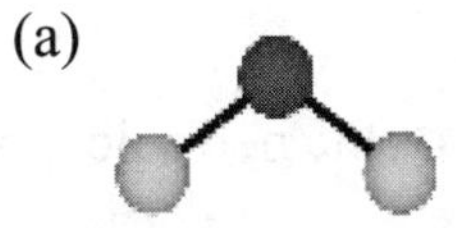

is OF_2;

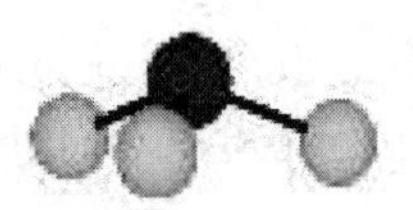

is NF_3;

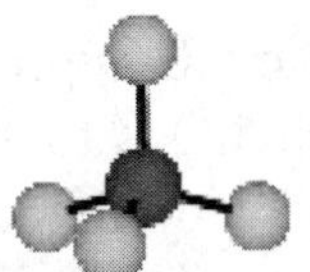

is CF_4 and SiF_4;

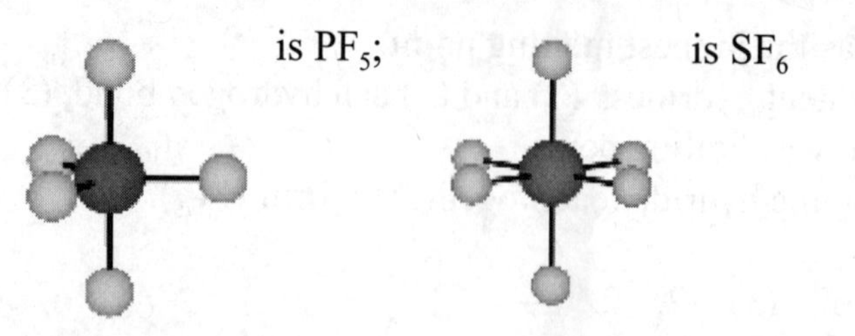

(b) The small N atom is limited to three nearest neighbors in NF_3, whereas the larger P atom can accommodate five nearest neighbors in PF_5. N uses its three unpaired electrons in bonding to three F atoms and has one lone pair, whereas P uses all five valence electrons in bonding to five F atoms. Both C and Si have four valence electrons and use sp^3 hybrid orbitals to bond to four F atoms.

22.46 (a) CO_2, Cl_2O_7, SO_3, N_2O_5

(b) :Ö=C=Ö:

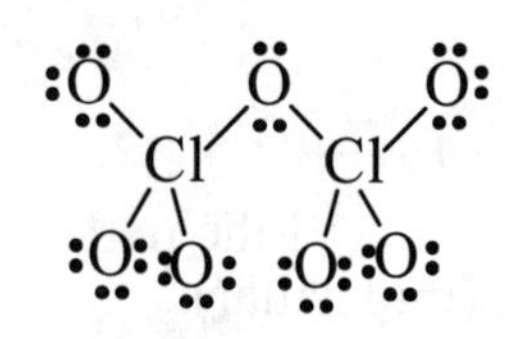

:Ö:
‖
S
:Ö: :Ö:

(resonance structures are needed)

:Ö   :Ö:
N—Ö—N
:Ö   :Ö:

(resonance structures are needed)

22.47 (a) SiO_4^{4-} (b) $Si_3O_{10}^{8-}$ (c) $Si_4O_{12}^{8-}$

Section Problems

General Properties and Periodic Trends (Sections 22.1 and 22.2)

22.48 (a) Cl (group 7A) is to the right of S (group 6A) in the same row of the periodic table. Cl has the higher ionization energy.
(b) Si is above Ge in group 4A. Si has the higher ionization energy.
(c) O (group 6A) is above and to the right of In (group 3A) in the periodic table. O has the higher ionization energy.

22.49 K < Al < P < F

22.50 (a) Al is below B in group 3A. Al has the larger atomic radius.
(b) P (group 5A) is to the left of S (group 6A) in the same row of the periodic table. P has the larger atomic radius.
(c) Pb (group 4A) is below and to the left of Br (group 7A) in the periodic table. Pb has the larger atomic radius.

22.51 O < S < As < Sn

22.52 (a) I (group 7A) is to the right of Te (group 6A) in the same row of the periodic table. I has the higher electronegativity.
(b) N is above P in group 5A. N has the higher electronegativity.
(c) F (group 7A) is above and to the right of In (group 3A) in the periodic table. F has the higher electronegativity.

22.53 Ge < P < N < O

22.54 (a) Sn is below Si in group 4A. Sn has more metallic character.
(b) Ge (group 4A) is to the left of Se (group 6A) in the same row of the periodic table. Ge has more metallic character.
(c) Bi (group 5A) is below and to the left of I (group 7A) in the periodic table. Bi has more metallic character.

22.55 (a) S is above Te in group 6A. S has more nonmetallic character.
(b) Cl (group 7A) is to the right of P (group 5A) in the same row of the periodic table. Cl has more nonmetallic character.
(c) Br (group 7A) is above and to the right of Bi (group 5A) in the periodic table. Br has more nonmetallic character.

22.56 In each case the more ionic compound is the one formed between a metal and nonmetal.
(a) CaH_2 (b) Ga_2O_3 (c) KCl (d) $AlCl_3$

22.57 In each case the more covalent molecule is the one formed between two nonmetals.
(a) PCl_3 (b) NO (c) NH_3 (d) SiO_2

22.58 Molecular (a) B_2H_6 (c) SO_3 (d) $GeCl_4$
Extended three-dimensional structure (b) $KAlSi_3O_8$

22.59 Molecular (b) P_4O_{10} (c) $SiCl_4$
Extended three-dimensional structure (a) KF (d) $CaMgSi_2O_6$

22.60 (a) Sn (b) Cl (c) Sn (d) Se (e) B

22.61 (a) F (b) Al (c) N (d) Si (e) S

22.62 The smaller B atom can bond to a maximum of four nearest neighbors, whereas the larger Al atom can accommodate more than four nearest neighbors.

22.63 The smaller C atom can bond to a maximum of four nearest neighbors, whereas the larger Ge atom can accommodate more than four nearest neighbors.

22.64 In O_2 a π bond is formed by 2p orbitals on each O. S does not form strong π bonds with its 3p orbitals, which leads to the S_8 ring structure with single bonds.

22.65 In N_2 π bonds are formed by 2p orbitals on each N. P does not form strong π bonds with its 3p orbitals, which leads to the P_4 tetrahedral structure with single bonds.

Group 1A: Hydrogen (Sections 22.3)

22.66 (a) $Zn(s) + 2\ H^+(aq) \rightarrow H_2(g) + Zn^{2+}(aq)$
(b) at 1000 °C, $H_2O(g) + C(s) \rightarrow CO(g) + H_2(g)$
(c) at 1100 °C with a Ni catalyst, $H_2O(g) + CH_4(g) \rightarrow CO(g) + 3\ H_2(g)$
(d) There are a number of possibilities. (b) and (c) above are two; electrolysis is another:
$2\ H_2O(l) \rightarrow 2\ H_2(g) + O_2(g)$

22.67 (a) $Fe(s) + 2\ H^+(aq) \rightarrow H_2(g) + Fe^{2+}(aq)$
(b) $Ca(s) + 2\ H_2O(l) \rightarrow H_2(g) + Ca^{2+}(aq) + 2\ OH^-(aq)$
(c) $2\ Al(s) + 6\ H^+(aq) \rightarrow 3\ H_2(g) + 2\ Al^{3+}(aq)$
(d) $C_2H_6(aq) + 2\ H_2O(l) \xrightarrow[\text{Catalyst}]{\text{Heat}} 2\ CO(g) + 5\ H_2(g)$

22.68 $CaH_2(s) + 2\ H_2O(l) \rightarrow 2\ H_2(g) + Ca^{2+}(aq) + 2\ OH^-(aq)$
CaH_2, 42.09; 25 °C = 298 K

$$PV = nRT; \quad n_{H_2} = \frac{PV}{RT} = \frac{(1.00\ \text{atm})(2.0 \times 10^5\ \text{L})}{\left(0.082\ 06\ \frac{\text{L}\cdot\text{atm}}{\text{K}\cdot\text{mol}}\right)(298\ \text{K})} = 8.18 \times 10^3\ \text{mol}\ H_2$$

$$8.18 \times 10^3\ \text{mol}\ H_2 \times \frac{1\ \text{mol}\ CaH_2}{2\ \text{mol}\ H_2} \times \frac{42.09\ \text{g}\ CaH_2}{1\ \text{mol}\ CaH_2} \times \frac{1\ \text{kg}}{1000\ \text{g}} = 1.7 \times 10^2\ \text{kg}\ CaH_2$$

22.69 PV = nRT

$$n_{H_2} = \frac{PV}{RT} = \frac{\left(740\ \text{mm Hg} \times \frac{1\ \text{atm}}{760\ \text{mm Hg}}\right)(1.99 \times 10^8\ \text{L})}{\left(0.082\ 06\ \frac{\text{L}\cdot\text{atm}}{\text{K}\cdot\text{mol}}\right)(293\ \text{K})} = 8.059 \times 10^6\ \text{mol}\ H_2$$

$n_C = n_{H_2} = 8.059 \times 10^6$ mol C
mass C = $(8.059 \times 10^6$ mol C)(12.011 g C/mol)(1 kg/1000 g) = 9.68×10^4 kg C

22.70 (a) MgH_2, H^- (b) PH_3, covalent (c) KH, H^- (d) HBr, covalent

22.71 (a) H_2Se, covalent (b) RbH, H^- (c) CaH_2, H^- (d) GeH_4, covalent

22.72 H_2S – covalent hydride, gas, weak acid in H_2O
NaH – ionic hydride, solid (salt like), reacts with H_2O to produce H_2
PdH_x – metallic (interstitial) hydride, solid, stores hydrogen

22.73 $TiH_{1.7}$ – an interstitial metal hydride, nonstoichiometric, solid, probably high melting
HCl – a covalent hydride, molecular, gas at 25 °C
CaH_2 – an ionic hydride, saltlike, white solid, high melting

---

22.74 (a) CH_4, covalent bonding (b) NaH, ionic bonding

22.75 (a) CaH_2, ionic bonding (b) NH_3, covalent bonding

22.76 (a) H—S̈e—H (with two lone pairs on Se), bent (b) H—Äs—H with a third H bonded to As (one lone pair on As), trigonal pyramidal

(c) SiH_4: Si bonded to four H atoms, tetrahedral

22.77 (a) GeH_4: Ge bonded to four H atoms, tetrahedral (b) H—S̈—H (with two lone pairs on S), bent

(c) H—N̈—H with a third H bonded to N (one lone pair on N), trigonal pyramidal

22.78 A nonstoichiometric compound is a compound whose atomic composition cannot be expressed as a ratio of small whole numbers. An example is PdH_x. The lack of stoichiometry results from the hydrogen occupying holes in the solid state structure.

22.79 Hydrogen atoms in interstitial hydrides are mobile because they occupy some of the holes between the larger metal atoms and can jump from the occupied holes to nearby unoccupied holes.

Group 1A and 2A: Alkali and Alkaline Earth Metals (Sections 22.4 and 22.5)

22.80 Predicted for Fr: melting point ≈ 23 °C; boiling point ≈ 650 °C; density ≈ 2 g/cm^3; atomic radius ≈ 275 pm

22.81 Group 1A and 2A metals react by losing one and two electrons, respectively. As you go down each group, the valence electrons are farther from the nucleus and more easily removed. This trend parallels chemical reactivity.

22.82 (a) $2\,K(s) + 2\,H_2O(l) \rightarrow 2\,K^+(aq) + 2\,OH^-(aq) + H_2(g)$
(b) $2\,K(s) + Br_2(l) \rightarrow 2\,KBr(s)$
(c) $K(s) + O_2(g) \rightarrow KO_2(s)$

22.83 (a) $Ca(s) + 2\,H_2O(l) \rightarrow Ca^{2+}(aq) + 2\,OH^-(aq) + H_2(g)$
(b) $Ca(s) + He(g) \rightarrow$ N. R.
(c) $Ca(s) + Br_2(l) \rightarrow CaBr_2(s)$
(d) $2\,Ca(s) + O_2(g) \rightarrow 2\,CaO(s)$

22.84 $2\ Mg(s) + O_2(g) \rightarrow 2\ MgO(s)$
$MgO(s) + H_2O(l) \rightarrow Mg(OH)_2(aq)$

22.85 $3\ BaO(s) + 2\ Al(l) \rightarrow Al_2O_3(s) + 3\ Ba(g)$

22.86 anode $Mg^{2+}(l) + 2\ e^- \rightarrow Mg(l)$
cathode $2\ Cl^-(l) \rightarrow Cl_2(g) + 2\ e^-$
overall $Mg^{2+}(l) + 2\ Cl^-(l) \rightarrow Mg(l) + Cl_2(g)$

22.87 $Mg^{2+}(l) + 2\ Cl^-(l) \rightarrow Mg(l) + Cl_2(g)$
$Mg^{2+}(l) + 2\ e^- \rightarrow Mg(l)$; $1\ A = 1\ C/s$; $10.0\ kg = 1.00 \times 10^4\ g$

$$\text{Charge} = 1.00 \times 10^4\ g\ Mg \times \frac{1\ mol\ Mg}{24.31\ g\ Mg} \times \frac{2\ mol\ e^-}{1\ mol\ Mg} \times \frac{96{,}500\ C}{1\ mol\ e^-} = 7.94 \times 10^7\ C$$

$$\text{Time} = \frac{7.94 \times 10^7\ C}{1.00 \times 10^4\ C/s} \times \frac{1\ h}{3600\ s} = 2.21\ h$$

$$1.00 \times 10^4\ g\ Mg \times \frac{1\ mol\ Mg}{24.31\ g\ Mg} \times \frac{1\ mol\ Cl_2}{1\ mol\ Mg} = 411\ mol\ Cl_2$$

$PV = nRT$

$$V = \frac{nRT}{P} = \frac{(411\ mol)\left(0.082\ 06\ \frac{L \cdot atm}{K \cdot mol}\right)(273.15\ K)}{1.00\ atm} = 9.21 \times 10^3\ L\ Cl_2$$

Group 3A: Boron (Section 22.6)

22.88 (a) Al (b) Tl (c) B

22.89 (a) Ga (b) B (c) Tl

22.90 +3 for B, Al, Ga and In; +1 for Tl

22.91 (a) Na +1, B +3, F −1 (b) Ga +3, Cl −1
(c) Tl +1, Cl −1 (d) B +3, H −1

22.92 Boron is a hard semiconductor with a high melting point. Boron forms only molecular compounds and does not form an aqueous B^{3+} ion. $B(OH)_3$ is an acid.

22.93 Boron is a semimetal, whereas all the other 3A elements are metals. Boron has a much smaller atomic radius and a higher electronegativity than the other group 3A elements.

22.94 (a) An electron deficient molecule is a molecule that doesn't have enough electrons to form a two-center, two-electron bond between each pair of bonded atoms. B_2H_6 is an electron deficient molecule.
(b) A three-center, two-electron bond has three atoms bonded together using just two electrons. The B–H–B bridging bond in B_2H_6 is a three-center, two-electron bond.

22.95 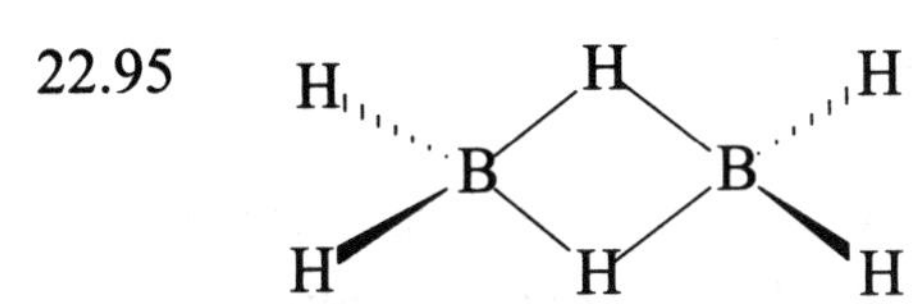

The terminal B–H bonds in diborane are ordinary two-center, two-electron bonds. The bridging B–H–B bonds in diborane are three-center, two-electron bonds. A three-center, two-electron bond is longer than a two-center, two-electron bond because it has less electron density between each pair of adjacent atoms.

Group 4A: Carbon and Silicon (Sections 22.7 and 22.8)

22.96 (a) Pb (b) C (c) Si (d) C

22.97 (a) Pb (b) Si (c) C (d) C

22.98 (a) $GeBr_4$, tetrahedral; Ge is sp^3 hybridized.
(b) CO_2, linear; C is sp hybridized.
(c) CO_3^{2-}, trigonal planar; C is sp^2 hybridized.
(d) $SnCl_3^-$, trigonal pyramidal; Sn is sp^3 hybridized.

22.99 (a) SiO_4^{4-}, tetrahedral; Si is sp^3 hybridized.
(b) CCl_4, tetrahedral; C is sp^3 hybridized.
(c) $SnCl_2$, bent; Sn is sp^2 hybridized.
(d) HCN, linear; C is sp hybridized.

22.100 Diamond is a very hard, high melting solid. It is an electrical insulator.
Diamond has a covalent network structure in which each C atom uses sp^3 hybrid orbitals to form a tetrahedral array of σ bonds. The interlocking, three-dimensional network of strong bonds makes diamond the hardest known substance with the highest melting point for an element. Because the valence electrons are localized in the σ bonds, diamond is an electrical insulator.

22.101 Graphite has a two-dimensional sheetlike structure in which each C atom uses sp^2 hybrid orbitals to form trigonal planar σ bonds to three neighboring C atoms. In addition, each C atom uses its remaining p orbital, perpendicular to the plane of the sheet, to form a π bond. Because each C atom must share its π bond with its three neighbors, the π electrons are delocalized and are free to move in the plane of the sheet. As a result, the electrical conductivity of graphite in a direction parallel to the sheets is about 10^{20} times greater than the conductivity of diamond. The conductivity of graphite perpendicular to the sheets of C atoms is lower because electrons must hop from one sheet to the next. The carbon sheets in graphite are separated by a distance of 335 pm and are held together by weak London dispersion forces. Consequently, the sheets can easily slide over one another, thus accounting for the slippery feel of graphite and its use as a lubricant.

22.102 Graphene is a two-dimensional array of hexagonally arranged carbon atoms just one atom thick, essentially one layer of graphite. Graphene is extremely strong and flexible, and is a superb conductor of electricity.

22.103 CaC_2, calcium carbide; C, –1

22.104 CO bonds to hemoglobin and prevents it from carrying O_2. CN^- bonds to cytochrome oxidase and interferes with the electron transfer associated with oxidative phosphorylation.

22.105 $SiO_2(l) + 2\ C(s) \rightarrow Si(l) + 2\ CO(g)$
(sand)
Purification of silicon for semiconductor devices:
$Si(s) + 2\ Cl_2(g) \rightarrow SiCl_4(l)$; $SiCl_4$ is purified by distillation.
$$SiCl_4(g) + 2\ H_2(g) \xrightarrow{\text{heat}} Si(s) + 4\ HCl(g)$$; Si is purified by zone refining.

22.106 Silicon and germanium are semimetals, and tin and lead are metals. Silicon is a hard, gray, semiconducting solid that melts at 1414 °C. It crystallizes in a diamondlike structure but does not form a graphitelike allotrope because of the relatively poor overlap of silicon p orbitals. Germanium is a relatively high-melting, brittle semiconductor that has the same crystal structure as diamond and silicon. Tin exists in two allotropic forms: the usual silvery white metallic form called white tin and a brittle, semiconducting form with the diamond structure called gray tin. Both white tin and lead are soft, malleable, low-melting metals. Only the metallic form occurs for lead.

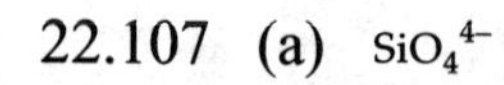
22.107 (a) SiO_4^{4-} (b) $Si_4O_{13}^{10-}$

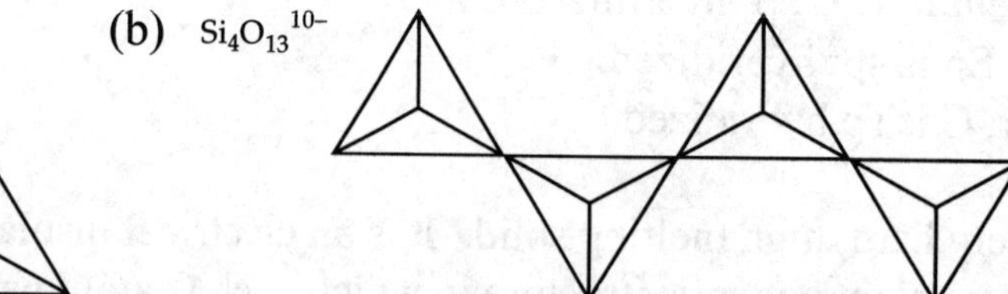

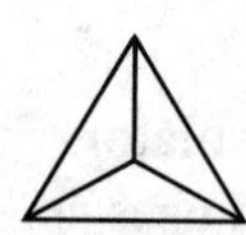

The charge on the anion is equal to the number of terminal O atoms.

22.108

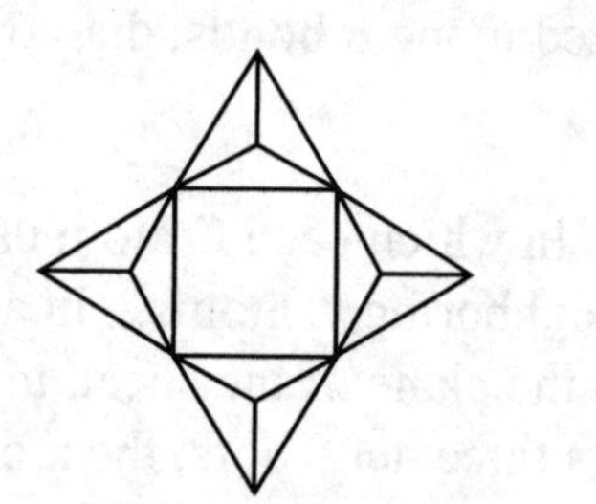
$Si_4O_{12}^{8-}$

22.109 (a) spodumene, $LiAlSi_2O_6$

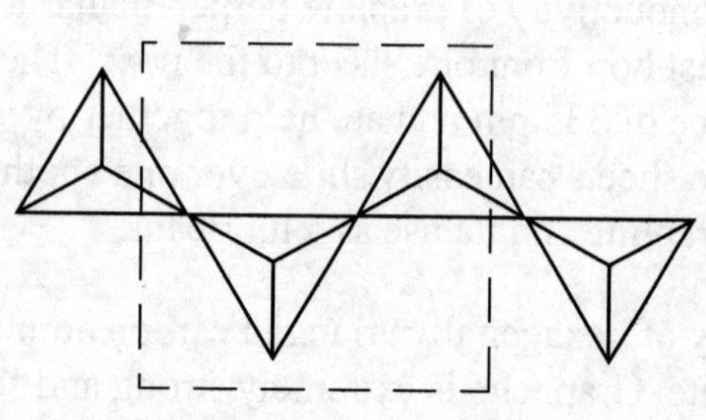
Repeating unit, $Si_2O_6^{4-}$

(b) thortveitite, $Sc_2Si_2O_7$

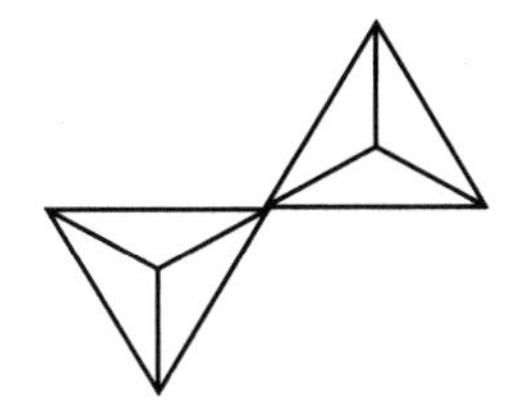

$Si_2O_7^{6-}$

Group 5A: Nitrogen and Phosphorus (Sections 22.9 and 22.10)

22.110 (a) P (b) Sb and Bi (c) N (d) Bi

22.111 (a) N (b) Bi (c) P (d) N

22.112 (a) N_2O, +1 (b) N_2H_4, −2 (c) Ca_3P_2, −3
(d) H_3PO_3, +3 (e) H_3AsO_4, +5

22.113 (a) NO, +2 (b) HNO_2, +3 (c) PH_3, −3
(d) P_4O_{10}, +5 (e) H_3PO_4, +5

22.114 **:N≡N:**
N_2 is unreactive because of the large amount of energy necessary to break the N≡N triple bond.

22.115 (a) NO_2^-, bent
(b) PH_3, trigonal pyramidal
(c) PF_5, trigonal bipyramidal
(d) PCl_4^+, tetrahedral

22.116 White phosphorus consists of tetrahedral P_4 molecules with 60° bond angles.

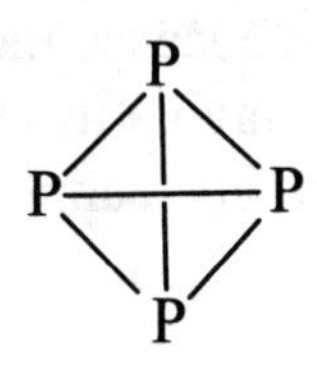

Red phosphorus is polymeric.
White phosphorus is reactive due to the considerable strain in the P_4 molecule.

22.117 (a) tetraphosphorus hexoxide, P_4O_6 (b) tetraphosphorus decoxide, P_4O_{10}

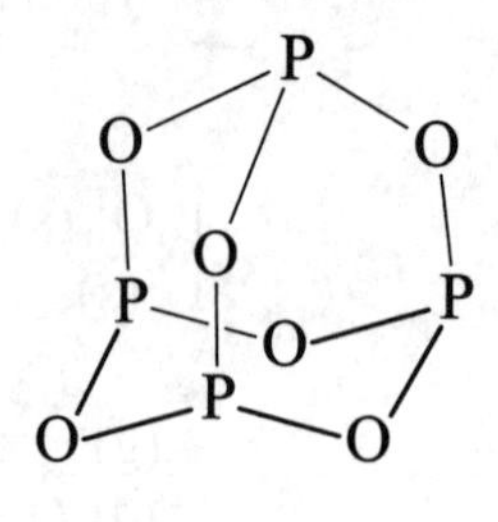

(c) phosphorous acid (d) phosphoric acid

22.118 (a) The structure for phosphorous acid is

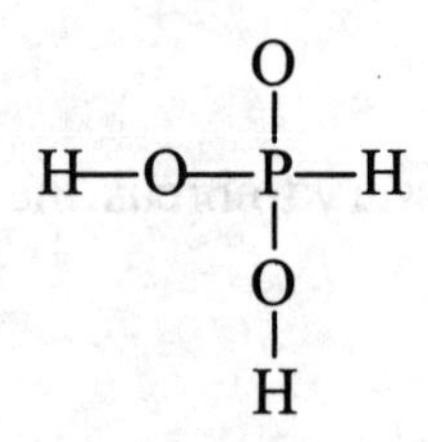

Only the two hydrogens bonded to oxygen are acidic.

(b) Nitrogen forms strong π bonds, and in N_2 the nitrogen atoms are triple bonded to each other. Phosphorus does not form strong pπ - pπ bonds, and so the P atoms are single bonded to each other in P_4.

22.119 (a) Nitric acid is a strong oxidizing agent, but phosphoric acid is not because the nitrogen atom is smaller and more electronegative than the phosphorus atom. This favors its reduction.

(b) P, As, and Sb can use d orbitals to become five-coordinate. Nitrogen cannot.

Group 6A: Oxygen and Sulfur (Sections 20.11 and 20.12)

22.120 (a) O (b) Te (c) Po (d) O

22.121 (a) Po (b) O (c) O (d) S

22.122 (a) O_2 is obtained in industry by the fractional distillation of liquid air.

(b) In the laboratory, O_2 is prepared by the thermal decomposition of $KClO_3(s)$.

$$2\ KClO_3(s) \xrightarrow[MnO_2]{heat} 2\ KCl(s) + 3\ O_2(g)$$

22.123 In nature, oxygen is found as $O_2(g)$ in the atmosphere, in the form of H_2O in the hydrosphere, and combined with other elements in the form of silicates, carbonates, oxides, and other oxygen containing minerals in the lithosphere.

22.124 (a) $4\,Li(s) + O_2(g) \rightarrow 2\,Li_2O(s)$ (b) $P_4(s) + 5\,O_2(g) \rightarrow P_4O_{10}(s)$
(c) $4\,Al(s) + 3\,O_2(g) \rightarrow 2\,Al_2O_3(s)$ (d) $Si(s) + O_2(g) \rightarrow SiO_2(s)$

22.125 (a) $2\,Ca(s) + O_2(g) \rightarrow 2\,CaO(s)$ (b) $C(s) + O_2(g) \rightarrow CO_2(g)$
(c) $4\,As(s) + 5\,O_2(g) \rightarrow 2\,As_2O_5(s)$ (d) $4\,B(s) + 3\,O_2(g) \rightarrow 2\,B_2O_3(s)$

22.126 :Ö::Ö: The electron dot structure shows an O=O double bond. It also shows all electrons paired. This is not consistent with the fact that O_2 is paramagnetic.

22.127 The highest occupied molecular orbital in O_2 is the doubly degenerate π^*_{2p} orbital which contains 2 unpaired electrons. The bond order is 2 because O_2 has four π_{2p}, two σ_{2p} bonding electrons, and two π^*_{2p} antibonding electrons. (See text, Figure 8.23.)

22.128 An element that forms an acidic oxide is more likely to form a covalent hydride. C and N are examples.

22.129 An element that forms an ionic hydride is more likely to form a basic oxide. Na and Ca are examples.

22.130 $Li_2O < BeO < B_2O_3 < CO_2 < N_2O_5$ (see Figure 22.16)

22.131 $P_4O_{10} < SiO_2 < GeO_2 < Ga_2O_3 < K_2O$ (see Figure 22.16)

22.132 $N_2O_5 < Al_2O_3 < K_2O < Cs_2O$ (see Figure 22.16)

22.133 $BaO < SnO_2 < SO_3 < Cl_2O_7$ (see Figure 22.16)

22.134 (a) CrO_3 (higher Cr oxidation state) (b) N_2O_5 (higher N oxidation state)
(c) SO_3 (higher S oxidation state)

22.135 (a) CrO (lower Cr oxidation state) (b) SnO (lower Sn oxidation state)
(c) As_2O_3 (lower As oxidation state)

22.136 (a) $Cl_2O_7(l) + H_2O(l) \rightarrow 2\,H^+(aq) + 2\,ClO_4^-(aq)$
(b) $K_2O(s) + H_2O(l) \rightarrow 2\,K^+(aq) + 2\,OH^-(aq)$
(c) $SO_3(l) + H_2O(l) \rightarrow H^+(aq) + HSO_4^-(aq)$

22.137 (a) $BaO(s) + H_2O(l) \rightarrow Ba^{2+}(aq) + 2\,OH^-(aq)$
(b) $Cs_2O(s) + H_2O(l) \rightarrow 2\,Cs^+(aq) + 2\,OH^-(aq)$
(c) $N_2O_5(s) + H_2O(l) \rightarrow 2\,H^+(aq) + 2\,NO_3^-(aq)$

22.138 (a) $ZnO(s) + 2 H^+(aq) \rightarrow Zn^{2+}(aq) + H_2O(l)$
(b) $ZnO(s) + 2 OH^-(aq) + H_2O(l) \rightarrow Zn(OH)_4^{2-}(aq)$

22.139 (a) $Ga_2O_3(s) + 3 H^+(aq) + 3 HSO_4^-(aq) \rightarrow 2 Ga^{3+}(aq) + 3 SO_4^{2-}(aq) + 3 H_2O(l)$
(b) $Ga_2O_3(s) + 2 OH^-(aq) + 3 H_2O(l) \rightarrow 2 Ga(OH)_4^-(aq)$

22.140 (a) rhombic sulfur – yellow crystalline solid (mp 113 °C) that contains crown-shaped S_8 rings.
(b) monoclinic sulfur – an allotrope of sulfur in which the S_8 rings pack differently in the crystal.
(c) plastic sulfur – when sulfur is cooled rapidly, the sulfur forms disordered, tangled chains, yielding an amorphous, rubbery material called plastic sulfur.
(d) Liquid sulfur between 160 and 195 °C becomes dark reddish-brown and very viscous forming long polymer chains (S_n, $n > 200,000$).

22.141 The dramatic increase in the viscosity of molten sulfur at 160-195 °C is due to opening of the S_8 rings, yielding S_8 chains that form long polymer chains, which become entangled. Above 200 °C, the polymer chains begin to fragment into smaller pieces, with a decrease in viscosity.

22.142 (a) $Zn(s) + 2 H_3O^+(aq) \rightarrow Zn^{2+}(aq) + H_2(g) + 2 H_2O(l)$
(b) $BaSO_3(s) + 2 H_3O^+(aq) \rightarrow H_2SO_3(aq) + Ba^{2+}(aq) + 2 H_2O(l)$
(c) $Cu(s) + 2 H_2SO_4(l) \rightarrow Cu^{2+}(aq) + SO_4^{2-}(aq) + SO_2(g) + 2 H_2O(l)$
(d) $H_2S(aq) + I_2(aq) \rightarrow S(s) + 2 H^+(aq) + 2 I^-(aq)$

22.143 (a) $ZnS(s) + 2 H_3O^+(aq) \rightarrow Zn^{2+}(aq) + H_2S(g) + 2 H_2O(l)$
(b) $H_2S(aq) + 2 Fe^{3+}(aq) \rightarrow S(s) + 2 Fe^{2+}(aq) + 2 H^+(aq)$
(c) $Fe(s) + 2 H_3O^+(aq) \rightarrow Fe^{2+}(aq) + H_2(g) + 2 H_2O(l)$
(d) $BaO(s) + H_3O^+(aq) + HSO_4^-(aq) \rightarrow BaSO_4(s) + 2 H_2O(l)$

22.144 (a) Acid strength increases as the number of O atoms increases.
(b) In comparison with S, O is much too electronegative to form compounds of O in the +4 oxidation state. Also, an S atom is large enough to accommodate four bond pairs and a lone pair in its valence shell, but an O atom is too small to do so.
(c) Each S is sp^3 hybridized with two lone pairs of electrons. The bond angles are therefore 109.5°. A planar ring would require bond angles of 135°.

22.145 (a) O is more electronegative than S because of its smaller size.
(b) S forms long S_n chains by forming single bonds because it does not effectively π-bond with itself. In contrast, O can effectively π-bond with itself.
(c) SO_3 has three bond pairs and no lone pairs, therefore the geometry is trigonal planar. SO_3^{2-} has three bond pairs and one lone pair, therefore the geometry is trigonal pyramidal.

Group 7A and 8A: Halogen and Noble Gases (Sections 22.13 and 22.14)

22.146 (a) At is in Group 7A. The trend going down the group is gas → liquid → solid. At, being at the bottom of the group, should be a solid.

(b) At is likely to react with Na just like the other halogens, yielding NaAt.

22.147 Group 7A nonmetals react by gaining an electron. The electron affinity generally decreases going down the group. This trend parallels chemical reactivity.

22.148 $MnO_2(s) + 2\ Br^-(aq) + 4\ H^+(aq) \rightarrow Mn^{2+}(aq) + 2\ H_2O(l) + Br_2(aq)$

22.149 $4\ HCl(g) + O_2(g) \rightarrow 2\ Cl_2(g) + 2\ H_2O(g)$

22.150 (a) Assume a 100.0 g sample. From the percent composition data, a 100.0 g sample contains 25.25 g Ti, and 74.75 g Cl.

$$25.25\text{ g Ti} \times \frac{1\text{ mol Ti}}{47.87\text{ g Ti}} = 0.5275\text{ mol Ti}$$

$$74.75\text{ g Cl} \times \frac{1\text{ mol Cl}}{35.45\text{ g Cl}} = 2.109\text{ mol Cl}$$

$Ti_{0.5275}Cl_{2.109}$; divide each subscript by the smaller, 0.5275.

$Ti_{0.5275\,/\,0.5275}Cl_{2.109\,/\,0.5275}$

The empirical and molecular formula is $TiCl_4$, titanium tetrachloride.

(b) $Ti(s) + 2\ Cl_2(g) \rightarrow TiCl_4(g)$

(c) $TiCl_4(l) + 2\ Mg(s) \rightarrow Ti(s) + 2\ MgCl_2(s)$

22.151 Assume a 100.0 g sample. From the percent composition data, a 100.0 g sample contains 49.44 g Nb, and 50.56 g F.

$$49.44\text{ g Nb} \times \frac{1\text{ mol Nb}}{92.91\text{ g Nb}} = 0.5321\text{ mol Nb}$$

$$50.56\text{ g F} \times \frac{1\text{ mol F}}{19.00\text{ g F}} = 2.661\text{ mol F}$$

$Nb_{0.5321}F_{2.661}$; divide each subscript by the smaller, 0.5321.

$Nb_{0.5321\,/\,0.5321}F_{2.661\,/\,0.5321}$

The empirical formula is NbF_5.

(b) $2\ Nb(s) + 5\ F_2(g) \rightarrow 2\ NbF_5(s)$

(c) $2\ NbF_5(s) + 5\ H_2(g) \rightarrow 2\ Nb(s) + 10\ HF(g)$

22.152 (a) $HBrO_3$, +5 (b) HIO, +1

22.153 (a) H_5IO_6, +7 (b) $HClO_2$, +3

22.154 (a) HIO_3 :Ö—Ï—Ö—H with :Ö: bonded below I — trigonal pyramidal

(b) ClO_2^- [:Ö—C̈l—Ö:]$^-$ — bent

(c) HOCl H—Ö—C̈l: — bent

(d) IO_6^{5-} 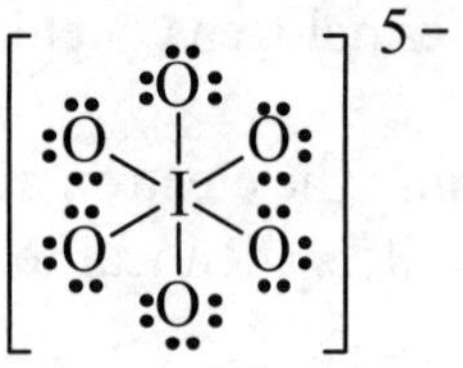   octahedral

22.155 (a) BrO_4^- 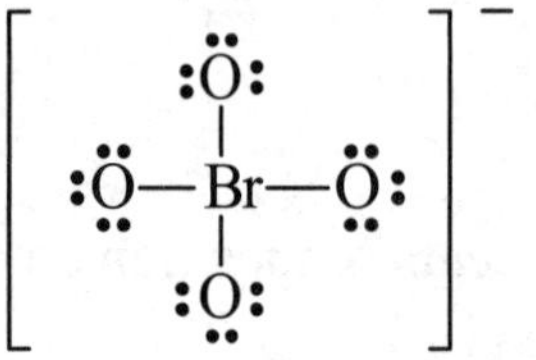   tetrahedral

(b) ClO_3^- 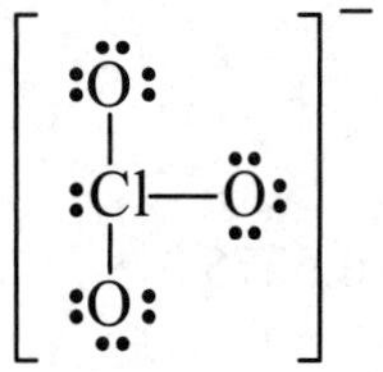   trigonal pyramidal

(c) HIO_4 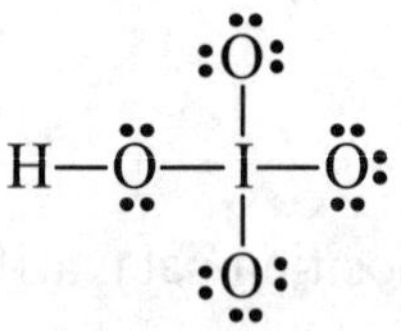   tetrahedral

(d) HOBr   H—O—Br   bent

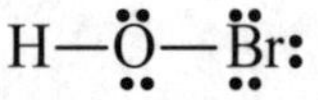

22.156 Oxygen atoms are highly electronegative. Increasing the number of oxygen atoms increases the polarity of the O–H bond and increases the acid strength.

22.157 Acid strength increases in the order HIO < HBrO < HClO because electronegativity increases in the order I < Br < Cl. The higher the electronegativity, the more polarized is the O–H bond.

Chapter Problems

22.158 (a) $Si_3O_{10}^{8-}$
(b) The charge on the anion is 8–. Because the Ca^{2+} to Cu^{2+} ratio is 1:1, there must be 2 Ca^{2+} and 2 Cu^{2+} ions in the formula for the mineral. There are also 2 waters. The formula of the mineral is: $Ca_2Cu_2Si_3O_{10} \cdot 2\,H_2O$

22.159 $Mg(s) + 2\,H_2SO_4(l) \rightarrow Mg^{2+}(aq) + SO_4^{2-}(aq) + SO_2(g) + 2\,H_2O(l)$

22.160 $I_2O_5(aq) + H_2O(l) \rightarrow 2\,HIO_3(aq)$; HIO_3 is iodic acid.

22.161 (a) LiCl is an ionic compound. PCl_3 is a covalent molecular compound.
The ionic compound, LiCl, has the higher melting point.

(b) Carbon forms strong π bonds with oxygen, and CO_2 is a covalent molecular compound with a low melting point. Silicon prefers to form single bonds with oxygen. SiO_2 is three dimensional extended structure with alternating silicon and oxygen singly bonded to each other. SiO_2 is a high melting solid.
(c) Nitrogen forms strong π bonds with oxygen and NO_2 is a covalent molecular compound with a low melting point. Phosphorus prefers to form single bonds with oxygen. P_4O_{10} is a larger covalent molecular compound than NO_2, with a higher melting point.

22.162 (a) Ga (b) In (c) Pb
Metals are better electrical conductors than nonmetals (S and P) or semimetals (B).

22.163 Both NH_3 and PH_3 are colorless gases at room temperature. Both have a trigonal pyramidal geometry. NH_3 can hydrogen bond, PH_3 cannot. Aqueous solutions of NH_3 are basic. Aqueous solutions of PH_3 are neutral.

22.164 C, Si, Ge and Sn have allotropes with the diamond structure.
Sn and Pb have metallic allotropes.
C (nonmetal), Si (semimetal), Ge (semimetal), Sn (semimetal and metal), Pb (metal)

22.165 (a) $H_3PO_4(aq) + H_2O(l) \rightleftarrows H_3O^+(aq) + H_2PO_4^-(aq)$
H_3PO_4 is a Brønsted-Lowry acid.
(b) $B(OH)_3(aq) + 2\ H_2O(l) \rightleftarrows B(OH)_4^-(aq) + H_3O^+(aq)$
$B(OH)_3$ is a Lewis acid.

22.166 (a) In diamond each C is covalently bonded to four additional C atoms in a rigid three-dimensional network solid. Graphite is a two-dimensional covalent network solid of carbon sheets that can slide over each other. Both are high melting because melting requires the breaking of C–C bonds.

22.167

```
        O     O
        |     |
H—O—S—O—S—O—H
        |     |
        O     O
```

22.168 Cl—S—S—Cl

22.169 The pipes have undergone a phase transition in places from metallic white tin to brittle grey tin, a transformation that can occur when white tin is kept below 13 °C for a long time.

22.170 NH_3, $K_b = 1.8 \times 10^{-5}$; N_2H_4, $K_b = 8.9 \times 10^{-7}$; NH_2OH, $K_b = 9.1 \times 10^{-9}$
The strongest base, NH_3, will react to the greatest extent with HNO_2.

22.171

```
H     H     H
  \  / \   /
   Al    B
  /  \ /   \
H     H     H
```

22.172 (a) C as diamond
(b) $Cl_2(g) + H_2O(l) \rightarrow HOCl(aq) + H^+(aq) + Cl^-(aq)$
(c) NO (d) NO_2 (e) BF_3 (f) Al_2O_3 (g) Si (h) HNO_3
(i) C as diamond, graphite, and fullerene.

22.173 The angle required by P_4 is 60°. The strain would not be reduced by using sp^3 hybrid orbitals because their angle is ~109°.

22.174 Carbon is a versatile element that can form millions of very stable compounds with elements such as N, O, and H. Biomolecules contain chains and rings with many C–C bonds. Si–Si bonds are much less stable and chains of Si atoms are uncommon. In addition, carbon can form very stable pπ-pπ multiple bonds. On the other hand, the chemistry of silicon (which cannot form stable pπ-pπ bonds) is dominated by structures based on the SiO_4^{4-} anion.

Multiconcept Problems

22.175 (a) ·N̈=Ö: [:Ö–Ö·]⁻ ⟷ [·Ö–Ö:]⁻ [:Ö=N̈–Ö–Ö:]⁻

The O–N–O bond angle should be ~120°.

(b)
σ^*_{2p} —
π^*_{2p} ↑ —
σ_{2p} ↑↓
π_{2p} ↑↓ ↑↓
σ^*_{2s} ↑↓
σ_{2s} ↑↓

The bond order is 2½ with one unpaired electron.

22.176 (a)

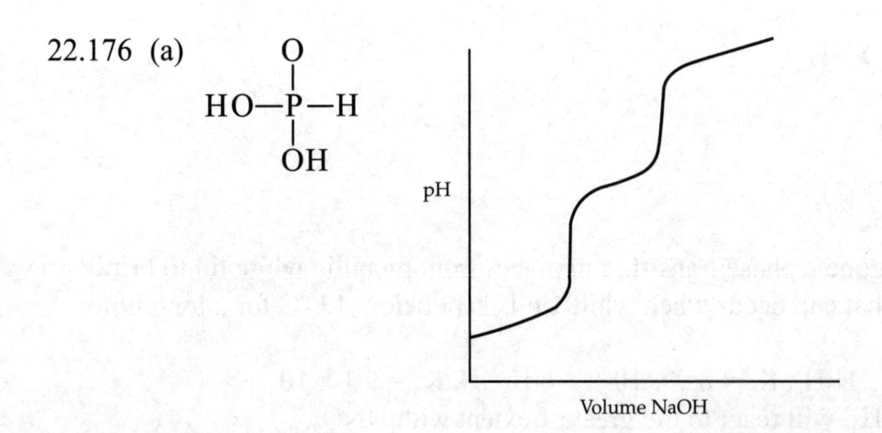

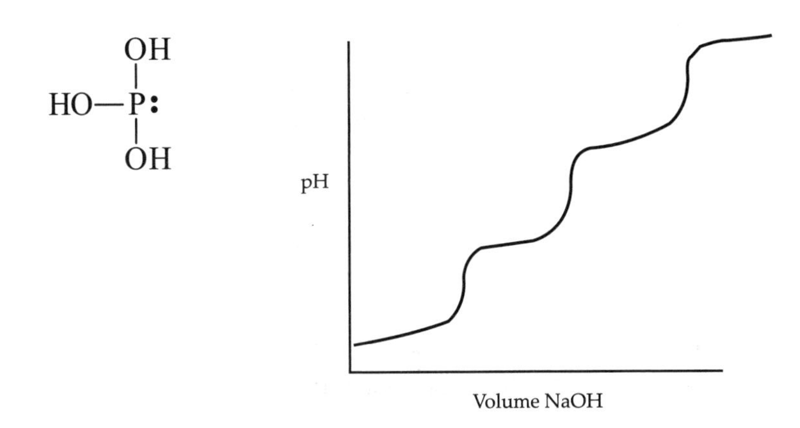

(b) $K_{a1} = 1.0 \times 10^{-2}$; $pK_{a1} = -\log(1.0 \times 10^{-2}) = 2.00$
$K_{a2} = 2.6 \times 10^{-7}$; $pK_{a2} = -\log(2.6 \times 10^{-7}) = 6.59$

At the first equivalence point, $pH = \dfrac{pK_{a1} + pK_{a2}}{2} = \dfrac{2.00 + 6.59}{2} = 4.29$

mmol HPO_3^{2-} = (30.00 mL)(0.1240 mmol/mL) = 3.72 mmol HPO_3^{2-}
volume NaOH to reach second equivalence point

$$= 3.72 \text{ mmol } HPO_3^{2-} \times \frac{2 \text{ mmol NaOH}}{1 \text{ mmol } HPO_3^{2-}} \times \frac{1.00 \text{ mL}}{0.1000 \text{ mmol NaOH}} = 74.40 \text{ mL}$$

At the second equivalence point only Na_2HPO_3, a basic salt, is in solution.

$$[HPO_3^{2-}] = \frac{3.72 \text{ mmol}}{30.00 \text{ mL} + 74.40 \text{ mL}} = 0.0356 \text{ mmol/mL} = 0.0356 \text{ M}$$

$$K_b = \frac{K_w}{K_{a2}} = \frac{1.0 \times 10^{-14}}{2.6 \times 10^{-7}} = 3.8 \times 10^{-8}$$

	$HPO_3^{2-}(aq)$ + $H_2O(l)$	⇄ $H_2PO_3^{2-}(aq)$	+ $OH^-(aq)$
initial (M)	0.0356	0	~0
change (M)	−x	+x	+x
equil (M)	0.0356 − x	x	x

$$K_b = \frac{[H_2PO_3^-][OH^-]}{[HPO_3^{2-}]} = 3.8 \times 10^{-8} = \frac{(x)(x)}{0.0356 - x} \approx \frac{x^2}{0.0356}$$

Solve for x.

$$x = [OH^-] = \sqrt{(3.8 \times 10^{-8})(0.0356)} = 3.68 \times 10^{-5} \text{ M}$$

$$[H_3O^+] = \frac{K_w}{[OH^-]} = \frac{1.0 \times 10^{-14}}{3.68 \times 10^{-5}} = 2.72 \times 10^{-10} \text{ M}$$

$pH = -\log[H_3O^+] = -\log(2.72 \times 10^{-10}) = 9.57$

22.177

$2\,In^+(aq) + 2\,e^- \rightarrow 2\,In(s)$	$E^\circ = -0.14$ V
$In^+(aq) \rightarrow In^{3+}(aq) + 2\,e^-$	$E^\circ = 0.44$ V
$3\,In^+(aq) \rightarrow In^{3+}(aq) + 2\,In(s)$	$E^\circ = 0.30$ V

$1 \text{ J} = 1 \text{ V} \cdot \text{C}$
$\Delta G^\circ = -nFE^\circ = -(2)(96{,}500 \text{ C})(0.30 \text{ V}) = -5.8 \times 10^4 \text{ J} = -58 \text{ kJ}$
Because $\Delta G^\circ < 0$, the disproportionation of In^+ is spontaneous.

$$\begin{array}{ll} 2\,Tl^+(aq) + 2\,e^- \rightarrow 2\,Tl(s) & E^\circ = -0.34 \text{ V} \\ \underline{Tl^+(aq) \rightarrow Tl^{3+}(aq) + 2\,e^-} & \underline{E^\circ = -1.25 \text{ V}} \\ 3\,Tl^+(aq) \rightarrow Tl^{3+}(aq) + 2\,Tl(s) & E^\circ = -1.59 \text{ V} \end{array}$$

$\Delta G^\circ = -nFE^\circ = -(2)(96{,}500 \text{ C})(-1.59 \text{ V}) = +3.07 \times 10^5 \text{ J} = +307 \text{ kJ}$
Because $\Delta G^\circ > 0$, the disproportionation of Tl^+ is nonspontaneous.

22.178 (a) $P_4(s) + 5\,O_2(g) \rightarrow P_4O_{10}(s)$
$P_4O_{10}(s) + 6\,H_2O(l) \rightarrow 4\,H_3PO_4(aq)$
(b) P_4, 123.90

$$\text{mol } H_3PO_4 = 5.00 \text{ g } P_4 \times \frac{1 \text{ mol } P_4}{123.90 \text{ g } P_4} \times \frac{1 \text{ mol } P_4O_{10}}{1 \text{ mol } P_4} \times \frac{4 \text{ mol } H_3PO_4}{1 \text{ mol } P_4O_{10}} = 0.1614 \text{ mol}$$

$$[H_3PO_4] = \frac{0.1614 \text{ mol}}{0.2500 \text{ L}} = 0.646 \text{ M}$$

For the dissociation of the first proton, the following equilibrium must be considered:

$$H_3PO_4(aq) + H_2O(l) \rightleftharpoons H_3O^+(aq) + H_2PO_4^-(aq)$$

	H_3PO_4	H_3O^+	$H_2PO_4^-$
initial (M)	0.646	~0	0
change (M)	–x	+x	+x
equil (M)	0.646 – x	x	x

$$K_{a1} = \frac{[H_3O^+][H_2PO_4^-]}{[H_3PO_4]} = 7.5 \times 10^{-3} = \frac{x^2}{0.646 - x}$$

$x^2 + (7.5 \times 10^{-3})x - (4.84 \times 10^{-3}) = 0$
Solve for x using the quadratic formula.

$$x = \frac{-(7.5 \times 10^{-3}) \pm \sqrt{(7.5 \times 10^{-3})^2 - (4)(1)(-4.84 \times 10^{-3})}}{2(1)} = \frac{(-7.5 \times 10^{-3}) \pm 0.139}{2}$$

$x = 0.0658$ and -0.0733; Of the two solutions for x, only the positive value of x has physical meaning, because x is the $[H_3O^+]$.
$x = 0.0658 \text{ M} = [H_2PO_4^-] = [H_3O^+]$
Only the dissociation of the first proton contributes a significant amount of H_3O^+.
$pH = -\log[H_3O^+] = -\log(0.0658) = 1.18$

(c) $3\,Ca^{2+}(aq) + 2\,H_3PO_4(aq) \rightarrow Ca_3(PO_4)_2(s) + 6\,H^+(aq)$
$Ca_3(PO_4)_2$, 310.18

$$\text{mass } Ca_3(PO_4)_2 = 0.1614 \text{ mol } H_3PO_4 \times \frac{1 \text{ mol } Ca_3(PO_4)_2}{2 \text{ mol } H_3PO_4} \times \frac{310.18 \text{ g } Ca_3(PO_4)_2}{1 \text{ mol } Ca_3(PO_4)_2} = 25.0 \text{ g}$$

(d) $Zn(s) + 2\,H^+(aq) \rightarrow H_2(g) + Zn^{2+}(aq)$; the gas is H_2.

$$\text{mol } H_2 = 0.1614 \text{ mol } H_3PO_4 \times \frac{6 \text{ mol } H^+}{2 \text{ mol } H_3PO_4} \times \frac{1 \text{ mol } H_2}{2 \text{ mol } H^+} = 0.242 \text{ mol } H_2$$

$PV = nRT$; 20 °C = 293 K

$$V = \frac{nRT}{P} = \frac{(0.242\ \text{mol})\left(0.082\ 06\ \frac{\text{L}\cdot\text{atm}}{\text{K}\cdot\text{mol}}\right)(293\ \text{K})}{\left(742\ \text{mm Hg} \times \frac{1.00\ \text{atm}}{760\ \text{mm Hg}}\right)} = 5.96\ \text{L}$$

22.179 $N_2O_4(g) \rightleftarrows 2\ NO_2(g)$

$$P_{Total} = 753\ \text{mm Hg} \times \frac{1.00\ \text{atm}}{760\ \text{mm Hg}} = 0.991\ \text{atm}$$

$$P_{Total} = P_{N_2O_4} + P_{NO_2} = 0.991\ \text{atm}$$

$$P_{N_2O_4} = 0.991\ \text{atm} - P_{NO_2}$$

$$K_p = \frac{(P_{NO_2})^2}{P_{N_2O_4}} = 0.113; \qquad K_p = \frac{(P_{NO_2})^2}{(0.991\ \text{atm} - P_{NO_2})} = 0.113$$

$$(P_{NO_2})^2 + 0.113(P_{NO_2}) - 0.112 = 0$$

Use the quadratic formula to solve for P_{NO_2}.

$$P_{NO_2} = \frac{-(0.113) \pm \sqrt{(0.113)^2 - (4)(1)(-0.112)}}{2(1)} = \frac{-0.113 \pm 0.679}{2}$$

$$P_{NO_2} = -0.396 \text{ and } 0.283$$

Of the two solutions for P_{NO_2}, only 0.283 has physical meaning because NO_2 can't have a negative partial pressure.

$P_{NO_2} = 0.283$ atm and $P_{N_2O_4} = 0.991\ \text{atm} - P_{NO_2} = 0.991 - 0.283 = 0.708$ atm

	$N_2O_4(g)$	$\rightarrow$	$2\ NO_2(g)$
before reaction (atm)	0.708		0.283
change (atm)	–0.708		+2(0.708)
after reaction (atm)	0		0.283 + 2(0.708) = 1.70 atm

$PV = nRT$; 25 °C = 298 K

$$n_{NO_2} = \frac{PV}{RT} = \frac{(1.70\ \text{atm})(0.5000\ \text{L})}{\left(0.082\ 06\ \frac{\text{L}\cdot\text{atm}}{\text{K}\cdot\text{mol}}\right)(298\ \text{K})} = 0.0348\ \text{mol}\ NO_2$$

(a)	$2\ NO_2(aq) + 2\ H_2O(l)$	$\rightarrow$	$HNO_2(aq)$	+ $H_3O^+(aq)$	+ $NO_3^-(aq)$
before reaction (mol)	0.0348		0	~0	0
change (mol)	–0.0348		+0.0348/2	+0.0348/2	+0.0348/2
after reaction (mol)	0		0.0174	0.0174	0.0174

(b) $[HNO_2] = [H_3O^+] = 0.0174 \text{ mol}/0.250 \text{ L} = 0.0696 \text{ M}$

	$HNO_2(aq)$ + $H_2O(l)$	⇄	$H_3O^+(aq)$	+ $NO_2^-(aq)$
initial (M)	0.0696		0.0696	0
change (M)	−x		+x	+x
equil (M)	0.0696 − x		0.0696 + x	x

$$K_a = \frac{[H_3O^+][NO_2^-]}{[HNO_2]} = 4.5 \text{ x } 10^{-4} = \frac{(0.0696 + x)x}{(0.0696 - x)}$$

$x^2 + 0.070\,05 - (3.132 \text{ x } 10^{-5}) = 0$

Use the quadratic formula to solve for x.

$$x = \frac{-(0.070\,05) \pm \sqrt{(0.070\,05)^2 - (4)(1)(-3.132 \text{ x } 10^{-5})}}{2(1)} = \frac{-0.070\,05 \pm 0.070\,94}{2}$$

$x = -0.0705$ and $4.45 \text{ x } 10^{-4}$

Of the two solutions for x only $4.45 \text{ x } 10^{-4}$ has physical meaning because −0.0705 leads to negative concentrations.

$[NO_2^-] = x = 4.45 \text{ x } 10^{-4} \text{ M} = 4.4 \text{ x } 10^{-4} \text{ M}$

$[H_3O^+] = 0.0696 + x = 0.0696 + 4.45 \text{ x } 10^{-4} = 0.0700 \text{ M}$

$pH = -\log[H_3O^+] = -\log(0.0700) = 1.15$

(c) Total solution molarity = $[NO_3^-] + [NO_2^-] + [H_3O^+] + [HNO_2]$
$= 0.0696 \text{ M} + (4.45 \text{ x } 10^{-4}) + 0.0700 \text{ M} + 0.0692 \text{ M} = 0.2092 \text{ M}$

$$\Pi = MRT = (0.2092 \text{ M})\left(0.082\,06 \frac{L \cdot atm}{K \cdot mol}\right)(298 \text{ K}) = 5.12 \text{ atm}$$

(d) mol H_3O^+ = (0.0700 mol/L)(0.250 L) = 0.0175 mol H_3O^+

mol HNO_2 = (0.0692 mol/L)(0.250 L) = 0.0173 mol HNO_2

Total mol of acid to neutralize = 0.0175 + 0.0173 = 0.0348 mol

$$\text{mass CaO} = 0.0348 \text{ mol acid x } \frac{1 \text{ mol CaO}}{2 \text{ mol acid}} \text{ x } \frac{56.08 \text{ g CaO}}{1 \text{ mol CaO}} = 0.976 \text{ g CaO}$$

23 Organic and Biological Chemistry

23.1 (a) 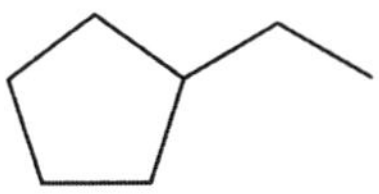 (b)

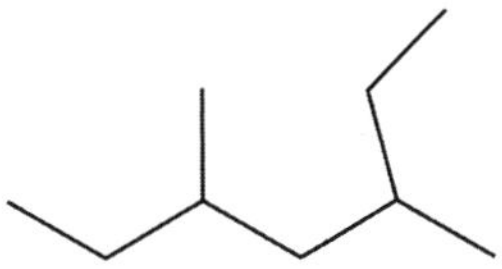

23.2 (a) C_8H_{16} (b) C_7H_{16}

23.3 Structures (a) and (c) are identical. They both contain a chain of six carbons with two $-CH_3$ branches at the fourth carbon and one $-CH_3$ branch at the second carbon. Structure (b) is different, having a chain of seven carbons.

23.4 $CH_3CH_2CH_2CH_2CH_2CH_3$

$CH_3CH(CH_3)CH_2CH_2CH_3$

$CH_3CH_2CH(CH_3)CH_2CH_3$

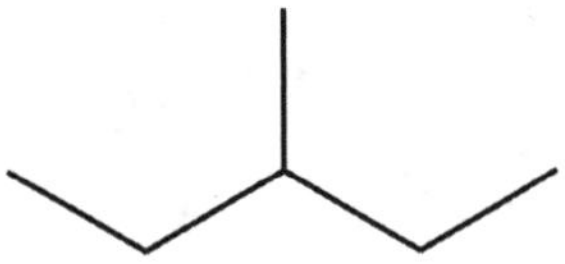

$CH_3C(CH_3)_2CH_2CH_3$

$CH_3CH(CH_3)CH(CH_3)CH_3$

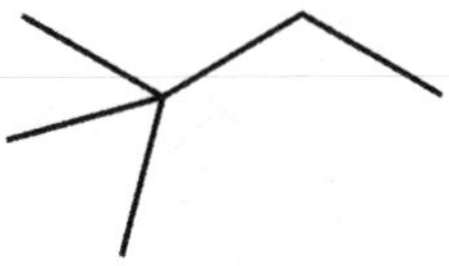

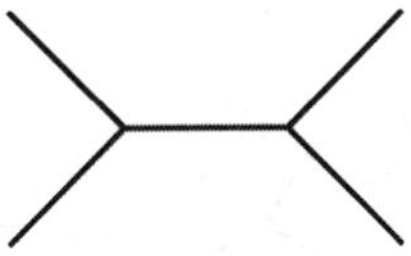

23.5 (a) (b)

23.6 (a) (b) (c) (d) $C_6H_{11}NH_2$

23.7 (a) (b) (c)

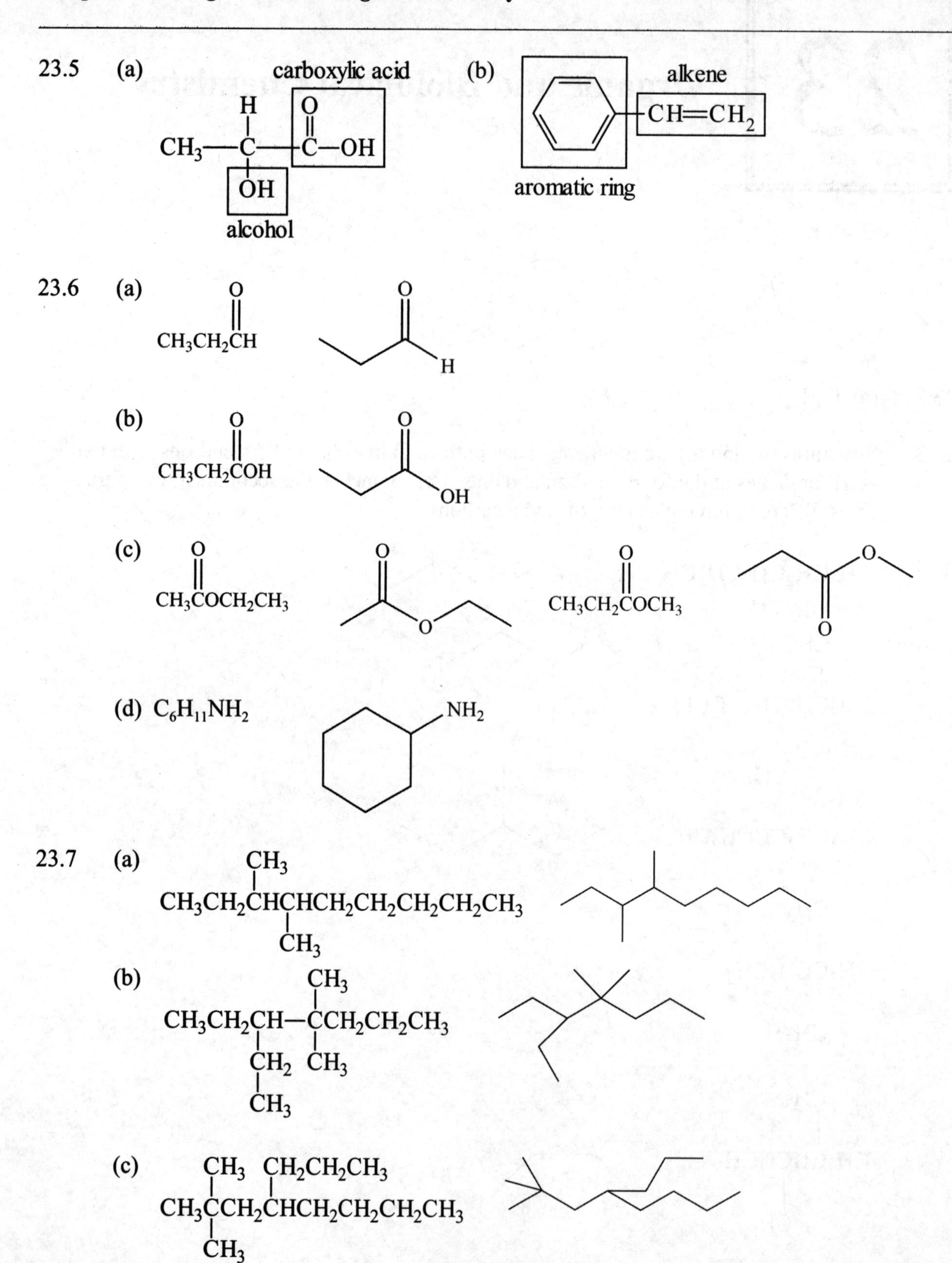

(d)

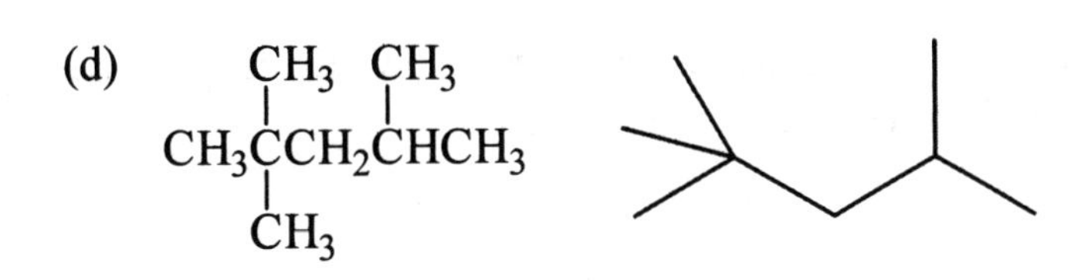

23.8 pentane

2-methylbutane

2,2-dimethypropane

23.9 (a) 3-methyl-1-butene

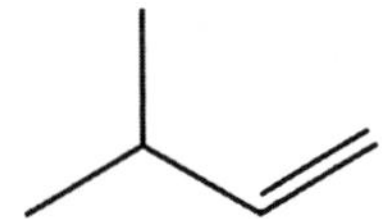

(b) 4-methyl-3-heptene

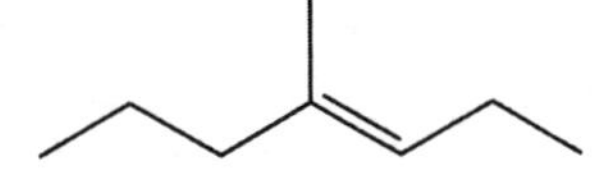

(c) 3-ethyl-1-hexyne

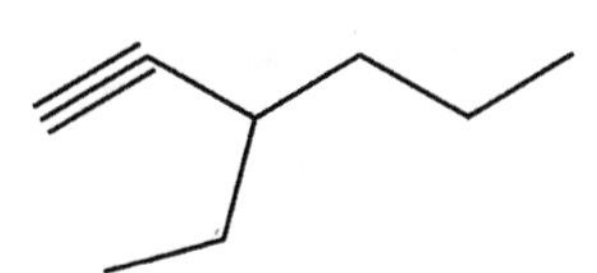

23.10 (a)

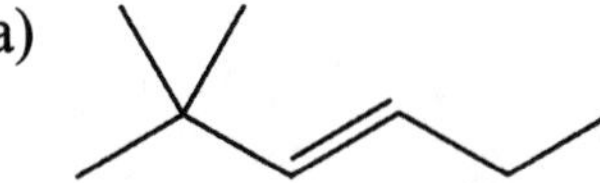

(b)

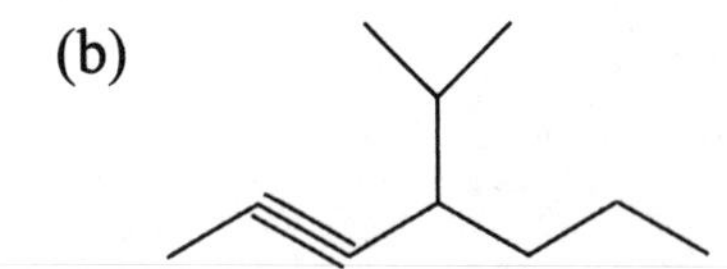

23.11 (a) aldopentose (b) ketotriose (c) aldotetrose

23.12

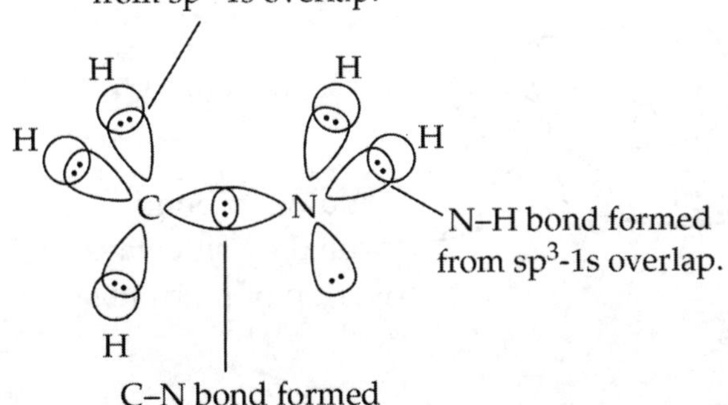

23.13 The molecule is unstable because the C–C bonds are relatively weak due to the less effective orbital overlap in a molecule with 60° bond angles.

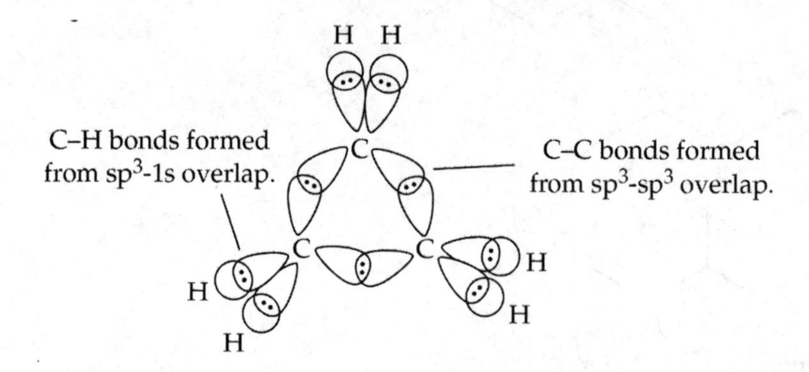
C–H bonds formed
from sp^3-1s overlap.
C–C bonds formed
from sp^3-sp^3 overlap.

23.14 (a)

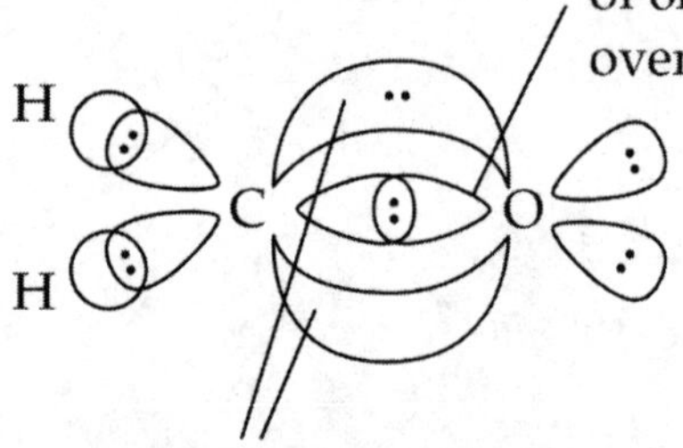
Carbon–Oxygen double bond consists
of one σ bond formed by head on
overlap of sp^2 orbitals ...
... and sideways overlap of
p orbitals

(b) No (c) Yes

23.15 (a) $H_2C{=}C{=}\ddot{O}:$

(b)

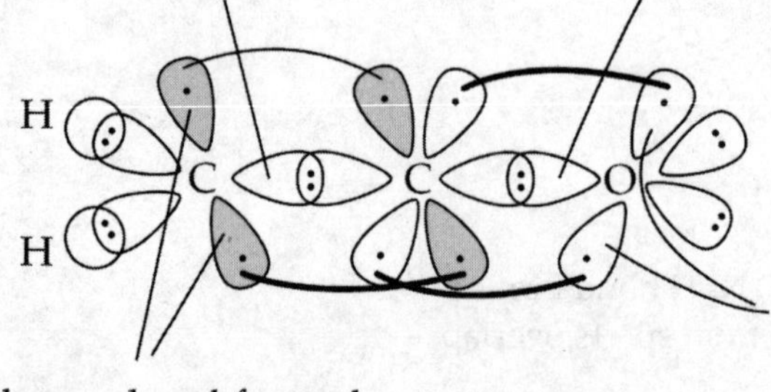
Carbon–Carbon double bond
consists of one σ bond
formed by head on overlap of
sp^2 and sp orbitals...
Carbon–Oxygen double bond
consists of one σ bond formed by
head on overlap of sp and sp^2 orbitals...
...and one π bond formed
by sideways overlap
of p orbitals
...and one π bond
formed by sideways
overlap of p orbitals

23.16 (a) Does not exhibit cis-trans isomerism.

(b) The cis isomer was shown in the problem.

The trans isomer is shown here:

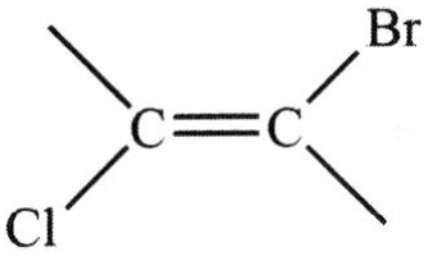

Br

C=C

Cl

(c) Does not exhibit cis-trans isomerism.

23.17 (a) Fumaric acid is a trans isomer.

(b) Maleic acid

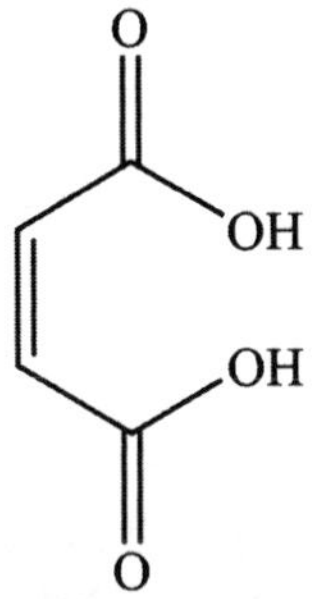

O

OH

OH

O

(c) No, because succinic acid does not have a carbon-carbon double bond.

23.18 (a) trans (b) cis

23.19 (a) 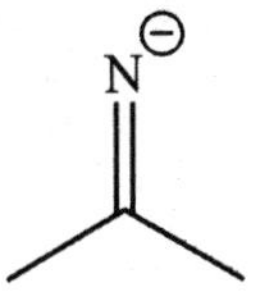 N⊖ (b)  ⊕ (c) 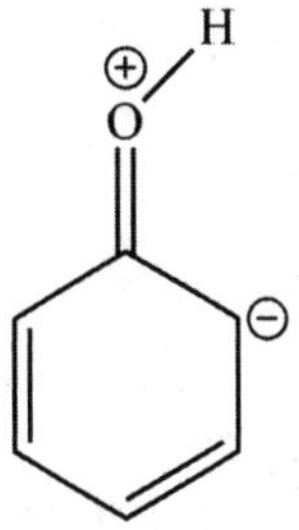 ⊕ H O ⊖

23.20 (a) 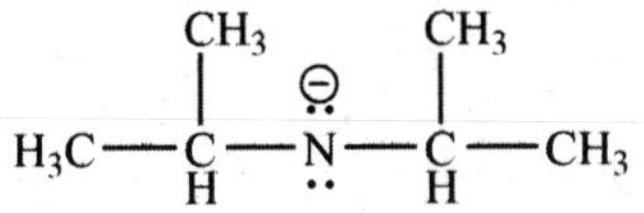 CH₃ CH₃ H₃C—C—N—C—CH₃ H H

(b) ⊖O ⊕ O⊖ N C ⊕ H_2C CH H_2C CH_2 C H_2

(c) H C⊖ HC CH HC—CH

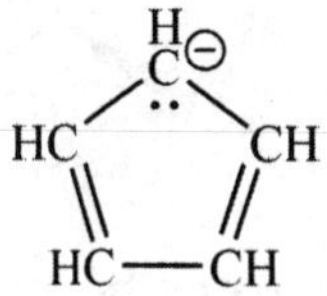

23.21 (a) 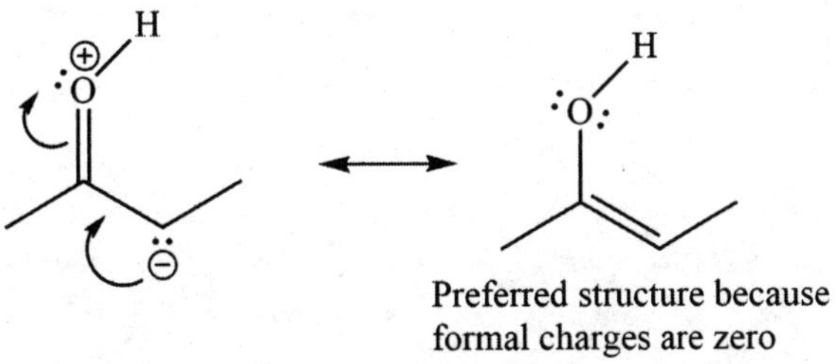 ⊕ H O ⊖ ⟷ H O

Preferred structure because formal charges are zero

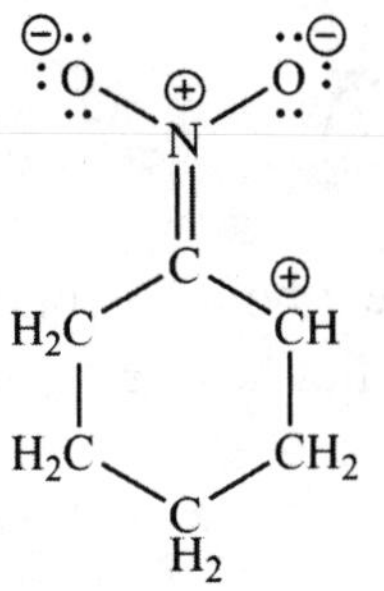

(b) 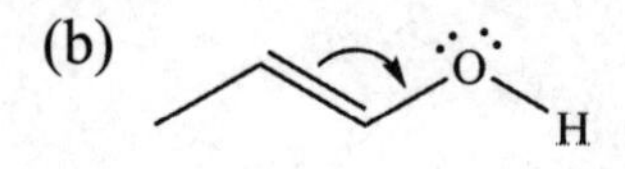

Moving electrons as indicated would result in an incorrect electron-dot structure because oxygen would have an expanded octet.

(c) 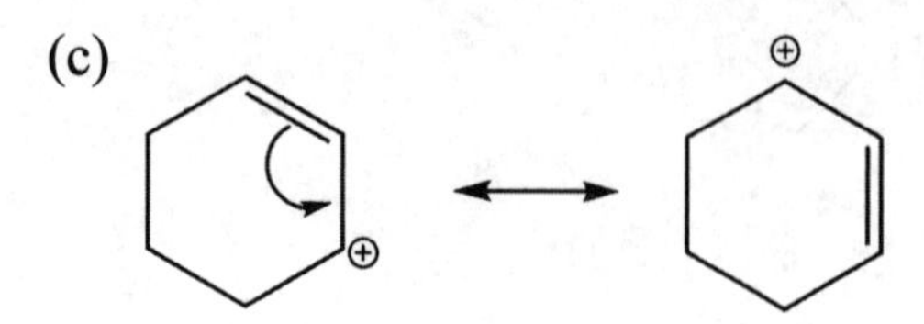

Both resonance structures are equivalent

(d)

Preferred resonance structure because negative formal charge is on more electronegative oxygen

23.22 (a) (b)

23.23 Not conjugated

Conjugated

23.24 In the triple bond one of the π bonds is perpendicular to the other π bonds in the molecule. Therefore, it cannot have sideways overlap with the p orbitals that make up the other π bonds.

23.25 (a) (b) 8

23.26

sp^3
HO

sp^2
HO

sp^3
OH

N

23.27

O

NH_2
delocalized

N
localized

The localized lone pair of electrons is in an sp^2 hybrid orbital that is in a plane perpendicular to the orbitals used for the delocalized electrons in the aromatic ring.

23.28 Val-Cys

$$H_2N\underset{\underset{\underset{CH_3}{|}}{CHCH_3}}{\underset{|}{CH}}\overset{O}{\overset{\|}{C}}NH\underset{CH_2SH}{\underset{|}{CH}}\overset{O}{\overset{\|}{C}}OH$$

Cys-Val

$$H_2N\underset{CH_2SH}{\underset{|}{CH}}\overset{O}{\overset{\|}{C}}NH\underset{\underset{\underset{CH_3}{|}}{CHCH_3}}{\underset{|}{CH}}\overset{O}{\overset{\|}{C}}OH$$

23.29 (a) 4 (b) Phe-nonpolar, Asn-polar, Trp-nonpolar, Ala-nonpolar

23.30 Proline is not aromatic because it does not have a ring of sp^2 hybridized atoms. Tyrosine is aromatic because it has a six-member ring of sp^2 hybridized carbon atoms that are part of a conjugated system. There are 6π electrons and satisfying the 4n + 2 rule for aromaticity.

23.31 (a)

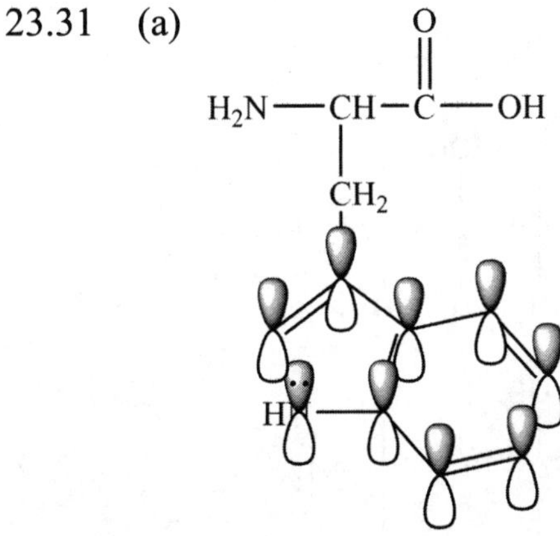

(b) All atoms in the two-ring system of the side chain of tryptophan are sp^2 hybridized and conjugated. Four double bonds contribute 8 π electrons and one lone pair contributes 2 π electrons for a total of 10 π electrons, satisfying the 4n + 2 rule.

23.32

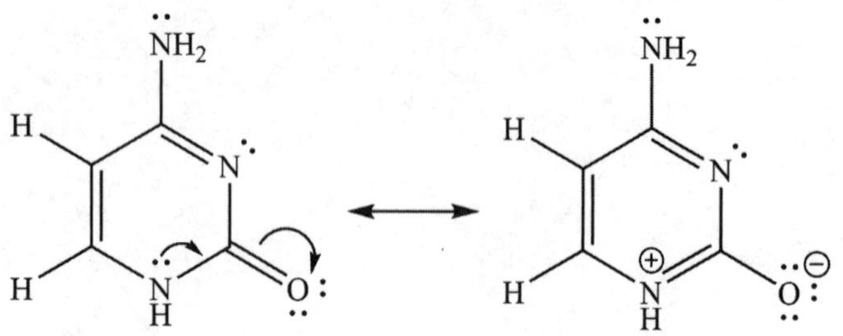

NH₂ H H N H N O ⟷ NH₂ H H N H N⊕ O⊖

23.33 (a) ester, alcohol, alkene

(b) $O = sp^2$, $C_a = sp^3$, $C_b = sp^2$, $C_c = sp^3$

(c) HO HO HO OH O O ⟷ HO HO HO OH O⊕ O⊖

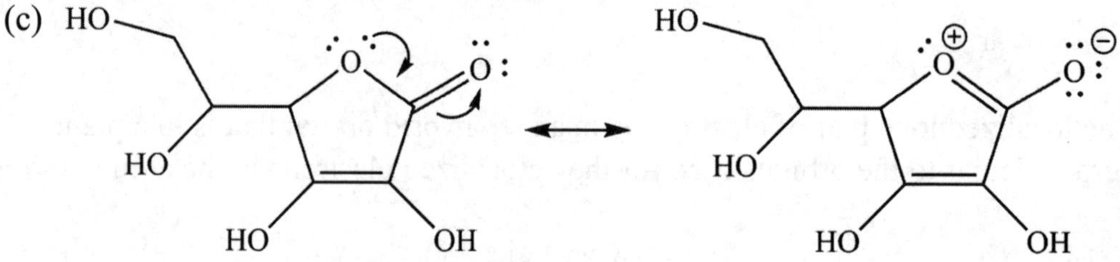

23.34 (a) carbonyl, amine, alkene

(b) O⊖ CH₃ H₃C N⊕ N ⊖O N N⊕ CH₃

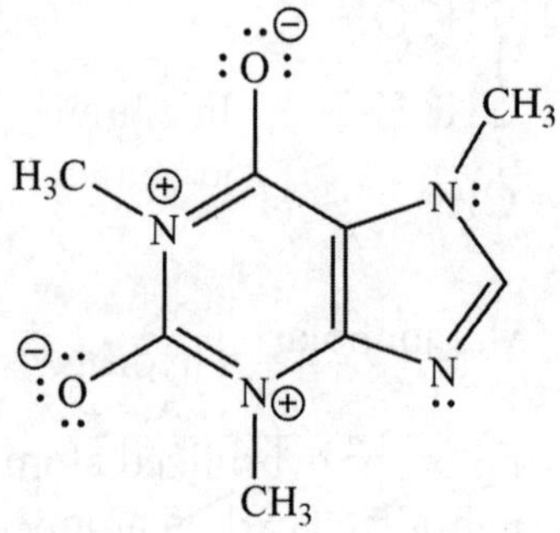

(c) sp^2

(d) 10 electrons in the conjugated system so it meets the 4n + 2 rule. Caffeine is an aromatic compound.

Conceptual Problems

23.35

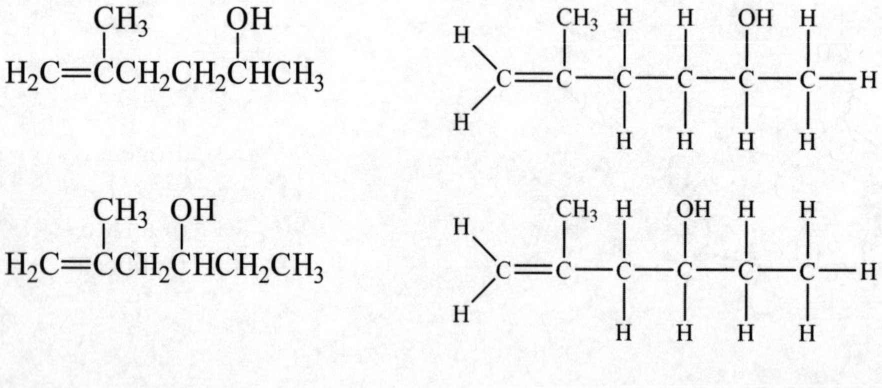

$$H_2C{=}\overset{CH_3}{\overset{|}{C}}CH_2CH_2\overset{OH}{\overset{|}{C}}HCH_3$$

$$H_2C{=}\overset{CH_3}{\overset{|}{C}}CH_2\overset{OH}{\overset{|}{C}}HCH_2CH_3$$

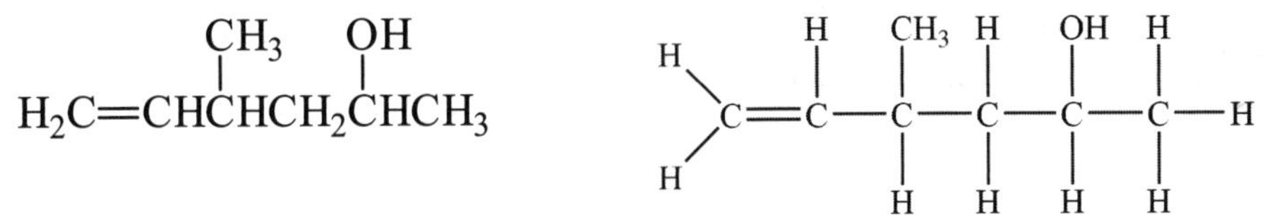

$$H_2C{=}CH\underset{}{\overset{CH_3}{\overset{|}{C}}}HCH_2\overset{OH}{\overset{|}{C}}HCH_3$$

23.36 (a) alkene, ketone, ether (b) alkene, amine, carboxylic acid

23.37 (a) 2,3-dimethylpentane (b) 2-methyl-2-hexene

23.38 (a) serine (b) methionine

23.39

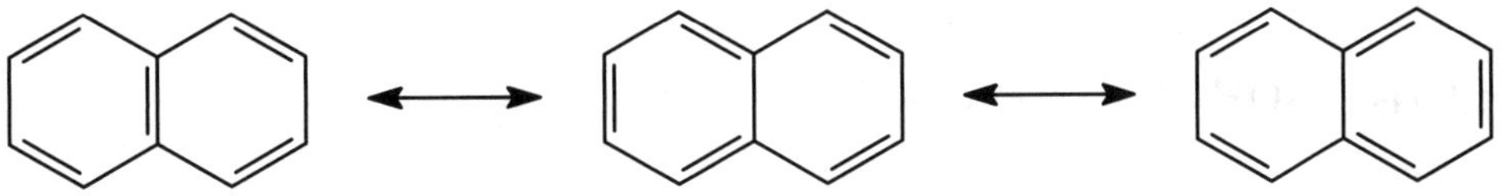

23.40 Ser-Val

23.41 (a) guanine (b) cytosine

Section Problems
Isomers, Functional Groups, and Naming (Sections 23.1–23.3)

23.42 In a straight-chain alkane, all the carbons are connected in a row. In a branched-chain alkane, there are branching connections of carbons along the carbon chain.

23.43 An alkane is a compound that contains only carbon and hydrogen and has only single bonds. An alkyl group is the part of an alkane that remains when a hydrogen is removed.

23.44 $CH_3CH_2CH_2CH_2CH_2CH_2CH_3$

23.45 $CH_3CH_2CH_2CH_2CH_3$

23.46 (a) No, because they contain different numbers of carbons and hydrogens.
(b) They are isomers of each other.
(c) No, they are identical.

23.47 (a) They are isomers of each other.
(b) They are isomers of each other.
(c) No, because they contain different numbers of carbons and hydrogens.

23.48 A functional group is a part of a larger molecule and is composed of an atom or group of atoms that has a characteristic chemical behavior. They are important because their chemistry controls the chemistry in molecules that contain them.

23.49 (a) $\rangle C{=}C \langle$ (b) $-\overset{|}{\underset{|}{C}}-O-H$ (c) $-\overset{|}{\underset{|}{C}}-\overset{O}{\overset{\|}{C}}-O-\overset{|}{\underset{|}{C}}-$ (d) $-\overset{|}{\underset{|}{C}}-\overset{|}{N}-$

23.50 (a) $CH_3CH_2\overset{O}{\overset{\|}{C}}CH_2CH_3$ (b) $CH_3CH_2CH_2\overset{O}{\overset{\|}{C}}OCH_2CH_3$ (c) $NH_2CH_2\overset{O}{\overset{\|}{C}}OH$

23.51 (a) NH_2 (b) (c) O (d) O, OH

23.52 $CH_3CH_2CH_2OH$ $\quad CH_3\overset{OH}{\overset{|}{C}}HCH_3$ $\quad CH_3CH_2OCH_3$

23.53 They have the same molecular formulas but different chemical structures. They are isomers.

23.54 ester, aromatic ring, and amine

23.55 (a) $\underline{C}H_3{=}CHCH_2CH_2OH$ Underlined carbon has five bonds.

(b) $CH_3CH_2CH{=}\underline{\overset{O}{\overset{\|}{C}}}CH_3$ Underlined carbon has five bonds.

(c) $CH_3CH_2C{\equiv}\underline{C}H_2CH_3$ Underlined carbon has six bonds.

23.56 (a) 4-ethyl-3-methyloctane
(b) 4-isopropyl-2-methylheptane
(c) 2,2,6-trimethylheptane
(d) 4-ethyl-4-methyloctane

23.57 2,2,4-trimethylpentane

23.58 (a) (b)

(c) (d)

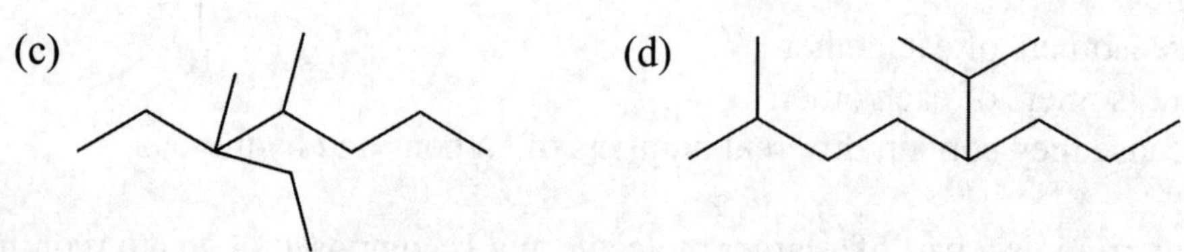

23.59 (a) (b) (c) (d)

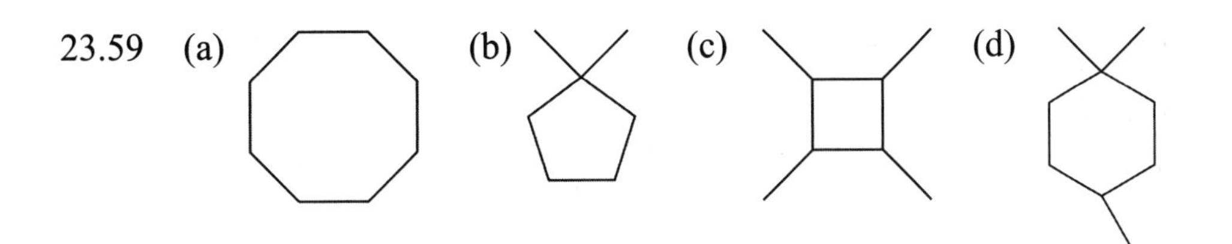

23.60 $CH_2{=}C{=}CHCH_2CH_3$ $CH_2{=}CHCH{=}CHCH_3$ $CH_2{=}CHCH_2CH{=}CH_2$

$CH_3CH{=}C{=}CHCH_3$ $CH_2{=}C(CH_3)CH{=}CH_2$ $CH_2{=}C{=}C(CH_3)CH_3$

23.61 C_3H_9 contains one more H than needed for an alkane.

23.62 (a) 4-methyl-2-pentene
(b) 3-methyl-1-pentene
(c) 1,2-dichlorobenzene, or o-dichlorobenzene
(d) 2-methyl-2-butene
(e) 7-methyl-3-octyne

23.63 (a) $CH_3C(H){=}C(H)CH_2CH_2CH_3$ (b) $CH_3CH(CH_3)CH{=}CHCH_2CH_3$ (c) $CH_2{=}C(CH_3)CH{=}CH_2$

23.64 (a) C_3H_4 (b) C_4H_8 (c) C_5H_6

23.65 (a) C_4H_{10} (b) C_6H_{10} (c) C_6H_6

23.66 (a) C_2H_6O (b) C_2H_7N

(c) $C_5H_{11}N$ (d) C_3H_8O

(e) 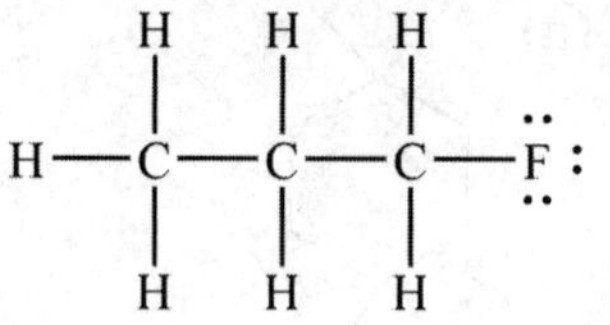 C_3H_7F (f) 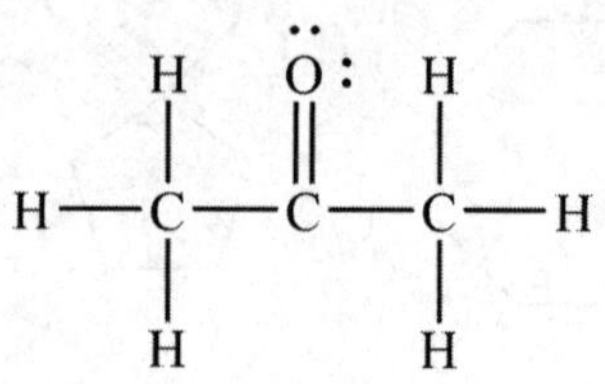 C_3H_6O

23.67 (a) 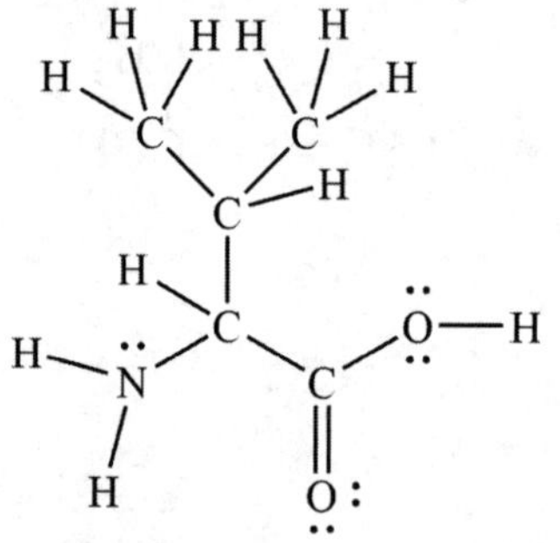 $C_5H_{11}NO_2$ (b) 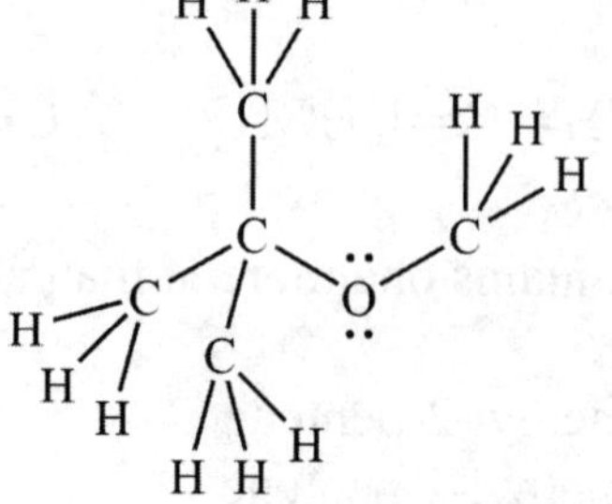 $C_5H_{12}O$

(c) 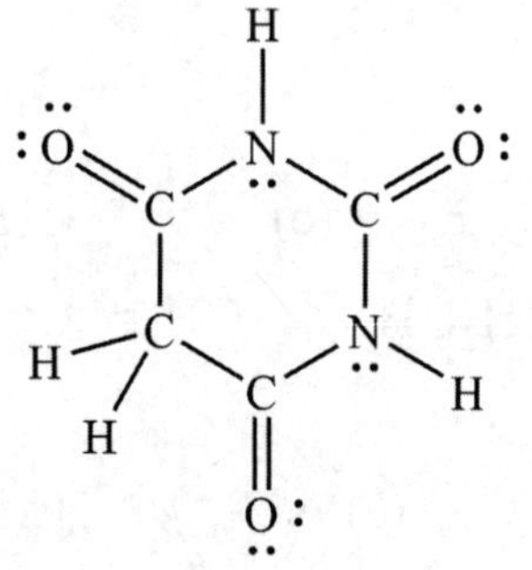 $C_4H_4N_2O_3$

23.68 (a) 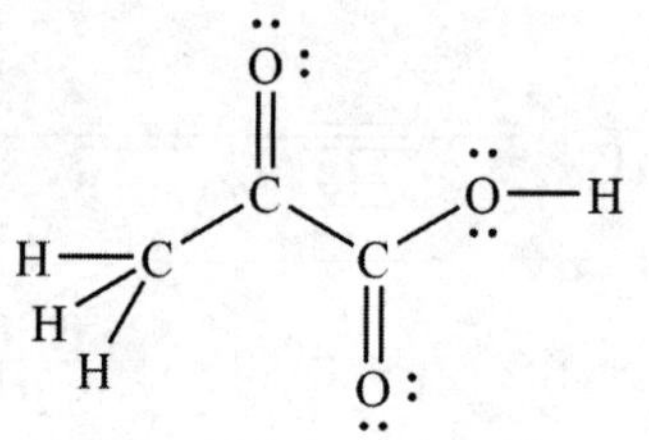 $C_3H_4O_3$ (b) 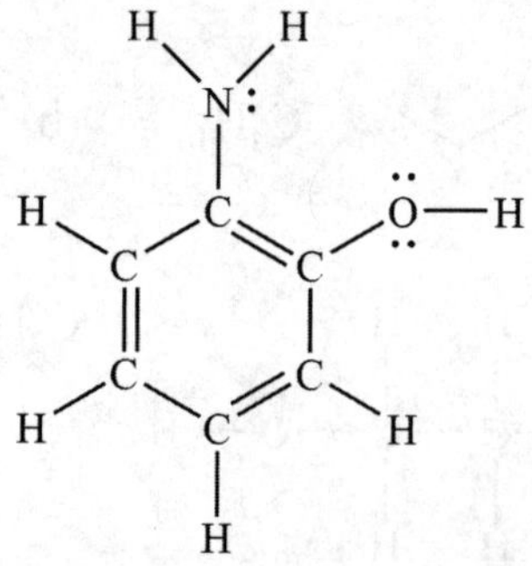 C_6H_7NO

(c) 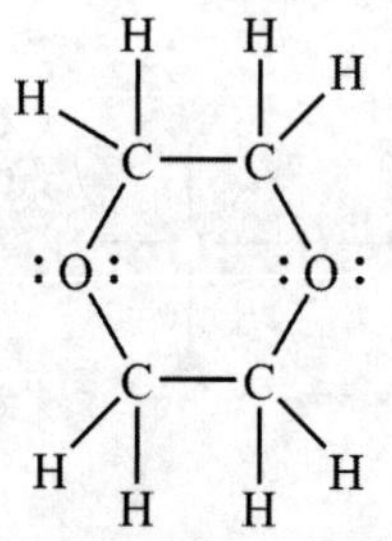 $C_4H_8O_2$

23.69 (a) $C_7H_6Cl_2O$ (b) C_5H_8O

(c) CH_4N_2O

Carbohydrates (Section 23.4)

23.70 An aldose contains the aldehyde functional group while a ketose contains the ketone functional group.

23.71 (a) aldopentose (b) aldohexose (c) ketohexose

23.72

$$\begin{array}{c} O \\ \| \\ HOCH_2CHCCH_2OH \\ | \\ OH \end{array}$$

23.73

$$\begin{array}{c} O \\ \| \\ HOCH_2CHCH_2CH \\ | \\ OH \end{array}$$

23.74 (a) constitutional isomers (b) anomers

23.75 (a) One has 4 carbons and the other has 3 carbons, they are not isomers.
(b) enantiomers

Valence Bond Theory, Cis-Trans Isomers (Section 23.5)

23.76 (a) All C–H bond angles are ~109.5°.

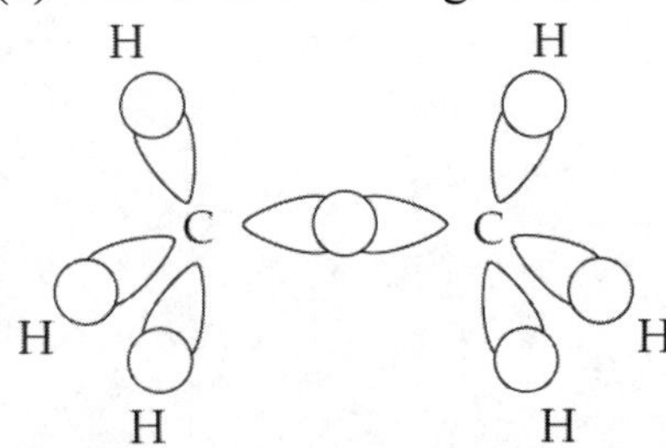

(b) Two C–H bond angles are ~120° and one is 90°.

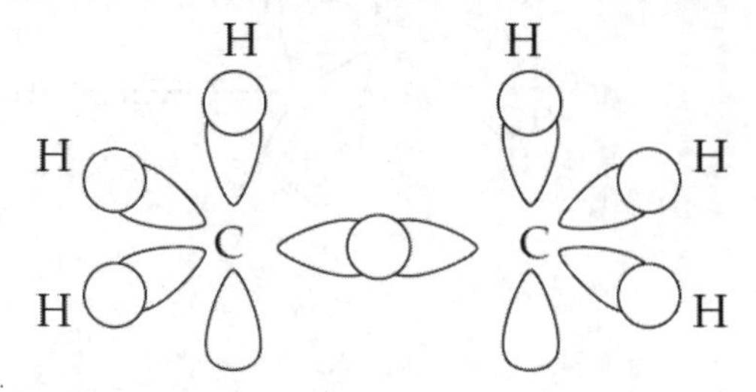

(c) In structure (b), the 90° bond angle introduces a larger repulsion and lower stability. Structure (a) is more favorable.

23.77 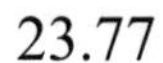

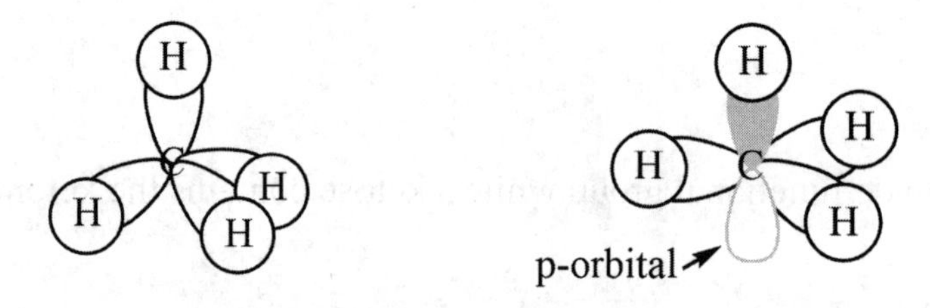

In the second structure, the 90° bond angles introduce a larger repulsion and lower stability. The first structure with 109.5° bond angles is more favorable.

23.78

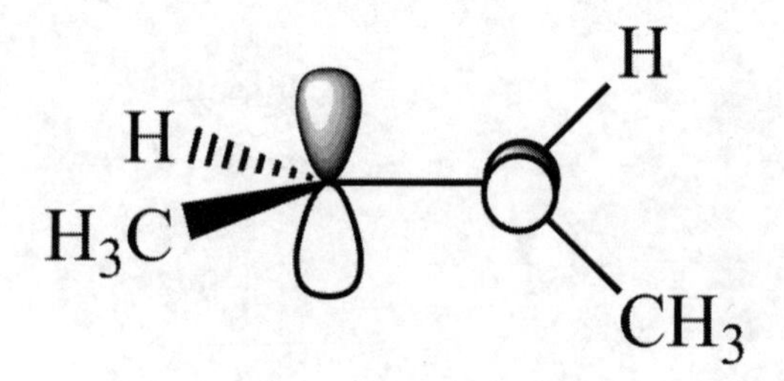

23.79

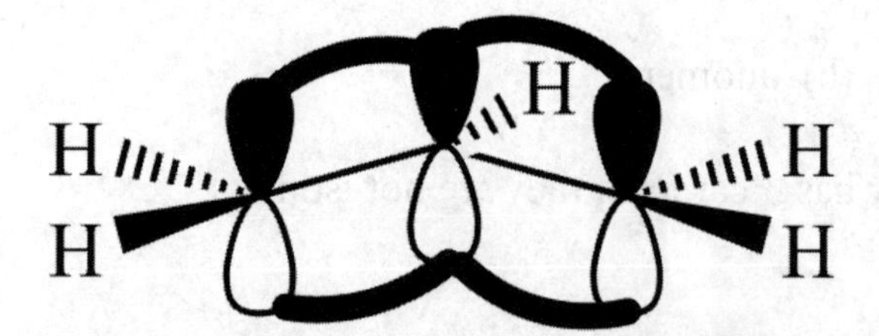

There is delocalization across the 3 carbons that are all sp^2 hybridized. This requires that all carbons and hydrogens be in the same plane.

23.80

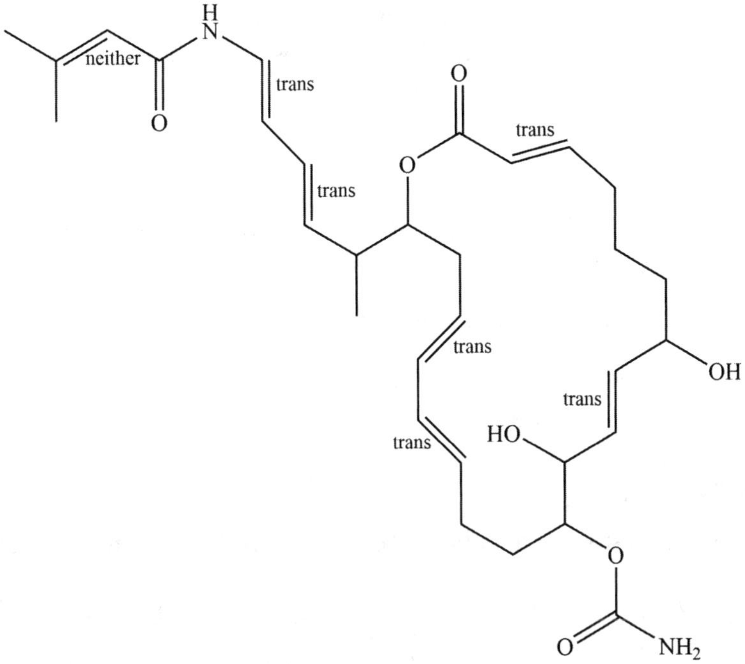

neither, trans, trans, trans, trans, trans, trans; H N; O; O; O; OH; HO; O; O; NH_2

23.81

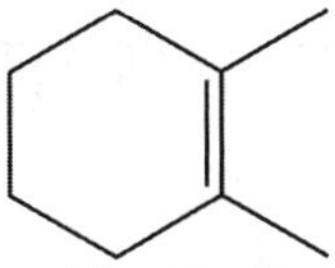

The six-membered ring lacks the flexibility of larger ring systems. It locks in the cis double bond.

23.82 Compounds (b) and (c) exhibit cis–trans isomerism.

(b) 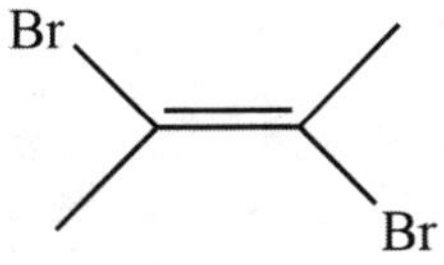 Br, Br — trans

(c) 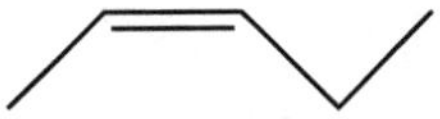 cis

23.83 Compounds (b) and (c) exhibit cis–trans isomerism.

(b) 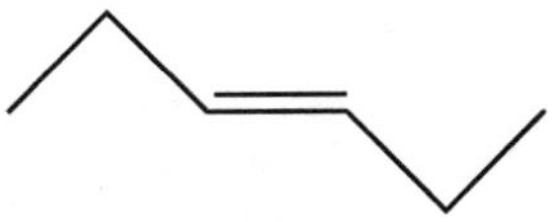 trans

(c) 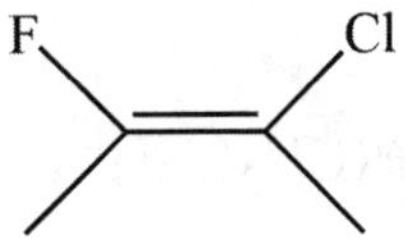 F, Cl — cis

23.84 (a) CH_2=$CHCH_2CH_2CH_2CH_3$ This compound cannot form cis-trans isomers.

(b) CH_3CH=$CHCH_2CH_2CH_3$ This compound can form cis-trans isomers because of the different groups on each double bond C.

(c) $CH_3CH_2CH{=}CHCH_2CH_3$ This compound can form cis-trans isomers because of the different groups on each double bond C.

23.85 (a)

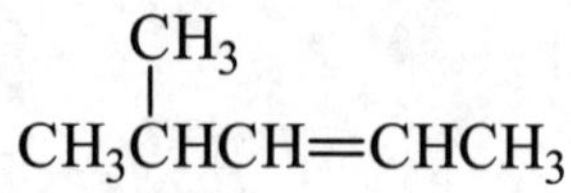

$$\begin{array}{l}\quad\;\; CH_3 \\ \quad\;\;\, | \\ CH_3CHCH{=}CHCH_3\end{array}$$

This compound can form cis-trans isomers because of the different groups on each double bond C.

(b)

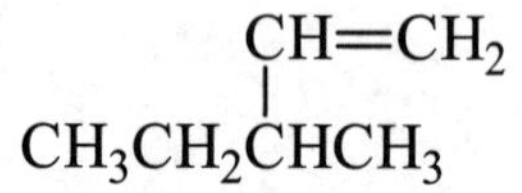

$$\begin{array}{l}\qquad\quad CH{=}CH_2 \\ \qquad\quad\; | \\ CH_3CH_2CHCH_3\end{array}$$

This compound cannot form cis-trans isomers.

(c)

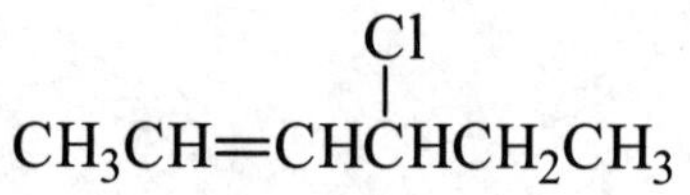

$$\begin{array}{l}\qquad\qquad\quad Cl \\ \qquad\qquad\quad\, | \\ CH_3CH{=}CHCHCH_2CH_3\end{array}$$

This compound can form cis-trans isomers because of the different groups on each double bond C.

Lipids (Section 23.6)

23.86 Long-chain carboxylic acids are called fatty acids. Fatty acids are usually unbranched and have an even number of carbon atoms in the range of 12-22.

23.87 All fats and oils are triesters of glycerol (1,2,3-propanetriol) with three long-chain carboxylic acids, hence they are called triacylglycerols, or triglycerides.

23.88

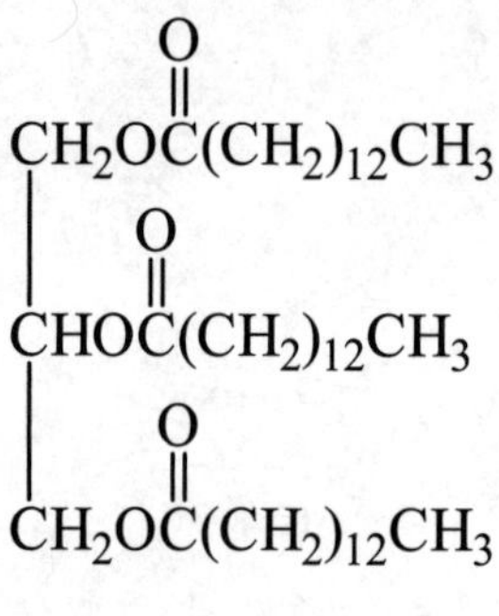

$$\begin{array}{l}\qquad\;\; O \\ \qquad\;\; \| \\ CH_2OC(CH_2)_{12}CH_3 \\ | \qquad O \\ | \qquad \| \\ CHOC(CH_2)_{12}CH_3 \\ | \qquad O \\ | \qquad \| \\ CH_2OC(CH_2)_{12}CH_3\end{array}$$

23.89

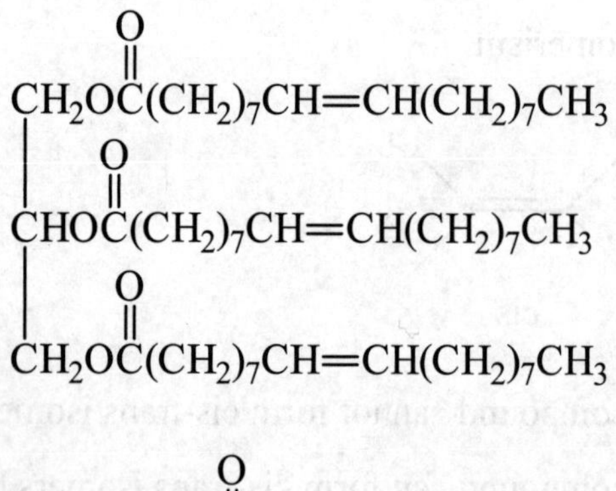

$$\begin{array}{l}\qquad\;\; O \\ \qquad\;\; \| \\ CH_2OC(CH_2)_7CH{=}CH(CH_2)_7CH_3 \\ | \qquad O \\ | \qquad \| \\ CHOC(CH_2)_7CH{=}CH(CH_2)_7CH_3 \\ | \qquad O \\ | \qquad \| \\ CH_2OC(CH_2)_7CH{=}CH(CH_2)_7CH_3\end{array}$$

23.90

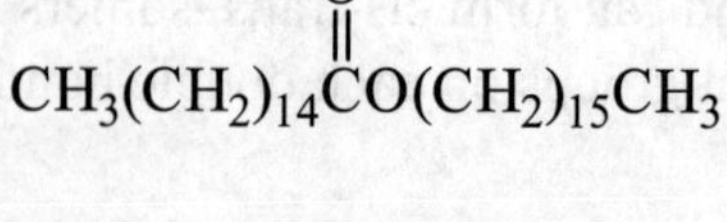

$$\begin{array}{l}\qquad\qquad\quad\; O \\ \qquad\qquad\quad\; \| \\ CH_3(CH_2)_{14}CO(CH_2)_{15}CH_3\end{array}$$

23.91

$$\begin{array}{l} \qquad\; O \\ \qquad\; \| \\ CH_2OC(CH_2)_{16}CH_3 \\ | \qquad O \\ | \qquad \| \\ CHOC(CH_2)_{16}CH_3 \\ | \qquad O \\ | \qquad \| \\ CH_2OC(CH_2)_{14}CH_3 \end{array} \qquad\qquad \begin{array}{l} \qquad\; O \\ \qquad\; \| \\ CH_2OC(CH_2)_{16}CH_3 \\ | \qquad O \\ | \qquad \| \\ CHOC(CH_2)_{14}CH_3 \\ | \qquad O \\ | \qquad \| \\ CH_2OC(CH_2)_{16}CH_3 \end{array}$$

The two fat molecules differ from each other depending on where the palmitic acid chain is located. In the first fat molecule, palmitic acid is on an end and in the second it is in the middle.

23.92 (a) 18:2 (ω-6) (b) 16:0

23.93

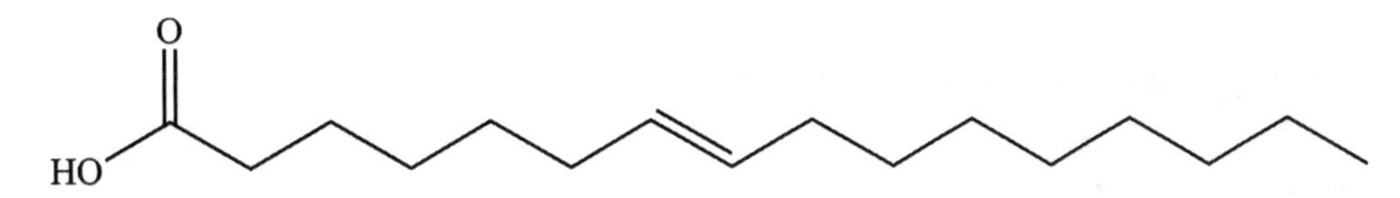

ω-9 fatty acids are nonessential.

23.94 Only (c) is an unsaturated fatty acid. It has double bonds in a cis configuration. This geometry substantially disrupts intermolecular forces, leading to oils (liquids) at room temperature.

23.95 Only (b) is a saturated fatty acid. Individual London dispersion interactions are relatively weak, however many interactions add up to strongly hold saturated fatty acid molecules together, making them solids at room temperature.

23.96 (a) true (b) true (c) false (d) true

23.97 (a) false (b) false (c) cannot be determined (d) true

23.98 None are possible. (a) has only 15 carbons. (b) is fully hydrogenated. (c) is not hydrogenated.

23.99 (b) is palmitic acid, a naturally occurring fatty acid. (a) and (c) could possibly be the product of partial hydrogenation.

Formal Charge and Resonance (Section 23.7)

23.100 (a)

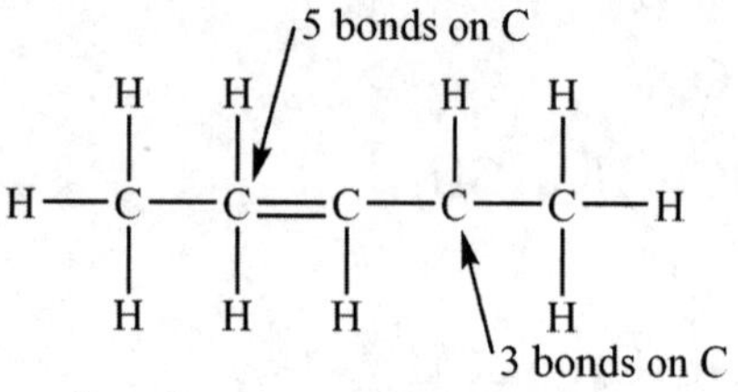

5 bonds on C

3 bonds on C

This structure is not valid.

(b)

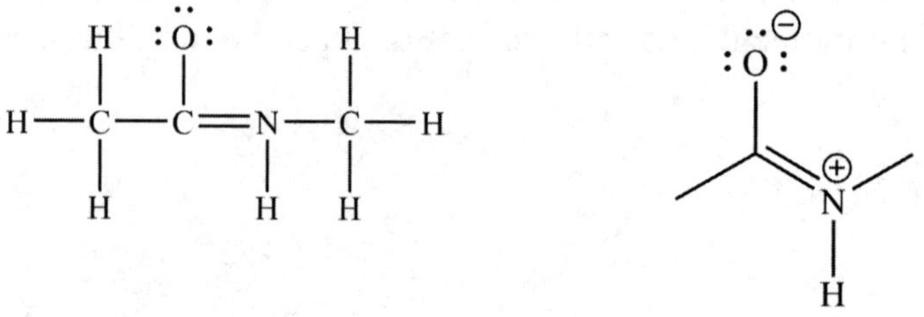

(c) incomplete octet

5 bonds on N & C

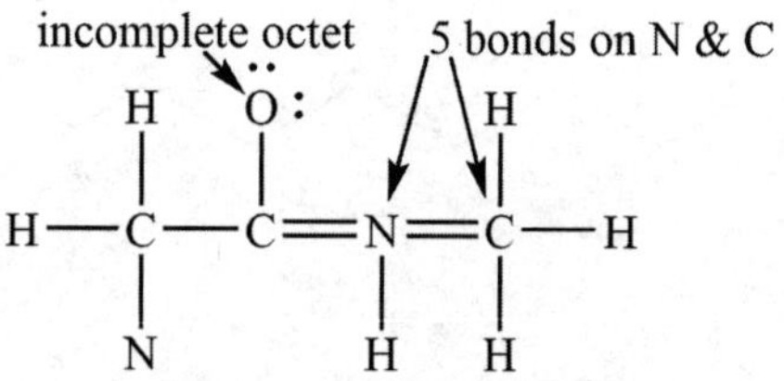

This structure is not valid.

23.101 (a)

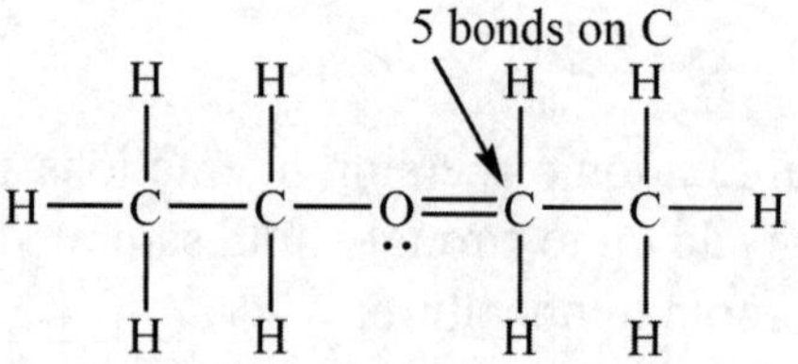

5 bonds on C

This structure is not valid.

(b)

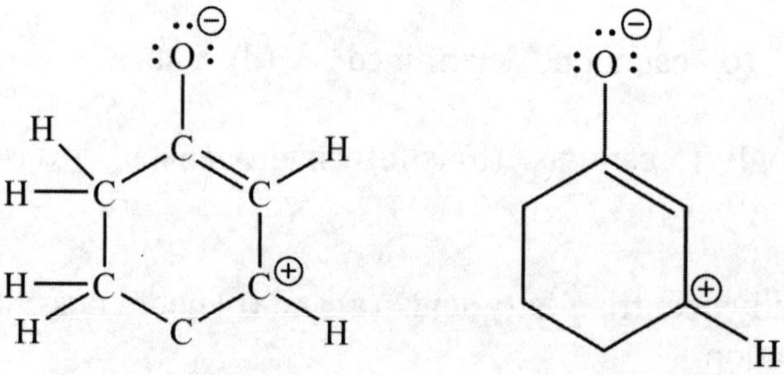

(c)

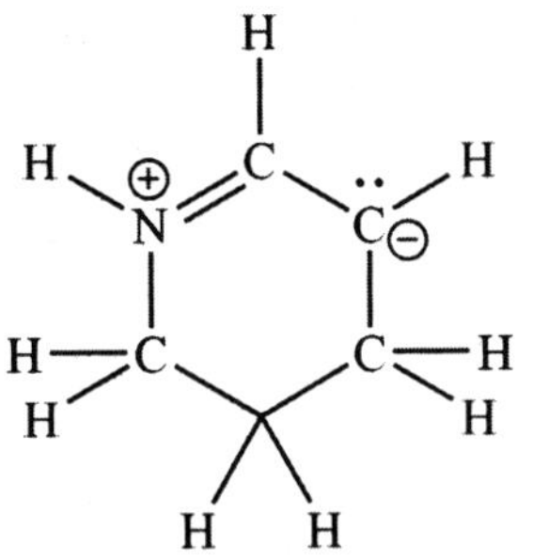

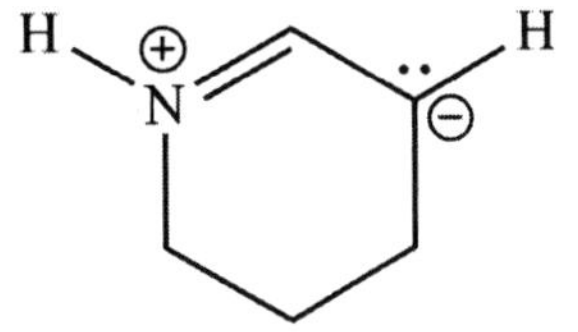

23.102 (a)

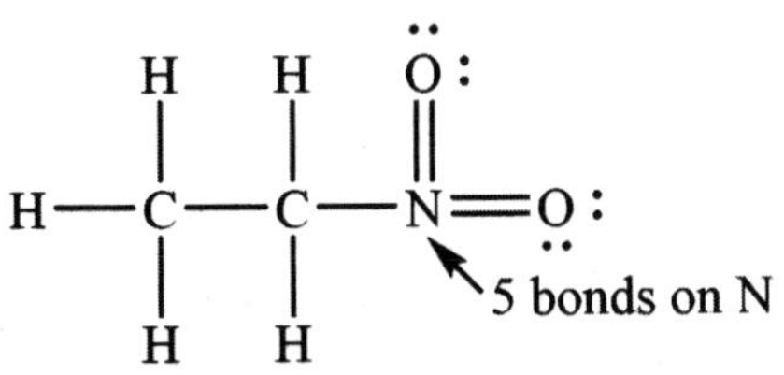

5 bonds on N

This structure is not valid.

(b)

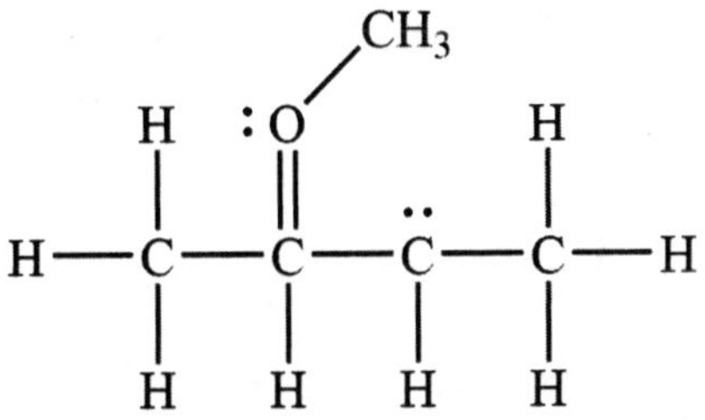

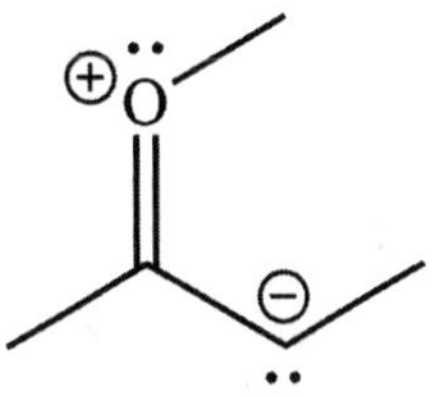

(c)

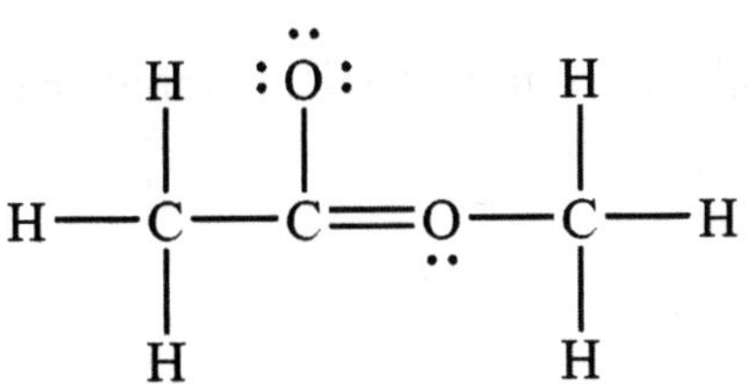

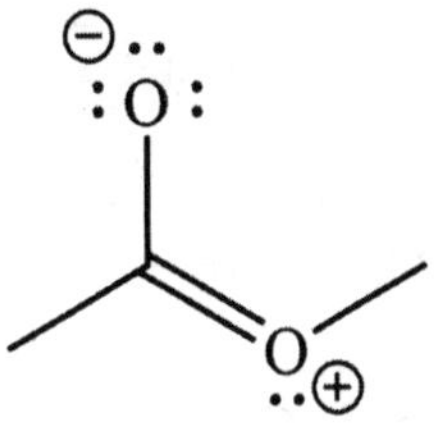

23.103 (a)

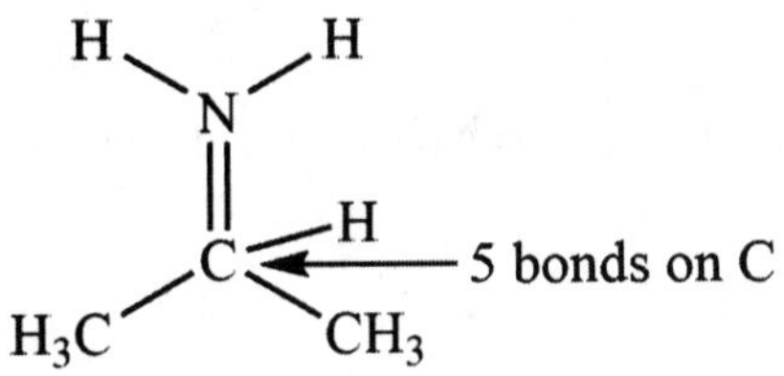

5 bonds on C

This structure is not valid.

(b)

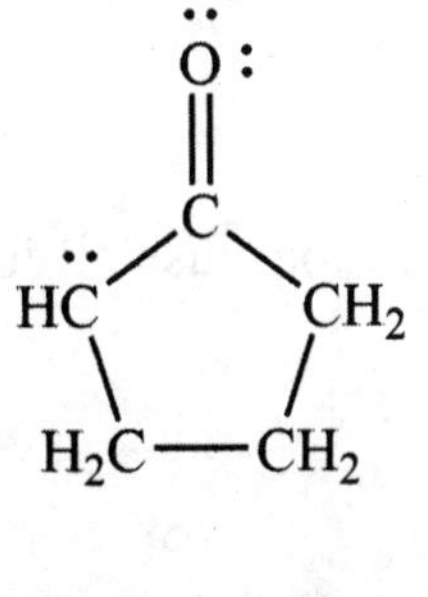

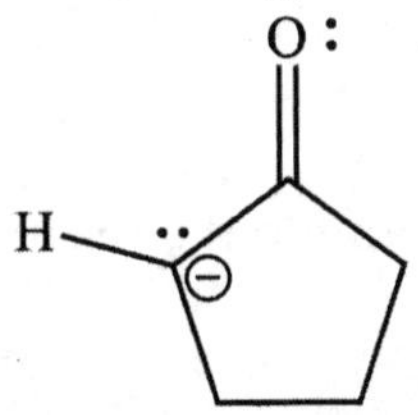

(c)

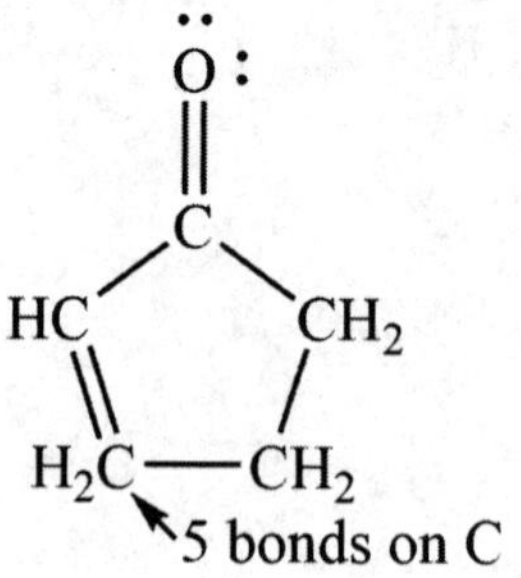

HC, C=O, CH₂, H₂C—CH₂

5 bonds on C

This structure is not valid.

23.104 (a)

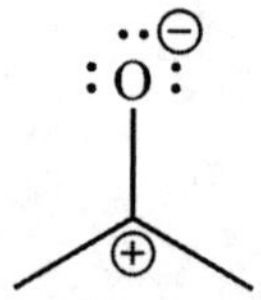

The original structure contributes more to the resonance hybrid because all formal charges are zero.

(b)

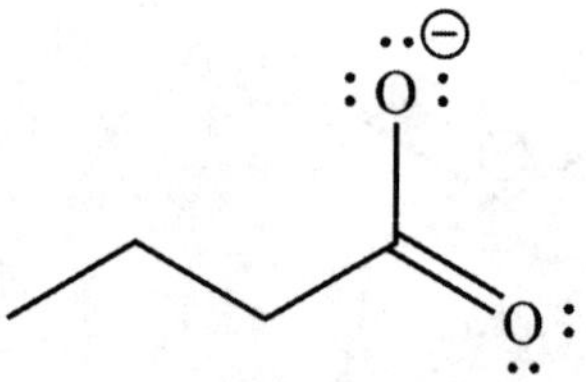

The two structures are identical and are equal contributors to the resonance hybrid.

23.105 (a)

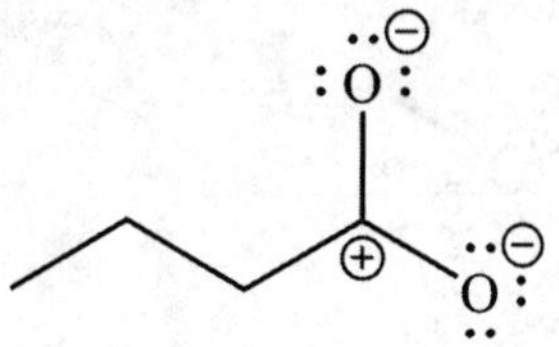

The original structure contributes more to the resonance hybrid because of fewer nonzero formal charges.

(b)

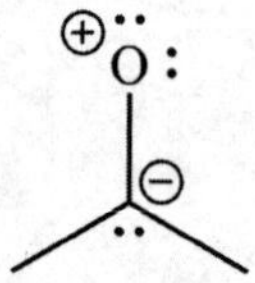

The original structure contributes more to the resonance hybrid because all formal charges are zero.

23.106 (a)

The original structure contributes more to the resonance hybrid because all formal charges are zero.

(b)

The original structure contributes more to the resonance hybrid because all formal charges are zero.

23.107 (a)

The original structure contributes more to the resonance hybrid because all formal charges are zero.

(b)

This structure contributes more to the resonance hybrid because all formal charges are zero.

23.108 They are resonance structures.

23.109 They are constitutional isomers.

23.110

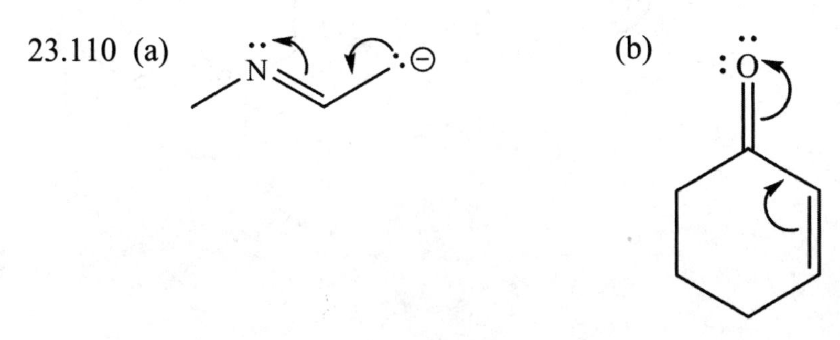

23.111 (a)

(b)

Conjugation Systems (Section 23.8)

23.112 (a)

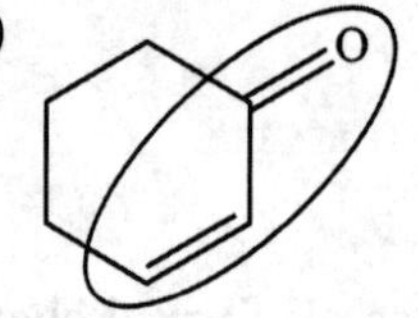

(b)

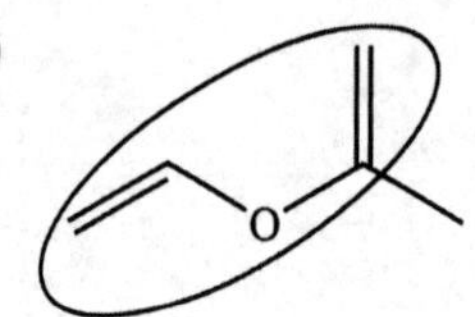

(c)

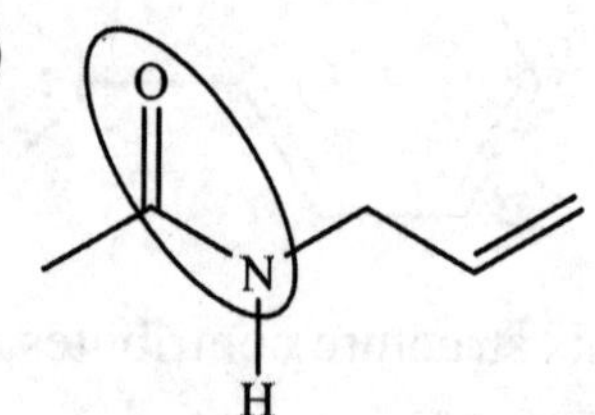

23.113 (a)

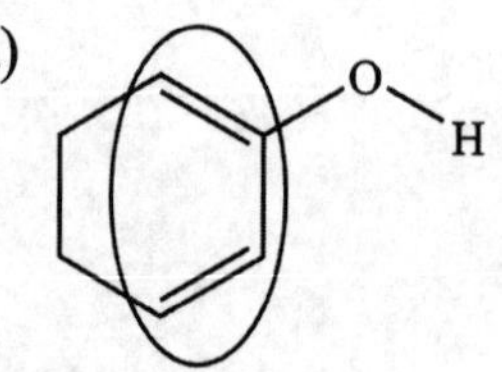

(b)

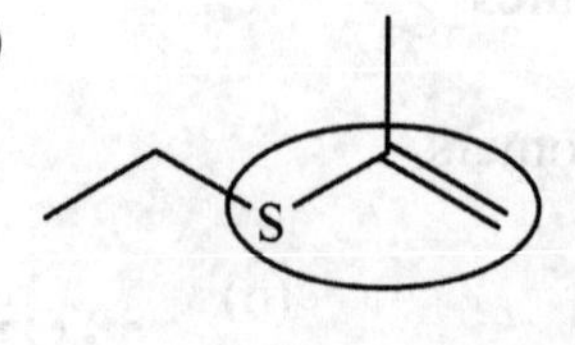

(c)

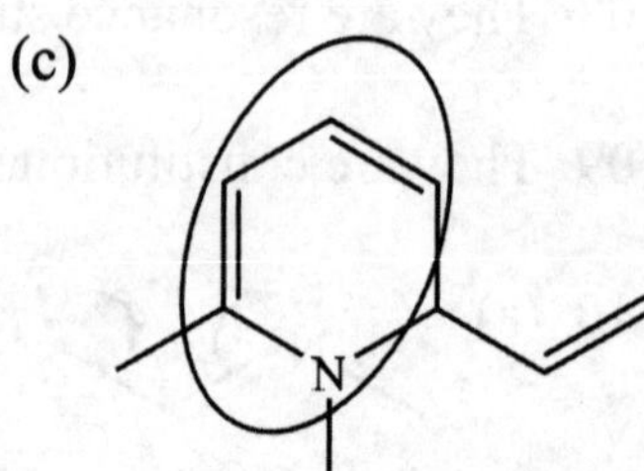

23.114

23.115

All four oxygens are sp^2 hybridized.

23.116

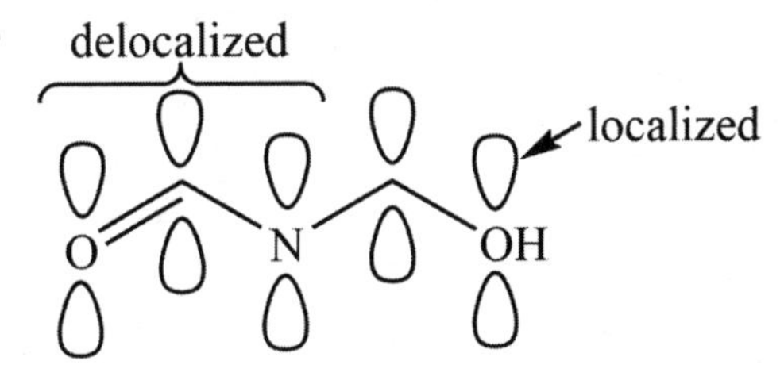

23.117

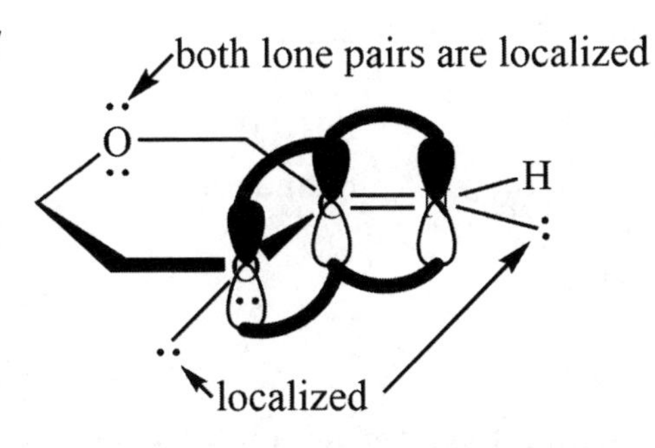

Proteins (Section 23.9)

23.118 (a) serine (b) threonine (c) proline (d) phenylalanine (e) cysteine

23.119 (a)

$$\underset{\text{valine}}{\underset{\text{NH}_2}{\overset{\text{CH}_3}{\text{CH}_3\text{CHCHCOOH}}}} \qquad \underset{\text{leucine}}{\underset{\text{NH}_2}{\overset{\text{CH}_3}{\text{CH}_3\text{CHCH}_2\text{CHCOOH}}}}$$

(b) $HOCH_2CHCOOH$ with NH_2 on the CH — serine

$CH_3CHCHCOOH$ with OH on C-3 and NH_2 on C-2 — threonine

$HO-C_6H_4-CH_2CHCOOH$ with NH_2 on the CH — tyrosine

serine threonine tyrosine

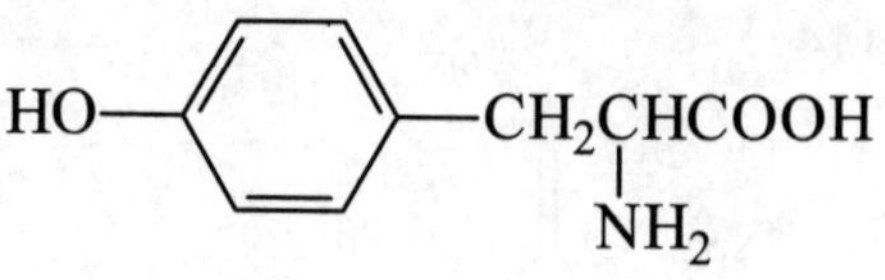

(c) $HSCH_2CHCOOH$ with NH_2 on the CH

cysteine

(d) $HO-C_6H_4-CH_2CHCOOH$ with NH_2 on the CH

H, N, NH_2, $CH_2CHCOOH$ (indole ring bearing $CH_2CHCOOH$ with NH_2 on the CH)

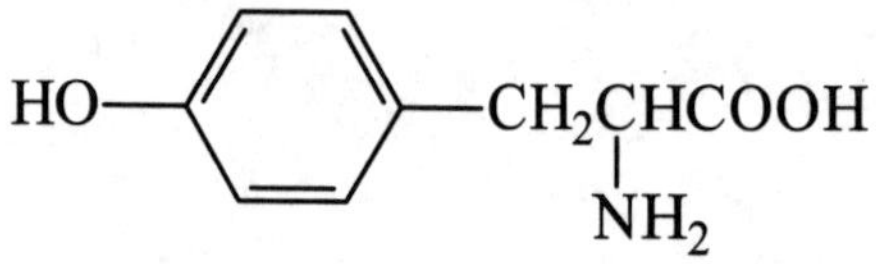

tyrosine 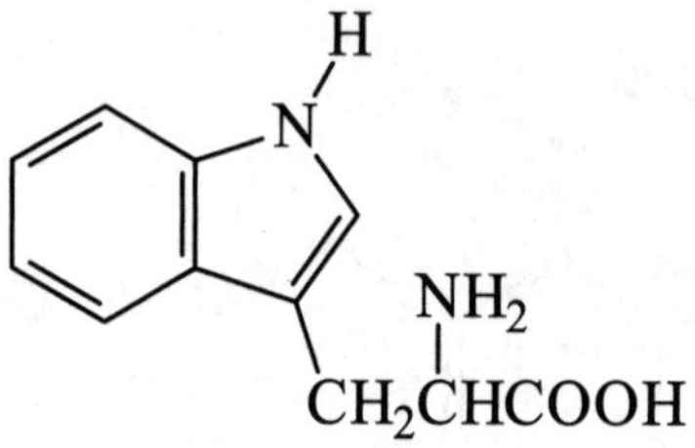 tryptophan

$C_6H_5-CH_2CHCOOH$ with NH_2 on the CH

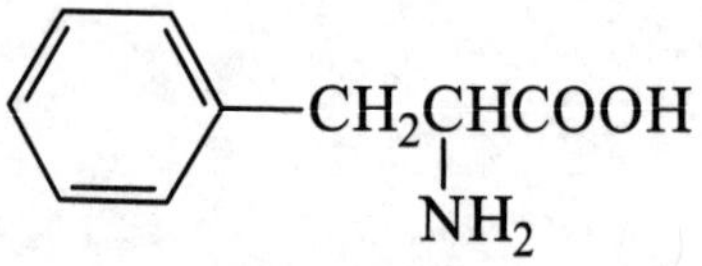

phenylalanine

23.120 Val-Ser-Phe-Met-Thr-Ala

23.121 Aspartic acid and phenylalanine are the amino acids in aspartame.
Digestion products:

$HOCCH_2CHCOH$ with C=O on each terminal carbon and NH_2 on the CH — aspartic acid

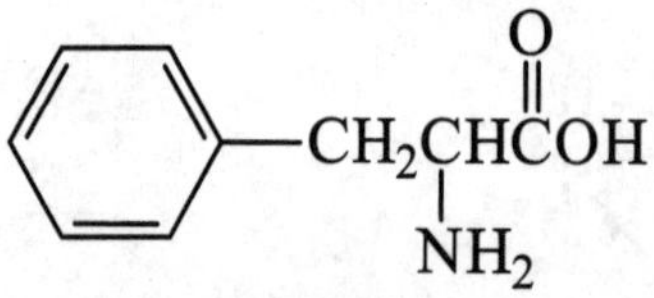

$C_6H_5-CH_2CHCOH$ with C=O on the carboxyl carbon and NH_2 on the CH — phenylalanine

CH_3OH — methanol

23.122 Met-Ile-Lys, Met-Lys-Ile, Ile-Met-Lys, Ile-Lys-Met, Lys-Met-Ile, Lys-Ile-Met

23.123 There are 24 tetrapeptides containing alanine, serine, leucine, and glutamic acid.
For example: Ala-Ser-Leu-Glu, Ser-Ala-Leu-Glu, Leu-Glu-Ala-Ser

Aromaticity and Molecular Orbital Theory (Section 23.10)

23.124 Only (b) is aromatic.

23.125 Only molecule (b) is aromatic.

23.126 There are 6 π-electrons, 2 each in the 2 π bonds and 2 in the p-orbital of the NH nitrogen.

23.127 There are 10 π-electrons, 2 each in the 5 π bonds.

Nucleic Acids (Section 23.11)

23.128 Just as proteins are polymers made of amino acid units, nucleic acids are polymers made up of nucleotide units linked together to form a long chain. Each nucleotide contains a phosphate group, an aldopentose sugar, and an amine base.

23.129 The sugar component in RNA is ribose, and the sugar in DNA is 2-deoxyribose (2-deoxy means that oxygen is missing from C2 of ribose).

23.130

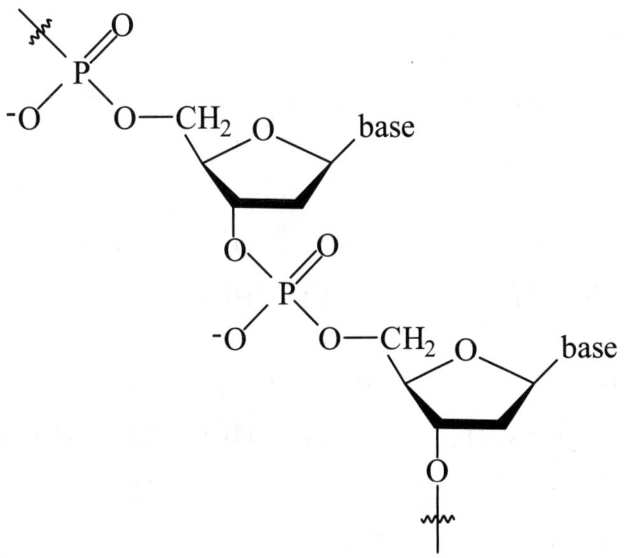

O
P
-O O—CH_2 O base
O O
P
-O O—CH_2 O base
O

23.131

amine base
CH_2OH O N
H H
H H
OH Y

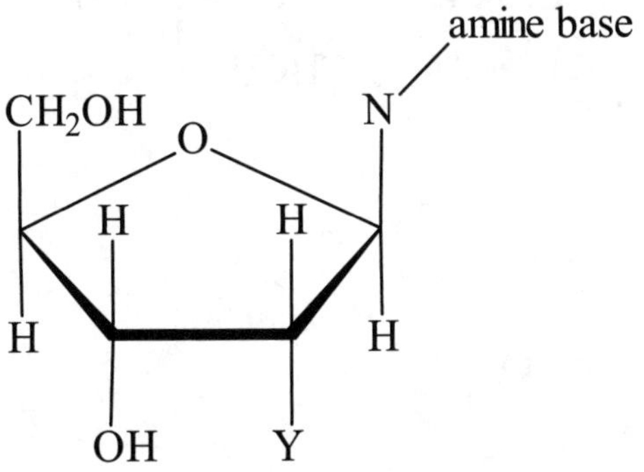

23.132 Original: T–A–C–C–G–A
Complement: A–T–G–G–C–T

23.133 32% A and T because they are complementary.
18% G and C because they are complementary.

23.134

23.135

All aromatic ring atoms are sp^2 hybridized.

Chapter Problems

23.136 (a)

$$\begin{array}{l}\quad CH_3 \\ \quad | \\ CH_3CHCH_2CH_2CH_2CH_2CH_3\end{array}$$

(b)

$$\begin{array}{l}\quad CH_3 \quad CH_2CH_3 \\ \quad | \qquad\quad | \\ CH_3CHCH_2CHCH_2CH_3\end{array}$$

(c)

$$\begin{array}{l}\qquad\quad CH_3 \; CH_2CH_3 \\ \qquad\quad | \qquad | \\ CH_3CH_2CH-CCH_2CH_2CH_2CH_3 \\ \qquad\qquad\quad | \\ \qquad\qquad\; CH_3\end{array}$$

(d)

$$\begin{array}{l}\quad CH_3 \quad CH_3 \\ \quad | \qquad\; | \\ CH_3CHCH_2CCH_2CH_2CH_3 \\ \qquad\qquad\; | \\ \qquad\quad\; CH_3\end{array}$$

(e) CH_3, CH_3 (1,1-disubstituted cyclopentane)

(f)

$$\begin{array}{l}\qquad\quad CH_3 \\ \qquad\quad | \\ CH_3CH_2CHCHCH_2CH_2CH_3 \\ \qquad\qquad\; | \\ \qquad\qquad CHCH_3 \\ \qquad\qquad\; | \\ \qquad\qquad CH_3\end{array}$$

23.137 (a) 2,3-dimethylhexane
(b) 4-isopropyloctane
(c) 4-ethyl-2,4-dimethylhexane
(d) 3,3-diethylpentane

23.138

$$\begin{array}{c}\qquad\qquad O \\ \qquad\qquad \| \\ CH_3(CH_2)_{18}CO(CH_2)_{31}CH_3\end{array}$$

23.139 Amine, ester, and arene

23.140 There are alkene and alkyne functional groups.

23.141 $C_9H_8O_4$

23.142 $C_{11}H_{12}N_2O_2$

23.143 $C_5H_6N_2O_2$

23.144 (a) ketone (b) aldehyde (c) ketone (d) amide (e) ester

23.145 (a)

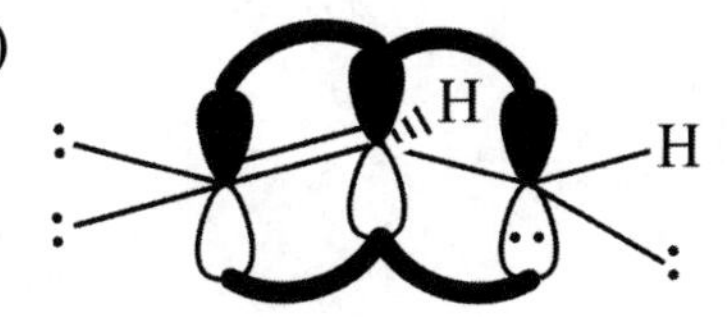

(b)

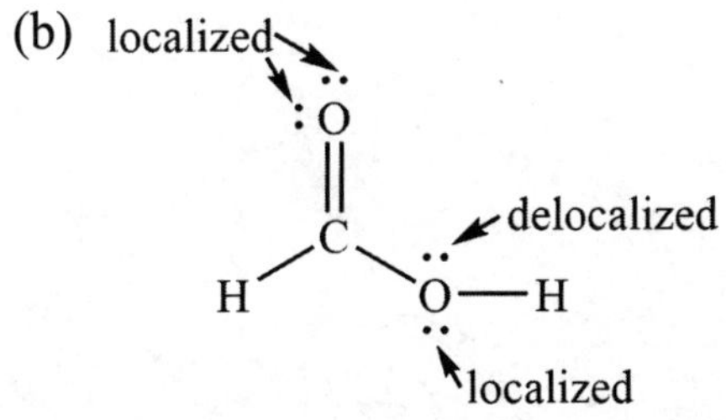

(c)

23.146 The bicyclic ring system is aromatic with 10 π electrons that satisfies the 4n +2 rule.

23.147 (a)

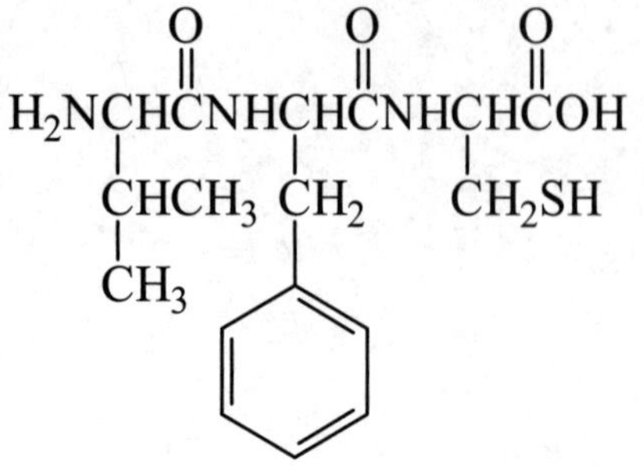

$$H_2NCH(CHCH_3CH_3)C(=O)NHCH(CH_2C_6H_5)C(=O)NHCH(CH_2SH)COOH$$

O O O
H₂NCHCNHCHCNHCHCOH
CHCH₃ CH₂ CH₂SH
CH₃

(b)

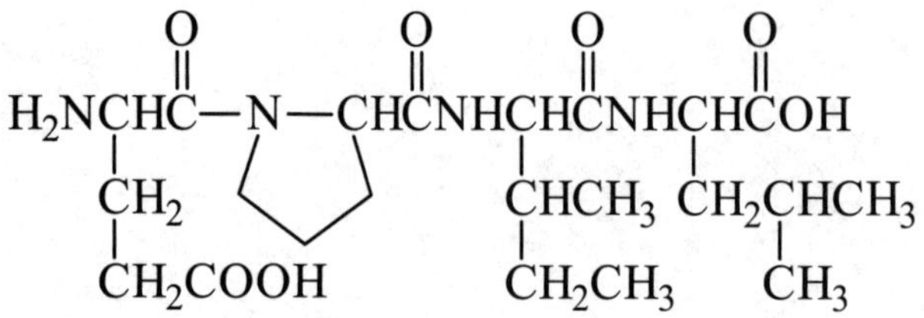

O O O O
$H_2NCHC{-}N{-}CHCNHCHCNHCHCOH$
CH_2 $CHCH_3$ CH_2CHCH_3
CH_2COOH CH_2CH_3 CH_3

23.148 (a) a fat

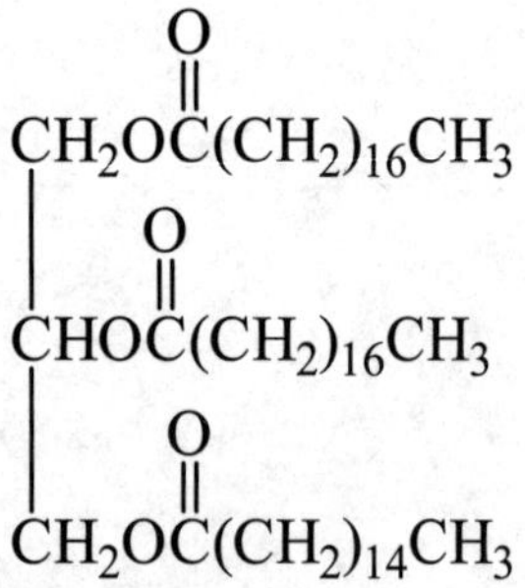

$$CH_2OC(=O)(CH_2)_{16}CH_3$$
$$CHOC(=O)(CH_2)_{16}CH_3$$
$$CH_2OC(=O)(CH_2)_{14}CH_3$$

(b) a vegetable oil

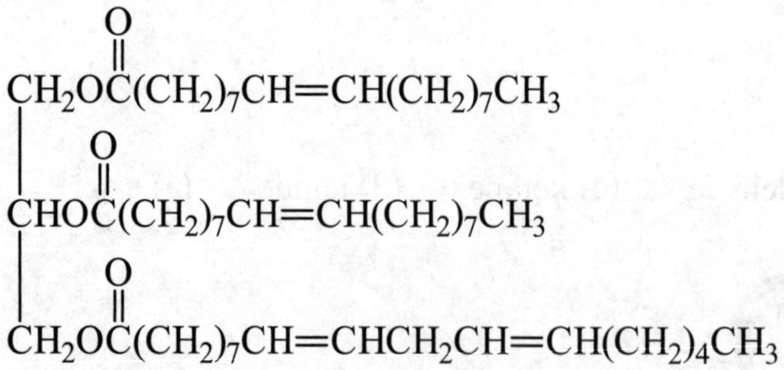

$$CH_2OC(=O)(CH_2)_7CH{=}CH(CH_2)_7CH_3$$
$$CHOC(=O)(CH_2)_7CH{=}CH(CH_2)_7CH_3$$
$$CH_2OC(=O)(CH_2)_7CH{=}CHCH_2CH{=}CH(CH_2)_4CH_3$$

(c) an aldotetrose

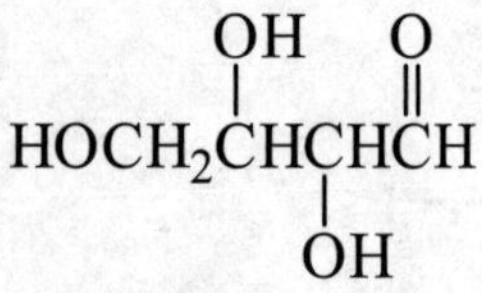

$$HOCH_2CH(OH)CH(OH)CH(=O)$$

23.149 Original: A–G–T–T–C–A–T–C–G
Complement: T–C–A–A–G–T–A–G–C

23.150

$$CH_3(CH_2)_{16}\overset{\overset{\displaystyle O}{||}}{C}O(CH_2)_{21}CH_3$$

23.151

oleic acid

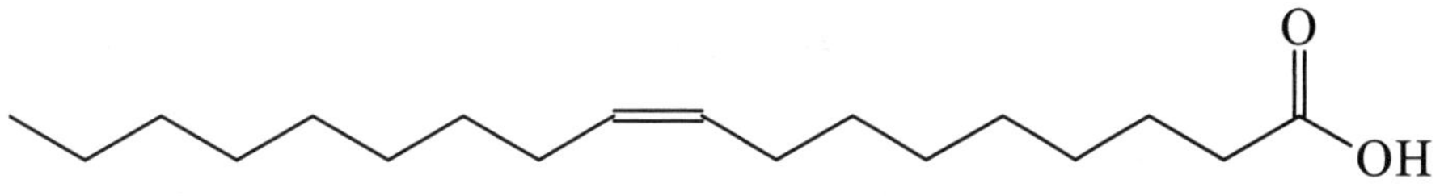

elaidic acid

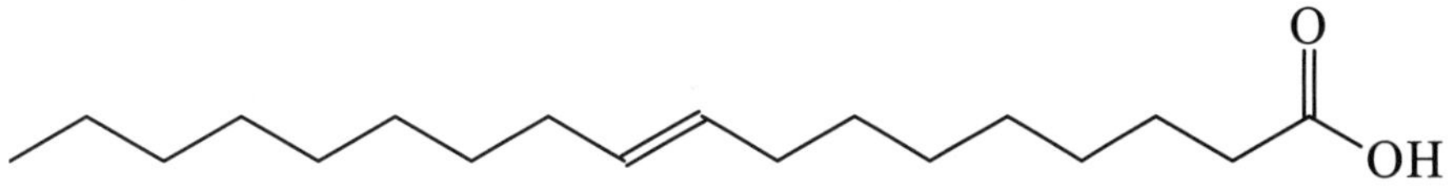

Multiconcept Problems

23.152 (a) Calculate the empirical formula. Assume a 100.0 g sample of fumaric acid.

$$41.4 \text{ g C x } \frac{1 \text{ mol C}}{12.01 \text{ g C}} = 3.45 \text{ mol C}$$

$$3.5 \text{ g H x } \frac{1 \text{ mol H}}{1.008 \text{ g H}} = 3.47 \text{ mol H}$$

$$55.1 \text{ g O x } \frac{1 \text{ mol O}}{16.00 \text{ g O}} = 3.44 \text{ mol O}$$

Because the mol amounts for the three elements are essentially the same, the empirical formula is CHO (29).

(b) Calculate the molar mass from the osmotic pressure.

$$\Pi = MRT; \quad M = \frac{\Pi}{RT} = \frac{\left(240.3 \text{ mm Hg x } \dfrac{1.00 \text{ atm}}{760 \text{ mm Hg}}\right)}{\left(0.082\,06 \dfrac{\text{L} \cdot \text{atm}}{\text{K} \cdot \text{mol}}\right)(298 \text{ K})} = 0.0129 \text{ M}$$

$(0.1000 \text{ L})(0.0129 \text{ mol/L}) = 1.29 \times 10^{-3}$ mol fumaric acid

$$\text{fumaric acid molar mass} = \frac{0.1500 \text{ g}}{1.29 \times 10^{-3} \text{ mol}} = 116 \text{ g/mol}$$

molecular mass = 116

(c) Determine the molecular formula. $\dfrac{\text{molar mass}}{\text{empirical formula mass}} = \dfrac{116}{29} = 4$

molecular formula = $C_{(1 \times 4)}H_{(1 \times 4)}O_{(1 \times 4)} = C_4H_4O_4$

From the titration, the number of carboxylic acid groups can be determined.

$$\text{mol } C_4H_4O_4 = 0.573 \text{ g x } \frac{1 \text{ mol } C_4H_4O_4}{116 \text{ g}} = 0.004\,94 \text{ mol } C_4H_4O_4$$

mol NaOH used = (0.0941 L)(0.105 mol/L) = 0.0099 mol NaOH

$$\frac{\text{mol NaOH}}{\text{mol } C_4H_4O_4} = \frac{0.0099 \text{ mol}}{0.004\,94 \text{ mol}} = 2$$

Because 2 mol of NaOH are required to titrate 1 mol $C_4H_4O_4$, $C_4H_4O_4$ is a diprotic acid. Because $C_4H_4O_4$ gives an addition product with HCl and a reduction product with H_2, it contains a double bond.

HO—C(=O)—CH=CH—C(=O)—OH HO—C(=O)—CH=CH—C(=O)—OH $H_2C{=}C(C(=O){-}OH)_2$

(d) The correct structure is HO—C(=O)—CH=CH—C(=O)—OH

23.153 $\frac{55.847 \text{ u Fe}}{\text{mol mass cyt c}} \times 100\% = 0.43\%$; mol mass cyt c $= \frac{55.847 \text{ u Fe}}{0.0043} = 13{,}000$

23.154 (a) $-CO_2H$ is the more acidic group because it has the smaller pK_a (larger K_a).

(b) At pH = 4.00 only HA and H_2A^+ are present in appreciable amounts.

$$pH = pK_a + \log\frac{[\text{Base}]}{[\text{Acid}]}$$

$$4.00 = 2.34 + \log\frac{[HA]}{[H_2A^+]};\quad 4.00 - 2.34 = \log\frac{[HA]}{[H_2A^+]};\quad 1.66 = \log\frac{[HA]}{[H_2A^+]}$$

$$\frac{[HA]}{[H_2A^+]} = 10^{1.66} = 45.7$$

Let [HA] = 100 – x and $[H_2A^+] = x$

$$\frac{100 - x}{x} = 45.7;\ 100 - x = 45.7x;\ 100 = 46.7x;\ x = \frac{100}{46.7} = 2.1$$

HA = 100 – x = 98% and H_2A^+ = x = 2.1%

(c) At pH = 8.50 only HA and A^- are present in appreciable amounts.

$$pH = pK_a + \log\frac{[\text{Base}]}{[\text{Acid}]}$$

$$8.50 = 9.69 + \log\frac{[A^-]}{[HA]};\quad 8.50 - 9.69 = \log\frac{[A^-]}{[HA]};\quad -1.19 = \log\frac{[A^-]}{[HA]}$$

$$\frac{[A^-]}{[HA]} = 10^{-1.19} = 0.0646$$

Let [HA] = 100 – x and $[A^-] = x$

$$\frac{x}{100 - x} = 0.0646;\ 0.0646(100 - x) = x;\ 6.46 - 0.0646x = x;\ 6.46 = 1.0646x$$

$$x = \frac{6.46}{1.0646} = 6.1;\qquad HA = 100 - x = 94\% \text{ and } A^- = x = 6.1\%$$

(d) The maximum amount of HA is found at a pH that is midway between the two pK_a values, 2.34 and 9.69. That pH = 6.01.

23.155 (a) $\text{grams } I_2 = (0.0250 \text{ L})(0.200 \text{ mol/L})\left(\frac{253.81 \text{ g } I_2}{1 \text{ mol } I_2}\right) = 1.27 \text{ g } I_2$

(b) $\text{mol } Na_2S_2O_3 = (0.08199 \text{ L})(0.100 \text{ mol/L}) = 8.20 \times 10^{-3} \text{ mol}$

$$\text{grams excess } I_2 = 8.20 \times 10^{-3} \text{ mol } Na_2S_2O_3 \times \frac{1 \text{ mol } I_2}{2 \text{ mol } Na_2S_2O_3} \times \frac{253.81 \text{ g } I_2}{1 \text{ mol } I_2} = 1.04 \text{ g } I_2$$

$\text{grams } I_2 \text{ reacted} = 1.27 \text{ g} - 1.04 \text{ g} = 0.23 \text{ g } I_2$

(c) $$\text{iodine number} = \frac{0.23 \text{ g } I_2}{0.500 \text{ g milkfat}} \times 100 = 46$$

(d) $$\text{mol milkfat} = 0.500 \text{ g milkfat} \times \frac{1 \text{ mol milkfat}}{800 \text{ g milkfat}} = 6.25 \times 10^{-4} \text{ mol}$$

$$\text{mol } I_2 \text{ reacted} = 0.23 \text{ g } I_2 \times \frac{1 \text{ mol } I_2}{253.81 \text{ g } I_2} = 9.06 \times 10^{-4} \text{ mol}$$

$$\text{number of double bonds per molecule} = \frac{9.06 \times 10^{-4} \text{ mol } I_2}{6.25 \times 10^{-4} \text{ mol milkfat}} = 1.4 \text{ double bonds}$$